# QUASI-HOPF ALGEBRAS

This is the first book to be dedicated entirely to Drinfeld's quasi-Hopf algebras. Ideal for graduate students and researchers in mathematics and mathematical physics, this treatment is largely self-contained, taking the reader from the basics, with complete proofs, to much more advanced topics, with almost complete proofs. Many of the proofs are based on general categorical results; the same approach can then be used in the study of other Hopf-type algebras, for example Turaev or Zunino Hopf algebras, Hom-Hopf algebras, Hopfish algebras, and in general any algebra for which the category of representations is monoidal.

Newcomers to the subject will appreciate the detailed introduction to (braided) monoidal categories, (co)algebras and the other tools they will need in this area. More advanced readers will benefit from having recent research gathered in one place, with open questions to inspire their own research.

**Encyclopedia of Mathematics and Its Applications**

This series is devoted to significant topics or themes that have wide application in mathematics or mathematical science and for which a detailed development of the abstract theory is less important than a thorough and concrete exploration of the implications and applications.

Books in the **Encyclopedia of Mathematics and Its Applications** cover their subjects comprehensively. Less important results may be summarized as exercises at the ends of chapters. For technicalities, readers can be referred to the bibliography, which is expected to be comprehensive. As a result, volumes are encyclopedic references or manageable guides to major subjects.

All the titles listed below can be obtained from good booksellers or from Cambridge University Press. For a complete series listing visit www.cambridge.org/mathematics.

122 S. Khrushchev *Orthogonal Polynomials and Continued Fractions*
123 H. Nagamochi and T. Ibaraki *Algorithmic Aspects of Graph Connectivity*
124 F. W. King *Hilbert Transforms I*
125 F. W. King *Hilbert Transforms II*
126 O. Calin and D.-C. Chang *Sub-Riemannian Geometry*
127 M. Grabisch *et al. Aggregation Functions*
128 L. W. Beineke and R. J. Wilson (eds.) with J. L. Gross and T. W. Tucker *Topics in Topological Graph Theory*
129 J. Berstel, D. Perrin and C. Reutenauer *Codes and Automata*
130 T. G. Faticoni *Modules over Endomorphism Rings*
131 H. Morimoto *Stochastic Control and Mathematical Modeling*
132 G. Schmidt *Relational Mathematics*
133 P. Kornerup and D. W. Matula *Finite Precision Number Systems and Arithmetic*
134 Y. Crama and P. L. Hammer (eds.) *Boolean Models and Methods in Mathematics, Computer Science, and Engineering*
135 V. Berthé and M. Rigo (eds.) *Combinatorics, Automata and Number Theory*
136 A. Kristály, V. D. Rădulescu and C. Varga *Variational Principles in Mathematical Physics, Geometry, and Economics*
137 J. Berstel and C. Reutenauer *Noncommutative Rational Series with Applications*
138 B. Courcelle and J. Engelfriet *Graph Structure and Monadic Second-Order Logic*
139 M. Fiedler *Matrices and Graphs in Geometry*
140 N. Vakil *Real Analysis through Modern Infinitesimals*
141 R. B. Paris *Hadamard Expansions and Hyperasymptotic Evaluation*
142 Y. Crama and P. L. Hammer *Boolean Functions*
143 A. Arapostathis, V. S. Borkar and M. K. Ghosh *Ergodic Control of Diffusion Processes*
144 N. Caspard, B. Leclerc and B. Monjardet *Finite Ordered Sets*
145 D. Z. Arov and H. Dym *Bitangential Direct and Inverse Problems for Systems of Integral and Differential Equations*
146 G. Dassios *Ellipsoidal Harmonics*
147 L. W. Beineke and R. J. Wilson (eds.) with O. R. Oellermann *Topics in Structural Graph Theory*
148 L. Berlyand, A. G. Kolpakov and A. Novikov *Introduction to the Network Approximation Method for Materials Modeling*
149 M. Baake and U. Grimm *Aperiodic Order I: A Mathematical Invitation*
150 J. Borwein *et al. Lattice Sums Then and Now*
151 R. Schneider *Convex Bodies: The Brunn–Minkowski Theory (Second Edition)*
152 G. Da Prato and J. Zabczyk *Stochastic Equations in Infinite Dimensions (Second Edition)*
153 D. Hofmann, G. J. Seal and W. Tholen (eds.) *Monoidal Topology*
154 M. Cabrera García and Á. Rodríguez Palacios *Non-Associative Normed Algebras I: The Vidav–Palmer and Gelfand–Naimark Theorems*
155 C. F. Dunkl and Y. Xu *Orthogonal Polynomials of Several Variables (Second Edition)*
156 L. W. Beineke and R. J. Wilson (eds.) with B. Toft *Topics in Chromatic Graph Theory*
157 T. Mora *Solving Polynomial Equation Systems III: Algebraic Solving*
158 T. Mora *Solving Polynomial Equation Systems IV: Buchberger Theory and Beyond*
159 V. Berthé and M. Rigo (eds.) *Combinatorics, Words and Symbolic Dynamics*
160. B. Rubin *Introduction to Radon Transforms: With Elements of Fractional Calculus and Harmonic Analysis*
161 M. Ghergu and S. D. Taliaferro *Isolated Singularities in Partial Differential Inequalities*
162 G. Molica Bisci, V. D. Radulescu and R. Servadei *Variational Methods for Nonlocal Fractional Problems*
163 S. Wagon *The Banach–Tarski Paradox (Second Edition)*
164 K. Broughan *Equivalents of the Riemann Hypothesis I: Arithmetic Equivalents*
165 K. Broughan *Equivalents of the Riemann Hypothesis II: Analytic Equivalents*
166 M. Baake and U. Grimm (eds.) *Aperiodic Order II: Crystallography and Almost Periodicity*
167 M. Cabrera García and Á. Rodríguez Palacios *Non-Associative Normed Algebras II: Representation Theory and the Zel'manov Approach*
168 A. Yu. Khrennikov, S. V. Kozyrev and W. A. Zúñiga-Galindo *Ultrametric Pseudodifferential Equations and Applications*
169 S. R. Finch *Mathematical Constants II*
170 J. Krajíček *Proof Complexity*
171 D. Bulacu, S. Caenepeel, F. Panaite and F. Van Oystaeyen *Quasi-Hopf Algebras*

Encyclopedia of Mathematics and its Applications

# *Quasi-Hopf Algebras*

## A Categorical Approach


DANIEL BULACU

*Universitatea din Bucureşti, Romania*

STEFAAN CAENEPEEL

*Vrije Universiteit Brussel, Belgium*

FLORIN PANAITE

*Institute of Mathematics of the Romanian Academy*

FREDDY VAN OYSTAEYEN

*Universiteit Antwerpen, Belgium*



CAMBRIDGE
UNIVERSITY PRESS

CAMBRIDGE
UNIVERSITY PRESS

University Printing House, Cambridge CB2 8BS, United Kingdom

One Liberty Plaza, 20th Floor, New York, NY 10006, USA

477 Williamstown Road, Port Melbourne, VIC 3207, Australia

314–321, 3rd Floor, Plot 3, Splendor Forum, Jasola District Centre,
New Delhi – 110025, India

79 Anson Road, #06–04/06, Singapore 079906

Cambridge University Press is part of the University of Cambridge.

It furthers the University's mission by disseminating knowledge in the pursuit of education, learning, and research at the highest international levels of excellence.

www.cambridge.org
Information on this title: www.cambridge.org/9781108427012
DOI: 10.1017/9781108582780



First published 2019

Printed and bound in Great Britain by Clays Ltd, Elcograf S.p.A.

*A catalogue record for this publication is available from the British Library*

*Library of Congress Cataloging-in-Publication Data*
Names: Bulacu, Daniel, 1973– author. | Caenepeel, Stefaan, 1956– author. |
Panaite, Florin, 1970– author. | Oystaeyen, F. Van, 1947– author.
Title: Quasi-Hopf algebras : a categorical approach / Daniel Bulacu
(Universitatea din Bucureti, Romania), Stefaan Caenepeel (Vrije
Universiteit, Amsterdam), Florin Panaite (Institute of Mathematics of the
Romanian Academy), Freddy van Oystaeyen (Universiteit Antwerpen, Belgium).
Description: Cambridge ; New York, NY : Cambridge University Press, [2019] |
Series: Encyclopedia of mathematics and its applications ; 171 | Includes
bibliographical references and index.
Identifiers: LCCN 2018034517 | ISBN 9781108427012 (hardback)
Subjects: LCSH: Hopf algebras. | Tensor products. | Tensor algebra.
Classification: LCC QA613.8 .B85 2019 | DDC 512/.55–dc23
LC record available at https://lccn.loc.gov/2018034517

ISBN 978-1-108-42701-2 Hardback

Dedicated to our wives
Adriana, Lieve, Cristina, Danielle.

# Contents

# Preface

Some basic ideas in mathematics are very generic and almost omnipresent. Let us just mention "operators on some structure," an idea going back to symmetry of geometric configurations, and also "duality." These ideas are also at the roots of the modern theory of quasi-Hopf algebras, which is the topic of this book.

Geometry is at the root of many developments in mathematics, and for our topic of interest we may go back to algebraic geometry and the theory of (affine) algebraic varieties, which may be seen as sets of solutions of polynomial equations in some affine space over some field. One then studies such varieties via the ring of functions on them with values in the base field; in fact one restricts attention to polynomial functions forming the coordinate ring of the variety. There one observes the fundamental duality between commutative (affine) algebra and the algebraic geometry of (affine) algebraic varieties, later better phrased in the more general scheme theory.

The other generic idea of operators acting on geometric structures led directly to actions or transformation groups and operator algebras. The idea of group actions and their invariants is deeply embedded in the philosophy of mathematics; for example, in the "Erlangen Program" of F. Klein, geometry was redefined as the study of properties invariant for actions of transformation groups. On the more algebraic side, actions of groups of automorphisms of fields were used by E. Galois to solve some problems about solutions to polynomial equations over a field. In the resulting Galois theory another duality appeared, namely the duality between subgroups of the Galois group of some field extension and the lattice of subfields of the field. This Galois duality originally was considered for finite-dimensional separable field extensions but it was extended to inseparable extensions by using derivations and higher derivations, leading to Lie algebra actions and their invariants. Thus, a more general Galois theory mixing Lie actions (of derivations) and group actions (of automorphisms) resulted, immediately leading to a Galois theory for Hopf algebra actions. Further extensions of the Galois theory were in the direction of continuous groups, later called Lie groups. So here the generic ideas of action and duality met, and Hopf algebras appeared naturally. But also the geometric line of development showed a similar phenomenon with the study of abelian varieties and algebraic groups. Roughly stated, an algebraic group is an algebraic variety with a group structure on its points;

interesting examples are matrix groups, that is, groups embedded in a matrix ring and having the structure of an algebraic variety, like $\mathrm{GL}_n(k)$ and $\mathrm{SL}_n(k)$, the general and special linear groups over the field $k$, respectively. The group structure on the variety translates into a structure of the coordinate ring given by a comultiplication, a counit and an antipode satisfying suitable conditions that turn it into a commutative Hopf algebra. Hopf algebras got their name because they appeared first in a celebrated paper by H. Hopf on algebraic topology. In fact the structure was discovered on the cohomology ring of an $H$-space; roughly stated, that is a topological space with a multiplication on it together with a special element such that left and right multiplication by this element defines a map which is homotopic to the identity map (so a kind of neutral element up to homotopy).

Group actions on vector spaces may be studied by looking at modules over the group algebra $k[G]$ of the acting group $G$ over the base field $k$; similarly, Lie algebra actions of a Lie algebra $\mathfrak{g}$ on a vector space may be studied by looking at the universal enveloping algebra of $\mathfrak{g}$ over $k$, say $U_k(\mathfrak{g})$. Now both $k[G]$ and $U_k(\mathfrak{g})$ are Hopf algebras but not commutative anymore; instead, they are cocommutative. So aspects of group actions and Lie algebra actions become unified in a theory of actions of arbitrary Hopf algebras on general algebras or vector spaces or modules, and this received extensive interest in ring theory.

Let us point out one important "generality" for general Hopf algebras: they need not be commutative or cocommutative, as many of the early examples of Hopf algebras were. In his famous address to the International Congress of Mathematicians in 1986, Drinfeld introduced the term "quantum group," roughly referring to a quasitriangular Hopf algebra, that is, a Hopf algebra endowed with a so-called R-matrix, satisfying certain axioms that represent a relaxation of the cocommutativity condition and implying the (equally famous) quantum Yang–Baxter equation. Drinfeld proved that any finite-dimensional Hopf algebra can be embedded in a quasitriangular one, called its quantum (or Drinfeld) double. There is a vast literature on quantum groups and many examples could be obtained from deforming well-known easier Hopf algebras. Combined with the restriction to special Hopf algebras it also makes sense to restrict to special categories of modules like so-called Yetter–Drinfeld modules, to name just one.

Essential for the transition from Hopf algebras to quasi-Hopf algebras was the concept of monoidal category, roughly stated a category with a product (called the "tensor product") generalizing the tensor product of vector spaces in a suitable way and satisfying natural conditions. For example, the category of sets is a monoidal category, the "tensor product" being the Cartesian product of sets. One of the axioms of a monoidal category is the so-called "associativity constraint," which for the categories of vector spaces and of sets is "trivial;" for instance, for vector spaces this boils down to saying that, if $U$, $V$, $W$ are vector spaces, then $(U \otimes V) \otimes W$ and $U \otimes (V \otimes W)$ can be identified in the usual (or "trivial") way.

One of the fundamental features of a Hopf algebra, $H$, is that its category of (left) representations is a monoidal category, with tensor product inherited from the category of vector spaces, and the tensor product of two left $H$-modules is again a left $H$-module via the comultiplication of $H$. The associativity constraint is, again, "trivial."

If one is not interested in an a priori given type of algebra but wants to make sure that there is a "product" on the category of its representations, then one finds the motivation for the introduction of quasi-Hopf algebras as Drinfeld did in his seminal paper [80]. Roughly, a quasi-Hopf algebra is an algebra for which its category of left modules is monoidal, but maybe with non-trivial associativity constraint. More precisely, what Drinfeld did was to weaken the coassociativity condition for a Hopf algebra so that the comultiplication is only coassociative up to conjugation by an invertible element of $H \otimes H \otimes H$ (which is a sort of 3-cocycle). Moreover, examples of quasi-Hopf algebras can be obtained by "twisting" the comultiplication of a Hopf algebra via a so-called "gauge transformation" (only if the gauge transformation is a sort of 2-cocycle is the twisted object again a Hopf algebra). After specialization to quantum groups, sometimes just taken to be non-commutative non-cocommutative Hopf algebras but usually with extra conditions like quasitriangularity, the generalization in terms of non-coassociativity became popular too and it found several applications as well. Again, the fundamental property is that the relaxation of coassociativity still makes the representation category into a monoidal category, and moreover the rigidity (i.e. the existence of dual objects) of the category of finite-dimensional representations of a Hopf algebra, owing to the presence of an antipode, is preserved by replacing the notion of an antipode by a suitable analogue. Categorically speaking, passing from the category of Hopf algebras to the one of quasi-Hopf algebras does not (in principle) really add to the complexity; in fact the latter is in some sense more manageable because of the presence of a kind of gauge group.

Monoidal categories were present, if hidden, in the classical ideas mentioned before and they have been very useful in obtaining a unified theory. One of the early facts that stimulated interest in monoidal categories stemmed from their applicability in rational conformal field theory (RCFT). The monoidal categories in RCFT could, by Tannaka–Krein reconstruction, be considered as module categories over some "Hopf-like" algebras. Back in 1984 Drinfeld and Jimbo introduced a quantum group by deforming a universal enveloping algebra $U(\mathfrak{g})$ for some Lie algebra $\mathfrak{g}$; in fact for every semisimple Lie algebra they constructed what was called afterwards the Drinfeld–Jimbo algebra. For the study of some categories of modules over the Drinfeld–Jimbo algebras, a relation with the so-called KZ-equations had to be used; these equations were introduced by Knizhnik and Zamolodchikov in 1984. The KZ-equations are linear differential equations satisfied by two-dimensional conformal field theories associated with affine Lie algebras. Such KZ-equations may be used to obtain a quantization of universal enveloping algebras, and Drinfeld used KZ-equations to construct a quasi-Hopf algebra for some Lie algebra $\mathfrak{g}$, say $Q_{\mathfrak{g}}$, so that some categories of modules over $Q_{\mathfrak{g}}$ are equivalent to similar ones over the Drinfeld–Jimbo algebra of the Lie algebra $\mathfrak{g}$. Further interesting applications of KZ-equations follow, for example, from the fact that their monodromy along closed paths yields a representation of the braid group. We refer to the specialized literature for more detail concerning applications in physics. We do the same for some deep relations with number theory in the sense of A. Grothendieck's "Esquisse."

In this book we aim to develop the theory of quasi-Hopf algebras from scratch, or almost, dealing mainly with algebraic methods. Knowledge of Hopf algebras will benefit the reader but we do introduce the necessary concepts. Using monoidal categories as the main tool makes for a rather abstract treatment of the material, but we hope the unifying effect of it will expose well the beautiful generalization from Hopf algebras to quasi-Hopf algebras; moreover, the categorical point of view also stays close to the applications in physics, as indicated by the foregoing remarks.

We now outline the content of the book (more historical and bibliographical remarks can be found in the Notes section at the end of each chapter).

In Chapters 1 and 2 we present the basic categorical concepts and tools needed for the rest of the book (monoidal, rigid and braided categories and algebras, coalgebras and Hopf algebras in such categories). We included detailed definitions and proofs; we do not assume that the reader has prior knowledge of these topics. In particular, we introduce the concepts of coalgebra, bialgebra and Hopf algebra in the usual sense (over a field), so we do not assume from the reader a knowledge of these concepts either.

In Chapter 3 we introduce the main objects of our study, quasi-bialgebras and quasi-Hopf algebras (as well as the dual concepts), present their basic properties and some classes of examples. We have two warnings for the reader: (1) the concept of quasi-bialgebra is introduced in Definition 3.4, but afterwards we make a reduction, and the axioms of a quasi-bialgebra that will be used from there on are the ones presented in equations (3.1.7)–(3.1.10); (2) unlike Drinfeld, we do not include the bijectivity of the antipode in the definition of a quasi-Hopf algebra, and we shall see in later chapters that the bijectivity is automatic in the finite-dimensional and the quasitriangular case.

In Chapter 4 we study "(co)actions" of quasi-bialgebras and quasi-Hopf algebras, namely we introduce the concepts of module (co)algebra and (bi)comodule algebra over a quasi-bialgebra, we give some examples and present some connections that exist between these structures.

In Chapter 5 we introduce various types of crossed products that appear in the context of quasi-Hopf algebras (smash products, diagonal crossed products, etc.), we study the relations between them and as an application we present a duality theorem for finite-dimensional quasi-Hopf algebras.

In Chapter 6 we introduce so-called quasi-Hopf bimodules over a quasi-Hopf algebra $H$, prove some structure theorems for them leading to the fact that their category is monoidally equivalent to the category of left $H$-modules and, as an application, we prove a structure theorem for quasi-Hopf comodule algebras.

In Chapter 7 we study finite-dimensional quasi-Hopf algebras, more precisely integrals and cointegrals for them. We use the machinery provided by Frobenius algebras, and we present some basic results about Frobenius, symmetric and Frobenius augmented algebras (so again we do not assume from the reader a knowledge of these topics). A consequence of the theory we develop is that the antipode of a finite-dimensional quasi-Hopf algebra is bijective. We end the chapter with a section containing a freeness result for quasi-Hopf algebras (for that section the reader is assumed to have some knowledge of module theory).

In Chapter 8 we introduce the four categories of Yetter–Drinfeld modules over a quasi-Hopf algebra, prove that they are all braided isomorphic and, when restricted to finite-dimensional objects, rigid. Then we introduce the quantum double of a finite-dimensional quasi-Hopf algebra (for the moment, only as a quasi-Hopf algebra), and two particular cases, objects denoted by $D^{\omega}(H)$ and $D^{\omega}(G)$ (for the latter, $G$ is a finite group and $D^{\omega}(G)$ is called the twisted quantum double of $G$). We end the chapter with some properties and examples of algebras in Yetter–Drinfeld categories.

In Chapter 9 we define so-called two-sided two-cosided Hopf modules over a quasi-Hopf algebra, prove that their category is monoidally equivalent to a category of Yetter–Drinfeld modules and use this equivalence to prove some structure theorems for bicomodule algebras and bimodule coalgebras. We characterize Hopf algebras within the category of two-sided two-cosided Hopf modules and use this to define biproduct quasi-Hopf algebras.

In Chapter 10 we study quasitriangular quasi-Hopf algebras, QT for short. We show that the antipode of a QT quasi-Hopf algebra is inner, hence bijective. We prove that the quantum double of a finite-dimensional quasi-Hopf algebra is a QT quasi-Hopf algebra and we characterize the quantum double of a QT finite-dimensional quasi-Hopf algebra as a certain biproduct quasi-Hopf algebra.

In Chapter 11 we introduce the concept of factorizable quasi-Hopf algebra, prove that the quantum double of a finite-dimensional quasi-Hopf algebra is factorizable, and describe the quantum double of a factorizable quasi-Hopf algebra. We prove also that any factorizable quasi-Hopf algebra is unimodular (i.e. the spaces of left and right integrals coincide).

In Chapter 12 we describe the integrals of a quantum double of a finite-dimensional quasi-Hopf algebra (reproving that it is unimodular). We define the quantum dimension of an object in a braided rigid category, apply this to the category of finite-dimensional modules over a quasi-Hopf algebra and compute the quantum dimension of a finite-dimensional quasi-Hopf algebra $H$ and of its quantum double $D(H)$ regarded as left $D(H)$-modules. We present a trace formula for quasi-Hopf algebras, and then we introduce the concept of involutory quasi-Hopf algebra.

In Chapter 13 we introduce the concepts of balanced and ribbon categories, leading to the concept of ribbon quasi-Hopf algebra, which is a QT quasi-Hopf algebra endowed with an element (called ribbon element) satisfying some axioms. In the final two sections, we present two classes of examples of ribbon quasi-Hopf algebras.

We have tried to make this book as self-contained as possible, providing as many details (in definitions and proofs) as we could. Owing to lack of space, we had to leave aside some other topics on quasi-Hopf algebras that would have deserved to be presented here (we intentionally left aside Drinfeld's theory of quantum enveloping algebras, because this is very well presented in C. Kassel's book [127]). In order to help the reader to get an idea of what else can be said about quasi-Hopf algebras, we have included in the bibliography a number of papers on (or related to) quasi-Hopf algebras that we did not cite or use in the book. We have also included some papers or books about Hopf algebras or category theory or other topics that we considered relevant for us or for the subject of the book.

This book is an outcome of the long-term scientific cooperation between the Noncommutative Algebra groups from Antwerp, Brussels and Bucharest. We wholeheartedly thank our colleagues from the University of Antwerp, the University of Brusssels, the University of Bucharest and the Institute of Mathematics of the Romanian Academy for the scientific discussions we had with them over the years.

Finally, the authors would like to thank Paul Taylor and Bodo Pareigis for sharing their "diagrams" programs, which were intensively used in this book.

# 1
# Monoidal and Braided Categories

In this chapter we introduce the basic categorical language that will be used throughout this book. We define the concepts of monoidal and braided monoidal category and prove that any monoidal category is monoidally equivalent to a strict one.

## 1.1 Monoidal Categories

Recall that a category $\mathscr{C}$ consists of the following:

- a collection $\mathrm{Ob}(\mathscr{C})$, whose elements are called the objects of $\mathscr{C}$; if $X$ is an object of $\mathscr{C}$, we write either $X \in \mathrm{Ob}(\mathscr{C})$ or simply $X \in \mathscr{C}$;
- for every two objects $X, Y \in \mathrm{Ob}(\mathscr{C})$, a set $\mathrm{Hom}_{\mathscr{C}}(X,Y)$, whose elements are denoted by $f : X \to Y$ and called the morphisms from $X$ to $Y$ in $\mathscr{C}$;
- for every object $X$ of $\mathscr{C}$, a specified morphism $\mathrm{Id}_X \in \mathrm{Hom}_{\mathscr{C}}(X,X)$, called the identity morphism of $X$;
- for every three objects $X,Y,Z$ of $\mathscr{C}$, a function

$$\circ : \mathrm{Hom}_{\mathscr{C}}(X,Y) \times \mathrm{Hom}_{\mathscr{C}}(Y,Z) \to \mathrm{Hom}_{\mathscr{C}}(X,Z),$$

  called the composition function, that maps a pair $(f,g)$ to $\circ(f,g) := g \circ f$, where $f : X \to Y$ and $g : Y \to Z$ are morphisms in $\mathscr{C}$.

  These data are subject to the following axioms:

(A) Associativity axiom: for all morphisms $f : X \to Y$, $g : Y \to Z$ and $h : Z \to T$ in $\mathscr{C}$ we have $(h \circ g) \circ f = h \circ (g \circ f)$.
(I) Identity axiom: $f \circ \mathrm{Id}_X = f = \mathrm{Id}_Y \circ f$, for every morphism $f : X \to Y$ in $\mathscr{C}$.

A morphism $f : X \to Y$ in $\mathscr{C}$ will also be denoted by $X \xrightarrow{f} Y$. Note that, when there is no danger of confusion, the composition of two morphisms $f : X \to Y$ and $g : Y \to Z$ in $\mathscr{C}$ will often be written as $gf$ instead of $g \circ f$.

A morphism $f : X \to Y$ in $\mathscr{C}$ is called an isomorphism if there exists a morphism $g : Y \to X$ in $\mathscr{C}$, called the inverse of $f$, such that $g \circ f = \mathrm{Id}_X$ and $f \circ g = \mathrm{Id}_Y$. Note that the inverse is unique.

If $X \in \mathrm{Ob}(\mathscr{C})$, we denote $\mathrm{End}_{\mathscr{C}}(X) := \mathrm{Hom}_{\mathscr{C}}(X,X)$.

A subcategory $\mathscr{D}$ of a category $\mathscr{C}$ is a collection of some objects and some morphisms of $\mathscr{C}$ in such a way that $\mathscr{D}$ becomes a category with composition and identities from $\mathscr{C}$. Furthermore, we say that $\mathscr{D}$ is a full subcategory when $\mathrm{Hom}_{\mathscr{D}}(X,Y) = \mathrm{Hom}_{\mathscr{C}}(X,Y)$, for all $X,Y \in \mathrm{Ob}(\mathscr{D})$.

Recall also that a functor $F$ between two categories $\mathscr{C}$ and $\mathscr{D}$ consists of:

- a map $\mathrm{Ob}(F) : \mathrm{Ob}(\mathscr{C}) \to \mathrm{Ob}(\mathscr{D})$; we will denote $\mathrm{Ob}(F)(X) = F(X)$, for all $X \in \mathrm{Ob}(\mathscr{C})$;
- a function

$$\mathrm{Hom}_F(X,Y) : \mathrm{Hom}_{\mathscr{C}}(X,Y) \to \mathrm{Hom}_{\mathscr{D}}(F(X),F(Y))$$

for any objects $X,Y$ of $\mathscr{C}$; we will denote $\mathrm{Hom}_F(X,Y)(f) = F(f)$, for any morphism $f : X \to Y$ in $\mathscr{C}$.

These data are subject to the following axioms:

(A1) Identities are preserved by $F$, that is, $F(\mathrm{Id}_X) = \mathrm{Id}_{F(X)}$, for all $X \in \mathscr{C}$.

(A2) Composition is preserved by $F$, i.e. $F(g \circ f) = F(g) \circ F(f)$, for any morphisms $f : X \to Y$ and $g : Y \to Z$ in $\mathscr{C}$.

If $F : \mathscr{C} \to \mathscr{D}$ and $G : \mathscr{D} \to \mathscr{E}$ are two functors then the pointwise composition defines a functor from $\mathscr{C}$ to $\mathscr{E}$. It will be denoted by $G \circ F$, or simply $GF$ when there is no danger of confusion.

If $\mathscr{C}$ is a category, there exists a functor $\mathrm{Id}_{\mathscr{C}} : \mathscr{C} \to \mathscr{C}$, called the identity functor on $\mathscr{C}$, which is the identity on both objects and morphisms in $\mathscr{C}$.

A functor $F : \mathscr{C} \to \mathscr{D}$ is called an isomorphism if there exists a functor $G : \mathscr{D} \to \mathscr{C}$ such that $FG = \mathrm{Id}_{\mathscr{D}}$ and $GF = \mathrm{Id}_{\mathscr{C}}$. Such a functor $G$, if it exists, is unique and is called the inverse of $F$. Two categories are isomorphic if there exists an isomorphism between them.

If $F : \mathscr{C} \to \mathscr{D}$ is a functor, we call the full image of $F$ (denoted $\mathrm{Im}(F)$) the full subcategory of $\mathscr{D}$ whose objects are $(F(X))_{X \in \mathrm{Ob}(\mathscr{C})}$.

A natural transformation $\mu$ between two functors $F, G : \mathscr{C} \to \mathscr{D}$ consists of a family of morphisms in $\mathscr{D}$, $\mu = (\mu_X : F(X) \to G(X))_{X \in \mathrm{Ob}(\mathscr{C})}$, having the property that $G(f) \circ \mu_X = \mu_Y \circ F(f)$, for any morphism $f : X \to Y$ in $\mathscr{C}$. If, moreover, $\mu_X$ is an isomorphism in $\mathscr{D}$, for all $X \in \mathrm{Ob}(\mathscr{C})$, then $\mu$ is called a natural isomorphism between $F$ and $G$.

Finally, if $\mathscr{C}, \mathscr{D}$ are categories then $\mathscr{C} \times \mathscr{D}$ is the category whose

- objects are pairs $(X,Y)$, where $X$ is an object of $\mathscr{C}$ and $Y$ is an object of $\mathscr{D}$;
- morphisms between $(X,Y)$ and $(X',Y')$ are pairs $(f,g)$ consisting of a morphism $f : X \to X'$ in $\mathscr{C}$ and a morphism $g : Y \to Y'$ in $\mathscr{D}$.

The identity morphisms and the composition functions in $\mathscr{C} \times \mathscr{D}$ are canonically defined in terms of those of $\mathscr{C}$ and $\mathscr{D}$. The new category $\mathscr{C} \times \mathscr{D}$ is called the product of $\mathscr{C}$ and $\mathscr{D}$.

We can now introduce the concept of monoidal category, which is roughly a category $\mathscr{C}$ endowed with an associative "tensor product" $\otimes : \mathscr{C} \times \mathscr{C} \to \mathscr{C}$, with a unit object $\underline{1}$ and coherence. Rigorously, we have the following:

**Definition 1.1** A monoidal category consists of a category $\mathscr{C}$ endowed with a functor $\otimes : \mathscr{C} \times \mathscr{C} \to \mathscr{C}$ (called the tensor product), a distinguished object $\underline{1} \in \mathscr{C}$ (called the unit object of $\mathscr{C}$) and natural isomorphisms ($X, Y, Z$ are arbitrary objects of $\mathscr{C}$)

$$\begin{aligned} &a_{X,Y,Z} : (X \otimes Y) \otimes Z \to X \otimes (Y \otimes Z) \ \text{(the associativity constraint)},\\ &l_X : \underline{1} \otimes X \to X \ \text{(the left unit constraint)},\\ &r_X : X \otimes \underline{1} \to X \ \text{(the right unit constraint)}, \end{aligned}$$

satisfying the so-called Pentagon Axiom and Triangle Axiom, namely for any objects $X, Y, Z, T \in \mathscr{C}$ the following diagrams are commutative:

$$\begin{array}{ccccc} ((X \otimes Y) \otimes Z) \otimes T & \xrightarrow{a_{X \otimes Y,Z,T}} & (X \otimes Y) \otimes (Z \otimes T) & \xrightarrow{a_{X,Y,Z \otimes T}} & X \otimes (Y \otimes (Z \otimes T)) \\ \downarrow{\scriptstyle a_{X,Y,Z} \otimes \mathrm{Id}_T} & & & & \uparrow{\scriptstyle \mathrm{Id}_X \otimes a_{Y,Z,T}} \\ (X \otimes (Y \otimes Z)) \otimes T & & \xrightarrow{\quad a_{X,Y \otimes Z,T} \quad} & & X \otimes ((Y \otimes Z) \otimes T), \end{array} \tag{1.1.1}$$

$$\begin{array}{ccc} (X \otimes \underline{1}) \otimes Y & \xrightarrow{a_{X,\underline{1},Y}} & X \otimes (\underline{1} \otimes Y) \\ & {\scriptstyle r_X \otimes \mathrm{Id}_Y} \searrow \quad \swarrow {\scriptstyle \mathrm{Id}_X \otimes l_Y} & \\ & X \otimes Y & \end{array} . \tag{1.1.2}$$

The monoidal category $(\mathscr{C}, \otimes, \underline{1}, a, l, r)$ is called strict if all the natural isomorphisms $a, l$ and $r$ are defined by identity morphisms in $\mathscr{C}$.

**Remark 1.2** Let $X \xrightarrow{f} Y \xrightarrow{g} Z$ and $X' \xrightarrow{f'} Y' \xrightarrow{g'} Z'$ be morphisms in $\mathscr{C}$. The fact that $\otimes : \mathscr{C} \times \mathscr{C} \to \mathscr{C}$ is a functor implies the following equality:

$$(g \circ f) \otimes (g' \circ f') = (g \otimes g') \circ (f \otimes f') : X \otimes X' \to Z \otimes Z'.$$

Also, for all objects $X, Y$ of $\mathscr{C}$ we have $\mathrm{Id}_{X \otimes Y} = \mathrm{Id}_X \otimes \mathrm{Id}_Y$.

If $\mathscr{C}$ is a monoidal category and $X, Y, Z, T$ are objects of $\mathscr{C}$, there are two different ways to go from $((X \otimes Y) \otimes Z) \otimes T$ to $X \otimes (Y \otimes (Z \otimes T))$. The Pentagon Axiom says that these two ways coincide. Then it is automatic that all the other consistency problems of this type are solved as well; see Remark 1.35 below.

**Proposition 1.3** *Let $(\mathscr{C}, \otimes, \underline{1}, a, l, r)$ be a monoidal category and consider the switch functor $\tau : \mathscr{C} \times \mathscr{C} \to \mathscr{C} \times \mathscr{C}$, defined by $\tau(X,Y) = (Y,X)$, for any $X, Y \in \mathscr{C}$, and $\tau(f,g) = (g,f)$, for any morphisms $X \xrightarrow{f} X'$ and $Y \xrightarrow{g} Y'$ in $\mathscr{C}$. Then*

$$\overline{\mathscr{C}} := (\mathscr{C},\ \overline{\otimes} := \otimes \circ \tau,\ \overline{a},\ \underline{1},\ \overline{l} := r,\ \overline{r} := l)$$

*is a monoidal category, where $\overline{a}_{X,Y,Z} := a^{-1}_{Z,Y,X}$, for all $X, Y, Z \in \mathscr{C}$.*

*In what follows $\overline{\mathscr{C}}$ will be called the reverse monoidal category associated to $\mathscr{C}$.*

*Proof* All the axioms for $\overline{\mathscr{C}}$ to be a monoidal category follow from those of $\mathscr{C}$ and the fact that $(g\circ f)^{-1}=f^{-1}\circ g^{-1}$, for any isomorphisms $X\xrightarrow{f}Y\xrightarrow{g}Z$ in $\mathscr{C}$. For example, the Pentagon Axiom for $\overline{\mathscr{C}}$ reduces to the commutativity of the diagram

$$\begin{array}{ccccc}
T\otimes(Z\otimes(Y\otimes X)) & \xrightarrow{a^{-1}_{T,Z,Y\otimes X}} & (T\otimes Z)\otimes(Y\otimes X) & \xrightarrow{a^{-1}_{T\otimes Z,Y,X}} & ((T\otimes Z)\otimes Y)\otimes X \\
{\scriptstyle \mathrm{Id}_T\otimes a^{-1}_{Z,Y,X}}\downarrow & & & & \uparrow{\scriptstyle a^{-1}_{T,Z,Y}\otimes \mathrm{Id}_X} \\
T\otimes((Z\otimes Y)\otimes X) & & \xrightarrow[a^{-1}_{T,Z\otimes Y,X}]{} & & (T\otimes(Z\otimes Y))\otimes X,
\end{array}$$

which holds because of (1.1.1). Similarly, for $\overline{a}$ as above, $\overline{l}=r$ and $\overline{r}=l$, the Triangle Axiom is satisfied because of (1.1.2). □

**Remark 1.4** Apart from $\overline{\mathscr{C}}$, to a monoidal category $\mathscr{C}$ we can associate a new one that will be denoted by $\mathscr{C}^{\mathrm{opp}}$ and called the opposite category associated to $\mathscr{C}$. As a category, $\mathscr{C}^{\mathrm{opp}}$ has the same objects as $\mathscr{C}$ and $\mathrm{Hom}_{\mathscr{C}^{\mathrm{opp}}}(X,Y)=\mathrm{Hom}_{\mathscr{C}}(Y,X)$, for any objects $X,Y$ of $\mathscr{C}$. If $f\in\mathrm{Hom}_{\mathscr{C}^{\mathrm{opp}}}(X,Y)$ and $g\in\mathrm{Hom}_{\mathscr{C}^{\mathrm{opp}}}(Y,Z)$ then the composition $\circ_{\mathrm{opp}}$ between $g$ and $f$ in $\mathscr{C}^{\mathrm{opp}}$ is $g\circ_{\mathrm{opp}}f=f\circ g$, the latest composition being in $\mathscr{C}$.

If $\mathscr{C}$ is monoidal then so is $\mathscr{C}^{\mathrm{opp}}$, with the monoidal structure induced by that of $\mathscr{C}$, namely $\mathscr{C}^{\mathrm{opp}}=(\mathscr{C}^{\mathrm{opp}},\otimes,\underline{1},a^{-1},l^{-1},r^{-1})$.

The Triangle Axiom in Definition 1.1 gives the compatibility between the left and right unit constraints. There also exist other compatibilities of this type:

**Proposition 1.5** *Let $(\mathscr{C},\otimes,\underline{1},a,l,r)$ be a monoidal category. Then the diagrams*

$$\begin{array}{ccc}
(X\otimes Y)\otimes\underline{1} & \xrightarrow{a_{X,Y,\underline{1}}} & X\otimes(Y\otimes\underline{1}) \\
 & {\scriptstyle r_{X\otimes Y}}\searrow \quad \swarrow{\scriptstyle \mathrm{Id}_X\otimes r_Y} & \\
 & X\otimes Y &
\end{array}$$

*and*

$$\begin{array}{ccc}
(\underline{1}\otimes X)\otimes Y & \xrightarrow{a_{\underline{1},X,Y}} & \underline{1}\otimes(X\otimes Y) \\
 & {\scriptstyle l_X\otimes \mathrm{Id}_Y}\searrow \quad \swarrow{\scriptstyle l_{X\otimes Y}} & \\
 & X\otimes Y &
\end{array}$$

*are commutative, for any objects $X,Y\in\mathscr{C}$. Moreover, we have that $l_{\underline{1}}=r_{\underline{1}}$.*

*Proof* Since $a$ is natural, the following diagrams are commutative:

$$\begin{array}{ccc}
(X\otimes(Y\otimes\underline{1}))\otimes T & \xrightarrow{a_{X,Y\otimes\underline{1},T}} & X\otimes((Y\otimes\underline{1})\otimes T) \\
{\scriptstyle (\mathrm{Id}_X\otimes r_Y)\otimes \mathrm{Id}_T}\downarrow & & \downarrow{\scriptstyle \mathrm{Id}_X\otimes(r_Y\otimes \mathrm{Id}_T)} \\
(X\otimes Y)\otimes T & \xrightarrow{a_{X,Y,T}} & X\otimes(Y\otimes T),
\end{array} \tag{1.1.3}$$

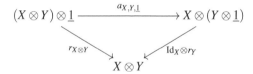

$$\begin{array}{ccc}
(X\otimes Y)\otimes(\underline{1}\otimes T) & \xrightarrow{a_{X,Y,\underline{1}\otimes T}} & X\otimes(Y\otimes(\underline{1}\otimes T)) \\
{\scriptstyle \mathrm{Id}_{X\otimes Y}\otimes l_T}\downarrow & & \downarrow{\scriptstyle \mathrm{Id}_X\otimes(\mathrm{Id}_Y\otimes l_T)} \\
(X\otimes Y)\otimes T & \xrightarrow{a_{X,Y,T}} & X\otimes(Y\otimes T),
\end{array} \tag{1.1.4}$$

for all $X,Y,T\in\mathscr{C}$. Then we have:

$$\begin{aligned}
& a_{X,Y,T}((\mathrm{Id}_X\otimes r_Y)\otimes\mathrm{Id}_T)(a_{X,Y,\underline{1}}\otimes\mathrm{Id}_T) \\
&\overset{(1.1.3)}{=} (\mathrm{Id}_X\otimes(r_Y\otimes\mathrm{Id}_T))a_{X,Y\otimes\underline{1},T}(a_{X,Y,\underline{1}}\otimes\mathrm{Id}_T) \\
&\overset{(1.1.2)}{=} (\mathrm{Id}_X\otimes(\mathrm{Id}_Y\otimes l_T))(\mathrm{Id}_X\otimes a_{Y,\underline{1},T})a_{X,Y\otimes\underline{1},T}(a_{X,Y,\underline{1}}\otimes\mathrm{Id}_T) \\
&\overset{(1.1.1)}{=} (\mathrm{Id}_X\otimes(\mathrm{Id}_Y\otimes l_T))a_{X,Y,\underline{1}\otimes T}a_{X\otimes Y,\underline{1},T} \\
&\overset{(1.1.4)}{=} a_{X,Y,T}(\mathrm{Id}_{X\otimes Y}\otimes l_T)a_{X\otimes Y,\underline{1},T} \\
&\overset{(1.1.2)}{=} a_{X,Y,T}(r_{X\otimes Y}\otimes\mathrm{Id}_T).
\end{aligned}$$

Using that $a_{X,Y,T}$ is an isomorphism we get, for all $X,Y,T\in\mathscr{C}$,

$$(\mathrm{Id}_X\otimes r_Y)a_{X,Y,\underline{1}}\otimes\mathrm{Id}_T = r_{X\otimes Y}\otimes\mathrm{Id}_T. \tag{1.1.5}$$

Now, by the naturality of $r$ the diagrams

$$\begin{array}{ccc}
((X\otimes Y)\otimes\underline{1})\otimes\underline{1} & \xrightarrow{r_{(X\otimes Y)\otimes\underline{1}}} & (X\otimes Y)\otimes\underline{1} \\
{\scriptstyle (\mathrm{Id}_X\otimes r_Y)a_{X,Y,\underline{1}}\otimes\mathrm{Id}_{\underline{1}}}\downarrow\downarrow{\scriptstyle r_{X\otimes Y}\otimes\mathrm{Id}_{\underline{1}}} & & {\scriptstyle (\mathrm{Id}_X\otimes r_Y)a_{X,Y,\underline{1}}}\downarrow\downarrow{\scriptstyle r_{X\otimes Y}} \\
(X\otimes Y)\otimes\underline{1} & \xrightarrow[r_{X\otimes Y}]{} & X\otimes Y
\end{array}$$

are commutative, so by (1.1.5) (with $T=\underline{1}$) we obtain that

$$(\mathrm{Id}_X\otimes r_Y)a_{X,Y,\underline{1}}r_{(X\otimes Y)\otimes\underline{1}} = r_{X\otimes Y}r_{(X\otimes Y)\otimes\underline{1}},$$

and therefore the first triangle in the proposition is commutative because $r_{(X\otimes Y)\otimes\underline{1}}$ is an isomorphism.

If we express the commutativity of the first triangle for $\overline{\mathscr{C}}$, the reverse monoidal category associated to $\mathscr{C}$ as in Proposition 1.3, we obtain the commutativity of the second triangle in the proposition.

So it remains to prove $l_{\underline{1}} = r_{\underline{1}}$. For this, note that the naturality of $r$ implies that

$$\begin{array}{ccc}
(X\otimes\underline{1})\otimes\underline{1} & \xrightarrow{r_{X\otimes\underline{1}}} & X\otimes\underline{1} \\
{\scriptstyle r_X\otimes\mathrm{Id}_{\underline{1}}}\downarrow & & \downarrow{\scriptstyle r_X} \\
X\otimes\underline{1} & \xrightarrow[r_X]{} & X
\end{array}$$

is commutative, for any $X\in\mathscr{C}$. Since $r_X$ is an isomorphism we deduce that

$$r_{X\otimes\underline{1}} = r_X\otimes\mathrm{Id}_{\underline{1}}. \tag{1.1.6}$$

Note that, by applying equation (1.1.6) in $\overline{\mathscr{C}}$, we obtain in $\mathscr{C}$ the relation

$$l_{\underline{1}\otimes X} = \mathrm{Id}_{\underline{1}} \otimes l_X. \tag{1.1.7}$$

Now, by (1.1.2) we have $r_{\underline{1}\otimes\underline{1}} = r_{\underline{1}} \otimes \mathrm{Id}_{\underline{1}} = (\mathrm{Id}_{\underline{1}} \otimes l_{\underline{1}})a_{\underline{1},\underline{1},\underline{1}}$, and by the commutativity of the first triangle in the proposition we get $r_{\underline{1}\otimes\underline{1}} = (\mathrm{Id}_{\underline{1}} \otimes r_{\underline{1}})a_{\underline{1},\underline{1},\underline{1}}$. Since $a_{\underline{1},\underline{1},\underline{1}}$ is an isomorphism we obtain $\mathrm{Id}_{\underline{1}} \otimes l_{\underline{1}} = \mathrm{Id}_{\underline{1}} \otimes r_{\underline{1}}$.

By the naturality of $l$ the diagrams

$$\begin{array}{ccc}
\underline{1}\otimes(\underline{1}\otimes\underline{1}) & \xrightarrow{\ l_{\underline{1}\otimes\underline{1}}\ } & \underline{1}\otimes\underline{1} \\
{\scriptstyle \mathrm{Id}_{\underline{1}}\otimes l_{\underline{1}}}\ \Big\downarrow\Big\downarrow\ {\scriptstyle \mathrm{Id}_{\underline{1}}\otimes r_{\underline{1}}} & & {\scriptstyle l_{\underline{1}}}\ \Big\downarrow\Big\downarrow\ {\scriptstyle r_{\underline{1}}} \\
\underline{1}\otimes\underline{1} & \xrightarrow[\ l_{\underline{1}}\ ]{} & \underline{1}
\end{array}$$

are commutative. Using that $l_{\underline{1}\otimes\underline{1}}$ is an isomorphism and $\mathrm{Id}_{\underline{1}} \otimes l_{\underline{1}} = \mathrm{Id}_{\underline{1}} \otimes r_{\underline{1}}$ we get $l_{\underline{1}} = r_{\underline{1}}$, and this finishes the proof. □

**Proposition 1.6** *Let $\underline{1}$ be the unit object of a monoidal category $\mathscr{C}$. Then* $\mathrm{End}_{\mathscr{C}}(\underline{1})$ *is a commutative monoid, and if we identify $\underline{1}\otimes\underline{1}$ with $\underline{1}$ via $l_{\underline{1}} = r_{\underline{1}}$ then the tensor product of two morphisms in* $\mathrm{End}_{\mathscr{C}}(\underline{1})$ *coincides with their composition.*

*Proof* It can be easily checked that the composition endows $\mathrm{End}_{\mathscr{C}}(\underline{1})$ with a monoid structure, the unit element being $\mathrm{Id}_{\underline{1}}$. Thus, we only need to show that

$$f\otimes g = r_{\underline{1}}^{-1}\circ(f\circ g)\circ r_{\underline{1}} = r_{\underline{1}}^{-1}\circ(g\circ f)\circ r_{\underline{1}},$$

for all $f,g \in \mathrm{End}_{\mathscr{C}}(\underline{1})$. To this end, note that the naturality of $l$ and $r$ imply the commutativity of the following diagrams:

$$\begin{array}{ccc}
\underline{1}\otimes\underline{1} & \xrightarrow{\ l_{\underline{1}}\ } & \underline{1} \\
{\scriptstyle \mathrm{Id}_{\underline{1}}\otimes g}\Big\downarrow & & \Big\downarrow{\scriptstyle g} \\
\underline{1}\otimes\underline{1} & \xrightarrow{\ l_{\underline{1}}\ } & \underline{1}
\end{array}
\qquad \text{and} \qquad
\begin{array}{ccc}
\underline{1}\otimes\underline{1} & \xrightarrow{\ r_{\underline{1}}\ } & \underline{1} \\
{\scriptstyle f\otimes \mathrm{Id}_{\underline{1}}}\Big\downarrow & & \Big\downarrow{\scriptstyle f} \\
\underline{1}\otimes\underline{1} & \xrightarrow{\ r_{\underline{1}}\ } & \underline{1}\,.
\end{array}$$

Now, since $l_{\underline{1}}$ and $r_{\underline{1}}$ are isomorphisms we obtain $\mathrm{Id}_{\underline{1}}\otimes g = l_{\underline{1}}^{-1}\circ g\circ l_{\underline{1}}$ and $f\otimes \mathrm{Id}_{\underline{1}} = r_{\underline{1}}^{-1}\circ f\circ r_{\underline{1}}$. Since $l_{\underline{1}} = r_{\underline{1}}$ it follows that

$$f\otimes g = (f\otimes \mathrm{Id}_{\underline{1}})\circ(\mathrm{Id}_{\underline{1}}\otimes g) = r_{\underline{1}}^{-1}\circ(f\circ g)\circ r_{\underline{1}},$$
$$f\otimes g = (\mathrm{Id}_{\underline{1}}\otimes g)\circ(f\otimes \mathrm{Id}_{\underline{1}}) = r_{\underline{1}}^{-1}\circ(g\circ f)\circ r_{\underline{1}}.$$

Thus, we proved the equalities $f\otimes g = r_{\underline{1}}^{-1}\circ(f\circ g)\circ r_{\underline{1}} = r_{\underline{1}}^{-1}\circ(g\circ f)\circ r_{\underline{1}}$. Note that by interchanging $f$ and $g$ in the above relation we also obtain $f\otimes g = g\otimes f$, for all $f,g \in \mathrm{End}_{\mathscr{C}}(\underline{1})$. □

## 1.2 Examples of Monoidal Categories

### 1.2.1 The Category of Sets

We denote the category of sets by Set, and by $\{*\}$ a fixed singleton, that is, a fixed set with one element. Furthermore, by $\times$ we denote the direct product of sets, that is, for any sets $X$ and $Y$, $X \times Y$ is the set of ordered pairs $(x,y)$ with $x \in X$ and $y \in Y$, and by $f \times f'$ the direct product of two functions $X \xrightarrow{f} Y$ and $X' \xrightarrow{f'} Y'$, that is, $f \times f' : X \times X' \to Y \times Y'$ is defined by $f \times f'(x,x') = (f(x), f'(x'))$, for all $x \in X$ and $x' \in X'$. If follows that $\times$ defines a functor from Set $\times$ Set to Set.

For any sets $X, Y$ and $Z$, we have canonical isomorphisms, defined for all $x \in X$, $y \in Y$ and $z \in Z$, by

$$a_{X,Y,Z} : (X \times Y) \times Z \to X \times (Y \times Z), \quad a_{X,Y,Z}((x,y),z) = (x,(y,z)),$$
$$l_X : \{*\} \times X \to X, \quad l_X(*,x) = x,$$
$$r_X : X \times \{*\} \to X, \quad r_X(x,*) = x.$$

The proof of the next result is straightforward, so it is left to the reader.

**Proposition 1.7** *With notation as above,* $(\underline{\mathrm{Set}}, \times, \{*\}, a, l, r)$ *is a monoidal category.*

### 1.2.2 The Category of Vector Spaces

One of the most important examples of a monoidal category for what follows is the category ${}_k\mathcal{M}$ of vector spaces over a base field $k$. The tensor product in ${}_k\mathcal{M}$ is the usual tensor product of vector spaces, the unit object $\underline{1}$ is the field $k$ itself, and the associativity and unit constraints are the natural isomorphisms (for all $X, Y, Z \in {}_k\mathcal{M}$)

$$a_{X,Y,Z} : (X \otimes Y) \otimes Z \to X \otimes (Y \otimes Z), \quad a_{X,Y,Z}((x \otimes y) \otimes z) = x \otimes (y \otimes z),$$
$$l_X : k \otimes X \to X, \quad l_X(\lambda \otimes x) = \lambda x,$$
$$r_X : X \otimes k \to X, \quad r_X(x \otimes \lambda) = \lambda x,$$

for all $\lambda \in k$, $x \in X$, $y \in Y$ and $z \in Z$. The above statement remains valid if we consider $k$ a commutative ring and take ${}_k\mathcal{M}$ equal to the category of modules over $k$.

### 1.2.3 The Category of Bimodules

We present now the noncommutative version of Subsection 1.2.2.

Let $k$ be a field (or, more generally, a commutative ring) and $R$ a $k$-algebra. Denote by ${}_R\mathcal{M}_R$ the category of $R$-bimodules and $R$-bimodule maps. Then ${}_R\mathcal{M}_R$ is monoidal with the following structure:

- The tensor product functor is $\otimes_R : {}_R\mathcal{M}_R \times {}_R\mathcal{M}_R \to {}_R\mathcal{M}_R$ defined as follows. On objects, we have $\otimes_R(M,N) := M \otimes_R N$, the tensor product over $R$ between $M$ and $N$. It becomes an $R$-bimodule in the canonical way: $r \cdot (m \otimes_R n) \cdot r' = rm \otimes_R nr'$, for all $r, r' \in R$, $m \in M$ and $n \in N$. If $f : M \to N$, $g : P \to Q$ are morphisms in ${}_R\mathscr{M}_R$ then the map $(f \otimes_R g)(m \otimes_R p) = f(m) \otimes_R g(p)$, for all $m \in M$ and $p \in P$, is a morphism in ${}_R\mathscr{M}_R$.

- The unit is $R$, considered as an $R$-bimodule via its multiplication.
- The associativity and unit constraints are defined as follows:

$$a_{X,Y,Z} : (X \otimes_R Y) \otimes_R Z \to X \otimes_R (Y \otimes_R Z),\ \ a_{X,Y,Z}((x \otimes_R y) \otimes_R z) = x \otimes_R (y \otimes_R z),$$
$$l_X : R \otimes_R X \to X,\ \ l_X(r \otimes_R x) = rx,$$
$$r_X : X \otimes_R R \to X,\ \ r_X(x \otimes_R r) = xr,$$

for all $r \in R$, $x \in X$, $y \in Y$ and $z \in Z$.

We leave it to the reader to check that this defines a monoidal structure on ${}_R\mathscr{M}_R$. Note that if $R = k$ then ${}_R\mathscr{M}_R$ coincides with ${}_k\mathscr{M}$ as a monoidal category.

### 1.2.4 The Category of $G$-graded Vector Spaces

Throughout this subsection $G$ is a group written multiplicatively and with neutral element $e$, $k$ is a field and $k^* = k \backslash \{0\}$.

**Definition 1.8** A $G$-graded vector space over $k$ is a $k$-vector space $V$ which decomposes into a direct sum of the form $V = \bigoplus_{g \in G} V_g$, where each $V_g$ is a $k$-vector space. For a given $g \in G$ the elements of $V_g$ are called homogeneous elements of degree $g$. If $v \in V$ is a homogeneous element then we denote the degree of $v$ by $\mid v \mid \in G$.

Let $W = \bigoplus_{g \in G} W_g$ be another $G$-graded vector space. Then a $k$-linear map $f : V \to W$ is called a $G$-graded morphism if it preserves the degree of homogeneous elements, that is, $f(V_g) \subseteq W_g$, for all $g \in G$.

$\mathrm{Vect}^G$ denotes the category of $G$-graded vector spaces and $G$-graded morphisms.

If $V = \bigoplus_{g \in G} V_g$ and $W = \bigoplus_{g \in G} W_g$ are $G$-graded vector spaces then $V \otimes W$ is also a $G$-graded vector space with the grading defined by

$$(V \otimes W)_g := \bigoplus_{\sigma\tau = g} V_\sigma \otimes W_\tau, \tag{1.2.1}$$

for all $g \in G$. Indeed, it is an elementary fact that in ${}_k\mathscr{M}$ the tensor product commutes with arbitrary direct sums. Hence

$$\bigoplus_{g \in G} (V \otimes W)_g = \bigoplus_{g \in G} \left( \bigoplus_{\sigma\tau = g} V_\sigma \otimes W_\tau \right) = \left( \bigoplus_{g \in G} V_g \right) \otimes \left( \bigoplus_{g' \in G} W_{g'} \right) = V \otimes W,$$

as required. Furthermore, if $f : V \to V'$ and $g : W \to W'$ are morphisms in $\mathrm{Vect}^G$ then $f \otimes g$ becomes a morphism in $\mathrm{Vect}^G$. Thus, the tensor product $\otimes$ of the category of $k$-vector spaces induces a tensor product on $\mathrm{Vect}^G$.

Also, $k$ can be viewed as a $G$-graded vector space via the trivial grading, that is, $k_e = k$ and $k_g = 0$, for all $G \ni g \neq e$. In this way the left and right unit constraints $l$ and $r$ of ${}_k\mathscr{M}$ become graded morphisms, that is, morphisms in $\mathrm{Vect}^G$.

Our next aim is to describe the monoidal structures of $\mathrm{Vect}^G$, somehow induced by the monoidal structure of ${}_k\mathcal{M}$. To this end we first need some group cohomology, with a particular emphasis on the third cohomology group of a group $G$ with coefficients in $k^*$, the group of units of a field $k$, viewed trivially as a $\mathbb{Z}[G]$-module. Here $\mathbb{Z}$ is the ring of integers and $\mathbb{Z}[G]$ is the group algebra associated to $G$ over the commutative ring $\mathbb{Z}$. More generally, for $G$ a (multiplicative) group with neutral element $e$ and $R$ a commutative ring we denote by $R[G]$ the free $R$-module with basis $\{g \mid g \in G\}$, so any element of $R[G]$ has the form $\sum_{g\in G} \alpha_g g$ with $(\alpha_g)_{g\in G}$ a family of elements of $R$ having only a finite number of non-zero elements. Then $R[G]$ with multiplication defined by $(\alpha_h h)(\beta_g g) = \alpha_h \beta_g hg$, extended by linearity, and unit $e$, is called the group algebra associated to $G$ over $R$. It is easy to see that $R[G]$ is a unital associative $R$-algebra, and that $R[G]$ is a $G$-graded vector space with grading defined by $R[G]_g = Rg$, for all $g \in G$.

Coming back to the survey on group cohomology, let $K^n(G,k^*)$ be the set of maps from $G^n$ to $k^*$. Then one can easily see that $K^n(G,k^*)$ is a group under pointwise multiplication. There exist maps $\Delta_n : K^n(G,k^*) \to K^{n+1}(G,k^*)$, which for $n \in \{2,3\}$ are given by the formulas

$$\Delta_2(g)(x,y,z) = g(y,z)g(xy,z)^{-1}g(x,yz)g(x,y)^{-1},$$
$$\Delta_3(f)(x,y,z,t) = f(y,z,t)f(xy,z,t)^{-1}f(x,yz,t)f(x,y,zt)^{-1}f(x,y,z).$$

It is known that $B^n(G,k^*) := \mathrm{Im}\Delta_{n-1} \subseteq Z^n(G,k^*) := \mathrm{Ker}(\Delta_n)$. The $n$th cohomology group is defined as $H^n(G,k^*) = Z^n(G,k^*)/B^n(G,k^*)$, and two elements of $H^n(G,k^*)$ are called cohomologous if they lie in the same equivalence class.

The elements of $Z^3(G,k^*)$ are called 3-cocycles, and the elements of $B^3(G,k^*)$ are called 3-coboundaries. We have the following.

**Definition 1.9** A 3-cocycle on $G$ with coefficients in $k^*$ is a map $\phi : G \times G \times G \to k^*$ such that

$$\phi(y,z,t)\phi(x,yz,t)\phi(x,y,z) = \phi(x,y,zt)\phi(xy,z,t), \tag{1.2.2}$$

for all $x,y,z,t \in G$. A 3-cocycle $\phi$ is called normalized if $\phi(x,e,y) = 1$, for all $x,y \in G$.

**Remarks 1.10** (1) If $\phi$ is a normalized 3-cocycle, then $\phi(e,y,z) = \phi(x,y,e) = 1$, for all $x,y,z \in G$.

Indeed, by taking $z = e$ in (1.2.2), we find that $\phi(x,y,e) = 1$. By taking $y = e$, we find that $\phi(e,z,t) = 1$.

(2) A coboundary $\Delta_2(g)$ is normalized if and only if $g(e,x) = g(z,e)$, for all $x,z \in G$.

As we shall see, $H^3(G,k^*)$ is completely determined by the normalized 3-cocycles.

**Lemma 1.11** *Every 3-cocycle $\phi$ is cohomologous to a normalized 3-cocycle.*

*Proof* By taking $y = z = e$ in (1.2.2) we find $\phi(x,e,t) = \phi(e,e,t)\phi(x,e,e)$. In particular, by taking $x = t = e$, it follows that $\phi(e,e,e) = 1$. Then we consider the map

$f\colon G\times G\to k^*$, $f(x,y)=\phi(e,e,y)^{-1}\phi(x,e,e)$, and compute:

$$\begin{aligned}\Delta_2(f)(x,e,y)&=f(e,y)f(x,y)^{-1}f(x,y)f(x,e)^{-1}\\&=\phi(e,e,y)^{-1}\phi(e,e,e)\phi(e,e,e)\phi(x,e,e)^{-1}=\phi(x,e,y)^{-1}.\end{aligned}$$

It then follows that $\phi\Delta_2(f)$ is normalized. □

Let $B^3_n(G,k^*)$ and $Z^3_n(G,k^*)$ be the subgroups of $B^3(G,k^*)$ and $Z^3(G,k^*)$ consisting of normalized elements. We have a well-defined group morphism

$$Z^3_n(G,k^*)/B^3_n(G,k^*)\ni\hat{\phi}\mapsto\overline{\phi}\in Z^3(G,k^*)/B^3(G,k^*)$$

which is surjective by Lemma 1.11. One can see that it is also injective, and therefore

$$H^3(G,k^*)=Z^3_n(G,k^*)/B^3_n(G,k^*).$$

**Example 1.12** If $k$ is a field of characteristic different from 2 and $C_2$ is the cyclic group of order 2 then $H^3(C_2,k^*)=C_2$. If $\mathrm{char}(k)=2$, then $H^3(C_2,k^*)=\{e\}$.

*Proof* Write $C_2=\{1,\sigma\}$. A straightforward computation shows that all normalized coboundaries are trivial. If $\phi$ is a normalized 3-cocycle, then the only value of $\phi(x,y,z)$ that is possibly different from 1 is $\phi(\sigma,\sigma,\sigma)$. By substituting $x=y=z=t=\sigma$ in (1.2.2), we find that $\phi(\sigma,\sigma,\sigma)=\pm1$. If $\phi(\sigma,\sigma,\sigma)=1$, then $\phi$ is trivial. The only possibly non-trivial normalized 3-cocycle is given by $\phi(\sigma,\sigma,\sigma)=-1$.

Consequently, if $\mathrm{char}(k)=2$ then any normalized 3-cocycle is trivial, and so $H^3(C_2,k^*)=\{e\}$. □

One can now provide the connection between $H^3(G,k^*)$ and some monoidal structures on $\mathrm{Vect}^G$.

**Proposition 1.13** *Let $G$ be a group, $k$ a field and $\mathrm{Vect}^G$ the category of $G$-graded $k$-vector spaces. There is a bijective correspondence between the monoidal structures on $\mathrm{Vect}^G$ of the form $(\mathrm{Vect}^G,\otimes,a,k,l,r)$ and the set of normalized 3-cocycles on $G$, where $\otimes$ is defined by (1.2.1) and $l,r$ are the constraints of the monoidal category ${}_k\mathcal{M}$ as defined in Subsection 1.2.2.*

*More precisely, any associativity constraint $a$ on $\mathrm{Vect}^G$ is completely determined by a normalized 3-cocycle $\phi\in H^3(G,k^*)$, in the sense that, for any $U,V,W\in\mathrm{Vect}^G$ and any homogeneous elements $u\in U$, $v\in V$ and $w\in W$, $a_{U,V,W}$ is the $k$-linear map*

$$a_{U,V,W}((u\otimes v)\otimes w)=\phi(|u|,|v|,|w|)u\otimes(v\otimes w).$$

*We denote by $\mathrm{Vect}^G_\phi$ the category $\mathrm{Vect}^G$ with monoidal structure determined by $\phi$.*

*Proof* If $\phi$ is a normalized 3-cocycle on $G$ then, clearly, the morphism $a_{U,V,W}$ defined above preserves the degree of homogeneous elements, so it is a morphism in $\mathrm{Vect}^G$. The Pentagon Axiom (1.1.1) follows now from (1.2.2), while the Triangle Axiom in (1.1.2) follows because $\phi$ is normalized. The details are straightforward, so they are left to the reader.

Conversely, suppose that $\mathrm{Vect}^G$ has a monoidal structure of the form mentioned in the statement. For any $U \in \mathrm{Vect}^G$ and any $k$-linear map $f : U \to k$ define $\theta_f : U \to k[G]$ by $\theta_f(u) = \sum_{x\in G} f(u_x)x$, for all $u \in U$, where $u = \sum_{x\in G} u_x$ is the decomposition of $u$ in homogeneous components. Obviously, $\theta_f$ is a graded morphism. Likewise we define $\theta_g$ and $\theta_h$, for any $V, W \in \mathrm{Vect}^G$ and all $g : V \to k$ and $h : W \to k$.

Now take $\varepsilon : k[G] \to k$ defined by $\varepsilon(g) = 1$, for all $g \in G$, extended by linearity. A simple computation shows that $(\varepsilon \otimes \varepsilon)(\theta_f \otimes \theta_g) = f \otimes g$.

We are now able to show that an associativity constraint $a$ of $\mathrm{Vect}^G$ is completely determined by $a_{k[G],k[G],k[G]}$. More precisely, for $a$, an associativity constraint on $\mathrm{Vect}^G$, define $\phi(x,y,z) := (\varepsilon \otimes (\varepsilon \otimes \varepsilon))a_{k[G],k[G],k[G]}((x \otimes y) \otimes z)$, for all $x,y,z \in G$. By the naturality of $a$, for any $U,V,W$ and $f,g,h$ as above, we have

$$(\theta_f \otimes (\theta_g \otimes \theta_h))a_{U,V,W}((u \otimes v) \otimes w) = a_{k[G],k[G],k[G]}((\theta_f \otimes \theta_g) \otimes \theta_h)((u \otimes v) \otimes w),$$

where $u \in U$, $v \in V$ and $w \in W$ are arbitrary elements.

Assume that $u,v,w$ are homogeneous of degrees $x,y$ and $z$, respectively, and write

$$a_{U,V,W}((u \otimes v) \otimes w) = \sum_i u_i \otimes (v_i \otimes w_i),$$

for some homogeneous elements $u_i \in U$, $v_i \in V$ and $w_i \in W$. We obtain that

$$\sum_i f(u_i)g(v_i)h(w_i) \mid u_i \mid \otimes (\mid v_i \mid \otimes \mid w_i \mid) = f(u)g(v)h(w)a_{k[G],k[G],k[G]}((x \otimes y) \otimes z).$$

By applying $\varepsilon \otimes (\varepsilon \otimes \varepsilon)$ on both sides of the above equality we obtain

$$\sum_i f(u_i)g(v_i)h(w_i) = \phi(x,y,z)f(u)g(v)h(w).$$

Since $f,g,h$ are arbitrary we get $a_{U,V,W}((u \otimes v) \otimes w) = \phi(\mid u \mid, \mid v \mid, \mid w \mid)u \otimes (v \otimes w)$, as stated. Note that the bijectivity of the associativity constraint implies that $\phi(x,y,z) \neq 0$, for all $x,y,z \in G$. It is clear now that $a$ satisfies the Pentagon and Triangle Axioms if and only if $\phi$ is a normalized 3-cocycle on $G$, and so we are done. □

**Remark 1.14** If $\phi$ is the trivial 3-cocycle on $G$ then the monoidal structure on $\mathrm{Vect}^G$ is entirely induced by the monoidal structure of ${}_k\mathcal{M}$ described in Subsection 1.2.2. In this case $\mathrm{Vect}^G$ is strict monoidal and the grading is relevant only in the definition of the tensor product of $\mathrm{Vect}^G$.

A non-strict monoidal structure on $\mathrm{Vect}^G$ when $G$ is cyclic of order 2 can be obtained by considering Example 1.12. Note that in this case we get the so-called category of super vector spaces.

**Example 1.15** (Super vector spaces) Let $k$ be a field, $\mathbb{Z}_2 = \{\overline{0}, \overline{1}\}$ the cyclic group of order 2, this time written additively, and consider the category $\mathrm{Vect}^{\mathbb{Z}_2}$ of $\mathbb{Z}_2$-graded $k$-vector spaces. It can be identified with the category whose objects are pairs $V = (V_{\overline{0}}, V_{\overline{1}})$ of $k$-vector spaces. A morphism from $(V_{\overline{0}}, V_{\overline{1}})$ to $(V'_{\overline{0}}, V'_{\overline{1}})$ in $\mathrm{Vect}^{\mathbb{Z}_2}$ is a pair $(f_{\overline{0}}, f_{\overline{1}})$ of $k$-linear maps with $f_{\overline{i}} : V_{\overline{i}} \to V'_{\overline{i}}$, $i \in \{0,1\}$.

Then $\mathrm{Vect}^{\mathbb{Z}_2}$ is monoidal with tensor product defined by

$$(V_{\overline{0}},V_{\overline{1}})\otimes(W_{\overline{0}},W_{\overline{1}})=\left((V_{\overline{0}}\otimes W_{\overline{0}})\oplus(V_{\overline{1}}\otimes W_{\overline{1}}),(V_{\overline{0}}\otimes W_{\overline{1}})\oplus(V_{\overline{1}}\otimes W_{\overline{0}})\right),$$
$$(f_{\overline{0}},f_{\overline{1}})\otimes(g_{\overline{0}},g_{\overline{1}})=\left((f_{\overline{0}}\otimes g_{\overline{0}})\oplus(f_{\overline{1}}\otimes g_{\overline{1}}),(f_{\overline{0}}\otimes g_{\overline{1}})\oplus(f_{\overline{1}}\otimes g_{\overline{0}})\right).$$

The associativity constraint is $(a_{V,W,Z})_{V,W,Z\in\mathrm{Vect}^{\mathbb{Z}_2}}$, defined on homogeneous elements $v\in V$, $w\in W$ and $z\in Z$ by

$$a_{V,W,Z}((v\otimes w)\otimes z)=\begin{cases}-v\otimes(w\otimes z) & \text{for } |v|,|w|,|z| \text{ all odd},\\ v\otimes(w\otimes z) & \text{otherwise.}\end{cases}$$

The unit object is $(k,0)$ and the left and right unit constrains of $\mathrm{Vect}^{\mathbb{Z}_2}$ are defined by the left and right unit constraints of ${}_k\mathcal{M}$.

As we have already mentioned, $\mathrm{Vect}^{\mathbb{Z}_2}$ with the monoidal structure described above is called the category of super vector spaces.

*Proof* In Proposition 1.13 take $G=\mathbb{Z}_2$ and $\phi$ the (possibly) non-trivial 3-cocycle constructed in Example 1.12, that is, $\phi$ is 1 everywhere except on $(\overline{1},\overline{1},\overline{1})$ where its value is $-1$. □

More generally, examples of monoidal structures on $\mathrm{Vect}^{\mathbb{Z}_n}$ are given by the following family of normalized 3-cocycles on $\mathbb{Z}_n$.

**Example 1.16** Let $\mathbb{Z}_n=\{\overline{0},\overline{1},\ldots,\overline{n-1}\}$ be the cyclic group of order $n\geq 2$ written additively and $q$ be an $n$th root of unity in $k$. Then

$$\phi_q(\overline{x},\overline{y},\overline{z})=\begin{cases}1 & \text{if } y+z<n\\ q^x & \text{if } y+z\geq n\end{cases},\quad \forall\, x,y,z\in\{0,1,\ldots,n-1\},$$

defines a normalized 3-cocycle on $\mathbb{Z}_n$. Moreover, $\phi_q$ is a coboundary if and only if $q=1$. Consequently, $\phi_q$ and $\phi_{q'}$ are cohomologous if and only if $q=q'$, whenever $q'$ is another $n$th root of unity in $k$.

*Proof* If $y=0$ then clearly $y+z<n$, and so $\phi_q(x,0,z)=1$. Clearly we have $\phi_q(\overline{0},\overline{y},\overline{z})=\phi_q(\overline{x},\overline{y},\overline{0})=1$, for all $x,y,z\in\{0,1,\ldots,n-1\}$. Thus we only have to show that, for all $x,y,z,t\in\{1,2,\ldots,n-1\}$,

$$\phi_q(\overline{y},\overline{z},\overline{t})\phi_q(\overline{x},\overline{y+z},\overline{t})\phi_q(\overline{x},\overline{y},\overline{z})=\phi_q(\overline{x},\overline{y},\overline{z+t})\phi_q(\overline{x+y},\overline{z},\overline{t}).\qquad(1.2.3)$$

To see this, consider the following situations:

1. If $y+z+t<n$ then $z+t<n$ and $y+z<n$, and so (1.2.3) reduces to $1=1$.

2. If $y+z+t\geq n$ then there are the following possibilities:

(i) If $z+t<n$ then (1.2.3) reduces to $\phi_q(\overline{x},\overline{y+z},\overline{t})\phi_q(\overline{x},\overline{y},\overline{z})=q^x$. It is clearly satisfied when $y+z<n$. In the case when $y+z\geq n$ we have $y+z=n+u$, for some $u\in\{0,\ldots,n-2\}$, and we will prove that $\phi_q(\overline{x},\overline{u},\overline{t})=1$. Indeed, $u+t=y+z+t-n<y<n$ and this implies the desired equality.

(ii) If $z+t\geq n$ we write $z+t=n+v$, for some $v\in\{0,\ldots,n-2\}$. Then (1.2.3) reduces to $\phi_q(\overline{x},\overline{y+z},\overline{t})\phi_q(\overline{x},\overline{y},\overline{z})=q^x\phi_q(\overline{x},\overline{y},\overline{v})$. To prove this, consider the following cases:

(ii.1) If $y+z<n$ then we have to prove that $\phi_q(\overline{x},\overline{y},\overline{v})=1$. This follows from $y+v=y+z+t-n<t<n$ and from the definition of $\phi_q$.

(ii.2) If $y+z\geq n$, there exists $w\in\{0,\ldots,n-2\}$ such that $y+z=w+n$, and we will prove that $\phi_q(\overline{x},\overline{w},\overline{t})=\phi_q(\overline{x},\overline{y},\overline{v})$. To this end observe that $w+t<n\Leftrightarrow y+z+t<2n$ and, similarly, $y+v<n\Leftrightarrow y+z+t<2n$, which implies that $w+t<n\Leftrightarrow y+v<n$.

Thus, we have proved that $\phi_q$ is a normalized 3-cocycle on $\mathbb{Z}_n$. Obviously, it is a coboundary if $q=1$. Conversely, if $\phi_q$ is a coboundary then there exists $g\in K^2(\mathbb{Z}_n,k^*)$ such that $g(\overline{0},\overline{x})=g(\overline{z},\overline{0})$ and

$$g(\overline{y},\overline{z})g(\overline{x+y},\overline{z})^{-1}g(\overline{x},\overline{y+z})g(\overline{x},\overline{y})^{-1}=\begin{cases}1 & \text{if } y+z<n\\ q^x & \text{if } y+z\geq n,\end{cases}$$

for all $x,y,z\in\{0,1,\ldots,n-1\}$. By taking in the above relation $x=1$, $y=n-1$ and $z=1$ we get that $q=g(\overline{n-1},\overline{1})g(\overline{1},\overline{n-1})^{-1}$. If we take $x=1$, $y=n-k$ and $z=1$, where $k\in\{2,\ldots,n-1\}$, we obtain $g(\overline{n-k+1},\overline{1})g(\overline{1},\overline{n-k+1})^{-1}=g(\overline{n-k},\overline{1})g(\overline{1},\overline{n-k})^{-1}$. We conclude that

$$q=g(\overline{n-1},\overline{1})g(\overline{1},\overline{n-1})^{-1}=g(\overline{n-2},\overline{1})g(\overline{1},\overline{n-2})^{-1}=\cdots=g(\overline{1},\overline{1})g(\overline{1},\overline{1})^{-1}=1,$$

as needed. Finally, since $\phi_q\phi_{q'}=\phi_{qq'}$, for any $q$ and $q'$ $n$th roots of unity in $k$, it follows that $\phi_q$ and $\phi_{q'}$ are cohomologous if and only if $\phi_{qq'^{-1}}$ is a coboundary, if and only if $q=q'$. This finishes the proof. □

### 1.2.5 The Category of Endo-functors

Let $\mathscr{C},\mathscr{D}$ be arbitrary categories and $[\mathscr{C},\mathscr{D}]$ the category whose objects are functors $F:\mathscr{C}\to\mathscr{D}$. A morphism in $[\mathscr{C},\mathscr{D}]$ between two functors $F,G:\mathscr{C}\to\mathscr{D}$ is a natural transformation $\mu:F\to G$. The composition $\nu\circ\mu$ of two natural transformations $F\overset{\mu}{\to}G\overset{\nu}{\to}H$ is the "vertical" composition of $\nu$ and $\mu$, that is, $(\nu\circ\mu)_X:=\nu_X\circ\mu_X$, for any object $X$ of $\mathscr{C}$. One can easily see that this composition is indeed a natural transformation, and that it is associative. Moreover, for any natural transformation $\mu:F\to G$ we have $1_G\circ\mu=\mu=\mu\circ 1_F$, where, in general, for a functor $T:\mathscr{C}\to\mathscr{D}$ we denote by $1_T:T\to T$ the identity natural transformation of $T$, $(1_T)_X=\mathrm{Id}_{T(X)}$, for any object $X$ of $\mathscr{C}$. Thus $[\mathscr{C},\mathscr{D}]$ is a category.

Consider now $\mathscr{C}=\mathscr{D}$. The composition of functors defines a tensor product $\otimes$ on $[\mathscr{C},\mathscr{C}]$ (which is called the category of endo-functors on $\mathscr{C}$), that is, $F\otimes G:=F\circ G$, and for $\mu:F\to G$ and $\mu':F'\to G'$ natural transformations, $\mu\otimes\mu':F\otimes F'=F\circ F'\to G\otimes G'=G\circ G'$ is the natural transformation defined, for any $X\in\mathscr{C}$, by the diagram

$$\begin{array}{ccc}
FF'(X) & \xrightarrow{F(\mu'_X)} & FG'(X)\\
{\scriptstyle \mu_{F'(X)}}\downarrow & \overset{(\mu\otimes\mu')_X}{\searrow} & \downarrow{\scriptstyle \mu_{G'(X)}}\\
GF'(X) & \xrightarrow[G(\mu'_X)]{} & GG'(X)\,.
\end{array}$$

Note that the above diagram is commutative because of the naturality of $\mu$. The natural transformation $\mu\otimes\mu'$ is called the "horizontal" composition or the Godement product of the natural transformations $\mu$ and $\mu'$. It can be defined for any given functors and natural transformations

$$\mathscr{B} \underset{G'}{\overset{F'}{\rightrightarrows}}\ \downarrow\mu'\ \mathscr{C} \underset{G}{\overset{F}{\rightrightarrows}}\ \downarrow\mu\ \mathscr{D}.$$

To show that $\mu\otimes\mu'$ is indeed a natural transformation consider the diagram

$$\begin{array}{ccccc}
FF'(X) & \xrightarrow{F(\mu'_X)} & FG'(X) & \xrightarrow{\mu_{G'(X)}} & GG'(X) \\
{\scriptstyle FF'(f)}\downarrow & & {\scriptstyle FG'(f)}\downarrow & & {\scriptstyle GG'(f)}\downarrow \\
FF'(Y) & \xrightarrow[F(\mu'_Y)]{} & FG'(Y) & \xrightarrow[\mu_{G'(Y)}]{} & GG'(Y)\ ,
\end{array}$$

where $X \xrightarrow{f} Y$ is an arbitrary morphism in $\mathscr{C}$. The left-handed square is commutative because $\mu'$ is natural and $F$ is a functor, the right-handed square also commutes because $\mu$ is natural and $G'(f) : G'(X) \to G'(Y)$ is a morphism in $\mathscr{C}$, and horizontally the compositions are $(\mu\otimes\mu')_X$ and $(\mu\otimes\mu')_Y$, respectively. These facts show that $\mu\otimes\mu'$ is natural, as claimed. Furthermore, the horizontal composition can be rewritten, using vertical composition, as

$$\mu\otimes\mu' = (1_G\otimes\mu')\circ(\mu\otimes 1_{F'}) = (\mu\otimes 1_{G'})\circ(1_F\otimes\mu').$$

To see this, note that for any object $T$ of $[\mathscr{C},\mathscr{C}]$ we have $(1_T\otimes\mu)_X = T(\mu_X)$ and $(\mu\otimes 1_T)_X = \mu_{T(X)}$, for all $X\in\mathscr{C}$.

The identity for the composition $\otimes$ is $1_{\mathscr{C}} : \mathrm{Id}_{\mathscr{C}} \to \mathrm{Id}_{\mathscr{C}}$, the identity natural transformation of the identity functor $\mathrm{Id}_{\mathscr{C}}$ to itself. This means that $\mu\otimes 1_{\mathscr{C}} = \mu = 1_{\mathscr{C}}\otimes\mu$, for any morphism $\mu : F\to G$ in $[\mathscr{C},\mathscr{C}]$.

**Proposition 1.17** $([\mathscr{C},\mathscr{C}],\otimes,\mathrm{Id}_{\mathscr{C}})$ *is a strict monoidal category.*

*Proof* Clearly, $(F\otimes G)\otimes H = F\otimes(G\otimes H)$ and $F\otimes\mathrm{Id}_{\mathscr{C}} = F = \mathrm{Id}_{\mathscr{C}}\otimes F$, for any functors $F,G,H\in[\mathscr{C},\mathscr{C}]$. Also, it can be easily checked that the composition $\otimes$ is associative. Thus, we only need to prove that for any given functors and natural transformations

$$\mathscr{C}\ \begin{array}{c} \xrightarrow{G} \\ \downarrow\mu' \\ \xrightarrow{G'} \\ \downarrow\nu' \\ \xrightarrow[G'']{} \end{array}\ \mathscr{C}\ \begin{array}{c} \xrightarrow{F} \\ \downarrow\mu \\ \xrightarrow{F'} \\ \downarrow\nu \\ \xrightarrow[F'']{} \end{array}\ \mathscr{C},$$

we have $(\nu\circ\mu)\otimes(\nu'\circ\mu') = (\nu\otimes\nu')\circ(\mu\otimes\mu')$. Indeed, we compute:

$$\big((\nu\circ\mu)\otimes(\nu'\circ\mu')\big)_X = F''((\nu'\circ\mu')_X)(\nu\circ\mu)_{G(X)}$$

$$
\begin{aligned}
&= F''(\nu'_X)F''(\mu'_X)\nu_{G(X)}\mu_{G(X)} \\
&= F''(\nu'_X)\nu_{G'(X)}F'(\mu'_X)\mu_{G(X)} \\
&= (\nu \otimes \nu')_X(\mu \otimes \mu')_X \\
&= ((\nu \otimes \nu') \circ (\mu \otimes \mu'))_X,
\end{aligned}
$$

for any object $X$ of $\mathscr{C}$, and this finishes the proof. □

### 1.2.6 A Strict Category Associated to a Monoidal Category

Let $(\mathscr{C}, \otimes, \underline{1}, a, l, r)$ be a monoidal category. The aim of this subsection is to associate to $\mathscr{C}$ a strict monoidal category, denoted by $\mathfrak{e}(\mathscr{C})$. The objects of $\mathfrak{e}(\mathscr{C})$ are pairs $(F, \rho^F)$, where $F : \mathscr{C} \to \mathscr{C}$ is a functor and $\rho^F$ is a family of natural isomorphisms $\rho^F_{X,Y} : F(X) \otimes Y \to F(X \otimes Y)$, indexed by $(X, Y) \in \mathscr{C} \times \mathscr{C}$. A morphism $(F, \rho^F) \xrightarrow{\mu} (G, \rho^G)$ in $\mathfrak{e}(\mathscr{C})$ consists of a natural transformation $\mu : F \to G$ such that, for any objects $X, Y$ of $\mathscr{C}$, the diagram below is commutative:

$$
\begin{array}{ccc}
F(X) \otimes Y & \xrightarrow{\rho^F_{X,Y}} & F(X \otimes Y) \\
\downarrow{\scriptstyle \mu_X \otimes \mathrm{Id}_Y} & & \downarrow{\scriptstyle \mu_{X \otimes Y}} \\
G(X) \otimes Y & \xrightarrow[\rho^G_{X,Y}]{} & G(X \otimes Y).
\end{array}
\tag{1.2.4}
$$

If $(F, \rho^F) \xrightarrow{\mu} (G, \rho^G) \xrightarrow{\mu'} (H, \rho^H)$ are morphisms in $\mathfrak{e}(\mathscr{C})$ then $\mu' \circ \mu : (F, \rho^F) \to (H, \rho^H)$ is clearly a morphism in $\mathfrak{e}(\mathscr{C})$. Moreover, the identity morphism associated to the object $(F, \rho^F)$ is $1_F : F \to F$, the identity natural transformation associated to $F$. One can easily see that $1_F$ is a morphism in $\mathfrak{e}(\mathscr{C})$.

The category $\mathfrak{e}(\mathscr{C})$ is strict monoidal. The tensor product is given by $(F, \rho^F) \otimes (G, \rho^G) = (FG, \rho^{FG})$, where, for all $X, Y \in \mathscr{C}$,

$$
\rho^{FG}_{X,Y} : FG(X) \otimes Y \xrightarrow{\rho^F_{G(X),Y}} F(G(X) \otimes Y) \xrightarrow{F(\rho^G_{X,Y})} FG(X \otimes Y),
$$

and the unit object is $(\mathrm{Id}_{\mathscr{C}}, \rho^{\mathrm{Id}_{\mathscr{C}}} = (\mathrm{Id}_{X \otimes Y})_{X,Y \in \mathscr{C}})$. To see this, note that

$$
\begin{aligned}
\rho^{(FG)H}_{X,Y} &= FG(\rho^H_{X,Y})\rho^{FG}_{H(X),Y} = FG(\rho^H_{X,Y})F(\rho^G_{H(X),Y})\rho^F_{GH(X),Y} \\
&= F(G(\rho^H_{X,Y})\rho^G_{H(X),Y})\rho^F_{GH(X),Y} = F(\rho^{GH}_{X,Y})\rho^F_{GH(X),Y} = \rho^{F(GH)}_{X,Y},
\end{aligned}
$$

for any $(F, \rho^F)$, $(G, \rho^G)$, $(H, \rho^H) \in \mathfrak{e}(\mathscr{C})$ and $X, Y \in \mathscr{C}$.

If $(F, \rho^F) \xrightarrow{\mu} (G, \rho^G)$, $(F', \rho^{F'}) \xrightarrow{\mu'} (G', \rho^{G'})$ are morphisms in $\mathfrak{e}(\mathscr{C})$ then $\mu \otimes \mu'$, the horizontal composition defined in Subsection 1.2.5, is a morphism in $\mathfrak{e}(\mathscr{C})$ from

$(FF', \rho^{FF'})$ to $(GG', \rho^{GG'})$. To see this, we use the commutativity of the diagrams

$$\begin{array}{ccccc}
F(F'(X)\otimes Y) & \xrightarrow{F(\mu'_X \otimes \mathrm{Id}_Y)} & F(G'(X)\otimes Y) & \xrightarrow{F(\rho^{G'}_{X,Y})} & FG'(X\otimes Y) \\
\Big\downarrow {\scriptstyle \mu_{F'(X)\otimes Y}} & & \Big\downarrow {\scriptstyle \mu_{G'(X)\otimes Y}} & & \Big\downarrow {\scriptstyle \mu_{G'(X\otimes Y)}} \\
G(F'(X)\otimes Y) & \xrightarrow[G(\mu'_X \otimes \mathrm{Id}_Y)]{} & G(G'(X)\otimes Y) & \xrightarrow[G(\rho^{G'}_{X,Y})]{} & GG'(X\otimes Y),
\end{array}$$

$$\begin{array}{ccc}
G(F'(X))\otimes Y & \xrightarrow{G(\mu'_X)\otimes \mathrm{Id}_Y} & G(G'(X))\otimes Y \\
\Big\downarrow {\scriptstyle \rho^G_{F'(X),Y}} & & \Big\downarrow {\scriptstyle \rho^G_{G'(X),Y}} \\
G(F'(X)\otimes Y) & \xrightarrow[G(\mu'_X \otimes \mathrm{Id}_Y)]{} & G(G'(X)\otimes Y),
\end{array}$$

which follow by using that $\mu$ and $\rho^G_{-,Y}$ are natural transformations. This fact allows to compute:

$$\begin{aligned}
(\mu\otimes\mu')_{X\otimes Y}\rho^{FF'}_{X,Y} &= (\mu\otimes\mu')_{X\otimes Y} F(\rho^{F'}_{X,Y})\rho^F_{F'(X),Y} \\
&= \mu_{G'(X\otimes Y)} F(\mu'_{X\otimes Y}\rho^{F'}_{X,Y})\rho^F_{F'(X),Y} \\
&= \mu_{G'(X\otimes Y)} F(\rho^{G'}_{X,Y}) F(\mu'_X \otimes \mathrm{Id}_Y)\rho^F_{F'(X),Y} \\
&= G(\rho^{G'}_{X,Y})\mu_{G'(X)\otimes Y} F(\mu'_X \otimes \mathrm{Id}_Y)\rho^F_{F'(X),Y} \\
&= G(\rho^{G'}_{X,Y}(\mu'_X \otimes \mathrm{Id}_Y))\mu_{F'(X)\otimes Y}\rho^F_{F'(X),Y} \\
&= G(\rho^{G'}_{X,Y}) G(\mu'_X \otimes \mathrm{Id}_Y)\rho^G_{F'(X),Y}(\mu_{F'(X)}\otimes \mathrm{Id}_Y) \\
&= G(\rho^{G'}_{X,Y})\rho^G_{G'(X),Y}(G(\mu'_X)\otimes \mathrm{Id}_Y)(\mu_{F'(X)}\otimes \mathrm{Id}_Y) \\
&= G(\rho^{G'}_{X,Y})\rho^G_{G'(X),Y}((\mu\otimes\mu')_X \otimes \mathrm{Id}_Y) \\
&= \rho^{GG'}_{X,Y}((\mu\otimes\mu')_X \otimes \mathrm{Id}_Y),
\end{aligned}$$

as needed.

## 1.3 Monoidal Functors

The aim of this section is to introduce and study functors between monoidal categories that behave well with respect to their monoidal structures. This will allow us to extend, for instance, the notion of equivalent categories to the monoidal setting.

**Definition 1.18** A functor $F : \mathscr{C} \to \mathscr{D}$ is called an equivalence of categories if there exists a functor $G : \mathscr{D} \to \mathscr{C}$ such that $FG$ is naturally isomorphic to $\mathrm{Id}_{\mathscr{D}}$ and $GF$ is naturally isomorphic to $\mathrm{Id}_{\mathscr{C}}$.

We say that $\mathscr{C}$ and $\mathscr{D}$ are equivalent categories if there exists a functor $F : \mathscr{C} \to \mathscr{D}$ which is an equivalence of categories.

We have a criterion for a functor $F : \mathscr{C} \to \mathscr{D}$ to be an equivalence of categories.

**Proposition 1.19** *$F : \mathscr{C} \to \mathscr{D}$ is an equivalence of categories if and only if*

- *$F$ is essentially surjective, that is, for any object $U$ of $\mathscr{D}$ there exists an object $X$ of $\mathscr{C}$ such that $U \cong F(X)$;*
- *$F$ is fully faithful, that is, for any $X, Y \in \mathscr{C}$ the following map is bijective:*

$$\mathrm{Hom}_{\mathscr{C}}(X,Y) \ni f \mapsto F(f) \in \mathrm{Hom}_{\mathscr{D}}(F(X),F(Y)).$$

*Proof* Suppose that $F : \mathscr{C} \to \mathscr{D}$ defines an equivalence of categories and let $G : \mathscr{D} \to \mathscr{C}$ be a functor such that $FG$ is naturally isomorphic to $\mathrm{Id}_{\mathscr{D}}$, say via $\mu : \mathrm{Id}_{\mathscr{D}} \to FG$, and $GF$ is naturally isomorphic to $\mathrm{Id}_{\mathscr{C}}$, say via $\nu : GF \to \mathrm{Id}_{\mathscr{C}}$. Then, for any object $U$ of $\mathscr{D}$, $\mu_U : U \to FG(U)$ defines an isomorphism in $\mathscr{D}$, and so $F$ is essentially surjective.

If $f, f' \in \mathrm{Hom}_{\mathscr{C}}(X,Y)$ such that $F(f) = F(f')$ then the naturality of $\nu$ implies

$$f\nu_X = \nu_Y GF(f) = \nu_Y GF(f') = f'\nu_X,$$

and therefore $f = f'$. Similarly, by using the naturality of $\mu$ we get that the map

$$\mathrm{Hom}_{\mathscr{D}}(U,V) \ni g \to G(g) \in \mathrm{Hom}_{\mathscr{C}}(G(U),G(V))$$

is injective, for any objects $U, V$ of $\mathscr{D}$.

Now let $g \in \mathrm{Hom}_{\mathscr{D}}(F(X),F(Y))$. By the naturality of $\nu$, for $f = \nu_Y G(g)\nu_X^{-1} : X \to Y$ it follows that

$$\nu_Y GF(f) = f\nu_X = \nu_Y G(g).$$

Thus $GF(f) = G(g)$, and so $F(f) = g$. This shows that $F$ is fully faithful, as needed.

Conversely, assume that $F$ is essentially surjective and fully faithful. Then, for any object $U$ of $\mathscr{D}$ we fix an object $G(U)$ of $\mathscr{C}$ and an isomorphism $\mu_U : U \to FG(U)$ in $\mathscr{D}$. Moreover, if $U \xrightarrow{g} V$ is a morphism in $\mathscr{D}$ then $\mu_V g \mu_U^{-1} \in \mathrm{Hom}_{\mathscr{D}}(FG(U),FG(V))$, and since $F$ is fully faithful there is a unique morphism $G(g) : G(U) \to G(V)$ in $\mathscr{C}$ such that $FG(g) = \mu_V g \mu_U^{-1}$. Hence, the correspondences $U \mapsto G(U)$ and $g \mapsto G(g)$ define a functor from $\mathscr{D}$ to $\mathscr{C}$ that turns the family of isomorphisms $\mu = (\mu_U : U \to FG(U))_{U \in \mathscr{D}}$ into a natural isomorphism from $\mathrm{Id}_{\mathscr{D}}$ to $FG$.

It remains to prove that $GF$ is naturally isomorphic to $\mathrm{Id}_{\mathscr{C}}$. For any $X \in \mathscr{C}$ consider $\mu_{F(X)}^{-1} : FGF(X) \to F(X)$. The functor $F$ is fully faithful, so there exists a unique morphism $\nu_X : GF(X) \to X$ in $\mathscr{C}$ such that $F(\nu_X) = \mu_{F(X)}^{-1}$. Actually, since $\mu_{F(X)}$ is an isomorphism in $\mathscr{D}$ it follows that $\nu_X$ is an isomorphism in $\mathscr{C}$. In addition, the naturality of $\mu$ implies that, for any $X \xrightarrow{f} Y$ in $\mathscr{C}$,

$$\begin{aligned} F(f\nu_X) &= F(f)F(\nu_X) = F(f)\mu_{F(X)}^{-1} = \mu_{F(Y)}^{-1} FGF(f) \\ &= F(\nu_Y)FGF(f) = F(\nu_Y GF(f)). \end{aligned}$$

From here we conclude that $f\nu_X = \nu_Y GF(f)$. Thus $\nu = (\nu_X : GF(X) \to X)_{X \in \mathscr{C}}$ is a natural isomorphism from $GF$ to $\mathrm{Id}_{\mathscr{C}}$, and this finishes the proof. □

**Remark 1.20** Let $F : \mathscr{C} \to \mathscr{D}$ be an equivalence of categories. If $G : \mathscr{D} \to \mathscr{C}$, $\mu : \mathrm{Id}_{\mathscr{D}} \to FG$ and $\nu : GF \to \mathrm{Id}_{\mathscr{C}}$ are as in the second part of the proof of Proposition 1.19 then, for all $X \in \mathscr{C}$ and $Y \in \mathscr{D}$, we have

$$F(\nu_X) = \mu^{-1}_{F(X)} \text{ and } G(\mu_Y) = \nu^{-1}_{G(Y)}. \tag{1.3.1}$$

Indeed, the first equality is just the definition of $\nu$. For the second one, we use the definition of $G$ to see that $FG(\mu_Y) = \mu_{FG(Y)} = F(\nu^{-1}_{G(Y)})$, and since $F$ is fully faithful it follows that $G(\mu_Y) = \nu^{-1}_{G(Y)}$, as desired.

**Proposition 1.21** *Let $\mathscr{C}$ be a monoidal category and $\mathfrak{e}(\mathscr{C})$ the strict monoidal category associated to $\mathscr{C}$; see Subsection 1.2.6. Then the functor $F : \mathscr{C} \to \mathfrak{e}(\mathscr{C})$,*

$$F(X) = (X \otimes -, a_{X,-,-}) \quad \text{and} \quad F(f) = f \otimes -$$

*is fully faithful. Consequently, $\mathscr{C}$ is equivalent to the full image of $F$.*

*Proof* Let $X, Y$ be objects of $\mathscr{C}$ and $X \overset{g}{\underset{f}{\rightrightarrows}} Y$ morphisms in $\mathscr{C}$ such that $F(f) = F(g)$. Then $f \otimes \mathrm{Id}_Z = g \otimes \mathrm{Id}_Z$, for any $Z \in \mathscr{C}$. Together with the naturality of $r$ this implies $f r_X = r_Y(f \otimes \mathrm{Id}_{\underline{1}}) = r_Y(g \otimes \mathrm{Id}_{\underline{1}}) = g r_X$. Since $r_X$ is an isomorphism we get $f = g$, and so the map $f \mapsto F(f)$ is injective. It is also surjective, that is, any natural transformation $\theta : F(X) \to F(Y)$ satisfying (1.2.4) has the form $\theta = (\theta_Z = f_\theta \otimes \mathrm{Id}_Z)_{Z \in \mathscr{C}}$, for some morphism $f_\theta : X \to Y$ in $\mathscr{C}$. Actually, $f_\theta : X \to Y$ is the morphism in $\mathscr{C}$ defined by the composition

$$f_\theta : X \xrightarrow{r_X^{-1}} X \otimes \underline{1} \xrightarrow{\theta_{\underline{1}}} Y \otimes \underline{1} \xrightarrow{r_Y} Y.$$

To see this observe that the three (small) rectangular diagrams below are commutative, for all $X, Y, Z \in \mathscr{C}$:

$$\begin{array}{ccccccc}
X \otimes Z & \xrightarrow{r_X^{-1} \otimes \mathrm{Id}_Z} & (X \otimes \underline{1}) \otimes Z & \xrightarrow{a_{X,\underline{1},Z}} & X \otimes (\underline{1} \otimes Z) & \xrightarrow{\mathrm{Id}_X \otimes l_Z} & X \otimes Z \\
\downarrow{\scriptstyle f_\theta \otimes \mathrm{Id}_Z} & & \downarrow{\scriptstyle \theta_{\underline{1}} \otimes \mathrm{Id}_Z} & & \downarrow{\scriptstyle \theta_{\underline{1} \otimes Z}} & & \downarrow{\scriptstyle \theta_Z} \\
Y \otimes Z & \xrightarrow[r_Y^{-1} \otimes \mathrm{Id}_Z]{} & (Y \otimes \underline{1}) \otimes Z & \xrightarrow[a_{Y,\underline{1},Z}]{} & Y \otimes (\underline{1} \otimes Z) & \xrightarrow[\mathrm{Id}_Y \otimes l_Z]{} & Y \otimes Z.
\end{array}$$

Indeed, the first one is commutative because of the definition of $f_\theta$, the second one is commutative because of (1.2.4), and the last one is commutative because of the naturality of $\theta$. Therefore, the exterior rectangular diagram is commutative.

According to (1.1.2) we have $a_{X,\underline{1},Z} = (\mathrm{Id}_X \otimes l_Z^{-1})(r_X \otimes \mathrm{Id}_Z)$, and consequently we obtain $(\mathrm{Id}_X \otimes l_Z) a_{X,\underline{1},Z}(r_X^{-1} \otimes \mathrm{Id}_Z) = \mathrm{Id}_{X \otimes Z}$. Similarly, $(\mathrm{Id}_Y \otimes l_Z) a_{Y,\underline{1},Z}(r_Y^{-1} \otimes \mathrm{Id}_Z) = \mathrm{Id}_{Y \otimes Z}$. Hence $\theta_Z = f_\theta \otimes \mathrm{Id}_Z = F(f_\theta)_Z$, for all $Z \in \mathscr{C}$, and this finishes the proof. □

An (op)monoidal functor between two monoidal categories is a functor that respects the two monoidal structures. More precisely:

**Definition 1.22** Let $(\mathscr{C},\otimes,\underline{1},a,l,r)$ and $(\mathscr{D},\Box,\underline{I},a',l',r')$ be monoidal categories and $F:\mathscr{C}\to\mathscr{D}$ a functor.

(i) $F$ is called monoidal if there exists a family of morphisms in $\mathscr{D}$

$$\varphi_2=(\varphi_{2,X,Y}:F(X)\Box F(Y)\to F(X\otimes Y))_{X,Y\in\mathscr{C}},$$

natural in $X$ and $Y$, and $\varphi_0:\underline{I}\to F(\underline{1})$ a morphism in $\mathscr{D}$ such that, for all $X,Y,Z\in\mathscr{C}$, the corresponding diagrams in (1.3.2) are commutative.

If $\varphi_0$ and $\varphi_2$ are defined by identity morphisms in $\mathscr{D}$ then we call the monoidal functor $(F,\varphi_0,\varphi_2)$ strict monoidal.

(ii) $F$ is called opmonoidal if there exists a family of morphisms in $\mathscr{D}$

$$\psi_2=(\psi_{2,X,Y}:F(X\otimes Y)\to F(X)\Box F(Y))_{X,Y\in\mathscr{C}},$$

natural in $X$ and $Y$, and $\psi_0:F(\underline{1})\to\underline{I}$ a morphism in $\mathscr{D}$ such that, for all $X,Y,Z\in\mathscr{C}$, the corresponding diagrams in (1.3.2) are commutative.

If $\psi_0$ and $\psi_2$ are defined by identity morphisms in $\mathscr{D}$ then the opmonoidal functor $(F,\psi_0,\psi_2)$ is actually strict monoidal.

(iii) $F$ is called a strong monoidal functor if it is monoidal and, moreover, $\varphi_0$ and $\varphi_2$ are defined by isomorphisms in $\mathscr{D}$. Equivalently, $F$ is strong monoidal if it is opmonoidal and, moreover, $\psi_0$ and $\psi_2$ are defined by isomorphisms in $\mathscr{D}$.

$$\begin{array}{ccc}
(F(X)\Box F(Y))\Box F(Z) & \xrightarrow{\quad a'_{F(X),F(Y),F(Z)}\quad} & F(X)\Box(F(Y)\Box F(Z)) \\
{\scriptstyle \varphi_{2,X,Y}\Box \mathrm{Id}_{F(Z)}}\downarrow\;\uparrow{\scriptstyle \psi_{2,X,Y}\Box \mathrm{Id}_{F(Z)}} & & {\scriptstyle \mathrm{Id}_{F(X)}\Box\psi_{2,Y,Z}}\uparrow\;\downarrow{\scriptstyle \mathrm{Id}_{F(X)}\Box\varphi_{2,Y,Z}} \\
F(X\otimes Y)\Box F(Z) & & F(X)\Box F(Y\otimes Z) \\
{\scriptstyle \varphi_{2,X\otimes Y,Z}}\downarrow\;\uparrow{\scriptstyle \psi_{2,X\otimes Y,Z}} & & {\scriptstyle \psi_{2,X,Y\otimes Z}}\uparrow\;\downarrow{\scriptstyle \varphi_{2,X,Y\otimes Z}} \\
F((X\otimes Y)\otimes Z) & \xrightarrow[\quad F(a_{X,Y,Z})\quad]{} & F(X\otimes(Y\otimes Z)),
\end{array}$$

$$\begin{array}{ccc}
\underline{I}\Box F(X) & \underset{\psi_0\Box\mathrm{Id}_{F(X)}}{\overset{\varphi_0\Box\mathrm{Id}_{F(X)}}{\rightleftarrows}} & F(\underline{1})\Box F(X) \\
{\scriptstyle l'_{F(X)}}\downarrow & & {\scriptstyle \psi_{2,\underline{1},X}}\uparrow\;\downarrow{\scriptstyle \varphi_{2,\underline{1},X}} \\
F(X) & \xleftarrow[\quad F(l_X)\quad]{} & F(\underline{1}\otimes X),
\end{array}
\qquad
\begin{array}{ccc}
F(X)\Box\underline{I} & \underset{\mathrm{Id}_{F(X)}\Box\psi_0}{\overset{\mathrm{Id}_{F(X)}\Box\varphi_0}{\rightleftarrows}} & F(X)\Box F(\underline{1}) \\
{\scriptstyle r'_{F(X)}}\downarrow & & {\scriptstyle \psi_{2,X,\underline{1}}}\uparrow\;\downarrow{\scriptstyle \varphi_{2,X,\underline{1}}} \\
F(X) & \xleftarrow[\quad F(r_X)\quad]{} & F(X\otimes\underline{1}).
\end{array}
\tag{1.3.2}$$

**Example 1.23** If $\mathscr{C}$ is a monoidal category then the identity functor $\mathrm{Id}_{\mathscr{C}}$ is a strict monoidal functor.

**Remark 1.24** Many times in what follows, unless otherwise specified, if $F:\mathscr{C}\to\mathscr{D}$ is an (op)monoidal functor, we will denote by the same symbols $\otimes$, $\underline{1}$, $a$, $l$, $r$ the tensor product, unit, etc. both in $\mathscr{C}$ and $\mathscr{D}$.

**Remark 1.25** If $\mathscr{C} \xrightarrow{(F,\varphi_0,\varphi_2)} \mathscr{D} \xrightarrow{(G,\psi_0,\psi_2)} \mathscr{E}$ are monoidal functors then so is $GF$ with

$$\xi_0 : \underline{1} \xrightarrow{\psi_0} G(\underline{I}) \xrightarrow{G(\varphi_0)} GF(\underline{1}) \text{ and}$$
$$\xi_{2,X,Y} : GF(X) \otimes GF(Y) \xrightarrow{\psi_{2,F(X),F(Y)}} G(F(X)\Box F(Y)) \xrightarrow{G(\varphi_{2,X,Y})} GF(X \otimes Y).$$

If $F, G$ are opmonoidal or strong monoidal functors, the same is true for $GF$.

**Definition 1.26** Let $\mathscr{C}$, $\mathscr{D}$ be monoidal categories. They are called monoidally isomorphic if there exist monoidal functors $(F,\varphi_0,\varphi_2) : \mathscr{C} \to \mathscr{D}$ and $(G,\psi_0,\psi_2) : \mathscr{D} \to \mathscr{C}$ such that $FG = \mathrm{Id}_{\mathscr{D}}$ and $GF = \mathrm{Id}_{\mathscr{C}}$ as monoidal functors (this implies that $F$ and $G$ are automatically strong monoidal).

**Proposition 1.27** *Let $\mathscr{C}$, $\mathscr{D}$ be monoidal categories. Then they are monoidally isomorphic if and only if there exists a strong monoidal functor $F : \mathscr{C} \to \mathscr{D}$ which is also an isomorphism of categories.*

*Proof* We only have to prove the converse. Let $G : \mathscr{D} \to \mathscr{C}$ be the inverse of the monoidal functor $(F,\varphi_0,\varphi_2)$; define $\psi_0 := G(\varphi_0^{-1})$ and $\psi_{2,U,V} := G(\varphi^{-1}_{2,G(U),G(V)})$, for all $U, V \in \mathscr{D}$. We leave it to the reader to check that $(G,\psi_0,\psi_2)$ is indeed a monoidal functor, and that $FG = \mathrm{Id}_{\mathscr{D}}$ and $GF = \mathrm{Id}_{\mathscr{C}}$ as monoidal functors. □

An example of a strong monoidal functor is the following.

**Proposition 1.28** *The functor $F$ defined in Proposition 1.21 is strong monoidal.*

*Proof* The commutativity of the diagram

$$\begin{array}{ccc} X \otimes Y & \xrightarrow{\mathrm{Id}_{X\otimes Y}} & X \otimes Y \\ {\scriptstyle l_X^{-1}\otimes \mathrm{Id}_Y}\downarrow & & \downarrow{\scriptstyle l^{-1}_{X\otimes Y}} \\ (\underline{1}\otimes X)\otimes Y & \xrightarrow[a_{\underline{1},X,Y}]{} & \underline{1}\otimes (X \otimes Y) \end{array}$$

is equivalent to the fact that the second diagram in the statement of Proposition 1.5 is commutative. So $\varphi_0 := l^{-1} : (\mathrm{Id}_{\mathscr{C}}, \rho^{\mathrm{Id}_{\mathscr{C}}} = (\mathrm{Id}_{X\otimes Y})_{X,Y\in\mathscr{C}}) \to F(\underline{1}) = (\underline{1}\otimes -, a_{\underline{1},-,-})$ is an isomorphism in $\mathfrak{e}(\mathscr{C})$.

For any $X, Y \in \mathscr{C}$ we have

$$F(X) \otimes F(Y) = (X \otimes (Y \otimes -), (\mathrm{Id}_X \otimes a_{Y,-,-})a_{X,Y\otimes -,-}).$$

Since for any $Z, T \in \mathscr{C}$ the commutativity of the diagram

$$\begin{array}{ccc} (X\otimes (Y\otimes Z))\otimes T & \xrightarrow{(\mathrm{Id}_X \otimes a_{Y,Z,T})a_{X,Y\otimes Z,T}} & X \otimes (Y \otimes (Z \otimes T)) \\ {\scriptstyle a^{-1}_{X,Y,Z}\otimes \mathrm{Id}_T}\downarrow & & \downarrow{\scriptstyle a^{-1}_{X,Y,Z\otimes T}} \\ ((X\otimes Y)\otimes Z)\otimes T & \xrightarrow{a_{X\otimes Y,Z,T}} & (X\otimes Y)\otimes (Z\otimes T) \end{array}$$

is equivalent to the commutativity of (1.1.1), it follows that

$$\varphi_{2,X,Y} := a^{-1}_{X,Y,-} : F(X)\otimes F(Y) \to F(X\otimes Y) = ((X\otimes Y)\otimes -, a_{X\otimes Y,-,-})$$

is an isomorphism in $\mathfrak{e}(\mathscr{C})$. We now prove that $(F, \varphi_0, (\varphi_{2,X,Y})_{X,Y\in\mathscr{C}})$ is monoidal.

The commutativity of the first diagram in Definition 1.22 comes out as

$$\varphi_{2,X,Y\otimes Z}(\mathrm{id}_{X\otimes -}\otimes \varphi_{2,Y,Z}) = (a_{X,Y,Z}\otimes -)\varphi_{2,X\otimes Y,Z}(\varphi_{2,X,Y}\otimes \mathrm{id}_{Z\otimes -}),$$

an equality between two natural transformations from $X\otimes (Y\otimes (Z\otimes -))$ to $(X\otimes (Y\otimes Z))\otimes -$. This equality holds since

$$\begin{aligned}(\varphi_{2,X,Y\otimes Z})_T(\mathrm{id}_{X\otimes -}\otimes \varphi_{2,Y,Z})_T &= a^{-1}_{X,Y\otimes Z,T}(\mathrm{Id}_X\otimes a^{-1}_{Y,Z,T})\\ &\overset{(1.1.1)}{=} (a_{X,Y,Z}\otimes \mathrm{Id}_T)a^{-1}_{X\otimes Y,Z,T}a^{-1}_{X,Y,Z\otimes T}\\ &= (a_{X,Y,Z}\otimes -)_T(\varphi_{2,X\otimes Y,Z})_T(\varphi_{2,X,Y}\otimes \mathrm{id}_{Z\otimes -})_T,\end{aligned}$$

for all $T\in\mathscr{C}$. Similarly, by using again the commutativity of the second triangle in Proposition 1.5 and the fact that $\mathfrak{e}(\mathscr{C})$ is strict monoidal we compute

$$F(l_X)_Y(\varphi_{2,\underline{1},X})_Y(\varphi_0\otimes \mathrm{id}_{X\otimes -})_Y = (l_X\otimes \mathrm{Id}_Y)a^{-1}_{\underline{1},X,Y}l^{-1}_{X\otimes Y} = \mathrm{Id}_{X\otimes Y} = (l_{X\otimes -})_Y,$$

for all $Y\in\mathscr{C}$. By (1.1.2) we have

$$F(r_X)_Y(\varphi_{2,X,\underline{1}})_Y(\mathrm{id}_{X\otimes -}\otimes \varphi_0)_Y = (r_X\otimes \mathrm{Id}_Y)a^{-1}_{X,\underline{1},Y}(\mathrm{Id}_X\otimes l_Y^{-1}) = \mathrm{Id}_{X\otimes Y} = (r_{X\otimes -})_Y,$$

for all $Y\in\mathscr{C}$, and therefore the other two (square) diagrams in Definition 1.22 are commutative as well. Thus the proof is finished. □

The notion of natural transformation extends to the monoidal setting as follows.

**Definition 1.29** Let $\mathscr{C}$, $\mathscr{D}$ be monoidal categories and $(F, \varphi_0^F, \varphi_2^F), (G, \varphi_0^G, \varphi_2^G) : \mathscr{C}\to\mathscr{D}$ monoidal functors. A monoidal natural transformation $\omega$ from $(F, \varphi_0^F, \varphi_2^F)$ to $(G, \varphi_0^G, \varphi_2^G)$ is a natural transformation $\omega : F\to G$ such that, for any objects $X, Y$ of $\mathscr{C}$, the following diagrams are commutative:

$$\begin{array}{ccc} F(X)\otimes F(Y) & \xrightarrow{\varphi^F_{2,X,Y}} & F(X\otimes Y)\\ {\scriptstyle \omega_X\otimes\omega_Y}\downarrow & & \downarrow{\scriptstyle \omega_{X\otimes Y}}\\ G(X)\otimes G(Y) & \xrightarrow[\varphi^G_{2,X,Y}]{} & G(X\otimes Y)\end{array} \qquad \text{and} \qquad \begin{array}{ccc} & \underline{1} & \\ {\scriptstyle \varphi_0^F}\swarrow & & \searrow{\scriptstyle \varphi_0^G}\\ F(\underline{1}) & \xrightarrow[\omega_{\underline{1}}]{} & G(\underline{1}).\end{array}$$

The transformation $\omega$ is called a monoidal natural isomorphism if $\omega$ is both a monoidal natural transformation and a natural isomorphism.

Reversing the appropriate arrows in the above diagrams leads to the definition of an opmonoidal natural transformation between two opmonoidal functors.

We are now able to define the concept of monoidal equivalence.

**Definition 1.30** Let $\mathscr{C}$, $\mathscr{D}$ be monoidal categories and $F : \mathscr{C} \to \mathscr{D}$ a monoidal (resp. opmonoidal) functor. We call $F$ a monoidal (resp. opmonoidal) equivalence if there exists a monoidal (resp. opmonoidal) functor $G : \mathscr{D} \to \mathscr{C}$ such that $FG$ is monoidally (resp. opmonoidally) naturally isomorphic to $\mathrm{Id}_{\mathscr{D}}$ and $GF$ is monoidally (resp. opmonoidally) naturally isomorphic to $\mathrm{Id}_{\mathscr{C}}$.

If $F$ and $G$ as above are both strong monoidal then $F$ is called a strong monoidal equivalence between $\mathscr{C}$ and $\mathscr{D}$.

If a functor $F : \mathscr{C} \to \mathscr{D}$ defines an (op)monoidal (resp. strong monoidal) equivalence between $\mathscr{C}$ and $\mathscr{D}$ we say that the categories $\mathscr{C}$ and $\mathscr{D}$ are (op)monoidally (resp. strong monoidally) equivalent.

In the strong monoidal case the above definition can be reformulated.

**Proposition 1.31** *A functor $F : \mathscr{C} \to \mathscr{D}$ defines a strong monoidal equivalence if and only if $F$ is strong monoidal and an equivalence of categories.*

*Proof* The direct implication is immediate. For the converse, let $G : \mathscr{D} \to \mathscr{C}$ be a functor as in Remark 1.20. If $\mu : \mathrm{Id}_{\mathscr{D}} \to FG$ and $\nu : GF \to \mathrm{Id}_{\mathscr{C}}$ are the natural isomorphisms satisfying (1.3.1), we show that $G$ admits a unique strong monoidal structure with respect to which $\mu$ and $\nu$ become monoidal transformations.

In what follows we denote by $(\varphi_2^F := (\varphi_{2,X,Y}^F)_{X,Y\in\mathscr{C}}, \varphi_0^F)$ the strong monoidal structure of $F$, and by $\widetilde{\varphi}_2^F$, $\widetilde{\varphi}_0^F$ the inverse morphisms of $\varphi_2^F$, $\varphi_0^F$, respectively.

Assume that $G$ admits a strong monoidal structure $(\varphi_2^G, \varphi_0^G)$ with respect to which $\nu$ becomes a monoidal natural transformation. Remark 1.25 implies that $GF$ is strong monoidal via $\varphi_2^{GF} = (\varphi_{2,X,Y}^{GF} = G(\varphi_{2,X,Y}^F)\varphi_{2,F(X),F(Y)}^G)_{X,Y\in\mathscr{C}}$ and $\varphi_0^{GF} = G(\varphi_0^F)\varphi_0^G$, and from here we get that

$$\nu_{X\otimes Y}^{-1}(\nu_X \otimes \nu_Y) = \varphi_{2,X,Y}^{GF} = G(\varphi_{2,X,Y}^F)\varphi_{2,F(X),F(Y)}^G$$

or, equivalently,

$$\varphi_{2,F(X),F(Y)}^G = G(\widetilde{\varphi}_{2,X,Y}^F)\nu_{X\otimes Y}^{-1}(\nu_X \otimes \nu_Y). \tag{1.3.3}$$

Let $U,V \in \mathscr{D}$. By the naturality of $\varphi_2^G$ the diagram

$$\begin{array}{ccc}
GFG(U)\otimes GFG(V) & \xrightarrow{\varphi^G_{2,FG(U),FG(V)}} & G(FG(U)\Box FG(V)) \\
{\scriptstyle G(\mu_U)\otimes G(\mu_V)}\uparrow & & \uparrow{\scriptstyle G(\mu_U\Box\mu_V)} \\
G(U)\otimes G(V) & \xrightarrow{\varphi^G_{2,U,V}} & G(U\Box V)
\end{array}$$

is commutative, and therefore

$$\begin{aligned}
\varphi_{2,U,V}^G &= G(\mu_U^{-1}\Box\mu_V^{-1})\varphi_{2,FG(U),FG(V)}^G(G(\mu_U)\otimes G(\mu_V)) \\
&\overset{(1.3.1),(1.3.3)}{=} G((\mu_U^{-1}\Box\mu_V^{-1})\widetilde{\varphi}_{2,G(U),G(V)}^F)\nu_{G(U)\otimes G(V)}^{-1},
\end{aligned} \tag{1.3.4}$$

for all $U,V \in \mathscr{D}$. Together with $\nu_{\underline{1}}\varphi_0^{GF} = \mathrm{Id}_{\underline{1}}$ or, equivalently, with $\varphi_0^G = G(\widetilde{\varphi}_0^F)\nu_{\underline{1}}^{-1}$, this guarantees the uniqueness of the stated strong monoidal strucure of $G$.

Conversely, define $\varphi_2^G$ as in (1.3.4) and $\varphi_0^G := G(\widetilde{\varphi}_0^F)\nu_{\underline{1}}^{-1}$. We prove that with this structure $G$ is strong monoidal and $\mu$, $\nu$ become monoidal natural transformations.

To show the commutativity of the diagrams in (1.3.2) we need the equality

$$F(\varphi^G_{2,U,V}) = \mu_{U\Box V}(\mu_U^{-1}\Box\mu_V^{-1})\widetilde{\varphi}^F_{2,G(U),G(V)}, \tag{1.3.5}$$

which holds for any $U,V\in\mathscr{D}$. Indeed, the naturality of $\mu$ applied to $\varphi^F_{2,G(U),G(V)}$ and respectively to $\mu_U\Box\mu_V$ gives the following commutative diagrams:

$$\begin{array}{ccc}
FG(FG(U)\Box FG(V)) & \xrightarrow{FG(\varphi^F_{2,G(U),G(V)})} & FGF(G(U)\otimes G(V)) \\
\uparrow{\scriptstyle \mu_{FG(U)\Box FG(V)}} & & \uparrow{\scriptstyle \mu_{F(G(U)\otimes G(V))}} \\
FG(U)\Box FG(V) & \xrightarrow{\varphi^F_{2,G(U),G(V)}} & F(G(U)\otimes G(V))\ ,
\end{array}$$

$$\begin{array}{ccc}
FG(U)\Box FG(V) & \xrightarrow{\mu_{FG(U)\Box FG(V)}} & FG(FG(U)\Box FG(V)) \\
\uparrow{\scriptstyle \mu_U\Box\mu_V} & & \uparrow{\scriptstyle FG(\mu_U\Box\mu_V)} \\
U\Box V & \xrightarrow{\mu_{U\Box V}} & FG(U\Box V).
\end{array}$$

Therefore,

$$\begin{aligned}
F(\varphi^G_{2,U,V}) &= FG(\mu_U^{-1}\Box\mu_V^{-1})FG(\widetilde{\varphi}^F_{2,G(U),G(V)})F(\nu^{-1}_{G(U)\otimes G(V)}) \\
&\overset{(1.3.1)}{=} FG(\mu_U^{-1}\Box\mu_V^{-1})FG(\widetilde{\varphi}^F_{2,G(U),G(V)})\mu_{F(G(U)\otimes G(V))} \\
&= FG(\mu_U^{-1}\Box\mu_V^{-1})\mu_{FG(U)\Box FG(V)}\widetilde{\varphi}^F_{2,G(U),G(V)} \\
&= \mu_{U\Box V}(\mu_U^{-1}\Box\mu_V^{-1})\widetilde{\varphi}^F_{2,G(U),G(V)},
\end{aligned}$$

for all $U,V\in\mathscr{D}$, as stated. We can compute now that

$$\begin{aligned}
&FG(a'_{U,V,W})F(\varphi^G_{2,U\Box V,W})F(\varphi^G_{2,U,V}\otimes \mathrm{Id}_{G(W)}) \\
&= FG(a'_{U,V,W})\mu_{(U\Box V)\Box W}(\mu^{-1}_{U\Box V}\Box\mu_W^{-1})\widetilde{\varphi}^F_{2,G(U\Box V),G(W)}F(\varphi^G_{2,U,V}\otimes \mathrm{Id}_{G(W)}) \\
&= FG(a'_{U,V,W})\mu_{(U\Box V)\Box W}(\mu^{-1}_{U\Box V}\Box\mu_W^{-1})(F(\varphi^G_{2,U,V})\Box\mathrm{Id}_{FG(W)})\widetilde{\varphi}^F_{2,G(U)\otimes G(V),G(W)} \\
&\overset{(1.3.5)}{=} FG(a'_{U,V,W})\mu_{(U\Box V)\Box W}((\mu_U^{-1}\Box\mu_V^{-1})\Box\mu_W^{-1}) \\
&\qquad (\widetilde{\varphi}^F_{2,G(U),G(V)}\Box\mathrm{Id}_{FG(W)})\widetilde{\varphi}^F_{2,G(U)\otimes G(V),G(W)} \\
&= \mu_{U\Box(V\Box W)}a'_{U,V,W}((\mu_U^{-1}\Box\mu_V^{-1})\Box\mu_W^{-1}) \\
&\qquad (\widetilde{\varphi}^F_{2,G(U),G(V)}\Box\mathrm{Id}_{FG(W)})\widetilde{\varphi}^F_{2,G(U)\otimes G(V),G(W)} \\
&= \mu_{U\Box(V\Box W)}(\mu_U^{-1}\Box(\mu_V^{-1}\Box\mu_W^{-1}))a'_{FG(U),FG(V),FG(W)} \\
&\qquad (\widetilde{\varphi}^F_{2,G(U),G(V)}\Box\mathrm{Id}_{FG(W)})\widetilde{\varphi}^F_{2,G(U)\otimes G(V),G(W)} \\
&= \mu_{U\Box(V\Box W)}(\mu_U^{-1}\Box(\mu_V^{-1}\Box\mu_W^{-1}))(\mathrm{Id}_{FG(U)}\Box\widetilde{\varphi}^F_{2,G(V),G(W)}) \\
&\qquad \widetilde{\varphi}^F_{2,G(U),G(V)\otimes G(W)}F(a_{G(U),G(V),G(W)}),
\end{aligned}$$

where we applied: in the second equality the naturality of $\varphi_2^F$ to the morphisms

$\varphi^G_{2,U,V}$ and $\mathrm{Id}_{G(W)}$, in the fourth equality the naturality of $\mu$ to the morphism $a'_{U,V,W}$, in the fifth equality the naturality of $a'$ to the morphisms $\mu_U^{-1}$, $\mu_V^{-1}$, $\mu_W^{-1}$, and in the last equality the fact that $\varphi^F_2$ closes commutatively the diagram (1.3.2).

Similar arguments allow us to calculate that

$$\begin{aligned}
&F(\varphi^G_{2,U,V\Box W})F(\mathrm{Id}_{G(U)}\otimes\varphi^G_{2,V,W})F(a_{G(U),G(V),G(W)})\\
&=\mu_{U\Box(V\Box W)}(\mu_U^{-1}\Box\mu_{V\Box W}^{-1})\widetilde{\varphi}^F_{2,G(U),G(V\Box W)}F(\mathrm{Id}_{G(U)}\otimes\varphi^G_{2,V,W})F(a_{G(U),G(V),G(W)})\\
&=\mu_{U\Box(V\Box W)}(\mu_U^{-1}\Box\mu_{V\Box W}^{-1})(\mathrm{Id}_{FG(U)}\Box F(\varphi^G_{2,V,W}))\\
&\qquad\widetilde{\varphi}^F_{2,G(U),G(V)\otimes G(W)}F(a_{G(U),G(V),G(W)})\\
&=\mu_{U\Box(V\Box W)}(\mu_U^{-1}\Box(\mu_V^{-1}\Box\mu_W^{-1}))(\mathrm{Id}_{FG(U)}\Box\widetilde{\varphi}^F_{2,G(V),G(W)})\\
&\qquad\widetilde{\varphi}^F_{2,G(U),G(V)\otimes G(W)}F(a_{G(U),G(V),G(W)}).
\end{aligned}$$

Thus, since $F$ is fully faithful it follows that our $\varphi^G_2$ closes commutatively the corresponding hexagonal diagram in (1.3.2). Together with $\varphi^G_0$ as above, it also closes commutatively the two square diagrams in (1.3.2), since $F$ is fully faithful:

$$\begin{aligned}
&FG(l'_U)F(\varphi^G_{2,\underline{1},U})F(\varphi^G_0\otimes\mathrm{Id}_{G(U)})\\
&=FG(l'_U)\mu_{\underline{1}\Box U}(\mu_{\underline{1}}^{-1}\Box\mu_U^{-1})\widetilde{\varphi}^F_{2,G(\underline{1}),G(U)}F(\varphi^G_0\otimes\mathrm{Id}_{G(U)})\\
&=FG(l'_U)\mu_{\underline{1}\Box U}(\mu_{\underline{1}}^{-1}F(\varphi^G_0)\Box\mu_U^{-1})\widetilde{\varphi}^F_{2,\underline{1},G(U)}\\
&=FG(l'_U)\mu_{\underline{1}\Box U}(\widetilde{\varphi}^F_0\Box\mu_U^{-1})\widetilde{\varphi}^F_{2,\underline{1},G(U)}\\
&=\mu_U l'_U(\mathrm{Id}_{\underline{1}}\Box\mu_U^{-1})(\widetilde{\varphi}^F_0\Box\mathrm{Id}_{FG(U)})\widetilde{\varphi}^F_{2,\underline{1},G(U)}\\
&=l'_{FG(U)}(\widetilde{\varphi}^F_0\Box\mathrm{Id}_{FG(U)})\widetilde{\varphi}^F_{2,\underline{1},G(U)}=F(l_{G(U)}),
\end{aligned}$$

and similarly $FG(r'_U)F(\varphi^G_{2,U,\underline{1}})F(\mathrm{Id}_{G(U)}\otimes\varphi^G_0)=F(r_{G(U)})$, for all $U\in\mathscr{D}$. The remaining details are left to the reader.

We check that $\mu,\nu$ are monoidal natural transformations. For all $X,Y\in\mathscr{C}$,

$$\begin{aligned}
F(\varphi^{GF}_{2,X,Y}) &= FG(\varphi^F_{2,X,Y})F(\varphi^G_{2,F(X),F(Y)})\\
&\overset{(1.3.5)}{=} FG(\varphi^F_{2,X,Y})\mu_{F(X)\Box F(Y)}(\mu_{F(X)}^{-1}\Box\mu_{F(Y)}^{-1})\widetilde{\varphi}^F_{2,GF(X),GF(Y)}\\
&\overset{(1.3.1)}{=} \mu_{F(X\otimes Y)}\varphi^F_{2,X,Y}(F(\nu_X)\Box F(\nu_Y))\widetilde{\varphi}^F_{2,GF(X),GF(Y)}\\
&= \mu_{F(X\otimes Y)}F(\nu_X\otimes\nu_Y)\overset{(1.3.1)}{=}F(\nu_{X\otimes Y}^{-1}(\nu_X\otimes\nu_Y)),
\end{aligned}$$

and since $F$ is fully faithful this proves that $\nu$ is a monoidal natural transformation (clearly, $\varphi^{GF}_0=\nu_{\underline{1}}^{-1}$ since $\varphi^G_0=G(\widetilde{\varphi}^F_0)\nu_{\underline{1}}^{-1}$). From the definitions we have

$$\varphi^{FG}_{2,U,V}=F(\varphi^G_{2,U,V})\varphi^F_{2,G(U),G(V)}=\mu_{U\Box V}(\mu_U^{-1}\Box\mu_V^{-1}),\ \ \forall\, U,\ V\in\mathscr{D}.$$

So $\mu$ is a monoidal natural transformation (it is easy to see that $\varphi^{FG}_0=\mu_{\underline{1}}$). □

**Proposition 1.32** *The functor $F:\mathscr{C}\to\mathfrak{e}(\mathscr{C})$ defined in Proposition 1.28 provides a strong monoidal equivalence between $\mathscr{C}$ and the full image of $F$.*

*Proof* This follows from Propositions 1.21, 1.28 and 1.31. □

## 1.4 Mac Lane's Strictification Theorem for Monoidal Categories

We will prove that any monoidal category $\mathscr{C}$ is strong monoidally equivalent to a strict category, $\mathscr{C}^{\rm str}$. The construction associating to a monoidal category a strict one which is strong monoidally equivalent to it is called strictification.

To this end, we first need some concepts and preliminary results.

Let $(\mathscr{C},\otimes,\underline{1},a,l,r)$ be a monoidal category. For any positive integer $m$, a word of length $m$ with objects of $\mathscr{C}$ is a sequence $S$ of the form $S=(X_1,\ldots,X_m)$, where $X_1,\ldots,X_m\in\mathscr{C}$. By convention, the word of length zero with objects of $\mathscr{C}$ is the empty sequence $\phi$.

To any non-empty word $S=(X_1,\ldots,X_m)$ with objects of $\mathscr{C}$ we associate an object $F(S)$ of $\mathscr{C}$ defined by

$$F(S)=((\cdots((X_1\otimes X_2)\otimes X_3)\otimes\cdots)\otimes X_{m-1})\otimes X_m.$$

If $S=\phi$ then $F(\phi)=\underline{1}$, the unit object of $\mathscr{C}$.

**Definition 1.33** Let $\mathscr{C}$ be a monoidal category. By $\mathscr{C}^{\rm str}$ we denote the category whose objects are words of finite length with objects of $\mathscr{C}$. If $S$ and $S'$ are objects of $\mathscr{C}^{\rm str}$ then we define ${\rm Hom}_{\mathscr{C}^{\rm str}}(S,S'):={\rm Hom}_{\mathscr{C}}(F(S),F(S'))$. The identities and composition of morphisms are taken from $\mathscr{C}$.

By the above definition it is immediate that $F:\mathscr{C}^{\rm str}\to\mathscr{C}$, $S\mapsto F(S)$, defines a functor ($F$ acts as identity on morphisms).

We define the monoidal structure of $\mathscr{C}^{\rm str}$. If $S=(X_1,\ldots,X_m)$ and $S'=(Y_1,\ldots,Y_n)$ are non-empty words of length $m$ and $n$, respectively, with objects of $\mathscr{C}$, we define the tensor product $S\square S'$ between $S$ and $S'$ as follows:

$$S\square S'=(X_1,\ldots,X_m,Y_1,\ldots,Y_n).$$

By convention, $\phi\square S=S\square\phi=S$, for any word $S$ of finite length with objects of $\mathscr{C}$. Clearly, $(S\square S')\square S''=S\square(S'\square S''):=S\square S'\square S''$, for any objects $S,S',S''$ of $\mathscr{C}^{\rm str}$.

First we define $\varphi_{2,\phi,\phi}:F(\phi)\otimes F(\phi)=\underline{1}\otimes\underline{1}\to\underline{1}=F(\phi)=F(\phi\square\phi)$, $\varphi_{2,\phi,\phi}=l_{\underline{1}}=r_{\underline{1}}$. For a non-empty word $S$ with objects of $\mathscr{C}$ define

$$\varphi_{2,\phi,S}:F(\phi)\otimes F(S)=\underline{1}\otimes F(S)\to F(S)=F(\phi\square S),\ \varphi_{2,\phi,S}=l_{F(S)},$$
$$\varphi_{2,S,\phi}:F(S)\otimes F(\phi)=F(S)\otimes\underline{1}\to F(S)=F(S\square\phi),\ \varphi_{2,S,\phi}=r_{F(S)},$$
$$\varphi_{2,S,(Y)}:F(S)\otimes F((Y))=F(S)\otimes Y\to F(S)\otimes Y=F(S\square(Y)),\ \varphi_{2,S,(Y)}={\rm Id}_{F(S)\otimes Y},$$

and, inductively, if $S'=(Y_1,\ldots,Y_m)$ is a word of length $m\geq 2$ with objects of $\mathscr{C}$ and $S'_*=(Y_1,\ldots,Y_{m-1})$ then

$$\varphi_{2,S,S'}:F(S)\otimes F(S')=F(S)\otimes(F(S'_*)\otimes Y_m)\xrightarrow{a^{-1}_{F(S),F(S'_*),Y_m}}(F(S)\otimes F(S'_*))\otimes Y_m$$
$$\xrightarrow{\varphi_{2,S,S'_*}\otimes{\rm Id}_{Y_m}}F(S\square S'_*)\otimes Y_m=F(S\square S'_*\square(Y_m))=F(S\square S').$$

By induction on the length of $S'$ it follows that $\varphi_2:=(\varphi_{2,S,S'}:F(S)\otimes F(S')\to$

$F(S\Box S'))_{S,S'\in\mathscr{C}^{\rm str}}$ is a natural isomorphism. Then for any morphisms $S\xrightarrow{f}T$ and $S'\xrightarrow{f'}T'$ in $\mathscr{C}^{\rm str}$, that is, for any morphisms $F(S)\xrightarrow{f}F(T)$ and $F(S')\xrightarrow{f'}F(T')$ in $\mathscr{C}$, define the tensor product morphism $S\Box S'\xrightarrow{f\Box f'}T\Box T'$ in $\mathscr{C}^{\rm str}$ as being the unique morphism that makes the diagram

$$\begin{array}{ccc} F(S)\otimes F(S') & \xrightarrow{\varphi_{2,S,S'}} & F(S\Box S') \\ {\scriptstyle f\otimes f'}\downarrow & & \downarrow{\scriptstyle f\Box f'} \\ F(T)\otimes F(T') & \xrightarrow{\varphi_{2,T,T'}} & F(T\Box T') \end{array}$$

commutative. It is clear at this point that $(\mathscr{C}^{\rm str},\Box,\phi)$ is a strict monoidal category.

**Theorem 1.34** *The categories $\mathscr{C}$ and $\mathscr{C}^{\rm str}$ are strong monoidally equivalent.*

*Proof* We show that $(F,\mathrm{Id}_{\underline{1}},\varphi_2):\mathscr{C}^{\rm str}\to\mathscr{C}$ defines a strong monoidal equivalence.

Since $X=F((X))$, for any object of $\mathscr{C}$, we get that $F$ is essentially surjective. Also, since $\mathrm{Hom}_{\mathscr{C}^{\rm str}}(S,S')=\mathrm{Hom}_{\mathscr{C}}(F(S),F(S'))$, for any objects $S,S'$ of $\mathscr{C}^{\rm str}$, we obtain that $F$ is fully faithful. Thus, according to Proposition 1.19, $F:\mathscr{C}^{\rm str}\to\mathscr{C}$ is an equivalence of categories.

We will prove that $(F,\mathrm{Id}_{\underline{1}},\varphi_2)$ is a monoidal functor. For this, we only need to check that the first diagram that appears in Definition 1.22, specialized for our situation, is commutative (the commutativity of the other two diagrams is immediate). This reduces, for any $S,S',S''\in\mathscr{C}^{\rm str}$, to

$$\varphi_{2,S,S'\Box S''}(\mathrm{Id}_{F(S)}\otimes\varphi_{2,S',S''})a_{F(S),F(S'),F(S'')}=\varphi_{2,S\Box S',S''}(\varphi_{2,S,S'}\otimes\mathrm{Id}_{F(S'')}). \quad (1.4.1)$$

We prove the relation in (1.4.1) by induction on the length of $S''$.

By the naturality of $r$, the diagram

$$\begin{array}{ccc} (F(S)\otimes F(S'))\otimes\underline{1} & \xrightarrow{r_{F(S)\otimes F(S')}} & F(S)\otimes F(S') \\ {\scriptstyle \varphi_{2,S,S'}\otimes\mathrm{Id}_{\underline{1}}}\downarrow & & \downarrow{\scriptstyle \varphi_{2,S,S'}} \\ F(S\Box S')\otimes\underline{1} & \xrightarrow[r_{F(S\Box S')}]{} & F(S\Box S') \end{array}$$

is commutative. This fact allows us to compute

$$\begin{aligned} \varphi_{2,S,S'}(\mathrm{Id}_{F(S)}\otimes\varphi_{2,S',\phi})a_{F(S),F(S'),F(\phi)} &= \varphi_{2,S,S'}(\mathrm{Id}_{F(S)}\otimes r_{F(S')})a_{F(S),F(S'),\underline{1}} \\ \text{(by Proposition 1.5)} &= \varphi_{2,S,S'}r_{F(S)\otimes F(S')} \\ &= r_{F(S\Box S')}(\varphi_{2,S,S'}\otimes\mathrm{Id}_{\underline{1}}) \\ &= \varphi_{2,S\Box S',\phi}(\varphi_{2,S,S'}\otimes\mathrm{Id}_{F(\phi)}), \end{aligned}$$

and this shows that (1.4.1) holds for $S''=\phi$.

Assume that (1.4.1) is valid for any $S''$ of length $p-1$ and let $S''=(Z_1,\ldots,Z_p)$ be a word of length $p \geq 1$ with objects of $\mathscr{C}$. If $S''_* = (Z_1, \dots, Z_{p-1})$ we have that

$$\begin{array}{ccc}
(F(S)\otimes(F(S')\otimes F(S''_*)))\otimes Z_p & \xrightarrow{a_{F(S),F(S')\otimes F(S''_*),Z_p}} & F(S)\otimes((F(S')\otimes F(S''_*))\otimes Z_p) \\
{\scriptstyle (\mathrm{Id}_{F(S)}\otimes\varphi_{2,S',S''_*})\otimes\mathrm{Id}_{Z_p}}\Big\downarrow & & \Big\downarrow{\scriptstyle \mathrm{Id}_{F(S)}\otimes(\varphi_{2,S',S''_*}\otimes\mathrm{Id}_{Z_p})} \\
(F(S)\otimes F(S'\square S''_*))\otimes Z_p & \xrightarrow[a_{F(S),F(S'\square S''_*),Z_p}]{} & F(S)\otimes(F(S'\square S''_*)\otimes Z_p),
\end{array}$$

$$\begin{array}{ccc}
((F(S)\otimes F(S'))\otimes F(S''_*))\otimes Z_p & \xrightarrow{a_{F(S)\otimes F(S'),F(S''_*),Z_p}} & (F(S)\otimes F(S'))\otimes(F(S''_*)\otimes Z_p) \\
{\scriptstyle (\varphi_{2,S,S'}\otimes\mathrm{Id}_{F(S''_*)})\otimes\mathrm{Id}_{Z_p}}\Big\downarrow & & \Big\downarrow{\scriptstyle \varphi_{2,S,S'}\otimes\mathrm{Id}_{F(S''_*)\otimes Z_p}} \\
(F(S\square S')\otimes F(S''_*))\otimes Z_p & \xrightarrow[a_{F(S\square S'),F(S''_*),Z_p}]{} & F(S\square S')\otimes(F(S''_*)\otimes Z_p)
\end{array}$$

are commutative, because of the naturality of $a$. Therefore

$$\begin{aligned}
&\varphi_{2,S,S'\square S''}(\mathrm{Id}_{F(S)}\otimes\varphi_{2,S',S''})a_{F(S),F(S'),F(S'')} \\
&= (\varphi_{2,S,S'\square S''_*}\otimes\mathrm{Id}_{Z_p})a^{-1}_{F(S),F(S'\square S''_*),Z_p} \\
&\quad (\mathrm{Id}_{F(S)}\otimes(\varphi_{2,S',S''_*}\otimes\mathrm{Id}_{Z_p}))(\mathrm{Id}_{F(S)}\otimes a^{-1}_{F(S'),F(S''_*),Z_p})a_{F(S),F(S'),F(S'')} \\
&= \left(\varphi_{2,S,S'\square S''_*}(\mathrm{Id}_{F(S)}\otimes\varphi_{2,S',S''_*})\otimes\mathrm{Id}_{Z_p}\right) a^{-1}_{F(S),F(S')\otimes F(S''_*),Z_p} \\
&\quad (\mathrm{Id}_{F(S)}\otimes a^{-1}_{F(S'),F(S''_*),Z_p})a_{F(S),F(S'),F(S'')} \\
&\overset{(1.1.1)}{=} \left(\varphi_{2,S,S'\square S''_*}(\mathrm{Id}_{F(S)}\otimes\varphi_{2,S',S''_*})a_{F(S),F(S'),F(S''_*)}\otimes\mathrm{Id}_{Z_p}\right) a^{-1}_{F(S)\otimes F(S'),F(S''_*),Z_p} \\
&= \left(\varphi_{2,S\square S',S''_*}(\varphi_{2,S,S'}\otimes\mathrm{Id}_{F(S''_*)})\otimes\mathrm{Id}_{Z_p}\right) a^{-1}_{F(S)\otimes F(S'),F(S''_*),Z_p} \\
&= (\varphi_{2,S\square S',S''_*}\otimes\mathrm{Id}_{Z_p})a^{-1}_{F(S\square S'),F(S''_*),Z_p}(\varphi_{2,S,S'}\otimes\mathrm{Id}_{F(S''_*)\otimes Z_p}) \\
&= \varphi_{2,S\square S',S''}(\varphi_{2,S,S'}\otimes\mathrm{Id}_{F(S'')}),
\end{aligned}$$

where in the first and the last equality we used the inductive definition of $\varphi_2$ and in the fourth equality the inductive hypothesis.

By using also the inductive definition of $\varphi_2$, it follows immediately that $F$ is actually a strong monoidal functor. Since $F$ is an equivalence of categories as well, we are in a position to apply Proposition 1.31 and we obtain that $F$ is a strong monoidal equivalence.

The equivalence inverse $G : \mathscr{C} \to \mathscr{C}^{\mathrm{str}}$ of $F$ is defined by $G(X) = (X)$, the word of length 1 defined by $X \in \mathscr{C}$, and $G(f) = f$, for any morphism $f$ in $\mathscr{C}$. Also by the proof of Proposition 1.31 it follows that the strong monoidal structure of $G$ is defined by $\psi_0 = \mathrm{Id}_{\underline{1}} : \phi \to (\underline{1})$ and

$$\psi_{2,X,Y} : G(X)\square G(Y) = (X,Y) \to G(X\otimes Y) = (X\otimes Y), \quad \psi_{2,X,Y} = \mathrm{Id}_{X\otimes Y},$$

for all $X, Y \in \mathscr{C}$. □

To any strong monoidal functor $(T, t_0, t_2) : \mathscr{C} \to \mathscr{D}$ we associate a strict monoidal functor $T^{\mathrm{str}} : \mathscr{C}^{\mathrm{str}} \to \mathscr{D}^{\mathrm{str}}$ as follows.

Let $(F, \mathrm{Id}_{\underline{1}}, \varphi_2) : \mathscr{C}^{\rm str} \to \mathscr{C}$ and $(G, \mathrm{Id}_{\underline{1}}, \psi_2) : \mathscr{D}^{\rm str} \to \mathscr{D}$ be the functors from Theorem 1.34 that provide the monoidal equivalences. If $S = (X_1, \ldots, X_m)$ is a non-empty word with objects of $\mathscr{C}$ define $T^{\rm str}(S) = T(S) := (T(X_1), \ldots, T(X_m)) \in \mathscr{D}^{\rm str}$. In addition, $T^{\rm str}(\phi) = \phi$. In order to define $T^{\rm str}$ on morphisms observe first that $T(F(S)) \cong G(T(S))$, for any $S \in \mathscr{C}^{\rm str}$. Indeed, if $S = (X_1, \ldots, X_m) \in \mathscr{C}^{\rm str}$ is a non-empty word and $S_* = (X_1, \ldots, X_{m-1})$ then

$$T(F(S)) = T(F(S_* \square (X_m))) = T(F(S_*) \otimes X_m) \cong T(F(S_*)) \otimes T(X_m),$$

because $T$ is a strong monoidal functor. Since $T^{\rm str}(\phi) = \phi$, by induction on the length of $S$ it follows now that $T(F(S)) \cong G(T(S))$, for any $S \in \mathscr{C}^{\rm str}$.

Let now $S \xrightarrow{f} S'$ be a morphism in $\mathscr{C}^{\rm str}$, that is, $f \in \mathrm{Hom}_{\mathscr{C}}(F(S), F(S'))$. We then define $T^{\rm str}(f) : T(S) \to T(S')$ in $\mathscr{D}^{\rm str}$ (i.e. $T^{\rm str}(f) : G(T(S)) \to G(T(S'))$ in $\mathscr{D}$) by requiring the commutativity of the diagram

$$\begin{array}{ccc} G(T(S)) & \xrightarrow{T^{\rm str}(f)} & G(T(S')) \\ \cong \downarrow & & \downarrow \cong \\ T(F(S)) & \xrightarrow[T(f)]{} & T(F(S')). \end{array}$$

In this way we have produced a strict monoidal functor $T^{\rm str} : \mathscr{C}^{\rm str} \to \mathscr{D}^{\rm str}$, as claimed.

**Remark 1.35** The strictification theorem $\mathscr{C} \mapsto \mathscr{C}^{\rm str}$ allows us also to obtain Mac Lane's coherence theorem, which states that in any monoidal category $\mathscr{C}$ the commutativity of a diagram built from identities, associativity, and left and right unit constraints of $\mathscr{C}$, by tensoring and composing, is equivalent to the Pentagon and Triangle Axioms. Moreover, the coherence theorem gives a unique natural transformation between tensor words of the same length, constructed out of the constraints $a$, $l$, $r$ of $\mathscr{C}$. By a tensor word of length $m$ in $\mathscr{C}$ we mean an expression like

$$((\cdots((X_1 \otimes X_2) \otimes X_3) \otimes \cdots) \otimes X_{m-1}) \otimes X_m, \tag{1.4.2}$$

built with $m$ objects $X_1, \ldots, X_m$ of $\mathscr{C}$; note that in (1.4.2) the order of the parentheses can vary and that the tensor word of length 0 is $\underline{1}$, the unit object of the category. Thus, in diagrammatic computations, this fact allows us to omit the parentheses in expressions such as (1.4.2). Also, any diagram in $\mathscr{C}$ produces a diagram in $\mathscr{C}^{\rm str}$, whose commutativity implies the commutativity of the original diagram in $\mathscr{C}$. Hence, in practice, it is enough to prove that a certain diagram commutes when it is interpreted in a strict monoidal category.

## 1.5 (Pre-)Braided Monoidal Categories

If $\mathscr{C}$ is a monoidal category and $X, Y$ are objects of $\mathscr{C}$ then $X \otimes Y$ and $Y \otimes X$ are not necessarily isomorphic objects of $\mathscr{C}$. The definition of a braiding on a monoidal category $\mathscr{C}$ requires the existence of isomorphisms between $X \otimes Y$ and $Y \otimes X$, for all objects $X, Y$ of $\mathscr{C}$ that are compatible with the monoidal structure of $\mathscr{C}$.

Recall that if $\mathscr{C}$ is a category then the switch functor $\tau : \mathscr{C} \times \mathscr{C} \to \mathscr{C} \times \mathscr{C}$ is defined by $\tau(X,Y) = (Y,X)$ and $\tau(f,g) = (g,f)$.

**Definition 1.36** A pre-braiding on a monoidal category $\mathscr{C}$ is a natural transformation $c : \otimes \to \otimes \circ \tau$ satisfying the so-called Hexagon Axiom, namely for any objects $X, Y, Z \in \mathscr{C}$ the following diagrams are commutative:

$$\begin{array}{ccccc} & (Y\otimes X)\otimes Z & \xrightarrow{a_{Y,X,Z}} & Y\otimes(X\otimes Z) & \\ {\scriptstyle c_{X,Y}\otimes \mathrm{Id}_Z}\nearrow & & & & \searrow{\scriptstyle \mathrm{Id}_Y\otimes c_{X,Z}} \\ (X\otimes Y)\otimes Z & & & & Y\otimes(Z\otimes X) \\ {\scriptstyle a_{X,Y,Z}}\searrow & & & & \nearrow{\scriptstyle a_{Y,Z,X}} \\ & X\otimes(Y\otimes Z) & \xrightarrow{c_{X,Y\otimes Z}} & (Y\otimes Z)\otimes X, & \end{array} \tag{1.5.1}$$

$$\begin{array}{ccccc} & X\otimes(Z\otimes Y) & \xrightarrow{a^{-1}_{X,Z,Y}} & (X\otimes Z)\otimes Y & \\ {\scriptstyle \mathrm{Id}_X\otimes c_{Y,Z}}\nearrow & & & & \searrow{\scriptstyle c_{X,Z}\otimes \mathrm{Id}_Y} \\ X\otimes(Y\otimes Z) & & & & (Z\otimes X)\otimes Y \\ {\scriptstyle a^{-1}_{X,Y,Z}}\searrow & & & & \nearrow{\scriptstyle a^{-1}_{Z,X,Y}} \\ & (X\otimes Y)\otimes Z & \xrightarrow{c_{X\otimes Y,Z}} & Z\otimes(X\otimes Y). & \end{array} \tag{1.5.2}$$

A pre-braiding $c$ is called a braiding if it is a natural isomorphism.

A (pre-)braided category is a pair $(\mathscr{C}, c)$ consisting of a monoidal category $\mathscr{C}$ and a (pre-)braiding $c$ on $\mathscr{C}$.

Assuming $\mathscr{C}$ strict, we have that $c$ is a pre-braiding on $\mathscr{C}$ if $c$ is a natural transformation $c : \otimes \to \otimes \circ \tau$, satisfying the conditions

(a) $c_{X,Y\otimes Z} =$ 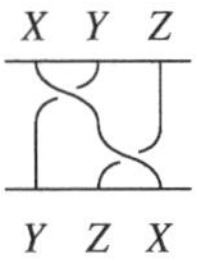

and (b) $c_{X\otimes Y,Z} =$ 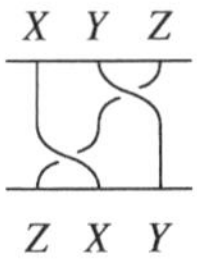

, (1.5.3)

where, from now on, for any two objects $X$ and $Y$ of $\mathscr{C}$, we will denote $c_{X,Y}$ by 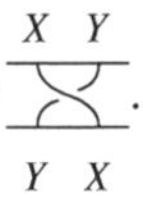

.

If $c$ is a braiding on $\mathscr{C}$, by 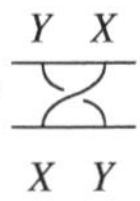

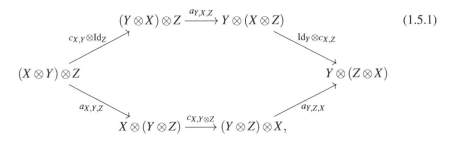

we denote $c^{-1}_{X,Y}$, the inverse of $c_{X,Y}$.

In what follows, for $\mathscr{C}$ a (strict) monoidal category, we will denote by

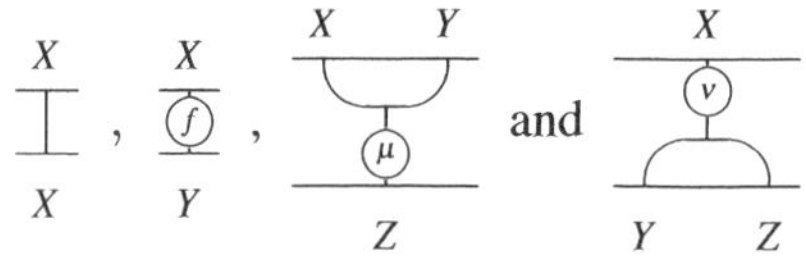

the following morphisms in $\mathscr{C}$: $\mathrm{Id}_X : X \to X$, $f : X \to Y$, $\mu : X \otimes Y \to Z$ and $\nu : X \to Y \otimes Z$, respectively.

In diagrammatic notation, the fact that $c$ is a natural transformation comes out as

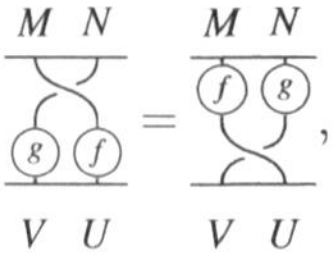

$\forall f : M \to U$ and $g : N \to V$ in $\mathscr{C}$. (1.5.4)

**Proposition 1.37** *For $(\mathscr{C}, c)$ a braided category, we have*

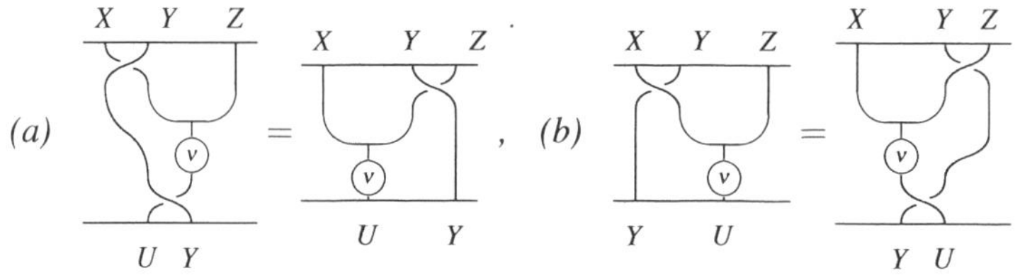

(1.5.5)

*for any morphism $\nu : X \otimes Z \to U$ in $\mathscr{C}$, and*

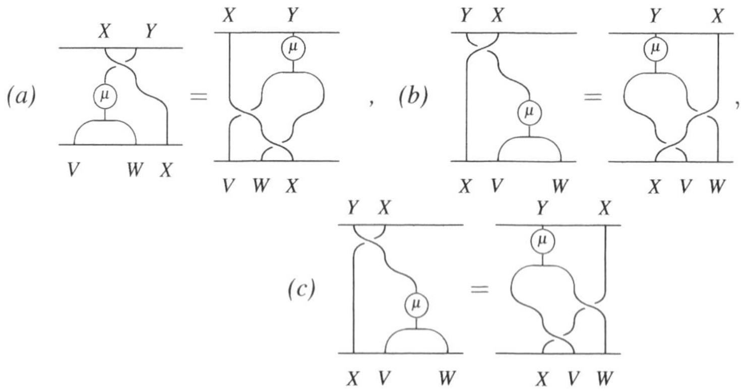

(1.5.6)

*for any morphism $\mu : Y \to V \otimes W$ in $\mathscr{C}$.*

*Proof* This follows from (1.5.4) and (1.5.3). For example, to prove the first equality in (1.5.5) we have to consider in (1.5.4) $f = \mathrm{Id}_Y$, $g = \nu$, and then use the relation (a) in (1.5.3). □

We next provide some examples of (pre-)braided categories.

**Example 1.38** For a braided category $\mathscr{C}$, let $\mathscr{C}^{\mathrm{in}}$ be equal to $\mathscr{C}$ as a monoidal category, equipped with the mirror-reversed braiding $\underline{c}_{X,Y} = c^{-1}_{Y,X}$. Then $(\mathscr{C}^{\mathrm{in}}, \underline{c})$ is also a braided category. To see this, note that (1.5.1) and (1.5.2) are obtained from (1.5.2) and (1.5.1), respectively, by replacing $c$ with $\underline{c}$.

**Definition 1.39** We call a braided category $(\mathscr{C}, c)$ symmetric if $\mathscr{C} = \mathscr{C}^{\text{in}}$, as braided categories. That is, $c_{Y,X}^{-1} = c_{X,Y}$, for any $X, Y \in \mathscr{C}$.

**Example 1.40** If $(\mathscr{C}, c)$ is (pre-)braided then the reverse monoidal category $\overline{\mathscr{C}}$ associated to $\mathscr{C}$ is (pre-)braided with $\overline{c}_{X,Y} = c_{Y,X} : X\overline{\otimes}Y = Y \otimes X \to X \otimes Y = Y\overline{\otimes}X$.

Indeed, it can be easily checked that (1.5.1) for $(\overline{\mathscr{C}}, \overline{c})$ follows from (1.5.2) for $(\mathscr{C}, c)$, and that (1.5.2) for $(\overline{\mathscr{C}}, \overline{c})$ follows from (1.5.1) for $(\mathscr{C}, c)$.

**Example 1.41** If $(\mathscr{C}, c)$ is braided then so is $\mathscr{C}^{\text{opp}}$, the opposite category associated to $\mathscr{C}$, with $c_{X,Y}^{\text{opp}} := c_{X,Y}^{-1}$.

**Example 1.42** The monoidal category $\underline{\text{Set}}$ of sets is symmetric with $c$ defined by $c_{X,Y} : X \times Y \to Y \times X$, $c_{X,Y}(x, y) = (y, x)$.

**Example 1.43** For $k$ a field the category ${}_k\mathscr{M}$ of vector spaces over $k$ is symmetric with $c_{X,Y}(x \otimes y) = y \otimes x$, for any $X, Y \in {}_k\mathscr{M}$ and $x \in X$, $y \in Y$.

We describe the braided structures on a category of $G$-graded vector spaces over a field $k$, $G$ being a multiplicative group with neutral element $e$.

**Proposition 1.44** *Let $G$ be a group, $\phi$ a normalized 3-cocycle on $G$ and $\text{Vect}_\phi^G$ the category of $G$-graded vector spaces endowed with the monoidal structure induced by $\phi$ as in Proposition 1.13. Then $\text{Vect}_\phi^G$ is braided if and only if $G$ is abelian and there exists $\mathscr{R} : G \times G \to k^*$ such that, for all $x, y, z \in G$, we have*

$$\mathscr{R}(xy, z)\phi(x, z, y) = \phi(x, y, z)\mathscr{R}(x, z)\phi(z, x, y)\mathscr{R}(y, z), \tag{1.5.7}$$

$$\phi(x, y, z)\mathscr{R}(x, yz)\phi(y, z, x) = \mathscr{R}(x, y)\phi(y, x, z)\mathscr{R}(x, z). \tag{1.5.8}$$

*Proof* Let $k[G]$ be the group algebra associated to $G$, and let $\varepsilon : k[G] \to k$ be defined by $\varepsilon(g) = 1$, for all $g \in G$, extended by linearity.

If $c$ is a braiding for $\text{Vect}_\phi^G$ we claim that the restriction of

$$\mathscr{R} := (\varepsilon \otimes \varepsilon)c_{k[G],k[G]} : k[G] \otimes k[G] \to k$$

at $G \times G$ (also denoted by $\mathscr{R}$) satisfies the two conditions in (1.5.7) and (1.5.8). To see this we first show that $c$ is completely determined by $\mathscr{R}$, in the sense that

$$c_{V,W}(v \otimes w) = \mathscr{R}(|v|, |w|)w \otimes v, \tag{1.5.9}$$

for any $V, W \in \text{Vect}^G$ and homogeneous elements $v \in V$ and $w \in W$.

Indeed, as in the proof of Proposition 1.13, for any $f : V \to k$ define $\theta_f : V \to k[G]$ by $\theta_f(v) = \sum_{x \in G} f(v_x)x$, for all $v \in V$, where $v = \sum_{x \in G} v_x$ is the decomposition of $v$ in homogeneous components. We have already remarked that $\theta_f$ is a morphism in $\text{Vect}^G$. Also, for $g : W \to k$ let $\theta_g : W \to k[G]$. We recall that $(\varepsilon \otimes \varepsilon)(\theta_g \otimes \theta_f) = g \otimes f$.

Now, by the naturality of $c$ we have $(\theta_g \otimes \theta_f)c_{V,W} = c_{k[G],k[G]}(\theta_f \otimes \theta_g)$, and composing both sides of this equality to the left with $\varepsilon \otimes \varepsilon$ we get that

$$f(v)g(w)\mathscr{R}(|v|,|w|) = (g \otimes f)c_{V,W}(v \otimes w),$$

for any homogeneous elements $v \in V$, $w \in W$. If we write $c_{V,W}(v \otimes w) = \sum_i w_i \otimes v_i$ we obtain $\sum_i g(w_i)f(v_i) = f(v)g(w)\mathscr{R}(|v|,|w|)$, for all $f : V \to k$, $g : W \to k$. Thus

$$c_{V,W}(v \otimes w) = \sum_i w_i \otimes v_i = \mathscr{R}(|v|,|w|)w \otimes v,$$

and this proves (1.5.9). If $|v| = x$ and $|w| = y$ then we must have $w \otimes v \in (W \otimes V)_{xy} = \bigoplus_{z \in G} W_z \otimes V_{z^{-1}xy}$. But $w \otimes v \in W_y \otimes V_x$ and this forces $xy = yx$. Hence $G$ must be abelian.

It is easy to see now that the two conditions on $\mathscr{R}$ in (1.5.7) and (1.5.8) are equivalent to the commutativity of the diagrams in (1.5.2) and (1.5.1), respectively, and that $c_{V,W}$ is an isomorphism if and only if $\mathscr{R}$ takes values in $k^*$.

Conversely, if $G$ is abelian and there is an $\mathscr{R}$ satisfying the two conditions above then it can be easily checked that $c$ defined by (1.5.9) is a braiding for $\mathrm{Vect}^G_\phi$. □

**Definition 1.45** Let $G$ be an abelian group and $k$ a field. A pair $(\phi, \mathscr{R})$ consisting of a normalized 3-cocycle $\phi$ on $G$ with values in $k^*$ and a map $\mathscr{R} : G \times G \to k^*$ satisfying (1.5.7) and (1.5.8) is called an abelian 3-cocycle on $G$ with values in $k^*$.

We denote by $\mathrm{Vect}^G_{(\phi,\mathscr{R})}$ the category $\mathrm{Vect}^G$ endowed with the braided structure given by the abelian 3-cocycle $(\phi, \mathscr{R})$.

**Remark 1.46** For an abelian group $G$, examples of pairs $(\phi, \mathscr{R})$ satisfying the conditions of Proposition 1.44 are given by the so-called normalized coboundary abelian 3-cocycles. More precisely, let $g \in K^2(G, k^*)$ such that $g(x,e) = g(e,z)$, for all $x, z \in G$, and define $\mathscr{R}_g : G \times G \to k^*$ by $\mathscr{R}_g(x,y) = g(x,y)^{-1}g(y,x)$, for all $x, y \in G$. Then the assumption on $G$ to be abelian ensures that the pair $(\Delta_2(g), \mathscr{R}_g)$ satisfies the equalities (1.5.7) and (1.5.8).

In what follows, a pair $(\Delta_2(g), \mathscr{R}_g)$ as above will be called a coboundary abelian 3-cocycle on $G$ with coefficients in $k^*$. Then $H^3_{\mathrm{ab}}(G, k^*)$ is the abelian group of abelian 3-cocycles modulo their coboundaries.

**Example 1.47** The strict monoidal category $\mathrm{Vect}^{\mathbb{Z}_2}$ is braided via the following structure. If $V, W \in \mathrm{Vect}^{\mathbb{Z}_2}$ we define $c_{V,W} : V \otimes W \to W \otimes V$ by

$$c_{V,W}(v \otimes w) = (-1)^{|v||w|} w \otimes v,$$

for homogeneous elements $v \in V$, $w \in W$. We denote this braided category by $\mathrm{Vect}^{\mathbb{Z}_2}_{-1}$.

*Proof* Define $\mathscr{R} : \mathbb{Z}_2 \times \mathbb{Z}_2 \to k^*$ by $\mathscr{R}(\bar{x}, \bar{y}) = (-1)^{xy}$, for all $x, y \in \{0, 1\}$. If $\phi$ is the trivial 3-cocycle on $\mathbb{Z}_2$, then it can be easily checked that $(\phi, \mathscr{R})$ is an abelian 3-cocycle on $\mathbb{Z}_2$. Moreover, one has $\mathrm{Vect}^{\mathbb{Z}_2}_{(\phi,\mathscr{R})} = \mathrm{Vect}^{\mathbb{Z}_2}_{-1}$, as braided categories. □

More generally, for $\mathrm{Vect}^{\mathbb{Z}_n}$ we have the following braided structures.

**Example 1.48** Let $G = \mathbb{Z}_n$ be the cyclic group of order $n \geq 2$ written additively and $\nu \in k$ such that $\nu^{n^2} = \nu^{2n} = 1$. If $\phi_{\nu^n}$ is the normalized 3-cocycle on $G$ defined in Example 1.16 and $\mathscr{R}_\nu : G \times G \to k^*$ is given by $\mathscr{R}_\nu(\overline{x}, \overline{y}) = \nu^{xy}$, for all $x, y \in \{0, \ldots, n-1\}$, then $(\phi_{\nu^n}, \mathscr{R}_\nu)$ endows the category of $\mathbb{Z}_n$-graded vector spaces with a braided structure.

*Proof* Let us first prove that the relations in (1.5.7) hold, that is,

$$\phi_{\nu^n}(\overline{x},\overline{z},\overline{y})\mathscr{R}_\nu(\overline{x+y},\overline{z}) = \mathscr{R}_\nu(\overline{x},\overline{z})\mathscr{R}_\nu(\overline{y},\overline{z})\phi_{\nu^n}(\overline{z},\overline{x},\overline{y})\phi_{\nu^n}(\overline{x},\overline{y},\overline{z}),$$

for all $x, y, z \in \{0, \ldots, n-1\}$. For this we consider the following cases:

1. If $x + y < n$ the above relation reduces to $\phi_{\nu^n}(\overline{x},\overline{z},\overline{y}) = \phi_{\nu^n}(\overline{x},\overline{y},\overline{z})$, and this follows directly from the definition of $\phi_{\nu^n}$.

2. If $x + y \geq n$ we write $x + y = u + n$, for some $0 \leq u \leq n-2$. We must show that

$$\phi_{\nu^n}(\overline{x},\overline{z},\overline{y})\nu^{uz} = \nu^{(x+y)z}\nu^{nz}\phi_{\nu^n}(\overline{x},\overline{y},\overline{z}).$$

By the definition of $\phi_{\nu^n}$ this is equivalent to $\nu^{uz} = \nu^{(x+y)z+nz}$, and the latter follows since $uz - (x+y)z - nz = -2nz$ and $\nu^{2n} = 1$.

The relation (1.5.8) can be proved in a similar way. This time we have to show

$$\phi_{\nu^n}(\overline{y},\overline{z},\overline{x})\phi_{\nu^n}(\overline{x},\overline{y},\overline{z})\mathscr{R}_\nu(\overline{x},\overline{y+z}) = \mathscr{R}_\nu(\overline{x},\overline{z})\mathscr{R}_\nu(\overline{x},\overline{y})\phi_{\nu^n}(\overline{y},\overline{x},\overline{z}),$$

for all $x, y, z \in \{0, \ldots, n-1\}$. To this end consider the following possibilities:

1′. If $y + z < n$ the above equality comes out as

$$\nu^{x(y+z)}\phi_{\nu^n}(\overline{y},\overline{z},\overline{x}) = \nu^{xz}\nu^{xy}\phi_{\nu^n}(\overline{y},\overline{x},\overline{z}),$$

which is clearly satisfied.

2′. If $y + z \geq n$ we write $y + z = v + n$, for some $0 \leq v \leq n-2$. The required equality reduces to $\nu^{nx}\nu^{xv} = \nu^{x(y+z)}$, and this holds since $nx + xv = x(y+z)$.

Thus, the category $\mathrm{Vect}^{\mathbb{Z}_n}$ is braided with the monoidal structure from Example 1.16 (where $q = \nu^n$), and braiding given by $c_{V,W}(v \otimes w) = \nu^{|v||w|}w \otimes v$, for any $\mathbb{Z}_n$-graded vector spaces $V$ and $W$ and homogeneous elements $v \in V$, $w \in W$. □

Coming back to the full generality, we next express the compatibilities between braidings and unit constraints.

**Proposition 1.49** *If $(\mathscr{C}, c)$ is a braided category then the diagrams*

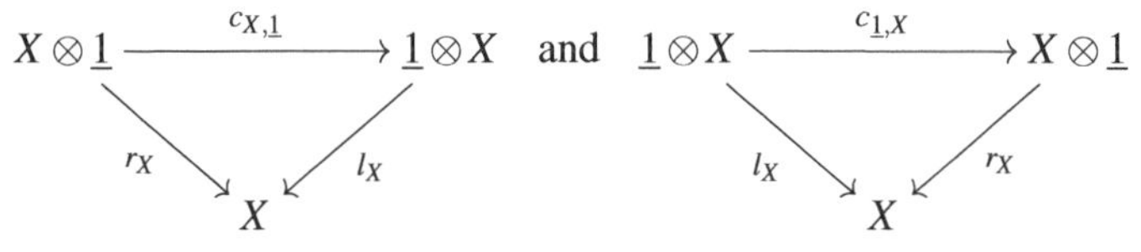


*are commutative, for any $X \in \mathscr{C}$.*

*Proof* It suffices to prove the commutativity of the first triangle. The second triangle can be viewed as the first one, but now in $(\overline{\mathscr{C}}, \overline{c})$ instead of $(\mathscr{C}, c)$.

If $Y$ is another object of $\mathscr{C}$ we have

$$
\begin{aligned}
c_{X,Y}(l_X \otimes \mathrm{Id}_Y)(c_{X,\underline{1}} \otimes \mathrm{Id}_Y) \\
&= c_{X,Y} l_{X \otimes Y} a_{\underline{1},X,Y}(c_{X,\underline{1}} \otimes \mathrm{Id}_Y) \\
&= l_{Y \otimes X}(\mathrm{Id}_{\underline{1}} \otimes c_{X,Y}) a_{\underline{1},X,Y}(c_{X,\underline{1}} \otimes \mathrm{Id}_Y) \\
&= l_{Y \otimes X} a_{\underline{1},Y,X} c_{X,\underline{1} \otimes Y} a_{X,\underline{1},Y} \\
&= (l_Y \otimes \mathrm{Id}_X) c_{X,\underline{1} \otimes Y} a_{X,\underline{1},Y} \\
&= c_{X,Y}(\mathrm{Id}_X \otimes l_Y) a_{X,\underline{1},Y} = c_{X,Y}(r_X \otimes \mathrm{Id}_Y),
\end{aligned}
$$

where in the first equality we used Proposition 1.5, in the second one the naturality of $l$, in the third one (1.5.1), in the fourth one again Proposition 1.5, in the fifth one the naturality of $c$, and in the last one (1.1.2). Thus, since $c_{X,Y}$ is an isomorphism, we get that $l_X c_{X,\underline{1}} \otimes \mathrm{Id}_Y = r_X \otimes \mathrm{Id}_Y$, for all $Y \in \mathscr{C}$. By the naturality of $r$ the two diagrams below are commutative

$$
\begin{array}{ccc}
(X \otimes \underline{1}) \otimes \underline{1} & \xrightarrow{\ r_{X \otimes \underline{1}}\ } & X \otimes \underline{1} \\
{\scriptstyle r_X \otimes \mathrm{Id}_{\underline{1}}} \Big\| \Big\downarrow {\scriptstyle l_X c_{X,\underline{1}} \otimes \mathrm{Id}_{\underline{1}}} & & {\scriptstyle r_X} \Big\| \Big\downarrow {\scriptstyle l_X c_{X,\underline{1}}} \\
X \otimes \underline{1} & \xrightarrow[\ \ r_X\ \ ]{} & X.
\end{array}
$$

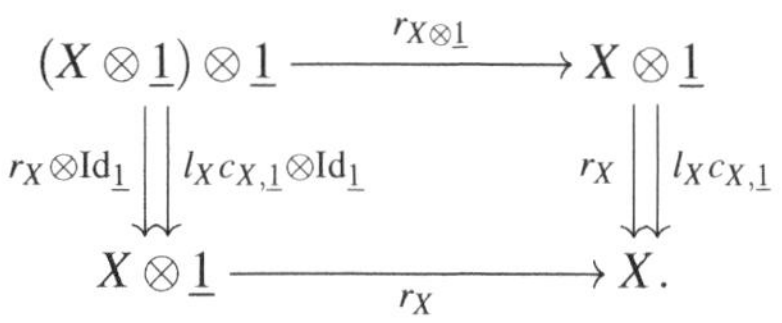

Therefore $r_X r_{X \otimes \underline{1}} = r_X(r_X \otimes \mathrm{Id}_{\underline{1}}) = r_X(l_X c_{X,\underline{1}} \otimes \mathrm{Id}_{\underline{1}}) = l_X c_{X,\underline{1}} r_{X \otimes \underline{1}}$. But $r_{X \otimes \underline{1}}$ is an isomorphism in $\mathscr{C}$, hence $r_X = l_X c_{X,\underline{1}}$. $\qquad\square$

**Corollary 1.50** *If $(\mathscr{C}, c)$ is a braided category then $c_{X,\underline{1}} = c_{\underline{1},X}^{-1}$, for all $X \in \mathscr{C}$.*

The next result is known as the categorical version of the Yang–Baxter equation.

**Proposition 1.51** *In any (pre-)braided category $(\mathscr{C}, c)$ the diagram*

$$
\begin{array}{c}
X \otimes (Y \otimes Z) \xrightarrow{\mathrm{Id}_X \otimes c_{Y,Z}} X \otimes (Z \otimes Y) \xrightarrow{a_{X,Z,Y}^{-1}} (X \otimes Z) \otimes Y \xrightarrow{c_{X,Z} \otimes \mathrm{Id}_Y} (Z \otimes X) \otimes Y
\end{array}
$$

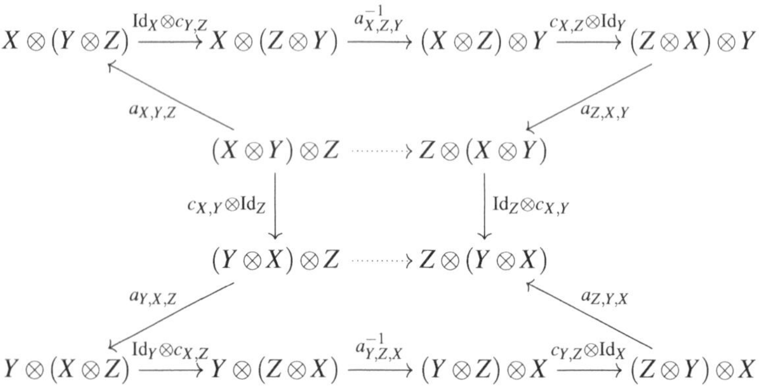

*is commutative, for any $X, Y, Z$ objects of $\mathscr{C}$.*

*Proof* As the picture suggests, we split the dodecagon diagram into two hexagons and a square. Then the upper dashed arrow is $c_{X \otimes Y, Z}$ and the lower dashed arrow is

$c_{Y\otimes X,Z}$, because of (1.5.2). Now, the square diagram is commutative because of the naturality of $c$, and the proof is finished. □

In diagrammatic notation, the categorical version of the Yang–Baxter equation can be expressed as

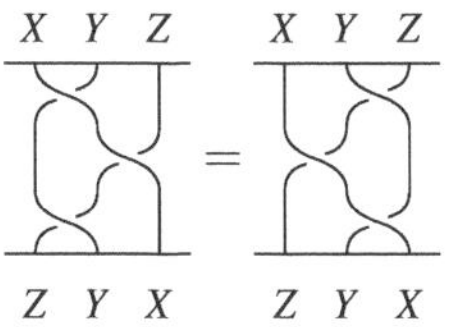


(1.5.10)

and holds for any objects $X, Y$ and $Z$ of $\mathscr{C}$.

We end this section by introducing the concept of a braided monoidal functor.

**Definition 1.52** Let $(\mathscr{C}, c)$ and $(\mathscr{D}, d)$ be (pre-)braided categories and $(F, \varphi_0, \varphi_2) : \mathscr{C} \to \mathscr{D}$ a strong monoidal functor. We call $F$ (pre-)braided monoidal if, for all objects $X, Y \in \mathscr{C}$, the following diagram is commutative:

$$\begin{array}{ccc} F(X)\otimes F(Y) & \xrightarrow{\varphi_{2,X,Y}} & F(X\otimes Y) \\ \Big\downarrow {\scriptstyle d_{F(X),F(Y)}} & & {\scriptstyle F(c_{X,Y})} \Big\downarrow \\ F(Y)\otimes F(X) & \xrightarrow{\varphi_{2,Y,X}} & F(Y\otimes X). \end{array} \tag{1.5.11}$$

**Example 1.53** If $\mathscr{C}$ is a (pre-)braided category then the identity functor $\mathrm{Id}_{\mathscr{C}}$ is a (pre-)braided monoidal functor.

**Remark 1.54** If $\mathscr{C} \xrightarrow{(F,\varphi_0,\varphi_2)} \mathscr{D} \xrightarrow{(G,\psi_0,\psi_2)} \mathscr{E}$ are (pre-)braided monoidal functors then so is $GF$, considered as a strong monoidal functor as in Remark 1.25.

**Definition 1.55** Let $\mathscr{C}$, $\mathscr{D}$ be (pre-)braided categories.

(i) $\mathscr{C}$ and $\mathscr{D}$ are called isomorphic as (pre-)braided categories if there exist two (pre-)braided monoidal functors $(F, \varphi_0, \varphi_2) : \mathscr{C} \to \mathscr{D}$ and $(G, \psi_0, \psi_2) : \mathscr{D} \to \mathscr{C}$ such that $FG = \mathrm{Id}_{\mathscr{D}}$ and $GF = \mathrm{Id}_{\mathscr{C}}$, as (pre-)braided monoidal functors.

(ii) $\mathscr{C}$ and $\mathscr{D}$ are called equivalent as (pre-)braided categories if there exist two (pre-)braided monoidal functors $(F, \varphi_0, \varphi_2) : \mathscr{C} \to \mathscr{D}$ and $(G, \psi_0, \psi_2) : \mathscr{D} \to \mathscr{C}$ such that $(F, G)$ gives a monoidal equivalence between $\mathscr{C}$ and $\mathscr{D}$. We also say that $F$ (and $G$ as well) is a braided equivalence functor.

Similar to the monoidal case, one can prove that two braided categories are isomorphic (resp. equivalent) as braided categories if and only if there exists a braided monoidal functor between them which is also an isomorphism (resp. equivalence) of categories. The case of the isomorphism is clear and we will prove below only the equivalence case.

**Proposition 1.56** *Let $(F, \varphi_0, \varphi_2) : \mathscr{C} \to \mathscr{D}$ be a strong monoidal functor that provides an equivalence of categories. If $\mathscr{C}$ is braided then there exists a unique braiding on $\mathscr{D}$ such that $F$ becomes a braided monoidal functor.*

*Proof* Let $G : \mathscr{D} \to \mathscr{C}$ be the equivalence inverse of $F$ and $\mu : \mathrm{Id}_{\mathscr{D}} \to FG$ and $\nu : GF \to \mathrm{Id}_{\mathscr{C}}$ the required natural monoidal isomorphisms. Assume, moreover, that

$$\mu^{-1}_{F(X)} = F(\nu_X) : FGF(X) \to F(X), \tag{1.5.12}$$

for all $X \in \mathscr{C}$, which is always possible in view of Remark 1.20 and the proof of Proposition 1.31.

Any braiding $c$ for $\mathscr{C}$ defines a braiding $d$ on $\mathscr{D}$ as follows. For any objects $U, V$ of $\mathscr{D}$ take $d_{U,V}$ to be the following composition:

$$\begin{array}{ccccc} U\square V & \xrightarrow{\mu_U\square\mu_V} & FG(U)\square FG(V) & \xrightarrow{\varphi_{2,G(U),G(V)}} & F(G(U)\otimes G(V)) \\ {\scriptstyle d_{U,V}}\Big\downarrow & & & & \Big\downarrow{\scriptstyle F(c_{G(U),G(V)})} \\ V\square U & \xleftarrow[\mu_V^{-1}\square\mu_U^{-1}]{} & FG(V)\square FG(U) & \xleftarrow[\varphi^{-1}_{2,G(V),G(U)}]{} & F(G(V)\otimes G(U)), \end{array} \tag{1.5.13}$$

where $\square$ is the tensor product of $\mathscr{D}$. Then $(\mathscr{D}, d = (d_{U,V})_{U,V\in\mathscr{D}})$ is a braided category and $F : (\mathscr{C}, c) \to (\mathscr{D}, d)$ becomes a braided monoidal functor.

Indeed, by the naturality of $\varphi_2$ and $c$, the three rectangle diagrams below are commutative:

$$\begin{array}{ccccc} F(X)\square F(Y) & \xrightarrow{F(\nu_X^{-1})\square F(\nu_Y^{-1})} & FGF(X)\square FGF(Y) & \xrightarrow{\varphi_{2,GF(X),GF(Y)}} & F(GF(X)\otimes GF(Y)) \\ & & {\scriptstyle F(\nu_X)\square F(\nu_Y)}\Big\downarrow & & \Big\downarrow{\scriptstyle F(\nu_X\otimes\nu_Y)} \\ & & F(X)\square F(Y) & \xrightarrow[\varphi_{2,X,Y}]{} & F(X\otimes Y)\ , \end{array}$$

$$\begin{array}{ccccc} F(GF(X)\otimes GF(Y)) & \xrightarrow{F(c_{GF(X),GF(Y)})} & F(GF(Y)\otimes GF(X)) & \xrightarrow{\varphi^{-1}_{2,GF(Y),GF(X)}} & FGF(Y)\square FGF(X) \\ {\scriptstyle F(\nu_X\otimes\nu_Y)}\Big\downarrow & & {\scriptstyle F(\nu_Y\otimes\nu_X)}\Big\downarrow & & \Big\downarrow{\scriptstyle F(\nu_Y)\square F(\nu_X)} \\ F(X\otimes Y) & \xrightarrow{F(c_{X,Y})} & F(Y\otimes X) & \xrightarrow[\varphi^{-1}_{2,Y,X}]{} & F(Y)\square F(X) \\ & & & & \Big\downarrow{\scriptstyle \varphi_{2,Y,X}} \\ & & & & F(Y\otimes X)\ . \end{array}$$

Together with (1.5.12) this guarantees that (1.5.11), specialized for our $F : (\mathscr{C}, c) \to (\mathscr{D}, d)$, is commutative. Since $c$ is a braiding for $\mathscr{C}$, it follows that $d$ is a braiding on $\mathscr{D}$, and consequently that $F : (\mathscr{C}, c) \to (\mathscr{D}, d)$ is a braided monoidal functor. We point out that the required commutativity of the two hexagonal diagrams corresponding to $d$ follows from those of $c$, by using that $\mu$ is a monoidal natural isomorphism, the naturality of $\varphi_2$ and its compatibility relations with the monoidal structures on $\mathscr{C}$ and $\mathscr{D}$, respectively. We leave the verification of the details to the reader.

We prove now the uniqueness part. Indeed, if $d'$ is a braiding on $\mathscr{D}$ making $F$ braided monoidal then

$$d'_{F(X),F(Y)} = \varphi^{-1}_{2,Y,X} F(c_{X,Y})\varphi_{2,X,Y}, \ \ \forall\, X,\ Y \in \mathscr{C}. \tag{1.5.14}$$

Since $d'$ is a natural transformation and $\mu_U$, $\mu_V$ are morphisms in $\mathscr{D}$, we get

$$\begin{aligned} d'_{U,V} &= (\mu_V^{-1}\Box\mu_U^{-1})d'_{FG(U),FG(V)}(\mu_U\Box\mu_V) \\ &\overset{(1.5.14)}{=} (\mu_V^{-1}\Box\mu_U^{-1})\varphi^{-1}_{2,G(V),G(U)}F(c_{G(U),G(V)})\varphi_{2,G(U),G(V)}(\mu_U\Box\mu_V) \overset{(1.5.13)}{=} d_{U,V}, \end{aligned}$$

for all $U,V \in \mathscr{D}$, as desired. □

**Corollary 1.57** *Let $(\mathscr{C},c)$ and $(\mathscr{D},d)$ be braided categories. A functor $F:\mathscr{C}\to\mathscr{D}$ is a braided equivalence if and only if $F$ is braided and an equivalence of categories.*

*Proof* We have only to prove that if $F$ is braided and an equivalence then $F$ has a monoidal equivalence inverse which is also a braided functor. Indeed, by the uniqueness of $d$ in Proposition 1.56 we have that $d$ is completely determined by (1.5.13). It is enough to show that, with respect to this braiding, the equivalence inverse $G$ of $F$ considered in the proof of Proposition 1.56 is a braided functor, that is, $G(d_{U,V})\varphi^G_{2,U,V} = \varphi^G_{2,V,U}c_{G(U),G(V)}$, for all $U,V \in \mathscr{D}$, where $\varphi^G_2$ is defined by the relations in (1.3.4). Note that $(F,G)$ provides a monoidal equivalence; see Proposition 1.31.

As $F$ is injective on morphisms, the last relation follows from

$$\begin{aligned} FG(d_{U,V})F(\varphi^G_{2,U,V}) &\overset{(1.3.5)}{=} FG(d_{U,V})\mu_{U\Box V}(\mu_U^{-1}\Box\mu_V^{-1})\varphi^{-1}_{2,G(U),G(V)} \\ &= \mu_{V\Box U}d_{U,V}(\mu_U^{-1}\Box\mu_V^{-1})\varphi^{-1}_{2,G(U),G(V)} \\ &\overset{(1.5.13)}{=} \mu_{V\Box U}(\mu_V^{-1}\Box\mu_U^{-1})\varphi^{-1}_{2,G(V),G(U)}F(c_{G(U),G(V)}) \\ &\overset{(1.3.5)}{=} F(\varphi^G_{2,V,U})F(c_{G(U),G(V)}), \end{aligned}$$

for all $U,V \in \mathscr{D}$, where in the second equality we used the fact that $\mu$ is a natural transformation and $d_{U,V}$ is a morphism in $\mathscr{D}$. □

By Proposition 1.56 and Corollary 1.57 we obtain the following.

**Corollary 1.58** *If $\mathscr{C}$ is a braided category then so is $\mathscr{C}^{\mathrm{str}}$. Thus any braided category is braided equivalent to a strict one.*

**Example 1.59** If $(\mathscr{C},c)$ is a braided category and $(\overline{\mathscr{C}},\overline{c})$ is the reverse braided category associated to it then $(\mathrm{Id}_{\mathscr{C}},\varphi_0 = \mathrm{Id}_{\underline{1}},\varphi_2 = (c_{Y,X})_{Y,X\in\mathscr{C}}) : (\mathscr{C},c) \to (\overline{\mathscr{C}},\overline{c})$ is a braided monoidal functor.

Indeed, for $(\mathrm{Id}_{\mathscr{C}},\varphi_0 = \mathrm{Id}_{\underline{1}},\varphi_2 = (c_{Y,X})_{Y,X\in\mathscr{C}})$ the first diagram in Definition 1.22 is commutative because of (1.5.1), (1.5.2) and Proposition 1.51, while the next two square diagrams in Definition 1.22 are commutative because of Proposition 1.49. Thus $(\mathrm{Id}_{\mathscr{C}},\varphi_0 = \mathrm{Id}_{\underline{1}},\varphi_2 = (c_{Y,X})_{Y,X\in\mathscr{C}})$ is a (strong) monoidal functor. It is also braided since the commutativity of the diagram in Definition 1.52 reduces to the condition $c_{X,Y}c_{Y,X} = c_{X,Y}c_{Y,X}$, for any $X,Y \in \mathscr{C}$.

## 1.6 Rigid Monoidal Categories

Throughout this section $(\mathscr{C},\otimes,\underline{1},a,l,r)$ is a monoidal category. Our aim is to define the concepts of left and right dual of an object of $\mathscr{C}$.

**Definition 1.60** Let $X,Y$ be objects of $\mathscr{C}$. A pairing between $Y$ and $X$ is a morphism $\varepsilon : Y\otimes X \to \underline{1}$ in $\mathscr{C}$.

A copairing in $\mathscr{C}$ between $Y$ and $X$ is a morphism $\eta : \underline{1} \to X\otimes Y$ in $\mathscr{C}$.

If $\varepsilon : Y\otimes X \to \underline{1}$ is a pairing between $Y$ and $X$ then $\tilde{\varepsilon} : \mathrm{Hom}_{\mathscr{C}}(Z,X\otimes T) \to \mathrm{Hom}_{\mathscr{C}}(Y\otimes Z,T)$ is the map sending $f\in \mathrm{Hom}_{\mathscr{C}}(Z,X\otimes T)$ to

$$\tilde{\varepsilon}(f) : Y\otimes Z \xrightarrow{\mathrm{Id}_Y\otimes f} Y\otimes(X\otimes T) \xrightarrow{a^{-1}_{Y,X,T}} (Y\otimes X)\otimes T \xrightarrow{\varepsilon\otimes\mathrm{Id}_T} \underline{1}\otimes T \xrightarrow{l_T} T.$$

**Definition 1.61** The pairing $\varepsilon : Y\otimes X \to \underline{1}$ is called exact if the associated map $\tilde{\varepsilon}$ is bijective, for all $Z,T$ objects of $\mathscr{C}$.

We next see that an exact pairing is a pairing for which there exists a copairing that is compatible with it in the following sense.

**Proposition 1.62** *A pairing $\varepsilon : Y\otimes X \to \underline{1}$ is exact if and only if there exists a copairing $\eta : \underline{1} \to X\otimes Y$ such that the following diagrams are commutative:*

$$\begin{array}{ccccc}
Y\otimes\underline{1} & & \xrightarrow{\quad\mathrm{Id}_Y\otimes\eta\quad} & & Y\otimes(X\otimes Y) \\
\downarrow{\scriptstyle r_Y} & & & & \downarrow{\scriptstyle a^{-1}_{Y,X,Y}} \\
Y & \xleftarrow{\ l_Y\ } & \underline{1}\otimes Y & \xleftarrow{\varepsilon\otimes\mathrm{Id}_Y} & (Y\otimes X)\otimes Y,
\end{array} \tag{1.6.1}$$

$$\begin{array}{ccccc}
\underline{1}\otimes X & & \xrightarrow{\quad\eta\otimes\mathrm{Id}_X\quad} & & (X\otimes Y)\otimes X \\
\downarrow{\scriptstyle l_X} & & & & \downarrow{\scriptstyle a_{X,Y,X}} \\
X & \xleftarrow{\ r_X\ } & X\otimes\underline{1} & \xleftarrow{\mathrm{Id}_X\otimes\varepsilon} & X\otimes(Y\otimes X).
\end{array} \tag{1.6.2}$$

*Moreover, if $\varepsilon$ is exact, the copairing $\eta$ for which the diagrams* (1.6.1) *and* (1.6.2) *are commutative is unique.*

*Proof* If $\varepsilon$ is exact then for $Z=\underline{1}$ and $T=Y$ let $\eta : \underline{1}\to X\otimes Y$ be the morphism in $\mathscr{C}$ satisfying $\tilde{\varepsilon}(\eta) = r_Y$; this is exactly the condition that the diagram in (1.6.1) is commutative (in particular, this proves the uniqueness part in the statement).

Let now $Z,T$ be arbitrary objects of $\mathscr{C}$. The morphism $\eta$ induces a map $\tilde{\eta} : \mathrm{Hom}_{\mathscr{C}}(Y\otimes Z,T) \to \mathrm{Hom}_{\mathscr{C}}(Z,X\otimes T)$, defined for any morphism $g : Y\otimes Z\to T$ by

$$\tilde{\eta}(g) : Z \xrightarrow{l_Z^{-1}} \underline{1}\otimes Z \xrightarrow{\eta\otimes\mathrm{Id}_Z} (X\otimes Y)\otimes Z \xrightarrow{a_{X,Y,Z}} X\otimes(Y\otimes Z) \xrightarrow{\mathrm{Id}_X\otimes g} X\otimes T. \tag{1.6.3}$$

We show that $\tilde{\varepsilon}$ and $\tilde{\eta}$ are inverses. It is enough to prove that $\tilde{\varepsilon}\tilde{\eta} = \mathrm{Id}_{\mathrm{Hom}_{\mathscr{C}}(Y\otimes Z,T)}$. Indeed, for any $Y\otimes Z \xrightarrow{\ g\ } T$ we have

$$\tilde{\varepsilon}\tilde{\eta}(g) = l_T(\varepsilon\otimes\mathrm{Id}_T)a^{-1}_{Y,X,T}(\mathrm{Id}_Y\otimes(\mathrm{Id}_X\otimes g)a_{X,Y,Z}(\eta\otimes\mathrm{Id}_Z)l_Z^{-1})$$

$$
\begin{aligned}
&= l_T(\varepsilon\otimes \mathrm{Id}_T)(\mathrm{Id}_{Y\otimes X}\otimes g)a^{-1}_{Y,X,Y\otimes Z}(\mathrm{Id}_Y\otimes a_{X,Y,Z}(\eta\otimes \mathrm{Id}_Z)l_Z^{-1})\\
&= l_T(\mathrm{Id}_{\underline{1}}\otimes g)(\varepsilon\otimes \mathrm{Id}_{Y\otimes Z})a^{-1}_{Y,X,Y\otimes Z}(\mathrm{Id}_Y\otimes a_{X,Y,Z}(\eta\otimes \mathrm{Id}_Z)l_Z^{-1})\\
&= g l_{Y\otimes Z}(\varepsilon\otimes \mathrm{Id}_{Y\otimes Z})a^{-1}_{Y,X,Y\otimes Z}(\mathrm{Id}_Y\otimes a_{X,Y,Z}(\eta\otimes \mathrm{Id}_Z)l_Z^{-1})\\
&= g l_{Y\otimes Z}(\varepsilon\otimes \mathrm{Id}_{Y\otimes Z})a_{Y\otimes X,Y,Z}(a^{-1}_{Y,X,Y}\otimes \mathrm{Id}_Z)a^{-1}_{Y,X\otimes Y,Z}(\mathrm{Id}_Y\otimes(\eta\otimes \mathrm{Id}_Z)l_Z^{-1})\\
&= g l_{Y\otimes Z}a_{\underline{1},Y,Z}((\varepsilon\otimes \mathrm{Id}_Y)\otimes \mathrm{Id}_Z)(a^{-1}_{Y,X,Y}\otimes \mathrm{Id}_Z)a^{-1}_{Y,X\otimes Y,Z}(\mathrm{Id}_Y\otimes(\eta\otimes \mathrm{Id}_Z)l_Z^{-1})\\
&= g(l_Y(\varepsilon\otimes \mathrm{Id}_Y)a^{-1}_{Y,X,Y}(\mathrm{Id}_Y\otimes\eta)\otimes \mathrm{Id}_Z)a^{-1}_{Y,\underline{1},Z}(\mathrm{Id}_Y\otimes l_Z^{-1})\\
&= g(l_Y(\varepsilon\otimes \mathrm{Id}_Y)a^{-1}_{Y,X,Y}(\mathrm{Id}_Y\otimes\eta)r_Y^{-1}\otimes \mathrm{Id}_Z)\\
&= g(\mathrm{Id}_Y\otimes \mathrm{Id}_Z)=g,
\end{aligned}
$$

where we used: the naturality of $a$ in the second equality, the naturality of $l$ in the fourth equality, (1.1.1) in the fifth equality, again the naturality of $a$ in the sixth equality, Proposition 1.5 and the naturality of $a$ in the seventh equality, (1.1.2) in the penultimate equality, and (1.6.1) in the last one.

Since $\tilde{\varepsilon}$ is bijective we also have $\tilde{\eta}\tilde{\varepsilon}=\mathrm{Id}_{\mathrm{Hom}_{\mathscr{C}}(Z,X\otimes T)}$. In particular, for $Z=X$, $T=\underline{1}$ and $f=r_X^{-1}:X\to X\otimes\underline{1}$ we have $\widetilde{\eta}\widetilde{\varepsilon}(f)=f$, which means

$$
r_X^{-1}=(\mathrm{Id}_X\otimes l_{\underline{1}}(\varepsilon\otimes \mathrm{Id}_{\underline{1}})a^{-1}_{Y,X,\underline{1}}(\mathrm{Id}_Y\otimes r_X^{-1}))a_{X,Y,X}(\eta\otimes \mathrm{Id}_X)l_X^{-1}.
$$

From Proposition 1.5 and since $r$ is a natural transformation we have

$$
l_{\underline{1}}(\varepsilon\otimes \mathrm{Id}_{\underline{1}})a^{-1}_{Y,X,\underline{1}}(\mathrm{Id}_Y\otimes r_X^{-1})=l_{\underline{1}}(\varepsilon\otimes \mathrm{Id}_{\underline{1}})r^{-1}_{Y\otimes X}=\varepsilon,
$$

and therefore $r_X^{-1}=(\mathrm{Id}_X\otimes\varepsilon)a_{X,Y,X}(\eta\otimes \mathrm{Id}_X)l_X^{-1}$, which means that the diagram in (1.6.2) is commutative.

Conversely, assume that there exists a copairing $\eta:\underline{1}\to X\otimes Y$ such that (1.6.1) and (1.6.2) hold, and define $\tilde{\eta}$ as in (1.6.3). From the above computation we know that $\tilde{\varepsilon}\tilde{\eta}=\mathrm{Id}_{\mathrm{Hom}_{\mathscr{C}}(Z,X\otimes T)}$. Thus we only have to show that $\tilde{\eta}\tilde{\varepsilon}=\mathrm{Id}_{\mathrm{Hom}_{\mathscr{C}}(Z,X\otimes T)}$.

For any $f\in\mathrm{Hom}_{\mathscr{C}}(Z,X\otimes T)$ we compute:

$$
\begin{aligned}
\tilde{\eta}\tilde{\varepsilon}(f)&=(\mathrm{Id}_X\otimes l_T(\varepsilon\otimes \mathrm{Id}_T)a^{-1}_{Y,X,T}(\mathrm{Id}_Y\otimes f))a_{X,Y,Z}(\eta\otimes \mathrm{Id}_Z)l_Z^{-1}\\
&=(\mathrm{Id}_X\otimes l_T(\varepsilon\otimes \mathrm{Id}_T)a^{-1}_{Y,X,T})a_{X,Y,X\otimes T}(\mathrm{Id}_{X\otimes Y}\otimes f)(\eta\otimes \mathrm{Id}_Z)l_Z^{-1}\\
&=(\mathrm{Id}_X\otimes l_T(\varepsilon\otimes \mathrm{Id}_T)a^{-1}_{Y,X,T})a_{X,Y,X\otimes T}(\eta\otimes \mathrm{Id}_{X\otimes T})(\mathrm{Id}_{\underline{1}}\otimes f)l_Z^{-1}\\
&=(\mathrm{Id}_X\otimes l_T(\varepsilon\otimes \mathrm{Id}_T))a_{X,Y\otimes X,T}(a_{X,Y,X}\otimes \mathrm{Id}_T)a^{-1}_{X\otimes Y,X,T}(\eta\otimes \mathrm{Id}_{X\otimes T})l^{-1}_{X\otimes T}f\\
&=(\mathrm{Id}_X\otimes l_T)a_{X,\underline{1},T}((\mathrm{Id}_X\otimes\varepsilon)a_{X,Y,X}\otimes \mathrm{Id}_T)a^{-1}_{X\otimes Y,X,T}(\eta\otimes \mathrm{Id}_{X\otimes T})l^{-1}_{X\otimes T}f\\
&=(r_X(\mathrm{Id}_X\otimes\varepsilon)a_{X,Y,X}(\eta\otimes \mathrm{Id}_X)\otimes \mathrm{Id}_T)a^{-1}_{\underline{1},X,T}l^{-1}_{X\otimes T}f\\
&=(r_X(\mathrm{Id}_X\otimes\varepsilon)a_{X,Y,X}(\eta\otimes \mathrm{Id}_X)l_X^{-1}\otimes \mathrm{Id}_T)f=(\mathrm{Id}_X\otimes \mathrm{Id}_T)f=f,
\end{aligned}
$$

as required. We used the naturality of $a$ in the second and fifth equality, the naturality of $l$ and (1.1.1) in the fourth equality, the naturality of $a$ and (1.1.2) in the sixth equality, Proposition 1.5 in the last but one equality, and (1.6.2) in the last one. □

**Definition 1.63** Let $\varepsilon:Y\otimes X\to\underline{1}$ be an exact pairing and $\eta:\underline{1}\to X\otimes Y$ the copairing satisfying (1.6.1) and (1.6.2). In this case we say that $(\eta,\varepsilon)$ is an adjunction between $Y$ and $X$, and denote this fact by $(\eta,\varepsilon) : Y \dashv X$. Moreover, we call $Y$ a left dual to $X$ and $X$ a right dual to $Y$.

A monoidal category is called left (resp. right) rigid if any object $X$ of $\mathscr{C}$ has a left (resp. right) dual. We call $\mathscr{C}$ rigid if it is left and right rigid.

If $\mathscr{C}$ is a left rigid monoidal category then the left dual of $X \in \mathscr{C}$ will be denoted by $X^*$, and the corresponding adjunction will be denoted by $(\mathrm{coev}_X, \mathrm{ev}_X) : X^* \dashv X$. Thus $\mathrm{ev}_X : X^* \otimes X \to \underline{1}$ and $\mathrm{coev}_X : \underline{1} \to X \otimes X^*$ are morphisms in $\mathscr{C}$ (called evaluation, respectively coevaluation) such that

$$r_X \circ (\mathrm{Id}_X \otimes \mathrm{ev}_X) \circ a_{X,X^*,X} \circ (\mathrm{coev}_X \otimes \mathrm{Id}_X) \circ l_X^{-1} = \mathrm{Id}_X, \tag{1.6.4}$$

$$l_{X^*} \circ (\mathrm{ev}_X \otimes \mathrm{Id}_{X^*}) \circ a^{-1}_{X^*,X,X^*} \circ (\mathrm{Id}_{X^*} \otimes \mathrm{coev}_X) \circ r^{-1}_{X^*} = \mathrm{Id}_{X^*}. \tag{1.6.5}$$

In this situation we denote $\mathrm{ev}_X =$ 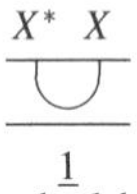

and $\mathrm{coev}_X =$ 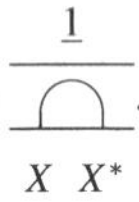

. Hence, when $\mathscr{C}$ is strict monoidal the following relations hold:

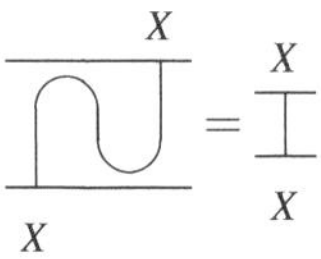

and
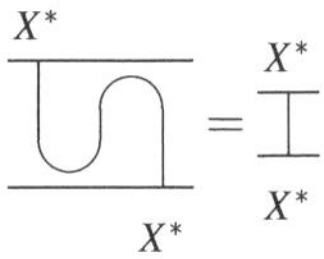

. (1.6.6)

Similarly, $\mathscr{C}$ is right rigid if for any $X \in \mathscr{C}$ there exist an object ${}^*X \in \mathscr{C}$ and morphisms $\mathrm{ev}'_X : X \otimes {}^*X \to \underline{1}$ and $\mathrm{coev}'_X : \underline{1} \to {}^*X \otimes X$ such that

$$l_X \circ (\mathrm{ev}'_X \otimes \mathrm{Id}_X) \circ a^{-1}_{X,{}^*X,X} \circ (\mathrm{Id}_X \otimes \mathrm{coev}'_X) \circ r_X^{-1} = \mathrm{Id}_X, \tag{1.6.7}$$

$$r_{{}^*X} \circ (\mathrm{Id}_{{}^*X} \otimes \mathrm{ev}'_X) \circ a_{{}^*X,X,{}^*X} \circ (\mathrm{coev}'_X \otimes \mathrm{Id}_{{}^*X}) \circ l^{-1}_{{}^*X} = \mathrm{Id}_{{}^*X}. \tag{1.6.8}$$

In what follows we will denote $\mathrm{ev}'_X :=$ 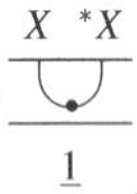

and $\mathrm{coev}'_X :=$ 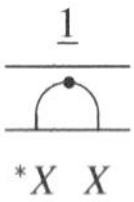

. Then if $\mathscr{C}$ is strict the relations (1.6.7) and (1.6.8) above can be written as

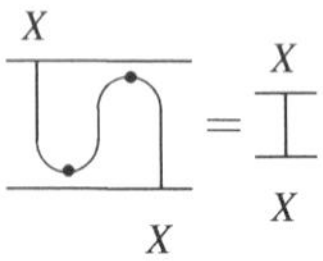

and
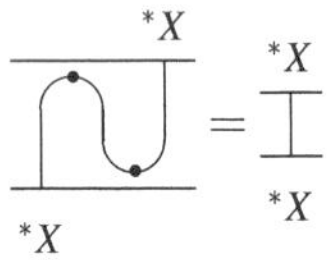

, (1.6.9)

respectively. Thus a right dual for $X$ in $\mathscr{C}$ is nothing else than a left dual for $X$ in $\overline{\mathscr{C}}$, the reverse monoidal category associated to $\mathscr{C}$.

**Example 1.64** For any field $k$ the category ${}_k\mathscr{M}^{\mathrm{fd}}$ of finite-dimensional $k$-vector spaces is rigid monoidal.

*Proof* For any $k$-vector space $V$ denote by $V^* = \mathrm{Hom}_k(V,k)$ the linear dual space of $V$. If, moreover, $V$ is finite dimensional then for a basis $\{v_i\}_i$ of $V$ consider its dual basis $\{v^i\}_i$ in $V^*$; thus

$$v^i(v_j) = \delta_{i,j}, \quad \sum_i v^*(v_i)v^i = v^* \quad \text{and} \quad \sum_i v^i(v)v_i = v,$$

for all $v^* \in V^*$, $v \in V$, where $\delta_{i,j}$ is the Kronecker delta symbol:

$$\delta_{i,j} = \begin{cases} 1 & \text{if } i = j \\ 0 & \text{otherwise.} \end{cases}$$

Then it can be easily seen that $V^*$ together with $\mathrm{ev}_V : V^* \otimes V \to k$, $\mathrm{ev}_V(v^* \otimes v) = v^*(v)$, for all $v^* \in V^*$ and $v \in V$, and with $\mathrm{coev}_V : k \to V \otimes V^*$ defined by $\mathrm{coev}_V(1) = \sum_i v_i \otimes v^i$, is a left dual object of $V$.

Similarly, a right dual for $V$ is the same vector space $V^*$ coming now with the linear maps $\mathrm{ev}'_V : V \otimes V^* \ni v \otimes v^* \mapsto v^*(v) \in k$ and $\mathrm{coev}'_V : k \to V^* \otimes V$ given by $\mathrm{coev}'_V(1) = \sum_i v^i \otimes v_i$. The verification of all these details is left to the reader. □

**Example 1.65** Let $G$ be a group and $\phi \in H^3(G, k^*)$ a normalized 3-cocycle on $G$ with coefficients in $k^*$, $k$ a field. Then $\mathrm{vect}_\phi^G$, the category of finite-dimensional $G$-graded vector spaces, endowed with the monoidal structure of $\mathrm{Vect}_\phi^G$ from Proposition 1.13, is a rigid monoidal category.

*Proof* If $V$ is a finite-dimensional $G$-graded vector space with a basis $\{{}_iv\}_i$ and dual basis $\{{}^iv\}_i$ in $V^*$, the left dual of $V$ is $V^* = \mathrm{Hom}_k(V, k)$, the vector space of $k$-linear maps from $V$ to $k$, with homogeneous component of degree $g \in G$ defined by

$$V_g^* = \left\{ \left( v \mapsto v^*(v^{g^{-1}}) \right) \mid v^* \in V^* \right\} = \left\{ v^* \in V^* \mid v^*_{|V_\sigma} = 0, \ \forall\, \sigma \neq g^{-1} \right\},$$

where by $v^{g^{-1}}$ we have denoted the component of degree $g^{-1}$ of $v \in V$. The evaluation and coevaluation maps are respectively given by

$$\begin{aligned} &\mathrm{ev}_V : V^* \otimes V \to k, \quad \mathrm{ev}_V(v^* \otimes v) = v^*(v), \quad \forall\, v^* \in V^*, \quad v \in V, \\ &\mathrm{coev}_V : k \to V \otimes V^*, \quad \mathrm{coev}_V(1) = \sum_{i; g \in G} \phi(g, g^{-1}, g)^{-1} ({}_iv)^g \otimes {}^iv. \end{aligned}$$

The right dual ${}^*V$ of $V$ coincides, as a $G$-graded vector space, with $V^*$ described above, but the evaluation and coevaluation maps are now respectively defined by

$$\begin{aligned} &\mathrm{ev}'_V : V \otimes {}^*V \to k, \quad \mathrm{ev}'_V(v \otimes {}^*v) = {}^*v(v), \quad \forall\, {}^*v \in {}^*V, \quad v \in V, \\ &\mathrm{coev}'_V : k \to {}^*V \otimes V, \quad \mathrm{coev}'_V(1) = \sum_{i; g \in G} \phi(g^{-1}, g, g^{-1})^{-1}\ {}^iv \otimes ({}_iv)^g. \end{aligned}$$

For the moment we leave the verification of all these details to the reader. Later on we will see that this result follows from Corollary 3.53. □

We show that the left/right dual object, if it exists, is unique up to isomorphism.

**Proposition 1.66** *In a left/right rigid monoidal category $\mathscr{C}$ a left/right dual of an object is unique up to an isomorphism in $\mathscr{C}$.*

*Proof* Let $Y$ and $X^*$ be left duals for $X$, with notation as before for $\varepsilon, \eta, \mathrm{ev}_X, \mathrm{coev}_X$. One can check directly that $g : X^* \to Y$ defined by the composition

$$X^* \xrightarrow{r_{X^*}^{-1}} X^* \otimes \underline{1} \xrightarrow{\mathrm{Id}_{X^*}\otimes\eta} X^* \otimes (X \otimes Y) \xrightarrow{a_{X^*,X,Y}^{-1}} (X^* \otimes X) \otimes Y \xrightarrow{\mathrm{ev}_X\otimes\mathrm{Id}_Y} \underline{1} \otimes Y \xrightarrow{l_Y} Y$$

is an isomorphism in $\mathscr{C}$ with inverse defined by the composition

$$Y \xrightarrow{r_Y^{-1}} Y \otimes \underline{1} \xrightarrow{\mathrm{Id}_Y\otimes\mathrm{coev}_X} Y \otimes (X \otimes X^*) \xrightarrow{a_{Y,X,X^*}^{-1}} (Y \otimes X) \otimes X^* \xrightarrow{\varepsilon\otimes\mathrm{Id}_{X^*}} \underline{1} \otimes X^* \xrightarrow{l_{X^*}} X^*.$$

For instance, if we assume $\mathscr{C}$ strict and denote $\varepsilon =$ 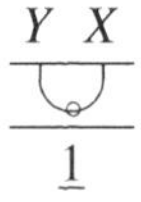 and $\eta =$ 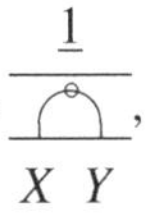, then

$g^{-1}g =$ 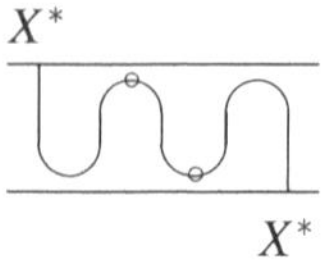 $= \mathrm{Id}_{X^*}$ and $gg^{-1} =$ 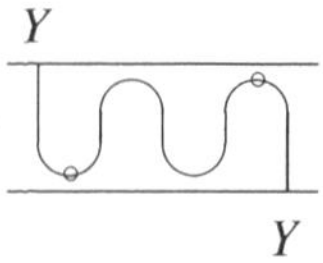 $= \mathrm{Id}_Y$.

The right-handed version is similar; the details are left to the reader. □

Strong monoidal functors preserve adjunctions, and consequently dual objects.

**Proposition 1.67** *For $(F, \varphi_0, \varphi_2) : \mathscr{C} \to \mathscr{D}$ a strong monoidal functor and $(\eta, \varepsilon) : Y \dashv X$ an adjunction in $\mathscr{C}$, $(\eta^F, \varepsilon^F) : F(Y) \dashv F(X)$ is an adjunction in $\mathscr{D}$, where*

$$\varepsilon^F : F(Y) \otimes F(X) \xrightarrow{\varphi_{2,Y,X}} F(Y \otimes X) \xrightarrow{F(\varepsilon)} F(\underline{1}) \xrightarrow{\varphi_0^{-1}} \underline{1},$$

$$\eta^F : \underline{1} \xrightarrow{\varphi_0} F(\underline{1}) \xrightarrow{F(\eta)} F(X \otimes Y) \xrightarrow{\varphi_{2,X,Y}^{-1}} F(X) \otimes F(Y).$$

*Consequently, if $\mathscr{C}$ is left rigid and $(G, \psi_0, \psi_2) : \mathscr{C} \to \mathscr{D}$ is a strong monoidal functor, any monoidal natural transformation $\omega : F \to G$ is a monoidal natural isomorphism.*

*Proof* By the commutativity of the diagrams in Definition 1.22 we have

$$\begin{aligned} & l_{F(Y)}(\varepsilon^F \otimes \mathrm{Id}_{F(Y)})a_{F(Y),F(X),F(Y)}^{-1}(\mathrm{Id}_{F(Y)} \otimes \eta^F)r_{F(Y)}^{-1} \\ &\quad = F(l_Y)\varphi_{2,\underline{1},Y}(F(\varepsilon)\varphi_{2,Y,X} \otimes \mathrm{Id}_{F(Y)}) \\ &\quad\quad a_{F(Y),F(X),F(Y)}^{-1}(\mathrm{Id}_{F(Y)} \otimes \varphi_{2,X,Y}^{-1}F(\eta))\varphi_{2,Y,\underline{1}}^{-1}F(r_Y^{-1}) \\ &\quad = F(l_Y)\varphi_{2,\underline{1},Y}(F(\varepsilon) \otimes F(\mathrm{Id}_Y))\varphi_{2,Y\otimes X,Y}^{-1}F(a_{Y,X,Y}^{-1}) \\ &\quad\quad \varphi_{2,Y,X\otimes Y}(F(\mathrm{Id}_Y) \otimes F(\eta))\varphi_{2,Y,\underline{1}}^{-1}F(r_Y^{-1}) \\ &\quad = F(l_Y(\varepsilon \otimes \mathrm{Id}_Y)a_{Y,X,Y}^{-1}(\mathrm{Id}_Y \otimes \eta)r_Y^{-1}) = F(\mathrm{Id}_Y) = \mathrm{Id}_{F(Y)}, \end{aligned}$$

where in the third equality we used the naturality of $\varphi_2$ twice. The corresponding equality for (1.6.2) can be proved in a similar manner, so it is left to the reader.

Assume now that $\mathscr{C}$ is left rigid and take $\omega : F \to G$ a monoidal natural transformation. We claim that, for any object $X$ of $\mathscr{C}$, $\omega_X$ is an isomorphism in $\mathscr{D}$ with

$$\omega_X^{-1} : G(X) \xrightarrow{l_{G(X)}^{-1}} \underline{1} \otimes G(X) \xrightarrow{\eta^F\otimes\mathrm{Id}_{G(X)}} (F(X) \otimes F(Y)) \otimes G(X) \xrightarrow{(\mathrm{Id}_{F(X)}\otimes\omega_Y)\otimes\mathrm{Id}_{G(X)}}$$

$$(F(X)\otimes G(Y))\otimes G(X) \overset{a_{F(X),G(Y),G(X)}}{\longrightarrow} F(X)\otimes(G(Y)\otimes G(X))$$
$$\overset{\mathrm{Id}_{F(X)}\otimes\varepsilon^G}{\longrightarrow} F(X)\otimes\underline{1} \overset{r_{F(X)}}{\longrightarrow} F(X),$$

where $(\eta,\varepsilon): Y \dashv X$ is an adjunction in $\mathscr{C}$. Indeed, we have

$$\begin{aligned}
\omega_X\omega_X^{-1} &= \omega_X r_{F(X)}(\mathrm{Id}_{F(X)}\otimes\varepsilon^G)a_{F(X),G(Y),G(X)}((\mathrm{Id}_{F(X)}\otimes\omega_Y)\eta^F\otimes\mathrm{Id}_{G(X)})l_{G(X)}^{-1}\\
&= r_{G(X)}(\omega_X\otimes\varepsilon^G)a_{F(X),G(Y),G(X)}((\mathrm{Id}_{F(X)}\otimes\omega_Y)\eta^F\otimes\mathrm{Id}_{G(X)})l_{G(X)}^{-1}\\
&= r_{G(X)}(\mathrm{Id}_{G(X)}\otimes\varepsilon^G)(\omega_X\otimes\mathrm{Id}_{G(Y)\otimes G(X)})a_{F(X),G(Y),G(X)}\\
&\qquad((\mathrm{Id}_{F(X)}\otimes\omega_Y)\eta^F\otimes\mathrm{Id}_{G(X)})l_{G(X)}^{-1}\\
&= r_{G(X)}(\mathrm{Id}_{G(X)}\otimes\varepsilon^G)a_{G(X),G(Y),G(X)}((\omega_X\otimes\omega_Y)\eta^F\otimes\mathrm{Id}_{G(X)})l_{G(X)}^{-1}\\
&= r_{G(X)}(\mathrm{Id}_{G(X)}\otimes\varepsilon^G)a_{G(X),G(Y),G(X)}(\psi_{2,X,Y}^{-1}\omega_{X\otimes Y}F(\eta)\varphi_0\otimes\mathrm{Id}_{G(X)})l_{G(X)}^{-1}\\
&= r_{G(X)}(\mathrm{Id}_{G(X)}\otimes\varepsilon^G)a_{G(X),G(Y),G(X)}(\psi_{2,X,Y}^{-1}G(\eta)\omega_{\underline{1}}\varphi_0\otimes\mathrm{Id}_{G(X)})l_{G(X)}^{-1}\\
&= r_{G(X)}(\mathrm{Id}_{G(X)}\otimes\varepsilon^G)a_{G(X),G(Y),G(X)}(\eta^G\otimes\mathrm{Id}_{G(X)})l_{G(X)}^{-1} \overset{(1.6.2)}{=} \mathrm{Id}_{G(X)},
\end{aligned}$$

where we also used: the naturality of $r$ in the second equality, the naturality of $a$ in the fourth equality, the commutativity of the square diagram in Definition 1.29 in the fifth equality, the naturality of $\omega$ in the sixth equality, the commutativity of the triangle diagram in Definition 1.29 and the definition of $\eta^G$ in the last but one equality. Similarly, $\omega_X^{-1}\omega_X = \mathrm{Id}_{F(X)}$. So the proof is finished. □

## 1.7 The Left and Right Dual Functors

Let $\mathscr{C}$ be a monoidal category which is left/right rigid. Using the existence of left/right dual objects in $\mathscr{C}$ we will construct strong monoidal functors $(-)^*, {}^*(-) : \mathscr{C} \to \overline{\mathscr{C}^{\mathrm{opp}}}$. Recall that $\overline{\mathscr{C}^{\mathrm{opp}}}$ is our notation for the reverse of the opposite category associated to $\mathscr{C}$, see Proposition 1.3 and Remark 1.4.

Let us start by defining explicitly the two functors $(-)^*, {}^*(-) : \mathscr{C} \to \mathscr{C}^{\mathrm{opp}}$.

**Proposition 1.68** *Let $\mathscr{C}$ be a left (resp. right) rigid monoidal category and for any object $X$ of $\mathscr{C}$ consider a left (resp. right) dual object $X^*$ (resp. ${}^*X$) for $X$. Then $(-)^* : \mathscr{C} \ni X \mapsto X^* \in \mathscr{C}^{\mathrm{opp}}$ (resp. ${}^*(-) : \mathscr{C} \ni X \mapsto {}^*X \in \mathscr{C}^{\mathrm{opp}}$) defines a functor.*

*Proof* We prove that $F := (-)^*$ is a functor. For $f : X \to Y$ morphism in $\mathscr{C}$, let

$$f^* := \Big(Y^* \overset{r_{Y^*}^{-1}}{\to} Y^*\otimes\underline{1} \overset{\mathrm{Id}_{Y^*}\otimes\mathrm{coev}_X}{\longrightarrow} Y^*\otimes(X\otimes X^*) \overset{\mathrm{Id}_{Y^*}\otimes(f\otimes\mathrm{Id}_{X^*})}{\longrightarrow} Y^*\otimes(Y\otimes X^*)$$
$$\overset{a_{Y^*,Y,X^*}^{-1}}{\longrightarrow} (Y^*\otimes Y)\otimes X^* \overset{\mathrm{ev}_Y\otimes\mathrm{Id}_{X^*}}{\longrightarrow} \underline{1}\otimes X^* \overset{l_{X^*}}{\to} X^*\Big).$$

If we take $F(f) = f^*$, $F$ is a functor. To see this, assume $\mathscr{C}$ strict monoidal. Then

$f^* =$ 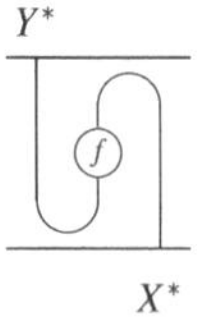

, and so $g^* \circ f^* =$ 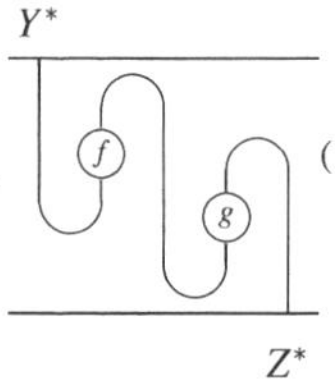

$\overset{(1.6.6)}{=}$ 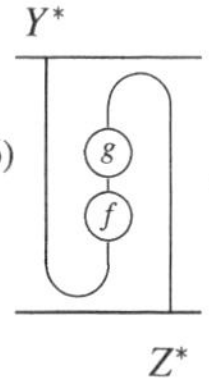

$= (f \circ g)^*$,

for any morphisms $Z \xrightarrow{g} X \xrightarrow{f} Y$ in $\mathscr{C}$. Likewise, $(\mathrm{Id}_X)^* = \mathrm{Id}_{X^*}$. Thus, $F$ is indeed a functor from $\mathscr{C}$ to $\mathscr{C}^{\mathrm{opp}}$, as claimed.

The right-handed version in the statement follows from the left-handed one by replacing $\mathscr{C}$ with the reverse category $\overline{\mathscr{C}}$; the details are left to the reader. □

We call $f^*$ the left transpose of $f$ in $\mathscr{C}$. In the right-handed version the (right) transpose of a morphism $f : X \to Y$ in $\mathscr{C}$ is ${}^*f :=$ 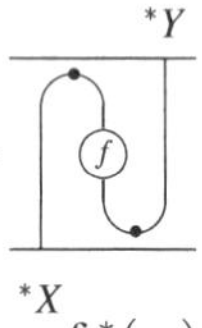

$: {}^*Y \to {}^*X$.

We focus now on the strong monoidal structures of ${}^*(-)$ and $(-)^*$.

**Proposition 1.69** *Let $\mathscr{C}$ be a monoidal category and $X, Y$ objects of $\mathscr{C}$ that admit left duals. Then $Y^* \otimes X^*$ is a left dual for $X \otimes Y$.*

*Proof* For simplicity assume that $\mathscr{C}$ is strict monoidal. If $\eta : \underline{1} \to X \otimes Y \otimes Y^* \otimes X^*$ and $\varepsilon : Y^* \otimes X^* \otimes X \otimes Y \to \underline{1}$ are defined by

$$\eta : \underline{1} \overset{\mathrm{coev}_X}{\to} X \otimes X^* \overset{\mathrm{Id}_X \otimes \mathrm{coev}_Y \otimes \mathrm{Id}_{X^*}}{\longrightarrow} X \otimes Y \otimes Y^* \otimes X^*,$$
$$\varepsilon : Y^* \otimes X^* \otimes X \otimes Y \overset{\mathrm{Id}_{Y^*} \otimes \mathrm{ev}_X \otimes \mathrm{Id}_Y}{\longrightarrow} Y^* \otimes Y \overset{\mathrm{ev}_Y}{\to} \underline{1},$$

then from the definition of a left dual it follows that $(\eta, \varepsilon) : Y^* \otimes X^* \dashv X \otimes Y$ is an adjunction in $\mathscr{C}$. Thus $Y^* \otimes X^*$ is a left dual for $X \otimes Y$. □

**Corollary 1.70** *Let $\mathscr{C}$ be a left rigid monoidal category. Then for any pair of objects $X, Y$ of $\mathscr{C}$ we have $(X \otimes Y)^* \cong Y^* \otimes X^*$ as objects of $\mathscr{C}$.*

*Proof* Since $\mathscr{C}$ is left rigid there exists an adjunction of the form $(\mathrm{coev}_{X \otimes Y}, \mathrm{ev}_{X \otimes Y}) : (X \otimes Y)^* \dashv X \otimes Y$ in $\mathscr{C}$. If $(\eta, \varepsilon) : Y^* \otimes X^* \dashv X \otimes Y$ is the adjunction considered in Proposition 1.69 then, by Proposition 1.66, we get that $(X \otimes Y)^* \cong Y^* \otimes X^*$ as objects of $\mathscr{C}$. Moreover, the isomorphism is produced by $\lambda_{X,Y} : (X \otimes Y)^* \to Y^* \otimes X^*$, which in the case when $\mathscr{C}$ is strict monoidal reads

$$\lambda_{X,Y} = (\mathrm{ev}_{X \otimes Y} \otimes \mathrm{Id}_{Y^* \otimes X^*}) \circ (\mathrm{Id}_{(X \otimes Y)^*} \otimes (\mathrm{Id}_X \otimes \mathrm{coev}_Y \otimes \mathrm{Id}_{X^*}) \circ \mathrm{coev}_X). \quad (1.7.1)$$

Its inverse is defined by $\lambda_{X,Y}^{-1} : Y^* \otimes X^* \to (X \otimes Y)^*$,

$$\lambda_{X,Y}^{-1} = (\mathrm{ev}_Y \circ (\mathrm{Id}_{Y^*} \otimes \mathrm{ev}_X \otimes \mathrm{Id}_Y) \otimes \mathrm{Id}_{(X \otimes Y)^*}) \circ (\mathrm{Id}_{Y^* \otimes X^*} \otimes \mathrm{coev}_{X \otimes Y}). \quad (1.7.2)$$

A direct computation shows also directly that $\lambda_{X,Y}$ and $\lambda_{X,Y}^{-1}$ are inverses. □

If $X_1,\dots,X_n$ are objects of a left rigid (strict) monoidal category, in diagrammatic notation we denote the evaluation and coevaluation morphisms of $X_1\otimes\cdots\otimes X_n$ by

$\mathrm{ev}_{X_1\otimes\cdots\otimes X_n} =$ 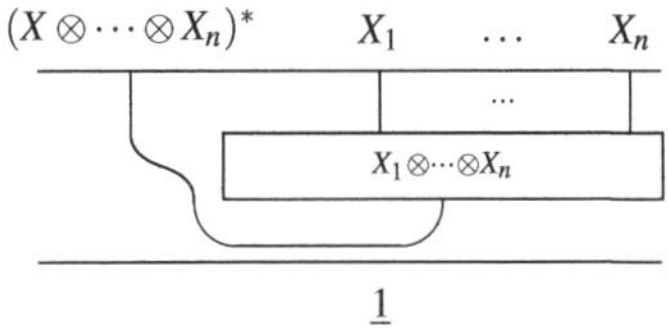 ,


$\mathrm{coev}_{X_1\otimes\cdots\otimes X_n} =$ 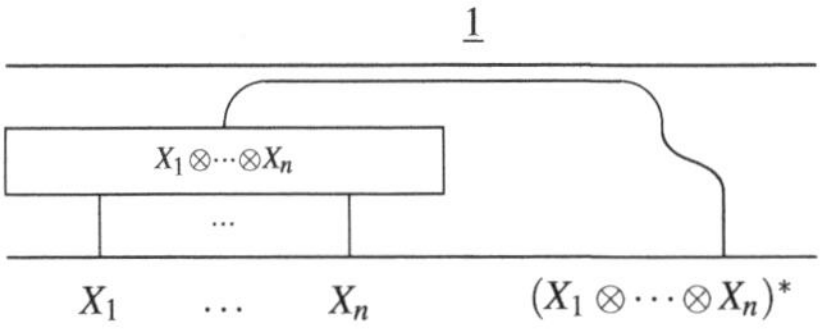 ,


respectively. We adopt a similar notation for the case when $\mathscr{C}$ is right rigid monoidal.

With this notation the morphism $\lambda$ in (1.7.1) and its inverse take the form

$\lambda_{X,Y} =$ 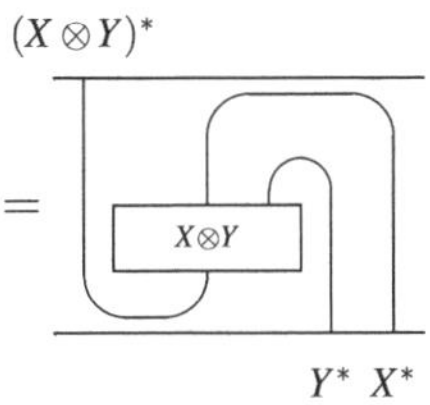 , $\lambda^{-1}_{X,Y} =$ 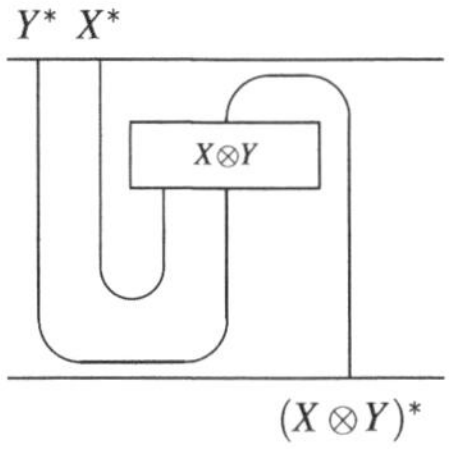 .


When $\mathscr{C}$ is right rigid there exists $\lambda'_{X,Y} : {}^*(X\otimes Y)\to {}^*Y\otimes {}^*X$, an isomorphism in $\mathscr{C}$:

$\lambda'_{X,Y} =$ 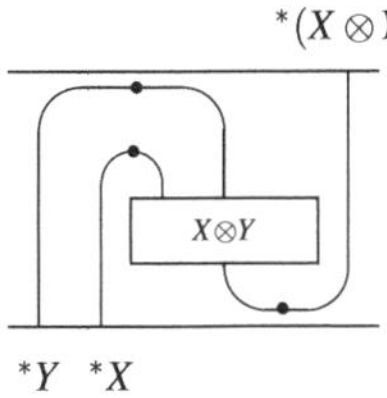 with inverse $\lambda'^{-1}_{X,Y} =$ 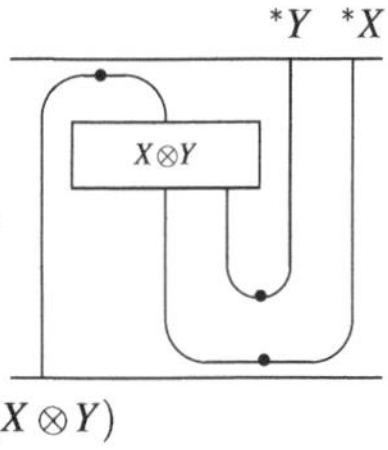 .


With the help of these isomorphisms in $\mathscr{C}$ the functors constructed in Proposition 1.68 become strong monoidal. To see this, observe first that

$$\overline{\mathscr{C}^{\mathrm{opp}}} = \overline{\mathscr{C}}^{\mathrm{opp}} = (\mathscr{C}^{\mathrm{opp}}, \otimes\circ\tau, \underline{1}, (a_{Z,Y,X})_{X,Y,Z\in\mathscr{C}}, r^{-1}, l^{-1})$$

for any monoidal category $(\mathscr{C},\otimes,\underline{1},a,l,r)$. Then we have the following.

**Proposition 1.71** *Let $\mathscr{C}$ be a left, respectively right, rigid monoidal category. Then the functor $(-)^*$, respectively ${}^*(-)$, defined in Proposition 1.68 is strong monoidal if it is viewed as a functor from $\mathscr{C}$ to $\overline{\mathscr{C}^{\mathrm{opp}}}$.*

*Proof* We prove only the statement for $(-)^*$; the proof for ${}^*(-)$ is similar.

By (1.1.2) and Proposition 1.5 it follows that $\underline{1}$ is a self-dual object of $\mathscr{C}$, that is, $\underline{1}$ is a left dual for itself. The adjunction $(\eta,\varepsilon) : \underline{1} \dashv \underline{1}$ is given by $\varepsilon = l_{\underline{1}} = r_{\underline{1}}$

and $\eta = l_{\underline{1}}^{-1} = r_{\underline{1}}^{-1}$, respectively. So there exists $\varphi_0 : \underline{1}^* \to \underline{1}$ an isomorphism in $\mathscr{C}$. Concretely, according to Proposition 1.66 we have

$$\begin{aligned}\varphi_0 &= l_{\underline{1}}(\mathrm{ev}_{\underline{1}} \otimes \mathrm{Id}_{\underline{1}})a^{-1}_{\underline{1}^*,\underline{1},\underline{1}}(\mathrm{Id}_{\underline{1}^*} \otimes l_{\underline{1}}^{-1})r_{\underline{1}^*}^{-1} \\ &= l_{\underline{1}}(\mathrm{ev}_{\underline{1}} \otimes \mathrm{Id}_{\underline{1}})(r_{\underline{1}^*}^{-1} \otimes \mathrm{Id}_{\underline{1}})r_{\underline{1}^*}^{-1} = l_{\underline{1}}r_{\underline{1}}^{-1}\mathrm{ev}_{\underline{1}}r_{\underline{1}^*}^{-1} = \mathrm{ev}_{\underline{1}}r_{\underline{1}^*}^{-1},\end{aligned}$$

where we used (1.1.2) in the second equality, the naturality of $r$ and (1.1.6) in the third equality and Proposition 1.5 in the fourth equality.

We claim that $((-)^*, \varphi_0, \varphi_2 = (\lambda_{X,Y})_{X,Y\in\mathscr{C}})$ is a strong monoidal functor from $\mathscr{C}$ to $\overline{\mathscr{C}^{\mathrm{opp}}}$. We assume, as usual, $\mathscr{C}$ strict monoidal and notice that the commutativity of the two square diagrams in Definition 1.22, specialized to our situation, reads

$$(\mathrm{Id}_{X^*} \otimes \varphi_0)\lambda_{\underline{1},X}l_X^* = r_{X^*}^{-1} \text{ and } (\varphi_0 \otimes \mathrm{Id}_{X^*})\lambda_{X,\underline{1}}r_X^* = l_{X^*}^{-1}.$$

We check only the second equality; the first one is similar. For this, note that

$$\begin{aligned}\lambda_{X,\underline{1}}r_X^* &= \lambda_{X,\underline{1}}l_{(X\otimes\underline{1})^*}(\mathrm{ev}_X \otimes \mathrm{Id}_{(X\otimes\underline{1})^*})a^{-1}_{X^*,X,(X\otimes\underline{1})^*} \\ &\quad \left(\mathrm{Id}_{X^*} \otimes (r_X \otimes \mathrm{Id}_{(X\otimes\underline{1})^*})\mathrm{coev}_{X\otimes\underline{1}}\right) r_{X^*}^{-1} \\ &= l_{\underline{1}^*\otimes X^*}(\mathrm{Id}_{\underline{1}} \otimes \lambda_{X,\underline{1}})(\mathrm{ev}_X \otimes \mathrm{Id}_{(X\otimes\underline{1})^*})a^{-1}_{X^*,X,(X\otimes\underline{1})^*} \\ &\quad \left(\mathrm{Id}_{X^*} \otimes (r_X \otimes \mathrm{Id}_{(X\otimes\underline{1})^*})\mathrm{coev}_{X\otimes\underline{1}}\right) r_{X^*}^{-1},\end{aligned}$$

because of the naturality of $l$, and therefore

$$\begin{aligned}&(\varphi_0 \otimes \mathrm{Id}_{X^*})\lambda_{X,\underline{1}}r_X^* \\ &\quad = l_{\underline{1}\otimes X^*}(\mathrm{Id}_{\underline{1}} \otimes (\varphi_0 \otimes \mathrm{Id}_{X^*})\lambda_{X,\underline{1}})(\mathrm{ev}_X \otimes \mathrm{Id}_{(X\otimes\underline{1})^*})a^{-1}_{X^*,X,(X\otimes\underline{1})^*} \\ &\quad\quad \left(\mathrm{Id}_{X^*} \otimes (r_X \otimes \mathrm{Id}_{(X\otimes\underline{1})^*})\mathrm{coev}_{X\otimes\underline{1}}\right) r_{X^*}^{-1} \\ &\quad = l_{\underline{1}\otimes X^*}(\mathrm{ev}_X \otimes \mathrm{Id}_{\underline{1}\otimes X^*})(\mathrm{Id}_{X^*\otimes X} \otimes (\varphi_0 \otimes \mathrm{Id}_{X^*})\lambda_{X,\underline{1}}) \\ &\quad\quad a^{-1}_{X^*,X,(X\otimes\underline{1})^*}\left(\mathrm{Id}_{X^*} \otimes (r_X \otimes \mathrm{Id}_{(X\otimes\underline{1})^*})\mathrm{coev}_{X\otimes\underline{1}}\right) r_{X^*}^{-1} \\ &\quad = l_{\underline{1}\otimes X^*}(\mathrm{ev}_X \otimes \mathrm{Id}_{\underline{1}\otimes X^*})a^{-1}_{X^*,X,\underline{1}\otimes X^*} \\ &\quad\quad \left(\mathrm{Id}_{X^*} \otimes (\mathrm{Id}_X \otimes (\varphi_0 \otimes \mathrm{Id}_{X^*})\lambda_{X,\underline{1}})(r_X \otimes \mathrm{Id}_{(X\otimes\underline{1})^*})\mathrm{coev}_{X\otimes\underline{1}}\right) r_{X^*}^{-1} \\ &\quad = l_{\underline{1}\otimes X^*}(\mathrm{ev}_X \otimes \mathrm{Id}_{\underline{1}\otimes X^*})a^{-1}_{X^*,X,\underline{1}\otimes X^*} \\ &\quad\quad (\mathrm{Id}_{X^*} \otimes (r_X \otimes \mathrm{Id}_{\underline{1}\otimes X^*})(\mathrm{Id}_{X\otimes\underline{1}} \otimes (\varphi_0 \otimes \mathrm{Id}_{X^*})\lambda_{X,\underline{1}})\mathrm{coev}_{X\otimes\underline{1}}) r_{X^*}^{-1}.\end{aligned}$$

We used the naturality of $l$ (resp. $a$) in the first (resp. third) equality.

Now, from the formula of $\lambda_{X,Y}^{-1}$, it follows that $\mathrm{coev}_{X\otimes Y} =$

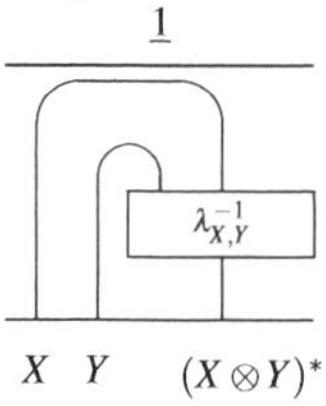


, for any objects $X$ and $Y$ of $\mathscr{C}$. When $\mathscr{C}$ is arbitrary the above formula takes the form

$$\mathrm{coev}_{X\otimes Y} = (\mathrm{Id}_{X\otimes Y} \otimes \lambda_{X,Y}^{-1})a_{X\otimes Y,Y^*,X^*}(a^{-1}_{X,Y,Y^*}(\mathrm{Id}_X \otimes \mathrm{coev}_Y)r_X^{-1} \otimes \mathrm{Id}_{X^*})\mathrm{coev}_X.$$

Thus the last computation continues as follows:

$$
\begin{aligned}
&(\varphi_0 \otimes \mathrm{Id}_{X^*})\lambda_{X,\underline{1}} r_X^* \\
&= l_{\underline{1}\otimes X^*}(\mathrm{ev}_X \otimes \mathrm{Id}_{\underline{1}\otimes X^*}) a^{-1}_{X^*,X,\underline{1}\otimes X^*} \Big( \mathrm{Id}_{X^*} \otimes (r_X \otimes \mathrm{Id}_{\underline{1}\otimes X^*})(\mathrm{Id}_{X\otimes \underline{1}} \otimes (\varphi_0 \otimes \mathrm{Id}_{X^*})) \\
&\qquad a_{X\otimes \underline{1},\underline{1}^*,X^*} \big(a^{-1}_{X,\underline{1},\underline{1}^*}(\mathrm{Id}_X \otimes \mathrm{coev}_{\underline{1}}) r_X^{-1} \otimes \mathrm{Id}_{X^*}\big)\mathrm{coev}_X \Big) r^{-1}_{X^*} \\
&= l_{\underline{1}\otimes X^*}(\mathrm{ev}_X \otimes \mathrm{Id}_{\underline{1}\otimes X^*}) a^{-1}_{X^*,X,\underline{1}\otimes X^*} \Big( \mathrm{Id}_{X^*} \otimes (r_X \otimes \mathrm{Id}_{\underline{1}\otimes X^*}) a_{X\otimes \underline{1},\underline{1},X^*} \\
&\qquad \Big( (\mathrm{Id}_{X\otimes \underline{1}} \otimes \varphi_0) a^{-1}_{X,\underline{1},\underline{1}^*}(\mathrm{Id}_X \otimes \mathrm{coev}_{\underline{1}}) r_X^{-1} \otimes \mathrm{Id}_{X^*}\Big)\mathrm{coev}_X \Big) r^{-1}_{X^*} \\
&= l_{\underline{1}\otimes X^*}(\mathrm{ev}_X \otimes \mathrm{Id}_{\underline{1}\otimes X^*}) a^{-1}_{X^*,X,\underline{1}\otimes X^*} \Big( \mathrm{Id}_{X^*} \otimes (r_X \otimes \mathrm{Id}_{\underline{1}\otimes X^*}) a_{X\otimes \underline{1},\underline{1},X^*} \\
&\qquad \Big( a^{-1}_{X,\underline{1},\underline{1}}(\mathrm{Id}_X \otimes (\mathrm{Id}_{\underline{1}} \otimes \varphi_0)\mathrm{coev}_{\underline{1}}) r_X^{-1} \otimes \mathrm{Id}_{X^*}\Big)\mathrm{coev}_X \Big) r^{-1}_{X^*} \\
&= l_{\underline{1}\otimes X^*}(\mathrm{ev}_X \otimes \mathrm{Id}_{\underline{1}\otimes X^*}) a^{-1}_{X^*,X,\underline{1}\otimes X^*} \big( \mathrm{Id}_{X^*} \otimes a_{X,\underline{1},X^*}\big((r_X \otimes \mathrm{Id}_{\underline{1}}) a^{-1}_{X,\underline{1},\underline{1}} \\
&\qquad (\mathrm{Id}_X \otimes (\mathrm{Id}_{\underline{1}} \otimes \varphi_0)\mathrm{coev}_{\underline{1}}) r_X^{-1} \otimes \mathrm{Id}_{X^*}\big)\mathrm{coev}_X \big) r^{-1}_{X^*} \\
&= l_{\underline{1}\otimes X^*}(\mathrm{ev}_X \otimes \mathrm{Id}_{\underline{1}\otimes X^*}) a^{-1}_{X^*,X,\underline{1}\otimes X^*} \big( \mathrm{Id}_{X^*} \otimes a_{X,\underline{1},X^*}\big((\mathrm{Id}_X \otimes l_{\underline{1}}(\mathrm{Id}_{\underline{1}} \otimes \mathrm{ev}_{\underline{1}}) \\
&\qquad a_{\underline{1},\underline{1}^*,\underline{1}} r^{-1}_{\underline{1}\otimes \underline{1}^*}\mathrm{coev}_{\underline{1}}) r_X^{-1} \otimes \mathrm{Id}_{X^*}\big)\mathrm{coev}_X \big) r^{-1}_{X^*} \\
&= l_{\underline{1}\otimes X^*}(\mathrm{ev}_X \otimes \mathrm{Id}_{\underline{1}\otimes X^*}) a^{-1}_{X^*,X,\underline{1}\otimes X^*} \big( \mathrm{Id}_{X^*} \otimes a_{X,\underline{1},X^*}(r_X^{-1} \otimes \mathrm{Id}_{X^*})\mathrm{coev}_X \big) r^{-1}_{X^*} \\
&= l_{\underline{1}\otimes X^*}(\mathrm{ev}_X \otimes \mathrm{Id}_{\underline{1}\otimes X^*}) a^{-1}_{X^*,X,\underline{1}\otimes X^*} \big( \mathrm{Id}_{X^*} \otimes (\mathrm{Id}_X \otimes l^{-1}_{X^*})\mathrm{coev}_X \big) r^{-1}_{X^*} \\
&= l_{\underline{1}\otimes X^*}(\mathrm{ev}_X \otimes \mathrm{Id}_{\underline{1}\otimes X^*})(\mathrm{Id}_{X^*\otimes X} \otimes l^{-1}_{X^*}) a^{-1}_{X^*,X,X^*}(\mathrm{Id}_{X^*} \otimes \mathrm{coev}_X) r^{-1}_{X^*} \\
&= l_{\underline{1}\otimes X^*}(\mathrm{Id}_{\underline{1}} \otimes l^{-1}_{X^*})(\mathrm{ev}_X \otimes \mathrm{Id}_{X^*}) a^{-1}_{X^*,X,X^*}(\mathrm{Id}_{X^*} \otimes \mathrm{coev}_X) r^{-1}_{X^*} \\
&= l_{\underline{1}\otimes X^*}(\mathrm{Id}_{\underline{1}} \otimes l^{-1}_{X^*}) l^{-1}_{X^*} = l^{-1}_{X^*},
\end{aligned}
$$

where we applied the naturality of $a$ in the second, third, fourth and eighth equality, (1.1.2) and Proposition 1.5 in the fifth equality, the naturality of $r$ and (1.6.4) in the sixth equality, (1.1.2) in the seventh equality, that $\otimes$ is a functor in the ninth equality, (1.6.5) in the penultimate equality, and the naturality of $l$ in the last one.

Finally, the commutativity of the first diagram in (1.3.2) is equivalent to

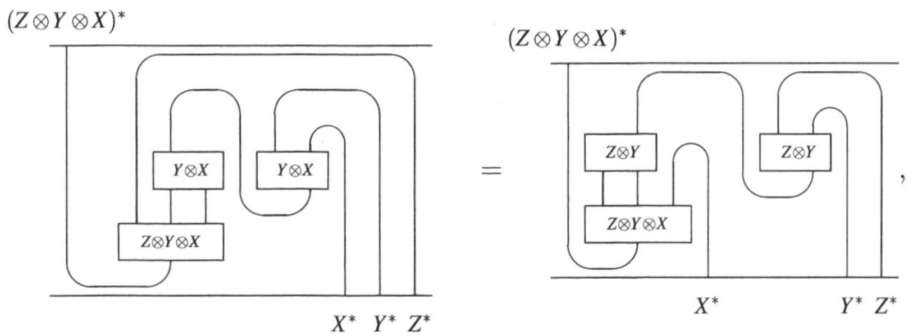


for all objects $X, Y, Z$ of $\mathscr{C}$. The above equality is true because of (1.6.6). □

**Definition 1.72** A monoidal category is called sovereign if it is left and right rigid such that the corresponding left and right dual functors $(-)^*, {}^*(-) : \mathscr{C} \to \overline{\mathscr{C}^{\mathrm{opp}}}$ are equal as strong monoidal functors.

The next result presents some canonical isomorphisms between an object and its "double" duals.

**Proposition 1.73** *Let $\mathscr{C}$ be a monoidal category, $X$ an object of $\mathscr{C}$ and assume that $X^*$ is a left, and ${}^*X$ is a right, dual object of $X$ in $\mathscr{C}$. If $X^*$ has a right dual object and ${}^*X$ has a left dual object in $\mathscr{C}$, then $X \cong ({}^*X)^* \cong {}^*(X^*)$.*

*Proof* Consider $\theta_X : X \to ({}^*X)^*$ defined by the following composition:

$$\theta_X : X \xrightarrow{r_X^{-1}} X \otimes \underline{1} \xrightarrow{\mathrm{Id}_X \otimes \mathrm{coev}_{{}^*X}} X \otimes ({}^*X \otimes ({}^*X)^*)$$
$$\xrightarrow{a^{-1}_{X,{}^*X,({}^*X)^*}} (X \otimes {}^*X) \otimes ({}^*X)^* \xrightarrow{\mathrm{ev}'_X \otimes \mathrm{Id}_{({}^*X)^*}} \underline{1} \otimes ({}^*X)^* \xrightarrow{l_{({}^*X)^*}} ({}^*X)^*. \tag{1.7.3}$$

We claim that $\theta_X$ is an isomorphism in $\mathscr{C}$ with inverse given by

$$\theta_X^{-1} : ({}^*X)^* \xrightarrow{r^{-1}_{({}^*X)^*}} ({}^*X)^* \otimes \underline{1} \xrightarrow{\mathrm{Id}_{({}^*X)^*} \otimes \mathrm{coev}'_X} ({}^*X)^* \otimes ({}^*X \otimes X)$$
$$\xrightarrow{a^{-1}_{({}^*X)^*,{}^*X,X}} (({}^*X)^* \otimes {}^*X) \otimes X \xrightarrow{\mathrm{ev}_{{}^*X} \otimes \mathrm{Id}_X} \underline{1} \otimes X \xrightarrow{l_X} X. \tag{1.7.4}$$

In graphical notation this means that $\theta_X =$ 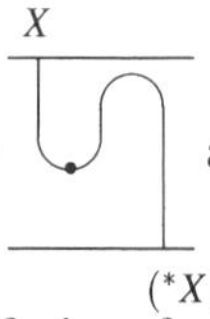

and $\theta_X^{-1} =$ 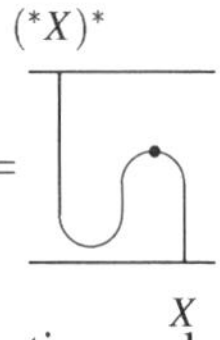

, where, as before, in order to avoid any possible confusion, for the evaluation and coevaluation morphisms of the left dual object of ${}^*X$ in $\mathscr{C}$ we kept the standard notation and in the notation for $\mathrm{ev}'_X$ and $\mathrm{coev}'_X$ we added a black dot. From (1.6.6) it is clear at this point that $\theta_X$ and $\theta_X^{-1}$ are indeed inverse to each other, as we claimed.

In the same way one can construct an isomorphism $\theta'_X : X \to {}^*(X^*)$. □

## 1.8 Braided Rigid Monoidal Categories

This section deals with braidings and duals. As we shall see, some conditions in the definition of a braided, or rigid, monoidal category are automatic. In this direction a first result is the following.

**Proposition 1.74** *Let $\mathscr{C}$ be a braided category. If $\mathscr{C}$ is left or right rigid monoidal then it is rigid monoidal.*

*Proof* As we have seen before, $\mathscr{C}$ is left rigid if and only if its reverse monoidal category $\overline{\mathscr{C}}$ is right rigid. Since $\overline{\mathscr{C}}$ is braided as well, it suffices to prove only the left-handed version in the statement.

Let $\mathscr{C}$ be a left rigid braided category. By Examples 1.59 the functor $(\mathrm{Id}_{\mathscr{C}}, \varphi_0 = \mathrm{Id}_{\underline{1}}, \varphi_2 = (c_{Y,X})_{Y,X \in \mathscr{C}}) : (\mathscr{C}, c) \to (\overline{\mathscr{C}}, \overline{c})$ is a braided monoidal functor. Thus, it carries left duals to left duals; see Proposition 1.67. In our situation this means that an adjunction $(\mathrm{coev}_X, \mathrm{ev}_X) : X^* \dashv X$ is mapped to an adjunction $(\mathrm{coev}'_X, \mathrm{ev}'_X) : X^* \dashv X$ in $\overline{\mathscr{C}}$, that is, to an adjunction $(\mathrm{coev}'_X, \mathrm{ev}'_X) : X \dashv X^*$ in $\mathscr{C}$. Thus, $X^*$ together with

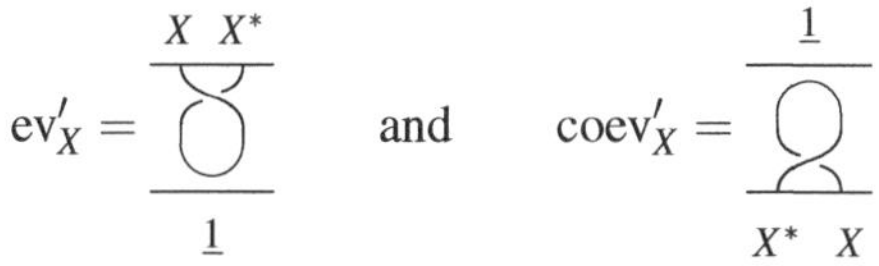

$\mathrm{ev}'_X =$ 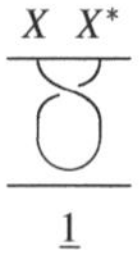 and $\mathrm{coev}'_X =$ 

is a right dual for $X$ in $\mathscr{C}$, so the proof is finished. □

**Remark 1.75** Since the definition of rigidity is independent of the choice of the braiding it follows that $X^*$, the left dual of an object $X$ of a left rigid braided category, is a right dual as well if it is considered together with $\mathrm{coev}''_X = c_{X,X^*} \circ \mathrm{coev}_X$ and $\mathrm{ev}''_X = \mathrm{ev}_X \circ c^{-1}_{X^*,X}$. Moreover, the isomorphism constructed in Proposition 1.66 between $X^*$ and $X^*$ viewed as a right dual of $X$ in the two different ways presented above is just $\mathrm{Id}_{X^*}$, and this is because

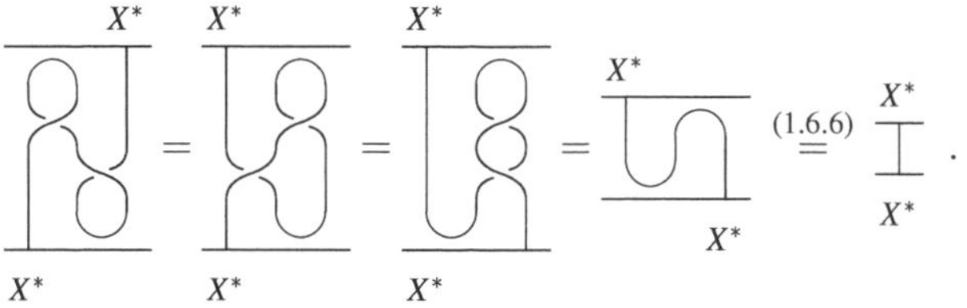

We used in the first two equalities the naturality of $c$ and Proposition 1.49.

**Corollary 1.76** *If $\mathscr{C}$ is a braided rigid monoidal category then $X^* \cong {}^*X$ for any object $X$ of $\mathscr{C}$. Consequently, $X^{**} := (X^*)^* \cong {}^*(X^*) \cong X \cong ({}^*X)^* \cong {}^*({}^*X) := {}^{**}X$.*

*Proof* This result follows from Propositions 1.66, 1.74 and 1.73. Note that $\Theta_X : X^* \to {}^*X$ and $\Theta_X^{-1} : {}^*X \to X^*$ defined by

$\Theta_X =$ 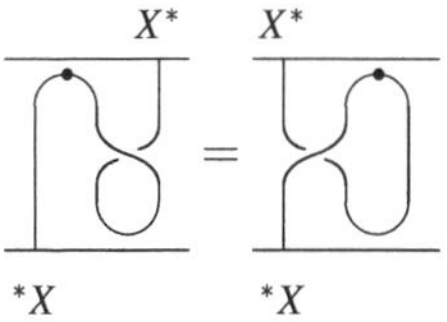 and $\Theta_X^{-1} =$ 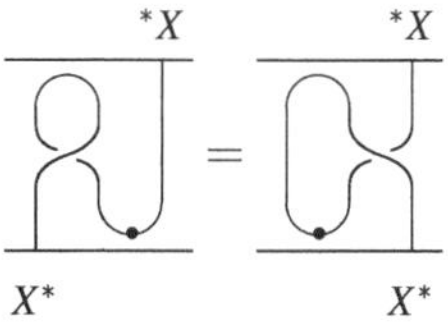

are inverses of each other. The black dots are used to distinguish the evaluation and coevaluation morphisms of the right dual of $X$ from the ones of the left dual of $X$. By the above remark, another pair of inverse isomorphisms can be obtained by replacing in the above definitions of $\Theta_X$ and $\Theta_X^{-1}$ the braiding of $\mathscr{C}$ with its inverse, that is, by thinking of everything in terms of $\mathscr{C}^{\mathrm{in}}$ instead of $\mathscr{C}$. We denote these by $\Theta'_X$ and $\Theta'^{-1}_X$, respectively. □

The next result shows that the condition of invertibility for a braiding in a left rigid monoidal category is redundant.

**Theorem 1.77** *A pre-braided left rigid monoidal category is braided, and so rigid monoidal as well.*

*Proof* Let $\mathscr{C}$ be a pre-braided left rigid monoidal category. If $c$ is the pre-braiding of $\mathscr{C}$ we claim that $c$ is invertible with inverse given, for $V$, $W$ objects of $\mathscr{C}$, by

$c^{-1}_{V,W} =$ 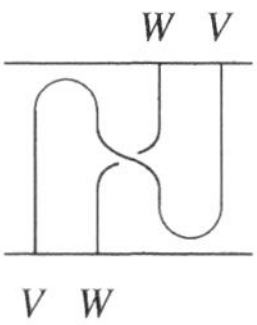 , (1.8.1)


where the evaluation and coevaluation morphisms are those of $V$. To prove this assertion observe first that

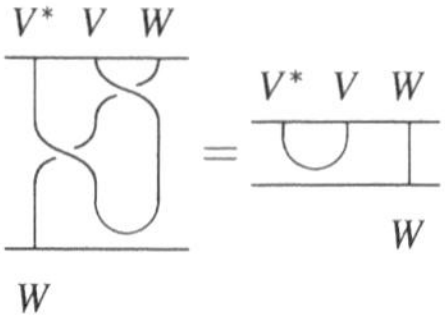


and

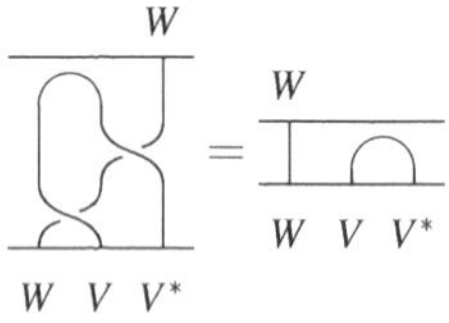


relations that follow from (1.5.3), the naturality of $c$ and Proposition 1.49.

We can compute now

$c^{-1}_{V,W} \circ c_{V,W} =$ 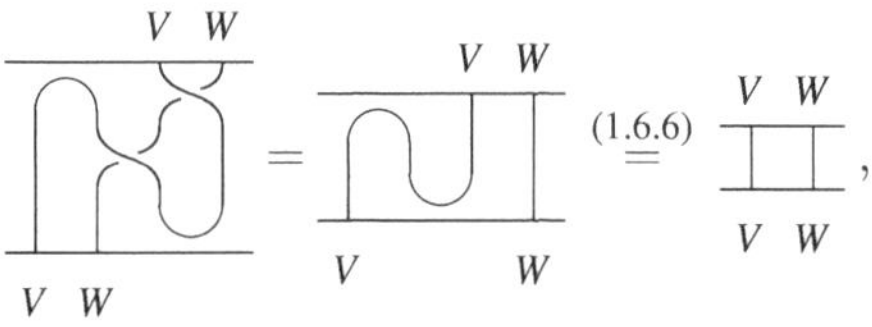 ,


$c_{V,W} \circ c^{-1}_{V,W} =$ 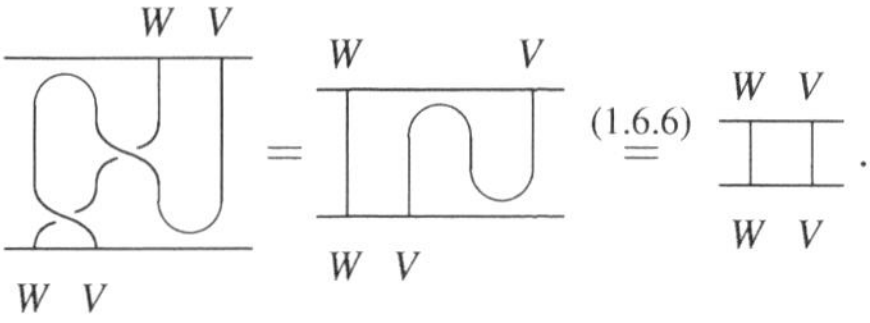 .


So our claim is proved. The second assertion follows from Proposition 1.74. □

The dual functors $(-)^*$ and ${}^*(-)$ behave well with respect to braidings.

**Proposition 1.78** *If $\mathscr{C}$ is a braided category then the strong monoidal functors from Proposition 1.71 are, moreover, braided monoidal if they are viewed as functors from $\mathscr{C}$ to $\overline{\mathscr{C}^{\mathrm{opp}}}^{\mathrm{in}}$.*

*Proof* The braiding of $\overline{\mathscr{C}^{\mathrm{opp}}}$ is $\overline{c^{\mathrm{opp}}_{X,Y}} = c^{\mathrm{opp}}_{Y,X} = c^{-1}_{Y,X} : X\overline{\otimes}Y = Y\otimes X \leftarrow X\otimes Y = Y\overline{\otimes}X$, and so the braiding of $\overline{\mathscr{C}^{\mathrm{opp}}}^{\mathrm{in}}$ coincides with that of $\mathscr{C}$. Now, the fact that $(-)^* : \mathscr{C} \to \overline{\mathscr{C}^{\mathrm{opp}}}^{\mathrm{in}}$ is braided means that

$$\begin{array}{ccc} Y^*\otimes X^* = X^*\overline{\otimes}Y^* & \xleftarrow{c_{X^*,Y^*}} & X^*\otimes Y^* = Y^*\overline{\otimes}X^* \\ \uparrow{\scriptstyle \lambda_{X,Y}} & & \uparrow{\scriptstyle \lambda_{Y,X}} \\ (X\otimes Y)^* & \xleftarrow[c^*_{X,Y}]{} & (Y\otimes X)^* \end{array}$$

is commutative. In diagrammatic language this comes out as

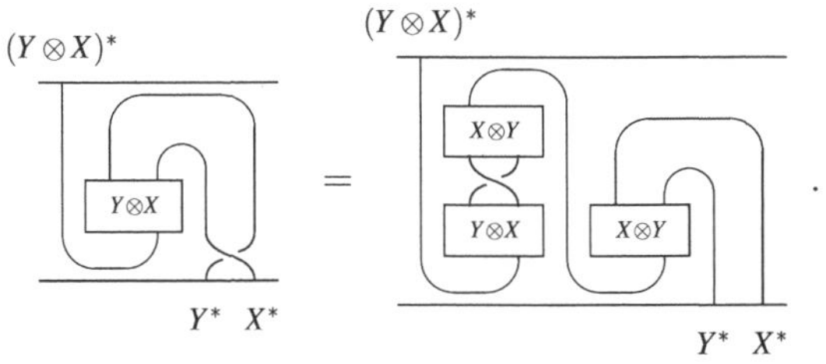


Now, by (1.6.6) the right-hand side of the above equality is equal to

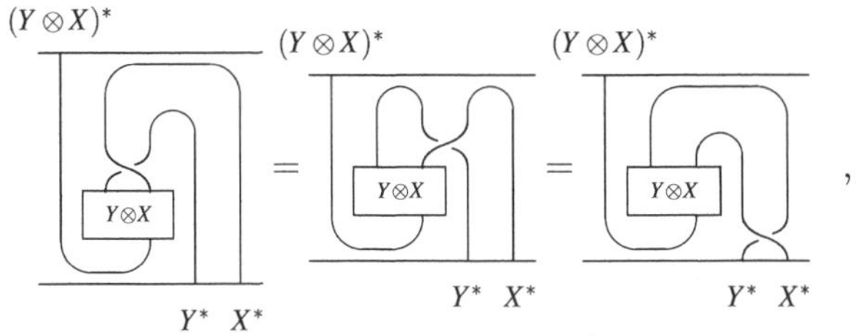


proving the desired equality. Note that we used 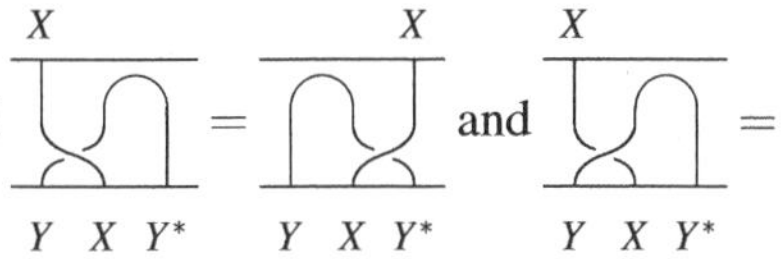


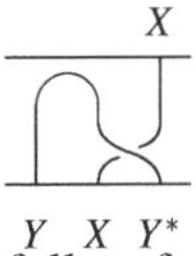


in the first and second equalities, respectively, and that these equalities follow from the naturality of $c$ and Proposition 1.49. Thus the proof is finished. □

We will need later the mate of the braiding of a left rigid braided category. The notion "mates under adjunction" can be regarded as a "halfway" dualizing process.

**Definition 1.79** Let $\mathscr{C}$ be a monoidal category and $X, Y, V, W$ objects of $\mathscr{C}$ such that $X$ and $W$ admit left dual objects. If $f : V \otimes W \to X \otimes Y$ is a morphism in $\mathscr{C}$ then the mate of $f$ is the morphism $f^\flat : X^* \otimes V \to Y \otimes W^*$ defined by the following composition:

$$
\begin{aligned}
f^\flat : X^* \otimes V &\xrightarrow{r^{-1}_{X^*\otimes V}} (X^* \otimes V) \otimes \underline{1} \xrightarrow{\mathrm{Id}_{X^*\otimes V}\otimes \mathrm{coev}_W} (X^* \otimes V) \otimes (W \otimes W^*)\\
&\xrightarrow{a_{X^*,V,W\otimes W^*}} X^* \otimes (V \otimes (W \otimes W^*)) \xrightarrow{\mathrm{Id}_{X^*}\otimes a^{-1}_{V,W,W^*}} X^* \otimes ((V \otimes W) \otimes W^*)\\
&\xrightarrow{\mathrm{Id}_{X^*}\otimes(f\otimes \mathrm{Id}_{W^*})} X^* \otimes ((X \otimes Y) \otimes W^*) \xrightarrow{\mathrm{Id}_{X^*}\otimes a_{X,Y,W^*}} X^* \otimes (X \otimes (Y \otimes W^*))\\
&\xrightarrow{a^{-1}_{X^*,X,Y\otimes W^*}} (X^* \otimes X) \otimes (Y \otimes W^*) \xrightarrow{\mathrm{ev}_X\otimes \mathrm{Id}_{Y\otimes W^*}} \underline{1} \otimes (Y \otimes W^*) \xrightarrow{l_{Y\otimes W^*}} Y \otimes W^*.
\end{aligned}
$$

When $V = Y = \underline{1}$, through the natural identifications, the mate of $f$ is the transpose morphism $f^*$. In general, assuming $\mathscr{C}$ is strict monoidal, the mate of $f$ is

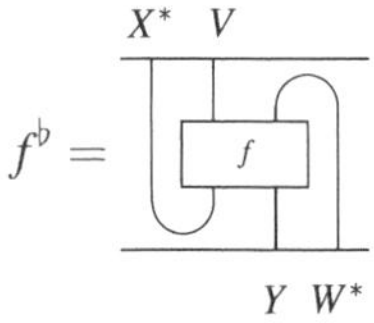


, where

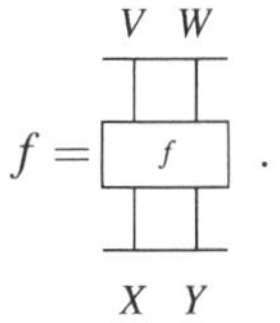


Moreover, the mate of a morphism can be characterized as follows.

**Proposition 1.80** *Let $\mathscr{C}$ be a monoidal category and $V$, $W$, $X$, $Y$ objects of $\mathscr{C}$ such that there exist $(\mathrm{coev}_W, \mathrm{ev}_W) : W^* \dashv W$ and $(\mathrm{coev}_X, \mathrm{ev}_X) : X^* \dashv X$ adjunctions in $\mathscr{C}$. If $f : V \otimes W \to X \otimes Y$ is a morphism in $\mathscr{C}$ then its mate $f^\flat$ is the unique morphism $f^\flat : X^* \otimes V \to Y \otimes W^*$ in $\mathscr{C}$ making one (hence also the other one) of the following two diagrams commutative:*

$$\begin{array}{ccccc}
V & \xrightarrow{\;r_V^{-1}\;} & V \otimes \underline{1} & \xrightarrow{\;a^{-1}_{V,W,W^*}(\mathrm{Id}_V \otimes \mathrm{coev}_W)\;} & (V \otimes W) \otimes W^* \\
\downarrow{\scriptstyle l_V^{-1}} & & & & \downarrow{\scriptstyle a_{X,Y,W^*}(f \otimes \mathrm{Id}_{W^*})} \\
\underline{1} \otimes V & \xrightarrow{\;a_{X,X^*,V}(\mathrm{coev}_X \otimes \mathrm{Id}_V)\;} & X \otimes (X^* \otimes V) & \xrightarrow{\;\mathrm{Id}_X \otimes f^\flat\;} & X \otimes (Y \otimes W^*),
\end{array}$$

$$\begin{array}{ccccc}
X^* \otimes (V \otimes W) & \xrightarrow{\;a^{-1}_{X^*,X,Y}(\mathrm{Id}_{X^*} \otimes f)\;} & (X^* \otimes X) \otimes Y & \xrightarrow{\;l_Y(\mathrm{ev}_X \otimes \mathrm{Id}_Y)\;} & Y \\
\downarrow{\scriptstyle a^{-1}_{X^*,V,W}} & & & & \uparrow{\scriptstyle r_Y} \\
(X^* \otimes V) \otimes W & \xrightarrow{\;f^\flat \otimes \mathrm{Id}_W\;} & (Y \otimes W^*) \otimes W & \xrightarrow{\;(\mathrm{Id}_Y \otimes \mathrm{ev}_W) a_{Y,W^*,W}\;} & Y \otimes \underline{1}.
\end{array}$$

*Proof* In diagrammatic notation we have to prove that $f^\flat$ is the unique morphism in $\mathscr{C}$ satisfying one of the following two equivalent identities:

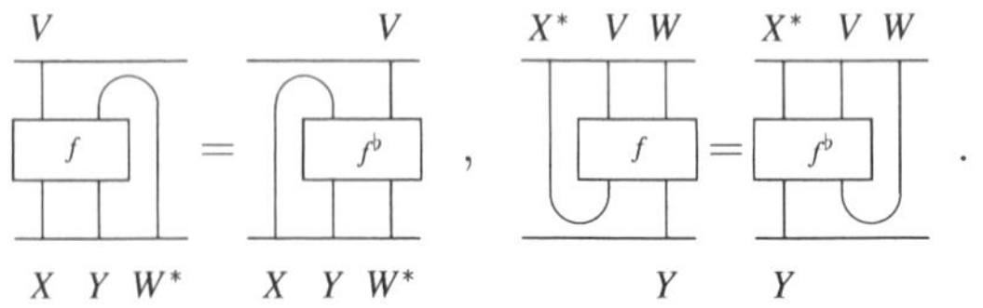


This is immediate if we use the relations in (1.6.6) in the definition of $f^\flat$. □

**Proposition 1.81** *Let $\mathscr{C}$ be a monoidal category and $V$, $W$, $X$, $Y$ objects of $\mathscr{C}$ such that $X$ and $W$ admit left dual objects. Then the correspondence*

$$\mathrm{Hom}_{\mathscr{C}}(V \otimes W, X \otimes Y) \ni f \mapsto f^\flat \in \mathrm{Hom}_{\mathscr{C}}(X^* \otimes V, Y \otimes W^*)$$

*is one to one. Consequently, we have $\mathrm{Hom}_{\mathscr{C}}(V \otimes W, X \otimes Y) \cong \mathrm{Hom}_{\mathscr{C}}(X^* \otimes V, Y \otimes W^*) \cong \mathrm{Hom}_{\mathscr{C}}((X^* \otimes V) \otimes W, Y) \cong \mathrm{Hom}_{\mathscr{C}}(V, (X \otimes Y) \otimes W^*)$.*

*Proof* We show that the inverse of $f \mapsto f^\flat$ is $g \mapsto g^\sharp$, where for any $g : X^* \otimes V \to Y \otimes W^*$ define $g^\sharp : V \otimes W \to X \otimes Y$ by $g^\sharp =$ 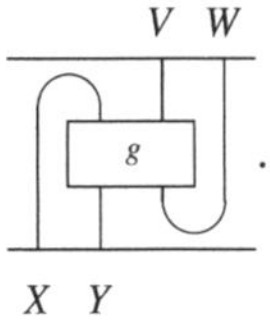

. To see this we compute

$(f^\flat)^\sharp =$ 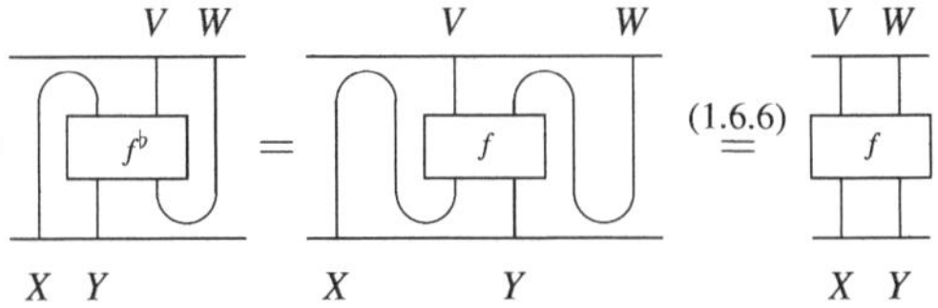

.

On the other hand we have

$(g^\sharp)^\flat =$ 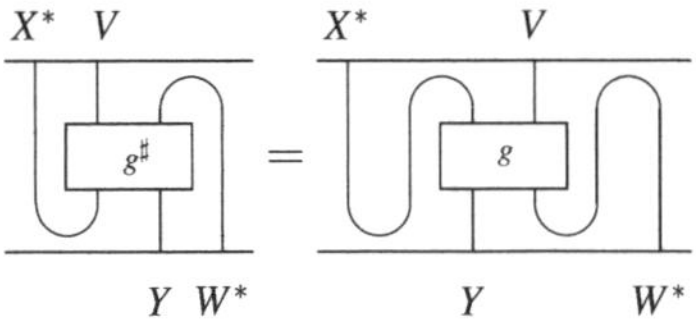

$= g,$

as required. If we specialize this isomorphism for $X = \underline{1}$, which always has a left dual, we get that $\mathrm{Hom}_{\mathscr{C}}(V \otimes W, Y) \cong \mathrm{Hom}_{\mathscr{C}}(V, Y \otimes W^*)$, for any other objects $Y$, $V$, $W$ of $\mathscr{C}$ with $W$ admitting a left dual object. It is clear now that we also have $\mathrm{Hom}_{\mathscr{C}}(V \otimes W, X \otimes Y) \cong \mathrm{Hom}_{\mathscr{C}}(V, (X \otimes Y) \otimes W^*)$ and $\mathrm{Hom}_{\mathscr{C}}(X^* \otimes V, Y \otimes W^*) \cong \mathrm{Hom}_{\mathscr{C}}((X^* \otimes V) \otimes W, Y)$, and this completes the proof. □

We finally compute the mate of the braiding of a left rigid braided category.

**Proposition 1.82** *If $\mathscr{C}$ is a left rigid braided category, then for all $X, Y$ objects of $\mathscr{C}$ the mate of the braiding $c_{X,Y} : X \otimes Y \to Y \otimes X$ is $c^{-1}_{X,Y^*} : Y^* \otimes X \to X \otimes Y^*$.*

*Proof* It is enough to show that $c_{X,Y^*} \circ c^\flat_{X,Y} = \mathrm{Id}_{Y^* \otimes X}$. For this we notice first that 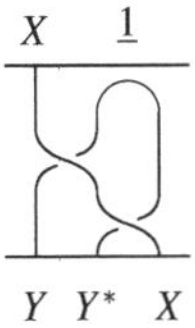

$=$ 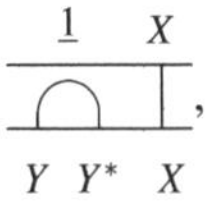

, an equality that follows from (1.5.3), the naturality of $c$ and Proposition 1.49. This fact allows to compute

$c_{X,Y^*} \circ c^\flat_{X,Y} =$ 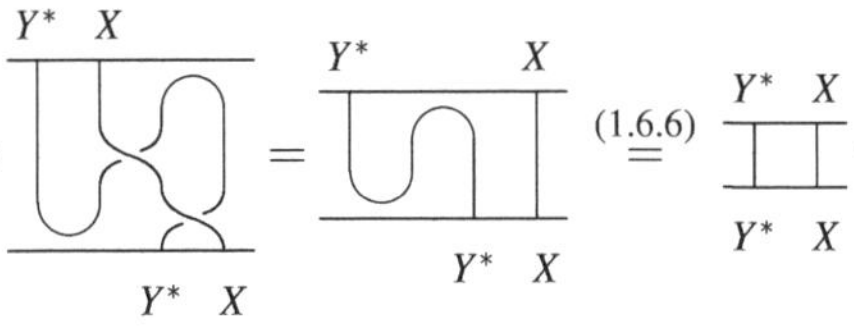

$= \mathrm{Id}_{Y^* \otimes X},$

as required. □

## 1.9 Notes

Monoidal categories were introduced by Bénabou in [30]; they are also known under the name of tensor categories. The notion of symmetric monoidal category goes back to Eilenberg and Kelly [84]. The concept of braided category was introduced by Joyal and Street in [119, 121]. The concept of rigidity abstracts the notion of duality, a concept intensively used for categories of vector spaces. The canonical isomorphisms constructed in the braided monoidal setting are natural generalizations of the well-known isomorphisms that exist in a category of vector spaces.

The concrete monoidal structures on some categories of graded vector spaces were taken from [119, 121], which were our main source of inspiration for many results of this chapter. We also took examples of normalized 3-cocycles from these papers and [5]. Note that $H^3(\mathbb{Z}_n, k^*)$ is completely determined by the $n$th roots of unity in $k$; see [192, Theorem 10.35] or [219, Theorem 6.2.2]. Thus $H^3(\mathbb{Z}_n, k^*)$ is completely determined by the cohomology classes defined by the cocycles $\phi_q$ from Example 1.16. Also, by [219, Example 6.1.4], $H^3(\mathbb{Z}, k^*) = \{e\}$, where $\mathbb{Z}$ is the infinite cyclic group written additively; this means that any normalized 3-cocycle on $\mathbb{Z}$ with coefficients in $k^*$ is a coboundary. This completes the description of the third cohomology group for any cyclic group.

Apart from the papers already mentioned above we used in the presentation of this chapter the books of C. Kassel [127], S. Majid [148] and S. Mac Lane [141]. Proofs for the coherence theorem of Mac Lane can be found in [141, VII.2], [95, Theorem 2.9.2] and [217, Theorem 3.17]. The fact that a diagram in a monoidal category $\mathscr{C}$ is commutative if it is commutative when it is considered in a strict monoidal category can be proved for instance by using so-called "cliques" in $\mathscr{C}$ and morphisms between them; as in the case of the coherence theorem, a detailed proof involves categorical techniques that are not related to the topic of this book, and this is why we skip it.

# 2
# Algebras and Coalgebras in Monoidal Categories

We define the notions of algebra and coalgebra, and of bialgebra and Hopf algebra, respectively, within a monoidal category and a (pre-)braided monoidal category, respectively. We also present several constructions associated to them.

## 2.1 Algebras in Monoidal Categories

Recall that an algebra over a field $k$ is a vector space $A$ over $k$ that also has a unital ring structure such that $\kappa(ab) = (\kappa a)b = a(\kappa b)$, for all $\kappa \in k$ and $a, b \in A$. Equivalently, an algebra over $k$, or a $k$-algebra, is a $k$-vector space for which there exist two $k$-linear maps $m_A : A \otimes A \to A$ and $\eta_A : k \to A$ making the diagrams below commutative:

$$\begin{array}{ccc} A\otimes A\otimes A & \xrightarrow{m_A\otimes \mathrm{Id}_A} & A\otimes A \\ {\scriptstyle \mathrm{Id}_A\otimes m_A}\downarrow & & \downarrow{\scriptstyle m_A} \\ A\otimes A & \xrightarrow[m_A]{} & A, \end{array}$$

$$\begin{array}{ccc} k\otimes A & \xrightarrow{\eta_A\otimes \mathrm{Id}_A} & A\otimes A, \\ & {\scriptstyle l_A}\searrow \quad \swarrow{\scriptstyle m_A} & \\ & A & \end{array} \qquad \begin{array}{ccc} A\otimes k & \xrightarrow{\mathrm{Id}_A\otimes \eta_A} & A\otimes A. \\ & {\scriptstyle r_A}\searrow \quad \swarrow{\scriptstyle m_A} & \\ & A & \end{array}$$

Here $l_A : k \otimes A \to A$, $r_A : A \otimes k \to A$ are the natural isomorphisms and all the unadorned tensor products are over $k$. Note that $m_A$ and $\eta_A$ are obtained from the unital ring structure of $A$ as follows: $m_A(a \otimes b) = ab$ and $\eta(\kappa) = \kappa 1_A$, for all $\kappa \in k$, where $1_A$ is the unit of $A$. Thus a $k$-algebra identifies with a triple $(A, m_A, \eta_A)$ as above.

Recall also that for two $k$-algebras $A$, $B$ an algebra morphism from $A$ to $B$ is a $k$-linear map $f : A \to B$ such that $f(1_A) = 1_B$ and $f(ab) = f(a)f(b)$, for all $a, b \in A$. If we consider $A, B$ as triples $(A, m_A, \eta_A)$ and $(B, m_B, \eta_B)$, respectively, then the definition of a $k$-algebra morphism $f$ from $A$ to $B$ can be rephrased as follows: $f$ is a $k$-linear morphism such that $fm_A = m_B(f \otimes f)$ and $f\eta_A = \eta_B$.

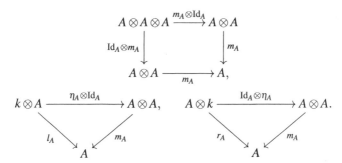

The diagrammatic reformulation of the definition of a $k$-algebra and a $k$-algebra morphism allows to define these concepts in any monoidal category, leading thus to the concepts of monoidal algebra and morphism of monoidal algebras.

**Definition 2.1** Let $\mathscr{C} = (\mathscr{C}, \otimes, \underline{1}, a, l, r)$ be a monoidal category. An algebra in $\mathscr{C}$ is a triple $(A, \underline{m}_A, \underline{\eta}_A)$, where $A \in \mathscr{C}$ and $\underline{m}_A : A \otimes A \to A$ (called the multiplication of $A$) and $\underline{\eta}_A : \underline{1} \to A$ (called the unit of $A$) are morphisms in $\mathscr{C}$ such that:

- $\underline{m}_A$ is associative up to the associativity constraint $a$ of $\mathscr{C}$, that is, the diagram

$$\begin{array}{ccccc}
(A \otimes A) \otimes A & \xrightarrow{\underline{m}_A \otimes \mathrm{Id}_A} & A \otimes A & \xrightarrow{\underline{m}_A} & A \\
\Big\downarrow{\scriptstyle a_{A,A,A}} & & & \nearrow{\scriptstyle \underline{m}_A} & \\
A \otimes (A \otimes A) & \xrightarrow[\mathrm{Id}_A \otimes \underline{m}_A]{} & A \otimes A & &
\end{array}$$

  is commutative;
- $\underline{\eta}_A$ is a unit for $\underline{m}_A$, that is, the following triangle diagrams are commutative:

$$\begin{array}{ccc}
\underline{1} \otimes A & \xrightarrow{\underline{\eta}_A \otimes \mathrm{Id}_A} & A \otimes A \\
& {\scriptstyle l_A} \searrow \quad \swarrow {\scriptstyle \underline{m}_A} & \\
& A &
\end{array}
\quad \text{and} \quad
\begin{array}{ccc}
A \otimes \underline{1} & \xrightarrow{\mathrm{Id}_A \otimes \underline{\eta}_A} & A \otimes A \\
& {\scriptstyle r_A} \searrow \quad \swarrow {\scriptstyle \underline{m}_A} & \\
& A. &
\end{array}$$

If $(A, \underline{m}_A, \underline{\eta}_A)$ and $(B, \underline{m}_B, \underline{\eta}_B)$ are algebras in $\mathscr{C}$ then an algebra morphism between $A$ and $B$ is a morphism $f : A \to B$ in $\mathscr{C}$ such that the diagrams

$$\begin{array}{ccc}
A \otimes A & \xrightarrow{f \otimes f} & B \otimes B \\
\Big\downarrow{\scriptstyle \underline{m}_A} & & \Big\downarrow{\scriptstyle \underline{m}_B} \\
A & \xrightarrow{f} & B
\end{array}
\quad \text{and} \quad
\begin{array}{ccc}
A & \xrightarrow{f} & B \\
& {\scriptstyle \underline{\eta}_A} \nwarrow \quad \nearrow {\scriptstyle \underline{\eta}_B} & \\
& \underline{1} &
\end{array}$$

are commutative. If, moreover, $f$ is an isomorphism in $\mathscr{C}$ then we say that $A$ and $B$ are isomorphic as algebras in $\mathscr{C}$.

Clearly, a $k$-algebra is nothing but an algebra in the category of $k$-vector spaces. In general, the meaning of an algebra in an arbitrary monoidal category $\mathscr{C}$ might be far from the classical one, as it is strongly connected with the monoidal structure of $\mathscr{C}$. Also, since the multiplication of $A$ is associative up to the associativity constraint $a$ of $\mathscr{C}$, it might happen that $A$ is not associative in the usual sense. For a good understanding of these ideas we next supply a list of examples.

**Examples 2.2** (1) The unit object of a monoidal category $\mathscr{C}$ is an algebra in $\mathscr{C}$ with multiplication $\underline{m}_{\underline{1}} = l_{\underline{1}} = r_{\underline{1}} : \underline{1} \otimes \underline{1} \to \underline{1}$ and unit $\underline{\eta}_{\underline{1}} = \mathrm{Id}_{\underline{1}}$.

(2) An algebra $A$ in $\underline{\mathrm{Set}}$ is a monoid, and a morphism of algebras in $\underline{\mathrm{Set}}$ is nothing but a morphism of monoids.

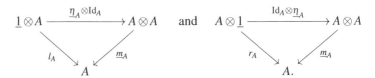

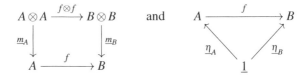

Indeed, $A$ is an algebra in $\underline{\mathrm{Set}}$ if and only if there exist functions $\underline{m}_A : A \times A \to A$ and $\underline{\eta}_A : \{*\} \to A$ such that the first three diagrams in Definition 2.1 are commutative. If we denote $\underline{m}_A(a,b) = ab$ and $\underline{\eta}_A(*) = e$, the commutativity of these diagrams reads $(ab)c = a(bc)$ and $ae = a = ea$, for all $a,b,c \in A$. Thus $A$ is a monoid.

It is clear that if $A, A'$ are algebras in $\underline{\text{Set}}$ then $f : A \to A'$ is an algebra morphism if and only if $f(ab) = f(a)f(b)$, for all $a,b \in A$, and $f(e) = e'$, where $e'$ is the neutral element of $A'$. Thus $f$ has to be a morphism of monoids.

(3) Let $[\mathscr{C},\mathscr{C}]$ be the category of endo-functors associated to a category $\mathscr{C}$, endowed with the monoidal structure from Proposition 1.17. An algebra in $[\mathscr{C},\mathscr{C}]$, called a monad in $\mathscr{C}$, is a triple $(T,\underline{m},\underline{\eta})$ where $T : \mathscr{C} \to \mathscr{C}$ is a functor and $\underline{m} : T \circ T \to T$ and $\underline{\eta} : \mathrm{Id}_{\mathscr{C}} \to T$ are natural transformations such that, for any object $X$ of $\mathscr{C}$, $\underline{m}_X \underline{m}_{T(X)} = \underline{m}_X T(\underline{m}_X)$ and $\underline{m}_X \underline{\eta}_{T(X)} = \underline{m}_X T(\underline{\eta}_X) = \mathrm{Id}_{T(X)}$.

Indeed, if $1_T$ is the identity natural transformation from $T$ to $T$, the two sets of conditions stated above are equivalent to $\underline{m} \circ (\underline{m} \otimes 1_T) = \underline{m} \circ (1_T \otimes \underline{m})$ and $\underline{m} \circ (\underline{\eta} \otimes 1_T) = \underline{m} \circ (1_T \otimes \underline{\eta}) = 1_T$, as natural transformations. These conditions are equivalent to the commutativity of the diagrams in Definition 2.1, specialized for our situation.

(4) Let $G$ be a group and $\phi \in H^3(G,k^*)$. If $\mathrm{Vect}_\phi^G$ is the category of $G$-graded vector spaces endowed with the monoidal structure from Proposition 1.13 then an algebra in $\mathrm{Vect}_\phi^G$ is a $G$-graded vector space $A$ together with a multiplication $\bullet$ and a usual unit $1_A \in A_e$ satisfying $A_x A_y \subseteq A_{xy}$, for all $x,y \in G$, and

$$(a \bullet b) \bullet c = \phi(|a|,|b|,|c|) a \bullet (b \bullet c), \tag{2.1.1}$$

for all homogeneous elements $a,b,c \in A$.

Indeed, let $A$ be an object of $\mathrm{Vect}_\phi^G$, that is, a $G$-graded $k$-vector space. Then $A$ has an algebra structure in $\mathrm{Vect}_\phi^G$ if and only if there exist morphisms $\underline{m}_A : A \otimes A \to A$ and $\underline{\eta}_A : k \to A$ in $\mathrm{Vect}_\phi^G$ that make the diagrams in Definition 2.1 commutative. Now, $\underline{m}_A$ preserves the degree of homogeneous elements if and only if $A_x A_y \subseteq A_{xy}$ and so does $\underline{\eta}_A$ if and only if $1_A := \underline{\eta}(1) \in A_e$. Moreover, those diagrams are commutative if and only if $1_A$ is a usual unit for $\underline{m}_A$ and $\underline{m}_A$ is associative in the sense of (2.1.1).

In what follows an algebra in $\mathrm{Vect}_\phi^G$ will be called a $G$-graded quasialgebra with reassociator $\phi$. If $\phi$ is trivial then we simply call such an algebra a $G$-graded algebra.

(5) Let $k$ be a commutative ring and $R$ a $k$-algebra. Then giving an algebra $A$ in ${}_R\mathscr{M}_R$ is equivalent to giving a $k$-algebra $A$ and a $k$-algebra morphism $i : R \to A$. Such a pair $(A,i)$ is called an $R$-ring.

To see this, we consider $A$ an algebra in ${}_R\mathscr{M}_R$. Thus $A$ is an $R$-bimodule and there exist $R$-bilinear morphisms $\underline{m}_A : A \otimes_R A \to A$ and $\underline{\eta}_A : R \to A$ such that $\underline{m}_A$ is associative and $\underline{\eta}_A$ is a unit for it. Otherwise stated, we have a $k$-linear map $m_A := \underline{m}_A q_{A,A}^R : A \otimes A \ni a \otimes b \mapsto ab \in A$ that is associative and unital in ${}_k\mathscr{M}$ and such that $r(ab) = (ra)b$, $(ar)b = a(rb)$ and $(ab)r = a(br)$, for all $a,b \in A$ and $r \in R$. Here $q_{A,A}^R : A \otimes A \to A \otimes_R A$ is the canonical projection. Furthermore, the fact that $\underline{\eta}_R$ is $R$-bilinear is equivalent to $r1_A = 1_A r$, for all $r \in R$, where $1_A := \underline{\eta}_A(1_R)$ is the unit for $m_A$. So $A$ becomes a $k$-algebra and $\underline{\eta}_A : R \to A$ turns into a $k$-algebra morphism since $\underline{\eta}_A(1_R) = 1_A$ and, for all $r,r' \in R$,

$$\underline{\eta}_A(rr') = (rr')1_A = r(r'1_A) = r(1_A r') = r(1_A(1_A r')) = (r1_A)(r'1_A) = \underline{\eta}_A(r)\underline{\eta}_A(r').$$

Conversely, if $A$ admits a $k$-algebra structure such that there exists a $k$-algebra morphism $i : R \to A$, then $A$ is an $R$-bimodule via $rar' = i(r)ai(r')$, for all $a \in A$ and $r, r' \in R$. Moreover, in this way the multiplication of $A$ becomes $R$-balanced and so we have a well-defined morphism $\underline{m}_A : A \otimes_R A \to A$ given by $\underline{m}_A(a \otimes_R b) = m_A(a \otimes b) = ab$, for all $a, b \in A$, where $m_A$ is the multiplication of $A$ in ${}_k\mathscr{M}$. It then follows that $(A, \underline{m}_A, i)$ is an algebra in ${}_R\mathscr{M}_R$, as required.

The two correspondences defined above are inverse to each other, so we are done.

A monoidal functor takes algebras to algebras.

**Proposition 2.3** *Let $(F, \varphi_0, \varphi_2) : \mathscr{C} \to \mathscr{D}$ be a monoidal functor and $A$ an algebra in $\mathscr{C}$. Then $F(A)$ has an algebra structure in $\mathscr{D}$. Moreover, if $A \xrightarrow{f} B$ is an algebra morphism in $\mathscr{C}$ then $F(A) \xrightarrow{F(f)} F(B)$ is an algebra morphism in $\mathscr{D}$.*

*Proof* We show that $F(A)$ with $\underline{m}_{F(A)}$ and $\underline{\eta}_{F(A)}$ given by

$$\underline{m}_{F(A)} : F(A) \otimes F(A) \xrightarrow{\varphi_{2,A,A}} F(A \otimes A) \xrightarrow{F(\underline{m}_A)} F(A),$$
$$\underline{\eta}_{F(A)} : \underline{1} \xrightarrow{\varphi_0} F(\underline{1}) \xrightarrow{F(\underline{\eta}_A)} F(A),$$

is an algebra in $\mathscr{D}$. For this, since $\varphi_2$ is a natural transformation, the diagrams

$$\begin{array}{ccc} F(A) \otimes F(A \otimes A) & \xrightarrow{\varphi_{2,A,A\otimes A}} & F(A \otimes (A \otimes A)) \\ \downarrow{\scriptstyle \mathrm{Id}_{F(A)} \otimes F(\underline{m}_A)} & & \downarrow{\scriptstyle F(\mathrm{Id}_A \otimes \underline{m}_A)} \\ F(A) \otimes F(A) & \xrightarrow{\varphi_{2,A,A}} & F(A \otimes A), \end{array} \qquad \begin{array}{ccc} F(\underline{1}) \otimes F(A) & \xrightarrow{\varphi_{2,\underline{1},A}} & F(\underline{1} \otimes A) \\ \downarrow{\scriptstyle F(\underline{\eta}_A) \otimes \mathrm{Id}_{F(A)}} & & \downarrow{\scriptstyle F(\underline{\eta}_A \otimes \mathrm{Id}_A)} \\ F(A) \otimes F(A) & \xrightarrow{\varphi_{2,A,A}} & F(A \otimes A) \end{array}$$

and

$$\begin{array}{ccc} F(A \otimes A) \otimes F(A) & \xrightarrow{\varphi_{2,A\otimes A,A}} & F((A \otimes A) \otimes A) \\ \downarrow{\scriptstyle F(\underline{m}_A) \otimes \mathrm{Id}_{F(A)}} & & \downarrow{\scriptstyle F(\underline{m}_A \otimes \mathrm{Id}_A)} \\ F(A) \otimes F(A) & \xrightarrow{\varphi_{2,A,A}} & F(A \otimes A), \end{array} \qquad \begin{array}{ccc} F(A) \otimes F(\underline{1}) & \xrightarrow{\varphi_{2,A,\underline{1}}} & F(A \otimes \underline{1}) \\ \downarrow{\scriptstyle \mathrm{Id}_{F(A)} \otimes F(\underline{\eta}_A)} & & \downarrow{\scriptstyle F(\mathrm{Id}_A \otimes \underline{\eta}_A)} \\ F(A) \otimes F(A) & \xrightarrow{\varphi_{2,A,A}} & F(A \otimes A) \end{array}$$

are all commutative. By using this, we compute:

$$\begin{aligned}
&\underline{m}_{F(A)}(\mathrm{Id}_{F(A)} \otimes \underline{m}_{F(A)})a_{F(A),F(A),F(A)} \\
&\qquad = F(\underline{m}_A)\varphi_{2,A,A}(\mathrm{Id}_{F(A)} \otimes F(\underline{m}_A)\varphi_{2,A,A})a_{F(A),F(A),F(A)} \\
&\qquad = F(\underline{m}_A(\mathrm{Id}_A \otimes \underline{m}_A))\varphi_{2,A,A\otimes A}(\mathrm{Id}_{F(A)} \otimes \varphi_{2,A,A})a_{F(A),F(A),F(A)} \\
&\qquad = F(\underline{m}_A(\mathrm{Id}_A \otimes \underline{m}_A)a_{A,A,A})\varphi_{2,A\otimes A,A}(\varphi_{2,A,A} \otimes \mathrm{Id}_{F(A)}) \\
&\qquad = F(\underline{m}_A)F(\underline{m}_A \otimes \mathrm{Id}_A)\varphi_{2,A\otimes A,A}(\varphi_{2,A,A} \otimes \mathrm{Id}_{F(A)}) \\
&\qquad = F(\underline{m}_A)\varphi_{2,A,A}(F(\underline{m}_A)\varphi_{2,A,A} \otimes \mathrm{Id}_{F(A)}) \\
&\qquad = \underline{m}_{F(A)}(\underline{m}_{F(A)} \otimes \mathrm{Id}_{F(A)}),
\end{aligned}$$

where in the third equality we used the commutativity of the first diagram in Definition 1.22 and in the fourth equality the associativity of $\underline{m}_A$.

The fact that $\underline{\eta}_{F(A)}$ is a unit for $\underline{m}_{F(A)}$ follows from the computations:

$$\begin{aligned}\underline{m}_{F(A)}(\underline{\eta}_{F(A)} \otimes \mathrm{Id}_{F(A)}) &= F(\underline{m}_A)\varphi_{2,A,A}(F(\underline{\eta}_A) \otimes \mathrm{Id}_{F(A)})(\varphi_0 \otimes \mathrm{Id}_{F(A)}) \\ &= F(\underline{m}_A(\underline{\eta}_A \otimes \mathrm{Id}_A))\varphi_{2,\underline{1},A}(\varphi_0 \otimes \mathrm{Id}_{F(A)}) \\ &= F(l_A)\varphi_{2,\underline{1},A}(\varphi_0 \otimes \mathrm{Id}_{F(A)}) = l_{F(A)}\end{aligned}$$

(in the last but one equality we used the fact that $\underline{\eta}_A$ is a unit for $\underline{m}_A$, and in the last equality we used the commutativity of the second diagram in Definition 1.22), and

$$\begin{aligned}\underline{m}_{F(A)}(\mathrm{Id}_{F(A)} \otimes \underline{\eta}_{F(A)}) &= F(\underline{m}_A)\varphi_{2,A,A}(\mathrm{Id}_{F(A)} \otimes F(\underline{\eta}_A))(\mathrm{Id}_{F(A)} \otimes \varphi_0) \\ &= F(\underline{m}_A(\mathrm{Id}_A \otimes \underline{\eta}_A))\varphi_{2,A,\underline{1}}(\mathrm{Id}_{F(A)} \otimes \varphi_0) \\ &= F(r_A)\varphi_{2,A,\underline{1}}(\mathrm{Id}_{F(A)} \otimes \varphi_0) = r_{F(A)},\end{aligned}$$

where, this time, in the last equality we used the commutativity of the third diagram in Definition 1.22.

Finally, let $A \xrightarrow{f} B$ be an algebra morphism in $\mathscr{C}$. Then

$$\begin{aligned}\underline{m}_{F(B)}(F(f) \otimes F(f)) &= F(\underline{m}_B)\varphi_{2,B,B}(F(f) \otimes F(f)) \\ &= F(\underline{m}_B(f \otimes f))\varphi_{2,A,A} \\ &= F(f)F(\underline{m}_A)\varphi_{2,A,A} = F(f)\underline{m}_{F(A)},\end{aligned}$$

and $F(f)\underline{\eta}_{F(A)} = F(f\underline{\eta}_A)\varphi_0 = F(\underline{\eta}_B)\varphi_0 = \underline{\eta}_{F(B)}$, which completes the proof. □

When $A$ is an algebra in a monoidal category $\mathscr{C}$ we will denote $\underline{m}_A$ by 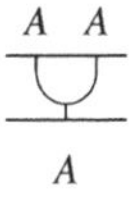 and the unit morphism $\underline{\eta}_A$ by 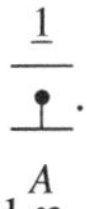. Then, in diagrammatic notation, the associativity and unit conditions for $\underline{m}_A$ and $\underline{\eta}_A$ take the form

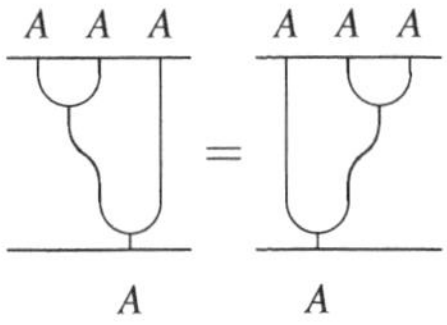
and
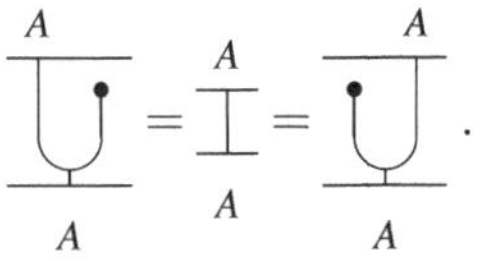 (2.1.2)

Note that, in this kind of computation, we will always assume that all the constraints of $\mathscr{C}$ are defined by identity morphisms, that is, that $\mathscr{C}$ is strict monoidal. Then the computations are still valid for an arbitrary monoidal category; see Remark 1.35.

In general, if $\mathscr{C}$ is a monoidal category, by  we denoted a morphism $f : X \to Y$

in $\mathscr{C}$. Thus, $f : A \to B$ is an algebra morphism in $\mathscr{C}$ if

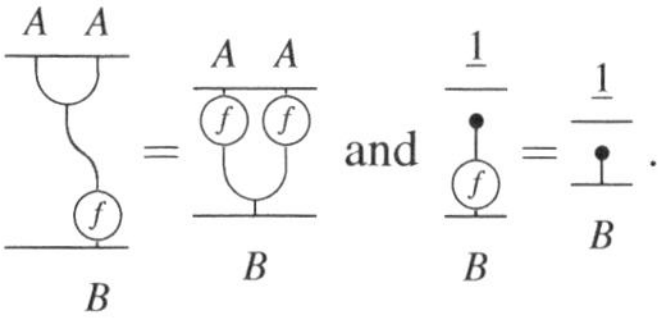


When $\mathscr{C}$ is braided, one can associate to an algebra $A$ in $\mathscr{C}$ another two (possibly different) algebra structures on $A$.

**Proposition 2.4** *Let $(\mathscr{C}, c)$ be a braided category and $(A, \underline{m}_A, \underline{\eta}_A)$ an algebra in $\mathscr{C}$. Then $A^{\mathrm{op}+} := (A, \underline{m}_A \circ c_{A,A}, \underline{\eta}_A)$ and $A^{\mathrm{op}-} := (A, \underline{m}_A \circ c^{-1}_{A,A}, \underline{\eta}_A)$ are algebras in $\mathscr{C}$. $A^{\mathrm{op}+}$ (resp. $A^{\mathrm{op}-}$) is called the c-opposite (resp. $c^{-1}$-opposite) algebra associated to $A$ in the category $(\mathscr{C}, c)$.*

*Proof* The multiplication $\underline{m}_{A^{\mathrm{op}+}}$ is associative since Proposition 1.37, (1.5.10) and the associativity of $\underline{m}_A$ imply that

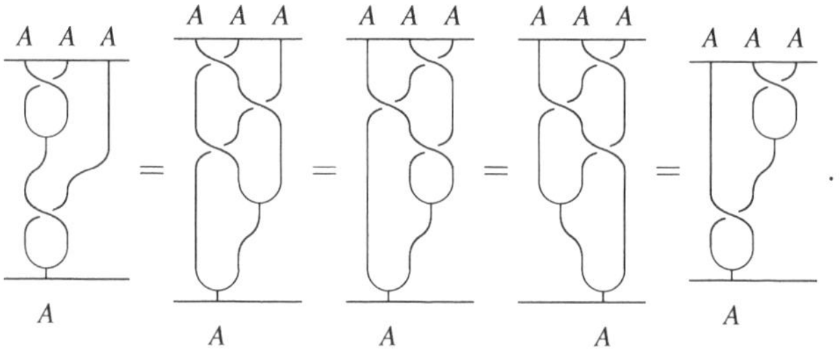


We also have that

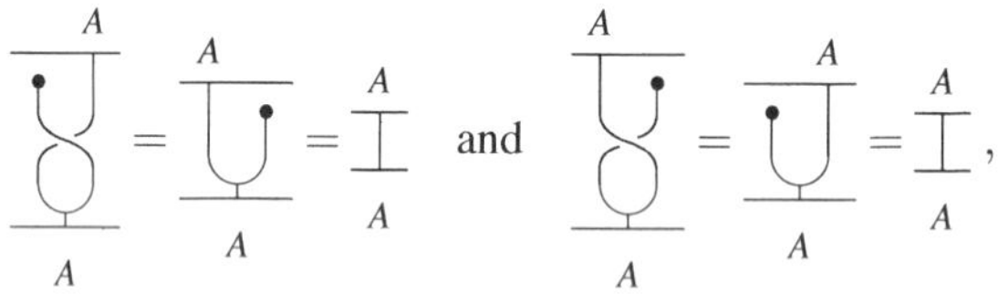


because of Proposition 1.49, and so $A^{\mathrm{op}+}$ is an algebra in $\mathscr{C}$. That $A^{\mathrm{op}-}$ is an algebra in $\mathscr{C}$ follows by applying the above arguments to $A$ viewed as an algebra in $\mathscr{C}^{\mathrm{in}}$. □

**Definition 2.5** An algebra $A$ in a braided category is called commutative (or braided commutative) if $A = A^{\mathrm{op}+}$ or, equivalently, if $A = A^{\mathrm{op}-}$, as algebras in $\mathscr{C}$.

We next see that the braiding $c$ of $\mathscr{C}$ or its inverse $c^{-1}$ gives rise to an algebra structure on the tensor product $A \otimes B$ of two algebras $A$, $B$. This is a consequence of the naturality of the braiding, as the cross product construction below shows.

If $A$ and $B$ are algebras in a monoidal category $\mathscr{C}$ and $\psi : B \otimes A \to A \otimes B$ is a morphism in $\mathscr{C}$ then one can introduce on $A \otimes B$ a multiplication given, in the case when $\mathscr{C}$ is strict, by the formula

$$\underline{m} = (\underline{m}_A \otimes \underline{m}_B) \circ (\mathrm{Id}_A \otimes \psi \otimes \mathrm{Id}_B). \tag{2.1.3}$$

By $A\#_\psi B$ we denote the object $A \otimes B$ endowed with the unit tensor product morphism and with the multiplication defined as in (2.1.3). Then $\underline{m}$, written as $\underline{m}_{A\#_\psi B}$ in what follows, takes the form

$\underline{m}_{A\#_\psi B} =$ 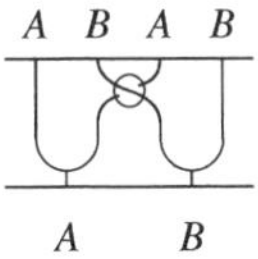

, where $\psi =$ 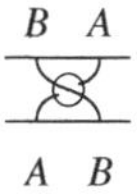

.

**Definition 2.6** If $A\#_\psi B$ is an algebra in $\mathscr{C}$ then we call it a cross product algebra of $A$ and $B$.

$A\#_\psi B$ is a cross product algebra under the following conditions on $\psi$.

**Theorem 2.7** *Let $\mathscr{C}$ be a monoidal category, $A$ and $B$ algebras in $\mathscr{C}$ and $\psi : B \otimes A \to A \otimes B$ a morphism in $\mathscr{C}$. Then the following are equivalent:*

*(i) $A\#_\psi B$ is a cross product algebra;*

*(ii) the following equalities hold:*

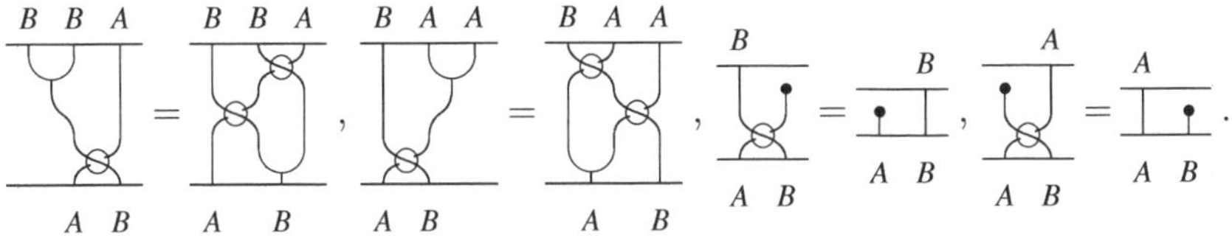


*If this is the case, $\psi$ is called a twisting morphism between $A$ and $B$.*

*Proof* We first prove that $\underline{\eta}_A \otimes \underline{\eta}_B$ is a right (resp. left) unit for $\underline{m}_{A\#_\psi B}$ defined in (2.1.3) if and only if the third (resp. fourth) equality in (ii) holds. Indeed, if $\underline{\eta}_A \otimes \underline{\eta}_B$ is a right unit for $\underline{m}_{A\#_\psi B}$ we have

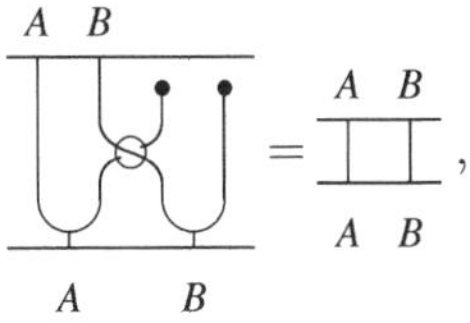

, so that 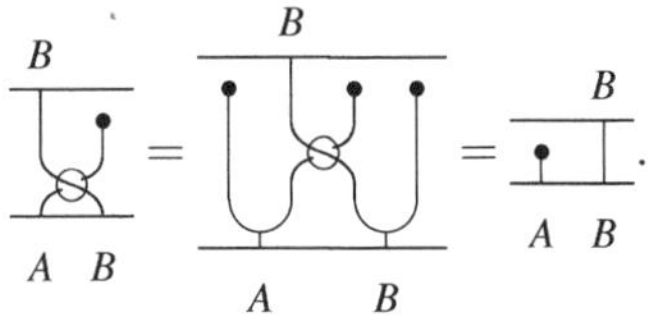


Conversely, if the third equality in (ii) holds then

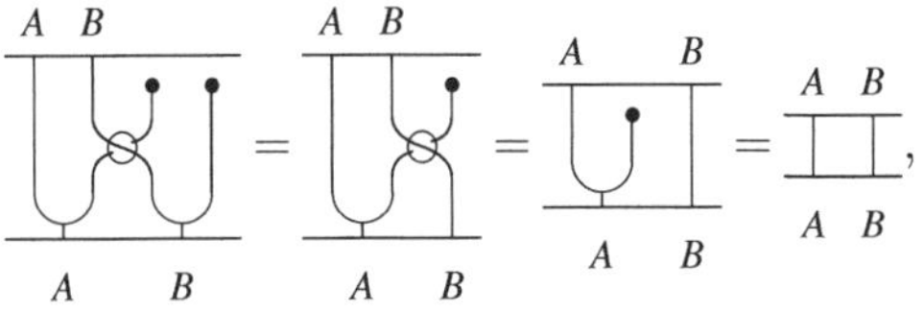


so $\underline{\eta}_A \otimes \underline{\eta}_B$ is a right unit for $\underline{m}_{A\#_\psi B}$. The left-handed case is similar.

$(i) \Leftrightarrow (ii)$. We have $\underline{m}_{A\#_\psi B}$ associative if and only if

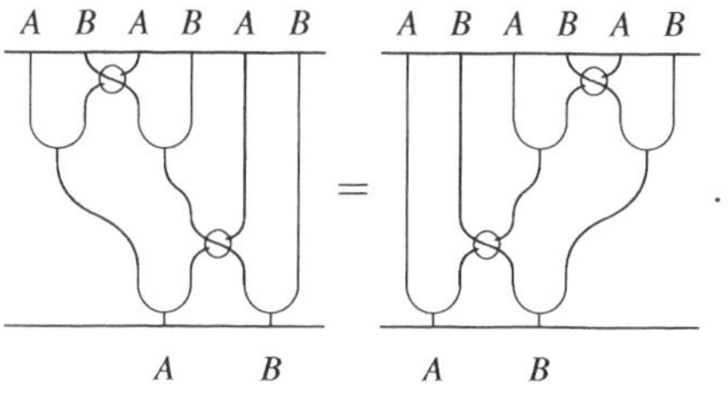


(2.1.4)

In view of the above results it is enough to show that (2.1.4) is equivalent to the first two equalities in (ii). Indeed, if (2.1.4) holds then by (2.1.2) and the unit conditions proved above we have

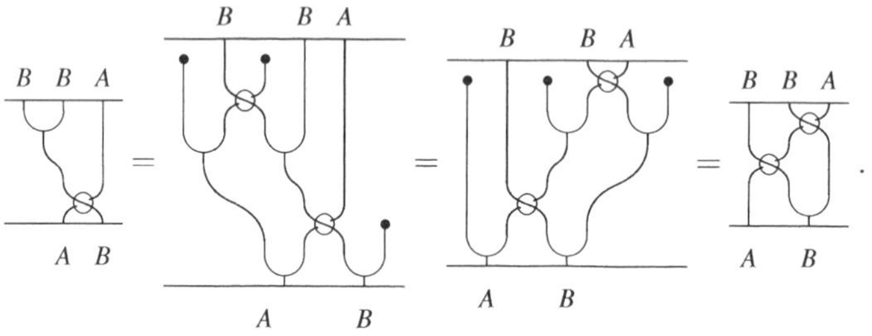


Similarly, one can prove the second equality in (ii).

Conversely, if the first two equalities in (ii) hold then

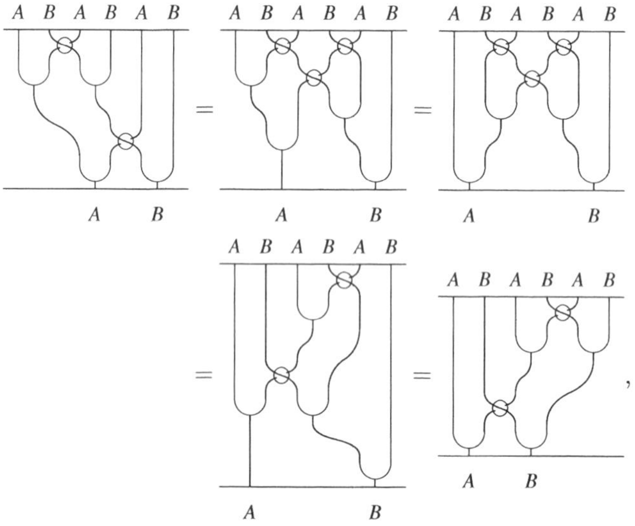


as required, where we used: in the first equality the first relation in (ii), in the second one the fact that the multiplication of $A$ is associative, in the third one the second equality in (ii), and in the last one the fact that the multiplication of $B$ is associative. This finishes the proof of the theorem. □

**Remark 2.8** If $A\#_\psi B$ is a cross product algebra in $\mathscr{C}$, the morphisms $i_A := \mathrm{Id}_A \otimes \underline{\eta}_B : A \to A\#_\psi B$ and $i_B := \underline{\eta}_A \otimes \mathrm{Id}_B : B \to A\#_\psi B$ are algebra morphisms in $\mathscr{C}$.

The cross product algebra has the following Universal Property.

**Proposition 2.9** *Let $A\#_\psi B$ be a cross product algebra in $\mathscr{C}$. If $(X, \underline{m}_X, \underline{\eta}_X)$ is an algebra in $\mathscr{C}$ and $u : A \to X$, $v : B \to X$ are morphisms of algebras in $\mathscr{C}$ such that*

$$\underline{m}_X(u \otimes v)\psi = \underline{m}_X(v \otimes u), \tag{2.1.5}$$

*then there exists a unique morphism $w : A\#_\psi B \to X$ of algebras in $\mathscr{C}$ such that $wi_A = u$ and $wi_B = v$. This morphism $w$ is given explicitly by $w = \underline{m}_X(u \otimes v)$.*

*Proof* Observe first that the algebra morphisms $i_A$ and $i_B$ in Remark 2.8 satisfy the condition in the statement, namely that $\underline{m}_{A\#_\psi B}(i_B \otimes i_A) = \underline{m}_{A\#_\psi B}(i_A \otimes i_B)\psi = \psi$.

Suppose there is an algebra map $w : A\#_\psi B \to X$ such that $wi_A = u$ and $wi_B = v$. Then $w = w\underline{m}_{A\#_\psi B}(i_A \otimes i_B) = \underline{m}_X(w \otimes w)(i_A \otimes i_B) = \underline{m}_X(wi_A \otimes wi_B) = \underline{m}_X(u \otimes v)$, and this shows the uniqueness of $w$.

Conversely, define $w = \underline{m}_X(u \otimes v)$. Then $w$ is multiplicative since

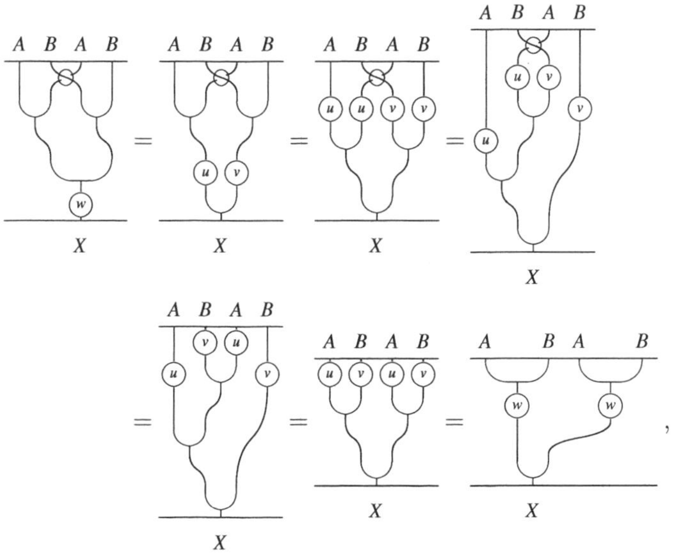


where we denoted $w =$ 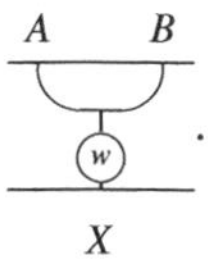

. We also have 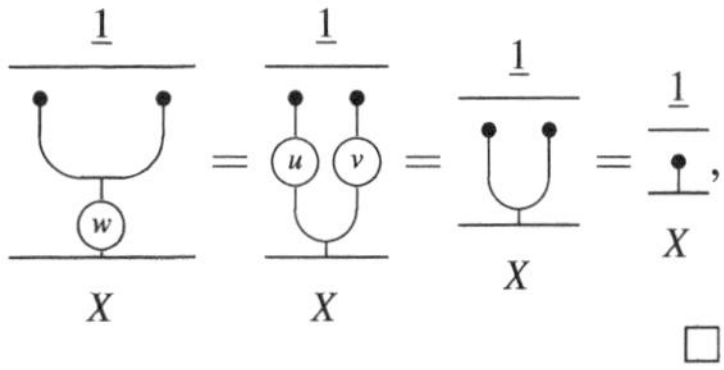


and this finishes the proof. □

The braiding $c$ of a braided category $\mathscr{C}$ and its inverse $c^{-1}$ satisfy the conditions in Theorem 2.7 (ii), because of Proposition 1.49 and Proposition 1.37. Thus $A\#_{c_{B,A}}B$ and $A\#_{c^{-1}_{A,B}}B$ are cross product algebras in $\mathscr{C}$.

**Definition 2.10** Let $A, B$ be two algebras in a braided category $(\mathscr{C}, c)$. We call the $c$-tensor product (resp. $c^{-1}$-tensor product) algebra between $A$ and $B$ in $\mathscr{C}$ the cross product algebra $A\#_{c_{B,A}}B$ (resp. $A\#_{c^{-1}_{A,B}}B$), and we denote it by $A\otimes_+ B$ (resp. $A\otimes_- B$).

Note that $A\otimes_- B$ is nothing but $A\otimes_+ B$ considered in $\mathscr{C}^{\text{in}}$ instead of $\mathscr{C}$. Also, $A\otimes_+ B$ is the object $A\otimes B$ of $\mathscr{C}$ endowed with the algebra structure in $\mathscr{C}$ given, if $\mathscr{C}$ is strict, by $(\underline{m}_A\otimes\underline{m}_B)\circ(\mathrm{Id}_A\otimes c_{B,A}\otimes\mathrm{Id}_B)$ and tensor product unit morphism. Similarly, $A\otimes_- B$ is an algebra in $\mathscr{C}$ with the multiplication given, in the strict case, by $(\underline{m}_A\otimes\underline{m}_B)\circ(\mathrm{Id}_A\otimes c^{-1}_{A,B}\otimes\mathrm{Id}_B)$ and tensor product unit morphism.

As a concrete example, let $G$ be an abelian group and endow the category of $G$-graded vector spaces $\mathrm{Vect}^G$ with a braided structure given by an abelian 3-cocycle $(\phi, \mathscr{R})$ on $G$ as in Proposition 1.44.

**Example 2.11** If $(\phi, \mathscr{R})$ is an abelian 3-cocycle on an abelian group $G$ and $A, A'$ are $G$-graded quasialgebras with reassociator $\phi$, then the multiplication of the $G$-graded quasialgebra $A\otimes_+ A'$ (with reassociator $\phi$) is given by

$$(a\otimes a')(b\otimes b') = \frac{\phi(|a|,|a'|,|b|)\phi(|a||b|,|a'|,|b'|)}{\phi(|a||a'|,|b|,|b'|)\phi(|a|,|b|,|a'|)}\mathscr{R}(|a'|,|b|)ab\otimes a'b',$$

for all homogeneous elements $a, b\in A$ and $a', b'\in A'$.

For instance, if $\mathrm{Vect}^{\mathbb{Z}_2}_{-1}$ is the braided category of super vector spaces (see Example 1.47), then for any two algebras $A, A'$ in $\mathrm{Vect}^{\mathbb{Z}_2}_{-1}$ the algebra structure of $A\otimes_+ A'$ in $\mathrm{Vect}^{\mathbb{Z}_2}_{-1}$ is determined by

$$(a\otimes a')(b\otimes b') = (-1)^{|a'||b|}ab\otimes a'b',$$

for all homogeneous elements $a, b\in A$ and $a', b'\in A'$. The unit of $A\otimes_+ A'$ is $1_A\otimes 1_{A'}$.

*Proof* Since the category $\mathrm{Vect}^G_{(\phi,\mathscr{R})}$ is not in general strict monoidal, the multiplication of the tensor product $G$-graded quasialgebra $A\otimes_+ A'$ is given by the following composition:

$$\begin{aligned}\underline{m}_{A\otimes_+ A'} : (A\otimes A')\otimes(A\otimes A') &\xrightarrow{a^{-1}_{A\otimes A',A,A'}} ((A\otimes A')\otimes A)\otimes A'\\ &\xrightarrow{a_{A,A',A}\otimes\mathrm{Id}_{A'}} (A\otimes(A'\otimes A))\otimes A' \xrightarrow{(\mathrm{Id}_A\otimes c_{A',A})\otimes\mathrm{Id}_{A'}} (A\otimes(A\otimes A'))\otimes A'\\ &\xrightarrow{a^{-1}_{A,A,A'}\otimes\mathrm{Id}_{A'}} ((A\otimes A)\otimes A')\otimes A' \xrightarrow{a_{A\otimes A,A',A'}} (A\otimes A)\otimes(A'\otimes A')\\ &\xrightarrow{\underline{m}_A\otimes\underline{m}_{A'}} A\otimes A'\ .\end{aligned}$$

If we keep in mind the monoidal structure of $\mathrm{Vect}^G_\phi$, a straightforward computation leads to the formula in the statement; we leave the verification to the reader.

The second assertion follows by specializing this result for $G=\mathbb{Z}_2$ and $(\phi, \mathscr{R})$ the abelian 3-cocycle on $\mathbb{Z}_2$ defined in Example 1.47. □

**Proposition 2.12** *Let $A, B, C$ be three algebras in a braided category $(\mathscr{C}, c)$. Then $(A\otimes_\pm B)\otimes_\pm C$ and $A\otimes_\pm(B\otimes_\pm C)$ are isomorphic as algebras in $\mathscr{C}$.*

*Proof* The isomorphism is produced by the associativity constraint of $\mathscr{C}$. If we assume $\mathscr{C}$ strict monoidal then the multiplications of $(A \otimes_+ B) \otimes_+ C$ and $A \otimes_+ (B \otimes_+ C)$ come out as

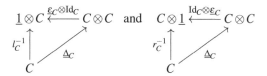

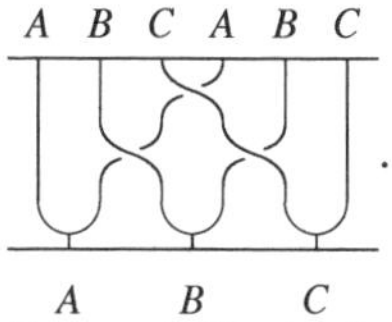

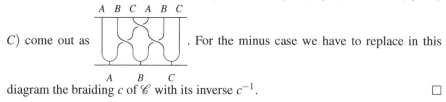

. For the minus case we have to replace in this diagram the braiding $c$ of $\mathscr{C}$ with its inverse $c^{-1}$. □

## 2.2 Coalgebras in Monoidal Categories

The concept of monoidal coalgebra, that is, of a coalgebra in a monoidal category $\mathscr{C}$, can be easily obtained by reversing the sense of the arrows in the definition of a monoidal algebra. In other words, a coalgebra in $\mathscr{C}$ is precisely an algebra in $\mathscr{C}^{\mathrm{opp}}$, the opposite monoidal category associated to $\mathscr{C}$. Consequently, results and constructions for monoidal algebras that have diagrammatic proofs or definitions produce "dual" results and constructions for coalgebras: we simply have to reverse the sense of the arrows in diagrams or, equivalently, to turn upside down diagrammatic computations. Furthermore, if the category is rigid monoidal, we shall see in the next section that the algebra and coalgebra concepts are actually equivalent. For these reasons in the remainder of this section we will omit most of the proofs.

**Definition 2.13** Let $(\mathscr{C}, \otimes, \underline{1}, a, l, r)$ be a monoidal category. A coalgebra in $\mathscr{C}$ is an algebra in $\mathscr{C}^{\mathrm{opp}}$, the opposite monoidal category associated to $\mathscr{C}$. Explicitly, a coalgebra in $\mathscr{C}$ is a triple $(C, \underline{\Delta}_C, \underline{\varepsilon}_C)$, where $\underline{\Delta}_C : C \to C \otimes C$ and $\underline{\varepsilon}_C : C \to \underline{1}$ are morphisms in $\mathscr{C}$ such that

- $\underline{\Delta}_C$, called the comultiplication of $C$, is coassociative in the sense that the diagram

$$\begin{array}{ccccc} (C\otimes C)\otimes C & \xleftarrow{\underline{\Delta}_C\otimes \mathrm{Id}_C} & C\otimes C & \xleftarrow{\underline{\Delta}_C} & C \\ \downarrow{\scriptstyle a_{C,C,C}} & & & \swarrow{\scriptstyle \underline{\Delta}_C} & \\ C\otimes (C\otimes C) & \xleftarrow{\mathrm{Id}_C\otimes \underline{\Delta}_C} & C\otimes C & & \end{array}$$

  is commutative;
- $\underline{\varepsilon}_C$, called the counit of $C$, is a counit for $\underline{\Delta}_C$; this means that the diagrams

$$\begin{array}{ccc} \underline{1}\otimes C & \xleftarrow{\underline{\varepsilon}_C\otimes \mathrm{Id}_C} & C\otimes C \\ {\scriptstyle l_C^{-1}}\uparrow & \nearrow{\scriptstyle \underline{\Delta}_C} & \\ C & & \end{array} \quad \text{and} \quad \begin{array}{ccc} C\otimes \underline{1} & \xleftarrow{\mathrm{Id}_C\otimes \underline{\varepsilon}_C} & C\otimes C \\ {\scriptstyle r_C^{-1}}\uparrow & \nearrow{\scriptstyle \underline{\Delta}_C} & \\ C & & \end{array}$$

  are commutative.

If $(C, \underline{\Delta}_C, \underline{\varepsilon}_C)$ and $(D, \underline{\Delta}_D, \underline{\varepsilon}_D)$ are coalgebras in $\mathscr{C}$ then a coalgebra morphism between $C$ and $D$ is a morphism $f : C \to D$ in $\mathscr{C}$ such that the diagrams

$$\begin{array}{ccc} C & \xrightarrow{\ f\ } & D \\ {\scriptstyle \underline{\Delta}_C}\downarrow & & \downarrow{\scriptstyle \underline{\Delta}_D} \\ C \otimes C & \xrightarrow{\ f \otimes f\ } & D \otimes D \end{array} \qquad \text{and} \qquad \begin{array}{ccc} C & \xrightarrow{\ f\ } & D \\ {\scriptstyle \underline{\varepsilon}_C}\downarrow & \swarrow{\scriptstyle \underline{\varepsilon}_D} & \\ \underline{1} & & \end{array}$$

are commutative (that is, $f : D \to C$ is an algebra morphism in $\mathscr{C}^{\text{opp}}$).

If, moreover, $f$ is an isomorphism in $\mathscr{C}$ we say that $C$ and $D$ are isomorphic coalgebras in $\mathscr{C}$.

As in the algebra case, the meaning of a coalgebra in a monoidal category differs from one category to another.

**Examples 2.14** (1) The unit object of a monoidal category $\mathscr{C}$ is a coalgebra in $\mathscr{C}$ via $\underline{\Delta}_{\underline{1}} = l_{\underline{1}}^{-1} = r_{\underline{1}}^{-1} : \underline{1} \to \underline{1} \otimes \underline{1}$ and $\underline{\varepsilon}_{\underline{1}} = \mathrm{Id}_{\underline{1}}$.

(2) Any set $X$ has a unique coalgebra structure in $\underline{\mathrm{Set}}$. In this way any function $f : X \to Y$ becomes a coalgebra morphism in $\underline{\mathrm{Set}}$.

To see this, suppose that $X$ has a coalgebra structure in $\underline{\mathrm{Set}}$ given by $\underline{\Delta}_X : X \to X \times X$ and $\underline{\varepsilon}_X : X \to \{*\}$. Then $\underline{\varepsilon}_X(x) = *$, for all $x \in X$.

For $x \in X$ write $\underline{\Delta}_X(x) = (y, z)$, for some $y, z \in X$. Since $\underline{\varepsilon}_X$ is a counit for $\underline{\Delta}_X$ it follows that $(*, z) = (*, x)$ and $(y, *) = (x, *)$, so $y = z = x$ and $\underline{\Delta}_X(x) = (x, x)$.

Conversely, it is easy to see that $(X, \underline{\Delta}_X, \underline{\varepsilon}_X)$ is a coalgebra in $\underline{\mathrm{Set}}$, where $\underline{\Delta}_X(x) = (x, x)$ and $\underline{\varepsilon}_X(x) = *$, for any $x \in X$.

(3) Let $\mathscr{C}$ be a category. A coalgebra in the monoidal category of endo-functors $[\mathscr{C}, \mathscr{C}]$ is called a comonad in $\mathscr{C}$. It consists of a triple $(U, \underline{\Delta}, \underline{\varepsilon})$, where $U : \mathscr{C} \to \mathscr{C}$ is a functor and $\underline{\Delta} : U \to U \circ U$ and $\underline{\varepsilon} : U \to \mathrm{Id}_{\mathscr{C}}$ are natural transformations such that

$$U(\underline{\Delta}_{\mathfrak{M}}) \circ \underline{\Delta}_{\mathfrak{M}} = \underline{\Delta}_{U(\mathfrak{M})} \circ \underline{\Delta}_{\mathfrak{M}}, \tag{2.2.1}$$

$$U(\underline{\varepsilon}_{\mathfrak{M}}) \circ \underline{\Delta}_{\mathfrak{M}} = \underline{\varepsilon}_{U(\mathfrak{M})} \circ \underline{\Delta}_{\mathfrak{M}} = \mathrm{Id}_{U(\mathfrak{M})}, \tag{2.2.2}$$

for any object $\mathfrak{M}$ of $\mathscr{C}$.

Indeed, since a coalgebra in $[\mathscr{C}, \mathscr{C}]$ is an algebra in $[\mathscr{C}, \mathscr{C}]^{\text{opp}}$ everything follows from Example 2.2 (3).

(4) Let $G$ be a group, $\phi$ an invertible normalized 3-cocycle with coefficients in a field $k$, and $\mathrm{Vect}_\phi^G$ the category of $G$-graded $k$-vector spaces endowed with the monoidal structure from Proposition 1.13. Then a coalgebra in $\mathrm{Vect}_\phi^G$ is a $G$-graded vector space $C = \bigoplus_{x \in G} C_x$ together with $k$-linear maps $\Delta : C \to C \otimes C$ and $\varepsilon : C \to k$ such that $(\mathrm{Id}_C \otimes \varepsilon)\Delta = \mathrm{Id}_C = (\varepsilon \otimes \mathrm{Id}_C)\Delta$, and

- $\Delta(C_x) \subseteq \bigoplus_{uv = x} C_u \otimes C_v$, for all $x \in G$;
- $\varepsilon(C_x) = 0$, for all $x \in G$, $x \neq e$;
- for any $c \in C_x$, if we write $(\Delta \otimes \mathrm{Id}_C)\Delta(c) = \sum_i c_i \otimes d_i \otimes e_i$ and $(\mathrm{Id}_C \otimes \Delta)\Delta(c) =$

$\sum_j c'_j \otimes d'_j \otimes e'_j$, with all $c_i$s , $d_i$s, etc. homogeneous elements, then

$$\sum_i \phi(|c_i|,|d_i|,|e_i|)c_i \otimes d_i \otimes e_i = \sum_j c'_j \otimes d'_j \otimes e'_j.$$

Indeed, all the conditions are reformulations of the following facts: $\varepsilon$ is a counit for $\Delta$, $\Delta$ and $\varepsilon$ must preserve the degree of homogeneous elements, and $\Delta$ has to be coassociative up to the associativity constraint of $\mathrm{Vect}_\phi^G$.

In what follows, we call a coalgebra $C$ in $\mathrm{Vect}_\phi^G$ a $G$-graded quasicoalgebra with reassociator $\phi$. When $\phi$ is trivial we simply say that $C$ is a $G$-graded coalgebra.

(5) Let $k$ be a commutative ring and $R$ a $k$-algebra. We call a coalgebra in ${}_R\mathscr{M}_R$ an $R$-coring. Explicitly, an $R$-coring is a triple $(C,\Delta_C,\varepsilon_C)$ consisting of an $R$-bimodule $C$ and $R$-bimodule morphisms $\Delta_C : C \to C \otimes_R C$ and $\varepsilon_C : C \to R$ such that $\Delta_C$ is coassociative up to the associativity constraint of ${}_R\mathscr{M}_R$ and $\varepsilon_C$ is a counit for it. If for $c \in C$ we denote $\Delta_C(c) = c_1 \otimes_R c_2$ (summation implicitly understood) then

$$\Delta_C(rc) = rc_1 \otimes_R c_2\ ,\ \Delta_C(cr) = c_1 \otimes_R c_2 r,$$
$$(c_1)_1 \otimes_R (c_1)_2 \otimes_R c_2 = c_1 \otimes_R (c_2)_1 \otimes_R (c_2)_2,$$
$$\varepsilon_C(rcr') = r\varepsilon(c)r' \quad \text{and} \quad c_1\varepsilon_C(c_2) = \varepsilon_C(c_1)c_2 = c,$$

for all $c \in C$ and $r, r' \in R$. When $R = k$ this reduces to the notion of $k$-coalgebra.

**Remark 2.15** For $C$ a $k$-coalgebra denote $\Delta_2 = (\Delta_C \otimes \mathrm{Id}_C)\Delta_C = (\mathrm{Id}_C \otimes \Delta_C)\Delta_C$, and $\Delta_n = (\Delta_C \otimes \mathrm{Id}_C^{\otimes(n-1)}) \circ \Delta_{n-1}$, for all $n \geq 2$. By induction on $n$ one can show that

$$\Delta_n = (\mathrm{Id}_C^{\otimes p} \otimes \Delta_C \otimes \mathrm{Id}_C^{\otimes(n-p-1)}) \circ \Delta_{n-1},$$

for all $n \geq 2$ and $1 \leq p \leq n-1$. This equality allows us to introduce the so-called sigma notation for coalgebras, also known as the Sweedler notation or as the Heyneman–Sweedler notation for coalgebras.

For $(C,\Delta_C,\varepsilon_C)$ a $k$-coalgebra and $c \in C$ we usually have $\Delta_C(c) = \sum_i x_i \otimes y_i$, for some families of elements $(x_i)_i$ and $(y_i)_i$ in $C$. Instead, we denote $\Delta_C(c) = c_1 \otimes c_2$, summation understood. Then the coassociativity property of $\Delta_C$ comes out as

$$(c_1)_1 \otimes (c_1)_2 \otimes c_2 = c_1 \otimes (c_2)_1 \otimes (c_2)_2 := c_1 \otimes c_2 \otimes c_3,$$

for all $c \in C$. Likewise, the fact that $\varepsilon_C$ is a counit for $\Delta_C$ can be written in sigma notation as $\varepsilon(c_1)c_2 = \varepsilon(c_2)c_1 = c$, for all $c \in C$.

In general, for $c \in C$ we denote $\Delta_n(c) = c_1 \otimes \cdots \otimes c_{n+1}$ and owing to the above formula the meaning of this notation is

$$(c_1)_1 \otimes (c_1)_2 \otimes c_2 \otimes \cdots \otimes c_n \text{ or } c_1 \otimes c_2 \otimes ((c_3)_1)_1 \otimes ((c_3)_1)_2 \otimes (c_3)_2 \otimes c_4 \otimes \cdots \otimes c_{n-1},$$

and so on. Similar notation will be used in what follows when we work with different types of coalgebras, if of course we are allowed to evaluate morphisms on elements. For instance see the $R$-coring case considered in Example 2.14 (5).

The counterpart of Proposition 2.3 is the following.

**Proposition 2.16** *Let $(F,\psi_0,\psi_2):\mathscr{C}\to\mathscr{D}$ be an opmonoidal functor and $C$ a coalgebra in $\mathscr{C}$. Then $F(C)$ is a coalgebra in $\mathscr{D}$. Moreover, if $C\xrightarrow{f}D$ is a coalgebra morphism in $\mathscr{C}$ then $F(C)\xrightarrow{F(f)}F(D)$ is a coalgebra morphism in $\mathscr{D}$.*

*Proof* One can easily see that $(F,\psi_0,\psi_2):\mathscr{C}^{\rm opp}\to\mathscr{D}^{\rm opp}$ is a monoidal functor. Thus, if $C$ is a coalgebra in $\mathscr{C}$ then $C$ is an algebra in $\mathscr{C}^{\rm opp}$ and so, by Proposition 2.3, we obtain that $F(C)$ is an algebra in $\mathscr{D}^{\rm opp}$, and therefore a coalgebra in $\mathscr{D}$.

For further use note that $F(C)$ becomes a coalgebra in $\mathscr{D}$ with

$$\underline{\Delta}_{F(C)}:F(C)\overset{F(\underline{\Delta}_C)}{\longrightarrow}F(C\otimes C)\overset{\psi_{2,C,C}}{\longrightarrow}F(C)\otimes F(C),$$
$$\underline{\varepsilon}_{F(C)}:F(C)\overset{F(\underline{\varepsilon}_C)}{\longrightarrow}F(\underline{1})\overset{\psi_0}{\longrightarrow}\underline{1}.$$

If $C\xrightarrow{f}D$ is a coalgebra morphism in $\mathscr{C}$ then $D\xrightarrow{f}C$ is an algebra morphism in $\mathscr{C}^{\rm opp}$, and therefore $F(D)\xrightarrow{F(f)}F(C)$ is an algebra morphism in $\mathscr{D}^{\rm opp}$; see Proposition 2.3. Hence $F(C)\xrightarrow{F(f)}F(D)$ is a coalgebra morphism in $\mathscr{D}$, as needed. □

For a coalgebra $(C,\underline{\Delta}_C,\underline{\varepsilon}_C)$ in a monoidal category $\mathscr{C}$ we denote the morphism $\underline{\Delta}_C$ by 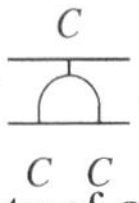 and the counit $\underline{\varepsilon}_C$ by . Then the coassociativity of $\underline{\Delta}_C$ and the counit property of $\underline{\varepsilon}_C$ can be written as follows:

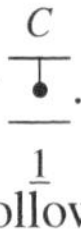

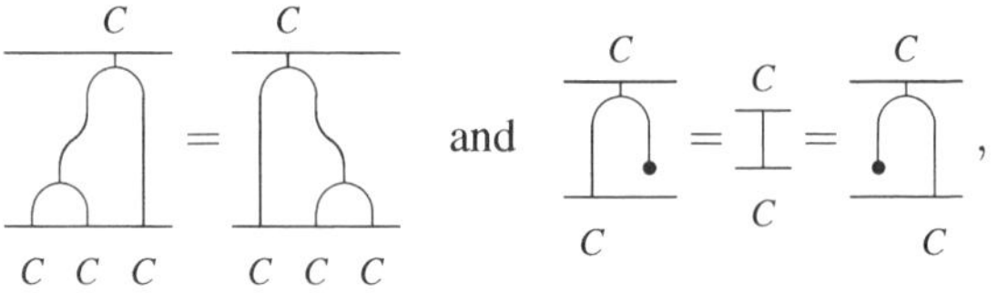

respectively, where again we used the convention that in diagrammatic computations the monoidal category $\mathscr{C}$ is always assumed strict monoidal. Note that, in general, $\overset{X}{\underset{X}{\rule[0pt]{0.5pt}{10pt}}}$  is the diagrammatic notation for the identity morphism of an object $X$ of $\mathscr{C}$.

Thus $f:C\to D$ is a coalgebra morphism in $\mathscr{C}$ if and only if

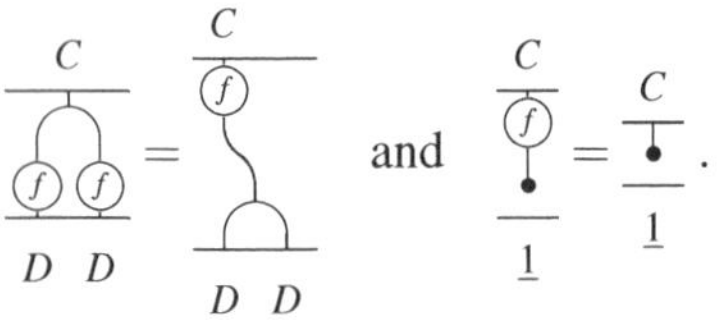

**Proposition 2.17** *If $(\mathscr{C},c)$ is a braided category and $(C,\underline{\Delta}_C,\underline{\varepsilon}_C)$ is a coalgebra in $\mathscr{C}$ then $C^{\rm cop+}:=(C,c_{C,C}\circ\underline{\Delta}_C,\underline{\varepsilon}_C)$ (resp. $C^{\rm cop-}:=(C,c^{-1}_{C,C}\circ\underline{\Delta}_C,\underline{\varepsilon}_C)$) is a coalgebra in $\mathscr{C}$ called the $c$-coopposite (resp. $c^{-1}$-coopposite) coalgebra associated to $C$.*

*Proof* Turn upside down the diagrammatic proof of Proposition 2.4. □

**Definition 2.18** A coalgebra $C$ in a braided category $\mathscr{C}$ is called cocommutative if $C = C^{\mathrm{cop}+}$ or equivalently $C = C^{\mathrm{cop}-}$, as coalgebras in $\mathscr{C}$. We also say that $C$ is braided cocommutative.

We end this section by presenting the cross product coalgebra construction. Once more, we will omit the proofs since they can be obtained from the corresponding proofs for algebras by turning the diagrams upside down.

If $C$, $D$ are coalgebras in a monoidal category $\mathscr{C}$ and $\psi : C\otimes D \to D\otimes C$ is a morphism in $\mathscr{C}$, we define on $C\otimes D$ a comultiplication which, for $\mathscr{C}$ strict, reads

$$\underline{\Delta} = (\mathrm{Id}_C\otimes\psi\otimes\mathrm{Id}_D)\circ(\underline{\Delta}_C\otimes\underline{\Delta}_D) =$$ 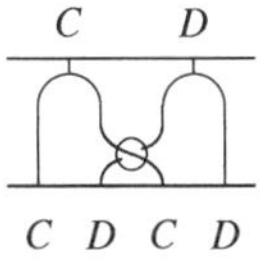

, where $\psi =$ 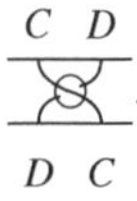

. (2.2.3)

We present necessary and sufficient conditions such that $C\otimes D$ with the comultiplication (2.2.3) and tensor product counit morphism is a coalgebra in $\mathscr{C}$. If this is the case, we call $C\otimes D$ a cross product coalgebra between $C$ and $D$, and denote it by $C\#^{\psi}D$. Moreover, $\underline{\Delta}$ defined in (2.2.3) will be denoted by $\underline{\Delta}_{C\#^{\psi}D}$.

**Theorem 2.19** *Let $\mathscr{C}$ be a monoidal category, $C$ and $D$ coalgebras in $\mathscr{C}$, and $\psi : C\otimes D\to D\otimes C$ a morphism in $\mathscr{C}$. Then the following are equivalent:*

*(i) $C\#^{\psi}D$ is a cross product coalgebra;*

*(ii) the following equalities hold:*

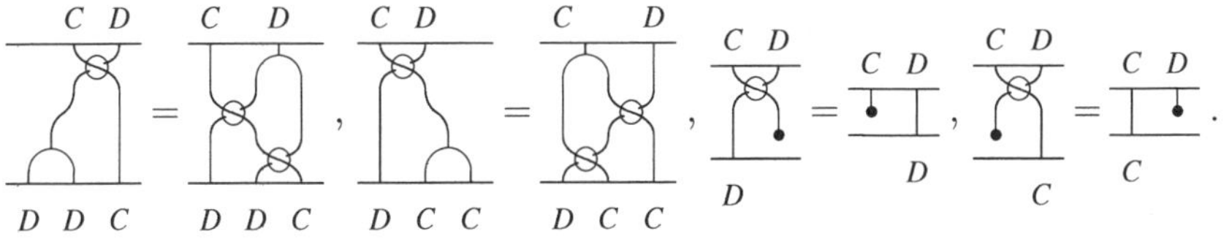


If $(\mathscr{C},c)$ is a braided category and $C,D$ are coalgebras in $\mathscr{C}$ then $\psi := c_{C,D} : C\otimes D\to D\otimes C$ satisfies the conditions in Theorem 2.19, and therefore $C\#^{c_{C,D}}D$ is a cross product coalgebra in $\mathscr{C}$. It will be called the $c$-tensor product coalgebra between $C$ and $D$ in $\mathscr{C}$, and will be denoted by $C\otimes^{+}D$. Thus $C\otimes^{+}D$ is the object $C\otimes D$ of $\mathscr{C}$ viewed as a coalgebra in $\mathscr{C}$ with comultiplication defined, if $\mathscr{C}$ is strict, by $(\mathrm{Id}_C\otimes c_{C,D}\otimes\mathrm{Id}_D)\circ(\underline{\Delta}_C\otimes\underline{\Delta}_D)$ and tensor product counit morphism.

Similarly, by $C\otimes^{-}D$ we denote the coalgebra in $\mathscr{C}$ with comultiplication given, if $\mathscr{C}$ is strict, by $(\mathrm{Id}_C\otimes c^{-1}_{D,C}\otimes\mathrm{Id}_D)\circ(\underline{\Delta}_C\otimes\underline{\Delta}_D)$ and tensor product counit morphism. We will call it the $c^{-1}$-tensor product coalgebra between $C$ and $D$; it coincides with $C\otimes^{+}D$ considered in $\mathscr{C}^{\mathrm{in}}$ instead of $\mathscr{C}$.

**Example 2.20** Let $(\phi,\mathscr{R})$ be an abelian 3-cocycle on an abelian group $G$, and $C,D$ two $G$-graded quasicoalgebras with reassociator $\phi$. Then

$$\Delta(c\otimes d) = \frac{\phi(|c_1|,|c_2|,|d_1|)\phi(|c_1||d_1|,|c_2|,|d_2|)}{\phi(|c_1||c_2|,|d_1|,|d_2|)\phi(|c_1|,|d_1|,|c_2|)} \mathscr{R}(|c_2|,|d_1|)(c_1\otimes d_1)\otimes(c_2\otimes d_2)$$

gives the comultiplication of the $c$-tensor product $G$-graded quasicoalgebra with reassociator $\phi$. Here $c \in C$, $d \in D$ are homogeneous elements and $\Delta_C(c) = c_1 \otimes c_2$ and $\Delta_D(d) = d_1 \otimes d_2$ (summations implicitly understood).

Consequently, if $C, D$ are coalgebras in the braided category of super vector spaces $\mathrm{Vect}_{-1}^{\mathbb{Z}_2}$, we have that their $c$-tensor product coalgebra is $C \otimes D$ endowed with the comultiplication given by $\Delta_{C\otimes^+ D}(c \otimes d) = (-1)^{|c_2||d_1|}(c_1 \otimes d_1) \otimes (c_2 \otimes d_2)$, for all homogeneous elements $c \in C$, $d \in D$. The counit is defined by $\varepsilon_{C\otimes^+ D}(c \otimes d) = \varepsilon_C(c)\varepsilon_D(d)$, for all $c \in C$, $d \in D$. Here we assumed that $\Delta_C(c)$ decomposes as a sum of homogeneous elements as $c_1 \otimes c_2$, and similarly for $\Delta_D(d)$.

*Proof* The $c$-tensor product coalgebra structure on $C \otimes D$ is given by

$$\begin{aligned} \Delta : C \otimes D &\xrightarrow{\Delta_C \otimes \Delta_D} (C \otimes C) \otimes (D \otimes D) \xrightarrow{a^{-1}_{C\otimes C, D, D}} ((C \otimes C) \otimes D) \otimes D \\ &\xrightarrow{a_{C,C,D} \otimes \mathrm{Id}_D} (C \otimes (C \otimes D)) \otimes D \xrightarrow{(\mathrm{Id}_C \otimes c_{C,D}) \otimes \mathrm{Id}_D} (C \otimes (D \otimes C)) \otimes D \\ &\xrightarrow{a^{-1}_{C,D,C} \otimes \mathrm{Id}_D} ((C \otimes D) \otimes C) \otimes D \xrightarrow{a_{C\otimes D, C, D}} (C \otimes D) \otimes (C \otimes D). \end{aligned}$$

The braided structure on $\mathrm{Vect}^G_{(\phi,\mathscr{R})}$ from Proposition 1.44 yields the formula for $\Delta$ in the statement.

The particular situation follows from the above arguments and the braided structure on $\mathrm{Vect}_{-1}^{\mathbb{Z}_2}$ described in Example 1.47. □

The next result can be regarded as the dual version of Proposition 2.12; the proof is left to the reader.

**Proposition 2.21** *For $C, D, E$ coalgebras in a braided category $\mathscr{C}$ we have that $(C \otimes^{\pm} D) \otimes^{\pm} E \cong C \otimes^{\pm} (D \otimes^{\pm} E)$ as coalgebras, the isomorphism being defined by the associativity constraint of $\mathscr{C}$.*

## 2.3 The Dual Coalgebra/Algebra of an Algebra/Coalgebra

Unless otherwise specified, throughout this section $\mathscr{C}$ is a monoidal category with left/right duality. The goal is to prove that in this case there exists an equivalence between the category of algebras and algebra morphisms in $\mathscr{C}$ and the category of coalgebras and coalgebra morphisms in $\mathscr{C}$. To this end we first show that when $\mathscr{C}$ is a monoidal category and $A$ is an algebra in $\mathscr{C}$ that admits a left or right dual object then the dual object admits a coalgebra structure in $\mathscr{C}$. An analogous result holds if we consider coalgebras instead of algebras.

**Proposition 2.22** *Let $\mathscr{C}$ be a monoidal category, $A$ an algebra in $\mathscr{C}$ and $C$ a coalgebra in $\mathscr{C}$. Then the following assertions hold:*

*(i) if $A$ has a left/right dual object then the dual has a coalgebra structure in $\mathscr{C}$;*

*(ii) if $C$ has a left/right dual object then the dual has an algebra structure in $\mathscr{C}$.*

*Proof* (i). We only prove the left-handed version, as the right-handed one is the reverse monoidal version of the left-handed one.

To see how the coalgebra structure on $A^*$ appears from the algebra structure of $A$, assume for the moment that $\mathscr{C}$ is left rigid. Then by Proposition 1.71 we have that the left dual functor $(-)^*$ is a strong monoidal functor from $\mathscr{C}$ to $\overline{\mathscr{C}^{\mathrm{opp}}} = \overline{\mathscr{C}}^{\mathrm{opp}}$. Thus, according to Proposition 2.3 it carries algebras to algebras. Since an algebra in $\overline{\mathscr{C}}^{\mathrm{opp}}$ is actually a coalgebra in $\overline{\mathscr{C}}$ we get that $A^*$ has a coalgebra structure in $\overline{\mathscr{C}}$, and this can be also viewed as a coalgebra structure in $\mathscr{C}$.

Working out the details, the coalgebra structure on $A^*$ (resp. on ${}^*A$ if $\mathscr{C}$ is right rigid) is obtained with the help of the isomorphism $\lambda$ (resp. $\lambda'$) constructed in Section 1.7. More precisely, if $(A, \underline{m}_A, \underline{\eta}_A)$ is an algebra in $\mathscr{C}$ with left dual $A^*$, we have that $A^\star$, which is $A^*$ equipped with

$$\underline{\Delta}_{A^\star} : A^* \xrightarrow{\underline{m}_A^*} (A\otimes A)^* \xrightarrow{\lambda^{-1}_{A,A}} A^*\otimes A^* \text{ and } \underline{\varepsilon}_{A^\star} : A^* \xrightarrow{r^{-1}_{A^*}} A^*\otimes \underline{1} \xrightarrow{\mathrm{Id}_{A^*}\otimes\underline{\eta}_A} A^*\otimes A \xrightarrow{\mathrm{ev}_A} \underline{1},$$

is a coalgebra in $\mathscr{C}$, where $\underline{m}_A^*$ is the transposed morphism associated to $\underline{m}_A$. In diagrammatic notation, by the definition of $\lambda^{-1}$ and (1.6.6) we have that

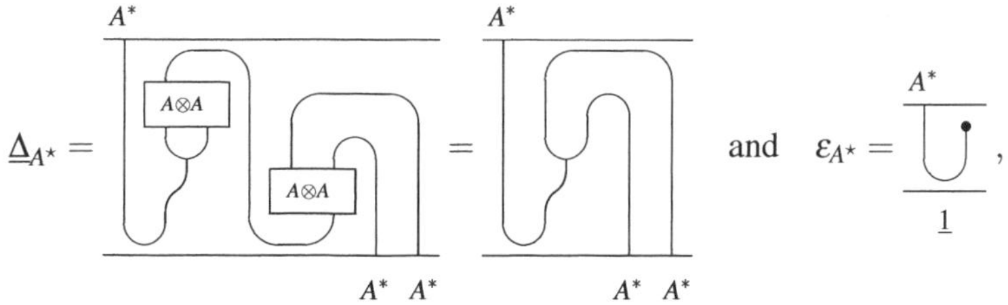


and so for defining $\underline{\Delta}_{A^\star}$ and $\underline{\varepsilon}_{A^\star}$ we only need the existence of the left dual object of $A$, as we assumed in the statement of the proposition.

So, when $\mathscr{C}$ is monoidal and left rigid, from the above comments it follows that $A^\star$ is a coalgebra in $\mathscr{C}$. Since we assumed that only $A$ (not all the objects of $\mathscr{C}$) admits a left dual object, for the completeness of the proof we need to include a direct check for the fact that $A^\star$ is indeed a coalgebra in $\mathscr{C}$.

From the definition, the coassociativity of $\underline{\Delta}_{A^\star}$ is equivalent to

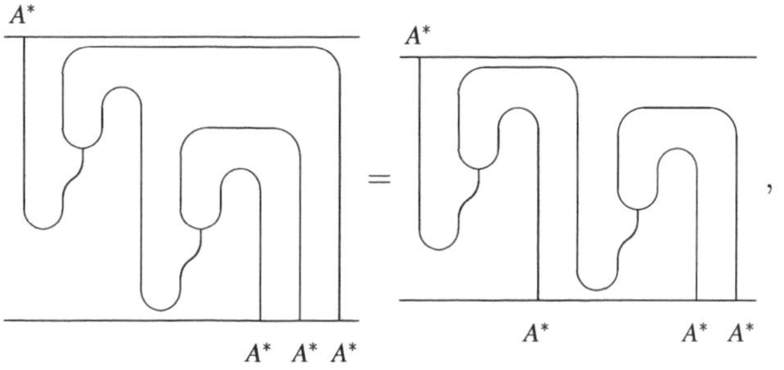

which follows because of (1.6.6) and the fact that $\underline{m}_A$ is associative. Now,

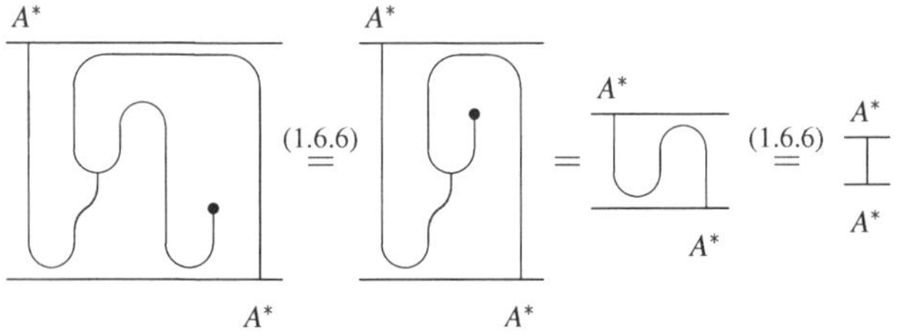


shows that $\underline{\varepsilon}_{A^\star}$ is a left counit for $\underline{\Delta}_{A^\star}$. That $\underline{\varepsilon}_{A^\star}$ is a right counit for $\underline{\Delta}_{A^\star}$ can be proved in a similar manner.

(ii) We omit this as it is the dual version of (i). Note only that when $C$ is a coalgebra in $\mathscr{C}$ such that $C$ has a left dual object $C^*$ in $\mathscr{C}$ then $C^*$ is an algebra in $\mathscr{C}$ via the following structure:

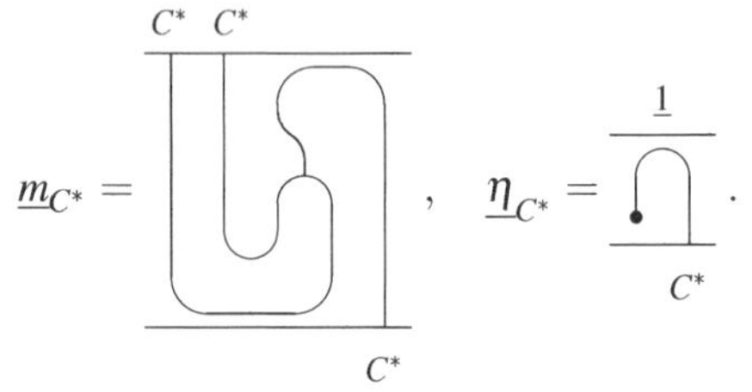


From now on by $C^\star$ we denote the above algebra structure on $C^*$ in $\mathscr{C}$. □

**Remark 2.23** The coalgebra structure in $\mathscr{C}$ on a right dual ${}^*\!A$ of an algebra $A$ in $\mathscr{C}$ is given by

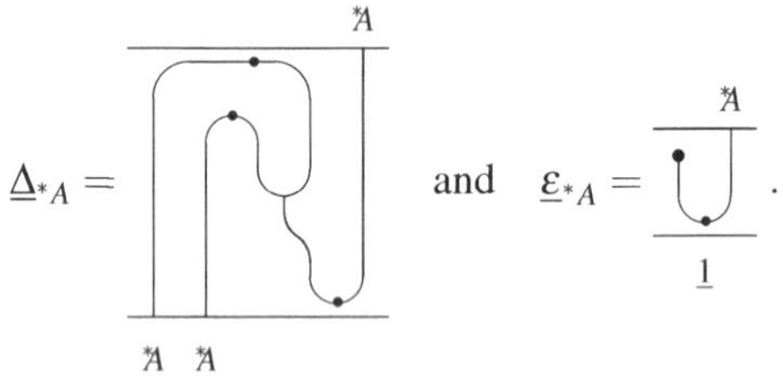


In what follows it will be denoted by ${}^\star\!A$. Likewise, if $C$ is a coalgebra in $\mathscr{C}$ admitting a right dual object ${}^*C$ in $\mathscr{C}$ then ${}^*C$ is an algebra in $\mathscr{C}$ via the structure given by

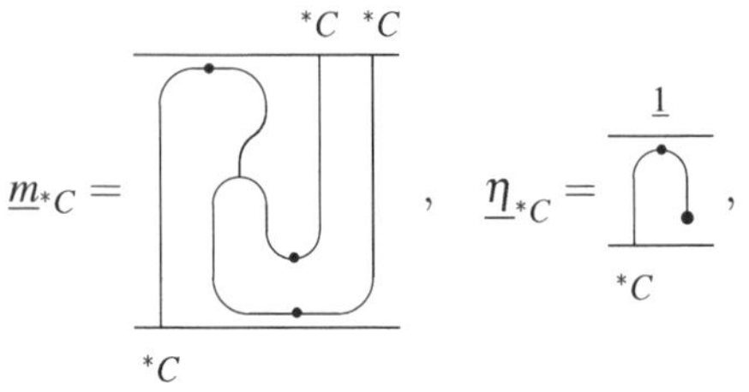


where, as before, the evaluation and coevaluation morphisms with a black dot are those associated to a right dual object. This algebra will be denoted by ${}^\star C$.

**Examples 2.24** (1) In Hopf algebra theory the coalgebra structure on $A^*$ presented above is the so-called co-opposite dual coalgebra of the algebra $A$. To obtain the dual coalgebra of the algebra $A$ we have to switch the tensor components of $A^* \otimes A^*$ with the help of the (symmetric) braiding of ${}_k\mathcal{M}$. This is why, in general, we keep the same terminology for the coalgebra structure on $A^*$ from Proposition 2.22. Namely, we call $A^\star$ the left co-opposite dual coalgebra of the algebra $A$. When $\mathscr{C}$ is braided then $A^{\star\mathrm{cop}+}$ is a coalgebra in $\mathscr{C}$ as well; see Proposition 2.17. We simply denote it by $A^*$ and call it the left dual coalgebra of the algebra $A$. Hence

$\underline{\Delta}_{A^*} =$ 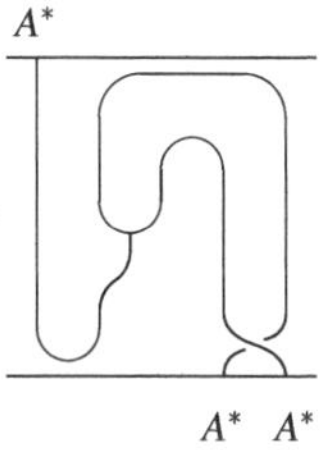 and $\varepsilon_{A^*} =$ 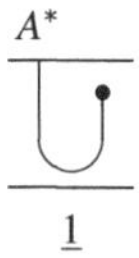 .

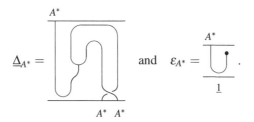

In the case of right duality we call ${}^\star A$ the right co-opposite dual coalgebra of the algebra $A$, and if, moreover, $\mathscr{C}$ is braided we call ${}^*A := ({}^\star A)^{\mathrm{cop}+}$ the right dual coalgebra of the algebra $A$.

(2) Assume again that $\mathscr{C}$ is the category of vector spaces over a field $k$, and that $C$ is a finite-dimensional coalgebra in ${}_k\mathcal{M}$. Then the algebra structure on $C^* = {}^*C$ is determined by

$$c^*d^*(c) = c^*(c_2)d^*(c_1), \quad 1_{C^*}(c) = \varepsilon(c), \ \forall\, c^*,\, d^* \in C^*, \ c \in C,$$

and remark that actually we do not need $C$ to be finite dimensional. Furthermore, in Hopf algebra theory this is precisely the opposite dual algebra structure of the coalgebra $C$. This is why, in general, for a coalgebra $C$ in a monoidal category we call the algebra structure on $C^*/{}^*C$ from Proposition 2.22 the left/right opposite dual algebra structure of the coalgebra $C$. As we have already mentioned it will be denoted by $C^\star/{}^\star C$. Furthermore, in the situation when $\mathscr{C}$ is braided we denote $C^{\star\mathrm{op}-}/({}^\star C)^{\mathrm{op}-}$ by $C^*/{}^*C$ and call it, simply, the left/right dual algebra structure of the coalgebra $C$.

The dual (co)algebra process can be iterated. For instance, if $C$ is a coalgebra then $C^\star$ is an algebra, and so we can consider either the coalgebra ${}^\star(C^\star)$ or $(C^\star)^\star$. We next see that the canonical isomorphisms between an object $X$ and its "double" duals behave well with respect to the initial (co)algebra structure of $X$. More precisely, we have the following results. For simplicity, we assume from the beginning that our category is rigid.

**Proposition 2.25** *Let $\mathscr{C}$ be a rigid monoidal category and $C$ a coalgebra in $\mathscr{C}$. Then the isomorphisms $\theta_C : C \to ({}^\star C)^\star$ and $\theta'_C : C \to {}^\star(C^\star)$ considered in Proposition 1.73 are coalgebra isomorphisms in $\mathscr{C}$.*

*Proof* Let us start by noting that the coalgebra structure of $({}^\star C)^\star$ coming from the left co-opposite dual coalgebra structure of the algebra ${}^{\star}C$ is given by

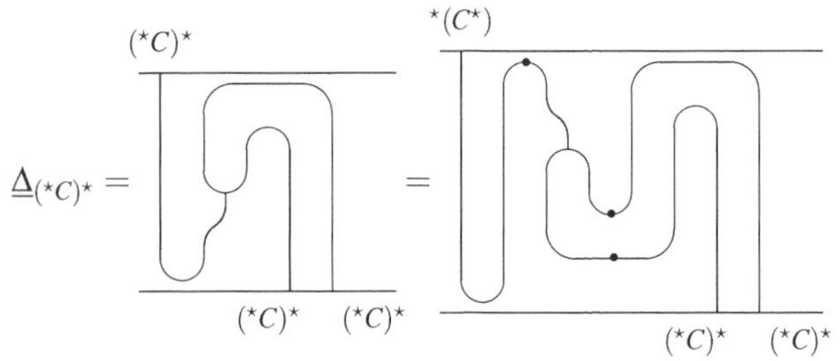


and 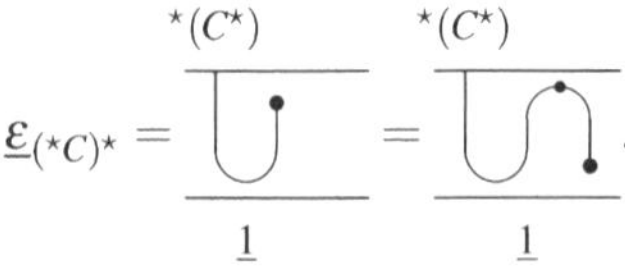

. Thus $\theta_C$ is a coalgebra morphism if and only if

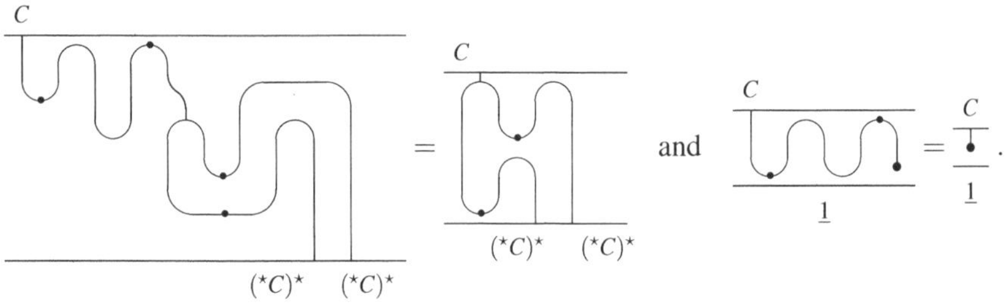


The two relations above follow easily by applying first (1.6.6) and then (1.6.9), so $\theta_C : C \to ({}^{\star}C)^{\star}$ is indeed a coalgebra isomorphism in $\mathscr{C}$. The assertion concerning $\theta'_C$ can be proved in a similar manner, so we leave it to the reader. □

We consider now the braided case.

**Proposition 2.26** *Let $\mathscr{C}$ be a rigid braided category and $C$ a coalgebra in $\mathscr{C}$. Then $C^{\star} \cong {}^{\star}C$, as algebras in $\mathscr{C}$. Consequently, $C^{\star\star} := (C^{\star})^{\star} \cong ({}^{\star}C)^{\star} \cong C \cong {}^{\star}(C^{\star}) \cong {}^{\star}({}^{\star}C) := {}^{\star\star}C$, as coalgebras in $\mathscr{C}$.*

*Proof* We will see that the isomorphism $\Theta'_C : C^{\star} \to {}^{\star}C$ defined in the proof of Corollary 1.76 is an algebra morphism, too. For this we have to show that

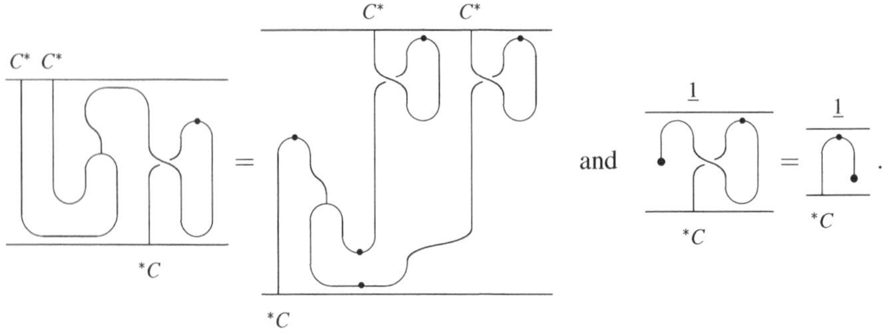

Indeed, by the naturality of the braiding and Proposition 1.49 we have

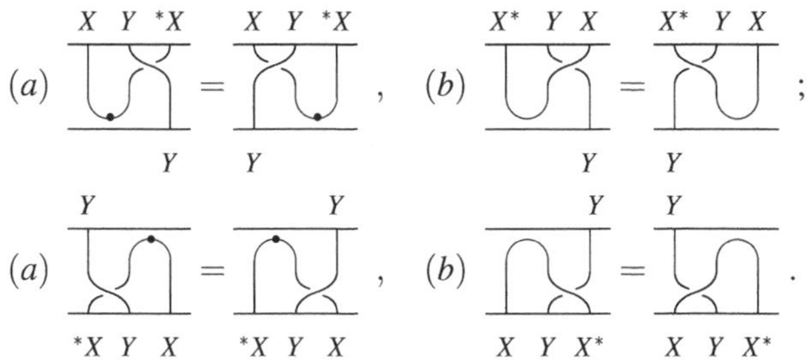


(2.3.1)

(2.3.2)

With the help of these relations we compute:

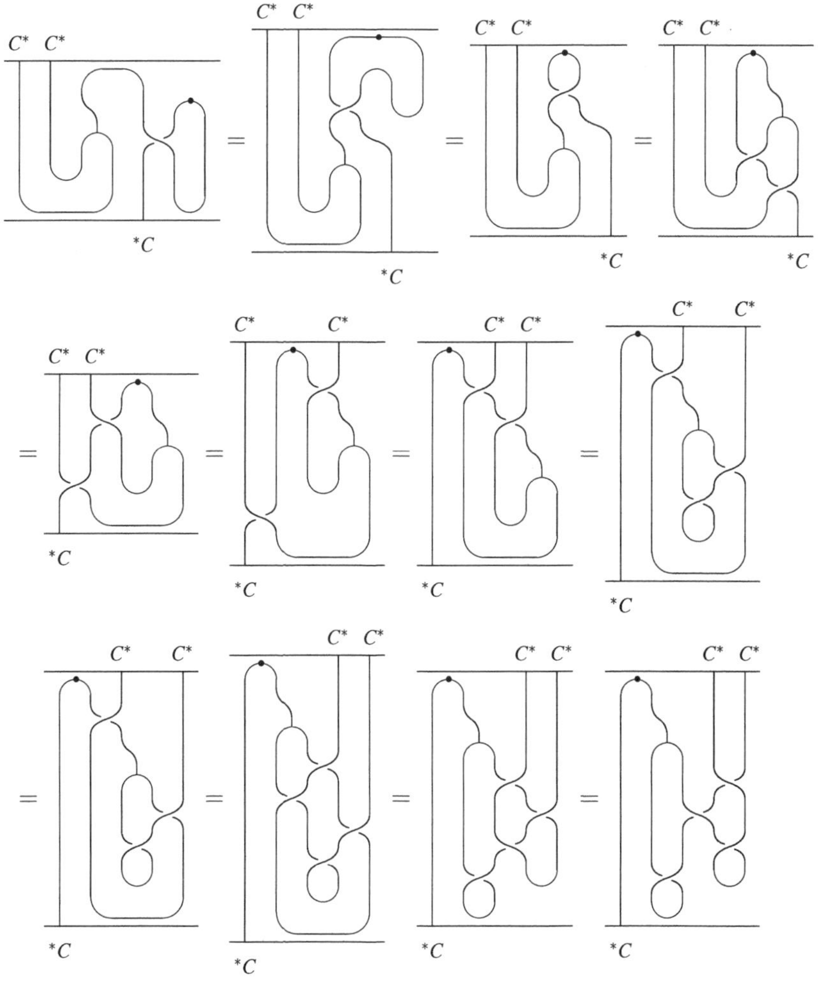

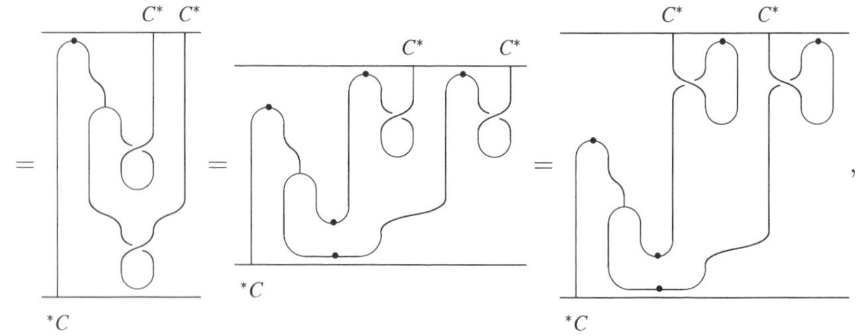


as required. More precisely, we used (2.3.2a) in the first equality, and a similar relation in the tenth equality; (1.6.6) in the second equality; Proposition 1.37 in the third, seventh, eighth, ninth and twelfth equalities; the equation (2.3.1b) in the fourth equality; the identity (2.3.2a) in the fifth and sixth equality; (1.5.10) in the eleventh equality; (1.6.9) in the last but one equality and the definition of $\Theta'_C$ from the proof of Corollary 1.76 in the last one.

That $\Theta'_C$ respects the counits follows from (2.3.2b), which implies

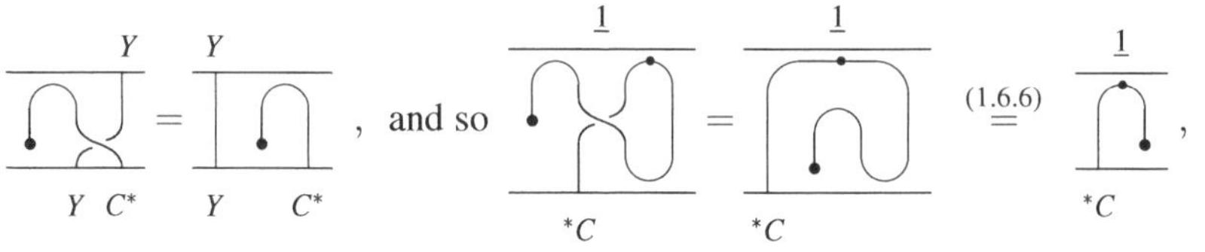


as required. Hence our proof is complete. □

By working with algebras instead of coalgebras one can prove the following results. As these are the dual situations the details will be skipped.

**Proposition 2.27** *Let $A$ be an algebra in a rigid monoidal category $\mathscr{C}$. Then the morphisms $\theta_A : A \to ({}^\star A)^\star$ and $\theta'_A : A \to {}^\star(A^\star)$ defined in the proof of Proposition 1.73 are algebra isomorphisms in $\mathscr{C}$.*

**Proposition 2.28** *For $A$ an algebra in a rigid braided category $\mathscr{C}$, the isomorphism $\Theta_A^{-1} : {}^\star A \to A^\star$ from the proof of Corollary 1.76 is a coalgebra isomorphism in $\mathscr{C}$. Consequently,*

$$A^{\star\star} := (A^\star)^\star \cong ({}^\star A)^\star \cong A \cong {}^\star(A^\star) \cong {}^\star({}^\star A) := {}^{\star\star}A,$$

*as algebras in $\mathscr{C}$.*

*Proof* This is the "upside down" version of Proposition 2.26. More precisely, it follows by turning upside down the diagrammatic computations performed in the proof of Proposition 2.26. This is because $\Theta_A^{-1}$ is a coalgebra morphism if and only if the equalities below hold:

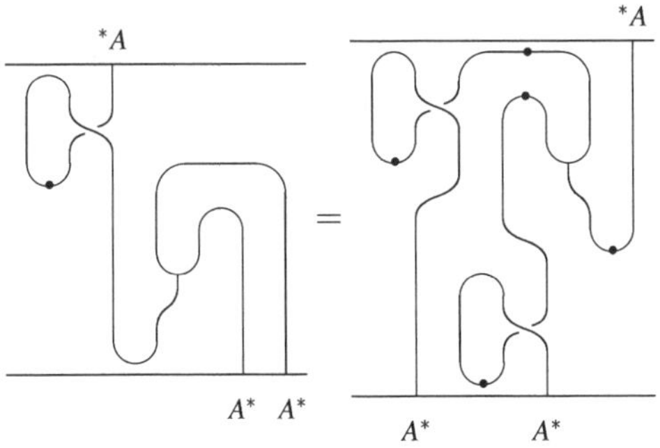


and

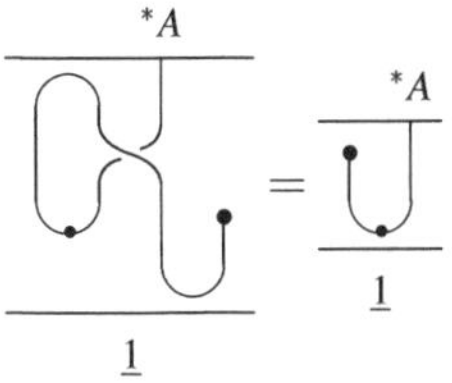


,

and these are precisely the ones proved in Proposition 2.26, turned upside down. □

If $\mathscr{C}$ is a monoidal category then by $\mathrm{Coalg}(\mathscr{C})$ we denote the subcategory of $\mathscr{C}$ whose objects are coalgebras and morphisms are coalgebra morphisms. Likewise, by $\mathrm{Alg}(\mathscr{C})$ we denote the subcategory of algebra objects and algebra morphisms in $\mathscr{C}$. The next result contributes to the duality formalism that we have already mentioned several times so far.

**Corollary 2.29** *If $\mathscr{C}$ is a rigid monoidal category then* $\mathrm{Coalg}(\mathscr{C})$ *and* $\mathrm{Alg}(\mathscr{C})$ *are equivalent categories.*

*Proof* We show that $\mathrm{Coalg}(\mathscr{C}) \underset{{}^*(-)}{\overset{(-)^*}{\rightleftarrows}} \mathrm{Alg}(\mathscr{C})$ is a pair of functors that produces the desired equivalence. For this, in view of the above results, it is enough to prove that $\theta = (\theta_X : X \to ({}^*X)^*)_{X\in\mathscr{C}}$ and $\theta' = (\theta'_X : X \to {}^*(X^*))_{X\in\mathscr{C}}$ are natural transformations. Indeed, for instance, if $X \xrightarrow{f} Y$ is a morphism in $\mathscr{C}$ then

$({}^*f)^* =$ 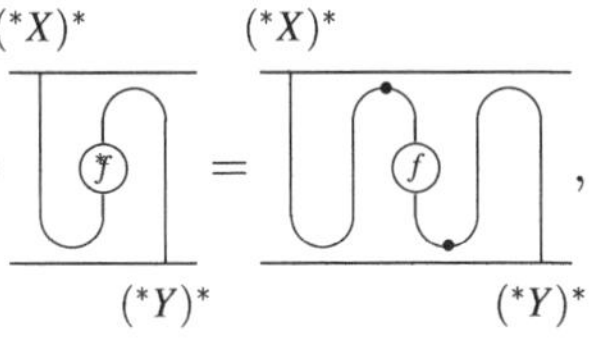


,

and so

$({}^*f)^* \circ \theta_X =$ 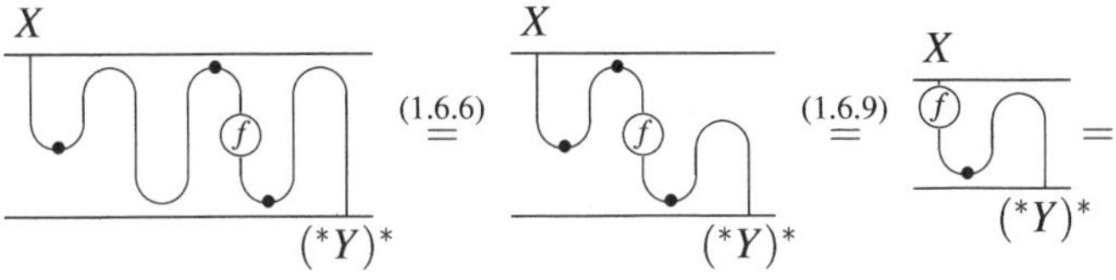


$= \theta_Y \circ f,$

as required. In a similar way one can show that $\theta'$ is a natural transformation; the details are left to the reader. □

## 2.4 Categories of Representations

To any algebra in a monoidal category $\mathscr{C}$ we associate the so-called category of representations (or modules) over it. Of course, this is inspired by the classical representation theory for $k$-algebras.

Let $\mathscr{C}$ be a monoidal category and $A$ an algebra in $\mathscr{C}$.

**Definition 2.30** A left $A$-module $M$ in $\mathscr{C}$ is an object $M \in \mathscr{C}$ together with a morphism $\underline{\mu}_M : A \otimes M \to M$ in $\mathscr{C}$ which is associative and unital, in the sense that

$$\begin{array}{ccccc} (A\otimes A)\otimes M & \xrightarrow{m_A\otimes \mathrm{Id}_M} & A\otimes M & \xrightarrow{\underline{\mu}_M} & M \\ {\scriptstyle a_{A,A,M}}\downarrow & & & & \uparrow{\scriptstyle \underline{\mu}_M} \\ A\otimes(A\otimes M) & & \xrightarrow{\mathrm{Id}_A\otimes \underline{\mu}_M} & & A\otimes M, \end{array} \qquad \begin{array}{ccc} A\otimes M & \xrightarrow{\underline{\mu}_M} & M \\ {\scriptstyle \underline{\eta}_A\otimes \mathrm{Id}_M}\uparrow & \nearrow{\scriptstyle l_M} & \\ \underline{1}\otimes M. & & \end{array}$$

If $(M, \underline{\mu}_M)$ and $(N, \underline{\mu}_N)$ are left $A$-modules then a morphism $f : M \to N$ in $\mathscr{C}$ is called left $A$-linear or a morphism of left $A$-modules if and only if $f\underline{\mu}_M = \underline{\mu}_N (\mathrm{Id}_A \otimes f)$.

We denote by ${}_A\mathscr{C}$ the category of left $A$-modules and left $A$-linear morphisms in $\mathscr{C}$.

In diagrammatic computations, 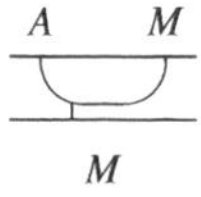 will be the short form of 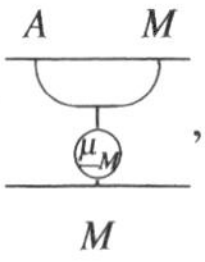, the left $A$-module structure morphism $\underline{\mu}_M$. Then we have

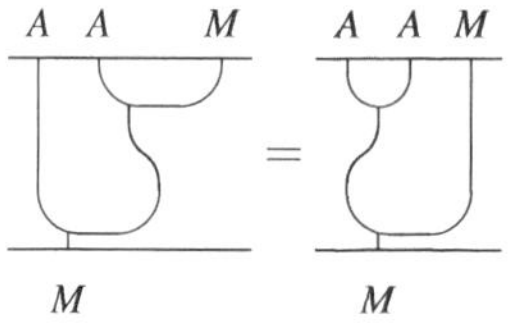
and
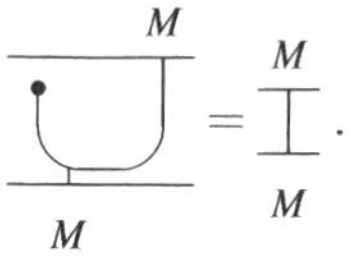
(2.4.1)

In a similar manner one can define $\mathscr{C}_A$, the category of right $A$-modules and right $A$-linear morphisms in $\mathscr{C}$. For simplicity, we denote by 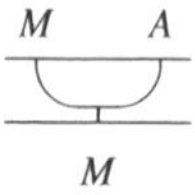 the morphism 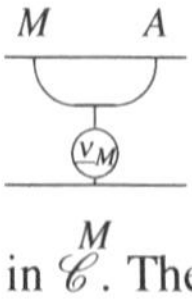 representing the right $A$-module structure morphism $\underline{\nu}_M : M \otimes A \to M$ of $M$ in $\mathscr{C}$. The equality below together with the unit property expresses that $M$ is a right $A$-module in the monoidal category $\mathscr{C}$:

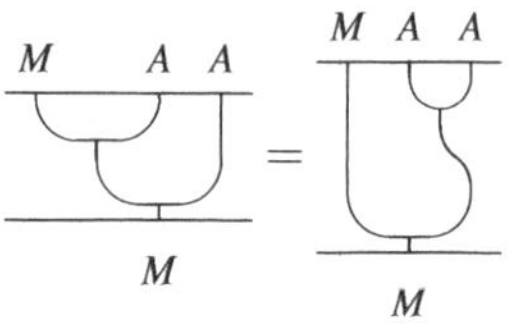
(2.4.2)

Actually, a right $A$-module in $\mathscr{C}$ is nothing but a left $A$-module in $\overline{\mathscr{C}}$, the reverse monoidal category associated to $\mathscr{C}$.

We now present some concrete examples of modules in monoidal categories.

**Examples 2.31** (1) Any algebra $A$ in a monoidal category is a left and right $A$-module via its multiplication $\underline{m}_A$.

(2) A left module in $\underline{\mathrm{Set}}$ is a left $G$-set with $G$ a monoid. A module morphism in $\underline{\mathrm{Set}}$ is a morphism of $G$-sets.

(3) If $\mathscr{C}$ is a category and $T = (T, \underline{m}, \underline{\eta})$ is a monad in $\mathscr{C}$ then a left $T$-module in $[\mathscr{C}, \mathscr{C}]$ is a functor $M \in [\mathscr{C}, \mathscr{C}]$ together with a natural transformation $\mu : T \circ M \to M$ such that $\mu_X \circ T(\mu_X) = \mu_X \circ \underline{m}_{T(X)}$ and $\mu_X \circ \underline{\eta}_{M(X)} = \mathrm{Id}_{M(X)}$, for any object $X$ of $\mathscr{C}$. A morphism of left $T$-modules in $[\mathscr{C}, \mathscr{C}]$ between $(M, \mu)$ and $(M', \mu')$ is a natural transformation $f : M \to M'$ satisfying $f_X \circ \mu_X = \mu'_X \circ T(f_X)$, for any $X \in \mathscr{C}$.

(4) Let $G$ be a group and $\phi \in H^3(G, k^*)$. If $A$ is a $G$-graded quasialgebra with reassociator $\phi$ then a left $A$-module in $\mathrm{Vect}_\phi^G$ is a $G$-graded vector space $M$ together with a $k$-linear map $A \otimes M \to M$ ($a \otimes m \mapsto am$ is the left action of $A$ on $M$) such that $A_x M_y \subseteq M_{xy}$, for all $x, y \in G$, $1_A m = m$ and $(aa')m = \phi(|a|, |a'|, |m|)a(a'm)$, for all homogeneous elements $a, a' \in A$ and $m \in M$.

When $\mathscr{C}$ is a braided category we have the following result.

**Proposition 2.32** *Let $\mathscr{C}$ be a braided category and $A$ an algebra in $\mathscr{C}$. Then ${}_A\mathscr{C}$ and $\mathscr{C}_{A^{\mathrm{op}+}}$ are isomorphic, where $A^{\mathrm{op}+}$ is the c-opposite algebra associated to $A$.*

*Proof* If $M$ is a left $A$-module then it becomes a right $A^{\mathrm{op}+}$-module with 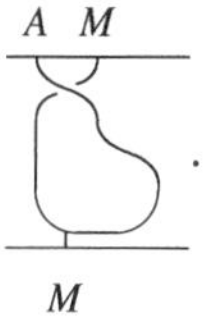

.

Indeed, by Proposition 1.37, $M \in {}_A\mathscr{C}$ and (1.5.10) we have

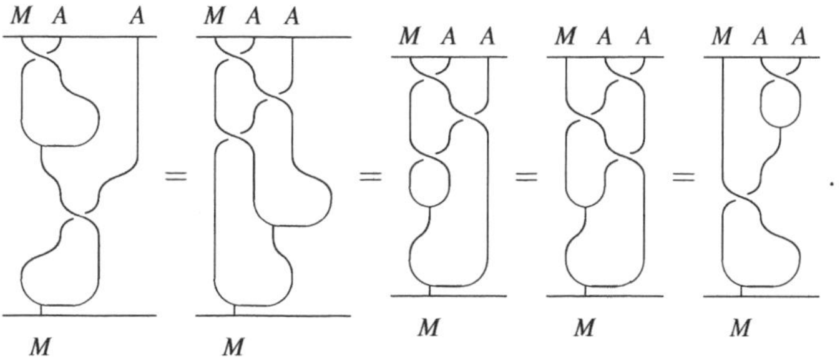


This, together with Proposition 1.49, implies that $M \in \mathscr{C}_{A^{\mathrm{op}+}}$, as claimed. Also, by the naturality of the braiding it follows that a left $A$-module morphism becomes a right $A^{\mathrm{op}+}$-module morphism. Thus we have defined a functor from ${}_A\mathscr{C}$ to $\mathscr{C}_{A^{\mathrm{op}+}}$.

Similarly, if $M \in \mathscr{C}_{A^{\mathrm{op}+}}$ then $M$ is a left $A$-module with 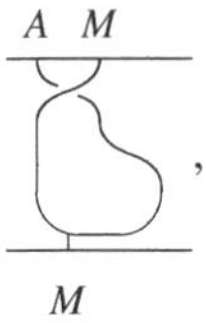 , and in this way a right $A^{\mathrm{op}+}$-module morphism turns into a left $A$-module morphism. Clearly, this functor is the inverse of the one defined above, so ${}_A\mathscr{C}$ and $\mathscr{C}_{A^{\mathrm{op}+}}$ are isomorphic. □

By working with dual objects, we get another way to pass from left/right to right/left modules.

**Proposition 2.33** *Let $\mathscr{C}$ be a monoidal category and $A$ an algebra in $\mathscr{C}$. If $M$ is a left $A$-module in $\mathscr{C}$ such that $M$ admits a left dual object $M^*$ in $\mathscr{C}$ then $M^*$ becomes a right $A$-module in $\mathscr{C}$ via the structure morphism*

$$\begin{aligned}\underline{\nu}_{M^*}: M^*\otimes A &\xrightarrow{r^{-1}_{M^*\otimes A}} (M^*\otimes A)\otimes \underline{1} \xrightarrow{\mathrm{Id}_{M^*\otimes A}\otimes \mathrm{coev}_M} (M^*\otimes A)\otimes (M\otimes M^*)\\ &\xrightarrow{a_{M^*,A,M\otimes M^*}} M^*\otimes (A\otimes (M\otimes M^*)) \xrightarrow{\mathrm{Id}_{M^*}\otimes a^{-1}_{A,M,M^*}} M^*\otimes ((A\otimes M)\otimes M^*)\\ &\xrightarrow{\mathrm{Id}_{M^*}\otimes(\underline{\mu}_M\otimes \mathrm{Id}_{M^*})} M^*\otimes (M\otimes M^*) \xrightarrow{a^{-1}_{M^*,M,M^*}} (M^*\otimes M)\otimes M^* \xrightarrow{\mathrm{ev}_M\otimes \mathrm{Id}_{M^*}} \underline{1}\otimes M^* \xrightarrow{l_{M^*}} M^*,\end{aligned} \tag{2.4.3}$$

*where $\underline{\mu}_M : A\otimes M \to M$ is the structure morphism of $M$ as a left $A$-module in $\mathscr{C}$.*

*Similarly, if $M$ is a right $A$-module and $M$ has a right dual object ${}^*M$ in $\mathscr{C}$ then ${}^*M$ with the structure given by*

$$\begin{aligned}\underline{\mu}_{{}^*M}: A\otimes {}^*M &\xrightarrow{l^{-1}_{A\otimes {}^*M}} \underline{1}\otimes (A\otimes {}^*M) \xrightarrow{\mathrm{coev}'_M\otimes \mathrm{Id}_{A\otimes {}^*M}} ({}^*M\otimes M)\otimes (A\otimes {}^*M)\\ &\xrightarrow{a_{{}^*M,M,A\otimes {}^*M}} {}^*M\otimes (M\otimes (A\otimes {}^*M)) \xrightarrow{\mathrm{Id}_{{}^*M}\otimes a^{-1}_{M,A,{}^*M}} {}^*M\otimes ((M\otimes A)\otimes {}^*M)\\ &\xrightarrow{\mathrm{Id}_{{}^*M}\otimes(\underline{\nu}_M\otimes \mathrm{Id}_{{}^*M})} {}^*M\otimes (M\otimes {}^*M) \xrightarrow{\mathrm{Id}_{{}^*M}\otimes \mathrm{ev}'_M} {}^*M\otimes \underline{1} \xrightarrow{r_{{}^*M}} M^*\end{aligned} \tag{2.4.4}$$

*becomes a left $A$-module in $\mathscr{C}$, where $\underline{\nu}_M : M\otimes A \to M$ is the morphism that defines on $M$ a right $A$-module structure in $\mathscr{C}$.*

*Proof* This time we prove only the second assertion; the first one follows in a similar manner (or it can be viewed as the reverse monoidal version of the second one).

Without loss of generality, we can assume that $\mathscr{C}$ is strict monoidal. Then we have

$\underline{\mu}_{{}^*M} =$ 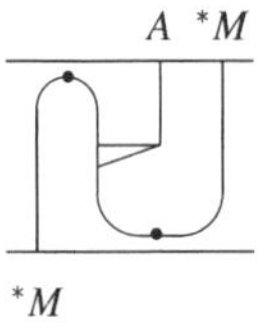 , where $\underline{\nu}_M :=$  .

Thus ${}^*M$ is a left $A$-module in $\mathscr{C}$ via $\mu_{{}^*M}$ since

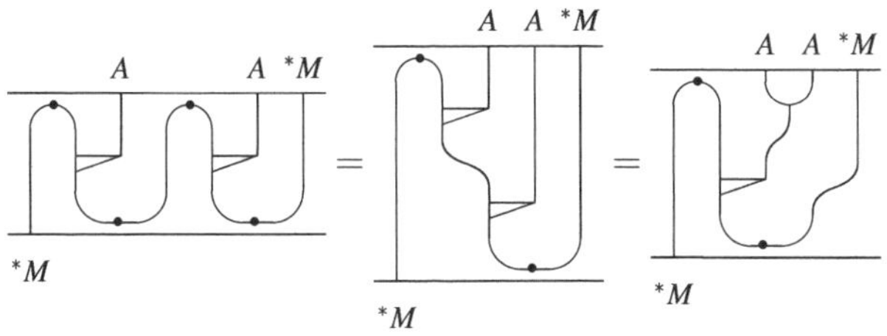


and 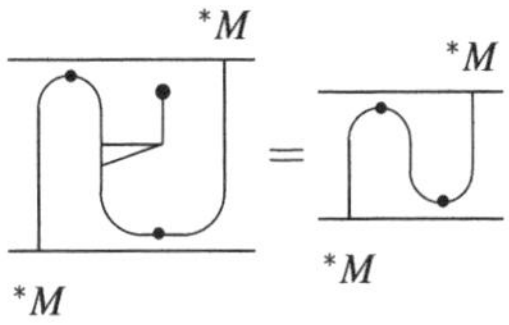

$= \mathrm{Id}_{{}^*M}$, as required. □

To any algebra morphism we can associate the functor called restriction of scalars; the proof of the next result is easy and left to the reader.

**Proposition 2.34** *Let $A, B$ be algebras in a monoidal category $\mathscr{C}$, $f : A \to B$ an algebra morphism in $\mathscr{C}$, and $M$ a left $B$-module in $\mathscr{C}$. Then $M$ is a left $A$-module in $\mathscr{C}$ via the structure morphism* 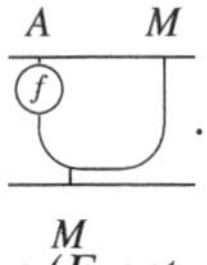

*. In this way we have a functor $F : {}_B\mathscr{C} \to {}_A\mathscr{C}$, called the restriction of scalars ($F$ acts as identity on morphisms).*

We now show that in a braided category the tensor product of two modules is still a module. This result will be used in Section 2.6 when we will study when a category of representations is a monoidal category as well.

**Proposition 2.35** *Let $\mathscr{C}$ be a braided category and $A, B$ algebras in $\mathscr{C}$. If $M \in {}_A\mathscr{C}$ and $N \in {}_B\mathscr{C}$ then $M \otimes N$ is a left $A \otimes_+ B$-module in $\mathscr{C}$ with*

$\underline{\mu}_{M\otimes N} :=$ 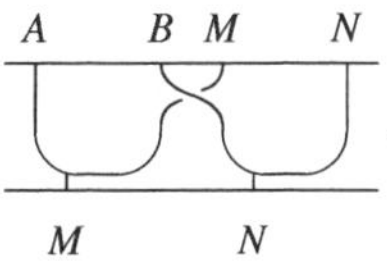

.

*If we replace the braiding in the above definition by its inverse then with this new structure $M \otimes N$ is a left $A \otimes_- B$-module in $\mathscr{C}$.*

*Proof* Consider only the first situation; the other one is obtained by considering $\mathscr{C}^{\mathrm{in}}$

instead of $\mathscr{C}$. We compute:

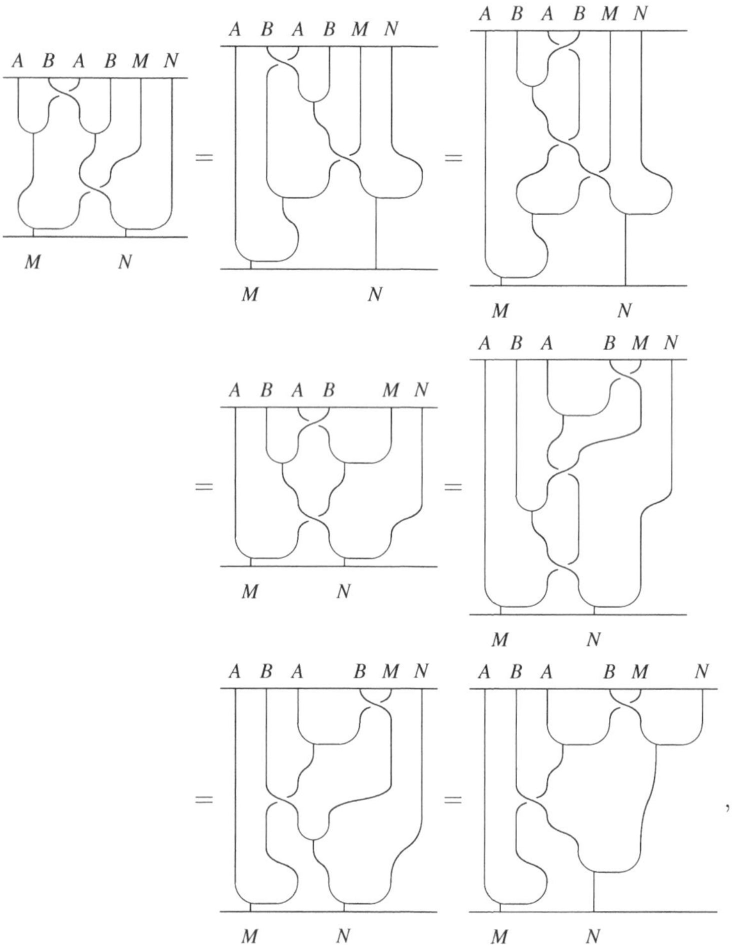


as required. We used (2.4.1) in the first and sixth equality and Proposition 1.37 in the second, third, fourth and fifth equalities. By using Proposition 1.49, one can easily check that $\underline{\eta}_A \otimes \underline{\eta}_B$ acts as identity on $M \otimes N$. □

## 2.5 Categories of Corepresentations

In this section we deal with the dual situation of Section 2.4. As we have seen earlier, a coalgebra in a monoidal category $\mathscr{C}$ is nothing else than an algebra in $\mathscr{C}^{\mathrm{opp}}$, the opposite category associated to $\mathscr{C}$. Thus, by duality, we can obtain the notion of a comodule over a coalgebra. Namely, a comodule in $\mathscr{C}$ is a module in $\mathscr{C}^{\mathrm{opp}}$.

**Definition 2.36** Let $\mathscr{C}$ be a monoidal category and $(C, \underline{\Delta}, \underline{\varepsilon})$ a coalgebra in $\mathscr{C}$. We say that $N \in \mathscr{C}$ together with a morphism $\underline{\rho}_N : N \to N \otimes C$ in $\mathscr{C}$ is a right $C$-comodule (or corepresentation) if $(\mathrm{Id}_N \otimes \underline{\varepsilon}_C) \circ \underline{\rho}_N = \mathrm{Id}_N$ and the following diagram

$$\begin{array}{ccccc} N & \xrightarrow{\underline{\rho}_N} & N \otimes C & \xrightarrow{\underline{\rho}_N \otimes \mathrm{Id}_C} & (N \otimes C) \otimes C \\ \downarrow{\scriptstyle \underline{\rho}_N} & & & & \downarrow{\scriptstyle a_{N,C,C}} \\ N \otimes C & & \xrightarrow{\mathrm{Id}_N \otimes \underline{\Delta}_C} & & N \otimes (C \otimes C) \end{array}$$

is commutative. If $(N, \underline{\rho}_N)$ and $(N', \underline{\rho}_{N'})$ are right $C$-comodules in $\mathscr{C}$ then a morphism $f : N \to N'$ in $\mathscr{C}$ is called right $C$-colinear, or a morphism of right $C$-comodules, if $(f \otimes \mathrm{Id}_C) \circ \underline{\rho}_N = \underline{\rho}_{N'} \circ f$. By $\mathscr{C}^C$ we denote the category of right $C$-comodules and right $C$-comodule morphisms in $\mathscr{C}$.

For a right $C$-comodule $N$ in $\mathscr{C}$ we denote its structure morphism $\underline{\rho}_N$ by 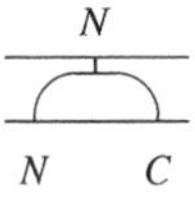
instead of 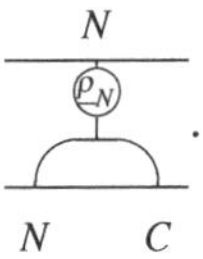 . Then we have

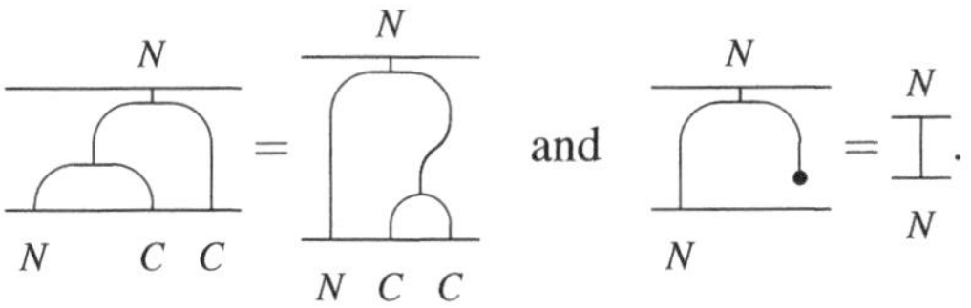 (2.5.1)

Likewise, we can define left $C$-comodules in a monoidal category $\mathscr{C}$. This time the structure morphism of a left $C$-comodule $N$ is of the form $\underline{\lambda}_N : N \to C \otimes N$. For simplicity, it will be denoted by 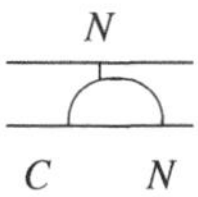 instead of 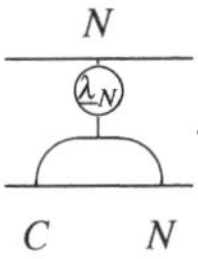 . Then 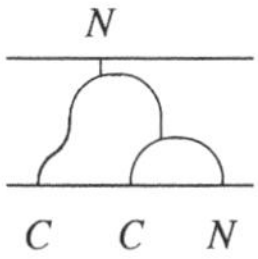 = 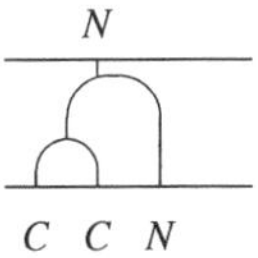 and 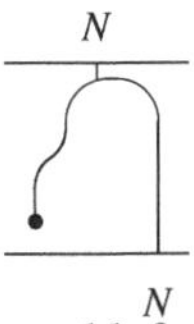 = . In what follows, by ${}^C\mathscr{C}$ we denote the category of left $C$-comodules and left $C$-comodule morphisms in $\mathscr{C}$.

Usually, we will work with left modules over algebras and right comodules over coalgebras. Any result that we obtain for right comodules has an analogue for left comodules, because a left $C$-comodule can be identified with a right $C$-comodule in $\overline{\mathscr{C}}$, the reverse monoidal category associated to $\mathscr{C}$. Furthermore, if $\mathscr{C}$ is braided we have the following result.

**Proposition 2.37** *Let $C$ be a coalgebra in a braided category $\mathscr{C}$. Then ${}^C\mathscr{C}$ and $\mathscr{C}^{C^{\mathrm{cop}+}}$ are isomorphic, where $C^{\mathrm{cop}+}$ is the c-coopposite coalgebra associated to $C$.*

*Proof* If $M \in {}^{C}\mathscr{C}$ then $M$ becomes a right $C^{\mathrm{cop}+}$-comodule via 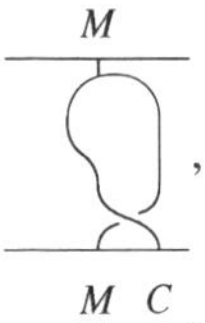

, and then a left $C$-comodule morphism becomes a right $C^{\mathrm{cop}+}$-comodule morphism in $\mathscr{C}$.

Similarly, a right $C^{\mathrm{cop}+}$-comodule becomes a left $C$-comodule with the structure morphism 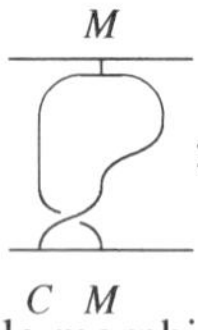

; in this way a right $C^{\mathrm{cop}+}$-comodule morphism turns into a left $C$-comodule morphism.

The two correspondences above define functors that are inverse to each other. Since the proof of the above assertions are the formal duals of the ones in Proposition 2.32 we leave the details to the reader. □

**Examples 2.38** (1) Any coalgebra $C$ in a monoidal category $\mathscr{C}$ is a left and right $C$-comodule via its comultiplication $\underline{\Delta}$.

(2) From Examples 2.14 (2) we know that any object $X$ in Set has a unique coalgebra structure in Set given by $\underline{\Delta}_X(x) = (x,x)$ and $\underline{\varepsilon}_X(x) = *$, for all $x \in X$. Thus, giving a right $X$-comodule structure on a set $N$ reduces to giving a map from $N$ to $X$, in the sense that any right $X$-comodule structure on $N$ has the form $\rho_N(n) = (n, f(n))$, for some map $f : N \to X$.

If $(N, f)$ and $(N', f')$ are right $X$-comodules in Set then a map $g : N \to N'$ is right $X$-colinear if and only if $f' \circ g = f$.

(3) If $(L, \underline{\Delta}, \underline{\varepsilon})$ is a comonad on a category $\mathscr{C}$ then a right $L$-comodule in $[\mathscr{C}, \mathscr{C}]$ is a functor $N : \mathscr{C} \to \mathscr{C}$ together with a natural transformation $\rho : N \to N \circ L$ such that $\rho_{L(X)} \circ \rho_X = N(\underline{\Delta}_X) \circ \rho_X$ and $N(\underline{\varepsilon}_X) \circ \rho_X = \mathrm{Id}_{N(X)}$, for all $X \in \mathscr{C}$.

A morphism $f : (N, \rho) \to (N', \rho')$ between two right $L$-comodules in $[\mathscr{C}, \mathscr{C}]$ is a natural transformation $f : N \to N'$ satisfying $\rho'_X \circ f_X = f_{L(X)} \circ \rho_X$, for all $X \in \mathscr{C}$.

(4) If $k$ is a field and $(C, \Delta_C, \varepsilon_C)$ is a $k$-coalgebra then a right $C$-comodule in ${}_k\mathscr{M}$ is a pair $(N, \rho_N)$, where $N$ is a $k$-vector space and $\rho_N : N \to N \otimes C$ is a $k$-linear map such that $(\rho_N \otimes \mathrm{Id}_C) \circ \rho_N = (\mathrm{Id}_N \otimes \Delta_C) \circ \rho_N$ and $(\mathrm{Id}_N \otimes \varepsilon_C) \circ \rho_N = \mathrm{Id}_N$. The sigma notation for a right $C$-comodule $N$ with structure map $\rho_N : N \to N \otimes C$ is $\rho(n) = n_{(0)} \otimes n_{(1)}$, for all $n \in N$. Then the conditions in the definition of a right $C$-comodule can now be written as

$$(n_{(0)})_{(0)} \otimes (n_{(0)})_{(1)} \otimes n_{(1)} = n_{(0)} \otimes (n_{(1)})_1 \otimes (n_{(1)})_2 := n_{(0)} \otimes n_{(1)} \otimes n_{(2)},$$

and $\varepsilon(n_{(1)}) n_{(0)} = n$, for all $n \in N$.

Also, $f : (N, \rho) \to (N', \rho')$ is a morphism of right $C$-comodules in ${}_k\mathscr{M}$ if $f$ is a $k$-linear map obeying $(f \otimes \mathrm{Id}_C) \circ \rho = \rho' \circ f$. In sigma notation this means that

$$f(n)_{(0)} \otimes f(n)_{(1)} = f(n_{(0)}) \otimes n_{(1)}, \;\; \forall\, n \in N.$$

The functor corestiction of scalars can be constructed as follows (it is the formal dual of the functor constructed in Proposition 2.34).

**Proposition 2.39** *Let $C$ and $D$ be coalgebras in a monoidal category and $f : C \to D$ a coalgebra morphism in $\mathscr{C}$. If $M$ is a right $C$-comodule then $M$ becomes a right $D$-comodule in $\mathscr{C}$ with* 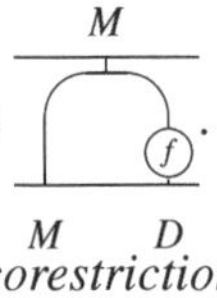

*. In this way we have a functor $F : \mathscr{C}^C \to \mathscr{C}^D$, which we call the functor corestriction of scalars ($F$ acts as identity of morphisms).*

We leave it to the reader to prove the dual version of Proposition 2.35.

**Proposition 2.40** *Let $C, D$ be coalgebras in a braided category $\mathscr{C}$, $M$ a right $C$-comodule and $N$ a right $D$-comodule in $\mathscr{C}$. Then $M \otimes N$ is a right $C \otimes^+ D$-comodule in $\mathscr{C}$ with* 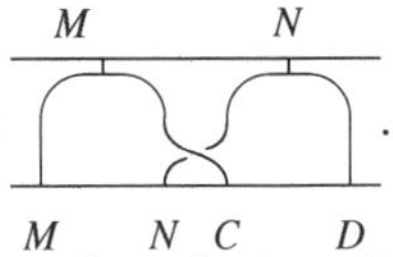

*. If we consider the inverse of the braiding instead of the braiding we then obtain a $C \otimes^- D$-comodule structure for $M \otimes N$ in $\mathscr{C}$.*

If the category $\mathscr{C}$ is left or right rigid then working with modules is equivalent to working with comodules over the co-opposite dual coalgebra.

**Proposition 2.41** *Let $\mathscr{C}$ be a monoidal category and $A$ an algebra in $\mathscr{C}$.*

*(i) If $\mathscr{C}$ is right rigid then the categories ${}_A\mathscr{C}$ and ${}^{*A}\mathscr{C}$ are isomorphic.*

*(ii) If $\mathscr{C}$ is left rigid then the categories $\mathscr{C}_A$ and $\mathscr{C}^{A^*}$ are isomorphic.*

*Consequently, if $\mathscr{C}$ is braided rigid then ${}_A\mathscr{C}$ and $\mathscr{C}^{*A}$ are isomorphic, and also $\mathscr{C}_A$ and ${}^{A^*}\mathscr{C}$ are isomorphic.*

*Proof* Since ${}^*A = ({}^*A)^{\mathrm{cop}+}$ the last assertions are consequences of the first ones and of Proposition 2.37.

We will prove only the isomorphism ${}_A\mathscr{C} \cong {}^{*A}\mathscr{C}$; the isomorphism $\mathscr{C}_A \cong \mathscr{C}^{A^*}$ can be proved in a similar manner. For this, we define first a functor $F : {}_A\mathscr{C} \to {}^{*A}\mathscr{C}$. $F$ acts as identity on objects and morphisms, and if $M$ is a left $A$-module in $\mathscr{C}$ then $F(M)$ becomes a left ${}^*A$-comodule in $\mathscr{C}$ with the structure morphism 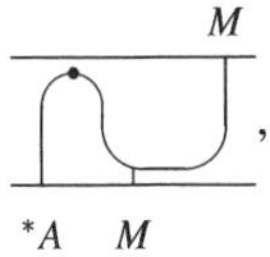

,

where 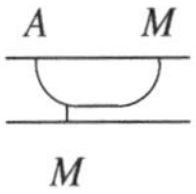

is the left $A$-module structure of $M$ in $\mathscr{C}$. Indeed, we have

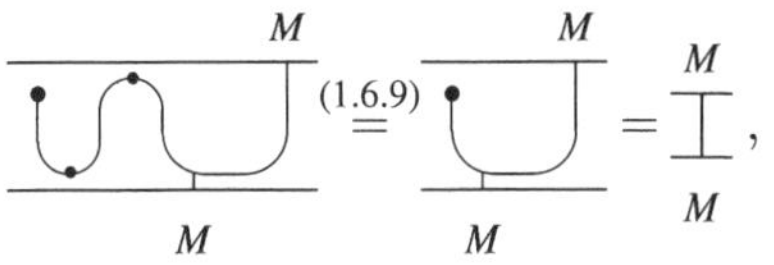

,

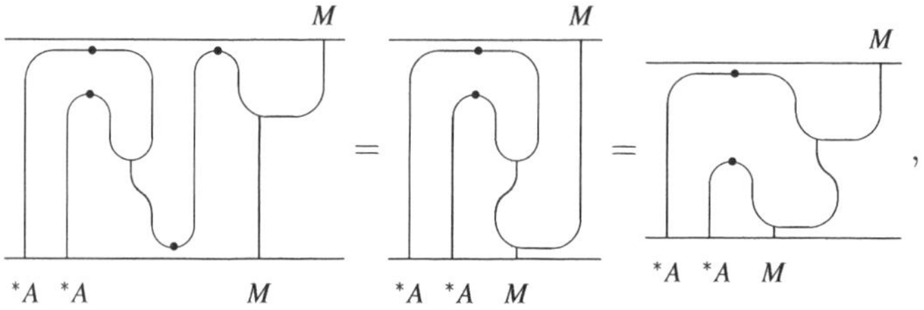


by (1.6.9) and the fact that $M \in {}_A\mathscr{C}$. One can easily check that in this way an $A$-linear morphism becomes a left ${}^{\star}A$-colinear morphism in $\mathscr{C}$, and so $F$ is indeed a functor.

For the other way around, if $M$ is a left ${}^{\star}A$-comodule in $\mathscr{C}$ then by $G(M)$ we denote the same object $M$ endowed with the left $A$-action 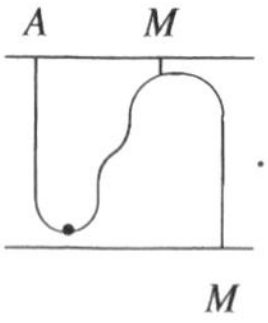

. We have

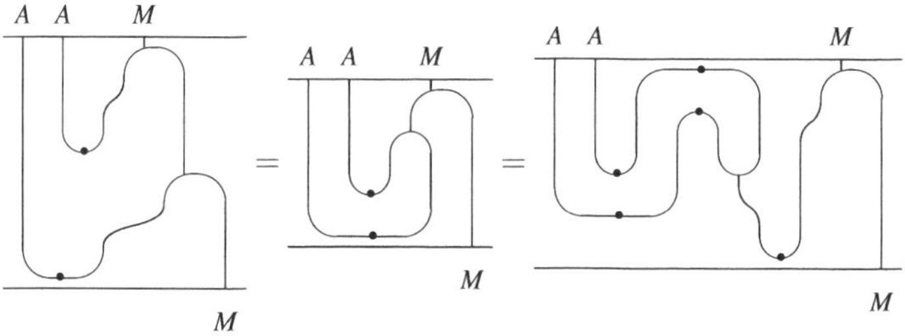


which is 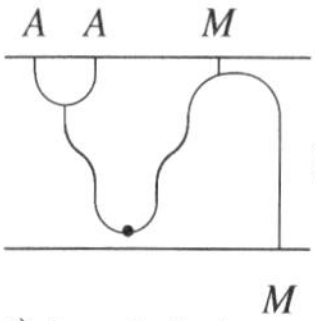

; see (1.6.9). Also, 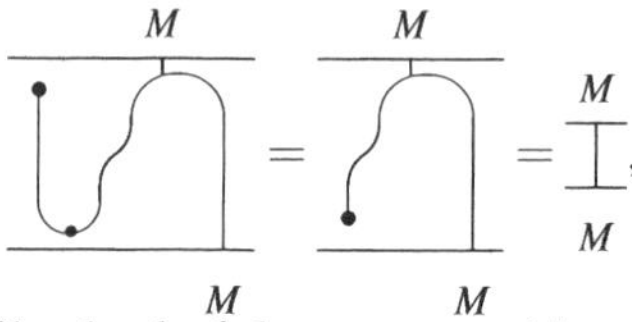

, and therefore $G(M)$ is a left $A$-module. It can be easily checked that any morphism $f : M \to N$ in ${}^{\star A}\mathscr{C}$ turns into a morphism from $G(M)$ to $G(N)$ in ${}_A\mathscr{C}$. Thus we have a well-defined functor $G : {}^{\star A}\mathscr{C} \to {}_A\mathscr{C}$. By (1.6.9) it follows that $F$ and $G$ define a pair of inverse functors, hence the categories ${}^{\star A}\mathscr{C}$ and ${}_A\mathscr{C}$ are isomorphic. □

The dual version of Proposition 2.41 is the following. Its proof is left to the reader.

**Proposition 2.42** *Let $\mathscr{C}$ be a monoidal category and $C$ a coalgebra in $\mathscr{C}$.*

*(i) If $\mathscr{C}$ is right rigid then the categories $\mathscr{C}^C$ and $\mathscr{C}_{\star C}$ are isomorphic.*

*(ii) If $\mathscr{C}$ is left rigid then the categories ${}^C\mathscr{C}$ and ${}_{C^\star}\mathscr{C}$ are isomorphic.*

*Consequently, if $\mathscr{C}$ is braided rigid then $\mathscr{C}^C$ and ${}_{\star C}\mathscr{C}$ are isomorphic, and also ${}^C\mathscr{C}$ and $\mathscr{C}_{C^\star}$ are isomorphic.*

Another result involving dual objects that we can prove is the following.

**Proposition 2.43** *Let $C$ be a coalgebra in a monoidal category $\mathscr{C}$.*

*(i) Let $(V, \underline{\lambda}_V)$ be a left $C$-comodule admitting a right dual ${}^*V$, with evaluation*

*and coevaluation morphisms* $\mathrm{ev}'_V$ *and* $\mathrm{coev}'_V$. *Then* ${}^*V$ *becomes a right* $C$*-comodule, with right* $C$*-coaction*

$$
\begin{aligned}
{}^*V &\xrightarrow{l^{-1}_{{}^*V}} \underline{1}\otimes {}^*V \xrightarrow{\mathrm{coev}'_V\otimes \mathrm{Id}_{{}^*V}} ({}^*V\otimes V)\otimes {}^*V \xrightarrow{(\mathrm{Id}_{{}^*V}\otimes \underline{\lambda}_V)\otimes \mathrm{Id}_{{}^*V}} ({}^*V\otimes(C\otimes V))\otimes {}^*V\\
&\xrightarrow{a^{-1}_{{}^*V,C,V}\otimes \mathrm{Id}_{{}^*V}} (({}^*V\otimes C)\otimes V)\otimes {}^*V \xrightarrow{a_{{}^*V\otimes C,V,{}^*V}} ({}^*V\otimes C)\otimes(V\otimes {}^*V)\\
&\xrightarrow{\mathrm{Id}_{{}^*V\otimes C}\otimes \mathrm{ev}'_V} ({}^*V\otimes C)\otimes \underline{1} \xrightarrow{r_{{}^*V\otimes C}} {}^*V\otimes C\,.
\end{aligned}
$$

*(ii) In a similar way, if* $(V,\underline{\rho}_V)$ *is a right* $C$*-comodule admitting a left dual* $V^*$*, with evaluation and coevaluation morphisms* $\mathrm{ev}_V$ *and* $\mathrm{coev}_V$*, then* $V^*$ *becomes a left* $C$*-comodule, with left* $C$*-coaction*

$$
\begin{aligned}
V^* &\xrightarrow{r^{-1}_{V^*}} V^*\otimes \underline{1} \xrightarrow{\mathrm{Id}_{V^*}\otimes \mathrm{coev}_V} V^*\otimes(V\otimes V^*) \xrightarrow{\mathrm{Id}_{V^*}\otimes(\underline{\rho}_V\otimes \mathrm{Id}_{V^*})} V^*\otimes((V\otimes C)\otimes V^*)\\
&\xrightarrow{\mathrm{Id}_{V^*}\otimes a_{V,C,V^*}} V^*\otimes(V\otimes(C\otimes V^*)) \xrightarrow{a^{-1}_{V^*,V,C\otimes V^*}} (V^*\otimes V)\otimes(C\otimes V^*)\\
&\xrightarrow{\mathrm{ev}_V\otimes \mathrm{Id}_{C\otimes V^*}} \underline{1}\otimes(C\otimes V^*) \xrightarrow{l_{C\otimes V^*}} C\otimes V^*\,.
\end{aligned}
$$

*Proof* Assume that $\mathscr{C}$ is strict monoidal, and then look through a mirror at the upside down version of the diagrammatic proof of Proposition 2.33. □

## 2.6 Braided Bialgebras

Let $(\mathscr{C},\otimes,\underline{1},a,l,r,c)$ be a pre-braided category and $H=(H,\underline{m},\underline{\eta})$ an algebra in $\mathscr{C}$ equipped with two algebra morphisms $\underline{\Delta}: H\to H\otimes_+ H$ and $\underline{\varepsilon}: H\to \underline{1}$ in $\mathscr{C}$. Here $\underline{1}$ has the algebra structure in $\mathscr{C}$ given by $l_{\underline{1}}=r_{\underline{1}}:\underline{1}\otimes\underline{1}\to\underline{1}$ and $\mathrm{Id}_{\underline{1}}$.

If $X,Y\in {}_H\mathscr{C}$ then, by Proposition 2.35, $X\otimes Y\in {}_{H\otimes_+ H}\mathscr{C}$. Since $\underline{\Delta}$ and $\underline{\varepsilon}$ are algebra morphisms in $\mathscr{C}$, by Proposition 2.34 we get that $X\otimes Y$ and $\underline{1}$ are objects of ${}_H\mathscr{C}$ via the structure morphisms given by

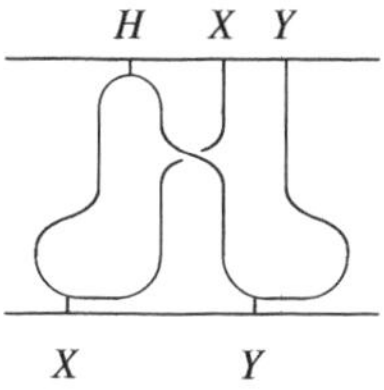

and $\underline{\varepsilon}:=$ 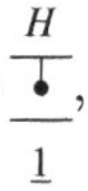

, (2.6.1)

respectively, where 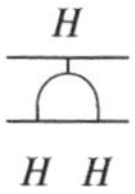

is, as usual, the notation for the morphism $\underline{\Delta}: H\to H\otimes H$.

Clearly, if $f$ and $g$ are morphisms in ${}_H\mathscr{C}$ then $f\otimes g$ is left $H$-linear, too. So the tensor product of $\mathscr{C}$ induces a functor from ${}_H\mathscr{C}\times {}_H\mathscr{C}$ to ${}_H\mathscr{C}$.

**Proposition 2.44** *With notation as above,* $({}_H\mathscr{C},\otimes,\underline{1},a,l,r)$ *is a monoidal category if and only if* $(H,\underline{\Delta},\underline{\varepsilon})$ *is a coalgebra in* $\mathscr{C}$.

*Proof* Without loss of generality we can assume that $\mathscr{C}$ is strict monoidal. We have that $({}_H\mathscr{C}, \otimes, \underline{1}, a, l, r)$ is a monoidal category if and only if $a$, $l$ and $r$ are defined by families of isomorphisms in ${}_H\mathscr{C}$. Now, according to (2.6.1), for $X, Y, Z \in {}_H\mathscr{C}$ we have that $a_{X,Y,Z}$ is an $H$-linear morphism if and only if

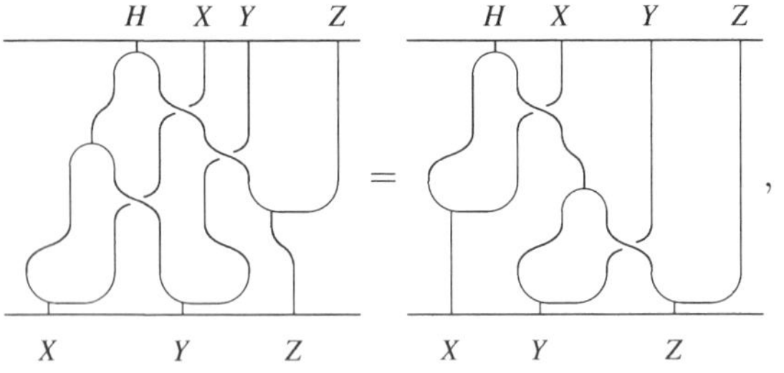


if and only if

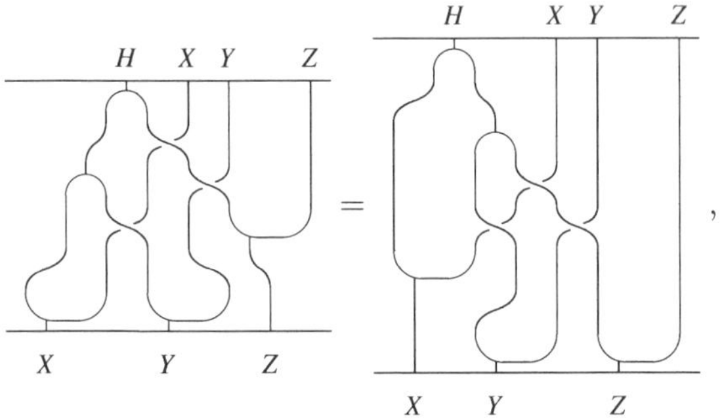


because of Proposition 1.37. It is clear now that the above equality is satisfied if $\underline{\Delta}$ is coassociative. Conversely, if the above equality holds for any $X, Y, Z \in {}_H\mathscr{C}$ then take $X = Y = Z = H$ and regard them as left $H$-modules via the multiplication $\underline{m}$ of $H$. Composing both sides of the resulting equality to the right with $\mathrm{Id}_H \otimes \underline{\eta}_H \otimes \underline{\eta}_H \otimes \underline{\eta}_H$, by Proposition 1.49 we get that $\underline{\Delta}$ is coassociative, as required.

Now, for an object $X \in {}_H\mathscr{C}$ we have that $l_X : \underline{1} \otimes X \to X$ is left $H$-linear if and only if 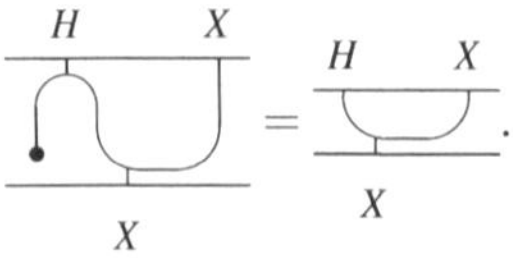

. By similar arguments as above the latest equality is equivalent to 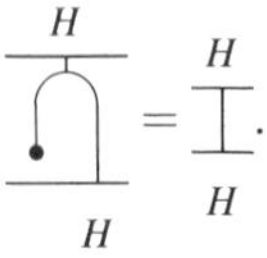

. In a similar manner one can show that $r_X : X \otimes \underline{1} \to X$ is left $H$-linear if and only if 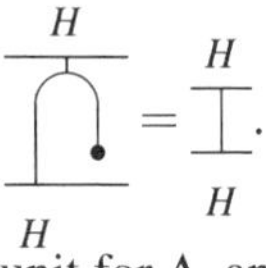

. Thus $l$ and $r$ of $\mathscr{C}$ induce unit constraints for ${}_H\mathscr{C}$ if and only if $\underline{\varepsilon}$ is a counit for $\underline{\Delta}$, and this finishes the proof. □

We can also consider the dual situation. Namely, consider a coalgebra $(H, \underline{\Delta}, \underline{\varepsilon})$ in a pre-braided category $\mathscr{C}$ and two coalgebra morphisms $\underline{m} : H \otimes^+ H \to H$ and $\underline{\eta}$ : $\underline{1} \to H$. If $X, Y \in \mathscr{C}^H$ then $X \otimes Y \in \mathscr{C}^{H \otimes^+ H}$, by Proposition 2.40. By Proposition 2.39 we obtain that $X \otimes Y \in \mathscr{C}^H$, and so $\otimes$ induces a tensor product on $\mathscr{C}^H$. In fact, $X \otimes Y$ is a right $H$-comodule via

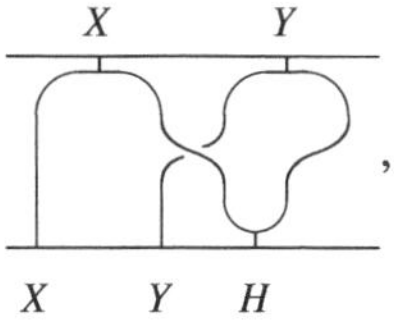


, where

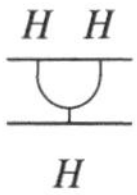


is the standard notation for $\underline{m}$. If $\underline{\eta} =$

then $\underline{1}$ is a right $H$-comodule via

. If we turn the diagrams in the proof of Proposition 2.44 upside down we obtain a proof for the following result.

**Proposition 2.45** *Let $(H, \underline{\Delta}, \underline{\varepsilon})$ be a coalgebra in a pre-braided category $\mathscr{C}$ and $\underline{m} : H \otimes^+ H \to H$ and $\underline{\eta} : \underline{1} \to H$ two coalgebra morphisms in $\mathscr{C}$. Then the category $\mathscr{C}^H$ equipped with the tensor product defined above, unit object $\underline{1}$ of $\mathscr{C}$ viewed as a right $H$-comodule via $\underline{\eta}$, and with the constraints a, l, r of $\mathscr{C}$, is a monoidal category if and only if $(H, \underline{m}, \underline{\eta})$ is an algebra in $\mathscr{C}$.*

The two results presented above can be unified by introducing the notion of bialgebra in a pre-braided category.

**Definition 2.46** A bialgebra $H = (H, \underline{m}, \underline{\eta}, \underline{\Delta}, \underline{\varepsilon})$ in a pre-braided category $\mathscr{C}$ is an algebra $(H, \underline{m}, \underline{\eta})$ and a coalgebra $(H, \underline{\Delta}, \underline{\varepsilon})$ in $\mathscr{C}$ such that $\underline{\Delta} : H \to H \otimes_+ H$ and $\underline{\varepsilon} : H \to \underline{1}$ are algebra morphisms.

A morphism of bialgebras in $\mathscr{C}$ is a morphism in $\mathscr{C}$ that is both an algebra and a coalgebra morphism.

In diagrammatic notation, the axioms for $\underline{\Delta}$ and $\underline{\varepsilon}$ to be algebra morphisms in $\mathscr{C}$ read as

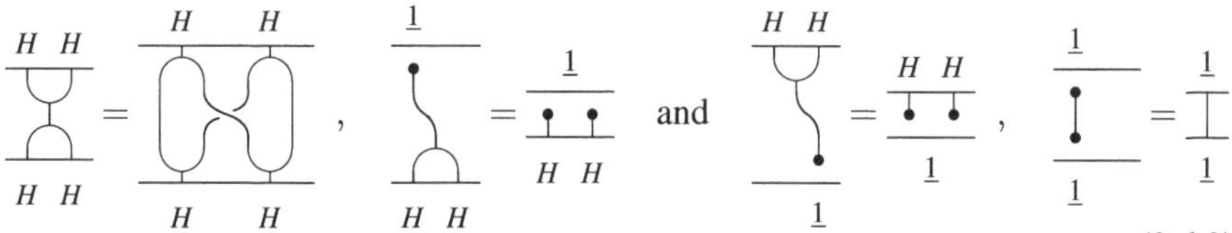


(2.6.2)

respectively. Having in mind the $c$-tensor product coalgebra construction we obtain:

**Corollary 2.47** *$H = (H, \underline{m}, \underline{\eta}, \underline{\Delta}, \underline{\varepsilon})$ is a bialgebra in $(\mathscr{C}, c)$ if and only if $(H, \underline{m}, \underline{\eta})$ is an algebra in $\mathscr{C}$, $(H, \underline{\Delta}, \underline{\varepsilon})$ is a coalgebra in $\mathscr{C}$, and $\underline{m} : H \otimes^+ H \to H$ and $\underline{\eta} : \underline{1} \to H$ are coalgebra morphisms in $\mathscr{C}$. Consequently, if $H$ is a bialgebra in $\mathscr{C}$ then both ${}_H\mathscr{C}$ and $\mathscr{C}^H$ are monoidal categories.*

A bialgebra in a pre-braided category will also be called a braided bialgebra. If $\mathscr{C}$ is braided then to a braided bialgebra we can associate another four braided bialgebras.

**Proposition 2.48** *Let $(B, \underline{m}, \underline{\eta}, \underline{\Delta}, \underline{\varepsilon})$ be a bialgebra in a braided category $\mathscr{C}$.*

*Denote by* $B^{\mathrm{op}-}$ *the object* $B$ *endowed with the* $c^{-1}$*-opposite multiplication associated to* $\underline{m}$ *and with the same coalgebra structure as that of* $B$*, and by* $B^{\mathrm{cop}-}$ *the object* $B$ *endowed with the algebra structure of* $B$ *and with the* $c^{-1}$*-coopposite coalgebra structure associated to* $\underline{\Delta}$*.*

*Similarly, denote by* $B^{\mathrm{op}+,\mathrm{cop}-}$ *the object* $B$ *with the algebra structure of* $B^{\mathrm{op}+}$ *and coalgebra structure of* $B^{\mathrm{cop}-}$*, and by* $B^{\mathrm{op}-,\mathrm{cop}+}$ *the object* $B$ *equipped with the algebra structure of* $B^{\mathrm{op}-}$ *and coalgebra structure of* $B^{\mathrm{cop}+}$*.*

*Then* $B^{\mathrm{op}-}$ *and* $B^{\mathrm{cop}-}$ *are bialgebras in* $\mathscr{C}^{\mathrm{in}}$*, and* $B^{\mathrm{op}+,\mathrm{cop}-}$ *and* $B^{\mathrm{op}-,\mathrm{cop}+}$ *are bialgebras in* $\mathscr{C}$*.*

*Proof* We use the fact that $B$ is a bialgebra in $\mathscr{C}$, Proposition 1.37 and the definition of the multiplication of $B^{\mathrm{op}-}$ to compute

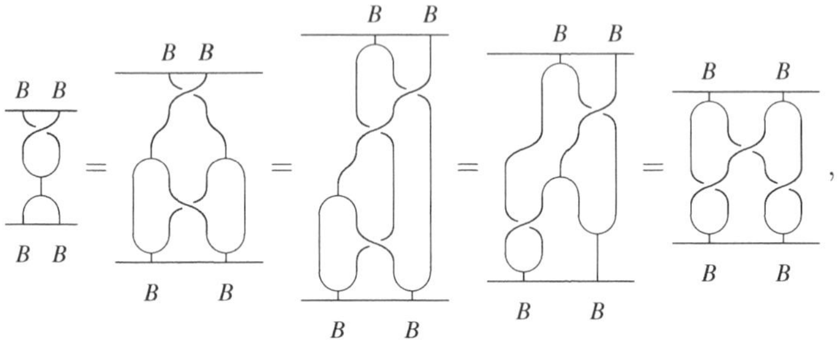


and this proves that $B^{\mathrm{op}-}$ is a bialgebra in $\mathscr{C}^{\mathrm{in}}$. If we turn the above computation upside down we get a proof for the fact that $B^{\mathrm{cop}-}$ is a bialgebra in $\mathscr{C}^{\mathrm{in}}$, too.

We next show that $B^{\mathrm{op}+,\mathrm{cop}-}$ is a bialgebra in $\mathscr{C}$. Actually, using the fact that $B^{\mathrm{cop}-}$ is a bialgebra in $\mathscr{C}^{\mathrm{in}}$ and arguments similar to the ones above we get that

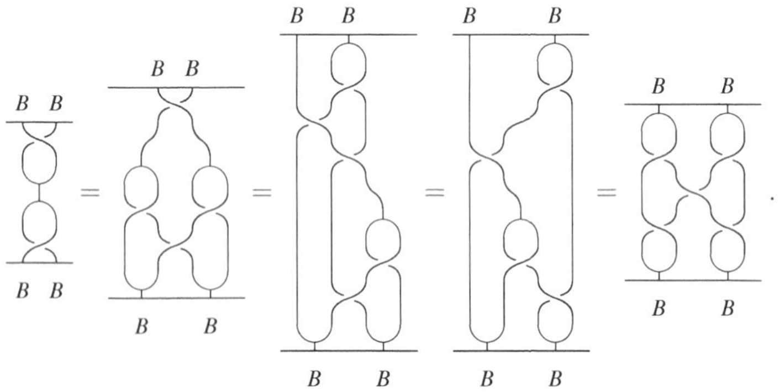


Once again, the proof for the fact that $B^{\mathrm{op}-,\mathrm{cop}+}$ is a bialgebra in $\mathscr{C}$ follows by turning the above diagrams upside down. □

We next supply a list of examples of braided bialgebras.

**Example 2.49** A bialgebra in $\underline{\mathrm{Set}}$ is a monoid. Indeed, we know from Example 2.2 (2) that an algebra in $\underline{\mathrm{Set}}$ is a monoid, and by Example 2.14 (2) that any set has a unique coalgebra structure in $\underline{\mathrm{Set}}$. Thus a bialgebra in $\underline{\mathrm{Set}}$ must have the form

$(H, \underline{m}, \underline{\eta}, \underline{\Delta}, \underline{\varepsilon})$, with $\underline{m}$ defined by the multiplication of the monoid $H$, $\underline{\eta}$ defined by the neutral element $e$ of $H$, and with $\underline{\Delta}(h) = (h,h)$ and $\underline{\varepsilon}(h) = \{*\}$, for all $h \in H$.

Now, $H \times H$ is an algebra in $\underline{\text{Set}}$ (i.e. a monoid) with multiplication $(h,g)(h',g') = (hh',gg')$ and unit $(e,e)$. From here we easily conclude that $\underline{\Delta}$ and $\underline{\varepsilon}$ are algebra morphisms, and so $(H, \underline{m}, \underline{\eta}, \underline{\Delta}, \underline{\varepsilon})$ is a bialgebra in $\underline{\text{Set}}$.

**Example 2.50** Let $k$ be a field and ${}_k\mathcal{M}$ the category of $k$-vector spaces. A bialgebra in ${}_k\mathcal{M}$, which we will call a $k$-bialgebra, is a $k$-algebra $H$ which is also a $k$-coalgebra such that, for all $h, g \in H$,

$$\Delta(hg) = \Delta(h)\Delta(g) = h_1g_1 \otimes h_2g_2 \quad \text{and} \quad \Delta(1_H) = 1_H \otimes 1_H,$$
$$\varepsilon(hg) = \varepsilon(h)\varepsilon(g) \quad \text{and} \quad \varepsilon(1_H) = 1.$$

**Example 2.51** Let $k$ be a field and $H = \frac{k[X]}{(X^2)} = k1 \oplus kx$, the $k$-algebra generated by 1 and $x$ with relation $x^2 = 0$. Then $H$ is a bialgebra within $\text{Vect}_{-1}^{\mathbb{Z}_2}$, the braided category of super vector spaces defined in Example 1.47, with structure given by $\Delta(x) = x \otimes 1 + 1 \otimes x$ and $\varepsilon(x) = 0$, extended by linearity and as algebra morphisms from $H$ to $H \otimes_+ H$ and from $H$ to $k$ in $\text{Vect}_{-1}^{\mathbb{Z}_2}$, respectively. Since $H_0 = k1$ and $H_1 = kx$ we have

$$\begin{aligned}\Delta(x)^2 &= (x \otimes 1 + 1 \otimes x)(x \otimes 1 + 1 \otimes x)\\ &= (-1)^{|1||x|}x^2 \otimes 1 + (-1)^{|1||1|}x \otimes x + (-1)^{|x||x|}x \otimes x + (-1)^{|x||1|}1 \otimes x^2 = 0,\end{aligned}$$

and so $\Delta$ is well defined; obviously, $\varepsilon$ is also well defined. It is easy to see that $\Delta$ is coassociative and that $\varepsilon$ is a counit for it. Thus $H$ is a bialgebra in $\text{Vect}_{-1}^{\mathbb{Z}_2}$, as stated. The above computation shows that $\Delta$ is not well defined in ${}_k\mathcal{M}$, unless $\text{char}(k) = 2$, so in general $H$ is not an ordinary $k$-bialgebra.

We point out that a bialgebra in $\text{Vect}_{-1}^{\mathbb{Z}_2}$ is usually called a super bialgebra.

A pre-braided monoidal functor carries bialgebras to bialgebras.

**Proposition 2.52** *Let $(F, \varphi_0, \varphi_2) : \mathscr{C} \to \mathscr{D}$ be a pre-braided monoidal functor between the pre-braided categories $(\mathscr{C}, c)$ and $(\mathscr{D}, d)$. If $H$ is a bialgebra in $\mathscr{C}$ then $F(H)$ is a bialgebra in $\mathscr{D}$.*

*Proof* Let $(H, \underline{m}_H, \underline{\eta}_H, \Delta_H, \underline{\varepsilon}_H)$ be the bialgebra in the statement. By Propositions 2.3 and 2.16, $F(H)$ has an algebra and a coalgebra structure in $\mathscr{D}$. We show that with this structures $F(H)$ is a bialgebra in $\mathscr{D}$. To this end observe first that

$$\begin{aligned}&(\varphi_{2,H,H} \otimes \varphi_{2,H,H})(\text{Id}_{F(H)} \otimes c_{F(H),F(H)} \otimes \text{Id}_{F(H)})(\varphi_{2,H,H}^{-1} \otimes \varphi_{2,H,H}^{-1})\\
&= (\varphi_{2,H,H} \otimes \varphi_{2,H,H})(\text{Id}_{F(H)} \otimes \varphi_{2,H,H}^{-1} \otimes \text{Id}_{F(H)})(F(\text{Id}_H) \otimes F(c_{H,H}) \otimes F(\text{Id}_H))\\
&\quad (\text{Id}_{F(H)} \otimes \varphi_{2,H,H} \otimes \text{Id}_{F(H)})(\varphi_{2,H,H}^{-1} \otimes \varphi_{2,H,H}^{-1})\\
&= (\varphi_{2,H,H} \otimes \varphi_{2,H,H})((\text{Id}_{F(H)} \otimes \varphi_{2,H,H}^{-1})\varphi_{2,H,H\otimes H}^{-1} \otimes \text{Id}_{F(H)})(F(\text{Id}_H \otimes c_{H,H})\\
&\quad \otimes F(\text{Id}_H))(\varphi_{2,H,H\otimes H}(\text{Id}_{F(H)} \otimes \varphi_{2,H,H}) \otimes \text{Id}_{F(H)})(\varphi_{2,H,H}^{-1} \otimes \varphi_{2,H,H}^{-1})\\
&= (\varphi_{2,H,H} \otimes \varphi_{2,H,H})((\varphi_{2,H,H}^{-1} \otimes \text{Id}_{F(H)})\varphi_{2,H\otimes H,H}^{-1} \otimes \text{Id}_{F(H)})(F(\text{Id}_H \otimes c_{H,H})\end{aligned}$$

$$\otimes F(\mathrm{Id}_H))(\varphi_{2,H\otimes H,H}(\varphi_{2,H,H}\otimes \mathrm{Id}_{F(H)})\otimes \mathrm{Id}_{F(H)})(\varphi_{2,H,H}^{-1}\otimes \varphi_{2,H,H}^{-1})$$
$$= (\mathrm{Id}_{F(H\otimes H)}\otimes \varphi_{2,H,H})(\varphi_{2,H\otimes H,H}^{-1}\otimes \mathrm{Id}_{F(H)})(F(\mathrm{Id}_H\otimes c_{H,H})\otimes F(\mathrm{Id}_H))$$
$$(\varphi_{2,H\otimes H,H}\otimes \mathrm{Id}_{F(H)})(\mathrm{Id}_{F(H\otimes H)}\otimes \varphi_{2,H,H}^{-1})$$
$$= (\mathrm{Id}_{F(H\otimes H)}\otimes \varphi_{2,H,H})(\varphi_{2,H\otimes H,H}^{-1}\otimes \mathrm{Id}_{F(H)})\varphi_{2,H\otimes H\otimes H,H}^{-1}$$
$$F(\mathrm{Id}_H\otimes c_{H,H}\otimes \mathrm{Id}_H)\varphi_{2,H\otimes H\otimes H,H}(\varphi_{2,H\otimes H,H}\otimes \mathrm{Id}_{F(H)})(\mathrm{Id}_{F(H\otimes H)}\otimes \varphi_{2,H,H}^{-1})$$
$$= \varphi_{2,H\otimes H,H\otimes H}^{-1}F(\mathrm{Id}_H\otimes c_{H,H}\otimes \mathrm{Id}_H)\varphi_{2,H\otimes H,H\otimes H},$$

where, for simplicity, we assumed that $\mathscr{C}$ is strict monoidal. We used the fact that $F$ is a pre-braided monoidal functor in the first equality, the naturality of $\varphi_2$ in the second and the fifth equality, and the commutativity of the first diagram in Definition 1.22 for the third and the sixth equality. This fact allows us to compute:

$$\begin{aligned}
\underline{\Delta}_{F(H)}\underline{m}_{F(H)} &= \varphi_{2,H,H}^{-1}F(\underline{\Delta}_H\underline{m}_H)\varphi_{2,H,H}\\
&= \varphi_{2,H,H}^{-1}F(\underline{m}_H\otimes \underline{m}_H)F(\mathrm{Id}_H\otimes c_{H,H}\otimes \mathrm{Id}_H)F(\underline{\Delta}_H\otimes \underline{\Delta}_H)\varphi_{2,H,H}\\
&= (F(\underline{m}_H)\otimes F(\underline{m}_H))\varphi_{2,H\otimes H,H\otimes H}^{-1}F(\mathrm{Id}_H\otimes c_{H,H}\otimes \mathrm{Id}_H)\\
&\quad \varphi_{2,H\otimes H,H\otimes H}(F(\underline{\Delta}_H)\otimes F(\underline{\Delta}_H))\\
&= (F(\underline{m}_H)\varphi_{2,H,H}\otimes F(\underline{m}_H)\varphi_{2,H,H})(\mathrm{Id}_{F(H)}\otimes c_{F(H),F(H)}\otimes \mathrm{Id}_{F(H)})\\
&\quad (\varphi_{2,H,H}^{-1}F(\underline{\Delta}_H)\otimes \varphi_{2,H,H}^{-1}F(\underline{\Delta}_H))\\
&= (\underline{m}_{F(H)}\otimes \underline{m}_{F(H)})(\mathrm{Id}_{F(H)}\otimes c_{F(H),F(H)}\otimes \mathrm{Id}_{F(H)})(\underline{\Delta}_{F(H)}\otimes \underline{\Delta}_{F(H)}),
\end{aligned}$$

as required, where we applied the naturality of $\varphi_2$ in the third equality and the relation obtained above in the fourth equality. We also have:

$$\begin{aligned}
\underline{\Delta}_{F(H)}\underline{\eta}_{F(H)} &= \varphi_{2,H,H}^{-1}F(\underline{\Delta}_H\underline{\eta}_H)\varphi_0\\
&= \varphi_{2,H,H}^{-1}F(\underline{\eta}_H\otimes \underline{\eta}_H)F(l_{\underline{1}}^{-1})\varphi_0\\
&= (F(\underline{\eta}_H)\otimes F(\underline{\eta}_H))\varphi_{2,\underline{1},\underline{1}}^{-1}F(l_{\underline{1}}^{-1})\varphi_0\\
&= (F(\underline{\eta}_H)\otimes F(\underline{\eta}_H))(\varphi_0\otimes \mathrm{Id}_{F(\underline{1})})l_{F(\underline{1})}^{-1}\varphi_0\\
&= (F(\underline{\eta}_H)\otimes F(\underline{\eta}_H))(\varphi_0\otimes \varphi_0)l_{\underline{1}}^{-1},
\end{aligned}$$

and so $\underline{\Delta}_{F(H)}$ is an algebra morphism. This time we used in the third equality the naturality of $\varphi_2$, in the fourth equality the commutativity of the first square diagram in Definition 1.22, and in the last equality the naturality of $l$.

Similarly, one can prove that $\underline{\varepsilon}_{F(H)}$ is an algebra morphism, for instance

$$\begin{aligned}
\underline{\varepsilon}_{F(H)}\underline{m}_{F(H)} &= \varphi_0^{-1}F(l_{\underline{1}})F(\underline{\varepsilon}_H\otimes \underline{\varepsilon}_H)\varphi_{2,H,H} = \varphi_0^{-1}F(l_{\underline{1}})\varphi_{2,\underline{1},\underline{1}}(F(\underline{\varepsilon}_H)\otimes F(\underline{\varepsilon}_H))\\
&= \varphi_0^{-1}l_{F(\underline{1})}(\varphi_0^{-1}\otimes \mathrm{Id}_{F(\underline{1})})(F(\underline{\varepsilon}_H)\otimes F(\underline{\varepsilon}_H)) = l_{\underline{1}}(\underline{\varepsilon}_{F(H)}\otimes \underline{\varepsilon}_{F(H)}).
\end{aligned}$$

The remaining details are left to the reader. □

Let $\mathscr{C}$ be a rigid braided category and $H$ an object of $\mathscr{C}$ that has both an algebra structure $(H,\underline{m}_H,\underline{\eta}_H)$, and a coalgebra structure $(H,\underline{\Delta}_H,\underline{\varepsilon}_H)$ in $\mathscr{C}$.

**Proposition 2.53** *With $\mathscr{C}, H$ as above we have that $\underline{\Delta}_H$ is multiplicative if and only if $\underline{\Delta}_{H^\star}$ is multiplicative, if and only if $\underline{\Delta}_{\star H}$ is multiplicative.*

*Proof* We use the left dual structures of $H$ to compute

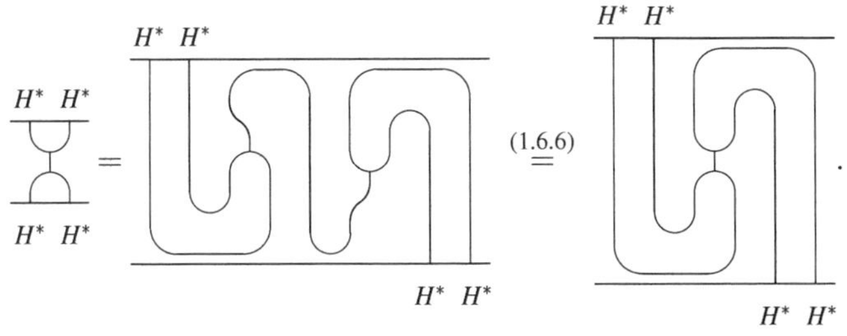


Similarly, $(\underline{m}_{H^*} \otimes \underline{m}_{H^*})(\mathrm{Id}_{H^*} \otimes c_{H^*,H^*} \otimes \mathrm{Id}_{H^*})(\underline{\Delta}_{H^*} \otimes \underline{\Delta}_{H^*})$ is equal to

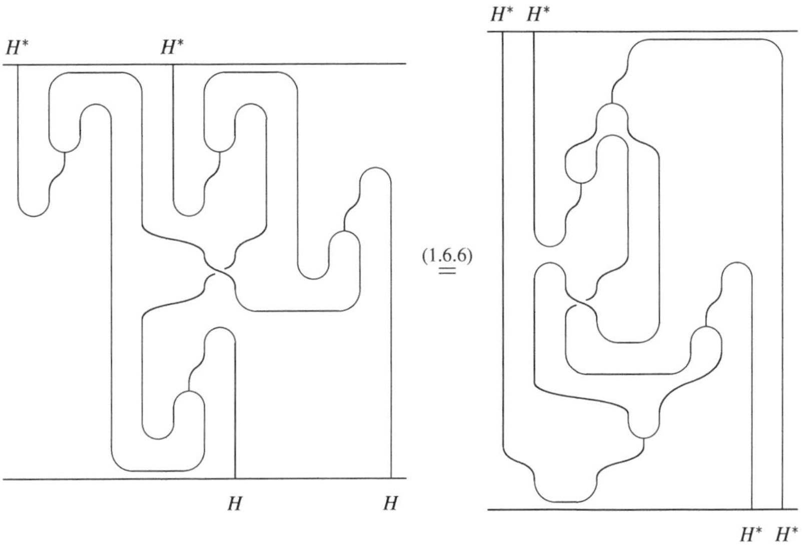


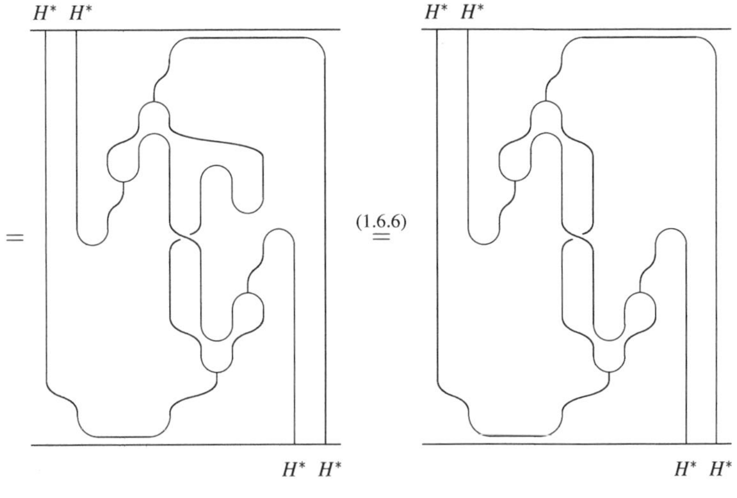

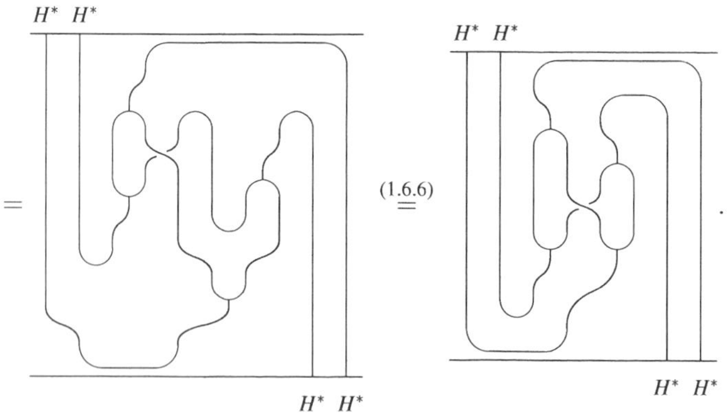


It is now clear that $\underline{\Delta}_{H^\star}$ is multiplicative if and only if $\underline{\Delta}_H$ is. Since a right dual in $\mathscr{C}$ is a left dual in the reverse monoidal category associated to $\mathscr{C}$, the second equivalence follows by applying the above arguments to $\overline{\mathscr{C}}$ instead of $\mathscr{C}$. □

From the previous result and the one below we can conclude that the bialgebra notion is selfdual, provided the category is rigid.

**Proposition 2.54** *If $\mathscr{C}$ is a rigid braided category and $H$ in $\mathscr{C}$ has both an algebra and a coalgebra structure, then $H$ is a bialgebra if and only if $H^\star$ is a bialgebra, if and only if $^\star H$ is a bialgebra.*

*Proof* We have proved in Proposition 2.53 that $\underline{\Delta}_H$ is multiplicative if and only if $\underline{\Delta}_{H^\star}$ is so. Furthermore,

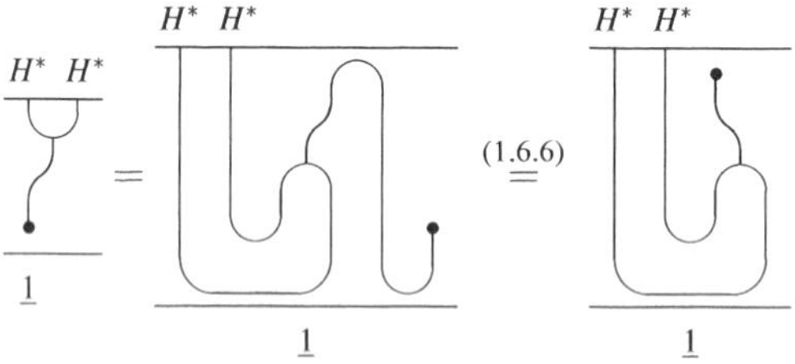


and 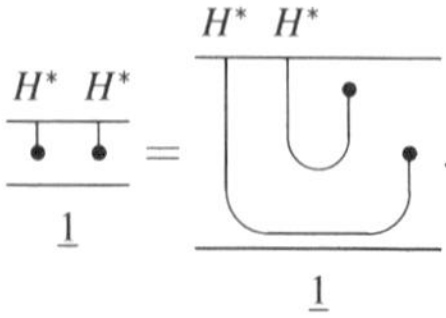

. From here we conclude that $\underline{\varepsilon}_{H^\star}$ is multiplicative if and only if 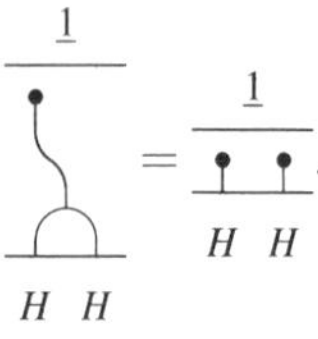

. In a similar way we compute that 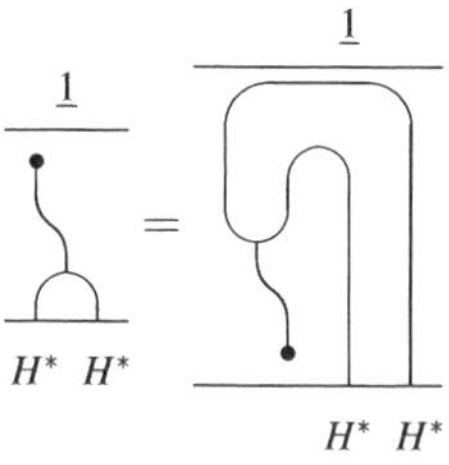

and

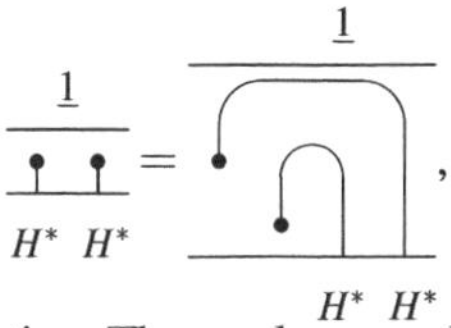


, and therefore $\underline{\Delta}_{H^\star} = \underline{\eta}_{H^\star} \otimes \underline{\eta}_{H^\star}$ if and only if $\underline{\varepsilon}_H$ is multiplicative. The reader can check that $\underline{\varepsilon}_{H^\star} \circ \underline{\eta}_{H^\star} = \mathrm{Id}_{\underline{1}}$ if and only if $\underline{\varepsilon}_H \circ \underline{\eta}_H = \mathrm{Id}_{\underline{1}}$. □

**Definition 2.55** Let $H$ be a bialgebra in a braided rigid category $\mathscr{C}$. We call $H^\star$ (resp. ${}^\star H$) the left (resp. right) op-cop dual braided bialgebra associated to $H$. Since $\mathscr{C}$ is braided, by Proposition 2.48 we have that $H^{\star \mathrm{op}-,\mathrm{cop}+}$ and $({}^\star H)^{\mathrm{op}-,\mathrm{cop}+}$ are bialgebras in $\mathscr{C}$, too. We will call $H^{\star \mathrm{op}-,\mathrm{cop}+}$ the left dual braided bialgebra of $H$, and $({}^\star H)^{\mathrm{op}-,\mathrm{cop}+}$ the right dual braided bialgebra of $H$. These two will be simply denoted by $H^*$, and ${}^*H$, respectively.

In the symmetric monoidal case the left and right duals are isomorphic as bialgebras in $\mathscr{C}$.

**Proposition 2.56** *If $\mathscr{C}$ is a symmetric category and $H$ is a bialgebra in $\mathscr{C}$ then ${}^\star H$ and $H^\star$, and so also ${}^*H$ and $H^*$, are isomorphic as braided bialgebras.*

*Proof* From Proposition 2.26 we know that $\Theta'_H : H^\star \to {}^\star H$ is an algebra isomorphism, and from Proposition 2.28 we have that $\Theta_H : H^\star \to {}^\star H$ is a coalgebra isomorphism. Moreover, $\Theta'_H$ is the morphism $\Theta_H$ regarded in $\mathscr{C}^{\mathrm{in}}$, so $\Theta'_H = \Theta_H$ when $\mathscr{C}$ is symmetric; so in this case $\Theta_H$ is a bialgebra isomorphism. □

## 2.7 Braided Hopf Algebras

We can achieve now the main goal of this chapter, namely to introduce the concept of a braided Hopf algebra. We first need the following result.

**Lemma 2.57** *Let $\mathscr{C}$ be a monoidal category, $(A, \underline{m}, \underline{\eta})$ an algebra in $\mathscr{C}$ and $(C, \underline{\Delta}, \underline{\varepsilon})$ a coalgebra in $\mathscr{C}$. Then $\mathrm{Hom}_{\mathscr{C}}(C, A)$ becomes an algebra in $\underline{\mathrm{Set}}$, that is, a monoid; see Example 2.2 (2). The multiplication is defined by $f * g = \underline{m}_A (f \otimes g) \underline{\Delta}_C$, for all $f, g \in \mathrm{Hom}_{\mathscr{C}}(C, A)$, and the unit is $\underline{\eta}_A \underline{\varepsilon}_C$. The multiplication $*$ is called the convolution product and the invertible elements in $\mathrm{Hom}_{\mathscr{C}}(C, A)$ are called convolution invertible.*

*Proof* The multiplication $*$ on $\mathrm{Hom}_{\mathscr{C}}(C, A)$ is associative since

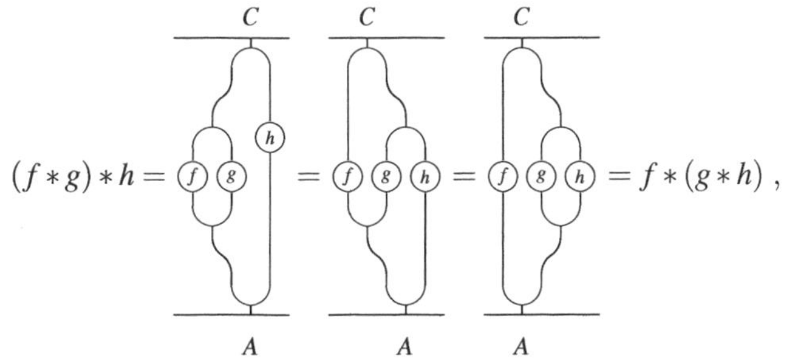

for all $f,g,h \in \mathrm{Hom}_{\mathscr{C}}(C,A)$. The following computations show that the morphism $\underline{\eta}_A\underline{\varepsilon}_C$ is a unit for the multiplication $*$:

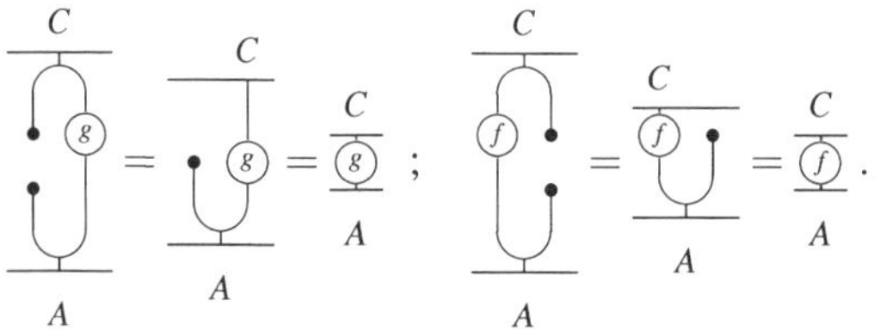


Note that in the above computations we used the coassociativity of $\underline{\Delta}$, the associativity of $\underline{m}$, and the unit and counit axioms, respectively. □

**Definition 2.58** Let $(\mathscr{C},c)$ be a pre-braided category. A bialgebra $H$ in $\mathscr{C}$ is called a Hopf algebra in $\mathscr{C}$ if the identity morphism $\mathrm{Id}_H$ is an invertible element in the monoid $\mathrm{Hom}_{\mathscr{C}}(H,H)$ considered in Lemma 2.57. The inverse of $\mathrm{Id}_H$ will be denoted by $\underline{S}$ and will be called the antipode of $H$.

**Remark 2.59** A Hopf algebra in a pre-braided category will be also called a braided Hopf algebra or a braided group. Thus, a braided bialgebra $H$ is a braided Hopf algebra if and only if there exists a morphism $\underline{S} : H \to H$ in $\mathscr{C}$ such that $\underline{S} * \mathrm{Id}_H = \mathrm{Id}_H * \underline{S} = \underline{\eta}_H\underline{\varepsilon}_H$, where $*$ was defined in Lemma 2.57. More precisely, $\underline{S}$ must satisfy

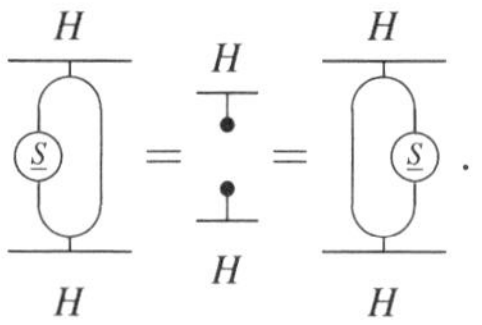


(2.7.1)

Note that the antipode of a braided Hopf algebra $H$ is uniquely determined by the equalities in (2.7.1).

We supply a list of examples of braided Hopf algebras. By using different techniques, more examples will be constructed later in this book, especially within the braided category of Yetter–Drinfeld modules over a quasi-Hopf algebra.

**Examples 2.60** (1) A Hopf algebra in $\underline{\mathrm{Set}}$ is a group.

(2) A Hopf algebra in a category of vector spaces ${}_k\mathscr{M}$ is a $k$-bialgebra $H$ together with a $k$-linear map $S : H \to H$ such that

$$S(h_1)h_2 = h_1S(h_2) = \varepsilon(h)1_H, \ \forall\, h \in H.$$

A Hopf algebra in ${}_k\mathscr{M}$ will be called a Hopf $k$-algebra.

(3) The $k$-super bialgebra described in Example 2.51 is a Hopf algebra in $\mathrm{Vect}^{\mathbb{Z}_2}_{-1}$ with antipode defined by $S(1) = 1$ and $S(x) = -x$, extended by linearity and as an anti-algebra morphism of $H$.

We call a Hopf algebra in $\mathrm{Vect}^{\mathbb{Z}_2}_{-1}$ a super Hopf algebra.

**Definition 2.61** If $H,K$ are Hopf algebras within a pre-braided category $\mathscr{C}$ then a morphism of Hopf algebras between $H$ and $K$ is a bialgebra morphism between $H$ and $K$ in $\mathscr{C}$.

A Hopf algebra morphism automatically respects the antipodes.

**Proposition 2.62** *If $H, K$ are Hopf algebras in $\mathscr{C}$ with antipodes $\underline{S}_H$ and $\underline{S}_K$, respectively, and $f : H \to K$ is a bialgebra morphism in $\mathscr{C}$, then $\underline{S}_K \circ f = f \circ \underline{S}_H$.*

*Proof* We will compute 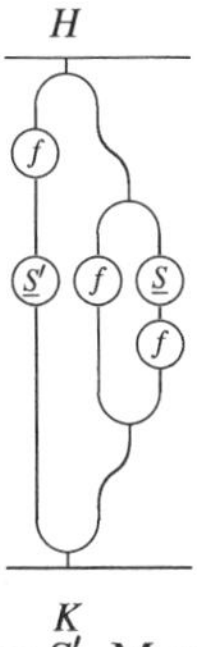

in two different ways, where, for simplicity, we denoted $\underline{S}_H$ by $\underline{S}$ and $\underline{S}_K$ by $\underline{S}'$. More precisely, on the one hand, we use the fact that $f$ is an algebra morphism and (2.7.1) to compute

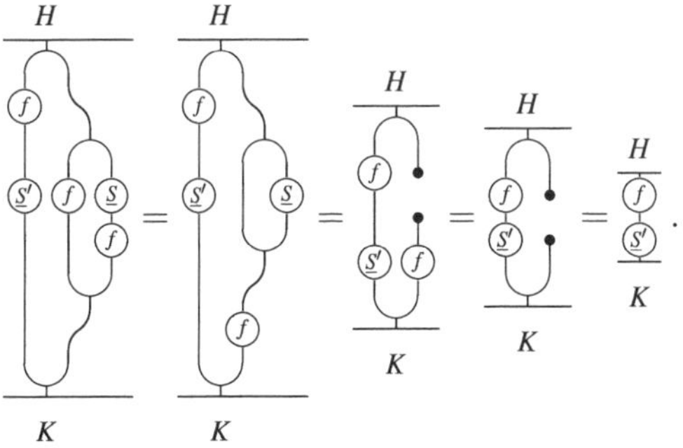


On the other hand, by using the fact that $\underline{\Delta}_H$ is coassociative, $f$ is a coalgebra morphism, $\underline{m}_K$ is associative and (2.7.1) we get that

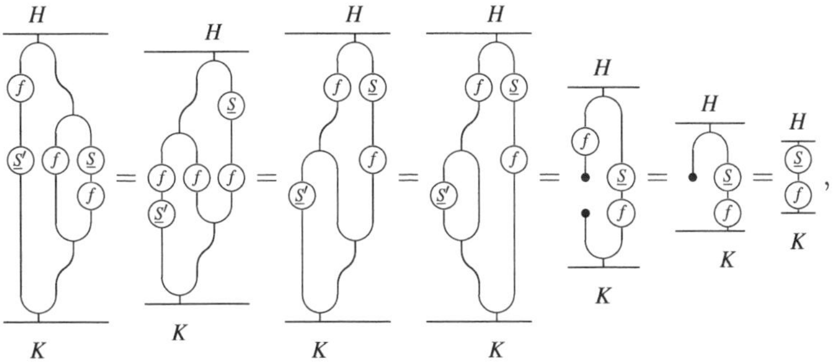


as desired. This finishes the proof. □

**Lemma 2.63** *Let $(F, \varphi_0, \varphi_2) : \mathscr{C} \to \mathscr{D}$ be a strong monoidal functor between two* *monoidal categories, $A$ an algebra in $\mathscr{C}$ and $C$ a coalgebra in $\mathscr{C}$. Then for $f,g \in \mathrm{Hom}_{\mathscr{C}}(C,A)$, we have $F(f*g) = F(f)*F(g)$.*

*Proof* From the naturality of $\varphi_2$, it follows that

$$F(f\otimes g)\varphi_{2,C,C} = \varphi_{2,A,A}(F(f)\otimes F(g)) : F(C)\otimes F(C)\to F(A\otimes A). \tag{2.7.2}$$

Then we easily compute that

$$\begin{aligned} F(f)*F(g) &= \underline{m}_{F(A)}\circ(F(f)\otimes F(g))\circ\Delta_{F(C)}\\ &= F(\underline{m}_A)\circ\varphi_{2,A,A}\circ(F(f)\otimes F(g))\circ\varphi_{2,C,C}^{-1}\circ F(\Delta_C)\\ &= F(\underline{m}_A\circ(f\otimes g)\circ\Delta_C) = F(f*g), \end{aligned}$$

as stated. □

A pre-braided monoidal functor behaves well with respect to the antipodes.

**Proposition 2.64** *Let $(F,\varphi_0,\varphi_2)$ be a pre-braided monoidal functor between the pre-braided categories $\mathscr{C}$ and $\mathscr{D}$. If $H$ is a Hopf algebra in $\mathscr{C}$, then $F(H)$ is a Hopf algebra in $\mathscr{D}$.*

*Proof* We have already seen that our result holds for bialgebras. If $\underline{S}$ is an antipode for $H$, then it follows from Lemma 2.63 that $F(\underline{S})$ is an antipode for $F(H)$. □

The next result says that the antipode of a braided Hopf algebra is an anti-algebra and an anti-coalgebra homomorphism of $H$.

**Proposition 2.65** *Let $(\mathscr{C},c)$ be a pre-braided category and $H$ a Hopf algebra in $\mathscr{C}$ with antipode $\underline{S}$. Then the following relations hold:*

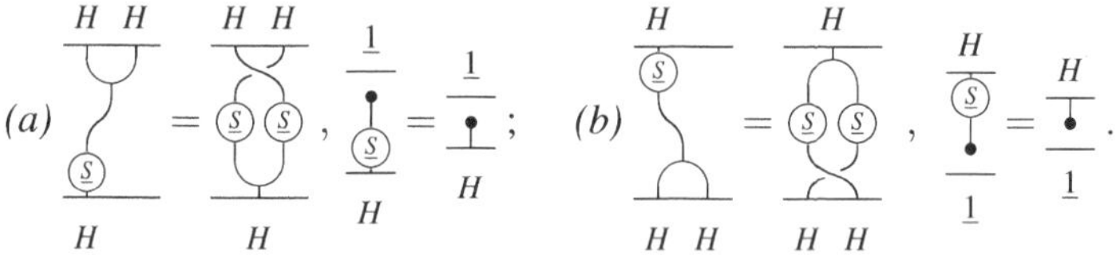

(2.7.3)

*Proof* We only prove the relations in (2.7.3a) (the ones in (2.7.3b) can be proved by turning the diagrams in the proof of (2.7.3a) upside down).

To prove the first equality in (2.7.3a) we compute:

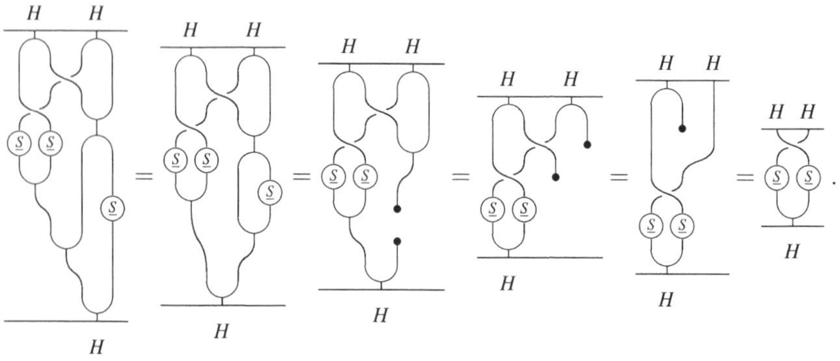

We used the associativity of $\underline{m}$ in the first equality, (2.6.2) in the second and third equalities, and Proposition 1.49 in the penultimate equality. But, on the other hand,

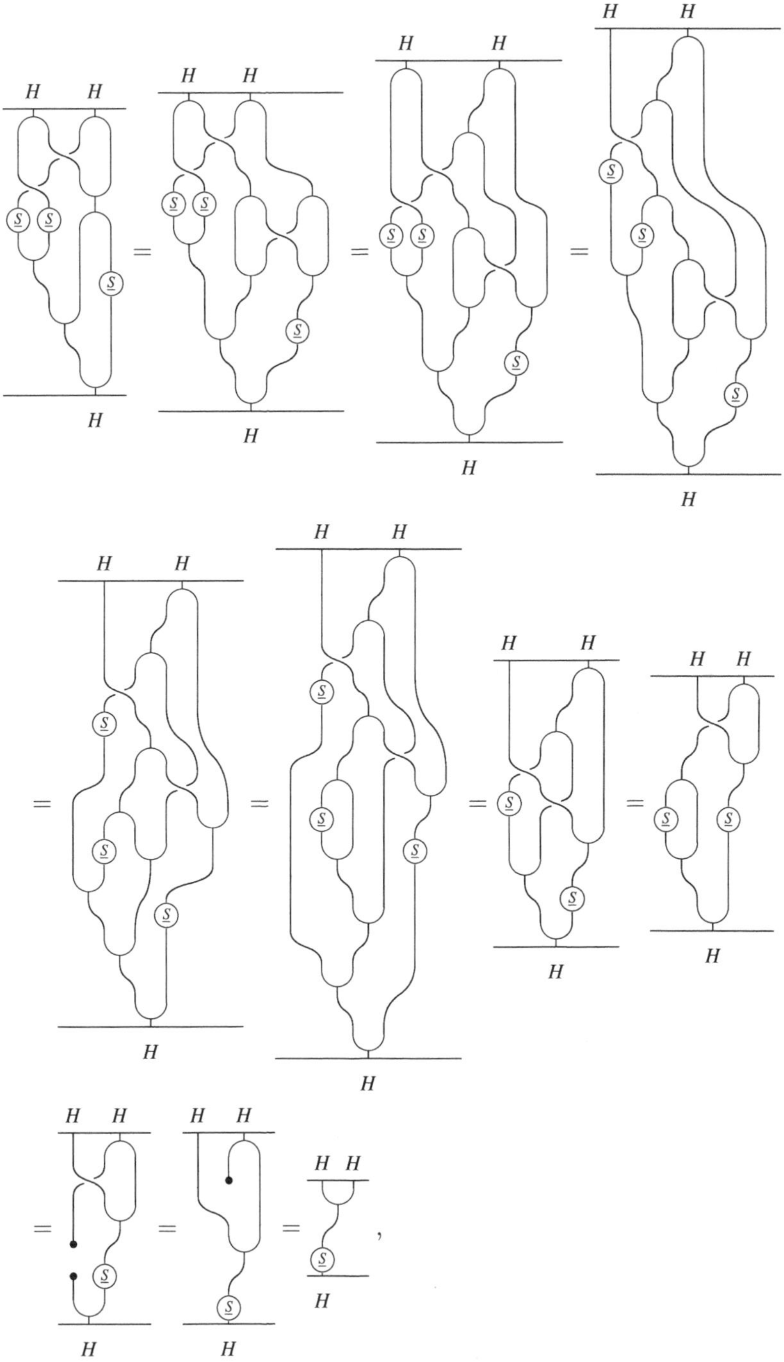

where this time we used: (2.6.2) in the first equality, the coassociativity of $\underline{\Delta}$ in the second and fourth equalities, Proposition 1.37 in the third and seventh equalities, the associativity of $\underline{m}$ in the fifth equality, (2.7.1) in the sixth and eighth equalities, Proposition 1.49 in the ninth equality, and the counit property in the last equality.

Comparing the two computations above we obtain the desired relation. The second relation in (2.7.3) follows, for instance, by "applying" the unit of $H$ to the upper parts of the two members of the first equality in (2.7.1). □

If $H$ is a Hopf algebra in a braided rigid category then its duals also have braided Hopf algebra structures. We denoted by $H^\star$ (resp. ${}^\star H$) the left (resp. right) dual of $H$ endowed with the left (resp. right) (co-)opposite (co)algebra structure dual to the underlying (co)algebra structure of $H$. A similar notation was used for $H^*/{}^*H$.

The result below can be seen as a completion of Proposition 2.54.

**Proposition 2.66** *Let $H$ be a Hopf algebra with antipode $\underline{S}$ in a braided rigid category $\mathscr{C}$. Then $H^\star$ and ${}^\star H$, and so also $H^*$ and ${}^*H$, are Hopf algebras in $\mathscr{C}$ with antipodes $\underline{S}^*$ and ${}^*\underline{S}$, respectively.*

*Proof* As usual, we prove only the statement related to $H^\star$. By the dual structure of $H^\star$ and the properties of $\underline{S}$ we have

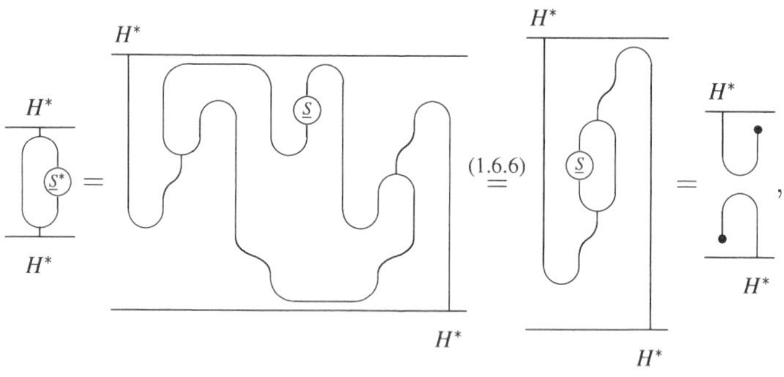


and similarly

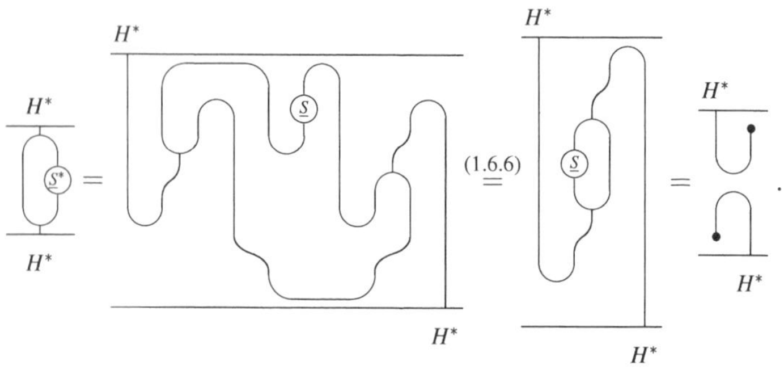


This finishes the proof of the proposition. □

## 2.8 Notes

The concept of monoidal algebra was introduced by Majid in [146] as a generalization to the monoidal setting of a reformulation at the level of commutative diagrams of the old notion of $k$-algebra. The cross product algebra construction was first considered in the particular case when $\mathscr{C}$ is a category of vector spaces. It was introduced independently and with different names in several papers, see for instance [69, 215, 67]. Later, Bespalov and Drabant generalized this construction to an arbitrary monoidal category; see [32, 33].

The notion of monoidal coalgebra is from [146], and the cross product coalgebra construction is from [32, 33]. As we have already explained, some of the results in this chapter are formal dualizations of properties valid for monoidal algebras.

The results in the sections about (co)representations are considered folklore.

Hopf algebras appeared in 1941 in a work in topology by H. Hopf, but Milnor and Moore founded the modern theory of Hopf algebras in [156]. For more information about Hopf algebra theory we recommend the books [1, 73, 157, 189, 209].

The concept of braided Hopf algebra was introduced by Majid in [146] as a natural generalization of the classical concept of Hopf algebra. The result stating that the notion of braided bialgebra is self-dual is taken from [212]. The observation that the dual (co)algebra structure makes sense in any monoidal category is taken from [148].

# 3

# Quasi-bialgebras and Quasi-Hopf Algebras

We introduce the concepts of quasi-bialgebra and quasi-Hopf algebra by using a categorical point of view. We present the basic properties of these objects and study their invariance under a twist. We also introduce the dual notions, called dual quasi-bialgebra and dual quasi-Hopf algebra.

## 3.1 Quasi-bialgebras

Throughout, from now on, by an algebra we mean a unital associative algebra over a field $k$, and any algebra or anti-algebra morphism is assumed to be unital. Usually we denote the multiplication of an algebra by $m_A : A \otimes A \ni a \otimes b \to ab \in A$ and the unit by $1_A$. If $A$ is an algebra and $M$ is a left (resp. right) $A$-module, unless otherwise specified, we will denote the action of $A$ on $M$ by $A \otimes M \ni a \otimes m \to a \cdot m \in M$ (resp. $M \otimes A \ni m \otimes a \to m \cdot a \in M$). A $k$-linear map will be simply called a linear map.

In this section we will construct non-strict monoidal structures on the category of representations of an ordinary $k$-algebra endowed with some additional structures. This will lead to the notion of quasi-bialgebra.

**Definition 3.1** Let $\mathscr{C}, \mathscr{D}$ be monoidal categories and $F : \mathscr{C} \to \mathscr{D}$ a functor between them. We call $F$ a quasi-monoidal functor if there exist a natural isomorphism $\varphi_2 = (\varphi_{2,X,Y} : F(X)\square F(Y) \to F(X \otimes Y))_{X,Y \in \mathscr{C}}$ and an isomorphism $\varphi_0 : F(\underline{1}) \to \underline{I}$ in $\mathscr{D}$ (without any further conditions). Here $\otimes$ and $\underline{1}$ stand for the tensor product and the unit object of $\mathscr{C}$, while $\square$ and $\underline{I}$ are the tensor product and the unit object of $\mathscr{D}$, respectively.

It is immediate that any strong monoidal functor is quasi-monoidal. This justifies our terminology and also the concept of quasi-bialgebra that we will introduce soon.

Throughout this section $k$ is a field and $H$ is a $k$-algebra with unit $1_H$.

We next investigate the monoidal structures on ${}_H\mathscr{M}$ for which the forgetful functor $F : {}_H\mathscr{M} \to {}_k\mathscr{M}$ is a quasi-monoidal functor.

**Lemma 3.2** *Let $H$ be a $k$-algebra. Then giving a monoidal structure on ${}_H\mathscr{M}$ such that the forgetful functor $F : {}_H\mathscr{M} \to {}_k\mathscr{M}$ is a quasi-monoidal functor is equivalent* *to giving a monoidal structure on ${}_H\mathcal{M}$ that comes by a restriction of the monoidal structure on ${}_k\mathcal{M}$ to ${}_H\mathcal{M}$. More precisely, this means that*

*(a) for any two left $H$-modules $X,Y$ the tensor product $X\otimes Y$ in ${}_k\mathcal{M}$ admits a left $H$-module structure;*

*(b) the tensor product in ${}_k\mathcal{M}$ of two left $H$-module morphisms is a morphism in ${}_H\mathcal{M}$, and so $\otimes$ induces a functor from ${}_H\mathcal{M}\times{}_H\mathcal{M}$ to ${}_H\mathcal{M}$;*

*(c) $k$, the unit object of ${}_k\mathcal{M}$, admits a left $H$-module structure;*

*(d) there exist functorial isomorphisms*

$$a=(a_{X,Y,Z}:(X\otimes Y)\otimes Z\to X\otimes(Y\otimes Z))_{X,Y,Z\,\in\,{}_H\mathcal{M}},$$

$$l=(l_X:k\otimes X\to X)_{X\,\in\,{}_H\mathcal{M}}\text{ and }r=(r_X:X\otimes k\to X)_{X\,\in\,{}_H\mathcal{M}}$$

*in ${}_H\mathcal{M}$ such that the Pentagon Axiom and the Triangle Axiom are satisfied.*

*Proof* Everything follows from the definition of a quasi-monoidal functor, since in our case it acts as identity on objects and morphisms. □

The next result is a reconstruction type theorem for quasi-bialgebras.

**Proposition 3.3** *Let $k$ be a field and $H$ a $k$-algebra. Then there exists a one-to-one correspondence between*

- *monoidal structures on ${}_H\mathcal{M}$ such that the forgetful functor $F:{}_H\mathcal{M}\to{}_k\mathcal{M}$ is a quasi-monoidal functor;*
- *5-tuples $(\Delta,\varepsilon,\Phi,l,r)$ consisting of two $k$-algebra maps $\Delta:H\to H\otimes H$ and $\varepsilon:H\to k$ and invertible elements $\Phi\in H\otimes H\otimes H$ and $l,r\in H$ such that, for all $h\in H$, the following relations hold:*

$$(\mathrm{Id}_H\otimes\Delta)(\Delta(h))=\Phi\Big((\Delta\otimes\mathrm{Id}_H)(\Delta(h))\Big)\Phi^{-1},$$

$$(\varepsilon\otimes\mathrm{Id}_H)(\Delta(h))=l^{-1}hl,\quad(\mathrm{Id}_H\otimes\varepsilon)(\Delta(h))=r^{-1}hr,$$

$$(\mathrm{Id}_H\otimes\mathrm{Id}_H\otimes\Delta)(\Phi)(\Delta\otimes\mathrm{Id}_H\otimes\mathrm{Id}_H)(\Phi)=(1_H\otimes\Phi)(\mathrm{Id}_H\otimes\Delta\otimes\mathrm{Id}_H)(\Phi)(\Phi\otimes 1_H),$$

$$(\mathrm{Id}_H\otimes\varepsilon\otimes\mathrm{Id}_H)(\Phi)=r\otimes l^{-1}.$$

*Proof* Assume that the strict monoidal structure on ${}_k\mathcal{M}$ induces a monoidal structure on ${}_H\mathcal{M}$, that is, the forgetful functor is a quasi-monoidal functor. In particular, this implies that we have a left $H$-module structure $\cdot:H\otimes(H\otimes H)\to H\otimes H$ on $H\otimes H$. If we define $\Delta:H\to H\otimes H$ by $\Delta(h)=h\cdot(1_H\otimes 1_H)$, for all $h\in H$, we claim that $\Delta$ is an algebra map. Indeed, it is clear that $\Delta(1_H)=1_H\otimes 1_H$. To see that $\Delta$ is multiplicative we proceed as follows. Let $X\in{}_H\mathcal{M}$ and fix $x\in X$. Then $\varphi_x:H\ni h\mapsto h\cdot x\in X$ is a left $H$-module morphism. Similarly, for $Y\in{}_H\mathcal{M}$ and $y\in Y$ define $\varphi_y:H\to Y$, a left $H$-module morphism. According to (a) and (b) in Lemma 3.2, we have that $\varphi_x\otimes\varphi_y:H\otimes H\to X\otimes Y$ is a left $H$-linear morphism, hence

$$(\varphi_x\otimes\varphi_y)(h\cdot(h'\otimes h''))=h\cdot(h'\cdot x\otimes h''\cdot y),$$

for all $h,h',h''\in H$. If we take $h'=h''=1_H$ and denote $\Delta(h)=h_1\otimes h_2$ (summation implicitly understood), we get that $h\cdot(x\otimes y)=h_1\cdot x\otimes h_2\cdot y$, for all $h\in H$, and so

$\Delta$ determines completely the left $H$-module structure on the tensor product $X \otimes Y$. From the condition $h \cdot (h' \cdot (1_H \otimes 1_H)) = (hh') \cdot (1_H \otimes 1_H)$, applied to all $h, h' \in H$, it follows immediately that $\Delta$ is multiplicative, assuming that $H \otimes H$ has the usual componentwise algebra structure.

We look now at the condition (c) in Lemma 3.2. We claim that giving a left $H$-module structure on $k$ is equivalent to giving an algebra map $\varepsilon : H \to k$. Indeed, if $\cdot : H \otimes k \to k$ gives a left $H$-module structure on $k$ then the map defined by $\varepsilon(h) = h \cdot 1_k$, for all $h \in H$, is an algebra morphism since $\varepsilon(hg) = (hg) \cdot 1_k = h \cdot (g \cdot 1_k) = \varepsilon(g) h \cdot 1_k = \varepsilon(h)\varepsilon(g)$ and $\varepsilon(1_H) = 1_H \cdot 1_k = 1_k$. For the converse, if $\varepsilon : H \to k$ is an algebra map then clearly $k$ is a left $H$-module via the structure defined by $h \cdot \kappa = \varepsilon(h)\kappa$, for all $h \in H$ and $\kappa \in k$. Thus (c) in Lemma 3.2 implies the existence of an algebra map $\varepsilon : H \to k$ such that $h \cdot \kappa = \varepsilon(h)\kappa$, for all $h \in H$ and $\kappa \in k$.

Thus, the monoidal structure on ${}_H\mathcal{M}$ is induced by a triple $(H, \Delta, \varepsilon)$ as follows. If $X$ and $Y$ are left $H$-modules then so is $X \otimes Y$ via the left diagonal $H$-action

$$h \cdot (x \otimes y) = h_1 \cdot x \otimes h_2 \cdot y, \tag{3.1.1}$$

where $\Delta(h) = h_1 \otimes h_2$ is the notation for the comultiplication $\Delta$ of $H$. The unit object is $k$ considered as a left $H$-module via $h \cdot \kappa = \varepsilon(h)\kappa$, for all $h \in H$, $\kappa \in k$.

We now look at the associativity and unit constraints of ${}_H\mathcal{M}$. For a left $H$-module $X$ and $x \in X$ consider again $\varphi_x : H \to X$ given by $\varphi_x(h) = h \cdot x$, for all $h \in H$. The map $\varphi_x$ is left $H$-linear, so by the naturality of $a$ the diagram

$$\begin{array}{ccc}
(H \otimes H) \otimes H & \xrightarrow{a_{H,H,H}} & H \otimes (H \otimes H) \\
{\scriptstyle (\varphi_x \otimes \varphi_y) \otimes \varphi_z} \downarrow & & \downarrow {\scriptstyle \varphi_x \otimes (\varphi_y \otimes \varphi_z)} \\
(X \otimes Y) \otimes Z & \xrightarrow{a_{X,Y,Z}} & X \otimes (Y \otimes Z)
\end{array}$$

is commutative, for all $X, Y, Z \in {}_H\mathcal{M}$, $x \in X$, $y \in Y$ and $z \in Z$. So, if we denote $\Phi := a_{H,H,H}((1_H \otimes 1_H) \otimes 1_H)$ we have

$$a_{X,Y,Z}((x \otimes y) \otimes z) = \Phi \cdot (x \otimes (y \otimes z)), \ \forall\, x \in X, \ y \in Y, \ z \in Z. \tag{3.1.2}$$

Furthermore, since $a$ is a natural isomorphism it follows that $\Phi$ is invertible, its inverse being defined by $a^{-1}_{H,H,H}(1_H \otimes (1_H \otimes 1_H))$.

With this description of $a$ we obtain that the left $H$-linearity of $a$ is equivalent to the first equality in the statement, and that (1.1.1) holds if and only if the third relation in the statement is verified.

The naturality of the unit constraints produces the commutative diagrams:

$$\begin{array}{ccc}
k \otimes H & \xrightarrow{l_H} & H \\
{\scriptstyle \mathrm{Id}_k \otimes \varphi_x} \downarrow & & \downarrow {\scriptstyle \varphi_x} \\
k \otimes X & \xrightarrow[l_X]{} & X
\end{array}
\quad \text{and} \quad
\begin{array}{ccc}
H \otimes k & \xrightarrow{r_H} & H \\
{\scriptstyle \varphi_x \otimes \mathrm{Id}_k} \downarrow & & \downarrow {\scriptstyle \varphi_x} \\
X \otimes k & \xrightarrow[r_X]{} & X.
\end{array}$$

If we denote $l := l_H(1_k \otimes 1_H)$ then, by writing the first diagram for $X = H$, with the left regular $H$-module structure, one can see that $l$ is invertible with $l^{-1} = l_H^{-1}(1_H)$. Also, the first diagram for arbitrary $X \in {}_H\mathscr{M}$ shows that $l_X(\kappa \otimes x) = \kappa l \cdot x$, for all $\kappa \in k$ and $x \in X$. Thus $l_X$ is left $H$-linear if and only if $\varepsilon(h_1)lh_2 \cdot x = hl \cdot x$, for all $h \in H$ and $x \in X$. By taking $X = H$ and $x = 1_H$ we get that $l_X$ is left $H$-linear if and only if $\varepsilon(h_1)lh_2 = hl$, for all $h \in H$. Similarly, using the commutativity of the second square diagram we obtain that $r_X$ is left $H$-linear if and only if $r\varepsilon(h_2)h_1 = hr$, for all $h \in H$, where $r := r_H(1_H \otimes 1_k)$ is invertible with $r^{-1} = r_H^{-1}(1_H)$. Thus the left and right unit constraints are left $H$-linear maps if and only if the second set of relations in the statement hold. Finally, one can easily see that (1.1.2) is equivalent to the fourth equality in the statement.

Conversely, if there exist $\Phi, l, r$ as above, then ${}_H\mathscr{M}$ is monoidal with the tensor product and unit object as in (3.1.1), with associativity constraint as in (3.1.2), left unit constraint defined by $l_X(\kappa \otimes x) = \kappa l \cdot x$, and right unit constraint defined by $r_X(x \otimes \kappa) = \kappa r \cdot x$, for all $X \in {}_H\mathscr{M}$, $x \in X$ and $\kappa \in k$. The verification of this fact is straightforward, so we leave it to the reader. □

**Definition 3.4** Let $H$ be a $k$-algebra, $\Delta : H \to H \otimes H$ and $\varepsilon : H \to k$ two algebra maps. We call $(H, \Delta, \varepsilon)$ a quasi-bialgebra if there exist invertible elements $\Phi \in H \otimes H \otimes H$ and $l, r \in H$ satisfying the conditions in Proposition 3.3. Hence, $(H, \Delta, \varepsilon)$ is a quasi-bialgebra if and only if its category of left modules is monoidal with tensor product given by $\Delta$ and unit object $k$ viewed as a left $H$-module via $\varepsilon$. The quasi-bialgebra with its entire structure is denoted by $(H, \Delta, \varepsilon, \Phi, l, r)$.

A $k$-bialgebra is a quasi-bialgebra for which $\Phi = 1_H \otimes 1_H \otimes 1_H$ and $l = r = 1_H$.

As before, for $k$-bialgebras we denote $\Delta(h) = h_1 \otimes h_2$ and

$$(\Delta \otimes \mathrm{Id}_H)(\Delta(h)) = (\mathrm{Id}_H \otimes \Delta)(\Delta(h)) = h_1 \otimes h_2 \otimes h_3 \text{ , etc.}$$

For quasi-bialgebras, since $\Delta$ is only quasi-coassociative we adopt the further convention (as above, the summation is implicitly understood):

$$(\Delta \otimes \mathrm{Id}_H)(\Delta(h)) = h_{(1,1)} \otimes h_{(1,2)} \otimes h_2, \quad (\mathrm{Id}_H \otimes \Delta)(\Delta(h)) = h_1 \otimes h_{(2,1)} \otimes h_{(2,2)},$$

for all $h \in H$. Furthermore, we will denote the tensor components of $\Phi$ by capital letters, and those of $\Phi^{-1}$ by lower case letters, namely

$$\Phi = X^1 \otimes X^2 \otimes X^3 = T^1 \otimes T^2 \otimes T^3 = V^1 \otimes V^2 \otimes V^3 = \cdots$$
$$\Phi^{-1} = x^1 \otimes x^2 \otimes x^3 = t^1 \otimes t^2 \otimes t^3 = v^1 \otimes v^2 \otimes v^3 = \cdots$$

We next show that any quasi-bialgebra is equivalent to a quasi-bialgebra for which $l = r = 1_H$. We first need to prove the following result.

**Proposition 3.5** *Let $H$ be a quasi-bialgebra, $F \in H \otimes H$ an invertible element. Let*

$$\Delta_F : H \to H \otimes H, \quad \Delta_F(h) = F\Delta(h)F^{-1}, \tag{3.1.3}$$
$$l_F = \varepsilon(G^1)lG^2, \quad r_F = \varepsilon(G^2)rG^1, \tag{3.1.4}$$
$$\Phi_F = (1_H \otimes F)(\mathrm{Id}_H \otimes \Delta)(F)\Phi(\Delta \otimes \mathrm{Id}_H)(F^{-1})(F^{-1} \otimes 1_H), \tag{3.1.5}$$

*where $F = F^1 \otimes F^2$, $F^{-1} = G^1 \otimes G^2$ is the formal notation for the tensor components of $F$ and $F^{-1}$, respectively. Then $H_F := (H, \Delta_F, \varepsilon, \Phi_F, l_F, r_F)$ is a quasi-bialgebra as well. Moreover, the categories ${}_H\mathscr{M}$ and ${}_{H_F}\mathscr{M}$ are monoidally isomorphic; more precisely, the identity functor from ${}_H\mathscr{M}$ to ${}_{H_F}\mathscr{M}$ is a strong monoidal functor.*

*Proof* Denote by $\mathscr{F}^1 \otimes \mathscr{F}^2 = \mathbb{F}^1 \otimes \mathbb{F}^2$ and $\mathscr{G}^1 \otimes \mathscr{G}^2 = \mathbb{G}^1 \otimes \mathbb{G}^2$ some more copies of $F$ and $F^{-1}$, respectively. We then have

$$\Phi_F = \mathbb{F}^1 X^1 G_1^1 \mathbb{G}^1 \otimes F^1 \mathbb{F}_1^2 X^2 G_2^1 \mathbb{G}^2 \otimes F^2 \mathbb{F}_2^2 X^3 G^2,$$

and so we compute:

$$\begin{aligned}
&\Phi_F(\Delta_F \otimes \mathrm{Id}_{H_F})(\Delta_F(h)) \\
&\quad = \Phi_F(\Delta_F(\mathscr{F}^1 h_1 \mathscr{G}^1) \otimes \mathscr{F}^2 h_2 \mathscr{G}^2) \\
&\quad = (\mathbb{F}^1 X^1 G_1^1 \otimes F^1 \mathbb{F}_1^2 X^2 G_2^1 \otimes F^2 \mathbb{F}_2^2 X^3 G^2)((\mathbb{G}^1 \otimes \mathbb{G}^2)\Delta_F(\mathscr{F}^1 h_1 \mathscr{G}^1) \otimes \mathscr{F}^2 h_2 \mathscr{G}^2) \\
&\quad = (\mathbb{F}^1 X^1 G_1^1 \otimes F^1 \mathbb{F}_1^2 X^2 G_2^1 \otimes F^2 \mathbb{F}_2^2 X^3 G^2)(\Delta(\mathscr{F}^1 h_1 \mathscr{G}^1)(\mathbb{G}^1 \otimes \mathbb{G}^2) \otimes \mathscr{F}^2 h_2 \mathscr{G}^2) \\
&\quad = (\mathbb{F}^1 X^1 \otimes F^1 \mathbb{F}_1^2 X^2 \otimes F^2 \mathbb{F}_2^2 X^3)(\Delta(h_1 \mathscr{G}^1)(\mathbb{G}^1 \otimes \mathbb{G}^2) \otimes h_2 \mathscr{G}^2) \\
&\quad = \mathbb{F}^1 h_1 X^1 \mathscr{G}_1^1 \mathbb{G}^1 \otimes F^1 \mathbb{F}_1^2 h_{(2,1)} X^2 \mathscr{G}_2^1 \mathbb{G}^2 \otimes F^2 \mathbb{F}_2^2 h_{(2,2)} X^3 \mathscr{G}^2 \\
&\quad = \mathbb{F}^1 h_1 X^1 \mathscr{G}_1^1 \mathbb{G}^1 \otimes \Delta_F(\mathbb{F}^2 h_2)(F^1 X^2 \mathscr{G}_2^1 \mathbb{G}^2 \otimes F^2 X^3 \mathscr{G}^2) \\
&\quad = (\mathrm{Id}_{H_F} \otimes \Delta_F)(\Delta_F(h))(\mathbb{F}^1 X^1 \mathscr{G}_1^1 \mathbb{G}^1 \otimes F^1 \mathbb{F}_1^2 X^2 \mathscr{G}_2^1 \mathbb{G}^2 \otimes F^2 \mathbb{F}_2^2 X^3 \mathscr{G}^2) \\
&\quad = (\mathrm{Id}_{H_F} \otimes \Delta_F)(\Delta_F(h))\Phi_F,
\end{aligned}$$

as required. Clearly $\varepsilon(F^1)F^2$ is invertible with inverse $\varepsilon(G^1)G^2$, and $\varepsilon(F^2)F^1$ is invertible with inverse $\varepsilon(G^2)G^1$. Therefore $l_F^{-1} = \varepsilon(F^1)F^2 l^{-1}$ and $r_F^{-1} = \varepsilon(F^2)F^1 r^{-1}$, and from here we compute

$$(\varepsilon \otimes \mathrm{Id}_{H_F})(\Delta_F(h)) = \varepsilon(F^1 h_1 G^1)F^2 h_2 G^2 = \varepsilon(F^1)F^2 l^{-1} h l \varepsilon(G^1)G^2 = l_F^{-1} h l_F$$

and similarly $(\mathrm{Id}_{H_F} \otimes \varepsilon)(\Delta_F(h)) = r_F^{-1} h r_F$, for all $h \in H$. We also have

$$\begin{aligned}
(\mathrm{Id}_{H_F} \otimes \varepsilon \otimes \mathrm{Id}_{H_F})(\Phi_F) &= \varepsilon(F^1 \mathbb{F}_1^2 X^2 G_2^1 \mathbb{G}^2)\mathbb{F}^1 X^1 G_1^1 \mathbb{G}^1 \otimes F^2 \mathbb{F}_2^2 X^3 G^2 \\
&= \varepsilon(F^1 X^2 \mathbb{G}^2)\mathbb{F}^1 X^1 r^{-1} G^1 r \mathbb{G}^1 \otimes F^2 l^{-1} \mathbb{F}^2 l X^3 G^2 \\
&= \varepsilon(F^1 \mathbb{G}^2)\mathbb{F}^1 G^1 r \mathbb{G}^1 \otimes F^2 l^{-1} \mathbb{F}^2 G^2 \\
&= \varepsilon(F^1)\varepsilon(\mathbb{G}^2) r \mathbb{G}^1 \otimes F^2 l^{-1} = r_F \otimes l_F^{-1}.
\end{aligned}$$

By the above formula for $\Phi_F$ we have

$$\begin{aligned}
(\mathrm{Id}_{H_F} \otimes \Delta_F \otimes \mathrm{Id}_{H_F})(\Phi_F) = {} & \mathbb{F}^1 X^1 G_1^1 \mathbb{G}^1 \otimes \mathscr{F}^1 F_1^1 \mathbb{F}_{(1,1)}^2 X_1^2 G_{(2,1)}^1 \mathbb{G}_1^2 \mathscr{G}^1 \\
& \otimes \mathscr{F}^2 F_2^1 \mathbb{F}_{(1,2)}^2 X_2^2 G_{(2,2)}^1 \mathbb{G}_2^2 \mathscr{G}^2 \otimes F^2 \mathbb{F}_2^2 X^3 G^2,
\end{aligned}$$

and this implies

$$\begin{aligned}
&(\mathrm{Id}_{H_F} \otimes \Delta_F \otimes \mathrm{Id}_{H_F})(\Phi_F)(\Phi_F \otimes 1_{H_F}) = \mathbb{F}^1 X^1 G_1^1 Y^1 \mathbb{G}_1^1 \mathscr{G}^1 \\
&\quad \otimes \mathscr{F}^1 F_1^1 \mathbb{F}_{(1,1)}^2 X_1^2 G_{(2,1)}^1 Y^2 \mathbb{G}_2^1 \mathscr{G}^2 \otimes \mathscr{F}^2 F_2^1 \mathbb{F}_{(1,2)}^2 X_2^2 G_{(2,2)}^1 Y^3 \mathbb{G}^2 \otimes F^2 \mathbb{F}_2^2 X^3 G^2.
\end{aligned}$$

We then compute:

$$
\begin{aligned}
&(1_{H_F}\otimes\Phi_F)(\mathrm{Id}_{H_F}\otimes\Delta_F\otimes\mathrm{Id}_{H_F})(\Phi_F)(\Phi_F\otimes 1_{H_F}) = \mathbb{F}^1X^1G_1^1Y^1\mathbb{G}_1^1\mathscr{G}^1\\
&\quad\otimes F^1Z^1\mathbb{F}^2_{(1,1)}X_1^2G^1_{(2,1)}Y^2\mathbb{G}_2^1\mathscr{G}^2\otimes\mathscr{F}^1F_1^2Z^2\mathbb{F}^2_{(1,2)}X_2^2G^1_{(2,2)}Y^3\mathbb{G}^2\otimes\mathscr{F}^2F_2^2Z^3\mathbb{F}_2^2X^3G^2\\
&= \mathbb{F}^1X^1Y^1G^1_{(1,1)}\mathbb{G}_1^1\mathscr{G}^1\otimes F^1\mathbb{F}_1^2Z^1X_1^2Y^2G^1_{(1,2)}\mathbb{G}_2^1\mathscr{G}^2\otimes\mathscr{F}^1F_1^2\mathbb{F}^2_{(2,1)}Z^2X_2^2Y^3G_2^1\mathbb{G}^2\\
&\quad\otimes\mathscr{F}^2F_2^2\mathbb{F}^2_{(2,2)}Z^3X^3G^2\\
&= \mathbb{F}^1X^1Y_1^1G^1_{(1,1)}\mathbb{G}_1^1\mathscr{G}^1\otimes F^1\mathbb{F}_1^2X^2Y_2^1G^1_{(1,2)}\mathbb{G}_2^1\mathscr{G}^2\otimes\mathscr{F}^1F_1^2\mathbb{F}^2_{(2,1)}X_1^3Y^2G_2^1\mathbb{G}^2\\
&\quad\otimes\mathscr{F}^2F_2^2\mathbb{F}^2_{(2,2)}X_2^3Y^3G^2\\
&= (\mathrm{Id}_{H_F}\otimes\mathrm{Id}_{H_F}\otimes\Delta_F)(\Phi_F)\Big(\mathbb{F}^1F_1^1Y_1^1G^1_{(1,1)}\mathbb{G}_1^1\mathscr{G}^1\otimes\mathbb{F}^2F_2^1Y_2^1G^1_{(1,2)}\mathbb{G}_2^1\mathscr{G}^2\\
&\qquad\qquad\otimes\mathscr{F}^1F_1^2Y^2G_2^1\mathbb{G}^2\otimes\mathscr{F}^2F_2^2Y^3G^2\Big)\\
&= (\mathrm{Id}_{H_F}\otimes\mathrm{Id}_{H_F}\otimes\Delta_F)(\Phi_F)(\Delta_F\otimes\mathrm{Id}_{H_F}\otimes\mathrm{Id}_{H_F})(\Phi_F).
\end{aligned}
$$

Hence $H_F$ is indeed a quasi-bialgebra.

We next show that the identity functor, viewed as a functor from ${}_H\mathscr{M}$ to ${}_{H_F}\mathscr{M}$, has a strong monoidal structure.

For $X, Y \in {}_H\mathscr{M}$ define $\varphi_{2,X,Y} : X\otimes Y \to X\otimes Y$ by

$$\varphi_{2,X,Y}(x\otimes y) = G^1\cdot x\otimes G^2\cdot y, \tag{3.1.6}$$

for all $x\in X$ and $y\in Y$. We have

$$
\begin{aligned}
\varphi_{2,X,Y}(h\cdot(x\otimes y)) &= \varphi_{2,X,Y}(F^1h_1G^1\cdot x\otimes F^2h_2G^2\cdot y)\\
&= h_1G^1\cdot x\otimes h_2G^2\cdot y\\
&= h\cdot\varphi_{2,X,Y}(x\otimes y),
\end{aligned}
$$

for all $h\in H$, and this shows that $\varphi_{2,X,Y}$ is left $H_F$-linear. It can be easily checked that $\varphi_2 = (\varphi_{2,X,Y})_{X,Y\,\in\,{}_H\mathscr{M}}$ is a family of natural isomorphisms and that the commutativity of the first diagram in Definition 1.22 reduces to the definition of $\Phi_F$. Also, if we take $\varphi_0 = \mathrm{Id}_k$ then the commutativity of the two square diagrams in Definition 1.22 follows from the definitions of $l_F$ and $r_F$, respectively. □

If $F\in H\otimes H$ is an invertible element, we call the quasi-bialgebra $H_F$ the twisting of $H$ by $F$. Motivated by the above results we introduce the following concepts.

**Definition 3.6** For $(H,\Delta,\varepsilon,\Phi,l,r)$ and $(H',\Delta',\varepsilon',\Phi',l',r')$ two quasi-bialgebras, a morphism between them is an algebra map $\chi: H\to H'$ such that

$$
\begin{aligned}
&(\chi\otimes\chi)\circ\Delta = \Delta'\circ\chi,\quad \varepsilon'\circ\chi = \varepsilon,\\
&(\chi\otimes\chi\otimes\chi)(\Phi) = \Phi',\quad \chi(l) = l' \quad\text{and}\ \chi(r) = r'.
\end{aligned}
$$

If, in addition, $\chi$ is an isomorphism we call it a quasi-bialgebra isomorphism.

Two quasi-bialgebras $H$ and $H'$ are called equivalent (or twist equivalent) if there exists an invertible element $F\in H'\otimes H'$ such that $H$ and $H'_F$ are isomorphic as quasi-bialgebras.

**Remark 3.7** If $H$ is a quasi-bialgebra and $F, F' \in H \otimes H$ are invertible elements, one can easily see that $(H_F)_{F'} = H_{FF'}$, and so $(H_F)_{F^{-1}} = (H_{F^{-1}})_F = H$. Consequently, the twist equivalence between quasi-bialgebras is an equivalence relation.

**Remark 3.8** If $H$ and $H'$ are isomorphic quasi-bialgebras, then ${}_H\mathcal{M}$ and ${}_{H'}\mathcal{M}$ are monoidally isomorphic (via the restriction of scalars functors).

**Corollary 3.9** *If $H$ and $H'$ are twist equivalent quasi-bialgebras then ${}_H\mathcal{M}$ and ${}_{H'}\mathcal{M}$ are monoidally isomorphic.*

**Lemma 3.10** *For any quasi-bialgebra $H$ there exists an invertible element $F \in H \otimes H$ for which $l_F = r_F = 1_H$.*

*Proof* Let us start by proving that $\varepsilon(l) = \varepsilon(r)$. For this, observe first that by applying $\varepsilon \otimes \varepsilon$ to both sides of $(\mathrm{Id}_H \otimes \varepsilon \otimes \mathrm{Id}_H)(\Phi) = r \otimes l^{-1}$ we get $\varepsilon(X^1)\varepsilon(X^2)\varepsilon(X^3) = \varepsilon(r)\varepsilon(l^{-1})$. On the other hand, it follows from $\varepsilon(h_1)h_2 = l^{-1}hl$ that $\varepsilon(h_1)\varepsilon(h_2) = \varepsilon(h)$, for any $h \in H$. Therefore, by applying $\varepsilon \otimes \varepsilon \otimes \varepsilon \otimes \varepsilon$ to both sides of

$$(\mathrm{Id}_H \otimes \mathrm{Id}_H \otimes \Delta)(\Phi)(\Delta \otimes \mathrm{Id}_H \otimes \mathrm{Id}_H)(\Phi) = (1_H \otimes \Phi)(\mathrm{Id}_H \otimes \Delta \otimes \mathrm{Id}_H)(\Phi)(\Phi \otimes 1_H)$$

and by using that $\varepsilon(X^1)\varepsilon(X^2)\varepsilon(X^3)$ is invertible in $k$ we obtain $\varepsilon(X^1)\varepsilon(X^2)\varepsilon(X^3) = 1$. Together with $\varepsilon(X^1)\varepsilon(X^2)\varepsilon(X^3) = \varepsilon(r)\varepsilon(l^{-1})$ this implies $\varepsilon(l) = \varepsilon(r)$, as stated.

Let us denote $c := \varepsilon(l) = \varepsilon(r)$. We show that $F = c^{-1}r \otimes l$ is the twist that we need. Note first that $c^{-1} = \varepsilon(l^{-1}) = \varepsilon(r^{-1})$. Now, since $F^{-1} = cr^{-1} \otimes l^{-1}$ we have

$$l_F = c\varepsilon(r^{-1})ll^{-1} = cc^{-1}1_H = 1_H \quad \text{and} \quad r_F = c\varepsilon(l^{-1})rr^{-1} = cc^{-1}1_H = 1_H.$$

So our proof is complete. □

**Corollary 3.11** *Any quasi-bialgebra is twist equivalent to a quasi-bialgebra of the form $(H, \Delta, \varepsilon, \Phi, l, r)$ for which $l = r = 1_H$.*

Motivated by these facts, we will modify Definition 3.4 and, from now on, by a quasi-bialgebra $H$ we mean a 4-tuple $(H, \Delta, \varepsilon, \Phi)$, where $H$ is an associative algebra with unit $1_H$, $\Phi$ is an invertible element in $H \otimes H \otimes H$ (called in what follows the reassociator of $H$) and $\Delta : H \to H \otimes H$ (called the comultiplication) and $\varepsilon : H \to k$ (called the counit) are algebra homomorphisms satisfying the identities

$$(\mathrm{Id}_H \otimes \Delta)(\Delta(h)) = \Phi(\Delta \otimes \mathrm{Id}_H)(\Delta(h))\Phi^{-1}, \tag{3.1.7}$$

$$(\mathrm{Id}_H \otimes \varepsilon)(\Delta(h)) = h, \quad (\varepsilon \otimes \mathrm{Id}_H)(\Delta(h)) = h, \tag{3.1.8}$$

for all $h \in H$, and $\Phi$ is a normalized 3-cocycle, in the sense that

$$(1_H \otimes \Phi)(\mathrm{Id}_H \otimes \Delta \otimes \mathrm{Id}_H)(\Phi)(\Phi \otimes 1_H) = (\mathrm{Id}_H \otimes \mathrm{Id}_H \otimes \Delta)(\Phi)(\Delta \otimes \mathrm{Id}_H \otimes \mathrm{Id}_H)(\Phi), \tag{3.1.9}$$

$$(\mathrm{Id}_H \otimes \varepsilon \otimes \mathrm{Id}_H)(\Phi) = 1_H \otimes 1_H. \tag{3.1.10}$$

One can easily see that the identities (3.1.8), (3.1.9) and (3.1.10) also imply that

$$(\varepsilon \otimes \mathrm{Id}_H \otimes \mathrm{Id}_H)(\Phi) = (\mathrm{Id}_H \otimes \mathrm{Id}_H \otimes \varepsilon)(\Phi) = 1_H \otimes 1_H. \tag{3.1.11}$$

Since we have assumed $l = r = 1_H$ in the definition of a quasi-bialgebra $H$, we need to assume also that $\varepsilon(F^1)F^2 = \varepsilon(F^2)F^1 = 1_H$ when we twist $H$ by an invertible element $F = F^1 \otimes F^2 \in H \otimes H$ (in order to obtain $l_F = r_F = 1_H$). We will call such an $F$ a twist or gauge transformation on $H$.

**Examples 3.12** (1) Any $k$-bialgebra $H$ is a quasi-bialgebra with $\Phi = 1_H \otimes 1_H \otimes 1_H$. In particular, if $F$ is a twist on $H$ then in general $H_F$ is no longer a $k$-bialgebra but it is a quasi-bialgebra.

(2) Together with a quasi-bialgebra $H = (H, \Delta, \varepsilon, \Phi)$ we also have $H^{\rm op}$, $H^{\rm cop}$ and $H^{\rm op,cop}$ as quasi-bialgebras, where "op" means opposite multiplication and "cop" means opposite comultiplication (i.e. $\Delta^{\rm cop}(h) = h_2 \otimes h_1$, for all $h \in H$). The quasi-bialgebra structures are obtained by putting $\Phi_{\rm op} = \Phi^{-1}$, $\Phi_{\rm cop} = (\Phi^{-1})^{321} := x^3 \otimes x^2 \otimes x^1$ and $\Phi_{\rm op,cop} = \Phi^{321} := X^3 \otimes X^2 \otimes X^1$, respectively.

More generally, if $t$ denotes a permutation of $\{1, 2, \ldots, n\}$ with inverse $t^{-1}$, for an element $\Omega^1 \otimes \cdots \otimes \Omega^n \in H^{\otimes n}$ we set

$$\Omega^{t(1)t(2)\cdots t(n)} = \Omega^{t^{-1}(1)} \otimes \cdots \otimes \Omega^{t^{-1}(n)}.$$

Non-trivial examples of quasi-bialgebras will be given in the next sections.

## 3.2 Quasi-Hopf Algebras

The categorical meaning of the next concept will be explained later on in this chapter.

**Definition 3.13** A quasi-bialgebra $H$ is called a quasi-Hopf algebra if there exist an anti-algebra endomorphism $S$ of $H$, called an antipode, and elements $\alpha, \beta \in H$ such that, for all $h \in H$, we have

$$S(h_1)\alpha h_2 = \varepsilon(h)\alpha \quad \text{and} \quad h_1 \beta S(h_2) = \varepsilon(h)\beta, \tag{3.2.1}$$

$$X^1 \beta S(X^2) \alpha X^3 = 1_H \quad \text{and} \quad S(x^1)\alpha x^2 \beta S(x^3) = 1_H. \tag{3.2.2}$$

Recall that a $k$-bialgebra is called a Hopf algebra if there exists a $k$-linear map $S : H \to H$ such that $S(h_1)h_2 = \varepsilon(h)1_H = h_1 S(h_2)$, for all $h \in H$. Note that the quasi-Hopf notion is more general than the Hopf algebra notion, as the next result explains.

**Proposition 3.14** *If $H$ is a Hopf algebra with antipode $S$ then $S$ is an anti-algebra and an anti-coalgebra morphism.*

*Proof* This follows from the more general result in Proposition 2.65. □

**Remark 3.15** Let $H$ be a quasi-Hopf algebra and $\mathrm{Alg}(H, k)$ the set of algebra maps from $H$ to $k$, endowed with the multiplication $\nu\xi = (\nu \otimes \xi) \circ \Delta$. One can easily see that this product is associative with unit $\varepsilon$. By (3.2.2) we obtain that for any $\nu \in \mathrm{Alg}(H, k)$ we have $\nu(\alpha) \neq 0 \neq \nu(\beta)$. By using this and (3.2.1), it follows that $\mathrm{Alg}(H, k)$ is a group, the inverse of an element $\nu \in \mathrm{Alg}(H, k)$ being the element $\nu \circ S$.

The observations below point out some differences that exist between Hopf algebras and quasi-Hopf algebras (proofs are straightforward and left to the reader).

**Remarks 3.16** (1) The axioms for a quasi-Hopf algebra immediately imply that $\varepsilon\circ S=\varepsilon$ and $\varepsilon(\alpha)\varepsilon(\beta)=1$, so, by rescaling $\alpha$ and $\beta$, we may (and will from now on) assume without loss of generality that $\varepsilon(\alpha)=\varepsilon(\beta)=1$.

(2) If $S,\alpha,\beta$ satisfy the conditions (3.2.1) and (3.2.2) then for any invertible element $u\in H$ the same conditions are satisfied by $\overline{S}$, $\overline{\alpha}$ and $\overline{\beta}$, where

$$\overline{S}(h)=uS(h)u^{-1},\ \forall\, h\in H,\ \overline{\alpha}=u\alpha \text{ and } \overline{\beta}=\beta u^{-1}. \tag{3.2.3}$$

So the antipode of a quasi-Hopf algebra is not unique.

(3) If $H$ is a $k$-bialgebra that admits a quasi-Hopf algebra structure then by (3.2.2) we obtain that $\alpha$ and $\beta$ are inverses of each other. If we take $u=\alpha^{-1}=\beta$ we then have $\overline{S}(h)=\alpha S(h)\alpha^{-1}$, for all $h\in H$, and $\overline{\alpha}=\overline{\beta}=1_H$, so $H$ is a Hopf algebra with antipode $\overline{S}$. Consequently, a $k$-bialgebra is a Hopf algebra if and only if it is a quasi-Hopf algebra.

(4) The quasi-Hopf notion is invariant under a twist. More precisely, if $F$ is a twist on $H$ then the quasi-bialgebra $H_F$ is, moreover, a quasi-Hopf algebra with $S_F=S$,

$$\alpha_F=S(G^1)\alpha G^2 \quad\text{and}\quad \beta_F=F^1\beta S(F^2), \tag{3.2.4}$$

where $F=F^1\otimes F^2$ and $F^{-1}=G^1\otimes G^2$.

(5) If $H=(H,\Delta,\varepsilon,\Phi,S,\alpha,\beta)$ is a quasi-Hopf algebra then $H^{\rm op,cop}$ is a quasi-Hopf algebra with $S_{\rm op,cop}=S$, $\alpha_{\rm op,cop}=\beta$ and $\beta_{\rm op,cop}=\alpha$. If $S$ is bijective then $H^{\rm op}$ and $H^{\rm cop}$ are quasi-Hopf algebras as well. This time the structures are obtained by putting $S_{\rm op}=S_{\rm cop}=S^{-1}$, $\alpha_{\rm op}=S^{-1}(\beta)$, $\beta_{\rm op}=S^{-1}(\alpha)$, $\alpha_{\rm cop}=S^{-1}(\alpha)$ and $\beta_{\rm cop}=S^{-1}(\beta)$, respectively.

The second remark above has also a converse.

**Proposition 3.17** *If two triples* $(S,\alpha,\beta)$ *and* $(\overline{S},\overline{\alpha},\overline{\beta})$ *satisfy* (3.2.1) *and* (3.2.2) *then there exists a unique invertible element* $u\in H$ *such that* (3.2.3) *holds.*

*Proof* If there exists $u$ satisfying (3.2.3) then

$$u=uS(x^1)\alpha x^2\beta S(x^3)=\overline{S}(x^1)u\alpha x^2\beta S(x^3)=\overline{S}(x^1)\overline{\alpha}x^2\beta S(x^3),$$

and this proves the uniqueness of $u$.

Set now $u=\overline{S}(x^1)\overline{\alpha}x^2\beta S(x^3)$. If we write (3.1.7) under the form

$$h_{(1,1)}x^1\otimes h_{(1,2)}x^2\otimes h_2x^3=x^1h_1\otimes x^2h_{(2,1)}\otimes x^3h_{(2,2)},\ \ \forall\, h\in H,$$

then we obtain

$$\overline{S}(h_{(1,1)}x^1)\overline{\alpha}h_{(1,2)}x^2\beta S(h_2x^3)=\overline{S}(x^1h_1)\overline{\alpha}x^2h_{(2,1)}\beta S(x^3h_{(2,2)}),$$

and this implies $uS(h)=\overline{S}(h)u$, for all $h\in H$.

Likewise, if we write (3.1.9) under the equivalent form

$$X^1Y_1^1x^1\otimes X^2Y_2^1x^2\otimes X_1^3Y^2x^3\otimes X_2^3Y^3=Y^1\otimes X^1Y_1^2\otimes X^2Y_2^2\otimes X^3Y^3,$$

we then have

$$\overline{S}(X^1Y_1^1x^1)\overline{\alpha}X^2Y_2^1x^2\beta S(X_1^3Y^2x^3)\alpha X_2^3Y^3=\overline{S}(Y^1)\overline{\alpha}X^1Y_1^2\beta S(X^2Y_2^2)\alpha X^3Y^3,$$

which can be rewritten as $\overline{S}(x^1)\overline{\alpha}x^2\beta S(x^3)\alpha=\overline{\alpha}X^1\beta S(X^2)\alpha X^3$, or as $u\alpha=\overline{\alpha}$. This formula for $H^{\mathrm{op,cop}}$ instead of $H$ becomes $\beta S(x^1)\alpha x^2\overline{\beta}\ \overline{S}(x^3)=\overline{\beta}$. So to end the proof it is enough to show that $u^{-1}:=S(x^1)\alpha x^2\overline{\beta}\ \overline{S}(x^3)$ is the inverse of $u$. On the one hand we have

$$uS(x^1)\alpha x^2\overline{\beta}\ \overline{S}(x^3)=\overline{S}(x^1)u\alpha x^2\overline{\beta}\ \overline{S}(x^3)=\overline{S}(x^1)\overline{\alpha}x^2\overline{\beta}\ \overline{S}(x^3)=1_H.$$

On the other hand, by using again (3.1.9), we compute:

$$\begin{aligned}\overline{\beta}u&=\overline{\beta}\ \overline{S}(x^1)\overline{\alpha}x^2\beta S(x^3)\\&=Y^1X^1y_1^1x^1\overline{\beta}\ \overline{S}(Y_1^2X^2y_2^1x^2)\overline{\alpha}Y_2^2X^3y^2x_1^3\beta S(Y^3y^3x_2^3)\\&=X^1\overline{\beta}\ \overline{S}(X^2)\overline{\alpha}X^3\beta=\beta,\end{aligned}$$

and this allows us to show that

$$S(x^1)\alpha x^2\overline{\beta}\ \overline{S}(x^3)u=S(x^1)\alpha x^2\overline{\beta}uS(x^3)=S(x^1)\alpha x^2\beta S(x^3)=1_H,$$

as desired. This completes the proof. □

**Definition 3.18** Let $H=(H,\Delta,\varepsilon,\Phi,\alpha,\beta,S)$ be a quasi-Hopf algebra. If $u\in H$ is invertible, we denote by $H_u$ the quasi-Hopf algebra $(H,\Delta,\varepsilon,\Phi,u\alpha,\beta u^{-1},S_u)$, where $S_u(h)=uS(h)u^{-1}$, for all $h\in H$.

For a quasi-Hopf algebra $H$ we show that the antipode is, up to conjugation by a twist, an anti-coalgebra morphism. In particular, we recover the result that the antipode is an ordinary anti-coalgebra morphism in the case when $H$ is a Hopf algebra.

To this end define $\gamma,\delta\in H\otimes H$ by

$$\begin{aligned}\gamma&=S(X^2x_2^1)\alpha X^3x^2\otimes S(X^1x_1^1)\alpha x^3\\&\overset{(3.1.9),(3.1.10),(3.2.1)}{=}S(x^1X^2)\alpha x^2X_1^3\otimes S(X^1)\alpha x^3X_2^3,\end{aligned}\tag{3.2.5}$$

$$\begin{aligned}\delta&=X_1^1x^1\beta S(X^3)\otimes X_2^1x^2\beta S(X^2x^3)\\&\overset{(3.1.9),(3.1.10),(3.2.1)}{=}x^1\beta S(x_2^3X^3)\otimes x^2X^1\beta S(x_1^3X^2).\end{aligned}\tag{3.2.6}$$

**Lemma 3.19** *In a quasi-Hopf algebra H the following relations hold:*

$$(S\otimes S)(\Delta^{\mathrm{cop}}(h_1))\gamma\Delta(h_2)=\varepsilon(h)\gamma,\tag{3.2.7}$$

$$\Delta(h_1)\delta(S\otimes S)(\Delta^{\mathrm{cop}}(h_2))=\varepsilon(h)\delta,\tag{3.2.8}$$

*for all $h\in H$, where $\Delta^{\mathrm{cop}}$ is the opposite comultiplication of H, and*

$$\Delta(X^1)\delta(S\otimes S)(\Delta^{\mathrm{cop}}(X^2))\gamma\Delta(X^3)=1_H\otimes 1_H,\tag{3.2.9}$$

$$(S\otimes S)(\Delta^{\mathrm{cop}}(x^1))\gamma\Delta(x^2)\delta(S\otimes S)(\Delta^{\mathrm{cop}}(x^3))=1_H\otimes 1_H.\tag{3.2.10}$$

*Proof* Set $\gamma=\gamma^1\otimes\gamma^2$ and $\delta=\delta^1\otimes\delta^2$. To check (3.2.7) we compute:

$$
\begin{aligned}
&(S\otimes S)(\Delta^{\rm cop}(h_1))\gamma\Delta(h_2)\\
&\quad = S(h_{(1,2)})\gamma^1(h_2)_1\otimes S(h_{(1,1)})\gamma^2(h_2)_2\\
&\quad = S(x^1X^2h_{(1,2)})\alpha x^2(X^3h_2)_1\otimes S(X^1h_{(1,1)})\alpha x^3(X^3h_2)_2\\
&\quad \overset{(3.1.7)}{=} S(x^1(h_2)_1X^2)\alpha x^2(h_2)_{(2,1)}X^3_1\otimes S(h_1X^1)\alpha x^3(h_2)_{(2,2)}X^3_2\\
&\quad \overset{(3.1.7)}{=} S((h_2)_{(1,1)}x^1X^2)\alpha(h_2)_{(1,2)}x^2X^3_1\otimes S(h_1X^1)\alpha(h_2)_2x^3X^3_2\\
&\quad \overset{(3.2.1)}{=} S(x^1X^2)\alpha x^2X^3_1\otimes S(h_1X^1)\alpha h_2x^3X^3_2\overset{(3.2.1)}{=}\varepsilon(h)\gamma.
\end{aligned}
$$

The formula in (3.2.8) can be proved in a similar manner, so we leave the details to the reader. Now (3.2.10) follows since

$$
\begin{aligned}
&(S\otimes S)(\Delta^{\rm cop}(x^1))\gamma\Delta(x^2)\delta(S\otimes S)(\Delta^{\rm cop}(x^3))\\
&\quad = S(x^1_2)\gamma^1x^2_1\delta^1S(x^3_2)\otimes S(x^1_1)\gamma^2x^2_2\delta^2S(x^3_1)\\
&\quad = S(y^1X^2x^1_2)\alpha y^2X^3_1x^2_1\delta^1S(x^3_2)\otimes S(X^1x^1_1)\alpha y^3X^3_2x^2_2\delta^2S(x^3_1)\\
&\quad \overset{(3.1.9)}{=} S(y^1z^2_1x^1X^2)\alpha y^2z^2_{(2,1)}x^2_1X^3_{(1,1)}\delta^1S(z^3_2x^3_2X^3_{(2,2)})\\
&\qquad\quad \otimes S(z^1X^1)\alpha y^3z^2_{(2,2)}x^2_2X^3_{(1,2)}\delta^2S(z^3_1x^3_1X^3_{(2,1)})\\
&\quad \overset{(3.2.8),(3.1.7)}{=} S(z^2_{(1,1)}y^1x^1)\alpha z^2_{(1,2)}y^2x^2_1\delta^1S(z^3_2x^3_2)\otimes S(z^1)\alpha z^2_2y^3x^2_2\delta^2S(z^3_1x^3_1)\\
&\quad \overset{(3.2.1)}{=} S(y^1x^1)\alpha y^2x^2_1\delta^1S(z^3_2x^3_2)\otimes S(z^1)\alpha z^2y^3x^2_2\delta^2S(z^3_1x^3_1)\\
&\quad = S(y^1x^1)\alpha y^2x^2_1t^1\beta S(z^3_2x^3_2t^3_2X^3)\\
&\qquad\quad \otimes S(z^1)\alpha z^2y^3x^2_2t^2X^1\beta S(z^3_1x^3_1t^3_1X^2)\\
&\quad \overset{(3.1.9),(3.1.11),(3.2.1)}{=} S(x^1)\alpha x^2\beta S(z^3_2x^3_{(2,2)}X^3)\otimes S(z^1)\alpha z^2x^3_1X^1\beta S(z^3_1x^3_{(2,1)}X^2)\\
&\quad \overset{(3.1.7),(3.2.1)}{=} S(x^1)\alpha x^2\beta S(z^3_2X^3x^3)\otimes S(z^1)\alpha z^2X^1\beta S(z^3_1X^2)\\
&\quad \overset{(3.2.2)}{=} S(z^3_2X^3)\otimes S(z^1)\alpha z^2X^1\beta S(z^3_1X^2)\\
&\quad \overset{(3.1.9)}{=} S(Y^3y^3)\otimes S(Y^1_1z^1y^1)\alpha Y^1_2z^2y^2_1\beta S(Y^2z^3y^2_2)\\
&\quad \overset{(3.2.1),(3.1.11),(3.1.10)}{=} 1_H\otimes S(z^1)\alpha z^2\beta S(z^3)\overset{(3.2.2)}{=}1_H\otimes 1_H,
\end{aligned}
$$

as desired. The formula in (3.2.9) can be proved in a similar way. □

Another preliminary result that we need is the following.

**Lemma 3.20** *Let $H$ be a quasi-Hopf algebra and $A$ a $k$-algebra. Suppose that there exist an algebra map $f:H\to A$, an anti-algebra map $g:H\to A$ and elements $\rho,\sigma\in A$ such that*

$$g(h_1)\rho f(h_2)=\varepsilon(h)\rho,\quad f(h_1)\sigma g(h_2)=\varepsilon(h)\sigma,\ \ \forall\, h\in H, \tag{3.2.11}$$

$$f(X^1)\sigma g(X^2)\rho f(X^3)=1_A,\quad g(x^1)\rho f(x^2)\sigma g(x^3)=1_A. \tag{3.2.12}$$

*If $\overline{g}:H\to A$ is another anti-algebra map and $\overline{\rho}$, $\overline{\sigma}\in A$ are such that* (3.2.11) *and* (3.2.12) *hold for $f$, $\overline{g}$, $\overline{\sigma}$ and $\overline{\rho}$ as well, then there exists a unique invertible element $F\in A$ such that $\overline{\rho}=F\rho$, $\overline{\sigma}=\sigma F^{-1}$ and $\overline{g}(h)=Fg(h)F^{-1}$, for all $h\in H$.*

*Proof* If it exists then $F$ is unique since

$$F = Fg(x^1)\rho f(x^2)\sigma g(x^3) = \overline{g}(x^1)F\rho f(x^2)\sigma g(x^3) = \overline{g}(x^1)\overline{\rho} f(x^2)\sigma g(x^3).$$

Now take $F = \overline{g}(x^1)\overline{\rho} f(x^2)\sigma g(x^3)$. By (3.1.7) we have

$$\overline{g}(h_{(1,1)}x^1)\overline{\rho} f(h_{(1,2)}x^2)\sigma g(h_2x^3) = \overline{g}(x^1h_1)\overline{\rho} f(x^2h_{(2,1)})\sigma g(x^3h_{(2,2)}),$$

and since $f$ is an algebra morphism and $g,\overline{g}$ are anti-algebra morphisms, by (3.2.11) we obtain $\overline{g}(x^1)\overline{\rho} f(x^2)\sigma g(x^3)g(h) = \overline{g}(h)\ \overline{g}(x^1)\overline{\rho} f(x^2)\sigma g(x^3)$, that is, $Fg(h) = \overline{g}(h)F$, for all $h \in H$. Moreover, by (3.1.9) we have

$$\begin{aligned}\overline{g}(X^1Y_1^1x^1)&\overline{\rho} f(X^2Y_2^1x^2)\sigma g(X_1^3Y^2x^3)\rho f(X_2^3Y^3)\\ &= \overline{g}(Y^1)\overline{\rho} f(X^1Y_1^2)\sigma g(X^2Y_2^2)\rho f(X^3Y^3),\end{aligned}$$

which implies $\overline{g}(x^1)\overline{\rho} f(x^2)\sigma g(x^3)\rho = \overline{\rho} f(X^1)\sigma g(X^2)\rho f(X^3)$, by (3.2.11), (3.1.10) and (3.1.11). So by (3.2.12) we deduce that $F\rho = \overline{\rho}$.

By using (3.1.9) again we have

$$\overline{\sigma}\,\overline{g}(x^1)\overline{\rho} f(x^2)\sigma g(x^3) = f(Y^1X^1y_1^1x^1)\overline{\sigma}\,\overline{g}(Y_1^2X^2y_2^1x^2)\,\overline{\rho} f(Y_2^2X^3y^2x_1^3)\sigma g(Y^3y^3x_2^3),$$

and so $\overline{\sigma}\,\overline{g}(x^1)\overline{\rho} f(x^2)\sigma g(x^3) = f(X^1)\overline{\sigma}\,\overline{g}(X^2)\overline{\rho} f(X^3)\sigma$, again because of (3.2.11), (3.1.10) and (3.1.11). Thus $\overline{\sigma}F = \sigma$, by (3.2.12).

We show that $F$ is invertible with inverse $F^{-1} := g(x^1)\rho f(x^2)\overline{\sigma}\,\overline{g}(x^3)$. Indeed,

$$\begin{aligned}FF^{-1} &= Fg(x^1)\rho f(x^2)\overline{\sigma}\,\overline{g}(x^3)\\ &= \overline{g}(x^1)F\rho f(x^2)\overline{\sigma}\,\overline{g}(x^3)\\ &= \overline{g}(x^1)\overline{\rho} f(x^2)\overline{\sigma}\,\overline{g}(x^3) \overset{(3.2.12)}{=} 1_A,\\ F^{-1}F &= g(x^1)\rho f(x^2)\overline{\sigma}\,\overline{g}(x^3)F\\ &= g(x^1)\rho f(x^2)\overline{\sigma}Fg(x^3)\\ &= g(x^1)\rho f(x^2)\sigma g(x^3) \overset{(3.2.12)}{=} 1_A,\end{aligned}$$

finishing the proof. □

We can prove now the main result of this section.

**Theorem 3.21** *For a quasi-Hopf algebra $H$ there exists a twist $F$ on $H$ such that*

$$F\Delta(S(h))F^{-1} = (S\otimes S)(\Delta^{\rm cop}(h)), \tag{3.2.13}$$

*for all $h \in H$. Furthermore,*

$$\gamma = F\Delta(\alpha) \quad \text{and} \quad \delta = \Delta(\beta)F^{-1}. \tag{3.2.14}$$

*Proof* We apply Lemma 3.20 to $A = H\otimes H$, $f = \Delta, g = \Delta\circ S : H \to H\otimes H$, $\rho = \Delta(\alpha)$, $\sigma = \Delta(\beta)$, $\overline{g} = (S\otimes S)\circ\Delta^{\rm cop} : H \to H\otimes H$, $\overline{\rho} = \gamma$ and $\overline{\sigma} = \delta$.

The conditions in (3.2.11) for $f, g, \sigma, \rho$ as above come out as

$$\Delta(S(h_1)\alpha h_2) = \varepsilon(h)\Delta(\alpha) \quad \text{and} \quad \Delta(h_1\beta S(h_2)) = \varepsilon(h)\Delta(\beta),\ \ \forall\, h \in H,$$

which are true because of (3.2.1). The relations (3.2.12) become

$$\Delta(X^1\beta S(X^2)\alpha X^3) = 1_H \otimes 1_H \ \text{ and } \ \Delta(S(x^1)\alpha x^2\beta S(x^3)) = 1_H \otimes 1_H,$$

which hold because of (3.2.2). Note now that the relations (3.2.11) for $f, \overline{g}, \overline{\sigma}, \overline{\rho}$ as above turn into (3.2.7) and (3.2.8), respectively, while the relations (3.2.12) for this setting turn into (3.2.9) and (3.2.10). Thus we are in the position to apply the above lemma. We obtain an invertible element

$$F = \overline{g}(x^1)\overline{\rho}f(x^2)\sigma g(x^3) = (S\otimes S)(\Delta^{\text{cop}}(x^1))\gamma\Delta(x^2\beta S(x^3)) \in H\otimes H, \tag{3.2.15}$$

with inverse

$$F^{-1} = g(x^1)\rho f(x^2)\overline{\sigma}\,\overline{g}(x^3) = \Delta(S(x^1)\alpha x^2)\delta(S\otimes S)(\Delta^{\text{cop}}(x^3)), \tag{3.2.16}$$

and satisfying the relations

$$\overline{\rho} = F\rho \Leftrightarrow \gamma = F\Delta(\alpha), \quad \overline{\sigma} = \sigma F^{-1} \Leftrightarrow \delta = \Delta(\beta)F^{-1} \ \text{ and}$$
$$\overline{g}(h) = Fg(h)F^{-1} \Leftrightarrow (S\otimes S)(\Delta^{\text{cop}}(h)) = F\Delta(S(h))F^{-1}, \ \forall\, h\in H.$$

These are precisely the desired relations. The equalities $\varepsilon(F^1)F^2 = \varepsilon(F^2)F^1 = 1_H$ are an immediate consequence of the axioms of a quasi-Hopf algebra. □

The twist defined in (3.2.15) will be called the Drinfeld twist associated to $H$, and will be usually denoted by $f$, $F$, $\mathscr{F}$, etc.

**Definition 3.22** If $(H,\Delta,\varepsilon,\Phi,S,\alpha,\beta)$ and $(H',\Delta',\varepsilon',\Phi',S',\alpha',\beta')$ are quasi-Hopf algebras then $\chi : H \to H'$ is a quasi-Hopf algebra morphism if it is a quasi-bialgebra morphism such that $\chi(\alpha) = \alpha'$, $\chi(\beta) = \beta'$ and $S'\circ\chi = \chi\circ S$. We say that $\chi$ is an isomorphism of quasi-Hopf algebras if, in addition, it is a bijective map (in which case its inverse $\chi^{-1}$ is also a quasi-Hopf algebra morphism).

For a Hopf algebra $H$ its antipode can be viewed as a Hopf algebra morphism from $H^{\text{op,cop}}$ to $H$. More generally, we have the following result.

**Proposition 3.23** *If $H$ is a quasi-Hopf algebra with antipode $S$, then $S : H^{\text{op,cop}} \to H_f$ is a quasi-Hopf algebra morphism, where $H_f$ is the twisting of the quasi-Hopf algebra $H$ by the Drinfeld twist $f$ defined in* (3.2.15).

*Proof* We have $S(\alpha_{\text{op,cop}}) = S(\beta)$ and $\alpha_f = S(g^1)\alpha g^2$, where $f^{-1} = g^1\otimes g^2$ is the inverse of $f = f^1\otimes f^2$ as in (3.2.16). It follows that $\alpha_f = S(\delta^1)\alpha\delta^2$, and using the definition of $\delta = \delta^1\otimes\delta^2$ from (3.2.6) we conclude that

$$\begin{aligned}\alpha_f &= S(X_1^1x^1\beta S(X^3))\alpha X_2^1x^2\beta S(X^2x^3)\\ &\overset{(3.2.1),(3.1.11)}{=} S(x^1\beta)\alpha x^2\beta S(x^3) \overset{(3.2.2)}{=} S(\beta) = S(\alpha_{\text{op,cop}}),\end{aligned}$$

as required. Likewise, one can prove that $S(\beta_{\text{op,cop}}) = S(\alpha) = \beta_f := f^1\beta S(f^2)$.

The map $S$ respects the multiplications and comultiplications of $H^{\text{op,cop}}$ and $H_f$ because it is, by definition, an anti-algebra endomorphism of $H$, and, moreover, the condition $(S\otimes S)\circ\Delta_{H^{\rm op,cop}} = \Delta_{H_f}\circ S$ is exactly the relation (3.2.13). Thus, we only have to check that $S$ respects the two reassociators, that is,

$$(1_H\otimes f)(\mathrm{Id}_H\otimes\Delta)(f)\Phi(\Delta\otimes\mathrm{Id}_H)(f^{-1})(f^{-1}\otimes 1_H) = (S\otimes S\otimes S)(X^3\otimes X^2\otimes X^1). \tag{3.2.17}$$

To prove this equality, with the help of Theorem 3.21, we compute:

$$\begin{aligned}
&(S\otimes S\otimes S)(\Phi^{321})(f\otimes 1_H)(\Delta\otimes\mathrm{Id}_H)(f)\\
&\quad= (S\otimes S\otimes S)(\Phi^{321})(f\otimes 1_H)(\Delta\circ S\otimes S)(\Delta^{\rm cop}(x^1))\\
&\qquad\qquad (\Delta\otimes\mathrm{Id}_H)(\gamma)(\Delta\otimes\mathrm{Id}_H)(\Delta(x^2\beta S(x^3)))\\
&\quad\overset{(3.2.13)}{=} (S\otimes S\otimes S)(\Phi^{321})(S\otimes S\otimes S)\Big((\Delta^{\rm cop}\otimes\mathrm{Id}_H)(\Delta^{\rm cop}(x^1))\Big)\\
&\qquad\qquad (f\otimes 1_H)(\Delta\otimes\mathrm{Id}_H)(\gamma)(\Delta\otimes\mathrm{Id}_H)(\Delta(x^2\beta S(x^3)))\\
&\quad\overset{(3.1.7)}{=} (S\otimes S\otimes S)\Big((\mathrm{Id}_H\otimes\Delta^{\rm cop})(\Delta^{\rm cop}(x^1))\Big)(S\otimes S\otimes S)(\Phi^{321})\\
&\qquad\qquad (f\otimes 1_H)(\Delta\otimes\mathrm{Id}_H)(\gamma)(\Delta\otimes\mathrm{Id}_H)(\Delta(x^2\beta S(x^3))),
\end{aligned}$$

and similarly

$$\begin{aligned}
&(1_H\otimes f)(\mathrm{Id}_H\otimes\Delta)(f)\Phi\\
&\quad= (1_H\otimes f)(S\otimes\Delta\circ S)(\Delta^{\rm cop}(x^1))(\mathrm{Id}_H\otimes\Delta)(\gamma)(\mathrm{Id}_H\otimes\Delta)(\Delta(x^2\beta S(x^3)))\Phi\\
&\quad\underset{(3.1.7)}{\overset{(3.2.13)}{=}} (S\otimes S\otimes S)\Big((\mathrm{Id}_H\otimes\Delta^{\rm cop})(\Delta^{\rm cop}(x^1))\Big)(1_H\otimes f)\\
&\qquad\qquad (\mathrm{Id}_H\otimes\Delta)(\gamma)\Phi(\Delta\otimes\mathrm{Id}_H)(\Delta(x^2\beta S(x^3))).
\end{aligned}$$

So, to prove (3.2.17), it suffices to show that

$$(S\otimes S\otimes S)(\Phi^{321})(f\otimes 1_H)(\Delta\otimes\mathrm{Id}_H)(\gamma) = (1_H\otimes f)(\mathrm{Id}_H\otimes\Delta)(\gamma)\Phi. \tag{3.2.18}$$

Indeed, we have:

$$\begin{aligned}
&(S\otimes S\otimes S)(\Phi^{321})(f\otimes 1_H)(\Delta\otimes\mathrm{Id}_H)(\gamma)\\
&\quad= (S\otimes S\otimes S)(\Phi^{321})\Big(f^1S(x^1X^2)_1\alpha_1x^2_1X^3_{(1,1)}\otimes f^2S(x^1X^2)_2\alpha_2x^2_2X^3_{(1,2)}\\
&\qquad\qquad \otimes S(X^1)\alpha x^3X^3_2\Big)\\
&\quad\overset{(3.2.13),(3.2.14)}{=} S(x^1_2X^2_2Y^3)\gamma^1x^2_1X^3_{(1,1)}\otimes S(x^1_1X^2_1Y^2)\gamma^2x^2_2X^3_{(1,2)}\otimes S(X^1Y^1)\alpha x^3X^3_2\\
&\quad= S(Z^2z^1_2x^1_2X^2_2Y^3)\alpha Z^3z^2x^2_1X^3_{(1,1)}\otimes S(Z^1z^1_1x^1_1X^2_1Y^2)\alpha z^3x^2_2X^3_{(1,2)}\\
&\qquad \otimes S(X^1Y^1)\alpha x^3X^3_2\\
&\quad\overset{(3.1.9)}{=} S(Z^2x^1_{(1,2)}z^1_2X^2_2Y^3)\alpha Z^3x^1_2z^2T^1X^3_{(1,1)}\\
&\qquad \otimes S(Z^1x^1_{(1,1)}z^1_1X^2_1Y^2)\alpha x^2z^3_1T^2X^3_{(1,2)}\otimes S(X^1Y^1)\alpha x^3z^3_2T^3X^3_2\\
&\quad\underset{(3.2.1)}{\overset{(3.1.7)}{=}} S(Z^2(z^1X^2)_2Y^3)\alpha Z^3z^2X^3_1T^1\otimes S(x^1Z^1(z^1X^2)_1Y^2)\alpha x^2(z^3X^3_2)_1T^2\\
&\qquad \otimes S(X^1Y^1)\alpha x^3(z^3X^3_2)_2T^3\\
&\quad\overset{(3.1.9)}{=} S(Z^2(X^2_1V^2z^1_2)_2Y^3)\alpha Z^3X^2_2V^3z^2T^1\otimes S(x^1Z^1(X^2_1V^2z^1_2)_1Y^2)\alpha
\end{aligned}$$

$$
\begin{aligned}
& x^2(X^3z^3)_1T^2\otimes S(X^1V^1z^1_1Y^1)\alpha x^3(X^3z^3)_2T^3\\
\overset{(3.1.7)}{\underset{(3.2.1)}{=}}\ & S(Z^2V^2_2Y^3z^1_2)\alpha Z^3V^3z^2T^1\otimes S(x^1X^2Z^1V^2_1Y^2z^1_{(1,2)})\alpha x^2X^3_1z^3_1T^2\\
& \otimes S(X^1V^1Y^1z^1_{(1,1)})\alpha x^3X^3_2z^3_2T^3\\
\overset{(3.1.9)}{\underset{(3.2.1)}{=}}\ & S(Z^2z^1_2)\alpha Z^3z^2T^1\otimes S(x^1X^2Z^1_2z^1_{(1,2)})\alpha x^2X^3_1z^3_1T^2\\
& \otimes S(X^1Z^1_1z^1_{(1,1)})\alpha x^3X^3_2z^3_2T^3\\
=\ & S(Z^2z^1_2)\alpha Z^3z^2T^1\otimes S(Z^1_2z^1_{(1,2)})\gamma^1z^3_1T^2\otimes S(Z^1_1z^1_{(1,1)})\gamma^2z^3_2T^3\\
\overset{(3.2.13),(3.2.14)}{=}\ & S(Z^2z^1_2)\alpha Z^3z^2T^1\otimes f\Delta(S(Z^1z^1_1)\alpha z^3)(T^2\otimes T^3)\\
=\ & (1_H\otimes f)(\gamma^1\otimes\Delta(\gamma^2))\Phi=(1_H\otimes f)(\mathrm{Id}_H\otimes\Delta)(\gamma)\Phi,
\end{aligned}
$$

as desired. So our proof is complete. □

We next point out that one of the two conditions in (3.2.2) is redundant.

**Proposition 3.24** *Let $H$ be a quasi-bialgebra and $(S,\alpha,\beta)$ a triple consisting of an anti-algebra endomorphism $S$ of $H$ and elements $\alpha,\beta\in H$ such that the conditions in* (3.2.1) *are satisfied. Denote $a:=X^1\beta S(X^2)\alpha X^3$ and $b:=S(x^1)\alpha x^2\beta S(x^3)$. Then $a$ is a central element of $H$ and $b$ commutes with the elements in the image of $S$. Consequently, if $b=1_H$ then $a=1_H$. The converse is also true when $S$ is surjective.*

*Proof* For all $h\in H$ we have

$$
ha\overset{(3.2.1)}{=}h_1X^1\beta S(h_{(2,1)}X^2)\alpha h_{(2,2)}X^3\overset{(3.1.7)}{=}X^1h_{(1,1)}\beta S(X^2h_{(1,2)})\alpha X^3h_2\overset{(3.2.1)}{=}ah.
$$

Similarly, we compute:

$$
\begin{aligned}
S(h)b&=S(x^1h)\alpha x^2\beta S(x^3)=S(x^1h_1)\alpha x^2h_{(2,1)}\beta S(x^3h_{(2,2)})\\
&=S(h_{(1,1)}x^1)\alpha h_{(1,2)}x^2\beta S(h_2x^3)=bS(h),
\end{aligned}
$$

for all $h\in H$. On the other hand, by using (3.1.9), we compute:

$$
\begin{aligned}
\alpha a&=\alpha X^1\beta S(X^2)\alpha X^3=S(X^1Y^1_1z^1y^1)\alpha X^2Y^1_2z^2y^2_1\beta S(X^3_1Y^2z^3y^2_2)\alpha X^3_2Y^3y^3\\
&=S(z^1)\alpha z^2\beta S(z^3)\alpha=b\alpha.
\end{aligned}
$$

Thus, for all $h,h'\in H$ we have $(S(h)\alpha h')a=S(h)\alpha h'a=S(h)\alpha ah'=S(h)b\alpha h'=b(S(h)\alpha h')$. Since $b$ has the form $\sum_i S(h_i)\alpha h'_i$ for some suitable families of elements $h_i,h'_i\in H$, we deduce that $ba=b^2$, and so if $b=1_H$ then clearly $a=1_H$, too.

If $S$ is surjective then $(h\alpha h')a=b(h\alpha h')$, for all $h,h'\in H$. Both $a$ and $b$ can be written as $\sum_i h_i\alpha h'_i$, for some $h_i,h'_i\in H$. Consequently, $a$ and $b$ are central elements of $H$ such that $a^2=b^2=ab$. It is clear now that $a=1_H$ if and only if $b=1_H$. □

We end this section by recording some formulas that will be intensively used from now on. They involve the following four elements of $H\otimes H$:

$$
p_R=x^1\otimes x^2\beta S(x^3),\quad q_R=X^1\otimes S^{-1}(\alpha X^3)X^2, \tag{3.2.19}
$$

$$
p_L=X^2S^{-1}(X^1\beta)\otimes X^3,\quad q_L=S(x^1)\alpha x^2\otimes x^3. \tag{3.2.20}
$$

In what follows we will denote these elements by

$$\begin{aligned}
p_R &= p^1 \otimes p^2 = P^1 \otimes P^2 = \cdots,\\
q_R &= q^1 \otimes q^2 = Q^1 \otimes Q^2 = \cdots,\\
p_L &= \tilde{p}^1 \otimes \tilde{p}^2 = \tilde{P}^1 \otimes \tilde{P}^2 = \cdots,\\
q_L &= \tilde{q}^1 \otimes \tilde{q}^2 = \tilde{Q}^1 \otimes \tilde{Q}^2 = \cdots.
\end{aligned}$$

Note that in the definitions of $q_R$ and $p_L$ it is implicitly understood that the antipode $S$ of $H$ is assumed to be bijective.

Note also that $p_R$ and $q_R$ in $H^{\mathrm{cop}}$ are exactly $(p_L)_{21} := \tilde{p}^2 \otimes \tilde{p}^1$ and $(q_L)_{21} := \tilde{q}^2 \otimes \tilde{q}^1$ in $H$, respectively, while in $H^{\mathrm{op}}$ they are exactly $q_R$ and $p_R$ in $H$, respectively.

The proof of the equations below is postponed until Section 4.3 where we will show that more general formulas hold. As before we denote by $f = f^1 \otimes f^2$ and $f^{-1} = g^1 \otimes g^2$ the Drinfeld twist and its inverse, defined in (3.2.15) and (3.2.16).

**Proposition 3.25** *If $H$ is a quasi-Hopf algebra with bijective antipode, then the following relations hold:*

$$\Delta(h_1)p_R[1_H \otimes S(h_2)] = p_R[h \otimes 1_H], \quad [1_H \otimes S^{-1}(h_2)]q_R\Delta(h_1) = [h \otimes 1_H]q_R, \tag{3.2.21}$$

$$\Delta(h_2)p_L[S^{-1}(h_1) \otimes 1_H] = p_L[1_H \otimes h], \quad [S(h_1) \otimes 1_H]q_L\Delta(h_2) = [1_H \otimes h]q_L, \tag{3.2.22}$$

$$\Delta(q^1)p_R[1_H \otimes S(q^2)] = 1_H \otimes 1_H, \quad [1_H \otimes S^{-1}(p^2)]q_R\Delta(p^1) = 1_H \otimes 1_H, \tag{3.2.23}$$

$$[S(\tilde{p}^1) \otimes 1_H]q_L\Delta(\tilde{p}^2) = 1_H \otimes 1_H, \quad \Delta(\tilde{q}^2)p_L[S^{-1}(\tilde{q}^1) \otimes 1_H] = 1_H \otimes 1_H, \tag{3.2.24}$$

$$\begin{aligned}
&X^1 p^1_1 P^1 \otimes X^2 p^1_2 P^2 \otimes X^3 p^2\\
&\qquad = x^1_1 p^1 \otimes x^1_{(1,2)} p^2_1 g^1 S(x^3) \otimes x^1_{(2,2)} p^2_2 g^2 S(x^2),
\end{aligned} \tag{3.2.25}$$

$$\begin{aligned}
&q^1 Q^1_1 x^1 \otimes q^2 Q^1_2 x^2 \otimes Q^2 x^3\\
&\qquad = q^1 X^1_1 \otimes S^{-1}(f^2 X^3) q^2_1 X^1_{(2,1)} \otimes S^{-1}(f^1 X^2) q^2_2 X^1_{(2,2)},
\end{aligned} \tag{3.2.26}$$

$$\begin{aligned}
&x^1 \tilde{p}^1 \otimes x^2 \tilde{p}^2_1 \tilde{P}^1 \otimes x^3 \tilde{p}^2_2 \tilde{P}^2\\
&\qquad = X^3_{(1,1)} \tilde{p}^1_1 S^{-1}(X^2 g^2) \otimes X^3_{(1,2)} \tilde{p}^1_2 S^{-1}(X^1 g^1) \otimes X^3_2 \tilde{p}^2,
\end{aligned} \tag{3.2.27}$$

$$\begin{aligned}
&\tilde{Q}^1 X^1 \otimes \tilde{q}^1 \tilde{Q}^2_1 X^2 \otimes \tilde{q}^2 \tilde{Q}^2_2 X^3\\
&\qquad = S(x^2) f^1 \tilde{q}^1_1 x^3_{(1,1)} \otimes S(x^1) f^2 \tilde{q}^1_2 x^3_{(1,2)} \otimes \tilde{q}^2 x^3_2.
\end{aligned} \tag{3.2.28}$$

Let $H$ be a quasi-Hopf algebra and $F \in H \otimes H$ a gauge transformation. We denote by $p^F_R$, $q^F_R$, $p^F_L$, $q^F_L$ the elements defined by (3.2.19) and (3.2.20) corresponding to the quasi-Hopf algebra $H_F$. Then, by using the explicit formulas for the structure of $H_F$ presented above, one can easily see that these elements are given by the following formulas (we denote $F = \mathscr{F}^1 \otimes \mathscr{F}^2$ and $F^{-1} = G^1 \otimes G^2$):

$$p^F_R = F\Delta(\mathscr{F}^1)p_R(1_H \otimes S(\mathscr{F}^2)), \tag{3.2.29}$$

$$q^F_R = (1_H \otimes S^{-1}(G^2))q_R\Delta(G^1)F^{-1}, \tag{3.2.30}$$

$$p^F_L = F\Delta(\mathscr{F}^2)p_L(S^{-1}(\mathscr{F}^1) \otimes 1_H), \tag{3.2.31}$$

$$q^F_L = (S(G^1) \otimes 1_H)q_L\Delta(G^2)F^{-1}. \tag{3.2.32}$$

## 3.3 Examples of Quasi-bialgebras and Quasi-Hopf Algebras

The aim of this section is to present some examples of quasi-Hopf algebras that are not twist equivalent to a Hopf algebra. A first candidate can be built out of a commutative bialgebra $H$, viewed as a quasi-bialgebra via a reassociator $\Phi$ which is not of the form $(1_H \otimes F)(\mathrm{Id}_H \otimes \Delta)(F)(\Delta \otimes \mathrm{Id}_H)(F^{-1})(F^{-1} \otimes 1_H)$, for some gauge transformation $F$ on $H$. Concretely, we have the following simple example.

**Example 3.26** For $k$ a field of characteristic different from 2 let $H(2)$ be the two-dimensional Hopf group algebra generated by the grouplike element $g$ such that $g^2 = 1$. It can be also viewed as a quasi-Hopf algebra with reassociator

$$\Phi = 1 \otimes 1 \otimes 1 - 2p_- \otimes p_- \otimes p_-,$$

and antipode defined by $S(g) = g$ and distinguished elements $\alpha = g$ and $\beta = 1$. Here $p_- = \frac{1}{2}(1-g)$.

*Proof* As a Hopf algebra $H(2)$ is $k[C_2]$, the group Hopf algebra associated to the cyclic group of order 2, $C_2$. If we assume $C_2 = \langle g \rangle$ then $k[C_2]$ is the $k$-algebra generated by $g$ with relation $g^2 = 1$, while the coalgebra structure is given by stating that $g$ is a grouplike element, that is,

$$\Delta(g) = g \otimes g \quad \text{and} \quad \varepsilon(g) = 1.$$

Since $H(2)$ is commutative it can be also considered as a quasi-bialgebra via any invertible element $\Phi \in H(2)^{\otimes 3}$ satisfying (3.1.9) and (3.1.10). Thus, to conclude that $H(2)$ is a quasi-bialgebra it is enough to check that $\Phi$ defined in the statement obeys these two conditions. In order to prove (3.1.9) first denote $p_+ = \frac{1}{2}(1+g)$ and observe that

$$\begin{aligned}\Delta(p_-) &= \frac{1}{2}(1 \otimes 1 - g \otimes g)\\ &= \frac{1}{2}((p_+ + p_-) \otimes (p_+ + p_-) - (p_+ - p_-) \otimes (p_+ - p_-))\\ &= p_+ \otimes p_- + p_- \otimes p_+.\end{aligned}$$

It can be easily checked that $\{p_+, p_-\}$ is a pair of orthogonal idempotents in $H(2)$, that is, $p_\pm^2 = p_\pm$ and $p_- p_+ = p_+ p_- = 0$. These remarks allow us to compute (for simplicity we denote $1 \otimes 1 \otimes 1 \otimes 1$ by 1):

$$\begin{aligned}&(1 \otimes \Phi)(\mathrm{Id} \otimes \Delta \otimes \mathrm{Id})(\Phi)(\Phi \otimes 1)\\ &= (1 - 2 \otimes p_- \otimes p_- \otimes p_-)(1 - 2p_- \otimes p_+ \otimes p_- \otimes p_- - 2p_- \otimes p_- \otimes p_+ \otimes p_-)\\ &\quad (1 - 2p_- \otimes p_- \otimes p_- \otimes 1)\\ &= (1 - 2p_- \otimes p_+ \otimes p_- \otimes p_- - 2p_- \otimes p_- \otimes p_+ \otimes p_- - 2 \otimes p_- \otimes p_- \otimes p_-)\\ &\quad (1 - 2p_- \otimes p_- \otimes p_- \otimes 1)\\ &= 1 - 2p_- \otimes p_+ \otimes p_- \otimes p_- - 2p_- \otimes p_- \otimes p_+ \otimes p_- - 2 \otimes p_- \otimes p_- \otimes p_-\\ &\quad - 2p_- \otimes p_- \otimes p_- \otimes 1 + 4p_- \otimes p_- \otimes p_- \otimes p_-\end{aligned}$$

$$\begin{aligned}
&= 1-2p_-\otimes g\otimes p_-\otimes p_- -2p_-\otimes p_-\otimes g\otimes p_- -2\otimes p_-\otimes p_-\otimes p_-\\
&\quad -2p_-\otimes p_-\otimes p_-\otimes 1\\
&= 1-2p_+\otimes p_-\otimes p_-\otimes p_- -2p_-\otimes p_+\otimes p_-\otimes p_- -2p_-\otimes p_-\otimes p_-\otimes p_+\\
&\quad -2p_-\otimes p_-\otimes p_+\otimes p_-\\
&= (1-2p_-\otimes p_-\otimes p_+\otimes p_- -2p_-\otimes p_-\otimes p_-\otimes p_+)\\
&\quad (1-2p_+\otimes p_-\otimes p_-\otimes p_- -2p_-\otimes p_+\otimes p_-\otimes p_-)\\
&= (\mathrm{Id}\otimes\mathrm{Id}\otimes\Delta)(\Phi)(\Delta\otimes\mathrm{Id}\otimes\mathrm{Id})(\Phi),
\end{aligned}$$

as required. Since $\varepsilon(p_-)=0$ it follows that (3.1.10) is satisfied as well. Thus $H(2)$ is a quasi-bialgebra. It is, moreover, a quasi-Hopf algebra, since $S(g)gg=g=\varepsilon(g)g$, $gS(g)=1=\varepsilon(g)1$,

$$X^1\beta S(X^2)\alpha X^3=\alpha-2p_-^2\alpha p_-=g-2p_-gp_-=g+2p_-=1$$

and

$$S(x^1)\alpha x^2\beta S(x^3)=\alpha-2p_-\alpha p_-^2=g-2p_-gp_-=g+2p_-=1,$$

where we used the fact that $\Phi^{-1}=\Phi$.

Assume now that there is a gauge transformation $F=F^1\otimes F^2$ on $H(2)$ such that

$$\Phi=(1\otimes F)(\mathrm{Id}\otimes\Delta)(F)(\Delta\otimes\mathrm{Id})(F^{-1})(F^{-1}\otimes 1).$$

Since $\{p_+,p_-\}$ is a basis of $H(2)$ as well we can write $F$ in the form

$$F=ap_+\otimes p_+ +bp_+\otimes p_- +cp_-\otimes p_+ +dp_-\otimes p_-,$$

for some scalars $a,b,c,d\in k$. Since we have $\varepsilon(p_-)=0$, $\varepsilon(p_+)=1$ and $\varepsilon(F^1)F^2=1=\varepsilon(F^2)F^1$ we get $a=b=c=1$, and so

$$F=p_+\otimes p_+ +p_+\otimes p_- +p_-\otimes p_+ +dp_-\otimes p_- =p_+\otimes 1+p_-\otimes p_+ +dp_-\otimes p_-,$$

for some $d\in k$. One can easily see that $F$ as above is invertible in $H(2)\otimes H(2)$ if and only if $d$ is non-zero, and in this case

$$F^{-1}=p_+\otimes 1+p_-\otimes p_+ +d^{-1}p_-\otimes p_-.$$

By similar computations we get $\Delta(p_+)=p_+\otimes p_+ +p_-\otimes p_-$, and therefore

$$\begin{aligned}
&(1\otimes F)(\mathrm{Id}\otimes\Delta)(F)\\
&\quad =(1\otimes p_+\otimes 1+1\otimes p_-\otimes p_+ +d\otimes p_-\otimes p_-)(p_+\otimes 1\otimes 1\\
&\qquad +p_-\otimes p_+\otimes p_+ +p_-\otimes p_-\otimes p_- +dp_-\otimes p_+\otimes p_- +dp_-\otimes p_-\otimes p_+)\\
&\quad =p_+\otimes p_+\otimes 1+p_-\otimes p_+\otimes p_+ +dp_-\otimes p_+\otimes p_- +p_+\otimes p_-\otimes p_+\\
&\qquad +dp_-\otimes p_-\otimes p_+ +dp_+\otimes p_-\otimes p_- +dp_-\otimes p_-\otimes p_-.
\end{aligned}$$

Likewise, we compute:

$$(\Delta\otimes\mathrm{Id})(F^{-1})(F^{-1}\otimes 1)$$

$$
\begin{aligned}
&= (p_+ \otimes p_+ \otimes 1 + p_- \otimes p_- \otimes 1 + p_+ \otimes p_- \otimes p_+ \\
&\qquad + p_- \otimes p_+ \otimes p_+ + d^{-1} p_+ \otimes p_- \otimes p_- + d^{-1} p_- \otimes p_+ \otimes p_-) \\
&\qquad (p_+ \otimes 1 \otimes 1 + p_- \otimes p_+ \otimes 1 + d^{-1} p_- \otimes p_- \otimes 1) \\
&= p_+ \otimes p_+ \otimes 1 + d^{-1} p_- \otimes p_- \otimes 1 + p_+ \otimes p_- \otimes p_+ + p_- \otimes p_+ \otimes p_+ \\
&\qquad + d^{-1} p_+ \otimes p_- \otimes p_- + d^{-1} p_- \otimes p_+ \otimes p_-.
\end{aligned}
$$

Hence

$$
\begin{aligned}
\Phi &= p_+ \otimes p_+ \otimes 1 + p_- \otimes p_+ \otimes p_+ + p_- \otimes p_+ \otimes p_- + p_+ \otimes p_- \otimes p_+ \\
&\qquad + p_- \otimes p_- \otimes p_+ + p_+ \otimes p_- \otimes p_- + p_- \otimes p_- \otimes p_- \\
&= p_+ \otimes p_+ \otimes 1 + p_- \otimes p_+ \otimes 1 + p_+ \otimes p_- \otimes 1 + p_- \otimes p_- \otimes 1 \\
&= p_+ \otimes 1 \otimes 1 + p_- \otimes 1 \otimes 1 = 1 \otimes 1 \otimes 1,
\end{aligned}
$$

which is clearly a contradiction. Thus $H(2)$ is a quasi-Hopf algebra that does not come from a Hopf algebra twisted by a gauge transformation. □

Infinite-dimensional quasi-Hopf algebras can be constructed as follows.

**Example 3.27** Let $k$ be a field that contains a primitive fourth root of unity $i$ (in particular the characteristic of $k$ is not 2), and consider $H_\pm(\infty)$ as being the $k$-algebra generated by $g$, $x$ with relations $g^2 = 1$ and $xg = -gx$. Then $H_\pm(\infty)$ are quasi-Hopf algebras with comultiplication defined by

$$
\begin{aligned}
&\Delta(g) = g \otimes g, \ \ \varepsilon(g) = 1, \\
&\Delta(x) = x \otimes (p_+ \pm i p_-) + 1 \otimes p_+ x + g \otimes p_- x, \ \ \varepsilon(x) = 0,
\end{aligned}
$$

reassociator given by $\Phi = 1 \otimes 1 \otimes 1 - 2p_- \otimes p_- \otimes p_-$, antipode determined by $S(g) = g$, $S(x) = -x(p_+ \pm i p_-)$, and distinguished elements $\alpha = g$ and $\beta = 1$. We denoted as before $p_\pm = \frac{1}{2}(1 \pm g)$.

*Proof* Observe that $H_\pm(\infty)$ contain $H(2)$, which is a quasi-Hopf algebra via the structure defined above. So to prove that $H_\pm(\infty)$ are quasi-bialgebras it is sufficient to show that

$$
\Phi(\Delta \otimes \mathrm{Id})(\Delta(x)) = (\mathrm{Id} \otimes \Delta)(\Delta(x))\Phi,
$$

and that $\Delta$, extended as an algebra morphism from $H_\pm(\infty)$ to $H_\pm(\infty) \otimes H_\pm(\infty)$, behaves well with respect to the relation $xg = -gx$ (the other details are immediate).

It is easy to see that

$$
\begin{aligned}
\Delta(p_+ x) &= p_+ x \otimes p_+ + p_+ \otimes p_+ x \pm i p_- x \otimes p_- - p_- \otimes p_- x, \\
\Delta(p_- x) &= \pm i p_+ x \otimes p_- + p_+ \otimes p_- x + p_- x \otimes p_+ + p_- \otimes p_+ x, \\
\Delta(p_+ \pm i p_-) &= p_+ \otimes (p_+ \pm i p_-) + p_- \otimes (p_- \pm i p_+),
\end{aligned}
$$

and so

$$
(\mathrm{Id} \otimes \Delta)(\Delta(x)) = x \otimes (p_+ \pm i p_-) \otimes p_+ \pm i x \otimes p_+ \otimes p_- + x \otimes p_- \otimes p_-
$$

$$+1\otimes p_+x\otimes p_+ + 1\otimes p_+\otimes p_+x \pm i1\otimes p_-x\otimes p_- - 1\otimes p_-\otimes p_-x \pm ig\otimes p_+x\otimes p_-$$
$$+g\otimes p_+\otimes p_-x + g\otimes p_-x\otimes p_+ + g\otimes p_-\otimes p_+x.$$

From here we compute:

$$\begin{aligned}
&(\mathrm{Id}\otimes\Delta)(\Delta(x))\Phi\\
&= (\mathrm{Id}\otimes\Delta)(\Delta(x)) - 2(\mathrm{Id}\otimes\Delta)(\Delta(x))p_-\otimes p_-\otimes p_-\\
&= (\mathrm{Id}\otimes\Delta)(\Delta(x)) - 2(p_+x\otimes p_-\otimes p_- \mp ip_-\otimes p_+x\otimes p_- - p_-\otimes p_-\otimes p_+x)\\
&= x\otimes(p_+ \pm ip_-)\otimes p_+ \pm ix\otimes p_+\otimes p_- - gx\otimes p_-\otimes p_- \pm i1\otimes x\otimes p_-\\
&\quad + 1\otimes p_-\otimes gx + 1\otimes p_+x\otimes p_+ + 1\otimes p_+\otimes p_+x + g\otimes p_+\otimes p_-x + g\otimes p_-x\otimes p_+\\
&= x\otimes(p_+ \pm ip_-)\otimes p_+ \pm ix\otimes p_+\otimes p_- - gx\otimes p_-\otimes p_- \pm i1\otimes x\otimes p_-\\
&\quad + 1\otimes p_+x\otimes p_+ + 1\otimes 1\otimes p_+x + g\otimes p_+\otimes p_-x - 1\otimes p_-\otimes p_-x + g\otimes p_-x\otimes p_+.
\end{aligned}$$

Now, we have

$$\begin{aligned}
(\Delta\otimes\mathrm{Id})(\Delta(x)) &= x\otimes(p_+ \pm ip_-)\otimes(p_+ \pm ip_-) + 1\otimes p_+x\otimes(p_+ \pm ip_-)\\
&\quad + g\otimes p_-x\otimes(p_+ \pm ip_-) + 1\otimes 1\otimes p_+x + g\otimes g\otimes p_-x,
\end{aligned}$$

which implies

$$\begin{aligned}
&\Phi(\Delta\otimes\mathrm{Id})(\Delta(x))\\
&= (\Delta\otimes\mathrm{Id})(\Delta(x)) - 2(p_-\otimes p_-\otimes p_-)(\Delta\otimes\mathrm{Id})(\Delta(x))\\
&= (\Delta\otimes\mathrm{Id})(\Delta(x)) + 2(p_-x\otimes p_-\otimes p_- \pm ip_-\otimes p_-x\otimes p_- - p_-\otimes p_-\otimes p_-x)\\
&= x\otimes(p_+ \pm ip_-)\otimes p_+ \pm ix\otimes p_+\otimes p_- + 1\otimes p_+x\otimes(p_+ \pm ip_-) + g\otimes p_-x\otimes p_+\\
&\quad + 1\otimes 1\otimes p_+x + g\otimes p_+\otimes p_-x - gx\otimes p_-\otimes p_- \pm i1\otimes p_-x\otimes p_- - 1\otimes p_-\otimes p_-x\\
&= x\otimes(p_+ \pm ip_-)\otimes p_+ \pm ix\otimes p_+\otimes p_- + 1\otimes p_+x\otimes p_+ + g\otimes p_-x\otimes p_+\\
&\quad + 1\otimes 1\otimes p_+x + g\otimes p_+\otimes p_-x - gx\otimes p_-\otimes p_- \pm i1\otimes x\otimes p_- - 1\otimes p_-\otimes p_-x.
\end{aligned}$$

By the two equalities above we conclude that $\Phi(\Delta\otimes\mathrm{Id})(\Delta(x)) = (\mathrm{Id}\otimes\Delta)(\Delta(x))\Phi$, so the first assertion is proved. To prove the second one we have to verify that $(g\otimes g)\Delta(x) = -\Delta(x)(g\otimes g)$. This follows easily from the equalities $gx = -xg$, $gp_+ = p_+$ and $gp_- = -p_-$, we leave the details to the reader.

Finally, $H_\pm(\infty)$ are quasi-Hopf algebras since

$$\begin{aligned}
S(x_1)gx_2 &= S(x)g(p_+ \pm ip_-) + gp_+x + S(g)gp_-x\\
&= -x(p_+ \pm ip_-)g(p_+ \pm ip_-) + p_+x + p_-x\\
&= -x(p_+ \mp ip_-)(p_+ \pm ip_-) + x\\
&= -x(p_+ + p_-) + x = -x + x = 0 = \varepsilon(x)g,
\end{aligned}$$

and similarly

$$\begin{aligned}
x_1S(x_2) &= xS(p_+ \pm ip_-) + S(x)S(p_+) + gS(x)S(p_-)\\
&= x(p_+ \pm ip_-) - x(p_+ \pm ip_-)p_+ - gx(p_+ \pm ip_-)p_-
\end{aligned}$$

$$= x(p_+ \pm ip_-) - xp_+ \mp igxp_-$$
$$= \pm ixp_- \pm ixgp_- = \pm ixp_- \mp ixp_- = 0 = \varepsilon(x)1.$$

The remaining relations hold because they hold in $H(2)$. □

Eight-dimensional quasi-Hopf algebras can be obtained from $H_\pm(\infty)$ by factorizing it through a quasi-Hopf ideal.

**Definition 3.28** If $H$ is a quasi-Hopf algebra then an ideal $I$ of $H$ is called a quasi-Hopf ideal if $\Delta(I) \subseteq I \otimes H + H \otimes I$, $\varepsilon(I) = 0$ and $S(I) \subseteq I$.

**Proposition 3.29** *If $I$ is a quasi-Hopf ideal of a quasi-Hopf algebra $H$ then the quotient algebra $H/I$ has a unique quasi-Hopf algebra structure such that the canonical surjection $p : H \to H/I$ becomes a quasi-Hopf algebra morphism.*

*Proof* From $\Delta(I) \subseteq I \otimes H + H \otimes I$ it follows that $(p \otimes p)\Delta(I) = 0$, and so, by the Universal Property of a quotient algebra, there exists a unique algebra morphism $\widehat{\Delta} : H/I \to H/I \otimes H/I$ such that $\widehat{\Delta} \circ p = (p \otimes p) \circ \Delta$. Similarly, from $\varepsilon(I) = 0$ it follows that there is a unique algebra morphism $\widehat{\varepsilon} : H/I \to k$ such that $\widehat{\varepsilon} \circ p = \varepsilon$, and $S(I) \subseteq I$ guarantees the existence and uniqueness of the anti-algebra morphism $\widehat{S} : H/I \to H/I$ obeying $\widehat{S} \circ p = p \circ S$. It is clear at this point that $(H/I, \widehat{\Delta}, \widehat{\varepsilon}, (p \otimes p \otimes p)(\Phi), \widehat{S}, p(\alpha), p(\beta))$ is the unique quasi-Hopf algebra structure on $H/I$ for which $p : H \to H/I$ becomes a quasi-Hopf algebra morphism. □

**Example 3.30** Consider $k$ a field that contains a primitive fourth root of unity $i$ and let $H_\pm(8)$ be the unital algebras generated by $g, x$ with relations $g^2 = 1$, $x^4 = 0$ and $gx = -xg$, and endowed with the (non-coassociative) comultiplication given by

$$\Delta(g) = g \otimes g, \quad \varepsilon(g) = 1,$$
$$\Delta(x) = x \otimes (p_+ \pm ip_-) + 1 \otimes p_+x + g \otimes p_-x, \quad \varepsilon(x) = 0,$$

where $p_\pm = \frac{1}{2}(1 \pm g)$. Then $H_\pm(8)$ are eight-dimensional quasi-Hopf algebras with reassociator $\Phi = 1 \otimes 1 \otimes 1 - 2p_- \otimes p_- \otimes p_-$, antipode defined by $S(g) = g$ and $S(x) = -x(p_+ \pm ip_-)$, and distinguished elements $\alpha = g$ and $\beta = 1$.

*Proof* By the above results all we have to check is the fact that the ideal generated by $x^4$ in $H_\pm(\infty)$ is a quasi-Hopf ideal. Then $H_\pm(8)$ are nothing else than the quotient quasi-Hopf algebras $H_\pm(\infty)/(x^4)$. Therefore, we must check that $\Delta(x^4) \in (x^4) \otimes H_\pm(\infty) + H_\pm(\infty) \otimes (x^4)$, $\varepsilon(x^4) = 0$ and $S(x^4) \in (x^4)$. Indeed, a straightforward computation ensures that

$$\Delta(x^2) = x^2 \otimes g + (1 \pm i)(p_+x \otimes p_+x + p_-x \otimes p_+x)$$
$$+ (1 \mp i)(p_+x \otimes p_-x - p_-x \otimes p_-x) + g \otimes x^2,$$

from which we obtain by calculation that $\Delta(x^4) = x^4 \otimes 1 + 1 \otimes x^4$. Since $\varepsilon$ is extended to the whole $H_\pm(\infty)$ as an algebra map and $\varepsilon(x) = 0$ we get that $\varepsilon(x^4) = 0$, and since

$$S(x^2) = x(p_+ \pm ip_-)x(p_+ \pm ip_-)$$

$$= x^2(p_- \pm ip_+)(p_+ \pm ip_-) = x^2(\pm ip_- \pm ip_+) = \pm ix^2$$

it follows that $S(x^4) = S(x^2)^2 = -x^4$. Thus our claim is proved. It is easy to see that $\{g^i x^j \mid 0 \le i \le 1,\ 0 \le j \le 3\}$ is a basis for $H_\pm(8)$, so $H_\pm(8)$ are eight-dimensional quasi-Hopf algebras. □

A 32-dimensional quasi-Hopf algebra can be constructed by amalgamating the quasi-Hopf algebras $H_+(8)$ and $H_-(8)$, along the quasi-Hopf algebra $H(2)$. To this end we first construct from $H_+(8)$ and $H_-(8)$ an infinite-dimensional quasi-Hopf algebra $H_{+-}(\infty)$ as follows.

The proof of the result below is based on the computations performed in Example 3.27, and we leave it to the reader.

**Example 3.31** Denote by $H_{+-}(\infty)$ the $k$-algebra generated by $g, x, y$ with relations $g^2 = 1, x^4 = y^4 = 0, gx = -xg$ and $gy = -yg$. Then $H_{+-}(\infty)$ is an infinite-dimensional quasi-Hopf algebra with structure determined by

$$\begin{aligned}
&\Delta(g) = g \otimes g, \ \ \varepsilon(g) = 1,\\
&\Delta(x) = x \otimes (p_+ + ip_-) + 1 \otimes p_+x + g \otimes p_-x, \ \ \varepsilon(x) = 0,\\
&\Delta(y) = y \otimes (p_+ - ip_-) + 1 \otimes p_+y + g \otimes p_-y, \ \ \varepsilon(y) = 0,\\
&S(g) = g, \quad S(x) = -x(p_+ + ip_-), \quad S(y) = -y(p_+ - ip_-).
\end{aligned}$$

Its reassociator is $\Phi = 1 \otimes 1 \otimes 1 - 2p_- \otimes p_- \otimes p_-$, and the two distinguished elements of $H_{+-}(\infty)$ are $\alpha = g$ and $\beta = 1$.

Factorizing $H_{+-}(\infty)$ through a certain quasi-Hopf ideal we obtain the following 32-dimensional quasi-Hopf algebra.

**Example 3.32** Assume again that $k$ is a field which contains a primitive fourth root of unity $i$. By $H(32)$ denote the unital $k$-algebra generated by $g$, $x$ and $y$ with relations

$$g^2 = 1, \ \ x^4 = y^4 = 0, \ \ gx = -xg, \ \ gy = -yg \ \text{ and } \ yx = ixy.$$

If we endow $H(32)$ with the comultiplication and counit defined by the formulas

$$\begin{aligned}
&\Delta(g) = g \otimes g, \ \ \varepsilon(g) = 1,\\
&\Delta(x) = x \otimes (p_+ + ip_-) + 1 \otimes p_+x + g \otimes p_-x, \ \ \varepsilon(x) = 0,\\
&\Delta(y) = y \otimes (p_+ - ip_-) + 1 \otimes p_+y + g \otimes p_-y, \ \ \varepsilon(y) = 0,
\end{aligned}$$

where $p_\pm = \frac{1}{2}(1 \pm g)$, then with these structures $H(32)$ is a 32-dimensional quasi-Hopf algebra with reassociator $\Phi = 1 \otimes 1 \otimes 1 - 2p_- \otimes p_- \otimes p_-$. Its antipode is determined by

$$S(g) = g, \ \ S(x) = -x(p_+ + ip_-) \ \text{ and } \ S(y) = -y(p_+ - ip_-),$$

and the distinguished elements are $\alpha = g$ and $\beta = 1$.

*Proof* We will prove that the ideal generated by $z := xy + iyx$ in $H_{+-}(\infty)$ is a quasi-Hopf ideal. Then the proof ends because of the identification of $H(32)$ with the quotient quasi-Hopf algebra $H_{+-}(\infty)/(z)$. To this end we compute:

$$\begin{aligned}\Delta(xy)&\\ &= [x\otimes(p_+ + ip_-) + 1\otimes p_+x + g\otimes p_-x][y\otimes(p_+ - ip_-) + 1\otimes p_+y + g\otimes p_-y]\\ &= xy\otimes 1 + x\otimes p_+y + ixg\otimes p_-y - iy\otimes p_+x + g\otimes p_+xy + gy\otimes p_-x + g\otimes p_-xy,\end{aligned}$$

and similarly

$$\begin{aligned}\Delta(yx) = yx\otimes 1 + y\otimes p_+x - iyg\otimes p_-x + ix\otimes p_+y&\\ + g\otimes p_+yx + gx\otimes p_-y + g\otimes p_-yx.&\end{aligned}$$

We conclude that

$$\begin{aligned}\Delta(z) &= \Delta(xy) + i\Delta(yx)\\ &= xy\otimes 1 + g\otimes p_+xy + g\otimes p_-xy + iyx\otimes 1 + ig\otimes p_+yx + ig\otimes p_-yx\\ &= z\otimes 1 + g\otimes p_-z + g\otimes p_+z\\ &= z\otimes 1 + g\otimes z \in (z)\otimes H + H\otimes(z),\end{aligned}$$

as required. By $\varepsilon(x) = \varepsilon(y) = 0$ we deduce $\varepsilon(z) = 0$, and since

$$\begin{aligned}S(z) &= S(y)S(x) + iS(x)S(y)\\ &= y(p_+ - ip_-)x(p_+ + ip_-) + ix(p_+ + ip_-)y(p_+ - ip_-)\\ &= yx(p_- - ip_+)(p_+ + ip_-) + ixy(p_- + ip_+)(p_+ - ip_-)\\ &= yx(ip_- - ip_+) - ixy(ip_- - ip_+)\\ &= z(p_- - p_+) = -zg \in (z),\end{aligned}$$

we obtain that $(z)$ is a quasi-Hopf ideal in $H_{+-}(\infty)$, as stated.

Finally, it can be easily checked that $\{g^jx^sy^t \mid 0\le j\le 1,\ 0\le s\le 3,\ 0\le t\le 3\}$ is a basis in $H(32)$, and so indeed $H(32)$ is 32-dimensional. □

Other classes of quasi-bialgebras and quasi-Hopf algebras will be presented in the forthcoming chapters.

## 3.4 The Rigid Monoidal Structure of ${}_H\mathcal{M}^{\rm fd}$ and $\mathcal{M}_H^{\rm fd}$

We will describe the rigid structures of the monoidal categories ${}_H\mathcal{M}^{\rm fd}$ and $\mathcal{M}_H^{\rm fd}$, the categories of finite-dimensional left and right representations, respectively, over the quasi-Hopf algebra $H$. Then we compute the canonical isomorphisms defined in Section 1.6 for the particular case when $\mathscr{C}$ is either ${}_H\mathcal{M}^{\rm fd}$ or $\mathcal{M}_H^{\rm fd}$.

In what follows, if $V$ is a finite-dimesional vector space, if $v^*\in V^*$ and $v\in V$, we will sometimes denote $v^*(v)$ by $\langle v^*, v\rangle$.

**Proposition 3.33** *Let $H$ be a quasi-Hopf algebra with antipode $S$ and distinguished elements $\alpha, \beta \in H$. Then ${}_H\mathcal{M}^{\rm fd}$ is left rigid and $\mathcal{M}_H^{\rm fd}$ is right rigid. When $S$ is bijective, both ${}_H\mathcal{M}^{\rm fd}$ and $\mathcal{M}_H^{\rm fd}$ are rigid monoidal categories.*

*Proof* Let $V$ be a finite-dimensional left $H$-module and consider $V^* = \mathrm{Hom}_k(V,k)$ with the left $H$-action $(h \cdot v^*)(v) = v^*(S(h) \cdot v)$, for all $v^* \in V^*$, $h \in H$ and $v \in V$. By definition $S$ is an anti-algebra homomorphism of $H$, so $V^*$ is a left $H$-module via this action. If we define the linear maps

$$\mathrm{ev}_V : V^* \otimes V \to k, \quad \mathrm{ev}_V(v^* \otimes v) = v^*(\alpha \cdot v), \ \forall\, v^* \in V^*,\ v \in V,$$
$$\mathrm{coev}_V : k \to V \otimes V^*, \quad \mathrm{coev}_V(1) = \sum_i \beta \cdot v_i \otimes v^i,$$

where $\{v_i\}_i$ and $\{v^i\}_i$ are dual bases in $V$ and $V^*$, we claim that $(\mathrm{coev}_V, \mathrm{ev}_V) : V^* \dashv V$ is an adjunction in ${}_H\mathcal{M}^{\rm fd}$. Indeed, $\mathrm{ev}_V$ is $H$-linear since

$$\begin{aligned}\mathrm{ev}_V(h \cdot (v^* \otimes v)) &= \mathrm{ev}_V(h_1 \cdot v^* \otimes h_2 \cdot v) = (h_1 \cdot v^*)(\alpha h_2 \cdot v)\\ &= v^*(S(h_1)\alpha h_2 \cdot v) \overset{(3.2.1)}{=} \varepsilon(h) v^*(\alpha \cdot v) = \varepsilon(h)\mathrm{ev}_V(v^* \otimes v),\end{aligned}$$

for all $v^* \in V^*$, $h \in H$ and $v \in V$. Similarly, using the second equality in (3.2.1), one can easily see that $\mathrm{coev}_V$ is left $H$-linear, too. Also, the left-hand side in (1.6.4) becomes, for all $v \in V$,

$$\begin{aligned}&r_V \circ (\mathrm{Id}_V \otimes \mathrm{ev}_V) \circ a_{V,V^*,V} \circ (\mathrm{coev}_V \otimes \mathrm{Id}_V) \circ l_V^{-1}(v)\\ &= \sum_i r_V \circ (\mathrm{Id}_V \otimes \mathrm{ev}_V) \circ a_{V,V^*,V}((\beta \cdot v_i \otimes v^i) \otimes v)\\ &= \sum_i r_V \circ (\mathrm{Id}_V \otimes \mathrm{ev}_V)(X^1\beta \cdot v_i \otimes (X^2 \cdot v^i \otimes X^3 \cdot v))\\ &= \sum_i (X^2 \cdot v^i)(\alpha X^3 \cdot v) X^1\beta \cdot v_i = \sum_i v^i(S(X^2)\alpha X^3 \cdot v) X^1 \beta \cdot v_i\\ &= X^1\beta S(X^2)\alpha X^3 \cdot v \overset{(3.2.2)}{=} 1_H \cdot v = v,\end{aligned}$$

as required. The second equality in (3.2.2) implies (1.6.5), so we are done.

Similarly, $\mathcal{M}_H^{\rm fd}$ is right rigid. For any object $V$ of $\mathcal{M}_H^{\rm fd}$ consider ${}^*V$ as being $\mathrm{Hom}_k(V,k)$ equipped with the right $H$-module structure $({}^*v \cdot h)(v) = {}^*v(v \cdot S(h))$, for all ${}^*v \in {}^*V$, $h \in H$ and $v \in V$. Then ${}^*V$ and the linear maps

$$\mathrm{ev}'_V : V \otimes {}^*V \to k, \quad \mathrm{ev}'_V(v \otimes {}^*v) = {}^*v(v \cdot \beta), \ \forall\, {}^*v \in {}^*V,\ v \in V,$$
$$\mathrm{coev}'_V : k \to {}^*V \otimes V, \quad \mathrm{coev}'_V(1) = \sum_i v^i \otimes v_i \cdot \alpha,$$

define a right dual for $V$ in $\mathcal{M}_H^{\rm fd}$; we leave the verification of the details to the reader.

When $S$ is bijective, any $V \in {}_H\mathcal{M}^{\rm fd}$ has a right dual as well. It is defined by ${}^*V := \mathrm{Hom}_k(V,k)$ viewed as a left $H$-module via $(h \cdot {}^*v)(v) = {}^*v(S^{-1}(h) \cdot v)$, for all ${}^*v \in {}^*V$, $h \in H$ and $v \in V$, and the linear maps

$$\mathrm{ev}'_V : V \otimes {}^*V \to k, \quad \mathrm{ev}'_V(v \otimes {}^*v) = {}^*v(S^{-1}(\alpha) \cdot v), \ \forall\, {}^*v \in {}^*V,\ v \in V,$$
$$\mathrm{coev}'_V : k \to {}^*V \otimes V, \quad \mathrm{coev}'_V(1) = \sum_i v^i \otimes S^{-1}(\beta) \cdot v_i.$$

Actually, (3.2.1) implies that $\mathrm{ev}'_V$ and $\mathrm{coev}'_V$ are left $H$-linear, while (3.2.2) implies (1.6.4) and (1.6.5).

In a similar manner it can be shown that any $V \in \mathcal{M}_H^{\rm fd}$ has a left dual, if $S$ is bijective. This time the left dual of $V$ is the same object $V^* = \mathrm{Hom}_k(V,k)$, viewed as a right $H$-module via $(v^* \cdot h)(v) = v^*(v \cdot S^{-1}(h))$, for all $v^* \in V^*$, $h \in H$, $v \in V$, with the two morphisms in $\mathcal{M}_H$, $\mathrm{ev}_V : V^* \otimes V \to k$ and $\mathrm{coev}_V : k \to V \otimes V^*$, defined by

$$\mathrm{ev}_V(v^* \otimes v) = v^*(v \cdot S^{-1}(\beta)) \text{ and } \mathrm{coev}_V(1) = \sum_i v_i \cdot S^{-1}(\alpha) \otimes v^i.$$

Once more we leave the details to the reader. □

The transpose of a morphism $f : V \to Y$ in ${}_H\mathcal{M}^{\rm fd}$ or $\mathcal{M}_H^{\rm fd}$ (in the sense of Proposition 1.68) coincides with the usual transpose of $f$ in the category of $k$-vector spaces.

**Proposition 3.34** *When $\mathscr{C}$ is the category of finite-dimensional left modules over a quasi-Hopf algebra, the maps $f^*$ and, if the antipode is bijective, ${}^*f$ coincide with the usual transpose map of $f$ in ${}_k\mathcal{M}$, that is, $f^*(y^*) = y^* \circ f$ and ${}^*f({}^*y) = {}^*y \circ f$, for all $y^* \in Y^*$, ${}^*y \in {}^*Y$, respectively. A similar result holds in the situation when $\mathscr{C}$ is the category of finite-dimensional right modules over a quasi-Hopf algebra.*

*Proof* Indeed, by the definition of $f^*$ and (3.2.2), we have that

$$\begin{aligned} f^*(y^*) &= \sum_i \langle x^1 \cdot y^*, \alpha x^2 \beta \cdot f(v_i)\rangle x^3 \cdot v^i \\ &= \sum_i \langle y^*, S(x^1)\alpha x^2 \beta S(x^3) \cdot f(v_i)\rangle v^i = y^* \circ f, \end{aligned}$$

where $\{v_i, v^i\}_i$ are dual bases in $V$ and $V^*$. Similarly one can show that ${}^*f({}^*y) = {}^*y \circ f$, for all ${}^*y \in {}^*Y$. The right-handed case can be proved in a similar manner. □

For ${}_H\mathcal{M}^{\rm fd}$ we compute explicitly the isomorphism $\lambda$ defined in (1.7.1) and its right-handed version $\lambda'$ as follows.

**Proposition 3.35** *Let $H$ be a quasi-Hopf algebra and $X, Y \in {}_H\mathcal{M}^{\rm fd}$. Consider $\{x_i\}_i$ and $\{x^i\}_i$ dual bases in $X$ and $X^*$, and $\{y_j\}_j$ and $\{y^j\}_j$ dual bases in $Y$ and $Y^*$. Then the maps $\lambda_{X,Y}$ and $\lambda^{-1}_{X,Y}$ are given by the formulas*

$$\begin{aligned} \lambda_{X,Y}(\mu) &= \sum_{i,j} \langle \mu, g^1 \cdot x_i \otimes g^2 \cdot y_j\rangle y^j \otimes x^i, \\ \lambda^{-1}_{X,Y}(y^* \otimes x^*)(x \otimes y) &= \langle x^*, f^1 \cdot x\rangle \langle y^*, f^2 \cdot y\rangle, \end{aligned} \tag{3.4.1}$$

*for all $\mu \in (X \otimes Y)^*$, $x^* \in X^*$, $y^* \in Y^*$, $x \in X$ and $y \in Y$.*

*If the antipode of $H$ is bijective then the maps $\lambda'_{X,Y}$ and $\lambda'^{-1}_{X,Y}$ are given by*

$$\begin{aligned} \lambda'_{X,Y}(\nu) &= \sum_{i,j} \langle \nu, S^{-1}(g^2) \cdot x_i \otimes S^{-1}(g^1) \cdot y_j\rangle y^j \otimes x^i, \\ \lambda'^{-1}_{X,Y}({}^*y \otimes {}^*x)(x \otimes y) &= \langle {}^*x, S^{-1}(f^2) \cdot x\rangle \langle {}^*y, S^{-1}(f^1) \cdot y\rangle, \end{aligned}$$

*for all $\nu \in {}^*(X \otimes Y)$, ${}^*x \in {}^*X$, ${}^*y \in {}^*Y$, $x \in X$ and $y \in Y$. Here $f = f^1 \otimes f^2$ is the Drinfeld twist from* (3.2.15) *and $g = g^1 \otimes g^2$ is its inverse defined in* (3.2.16).

*Proof* Let $p_R = p^1 \otimes p^2$ be the element defined in (3.2.19). We compute, for all $y^* \in Y^*, x^* \in X^*, x \in X$ and $y \in Y$:

$$\begin{aligned}
&\lambda_{X,Y}^{-1}(y^* \otimes x^*)(x \otimes y)\\
&\overset{(1.7.2)}{=} \langle x^*, S(x^1X^2y_2^1)\alpha x^2(X^3y^2\beta S(y^3))_1 \cdot x\rangle\\
&\qquad\qquad \langle y^*, S(X^1y_1^1)\alpha x^3(X^3y^2\beta S(y^3))_2 \cdot y\rangle\\
&\overset{(3.2.5)}{=} \langle x^*, S(y_2^1)\gamma^1(y^2\beta S(y^3))_1 \cdot x\rangle\langle y^*, S(y_1^1)\gamma^2(y^2\beta S(y^3))_2 \cdot y\rangle\\
&\overset{(3.2.19)}{=} \langle x^*, f^1S(p^1)_1g^1\gamma^1p_1^2 \cdot x\rangle\langle y^*, f^2S(p^1)_2g^2\gamma^2p_2^2 \cdot y\rangle\\
&= \langle x^*, f^1(S(p^1)\alpha p^2)_1 \cdot x\rangle\langle y^*, f^2(S(p^1)\alpha p^2)_2 \cdot y\rangle\\
&\overset{(3.2.19),(3.2.2)}{=} \langle x^*, f^1 \cdot x\rangle\langle y^*, f^2 \cdot y\rangle,
\end{aligned}$$

where we also used the relations in Theorem 3.21 in the third and fourth equality. By using (1.7.1) or by a direct computation it is easy to see that

$$\lambda_{X,Y}(\mu) = \sum_{i,j}\langle \mu, g^1 \cdot x_i \otimes g^2 \cdot y_j\rangle y^j \otimes x^i, \text{ for all } \mu \in (X \otimes Y)^*.$$

The assertion concerning the morphism $\lambda'_{X,Y}$ can be proved in a similar way; we leave the details to the reader. □

Note that the map $\theta_M$ defined in (1.7.3) specializes for $\mathscr{C} = {}_H\mathscr{M}^{\mathrm{fd}}$ as

$$\theta_M(m) = \sum_{i=1}^{n}\langle {}^im, m\rangle {}^{*i}m \tag{3.4.2}$$

for all $m \in M$, where, if $({}_im)_{i=\overline{1,n}}$ is a basis of $M \in {}_H\mathscr{M}^{\mathrm{fd}}$ with dual basis $({}^im)_{i=\overline{1,n}}$ in $M^*$, then ${}^{*i}m$ is the image of ${}_im$ under the canonical map $M \ni m \to (m^* \mapsto m^*(m)) \in M^{**}$. Furthermore, in this case the morphism $\theta'_M$ is defined by the same formula (3.4.2) as $\theta_M$. Similar formulas can be obtained for $\mathscr{C} = \mathscr{M}_H^{\mathrm{fd}}$.

## 3.5 The Reconstruction Theorem for Quasi-Hopf Algebras

We proved in Proposition 3.3 a reconstruction type theorem for quasi-bialgebras. The purpose of this section is to provide a similar result but now for quasi-Hopf algebras. For this, we must endow a quasi-monoidal functor with an extra property. Keeping in mind the statement of Proposition 3.33, this new property comes out naturally as follows:

**Definition 3.36** Let $F : \mathscr{C} \to \mathscr{D}$ be a quasi-monoidal functor between two left rigid monoidal categories. We say that $F$ is a left rigid quasi-monoidal functor if there exist isomorphisms $F(X^\vee) \cong F(X)^*$, natural in $X \in \mathrm{Ob}(\mathscr{C})$. Here $X^\vee$ is the left dual of $X$ in $\mathscr{C}$ and $F(X)^*$ is the left dual of $F(X)$ in $\mathscr{D}$.

Let $H$ be a quasi-bialgebra and denote by ${}_H\mathscr{M}^{\mathrm{fd}}$ the category of finite-dimensional left representations over $H$. If ${}_k\mathscr{M}^{\mathrm{fd}}$ is the category of finite-dimensional $k$-vector spaces it follows that the forgetful functor $F : {}_H\mathcal{M}^{\rm fd} \to {}_k\mathcal{M}^{\rm fd}$ is a quasi-monoidal functor. We next see when $F$ is a left rigid quasi-monoidal functor.

**Lemma 3.37** *Let $H$ be a quasi-bialgebra over $k$. Then the forgetful functor $F : {}_H\mathcal{M}^{\rm fd} \to {}_k\mathcal{M}^{\rm fd}$ is a left rigid quasi-monoidal functor if and only if:*

(LR1) *For any $V \in {}_H\mathcal{M}^{\rm fd}$, the left dual $V^*$ of $V$ in ${}_k\mathcal{M}^{\rm fd}$ admits a left $H$-module structure with respect to which there exists an adjunction $({\rm coev}_V^H, {\rm ev}_V^H) : V^* \dashv V$ in ${}_H\mathcal{M}^{\rm fd}$, that is, $V^*$ with a suitable structure becomes a left dual for $V$ in ${}_H\mathcal{M}^{\rm fd}$.*

(LR2) *For any morphism $f : V \to W$ in ${}_H\mathcal{M}^{\rm fd}$ we have $f^{\surd} = f^*$, where $f^{\surd}$ is the left transpose of $f$ in ${}_H\mathcal{M}^{\rm fd}$ and $f^*$ is the left transpose of $f$ in ${}_k\mathcal{M}^{\rm fd}$.*

*Proof* The functor $F$ acts as identity on objects and morphisms. So $F(V^{\surd}) \cong F(V)^*$ reduces to the choice of $V^{\surd}$ as being $V^*$, with additional structures as in (LR1) that come from those of $V^{\surd}$ (recall that the left dual of an object is unique up to an isomorphism; see Proposition 1.66). Furthermore, given this choice we must require $f^{\surd} = f^*$, for any $f : V \to W$ in ${}_H\mathcal{M}^{\rm fd}$, and this is because the $k$-linear isomorphisms $V^{\surd} \cong V^*$ are natural in $V \in {}_H\mathcal{M}^{\rm fd}$. Hence our proof is complete. □

We now present the reconstruction theorem for quasi-Hopf algebras. Note that in the proof we need the $k$-algebra $H$ to be an object in ${}_H\mathcal{M}^{\rm fd}$, so we have to assume from the beginning that $H$ is finite dimensional.

**Theorem 3.38** *Let $H$ be a finite-dimensional algebra over a field $k$. Then there exists a bijective correspondence between*

- *quasi-Hopf algebra structures on $H$;*
- *left rigid monoidal stuctures on ${}_H\mathcal{M}^{\rm fd}$ for which the functor $F : {}_H\mathcal{M}^{\rm fd} \to {}_k\mathcal{M}^{\rm fd}$ that forgets the left $H$-action is a left rigid quasi-monoidal functor.*

*Proof* By Proposition 3.33 and Proposition 3.34 it follows that for any quasi-Hopf algebra $H$ (not necessarily finite dimensional) the forgetful functor $F : {}_H\mathcal{M}^{\rm fd} \to {}_k\mathcal{M}^{\rm fd}$ is a left rigid quasi-monoidal functor. Thus by Proposition 3.3, Corollary 3.11 and Proposition 3.5 we only have to show that a finite-dimensional quasi-bialgebra $H$ is a quasi-Hopf algebra, provided that the forgetful functor $F : {}_H\mathcal{M}^{\rm fd} \to {}_k\mathcal{M}^{\rm fd}$ is a left rigid quasi-monoidal functor.

For this, let $H$ be a finite-dimensional quasi-bialgebra and $\{h_i\}_i$ a basis in $H$ with dual basis $\{h^i\}_i$ in $H^*$. Regard $H \in {}_H\mathcal{M}^{\rm fd}$ via its multiplication and denote by $\cdot : H \otimes H^* \to H^*$ the left $H$-module structure of $H^*$, and by $(\iota, \varepsilon) : H^* \dashv H$ the adjunction in ${}_H\mathcal{M}^{\rm fd}$ as in (LR1). We claim that this data defines completely the left rigid monoidal structure on ${}_H\mathcal{M}^{\rm fd}$ as well as the quasi-Hopf algebra structure on $H$, provided that the forgetful functor $F : {}_H\mathcal{M}^{\rm fd} \to {}_k\mathcal{M}^{\rm fd}$ is left rigid quasi-monoidal.

Indeed, consider $S : H \to H$ given by

$$S(h) = \sum_i (h \cdot h^i)(1_H) h_i, \ \ \forall\, h \in H. \tag{3.5.1}$$

Furthermore, for $V \in {}_H\mathscr{M}^{\mathrm{fd}}$ and a fixed element $v \in V$, define $\varphi_v : H \to V$ by $\varphi_v(h) = h \cdot v$, for all $h \in H$, a left $H$-linear morphism. Then $\varphi_v^* = \varphi_v^{\surd} : V^* \to H^*$ is a left $H$-linear morphism, that is, $\varphi_v^*(h \cdot v^*) = h \cdot \varphi_v^*(v^*)$, for all $h \in H$ and $v^* \in V^*$. This is clearly equivalent to

$$(h \cdot v^*) \circ \varphi_v = h \cdot \varphi_v^*(v^*) = h \cdot \sum_i \varphi_v^*(v^*)(h_i)h^i = \sum_i h \cdot v^*(\varphi_v(h_i))h^i = \sum_i v^*(h_i \cdot v)h \cdot h^i,$$

as elements in $H^*$, for all $h \in H$. Evaluating both sides on $1_H$ we get

$$(h \cdot v^*)(v) = \sum_i v^*(h_i \cdot v)(h \cdot h^i)(1_H) = v^*(S(h) \cdot v), \ \ \forall\, h \in H, \tag{3.5.2}$$

and this describes completely the left $H$-module structure of $V^*$, for any $V \in {}_H\mathscr{M}^{\mathrm{fd}}$. Since the definition of $S$ depends only on $H$ and the structure in (3.5.2) turns any $V^*$ into a left $H$-module, it follows that $S : H \to H$ is an anti-algebra morphism.

It remains to define the distinguished elements $\alpha, \beta \in H$ that together with $S$ defined by (3.5.1) obey the relations (3.2.1)–(3.2.2). To this end, we write $\iota(1_H) = \sum_j x_j \otimes x^j \in H \otimes H^*$ and define

$$\alpha = \sum_i \varepsilon(h^i \otimes 1_H)h_i \quad \text{and} \quad \beta = \sum_j x^j(1_H)x_j. \tag{3.5.3}$$

We next see that $\alpha, \beta$ determine completely the adjunction $(\iota_V, \varepsilon_V) : V^* \dashv V$ in (LR1) that behaves well with respect to the left $H$-module structures in (3.5.2). This claim follows from (LR2) since for any morphism $f : V \to W$ in ${}_H\mathscr{M}^{\mathrm{fd}}$ we have

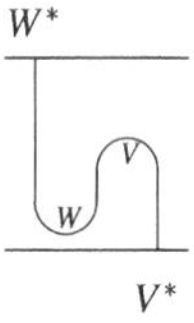

$= f^*$, and this implies
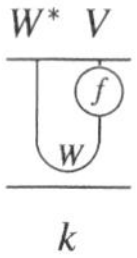

$=$
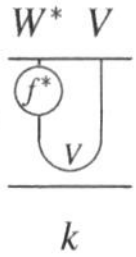

and
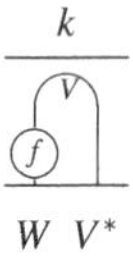

$=$
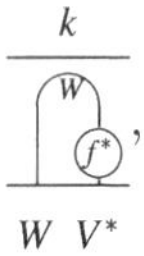

, (3.5.4)

where, as the notation suggests, the evaluation and coevaluation morphisms with the letter $V$ or $W$ nearby is our diagrammatic notation for the evaluation and coevaluation morphisms corresponding to $V$ and $W$, respectively, in ${}_H\mathscr{M}^{\mathrm{fd}}$.

For $V \in {}_H\mathscr{M}^{\mathrm{fd}}$ and fixed arbitrary $v \in V$ consider again $\varphi_v : H \to V$, the left $H$-linear morphism defined above. By the equality in (3.5.4) involving the evaluation morphisms, specialized for $\varphi_v$, we get that $\varepsilon_V(v^* \otimes h \cdot v) = \varepsilon(v^* \circ \varphi_v \otimes h)$, for all $v^* \in V^*$ and $h \in H$. By taking $h = 1_H$ we obtain that

$$\begin{aligned} \varepsilon_V(v^* \otimes v) &= \varepsilon(v^* \circ \varphi_v \otimes 1_H) \\ &= \sum_i v^*(\varphi_v(h_i))\varepsilon(h^i \otimes 1_H) \\ &\overset{(3.5.3)}{=} v^*(\alpha \cdot v), \end{aligned}$$

for all $v^* \in V^*$ and $v \in V$, so $\varepsilon_V$ is completely determined by $\alpha$.

Similarly, by taking $f = \varphi_v$ in the equality of (3.5.4) that involves the coevaluation morphisms we deduce that $(\varphi_v \otimes \mathrm{Id}_{H^*})\iota = (\mathrm{Id}_V \otimes \varphi_v^*)\iota_V$, as morphisms from $k$ to $H \otimes H^*$. If we write $\iota_V(1_k) := \sum_l y_l \otimes y^l \in V \otimes V^*$ we get

$$\sum_i \varphi_v(x_i) \otimes x^i = \sum_l y_l \otimes y^l \circ \varphi_v \in V \otimes H^*,$$

and therefore

$$\sum_l y^l(v) y_l = \sum_i x^i(1_H) x_i \cdot v \overset{(3.5.3)}{=} \beta \cdot v,$$

for all $v \in V$. It follows that $\iota_V(1_k) = \sum_s \beta \cdot v_s \otimes v^s$, where $\{v_s, v^s\}_s$ are dual bases in $V$ and $V^*$. So $\beta$ describes completely the coevaluation morphisms $\iota_V$, $V \in {}_H\mathscr{M}^{\mathrm{fd}}$.

It is clear at this moment that $\varepsilon_V$ and $\iota_V$ are $H$-linear morphisms, for all $V \in {}_H\mathscr{M}^{\mathrm{fd}}$, if and only if (3.2.1) are satisfied, and that $(\iota_V, \varepsilon_V)$ obeys (1.6.6) if and only if (3.2.2) are fulfilled. Otherwise stated, the triple $(S, \alpha, \beta)$ defines an antipode for the quasi-bialgebra $H$, and therefore $H$ is a quasi-Hopf algebra. □

## 3.6 Sovereign Quasi-Hopf Algebras

Let $H$ be a quasi-Hopf algebra. In this section we investigate when ${}_H\mathscr{M}^{\mathrm{fd}}$ is a sovereign monoidal category, in the sense of Definition 1.72. It will come up that sovereign structures on ${}_H\mathscr{M}^{\mathrm{fd}}$ are in a one-to-one correspondence with the pivotal structures, and that the latter are completely determined by certain elements of $H$.

If $\mathscr{C}$ is a monoidal category with left duality, then the functor $(-)^{**} := ((-)^*)^* : \mathscr{C} \to \mathscr{C}$ is strong monoidal: we have an isomorphism $\phi_0 : \underline{1} \to \underline{1}^{**}$, and for $V, W \in \mathscr{C}$, we have the following family of isomorphisms in $\mathscr{C}$:

$$\phi_{V,W} : V^{**} \otimes W^{**} \overset{\lambda^{-1}_{W^*,V^*}}{\longrightarrow} (W^* \otimes V^*)^* \overset{(\lambda_{V,W})^*}{\longrightarrow} (V \otimes W)^{**}.$$

$\lambda_{V,W} : (V \otimes W)^* \to W^* \otimes V^*$ is the isomorphism described in (1.7.1), $(\lambda_{V,W})^*$ is the transpose in $\mathscr{C}$ of $\lambda_{V,W}$, and $\lambda^{-1}_{W^*,V^*}$ is the inverse of $\lambda_{W^*,V^*}$.

**Definition 3.39** Let $\mathscr{C}$ be a left rigid monoidal category. A pivotal structure on $\mathscr{C}$ is a monoidal natural isomorphism $i$ between the strong monoidal functors $\mathrm{Id}_{\mathscr{C}}$ and $(-)^{**}$. This means that $i$ is a natural isomorphism satisfying the coherence conditions

$$\phi_{V,W} \circ (i_V \otimes i_W) = i_{V \otimes W}, \quad \phi_0 = i_{\underline{1}}. \tag{3.6.1}$$

Such a pair $(\mathscr{C}, i)$ is called pivotal category.

**Theorem 3.40** *Let $\mathscr{C}$ be a left rigid monoidal category. Then $\mathscr{C}$ admits a pivotal structure if and only if it is sovereign.*

*Proof* Let $\mathscr{C}$ be a sovereign category (in particular, it is also right rigid). By Proposition 1.73, for all $X \in \mathscr{C}$ we have an isomorphism $\theta_X : X \to ({}^*X)^*$. Since $\mathscr{C}$ is sovereign we have $(-)^* = {}^*(-)$, so $({}^*X)^* = X^{**}$. Thus we have $\theta_X : X \to X^{**}$. One can easily see that $\theta = (\theta_X)_{X\in \mathrm{Ob}(\mathscr{C})} : \mathrm{Id}_{\mathscr{C}} \to (-)^{**}$ is a natural isomorphism. It is, moreover, a natural monoidal transformation, and so provides a pivotal structure on $\mathscr{C}$. To see this, recall that $\theta_X = (\mathrm{ev}'_X \otimes \mathrm{Id}_{X^{**}})(\mathrm{Id}_X \otimes \mathrm{coev}_{X^*})$, for all $X \in \mathscr{C}$, and thus

$$\theta_{V\otimes W} =$$

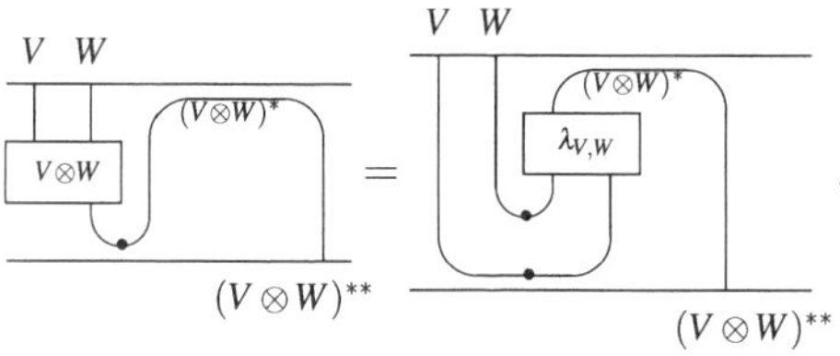


$$, \quad \forall\, V,\, W \in \mathscr{C},$$

because $\lambda = \lambda'$ since $\mathscr{C}$ is sovereign. On the other hand,

$$\phi_{V,W} =$$

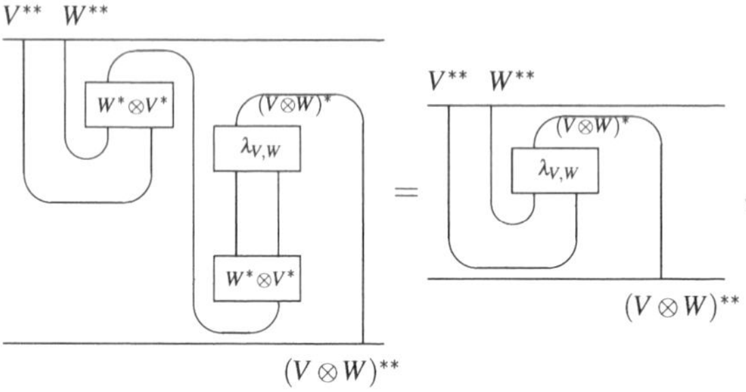


$$, \tag{3.6.2}$$

and therefore

$$\phi_{V,W}(\theta_V \otimes \theta_W) =$$

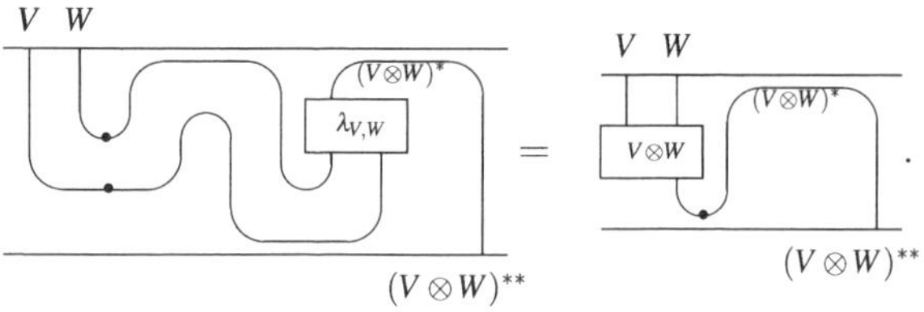


So $\theta$ is a pivotal structure on $\mathscr{C}$.

Conversely, let $i$ be a pivotal structure on $\mathscr{C}$, and for all $V \in \mathscr{C}$ define

$$\mathrm{ev}'_V : V \otimes V^* \overset{i_V \otimes \mathrm{Id}_{V^*}}{\longrightarrow} V^{**} \otimes V^* \overset{\mathrm{ev}_{V^*}}{\longrightarrow} \underline{1},$$

$$\mathrm{coev}'_V : \underline{1} \overset{\mathrm{coev}_{V^*}}{\longrightarrow} V^* \otimes V^{**} \overset{\mathrm{Id}_{V^*} \otimes i_V^{-1}}{\longrightarrow} V^* \otimes V.$$

It is immediate that $(V^*, \mathrm{ev}'_V, \mathrm{coev}'_V)$ is a right dual for $V$ in $\mathscr{C}$, and so $\mathscr{C}$ is right rigid, too. With respect to this right duality we have

$$\lambda'_{V,W} =$$

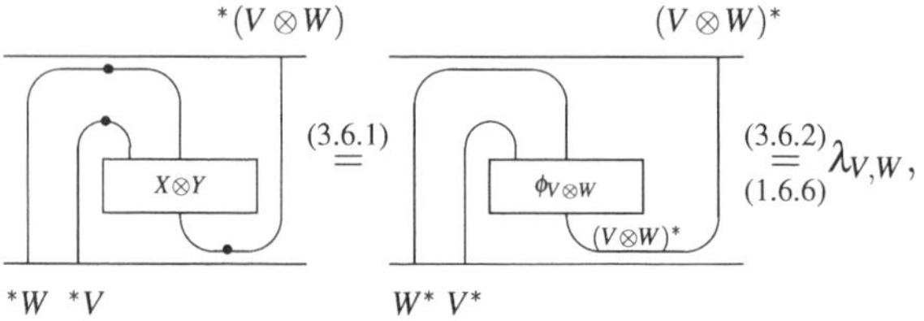


for all $V, W \in \mathscr{C}$. So we have shown that the left and right duality functors coincide as monoidal functors, and therefore $\mathscr{C}$ is sovereign. This ends the proof. □

We can give now a description of all pivotal/sovereign structures on ${}_H\mathscr{M}^{\mathrm{fd}}$.

**Proposition 3.41** *Let $H$ be quasi-Hopf algebra with bijective antipode over a field $k$. Then we have a bijective correspondence between*
*(i) pivotal structures $i$ on $\mathscr{C} = {}_H\mathscr{M}^{\mathrm{fd}}$;*
*(ii) sovereign structures on $\mathscr{C} = {}_H\mathscr{M}^{\mathrm{fd}}$;*
*(iii) invertible elements $\mathfrak{g}_i \in H$ satisfying*

$$S^2(h) = \mathfrak{g}_i^{-1} h \mathfrak{g}_i, \ \forall\, h \in H, \tag{3.6.3}$$

$$\Delta(\mathfrak{g}_i) = (\mathfrak{g}_i \otimes \mathfrak{g}_i)(S \otimes S)(f_{21}^{-1})f, \tag{3.6.4}$$

*where $f = f^1 \otimes f^2$ is the Drinfeld twist defined in* (3.2.15) *and $f_{21} = f^2 \otimes f^1$.*

*Proof* The bijection between (i) and (ii) is established by Theorem 3.40.

Let $V \in {}_H\mathscr{M}^{\mathrm{fd}}$, with $H$-action denoted by $H \otimes V \ni h \otimes v \to hv \in V$. We have a $k$-linear isomorphism $V \to V^{**}$, and $V^{**}$ can be regarded as $V$ with newly defined left $H$-action $h \cdot v = S^2(h)v$.

Let us check that there is a bijective correspondence between natural isomorphisms between the functors $\mathrm{Id}_{\mathscr{C}}$ and $(-)^{**}$ and invertible elements $\mathfrak{g}_i \in H$ satisfying (3.6.3). Indeed, let $i: \mathrm{Id}_{\mathscr{C}} \to (-)^{**}$ be a natural isomorphism, and let $\mathfrak{g}_i = S^2(i_H^{-1}(1_H))$, $\mathfrak{k}_i = i_H(1_H)$. For all $h \in H$, we have $i_H(h) = S^2(h)\mathfrak{k}_i$ and $i_H^{-1}(h) = S^{-2}(h\mathfrak{g}_i)$. In particular, $1_H = i_H^{-1}(\mathfrak{k}_i) = S^{-2}(\mathfrak{k}_i\mathfrak{g}_i)$ and $1_H = i_H(S^{-2}(\mathfrak{g}_i)) = \mathfrak{g}_i\mathfrak{k}_i$, hence $\mathfrak{k}_i = \mathfrak{g}_i^{-1}$. Now take $V \in {}_H\mathscr{M}^{\mathrm{fd}}$, fix $v \in V$ and define $f: H \ni h \mapsto hv \in V$, a morphism in ${}_H\mathscr{M}^{\mathrm{fd}}$. From the naturality of $i$, we deduce that $i_V(v) = (i_V \circ f)(1_H) = (f \circ i_H)(1_H) = \mathfrak{g}_i^{-1}v$. This means that $i$ is completely determined by $\mathfrak{g}_i$:

$$i_V(v) = \mathfrak{g}_i^{-1}v. \tag{3.6.5}$$

Now take $V = H$ and $v = h$. Equation (3.6.5) tells us that $S^2(h)\mathfrak{g}_i^{-1} = i_H(h) = \mathfrak{g}_i^{-1}h$, and it follows that (3.6.3) is satisfied.

For $H$ a quasi-Hopf algebra and $\mathscr{C} = {}_H\mathscr{M}^{\mathrm{fd}}$ the isomorphisms $\lambda_{V,W}$ were computed in Proposition 3.35, namely

$$\lambda_{V,W}(\psi) = \sum_{i,j} \psi(g^1 v_i \otimes g^2 w_j) w^j \otimes v^i,$$

for all $\psi \in (V \otimes W)^*$, where $\{v_i\}_i$ and $\{v^i\}_i$ are dual bases of $V$ and $V^*$, $\{w_j\}_j$ and $\{w^j\}_j$ are dual bases of $W$ and $W^*$, and $f^{-1} = g^1 \otimes g^2$ is the inverse of the Drinfeld twist. Also by Proposition 3.35 we know that

$$\lambda_{V,W}^{-1}(w^* \otimes v^*)(v \otimes w) = \langle v^*, f^1 v\rangle\langle w^*, f^2 w\rangle,$$

for all $v \in V$, $w \in W$, $v^* \in V^*$ and $w^* \in W^*$, where $f = f^1 \otimes f^2$ is the Drinfeld twist defined in (3.2.15).

It is now easy to establish that the first condition from (3.6.1) is equivalent to the following equivalent conditions:

$$\phi_{V,W}(i_V(v) \otimes i_W(w)) = i_{V\otimes W}(v \otimes w)$$

$$\Leftrightarrow \lambda^{-1}_{W^*,V^*}(i_V(v)\otimes i_W(w))\circ\lambda_{V,W}=i_{V\otimes W}(v\otimes w)$$
$$\Leftrightarrow \lambda^{-1}_{W^*,V^*}(i_V(v)\otimes i_W(w))(w^*\otimes v^*)=i_{V\otimes W}(v\otimes w)(\lambda^{-1}_{V,W}(w^*\otimes v^*))$$
$$\Leftrightarrow i_W(w)(f^1\cdot w^*)i_V(v)(f^2\cdot v^*)=\lambda^{-1}_{V,W}(w^*\otimes v^*)(\mathfrak{g}_i^{-1}(v\otimes w))$$
$$\Leftrightarrow v^*(S(f^2)\mathfrak{g}_i^{-1}v)w^*(S(f^1)\mathfrak{g}_i^{-1}w)=v^*(f^1(\mathfrak{g}_i^{-1})_1v)w^*(f^2(\mathfrak{g}_i^{-1})_2w),$$

for all $V,W\in\mathscr{C}$, $v\in V$, $w\in W$, $v^*\in V^*$ and $w^*\in W^*$. Since $H$ is an object of $\mathscr{C}$, this last condition is equivalent to

$$\Delta(\mathfrak{g}_i^{-1})=f^{-1}(S\otimes S)(f_{21})(\mathfrak{g}_i^{-1}\otimes\mathfrak{g}_i^{-1}),$$

which is equivalent to (3.6.4). Now take $x\in k$. It follows from (3.6.5) that $i_k(x)=\varepsilon(g_i^{-1})x$. Since $\phi_0:\ k\to k^{**}$ is the identity, we see that the second condition from (3.6.1) is equivalent to $\varepsilon(\mathfrak{g}_i)=1$. If (3.6.4) is satisfied, then $\varepsilon(\mathfrak{g}_i)^2=\varepsilon(\mathfrak{g}_i)$ (apply $\varepsilon\otimes\varepsilon$ to (3.6.4)), and it follows that $\varepsilon(\mathfrak{g}_i)=1$. This completes our proof. □

**Definition 3.42** A sovereign quasi-Hopf algebra is a quasi-Hopf algebra $H$ for which there exists an invertible element $\mathfrak{g}\in H$ satisfying (3.6.3) and (3.6.4).

**Remark 3.43** If $H$ is a sovereign quasi-Hopf algebra via an invertible element $\mathfrak{g}$ of $H$ obeying (3.6.3) and (3.6.4), then ${}_H\mathscr{M}^{\rm fd}$ is sovereign with the following rigid structure.

If $V\in{}_H\mathscr{M}^{\rm fd}$, with $H$-action denoted by $H\otimes V\ni h\otimes v\to hv\in V$, then the left dual of $V$ is $V^*$ with $H$-module structure and $\mathrm{ev}_V$ and $\mathrm{coev}_V$ as in Proposition 3.33. The right dual of $V$ is again $V^*$ considered as a left $H$-module via the antipode $S$ of $H$, and with the evaluation and coevaluation morphisms given by

$$\mathrm{ev}'_V:V\otimes V^*\ni v\otimes v^*\mapsto v^*(\mathfrak{g}^{-1}S^{-1}(\alpha)v)=v^*(S(\alpha)\mathfrak{g}^{-1}v)\in k,$$
$$\mathrm{coev}'_V:k\ni 1_k\mapsto v^i\otimes S^{-1}(\beta)\mathfrak{g}v_i=v^i\otimes\mathfrak{g}S(\beta)v_i\in V^*\otimes V,$$

summation implicitly understood, where $\{v_i,v^i\}_i$ are dual bases in $V$ and $V^*$.

**Example 3.44** The two-dimensional quasi-Hopf algebra $H(2)$ is sovereign via $\mathfrak{g}=g$, the generator of $C_2$.

*Proof* From the structure of $H(2)$ in Example 3.26 one can see that the Drinfeld twist of it is $f=g\otimes p_-+1\otimes p_+$. Since $f^2=1$, it follows that $f^{-1}=f$. But

$$\begin{aligned}f_{21}&=p_-\otimes g+p_+\otimes 1\\&=\frac{1}{2}(1\otimes g-g\otimes g+1\otimes 1+g\otimes 1)\\&=g\otimes p_-+1\otimes p_+=f,\end{aligned}$$

so $f^{-1}(S\otimes S)(f_{21})=1$, since $S$ is the identity of $H(2)$. Now everything follows from the commutativity of $H(2)$ and the fact that $g^{-1}=g$ is a grouplike element. □

## 3.7 Dual Quasi-Hopf Algebras

The notion of dual quasi-bialgebra can be regarded as the formal dual of the notion of quasi-bialgebra. So most of the results below will be presented without proof.

In what follows, by a $k$-coalgebra we mean a coassociative counital coalgebra $C$ with counit $\varepsilon : C \to k$ and comultiplication $\Delta : C \ni c \to c_1 \otimes c_2 \in C \otimes C$.

From a categorical point of view, dual quasi-bialgebras can be introduced as follows. Consider $H$ a $k$-coalgebra and denote by $\mathcal{M}^H$ the category of right $H$-comodules and right $H$-colinear maps in ${}_k\mathcal{M}$, the monoidal category of $k$-vector spaces.

**Definition 3.45** A dual quasi-bialgebra over a field $k$ is a $k$-coalgebra $H$ for which the monoidal structure of ${}_k\mathcal{M}$ induces a monoidal structure on $\mathcal{M}^H$, that is:

(a) for any two right $H$-comodules $X, Y$ the tensor product $X \otimes Y$ in ${}_k\mathcal{M}$ admits a right $H$-comodule structure;

(b) the tensor product in ${}_k\mathcal{M}$ of two right $H$-comodule morphisms is a morphism in $\mathcal{M}^H$, and so $\otimes$ induces a functor from $\mathcal{M}^H \times \mathcal{M}^H$ to $\mathcal{M}^H$;

(c) $k$, the unit object of ${}_k\mathcal{M}$, admits a right $H$-comodule structure;

(d) there exist functorial isomorphisms

$$a = (a_{X,Y,Z} : (X \otimes Y) \otimes Z \to X \otimes (Y \otimes Z))_{X,Y,Z \in \mathcal{M}^H},$$
$$l = (l_X : k \otimes X \to X)_{X \in \mathcal{M}^H} \text{ and } r = (r_X : X \otimes k \to X)_{X \in \mathcal{M}^H} \text{ in } \mathcal{M}^H$$

such that the Pentagon Axiom and the Triangle Axiom are satisfied.

After a reduction dual to the one in Corollary 3.11 we arrive at the following explicit characterization of a dual quasi-bialgebra.

**Proposition 3.46** *A dual quasi-bialgebra is a 4-tuple $(H, m, \eta, \varphi)$ consisting of a $k$-coalgebra $(H, \Delta, \varepsilon)$, two coalgebra morphisms $m : H \otimes H \to H$ (called multiplication) and $\eta : k \to H$ (called unit), with notation $\eta(1_k) = 1$, and an invertible element $\varphi : H \otimes H \otimes H \to k$ in the convolution algebra* $\mathrm{Hom}_k(H^{\otimes 3}, k)$, *called reassociator, such that the following conditions are satisfied:*

$$h_1(g_1 l_1)\varphi(h_2, g_2, l_2) = \varphi(h_1, g_1, l_1)(h_2 g_2) l_2, \tag{3.7.1}$$
$$1h = h1 = h, \tag{3.7.2}$$
$$\varphi(h_1, g_1, l_1 m_1)\varphi(h_2 g_2, l_2, m_2) = \varphi(g_1, l_1, m_1)\varphi(h_1, g_2 l_2, m_2)\varphi(h_2, g_3, l_3), \tag{3.7.3}$$
$$\varphi(h, 1, g) = \varepsilon(h)\varepsilon(g), \tag{3.7.4}$$

*for all $h, g, l, m \in H$, where we denoted $m(h \otimes g)$ by $hg$.*

It is easy to see that (3.7.3) and (3.7.4) also imply the identities:

$$\varphi(1, h, g) = \varphi(h, g, 1) = \varepsilon(h)\varepsilon(g), \ \ \forall\, h,\ g \in H. \tag{3.7.5}$$

Let $H$ be a dual quasi-bialgebra. For $M \in \mathcal{M}^H$ we denote by $\rho_M : M \ni m \to m_{(0)} \otimes m_{(1)} \in M \otimes H$ the right $H$-comodule structure of $M$. The category $\mathcal{M}^H$ is monoidal.

The tensor product is given via $m$, that is, for any $M,N \in \mathcal{M}^H$, $M \otimes N \in \mathcal{M}^H$ via the structure map

$$\rho_{M\otimes N}(m\otimes n) = m_{(0)} \otimes n_{(0)} \otimes m_{(1)}n_{(1)}. \tag{3.7.6}$$

The associativity constraint $a_{X,Y,Z} : (X\otimes Y)\otimes Z \to X\otimes(Y\otimes Z)$ of $\mathcal{M}^H$ is

$$a_{X,Y,Z}((x\otimes y)\otimes z) = \varphi(x_{(1)},y_{(1)},z_{(1)})x_{(0)} \otimes (y_{(0)} \otimes z_{(0)}), \tag{3.7.7}$$

for all $X,Y,Z \in \mathcal{M}^H$, $x \in X$, $y \in Y$ and $z \in Z$. The unit is $k$ as a trivial right $H$-comodule, and the left and right unit constraints are the usual ones.

If $H$ is a dual quasi-bialgebra then a twist or a gauge transformation on $H$ is an invertible element $\tau$ in the convolution algebra $\mathrm{Hom}_k(H^{\otimes 2},k)$, satisfying $\tau(1,h) = \tau(h,1) = \varepsilon(h)$, for all $h \in H$.

The notion of dual quasi-bialgebra is invariant under a twist as follows.

**Proposition 3.47** *Let $H$ be a dual quasi-bialgebra and $\tau$ a twist on $H$ with convolution inverse $\tau^{-1}$. Then $H$ with the same comultiplication, counit and unit as those of $H$, and with the new product and new reassociator given by*

$$\begin{aligned} &h \cdot_\tau g = \tau(h_1,g_1)h_2g_2\tau^{-1}(h_3,g_3),\\ &\varphi_\tau(h,g,l) = \tau(g_1,l_1)\tau(h_1,g_2l_2)\varphi(h_2,g_3,l_3)\tau^{-1}(h_3g_4,l_4)\tau^{-1}(h_4,g_5), \end{aligned}$$

*for all $h,g,l \in H$, is a dual quasi-bialgebra as well. It will be denoted by $H_\tau$ and called the twisting of $H$ by $\tau$.*

A dual quasi-bialgebra $H$ is called a dual quasi-Hopf algebra if, in addition, there exist an anti-coalgebra morphism $S : H \to H$ and $\alpha,\beta \in H^*$ such that, for all $h \in H$:

$$\begin{aligned} &S(h_1)\alpha(h_2)h_3 = \alpha(h)1, \quad h_1\beta(h_2)S(h_3) = \beta(h)1,\\ &\varphi(h_1\beta(h_2),S(h_3),\alpha(h_4)h_5) = \varphi^{-1}(S(h_1),\alpha(h_2)h_3,\beta(h_4)S(h_5)) = \varepsilon(h). \end{aligned}$$

It follows from the axioms that $S(1) = 1$ and $\alpha(1)\beta(1) = 1$, so we can assume that $\alpha(1) = \beta(1) = 1$. Also, if $\tau$ is a twist on $H$ then $H_\tau$ is a dual quasi-Hopf algebra with the same antipode as $H$ and distinguished elements

$$\alpha_\tau(h) = \tau^{-1}(S(h_1),h_3)\alpha(h_2) \text{ and } \beta_\tau(h) = \tau(h_1,S(h_3))\beta(h_2), \ \forall\, h \in H.$$

We now justify why we called such objects dual quasi-Hopf algebras.

Let $H$ be a finite-dimensional quasi-bialgebra and let $\{e_i\}_i$ be a basis of $H$ and $\{e^i\}_i$ the corresponding dual basis of $H^*$, the linear dual of $H$. Since $H$ is a $k$-algebra we get that $H^*$ is a $k$-coalgebra with comultiplication

$$\Delta_{H^*}(h^*) = \sum_{i,j} h^*(e_ie_j)e^i \otimes e^j, \ \forall\, h^* \in H^*,$$

and counit $\varepsilon_{H^*}(h^*) = h^*(1_H)$. $H^*$ is also an $H$-bimodule, by

$$\langle h \rightharpoonup h^*, h'\rangle = h^*(h'h) \text{ and } \langle h^* \leftharpoonup h, h'\rangle = h^*(hh'), \ \forall\, h^* \in H^*,\ h \in H.$$

Furthermore, the comultiplication of $H$ defines a multiplication on $H^*$. Namely,

$h^*g^*(h) := h^*(h_1)g^*(h_2)$, for all $h^*, g^* \in H^*$ and $h \in H$. It is not associative, but since $\Delta$ is coassociative up to conjugation by $\Phi$ we obtain, for all $h^*$, $g^*$, $l^* \in H^*$,

$$(h^*g^*)l^* = (X^1 \rightharpoonup h^* \leftharpoonup x^1)[(X^2 \rightharpoonup g^* \rightharpoonup x^2)(X^3 \rightharpoonup l^* \leftharpoonup x^3)].$$

**Proposition 3.48** *Let $H$ be a finite-dimensional quasi-bialgebra with reassociator $\Phi$. Then $H^*$ with the dual algebra and coalgebra structures of $H$ is a dual quasi-bialgebra with reassociator $\varphi$ defined for all $h^*, g^*, l^* \in H^*$ by $\varphi(h^*, g^*, l^*) = h^*(X^1)g^*(X^2)l^*(X^3)$. Furthermore, if $H$ is a quasi-Hopf algebra with antipode $S$ and distinguished elements $\alpha, \beta \in H$ then $H^*$ is a dual quasi-Hopf algebra with antipode $S^*$ and distinguished elements $\alpha^*, \beta^* \in H^{**}$, where $\alpha^*(h^*) = h^*(\alpha)$ and $\beta^*(h^*) = h^*(\beta)$, for all $h^* \in H^*$.*

*Proof* Let us start by noting that

$$\Delta_{H^*}(h^*) = h_1^* \otimes h_2^* \Longleftrightarrow h^*(hg) = h_1^*(h)h_2^*(g), \ \ \forall\, h,\ g \in H.$$

Thus

$$\begin{aligned} h_1^*(g_1^*l_1^*)\varphi(h_2^*, g_2^*, l_2^*) &= \varphi(h_1^*, g_1^*, l_1^*)(h_2^*g_2^*)l_2^* \\ \Leftrightarrow h_1^*(h_1)g_1^*(h_{(2,1)})l_1^*(h_{(2,2)})&h_2^*(X^1)g_2^*(X^2)l_2^*(X^3) \\ &= h_1^*(X^1)g_1^*(X^2)l_1^*(X^3)h_1^*(h_{(1,1)})g_1^*(h_{(1,2)})l_1^*(h_2) \\ \Leftrightarrow h^*(h_1X^1)g^*(h_{(2,1)}X^2)l^*(h_{(2,2)}X^3) &= h^*(X^1h_{(1,1)})g^*(X^2h_{(1,2)})l^*(X^3h_2), \end{aligned}$$

for all $h \in H$, and the last equality is true because of (3.1.7). $\Delta$ is multiplicative and therefore $\Delta_{H^*}$ is multiplicative, too. Since $\Delta$ respects the units we get that $\varepsilon_{H^*}$ is multiplicative, and since $\varepsilon$ is multiplicative it follows that $\Delta_{H^*}$ respects the units. Finally, one can easily see that $\varepsilon_{H^*}(1_{H^*}) = 1_{H^*}(1_H) = \varepsilon(1_H) = 1_k$, thus $H^*$ is a dual quasi-bialgebra.

If $H$ is a quasi-Hopf algebra then arguments similar to the ones above imply that $H^*$ is a dual quasi-Hopf algebra. For instance,

$$\begin{aligned} \langle S^*(h_1^*)\alpha^*(h_2^*)h_3^*, h\rangle &= h_2^*(\alpha)h_1^*(S(h_1))h_3^*(h_2) \\ &= h^*(S(h_1)\alpha h_2) \\ &= \varepsilon(h)h^*(\alpha) = \alpha^*(h^*)\langle 1_{H^*}, h\rangle, \end{aligned}$$

for all $h^* \in H^*$, as required. The remaining details are left to the reader. □

Note that if $H$ is a finite-dimensional dual quasi-bialgebra with reassociator $\varphi$ then $H^*$ is a quasi-Hopf algebra with the dual structure of $H$ and reassociator

$$\Phi = \sum_{i,j,s} \varphi(e_i, e_j, e_s)e^i \otimes e^j \otimes e^s \in H^* \otimes H^* \otimes H^*.$$

If, moreover, $H$ is a dual quasi-Hopf algebra with antipode $S$ and distinguished elements $\alpha, \beta \in H^*$ then $H^*$ is a quasi-Hopf algebra with antipode $S^*$ and distinguished elements $\alpha^*, \beta^* \in H^*$ defined by $\alpha^* = \alpha$ and $\beta^* = \beta$.

If we iterate the procedure we then obtain that $H \cong H^{**}$ as (dual) quasi-bialgebras or (dual) quasi-Hopf algebras. In both cases the isomorphism is produced by $\theta_H : H \to H^{**}$ defined by $\theta_H(h)(h^*) = h^*(h)$, for all $h \in H$ and $h^* \in H^*$.

As the reader probably expects, the antipode of a dual-quasi Hopf algebra is an anti-algebra morphism in the following sense.

**Proposition 3.49** *For a dual quasi-Hopf algebra $H$ the antipode is an anti-algebra morphism, up to a conjugation by a twist. More precisely, define $\gamma, \delta \in (H \otimes H)^*$ by*

$$\gamma(h,g) = \varphi(S(g_2), S(h_2), h_4)\alpha(h_3)\varphi^{-1}(S(g_1)S(h_1), h_5, g_4)\alpha(g_3),$$
$$\delta(h,g) = \varphi(h_1g_1, S(g_5), S(h_4))\beta(h_3)\varphi^{-1}(h_2, g_2, S(g_4))\beta(g_3),$$

*for all $h, g \in H$. If we define $f, f^{-1} \in (H \otimes H)^*$ by*

$$f(h,g) = \varphi^{-1}(S(g_1)S(h_1), h_3g_3, S(h_5g_5))\beta(h_4g_4)\gamma(h_2, g_2),$$
$$f^{-1}(h,g) = \varphi^{-1}(S(h_1g_1), h_3g_3, S(g_5)S(h_5))\alpha(h_2g_2)\delta(h_4, g_4),$$

*then $f$ and $f^{-1}$ are inverses in the convolution algebra* $\mathrm{Hom}_k(H^{\otimes 2}, k)$ *and*

$$f(h_1, g_1)S(h_2g_2)f^{-1}(h_3, g_3) = S(g)S(h),$$

*for all $h, g \in H$. Moreover, the following relations hold:*

$$\gamma(h,g) = f(h_1, g_1)\alpha(h_2g_2) \quad \text{and} \quad \delta(h,g) = \beta(h_1g_1)f^{-1}(h_2, g_2).$$

As we have seen, examples of dual quasi-Hopf algebras can be obtained from finite-dimensional quasi-Hopf algebras by dualizing. Another class of examples can be constructed as follows. Notice that $H(2)$ considered in Example 3.26 is the dual of $k_\phi[C_2]$ below, where $C_2 = \langle g \rangle$ is the cyclic group of order 2 and $\phi$ is the unique non-trivial 3-cocycle on $C_2$ determined by $\phi(g^u, g^v, g^s) = (-1)^{uvs}$, $0 \le u, v, s \le 1$.

**Example 3.50** Let $G$ be a group and $H = k[G]$ the group algebra associated to $G$. With the coproduct and counit

$$\Delta(g) = g \otimes g \;\text{ and }\; \varepsilon(g) = 1, \;\; \forall\, g \in G,$$

extended by linearity, $H$ has a Hopf algebra structure, its antipode $S$ being defined by $S(g) = g^{-1}$, for all $g \in G$. However, since $H$ is cocommutative (i.e. $\Delta^{\mathrm{cop}} = \Delta$), for any normalized 3-cocycle $\phi$ on $G$, that is, for any map $\phi : G \times G \times G \to k^*$ such that (1.2.2) holds and $\phi(x, e, y) = 1$, for all $x, y \in G$, the group Hopf algebra $k[G]$ can be also viewed as a dual quasi-Hopf algebra with reassociator $\varphi = \phi$, extended by linearity, antipode $S$ as above and distinguished elements

$$\alpha = \varepsilon \;\text{ and }\; \beta(g) = \phi(g, g^{-1}, g)^{-1}, \;\; \forall\, g \in G.$$

We denote this dual quasi-Hopf algebra structure on $k[G]$ by $k_\phi[G]$.

*Proof* This is straightforward, so we leave it to the reader. □

In the next result we show that a monoidal category of graded vector spaces is nothing else than a category of comodules over a certain dual quasi-Hopf algebra.

**Proposition 3.51** *Let $G$ be a group, $\phi \in H^3(G,k^*)$ a normalized 3-cocycle and $k_\phi[G]$ the dual quasi-Hopf algebra described in Example 3.50. Then the categories $\mathrm{Vect}_\phi^G$ and $\mathscr{M}^{k_\phi[G]}$ are monoidally isomorphic.*

*Proof* We first identify $\mathrm{Vect}^G$ and $\mathscr{M}^{k_\phi[G]}$ as usual categories.

If $X$ is a right $k_\phi[G]$-comodule with structure map $\rho : X \to X \otimes k_\phi[G]$ then for any $x \in X$ we write $\rho(x) = \sum_{g\in G} x^g \otimes g$, for some elements $x^g \in X$, $g \in G$, and this is possible because the elements of $G$ define a basis for $k_\phi[G]$. From the coassociativity of $\rho$ we obtain that

$$\sum_{g,h\in G} (x^g)^h \otimes h \otimes g = \sum_{g\in G} x^g \otimes g \otimes g,$$

and so we must have $(x^g)^h = \delta_{g,h}x^g$, for all $h,g \in G$. If for any $g \in G$ we define $X_g = \{x^g \mid x \in X\}$, then the latter equation shows that the sum $\sum_{g\in G} X_g$ is direct. But, on the other hand, since $(\mathrm{Id}_X \otimes \varepsilon)\rho = \mathrm{Id}_X$ we get that $x = \sum_{g\in G} x^g$, for all $x \in X$, and therefore $X = \bigoplus_{g\in G} X_g$. Thus $X$ is a $G$-graded vector space.

Conversely, if $X$ is a $G$-graded vector space we define $\rho^X : X \to X \otimes k_\phi[G]$ by $\rho^X(x) = \sum_{g\in G} x_g \otimes g$, where $x = \sum_{g\in G} x_g$ is the decomposition of $x$ in homogeneous components. It can be easily seen that with this structure $X$ becomes a right $k_\phi[G]$-comodule, and the two correspondences described above provide inverse bijections.

Now, recall that for the monoidal category of $G$-graded vector spaces $\mathrm{Vect}_\phi^G$, the associativity constraint is defined by

$$a_{X,Y,Z}((x\otimes y)\otimes z) = \phi(|x|,|y|,|z|)x\otimes(y\otimes z),$$

for all $X,Y,Z \in \mathrm{Vect}_\phi^G$ and homogeneous elements $x \in X$, $y \in Y$, $z \in Z$. If we regard $X$ as a right $k_\phi[G]$-comodule then $\rho^X(x) = x \otimes |x|$, thus $a_{X,Y,Z}$ can be rewritten as

$$a_{X,Y,Z}((x\otimes y)\otimes z) = \phi(x_{(1)}, y_{(1)}, z_{(1)})x_{(0)} \otimes (y_{(0)} \otimes z_{(0)}).$$

But this family of isomorphisms defines the associativity constraint of $\mathscr{M}^{k_\phi[G]}$, and therefore $\mathrm{Vect}_\phi^G$ and $\mathscr{M}^{k_\phi[G]}$ identify as monoidal categories, as claimed. □

The category of corepresentations over a dual quasi-Hopf algebra $H$ is left and/or right rigid, depending on the bijectivity of the antipode $S$ of $H$.

**Proposition 3.52** *Let $H$ be a dual quasi-Hopf algebra with antipode $S$ and distinguished elements $\alpha, \beta \in H^*$, and denote by $\mathscr{M}^H_{\mathrm{fd}}$ and ${}^H\mathscr{M}_{\mathrm{fd}}$ the category of right and left $H$-comodules, respectively, that are finite dimensional. Then $\mathscr{M}^H_{\mathrm{fd}}$ is left rigid and ${}^H\mathscr{M}_{\mathrm{fd}}$ is right rigid. Furthermore, if $S$ is bijective then $\mathscr{M}^H_{\mathrm{fd}}$ and ${}^H\mathscr{M}_{\mathrm{fd}}$ are both rigid monoidal categories.*

*Proof* By duality reasons we sketch the proof by indicating only the left/right dual object associated to a finite-dimensional $H$-comodule.

The left dual of $V \in \mathscr{M}^H_{\rm fd}$ is $V^* = \mathrm{Hom}_k(V,k)$ with the right $H$-comodule structure

$$\rho_{V^*}(v^*) = \langle v^*, {}_iv_{(0)}\rangle {}^iv \otimes S({}_iv_{(1)}),$$

for all $v^* \in V^*$, where $\{{}_iv\}_i$ is a basis in $V$ with dual basis $\{{}^iv\}_i$ in $V^*$, and $\rho_V(v) = v_{(0)} \otimes v_{(1)} \in V \otimes H$ is the right $H$-comodule structure of $V$. The evaluation and coevaluation maps are defined by

$$\mathrm{ev}_V : V^* \otimes V \to k, \quad \mathrm{ev}_V(v^* \otimes v) = \alpha(v_{(1)})v^*(v_{(0)}),$$
$$\mathrm{coev}_V : k \to V \otimes V^*, \quad \mathrm{coev}_V(1) = \beta({}_iv_{(1)}){}_iv_{(0)} \otimes {}^iv,$$

for all $v^* \in V^*$ and $v \in V$. If $S$ is bijective with inverse $S^{-1}$ then ${}^*V = \mathrm{Hom}_k(V,k)$ is a right $H$-comodule via the structure

$$\rho_{{}^*V}({}^*v) = \langle {}^*v, {}_iv_{(0)}\rangle {}^iv \otimes S^{-1}({}_iv_{(1)}),$$

for all ${}^*v \in {}^*V$, and together with the maps

$$\mathrm{ev}'_V : V \otimes {}^*V \to k, \quad \mathrm{ev}'_V(v \otimes {}^*v) = \alpha(S^{-1}(v_{(1)}))v^*(v_{(0)}),$$
$$\mathrm{coev}'_V : k \to {}^*V \otimes V, \quad \mathrm{coev}'_V(1) = \beta(S^{-1}({}_iv_{(1)})){}^iv \otimes {}_iv_{(0)}$$

define a right dual for $V$ in $\mathscr{M}^H_{\rm fd}$.

Consider now the category ${}^H\mathscr{M}_{\rm fd}$. It is always right rigid since any object $V$ of it admits as a right dual the object ${}^*V := \mathrm{Hom}_k(V,k)$, considered a left $H$-comodule via the structure morphism

$$\lambda_{{}^*V} : {}^*V \to H \otimes {}^*V, \quad \lambda_V({}^*v) = \langle {}^*v, {}_iv_{(0)}\rangle S({}_iv_{(-1)}) \otimes {}^iv,$$

where $\lambda_V(v) = v_{(-1)} \otimes v_{(0)}$ is the left $H$-comodule structure of $V$. The evaluation and coevaluation maps are defined by

$$\mathrm{ev}'_V : V \otimes {}^*V \to k, \quad \mathrm{ev}'_V(v \otimes {}^*v) = \beta(v_{(-1)}){}^*v(v_{(0)}),$$
$$\mathrm{coev}'_V : k \to {}^*V \otimes V, \quad \mathrm{coev}'_V(1) = \sum_i \alpha({}_iv_{(-1)}){}^iv \otimes {}_iv_{(0)}.$$

Assuming $S$ is bijective we get that ${}^H\mathscr{M}_{\rm fd}$ is rigid monoidal. For $V \in {}^H\mathscr{M}_{\rm fd}$ we have that $V^* = \mathrm{Hom}_k(V,k)$ is a left $H$-comodule via the structure morphism given by

$$\lambda_{V^*}(v^*) = \langle {}^*v, {}_iv_{(0)}\rangle S^{-1}({}_iv_{(-1)}) \otimes {}^iv,$$

for all $v^* \in V^*$, and a left dual of $V$ in ${}^H\mathscr{M}_{\rm fd}$ if it is considered together with the maps

$$\mathrm{ev}_V : V^* \otimes V \to k, \quad \mathrm{ev}_V(v^* \otimes v) = \beta(S^{-1}(v_{(-1)}))v^*(v_{(0)}),$$
$$\mathrm{coev}_V : k \to V \otimes V^*, \quad \mathrm{coev}_V(1) = \sum_i \alpha(S^{-1}({}_iv_{(-1)})){}_iv_{(0)} \otimes {}^iv.$$

So the proof is finished. □

**Corollary 3.53** *Let $G$ be a group, $k$ a field and $\phi \in H^3(G,k^*)$ a normalized 3-cocycle. Then* $\mathrm{vect}^G_\phi$*, the category of finite-dimensional vector spaces equipped with the monoidal structure of* $\mathrm{Vect}^G_\phi$*, is rigid monoidal.*

*Proof* This follows immediately from Proposition 3.51 and Proposition 3.52. The explicit structure of the dual objects is exactly the one stated in Example 1.65, we leave the details to the reader. □

The dual version of Proposition 3.35 says that the isomorphism $\lambda_{X,Y} : (X \otimes Y)^* \to Y^* \otimes X^*$ in $\mathscr{M}^H_{\rm fd}$, where $H$ is a dual quasi-Hopf algebra, is given by

$$\lambda_{X,Y}(\mu) = f^{-1}({}_ix_{(1)}, {}_jy_{(1)})\mu({}_ix_{(0)} \otimes {}_jy_{(0)})^jy \otimes {}^ix,$$

where $f^{-1}$ is the inverse of the Drinfeld twist defined in Proposition 3.49, and $\{{}_ix, {}^ix\}_i$ and $\{{}_jy, {}^jy\}_j$ are dual bases in $X$ and $X^*$, and in $Y$ and $Y^*$, respectively. Its inverse is given by

$$\lambda^{-1}_{X,Y}(y^* \otimes x^*)(x \otimes y) = f(x_{(1)}, y_{(1)})x^*(x_{(0)})y^*(y_{(0)}), \tag{3.7.8}$$

for all $x^* \in X^*$, $y^* \in Y^*$, $x \in X$ and $y \in Y$, where $f$ is the convolution inverse of $f^{-1}$.

We leave it to the reader to compute the right-handed version $\lambda'$ of $\lambda$, and $\lambda'^{-1}$.

## 3.8 Further Examples of (Dual) Quasi-Hopf Algebras

By Example 3.50 we can construct examples of dual quasi-Hopf algebras starting with a pair $(G, \phi)$ consisting of a group $G$ and a normalized 3-cocycle $\phi$ on it. When $G$ is finite, by duality, we also get examples of quasi-Hopf algebras; see the dual version of Proposition 3.48.

The goal of this section is to provide concrete examples of (dual) quasi-Hopf algebras of the type mentioned above. In other words, we will present concrete examples of pairs $(G, \phi)$ as above.

**Proposition 3.54** *Let $C_n = \langle\sigma\rangle$ be the cyclic group of order $n$ and $k$ a field that contains an $n$th root of unity $q$. Then $\phi(\sigma^a, \sigma^b, \sigma^c) := q^{abc}$ is a normalized 3-cocycle on $C_n$. It is a coboundary if and only if $q^{\frac{n(n-1)}{2}} = 1$.*

*Proof* The fact that $q$ is an $n$th root of unity in $k$ implies that $\phi$ is well defined. Indeed, if $a = a' + n$, $b = b' + n$ and $c = c' + n$ then $abc = n^3 + (a' + b' + c')n^2 + (a'b' + a'c' + b'c')n + a'b'c'$, and so $\phi(\sigma^a, \sigma^b, \sigma^c) = q^{abc} = q^{a'b'c'} = \phi(\sigma^{a'}, \sigma^{b'}, \sigma^{c'})$.

Now, the 3-cocycle condition follows because of the equality $bcd + a(b+c)d + abc = ab(c+d) + (a+b)cd$ in $\mathbb{Z}$. It is immediate that $\phi$ is normalized.

Suppose now that $\phi$ is a coboundary; take $g \in K^2(C_n, k^*)$ such that

$$g(\sigma^b, \sigma^c)g(\sigma^{a+b}, \sigma^c)^{-1}g(\sigma^a, \sigma^{b+c})g(\sigma^a, \sigma^b)^{-1} = q^{abc}, \ \forall\, a, b, c \in \mathbb{Z}, \tag{3.8.1}$$

and denote $\beta := g(\sigma^a, 1) = g(1, \sigma^b)$ and $\alpha_c := g(\sigma, \sigma^c)$, for $a, b, c \in \{1, \ldots, n-1\}$. By taking $a = 1$, $b = k$ and $c = n-1$ in (3.8.1) we obtain

$$g(\sigma^{k+1}, \sigma^{n-1}) = q^k g(\sigma^k, \sigma^{n-1})g(\sigma, \sigma^{k-1})g(\sigma, \sigma^k)^{-1}, \ \forall\, k \in \mathbb{Z}.$$

By mathematical induction it follows that $g(\sigma^k,\sigma^{n-1}) = q^{\frac{k(k-1)}{2}}\alpha_{n-1}\beta\alpha_{k-1}^{-1}$, for all $2 \le k \le n-1$. We then have

$$\beta = g(1,\sigma^{n-1}) = g(\sigma^{(n-1)+1},\sigma^{n-1}) = q^{n-1}g(\sigma^{n-1},\sigma^{n-1})g(\sigma,\sigma^{n-2})g(\sigma,\sigma^{n-1})^{-1}$$
$$= q^{n-1}q^{\frac{(n-1)(n-2)}{2}}\alpha_{n-1}\beta\alpha_{n-2}^{-1}\alpha_{n-2}\alpha_{n-1}^{-1} = q^{\frac{n(n-1)}{2}}\beta,$$

from which we conclude that $q^{\frac{n(n-1)}{2}} = 1$, as stated.

Conversely, if $q^{\frac{n(n-1)}{2}} = 1$ we consider the functions $f : \mathbb{Z}\times\mathbb{Z}\to\mathbb{Z}$ given by $f(x,y) = -\frac{(x-1)xy}{2}$, for all $x,y\in\mathbb{Z}$, and $g : C_n\times C_n\to k$ defined by $g(\sigma^a,\sigma^b) = q^{f(a,b)}$, for all $a,b\in\{0,1,\ldots,n-1\}$. If $a = a'+n$ and $b = b'+n$ then

$$f(a,b) - f(a',b') = -a'n^2 - \left(\frac{a'(a'-1)}{2} + a'b'\right)n - \frac{n^2(n-1)}{2} - \frac{n(n-1)}{2}b'.$$

This relation together with $q^n = 1$ and $q^{\frac{n(n-1)}{2}} = 1$ implies that $g$ is well defined.

Finally, a straightforward computation ensures that

$$f(y,z) - f(x+y,z) + f(x,y+z) - f(x,y) = xyz, \ \ \forall\, x,y,z\in\mathbb{Z},$$

and this shows that $\Delta_2(g) = \phi$, that is, that $\phi$ is a coboundary. □

We move now to the situation when $G$ is not a cyclic group.

**Proposition 3.55** *If $G = \mathbb{Z}_2\times\mathbb{Z}_2\times\mathbb{Z}_2$ then $\phi(\vec{x},\vec{y},\vec{z}) := (-1)^{(\vec{x}\times\vec{y})\cdot\vec{z}}$, for all $\vec{x},\vec{y},\vec{z}\in G$, is a normalized coboundary 3-cocycle on $G$. Here $(\vec{x}\times\vec{y})\cdot\vec{z}$ is the mixed double product of $\vec{x},\vec{y},\vec{y}\in G$, where we used a vector notation for the elements of $G$.*

*Proof* Let $A$ a be a square $3\times 3$-matrix whose rows are given by the components of three vectors in $G$. Then the determinant of $A$ coincides with the mixed double product of those three vectors. Hence, $(\vec{x},\vec{y},\vec{z}) := (\vec{x}\times\vec{y})\cdot\vec{z}$ is $\mathbb{Z}$-linear in each argument. From here we can easily see that

$$(\vec{y},\vec{z},\vec{w}) + (\vec{x},\vec{y}+\vec{z},\vec{w}) + (\vec{x},\vec{y},\vec{z}) = (\vec{x},\vec{y},\vec{z}+\vec{w}) + (\vec{x}+\vec{y},\vec{z},\vec{w}),$$

for all $\vec{x},\vec{y},\vec{z},\vec{w}\in G$, and this proves that $\phi$ is a 3-cocycle on $G$. Since the mixed double product is zero when a vector is zero it follows that $\phi$ is normalized.

To show that $\phi$ is a coboundary, define $f : G\times G\to\mathbb{Z}$ by

$$f(\vec{x},\vec{y}) = \sum_{1\le i\le j\le 3} x_iy_j + y_1x_2x_3 + x_1y_2x_3 + x_1x_2y_3, \ \ \forall\,\vec{x},\vec{y}\in G,$$

where we also made use of the multiplication of $\mathbb{Z}_2$. Taking into account that $-x = x$ in $\mathbb{Z}_2$, by a direct computation one can easily see that

$$f(\vec{y},\vec{z}) - f(\vec{x}+\vec{y},\vec{z}) + f(\vec{x},\vec{y}+\vec{z}) - f(\vec{x},\vec{y})$$
$$= -z_1(x_2y_3 + y_2x_3) - z_2(x_1y_3 + y_1x_3) - z_3(x_1y_2 + y_1x_2)$$
$$= z_1(x_2y_3 - y_2x_3) - z_2(x_1y_3 - y_1x_3) + z_3(x_1y_2 - y_1x_2) = (\vec{x}\times\vec{y})\cdot\vec{z}.$$

Thus $g\in K^2(G,k^*)$, given by $g(\vec{x},\vec{y}) = (-1)^{f(\vec{x},\vec{y})}$, for all $\vec{x},\vec{y}\in G$, satisfies $\Delta_2(g) = \phi$, and from here we conclude that $\phi$ is a coboundary. □

**Remark 3.56** If $G$ is a direct product of a finite number of copies of $\mathbb{Z}_n$ and $q$ is an $n$th root of unity in $k$ then any $\mathbb{Z}$-trilinear map $(\ ,\ ,\ ) : G \times G \times G \to \mathbb{Z}$ defines a normalized 3-cocycle on $G$, namely $\phi(\vec{x},\vec{y},\vec{z}) := q^{(\vec{x},\vec{y},\vec{z})}$, for all $\vec{x},\vec{y},\vec{z} \in G$, where we used the same vector notation for the elements of $G$ as above.

As we mentioned at the beginning of this section, classes of quasi-Hopf algebras can be obtained from groups and normalized 3-cocycles $\phi$ on them, by considering the dual structure of the dual quasi-Hopf algebra $k_\phi[G]$. A convenient situation occurs when $G$ is a finite abelian group.

Let $k[G]$ be the group algebra of $G$ over $k$. As we mentioned before, it has a Hopf algebra structure given by the comultiplication $\Delta(g) = g \otimes g$ and counit $\varepsilon(g) = 1$, for all $g \in G$, extended by linearity and as algebra maps. The antipode $S$ maps $g \in G$ to its inverse $g^{-1}$. Obviously $k[G]$ is a cocommutative $k$-coalgebra.

**Proposition 3.57** *Let $k$ be a field that contains a primitive nth root of unity. Then for any finite abelian group $C$ of order $n$ we have $k[C] \cong k[C]^*$ as Hopf algebras.*

*Proof* Consider first the situation when $C = C_n$, the cyclic group of order $n$. Since $k$ contains a primitive $n$th root of unity, say $\xi$, we deduce that the characteristic of $k$ does not divide $n$ (this follows easily from $n = (1-\xi)(1-\xi^2)\cdots(1-\xi^{n-1})$). Suppose $C = \langle c \rangle$, written multiplicatively, and let $\{P_1, P_c, \ldots, P_{c^{n-1}}\}$ be the basis of $k[C_n]^*$ dual to the basis $\{1, c, \ldots, c^{n-1}\}$ of $k[C_n]$.

Define $f \in k[C_n]^*$ by $f(c^i) = \xi^i$, for all $0 \le i \le n-1$. Then $f$ is a well-defined algebra map and $f^j(c^s) = \xi^{js}$, for all $0 \le s \le n-1$. We now claim that $\Psi : k[C_n] \ni c^j \mapsto f^j \in k[C_n]^*$, extended by linearity, is a Hopf algebra isomorphism. To see this we observe first that $f^j = \sum_{s=0}^{n-1} \xi^{js} P_{c^s}$, for all $0 \le j \le n-1$, and then we compute

$$\begin{aligned}\Delta_{k[C_n]^*}(\Psi(c^j)) = \Delta_{k[C_n]^*}(f^j) &= \sum_{s,t=0}^{n-1} f^j(c^{s+t}) P_{c^s} \otimes P_{c^t} = \sum_{s,t=0}^{n-1} \xi^{(s+t)j} P_{c^s} \otimes P_{c^t} \\ &= \sum_{s=0}^{n-1} \xi^{sj} P_{c^s} \otimes \sum_{t=0}^{n-1} \xi^{tj} P_{c^t} = f^j \otimes f^j = (\Psi \otimes \Psi)\Delta_{k[C_n]}(c^j)\end{aligned}$$

and $\varepsilon_{k[C_n]^*}(\Psi(c^j)) = \varepsilon_{k[C_n]^*}(f^j) = f^j(1) = 1 = \varepsilon_{k[C_n]}(c^j)$, for all $0 \le j \le n-1$. Thus, we have proved that $\Psi$ is a coalgebra morphism. It can be easily checked that $\Psi$ is an algebra morphism as well, and so it is a Hopf algebra morphism.

To show that $\Psi$ is an isomorphism we prove that $\{\varepsilon, f, \ldots, f^{n-1}\}$ is a basis of $k[C_n]^*$. Counting the dimensions it suffices to verify that $\{\varepsilon, f, \ldots, f^{n-1}\}$ is an independent linear system in $k[C_n]^*$. Indeed, if $\alpha_0, \ldots, \alpha_{n-1} \in k$ are such that $\sum_{j=0}^{n-1} \alpha_j f^j = 0$ then $\sum_{j=0}^{n-1} \alpha_j \xi^{js} = 0$, for all $0 \le s \le n-1$. In this way we get an $n \times n$ homogeneous linear system with indeterminates $\alpha_0, \cdots, \alpha_{n-1}$ whose discriminant is the Vandermonde discriminant of $1, \xi, \ldots, \xi^{n-1}$, usually denoted by $V(1, \xi, \ldots, \xi^{n-1})$.

Since $\xi$ is a primitive $n$th root of unity in $k$ we deduce that $V(1,\xi,\dots,\xi^{n-1})\neq 0$, therefore $\alpha_0=\cdots=\alpha_{n-1}=0$, as needed.

For further use note that the inverse of $\Psi$ is given by

$$\Psi^{-1}(P_{c^j})=\frac{1}{n}\sum_{s=0}^{n-1}\xi^{(n-s)j}c^s,\ \forall\, 0\leq j\leq n-1.$$

Indeed, for all $0\leq t\leq n-1$,

$$\begin{aligned}\frac{1}{n}&\sum_{s=0}^{n-1}\xi^{(n-s)j}f^s(c^t)\\ &=\frac{1}{n}\left(1+\xi^{t-j}+(\xi^{t-j})^2+\cdots+(\xi^{t-j})^{n-1}\right)=\frac{1}{n}\begin{cases} n & \text{if } t=j\\ \frac{1-(\xi^{t-j})^n}{1-\xi^{t-j}} & \text{if } t\neq j\end{cases}\\ &=\delta_{j,t}.\end{aligned}$$

So $\Psi\Psi^{-1}(P_{c^j})=P_{c^j}$, for all $0\leq j\leq n-1$, thus $\Psi$ and $\Psi^{-1}$ are bijective inverses.

Now take $C$ an arbitrary finite abelian group of order $n$. If we write $C$ as a direct product of finite cyclic groups then the Hopf algebra isomorphism $k[C]\cong k[C]^*$ follows from the first part of the proof and the fact that $k[G\times H]\ni(g,h)\mapsto g\otimes h\in k[G]\otimes k[H]$, extended by linearity, is an isomorphism of Hopf algebras, for any groups $G$ and $H$. Here $k[G]\otimes k[H]$ has the Hopf algebra structure given by the tensor product algebra and tensor product coalgebra structures. Actually, we have

$$k[G\times H]^*\cong(k[G]\otimes k[H])^*\cong k[G]^*\otimes k[H]^*\cong k[G]\otimes k[H]\cong k[G\times H]$$

as Hopf algebras, and then the general case follows by mathematical induction. Note that the second Hopf algebra isomorphism used above is given by

$$(k[G]\otimes k[H])^*\ni\mu\mapsto\sum_{g\in G,h\in H}\mu(g\otimes h)P_g\otimes P_h\in k[G]^*\otimes k[H]^*$$

and its inverse is $P_g\otimes P_h\mapsto\left(g'\otimes h'\mapsto\delta_{g,g'}\delta_{h,h'}\right)$. □

When $G$ is abelian, $k[G]$ is a commutative $k$-algebra. Thus in the abelian case a quasi-Hopf algebra structure on $k[G]$ is completely determined by a reassociator, that is, an invertible element $\Phi\in k[G]\otimes k[G]\otimes k[G]$ satisfying (3.1.9) and (3.1.10). If $G$ is, moreover, finite, as we have seen before, finding all the associators is equivalent to finding all the elements of $H^3(G,k^*)$. Next we point out how we can compute such elements $\Phi$ in some particular cases.

**Proposition 3.58** *Let $C_n=\langle c\rangle$ be the cyclic group of order $n$ written multiplicatively and $k$ a field that contains a primitive $n$th root of unity, say $\xi$. Then for any $k\in\{0,1,\dots,n-1\}$ the elements $\Phi_k\in k[C_n]\otimes k[C_n]\otimes k[C_n]$ of the form*

$$\Phi_k=1\otimes 1\otimes 1-\frac{1}{n^2}(1-c^k)\otimes\sum_{i,j=0}^{n-1}(1-n\delta_{i,j})(\xi^j-n\delta_{j,0})c^i\otimes c^j,$$

*are normalized 3-cocycles on $k[G]$. In particular, by considering $n=2$ we get that*

*$k[C_2]$ is a quasi-Hopf algebra with the non-trivial reassociator $\Phi_1 = 1\otimes 1\otimes 1 - 2p_-\otimes p_-\otimes p_-$, where $p_- = \frac{1}{2}(1-c)$.*

*Proof* Let $\Psi$ be the isomorphism from $k[C_n]$ to $k[C_n]^*$ defined in the proof of Proposition 3.57. Then we have

$$\sum_{j=0}^{n-1} q^j\Psi^{-1}(P_{c^j}) = \frac{1}{n}\sum_{s=0}^{n-1}\left(\sum_{j=0}^{n-1}\xi^{kj}(\xi^{n-s})^j\right)c^s$$
$$= \frac{1}{n}\sum_{s=0}^{n-1}\left(\sum_{j=0}^{n-1}(\xi^{k-s})^j\right)c^s = \frac{1}{n}\sum_{j=0}^{n-1} n\delta_{k,s}c^s = c^k.$$

Thus, according to the above comments and the definition of $\phi_q$ given in Example 1.16, we have that all the elements $\Phi\in k[C_n]\otimes k[C_n]\otimes k[C_n]$ of the form

$$\begin{aligned}
\Phi_k := &\sum_{u,v,s=0}^{n-1}\varphi_q(c^u,c^v,c^s)\Psi^{-1}(P_{c^u})\otimes\Psi^{-1}(P_{c^v})\otimes\Psi^{-1}(P_{c^s})\\
&= \sum_{u=0}^{n-1}\Psi^{-1}(P_{c^u})\otimes\sum_{v+s<n}\Psi^{-1}(P_{c^v})\otimes\Psi^{-1}(P_{c^s})\\
&\quad+\sum_{u=1}^{n-1}q^u\Psi^{-1}(P_{c^u})\otimes\sum_{v+s\geq n}\Psi^{-1}(P_{c^v})\otimes\Psi^{-1}(P_{c^s})\\
&= 1\otimes\sum_{v+s<n}\Psi^{-1}(P_{c^v})\otimes\Psi^{-1}(P_{c^s}) + c^k\otimes\sum_{v+s\geq n}\Psi^{-1}(P_{c^v})\otimes\Psi^{-1}(P_{c^s})\\
&= 1\otimes 1\otimes 1-(1-c^k)\otimes\sum_{v+s\geq n}\Psi^{-1}(P_{c^v})\otimes\Psi^{-1}(P_{c^s})\\
&= 1\otimes 1\otimes 1-(1-c^k)\otimes\sum_{v=1}^{n-1}\sum_{t=0}^{n-2}\Psi^{-1}(P_{c^v})\otimes\Psi^{-1}(P_{c^{n+t-v}})\\
&= 1\otimes 1\otimes 1-\frac{1}{n^2}(1-c^k)\otimes\sum_{i,j=0}^{n-1}\sum_{v=1}^{n-1}\sum_{t=0}^{n-2}\xi^{(n-i)v+(n-j)(n+t-v)}c^i\otimes c^j\\
&= 1\otimes 1\otimes 1-\frac{1}{n^2}(1-c^k)\otimes\sum_{i,j=0}^{n-1}\left(\sum_{v=1}^{n-1}(\xi^{j-i})^v\right)c^i\otimes\left(\sum_{t=0}^{n-2}(\xi^{-j})^t\right)c^j\\
&= 1\otimes 1\otimes 1-\frac{1}{n^2}(1-c^k)\otimes\sum_{i,j=0}^{n-1}(1-n\delta_{i,j})(\xi^j-n\delta_{j,0})c^i\otimes c^j,
\end{aligned}$$

are normalized 3-cocycles, as we stated. In the case when $n=2$ we compute

$$\begin{aligned}
\sum_{i,j=0}^{n}(1-2\delta_{i,j})((-1)^j-2\delta_{j,0})c^i\otimes c^j &= 1\otimes 1-1\otimes c-c\otimes 1+c\otimes c\\
&= (1-c)\otimes(1-c),
\end{aligned}$$

to deduce that $\Phi_1 = 1\otimes 1\otimes 1-2p_-\otimes p_-\otimes p_-$, as we claimed. Note that this is precisely the cocycle that endows $k[C_2]$ with the quasi-bialgebra structure described in Example 3.26. □

## 3.9 Notes

Quasi-bialgebras and quasi-Hopf algebras were introduced by Drinfeld [79, 80], and used to give an alternative proof for Kohno's theorem relating the monodromy of the Knizhnik–Zamolodchikov equations to a representation of the braid group arising from a quantum group. The dual concepts were studied in [148, 175, 41]. Both notions generalize the classical notion of Hopf algebra introduced by Hopf in 1941, and explained in a modern language by Milnor and More [156].

In the presentation of this chapter we used the books of Kassel [127] and Majid [148], and the papers of Drinfeld [79, 80]. Unlike these authors, we did not include the bijectivity of the antipode in the definition of a quasi-Hopf algebra. As we shall see later, this bijectivity is automatic in the case of a finite-dimensional or a quasitriangular quasi-Hopf algebra. Examples of quasi-bialgebras and quasi-Hopf algebras were taken from [91]. We mention that more examples can be obtained by using the explicit description of $H^3(\mathbb{Z}_2 \times \mathbb{Z}_2, k^*)$ from [53].The reconstruction type theorem in Section 3.5 is taken from [64],while the content of Section 3.6 is from [65].

# 4

# Module (Co)Algebras and (Bi)Comodule Algebras

We define the notions of module (co)algebra, respectively of (bi)comodule algebra over a quasi-bialgebra by using certain categorical points of view or by generalizing the axioms of a quasi-bialgebra, respectively. Then we give concrete classes of examples and the connections that exist between these structures.

## 4.1 Module Algebras over Quasi-bialgebras

Throughout this section $H$ is a quasi-bialgebra or a quasi-Hopf algebra with comultiplication $\Delta$, counit $\varepsilon$ and reassociator $\Phi$.

Recall from Chapter 3 that the category ${}_H\mathcal{M}$ of left $H$-modules is a monoidal category with tensor product $\otimes$ given via $\Delta$, associativity constraint

$$a = (a_{U,V,W} : (U \otimes V) \otimes W \to U \otimes (V \otimes W))_{U,V,W \,\in\, {}_H\mathcal{M}},$$
$$a_{U,V,W}((u \otimes v) \otimes w) = \Phi \cdot (u \otimes (v \otimes w)),$$

for all $u \in U$, $v \in V$ and $w \in W$, unit $k$ as a trivial $H$-module and the usual left and right unit constraints. So we can consider algebras within ${}_H\mathcal{M}$.

**Definition 4.1** Let $H$ be a quasi-bialgebra. We say that a $k$-vector space $A$ is a left $H$-module algebra if it is an algebra in the monoidal category ${}_H\mathcal{M}$, that is, $A$ has a multiplication (denoted by $a \otimes a' \mapsto aa'$) and a usual unit $1_A$ satisfying the conditions:

$$(aa')a'' = (X^1 \cdot a)[(X^2 \cdot a')(X^3 \cdot a'')], \tag{4.1.1}$$
$$h \cdot (aa') = (h_1 \cdot a)(h_2 \cdot a'), \tag{4.1.2}$$
$$h \cdot 1_A = \varepsilon(h) 1_A, \tag{4.1.3}$$

for all $a, a', a'' \in A$, $h \in H$, where $h \otimes a \to h \cdot a$ is the left $H$-module structure of $A$.

Likewise, the category $\mathcal{M}_H$ of right $H$-modules is monoidal. The tensor product, the unit object and the left and right unit constraints are similar to those of ${}_H\mathcal{M}$, while the associativity constraint is given by

$$\mathbf{a} = (\mathbf{a}_{U,V,W} : (U \otimes V) \otimes W \to U \otimes (V \otimes W))_{U,V,W \,\in\, \mathcal{M}_H},$$

$$\mathbf{a}_{U,V,W}((u\otimes v)\otimes w)=(u\otimes(v\otimes w))\cdot\Phi^{-1},$$

for all $u\in U$, $v\in V$ and $w\in W$. This leads to the notion of right module algebra over a quasi-bialgebra.

**Definition 4.2** Let $H$ be a quasi-bialgebra. We say that a $k$-linear space $B$ is a right $H$-module algebra if $B$ is an algebra in the monoidal category $\mathscr{M}_H$, that is, $B$ has a multiplication (denoted by $b\otimes b'\mapsto bb'$) and a usual unit $1_B$ satisfying the conditions:

$$(bb')b''=(b\cdot x^1)[(b'\cdot x^2)(b''\cdot x^3)], \tag{4.1.4}$$

$$(bb')\cdot h=(b\cdot h_1)(b'\cdot h_2), \tag{4.1.5}$$

$$1_B\cdot h=\varepsilon(h)1_B, \tag{4.1.6}$$

for all $b,b',b''\in B$, $h\in H$, where $b\otimes h\to b\cdot h$ is the right $H$-module structure of $B$.

If $H$ is a quasi-bialgebra then so is $H^{\rm op}$, where "op" means the opposite multiplication. The canonical identification $\mathscr{M}_H\equiv{}_{H^{\rm op}}\mathscr{M}$ is an isomorphism of monoidal categories. Thus, giving an algebra in $\mathscr{M}_H$ is equivalent to giving an algebra in ${}_{H^{\rm op}}\mathscr{M}$. This is why we will restrict ourselves to the study of left $H$-module algebras only.

We present a class of examples of module algebras over quasi-Hopf algebras.

**Proposition 4.3** *Let $H$ be a quasi-Hopf algebra, $A$ an associative unital algebra and $f:H\to A$ an algebra map. If we define on $A$ a new multiplication by*

$$a\circ b=f(X^1)af(S(x^1X^2)\alpha x^2X^3_1)bf(S(x^3X^3_2)),\ \ \forall\, a,b\in A, \tag{4.1.7}$$

*and we denote this new structure on $A$ by $A^f$, then $A^f$ becomes a left $H$-module algebra with unit $f(\beta)$ and with the left adjoint action induced by $f$, that is, $h\triangleright_f a=f(h_1)af(S(h_2))$, for all $h\in H$ and $a\in A$.*

*Proof* It is easy to see that $A^f$ is a left $H$-module. Now, we prove that $A^f$ is an algebra in ${}_H\mathscr{M}$, that is, the following relations hold, for all $a,b,c\in A$ and $h\in H$:

$$(i)\ \ (a\circ b)\circ c=(X^1\triangleright_f a)\circ[(X^2\triangleright_f b)\circ(X^3\triangleright_f c)],$$

$$(ii)\ \ a\circ f(\beta)=f(\beta)\circ a=a,$$

$$(iii)\ \ h\triangleright_f(a\circ b)=(h_1\triangleright_f a)\circ(h_2\triangleright_f b)\ \text{and}\ h\triangleright_f f(\beta)=\varepsilon(h)f(\beta).$$

We begin by noting that the relation (3.1.9) implies

$$a\circ b=f(X^1x^1_1)af(S(X^2x^1_2)\alpha X^3x^2)bf(S(x^3)). \tag{4.1.8}$$

Now, for (i) we have:

$$\begin{aligned}
&(X^1\triangleright_f a)\circ[(X^2\triangleright_f b)\circ(X^3\triangleright_f c)]\\
&\overset{(4.1.7)}{=}\ (X^1\triangleright_f a)\circ[f(Y^1X^2_1)bf(S(y^1Y^2X^2_2)\alpha y^2Y^3_1X^3_1)cf(S(y^3Y^3_2X^3_2))]\\
&\overset{(3.1.9)}{=}\ (X^1Y^1_1z^1\triangleright_f a)\circ[f(X^2Y^1_2z^2)bf(S(y^1X^3_1Y^2z^3)\alpha y^2X^3_{(2,1)})(Y^3\triangleright_f c)\\
&\qquad\qquad f(S(y^3X^3_{(2,2)}))]\\
&\overset{(3.1.7),\,(3.2.1)}{=}\ (X^1Y^1_1z^1\triangleright_f a)\circ[f(X^2Y^1_2z^2)bf(S(y^1Y^2z^3)\alpha y^2Y^3_1)cf(S(X^3y^3Y^3_2))]
\end{aligned}$$

$$
\begin{aligned}
&\overset{(4.1.8)}{=} f(Z^1)(x^1X^1Y_1^1z^1 \triangleright_f a)f(S(Z^2)\alpha Z^3x^2X^2Y_2^1z^2)bf(S(y^1Y^2z^3)\alpha y^2Y_1^3)\\
&\qquad cf(S(x^3X^3y^3Y_2^3))\\
&= f(Z^1Y^1_{(1,1)})(z^1 \triangleright_f a)f(S(Z^2Y^1_{(1,2)})\alpha Z^3Y_2^1z^2)bf(S(y^1Y^2z^3)\alpha y^2Y_1^3)\\
&\qquad cf(S(y^3Y_2^3))\\
&\overset{(3.1.7)}{=} f(Y_1^1X^1)(z^1 \triangleright_f a)f(S(Y^1_{(2,1)}X^2)\alpha Y^1_{(2,2)}X^3z^2)bf(S(y^1Y^2z^3)\alpha y^2Y_1^3)\\
&\qquad cf(S(y^3Y_2^3))\\
&\overset{(3.2.1),(3.1.8)}{=} f(Y^1X^1z_1^1)af(S(X^2z_2^1)\alpha X^3z^2)bf(S(y^1Y^2z^3)\alpha y^2Y_1^3)cf(S(y^3Y_2^3))\\
&\overset{(3.1.9)}{=} f(Y^1x^1X^1)af(S(x_1^2z^1X^2)\alpha x_2^2z^2X_1^3)bf(S(y^1Y^2x^3z^3X_2^3)\alpha y^2Y_1^3)\\
&\qquad cf(S(y^3Y_2^3))\\
&\overset{(3.2.1),(3.1.10)}{=} f(Y^1X^1)af(S(z^1X^2)\alpha z^2X_1^3)bf(S(y^1Y^2z^3X_2^3)\alpha y^2Y_1^3)cf(S(y^3Y_2^3))\\
&\overset{(4.1.7)}{=} [f(X^1)af(S(z^1X^2)\alpha z^2X_1^3)bf(S(z^3X_2^3))]\circ c\\
&= (a\circ b)\circ c.
\end{aligned}
$$

For (ii) we use the definitions. In fact, for all $a \in A$ we have:

$$
\begin{aligned}
a\circ f(\beta) &= f(X^1)af(S(x^1X^2)\alpha x^2X_1^3\beta S(x^3X_2^3))\\
&\overset{(3.2.1),(3.1.11)}{=} af(S(x^1)\alpha x^2\beta S(x^3))\overset{(3.2.2)}{=} a.
\end{aligned}
$$

Similarly, by using (4.1.8) we can obtain $f(\beta)\circ a = a$ for all $a \in A$, therefore $f(\beta)$ is the unit for $A^f$. Finally, we have to prove the relations in (iii). Indeed, by using (3.1.7), for all $a,b \in A$ and $h \in H$ we have:

$$
\begin{aligned}
(h_1 \triangleright_f a)\circ(h_2 \triangleright_f b) &= [f(h_{(1,1)})af(S(h_{(1,2)}))]\circ[f(h_{(2,1)})bf(S(h_{(2,2)}))]\\
&= f(X^1h_{(1,1)})af(S(X^2h_{(1,2)})S(x^1)\alpha x^2(X^3h_2)_1)b\\
&\qquad f(S((X^3h_2)_2)S(x^3))\\
&= f(h_1X^1)af(S(X^2)S(x^1h_{(2,1)})\alpha x^2h_{(2,2,1)}X_1^3)b\\
&\qquad f(S(X_2^3)S(x^3h_{(2,2,2)}))\\
&= f(h_1X^1)af(S(x^1X^2)S(h_{(2,1,1)})\alpha h_{(2,1,2)}x^2X_1^3)b\\
&\qquad f(S(x^3X_2^3)S(h_{(2,2)}))\\
&\overset{(3.2.1)}{=} f(h_1X^1)af(S(x^1X^2)\alpha x^2X_1^3)bf(S(x^3X_2^3)S(h_2))\\
&= f(h_1)(a\circ b)f(S(h_2)) = h\triangleright_f(a\circ b).
\end{aligned}
$$

Note that $h\triangleright_f f(\beta) = \varepsilon(h)f(\beta)$ is just (3.2.1). □

**Definition 4.4** If $H$ is a quasi-Hopf algebra, we can take $A = H$ and $f = \mathrm{Id}_H$ in Proposition 4.3, obtaining the left $H$-module algebra $H^{\mathrm{Id}_H}$, which will be denoted in what follows by $H_0$. Its multiplication is defined by

$$
g\circ h = X^1gS(x^1X^2)\alpha x^2X_1^3hS(x^3X_2^3),\ \forall\, g,h \in H, \tag{4.1.9}
$$

its unit is the element $\beta$ and the left action of $H$ on $H_0$ is defined by $h\triangleright g = h_1gS(h_2)$, for all $h \in H$ and $g \in H_0$.

**Remark 4.5** Let $H$ be a quasi-Hopf algebra and $A$, $B$ two left $H$-module algebras. Denote by $\mathrm{Alg}(A,B)$ the set of multiplicative and unital maps from $A$ to $B$ and by $\mathrm{Alg}_H(A,B)$ the set of elements $f \in \mathrm{Alg}(A,B)$ that are also $H$-module maps. An element in $\mathrm{Alg}_H(A,B)$ is called a morphism of left $H$-module algebras from $A$ to $B$.

In the hypotheses and notation of Proposition 4.3, it is easy to see that the map $f$ belongs to $\mathrm{Alg}_H(H_0, A^f)$.

We now give a class of examples of module algebras over quasi-bialgebras. To this end we first prove the following general result.

**Proposition 4.6** *Let $H$ be a dual quasi-bialgebra and denote by $\mathscr{M}^H$ its category of right comodules. If $H$ is finite dimensional then $\mathscr{M}^H$ is monoidally isomorphic to the category ${}_{H^*}\mathscr{M}$, where $H^*$ is the quasi-bialgebra dual to $H$. Consequently, giving an algebra in $\mathscr{M}^H$ is equivalent to giving a left $H^*$-module algebra.*

*Proof* We recall that the quasi-bialgebra structure of $H^*$ is given by the dual structure of $H$ and the reassociator

$$\Phi := \sum_{i,j,s} \varphi(e_i, e_j, e_s) e^i \otimes e^j \otimes e^s,$$

where $\{e_i, e^i\}_i$ are dual bases in $H$ and $H^*$.

That $\mathscr{M}^H$ and ${}_{H^*}\mathscr{M}$ are isomorphic follows from Proposition 2.42 applied to $\mathscr{C} = {}_k\mathscr{M}$ and $C = H$. Explicitly, the isomorphism is produced by the functor $F : \mathscr{M}^H \to {}_{H^*}\mathscr{M}$ defined as follows. $F$ acts as identity on objects and morphisms. If $M$ is a right $H$-comodule via the structure map $\rho : M \ni m \mapsto m_{(0)} \otimes m_{(1)} \in M \otimes H$ then $F(M) = M$ becomes a left $H^*$-module via $h^* \cdot m = h^*(m_{(1)}) m_{(0)}$, for all $m \in M$ and $h^* \in H^*$.

The inverse of $F$ is $G : {}_{H^*}\mathscr{M} \to \mathscr{M}^H$ defined as follows. $G$ acts as identity on objects and morphisms. If $N$ is a left $H^*$-module then $G(N)$ is the $k$-vector space $N$ endowed with the right $H$-comodule structure given by

$$\rho : N \ni n \mapsto e^i \cdot n \otimes e_i \in N \otimes H.$$

It remains to show that $F$ is a strong monoidal functor. Actually, we claim that $F$ is a strict monoidal functor. Indeed, since the right $H$-comodule structure on the tensor product $M \otimes M'$ of two right $H$-comodules $M$ and $M'$ is given by

$$M \otimes M' \ni m \otimes m' \mapsto m_{(0)} \otimes m'_{(0)} \otimes m_{(1)} m'_{(1)} \in M \otimes M' \otimes H,$$

it follows that the induced left $H^*$-module structure on $M \otimes M'$ is determined by

$$\begin{aligned} h^* \cdot (m \otimes m') &= h^*(m_{(1)} m'_{(1)}) m_{(0)} \otimes m'_{(0)} \\ &= h_1^*(m_{(1)}) h_2^*(m'_{(1)}) m_{(0)} \otimes m'_{(0)} \\ &= h_1^* \cdot m \otimes h_2^* \cdot m', \end{aligned}$$

for all $h^* \in H^*$, $m \in M$ and $m' \in M'$. This shows that the identity morphism from $F(M) \otimes F(M')$ to $F(M \otimes M')$ is a left $H^*$-module isomorphism, and therefore $F$ is a strict monoidal functor, as stated.

The last assertion in the statement follows from Proposition 2.3. □

Now consider the following context: $G$ is a group with neutral element $e$, $k$ is a field and $F$ is a 2-cochain on $G$, that is, a map $F : G\times G \to k^*$ obeying $F(e,x) = F(y,e) = 1$, for all $x,y \in G$. We deform the multiplication of the group algebra $k[G]$ in such a way that with respect to this new multiplication the $k$-vector space $k[G]$ becomes an algebra within the category $\mathrm{Vect}^G$, endowed with the monoidal structure given by the coboundary normalized 3-cocycle $\phi := \Delta_2(F^{-1})$. Here $F^{-1}$ is the pointwise multiplication inverse of $F$.

More precisely, we will prove the following.

**Proposition 4.7** *Let $G$, $k$ and $F$ be as above and denote by $k_F[G]$ the $k$-vector space $k[G]$ endowed with the multiplication*

$$x \bullet y = F(x,y)xy, \quad \forall\, x,y \in G,$$

*extended by linearity. Then $k_F[G]$ is an algebra within $\mathrm{Vect}^G_{\Delta_2(F^{-1})}$. Furthermore, if $G$ is abelian then $k_F[G]$ is a braided commutative algebra within $\mathrm{Vect}^G_{\Delta_2(F^{-1}),\mathscr{R}_{F^{-1}}}$.*

*Proof* Denote $\phi := \Delta_2(F^{-1})$. Then $\phi$ is given by

$$\phi(x,y,z) := \frac{F(x,y)F(xy,z)}{F(y,z)F(x,yz)}, \quad \forall\, x,y,z \in G,$$

and is a normalized 3-cocycle on $G$. Since $k_F[G] = k[G]$ as $k$-vector spaces it follows that $k_F[G]$ is a $G$-graded vector space with the same grading as that of $k[G]$; namely, $(k_F[G])_g = kg$, for all $g \in G$. It is a $G$-graded quasialgebra with reassociator $\phi$ since

$$\begin{aligned}(x\bullet y)\bullet z &= F(x,y)(xy)\bullet z = F(x,y)F(xy,z)(xy)z = \phi(x,y,z)F(y,z)F(x,yz)x(yz)\\ &= \phi(x,y,z)F(y,z)x\bullet(yz) = \phi(x,y,z)x\bullet(y\bullet z),\end{aligned}$$

for all $x,y,z \in G$, so that the relations in (2.1.1) are satisfied.

If $G$ is abelian, $\mathrm{Vect}^G_{\Delta_2(F^{-1})}$ is braided monoidal with the braiding $c$ defined by $\mathscr{R}_{F^{-1}}$ from Remark 1.46, that is, $\mathscr{R}_{F^{-1}}(x,y) = \frac{F(x,y)}{F(y,x)}$, for all $x,y \in G$. It is clear that

$$\bullet \circ c_{k_F[G],k_F[G]}(x,y) = \frac{F(x,y)}{F(y,x)}y\bullet x = F(x,y)yx = F(x,y)xy = x\bullet y,$$

for all $x,y \in G$, and this shows that $k_F[G]$ is a braided commutative algebra. □

Let $k_\phi[G]$ be the dual quasi-Hopf algebra constructed in Example 3.50 from the data $(k,G,\phi := \Delta_2(F^{-1}))$. By Proposition 3.51 it follows that $k_F[G]$ is an algebra within $\mathscr{M}^{k_\phi[G]}$. Thus by Proposition 4.6 we get the following class of examples of module algebras over quasi-bialgebras.

**Corollary 4.8** *Let $G$ be a finite group and $F$ a 2-cochain on $G$. Then $k_F[G]$ is a left $k_\phi[G]^*$-module algebra.*

A concrete example of module algebra as in Corollary 4.8 can be deduced from Proposition 3.54. Note that another one can be given by using Example 3.55.

**Example 4.9** Let $k$ be a field that contains a primitive $n$th root of unity $q$ such that $q^{\frac{n(n-1)}{2}} = 1$. Let $C_n = \langle \sigma \rangle$ be the cyclic group of order $n$ written multiplicatively and $k[C_n]$ the free $k$-vector space with basis $C_n$. Then the following assertions hold.

(*i*) $k[C_n]$ equipped with the group algebra structure, with the coalgebra structure given by $\Delta(\sigma^a) = \sigma^a \otimes \sigma^a$ and $\varepsilon(\sigma^a) = 1$, for all $0 \leq a \leq n-1$, extended as algebra maps and by linearity, and with the reassociator

$$\Phi = \frac{1}{n^2} \sum_{a,b,c,d=0}^{n-1} q^{-ac-bd} \sigma^{ab} \otimes \sigma^c \otimes \sigma^d$$

is a quasi-bialgebra that we denote by $H$.

(*ii*) Let $A$ be the $k$-vector space $k[C_n]$ endowed with the multiplication

$$\sigma^a \bullet \sigma^b = q^{\frac{(a-1)ab}{2}} \sigma^{a+b}, \ \forall\, 0 \leq a, b \leq n-1,$$

and unit $e$, the neutral element of $C_n$. Then $A$ is a left $H$-module algebra.

*Proof* Indeed, from Proposition 3.54 we know that $\phi : C_n \times C_n \times C_n \to k^*$ given by $\phi(\sigma^a, \sigma^b, \sigma^c) = q^{abc}$, for all $0 \leq a, b, c \leq n-1$, is a coboundary normalized 3-cocycle of $C_n$. More precisely, if $F : C_n \times C_n \to k^*$ is defined by $F(\sigma^a, \sigma^b) = q^{\frac{(a-1)ab}{2}}$, for all $0 \leq a, b \leq n-1$, then $\phi = \Delta_2(F^{-1})$. Thus $k_F[C_n]$ is a left $k_\phi[C_n]^*$-module algebra; see Corollary 4.8. Recall that the multiplication on $k_F[C_n]$ is given by

$$\sigma^a \bullet \sigma^b = F(\sigma^a, \sigma^b)\sigma^{a+b} = q^{\frac{(a-1)ab}{2}} \sigma^{a+b}, \ \forall\, 0 \leq a,\ b \leq n-1,$$

and that the unit of $k_F[G]$ is $e$. Hence $k_F[C_n] = A$ as algebras and so the proof ends if we show that $k_\phi[C_n]^* \cong H$ as a quasi-bialgebra.

From the proof of Proposition 3.57 we have that $\Psi : k[C_n] \to k[C_n]^*$ given by $\Psi(\sigma^a)(\sigma^b) = q^{ab}$ is a Hopf algebra isomorphism. Its inverse is

$$\Psi^{-1} : k[C_n]^* \to k[C_n], \quad \Psi^{-1}(P_{\sigma^a}) = \frac{1}{n} \sum_{s=0}^{n-1} q^{(n-s)a} \sigma^s,$$

extended by linearity, where $\{P_{\sigma^a}\}_a$ is the basis in $k[C_n]^*$ dual to the basis $\{\sigma^a | 0 \leq a \leq n-1\}$ of $k[C_n]$. Since $k_\phi[C_n]$ is a dual quasi-bialgebra it follows that $k[C_n]$ admits a unique quasi-bialgebra structure such that $\Psi$ becomes a quasi-bialgebra isomorphism. It is clear that this quasi-bialgebra structure on $k[C_n]$ is given by the group algebra and coalgebra structure of $k[C_n]$ and the reassociator

$$\begin{aligned}
\Phi &= \sum_{a,b,c=0}^{n-1} \phi(\sigma^a, \sigma^b, \sigma^c) \Psi^{-1}(P_{\sigma^a}) \otimes \Psi^{-1}(P_{\sigma^b}) \otimes \Psi^{-1}(P_{\sigma^c}) \\
&= \frac{1}{n^3} \sum_{a,b,c,i,j,s=0}^{n-1} q^{abc-ai-bj-cs} \sigma^i \otimes \sigma^j \otimes \sigma^s \\
&= \frac{1}{n^3} \sum_{b,c,i,j,s=0}^{n-1} \left( \sum_{a=0}^{n-1} (q^{bc-i})^a \right) q^{-bj-cs} \sigma^i \otimes \sigma^j \otimes \sigma^s \\
&= \frac{1}{n^2} \sum_{b,c,i,j,s=0}^{n-1} \delta_{bc,i} q^{-bj-cs} \sigma^i \otimes \sigma^j \otimes \sigma^s
\end{aligned}$$

$$= \frac{1}{n^2} \sum_{a,b,c,d=0}^{n-1} q^{-ac-bd} \sigma^{ab} \otimes \sigma^c \otimes \sigma^d.$$

In other words $k[C_n]$ endowed with the quasi-bialgebra structure that turns $\Psi$ into a quasi-bialgebra isomorphism is equal to $H$ from the statement. □

**Remark 4.10** A coboundary 3-cocycle $\phi = \Delta_2(F^{-1})$ is trivial if and only if

$$F(x,y)F(xy,z) = F(y,z)F(x,yz), \ \forall\, x,y \in G. \tag{4.1.10}$$

So $\phi$ is trivial if and only if $F$ is a normalized 2-cocycle on $G$ with coefficients in $k^*$.

**Definition 4.11** Two normalized 2-cocycles $F_1, F_2$ on a group $G$ are said to be cohomologous if there exists $s : G \to k^*$ such that $s(e) = 1$ and

$$F_1(x,y) = F_2(x,y)s(x)s(y)s(xy)^{-1},$$

for all $x,y \in G$, where $e$ is the neutral element of $G$.

**Proposition 4.12** *If $F_1, F_2$ are cohomologous 2-cocycles on $G$ then $k_{F_1}[G] \cong k_{F_2}[G]$ as a $k$-algebra.*

*Proof* The algebras $k_{F_1}[G]$ and $k_{F_2}[G]$ are unital associative $k$-algebras since $F_1$, $F_2$ are normalized 2-cocycles on $G$.

By our assumption there exists $s : G \to k^*$ satisfying the two conditions above. We claim that $\varphi : k_{F_1}[G] \to k_{F_2}[G]$ defined by $\varphi(x) = s(x)x$, for all $x \in G$, and extended by linearity, is a $k$-algebra isomorphism. Indeed, $\varphi(e) = s(e)e = e$ and

$$\begin{aligned}\varphi(x \bullet y) &= F_1(x,y)\varphi(xy) = F_1(x,y)s(xy)xy \\ &= F_2(x,y)s(x)s(y)xy = s(x)s(y)x \bullet y = \varphi(x) \bullet \varphi(y),\end{aligned}$$

for all $x,y \in G$. $\varphi$ is an isomorphism because $s$ is pointwise invertible, and so $k_{F_1}[G] \cong k_{F_2}[G]$ as a $k$-algebra. □

We end this section with the notion of bimodule algebra over a quasi-bialgebra.

As we mentioned before, if $H$ is a quasi-bialgebra then so is $H^{\rm op}$, where "op" means the opposite multiplication. The reassociator of $H^{\rm op}$ is $\Phi_{\rm op} = \Phi^{-1}$. Hence $H \otimes H^{\rm op}$ is a quasi-bialgebra with reassociator

$$\Phi_{H\otimes H^{\rm op}} = (X^1 \otimes x^1) \otimes (X^2 \otimes x^2) \otimes (X^3 \otimes x^3). \tag{4.1.11}$$

If we identify left $H \otimes H^{\rm op}$-modules with $H$-bimodules, then the category of $H$-bimodules, ${}_H\mathscr{M}_H$, is monoidal. More precisely, we have the following monoidal structure on ${}_H\mathscr{M}_H$.

**Lemma 4.13** *For a quasi-bialgebra $H$ the category ${}_H\mathscr{M}_H$ of bimodules over $H$ is monoidal with the following structure:*

- *for $M,N \in {}_H\mathscr{M}_H$ the tensor product $M \otimes N$ is an $H$-bimodule via the $H$-actions given by $h \cdot (m \otimes n) \cdot h' = h_1 \cdot m \cdot h'_1 \otimes h_2 \cdot n \cdot h'_2$, for all $m \in M$, $n \in N$ and $h,h' \in H$;*

- *the associativity constraint* $a_{M,N,P} : (M \otimes N) \otimes P \to M \otimes (N \otimes P)$ *is given by*

$$a_{M,N,P}((m \otimes n) \otimes p) = X^1 \cdot m \cdot x^1 \otimes (X^2 \cdot n \cdot x^2 \otimes X^3 \cdot p \cdot x^3); \tag{4.1.12}$$

- *the unit object is* $k$ *regarded as an* $H$*-bimodule via the counit* $\varepsilon$ *of* $H$;
- *the left and right unit constraints are given by the natural isomorphisms* $k \otimes M \cong M \cong M \otimes k$.

Therefore, we can define algebras in the category of $H$-bimodules. Such an algebra will be called an $H$-bimodule algebra. More precisely, we have the following definition.

**Definition 4.14** Let $H$ be a quasi-bialgebra. A $k$-vector space $\mathscr{A}$ is an $H$-bimodule algebra if $\mathscr{A}$ is an $H$-bimodule (denote the actions by $h \cdot \varphi$ and $\varphi \cdot h$, for $h \in H$ and $\varphi \in \mathscr{A}$) which has a multiplication and a usual unit $1_{\mathscr{A}}$ such that for all $\varphi, \varphi', \varphi'' \in \mathscr{A}$ and $h \in H$ the following relations hold:

$$(\varphi\varphi')\varphi'' = (X^1 \cdot \varphi \cdot x^1)[(X^2 \cdot \varphi' \cdot x^2)(X^3 \cdot \varphi'' \cdot x^3)], \tag{4.1.13}$$

$$h \cdot (\varphi\varphi') = (h_1 \cdot \varphi)(h_2 \cdot \varphi'), \ \ (\varphi\varphi') \cdot h = (\varphi \cdot h_1)(\varphi' \cdot h_2), \tag{4.1.14}$$

$$h \cdot 1_{\mathscr{A}} = \varepsilon(h)1_{\mathscr{A}}, \ \ 1_{\mathscr{A}} \cdot h = \varepsilon(h)1_{\mathscr{A}}. \tag{4.1.15}$$

To any quasi-bialgebra $H$ we can associate an $H$-bimodule algebra as follows.

**Example 4.15** Let $H$ be a quasi-bialgebra. Then $H^*$, the linear dual of $H$, is an $H$-bimodule via the $H$-actions

$$\langle h \rightharpoonup \varphi, h' \rangle = \varphi(h'h), \ \ \langle \varphi \leftharpoonup h, h' \rangle = \varphi(hh'), \tag{4.1.16}$$

for all $\varphi \in H^*$ and $h, h' \in H$. The convolution $\langle \varphi\psi, h \rangle = \varphi(h_1)\psi(h_2)$, $\varphi, \psi \in H^*$, $h \in H$, is a multiplication on $H^*$; it is not in general associative, but with this multiplication $H^*$ becomes an $H$-bimodule algebra.

*Proof* Follows easily from the axioms of a quasi-bialgebra. □

## 4.2 Module Coalgebras over Quasi-bialgebras

Let $H$ be a quasi-bialgebra. In the previous section we considered algebras within the monoidal categories ${}_H\mathscr{M}$, $\mathscr{M}_H$ and ${}_H\mathscr{M}_H$, and this led to the notions of left and right $H$-module algebra and to that of an $H$-bimodule algebra, respectively. In this section we will deal with the dual situation. Namely, instead of algebras we will consider coalgebras in the above mentioned monoidal categories.

**Definition 4.16** Let $H$ be a quasi-bialgebra. We call a coalgebra $C$ in ${}_H\mathscr{M}$ a left $H$-module coalgebra. So $C$ is a left $H$-module (denote by $\cdot$ its left $H$-module structure) with two morphisms $\underline{\Delta} : C \to C \otimes C$ and $\underline{\varepsilon} : C \to k$ in ${}_H\mathscr{M}$ such that $\underline{\varepsilon}$ is a counit for $\underline{\Delta}$ and $\underline{\Delta}$ is coassociative in ${}_H\mathscr{M}$. Explicitly, we have

$$\Phi \cdot (\underline{\Delta} \otimes \mathrm{Id}_C)(\underline{\Delta}(c)) = (\mathrm{Id}_C \otimes \underline{\Delta})(\underline{\Delta}(c)), \ \ \forall\, c \in C, \tag{4.2.1}$$

$$\underline{\varepsilon}(c_{\underline{1}})c_{\underline{2}} = \underline{\varepsilon}(c_{\underline{2}})c_{\underline{1}} = c, \ \forall\, c \in C, \tag{4.2.2}$$

$$\underline{\Delta}(h \cdot c) = h_1 \cdot c_{\underline{1}} \otimes h_2 \cdot c_{\underline{2}}, \ \forall\, c \in C, h \in H, \tag{4.2.3}$$

$$\underline{\varepsilon}(h \cdot c) = \varepsilon(h)\underline{\varepsilon}(c), \ \forall\, c \in C, h \in H, \tag{4.2.4}$$

where we used the Sweedler-type notation

$$\underline{\Delta}(c) = c_{\underline{1}} \otimes c_{\underline{2}}, \qquad (\underline{\Delta} \otimes \mathrm{Id}_C)(\underline{\Delta}(c)) = c_{(\underline{1},\underline{1})} \otimes c_{(\underline{1},\underline{2})} \otimes c_{\underline{2}}, \ \text{ etc.}$$

Likewise, a right $H$-module coalgebra is a coalgebra $C$ in $\mathscr{M}_H$. So $C$ is a right $H$-module (denote by $\cdot$ the right $H$-action on $C$) equipped with two morphisms $\underline{\Delta}: C \to C \otimes C$ and $\underline{\varepsilon}: C \to k$ in $\mathscr{M}_H$ such that $\underline{\varepsilon}$ is a counit for $\underline{\Delta}$ and $\underline{\Delta}$ is coassociative in $\mathscr{M}_H$. Explicitly, this means

$$(\underline{\Delta} \otimes \mathrm{Id}_C)(\underline{\Delta}(c)) \cdot \Phi^{-1} = (\mathrm{Id}_C \otimes \underline{\Delta})(\underline{\Delta}(c)), \ \forall\, c \in C, \tag{4.2.5}$$

$$\underline{\varepsilon}(c_{\underline{1}})c_{\underline{2}} = \underline{\varepsilon}(c_{\underline{2}})c_{\underline{1}} = c, \ \forall\, c \in C, \tag{4.2.6}$$

$$\underline{\Delta}(c \cdot h) = c_{\underline{1}} \cdot h_1 \otimes c_{\underline{2}} \cdot h_2, \ \forall\, c \in C, h \in H, \tag{4.2.7}$$

$$\underline{\varepsilon}(c \cdot h) = \underline{\varepsilon}(c)\varepsilon(h), \ \forall\, c \in C, h \in H. \tag{4.2.8}$$

It is immediate that a coalgebra in ${}_H\mathscr{M}$ is nothing else than a coalgebra in $\mathscr{M}_{H^{\mathrm{op}}}$ and vice versa. Hence, as in the algebra case, in most situations we deal only with the left-handed version of this concept.

If $H$ is an ordinary bialgebra then $H$ is a left and right $H$-module coalgebra; the $H$-module structure on $H$ is given by the multiplication of $H$ and the coalgebra structure $(\underline{\Delta}, \underline{\varepsilon})$ within the category of $H$-representations coincides with the coalgebra structure of $H$. We do not have a similar result for quasi-bialgebras, since for a quasi-bialgebra $H$ the coassociativity is controlled by an inner automorphism defined by $\Phi$ while for a coalgebra in ${}_H\mathscr{M}$ the coassociativity is up to multiplication by $\Phi$ on one side; see (3.1.7) and (4.2.1), respectively. However, a first class of examples of module coalgebras over a quasi-bialgebra can be obtained as follows.

**Proposition 4.17** *Let $F \in H \otimes H$ be a gauge transformation on a quasi-bialgebra $H$ and $(C, \underline{\Delta}, \underline{\varepsilon})$ a left $H$-module coalgebra. If we define on $C$ a new comultiplication by $\underline{\Delta}_F(c) = F\underline{\Delta}(c)$ and denote by $C^F$ the resulting structure, then $C^F$ becomes a left $H_F$-module coalgebra. The $H_F$-action on $C^F$ is the same as the $H$-action on $C$ and the counit of $C^F$ is $\underline{\varepsilon}$.*

*Proof* By Proposition 3.5 the categories ${}_H\mathscr{M}$ and ${}_{H_F}\mathscr{M}$ are monoidally isomorphic. The monoidal isomorphism is produced by the identity functor Id that is strong monoidal with the structure defined in (3.1.6). Therefore it carries coalgebras to coalgebras; see Proposition 2.16. Furthermore, from the proof of Proposition 2.16 we get easily that $\mathrm{Id}(C) = C^F$ as coalgebras in ${}_{H_F}\mathscr{M}$. □

Thus, if we start with $H$ a bialgebra and $F$ a gauge transformation on $H$, we can construct an $H_F$-module coalgebra $H^F$ but in the quasi-Hopf sense now. On the other hand, we have seen that to any quasi-Hopf algebra $H$ we can associate the $H$-module algebra $H_0$, obtained from $H$ by deforming its multiplication. So it is natural to see if a similar procedure exists in the module coalgebra case, too. As we next see, if we try to deform "canonically" the comultiplication of a quasi-bialgebra $H$ in such a way that it becomes a module coalgebra with this new comultiplication, then $H$ must come from a bialgebra deformed by a gauge transformation.

**Proposition 4.18** *Let $H$ be a quasi-bialgebra and $F = F^1 \otimes F^2 \in H \otimes H$ a gauge transformation on $H$. Define $\underline{\Delta} : H \to H \otimes H$ by $\underline{\Delta}(h) = h_1 F^1 \otimes h_2 F^2$, for all $h \in H$, where $\Delta(h) = h_1 \otimes h_2$. If we regard $H$ as a left $H$-module via its multiplication then $(H, \underline{\Delta}, \varepsilon)$ is a left $H$-module coalgebra if and only if $H_{F^{-1}}$ is an ordinary bialgebra, that is, $H$ is a quasi-bialgebra obtained by twisting from an ordinary bialgebra.*

*Proof* Clearly $\underline{\Delta}$ is left $H$-linear (this forces us to multiply $\Delta(h)$ by $F$ on the right and not on the left), and so is $\varepsilon$. Thus $(H, \underline{\Delta}, \varepsilon)$ is a left $H$-module coalgebra if and only if $\Phi(\underline{\Delta} \otimes \mathrm{Id}_H)\underline{\Delta}(h) = (\mathrm{Id}_H \otimes \underline{\Delta})\underline{\Delta}(h)$, for all $h \in H$. By (3.1.7) this means

$$h_1 X^1 F_1^1 \mathscr{F}^1 \otimes h_{(2,1)} X^2 F_2^1 \mathscr{F}^2 \otimes h_{(2,2)} X^3 F^2 = h_1 F^1 \otimes h_{(2,1)} F_1^2 \mathscr{F}^1 \otimes h_{(2,2)} F_2^2 \mathscr{F}^2,$$

for all $h \in H$, where $\mathscr{F}^1 \otimes \mathscr{F}^2$ is a second copy of $F$. Thus $(H, \underline{\Delta}, \varepsilon)$ is a left $H$-module coalgebra if and only if

$$X^1 F_1^1 \mathscr{F}^1 \otimes X^2 F_2^1 \mathscr{F}^2 \otimes X^3 F^2 = F^1 \otimes F_1^2 \mathscr{F}^1 \otimes F_2^2 \mathscr{F}^2,$$

or, equivalently, if and only if $\Phi_{F^{-1}} = 1 \otimes 1 \otimes 1$; see (3.1.5). So, we have shown that $H_{F^{-1}}$ is an ordinary $k$-bialgebra. Now everything follows since $H = (H_{F^{-1}})_F$. □

By Proposition 4.17 we can construct module coalgebras out of a finite group $G$ and a 2-cochain $F : G \times G \to k^*$ on $G$. Owing to the occurrence of $|G|$, the order of $G$, and of the multiple identifications that we need to perform, for completeness we prefer to supply a direct proof for the result below.

**Proposition 4.19** *Let $G$ be a finite group and $F$ a 2-cochain on $G$ with coefficients in $k^*$, where $k$ is a field such that $|G| \neq 0$ in $k$. If we denote by $k^F[G]$ the $G$-graded $k$-vector space $k[G]$, endowed with the comultiplication $\Delta_F$ and the counit $\varepsilon_F$ given by*

$$\Delta_F(x) = \frac{1}{|G|} \sum_{u \in G} F(u, u^{-1}x)^{-1} u \otimes u^{-1}x \ \text{ and } \ \varepsilon_F(x) = |G| \delta_{x,e},$$

*respectively, for all $x \in G$, where $\delta_{x,e}$ is the Kronecker delta, then $k^F[G]$ is a $G$-graded quasicoalgebra with reassociator $\Delta_2(F^{-1})$. When $G$ is abelian $k^F[G]$ is, moreover, a cocommutative coalgebra in $\mathrm{Vect}^G_{(\Delta_2(F^{-1}), \mathscr{R}_{F^{-1}})}$.*

*Proof* That $\varepsilon_F$ is a counit for $\Delta_F$ follows directly from their definitions and the fact that $F(e,x) = F(y,e) = 1$, for all $x, y \in G$. The comultiplication $\Delta_F$ is coassociative up to the reassociator $\phi := \Delta_2(F^{-1})$, $\phi(x,y,z) = \frac{F(x,y)F(xy,z)}{F(y,z)F(x,yz)}$, since

$$x \quad \overset{\Delta_F}{\to} \quad \frac{1}{|G|} \sum_{u \in G} F(u, u^{-1}x)^{-1} u \otimes u^{-1}x$$

$$\begin{aligned}
&\xrightarrow{\Delta_F\otimes \mathrm{Id}_{k^F[G]}} && \frac{1}{|G|^2}\sum_{u,v\in G} F(u,u^{-1}x)^{-1}F(v,v^{-1}u)^{-1}((v\otimes v^{-1}u)\otimes u^{-1}x)\\
&\xrightarrow{a_{k^F[G],k^F[G],k^F[G]}} && \frac{1}{|G|^2}\sum_{u,v\in G} F(v^{-1}u,u^{-1}x)^{-1}F(v,v^{-1}x)^{-1}v\otimes(v^{-1}u\otimes u^{-1}x)\\
&= && \frac{1}{|G|^2}\sum_{\theta,v\in G} F(v,v^{-1}x)^{-1}F(\theta,\theta^{-1}v^{-1}x)^{-1}v\otimes(\theta\otimes\theta^{-1}v^{-1}x)\\
&= && (\mathrm{Id}_{k^F[G]}\otimes\Delta_F)\Delta_F(x),
\end{aligned}$$

for all $x\in G$.

Now assume that $G$ is abelian. In order to see that $k^F[G]$ is braided cocommutative recall first that $\mathrm{Vect}^G_{\Delta_2(F^{-1})}$ is braided monoidal with

$$c_{M,N}(m\otimes n)=\mathscr{R}_{F^{-1}}(|m|,|n|)n\otimes m=\frac{F(|m|,|n|)}{F(|n|,|m|)}n\otimes m,$$

for any $M,N\in\mathrm{Vect}^G$ and homogeneous elements $m\in M$ and $n\in N$, respectively. We then compute

$$\begin{aligned}
x &\xrightarrow{\Delta_F} && \frac{1}{|G|}\sum_{u\in G} F(u,u^{-1}x)^{-1}u\otimes u^{-1}x\\
&\xrightarrow{c_{k^F[G],k^F[G]}} && \frac{1}{|G|}\sum_{u\in G} F(u^{-1}x,u)^{-1}u^{-1}x\otimes u=\Delta_F(x),
\end{aligned}$$

for all $x\in G$, and this finishes the proof. □

With the help of Proposition 4.19 we can give the following example of a module coalgebra over a quasi-Hopf algebra.

**Example 4.20** Let $C_n=\langle\sigma\rangle$ be the cyclic group of order $n$ and $k$ a field that contains an $n$th root of unity $q$ satisfying $q^{\frac{(n-1)n}{2}}=1$. If $H$ is the quasi-bialgebra built on the Hopf group algebra $k[C_n]$ as in Example 4.9, then $k^F[C_n]$ is a left $H$-module coalgebra, where $F: C_n\times C_n\to k^*$ is given by $F(\sigma^a,\sigma^b)=q^{\frac{(a-1)ab}{2}}$, for all $0\le a,b\le n-1$.

*Proof* By the proof of Example 4.9 we know that $\mathscr{M}^{k_\phi[C_n]}$ and ${}_H\mathscr{M}$ are monoidally isomorphic, where $\phi=\Delta_2(F^{-1})$. Since $k^F[C_n]$ is a coalgebra in $\mathscr{M}^{k_\phi[C_n]}$ it follows that $k^F[C_n]$ is a coalgebra in ${}_H\mathscr{M}$, too. □

Examples of module coalgebras can also be obtained by considering duals of module algebras.

**Proposition 4.21** *Let $H$ be a quasi-Hopf algebra and $A$ a finite-dimensional left $H$-module algebra. Then $A^*$, the linear dual of $A$, is a left $H$-module coalgebra with the following structure:*

$$\begin{aligned}
(h\cdot a^*)(a)=a^*(S(h)\cdot a),\quad \underline{\varepsilon}(a^*)=a^*(1_A),\\
\underline{\Delta}(a^*)=a^*((g^1\cdot a_i)\diamond(g^2\cdot a_j))a^j\otimes a_i,
\end{aligned}$$

*for all $a^* \in A^*$, $a \in A$ and $h \in H$, where $\{a_i, a^i\}_i$ are dual bases in $A$ and $A^*$, $1_A$ is the unit of $A$ and $\diamond$ stands for multiplication on $A$.*

*Proof* $A$ is an algebra in ${}_H\mathscr{M}$ that has $A^*$ as a left dual object in ${}_H\mathscr{M}$; see the proof of Proposition 3.33 for the full structure of $A^*$ as a left dual of $A$ in ${}_H\mathscr{M}$. Hence by Proposition 2.22 (i) and its proof we get that $A^*$ is a coalgebra in ${}_H\mathscr{M}$, that is, a left $H$-module coalgebra, via the following structure: $A^*$ is a left $H$-module with $(h \cdot a^*)(a) = a^*(S(h) \cdot a)$, for all $a^* \in A^*$, $a \in A$ and $h \in H$, and a coalgebra in ${}_H\mathscr{M}$ with counit and comultiplication defined by $\underline{\varepsilon}(a^*) = \mathrm{ev}_A(a^* \otimes 1_A) = a^*(\alpha \cdot 1_A) = \varepsilon(\alpha)a^*(1_A) = a^*(1_A)$ and

$$\underline{\Delta}_{A^*} : A^* \xrightarrow{\underline{m}_A^*} (A \otimes A)^* \xrightarrow{\lambda_{A,A}^{-1}} A^* \otimes A^*,$$

respectively, where $\lambda^{-1}$ is the isomorphism described in Proposition 3.35 and $\underline{m}_A^*$ is the transposed morphism associated to $\underline{m}_A$. Now by Proposition 3.34 and (3.4.1) it follows that

$$\underline{\Delta}(a^*) = \lambda_{A,A}^{-1}\underline{m}_A^*(a^*) = \underline{m}_A^*(a^*)(g^1 \cdot a_i \otimes g^2 \cdot a_j)a^j \otimes a_i = a^*((g^1 \cdot a_i) \diamond (g^2 \cdot a_j))a^j \otimes a_i,$$

for all $a^* \in A^*$, and this ends the proof. □

If $G$ is a finite group and $F$ is a 2-cochain on it then the left co-opposite dual coalgebra of $k_F[G]$ is isomorphic to $k^F[G]$.

**Proposition 4.22** *Let $G$ be a finite group and $F : G \times G \to k^*$ a 2-cochain on $G$. If $k_F[G]$ is the $G$-graded quasialgebra with reassociator $\Delta_2(F^{-1})$ built on $k[G]$ then $k_F[G]^\star \cong k^F[G]$ as monoidal coalgebras, where $k^F[G]$ is the $G$-graded quasicoalgebra with reassociator $\Delta_2(F^{-1})$ constructed from $k[G]$ and $F$. If $G$ is, moreover, abelian then $k_F[G]^\star = k_F[G]^* \cong k^F[G]$ as $G$-graded quasicoalgebras with reassociator $\Delta_2(F^{-1})$.*

*Proof* Recall that $k_F[G]$, respectively $k^F[G]$, is a monoidal algebra, respectively coalgebra, within $\mathrm{Vect}^G_{\Delta_2(F^{-1})}$.

If we take $H = k_\phi[G]$ with $\phi := \Delta_2(F^{-1})$ we have

$$\phi(x,y,z) = \frac{F(xy,z)F(x,y)}{F(y,z)F(x,yz)} \quad \text{and} \quad \beta(x) = \phi(x,x^{-1},x)^{-1} = \frac{F(x^{-1},x)}{F(x,x^{-1})},$$

for all $x, y, z \in G$. Thus, by the definition of $\delta$ in Proposition 3.49 we get that

$$\begin{aligned}\delta(x,y) &= \phi(xy,y^{-1},x^{-1})\beta(x)\phi^{-1}(x,y,y^{-1})\beta(y)\\ &= \frac{F(x,x^{-1})F(xy,y^{-1})}{F(y^{-1},x^{-1})F(xy,y^{-1}x^{-1})} \cdot \frac{F(x^{-1},x)}{F(x,x^{-1})} \cdot \frac{F(y,y^{-1})}{F(xy,y^{-1})F(x,y)} \cdot \frac{F(y^{-1},y)}{F(y,y^{-1})}\\ &= \frac{F(x^{-1},x)F(y^{-1},y)}{F(y^{-1},x^{-1})F(xy,y^{-1}x^{-1})F(x,y)},\end{aligned}$$

and therefore

$$f^{-1}(x,y) = \phi^{-1}((xy)^{-1},xy,y^{-1}x^{-1})\delta(x,y)$$

$$= \frac{F(xy,y^{-1}x^{-1})}{F(y^{-1}x^{-1},xy)} \cdot \frac{F(x^{-1},x)F(y^{-1},y)}{F(y^{-1},x^{-1})F(xy,y^{-1}x^{-1})F(x,y)}$$
$$= \frac{F(x^{-1},x)F(y^{-1},y)}{F(y^{-1}x^{-1},xy)F(y^{-1},x^{-1})F(x,y)},$$

for all $x,y \in G$. If $\{P_\theta \mid \theta \in G\}$ is the basis in $k_F[G]^*$ dual to the basis $\{\theta \mid \theta \in G\}$ of $k_F[G]$ then the comultiplication of the left co-opposite dual coalgebra structure of the algebra $k_F[G]$ is given, for all $\theta \in G$, by

$$\begin{aligned}
\underline{\Delta}_{k_F[G]^\star}(P_\theta) &= \lambda_{k_F[G],k_F[G]} \circ \underline{m}^*_{k_F[G]}(P_\theta) = \lambda_{k_F[G],k_F[G]}(P_\theta \circ \underline{m}_{k_F[G]}) \\
&= \sum_{\sigma,\tau\in G} f^{-1}(\sigma,\tau)P_\theta(\sigma\bullet\tau)P_\tau \otimes P_\sigma \\
&= \sum_{\sigma,\tau\in G} F(\sigma,\tau)\frac{F(\sigma^{-1},\sigma)F(\tau^{-1},\tau)}{F(\tau^{-1}\sigma^{-1},\sigma\tau)F(\tau^{-1},\sigma^{-1})F(\sigma,\tau)}P_\theta(\sigma\tau)P_\tau\otimes P_\sigma \\
&= F(\theta^{-1},\theta)^{-1}\sum_{\sigma\tau=\theta}\frac{F(\sigma^{-1},\sigma)F(\tau^{-1},\tau)}{F(\tau^{-1},\sigma^{-1})}P_\tau\otimes P_\sigma.
\end{aligned}$$

Its counit is $\underline{\varepsilon}_{k_F[G]^\star}(P_\theta) = \mathrm{ev}_{k_F[G]}(P_\theta \otimes e) = P_\theta(e) = \delta_{\theta,e}$, where $e$ is the neutral element of $G$.

We now claim that $\Xi : k^F[G] \to k_F[G]^*$ defined by $\Xi(\theta) = |G|F(\theta,\theta^{-1})P_{\theta^{-1}}$, for all $\theta \in G$ and extended by linearity, is an isomorphism of $G$-graded quasicoalgebras with reassociator $\Delta_2(F^{-1})$. To see this, we compute:

$$\begin{aligned}
(\Xi\otimes\Xi)\Delta_F(\theta) &= \frac{1}{|G|}\sum_{\sigma\tau=\theta}(\Xi\otimes\Xi)(F(\sigma,\tau)^{-1}\sigma\otimes\tau) \\
&= \frac{|G|^2}{|G|}\sum_{\sigma\tau=\theta}\frac{F(\sigma,\sigma^{-1})F(\tau,\tau^{-1})}{F(\sigma,\tau)}P_{\sigma^{-1}}\otimes P_{\tau^{-1}} \\
&= |G|\sum_{\tau^{-1}\sigma^{-1}=\theta^{-1}}\frac{F(\tau,\tau^{-1})F(\sigma,\sigma^{-1})}{F(\sigma,\tau)}P_{\sigma^{-1}}\otimes P_{\tau^{-1}} \\
&= |G|F(\theta,\theta^{-1})\underline{\Delta}_{k_F[G]^\star}(P_{\theta^{-1}}) = \underline{\Delta}_{k_F[G]^\star}(\Xi(P_\theta)),
\end{aligned}$$

and $\underline{\varepsilon}_{k_F[G]^\star}\circ\Xi(\theta) = |G|F(\theta,\theta^{-1})\underline{\varepsilon}_{k_F[G]^\star}(P_{\theta^{-1}}) = |G|\delta_{\theta,e} = \varepsilon_F(\theta)$, for all $\theta \in G$, as needed.

Since $P_{\theta^{-1}}$ has degree $\theta$ in $k_F[G]^*$ it follows that $\Xi$ is a morphism in $\mathrm{Vect}^G_{\Delta_2(F^{-1})}$. Moreover, one can easily see that it is an isomorphism with inverse given by

$$\Xi^{-1}(P_\theta) = \frac{1}{|G|F(\theta^{-1},\theta)}\theta^{-1},\ \forall\,\theta\in G.$$

Finally, if $G$ is abelian then the category $\mathrm{Vect}^G_{(\Delta_2(F^{-1}),\mathscr{R}_{F^{-1}})}$ is symmetric, where $\mathscr{R}_{F^{-1}}(x,y) = \frac{F(x,y)}{F(y,x)}$, for all $x,y \in G$. In this case $k_F[G]$ is braided commutative as a monoidal algebra and then $k_F[G]^\star$ is braided cocommutative as a monoidal coalgebra. Therefore $k_F[G]^* = k_F[G]^\star$ as coalgebras in $\mathrm{Vect}^G_{\Delta_2(F^{-1})}$. Note that a direct proof for this equality can be also done by computing, for all $\theta \in G$,

$$\begin{aligned}\Delta_{k_F[G]^*}(P_\theta) &= c_{k_F[G]^*,k_F[G]^*}\circ\underline{\Delta}_{k_F[G]^*}(P_\theta)\\ &= \frac{1}{F(\theta^{-1},\theta)}\sum_{\sigma\tau=\theta}\frac{F(\sigma^{-1},\sigma)F(\tau^{-1},\tau)}{F(\tau^{-1},\sigma^{-1})}c_{k_F[G]^*,k_F[G]^*}(P_\tau\otimes P_\sigma)\\ &= \frac{1}{F(\theta^{-1},\theta)}\sum_{\sigma\tau=\theta}\frac{F(\sigma^{-1},\sigma)F(\tau^{-1},\tau)}{F(\tau^{-1},\sigma^{-1})}\cdot\frac{F(\tau^{-1},\sigma^{-1})}{F(\sigma^{-1},\tau^{-1})}P_\sigma\otimes P_\tau\\ &= \frac{1}{F(\theta^{-1},\theta)}\sum_{\tau\sigma=\theta}\frac{F(\sigma^{-1},\sigma)F(\tau^{-1},\tau)}{F(\sigma^{-1},\tau^{-1})}P_\sigma\otimes P_\tau = \underline{\Delta}_{k_F[G]^\star}(P_\theta),\end{aligned}$$

as needed. In the above computation we used that $G$ is abelian and the fact that, in general, $P_x$ has degree $x^{-1}$ in $k_F[G]^*$, for any $x \in G$. So our proof is complete. □

We focus now on the algebra version of the above result. Recall from Section 2.3 that if $C$ is a coalgebra in $\mathscr{M}^{k_\phi[G]}$ then $C^*$ admits an algebra structure in $\mathscr{M}^{k_\phi[G]}$. The multiplication is given by

$$C^*\otimes C^* \xrightarrow{\lambda_{C,C}^{-1}} (C\otimes C)^* \xrightarrow{\underline{\Delta}^*} C^*,$$

where $\underline{\Delta}^*$ is the transpose of $\underline{\Delta}$ and $\lambda_{X,Y}^{-1} : Y^*\otimes X^* \to (X\otimes Y)^*$ is the bijective inverse of $\lambda_{X,Y}$ defined above; see (3.7.8). The unit of $C^*$ is

$$k \xrightarrow{\mathrm{coev}_C} C\otimes C^* \xrightarrow{\underline{\varepsilon}\otimes \mathrm{Id}_{C^*}} k\otimes C^* \cong C^*,$$

where $\underline{\varepsilon}_C$ is the counit of $C$. This monoidal algebra structure on $C^*$ was denoted by $C^\star$ and was called the left opposite dual algebra of the coalgebra $C$. When $\mathscr{M}^{k_\phi[G]}$ has a braided structure produced by the braiding $c$ then $C^*$ is also a unital algebra in $\mathscr{M}^{k_\phi[G]}$ with the same unit as that of $C^\star$ and multiplication $c_{C^*,C^*}^{-1}\underline{m}_{C^\star}$. In this case we said that $C^*$ is the left dual algebra of the coalgebra $C$, and we denoted it by $C^*$.

**Proposition 4.23** *Let $G$ be a finite group and $F : G\times G \to k^*$ a 2-cochain on $G$. Then $k^F[G]^\star \cong k_F[G]$ as $G$-graded quasialgebras with reassociator $\Delta_2(F^{-1})$. Consequently, if $G$ is, moreover, abelian then $k^F[G]^\star = k^F[G]^* \cong k_F[G]$ as monoidal algebras within* $\mathrm{Vect}^G_{(\Delta_2(F^{-1}),\mathscr{R}_{F^{-1}})}$.

*Proof* We show that the map $\Xi$ defined in the proof of Proposition 4.22, regarded now as a map from $k_F[G]$ to $k^F[G]^\star$, is an isomorphism of $G$-graded quasialgebras with reassociator $\Delta_2(F^{-1})$. To this end we first compute the left opposite dual algebra structure of the coalgebra $k^F[G]$. The multiplication $\diamond$ of $k^F[G]^\star$ is given by

$$\begin{aligned}(P_g\diamond P_h)(\theta) &= \lambda^{-1}_{k^F[G],k^F[G]}(P_g\otimes P_h)(\Delta_F(\theta))\\ &= \frac{1}{|G|}\sum_{\sigma\tau=\theta}F(\sigma,\tau)^{-1}\lambda^{-1}_{k^F[G],k^F[G]}(P_g\otimes P_h)(\sigma\otimes\tau)\\ &= \frac{1}{|G|}\sum_{\sigma\tau=\theta}F(\sigma,\tau)^{-1}f(\sigma,\tau)P_h(\sigma)P_g(\tau)\end{aligned}$$

$$= \frac{1}{|G|} \cdot \frac{1}{F(h,g)} \cdot \frac{F(g^{-1}h^{-1},hg)F(g^{-1},h^{-1})F(h,g)}{F(h^{-1},h)F(g^{-1},g)} \delta_{hg,\theta}$$
$$= \frac{1}{|G|} \cdot \frac{F(g^{-1}h^{-1},hg)F(g^{-1},h^{-1})}{F(h^{-1},h)F(g^{-1},g)} P_{hg}(\theta),$$

and so

$$P_g \diamond P_h = \frac{1}{|G|} \cdot \frac{F(g^{-1}h^{-1},hg)F(g^{-1},h^{-1})}{F(h^{-1},h)F(g^{-1},g)} P_{hg}, \ \forall\, g,\ h \in G.$$

The unit of $k^F[G]^\star$ is $\sum_{g\in G}\beta(g)\varepsilon_F(g)P_g = \beta(e)|G|P_e = |G|P_e$, where the notation is as in the proof of Proposition 4.22.

We are now in the position to show that $\Xi : k_F[G] \to k^F[G]^\star$,

$$\Xi(\theta) = |G|F(\theta,\theta^{-1})P_{\theta^{-1}}, \ \forall\, \theta \in G,$$

is an isomorphism of $G$-graded quasialgebras with reassociator $\Delta_2(F^{-1})$. We have already seen that $\Xi$ is an isomorphism in $\mathrm{Vect}^G_{\Delta_2(F^{-1})}$, hence we only have to prove that $\Xi$ is an algebra morphism. We compute:

$$\begin{aligned}
\Xi(\sigma)\diamond\Xi(\tau) &= |G|^2F(\sigma,\sigma^{-1})F(\tau,\tau^{-1})P_{\sigma^{-1}}\diamond P_{\tau^{-1}}\\
&= |G|F(\sigma,\sigma^{-1})F(\tau,\tau^{-1})\cdot\frac{F(\sigma\tau,\tau^{-1}\sigma^{-1})F(\sigma,\tau)}{F(\tau,\tau^{-1})F(\sigma,\sigma^{-1})}P_{\tau^{-1}\sigma^{-1}}\\
&= |G|F(\sigma\tau,\tau^{-1}\sigma^{-1})F(\sigma,\tau)P_{(\sigma\tau)^{-1}}\\
&= F(\sigma,\tau)\Xi(\sigma\tau) = \Xi(\sigma\bullet\tau),
\end{aligned}$$

for all $\sigma, \tau \in G$, and $\Xi(e) = |G|P_e$, as required.

Finally, when $G$ is abelian one can easily verify that the multiplication $\diamond'$ of $k^F[G]^*$ coincides with that of $k^F[G]^\star$. Actually,

$$\begin{aligned}
P_g\diamond' P_h &= \frac{F(g^{-1},h^{-1})}{F(h^{-1},g^{-1})}P_h\diamond P_g = \frac{1}{|G|}\cdot\frac{F((gh)^{-1},gh)F(g^{-1},h^{-1})}{F(g^{-1},g)F(h^{-1},h)}P_{gh}\\
&= \frac{1}{|G|}\cdot\frac{F((hg)^{-1},hg)F(g^{-1},h^{-1})}{F(g^{-1},g)F(h^{-1},h)}P_{hg} = P_g\diamond P_h,
\end{aligned}$$

for all $g, h \in G$. As an alternative proof, if $G$ is abelian then $k^F[G]$ is a braided cocommutative monoidal coalgebra and this implies that $k^F[G]^\star$ is a braided commutative monoidal algebra, and so $k^F[G]^\star = k^F[G]^*$ as coalgebras within $\mathrm{Vect}^G_{\Delta_2(F^{-1})}$. □

We end this section by recording the notion of an $H$-bimodule coalgebra.

**Definition 4.24** Let $H$ be a quasi-bialgebra. An $H$-bimodule coalgebra is a coalgebra in the category of $H$-bimodules. More precisely, an $H$-bimodule coalgebra $C$ is an $H$-bimodule (denote the actions by $h\cdot c$ and $c\cdot h$) with a comultiplication $\underline{\Delta} : C \to C\otimes C$ and a counit $\underline{\varepsilon} : C \to k$ satisfying the relations, for all $c \in C$, $h \in H$:

$$\Phi\cdot(\underline{\Delta}\otimes\mathrm{Id}_C)(\underline{\Delta}(c))\cdot\Phi^{-1} = (\mathrm{Id}_C\otimes\underline{\Delta})(\underline{\Delta}(c)), \tag{4.2.9}$$
$$\underline{\Delta}(h\cdot c) = h_1\cdot c_{\underline{1}}\otimes h_2\cdot c_{\underline{2}}, \qquad \underline{\Delta}(c\cdot h) = c_{\underline{1}}\cdot h_1\otimes c_{\underline{2}}\cdot h_2, \tag{4.2.10}$$

$$(\underline{\varepsilon}\otimes \mathrm{Id}_C)\circ\underline{\Delta} = (\mathrm{Id}_C\otimes\underline{\varepsilon})\circ\underline{\Delta} = \mathrm{Id}_C, \tag{4.2.11}$$
$$\underline{\varepsilon}(h\cdot c) = \varepsilon(h)\underline{\varepsilon}(c), \qquad \underline{\varepsilon}(c\cdot h) = \underline{\varepsilon}(c)\varepsilon(h), \tag{4.2.12}$$

where we used the same Sweedler-type notation as before.

We have explained that $H$ itself cannot be viewed as a left or right $H$-module coalgebra, but it has a natural $H$-bimodule coalgebra structure: $H\in {}_H\mathcal{M}_H$ via multiplication and is a coalgebra within ${}_H\mathcal{M}_H$ via its comultiplication.

## 4.3 Comodule Algebras over Quasi-bialgebras

The category of $H$-modules is monoidal, and an $H$-module (co)algebra is nothing else than a (co)algebra in this category. This categorical definition cannot be used to introduce H-comodule (co)algebras, since we do not have H-comodules. Instead, to introduce the concept of comodule algebra over a quasi-bialgebra one has to generalize the property of the comultiplication $\Delta$ of $H$ to an arbitrary $H$-coaction $\rho:\mathfrak{A}\to\mathfrak{A}\otimes H$ on a $k$-algebra $\mathfrak{A}$. The basic idea here is to relate $(\rho\otimes\mathrm{Id}_H)\rho$ and $(\mathrm{Id}_{\mathfrak{A}}\otimes\Delta)\rho$ by an inner automorphism defined by a reassociator $\Phi_\rho\in\mathfrak{A}\otimes H\otimes H$ that is required to satisfy a Pentagon Axiom and a Triangle Axiom, similar to those of the reassociator $\Phi$ of $H$.

**Definition 4.25** Let $H$ be a quasi-bialgebra. An algebra $\mathfrak{A}$ is called a right $H$-comodule algebra if there exist an algebra morphism $\rho:\mathfrak{A}\to\mathfrak{A}\otimes H$ and an invertible element $\Phi_\rho\in\mathfrak{A}\otimes H\otimes H$ such that:

$$\Phi_\rho(\rho\otimes\mathrm{Id}_H)(\rho(\mathfrak{a})) = (\mathrm{Id}_{\mathfrak{A}}\otimes\Delta)(\rho(\mathfrak{a}))\Phi_\rho,\ \ \forall\, \mathfrak{a}\in\mathfrak{A}, \tag{4.3.1}$$
$$(1_{\mathfrak{A}}\otimes\Phi)(\mathrm{Id}_{\mathfrak{A}}\otimes\Delta\otimes\mathrm{Id}_H)(\Phi_\rho)(\Phi_\rho\otimes 1_H)$$
$$= (\mathrm{Id}_{\mathfrak{A}}\otimes\mathrm{Id}_H\otimes\Delta)(\Phi_\rho)(\rho\otimes\mathrm{Id}_H\otimes\mathrm{Id}_H)(\Phi_\rho), \tag{4.3.2}$$
$$(\mathrm{Id}_{\mathfrak{A}}\otimes\varepsilon)\circ\rho = \mathrm{Id}_{\mathfrak{A}}, \tag{4.3.3}$$
$$(\mathrm{Id}_{\mathfrak{A}}\otimes\varepsilon\otimes\mathrm{Id}_H)(\Phi_\rho) = (\mathrm{Id}_{\mathfrak{A}}\otimes\mathrm{Id}_H\otimes\varepsilon)(\Phi_\rho) = 1_{\mathfrak{A}}\otimes 1_H. \tag{4.3.4}$$

Similarly, an algebra $\mathfrak{B}$ is called a left $H$-comodule algebra if there exist an algebra morphism $\lambda:\mathfrak{B}\to H\otimes\mathfrak{B}$ and an invertible element $\Phi_\lambda\in H\otimes H\otimes\mathfrak{B}$ such that:

$$(\mathrm{Id}_H\otimes\lambda)(\lambda(\mathfrak{b}))\Phi_\lambda = \Phi_\lambda(\Delta\otimes\mathrm{Id}_{\mathfrak{B}})(\lambda(\mathfrak{b})),\ \ \forall\, \mathfrak{b}\in\mathfrak{B}, \tag{4.3.5}$$
$$(1_H\otimes\Phi_\lambda)(\mathrm{Id}_H\otimes\Delta\otimes\mathrm{Id}_{\mathfrak{B}})(\Phi_\lambda)(\Phi\otimes 1_{\mathfrak{B}})$$
$$= (\mathrm{Id}_H\otimes\mathrm{Id}_H\otimes\lambda)(\Phi_\lambda)(\Delta\otimes\mathrm{Id}_H\otimes\mathrm{Id}_{\mathfrak{B}})(\Phi_\lambda), \tag{4.3.6}$$
$$(\varepsilon\otimes\mathrm{Id}_{\mathfrak{B}})\circ\lambda = \mathrm{Id}_{\mathfrak{B}}, \tag{4.3.7}$$
$$(\mathrm{Id}_H\otimes\varepsilon\otimes\mathrm{Id}_{\mathfrak{B}})(\Phi_\lambda) = (\varepsilon\otimes\mathrm{Id}_H\otimes\mathrm{Id}_{\mathfrak{B}})(\Phi_\lambda) = 1_H\otimes 1_{\mathfrak{B}}. \tag{4.3.8}$$

When $H$ is a quasi-bialgebra, particular examples of left and right $H$-comodule algebras are given by $\mathfrak{A}=\mathfrak{B}=H$ and $\rho=\lambda=\Delta$, $\Phi_\rho=\Phi_\lambda=\Phi$.

For a right $H$-comodule algebra $(\mathfrak{A}, \rho, \Phi_\rho)$ we will denote, for all $\mathfrak{a} \in \mathfrak{A}$,

$$\rho(\mathfrak{a}) = \mathfrak{a}_{\langle 0\rangle} \otimes \mathfrak{a}_{\langle 1\rangle}, \quad (\rho \otimes \mathrm{Id}_H)(\rho(\mathfrak{a})) = \mathfrak{a}_{\langle 0,0\rangle} \otimes \mathfrak{a}_{\langle 0,1\rangle} \otimes \mathfrak{a}_{\langle 1\rangle} \text{ etc.}$$

Similarly, for a left $H$-comodule algebra $(\mathfrak{B}, \lambda, \Phi_\lambda)$, if $\mathfrak{b} \in \mathfrak{B}$ then we will denote

$$\lambda(\mathfrak{b}) = \mathfrak{b}_{[-1]} \otimes \mathfrak{b}_{[0]}, \quad (\mathrm{Id}_H \otimes \lambda)(\lambda(\mathfrak{b})) = \mathfrak{b}_{[-1]} \otimes \mathfrak{b}_{[0,-1]} \otimes \mathfrak{b}_{[0,0]} \text{ etc.}$$

In analogy with the notation for the reassociator $\Phi$ of $H$, we will write

$$\Phi_\rho = \tilde{X}^1_\rho \otimes \tilde{X}^2_\rho \otimes \tilde{X}^3_\rho = \tilde{Y}^1_\rho \otimes \tilde{Y}^2_\rho \otimes \tilde{Y}^3_\rho = \cdots$$
$$\Phi_\rho^{-1} = \tilde{x}^1_\rho \otimes \tilde{x}^2_\rho \otimes \tilde{x}^3_\rho = \tilde{y}^1_\rho \otimes \tilde{y}^2_\rho \otimes \tilde{y}^3_\rho = \cdots$$

and similarly for the element $\Phi_\lambda$ of a left $H$-comodule algebra $\mathfrak{B}$. When there is no danger of confusion we will omit the subscripts $\rho$ or $\lambda$ for the tensor components of the elements $\Phi_\rho$, $\Phi_\lambda$ or for the tensor components of the elements $\Phi_\rho^{-1}$, $\Phi_\lambda^{-1}$.

**Definition 4.26** (i) Let $H$ be a quasi-bialgebra and $(\mathfrak{A}, \rho, \Phi_\rho)$, $(\mathfrak{D}, \theta, \Phi_\theta)$ two right $H$-comodule algebras. A morphism of right $H$-comodule algebras $f : \mathfrak{A} \to \mathfrak{D}$ is an algebra map such that $\Phi_\theta = (f \otimes \mathrm{Id}_H \otimes \mathrm{Id}_H)(\Phi_\rho)$ and $\theta \circ f = (f \otimes \mathrm{Id}_H) \circ \rho$.

(ii) Let $H$ be a quasi-bialgebra and $(\mathfrak{B}, \lambda, \Phi_\lambda)$, $(\mathfrak{C}, \mu, \Phi_\mu)$ two left $H$-comodule algebras. A morphism of left $H$-comodule algebras from $\mathfrak{B}$ to $\mathfrak{C}$ is an algebra map $g : \mathfrak{B} \to \mathfrak{C}$ such that $\Phi_\mu = (\mathrm{Id}_H \otimes \mathrm{Id}_H \otimes g)(\Phi_\lambda)$ and $\mu \circ g = (\mathrm{Id}_H \otimes g) \circ \lambda$.

We have seen that the comultiplication of a quasi-bialgebra $H$ may be twisted by a twist $F \in H \otimes H$. A similar process may also be performed for comodule algebras.

**Proposition 4.27** *Let $H$ be a quasi-bialgebra, $\mathfrak{A}$ a right $H$-comodule algebra and $\mathfrak{B}$ a left $H$-comodule algebra, with notation as before.*

*(i) Let $U \in \mathfrak{A} \otimes H$ be an invertible element such that $(\mathrm{Id}_{\mathfrak{A}} \otimes \varepsilon)(U) = 1_{\mathfrak{A}}$. If we define the linear map $\rho' : \mathfrak{A} \to \mathfrak{A} \otimes H$, $\rho'(\mathfrak{a}) = U\rho(\mathfrak{a})U^{-1}$, for all $\mathfrak{a} \in \mathfrak{A}$, then this is a new right $H$-comodule algebra structure on $\mathfrak{A}$, with*

$$\Phi_{\rho'} = (\mathrm{Id}_{\mathfrak{A}} \otimes \Delta)(U)\Phi_\rho(\rho \otimes \mathrm{Id}_H)(U^{-1})(U^{-1} \otimes 1_H),$$

*which will be denoted by $\mathfrak{A}'$ and we will say that $\mathfrak{A}$ and $\mathfrak{A}'$ are "twist equivalent".*

*(ii) Let $U \in H \otimes \mathfrak{B}$ be an invertible element such that $(\varepsilon \otimes \mathrm{Id}_{\mathfrak{B}})(U) = 1_{\mathfrak{B}}$. If we define the linear map $\lambda' : \mathfrak{B} \to H \otimes \mathfrak{B}$, $\lambda'(\mathfrak{b}) = U\lambda(\mathfrak{b})U^{-1}$, for all $\mathfrak{b} \in \mathfrak{B}$, then this is a new left $H$-comodule algebra structure on $\mathfrak{B}$, with*

$$\Phi_{\lambda'} = (1_H \otimes U)(\mathrm{Id}_H \otimes \lambda)(U)\Phi_\lambda(\Delta \otimes \mathrm{Id}_{\mathfrak{B}})(U^{-1}),$$

*which will be denoted by $\mathfrak{B}'$ and we will say that $\mathfrak{B}$ and $\mathfrak{B}'$ are "twist equivalent".*

*Proof* This is a straightforward computation that is left to the reader. □

**Remark 4.28** The twist equivalences defined above for right and left $H$-comodule algebras are equivalence relations.

The formulas below will be intensively used in what follows. If we specialize them for $\mathfrak{A} = H$ we obtain the relations (3.2.21), (3.2.23), (3.2.25) and (3.2.26) in Proposition 3.25.

**Proposition 4.29** *Let $H$ be a quasi-Hopf algebra with bijective antipode and $\mathfrak{A}$ a right $H$-comodule algebra. Define the elements $\tilde{p}_\rho, \tilde{q}_\rho \in \mathfrak{A}\otimes H$ as follows:*

$$\tilde{p}_\rho = \tilde{p}^1_\rho \otimes \tilde{p}^2_\rho = \tilde{x}^1_\rho \otimes \tilde{x}^2_\rho \beta S(\tilde{x}^3_\rho), \quad \tilde{q}_\rho = \tilde{q}^1_\rho \otimes \tilde{q}^2_\rho = \tilde{X}^1_\rho \otimes S^{-1}(\alpha \tilde{X}^3_\rho)\tilde{X}^2_\rho. \tag{4.3.9}$$

*Then, for all $\mathfrak{a} \in \mathfrak{A}$, the following relations hold:*

$$\rho(\mathfrak{a}_{\langle 0\rangle})\tilde{p}_\rho[1_{\mathfrak{A}} \otimes S(\mathfrak{a}_{\langle 1\rangle})] = \tilde{p}_\rho[\mathfrak{a}\otimes 1_H], \tag{4.3.10}$$
$$[1_{\mathfrak{A}} \otimes S^{-1}(\mathfrak{a}_{\langle 1\rangle})]\tilde{q}_\rho \rho(\mathfrak{a}_{\langle 0\rangle}) = [\mathfrak{a}\otimes 1_H]\tilde{q}_\rho, \tag{4.3.11}$$
$$\rho(\tilde{q}^1_\rho)\tilde{p}_\rho[1_{\mathfrak{A}} \otimes S(\tilde{q}^2_\rho)] = 1_{\mathfrak{A}} \otimes 1_H, \tag{4.3.12}$$
$$[1_{\mathfrak{A}} \otimes S^{-1}(\tilde{p}^2_\rho)]\tilde{q}_\rho \rho(\tilde{p}^1_\rho) = 1_{\mathfrak{A}} \otimes 1_H, \tag{4.3.13}$$
$$\begin{aligned}&\Phi_\rho(\rho\otimes \mathrm{Id}_H)(\tilde{p}_\rho)(\tilde{p}_\rho \otimes 1_H)\\ &\quad = (\mathrm{Id}_{\mathfrak{A}} \otimes \Delta)(\rho(\tilde{x}^1_\rho)\tilde{p}_\rho)(1_{\mathfrak{A}} \otimes g^1 S(\tilde{x}^3_\rho) \otimes g^2 S(\tilde{x}^2_\rho)),\end{aligned} \tag{4.3.14}$$
$$\begin{aligned}&(\tilde{q}_\rho \otimes 1_H)(\rho\otimes \mathrm{Id}_H)(\tilde{q}_\rho)\Phi_\rho^{-1}\\ &\quad = [1_{\mathfrak{A}} \otimes S^{-1}(f^2\tilde{X}^3_\rho) \otimes S^{-1}(f^1\tilde{X}^2_\rho)](\mathrm{Id}_{\mathfrak{A}} \otimes \Delta)(\tilde{q}_\rho \rho(\tilde{X}^1_\rho)),\end{aligned} \tag{4.3.15}$$

*where $f = f^1 \otimes f^2$ is the element defined in* (3.2.15) *and $f^{-1} = g^1 \otimes g^2$.*

*Proof* Note first that, by considering $(\rho, \Phi_\rho^{-1})$ as a right $H^{\mathrm{op}}$-coaction on $\mathfrak{A}^{\mathrm{op}}$, the roles of $\tilde{p}_\rho$ and $\tilde{q}_\rho$ interchange, so it is enough to prove the relations (4.3.11), (4.3.13) and (4.3.15). We compute:

$$\begin{aligned}[1_{\mathfrak{A}} \otimes S^{-1}(\mathfrak{a}_{\langle 1\rangle})]\tilde{q}_\rho \rho(\mathfrak{a}_{\langle 0\rangle}) &= \tilde{X}^1_\rho \mathfrak{a}_{\langle 0,0\rangle} \otimes S^{-1}(\alpha \tilde{X}^3_\rho \mathfrak{a}_{\langle 1\rangle})\tilde{X}^2_\rho \mathfrak{a}_{\langle 0,1\rangle}\\ &\overset{(4.3.1)}{=} \mathfrak{a}_{\langle 0\rangle}\tilde{X}^1_\rho \otimes S^{-1}(\alpha \mathfrak{a}_{\langle 1\rangle_2}\tilde{X}^3_\rho)\mathfrak{a}_{\langle 1\rangle_1}\tilde{X}^2_\rho\\ &\overset{(3.2.1),(4.3.3)}{=} \mathfrak{a}\tilde{X}^1_\rho \otimes S^{-1}(\alpha\tilde{X}^3_\rho)\tilde{X}^2_\rho = [\mathfrak{a}\otimes 1_H]\tilde{q}_\rho,\end{aligned}$$

thus proving (4.3.11). Then:

$$\begin{aligned}&[1_{\mathfrak{A}} \otimes S^{-1}(\tilde{p}^2_\rho)]\tilde{q}_\rho \rho(\tilde{p}^1_\rho)\\ &\quad = \tilde{X}^1_\rho(\tilde{x}^1_\rho)_{\langle 0\rangle} \otimes \tilde{x}^3_\rho S^{-1}(\alpha\tilde{X}^3_\rho \tilde{x}^2_\rho \beta)\tilde{X}^2_\rho(\tilde{x}^1_\rho)_{\langle 1\rangle}\\ &\quad \overset{(4.3.2)}{=} \tilde{x}^1_\rho\tilde{X}^1_\rho \otimes \tilde{x}^3_\rho x^3(\tilde{X}^3_\rho)_2 S^{-1}(\alpha(\tilde{x}^2_\rho)_2 x^2(\tilde{X}^3_\rho)_1\beta)(\tilde{x}^2_\rho)_1 x^1\tilde{X}^2_\rho\\ &\quad \overset{(3.2.1),(4.3.4)}{=} 1_{\mathfrak{A}} \otimes x^3 S^{-1}(\alpha x^2\beta)x^1 \overset{(3.2.2)}{=} 1_{\mathfrak{A}} \otimes 1_H,\end{aligned}$$

which proves (4.3.13). The proof of (4.3.15) is more complicated. First, we compute:

$$\begin{aligned}&[1_{\mathfrak{A}} \otimes S^{-1}(f^2)\otimes S^{-1}(f^1)](\mathrm{Id}_{\mathfrak{A}} \otimes \Delta)(\tilde{q}_\rho)\\ &= [1_{\mathfrak{A}} \otimes S^{-1}(f^2)\otimes S^{-1}(f^1)][\tilde{Y}^1_\rho \otimes \Delta(S^{-1}(\alpha\tilde{Y}^3_\rho)\tilde{Y}^2_\rho)]\\ &\overset{(3.2.13)}{=} \tilde{Y}^1_\rho \otimes S^{-1}(f^2(\alpha\tilde{Y}^3_\rho)_2)(\tilde{Y}^2_\rho)_1 \otimes S^{-1}(f^1(\alpha\tilde{Y}^3_\rho)_1)(\tilde{Y}^2_\rho)_2\\ &\overset{(3.2.14)}{=} \tilde{Y}^1_\rho \otimes S^{-1}(\gamma^2(\tilde{Y}^3_\rho)_2)(\tilde{Y}^2_\rho)_1 \otimes S^{-1}(\gamma^1(\tilde{Y}^3_\rho)_1)(\tilde{Y}^2_\rho)_2\end{aligned}$$

$$\begin{aligned}
&\overset{(3.2.5)}{=} \tilde{Y}^1_\rho \otimes S^{-1}(\alpha x^3(\tilde{Y}^3_\rho)_2)X^1x^1_1(\tilde{Y}^2_\rho)_1 \otimes S^{-1}(\alpha X^3x^2(\tilde{Y}^3_\rho)_1)X^2x^1_2(\tilde{Y}^2_\rho)_2\\
&\overset{(3.1.9)}{=} \tilde{Y}^1_\rho \otimes S^{-1}(\alpha x^3y^3X^3_2(\tilde{Y}^3_\rho)_2)x^1X^1(\tilde{Y}^2_\rho)_1 \otimes S^{-1}(\alpha x^2_2y^2X^3_1(\tilde{Y}^3_\rho)_1)x^2_1y^1X^2(\tilde{Y}^2_\rho)_2\\
&\overset{(3.2.1)}{=} \tilde{Y}^1_\rho \otimes S^{-1}(\alpha y^3X^3_2(\tilde{Y}^3_\rho)_2)X^1(\tilde{Y}^2_\rho)_1 \otimes S^{-1}(\alpha y^2X^3_1(\tilde{Y}^3_\rho)_1)y^1X^2(\tilde{Y}^2_\rho)_2.
\end{aligned}$$

Now define the linear map $\omega : \mathfrak{A} \otimes H^{\otimes 4} \to \mathfrak{A} \otimes H^{\otimes 2}$ by

$$\omega(\mathfrak{a} \otimes h \otimes h' \otimes g \otimes g') = \mathfrak{a} \otimes S^{-1}(\alpha g')h \otimes S^{-1}(\alpha g)h',$$

and consider the element $Y \in \mathfrak{A} \otimes H^{\otimes 4}$ defined by

$$\begin{aligned}
Y := {}&(1_{\mathfrak{A}} \otimes 1_H \otimes \Phi^{-1})(\mathrm{Id}_{\mathfrak{A}} \otimes \mathrm{Id}_H \otimes \mathrm{Id}_H \otimes \Delta)((1_{\mathfrak{A}} \otimes \Phi)(\mathrm{Id}_{\mathfrak{A}} \otimes \Delta \otimes \mathrm{Id}_H)(\Phi_\rho))\\
&((\mathrm{Id}_{\mathfrak{A}} \otimes \Delta) \circ \rho \otimes \mathrm{Id}_H \otimes \mathrm{Id}_H)(\Phi_\rho).
\end{aligned}$$

In view of the above computation, one can easily see that

$$[1_{\mathfrak{A}} \otimes S^{-1}(f^2\tilde{X}^3_\rho) \otimes S^{-1}(f^1\tilde{X}^2_\rho)](\mathrm{Id}_{\mathfrak{A}} \otimes \Delta)(\tilde{q}_\rho\rho(\tilde{X}^1_\rho)) = \omega(Y).$$

Now we compute:

$$\begin{aligned}
Y &\overset{(4.3.2)}{=} (1_{\mathfrak{A}} \otimes 1_H \otimes \Phi^{-1})(\mathrm{Id}_{\mathfrak{A}} \otimes \mathrm{Id}_H \otimes \mathrm{Id}_H \otimes \Delta)((\mathrm{Id}_{\mathfrak{A}} \otimes \mathrm{Id}_H \otimes \Delta)(\Phi_\rho)\\
&\qquad (\rho \otimes \mathrm{Id}_H \otimes \mathrm{Id}_H)(\Phi_\rho)(\Phi_\rho^{-1} \otimes 1_H))((\mathrm{Id}_{\mathfrak{A}} \otimes \Delta) \circ \rho \otimes \mathrm{Id}_H \otimes \mathrm{Id}_H)(\Phi_\rho)\\
&\overset{(3.1.7)}{=} (\mathrm{Id}_{\mathfrak{A}} \otimes \mathrm{Id}_H \otimes (\Delta \otimes \mathrm{Id}_H) \circ \Delta)(\Phi_\rho)(1_{\mathfrak{A}} \otimes 1_H \otimes \Phi^{-1})(\mathrm{Id}_{\mathfrak{A}} \otimes \mathrm{Id}_H \otimes \mathrm{Id}_H \otimes \Delta)\\
&\qquad ((\rho \otimes \mathrm{Id}_H \otimes \mathrm{Id}_H)(\Phi_\rho)(\Phi_\rho^{-1} \otimes 1_H))((\mathrm{Id}_{\mathfrak{A}} \otimes \Delta) \circ \rho \otimes \mathrm{Id}_H \otimes \mathrm{Id}_H)(\Phi_\rho)\\
&\overset{(4.3.1)}{=} (\mathrm{Id}_{\mathfrak{A}} \otimes \mathrm{Id}_H \otimes (\Delta \otimes \mathrm{Id}_H) \circ \Delta)(\Phi_\rho)(1_{\mathfrak{A}} \otimes 1_H \otimes \Phi^{-1})(\mathrm{Id}_{\mathfrak{A}} \otimes \mathrm{Id}_H \otimes \mathrm{Id}_H \otimes \Delta)\\
&\qquad (\rho \otimes \mathrm{Id}_H \otimes \mathrm{Id}_H)(\Phi_\rho)(\mathrm{Id}_{\mathfrak{A}} \otimes \mathrm{Id}_H \otimes \mathrm{Id}_H \otimes \Delta)(\Phi_\rho^{-1} \otimes 1_H)(\Phi_\rho \otimes 1_H \otimes 1_H)\\
&\qquad ((\rho \otimes \mathrm{Id}_H) \circ \rho \otimes \mathrm{Id}_H \otimes \mathrm{Id}_H)(\Phi_\rho)(\Phi_\rho^{-1} \otimes 1_H \otimes 1_H)\\
&= (\mathrm{Id}_{\mathfrak{A}} \otimes \mathrm{Id}_H \otimes (\Delta \otimes \mathrm{Id}_H) \circ \Delta)(\Phi_\rho)(\rho \otimes \mathrm{Id}_H \otimes \mathrm{Id}_H \otimes \mathrm{Id}_H)\\
&\qquad ((1_{\mathfrak{A}} \otimes \Phi^{-1})(\mathrm{Id}_{\mathfrak{A}} \otimes \mathrm{Id}_H \otimes \Delta)(\Phi_\rho)(\rho \otimes \mathrm{Id}_H \otimes \mathrm{Id}_H)(\Phi_\rho))(\Phi_\rho^{-1} \otimes 1_H \otimes 1_H)\\
&\overset{(4.3.2)}{=} (\mathrm{Id}_{\mathfrak{A}} \otimes \mathrm{Id}_H \otimes (\Delta \otimes \mathrm{Id}_H) \circ \Delta)(\Phi_\rho)(\rho \otimes \Delta \otimes \mathrm{Id}_H)(\Phi_\rho)\\
&\qquad [(\rho \otimes \mathrm{Id}_H \otimes \mathrm{Id}_H)(\Phi_\rho) \otimes 1_H](\Phi_\rho^{-1} \otimes 1_H \otimes 1_H)\\
&= \tilde{X}^1_\rho(\tilde{Y}^1_\rho)_{\langle 0\rangle}(\tilde{Z}^1_\rho)_{\langle 0\rangle}\tilde{x}^1_\rho \otimes \tilde{X}^2_\rho(\tilde{Y}^1_\rho)_{\langle 1\rangle}(\tilde{Z}^1_\rho)_{\langle 1\rangle}\tilde{x}^2_\rho \otimes (\tilde{X}^3_\rho)_{(1,1)}(\tilde{Y}^2_\rho)_1\tilde{Z}^2_\rho\tilde{x}^3_\rho\\
&\qquad \otimes(\tilde{X}^3_\rho)_{(1,2)}(\tilde{Y}^2_\rho)_2\tilde{Z}^3_\rho \otimes (\tilde{X}^3_\rho)_2\tilde{Y}^3_\rho.
\end{aligned}$$

With this new formula for $Y$, we can compute the right-hand side of (4.3.15):

$$\begin{aligned}
\omega(Y) &= \tilde{X}^1_\rho(\tilde{Y}^1_\rho)_{\langle 0\rangle}(\tilde{Z}^1_\rho)_{\langle 0\rangle}\tilde{x}^1_\rho \otimes S^{-1}(\alpha(\tilde{X}^3_\rho)_2\tilde{Y}^3_\rho)\tilde{X}^2_\rho(\tilde{Y}^1_\rho)_{\langle 1\rangle}(\tilde{Z}^1_\rho)_{\langle 1\rangle}\tilde{x}^2_\rho\\
&\qquad \otimes S^{-1}(\alpha(\tilde{X}^3_\rho)_{(1,2)}(\tilde{Y}^2_\rho)_2\tilde{Z}^3_\rho)(\tilde{X}^3_\rho)_{(1,1)}(\tilde{Y}^2_\rho)_1\tilde{Z}^2_\rho\tilde{x}^3_\rho\\
&\overset{(3.2.1),(4.3.4)}{=} \tilde{X}^1_\rho(\tilde{Z}^1_\rho)_{\langle 0\rangle}\tilde{x}^1_\rho \otimes S^{-1}(\alpha\tilde{X}^3_\rho)\tilde{X}^2_\rho(\tilde{Z}^1_\rho)_{\langle 1\rangle}\tilde{x}^2_\rho \otimes S^{-1}(\alpha\tilde{Z}^3_\rho)\tilde{Z}^2_\rho\tilde{x}^3_\rho\\
&= (\tilde{X}^1_\rho \otimes S^{-1}(\alpha\tilde{X}^3_\rho)\tilde{X}^2_\rho \otimes 1_H)((\tilde{Z}^1_\rho)_{\langle 0\rangle} \otimes (\tilde{Z}^1_\rho)_{\langle 1\rangle} \otimes S^{-1}(\alpha\tilde{Z}^3_\rho)\tilde{Z}^2_\rho)\\
&\qquad (\tilde{x}^1_\rho \otimes \tilde{x}^2_\rho \otimes \tilde{x}^3_\rho)\\
&= (\tilde{q}_\rho \otimes 1_H)(\rho \otimes \mathrm{Id}_H)(\tilde{q}_\rho)\Phi_\rho^{-1},
\end{aligned}$$

which is exactly the left-hand side of (4.3.15). □

We have a similar result for left $H$-comodule algebras, which is equivalent to Proposition 4.29, since $(\rho, \Phi_\rho)$ is a right $H$-coaction if and only if $(\rho^{\rm op}, (\Phi_\rho^{-1})^{321})$ is a left $H^{\rm cop}$-coaction, where we denoted $\rho^{\rm op} := \tau\rho$, $\tau$ being the switch map. Note that, when we specialize the next formulas for $\mathfrak{B} = H$, we get the relations (3.2.22), (3.2.24), (3.2.27) and (3.2.28) in Proposition 3.25.

**Proposition 4.30** *Let $H$ be a quasi-Hopf algebra with bijective antipode and $\mathfrak{B}$ a left $H$-comodule algebra. Define the elements $\tilde{p}_\lambda, \tilde{q}_\lambda \in H \otimes \mathfrak{B}$ as follows:*

$$\tilde{p}_\lambda = \tilde{p}_\lambda^1 \otimes \tilde{p}_\lambda^2 = \tilde{X}_\lambda^2 S^{-1}(\tilde{X}_\lambda^1 \beta) \otimes \tilde{X}_\lambda^3, \quad \tilde{q}_\lambda = \tilde{q}_\lambda^1 \otimes \tilde{q}_\lambda^2 = S(\tilde{x}_\lambda^1)\alpha \tilde{x}_\lambda^2 \otimes \tilde{x}_\lambda^3. \tag{4.3.16}$$

*Then, for all $\mathfrak{b} \in \mathfrak{B}$, the following relations hold:*

$$\lambda(\mathfrak{b}_{[0]})\tilde{p}_\lambda[S^{-1}(\mathfrak{b}_{[-1]}) \otimes 1_{\mathfrak{B}}] = \tilde{p}_\lambda[1_H \otimes \mathfrak{b}], \tag{4.3.17}$$

$$[S(\mathfrak{b}_{[-1]}) \otimes 1_{\mathfrak{B}}]\tilde{q}_\lambda \lambda(\mathfrak{b}_{[0]}) = [1_H \otimes \mathfrak{b}]\tilde{q}_\lambda, \tag{4.3.18}$$

$$\lambda(\tilde{q}_\lambda^2)\tilde{p}_\lambda[S^{-1}(\tilde{q}_\lambda^1) \otimes 1_{\mathfrak{B}}] = 1_H \otimes 1_{\mathfrak{B}}, \tag{4.3.19}$$

$$[S(\tilde{p}_\lambda^1) \otimes 1_{\mathfrak{B}}]\tilde{q}_\lambda \lambda(\tilde{p}_\lambda^2) = 1_H \otimes 1_{\mathfrak{B}}, \tag{4.3.20}$$

$$\begin{aligned}&\Phi_\lambda^{-1}(\mathrm{Id}_H \otimes \lambda)(\tilde{p}_\lambda)(1_H \otimes \tilde{p}_\lambda)\\ &\quad = (\Delta \otimes \mathrm{Id}_{\mathfrak{B}})(\lambda(\tilde{X}_\lambda^3)\tilde{p}_\lambda)[S^{-1}(\tilde{X}_\lambda^2 g^2) \otimes S^{-1}(\tilde{X}_\lambda^1 g^1) \otimes 1_{\mathfrak{B}}],\end{aligned} \tag{4.3.21}$$

$$\begin{aligned}&(1_H \otimes \tilde{q}_\lambda)(\mathrm{Id}_H \otimes \lambda)(\tilde{q}_\lambda)\Phi_\lambda\\ &\quad = [S(\tilde{x}_\lambda^2)f^1 \otimes S(\tilde{x}_\lambda^1)f^2 \otimes 1_{\mathfrak{B}}](\Delta \otimes \mathrm{Id}_{\mathfrak{B}})(\tilde{q}_\lambda \lambda(\tilde{x}_\lambda^3)),\end{aligned} \tag{4.3.22}$$

*where $f = f^1 \otimes f^2$ is the element defined in* (3.2.15) *and $f^{-1} = g^1 \otimes g^2$.*

When $H$ is a quasi-Hopf algebra, apart from the "cop"-machinery we have a second way to obtain right $H$-comodule algebras from left-handed ones.

**Proposition 4.31** *Let $H$ be a quasi-Hopf algebra with bijective antipode, and let $(\mathfrak{B}, \lambda, \Phi_\lambda)$ be a left $H$-comodule algebra. Then $\mathfrak{B}$ is a right $H^{\rm op}$-comodule algebra with structure*

$$\rho : \mathfrak{B} \to \mathfrak{B} \otimes H^{\rm op}, \ \rho(\mathfrak{b}) = \mathfrak{b}_{[0]} \otimes S^{-1}(\mathfrak{b}_{[-1]}), \tag{4.3.23}$$

$$\Phi_\rho = \tilde{x}_\lambda^3 \otimes S^{-1}(f^2 \tilde{x}_\lambda^2) \otimes S^{-1}(f^1 \tilde{x}_\lambda^1) \in \mathfrak{B} \otimes H^{\rm op} \otimes H^{\rm op}, \tag{4.3.24}$$

*where $f = f^1 \otimes f^2$ is the Drinfeld twist defined in* (3.2.15)*, and the multiplications in* (4.3.24) *are made in $H$, not in $H^{\rm op}$.*

*Proof* The relation (4.3.1) follows easily by applying (4.3.5) and (3.2.13), and the relations (4.3.3), (4.3.4) are trivial. We now prove (4.3.2). Since $\Phi_{\rm op} = \Phi^{-1}$ we have

$$\begin{aligned}&\tilde{X}_\rho^1 \tilde{Y}_\rho^1 \otimes x^1 \cdot_{\rm op} (\tilde{X}_\rho^2)_1 \cdot_{\rm op} \tilde{Y}_\rho^2 \otimes x^2 \cdot_{\rm op} (\tilde{X}_\rho^2)_2 \cdot_{\rm op} \tilde{Y}_\rho^3 \otimes x^3 \cdot_{\rm op} \tilde{X}_\rho^3\\ &= \quad \tilde{x}_\lambda^3 \tilde{y}_\lambda^3 \otimes S^{-1}(F^2 \tilde{y}_\lambda^2) S^{-1}(f^2 \tilde{x}_\lambda^2)_1 x^1\\ &\qquad \otimes S^{-1}(F^1 \tilde{y}_\lambda^1) S^{-1}(f^2 \tilde{x}_\lambda^2)_2 x^2 \otimes S^{-1}(f^1 \tilde{x}_\lambda^1) x^3\\ &\overset{(3.2.13)}{=} \quad \tilde{x}_\lambda^3 \tilde{y}_\lambda^3 \otimes S^{-1}(S(x^1) F^2 f_2^2 (\tilde{x}_\lambda^2)_2 \tilde{y}_\lambda^2)\\ &\qquad \otimes S^{-1}(S(x^2) F^1 f_1^2 (\tilde{x}_\lambda^2)_1 \tilde{y}_\lambda^1) \otimes S^{-1}(S(x^3) f^1 \tilde{x}_\lambda^1)\end{aligned}$$

$$
\begin{aligned}
&\overset{(3.2.17)}{=} \tilde{x}^3_\lambda \tilde{y}^3_\lambda \otimes S^{-1}(F^2 x^3 (\tilde{x}^2_\lambda)_2 \tilde{y}^2_\lambda) \\
&\quad \otimes S^{-1}(f^2 F^1_2 x^2 (\tilde{x}^2_\lambda)_1 \tilde{y}^1_\lambda) \otimes S^{-1}(f^1 F^1_1 x^1 \tilde{x}^1_\lambda) \\
&\overset{(4.3.6),(3.2.13)}{=} \tilde{y}^3_\lambda (\tilde{x}^3_\lambda)_{[0]} \otimes S^{-1}(F^2 \tilde{y}^2_\lambda) \cdot_{\mathrm{op}} S^{-1}((\tilde{x}^3_\lambda)_{[-1]}) \\
&\quad \otimes S^{-1}(F^1 \tilde{y}^1_\lambda)_1 \cdot_{\mathrm{op}} S^{-1}(f^2 \tilde{x}^2_\lambda) \otimes S^{-1}(F^1 \tilde{y}^1_\lambda)_2 \cdot_{\mathrm{op}} S^{-1}(f^1 \tilde{x}^1_\lambda) \\
&\overset{(4.3.23),(4.3.24)}{=} \tilde{X}^1_\rho (\tilde{Y}^1_\rho)_{\langle 0\rangle} \otimes \tilde{X}^2_\rho \cdot_{\mathrm{op}} (\tilde{Y}^1_\rho)_{\langle 1\rangle} \otimes (\tilde{X}^3_\rho)_1 \cdot_{\mathrm{op}} \tilde{Y}^2_\rho \otimes (\tilde{X}^3_\rho)_2 \cdot_{\mathrm{op}} \tilde{Y}^3_\rho,
\end{aligned}
$$

where $\cdot_{\mathrm{op}}$ is multiplication on $H^{\mathrm{op}}$ and $F^1 \otimes F^2$ is another copy of $f$. □

As we explained at the beginning of this section, we cannot introduce the notion of comodule algebra by using a categorical point of view, but we shall see that one can restate the definition of a comodule algebra in terms of monoidal categories.

If $H$ is a quasi-bialgebra and $\mathfrak{A}$ is an algebra then we define $\mathscr{C} := {}_{\mathfrak{A}\otimes H}\mathscr{M}_{\mathfrak{A}}$. Thus, an object of $\mathscr{C}$ is a $\mathfrak{A}$-bimodule and an $(H,\mathfrak{A})$-bimodule such that $h(\mathfrak{a}m) = \mathfrak{a}(hm)$, for all $\mathfrak{a} \in \mathfrak{A}$, $h \in H$ and $m \in M$, where we denoted by juxtaposition the actions of $H$ and $\mathfrak{A}$ on an object of $\mathscr{C}$. Morphisms are left $H$-linear maps which are also $\mathfrak{A}$-bimodule maps. We claim that $\mathscr{C}$ is a monoidal category. Indeed, it is not hard to see that $\mathscr{C}$ becomes a monoidal category with tensor product $\otimes_{\mathfrak{A}}$ given via $\Delta$, that is,

$$(\mathfrak{a} \otimes h)(m \otimes_{\mathfrak{A}} n)\mathfrak{a}' := \mathfrak{a} h_1 m \otimes_{\mathfrak{A}} h_2 n \mathfrak{a}',$$

for all $M, N \in \mathscr{C}$, $m \in M$, $n \in N$, $\mathfrak{a}, \mathfrak{a}' \in \mathfrak{A}$ and $h \in H$, associativity constraint

$$
\begin{aligned}
\underline{a}_{M,N,P} &: (M \otimes_{\mathfrak{A}} N) \otimes_{\mathfrak{A}} P \to M \otimes_{\mathfrak{A}} (N \otimes_{\mathfrak{A}} P), \\
\underline{a}_{M,N,P}((m \otimes_{\mathfrak{A}} n) \otimes_{\mathfrak{A}} p) &= X^1 m \otimes_{\mathfrak{A}} (X^2 n \otimes_{\mathfrak{A}} X^3 p),
\end{aligned}
$$

unit $\mathfrak{A}$ as a trivial left $H$-module, and the usual left and right unit constraints.

**Proposition 4.32** *Let $H$ be a quasi-bialgebra and $\mathfrak{A}$ an algebra. If $\mathfrak{A} \otimes H$ is viewed in the canonical way as an object in ${}_{\mathfrak{A}\otimes H}\mathscr{M}$, then $\mathfrak{A} \otimes H$ has a coalgebra structure $(\mathfrak{A} \otimes H, \underline{\Delta}, \underline{\varepsilon})$ in the monoidal category $\mathscr{C} = {}_{\mathfrak{A}\otimes H}\mathscr{M}_{\mathfrak{A}}$ such that $\underline{\Delta}(1_{\mathfrak{A}} \otimes 1_H)$ is invertible and $\underline{\varepsilon}(1_{\mathfrak{A}} \otimes 1_H) = 1_{\mathfrak{A}}$ if and only if $\mathfrak{A}$ is a right $H$-comodule algebra.*

*Proof* Suppose that $\mathfrak{A} \otimes H$ is an object of $\mathscr{C}$, and that there exists a coalgebra structure $(\mathfrak{A} \otimes H, \underline{\Delta}, \underline{\varepsilon})$ on $\mathfrak{A} \otimes H$ in the monoidal category $\mathscr{C}$ such that $\underline{\Delta}(1_{\mathfrak{A}} \otimes 1_H)$ is invertible and $\underline{\varepsilon}(1_{\mathfrak{A}} \otimes 1_H) = 1_{\mathfrak{A}}$. Then we define

$$\mathfrak{A} \ni \mathfrak{a} \mapsto \rho(\mathfrak{a}) = \mathfrak{a}_{\langle 0\rangle} \otimes \mathfrak{a}_{\langle 1\rangle} := (1_{\mathfrak{A}} \otimes 1_H)\mathfrak{a} \in \mathfrak{A} \otimes H,$$

and denote $\underline{\Delta}(1_{\mathfrak{A}} \otimes 1_H) := (\tilde{X}^1 \otimes \tilde{X}^2) \otimes_{\mathfrak{A}} (1_{\mathfrak{A}} \otimes \tilde{X}^3)$. Since $\mathfrak{A} \otimes H$ is a right $\mathfrak{A}$-module it follows that $\rho$ is an algebra map. Also, since $\underline{\Delta}(1_{\mathfrak{A}} \otimes 1_H)$ is invertible we obtain that $\Phi_\rho := \tilde{X}^1 \otimes \tilde{X}^2 \otimes \tilde{X}^3$ is an invertible element in $\mathfrak{A} \otimes H \otimes H$. By using the fact that $\underline{\Delta}$ and $\underline{\varepsilon}$ are morphisms in $\mathscr{C}$, and that $\underline{\varepsilon}(1_{\mathfrak{A}} \otimes 1_H) = 1_{\mathfrak{A}}$, one can see that

$$\underline{\Delta}(\mathfrak{a} \otimes h) = (\mathfrak{a}\tilde{X}^1 \otimes h_1 \tilde{X}^2) \otimes_{\mathfrak{A}} (1_{\mathfrak{A}} \otimes h_2 \tilde{X}^3) \quad \text{and} \quad \underline{\varepsilon}(\mathfrak{a} \otimes h) = \varepsilon(h)\mathfrak{a},$$

for all $\mathfrak{a} \in \mathfrak{A}$, $h \in H$. Now, (4.3.1) and (4.3.2) follow because $\underline{\Delta}((1_{\mathfrak{A}} \otimes 1_H)\mathfrak{a}) = \underline{\Delta}(1_{\mathfrak{A}} \otimes 1_H)\mathfrak{a}$ and $\Phi(\underline{\Delta} \otimes \mathrm{Id})\underline{\Delta}(\mathfrak{a} \otimes h) = (\mathrm{Id} \otimes \underline{\Delta})\underline{\Delta}(\mathfrak{a} \otimes h)$ for all $\mathfrak{a} \in \mathfrak{A}$ and $h \in H$.

Then it is easy to see that $\underline{\varepsilon}((1_{\mathfrak{A}} \otimes 1_H)\mathfrak{a}) = \mathfrak{a}$ implies (4.3.3), and the fact that $\underline{\varepsilon}$ is the counit for $\underline{\Delta}$ implies (4.3.4). We leave the details to the reader.

Conversely, suppose that $(\mathfrak{A}, \rho, \Phi_\rho)$ is a right $H$-comodule algebra and let $\mathfrak{C} = \mathfrak{A} \otimes H$. If we define

$$(\mathfrak{a} \otimes h)(\mathfrak{a}' \otimes h')\mathfrak{a}'' := \mathfrak{a}\mathfrak{a}'\mathfrak{a}''_{\langle 0 \rangle} \otimes hh'\mathfrak{a}''_{\langle 1 \rangle}, \tag{4.3.25}$$

for all $\mathfrak{a}, \mathfrak{a}', \mathfrak{a}'' \in \mathfrak{A}$ and $h, h' \in H$, then one can easily check that with this structure $\mathfrak{C} \in \mathscr{C}$. Moreover, we claim that $\mathfrak{C}$ with the structure given by

$$\underline{\Delta}_{\mathfrak{C}}(\mathfrak{a} \otimes h) := (\mathfrak{a}\tilde{X}^1 \otimes h_1\tilde{X}^2) \otimes_{\mathfrak{A}} (1_{\mathfrak{A}} \otimes h_2\tilde{X}^3), \tag{4.3.26}$$

$$\underline{\varepsilon}_{\mathfrak{C}}(\mathfrak{a} \otimes h) := \varepsilon(h)\mathfrak{a}, \tag{4.3.27}$$

for all $\mathfrak{a} \in \mathfrak{A}$ and $h \in H$, becomes a coalgebra in $\mathscr{C}$. Indeed, the fact that $\underline{\Delta}_{\mathfrak{C}}$ and $\underline{\varepsilon}_{\mathfrak{C}}$ are morphisms in $\mathscr{C}$ and that $\underline{\varepsilon}_{\mathfrak{C}}$ is the counit for $\underline{\Delta}_{\mathfrak{C}}$ follow from straightforward computations (which are left to the reader). We only show that the comultiplication $\underline{\Delta}_{\mathfrak{C}}$ is coassociative up to the associativity constraint of $\mathscr{C}$. Indeed, we compute:

$$\begin{aligned}
&(\underline{\Delta}_{\mathfrak{C}} \otimes_{\mathfrak{A}} \mathrm{Id})(\underline{\Delta}_{\mathfrak{C}}(\mathfrak{a} \otimes h)) \\
&= \underline{\Delta}_{\mathfrak{C}}(\mathfrak{a}\tilde{X}^1 \otimes h_1\tilde{X}^2) \otimes_{\mathfrak{A}} (1_{\mathfrak{A}} \otimes h_2\tilde{X}^3) \\
&= (\mathfrak{a}\tilde{X}^1\tilde{Y}^1 \otimes h_{(1,1)}\tilde{X}^2_1\tilde{Y}^2) \otimes_{\mathfrak{A}} (1_{\mathfrak{A}} \otimes h_{(1,2)}\tilde{X}^2_2\tilde{Y}^3) \otimes_{\mathfrak{A}} (1_{\mathfrak{A}} \otimes h_2\tilde{X}^3) \\
&\overset{(4.3.2)}{=} (\mathfrak{a}\tilde{X}^1\tilde{Y}^1_{\langle 0 \rangle} \otimes h_{(1,1)}x^1\tilde{X}^2\tilde{Y}^1_{\langle 1 \rangle}) \otimes_{\mathfrak{A}} (1_{\mathfrak{A}} \otimes h_{(1,2)}x^2\tilde{X}^3_1\tilde{Y}^2) \otimes_{\mathfrak{A}} (1_{\mathfrak{A}} \otimes h_2x^3\tilde{X}^3_2\tilde{Y}^3) \\
&\overset{(3.1.7)}{=} x^1(\mathfrak{a}\tilde{X}^1 \otimes h_1\tilde{X}^2)\tilde{Y}^1 \otimes_{\mathfrak{A}} x^2(1_{\mathfrak{A}} \otimes h_{(2,1)}\tilde{X}^3_1\tilde{Y}^2) \otimes_{\mathfrak{A}} x^3(1_{\mathfrak{A}} \otimes h_{(2,2)}\tilde{X}^3_2\tilde{Y}^3) \\
&= \Phi^{-1}(\mathfrak{a}\tilde{X}^1 \otimes h_1\tilde{X}^2) \otimes_{\mathfrak{A}} (\tilde{Y}^1 \otimes h_{(2,1)}\tilde{X}^3_1\tilde{Y}^2) \otimes_{\mathfrak{A}} (1_{\mathfrak{A}} \otimes h_{(2,2)}\tilde{X}^3_2\tilde{Y}^3) \\
&= \Phi^{-1}(\mathfrak{a}\tilde{X}^1 \otimes h_1\tilde{X}^2) \otimes_{\mathfrak{A}} \underline{\Delta}_{\mathfrak{C}}(1_{\mathfrak{A}} \otimes h_2\tilde{X}^3) \\
&= \Phi^{-1}(\mathrm{Id} \otimes_{\mathfrak{A}} \underline{\Delta}_{\mathfrak{C}})(\underline{\Delta}_{\mathfrak{C}}(\mathfrak{a} \otimes h)),
\end{aligned}$$

for all $\mathfrak{a} \in \mathfrak{A}$ and $h \in H$, as required. □

One can also formulate and prove a left-handed version of Proposition 4.32.

## 4.4 Bicomodule Algebras and Two-sided Coactions

In this section we introduce and study the concepts of bicomodule algebra and two-sided coaction for a quasi-bialgebra and the relations between these concepts.

**Definition 4.33** Let $H$ be a quasi-bialgebra and $\mathbb{A}$ an algebra. By an $H$-bicomodule algebra structure on $\mathbb{A}$ we mean a 5-tuple $(\lambda, \rho, \Phi_\lambda, \Phi_\rho, \Phi_{\lambda,\rho})$, where $\lambda$ and $\rho$ are left and right $H$-coactions on $\mathbb{A}$, respectively, and $\Phi_\lambda \in H \otimes H \otimes \mathbb{A}$, $\Phi_\rho \in \mathbb{A} \otimes H \otimes H$ and $\Phi_{\lambda,\rho} \in H \otimes \mathbb{A} \otimes H$ are invertible elements, such that:

(i) $(\mathbb{A}, \lambda, \Phi_\lambda)$ is a left $H$-comodule algebra;

(ii) $(\mathbb{A}, \rho, \Phi_\rho)$ is a right $H$-comodule algebra;

(iii) the following compatibility relations hold:

$$\Phi_{\lambda,\rho}(\lambda\otimes \mathrm{Id}_H)(\rho(u)) = (\mathrm{Id}_H\otimes\rho)(\lambda(u))\Phi_{\lambda,\rho},\quad \forall\, u\in\mathbb{A}, \tag{4.4.1}$$

$$(1_H\otimes\Phi_{\lambda,\rho})(\mathrm{Id}_H\otimes\lambda\otimes\mathrm{Id}_H)(\Phi_{\lambda,\rho})(\Phi_\lambda\otimes 1_H)$$
$$= (\mathrm{Id}_H\otimes\mathrm{Id}_H\otimes\rho)(\Phi_\lambda)(\Delta\otimes\mathrm{Id}_{\mathbb{A}}\otimes\mathrm{Id}_H)(\Phi_{\lambda,\rho}), \tag{4.4.2}$$

$$(1_H\otimes\Phi_\rho)(\mathrm{Id}_H\otimes\rho\otimes\mathrm{Id}_H)(\Phi_{\lambda,\rho})(\Phi_{\lambda,\rho}\otimes 1_H)$$
$$= (\mathrm{Id}_H\otimes\mathrm{Id}_{\mathbb{A}}\otimes\Delta)(\Phi_{\lambda,\rho})(\lambda\otimes\mathrm{Id}_H\otimes\mathrm{Id}_H)(\Phi_\rho). \tag{4.4.3}$$

It is easy to see that, if $\mathbb{A}$ is a bicomodule algebra then, in addition, we have

$$(\mathrm{Id}_H\otimes\mathrm{Id}_{\mathbb{A}}\otimes\varepsilon)(\Phi_{\lambda,\rho}) = 1_H\otimes 1_{\mathbb{A}},\qquad (\varepsilon\otimes\mathrm{Id}_{\mathbb{A}}\otimes\mathrm{Id}_H)(\Phi_{\lambda,\rho}) = 1_{\mathbb{A}}\otimes 1_H. \tag{4.4.4}$$

A first example of a bicomodule algebra over $H$ is $\mathbb{A}=H$, $\lambda=\rho=\Delta$ and $\Phi_\lambda = \Phi_\rho = \Phi_{\lambda,\rho} = \Phi$. Also, it is not hard to see that if $(\mathbb{A},\lambda,\rho,\Phi_\lambda,\Phi_\rho,\Phi_{\lambda,\rho})$ is an $H$-bicomodule algebra then

- $(\mathbb{A},\tau\circ\rho,\tau\circ\lambda,(\Phi_\rho^{-1})^{321},(\Phi_\lambda^{-1})^{321},(\Phi_{\lambda,\rho}^{-1})^{321})$ is an $H^{\mathrm{cop}}$-bicomodule algebra;
- $(\mathbb{A}^{\mathrm{op}},\tau\circ\rho,\tau\circ\lambda,\Phi_\rho^{321},\Phi_\lambda^{321},\Phi_{\lambda,\rho}^{321})$ is an $H^{\mathrm{op,cop}}$-bicomodule algebra;
- $(\mathbb{A}^{\mathrm{op}},\lambda,\rho,\Phi_\lambda^{-1},\Phi_\rho^{-1},\Phi_{\lambda,\rho}^{-1})$ is an $H^{\mathrm{op}}$-bicomodule algebra,

where $\tau$ is a generic notation for the switch map (for $X,Y$ vector spaces, $\tau_{X,Y}: X\otimes Y\to Y\otimes X$ is given by $\tau_{X,Y}(x\otimes y) = y\otimes x$, for all $x\in X$ and $y\in Y$).

For the left and right comodule algebra structures of $\mathbb{A}$ we will use notation as above. For simplicity we denote

$$\Phi_{\lambda,\rho} = \Theta^1\otimes\Theta^2\otimes\Theta^3 = \tilde{\Theta}^1\otimes\tilde{\Theta}^2\otimes\tilde{\Theta}^3 = \overline{\Theta}^1\otimes\overline{\Theta}^2\otimes\overline{\Theta}^3,$$
$$\Phi_{\lambda,\rho}^{-1} = \theta^1\otimes\theta^2\otimes\theta^3 = \tilde{\theta}^1\otimes\tilde{\theta}^2\otimes\tilde{\theta}^3 = \overline{\theta}^1\otimes\overline{\theta}^2\otimes\overline{\theta}^3.$$

By twisting a bicomodule algebra we obtain a bicomodule algebra as well.

**Proposition 4.34** *Let $H$ be a quasi-bialgebra and $\mathbb{A}$ an algebra such that $(\mathbb{A},\rho,\Phi_\rho)$ is a right $H$-comodule algebra and $(\mathbb{A},\lambda,\Phi_\lambda)$ is a left $H$-comodule algebra. Let $\Phi_{\lambda,\rho}\in H\otimes\mathbb{A}\otimes H$ be an invertible element and let $U_\rho\in\mathbb{A}\otimes H$ and $U_\lambda\in H\otimes\mathbb{A}$ be invertible elements such that $(\mathrm{Id}_{\mathbb{A}}\otimes\varepsilon)(U_\rho) = 1_{\mathbb{A}} = (\varepsilon\otimes\mathrm{Id}_{\mathbb{A}})(U_\lambda)$. Consider the right (resp. left) $H$-comodule algebra $(\mathbb{A},\rho',\Phi_{\rho'})$ (resp. $(\mathbb{A},\lambda',\Phi_{\lambda'})$) constructed as in Proposition 4.27, part (i) (resp. part (ii)). Define the element*

$$\Phi_{\lambda',\rho'} = (1_H\otimes U_\rho)(\mathrm{Id}_H\otimes\rho)(U_\lambda)\Phi_{\lambda,\rho}(\lambda\otimes\mathrm{Id}_H)(U_\rho^{-1})(U_\lambda^{-1}\otimes 1_H)\in H\otimes\mathbb{A}\otimes H.$$

*Then $(\mathbb{A},\lambda,\rho,\Phi_\lambda,\Phi_\rho,\Phi_{\lambda,\rho})$ is a bicomodule algebra over $H$ if and only if $(\mathbb{A},\lambda',\rho',\Phi_{\lambda'},\Phi_{\rho'},\Phi_{\lambda',\rho'})$ also is. If this is the case, these two $H$-bicomodule algebras are called "twist equivalent", and this twist equivalence is an equivalence relation.*

*Proof* This is a straightforward computation left to the reader. □

Bicomodule algebras are closely connected to so-called two-sided coactions.

**Definition 4.35** Let $H$ be a quasi-bialgebra and $\mathbb{A}$ an algebra. A two-sided coaction of $H$ on $\mathbb{A}$ is a pair $(\delta, \Psi)$, where $\delta : \mathbb{A} \to H \otimes \mathbb{A} \otimes H$ is an algebra map and $\Psi \in H^{\otimes 2} \otimes \mathbb{A} \otimes H^{\otimes 2}$ is an invertible element such that the following relations hold:

$$(\mathrm{Id}_H \otimes \delta \otimes \mathrm{Id}_H)(\delta(u))\Psi = \Psi(\Delta \otimes \mathrm{Id}_{\mathbb{A}} \otimes \Delta)(\delta(u)), \ \forall\, u \in \mathbb{A}, \tag{4.4.5}$$

$$\begin{aligned}&(1_H \otimes \Psi \otimes 1_H)(\mathrm{Id}_H \otimes \Delta \otimes \mathrm{Id}_{\mathbb{A}} \otimes \Delta \otimes \mathrm{Id}_H)(\Psi)(\Phi \otimes 1_{\mathbb{A}} \otimes \Phi^{-1})\\ &\quad = (\mathrm{Id}_H \otimes \mathrm{Id}_H \otimes \delta \otimes \mathrm{Id}_H \otimes \mathrm{Id}_H)(\Psi)(\Delta \otimes \mathrm{Id}_H \otimes \mathrm{Id}_{\mathbb{A}} \otimes \mathrm{Id}_H \otimes \Delta)(\Psi),\end{aligned} \tag{4.4.6}$$

$$(\varepsilon \otimes \mathrm{Id}_{\mathbb{A}} \otimes \varepsilon) \circ \delta = \mathrm{Id}_{\mathbb{A}}, \tag{4.4.7}$$

$$\begin{aligned}&(\mathrm{Id}_H \otimes \varepsilon \otimes \mathrm{Id}_{\mathbb{A}} \otimes \varepsilon \otimes \mathrm{Id}_H)(\Psi)\\ &\quad = (\varepsilon \otimes \mathrm{Id}_H \otimes \mathrm{Id}_{\mathbb{A}} \otimes \mathrm{Id}_H \otimes \varepsilon)(\Psi) = 1_H \otimes 1_{\mathbb{A}} \otimes 1_H.\end{aligned} \tag{4.4.8}$$

An example of a two-sided coaction of a quasi-bialgebra $H$ may be obtained by taking $\mathbb{A} = H$, $\delta = (\Delta \otimes \mathrm{Id}_H) \circ \Delta$ and

$$\Psi = [(\mathrm{Id}_H \otimes \Delta \otimes \mathrm{Id}_H)(\Phi) \otimes 1_H][\Phi \otimes 1_H \otimes 1_H][(\delta \otimes \mathrm{Id}_H \otimes \mathrm{Id}_H)(\Phi^{-1})].$$

Another example may be obtained by taking $\mathbb{A} = H$, $\delta' = (\mathrm{Id}_H \otimes \Delta) \circ \Delta$ and

$$\Psi' = [1_H \otimes (\mathrm{Id}_H \otimes \Delta \otimes \mathrm{Id}_H)(\Phi^{-1})][1_H \otimes 1_H \otimes \Phi^{-1}][(\mathrm{Id}_H \otimes \mathrm{Id}_H \otimes \delta')(\Phi)].$$

We show now that we can twist two-sided coactions as well.

**Proposition 4.36** *Let $H$ be a quasi-bialgebra, $\mathbb{A}$ an algebra and $(\delta, \Psi)$ a two-sided coaction of $H$ on $\mathbb{A}$. Let $U \in H \otimes \mathbb{A} \otimes H$ be an invertible element such that $(\varepsilon \otimes \mathrm{Id}_{\mathbb{A}} \otimes \varepsilon)(U) = 1_{\mathbb{A}}$. Define $\delta' : \mathbb{A} \to H \otimes \mathbb{A} \otimes H$ and $\Psi' \in H^{\otimes 2} \otimes \mathbb{A} \otimes H^{\otimes 2}$ by:*

$$\begin{aligned}&\delta'(u) = U\delta(u)U^{-1}, \ \forall\, u \in \mathbb{A},\\ &\Psi' = (1_H \otimes U \otimes 1_H)(\mathrm{Id}_H \otimes \delta \otimes \mathrm{Id}_H)(U)\Psi(\Delta \otimes \mathrm{Id}_{\mathbb{A}} \otimes \Delta)(U^{-1}).\end{aligned}$$

*Then $(\delta', \Psi')$ is a two-sided coaction of $H$ on $\mathbb{A}$, called "twist equivalent" with $(\delta, \Psi)$, and this twisting procedure provides an equivalence relation between two-sided coactions of $H$ on $\mathbb{A}$.*

*Proof* This is a straightforward computation, left to the reader. □

We will later need some additional relations concerning two-sided coactions.

**Lemma 4.37** *Let $H$ be a quasi-bialgebra and $(\delta, \Psi)$ a two-sided coaction of $H$ on an algebra $\mathbb{A}$. We denote $\lambda := (\mathrm{Id}_H \otimes \mathrm{Id}_{\mathbb{A}} \otimes \varepsilon) \circ \delta$ and $\rho := (\varepsilon \otimes \mathrm{Id}_{\mathbb{A}} \otimes \mathrm{Id}_H) \circ \delta$. Then the following relations are satisfied:*

$$\begin{aligned}&[(\mathrm{Id}_H^{\otimes 2} \otimes \mathrm{Id}_{\mathbb{A}} \otimes \varepsilon \otimes \mathrm{Id}_H)(\Psi) \otimes 1_H][(\varepsilon \otimes \Delta \otimes \mathrm{Id}_{\mathbb{A}} \otimes \mathrm{Id}_H^{\otimes 2})(\Psi)]\\ &\qquad = [(\varepsilon \otimes \mathrm{Id}_H \otimes \lambda \otimes \mathrm{Id}_H^{\otimes 2})(\Psi)][(\mathrm{Id}_H^{\otimes 2} \otimes \mathrm{Id}_{\mathbb{A}} \otimes \varepsilon \otimes \Delta)(\Psi)],\end{aligned} \tag{4.4.9}$$

$$\begin{aligned}&[(\mathrm{Id}_H^{\otimes 2} \otimes \mathrm{Id}_{\mathbb{A}} \otimes \mathrm{Id}_H \otimes \varepsilon)(\Psi) \otimes 1_H][(\varepsilon \otimes \Delta \otimes \mathrm{Id}_{\mathbb{A}} \otimes \mathrm{Id}_H^{\otimes 2})(\Psi)]\\ &\qquad = [(\varepsilon \otimes \mathrm{Id}_H \otimes \delta \otimes \varepsilon \otimes \mathrm{Id}_H)(\Psi)]\Psi,\end{aligned} \tag{4.4.10}$$

$$\begin{aligned}&[1_H \otimes (\varepsilon \otimes \mathrm{Id}_H \otimes \mathrm{Id}_{\mathbb{A}} \otimes \varepsilon \otimes \mathrm{Id}_H)(\Psi) \otimes 1_H]\Psi\\ &\qquad = [(\mathrm{Id}_H \otimes \varepsilon \otimes \lambda \otimes \mathrm{Id}_H^{\otimes 2})(\Psi)][(\mathrm{Id}_H^{\otimes 2} \otimes \mathrm{Id}_{\mathbb{A}} \otimes \varepsilon \otimes \Delta)(\Psi)],\end{aligned} \tag{4.4.11}$$

$$[1_H \otimes (\mathrm{Id}_H \otimes \varepsilon \otimes \mathrm{Id}_{\mathbb{A}} \otimes \mathrm{Id}_H^{\otimes 2})(\Psi)][(\mathrm{Id}_H^{\otimes 2} \otimes \mathrm{Id}_{\mathbb{A}} \otimes \Delta \otimes \varepsilon)(\Psi)]$$
$$= [(\mathrm{Id}_H^{\otimes 2} \otimes \rho \otimes \mathrm{Id}_H \otimes \varepsilon)(\Psi)][(\Delta \otimes \varepsilon \otimes \mathrm{Id}_{\mathbb{A}} \otimes \mathrm{Id}_H^{\otimes 2})(\Psi)], \quad (4.4.12)$$
$$[1_H \otimes (\varepsilon \otimes \mathrm{Id}_H \otimes \mathrm{Id}_{\mathbb{A}} \otimes \mathrm{Id}_H^{\otimes 2})(\Psi)][(\mathrm{Id}_H^{\otimes 2} \otimes \mathrm{Id}_{\mathbb{A}} \otimes \Delta \otimes \varepsilon)(\Psi)]$$
$$= [(\mathrm{Id}_H \otimes \varepsilon \otimes \delta \otimes \mathrm{Id}_H \otimes \varepsilon)(\Psi)]\Psi, \quad (4.4.13)$$
$$[1_H \otimes (\mathrm{Id}_H \otimes \varepsilon \otimes \mathrm{Id}_{\mathbb{A}} \otimes \mathrm{Id}_H \otimes \varepsilon)(\Psi) \otimes 1_H]\Psi$$
$$= [(\mathrm{Id}_H^{\otimes 2} \otimes \rho \otimes \varepsilon \otimes \mathrm{Id}_H)(\Psi)][(\Delta \otimes \varepsilon \otimes \mathrm{Id}_{\mathbb{A}} \otimes \mathrm{Id}_H^{\otimes 2})(\Psi)]. \quad (4.4.14)$$

*Proof* For $i = 1,2,3$, consider the maps $\varepsilon_i : H^{\otimes 3} \to H^{\otimes 2}$, given by acting with $\varepsilon$ on the $i$th tensor factor. By (3.1.10) and (3.1.11) we know that $\varepsilon_i(\Phi) = 1_H \otimes 1_H$, for $i = 1,2,3$. Then the relations (4.4.9)–(4.4.14) follow by using this after applying maps of the type $\varepsilon_i \otimes \mathrm{Id}_{\mathbb{A}} \otimes \varepsilon_j$, with $1 \le i, j \le 3$, to both sides of the relation (4.4.6); we leave the verification of the details to the reader. □

Let $H$ be a quasi-Hopf algebra with bijective antipode, $\mathbb{A}$ an algebra and $(\delta, \Psi)$ a two-sided coaction of $H$ on $\mathbb{A}$. We will use the following notation: $\delta(u) := u_{(-1)} \otimes u_{(0)} \otimes u_{(1)}$, for all $u \in \mathbb{A}$, $\Psi = \Psi^1 \otimes \cdots \otimes \Psi^5$, $\Psi^{-1} = \overline{\Psi}^1 \otimes \cdots \otimes \overline{\Psi}^5$. We associate to $(\delta, \Psi)$ the elements $p_\delta, q_\delta \in H \otimes \mathbb{A} \otimes H$ as follows:

$$p_\delta = p_\delta^1 \otimes p_\delta^2 \otimes p_\delta^3 = \Psi^2 S^{-1}(\Psi^1 \beta) \otimes \Psi^3 \otimes \Psi^4 \beta S(\Psi^5), \quad (4.4.15)$$
$$q_\delta = q_\delta^1 \otimes q_\delta^2 \otimes q_\delta^3 = S(\overline{\Psi}^1)\alpha\overline{\Psi}^2 \otimes \overline{\Psi}^3 \otimes S^{-1}(\alpha\overline{\Psi}^5)\overline{\Psi}^4. \quad (4.4.16)$$

**Lemma 4.38** *We have the following relations, for all $u \in \mathbb{A}$:*

$$p_\delta(1_H \otimes u \otimes 1_H) = \delta(u_{(0)}) p_\delta [S^{-1}(u_{(-1)}) \otimes 1_{\mathbb{A}} \otimes S(u_{(1)})], \quad (4.4.17)$$
$$(1_H \otimes u \otimes 1_H) q_\delta = [S(u_{(-1)}) \otimes 1_{\mathbb{A}} \otimes S^{-1}(u_{(1)})] q_\delta \delta(u_{(0)}), \quad (4.4.18)$$
$$\delta(q_\delta^2) p_\delta [S^{-1}(q_\delta^1) \otimes 1_{\mathbb{A}} \otimes S(q_\delta^3)] = 1_H \otimes 1_{\mathbb{A}} \otimes 1_H, \quad (4.4.19)$$
$$[S(p_\delta^1) \otimes 1_{\mathbb{A}} \otimes S^{-1}(p_\delta^3)] q_\delta \delta(p_\delta^2) = 1_H \otimes 1_{\mathbb{A}} \otimes 1_H, \quad (4.4.20)$$
$$\Psi^{-1}(\mathrm{Id}_H \otimes \delta \otimes \mathrm{Id}_H)(p_\delta)[1_H \otimes p_\delta \otimes 1_H]$$
$$= (\Delta \otimes \mathrm{Id}_{\mathbb{A}} \otimes \Delta)(\delta(\Psi^3) p_\rho)[S^{-1}(g^2) \otimes S^{-1}(g^1) \otimes 1_{\mathbb{A}} \otimes G^1 \otimes G^2]$$
$$[S^{-1}(\Psi^2) \otimes S^{-1}(\Psi^1) \otimes 1_{\mathbb{A}} \otimes S(\Psi^5) \otimes S(\Psi^4)], \quad (4.4.21)$$
$$[1_H \otimes q_\delta \otimes 1_H](\mathrm{Id}_H \otimes \delta \otimes \mathrm{Id}_H)(q_\delta)\Psi$$
$$= [S(\overline{\Psi}^2) f^1 \otimes S(\overline{\Psi}^1) f^2 \otimes 1_{\mathbb{A}} \otimes S^{-1}(F^2 \overline{\Psi}^5) \otimes S^{-1}(F^1 \overline{\Psi}^4)]$$
$$(\Delta \otimes \mathrm{Id}_{\mathbb{A}} \otimes \Delta)(q_\delta \delta(\overline{\Psi}^3)), \quad (4.4.22)$$

*where $f = f^1 \otimes f^2 = F^1 \otimes F^2$ is the Drinfeld twist defined in* (3.2.15) *and $f^{-1} = g^1 \otimes g^2 = G^1 \otimes G^2$.*

*Proof* This follows immediately from Proposition 4.30, because, after a permutation of tensor factors $\mathrm{Id}_H \otimes \tau_{\mathbb{A},H} : H \otimes \mathbb{A} \otimes H \to H \otimes H \otimes \mathbb{A}$, a two-sided $H$-coaction becomes a left $(H \otimes H^{\mathrm{cop}})$-comodule algebra. □

**Remark 4.39** The definitions of $q_\delta$ and of a two-sided coaction imply

$$q_\delta^1 \Psi^1 \otimes (q_\delta^2)_{(-1)} \Psi^2 \otimes (q_\delta^2)_{(0)} \Psi^3 \otimes (q_\delta^2)_{(1)} \Psi^4 \otimes q_\delta^3 \Psi^5$$

$$= S(\overline{\Psi}^1)q_L^1\overline{\Psi}^2_1 \otimes q_L^2\overline{\Psi}^2_2 \otimes \overline{\Psi}^3 \otimes q_R^1\overline{\Psi}^4_1 \otimes S^{-1}(\overline{\Psi}^5)q_R^2\overline{\Psi}^4_2, \tag{4.4.23}$$

where $q_L = q_L^1 \otimes q_L^2 = S(x^1)\alpha x^2 \otimes x^3$ and $q_R = q_R^1 \otimes q_R^2 = X^1 \otimes S^{-1}(\alpha X^3)X^2$.

We investigate the relation between bicomodule algebras and two-sided coactions.

**Proposition 4.40** *Let $H$ be a quasi-bialgebra and $\mathbb{A}$ an algebra such that $(\mathbb{A}, \lambda, \Phi_\lambda)$ is a left $H$-comodule algebra and $(\mathbb{A}, \rho, \Phi_\rho)$ is a right $H$-comodule algebra. Let $\Phi_{\lambda,\rho} \in H \otimes \mathbb{A} \otimes H$ be an invertible element. We define*

$$\begin{aligned} &\delta_l = (\lambda \otimes \mathrm{Id}_H) \circ \rho, \\ &\Psi_l = (\mathrm{Id}_H \otimes \lambda \otimes \mathrm{Id}_H^{\otimes 2})\Big((\Phi_{\lambda,\rho} \otimes 1_H)(\lambda \otimes \mathrm{Id}_H^{\otimes 2})(\Phi_\rho^{-1})\Big)[\Phi_\lambda \otimes 1_H^{\otimes 2}], \end{aligned} \tag{4.4.24}$$

$$\begin{aligned} &\delta_r = (\mathrm{Id}_H \otimes \rho) \circ \lambda, \\ &\Psi_r = (\mathrm{Id}_H^{\otimes 2} \otimes \rho \otimes \mathrm{Id}_H)\Big((1_H \otimes \Phi_{\lambda,\rho}^{-1})(\mathrm{Id}_H^{\otimes 2} \otimes \rho)(\Phi_\lambda)\Big)[1_H^{\otimes 2} \otimes \Phi_\rho^{-1}]. \end{aligned} \tag{4.4.25}$$

*(i) Consider the following conditions:*

*(1) $(\mathbb{A}, \lambda, \rho, \Phi_\lambda, \Phi_\rho, \Phi_{\lambda,\rho})$ is an $H$-bicomodule algebra;*

*(2) $(\delta_l, \Psi_l)$ is a two-sided coaction of $H$ on $\mathbb{A}$;*

*(3) $(\delta_r, \Psi_r)$ is a two-sided coaction of $H$ on $\mathbb{A}$.*

*Then $(1) \Leftrightarrow (2) \Leftrightarrow (3)$, and under these conditions $\Phi_{\lambda,\rho}$ provides a twist equivalence from $(\delta_l, \Psi_l)$ to $(\delta_r, \Psi_r)$.*

*(ii) Under the conditions of (i), let $(\mathbb{A}, \lambda', \rho', \Phi_{\lambda'}, \Phi_{\rho'}, \Phi_{\lambda',\rho'})$ be the $H$-bicomodule algebra obtained from $(\mathbb{A}, \lambda, \rho, \Phi_\lambda, \Phi_\rho, \Phi_{\lambda,\rho})$ as in Proposition 4.34 by twisting with some elements $U_\lambda \in H \otimes \mathbb{A}$ and $U_\rho \in \mathbb{A} \otimes H$. Let $(\delta_l', \Psi_l')$ and $(\delta_r', \Psi_r')$ be the associated two-sided coactions. Then $(U_\lambda \otimes 1_H)(\lambda \otimes \mathrm{Id}_H)(U_\rho)$ is a twist from $(\delta_l, \Psi_l)$ to $(\delta_l', \Psi_l')$ and $(1_H \otimes U_\rho)(\mathrm{Id}_H \otimes \rho)(U_\lambda)$ is a twist from $(\delta_r, \Psi_r)$ to $(\delta_r', \Psi_r')$.*

*Thus, we have a map from twist equivalence classes of $H$-bicomodule algebra structures on $\mathbb{A}$ to twist equivalence classes of two-sided coactions of $H$ on $\mathbb{A}$.*

*Proof* The proofs of the implications $(1) \Rightarrow (2)$ and $(1) \Rightarrow (3)$, as well as the proof of (ii), follow by tedious but straightforward computations that are left to the reader. We prove only the implication $(2) \Rightarrow (1)$, while $(3) \Rightarrow (1)$ is analogous and also left to the reader. We assume thus that $(\delta_l, \Psi_l)$ is a two-sided coaction. We denote $\Phi_{\lambda,\rho} = \Theta^1 \otimes \Theta^2 \otimes \Theta^3$. With this notation, we have, for all $u \in \mathbb{A}$:

$$\begin{aligned} \delta_l(u) &= u_{\langle 0\rangle_{[-1]}} \otimes u_{\langle 0\rangle_{[0]}} \otimes u_{\langle 1\rangle}, \\ \Psi_l &= \Theta^1 \tilde{x}^1_{\rho_{[-1]}} \tilde{X}^1_\lambda \otimes \Theta^2_{[-1]} \tilde{x}^1_{\rho_{[0]_{[-1]}}} \tilde{X}^2_\lambda \otimes \Theta^2_{[0]} \tilde{x}^1_{\rho_{[0]_{[0]}}} \tilde{X}^3_\lambda \otimes \Theta^3 \tilde{x}^2_\rho \otimes \tilde{x}^3_\rho. \end{aligned}$$

By using these formulas and (4.3.4), (4.3.7) and (4.3.8) we immediately obtain:

$$(\mathrm{Id}_H^{\otimes 2} \otimes \mathrm{Id}_{\mathbb{A}} \otimes \mathrm{Id}_H \otimes \varepsilon)(\Psi_l) = (\mathrm{Id}_H \otimes \lambda \otimes \mathrm{Id}_H)(\Phi_{\lambda,\rho})(\Phi_\lambda \otimes 1_H), \tag{4.4.26}$$

$$(\mathrm{Id}_H \otimes \varepsilon \otimes \mathrm{Id}_{\mathbb{A}} \otimes \mathrm{Id}_H \otimes \varepsilon)(\Psi_l) = \Phi_{\lambda,\rho}. \tag{4.4.27}$$

Since $\Psi_l$ is assumed to satisfy (4.4.8), from (4.4.27) we obtain that $(\varepsilon \otimes \mathrm{Id}_{\mathbb{A}} \otimes$

$\mathrm{Id}_H)(\Phi_{\lambda,\rho}) = 1_{\mathbb{A}} \otimes 1_H$ and $(\mathrm{Id}_H \otimes \mathrm{Id}_{\mathbb{A}} \otimes \varepsilon)(\Phi_{\lambda,\rho}) = 1_H \otimes 1_{\mathbb{A}}$. Thus we also have:

$$(\varepsilon \otimes \mathrm{Id}_H \otimes \mathrm{Id}_{\mathbb{A}} \otimes \varepsilon \otimes \mathrm{Id}_H)(\Psi_l) = 1_H \otimes 1_{\mathbb{A}} \otimes 1_H, \tag{4.4.28}$$
$$(\mathrm{Id}_H^{\otimes 2} \otimes \mathrm{Id}_{\mathbb{A}} \otimes \varepsilon^{\otimes 2})(\Psi_l) = \Phi_\lambda, \tag{4.4.29}$$
$$(\varepsilon^{\otimes 2} \otimes \mathrm{Id}_{\mathbb{A}} \otimes \mathrm{Id}_H^{\otimes 2})(\Psi_l) = \Phi_\rho^{-1}. \tag{4.4.30}$$

We want to prove (4.4.1), that is, $\Phi_{\lambda,\rho}\delta_l(u) = \delta_r(u)\Phi_{\lambda,\rho}$, for all $u \in \mathbb{A}$. One can easily check that we have the following relation:

$$\delta_r = (\mathrm{Id}_H \otimes \varepsilon \otimes \mathrm{Id}_{\mathbb{A}} \otimes \mathrm{Id}_H \otimes \varepsilon) \circ \delta_l^{(2)},$$

where we denote by $\delta_l^{(2)}$ the map $\delta_l^{(2)} = (\mathrm{Id}_H \otimes \delta_l \otimes \mathrm{Id}_H) \circ \delta_l$. Thus, it is enough to prove that we have, for all $u \in \mathbb{A}$:

$$(\mathrm{Id}_H \otimes \varepsilon \otimes \mathrm{Id}_{\mathbb{A}} \otimes \mathrm{Id}_H \otimes \varepsilon) \circ \delta_l^{(2)}(u) = \Phi_{\lambda,\rho}\delta_l(u)\Phi_{\lambda,\rho}^{-1}.$$

By (4.4.5) for $(\delta_l, \Psi_l)$, we have $\delta_l^{(2)}(u) = \Psi_l(\Delta \otimes \mathrm{Id}_{\mathbb{A}} \otimes \Delta)(\delta_l(u))\Psi_l^{-1}$. Thus, by applying (4.4.27), we have:

$$\begin{aligned}
&(\mathrm{Id}_H \otimes \varepsilon \otimes \mathrm{Id}_{\mathbb{A}} \otimes \mathrm{Id}_H \otimes \varepsilon)(\delta_l^{(2)}(u)) \\
&\quad = (\mathrm{Id}_H \otimes \varepsilon \otimes \mathrm{Id}_{\mathbb{A}} \otimes \mathrm{Id}_H \otimes \varepsilon)(\Psi_l(\Delta \otimes \mathrm{Id}_{\mathbb{A}} \otimes \Delta)(\delta_l(u))\Psi_l^{-1}) \\
&\quad = \Phi_{\lambda,\rho}[(\mathrm{Id}_H \otimes \varepsilon \otimes \mathrm{Id}_{\mathbb{A}} \otimes \mathrm{Id}_H \otimes \varepsilon) \circ (\Delta \otimes \mathrm{Id}_{\mathbb{A}} \otimes \Delta)(\delta_l(u))]\Phi_{\lambda,\rho}^{-1} \\
&\quad = \Phi_{\lambda,\rho}\delta_l(u)\Phi_{\lambda,\rho}^{-1}. \quad \text{QED}
\end{aligned}$$

To prove (4.4.2), one has to apply $\mathrm{Id}_H^{\otimes 2} \otimes \mathrm{Id}_{\mathbb{A}} \otimes \mathrm{Id}_H \otimes \varepsilon$ to (4.4.12), and then (4.4.2) follows immediately by also using (4.4.26), (4.4.27) and (4.4.29).

Now we prove (4.4.3). Note first that by using (4.4.11), (4.4.28) and (4.4.30) we immediately obtain

$$(\lambda \otimes \mathrm{Id}_H^{\otimes 2})(\Phi_\rho^{-1}) = (\varepsilon \otimes \mathrm{Id}_H \otimes \mathrm{Id}_{\mathbb{A}} \otimes \mathrm{Id}_H^{\otimes 2})(\Psi_l). \tag{4.4.31}$$

By applying $\mathrm{Id}_H \otimes \varepsilon \otimes \mathrm{Id}_{\mathbb{A}} \otimes \mathrm{Id}_H^{\otimes 2}$ to (4.4.10) and also using (4.4.28) we obtain:

$$\begin{aligned}
&[(\mathrm{Id}_H \otimes \varepsilon \otimes \mathrm{Id}_{\mathbb{A}} \otimes \mathrm{Id}_H \otimes \varepsilon)(\Psi_l) \otimes 1_H][(\varepsilon \otimes \mathrm{Id}_H \otimes \mathrm{Id}_{\mathbb{A}} \otimes \mathrm{Id}_H^{\otimes 2})(\Psi_l)] \\
&\qquad = (\mathrm{Id}_H \otimes \varepsilon \otimes \mathrm{Id}_{\mathbb{A}} \otimes \mathrm{Id}_H^{\otimes 2})(\Psi_l).
\end{aligned} \tag{4.4.32}$$

By applying $\mathrm{Id}_H \otimes \varepsilon \otimes \mathrm{Id}_{\mathbb{A}} \otimes \mathrm{Id}_H^{\otimes 2}$ to (4.4.12) we obtain:

$$\begin{aligned}
&[1_H \otimes (\varepsilon^{\otimes 2} \otimes \mathrm{Id}_{\mathbb{A}} \otimes \mathrm{Id}_H^{\otimes 2})(\Psi_l)][(\mathrm{Id}_H \otimes \varepsilon \otimes \mathrm{Id}_{\mathbb{A}} \otimes \Delta \otimes \varepsilon)(\Psi_l)] \\
&\quad = [(\mathrm{Id}_H \otimes \varepsilon \otimes \rho \otimes \mathrm{Id}_H \otimes \varepsilon)(\Psi_l)][(\mathrm{Id}_H \otimes \varepsilon \otimes \mathrm{Id}_{\mathbb{A}} \otimes \mathrm{Id}_H^{\otimes 2})(\Psi_l)].
\end{aligned} \tag{4.4.33}$$

Now we compute:

$$\begin{aligned}
&(\mathrm{Id}_H \otimes \rho \otimes \mathrm{Id}_H)(\Phi_{\lambda,\rho})(\Phi_{\lambda,\rho} \otimes 1_H)(\lambda \otimes \mathrm{Id}_H^{\otimes 2})(\Phi_\rho^{-1}) \\
&\overset{(4.4.31)}{\underset{(4.4.27)}{=}} [(\mathrm{Id}_H \otimes \varepsilon \otimes \rho \otimes \mathrm{Id}_H \otimes \varepsilon)(\Psi_l)][(\mathrm{Id}_H \otimes \varepsilon \otimes \mathrm{Id}_{\mathbb{A}} \otimes \mathrm{Id}_H \otimes \varepsilon)(\Psi_l) \otimes 1_H] \\
&\qquad [(\varepsilon \otimes \mathrm{Id}_H \otimes \mathrm{Id}_{\mathbb{A}} \otimes \mathrm{Id}_H^{\otimes 2})(\Psi_l)] \\
&\overset{(4.4.32)}{=} [(\mathrm{Id}_H \otimes \varepsilon \otimes \rho \otimes \mathrm{Id}_H \otimes \varepsilon)(\Psi_l)][(\mathrm{Id}_H \otimes \varepsilon \otimes \mathrm{Id}_{\mathbb{A}} \otimes \mathrm{Id}_H^{\otimes 2})(\Psi_l)]
\end{aligned}$$

$$\overset{(4.4.33)}{=} [1_H \otimes (\varepsilon^{\otimes 2} \otimes \mathrm{Id}_{\mathbb{A}} \otimes \mathrm{Id}_H^{\otimes 2})(\Psi_l)][(\mathrm{Id}_H \otimes \varepsilon \otimes \mathrm{Id}_{\mathbb{A}} \otimes \Delta \otimes \varepsilon)(\Psi_l)]$$
$$\overset{(4.4.27),(4.4.30)}{=} (1_H \otimes \Phi_\rho^{-1})[(\mathrm{Id}_H \otimes \mathrm{Id}_{\mathbb{A}} \otimes \Delta)(\Phi_{\lambda,\rho})],$$

finishing the proof of (4.4.3); thus, we have shown that $(\mathbb{A}, \lambda, \rho, \Phi_\lambda, \Phi_\rho, \Phi_{\lambda,\rho})$ is indeed an $H$-bicomodule algebra.

We prove now that, under the conditions (1)–(3), the element $U := \Phi_{\lambda,\rho}$ is a twist from $(\delta_l, \Psi_l)$ to $(\delta_r, \Psi_r)$, that is, we need to prove that

$$\delta_r(u) = U\delta_l(u)U^{-1}, \ \forall\, u \in \mathbb{A},$$
$$\Psi_r = (1_H \otimes U \otimes 1_H)(\mathrm{Id}_H \otimes \delta_l \otimes \mathrm{Id}_H)(U)\Psi_l(\Delta \otimes \mathrm{Id}_{\mathbb{A}} \otimes \Delta)(U^{-1}).$$

The first relation obviously holds, being equivalent to (4.4.1), so we prove the second one. Note first that, by using (4.4.3), the formula for $\Psi_l$ given in (4.4.24) becomes

$$\Psi_l = (\mathrm{Id}_H \otimes \delta_l \otimes \mathrm{Id}_H)(\Phi_{\lambda,\rho}^{-1})[1_H \otimes (\lambda \otimes \mathrm{Id}_H^{\otimes 2})(\Phi_\rho^{-1})]$$
$$(\mathrm{Id}_H \otimes \lambda \otimes \Delta)(\Phi_{\lambda,\rho})(\Phi_\lambda \otimes 1_H \otimes 1_H). \tag{4.4.34}$$

Now we compute:

$$(1_H \otimes U \otimes 1_H)(\mathrm{Id}_H \otimes \delta_l \otimes \mathrm{Id}_H)(U)\Psi_l$$
$$\overset{(4.4.34)}{=} [1_H \otimes (\Phi_{\lambda,\rho} \otimes 1_H)(\lambda \otimes \mathrm{Id}_H^{\otimes 2})(\Phi_\rho^{-1})]$$
$$(\mathrm{Id}_H^{\otimes 2} \otimes \mathrm{Id}_{\mathbb{A}} \otimes \Delta)((\mathrm{Id}_H \otimes \lambda \otimes \mathrm{Id}_H)(\Phi_{\lambda,\rho})(\Phi_\lambda \otimes 1_H))$$
$$\overset{(4.4.3)}{=} [1_H \otimes (\mathrm{Id}_H \otimes \rho \otimes \mathrm{Id}_H)(\Phi_{\lambda,\rho}^{-1})][1_H \otimes 1_H \otimes \Phi_\rho^{-1}]$$
$$(\mathrm{Id}_H^{\otimes 2} \otimes \mathrm{Id}_{\mathbb{A}} \otimes \Delta)((1_H \otimes \Phi_{\lambda,\rho})(\mathrm{Id}_H \otimes \lambda \otimes \mathrm{Id}_H)(\Phi_{\lambda,\rho})(\Phi_\lambda \otimes 1_H))$$
$$\overset{(4.4.2)}{=} [1_H \otimes (\mathrm{Id}_H \otimes \rho \otimes \mathrm{Id}_H)(\Phi_{\lambda,\rho}^{-1})][1_H \otimes 1_H \otimes \Phi_\rho^{-1}]$$
$$(\mathrm{Id}_H^{\otimes 2} \otimes (\mathrm{Id}_{\mathbb{A}} \otimes \Delta) \circ \rho)(\Phi_\lambda)(\Delta \otimes \mathrm{Id}_{\mathbb{A}} \otimes \Delta)(\Phi_{\lambda,\rho})$$
$$\overset{(4.3.1)}{=} [1_H \otimes (\mathrm{Id}_H \otimes \rho \otimes \mathrm{Id}_H)(\Phi_{\lambda,\rho}^{-1})](\mathrm{Id}_H^{\otimes 2} \otimes (\rho \otimes \mathrm{Id}_H) \circ \rho)(\Phi_\lambda)$$
$$[1_H \otimes 1_H \otimes \Phi_\rho^{-1}](\Delta \otimes \mathrm{Id}_{\mathbb{A}} \otimes \Delta)(\Phi_{\lambda,\rho})$$
$$= \Psi_r(\Delta \otimes \mathrm{Id}_{\mathbb{A}} \otimes \Delta)(\Phi_{\lambda,\rho}),$$

finishing the proof. □

We are now ready to prove that twist equivalence classes of $H$-bicomodule algebras are in bijection with twist equivalence classes of two-sided coactions of $H$.

**Proposition 4.41** *Let $H$ be a quasi-bialgebra, $\mathbb{A}$ an algebra and $(\delta, \Psi)$ a two-sided coaction of $H$ on $\mathbb{A}$. Define the following maps and elements:*

$$\lambda : \mathbb{A} \to H \otimes \mathbb{A}, \ \ \lambda := (\mathrm{Id}_H \otimes \mathrm{Id}_{\mathbb{A}} \otimes \varepsilon) \circ \delta,$$
$$\Phi_\lambda := (\mathrm{Id}_H^{\otimes 2} \otimes \mathrm{Id}_{\mathbb{A}} \otimes \varepsilon^{\otimes 2})(\Psi) \in H \otimes H \otimes \mathbb{A},$$
$$\rho : \mathbb{A} \to \mathbb{A} \otimes H, \ \ \rho := (\varepsilon \otimes \mathrm{Id}_{\mathbb{A}} \otimes \mathrm{Id}_H) \circ \delta,$$
$$\Phi_\rho := (\varepsilon^{\otimes 2} \otimes \mathrm{Id}_{\mathbb{A}} \otimes \mathrm{Id}_H^{\otimes 2})(\Psi^{-1}) \in \mathbb{A} \otimes H \otimes H,$$
$$U_l := (\varepsilon \otimes \mathrm{Id}_H \otimes \mathrm{Id}_{\mathbb{A}} \otimes \varepsilon \otimes \mathrm{Id}_H)(\Psi) \in H \otimes \mathbb{A} \otimes H,$$

$$U_r := (\mathrm{Id}_H \otimes \varepsilon \otimes \mathrm{Id}_{\mathbb{A}} \otimes \mathrm{Id}_H \otimes \varepsilon)(\Psi) \in H \otimes \mathbb{A} \otimes H,$$
$$\Phi_{\lambda,\rho} := U_r U_l^{-1} \in H \otimes \mathbb{A} \otimes H.$$

*Then:*

*(i)* $(\mathbb{A}, \lambda, \Phi_\lambda)$ *(resp.* $(\mathbb{A}, \rho, \Phi_\rho)$*) is a left (resp. right) H-comodule algebra.*

*(ii) To the left and right H-comodule algebras* $(\mathbb{A}, \lambda, \Phi_\lambda)$ *and* $(\mathbb{A}, \rho, \Phi_\rho)$ *and the element* $\Phi_{\lambda,\rho}$ *we associate the linear maps and elements* $(\delta_l, \Psi_l)$ *and* $(\delta_r, \Psi_r)$ *defined by* (4.4.24) *and* (4.4.25). *Then:*

  *(1)* $(\mathbb{A}, \lambda, \rho, \Phi_\lambda, \Phi_\rho, \Phi_{\lambda,\rho})$ *is an H-bicomodule algebra;*

  *(2)* $U_l$ *provides a twist equivalence from* $(\delta, \Psi)$ *to* $(\delta_l, \Psi_l)$ *and* $U_r$ *provides a twist equivalence from* $(\delta, \Psi)$ *to* $(\delta_r, \Psi_r)$*;*

  *(3) if* $(\delta', \Psi')$ *is twist equivalent to* $(\delta, \Psi)$ *then the H-bicomodule algebra* $(\mathbb{A}, \lambda', \rho', \Phi_{\lambda'}, \Phi_{\rho'}, \Phi_{\lambda',\rho'})$ *is twist equivalent to* $(\mathbb{A}, \lambda, \rho, \Phi_\lambda, \Phi_\rho, \Phi_{\lambda,\rho})$.

*Thus, we have a well-defined map from twist equivalence classes of two-sided coactions of H on* $\mathbb{A}$ *to twist equivalence classes of H-bicomodule algebras, which is the bijective inverse of the map defined in Proposition 4.40.*

*Proof* Parts (i) and (ii)(3) are straightforward and left to the reader. Then, (1) follows from (2) by using part (i) of Proposition 4.40, since the twist equivalence in (2) already guarantees that $(\delta_l, \Psi_l)$ and $(\delta_r, \Psi_r)$ are two-sided coactions. Thus, we only have to prove (2). We prove the statement about $U_l$; the one about $U_r$ is similar and left to the reader. We denote $\delta^{(2)} := (\mathrm{Id}_H \otimes \delta \otimes \mathrm{Id}_H) \circ \delta : \mathbb{A} \to H^{\otimes 2} \otimes \mathbb{A} \otimes H^{\otimes 2}$. It is very easy to see that the following relations hold:

$$(\lambda \otimes \mathrm{Id}_H) \circ \rho = (\varepsilon \otimes \mathrm{Id}_H \otimes \mathrm{Id}_{\mathbb{A}} \otimes \varepsilon \otimes \mathrm{Id}_H) \circ \delta^{(2)}, \tag{4.4.35}$$
$$(\mathrm{Id}_H \otimes \rho) \circ \lambda = (\mathrm{Id}_H \otimes \varepsilon \otimes \mathrm{Id}_{\mathbb{A}} \otimes \mathrm{Id}_H \otimes \varepsilon) \circ \delta^{(2)}. \tag{4.4.36}$$

These two relations together with (4.4.5) imply immediately that $\delta_l(u) = U_l \delta(u) U_l^{-1}$ and $\delta_r(u) = U_r \delta(u) U_r^{-1}$, for all $u \in \mathbb{A}$. By (4.4.8) we have that $(\varepsilon \otimes \mathrm{Id}_{\mathbb{A}} \otimes \varepsilon)(U_l) = 1_{\mathbb{A}}$. Thus, the only thing left to prove is the relation

$$(1_H \otimes U_l \otimes 1_H)(\mathrm{Id}_H \otimes \delta \otimes \mathrm{Id}_H)(U_l)\Psi = \Psi_l(\Delta \otimes \mathrm{Id}_{\mathbb{A}} \otimes \Delta)(U_l). \tag{4.4.37}$$

We need first some supplementary relations. By applying $\mathrm{Id}_H^{\otimes 2} \otimes \mathrm{Id}_{\mathbb{A}} \otimes \mathrm{Id}_H \otimes \varepsilon$ to (4.4.11) we obtain:

$$(1_H \otimes U_l)(\mathrm{Id}_H^{\otimes 2} \otimes \mathrm{Id}_{\mathbb{A}} \otimes \mathrm{Id}_H \otimes \varepsilon)(\Psi) = [(\mathrm{Id}_H \otimes \varepsilon \otimes \lambda \otimes \mathrm{Id}_H \otimes \varepsilon)(\Psi)]$$
$$[(\mathrm{Id}_H^{\otimes 2} \otimes \mathrm{Id}_{\mathbb{A}} \otimes \varepsilon \otimes \mathrm{Id}_H)(\Psi)]. \tag{4.4.38}$$

By applying $\varepsilon \otimes \Delta \otimes \mathrm{Id}_{\mathbb{A}} \otimes \mathrm{Id}_H^{\otimes 2}$ to (4.4.9) we obtain:

$$[(\varepsilon \otimes \Delta \otimes \mathrm{Id}_{\mathbb{A}} \otimes \varepsilon \otimes \mathrm{Id}_H)(\Psi) \otimes 1_H][(\varepsilon \otimes \Delta \otimes \mathrm{Id}_{\mathbb{A}} \otimes \mathrm{Id}_H^{\otimes 2})(\Psi)]$$
$$= [(\varepsilon^{\otimes 2} \otimes (\Delta \otimes \mathrm{Id}_{\mathbb{A}}) \circ \lambda \otimes \mathrm{Id}_H^{\otimes 2})(\Psi)][(\varepsilon \otimes \Delta \otimes \mathrm{Id}_{\mathbb{A}} \otimes \varepsilon \otimes \Delta)(\Psi)]. \tag{4.4.39}$$

By applying $\mathrm{Id}_H^{\otimes 2} \otimes \mathrm{Id}_{\mathbb{A}} \otimes \varepsilon \otimes \mathrm{Id}_H$ to (4.4.9) we obtain:

$$[(\mathrm{Id}_H^{\otimes 2} \otimes \mathrm{Id}_{\mathbb{A}} \otimes \varepsilon^{\otimes 2})(\Psi) \otimes 1_H][(\varepsilon \otimes \Delta \otimes \mathrm{Id}_{\mathbb{A}} \otimes \varepsilon \otimes \mathrm{Id}_H)(\Psi)]$$

$$= [(\varepsilon\otimes \mathrm{Id}_H\otimes \lambda\otimes \varepsilon\otimes \mathrm{Id}_H)(\Psi)][(\mathrm{Id}_H^{\otimes 2}\otimes \mathrm{Id}_{\mathbb{A}}\otimes \varepsilon\otimes \mathrm{Id}_H)(\Psi)]. \tag{4.4.40}$$

By using the definitions of $\Psi_l, \Phi_\lambda, \Phi_\rho, \Phi_{\lambda,\rho}$ and (4.3.5), we get:

$$\begin{aligned}\Psi_l = {} & [(\mathrm{Id}_H\otimes \varepsilon\otimes \lambda\otimes \mathrm{Id}_H\otimes \varepsilon)(\Psi)\otimes 1_H]\\ & [(\varepsilon\otimes \mathrm{Id}_H\otimes \lambda\otimes \varepsilon\otimes \mathrm{Id}_H)(\Psi^{-1})\otimes 1_H]\\ & [(\mathrm{Id}_H^{\otimes 2}\otimes \mathrm{Id}_{\mathbb{A}}\otimes \varepsilon^{\otimes 2})(\Psi)\otimes 1_H\otimes 1_H][(\varepsilon^{\otimes 2}\otimes (\Delta\otimes \mathrm{Id}_{\mathbb{A}})\circ\lambda\otimes \mathrm{Id}_H^{\otimes 2})(\Psi)].\end{aligned} \tag{4.4.41}$$

We compute the left-hand side of (4.4.37):

$$\begin{aligned}&(1_H\otimes U_l\otimes 1_H)(\mathrm{Id}_H\otimes \delta\otimes \mathrm{Id}_H)(U_l)\Psi\\ &\overset{(4.4.10)}{=} [1_H\otimes U_l\otimes 1_H][(\mathrm{Id}_H^{\otimes 2}\otimes \mathrm{Id}_{\mathbb{A}}\otimes \mathrm{Id}_H\otimes \varepsilon)(\Psi)\otimes 1_H]\\ &\qquad [(\varepsilon\otimes \Delta\otimes \mathrm{Id}_{\mathbb{A}}\otimes \mathrm{Id}_H^{\otimes 2})(\Psi)]\\ &\overset{(4.4.38)}{=} [(\mathrm{Id}_H\otimes \varepsilon\otimes \lambda\otimes \mathrm{Id}_H\otimes \varepsilon)(\Psi)\otimes 1_H][(\mathrm{Id}_H^{\otimes 2}\otimes \mathrm{Id}_{\mathbb{A}}\otimes \varepsilon\otimes \mathrm{Id}_H)(\Psi)\otimes 1_H]\\ &\qquad [(\varepsilon\otimes \Delta\otimes \mathrm{Id}_{\mathbb{A}}\otimes \mathrm{Id}_H^{\otimes 2})(\Psi)].\end{aligned}$$

Now we compute the right-hand side of (4.4.37):

$$\begin{aligned}&\Psi_l(\Delta\otimes \mathrm{Id}_{\mathbb{A}}\otimes \Delta)(U_l)\\ &\overset{(4.4.41)}{=} [(\mathrm{Id}_H\otimes \varepsilon\otimes \lambda\otimes \mathrm{Id}_H\otimes \varepsilon)(\Psi)\otimes 1_H][(\varepsilon\otimes \mathrm{Id}_H\otimes \lambda\otimes \varepsilon\otimes \mathrm{Id}_H)(\Psi^{-1})\otimes 1_H]\\ &\qquad [(\mathrm{Id}_H^{\otimes 2}\otimes \mathrm{Id}_{\mathbb{A}}\otimes \varepsilon^{\otimes 2})(\Psi)\otimes 1_H\otimes 1_H][(\varepsilon^{\otimes 2}\otimes (\Delta\otimes \mathrm{Id}_{\mathbb{A}})\circ\lambda\otimes \mathrm{Id}_H^{\otimes 2})(\Psi)]\\ &\qquad [(\varepsilon\otimes \Delta\otimes \mathrm{Id}_{\mathbb{A}}\otimes \varepsilon\otimes \Delta)(\Psi)]\\ &\overset{(4.4.39)}{=} [(\mathrm{Id}_H\otimes \varepsilon\otimes \lambda\otimes \mathrm{Id}_H\otimes \varepsilon)(\Psi)\otimes 1_H][(\varepsilon\otimes \mathrm{Id}_H\otimes \lambda\otimes \varepsilon\otimes \mathrm{Id}_H)(\Psi^{-1})\otimes 1_H]\\ &\qquad [(\mathrm{Id}_H^{\otimes 2}\otimes \mathrm{Id}_{\mathbb{A}}\otimes \varepsilon^{\otimes 2})(\Psi)\otimes 1_H\otimes 1_H][(\varepsilon\otimes \Delta\otimes \mathrm{Id}_{\mathbb{A}}\otimes \varepsilon\otimes \mathrm{Id}_H)(\Psi)\otimes 1_H]\\ &\qquad [(\varepsilon\otimes \Delta\otimes \mathrm{Id}_{\mathbb{A}}\otimes \mathrm{Id}_H^{\otimes 2})(\Psi)]\\ &\overset{(4.4.40)}{=} [(\mathrm{Id}_H\otimes \varepsilon\otimes \lambda\otimes \mathrm{Id}_H\otimes \varepsilon)(\Psi)\otimes 1_H][(\mathrm{Id}_H^{\otimes 2}\otimes \mathrm{Id}_{\mathbb{A}}\otimes \varepsilon\otimes \mathrm{Id}_H)(\Psi)\otimes 1_H]\\ &\qquad [(\varepsilon\otimes \Delta\otimes \mathrm{Id}_{\mathbb{A}}\otimes \mathrm{Id}_H^{\otimes 2})(\Psi)],\end{aligned}$$

which is equal to the left-hand side, and the proof is finished. □

## 4.5 Notes

The concept of module algebra over a quasi-bialgebra was introduced in [59]; the rest of the content of Section 4.1 is taken from [59] and [6]. The deformation of the group coalgebra into a module coalgebra over a quasi-Hopf algebra is taken from [42], and that its duals are isomorphic to a group algebra deformed by a 2-cochain is taken from [43]. The concepts of left and right comodule algebra over a quasi-bialgebra have been introduced in [107], as well as the concept of bicomodule algebra, but under a different name ("quasi-commuting pair of coactions"). Their properties, presented in Section 4.3 and Section 4.4, have been found in [107]. The monoidal characterization of a comodule algebra was given in [46].

# 5

# Crossed Products

The main purpose of this chapter is to introduce and study several kinds of crossed products corresponding to quasi-bialgebras and quasi-Hopf algebras (smash and quasi-smash products, diagonal crossed products, L–R-smash products). These constructions will also be used in subsequent chapters. The chapter ends with a duality theorem for finite-dimensional quasi-Hopf algebras.

## 5.1 Smash Products

For $H$ a quasi-bialgebra and $A$ an algebra within ${}_H\mathcal{M}$, we show that the category of representations of $A$ inside ${}_H\mathcal{M}$ can be identified with the category of modules over a certain associative algebra, namely the smash product between $A$ and $H$.

**Proposition 5.1** *Let $H$ be a quasi-bialgebra and $A$ a left $H$-module algebra. Define a multiplication on $A\otimes H$ by*

$$(a\#h)(b\#g)=(x^1\cdot a)(x^2h_1\cdot b)\#x^3h_2g, \tag{5.1.1}$$

*for all $a,b\in A$ and $h,g\in H$, where we write $a\#h$ for $a\otimes h$. If we denote this structure by $A\#H$, then $A\#H$ is an associative algebra with unit $1_A\#1_H$ and the map $j:H\to A\#H$, $j(h)=1_A\#h$, is an algebra homomorphism. The algebra $A\#H$ is called the smash product of $A$ and $H$.*

*Proof* For all $a,b,c\in A$ and $h,g,l\in H$ we have:

$$\begin{aligned}
[(a\#h)(b\#g)](c\#l) &= [(x^1\cdot a)(x^2h_1\cdot b)\#x^3h_2g)](c\#l)\\
&= [y^1\cdot((x^1\cdot a)(x^2h_1\cdot b))](y^2x_1^3h_{(2,1)}g_1\cdot c)\#y^3x_2^3h_{(2,2)}g_2l\\
&= \{(y_1^1x^1\cdot a)(y_2^1x^2h_1\cdot b)\}(y^2x_1^3h_{(2,1)}g_1\cdot c)\#y^3x_2^3h_{(2,2)}g_2l\\
&\overset{(3.1.9)}{=} \{(z^1y^1\cdot a)(z^2y_1^2x^1h_1\cdot b)\}(z^3y_2^2x^2h_{(2,1)}g_1\cdot c)\#y^3x^3h_{(2,2)}g_2l\\
&\overset{(3.1.7)}{=} \{(z^1y^1\cdot a)(z^2y_1^2h_{(1,1)}x^1\cdot b)\}(z^3y_2^2h_{(1,2)}x^2g_1\cdot c)\#y^3h_2x^3g_2l\\
&\overset{(4.1.1)}{=} \{X^1\cdot(z^1y^1\cdot a)\}\{[X^2\cdot(z^2y_1^2h_{(1,1)}x^1\cdot b)]\\
&\qquad [X^3\cdot(z^3y_2^2h_{(1,2)}x^2g_1\cdot c)]\}\#y^3h_2x^3g_2l
\end{aligned}$$

$$
\begin{aligned}
&= (y^1\cdot a)\{[y_1^2h_{(1,1)}x^1\cdot b][y_2^2h_{(1,2)}x^2g_1\cdot c]\}\#y^3h_2x^3g_2l\\
&= (y^1\cdot a)\{[y_1^2h_{(1,1)}\cdot(x^1\cdot b)][y_2^2h_{(1,2)}\cdot(x^2g_1\cdot c)]\}\#y^3h_2x^3g_2l\\
&= (y^1\cdot a)\{y^2h_1\cdot[(x^1\cdot b)(x^2g_1\cdot c)]\}\#y^3h_2x^3g_2l\\
&= (a\#h)[(x^1\cdot b)(x^2g_1\cdot c)\#x^3g_2l]\\
&= (a\#h)[(b\#g)(c\#l)].
\end{aligned}
$$

By (3.1.10), (3.1.11) and (4.1.3) if follows that $1_A\#1_H$ is the unit for $A\#H$. Finally, by again using (3.1.10) and (4.1.3), one can easily see that $j$ is an algebra map. □

**Example 5.2** For any quasi-bialgebra $H$ and any algebra $A$, we have the trivial action of $H$ on $A$ given by $h\cdot a=\varepsilon(h)a$ for all $h\in H$ and $a\in A$. In this case $A\#H\equiv A\otimes H$ as algebras.

**Example 5.3** Let $H$ be a quasi-Hopf algebra and $H_0$ the left $H$-module algebra introduced in Definition 4.4. Then we can define the smash product $H_0\#H$; it is easy to see that its multiplication is given, for all $g,g',h,h'\in H$, by:

$$(g\#h)(g'\#h')=X^1(y^1\triangleright g)S(x^1X^2)\alpha x^2(X^3y^2h_1\triangleright g')S(x^3)\#y^3h_2h'.$$

**Remarks 5.4** (1) In general, the map $i: A\to A\#H$, $i(a)=a\#1_H$, for all $a\in A$, is not multiplicative. A necessary and sufficient condition for $i$ to be multiplicative is:

$$(x^1\cdot a)(x^2\cdot b)\otimes x^3=ab\otimes 1_H, \tag{5.1.2}$$

for all $a,b\in A$. One can see immediately that if (5.1.2) holds then $A$ is an associative algebra. In this case we obtain

$$
\begin{aligned}
(a\#h)(b\#g) &= (x^1\cdot a)(x^2h_1\cdot b)\#x^3h_2g\\
&= a(h_1\cdot b)\#h_2g,
\end{aligned}
$$

for all $a,b\in A$ and $h,g\in H$. Note that the relation (5.1.2) implies (4.1.1) for any associative algebra $A$.

(2) Note that $i(a)j(h)=(x^1\cdot a)(x^2\cdot 1_A)\#x^3h=a\#h$, for all $a\in A$ and $h\in H$.

For a quasi-bialgebra $H$ and a left $H$-module algebra $A$, we aim to describe the category of left modules over $A\#H$ more precisely. First we need the following:

**Definition 5.5** We say that $M$, a $k$-linear space, is a left $A,H$-module if:

(i) $M$ is a left $H$-module with action denoted by $h\otimes m\mapsto h\cdot m$;

(ii) $A$ acts weakly on $M$ to the left, that is, there exists a $k$-linear map $A\otimes M\to M$, denoted by $a\otimes m\mapsto a\triangleright m$, such that $1_A\triangleright m=m$ for all $m\in M$;

(iii) the following compatibility relations hold:

$$a\triangleright(b\triangleright m)=[(x^1\cdot a)(x^2\cdot b)]\triangleright(x^3\cdot m),\ \forall\, a,b\in A, m\in M, \tag{5.1.3}$$

$$h\cdot(a\triangleright m)=(h_1\cdot a)\triangleright(h_2\cdot m),\ \forall\, h\in H, a\in A, m\in M. \tag{5.1.4}$$

The category of all left $A,H$-modules, with morphisms being the maps that are $H$-linear and preserve the weak $A$-action, will be denoted by ${}_{A,H}\mathcal{M}$.

**Remark 5.6** One can easily see that ${}_{A,H}\mathcal{M}$ coincides with the category of left $A$-modules inside the monoidal category of left $H$-modules.

**Proposition 5.7** *Let $H$ be a quasi-bialgebra and $A$ a left $H$-module algebra. Then the category ${}_{A,H}\mathcal{M}$ is isomorphic to the category ${}_{A\#H}\mathcal{M}$ of left $A\#H$-modules.*

*Proof* We only define the isomorphisms and leave verifications of some details to the reader. Let $M \in {}_{A\#H}\mathcal{M}$ with $A\#H$-module structure given by $(a\#h)\otimes m \mapsto (a\#h)m$. Since $j : H \to A\#H$ is an algebra map, $M$ becomes a left $H$-module by $h\otimes m \mapsto j(h)m$. It is clear that $A$ acts weakly on $M$ by $a\otimes m \mapsto i(a)m$. Then one can check that (5.1.3) and (5.1.4) hold, and hence $M \in {}_{A,H}\mathcal{M}$.

Conversely, if $M \in {}_{A,H}\mathcal{M}$, with notation as in Definition 5.5, we define

$$(A\#H)\otimes M \to M, \quad (a\#h)\otimes m \mapsto a\triangleright(h\cdot m),$$

for $a \in A$, $h \in H$, $m \in M$. Then with this structure $M$ becomes a left $A\#H$-module.

The above correspondences define two functors (which act as the identity on morphisms) that provide category isomorphisms, one inverse to each other. □

**Definition 5.8** If $(\mathfrak{A},\rho,\Phi_\rho)$ is a right $H$-comodule algebra, we define the subalgebra of coinvariants by $\mathfrak{A}^{\mathrm{co}(H)} = \{\mathfrak{a} \in \mathfrak{A} \mid \rho(\mathfrak{a}) = \mathfrak{a}\otimes 1_H\}$.

**Proposition 5.9** *Let $H$ be a quasi-bialgebra and $A$ a left $H$-module algebra. Define the following linear map and element:*

$$\rho : A\#H \to (A\#H)\otimes H, \ \rho(a\#h) = (x^1\cdot a\#x^2h_1)\otimes x^3h_2,$$
$$\Phi_\rho = (1_A\#X^1)\otimes X^2\otimes X^3 \in (A\#H)\otimes H\otimes H.$$

*Then $(A\#H,\rho,\Phi_\rho)$ is a right $H$-comodule algebra. Moreover, the subalgebra of coinvariants is*

$$(A\#H)^{\mathrm{co}(H)} = \{a\#1_H \mid a \in A \ \textit{satisfying} \ a\#1_H\otimes 1_H = x^1\cdot a\#x^2\otimes x^3\}.$$

*Proof* We begin by showing that $\rho$ is an algebra morphism. Indeed, $\rho(1_A\#1_H) = 1_A\#1_H\otimes 1_H$ and for all $a,b \in A$ and $h,g \in H$ we have:

$$\begin{aligned}
\rho((a\#h)(b\#g)) &= \rho((x^1\cdot a)(x^2h_1\cdot b)\#x^3h_2g)\\
&= (y_1^1x^1\cdot a)(y_2^1x^2h_1\cdot b)\#y^2x_1^3h_{(2,1)}g^1\otimes y^3x_2^3h_{(2,2)}g_2\\
&\overset{(3.1.9)}{=} (z^1y^1\cdot a)(z^2y_1^2x^1h_1\cdot b)\#z^3y_2^2x^2h_{(2,1)}g^1\otimes y^3x^3h_{(2,2)}g_2\\
&\overset{(3.1.7)}{=} (z^1y^1\cdot a)(z^2y_1^2h_{(1,1)}x^1\cdot b)\#z^3y_2^2h_{(1,2)}x^2g_1\otimes y^3h_2x^3g_2\\
&= (y^1\cdot a\#y^2h_1\otimes y^3h_2)(x^1\cdot b\#x^2g_1\otimes x^3g_2)\\
&= \rho(a\#h)\rho(b\#g).
\end{aligned}$$

Now we prove the relation (4.3.1):

$$\begin{aligned}
&\Phi_\rho(\rho\otimes \mathrm{Id}_H)(\rho(a\#h))\\
&\quad = (1_A\#X^1\otimes X^2\otimes X^3)(\rho\otimes \mathrm{Id}_H)(x^1\cdot a\#x^2h_1\otimes x^3h_2)
\end{aligned}$$

$$\begin{aligned}
&= (1_A\#X^1\otimes X^2\otimes X^3)(y^1x^1\cdot a\#y^2x_1^2h_{(1,1)}y^3x_2^2h_{(1,2)}\otimes x^3h_2)\\
&= X_1^1y^1x^1\cdot a\#X_2^1y^2x_1^2h_{(1,1)}\otimes X^2y^3x_2^2h_{(1,2)}\otimes X^3x^3h_2\\
&\overset{(3.1.9)}{=} x^1\cdot a\#x^2X^1h_{(1,1)}\otimes x_1^3X^2h_{(1,2)}\otimes x_2^3X^3h_2\\
&\overset{(3.1.7)}{=} x^1\cdot a\#x^2h_1X^1\otimes x_1^3h_{(2,1)}X^2\otimes x_2^3h_{(2,2)}X^3\\
&= (x^1\cdot a\#x^2h_1\otimes x_1^3h_{(2,1)}\otimes x_2^3h_{(2,2)})(1_A\#X^1\otimes X^2\otimes X^3)\\
&= (\mathrm{Id}_{A\#H}\otimes\Delta)(x^1\cdot a\#x^2h_1\otimes x^3h_2)(1_A\#X^1\otimes X^2\otimes X^3)\\
&= (\mathrm{Id}_{A\#H}\otimes\Delta)(\rho(a\#h))\Phi_\rho.
\end{aligned}$$

Finally, the relation (4.3.2) is just (3.1.9), and (4.3.3), (4.3.4) are trivial. Also, it is easy to see that the coinvariants are exactly the elements of the form $a\#1_H$ with $a\#1_H\otimes 1_H = x^1\cdot a\#x^2\otimes x^3$. □

The smash product is invariant under twisting in the following sense:

**Proposition 5.10** *Let $H$ be a quasi-bialgebra, $A$ a left $H$-module algebra and $F\in H\otimes H$ a gauge transformation, with notation $F=F^1\otimes F^2=f^1\otimes f^2=\mathbf{F}^1\otimes\mathbf{F}^2=\mathscr{F}^1\otimes\mathscr{F}^2$ and $F^{-1}=G^1\otimes G^2=g^1\otimes g^2=\mathbf{G}^1\otimes\mathbf{G}^2=\mathscr{G}^1\otimes\mathscr{G}^2$. If we introduce on $A$ a new multiplication*

$$a\diamond b=(G^1\cdot a)(G^2\cdot b),\ \ \forall\, a,b\in A,$$

*and we denote by $A_{F^{-1}}$ the resulting structure, then $A_{F^{-1}}$ becomes a left $H_F$-module algebra, with the same unit and $H$-action as for $A$. Moreover, the linear map*

$$\pi: A\#H\to A_{F^{-1}}\#H_F,\ \ \pi(a\#h)=F^1\cdot a\#F^2h,$$

*is an algebra isomorphism.*

*Proof* By Proposition 3.5 the categories ${}_H\mathscr{M}$ and ${}_{H_F}\mathscr{M}$ are monoidally isomorphic. Moreover, the monoidal isomorphism is produced by the identity functor. Thus, by Proposition 2.3, to $A$ we can associate an algebra in ${}_{H_F}\mathscr{M}$. A simple inspection shows that this algebra is precisely $A_{F^{-1}}$ defined in the statement.

We prove now that the map $\pi$ is an algebra isomorphism. We denote

$$\Phi_F^{-1}:=\tilde{x}^1\otimes\tilde{x}^2\otimes\tilde{x}^3=F^1f_1^1x^1G^1\otimes F^2f_2^1x^2G_1^2g^1\otimes f^2x^3G_2^2g^2.$$

The multiplication of $A_{F^{-1}}\#H_F$ is:

$$\begin{aligned}
(a\#h)(a'\#h') &= (\tilde{x}^1\cdot a)\diamond(\tilde{x}^2h_{(1)}\cdot a')\#\tilde{x}^3h_{(2)}h'\\
&= (F^1f_1^1x^1G^1\cdot a)\diamond(F^2f_2^1x^2G_1^2g^1\mathbf{F}^1h_1\mathbf{G}^1\cdot a')\#f^2x^3G_2^2g^2\mathbf{F}^2h_2\mathbf{G}^2h'\\
&= (f_1^1x^1G^1\cdot a)(f_2^1x^2G_1^2h_1\mathbf{G}^1\cdot a')\#f^2x^3G_2^2h_2\mathbf{G}^2h',
\end{aligned}$$

so we obtain

$$\begin{aligned}
\pi(a\#h)\pi(a'\#h') &= (F^1\cdot a\#F^2h)(\mathscr{F}^1\cdot a'\#\mathscr{F}^2h')\\
&= (f_1^1x^1G^1F^1\cdot a)(f_2^1x^2G_1^2F_1^2h_1\mathbf{G}^1\mathscr{F}^1\cdot a')\#f^2x^3G_2^2F_2^2h_2\mathbf{G}^2\mathscr{F}^2h'\\
&= f^1\cdot[(x^1\cdot a)(x^2h_1\cdot a')]\#f^2x^3h_2h'
\end{aligned}$$

$$= \pi((a\#h)(a'\#h')).$$

It is easy to see that $\pi(1_A\#1_H) = 1_A\#1_H$ and that the inverse of $\pi$ is $\pi^{-1}(a\#h) = G^1 \cdot a\#G^2h$, for all $a \in A$, $h \in H$. □

**Lemma 5.11** *Let $H$ be a quasi-Hopf algebra, $A$ a left $H$-module algebra and $(A\#H)^j$ the left $H$-module algebra obtained by using Proposition 4.3 for the canonical map $j : H \to A\#H$. If we define the map*

$$i_0 : A \to A\#H, \quad i_0(a) = x^1 \cdot a\#x^2\beta S(x^3), \quad \forall\, a \in A, \tag{5.1.5}$$

*then $i_0 \in Alg_H(A, (A\#H)^j)$, where we used notation as in Remark 4.5.*

*Proof* Let $\gamma, \delta, f \in H \otimes H$ be the elements defined by the formulas (3.2.5), (3.2.6) and (3.2.15). If we denote by $\gamma = \gamma^1 \otimes \gamma^2$, $\delta = \delta^1 \otimes \delta^2$, $f = f^1 \otimes f^2$ and $f^{-1} = g^1 \otimes g^2$ then, for all $a, b \in A$, we compute:

$$\begin{aligned}
& i_0(a) \circ i_0(b) \\
&= (1_A\#X^1)(y^1 \cdot a\#y^2\beta S(y^3))\, j(S(x^1X^2)\alpha x^2X_1^3) \\
&\quad (z^1 \cdot b\#z^2\beta S(z^3))(1_A\#S(x^2X_2^3)) \\
&= (X_1^1y^1 \cdot a\#X_2^1y^2\beta S(x^1X^2y^3)\alpha x^2X_1^3)(z^1 \cdot b\#z^2\beta S(x^3X_2^3z^3)) \\
&\overset{(3.1.9)}{=} (y^1Y^1 \cdot a\#y^2X^1Y_1^2\beta S(x^1y_1^3X^2Y_2^2)\alpha x^2y_{(2,1)}^3X_1^3Y_1^3) \\
&\quad (z^1 \cdot b\#z^2\beta S(x^3y_{(2,2)}^3X_2^3Y_2^3z^3)) \\
&\overset{(3.2.1),(3.1.7)}{=} (y^1 \cdot a\#y^2X^1\beta S(x^1X^2)\alpha x^2X_1^3)(z^1 \cdot b\#z^2\beta S(y^3x^3X_2^3z^3)) \\
&= (y^1 \cdot a\#y^2X^1\beta S(x^1X^2)\alpha x^2)(1_A\#X_1^3)(z^1 \cdot b\#z^2\beta S(y^3x^3X_2^3z^3)) \\
&\overset{(3.1.7)}{=} (y^1 \cdot a\#y^2X^1\beta S(x^1X^2)\alpha x^2)(z^1X_1^3 \cdot b\#z^2X_{(2,1)}^3\beta S(y^3x^3z^3X_{(2,2)}^3) \\
&= (y^1 \cdot a\#y^2X^1\beta S(x^1X^2)\alpha x^2)(z^1X^3 \cdot b\#z^2\beta S(y^3x^3z^3)) \\
&\overset{(3.2.14)}{=} (t^1y^1 \cdot a)(t^2y_1^2X_1^1\delta^1f^1S(x^1X^2)_1g^1\gamma^1x_1^2z^1X^3 \cdot b) \\
&\quad \#t^3y_2^2X_2^1\delta^2f^2S(x^1X^2)_2g^2\gamma^2x_2^2z^2\beta S(y^3x^3z^3) \\
&\overset{(3.2.13)}{=} (t^1y^1 \cdot a)(t^2y_1^2X_1^1\delta^1S(x_2^1X_2^2)\gamma^1x_1^2z^1X^3 \cdot b) \\
&\quad \#t^3y_2^2X_2^1\delta^2S(x_1^1X_1^2)\gamma^2x_2^2z^2\beta S(y^3x^3z^3) \\
&= (t^1y^1 \cdot a)(t^2y_1^2X_1^1u^1\beta S(v^1V^2x_2^1X_2^2u_2^3U^3)\alpha v^2V_1^3x_1^2z^1X^3 \cdot b) \\
&\quad \#t^3y_2^2X_2^1u^2U^1\beta S(V^1x_1^1X_1^2u_1^3U^2)\alpha v^3V_2^3x_2^2z^2\beta S(y^3x^3z^3) \\
&\overset{(3.1.9)}{=} (t^1y^1 \cdot a)(t^2y_1^2u^1Y^1\beta S(v^1x_1^2w^1V^2u_{(1,2)}^3X_2^2Y_{(2,2)}^2U^3)\alpha v^2(x_2^2w^2)_1 \\
&\qquad\qquad V_{(1,1)}^3z^1u_2^3X^3Y^3 \cdot b)\#t^3y_2^2u^2X^1Y_1^2U^1\beta S(x^1V^1u_{(1,1)}^3X_1^2Y_{(2,1)}^2U^2) \\
&\quad \alpha v^3(x_2^2w^2)_2V_{(1,2)}^3z^2\beta S(y^3x^3w^3V_2^3z^3) \\
&\overset{(3.1.7),\,(3.2.1)}{=} (t^1y^1 \cdot a)(t^2y_1^2u^1Y^1\beta S(v^1x_1^2w^1V^2u_{(1,2)}^3X_2^2U^3Y^2)\alpha v^2x_{(2,1)}^2w_1^2z^1 \\
&\qquad\qquad V^3u_2^3X^3Y^3 \cdot b)\#t^3y_2^2u^2X^1U^1\beta S(x^1V^1u_{(1,1)}^3X_1^2U^2)\alpha v^3x_{(2,2)}^2w_2^2z^2 \\
&\quad \beta S(y^3x^3w^3z^3)
\end{aligned}$$

$$
\begin{aligned}
&\overset{(3.1.7),(3.2.1)}{=} (t^1y^1\cdot a)(t^2y_1^2u^1Y^1\beta S(v^1w^1V^2u^3_{(1,2)}X_2^2U^3Y^2)\alpha v^2w_1^2z^1V^3u_2^3\\
&\qquad X^3Y^3\cdot b)\#t^3y_2^2u^2X^1U^1\beta S(x^1V^1u^3_{(1,1)}X_1^2U^2)\alpha x^2v^3w_2^2z^2\beta\\
&\quad S(y^3x^3w^3z^3)\\
&\overset{(3.1.9)}{=} (t^1y^1\cdot a)(t^2y_1^2u^1Y^1\beta S(v_1^1w^1V^2u^3_{(1,2)}X_2^2U^3Y^2)\alpha v_2^1w^2V^3u_2^3\\
&\qquad X^3V^3\cdot b)\#t^3y_2^2u^2X^1U^1\beta S(x^1V^1u^3_{(1,1)}X_1^2U^2)\alpha x^2v^2w_1^3\beta\\
&\quad S(y^3x^3v^3w_2^3)\\
&\overset{(3.2.1),(3.1.11)}{=} (t^1y^1\cdot a)(t^2y_1^2u^1Y^1\beta S(V^2X_2^2U^3Y^2)\alpha V^3X^3Y^3\cdot b)\\
&\quad \#t^3y_2^2u^2X^1U^1\beta S(x^1u^3V^1X_1^2U^2)\alpha x^2\beta S(y^3x^3)\\
&\overset{(3.1.9)}{=} (t^1y^1\cdot a)(t^2y_1^2u^1Y^1\beta S(U_1^3V^2Y^2)\alpha U_2^3V^3Y^3\cdot b)\#t^3y_2^2u^2U^1\\
&\quad V_1^1\beta S(x^1u^3U^2V_2^1)\alpha x^2\beta S(y^3x^3)\\
&\overset{(3.2.1),(3.1.11)}{=} (t^1y^1\cdot a)(t^2y_1^2u^1Y^1\beta S(Y^2)\alpha Y^3\cdot b)\#t^3y_2^2u^2\beta S(x^1u^3)\alpha x^2\beta S(y^3x^3)\\
&\overset{(3.2.2)}{=} (t^1y^1\cdot a)(t^2y_1^2u^1\cdot b)\#t^3y_2^2u^2\beta S(u^3)S(x^1)\alpha x^2\beta S(x^3)S(y^3)\\
&\overset{(3.2.2)}{=} (t^1y^1\cdot a)(t^2y_1^2u^1\cdot b)\#t^3y_2^2u^2\beta S(y^3u^3)\\
&\overset{(3.1.9)}{=} (x_1^1y^1\cdot a)(x_2^1y^2\cdot b)\#x^2y_1^3\beta S(x^3y_2^3)\\
&\overset{(3.2.1),(4.1.2)}{=} x^1\cdot(ab)\#x^2\beta S(x^3)=i_0(ab).
\end{aligned}
$$

Obviously $i_0(1_A)=1_A\#\beta=j(\beta)$, which is the unit of $(A\#H)^j$. Finally, we have to prove that $i_0$ is $H$-linear. For all $h\in H$ and $a\in A$ we have:

$$
\begin{aligned}
h\triangleright_j i_0(a) &= j(h_1)i_0(a)j(S(h_2))\\
&= h_{(1,1)}x^1\cdot a\#h_{(1,2)}x^2\beta S(h_2x^3)\\
&\overset{(3.1.7)}{=} x^1h_1\cdot a\#x^2h_{(2,1)}\beta S(x^3h_{(2,2)})\\
&\overset{(3.2.1)}{=} x^1h\cdot a\#x^2\beta S(x^3)=i_0(h\cdot a),
\end{aligned}
$$

finishing the proof. □

We have a Universal Property for smash products over quasi-Hopf algebras:

**Theorem 5.12** *Let $H$ be a quasi-Hopf algebra, $A$ a left $H$-module algebra and the maps $i_0:A\to A\#H$ and $j:H\to A\#H$ as defined above. Then, for any algebra $B$, and for any algebra map $v:H\to B$ and any $u\in Alg_H(A,B^v)$ (where $B^v$ means that $B$ is a left $H$-module algebra via $v$ as in Proposition 4.3), there exists a unique algebra map $u\#v:A\#H\to B$ such that $(u\#v)\circ i_0=u$ and $(u\#v)\circ j=v$.*

*Proof* Let us start by noting that $i_0(X^1\cdot a)\circ j(X^2hS(X^3))=a\#h$ in $(A\#H)^j$, for all $a\in A$ and $h\in H$. Indeed, we have:

$$
\begin{aligned}
i_0(a)\circ j(h) &= j(X^1)(y^1\cdot a\#y^2\beta S(y^3))j(S(x^1X^2)\alpha x^2X_1^3hS(x^3X_2^3))\\
&= X_1^1y^1\cdot a\#X_2^1y^2\beta S(x^1X^2y^3)\alpha x^2X_1^3hS(x^3X_2^3)\\
&\overset{(3.1.9)}{=} z^1Y^1\cdot a\#z^2X^1Y_1^2\beta S(x^1z_1^3X^2Y_2^2)\alpha x^2z^3_{(2,1)}X_1^3Y_1^3h\\
&\quad S(x^3z^3_{(2,2)}X_2^3Y_2^3)
\end{aligned}
$$

$$
\begin{aligned}
&\overset{(3.2.1),(3.1.7)}{=} z^1\cdot a\#z^2X^1\beta S(z^3_{(1,1)}x^1X^2)\alpha z^3_{(1,2)}x^2X^3_1hS(z^3_2x^3X^3_2)\\
&\overset{(3.2.1),(3.1.11)}{=} z^1\cdot a\#z^2X^1\beta S(x^1X^2)\alpha x^2X^3_1hS(z^3x^3X^3_2)\\
&= z^1\cdot a\#z^2(\beta\circ h)S(z^3)=z^1\cdot a\#z^2hS(z^3),
\end{aligned}
$$

and therefore $i_0(X^1\cdot a)\circ j(X^2hS(X^3))=a\#h$ for all $h\in H$ and $a\in A$.

In particular, we obtain the following useful relation in $(A\#H)^j$, for $a\in A$, $h\in H$:

$$
i_0(a)\circ j(h)=x^1\cdot a\#x^2hS(x^3). \tag{5.1.6}
$$

Now, let $v,u$ be as in the hypothesis, that is,

$$
\begin{aligned}
u(ab)&=v(X^1)u(a)v(S(x^1X^2)\alpha x^2X^3_1)u(b)v(S(x^3X^3_2)),\\
u(h\cdot a)&=v(h_1)u(a)v(S(h_2)),\\
u(1_A)&=v(\beta),
\end{aligned}
$$

for all $a,b\in A$ and $h\in H$. If there exists an algebra map $w: A\#H\to B$ such that $w\circ i_0=u$ and $w\circ j=v$, then for all $a\in A$, $h\in H$ we have:

$$
\begin{aligned}
w(a\#h)&=w(i_0(Y^1\cdot a)\circ j(Y^2hS(Y^3))\\
&=w(j(X^1)i_0(Y^1\cdot a)j(S(x^1X^2)\alpha x^2X^3_1Y^2hS(x^3X^3_2Y^3))\\
&=v(X^1)u(Y^1\cdot a)v(S(x^1X^2)\alpha x^2X^3_1Y^2hS(x^2X^3_2Y^3))\\
&=v(X^1Y^1_1)u(a)v(S(x^1X^2Y^1_2)\alpha x^2X^3_1Y^2hS(x^3X^3_2Y^3))\\
&\overset{(3.1.9)}{=}v(Y^1Z^1)u(a)v(S(Y^2_1Z^2)\alpha Y^2_2Z^3hS(Y^3))\\
&\overset{(3.2.1)}{=}v(Z^1)u(a)v(S(Z^2)\alpha Z^3h),
\end{aligned}
$$

and this shows the uniqueness of $u\#v$. We now prove the existence of $u\#v$. Define the linear map

$$
w: A\#H\to B,\quad w(a\#h)=v(X^1)u(a)v(S(X^2)\alpha X^3h), \tag{5.1.7}
$$

for all $a\in A$, $h\in H$. We have to prove that $w$ is an algebra map and $w\circ i_0=u$, $w\circ j=v$. For all $a,b\in A$, $h,g\in H$ we have:

$$
\begin{aligned}
w((a\#h)(b\#g))&=w((x^1\cdot a)(x^2h_1\cdot b)\#x^3h_2g)\\
&=v(X^1)u((x^1\cdot a)(x^2h_1\cdot b))v(S(X^2)\alpha X^3x^3h_2g)\\
&=v(X^1Y^1)u(x^1\cdot a)v(S(y^1Y^2)\alpha y^2Y^3_1)u(x^2h_1\cdot b)\\
&\quad v(S(X^2y^3Y^3_2)\alpha X^3x^3h_2g)\\
&=v(X^1Y^1x^1_1)u(a)v(S(y^1Y^2x^1_2)\alpha y^2Y^3_1x^2_1h_{(1,1)})u(b)\\
&\quad v(S(X^2y^3Y^3_2x^2_2h_{(1,2)})\alpha X^3x^3h_2g)\\
&\overset{(3.1.9)}{=}v(X^1t^1Y^1)u(a)v(S(y^1t^2_1z^1Y^2)\alpha y^2t^2_{(2,1)}z^2_1Y^3_{(1,1)}h_{(1,1)})\\
&\quad u(b)v(S(X^2y^3t^2_{(2,2)}z^2_2Y^3_{(1,2)}h_{(1,2)})\alpha X^3t^3z^3Y^3_2h_2g)\\
&\overset{(3.1.9)}{=}v(X^1t^1Y^1)u(a)v(S(y^1z^1Y^2)\alpha y^2z^2_1Y^3_{(1,1)}h_{(1,1)})\\
&\quad u(b)v(S(X^2t^2y^3z^2_2Y^3_{(1,2)}h_{(1,2)})\alpha X^3t^3z^3Y^3_2h_2g)
\end{aligned}
$$

$$\begin{aligned}
&\overset{(3.1.9)}{=} v(Y^1)u(a)v(S(y_1^1z^1Y^2)\alpha y_2^1z^2X^1(Y^3h)_{(1,1)}u(b)\\
&\qquad v(S(y^2z_1^3X^2(Y^3h)_{(1,2)}\alpha y^3z_2^3X^3(Y^3h)_2g)\\
&\overset{(3.1.7),(3.2.1)}{=} v(Y^1)u(a)v(S(Y^2)\alpha(Y^3h)_1X^1)u(b)\\
&\qquad v(S((Y^3h)_{(2,1)}X^2)\alpha(Y^3h)_{(2,2)}X^3g)\\
&\overset{(3.2.1)}{=} v(Y^1)u(a)v(S(Y^2)\alpha Y^3hX^1)u(b)v(S(X^2)\alpha X^3g)\\
&= w(a\#h)w(b\#g).
\end{aligned}$$

Also, $w(1_A\#1_H)=v(X^1\beta S(X^2)\alpha X^3)=v(1_H)=1_B$, therefore $w$ is an algebra homomorphism. Finally, observe that $(w\circ j)(h)=v(X^1\beta S(X^2)\alpha X^3h)=v(h)$, and

$$\begin{aligned}
w\circ i_0(a) &= w(x^1\cdot a\#x^2\beta S(x^3))\\
&= v(X^1)u(x^1\cdot a)v(S(X^2)\alpha X^3x^2\beta S(x^3))\\
&= v(X^1x_1^1)u(a)v(S(X^2x_2^1)\alpha X^3x^2\beta S(x^3))\\
&\overset{(3.1.9)}{=} v(y^1X^1)u(a)v(S(y_1^2x^1X^2)\alpha y_2^2x^2X_1^3\beta S(y^3x^3X_2^3))\\
&\overset{(3.2.1)}{=} u(a)v(S(x^1)\alpha x^2\beta S(x^3))\overset{(3.2.2)}{=}u(a),
\end{aligned}$$

and this finishes the proof. □

**Proposition 5.13** *Let $H$ be a quasi-Hopf algebra, $A$ a linear space with a multiplication and a unit, which is a left $H$-module such that* (4.1.2), (5.1.2) *hold. Then:*

*(i) $(h\cdot a)(\beta\cdot b)=h_1\cdot(a(\beta S(h_2)\cdot b))$, for all $h\in H$ and $a,b\in A$;*

*(ii) the relation* (4.1.3) *holds if and only if $\beta\cdot 1_A=\varepsilon(\beta)1_A$.*

*Proof* (i) Let $h\in H$ and $a,b\in A$. By using (3.1.8) and (3.2.1) we obtain $h\otimes\beta=h_1\otimes h_{(2,1)}\beta S(h_{(2,2)})$. Therefore, we have:

$$\begin{aligned}
(h\cdot a)(\beta\cdot b) &\overset{(5.1.2)}{=} (x^1\cdot(h\cdot a))(x^2\cdot(\beta S(x^3)\cdot b))\\
&= (x^1h_1\cdot a)(x^2h_{(2,1)}\beta S(h_{(2,2)})S(x^3)\cdot b)\\
&= (x^1h_1\cdot a)(x^2h_{(2,1)}\beta S(x^3h_{(2,2)})\cdot b)\\
&\overset{(3.1.7)}{=} (h_{(1,1)}x^1\cdot a)(h_{(1,2)}x^2\beta S(h_2x^3)\cdot b)\\
&= (h_{(1,1)}\cdot(x^1\cdot a))(h_{(1,2)}\cdot(x^2\beta S(h_2x^3)\cdot b))\\
&\overset{(4.1.2)}{=} h_1\cdot[(x^1\cdot a)(x^2\cdot(\beta S(h_2x^3)\cdot b))]\\
&\overset{(5.1.2)}{=} h_1\cdot(a(\beta S(h_2)\cdot b)).
\end{aligned}$$

(ii) This follows from (i) by taking $a=b=1_A$ and by using the fact that $\varepsilon(\beta)=1$. □

For completeness and further use, we introduce right-handed versions of some of the above constructions and results (proofs are left to the reader).

**Definition 5.14** Let $H$ be a quasi-bialgebra and $B$ a right $H$-module algebra. We define the (right-handed) smash product $H\#B$ as follows: as vector space $H\#B$ is $H\otimes B$ (elements $h\otimes b$ will be written $h\#b$) with multiplication:

$$(h\#b)(h'\#b')=hh'_1x^1\#(b\cdot h'_2x^2)(b'\cdot x^3), \tag{5.1.8}$$

for all $b, b' \in B$, $h, h' \in H$. This $H\#B$ is an associative algebra with unit $1_H\#1_B$ and it becomes a left $H$-comodule algebra, with structure

$$\lambda : H\#B \to H \otimes (H\#B), \ \ \lambda(h\#b) = h_1x^1 \otimes (h_2x^2\#b \cdot x^3), \ \ \forall\, h \in H, b \in B,$$
$$\Phi_\lambda = X^1 \otimes X^2 \otimes (X^3\#1_B) \in H \otimes H \otimes (H\#B).$$

## 5.2 Quasi-smash Products and Generalized Smash Products

We construct a class of examples of module algebras by using a certain type of crossed product between a comodule algebra and a bimodule algebra.

**Proposition 5.15** *Let $H$ be a quasi-bialgebra, $\mathfrak{A}$ a right $H$-comodule algebra and $\mathscr{A}$ an $H$-bimodule algebra. Define a multiplication on $\mathfrak{A} \otimes \mathscr{A}$ by*

$$(\mathfrak{a}\overline{\#}\varphi)(\mathfrak{a}'\overline{\#}\varphi') = \mathfrak{a}\mathfrak{a}'_{\langle 0\rangle}\tilde{x}^1_\rho\overline{\#}(\varphi\cdot\mathfrak{a}'_{\langle 1\rangle}\tilde{x}^2_\rho)(\varphi'\cdot\tilde{x}^3_\rho), \ \ \forall\, \mathfrak{a}, \mathfrak{a}' \in \mathfrak{A}, \varphi, \ \varphi' \in \mathscr{A}, \quad (5.2.1)$$

*where we write $\mathfrak{a}\,\overline{\#}\,\varphi$ for $\mathfrak{a} \otimes \varphi$, and denote this structure by $\mathfrak{A}\,\overline{\#}\,\mathscr{A}$. Then $\mathfrak{A}\,\overline{\#}\,\mathscr{A}$ becomes a left $H$-module algebra with unit $1_{\mathfrak{A}}\,\overline{\#}\,1_{\mathscr{A}}$ and with left $H$-action*

$$h \cdot (\mathfrak{a}\,\overline{\#}\,\varphi) = \mathfrak{a}\,\overline{\#}\,h \cdot \varphi, \ \ \forall\, \mathfrak{a} \in \mathfrak{A}, h \in H, \varphi \in \mathscr{A}.$$

*We call $\mathfrak{A}\,\overline{\#}\,\mathscr{A}$ the quasi-smash product of $\mathfrak{A}$ and $\mathscr{A}$.*

*Proof* Clearly $\mathfrak{A}\,\overline{\#}\,\mathscr{A}$ is a left $H$-module. We prove that $\mathfrak{A}\,\overline{\#}\,\mathscr{A}$ is an algebra in ${}_H\mathscr{M}$ with unit $1_{\mathfrak{A}}\,\overline{\#}\,1_{\mathscr{A}}$. For all $\mathfrak{a}, \mathfrak{a}', \mathfrak{a}'' \in \mathfrak{A}$ and $\varphi, \psi, \chi \in \mathscr{A}$ we have:

$$\begin{aligned}
&[X^1\cdot(\mathfrak{a}\,\overline{\#}\,\varphi)]\{[X^2\cdot(\mathfrak{a}'\,\overline{\#}\,\psi)][X^3\cdot(\mathfrak{a}''\,\overline{\#}\,\chi)]\}\\
&\quad = \quad (\mathfrak{a}\,\overline{\#}\,X^1\cdot\varphi)[(\mathfrak{a}'\,\overline{\#}\,X^2\cdot\psi)(\mathfrak{a}''\,\overline{\#}\,X^3\cdot\chi)]\\
&\quad = \quad (\mathfrak{a}\,\overline{\#}\,X^1\cdot\varphi)[\mathfrak{a}'\mathfrak{a}''_{\langle 0\rangle}\tilde{x}^1_\rho\,\overline{\#}\,(X^2\cdot\psi\cdot\mathfrak{a}''_{\langle 1\rangle}\tilde{x}^2_\rho)(X^3\cdot\chi\cdot\tilde{x}^3_\rho)]\\
&\quad = \quad \mathfrak{a}\mathfrak{a}'_{\langle 0\rangle}\mathfrak{a}''_{\langle 0,0\rangle}(\tilde{x}^1_\rho)_{\langle 0\rangle}\tilde{y}^1_\rho\,\overline{\#}\,(X^1\cdot\varphi\cdot\mathfrak{a}'_{\langle 1\rangle}\mathfrak{a}''_{\langle 0,1\rangle}(\tilde{x}^1_\rho)_{\langle 1\rangle}\tilde{y}^2_\rho)\\
&\qquad\qquad [(X^2\cdot\psi\cdot\mathfrak{a}''_{\langle 1\rangle}\tilde{x}^2_\rho(\tilde{y}^3_\rho)_1)(X^3\cdot\chi\cdot\tilde{x}^3_\rho(\tilde{y}^3_\rho)_2)]\\
&\overset{(4.1.13),(4.3.2)}{=} \mathfrak{a}\mathfrak{a}'_{\langle 0\rangle}\mathfrak{a}''_{\langle 0,0\rangle}\tilde{x}^1_\rho\tilde{y}^1_\rho\,\overline{\#}\,[(\varphi\cdot\mathfrak{a}'_{\langle 1\rangle}\mathfrak{a}''_{\langle 0,1\rangle}\tilde{x}^2_\rho(\tilde{y}^2_\rho)_1)(\psi\cdot\mathfrak{a}''_{\langle 1\rangle}\tilde{x}^3_\rho(\tilde{y}^2_\rho)_2)](\chi\cdot\tilde{y}^3_\rho)\\
&\quad \overset{(4.3.1)}{=} \mathfrak{a}\mathfrak{a}'_{\langle 0\rangle}\tilde{x}^1_\rho\mathfrak{a}''_{\langle 0\rangle}\tilde{y}^1_\rho\,\overline{\#}\,\{[(\varphi\cdot\mathfrak{a}'_{\langle 1\rangle}\tilde{x}^2_\rho)(\psi\cdot\tilde{x}^3_\rho)]\cdot\mathfrak{a}''_{\langle 1\rangle}\tilde{y}^2_\rho\}(\chi\cdot\tilde{y}^3_\rho)\\
&\quad = \quad [\mathfrak{a}\mathfrak{a}'_{\langle 0\rangle}\tilde{x}^1_\rho\,\overline{\#}\,(\varphi\cdot\mathfrak{a}'_{\langle 1\rangle}\tilde{x}^2_\rho)(\psi\cdot\tilde{x}^3_\rho)](\mathfrak{a}''\,\overline{\#}\,\chi)\\
&\quad = \quad [(\mathfrak{a}\,\overline{\#}\,\varphi)(\mathfrak{a}'\,\overline{\#}\,\psi)](\mathfrak{a}''\,\overline{\#}\,\chi).
\end{aligned}$$

It is easy to see that $h\cdot(1_{\mathfrak{A}}\,\overline{\#}\,1_{\mathscr{A}}) = \varepsilon(h)1_{\mathfrak{A}}\,\overline{\#}\,1_{\mathscr{A}}$ for all $h \in H$, and that $1_{\mathfrak{A}}\,\overline{\#}\,1_{\mathscr{A}}$ is the unit of $\mathfrak{A}\,\overline{\#}\,\mathscr{A}$. Finally, for all $h \in H$, $\mathfrak{a}, \mathfrak{a}' \in \mathfrak{A}$ and $\varphi, \psi \in \mathscr{A}$, we compute:

$$\begin{aligned}
[h_1\cdot(\mathfrak{a}\,\overline{\#}\,\varphi)][h_2\cdot(\mathfrak{a}'\,\overline{\#}\,\psi)] &= (\mathfrak{a}\,\overline{\#}\,h_1\cdot\varphi)(\mathfrak{a}'\,\overline{\#}\,h_2\cdot\psi)\\
&= \mathfrak{a}\mathfrak{a}'_{\langle 0\rangle}\tilde{x}^1_\rho\,\overline{\#}\,(h_1\cdot\varphi\cdot\mathfrak{a}'_{\langle 1\rangle}\tilde{x}^2_\rho)(h_2\cdot\psi\cdot\tilde{x}^3_\rho)\\
&= \mathfrak{a}\mathfrak{a}'_{\langle 0\rangle}\tilde{x}^1_\rho\,\overline{\#}\,h\cdot[(\varphi\cdot\mathfrak{a}'_{\langle 1\rangle}\tilde{x}^2_\rho)(\psi\cdot\tilde{x}^3_\rho)]\\
&= h\cdot[(\mathfrak{a}\,\overline{\#}\,\varphi)(\mathfrak{a}'\,\overline{\#}\,\psi)],
\end{aligned}$$

finishing the proof. □

**Corollary 5.16** *Let $H$ be a quasi-bialgebra, $B$ a right $H$-module algebra and $\mathfrak{A}$ a right $H$-comodule algebra. We denote by $\mathfrak{A} \rtimes B$ the k-vector space $\mathfrak{A} \otimes B$ with the newly defined multiplication*

$$(\mathfrak{a} \rtimes b)(\mathfrak{a}' \rtimes b') = \mathfrak{a}\mathfrak{a}'_{\langle 0\rangle}\tilde{x}^1_\rho \rtimes (b \cdot \mathfrak{a}'_{\langle 1\rangle}\tilde{x}^2_\rho)(b' \cdot \tilde{x}^3_\rho), \tag{5.2.2}$$

*for all $\mathfrak{a}, \mathfrak{a}' \in \mathfrak{A}$ and $b, b' \in B$. Then $\mathfrak{A} \rtimes B$ is an associative algebra with unit $1_{\mathfrak{A}} \rtimes 1_B$, called the generalized smash product of $\mathfrak{A}$ and $B$.*

*Proof* If we regard $B$ as an $H$-bimodule algebra with trivial left $H$-action, then obviously $\mathfrak{A} \rtimes B$ coincides with $\mathfrak{A} \overline{\#} B$, which is a left $H$-module algebra with trivial left $H$-action, namely an associative algebra. □

The above constructions have left-handed versions (proofs are left to the reader).

**Proposition 5.17** *Let $H$ be a quasi-bialgebra, $\mathfrak{B}$ a left $H$-comodule algebra and $\mathscr{A}$ an $H$-bimodule algebra. Define a multiplication on $\mathscr{A} \otimes \mathfrak{B}$ by*

$$(\varphi \overline{\#} \mathfrak{b})(\varphi' \overline{\#} \mathfrak{b}') = (\tilde{x}^1_\lambda \cdot \varphi)(\tilde{x}^2_\lambda \mathfrak{b}_{[-1]} \cdot \varphi') \overline{\#} \tilde{x}^3_\lambda \mathfrak{b}_{[0]}\mathfrak{b}', \quad \forall\, \varphi, \varphi' \in \mathscr{A}, \mathfrak{b}, \mathfrak{b}' \in \mathfrak{B}, \tag{5.2.3}$$

*where we write $\varphi \overline{\#} \mathfrak{b}$ for $\varphi \otimes \mathfrak{b}$, and denote this structure by $\mathscr{A} \overline{\#} \mathfrak{B}$. Then $\mathscr{A} \overline{\#} \mathfrak{B}$ becomes a right $H$-module algebra with unit $1_{\mathscr{A}} \overline{\#} 1_{\mathfrak{B}}$ and with right $H$-action*

$$(\varphi \overline{\#} \mathfrak{b}) \cdot h = \varphi \cdot h \overline{\#} \mathfrak{b}, \quad \forall\, \varphi \in \mathscr{A}, h \in H, \mathfrak{b} \in \mathfrak{B}.$$

*We call $\mathscr{A} \overline{\#} \mathfrak{B}$ the quasi-smash product of $\mathscr{A}$ and $\mathfrak{B}$.*

**Corollary 5.18** *Let $H$ be a quasi-bialgebra, $A$ a left $H$-module algebra and $\mathfrak{B}$ a left $H$-comodule algebra. Denote by $A \ltimes \mathfrak{B}$ the k-vector space $A \otimes \mathfrak{B}$ with multiplication*

$$(a \ltimes \mathfrak{b})(a' \ltimes \mathfrak{b}') = (\tilde{x}^1_\lambda \cdot a)(\tilde{x}^2_\lambda \mathfrak{b}_{[-1]} \cdot a') \ltimes \tilde{x}^3_\lambda \mathfrak{b}_{[0]}\mathfrak{b}', \tag{5.2.4}$$

*for all $a, a' \in A$ and $\mathfrak{b}, \mathfrak{b}' \in \mathfrak{B}$. Then $A \ltimes \mathfrak{B}$ is an associative algebra with unit $1_A \ltimes 1_{\mathfrak{B}}$, called the generalized smash product of $A$ and $\mathfrak{B}$.*

**Remark 5.19** If $H$ is a quasi-bialgebra, $A$ a left $H$-module algebra and $B$ a right $H$-module algebra, then $A \ltimes H = A\#H$ and $H \rtimes B = H\#B$.

The proof of the next result is similar to the proof of Proposition 5.9 and is left to the reader.

**Proposition 5.20** *Let $H$ be a quasi-bialgebra, $A$ a left $H$-module algebra, $B$ a right $H$-module algebra and $\mathbb{A}$ an $H$-bicomodule algebra. Then $A \ltimes \mathbb{A}$ becomes a right $H$-comodule algebra, with structure defined for all $a \in A$ and $u \in \mathbb{A}$ by*

$$\rho : A \ltimes \mathbb{A} \to (A \ltimes \mathbb{A}) \otimes H, \qquad \rho(a \ltimes u) = (\theta^1 \cdot a \ltimes \theta^2 u_{\langle 0\rangle}) \otimes \theta^3 u_{\langle 1\rangle},$$
$$\Phi_\rho = (1_A \ltimes \tilde{X}^1_\rho) \otimes \tilde{X}^2_\rho \otimes \tilde{X}^3_\rho \in (A \ltimes \mathbb{A}) \otimes H \otimes H,$$

*and $\mathbb{A} \rtimes B$ becomes a left $H$-comodule algebra, with structure defined for all $u \in \mathbb{A}$, $b \in B$ by*

$$\lambda : \mathbb{A} \rtimes B \to H \otimes (\mathbb{A} \rtimes B), \quad \lambda(u \rtimes b) = u_{[-1]}\theta^1 \otimes (u_{[0]}\theta^2 \rtimes b \cdot \theta^3),$$
$$\Phi_\lambda = \tilde{X}^1_\lambda \otimes \tilde{X}^2_\lambda \otimes (\tilde{X}^3_\lambda \rtimes 1_B) \in H \otimes H \otimes (\mathbb{A} \rtimes B).$$

**Proposition 5.21** *Let $H$ be a finite-dimensional quasi-Hopf algebra with bijective antipode. Define the linear map*

$$\mu : H \overline{\#} H^* \to \mathrm{End}_k(H), \quad \mu(h \overline{\#} \varphi)(h') = \varphi(h'_2\tilde{p}^2)hh'_1\tilde{p}^1,$$

*for all $h, h' \in H$ and $\varphi \in H^*$, where $p_L = \tilde{p}^1 \otimes \tilde{p}^2$ is the element defined by* (3.2.20). *Then $\mu$ is a bijection, and therefore there exists a unique left $H$-module algebra structure on $\mathrm{End}_k(H)$ such that $\mu$ becomes an $H$-module algebra isomorphism. The multiplication, unit and left $H$-module structure of $\mathrm{End}_k(H)$ are given by*

$$(u\overline{\circ}v)(h) = u(v(hx^3X^3_2)S^{-1}(S(x^1X^2)\alpha x^2X^3_1))S^{-1}(X^1), \tag{5.2.5}$$
$$\mathbf{1}_{\mathrm{End}_k(H)}(h) = hS^{-1}(\beta), \qquad (h \cdot u)(h') = u(h'h_2)S^{-1}(h_1), \tag{5.2.6}$$

*for all $u, v \in \mathrm{End}_k(H)$ and $h, h' \in H$. We denote by $END(H)$ this left $H$-module algebra structure of $\mathrm{End}_k(H)$.*

*Proof* Let $\{e_i\}_{i=\overline{1,n}}$ be a basis of $H$ and $\{e^i\}_{i=\overline{1,n}}$ the corresponding dual basis of $H^*$. We claim that the inverse of $\mu$ is the map $\mu^{-1} : \mathrm{End}_k(H) \to H \overline{\#} H^*$ given by

$$\mu^{-1}(u) = \sum_{i=1}^n u(\tilde{q}^2(e_i)_2)S^{-1}(\tilde{q}^1(e_i)_1) \overline{\#} e^i, \quad \forall\, u \in \mathrm{End}_k(H),$$

where $q_L = \tilde{q}^1 \otimes \tilde{q}^2$ is defined by (3.2.20). Indeed, for $h \in H$, $\varphi \in H^*$ we have:

$$\begin{aligned}
(\mu^{-1} \circ \mu)(h \overline{\#} \varphi) &= \sum_{i=1}^n \mu(h \overline{\#} \varphi)(\tilde{q}^2(e_i)_2)S^{-1}(\tilde{q}^1(e_i)_1) \overline{\#} e^i \\
&= \sum_{i=1}^n \varphi(\tilde{q}^2_2(e_i)_{(2,2)}\tilde{p}^2)h\tilde{q}^2_1(e_i)_{(2,1)}\tilde{p}^1S^{-1}(\tilde{q}^1(e_i)_1) \overline{\#} e^i \\
&\overset{(3.2.22)}{=} \sum_{i=1}^n \varphi(\tilde{q}^2_2\tilde{p}^2e_i)h\tilde{q}^2_1\tilde{p}^1S^{-1}(\tilde{q}^1) \overline{\#} e^i \\
&\overset{(3.2.24)}{=} \sum_{i=1}^n \varphi(e_i)h \overline{\#} e^i = h \overline{\#} \varphi.
\end{aligned}$$

Similarly, for $u \in \mathrm{End}_k(H)$ and $h \in H$ we have $(\mu \circ \mu^{-1})(u)(h) = u(h)$. By using the bijection $\mu$, we transfer the $H$-module algebra structure from $H \overline{\#} H^*$ to $\mathrm{End}_k(H)$. First we compute the transferred multiplication $\overline{\circ}$: for all $u, v \in \mathrm{End}_k(H)$, we find

$$\begin{aligned}
u\overline{\circ}v &= \mu(\mu^{-1}(u)\mu^{-1}(v)) \\
&= \sum_{i,j=1}^n \mu((u(\tilde{q}^2(e_i)_2)S^{-1}(\tilde{q}^1(e_i)_1) \overline{\#} e^i)(v(\tilde{Q}^2(e_j)_2)S^{-1}(\tilde{Q}^1(e_j)_1) \overline{\#} e^j)) \\
&= \sum_{i,j=1}^n \mu\Big(u(\tilde{q}^2(e_i)_2)S^{-1}(\tilde{q}^1(e_i)_1)[v(\tilde{Q}^2(e_j)_2)S^{-1}(\tilde{Q}^1(e_j)_1)]_1x^1
\end{aligned}$$

$$\overline{\#}\,(e^i \leftharpoonup [v(\tilde{Q}^2(e_j)_2)S^{-1}(\tilde{Q}^1(e_j)_1)]_2x^2)(e^j \leftharpoonup x^3)\Big),$$

where $\tilde{Q}^1 \otimes \tilde{Q}^2$ is another copy of $q_L$ and $\leftharpoonup$ is the right $H$-action on $H^*$ defined by (4.1.16). Note that (3.1.9) and (3.2.20) imply

$$S(x^1)\tilde{q}^1x_1^2 \otimes \tilde{q}^2x_2^2 \otimes x^3 = \tilde{q}^1X^1 \otimes \tilde{q}_1^2X^2 \otimes \tilde{q}_2^2X^3. \tag{5.2.7}$$

By using the above arguments, a long but straightforward computation shows that

$$(u\bar{\circ}v)(h) = u(v(hx^3X_2^3)S^{-1}(S(x^1X^2)\alpha x^2X_1^3))S^{-1}(X^1),$$

for all $h \in H$. Thus, we have obtained (5.2.5). Similar computations show that the transferred unit and the $H$-action on $\mathrm{End}_k(H)$ are given by (5.2.6). □

## 5.3 Endomorphism $H$-module Algebras

If $H$ is a quasi-Hopf algebra with bijective antipode, $M$ is a left $H$-module and $A$ is a left $H$-module algebra, we show that setting a left $A,H$-module structure on $M$ is equivalent to giving a morphism of left $H$-module algebras from $A$ to a certain deformation of $\mathrm{End}(M)$.

We begin with a lemma of independent interest.

**Lemma 5.22** *Let $H$ be a quasi-Hopf algebra, $B,C$ associative algebras, $\eta : B \to C$, $j : H \to B$, $v : H \to C$ algebra maps such that $\eta \circ j = v$. Then the map $\eta : B^j \to C^v$ is a morphism of left $H$-module algebras (notation as in Proposition 4.3).*

*Proof* This follows by a direct computation, using the formula (4.1.7). □

Let $H$ be a quasi-Hopf algebra and $M$ a left $H$-module, with action denoted by $h \otimes m \mapsto h \cdot m$. Consider the (usual) associative algebra $\mathrm{End}(M)$ of $k$-linear endomorphisms of $M$ (with composition) and define $v : H \to \mathrm{End}(M)$, $v(h)(m) = h \cdot m$, which is an algebra map, so we can consider the left $H$-module algebra $\mathrm{End}(M)^v$, whose multiplication, unit and $H$-action are given by

$$(u \circ u')(m) = X^1 \cdot u(S(x^1X^2)\alpha x^2X_1^3 \cdot u'(S(x^3X_2^3) \cdot m)), \tag{5.3.1}$$

$$1_{\mathrm{End}(M)^v}(m) = v(\beta)(m) = \beta \cdot m, \tag{5.3.2}$$

$$(h \triangleright_v u)(m) = h_1 \cdot u(S(h_2) \cdot m), \tag{5.3.3}$$

for all $h \in H$, $u, u' \in \mathrm{End}(M)^v$, $m \in M$.

Suppose we have also a left $H$-module algebra $A$ and the antipode of $H$ is bijective.

**Theorem 5.23** *Setting a structure of a left $A,H$-module on $M$ is equivalent to giving a morphism of left $H$-module algebras $\varphi : A \to \mathrm{End}(M)^v$. The correspondence is given as follows: if $M$ is a left $A,H$-module (with $A$-action denoted by $a \otimes m \mapsto a \triangleright m$) then the map $\varphi : A \to \mathrm{End}(M)^v$ is given by*

$$\varphi(a)(m) = (p^1 \cdot a) \triangleright (p^2 \cdot m), \ \ \forall\, a \in A,\ m \in M, \tag{5.3.4}$$

*where $p_R = p^1 \otimes p^2 = x^1 \otimes x^2\beta S(x^3)$. Conversely, if $\varphi : A \to \mathrm{End}(M)^\nu$ is a morphism of left $H$-module algebras, then $M$ becomes a left $A,H$-module, with $A$-action*

$$a \triangleright m = q^1 \cdot \varphi(a)(S(q^2) \cdot m), \quad \forall\, a \in A,\ m \in M, \tag{5.3.5}$$

*where $q_R = q^1 \otimes q^2 = X^1 \otimes S^{-1}(\alpha X^3)X^2$, and the $H$-action is the original $H$-module structure of $M$.*

*Proof* Suppose first that $M$ is a left $A,H$-module, with $A$-action $a \otimes m \mapsto a \triangleright m$. By Proposition 5.7, this is equivalent to $M$ being a left $A\#H$-module, with structure

$$(a\#h) \cdot m = a \triangleright (h \cdot m), \quad \forall\, a \in A,\ h \in H,\ m \in M.$$

So, by considering the usual associative algebra $\mathrm{End}(M)$, we obtain an algebra map $\eta : A\#H \to \mathrm{End}(M)$, $\eta(a\#h)(m) = (a\#h) \cdot m$. We also have the canonical algebra map $j : H \to A\#H$, $j(h) = 1_A\#h$; since we obviously have that $\eta \circ j = \nu$, we can apply Lemma 5.22 and obtain that the map $\eta : (A\#H)^j \to \mathrm{End}(M)^\nu$ is a morphism of left $H$-module algebras. By Lemma 5.11, the map $i_0 : A \to (A\#H)^j$, $i_0(a) = p^1 \cdot a\#p^2$, where $p_R = p^1 \otimes p^2 = x^1 \otimes x^2\beta S(x^3)$, is a morphism of left $H$-module algebras, so the composition $\varphi = \eta \circ i_0 : A \to \mathrm{End}(M)^\nu$ is also a morphism of left $H$-module algebras, and one can easily check that it is given by $\varphi(a)(m) = (p^1 \cdot a) \triangleright (p^2 \cdot m)$, for all $a \in A$, $m \in M$.

Conversely, let $\varphi : A \to \mathrm{End}(M)^\nu$ be a morphism of left $H$-module algebras; by applying the Universal Property of the smash product $A\#H$ (Theorem 5.12) for $B = \mathrm{End}(M)$, we obtain the algebra map $\varphi\#\nu : A\#H \to \mathrm{End}(M)$, which (by using the formula (5.1.7)) can be expressed as follows:

$$(\varphi\#\nu)(a\#h)(m) = q^1 \cdot \varphi(a)(S(q^2)h \cdot m), \quad \forall\, a \in A,\ h \in H,\ m \in M.$$

Hence, $M$ becomes a left $A\#H$-module (i.e. a left $A,H$-module) with action

$$(a\#h) \cdot m = q^1 \cdot \varphi(a)(S(q^2)h \cdot m), \quad \forall\, a \in A,\ h \in H,\ m \in M.$$

In particular, the $A$-action is given by

$$a \triangleright m = (a\#1_H) \cdot m = q^1 \cdot \varphi(a)(S(q^2) \cdot m), \quad \forall\, a \in A,\ m \in M,$$

and, by using the fact that $q^1\beta S(q^2) = 1_H$ (which follows from (3.2.2)), we obtain that the $H$-action is given by

$$(1_A\#h) \cdot m = q^1\beta S(q^2)h \cdot m = h \cdot m, \quad \forall\, h \in H,\ m \in M.$$

The only thing left to prove is that the two correspondences are inverse to each other. If $M$ is an $A,H$-module with $A$-action denoted by $\triangleright$, $\varphi$ is the associated map $\varphi : A \to \mathrm{End}(M)^\nu$ and $\triangleright'$ is the $A$-action associated to $\varphi$, we have

$$\begin{aligned} a \triangleright' m &= q^1 \cdot ((p^1 \cdot a) \triangleright (p^2 S(q^2) \cdot m)) \\ &\overset{(5.1.4)}{=} (q^1_1 p^1 \cdot a) \triangleright (q^1_2 p^2 S(q^2) \cdot m) \overset{(3.2.23)}{=} a \triangleright m, \end{aligned}$$

for all $a \in A$ and $m \in M$. Conversely, if $\varphi : A \to \mathrm{End}(M)^\nu$ is a left $H$-module algebra map, $\triangleright$ is the $A$-action obtained from $\varphi$ and $\varphi'$ is the map obtained from this $A,H$-module structure on $M$, we have (for all $a \in A$ and $m \in M$):

$$\begin{aligned}
\varphi'(a)(m) &= (p^1 \cdot a) \triangleright (p^2 \cdot m)\\
&= q^1 \cdot \varphi(p^1 \cdot a)(S(q^2)p^2 \cdot m)\\
&= q^1 \cdot ((p^1 \triangleright_\nu \varphi(a))(S(q^2)p^2 \cdot m))\\
&\overset{(5.3.3)}{=} q^1 \cdot (p_1^1 \cdot \varphi(a)(S(p_2^1)S(q^2)p^2 \cdot m))\\
&= q^1 p_1^1 \cdot \varphi(a)(S(S^{-1}(p^2)q^2 p_2^1) \cdot m) \overset{(3.2.23)}{=} \varphi(a)(m),
\end{aligned}$$

finishing the proof. □

By taking $A = \mathrm{End}(M)^\nu$ and $\varphi = \mathrm{Id}$ in Theorem 5.23, we obtain:

**Corollary 5.24** *If $H$ is a quasi-Hopf algebra with bijective antipode and $M$ is a left $H$-module, then $M$ becomes a left $\mathrm{End}(M)^\nu, H$-module (i.e. a left $\mathrm{End}(M)^\nu \# H$-module), with $\mathrm{End}(M)^\nu$-action given, for all $u \in \mathrm{End}(M)^\nu$ and $m \in M$, by*

$$u \triangleright m = q^1 \cdot u(S(q^2) \cdot m). \tag{5.3.6}$$

We now study the behavior of the construction $\mathrm{End}(M)^\nu$ under twisting. Let $H$ be a quasi-Hopf algebra, $F \in H \otimes H$ a gauge transformation and $M$ a left $H$-module. Then $M$ is also a left $H_F$-module, with the same $H$-action. Denote by $\nu : H \to \mathrm{End}(M)$ and $\nu_F : H_F \to \mathrm{End}(M)$ the corresponding algebra maps, and consider the $H$-module algebra $\mathrm{End}(M)^\nu$ and the $H_F$-module algebra $\mathrm{End}(M)^{\nu_F}$; we also consider the $H_F$-module algebra $\mathrm{End}(M)^\nu_{F^{-1}}$. We will prove that $\mathrm{End}(M)^{\nu_F}$ and $\mathrm{End}(M)^\nu_{F^{-1}}$ are isomorphic as left $H_F$-module algebras.

Actually, we will prove something more general. Let $H$ be a quasi-Hopf algebra, $F \in H \otimes H$ a gauge transformation, $B$ an algebra and $\nu : H \to B$ an algebra map, which will be denoted by $\nu_F$ when it is considered as a map from $H_F$ to $B$.

**Proposition 5.25** *The map $\psi : B^\nu_{F^{-1}} \to B^{\nu_F}$, $\psi(b) = \nu(F^1)b\nu(S(F^2))$, for all $b \in B$, is an isomorphism of left $H_F$-module algebras.*

*Proof* The map $\psi$ is obviously bijective, with inverse $\psi^{-1}(b) = \nu(G^1)b\nu(S(G^2))$, for all $b \in B$, where $F^{-1} = G^1 \otimes G^2$. Then one checks by a direct computation that $\psi$ is a morphism of left $H_F$-module algebras, by using the formulas for $\Delta_F$, $\Phi_F$, $\alpha_F$, $\beta_F$ and for the multiplications, units and actions in $B^{\nu_F}$ and $B^\nu_{F^{-1}}$. □

By taking $B = \mathrm{End}(M)$, where $M$ is a left $H$-module, we obtain:

**Corollary 5.26** $\mathrm{End}(M)^{\nu_F} \cong \mathrm{End}(M)^\nu_{F^{-1}}$ *as left $H_F$-module algebras.*

By taking $B = H$, $\nu = \mathrm{Id}_H$ in Proposition 5.25, we obtain:

**Corollary 5.27** *$(H_0)_{F^{-1}} \cong (H_F)_0$ as left $H_F$-module algebras, with an isomorphism given by $\psi : (H_0)_{F^{-1}} \to (H_F)_0$, $\psi(h) = F^1 h S(F^2)$ for all $h \in H$, where $F = F^1 \otimes F^2$ is a gauge transformation on $H$ and $H_0$ is the left $H$-module algebra that appears in Definition 4.4.*

**Proposition 5.28** *Let $H$ be a finite-dimensional quasi-Hopf algebra with bijective antipode. Denote by $M$ the vector space $H$ viewed as a left $H$-module with action $h \cdot m = mS^{-1}(h)$ for all $m, h \in H$, and consider the left $H$-module algebra* $\mathrm{End}(M)^{\nu}$ *constructed at the beginning of this section and the left $H$-module algebra $END(H)$ as in Proposition 5.21. Then $END(H) = \mathrm{End}(M)^{\nu}$ as left $H$-module algebras.*

*Proof* This is a straightforward verification, using the formulas (5.2.5) and (5.2.6). □

## 5.4 Two-sided Smash and Crossed Products

We introduce some new types of crossed products, which will turn out to be iterated (quasi-)smash products.

**Proposition 5.29** *Let $H$ be a quasi-bialgebra, $\mathfrak{A}$ a right $H$-comodule algebra, $\mathfrak{B}$ a left $H$-comodule algebra and $\mathscr{A}$ an $H$-bimodule algebra. On $\mathfrak{A} \otimes \mathscr{A} \otimes \mathfrak{B}$ define a multiplication by*

$$
\begin{aligned}
&(\mathfrak{a} \mathrel{>\!\!\!\lhd} \varphi \ltimes \mathfrak{b})(\mathfrak{a}' \mathrel{>\!\!\!\lhd} \varphi' \ltimes \mathfrak{b}') \\
&\quad = \mathfrak{a}\mathfrak{a}'_{\langle 0\rangle}\tilde{x}^1_\rho \mathrel{>\!\!\!\lhd} (\tilde{x}^1_\lambda \cdot \varphi \cdot \mathfrak{a}'_{\langle 1\rangle}\tilde{x}^2_\rho)(\tilde{x}^2_\lambda \mathfrak{b}_{[-1]} \cdot \varphi' \cdot \tilde{x}^3_\rho) \ltimes \tilde{x}^3_\lambda \mathfrak{b}_{[0]}\mathfrak{b}',
\end{aligned}
\tag{5.4.1}
$$

*for all $\mathfrak{a}, \mathfrak{a}' \in \mathfrak{A}$, $\mathfrak{b}, \mathfrak{b}' \in \mathfrak{B}$ and $\varphi, \varphi' \in \mathscr{A}$, where we write $\mathfrak{a} \mathrel{>\!\!\!\lhd} \varphi \ltimes \mathfrak{b}$ for $\mathfrak{a} \otimes \varphi \otimes \mathfrak{b}$. Then this multiplication yields an associative algebra with unit $1_{\mathfrak{A}} \mathrel{>\!\!\!\lhd} 1_{\mathscr{A}} \ltimes 1_{\mathfrak{B}}$, denoted by $\mathfrak{A} \mathrel{>\!\!\!\lhd} \mathscr{A} \ltimes \mathfrak{B}$ and called the two-sided crossed product.*

*Proof* We check the associativity of the multiplication:

$$
\begin{aligned}
&[(\mathfrak{a} \mathrel{>\!\!\!\lhd} \varphi \ltimes \mathfrak{b})(\mathfrak{a}' \mathrel{>\!\!\!\lhd} \varphi' \ltimes \mathfrak{b}')](\mathfrak{a}'' \mathrel{>\!\!\!\lhd} \varphi'' \ltimes \mathfrak{b}'') \\
&= (\mathfrak{a}\mathfrak{a}'_{\langle 0\rangle}\tilde{x}^1_\rho \mathrel{>\!\!\!\lhd} (\tilde{x}^1_\lambda \cdot \varphi \cdot \mathfrak{a}'_{\langle 1\rangle}\tilde{x}^2_\rho)(\tilde{x}^2_\lambda \mathfrak{b}_{[-1]} \cdot \varphi' \cdot \tilde{x}^3_\rho) \ltimes \tilde{x}^3_\lambda \mathfrak{b}_{[0]}\mathfrak{b}')(\mathfrak{a}'' \mathrel{>\!\!\!\lhd} \varphi'' \ltimes \mathfrak{b}'') \\
&= \mathfrak{a}\mathfrak{a}'_{\langle 0\rangle}\tilde{x}^1_\rho \mathfrak{a}''_{\langle 0\rangle}\tilde{y}^1_\rho \mathrel{>\!\!\!\lhd} [((\tilde{y}^1_\lambda)_1\tilde{x}^1_\lambda \cdot \varphi \cdot \mathfrak{a}'_{\langle 1\rangle}\tilde{x}^2_\rho \mathfrak{a}''_{\langle 1\rangle_1}(\tilde{y}^2_\rho)_1)((\tilde{y}^1_\lambda)_2\tilde{x}^2_\lambda \mathfrak{b}_{[-1]} \cdot \varphi' \\
&\qquad \cdot \tilde{x}^3_\rho \mathfrak{a}''_{\langle 1\rangle_2}(\tilde{y}^2_\rho)_2)](\tilde{y}^2_\lambda(\tilde{x}^3_\lambda)_{[-1]}\mathfrak{b}_{[0,-1]}\mathfrak{b}'_{[-1]} \cdot \varphi'' \cdot \tilde{y}^3_\rho) \ltimes \tilde{y}^3_\lambda(\tilde{x}^3_\lambda)_{[0]}\mathfrak{b}_{[0,0]}\mathfrak{b}'_{[0]}\mathfrak{b}'' \\
&\overset{(4.3.1)}{=} \mathfrak{a}\mathfrak{a}'_{\langle 0\rangle}\mathfrak{a}''_{\langle 0,0\rangle}\tilde{x}^1_\rho\tilde{y}^1_\rho \mathrel{>\!\!\!\lhd} [((\tilde{y}^1_\lambda)_1\tilde{x}^1_\lambda \cdot \varphi \cdot \mathfrak{a}'_{\langle 1\rangle}\mathfrak{a}''_{\langle 0,1\rangle}\tilde{x}^2_\rho(\tilde{y}^2_\rho)_1)((\tilde{y}^1_\lambda)_2\tilde{x}^2_\lambda \mathfrak{b}_{[-1]} \cdot \varphi' \\
&\qquad \cdot \mathfrak{a}''_{\langle 1\rangle}\tilde{x}^3_\rho(\tilde{y}^2_\rho)_2)](\tilde{y}^2_\lambda(\tilde{x}^3_\lambda)_{[-1]}\mathfrak{b}_{[0,-1]}\mathfrak{b}'_{[-1]} \cdot \varphi'' \cdot \tilde{y}^3_\rho) \ltimes \tilde{y}^3_\lambda(\tilde{x}^3_\lambda)_{[0]}\mathfrak{b}_{[0,0]}\mathfrak{b}'_{[0]}\mathfrak{b}'' \\
&\overset{(4.1.13)}{=} \mathfrak{a}\mathfrak{a}'_{\langle 0\rangle}\mathfrak{a}''_{\langle 0,0\rangle}\tilde{x}^1_\rho\tilde{y}^1_\rho \mathrel{>\!\!\!\lhd} (X^1(\tilde{y}^1_\lambda)_1\tilde{x}^1_\lambda \cdot \varphi \cdot \mathfrak{a}'_{\langle 1\rangle}\mathfrak{a}''_{\langle 0,1\rangle}\tilde{x}^2_\rho(\tilde{y}^2_\rho)_1x^1) \\
&\qquad [(X^2(\tilde{y}^1_\lambda)_2\tilde{x}^2_\lambda \mathfrak{b}_{[-1]} \cdot \varphi' \cdot \mathfrak{a}''_{\langle 1\rangle}\tilde{x}^3_\rho(\tilde{y}^2_\rho)_2x^2)(X^3\tilde{y}^2_\lambda(\tilde{x}^3_\lambda)_{[-1]}\mathfrak{b}_{[0,-1]}\mathfrak{b}'_{[-1]} \cdot \varphi'' \cdot \tilde{y}^3_\rho x^3)] \\
&\qquad \ltimes \tilde{y}^3_\lambda(\tilde{x}^3_\lambda)_{[0]}\mathfrak{b}_{[0,0]}\mathfrak{b}'_{[0]}\mathfrak{b}'' \\
&\overset{(4.3.6)}{=} \mathfrak{a}\mathfrak{a}'_{\langle 0\rangle}\mathfrak{a}''_{\langle 0,0\rangle}\tilde{x}^1_\rho\tilde{y}^1_\rho \mathrel{>\!\!\!\lhd} (\tilde{x}^1_\lambda \cdot \varphi \cdot \mathfrak{a}'_{\langle 1\rangle}\mathfrak{a}''_{\langle 0,1\rangle}\tilde{x}^2_\rho(\tilde{y}^2_\rho)_1x^1)[((\tilde{x}^2_\lambda)_1\tilde{y}^1_\lambda \mathfrak{b}_{[-1]} \cdot \varphi' \\
&\qquad \cdot \mathfrak{a}''_{\langle 1\rangle}\tilde{x}^3_\rho(\tilde{y}^2_\rho)_2x^2)((\tilde{x}^2_\lambda)_2\tilde{y}^2_\lambda \mathfrak{b}_{[0,-1]}\mathfrak{b}'_{[-1]} \cdot \varphi'' \cdot \tilde{y}^3_\rho x^3)] \ltimes \tilde{x}^3_\lambda\tilde{y}^3_\lambda \mathfrak{b}_{[0,0]}\mathfrak{b}'_{[0]}\mathfrak{b}'' \\
&\overset{(4.3.5)}{=} \mathfrak{a}\mathfrak{a}'_{\langle 0\rangle}\mathfrak{a}''_{\langle 0,0\rangle}\tilde{x}^1_\rho\tilde{y}^1_\rho \mathrel{>\!\!\!\lhd} (\tilde{x}^1_\lambda \cdot \varphi \cdot \mathfrak{a}'_{\langle 1\rangle}\mathfrak{a}''_{\langle 0,1\rangle}\tilde{x}^2_\rho(\tilde{y}^2_\rho)_1x^1)[((\tilde{x}^2_\lambda)_1\mathfrak{b}_{[-1]_1}\tilde{y}^1_\lambda \cdot \varphi' \\
&\qquad \cdot \mathfrak{a}''_{\langle 1\rangle}\tilde{x}^3_\rho(\tilde{y}^2_\rho)_2x^2)((\tilde{x}^2_\lambda)_2\mathfrak{b}_{[-1]_2}\tilde{y}^2_\lambda \mathfrak{b}'_{[-1]} \cdot \varphi'' \cdot \tilde{y}^3_\rho x^3)] \ltimes \tilde{x}^3_\lambda \mathfrak{b}_{[0]}\tilde{y}^3_\lambda \mathfrak{b}'_{[0]}\mathfrak{b}'' \\
&\overset{(4.3.2)}{=} \mathfrak{a}\mathfrak{a}'_{\langle 0\rangle}\mathfrak{a}''_{\langle 0,0\rangle}(\tilde{y}^1_\rho)_{\langle 0\rangle}\tilde{x}^1_\rho \mathrel{>\!\!\!\lhd} (\tilde{x}^1_\lambda \cdot \varphi \cdot \mathfrak{a}'_{\langle 1\rangle}\mathfrak{a}''_{\langle 0,1\rangle}(\tilde{y}^1_\rho)_{\langle 1\rangle}\tilde{x}^2_\rho)[((\tilde{x}^2_\lambda)_1\mathfrak{b}_{[-1]_1}\tilde{y}^1_\lambda \cdot \varphi'
\end{aligned}
$$

$$
\begin{aligned}
&\cdot\mathfrak{a}''_{\langle 1\rangle}\tilde{y}^2_\rho(\tilde{x}^3_\rho)_1)((\tilde{x}^2_\lambda)_2\mathfrak{b}_{[-1]_2}\tilde{y}^2_\lambda\mathfrak{b}'_{[-1]}\cdot\varphi''\cdot\tilde{y}^3_\rho(\tilde{x}^3_\rho)_2)]\ \triangleright\!\!<\ \tilde{x}^3_\lambda\mathfrak{b}_{[0]}\tilde{y}^3_\lambda\mathfrak{b}'_{[0]}\mathfrak{b}''\\
=\ &(\mathfrak{a}>\!\!\triangleleft\,\varphi\ \triangleright\!\!<\ \mathfrak{b})[\mathfrak{a}'\mathfrak{a}''_{\langle 0\rangle}\tilde{x}^1_\rho>\!\!\triangleleft(\tilde{x}^1_\lambda\cdot\varphi'\cdot\mathfrak{a}''_{\langle 1\rangle}\tilde{x}^2_\rho)(\tilde{x}^2_\lambda\mathfrak{b}'_{[-1]}\cdot\varphi''\cdot\tilde{x}^3_\rho)\ \triangleright\!\!<\ \tilde{x}^3_\lambda\mathfrak{b}'_{[0]}\mathfrak{b}'']\\
=\ &(\mathfrak{a}>\!\!\triangleleft\,\varphi\ \triangleright\!\!<\ \mathfrak{b})[(\mathfrak{a}'>\!\!\triangleleft\,\varphi'\ \triangleright\!\!<\ \mathfrak{b}')(\mathfrak{a}''>\!\!\triangleleft\,\varphi''\ \triangleright\!\!<\ \mathfrak{b}'')].
\end{aligned}
$$

The fact that $1_{\mathfrak{A}}>\!\!\triangleleft 1_{\mathscr{A}}\ \triangleright\!\!<\ 1_{\mathfrak{B}}$ is the unit is easy to check and left to the reader. □

**Proposition 5.30** *Let $H$ be a quasi-bialgebra, $A$ a left $H$-module algebra, $B$ a right $H$-module algebra and $\mathbb{A}$ an $H$-bicomodule algebra. Define on $A\otimes\mathbb{A}\otimes B$ a multiplication by*

$$
\begin{aligned}
&(a\blacktriangleright\!\!<u>\!\!\blacktriangleleft b)(a'\blacktriangleright\!\!<u'>\!\!\blacktriangleleft b')\\
&=(\tilde{x}^1_\lambda\cdot a)(\tilde{x}^2_\lambda u_{[-1]}\theta^1\cdot a')\blacktriangleright\!\!<\tilde{x}^3_\lambda u_{[0]}\theta^2u'_{\langle 0\rangle}\tilde{x}^1_\rho>\!\!\blacktriangleleft(b\cdot\theta^3u'_{\langle 1\rangle}\tilde{x}^2_\rho)(b'\cdot\tilde{x}^3_\rho), \quad (5.4.2)
\end{aligned}
$$

*for all $a,a'\in A$, $u,u'\in\mathbb{A}$, $b,b'\in B$ (where we write $a\blacktriangleright\!\!<u>\!\!\blacktriangleleft b$ for $a\otimes u\otimes b$), and denote this structure on $A\otimes\mathbb{A}\otimes B$ by $A\blacktriangleright\!\!<\mathbb{A}>\!\!\blacktriangleleft B$. Then $A\blacktriangleright\!\!<\mathbb{A}>\!\!\blacktriangleleft B$ is an associative algebra with unit $1_A\blacktriangleright\!\!<1_{\mathbb{A}}>\!\!\blacktriangleleft 1_B$, called the two-sided generalized smash product.*

*Proof* For all $a,a',a''\in A$, $u,u',u''\in\mathbb{A}$ and $b,b',b''\in B$ we compute:

$$
\begin{aligned}
&[(a\blacktriangleright\!\!<u>\!\!\blacktriangleleft b)(a'\blacktriangleright\!\!<u'>\!\!\blacktriangleleft b')](a''\blacktriangleright\!\!<u''>\!\!\blacktriangleleft b'')\\
\overset{(5.4.2)}{=}\ &\{\tilde{y}^1_\lambda\cdot[(\tilde{x}^1_\lambda\cdot a)(\tilde{x}^2_\lambda u_{[-1]}\theta^1\cdot a')]\}[\tilde{y}^2_\lambda(\tilde{x}^3_\lambda)_{[-1]}u_{[0,-1]}\theta^2_{[-1]}\\
&u'_{\langle 0\rangle_{[-1]}}(\tilde{x}^1_\rho)_{[-1]}\overline{\theta}^1\cdot a'']\blacktriangleright\!\!<\tilde{y}^3_\lambda(\tilde{x}^3_\lambda)_{[0]}u_{[0,0]}\theta^2_{[0]}u'_{\langle 0\rangle_{[0]}}(\tilde{x}^1_\rho)_{[0]}\overline{\theta}^2\\
&u''_{\langle 0\rangle}\tilde{y}^1_\rho>\!\!\blacktriangleleft\{[(b\cdot\theta^3u'_{\langle 1\rangle}\tilde{x}^2_\rho)(b'\cdot\tilde{x}^3_\rho)]\cdot\overline{\theta}^3u''_{\langle 1\rangle}\tilde{y}^2_\rho\}(b''\cdot\tilde{y}^3_\rho)\\
\underset{(4.1.4)}{\overset{(4.1.1)}{=}}\ &[(X^1(\tilde{y}^1_\lambda)_1\tilde{x}^1_\lambda\cdot a]\{[X^2(\tilde{y}^1_\lambda)_2\tilde{x}^2_\lambda u_{[-1]}\theta^1\cdot a']\\
&\qquad[X^3\tilde{y}^2_\lambda(\tilde{x}^3_\lambda)_{[-1]}u_{[0,-1]}\theta^2_{[-1]}u'_{\langle 0\rangle_{[-1]}}(\tilde{x}^1_\rho)_{[-1]}\overline{\theta}^1\cdot a'']\}\\
&\blacktriangleright\!\!<\tilde{y}^3_\lambda(\tilde{x}^3_\lambda)_{[0]}u_{[0,0]}\theta^2_{[0]}u'_{\langle 0\rangle_{[0]}}(\tilde{x}^1_\rho)_{[0]}\overline{\theta}^2u''_{\langle 0\rangle}\tilde{y}^1_\rho\\
&>\!\!\blacktriangleleft[b\cdot\theta^3u'_{\langle 1\rangle}\tilde{x}^2_\rho\overline{\theta}^3_1u''_{\langle 1\rangle_1}(\tilde{y}^2_\rho)_1x^1]\{[(b'\cdot\tilde{x}^3_\rho\overline{\theta}^3_2u''_{\langle 1\rangle_2}(\tilde{y}^2_\rho)_2x^2](b''\cdot\tilde{y}^3_\rho x^3)\}\\
\overset{(4.3.6)}{=}\ &(\tilde{y}^1_\lambda\cdot a)\{[(\tilde{y}^2_\lambda)_1\tilde{x}^1_\lambda u_{[-1]}\theta^1\cdot a'][(\tilde{y}^2_\lambda)_2\tilde{x}^2_\lambda u_{[0,-1]}\theta^2_{[-1]}u'_{\langle 0\rangle_{[-1]}}\\
&\qquad(\tilde{x}^1_\rho)_{[-1]}\overline{\theta}^1\cdot a'']\}\blacktriangleright\!\!<\tilde{y}^3_\lambda\tilde{x}^3_\lambda u_{[0,0]}\theta^2_{[0]}u'_{\langle 0\rangle_{[0]}}(\tilde{x}^1_\rho)_{[0]}\overline{\theta}^2u''_{\langle 0\rangle}\tilde{y}^1_\rho\\
&>\!\!\blacktriangleleft[b\cdot\theta^3u'_{\langle 1\rangle}\tilde{x}^2_\rho\overline{\theta}^3_1u''_{\langle 1\rangle_1}(\tilde{y}^2_\rho)_1x^1]\{[b'\cdot\tilde{x}^3_\rho\overline{\theta}^3_2u''_{\langle 1\rangle_2}(\tilde{y}^2_\rho)_2x^2](b''\cdot\tilde{y}^3_\rho x^3)\}\\
\underset{(4.4.2)}{\overset{(4.3.5)}{=}}\ &(\tilde{y}^1_\lambda\cdot a)\{\tilde{y}^2_\lambda u_{[-1]}\theta^1\cdot[(\tilde{x}^1_\lambda\cdot a')(\tilde{x}^2_\lambda\Theta^1u'_{\langle 0\rangle_{[-1]}}(\tilde{x}^1_\rho)_{[-1]}\overline{\theta}^1\cdot a'')]\}\\
&\blacktriangleright\!\!<\tilde{y}^3_\lambda u_{[0]}\theta^2(\tilde{x}^3_\lambda)_{\langle 0\rangle}\Theta^2u'_{\langle 0\rangle_{[0]}}(\tilde{x}^1_\rho)_{[0]}\overline{\theta}^2u''_{\langle 0\rangle}\tilde{y}^1_\rho>\!\!\blacktriangleleft[b\cdot\theta^3(\tilde{x}^3_\lambda)_{\langle 1\rangle}\\
&\Theta^3u'_{\langle 1\rangle}\tilde{x}^2_\rho\overline{\theta}^3_1u''_{\langle 1\rangle_1}(\tilde{y}^2_\rho))_1x^1]\{[b'\cdot\tilde{x}^3_\rho\overline{\theta}^3_2u''_{\langle 1\rangle_2}(\tilde{y}^2_\rho)_2x^2](b''\cdot\tilde{y}^3_\rho x^3)\}\\
\underset{(4.3.1)}{\overset{(4.4.1),(4.4.3)}{=}}\ &(\tilde{y}^1_\lambda\cdot a)\{\tilde{y}^2_\lambda u_{[-1]}\theta^1\cdot[((\tilde{x}^1_\lambda\cdot a')(\tilde{x}^2_\lambda u'_{[-1]}\overline{\theta}^1\cdot a'')]\}\blacktriangleright\!\!<\tilde{y}^3_\lambda u_{[0]}\theta^2\\
&(\tilde{x}^3_\lambda)_{\langle 0\rangle}u'_{[0]_{\langle 0\rangle}}\overline{\theta}^2_{\langle 0\rangle}u''_{\langle 0,0\rangle}\tilde{x}^1_\rho\tilde{y}^1_\rho>\!\!\blacktriangleleft[b\cdot\theta^3(\tilde{x}^3_\lambda)_{\langle 1\rangle}u'_{[0]_{\langle 1\rangle}}\overline{\theta}^2_{\langle 1\rangle}\\
&u''_{\langle 0,1\rangle}\tilde{x}^2_\rho(\tilde{y}^2_\rho)_1x^1]\{[b'\cdot\overline{\theta}^3u''_{\langle 1\rangle}\tilde{x}^3_\rho(\tilde{y}^2_\rho)_2x^2](b''\cdot\tilde{y}^3_\rho x^3)\}
\end{aligned}
$$

$$
\begin{aligned}
&\overset{(4.3.2)}{\underset{(4.1.5)}{=}} (\tilde{y}_\lambda^1 \cdot a)\{\tilde{y}_\lambda^2 u_{[-1]}\theta^1 \cdot [(\tilde{x}_\lambda^1 \cdot a')(\tilde{x}_\lambda^2 u'_{[-1]}\overline{\theta}^1 \cdot a'')]\} \blacktriangleright\!\!< \tilde{y}_\lambda^3 u_{[0]}\theta^2 \\
&\quad (\tilde{x}_\lambda^3 u'_{[0]}\overline{\theta}^2 u''_{\langle 0\rangle}\tilde{y}_\rho^1)_{\langle 0\rangle}\tilde{x}_\rho^1 >\!\!\blacktriangleleft [b \cdot \theta^3(\tilde{x}_\lambda^3 u'_{[0]}\overline{\theta}^2 u''_{\langle 0\rangle}\tilde{y}_\rho^1)_{\langle 1\rangle}\tilde{x}_\rho^2] \\
&\quad \{[(b' \cdot \overline{\theta}^3 u''_{\langle 1\rangle}\tilde{y}_\rho^2)(b'' \cdot \tilde{y}_\rho^3)] \cdot \tilde{x}_\rho^3\} \\
&\overset{(5.4.2)}{=} (a \blacktriangleright\!\!< u >\!\!\blacktriangleleft b)[(\tilde{x}_\lambda^1 \cdot a')(\tilde{x}_\lambda^2 u_{[-1]}\overline{\theta}^1 \cdot a'') \blacktriangleright\!\!< \tilde{x}_\lambda^3 u'_{[0]}\overline{\theta}^2 u''_{\langle 0\rangle}\tilde{y}_\rho^1 \\
&\quad >\!\!\blacktriangleleft (b' \cdot \overline{\theta}^3 u''_{\langle 1\rangle}\tilde{y}_\rho^2)(b'' \cdot \tilde{y}_\rho^3)] \\
&\overset{(5.4.2)}{=} (a \blacktriangleright\!\!< u >\!\!\blacktriangleleft b)[(a' \blacktriangleright\!\!< u' >\!\!\blacktriangleleft b')(a'' \blacktriangleright\!\!< u'' >\!\!\blacktriangleleft b'')].
\end{aligned}
$$

Finally, by (4.3.3), (4.3.4), (4.3.7), (4.3.8) and (4.4.4) it follows that $1_A \blacktriangleright\!\!< 1_{\mathbb{A}} >\!\!\blacktriangleleft 1_B$ is the unit of $A \blacktriangleright\!\!< \mathbb{A} >\!\!\blacktriangleleft B$. □

**Remarks 5.31** (i) The two-sided crossed product $\mathfrak{A} \rtimes \mathscr{A} \ltimes \mathfrak{B}$ cannot be particularized for $\mathfrak{A} = k$ or $\mathfrak{B} = k$ because, in general, $k$ is not a right or left $H$-comodule algebra. For the algebra $A \blacktriangleright\!\!< \mathbb{A} >\!\!\blacktriangleleft B$, we can take $A = k$ or $B = k$. In these cases we obtain the right or left generalized smash products $\mathbb{A} >\!\!\blacktriangleleft B$ and $A \blacktriangleright\!\!< \mathbb{A}$, respectively.

(ii) Let $\mathbb{A} = H$. In this particular case we will denote the algebra $A \blacktriangleright\!\!< H >\!\!\blacktriangleleft B$ by $A\#H\#B$ (the elements will be written $a\#h\#b$, $a \in A$, $h \in H$, $b \in B$) and will call it the two-sided smash product. This terminology is based on the fact that when we take $A = k$ or $B = k$ the resulting algebra is the right or left version of the smash product algebra. Note that the multiplication of $A\#H\#B$ is defined, for all $a, a' \in A$, $h, h' \in H$, $b, b' \in B$, by

$$(a\#h\#b)(a'\#h'\#b') = (x^1 \cdot a)(x^2 h_1 y^1 \cdot a')\#x^3 h_2 y^2 h'_1 z^1\#(b \cdot y^3 h'_2 z^2)(b' \cdot z^3).$$

It follows that the canonical maps $i : A\#H \to A\#H\#B$ and $j : H\#B \to A\#H\#B$, $i(a\#h) = a\#h\#1_B$ and $j(h\#b) = 1_A\#h\#b$, are algebra morphisms.

Suppose again that $H$ is a quasi-bialgebra, $A$ is a left $H$-module algebra and $F = F^1 \otimes F^2 \in H \otimes H$ is a gauge transformation with inverse $F^{-1} = G^1 \otimes G^2$. Suppose now that we also have a right $H$-module algebra $B$. If we introduce on $B$ another multiplication, by $b \star b' = (b \cdot F^1)(b' \cdot F^2)$ for all $b, b' \in B$, and denote this structure by ${}_FB$, then ${}_FB$ becomes a right $H_F$-module algebra with the same unit and right $H$-action as for $B$. We have the following type of invariance under twisting for two-sided smash products:

**Proposition 5.32** *With notation as before, we have an algebra isomorphism*

$$\varphi : A\#H\#B \cong A_{F^{-1}}\#H_F\#_FB,$$
$$\varphi(a\#h\#b) = F^1 \cdot a\#F^2 h G^1\#b \cdot G^2, \ \ \forall\, a \in A,\ h \in H,\ b \in B.$$

*In particular, by taking $A = k$, we have an algebra isomorphism*

$$H\#B \cong H_F\#_FB.$$

*Proof* This follows by a direct computation, similar to the one in Proposition 5.10. □

We now prove that the two-sided generalized smash product can be written (in two ways) as an iterated generalized smash product.

**Proposition 5.33** *Let $H$ be a quasi-bialgebra, $A$ a left $H$-module algebra, $B$ a right $H$-module algebra and $\mathbb{A}$ an $H$-bicomodule algebra. Consider the right and left $H$-comodule algebras $A \blacktriangleright\!\!< \mathbb{A}$ and $\mathbb{A} >\!\!\blacktriangleleft B$ as in Proposition 5.20. Then we have algebra isomorphisms*

$$A \blacktriangleright\!\!< \mathbb{A} >\!\!\blacktriangleleft B \equiv (A \blacktriangleright\!\!< \mathbb{A}) >\!\!\blacktriangleleft B, \quad A \blacktriangleright\!\!< \mathbb{A} >\!\!\blacktriangleleft B \equiv A \blacktriangleright\!\!< (\mathbb{A} >\!\!\blacktriangleleft B),$$

*given by the trivial identifications. In particular, we have*

$$A\#H\#B \equiv (A\#H) >\!\!\blacktriangleleft B, \quad A\#H\#B \equiv A \blacktriangleright\!\!< (H\#B).$$

*Proof* We will prove the first isomorphism, the second is similar. We compute the multiplication in $(A \blacktriangleright\!\!< \mathbb{A}) >\!\!\blacktriangleleft B$. For $a, a' \in A$, $b, b' \in B$ and $u, u' \in \mathbb{A}$ we have:

$$\begin{aligned}
&((a \blacktriangleright\!\!< u) >\!\!\blacktriangleleft b)((a' \blacktriangleright\!\!< u') >\!\!\blacktriangleleft b') \\
&\quad = (a \blacktriangleright\!\!< u)(a' \blacktriangleright\!\!< u')_{\langle 0\rangle}(1_A \blacktriangleright\!\!< \tilde{x}^1_\rho) >\!\!\blacktriangleleft (b \cdot (a' \blacktriangleright\!\!< u')_{\langle 1\rangle}\tilde{x}^2_\rho)(b' \cdot \tilde{x}^3_\rho) \\
&\quad = (a \blacktriangleright\!\!< u)(\theta^1 \cdot a' \blacktriangleright\!\!< \theta^2 u'_{\langle 0\rangle}\tilde{x}^1_\rho) >\!\!\blacktriangleleft (b \cdot \theta^3 u'_{\langle 1\rangle}\tilde{x}^2_\rho)(b' \cdot \tilde{x}^3_\rho) \\
&\quad = ((\tilde{x}^1_\lambda \cdot a)(\tilde{x}^2_\lambda u_{[-1]}\theta^1 \cdot a') \blacktriangleright\!\!< \tilde{x}^3_\lambda u_{[0]}\theta^2 u'_{\langle 0\rangle}\tilde{x}^1_\rho) >\!\!\blacktriangleleft (b \cdot \theta^3 u'_{\langle 1\rangle}\tilde{x}^2_\rho)(b' \cdot \tilde{x}^3_\rho).
\end{aligned}$$

Via the trivial identification, this is exactly the multiplication of $A \blacktriangleright\!\!< \mathbb{A} >\!\!\blacktriangleleft B$. □

**Proposition 5.34** *Let $H$ be a quasi-bialgebra, $\mathfrak{A}$ a right $H$-comodule algebra, $\mathfrak{B}$ a left $H$-comodule algebra and $\mathscr{A}$ an $H$-bimodule algebra. Consider the left and right $H$-module algebras $\mathfrak{A} \,\overline{\#}\, \mathscr{A}$ and $\mathscr{A} \,\overline{\#}\, \mathfrak{B}$ as in Propositions 5.15 and 5.17. Then we have algebra isomorphisms*

$$\mathfrak{A} >\!\!\triangleleft \mathscr{A} \triangleright\!\!< \mathfrak{B} \equiv (\mathfrak{A} \,\overline{\#}\, \mathscr{A}) \blacktriangleright\!\!< \mathfrak{B}, \quad \mathfrak{A} >\!\!\triangleleft \mathscr{A} \triangleright\!\!< \mathfrak{B} \equiv \mathfrak{A} >\!\!\blacktriangleleft (\mathscr{A} \,\overline{\#}\, \mathfrak{B}),$$

*obtained from the trivial identifications.*

*Proof* This follows by direct computations. □

**Theorem 5.35** *Let $H$ be a quasi-bialgebra, $\mathscr{A}$ an $H$-bimodule algebra, $\mathfrak{A}$ a right $H$-comodule algebra, $\mathbb{B}$ an $H$-bicomodule algebra and $\mathfrak{C}$ a left $H$-comodule algebra. Then:*

*(i) $\mathfrak{A} >\!\!\triangleleft \mathscr{A} \triangleright\!\!< \mathbb{B}$ admits a right $H$-comodule algebra structure;*
*(ii) $\mathbb{B} >\!\!\triangleleft \mathscr{A} \triangleright\!\!< \mathfrak{C}$ admits a left $H$-comodule algebra structure;*
*(iii) there is an algebra isomorphism (given by the trivial identification)*

$$(\mathfrak{A} >\!\!\triangleleft \mathscr{A} \triangleright\!\!< \mathbb{B}) >\!\!\triangleleft \mathscr{A} \triangleright\!\!< \mathfrak{C} \equiv \mathfrak{A} >\!\!\triangleleft \mathscr{A} \triangleright\!\!< (\mathbb{B} >\!\!\triangleleft \mathscr{A} \triangleright\!\!< \mathfrak{C}).$$

*Proof* By writing $\mathfrak{A} >\!\!\triangleleft \mathscr{A} \triangleright\!\!< \mathbb{B}$ as $(\mathfrak{A} \,\overline{\#}\, \mathscr{A}) \blacktriangleright\!\!< \mathbb{B}$, we obtain that this is a right $H$-comodule algebra (being a generalized smash product between a left $H$-module algebra and an $H$-bicomodule algebra), and we can explicitly write its structure:

$$\rho : \mathfrak{A} >\!\!\triangleleft \mathscr{A} \triangleright\!\!< \mathbb{B} \equiv (\mathfrak{A} \,\overline{\#}\, \mathscr{A}) \blacktriangleright\!\!< \mathbb{B} \to ((\mathfrak{A} \,\overline{\#}\, \mathscr{A}) \blacktriangleright\!\!< \mathbb{B}) \otimes H \equiv (\mathfrak{A} >\!\!\triangleleft \mathscr{A} \triangleright\!\!< \mathbb{B}) \otimes H,$$

$$\rho(\mathfrak{a} \bowtie \varphi \bowtie \mathfrak{b}) = (\mathfrak{a} \bowtie \theta^1 \cdot \varphi \bowtie \theta^2\mathfrak{b}_{\langle 0\rangle}) \otimes \theta^3\mathfrak{b}_{\langle 1\rangle}, \ \forall\, \mathfrak{a} \in \mathfrak{A},\ \varphi \in \mathscr{A},\ \mathfrak{b} \in \mathbb{B},$$
$$\Phi_\rho = (1_{\mathfrak{A}} \bowtie 1_{\mathscr{A}} \bowtie \tilde{X}^1_\rho) \otimes \tilde{X}^2_\rho \otimes \tilde{X}^3_\rho \in (\mathfrak{A} \bowtie \mathscr{A} \bowtie \mathbb{B}) \otimes H \otimes H.$$

Similarly, by writing $\mathbb{B} \bowtie \mathscr{A} \bowtie \mathfrak{C}$ as $\mathbb{B} \blacktriangleright\!\!\blacktriangleleft (\mathscr{A} \,\overline{\#}\, \mathfrak{C})$, we obtain that this is a left $H$-comodule algebra, with structure:

$$\lambda : \mathbb{B} \bowtie \mathscr{A} \bowtie \mathfrak{C} \equiv \mathbb{B} \blacktriangleright\!\!\blacktriangleleft (\mathscr{A} \,\overline{\#}\, \mathfrak{C}) \to H \otimes (\mathbb{B} \blacktriangleright\!\!\blacktriangleleft (\mathscr{A} \,\overline{\#}\, \mathfrak{C})) \equiv H \otimes (\mathbb{B} \bowtie \mathscr{A} \bowtie \mathfrak{C}),$$
$$\lambda(\mathfrak{b} \bowtie \varphi \bowtie \mathfrak{c}) = \mathfrak{b}_{[-1]}\theta^1 \otimes (\mathfrak{b}_{[0]}\theta^2 \bowtie \varphi \cdot \theta^3 \bowtie \mathfrak{c}), \ \forall\, \mathfrak{b} \in \mathbb{B},\ \varphi \in \mathscr{A},\ \mathfrak{c} \in \mathfrak{C},$$
$$\Phi_\lambda = \tilde{X}^1_\lambda \otimes \tilde{X}^2_\lambda \otimes (\tilde{X}^3_\lambda \bowtie 1_{\mathscr{A}} \bowtie 1_{\mathfrak{C}}) \in H \otimes H \otimes (\mathbb{B} \bowtie \mathscr{A} \bowtie \mathfrak{C}).$$

To prove (iii), we will use the identifications appearing in Propositions 5.33 and 5.34:

$$(\mathfrak{A} \bowtie \mathscr{A} \bowtie \mathbb{B}) \bowtie \mathscr{A} \bowtie \mathfrak{C} \equiv ((\mathfrak{A} \,\overline{\#}\, \mathscr{A}) \blacktriangleright\!\!< \mathbb{B}) \bowtie \mathscr{A} \bowtie \mathfrak{C}$$
$$\equiv ((\mathfrak{A} \,\overline{\#}\, \mathscr{A}) \blacktriangleright\!\!< \mathbb{B}) \blacktriangleright\!\!\blacktriangleleft (\mathscr{A} \,\overline{\#}\, \mathfrak{C}) \equiv (\mathfrak{A} \,\overline{\#}\, \mathscr{A}) \blacktriangleright\!\!< \mathbb{B} \blacktriangleright\!\!\blacktriangleleft (\mathscr{A} \,\overline{\#}\, \mathfrak{C}),$$

and

$$\mathfrak{A} \bowtie \mathscr{A} \bowtie (\mathbb{B} \bowtie \mathscr{A} \bowtie \mathfrak{C}) \equiv \mathfrak{A} \bowtie \mathscr{A} \bowtie (\mathbb{B} \blacktriangleright\!\!\blacktriangleleft (\mathscr{A} \,\overline{\#}\, \mathfrak{C}))$$
$$\equiv (\mathfrak{A} \,\overline{\#}\, \mathscr{A}) \blacktriangleright\!\!< (\mathbb{B} \blacktriangleright\!\!\blacktriangleleft (\mathscr{A} \,\overline{\#}\, \mathfrak{C})) \equiv (\mathfrak{A} \,\overline{\#}\, \mathscr{A}) \blacktriangleright\!\!< \mathbb{B} \blacktriangleright\!\!\blacktriangleleft (\mathscr{A} \,\overline{\#}\, \mathfrak{C}).$$

So, we have proved that the two iterated two-sided crossed products that appear in (iii) are both isomorphic as algebras (via the trivial identifications) to the two-sided generalized smash product $(\mathfrak{A} \,\overline{\#}\, \mathscr{A}) \blacktriangleright\!\!< \mathbb{B} \blacktriangleright\!\!\blacktriangleleft (\mathscr{A} \,\overline{\#}\, \mathfrak{C})$. □

By using the same results, we can obtain another relation between the two-sided crossed product and the two-sided generalized smash product. Namely, let $H$ be a quasi-bialgebra, $\mathscr{A}$ an $H$-bimodule algebra, $A$ a left $H$-module algebra, $B$ a right $H$-module algebra and $\mathbb{A}$ and $\mathbb{B}$ two $H$-bicomodule algebras. As we have seen before, $A \blacktriangleright\!\!< \mathbb{A}$ (resp. $\mathbb{B} \blacktriangleright\!\!\blacktriangleleft B$) becomes a right (resp. left) $H$-comodule algebra, so we can consider the two-sided crossed product $(A \blacktriangleright\!\!< \mathbb{A}) \bowtie \mathscr{A} \bowtie (\mathbb{B} \blacktriangleright\!\!\blacktriangleleft B)$. On the other hand, by Theorem 5.35, $\mathbb{A} \bowtie \mathscr{A} \bowtie \mathbb{B}$ becomes a right $H$-comodule algebra and a left $H$-comodule algebra, but actually, by using the explicit formulas for its structures that we gave, one can prove that it is even an $H$-bicomodule algebra, with $\Phi_{\lambda,\rho} = 1_H \otimes (1_{\mathbb{A}} \bowtie 1_{\mathscr{A}} \bowtie 1_{\mathbb{B}}) \otimes 1_H$, so we can consider the two-sided generalized smash product $A \blacktriangleright\!\!< (\mathbb{A} \bowtie \mathscr{A} \bowtie \mathbb{B}) \blacktriangleright\!\!\blacktriangleleft B$.

**Proposition 5.36** *We have an algebra isomorphism*

$$(A \blacktriangleright\!\!< \mathbb{A}) \bowtie \mathscr{A} \bowtie (\mathbb{B} \blacktriangleright\!\!\blacktriangleleft B) \equiv A \blacktriangleright\!\!< (\mathbb{A} \bowtie \mathscr{A} \bowtie \mathbb{B}) \blacktriangleright\!\!\blacktriangleleft B$$

*obtained from the trivial identification. In particular, we have*

$$(A\#H) \bowtie H^* \bowtie (H\#B) \equiv A \blacktriangleright\!\!< (H \bowtie H^* \bowtie H) \blacktriangleright\!\!\blacktriangleleft B.$$

*Proof* This can be proved by computing explicitly the multiplication rules in the two algebras and noting that they coincide. Alternatively, we provide a conceptual proof, by a sequence of identifications using the above results. We compute:

$$A \blacktriangleright\!\!< (\mathbb{A} \bowtie \mathscr{A} \bowtie \mathbb{B}) \blacktriangleright\!\!\blacktriangleleft B \equiv A \blacktriangleright\!\!< ((\mathbb{A} \bowtie \mathscr{A} \bowtie \mathbb{B}) \blacktriangleright\!\!\blacktriangleleft B)$$

$$
\begin{aligned}
&\equiv A \blacktriangleright\!\!< (((\mathbb{A}\,\overline{\#}\,\mathscr{A}) \blacktriangleright\!\!< \mathbb{B}) >\!\!\blacktriangleleft B) \equiv A \blacktriangleright\!\!< ((\mathbb{A}\,\overline{\#}\,\mathscr{A}) \blacktriangleright\!\!< (\mathbb{B} >\!\!\blacktriangleleft B)) \\
&\equiv A \blacktriangleright\!\!< (\mathbb{A} \rtimes \mathscr{A} \ltimes (\mathbb{B} >\!\!\blacktriangleleft B)) \equiv A \blacktriangleright\!\!< (\mathbb{A} >\!\!\blacktriangleleft (\mathscr{A}\,\overline{\#}\,(\mathbb{B} >\!\!\blacktriangleleft B))) \\
&\equiv (A \blacktriangleright\!\!< \mathbb{A}) >\!\!\blacktriangleleft (\mathscr{A}\,\overline{\#}\,(\mathbb{B} >\!\!\blacktriangleleft B)) \equiv (A \blacktriangleright\!\!< \mathbb{A}) \rtimes \mathscr{A} \ltimes (\mathbb{B} >\!\!\blacktriangleleft B),
\end{aligned}
$$

where the fourth and the fifth identities hold since the left $H$-comodule algebra structures on $(\mathbb{A} \rtimes \mathscr{A} \ltimes \mathbb{B}) >\!\!\blacktriangleleft B$, $\mathbb{A} \rtimes \mathscr{A} \ltimes (\mathbb{B} >\!\!\blacktriangleleft B)$ and $\mathbb{A} >\!\!\blacktriangleleft (\mathscr{A}\,\overline{\#}\,(\mathbb{B} >\!\!\blacktriangleleft B))$ coincide (via the trivial identifications). □

## 5.5 $H^*$-Hopf Bimodules

The aim of this section is to prove that a suitably defined category ${}_{H^*}^{H^*}\mathscr{M}_{H^*}^{H^*}$ of $H^*$-Hopf bimodules, where $H$ is a finite-dimensional quasi-Hopf algebra with bijective antipode, is isomorphic to a certain category of modules.

Let $H$ be a finite-dimensional quasi-bialgebra and $A$ a left $H$-module algebra. We define the category $\mathscr{M}_A^{H^*}$, whose objects are vector spaces $M$ such that $M$ is a right $H^*$-comodule (i.e. $M$ is a left $H$-module, with action denoted by $h \otimes m \mapsto h \triangleright m$) and $A$ acts on $M$ to the right (denote this action by $m \otimes a \mapsto m \cdot a$) such that $m \cdot 1_A = m$ for all $m \in M$, and the following relations hold, for all $a, a' \in A$, $m \in M$, $h \in H$:

$$(m \cdot a) \cdot a' = (X^1 \triangleright m) \cdot [(X^2 \cdot a)(X^3 \cdot a')], \tag{5.5.1}$$

$$h \triangleright (m \cdot a) = (h_1 \triangleright m) \cdot (h_2 \cdot a). \tag{5.5.2}$$

Similarly, the category ${}_A\mathscr{M}^{H^*}$ consists of vector spaces $M$ such that $M$ is a right $H^*$-comodule (i.e. a left $H$-module, with action denoted also by $\triangleright$) and $A$ acts on $M$ to the left (denote this action by $a \otimes m \mapsto a \cdot m$) such that $1_A \cdot m = m$ for all $m \in M$, and the following relations hold:

$$a \cdot (a' \cdot m) = [(x^1 \cdot a)(x^2 \cdot a')] \cdot (x^3 \triangleright m), \tag{5.5.3}$$

$$h \triangleright (a \cdot m) = (h_1 \cdot a) \cdot (h_2 \triangleright m), \tag{5.5.4}$$

for all $a, a' \in A$, $m \in M$, $h \in H$. Clearly, the category ${}_A\mathscr{M}^{H^*}$ coincides with the category ${}_{A,H}\mathscr{M}$ from Definition 5.5; by Proposition 5.7, we have that ${}_A\mathscr{M}^{H^*} \cong {}_{A\#H}\mathscr{M}$.

We need a description of $\mathscr{M}_A^{H^*}$ as a category of left modules over a right-handed smash product.

**Proposition 5.37** *Let $H$ be a quasi-Hopf algebra and $A$ a left $H$-module algebra. Define on $A$ a new multiplication, by putting*

$$a \star a' = (g^1 \cdot a')(g^2 \cdot a), \ \ \forall\, a,\ a' \in A, \tag{5.5.5}$$

*where $f^{-1} = g^1 \otimes g^2$ is given by* (3.2.16)*, and denote this new structure by $\overline{A}$. Then $\overline{A}$ becomes a right $H$-module algebra, with the same unit as $A$ and right $H$-action given by $a \cdot h = S(h) \cdot a$, for all $a \in A$, $h \in H$.*

*Proof* This is a straightforward computation, using (3.2.13) and (3.2.17). □

**Definition 5.38** Let $H$ be a quasi-bialgebra and $B$ a right $H$-module algebra. We say that $M$, a $k$-linear space, is a left $H,B$-module if

(i) $M$ is a left $H$-module with action denoted by $h\otimes m \mapsto h\triangleright m$;

(ii) $B$ acts weakly on $M$ from the left, that is, there exists a $k$-linear map $B\otimes M \to M$, denoted by $b\otimes m \mapsto b\cdot m$, such that $1_B\cdot m = m$ for all $m\in M$;

(iii) the following compatibility conditions hold:

$$b\cdot(b'\cdot m) = x^1\triangleright([(b\cdot x^2)(b'\cdot x^3)]\cdot m), \tag{5.5.6}$$

$$b\cdot(h\triangleright m) = h_1\triangleright[(b\cdot h_2)\cdot m], \tag{5.5.7}$$

for all $b,b'\in B$, $h\in H$, $m\in M$. The category of all left $H,B$-modules, with morphisms being the $H$-linear maps that preserve the $B$-action, will be denoted by ${}_{H,B}\mathcal{M}$.

**Proposition 5.39** *If $H$, $B$ are as above, the categories ${}_{H,B}\mathcal{M}$ and ${}_{H\#B}\mathcal{M}$ are isomorphic. The isomorphism is given as follows. If $M\in {}_{H\#B}\mathcal{M}$, define $h\triangleright m = (h\#1_B)\cdot m$ and $b\cdot m = (1_H\#b)\cdot m$. Conversely, if $M\in {}_{H,B}\mathcal{M}$, define $(h\#b)\cdot m = h\triangleright(b\cdot m)$.*

*Proof* This is a straightforward computation. □

**Proposition 5.40** *If $H$ is a finite-dimensional quasi-Hopf algebra with bijective antipode and $A$ is a left $H$-module algebra, then $\mathcal{M}_A^{H^*}$ is isomorphic to ${}_{H\#\overline{A}}\mathcal{M}$, where $\overline{A}$ is the right $H$-module algebra constructed in Proposition 5.37. The correspondences are given as follows (we fix $\{e_i\}_i$ a basis in $H$ with $\{e^i\}_i$ the dual basis in $H^*$):*

- *If $M\in {}_{H\#\overline{A}}\mathcal{M}$, then $M$ becomes an object in $\mathcal{M}_A^{H^*}$ with the following structures (we denote by $h\otimes m\mapsto h\triangleright m$ the left $H$-module structure of $M$ and by $a\otimes m\mapsto a\star m$ the weak left $\overline{A}$-action on $M$ arising from Proposition 5.39):*

$$M\to M\otimes H^*, \quad m\mapsto \sum_{i=1}^{n} e_i\triangleright m\otimes e^i, \ \forall\, m\in M,$$

$$M\otimes A\to M, \quad m\otimes a\mapsto m\cdot a = q^1\triangleright((S(q^2)\cdot a)\star m),$$

  *where $q_R = q^1\otimes q^2 = X^1\otimes S^{-1}(\alpha X^3)X^2\in H\otimes H$.*
- *Conversely, if $M\in\mathcal{M}_A^{H^*}$, if we denote the $H^*$-comodule structure of $M$ by $M\to M\otimes H^*$, $m\mapsto m_{(0)}\otimes m_{(1)}$, and the weak right $A$-action on $M$ by $m\otimes a\mapsto ma$, then $M$ becomes an object in ${}_{H\#\overline{A}}\mathcal{M}$ with the following structures (again via Proposition 5.39): $M$ is a left $H$-module with action $h\triangleright m = m_{(1)}(h)m_{(0)}$, and the weak left $\overline{A}$-action on $M$ is given by*

$$a\to m = (p^1\triangleright m)(p^2\cdot a), \ \forall\, a\in\overline{A},\ m\in M,$$

  *where $p_R = p^1\otimes p^2 = x^1\otimes x^2\beta S(x^3)\in H\otimes H$.*

*Proof* Assume first that $M\in {}_{H\#\overline{A}}\mathcal{M}$; then we have, by Propositions 5.39 and 5.37:

$$a\star(a'\star m) = x^1\triangleright([(g^1S(x^3)\cdot a')(g^2S(x^2)\cdot a)]\star m), \tag{5.5.8}$$

$$a\star(h\triangleright m) = h_1\triangleright[(S(h_2)\cdot a)\star m], \tag{5.5.9}$$

for all $a, a' \in A$, $h \in H$, $m \in M$. We have to prove that $M \in \mathscr{M}_A^{H^*}$. To prove (5.5.1), we compute (denoting by $Q^1 \otimes Q^2$ another copy of $q_R$):

$$\begin{aligned}
(m\cdot a)\cdot a' &= Q^1 \triangleright [(S(Q^2)\cdot a')\star(q^1 \triangleright [(S(q^2)\cdot a)\star m])] \\
&\overset{(5.5.9)}{=} Q^1 q_1^1 \triangleright [(S(q_2^1)S(Q^2)\cdot a')\star((S(q^2)\cdot a)\star m)] \\
&\overset{(5.5.8)}{=} Q^1 q_1^1 x^1 \triangleright [((g^1 S(x^3)S(q^2)\cdot a)(g^2 S(x^2)S(Q^2 q_2^1)\cdot a'))\star m] \\
&\overset{(3.2.26)}{=} q^1 X_1^1 \triangleright [((g^1 S(X^1_{(2,2)})S(q_2^2)f^1 X^2\cdot a) \\
&\qquad (g^2 S(X^1_{(2,1)})S(q_1^2)f^2 X^3\cdot a'))\star m] \\
&\overset{(3.2.13)}{=} q^1 X_1^1 \triangleright [((S(q^2 X_2^1)_1 X^2\cdot a)(S(q^2 X_2^1)_2 X^3\cdot a'))\star m] \\
&= q^1 X_1^1 \triangleright [(S(q^2 X_2^1)\cdot((X^2\cdot a)(X^3\cdot a')))\star m] \\
&= q^1 \triangleright [X_1^1 \triangleright [(S(X_2^1)\cdot(S(q^2)\cdot((X^2\cdot a)(X^3\cdot a'))))\star m]] \\
&\overset{(5.5.9)}{=} q^1 \triangleright [(S(q^2)\cdot((X^2\cdot a)(X^3\cdot a')))\star(X^1\triangleright m)] \\
&= (X^1 \triangleright m)\cdot((X^2\cdot a)(X^3\cdot a')). \text{ QED}
\end{aligned}$$

To prove (5.5.2), we compute:

$$\begin{aligned}
(h_1 \triangleright m)\cdot(h_2\cdot a) &= q^1 \triangleright ((S(q^2)h_2\cdot a)\star(h_1\triangleright m)) \\
&\overset{(5.5.9)}{=} q^1 h_{(1,1)} \triangleright ((S(h_{(1,2)})S(q^2)h_2\cdot a)\star m) \\
&\overset{(3.2.21)}{=} hq^1 \triangleright ((S(q^2)\cdot a)\star m) = h\triangleright(m\cdot a).
\end{aligned}$$

Obviously $m\cdot 1_A = m$, for all $m \in M$, hence indeed $M \in \mathscr{M}_A^{H^*}$.

Conversely, assume that $M \in \mathscr{M}_A^{H^*}$, that is,

$$(ma)a' = (X^1 \triangleright m)[(X^2\cdot a)(X^3\cdot a')], \tag{5.5.10}$$

$$h\triangleright(ma) = (h_1\triangleright m)(h_2\cdot a), \tag{5.5.11}$$

for all $m \in M$, $a, a' \in A$, $h \in H$, and we have to prove that

$$a \rightarrow (a' \rightarrow m) = x^1 \triangleright ([(g^1 S(x^3)\cdot a')(g^2 S(x^2)\cdot a)] \rightarrow m), \tag{5.5.12}$$

$$a \rightarrow (h\triangleright m) = h_1 \triangleright [(S(h_2)\cdot a) \rightarrow m], \tag{5.5.13}$$

for all $a, a' \in A$, $h \in H$, $m \in M$.

To prove (5.5.12), we compute (denoting by $P^1 \otimes P^2$ another copy of $p_R$):

$$\begin{aligned}
a \rightarrow (a' \rightarrow m) &= (p^1 \triangleright [(P^1 \triangleright m)(P^2\cdot a')])(p^2\cdot a) \\
&\overset{(5.5.11)}{=} [(p_1^1 P^1 \triangleright m)(p_2^1 P^2\cdot a')](p^2\cdot a) \\
&\overset{(5.5.10)}{=} (X^1 p_1^1 P^1 \triangleright m)[(X^2 p_2^1 P^2\cdot a')(X^3 p^2\cdot a)] \\
&\overset{(3.2.25)}{=} (x_1^1 p^1 \triangleright m)[(x^1_{(2,1)} p_1^2 g^1 S(x^3)\cdot a')(x^1_{(2,2)} p_2^2 g^2 S(x^2)\cdot a)] \\
&\overset{(5.5.11)}{=} x^1 \triangleright [(p^1 \triangleright m)[(p_1^2 g^1 S(x^3)\cdot a')(p_2^2 g^2 S(x^2)\cdot a)]] \\
&= x^1 \triangleright [((g^1 S(x^3)\cdot a')(g^2 S(x^2)\cdot a)) \rightarrow m]. \text{ QED}
\end{aligned}$$

To prove (5.5.13), we compute:

$$h_1 \triangleright [(S(h_2)\cdot a) \rightarrow m] = h_1 \triangleright [(p^1 \triangleright m)(p^2 S(h_2)\cdot a)]$$

$$\overset{(5.5.11)}{=} (h_{(1,1)}p^1 \triangleright m)(h_{(1,2)}p^2 S(h_2) \cdot a)$$
$$\overset{(3.2.21)}{=} (p^1 h \triangleright m)(p^2 \cdot a) = a \rightarrow (h \triangleright m).$$

Obviously $1_A \rightarrow m = m$, for all $m \in M$, hence indeed $M \in {}_{H\#\overline{A}}\mathcal{M}$.

To show that $\mathcal{M}_A^{H^*} \cong {}_{H\#\overline{A}}\mathcal{M}$, the only things left to prove are the following:

(1) If $M \in {}_{H\#\overline{A}}\mathcal{M}$, then $a \rightarrow m = a \star m$, for all $a \in A$, $m \in M$.
(2) If $M \in \mathcal{M}_A^{H^*}$, then $m \cdot a = ma$, for all $a \in A$, $m \in M$.

To prove (1), we compute:

$$\begin{aligned} a \rightarrow m &= (p^1 \triangleright m) \cdot (p^2 \cdot a) \\ &= q^1 \triangleright [(S(q^2)p^2 \cdot a) \star (p^1 \triangleright m)] \\ &\overset{(5.5.9)}{=} q^1 p_1^1 \triangleright [(S(p_2^1)S(q^2)p^2 \cdot a) \star m] \overset{(3.2.23)}{=} a \star m. \end{aligned}$$

To prove (2), we compute:

$$\begin{aligned} m \cdot a &= q^1 \triangleright [(S(q^2) \cdot a) \rightarrow m] \\ &= q^1 \triangleright [(p^1 \triangleright m)(p^2 S(q^2) \cdot a)] \\ &\overset{(5.5.11)}{=} (q_1^1 p^1 \triangleright m)(q_2^1 p^2 S(q^2) \cdot a) \overset{(3.2.23)}{=} ma, \end{aligned}$$

and the proof is finished. □

We also need the description of left modules over a two-sided smash product.

**Definition 5.41** Let $H$ be a quasi-bialgebra, $A$ a left $H$-module algebra and $B$ a right $H$-module algebra. Define the category ${}_{A,H,B}\mathcal{M}$ as follows: an object in this category is a left $H$-module $M$, with action denoted by $h \otimes m \mapsto h \triangleright m$, and we have left weak actions of $A$ and $B$ on $M$, denoted by $a \otimes m \mapsto a \cdot m$ and $b \otimes m \mapsto b \cdot m$, such that:

(i) $M \in {}_{A\#H}\mathcal{M}$, that is, the relations (5.5.3) and (5.5.4) hold;
(ii) $M \in {}_{H\#B}\mathcal{M}$, that is, the relations (5.5.6) and (5.5.7) hold;
(iii) the following compatibility condition holds:

$$b \cdot (a \cdot m) = (y^1 \cdot a) \cdot [y^2 \triangleright ((b \cdot y^3) \cdot m)], \tag{5.5.14}$$

for all $a \in A$, $b \in B$, $m \in M$. The morphisms in this category are the $H$-linear maps compatible with the two weak actions.

**Proposition 5.42** *If $H$, $A$, $B$ are as above, then ${}_{A\#H\#B}\mathcal{M} \cong {}_{A,H,B}\mathcal{M}$, the isomorphism being given as follows:*

- *If $M \in {}_{A\#H\#B}\mathcal{M}$, define $a \cdot m = (a\#1_H\#1_B) \cdot m$, $h \triangleright m = (1_A\#h\#1_B) \cdot m$, $b \cdot m = (1_A\#1_H\#b) \cdot m$;*
- *Conversely, if $M \in {}_{A,H,B}\mathcal{M}$, define $(a\#h\#b) \cdot m = a \cdot (h \triangleright (b \cdot m))$.*

*Proof* Straightforward computation, using the formula for the multiplication in $A\#H\#B$. Let us point out how the condition (5.5.14) occurs:

$$b \cdot (a \cdot m) = (1_A\#1_H\#b) \cdot ((a\#1_H\#1_B) \cdot m)$$

$$= [(1_A\#1_H\#b)(a\#1_H\#1_B)]\cdot m$$
$$= (y^1\cdot a\#y^2\#b\cdot y^3)\cdot m = (y^1\cdot a)\cdot(y^2\triangleright((b\cdot y^3)\cdot m)),$$

which is exactly (5.5.14). □

Let $H$ be a finite-dimensional quasi-bialgebra and $A$, $D$ two left $H$-module algebras. It is obvious that ${}_A\mathcal{M}^{H^*}$ coincides with the category of left $A$-modules within the monoidal category ${}_H\mathcal{M}$, and similarly $\mathcal{M}_D^{H^*}$ coincides with the category of right $D$-modules within ${}_H\mathcal{M}$. Hence, we can introduce the following new category:

**Definition 5.43** If $H$, $A$, $D$ are as above, define ${}_A\mathcal{M}_D^{H^*}$ as the category of $A$–$D$-bimodules within the monoidal category ${}_H\mathcal{M}$, that is, $M\in{}_A\mathcal{M}_D^{H^*}$ if and only if $M\in{}_A\mathcal{M}^{H^*}$, $M\in\mathcal{M}_D^{H^*}$ and the following relation holds, for all $a\in A$, $m\in M$, $d\in D$:

$$(a\cdot m)\cdot d = (X^1\cdot a)\cdot[(X^2\triangleright m)\cdot(X^3\cdot d)], \tag{5.5.15}$$

where $a\otimes m\mapsto a\cdot m$ and $m\otimes d\mapsto m\cdot d$ are the weak actions.

**Proposition 5.44** *Let $H$ be a finite-dimensional quasi-Hopf algebra with bijective antipode and $A$, $D$ two left $H$-module algebras. Then we have an isomorphism of categories ${}_A\mathcal{M}_D^{H^*}\cong{}_{A\#H\#\overline{D}}\mathcal{M}$, where the algebra $\overline{D}\in\mathcal{M}_H$ is as in Proposition 5.37.*

*Proof* Since ${}_A\mathcal{M}^{H^*}\cong{}_{A\#H}\mathcal{M}$ and $\mathcal{M}_D^{H^*}\cong{}_{H\#\overline{D}}\mathcal{M}$, the only thing left to prove is that the compatibility (5.5.14) in ${}_{A,H,\overline{D}}\mathcal{M}$ is equivalent to the compatibility (5.5.15) in ${}_A\mathcal{M}_D^{H^*}$. Let us first note the following easy consequences of (3.1.9), (3.2.1):

$$X^1p_1^1\otimes X^2p_2^1\otimes X^3p^2 = y^1\otimes y_1^2p^1\otimes y_2^2p^2S(y^3), \tag{5.5.16}$$
$$q_1^1y^1\otimes q_2^1y^2\otimes S(q^2y^3) = X^1\otimes q^1X_1^2\otimes S(q^2X_2^2)X^3, \tag{5.5.17}$$

where $p_R = p^1\otimes p^2$ and $q_R = q^1\otimes q^2$ are again the standard elements in $H\otimes H$.

Now let $M\in{}_A\mathcal{M}_D^{H^*}$, with right $D$-action on $M$ denoted by $m\otimes d\mapsto m\cdot d$. Then, by Proposition 5.40, the weak left $\overline{D}$-action on $M$ is given by $d\to m = (p^1\triangleright m)\cdot(p^2\cdot d)$. We check (5.5.14); we compute:

$$\begin{aligned}
d\to(a\cdot m) &= (p^1\triangleright(a\cdot m))\cdot(p^2\cdot d)\\
&\overset{(5.5.4)}{=} [(p_1^1\cdot a)\cdot(p_2^1\triangleright m)]\cdot(p^2\cdot d)\\
&\overset{(5.5.15)}{=} (X^1p_1^1\cdot a)\cdot[(X^2p_2^1\triangleright m)\cdot(X^3p^2\cdot d)]\\
&\overset{(5.5.16)}{=} (y^1\cdot a)\cdot[(y_1^2p^1\triangleright m)\cdot(y_2^2p^2S(y^3)\cdot d)]\\
&\overset{(5.5.2)}{=} (y^1\cdot a)\cdot[y^2\triangleright((p^1\triangleright m)\cdot(p^2S(y^3)\cdot d))]\\
&= (y^1\cdot a)\cdot[y^2\triangleright((S(y^3)\cdot d)\to m)]\\
&= (y^1\cdot a)\cdot[y^2\triangleright((d\cdot y^3)\to m)].
\end{aligned}$$

Conversely, assume that $M\in{}_{A\#H\#\overline{D}}\mathcal{M}$, and denote the actions of $A$, $H$, $\overline{D}$ on $M$ by $a\cdot m$, $h\triangleright m$, $d\cdot m$, respectively. Then, by Proposition 5.40, the right $D$-action on $M$ is given by $m\cdot d = q^1\triangleright((S(q^2)\cdot d)\cdot m)$. To check (5.5.15), we compute:

$$(a\cdot m)\cdot d = q^1\triangleright[(S(q^2)\cdot d)\cdot(a\cdot m)]$$

$$
\begin{aligned}
&\overset{(5.5.14)}{=} q^1 \triangleright [(y^1 \cdot a) \cdot (y^2 \triangleright ((S(q^2) \cdot d \cdot y^3) \cdot m))] \\
&= q^1 \triangleright [(y^1 \cdot a) \cdot (y^2 \triangleright ((S(q^2 y^3) \cdot d) \cdot m))] \\
&\overset{(5.5.4)}{=} (q_1^1 y^1 \cdot a) \cdot [q_2^1 y^2 \triangleright ((S(q^2 y^3) \cdot d) \cdot m)] \\
&\overset{(5.5.17)}{=} (X^1 \cdot a) \cdot [q^1 X_1^2 \triangleright ((S(q^2 X_2^2) X^3 \cdot d) \cdot m)] \\
&= (X^1 \cdot a) \cdot [q^1 X_1^2 \triangleright ((S(q^2) X^3 \cdot d \cdot X_2^2) \cdot m)] \\
&\overset{(5.5.7)}{=} (X^1 \cdot a) \cdot [q^1 \triangleright ((S(q^2) X^3 \cdot d) \cdot (X^2 \triangleright m))] \\
&= (X^1 \cdot a) \cdot [(X^2 \triangleright m) \cdot (X^3 \cdot d)],
\end{aligned}
$$

and the proof is finished. □

Let $H$ be a finite-dimensional quasi-bialgebra and $\mathscr{A}$, $\mathscr{D}$ two $H$-bimodule algebras. Define the category ${}^{H^*}_{\mathscr{A}}\mathscr{M}^{H^*}_{\mathscr{D}}$ as the category of $\mathscr{A}$–$\mathscr{D}$-bimodules within the monoidal category ${}_H\mathscr{M}_H$. By regarding $\mathscr{A}$ and $\mathscr{D}$ as left module algebras over $H \otimes H^{\mathrm{op}}$, it is easy to see that ${}^{H^*}_{\mathscr{A}}\mathscr{M}^{H^*}_{\mathscr{D}} \cong {}_{\mathscr{A}}\mathscr{M}^{(H\otimes H^{\mathrm{op}})^*}_{\mathscr{D}}$. Hence, as a consequence of Proposition 5.44, we finally obtain:

**Theorem 5.45** *If $H$ is a finite-dimensional quasi-Hopf algebra with bijective antipode and $\mathscr{A}$, $\mathscr{D}$ are two $H$-bimodule algebras, we have an isomorphism of categories ${}^{H^*}_{\mathscr{A}}\mathscr{M}^{H^*}_{\mathscr{D}} \cong {}_{\mathscr{A}\#(H\otimes H^{\mathrm{op}})\#\overline{\mathscr{D}}}\mathscr{M}$. In particular, ${}^{H^*}_{H^*}\mathscr{M}^{H^*}_{H^*} \cong {}_{H^*\#(H\otimes H^{\mathrm{op}})\#\overline{H^*}}\mathscr{M}$.*

## 5.6 Diagonal Crossed Products

Let $H$ be a quasi-Hopf algebra with bijective antipode, $\mathscr{A}$ an $H$-bimodule algebra and $(\delta, \Psi)$ a two-sided coaction of $H$ on an algebra $\mathbb{A}$. Define the following elements in $H^{\otimes 2} \otimes \mathbb{A} \otimes H^{\otimes 2}$:

$$\Omega_\delta = \Omega_\delta^1 \otimes \cdots \otimes \Omega_\delta^5 = \overline{\Psi}^1 \otimes \overline{\Psi}^2 \otimes \overline{\Psi}^3 \otimes S^{-1}(f^1 \overline{\Psi}^4) \otimes S^{-1}(f^2 \overline{\Psi}^5), \tag{5.6.1}$$

$$\Omega'_\delta = \Omega'^1_\delta \otimes \cdots \otimes \Omega'^5_\delta = S^{-1}(\Psi^1 g^1) \otimes S^{-1}(\Psi^2 g^2) \otimes \Psi^3 \otimes \Psi^4 \otimes \Psi^5. \tag{5.6.2}$$

Here $f = f^1 \otimes f^2$ is the twist defined in (3.2.15) and $f^{-1} = g^1 \otimes g^2$ is its inverse.

We denote by $\mathscr{A} \bowtie_\delta \mathbb{A}$ and $\mathbb{A} \bowtie_\delta \mathscr{A}$ the $k$-vector spaces $\mathscr{A} \otimes \mathbb{A}$ and $\mathbb{A} \otimes \mathscr{A}$, respectively, furnished with the multiplications given by:

$$
\begin{aligned}
&(\varphi \bowtie_\delta u)(\varphi' \bowtie_\delta u') \\
&\quad = (\Omega_\delta^1 \cdot \varphi \cdot \Omega_\delta^5)(\Omega_\delta^2 u_{(-1)} \cdot \varphi' \cdot S^{-1}(u_{(1)}) \Omega_\delta^4) \bowtie_\delta \Omega_\delta^3 u_{(0)} u',
\end{aligned} \tag{5.6.3}
$$

$$
\begin{aligned}
&(u \bowtie_\delta \varphi)(u' \bowtie_\delta \varphi') \\
&\quad = u u'_{(0)} \Omega'^3_\delta \bowtie_\delta (\Omega'^2_\delta S^{-1}(u'_{(-1)}) \cdot \varphi \cdot u'_{(1)} \Omega'^4_\delta)(\Omega'^1_\delta \cdot \varphi' \cdot \Omega'^5_\delta),
\end{aligned} \tag{5.6.4}
$$

respectively, for all $u, u' \in \mathbb{A}$ and $\varphi, \varphi' \in \mathscr{A}$, where we write $\varphi \bowtie_\delta u$ and $u \bowtie_\delta \varphi$ in place of $\varphi \otimes u$ and $u \otimes \varphi$, respectively, to distinguish the new algebraic structures, and where $\Omega_\delta = \Omega_\delta^1 \otimes \cdots \otimes \Omega_\delta^5$ and $\Omega'_\delta = \Omega'^1_\delta \otimes \cdots \otimes \Omega'^5_\delta$ are the elements defined by (5.6.1) and (5.6.2), respectively. We call $\mathscr{A} \bowtie_\delta \mathbb{A}$ and $\mathbb{A} \bowtie_\delta \mathscr{A}$ the left and right generalized diagonal crossed products, respectively, between $\mathscr{A}$ and $\mathbb{A}$.

The following (technical) lemma, expressing some relations fulfilled by the elements $\Omega_\delta$ and $\Omega'_\delta$, will be essential in what follows. It will help to prove that the generalized diagonal crossed products defined above are associative algebras, and moreover it will allow us to regard an $H$-bicomodule algebra $\mathbb{A}$, in two ways, as a left $H\otimes H^{\rm op}$-comodule algebra.

**Lemma 5.46** *Let $H$ be a quasi-Hopf algebra with bijective antipode, $\mathbb{A}$ an algebra and $(\delta,\Psi)$ a two-sided coaction of $H$ on $\mathbb{A}$.*

*(a) Let $\Omega_\delta=\Omega_\delta^1\otimes\cdots\otimes\Omega_\delta^5=\overline{\Omega}_\delta^1\otimes\cdots\otimes\overline{\Omega}_\delta^5$ be the element defined by* (5.6.1). *Then for all $u\in\mathbb{A}$ the following relations hold:*

$$\begin{aligned}
&\Omega_\delta^1u_{(-1)}\otimes\Omega_\delta^2u_{(0,-1)}\otimes\Omega_\delta^3u_{(0,0)}\otimes S^{-1}(u_{(0,1)})\Omega_\delta^4\otimes S^{-1}(u_{(1)})\Omega_\delta^5\\
&\qquad=u_{(-1)_1}\Omega_\delta^1\otimes u_{(-1)_2}\Omega_\delta^2\otimes u_{(0)}\Omega_\delta^3\otimes\Omega_\delta^4S^{-1}(u_{(1)})_2\otimes\Omega_\delta^5S^{-1}(u_{(1)})_1,
\end{aligned}\tag{5.6.5}$$

$$\begin{aligned}
&X^1(\overline{\Omega}_\delta^1)_1\Omega_\delta^1\otimes X^2(\overline{\Omega}_\delta^1)_2\Omega_\delta^2\otimes X^3\overline{\Omega}_\delta^2(\Omega_\delta^3)_{(-1)}\otimes\overline{\Omega}_\delta^3\Omega^3_{(0)}\otimes S^{-1}((\Omega_\delta^3)_{(1)})\overline{\Omega}_\delta^4x^3\\
&\qquad\otimes\Omega_\delta^4(\overline{\Omega}_\delta^5)_2x^2\otimes\Omega_\delta^5(\overline{\Omega}_\delta^5)_1x^1=\overline{\Omega}_\delta^1\otimes(\overline{\Omega}_\delta^2)_1\Omega_\delta^1\otimes(\overline{\Omega}_\delta^2)_2\Omega_\delta^2\\
&\qquad\qquad\otimes\overline{\Omega}_\delta^3\Omega_\delta^3\otimes\Omega_\delta^4(\overline{\Omega}_\delta^4)_2\otimes\Omega_\delta^5(\overline{\Omega}_\delta^4)_1\otimes\overline{\Omega}_\delta^5.
\end{aligned}\tag{5.6.6}$$

*(b) Let $\Omega'_\delta=\Omega_\delta'^1\otimes\cdots\otimes\Omega_\delta'^5=\overline{\Omega}_\delta'^1\otimes\cdots\otimes\overline{\Omega}_\delta'^5$ be the element defined by* (5.6.2). *Then for all $u\in\mathbb{A}$ the following relations hold:*

$$\begin{aligned}
&\Omega_\delta'^1S^{-1}(u_{(-1)})\otimes\Omega_\delta'^2S^{-1}(u_{(0,-1)})\otimes u_{(0,0)}\Omega_\delta'^3\otimes u_{(0,1)}\Omega_\delta'^4\otimes u_{(1)}\Omega_\delta'^5\\
&\qquad=S^{-1}(u_{(-1)})_2\Omega_\delta'^1\otimes S^{-1}(u_{(-1)})_1\Omega_\delta'^2\otimes\Omega_\delta'^3u_{(0)}\otimes\Omega_\delta'^4u_{(1)_1}\otimes\Omega_\delta'^5u_{(1)_2},
\end{aligned}\tag{5.6.7}$$

$$\begin{aligned}
&X^3\overline{\Omega}_\delta'^1\otimes X^2(\overline{\Omega}_\delta'^2)_2\Omega_\delta'^1\otimes X^1(\overline{\Omega}_\delta'^2)_1\Omega_\delta'^2\otimes\Omega_\delta'^3\overline{\Omega}_\delta'^3\otimes\Omega_\delta'^4(\overline{\Omega}_\delta'^4)_1x^1\\
&\qquad\otimes\Omega_\delta'^5(\overline{\Omega}_\delta'^4)_2x^2\otimes\overline{\Omega}_\delta'^5x^3=(\overline{\Omega}_\delta'^1)_1\Omega_\delta'^1\otimes(\overline{\Omega}_\delta'^1)_2\Omega_\delta'^2\otimes\overline{\Omega}_\delta'^2S^{-1}((\Omega_\delta'^3)_{(-1)})\\
&\qquad\qquad\otimes(\Omega_\delta'^3)_{(0)}\overline{\Omega}_\delta'^3\otimes(\Omega_\delta'^3)_{(1)}\overline{\Omega}_\delta'^4\otimes\Omega_\delta'^4(\overline{\Omega}_\delta'^5)_1\otimes\Omega_\delta'^5(\overline{\Omega}_\delta'^5)_2.
\end{aligned}\tag{5.6.8}$$

*Proof* We will prove only (a), (b) being similar. The relation (5.6.5) follows easily by applying (5.6.1), (4.4.5) and (3.2.13); the details are left to the reader. We now prove (5.6.6). We compute:

$$\begin{aligned}
&X^1(\overline{\Omega}_\delta^1)_1\Omega_\delta^1\otimes X^2(\overline{\Omega}_\delta^1)_2\Omega_\delta^2\otimes X^3\overline{\Omega}_\delta^2(\Omega_\delta^3)_{(-1)}\otimes\overline{\Omega}_\delta^3\Omega^3_{(0)}\\
&\qquad\otimes S^{-1}((\Omega_\delta^3)_{(1)})\overline{\Omega}_\delta^4x^3\otimes\Omega_\delta^4(\overline{\Omega}_\delta^5)_2x^2\otimes\Omega_\delta^5(\overline{\Omega}_\delta^5)_1x^1\\
&\overset{(5.6.1)}{\underset{(3.2.13)}{=}}X^1\overline{\Psi}_1^1\overline{\Upsilon}^1\otimes X^2\overline{\Psi}_2^1\overline{\Upsilon}^2\otimes X^3\overline{\Psi}^2\overline{\Upsilon}^3_{(-1)}\otimes\overline{\Psi}^3\overline{\Upsilon}^3_{(0)}\otimes S^{-1}(f^1\overline{\Psi}^4\overline{\Upsilon}^3_{(1)})x^3\\
&\qquad\otimes S^{-1}(F^1f_1^2\overline{\Psi}_1^5\overline{\Upsilon}^4)x^2\otimes S^{-1}(F^2f_2^2\overline{\Psi}_2^5\overline{\Upsilon}^5)x^1\\
&\overset{(4.4.6)}{=}\overline{\Upsilon}^1\otimes\overline{\Upsilon}_1^2\overline{\Psi}^1\otimes\overline{\Upsilon}_2^2\overline{\Psi}^2\otimes\overline{\Upsilon}^3\overline{\Psi}^3\otimes S^{-1}(S(x^3)f^1X^1\overline{\Upsilon}_1^4\overline{\Psi}^4)\\
&\qquad\otimes S^{-1}(S(x^2)F^1f_1^2X^2\overline{\Upsilon}_2^4\overline{\Psi}^5)\otimes S^{-1}(S(x^1)F^2f_2^2X^3\overline{\Upsilon}^5)\\
&\overset{(3.1.5),(3.2.17)}{\underset{(3.2.13)}{=}}\overline{\Upsilon}^1\otimes\overline{\Upsilon}_1^2\overline{\Psi}^1\otimes\overline{\Upsilon}_2^2\overline{\Psi}^2\otimes\overline{\Upsilon}^3\overline{\Psi}^3\otimes S^{-1}(f^1\overline{\Psi}^4)S^{-1}(F^1\overline{\Upsilon}^4)_2\\
&\qquad\otimes S^{-1}(f^2\overline{\Psi}^5)S^{-1}(F^1\overline{\Upsilon}^4)_1\otimes S^{-1}(F^2\overline{\Upsilon}^5)\\
&\overset{(5.6.1)}{=}\overline{\Omega}_\delta^1\otimes(\overline{\Omega}_\delta^2)_1\Omega_\delta^1\otimes(\overline{\Omega}_\delta^2)_2\Omega_\delta^2\otimes\overline{\Omega}_\delta^3\Omega_\delta^3\otimes\Omega_\delta^4(\overline{\Omega}_\delta^4)_2\otimes\Omega_\delta^5(\overline{\Omega}_\delta^4)_1\otimes\overline{\Omega}_\delta^5,
\end{aligned}$$

as claimed. We denoted by $\overline{\Upsilon}^1 \otimes \cdots \otimes \overline{\Upsilon}^5$ another copy of $\Psi^{-1}$ and by $F^1 \otimes F^2$ another copy of the Drinfeld twist $f$ defined in (3.2.15). □

Let $\mathbb{A}$ be an $H$-bicomodule algebra and let $(\delta, \Psi) = (\delta_{l/r}, \Psi_{l/r})$ be the two-sided coactions defined by (4.4.24) and (4.4.25), respectively. For simplicity we denote $\Omega = \Omega_{\delta_l}$, $\omega = \Omega_{\delta_r}$, $\Omega' = \Omega'_{\delta_l}$ and $\omega' = \Omega'_{\delta_r}$. Concretely, the elements $\Omega, \omega \in H^{\otimes 2} \otimes \mathbb{A} \otimes H^{\otimes 2}$ become

$$\begin{aligned}\Omega = {} & (\tilde{X}^1_\rho)_{[-1]_1} \tilde{x}^1_\lambda \theta^1 \otimes (\tilde{X}^1_\rho)_{[-1]_2} \tilde{x}^2_\lambda \theta^2_{[-1]} \\ & \otimes (\tilde{X}^1_\rho)_{[0]} \tilde{x}^3_\lambda \theta^2_{[0]} \otimes S^{-1}(f^1 \tilde{X}^2_\rho \theta^3) \otimes S^{-1}(f^2 \tilde{X}^3_\rho),\end{aligned} \tag{5.6.9}$$

$$\begin{aligned}\omega = {} & \tilde{x}^1_\lambda \otimes \tilde{x}^2_\lambda \Theta^1 \otimes (\tilde{x}^3_\lambda)_{\langle 0\rangle} \tilde{X}^1_\rho \Theta^2_{\langle 0\rangle} \\ & \otimes S^{-1}(f^1 (\tilde{x}^3_\lambda)_{\langle 1\rangle_1} \tilde{X}^2_\rho \Theta^2_{\langle 1\rangle}) \otimes S^{-1}(f^2 (\tilde{x}^3_\lambda)_{\langle 1\rangle_2} \tilde{X}^3_\rho \Theta^3),\end{aligned} \tag{5.6.10}$$

where $\Phi_\rho = \tilde{X}^1_\rho \otimes \tilde{X}^2_\rho \otimes \tilde{X}^3_\rho$, $\Phi_\lambda^{-1} = \tilde{x}^1_\lambda \otimes \tilde{x}^2_\lambda \otimes \tilde{x}^3_\lambda$, $\Phi_{\lambda,\rho} = \Theta^1 \otimes \Theta^2 \otimes \Theta^3$, $\Phi_{\lambda,\rho}^{-1} = \theta^1 \otimes \theta^2 \otimes \theta^3$ and $f = f^1 \otimes f^2$ is the twist defined in (3.2.15).

For further use we record the fact that the formulas in Lemma 5.46 (a) specialize to $(\delta_{l/r}, \Psi_{l/r})$ as follows (for all $u \in \mathbb{A}$):

$$\begin{aligned}& \Omega^1 u_{\langle 0\rangle_{[-1]}} \otimes \Omega^2 u_{\langle 0\rangle_{[0]_{\langle 0\rangle_{[-1]}}}} \otimes \Omega^3 u_{\langle 0\rangle_{[0]_{\langle 0\rangle_{[0]}}}} \otimes S^{-1}(u_{\langle 0\rangle_{[0]_{\langle 1\rangle}}}) \Omega^4 \otimes S^{-1}(u_{\langle 1\rangle}) \Omega^5 \\ & \quad = u_{\langle 0\rangle_{[-1]_1}} \Omega^1 \otimes u_{\langle 0\rangle_{[-1]_2}} \Omega^2 \otimes u_{\langle 0\rangle_{[0]}} \Omega^3 \otimes \Omega^4 S^{-1}(u_{\langle 1\rangle})_2 \otimes \Omega^5 S^{-1}(u_{\langle 1\rangle})_1,\end{aligned} \tag{5.6.11}$$

$$\begin{aligned}& X^1 \overline{\Omega}^1_1 \Omega^1 \otimes X^2 \overline{\Omega}^1_2 \Omega^2 \otimes X^3 \overline{\Omega}^2 \Omega^3_{\langle 0\rangle_{[-1]}} \otimes \overline{\Omega}^3 \Omega^3_{\langle 0\rangle_{[0]}} \otimes S^{-1}(\Omega^3_{\langle 1\rangle}) \overline{\Omega}^4 x^3 \otimes \Omega^4 \overline{\Omega}^5_2 x^2 \\ & \quad \otimes \Omega^5 \overline{\Omega}^5_1 x^1 = \overline{\Omega}^1 \otimes \overline{\Omega}^2_1 \Omega^1 \otimes \overline{\Omega}^2_2 \Omega^2 \otimes \overline{\Omega}^3 \Omega^3 \otimes \Omega^4 \overline{\Omega}^4_2 \otimes \Omega^5 \overline{\Omega}^4_1 \otimes \overline{\Omega}^5,\end{aligned} \tag{5.6.12}$$

and

$$\begin{aligned}& \omega^1 u_{[-1]} \otimes \omega^2 u_{[0]_{\langle 0\rangle_{[-1]}}} \otimes \omega^3 u_{[0]_{\langle 0\rangle_{[0]_{\langle 0\rangle}}}} \otimes S^{-1}(u_{[0]_{\langle 0\rangle_{[0]_{\langle 1\rangle}}}}) \omega^4 \otimes S^{-1}(u_{[0]_{\langle 1\rangle}}) \omega^5 \\ & \quad = u_{[-1]_1} \omega^1 \otimes u_{[-1]_2} \omega^2 \otimes u_{[0]_{\langle 0\rangle}} \omega^3 \otimes \omega^4 S^{-1}(u_{[0]_{\langle 1\rangle}})_2 \otimes \omega^5 S^{-1}(u_{[0]_{\langle 1\rangle}})_1,\end{aligned} \tag{5.6.13}$$

$$\begin{aligned}& \overline{\omega}^1_1 \omega^1 \otimes \overline{\omega}^1_2 \omega^2 \otimes \overline{\omega}^2 \omega^3_{[-1]} \otimes \overline{\omega}^3 \omega^3_{[0]_{\langle 0\rangle}} \otimes S^{-1}(\omega^3_{[0]_{\langle 1\rangle}}) \overline{\omega}^4 \otimes \omega^4 \overline{\omega}^5_2 \otimes \omega^5 \overline{\omega}^5_1 \\ & \quad = x^1 \overline{\omega}^1 \otimes x^2 \overline{\omega}^2_1 \omega^1 \otimes x^3 \overline{\omega}^2_2 \omega^2 \otimes \overline{\omega}^3 \omega^3 \otimes \omega^4 \overline{\omega}^4_2 X^3 \otimes \omega^5 \overline{\omega}^4_1 X^2 \otimes \overline{\omega}^5 X^1,\end{aligned} \tag{5.6.14}$$

respectively, where we denoted by $\Omega = \Omega^1 \otimes \cdots \otimes \Omega^5 = \overline{\Omega}^1 \otimes \cdots \otimes \overline{\Omega}^5$ the element defined in (5.6.9) and by $\omega = \omega^1 \otimes \cdots \otimes \omega^5 = \overline{\omega}^1 \otimes \cdots \otimes \overline{\omega}^5$ the element defined in (5.6.10).

If $(\mathbb{A}, \lambda, \rho, \Phi_\lambda, \Phi_\rho, \Phi_{\lambda,\rho})$ is an $H$-bicomodule algebra then, as we mentioned before, $\mathbb{A}^{\rm op,cop} := (\mathbb{A}^{\rm op}, \tau_{\mathbb{A},H} \circ \rho, \tau_{H,\mathbb{A}} \circ \lambda, \Phi_\rho^{321}, \Phi_\lambda^{321}, \Phi_{\lambda,\rho}^{321})$ is an $H^{\rm op,cop}$-bicomodule algebra. Moreover, in $H^{\rm op,cop}$ we have that the Drinfeld twist (defined for an arbitrary quasi-Hopf algebra in (3.2.15)) is given by $f_{\rm op,cop} = f_{21}^{-1} = g^2 \otimes g^1$, where $f$ is the Drinfeld twist of $H$. Now, if we denote by $\Omega_{\rm op,cop}$ and $\omega_{\rm op,cop}$ the elements $\Omega_{\delta_{l/r}}$ corresponding to the $H^{\rm op,cop}$-bicomodule algebra $\mathbb{A}^{\rm op,cop}$, then one checks that

$$\Omega' = (\omega_{\rm op,cop})^{54321} \quad \text{and} \quad \omega' = (\Omega_{\rm op,cop})^{54321},$$

so we restrict to the study of the elements $\Omega$, $\omega$ and their associated constructions.

If $H$ is a quasi-Hopf algebra with bijective antipode, $\mathscr{A}$ an $H$-bimodule algebra and $(\mathbb{A}, \lambda, \rho, \Phi_\lambda, \Phi_\rho, \Phi_{\lambda,\rho})$ an $H$-bicomodule algebra, we will denote $\mathscr{A} \bowtie_{\delta_l} \mathbb{A} = \mathscr{A} \bowtie \mathbb{A}$, $\mathscr{A} \bowtie_{\delta_r} \mathbb{A} = \mathscr{A} \blacktriangleright\!\!\blacktriangleleft \mathbb{A}$, $\mathbb{A} \bowtie_{\delta_l} \mathscr{A} = \mathbb{A} \bowtie \mathscr{A}$ and $\mathbb{A} \bowtie_{\delta_r} \mathscr{A} = \mathbb{A} \blacktriangleright\!\!\blacktriangleleft \mathscr{A}$, where $\delta_l$ and $\delta_r$ are the two-sided coactions given by (4.4.24) and (4.4.25). We call the first two constructions left diagonal crossed products and the last two right diagonal crossed products. For example, the multiplications in $\mathscr{A} \bowtie \mathbb{A}$ and $\mathscr{A} \blacktriangleright\!\!\blacktriangleleft \mathbb{A}$ are given by

$$
\begin{aligned}
&(\varphi \bowtie u)(\varphi' \bowtie u') \\
&\qquad = (\Omega^1 \cdot \varphi \cdot \Omega^5)(\Omega^2 u_{\langle 0\rangle_{[-1]}} \cdot \varphi' \cdot S^{-1}(u_{\langle 1\rangle})\Omega^4) \bowtie \Omega^3 u_{\langle 0\rangle_{[0]}} u', \qquad (5.6.15) \\
&(\varphi \blacktriangleright\!\!\blacktriangleleft u)(\varphi' \blacktriangleright\!\!\blacktriangleleft u') \\
&\qquad = (\omega^1 \cdot \varphi \cdot \omega^5)(\omega^2 u_{[-1]} \cdot \varphi' \cdot S^{-1}(u_{[0]_{\langle 1\rangle}})\omega^4) \blacktriangleright\!\!\blacktriangleleft \omega^3 u_{[0]_{\langle 0\rangle}} u', \qquad (5.6.16)
\end{aligned}
$$

respectively, for all $\varphi, \varphi' \in \mathscr{A}$ and $u, u' \in \mathbb{A}$, where we write $\varphi \bowtie u$ and $\varphi \blacktriangleright\!\!\blacktriangleleft u$ instead of $\varphi \otimes u$ to distinguish the new algebraic structures.

We are now ready to show that the generalized diagonal crossed products are unital associative algebras.

**Proposition 5.47** *Let $H$ be a quasi-Hopf algebra with bijective antipode, $\mathbb{A}$ an algebra and $(\delta, \Psi)$ a two-sided coaction of $H$ on $\mathbb{A}$. Consider $\mathscr{A} \bowtie_\delta \mathbb{A}$ and $\mathbb{A} \bowtie_\delta \mathscr{A}$, the vector spaces $\mathscr{A} \otimes \mathbb{A}$ and $\mathbb{A} \otimes \mathscr{A}$, respectively, with the multiplications defined in* (5.6.3) *and* (5.6.4), *respectively. Then these products define on $\mathscr{A} \bowtie_\delta \mathbb{A}$ and $\mathbb{A} \bowtie_\delta \mathscr{A}$ two associative algebra structures with unit $1_{\mathscr{A}} \bowtie_\delta 1_{\mathbb{A}}$ (resp. $1_{\mathbb{A}} \bowtie_\delta 1_{\mathscr{A}}$), containing $\mathbb{A} \equiv 1_{\mathscr{A}} \bowtie_\delta \mathbb{A}$ (resp. $\mathbb{A} \equiv \mathbb{A} \bowtie_\delta 1_{\mathscr{A}}$) as unital subalgebra.*

*Consequently, if $\mathbb{A}$ is an $H$-bicomodule algebra and $\mathscr{A}$ is an $H$-bimodule algebra then $\mathscr{A} \bowtie \mathbb{A}$, $\mathscr{A} \blacktriangleright\!\!\blacktriangleleft \mathbb{A}$, $\mathbb{A} \bowtie \mathscr{A}$ and $\mathbb{A} \blacktriangleright\!\!\blacktriangleleft \mathscr{A}$ are associative unital algebras containing $\mathbb{A}$ as unital subalgebra.*

*Proof* We will give the proof only for $\mathscr{A} \bowtie_\delta \mathbb{A}$, the one for $\mathbb{A} \bowtie_\delta \mathscr{A}$ being similar (it uses the relations satisfied by $\Omega'_\delta$, instead of the ones satisfied by $\Omega_\delta$). For $\varphi, \varphi', \varphi'' \in \mathscr{A}$ and $u, u', u'' \in \mathbb{A}$ we compute:

$$
\begin{aligned}
&(\varphi \bowtie_\delta u)[(\varphi' \bowtie_\delta u')(\varphi'' \bowtie_\delta u'')] \\
&\overset{(5.6.3)}{=} (\varphi \bowtie_\delta u)[(\Omega_\delta^1 \cdot \varphi' \cdot \Omega_\delta^5)(\Omega_\delta^2 u'_{(-1)} \cdot \varphi'' \cdot S^{-1}(u'_{(-1)})\Omega_\delta^4) \bowtie_\delta \Omega_\delta^3 u'_{(0)} u''] \\
&\underset{(4.1.14)}{\overset{(5.6.3)}{=}} (\overline{\Omega}_\delta^1 \cdot \varphi \cdot \overline{\Omega}_\delta^5)[((\overline{\Omega}_\delta^2)_1 u_{(-1)_1} \Omega_\delta^1 \cdot \varphi' \cdot \Omega_\delta^5 S^{-1}(u_{(-1)})_1 (\overline{\Omega}_\delta^4)_1)((\overline{\Omega}_\delta^2)_2 u_{(-1)_2} \\
&\qquad \Omega_\delta^2 u'_{(-1)} \cdot \varphi'' \cdot S^{-1}(u'_{(1)})\Omega_\delta^4 S^{-1}(u_{(1)})_2 (\overline{\Omega}_\delta^4)_2)] \bowtie_\delta \overline{\Omega}_\delta^3 u_{(0)} \Omega_\delta^3 u'_{(0)} u'' \\
&\overset{(5.6.5)}{=} (\overline{\Omega}_\delta^1 \cdot \varphi \cdot \overline{\Omega}_\delta^5)[(\overline{\Omega}_\delta^2)_1 \Omega_\delta^1 u_{(-1)} \cdot \varphi' \cdot S^{-1}(u_{(1)})\Omega_\delta^5 (\overline{\Omega}_\delta^4)_1) \\
&\qquad ((\overline{\Omega}_\delta^2)_2 \Omega_\delta^2 u_{(0,-1)} u'_{(-1)} \cdot \varphi'' \cdot S^{-1}(u_{(0,1)} u'_{(1)})\Omega_\delta^4 (\overline{\Omega}_\delta^4)_2)] \\
&\qquad \bowtie_\delta \overline{\Omega}_\delta^3 \Omega_\delta^3 u_{(0,0)} u'_{(0)} u'' \\
&\overset{(5.6.6),(4.1.13)}{=} [((\overline{\Omega}_\delta^1)_1 \Omega_\delta^1 \cdot \varphi \cdot \Omega_\delta^5 (\overline{\Omega}_\delta^5)_1)((\overline{\Omega}_\delta^1)_2 \Omega_\delta^2 u_{(-1)} \cdot \varphi' \cdot S^{-1}(u_{(1)})\Omega_\delta^4 (\overline{\Omega}_\delta^5)_2)]
\end{aligned}
$$

$$
\begin{aligned}
&(\overline{\Omega}^2_\delta(\Omega^3_\delta)_{(-1)}u_{(0,-1)}u'_{(-1)}\cdot\varphi''\cdot S^{-1}((\Omega^3_\delta)_{(1)}u_{(0,1)}u'_{(1)})\overline{\Omega}^4_\delta)\\
&\bowtie_\delta \overline{\Omega}^3_\delta(\Omega^3_\delta)_{(0)}u_{(0,0)}u'_{(0)}u''\\
\overset{(4.1.14),(5.6.3)}{=}\ &[(\Omega^1_\delta\cdot\varphi\cdot\Omega^5_\delta)(\Omega^2_\delta u_{(-1)}\cdot\varphi'\cdot S^{-1}(u_{(1)})\Omega^4_\delta)\bowtie_\delta \Omega^3_\delta u_{(0)}u'](\varphi''\bowtie_\delta u'')\\
\overset{(5.6.3)}{=}\ &[(\varphi\bowtie_\delta u)(\varphi'\bowtie_\delta u')](\varphi''\bowtie_\delta u'').
\end{aligned}
$$

The fact that $1_{\mathscr{A}}\bowtie_\delta 1_{\mathbb{A}}$ is the unit follows easily from the (co)unit axioms. □

**Remark 5.48** In the algebras $\mathscr{A}\bowtie_\delta\mathbb{A}$ and $\mathbb{A}\bowtie_\delta\mathscr{A}$ we have, for all $\varphi\in\mathscr{A}$ and $u\in\mathbb{A}$, $(\varphi\bowtie_\delta 1_{\mathbb{A}})(1_{\mathscr{A}}\bowtie_\delta u)=\varphi\bowtie_\delta u$ and $(u\bowtie_\delta 1_{\mathscr{A}})(1_{\mathbb{A}}\bowtie_\delta\varphi)=u\bowtie_\delta\varphi$.

**Examples 5.49** (i) We know that, if $H$ is a quasi-Hopf algebra with bijective antipode, then $H^*$ is an $H$-bimodule algebra, hence it makes sense to consider the algebras $H^*\bowtie_\delta\mathbb{A}$ and $\mathbb{A}\bowtie_\delta H^*$, where $\mathbb{A}$ is an algebra and $(\delta,\Psi)$ is a two-sided coaction of $H$ on $\mathbb{A}$.

(ii) Let $A$ be a left $H$-module algebra. Then $A$ becomes an $H$-bimodule algebra, with right $H$-action given via $\varepsilon$. In this particular case $A\bowtie H$ and $A\blacktriangleright\!\!\blacktriangleleft H$ both coincide to the smash product algebra $A\#H$. Moreover, if we replace the quasi-Hopf algebra $H$ by an arbitrary $H$-bicomodule algebra $\mathbb{A}$, then $A\bowtie\mathbb{A}$ and $A\blacktriangleright\!\!\blacktriangleleft\mathbb{A}$ coincide with the generalized smash product algebra $A\blacktriangleright\!\!<\mathbb{A}$. Therefore, the diagonal crossed products may be viewed as a generalization of the (generalized) smash product.

(iii) As we know, $H$ itself is an $H$-bicomodule algebra. So, in this case, the multiplications of the diagonal crossed products $\mathscr{A}\bowtie H$ and $\mathscr{A}\blacktriangleright\!\!\blacktriangleleft H$ specialize to

$$
\begin{aligned}
&(\varphi\bowtie h)(\varphi'\bowtie h')\\
&\quad=(\Omega^1\cdot\varphi\cdot\Omega^5)(\Omega^2h_{(1,1)}\cdot\varphi'\cdot S^{-1}(h_2)\Omega^4)\bowtie\Omega^3h_{(1,2)}h', \qquad (5.6.17)\\
&(\varphi\blacktriangleright\!\!\blacktriangleleft h)(\varphi'\blacktriangleright\!\!\blacktriangleleft h')\\
&\quad=(\omega^1\cdot\varphi\cdot\omega^5)(\omega^2h_1\cdot\varphi'\cdot S^{-1}(h_{(2,2)})\omega^4)\blacktriangleright\!\!\blacktriangleleft\omega^3h_{(2,1)}h', \qquad (5.6.18)
\end{aligned}
$$

for all $\varphi,\varphi'\in\mathscr{A}$ and $h,h'\in H$, where $\Omega=\Omega^1\otimes\cdots\otimes\Omega^5\in H^{\otimes 5}$, $\omega=\omega^1\otimes\ldots\otimes\omega^5\in H^{\otimes 5}$ are now given by

$$\Omega=X^1_{(1,1)}x^1y^1\otimes X^1_{(1,2)}x^2y^2_1\otimes X^1_2x^3y^2_2\otimes S^{-1}(f^1X^2y^3)\otimes S^{-1}(f^2X^3), \qquad (5.6.19)$$

$$\omega=x^1\otimes x^2Y^1\otimes x^3_1X^1Y^2_1\otimes S^{-1}(f^1x^3_{(2,1)}X^2Y^2_2)\otimes S^{-1}(f^2x^3_{(2,2)}X^3Y^3), \qquad (5.6.20)$$

and $f=f^1\otimes f^2$ is the twist defined in (3.2.15).

Let $H$ be a quasi-Hopf algebra with bijective antipode. For an $H$-bicomodule algebra $\mathbb{A}$ and an $H$-bimodule algebra $\mathscr{A}$ the multiplications of the right diagonal crossed products $\mathbb{A}\bowtie\mathscr{A}$ and $\mathbb{A}\blacktriangleright\!\!\blacktriangleleft\mathscr{A}$ are the following: if $\Omega'=\Omega'^1\otimes\cdots\otimes\Omega'^5$ and $\omega'=\omega'^1\otimes\cdots\otimes\omega'^5$ we have

$$
\begin{aligned}
&(u\bowtie\varphi)(u'\bowtie\varphi')\\
&\quad=uu'_{\langle 0\rangle_{[0]}}\Omega'^3\bowtie(\Omega'^2S^{-1}(u'_{\langle 0\rangle_{[-1]}})\cdot\varphi\cdot u'_{\langle 1\rangle}\Omega'^4)(\Omega'^1\cdot\varphi'\cdot\Omega'^5), \qquad (5.6.21)
\end{aligned}
$$

$$(u \blacktriangleright\!\!\blacktriangleleft \varphi)(u' \blacktriangleright\!\!\blacktriangleleft \varphi')$$
$$= uu'_{[0]_{\langle 0\rangle}}\omega'^3 \blacktriangleright\!\!\blacktriangleleft (\omega'^2 S^{-1}(u'_{[-1]})\cdot\varphi\cdot u'_{[0]_{\langle 1\rangle}}\omega'^4)(\omega'^1\cdot\varphi'\cdot\omega'^5), \quad (5.6.22)$$

for all $u,u' \in \mathbb{A}$ and $\varphi,\varphi' \in \mathscr{A}$. We know from Proposition 5.47 that $\mathbb{A} \bowtie \mathscr{A}$ and $\mathbb{A} \blacktriangleright\!\!\blacktriangleleft \mathscr{A}$ are associative algebras with unit $1_{\mathbb{A}} \bowtie 1_{\mathscr{A}}$ and $1_{\mathbb{A}} \blacktriangleright\!\!\blacktriangleleft 1_{\mathscr{A}}$, respectively, containing $\mathbb{A}$ as unital subalgebra. In fact, under the trivial permutation of tensor factors we have that

$$\mathbb{A} \bowtie \mathscr{A} \equiv (\mathscr{A}^{\rm op} \blacktriangleright\!\!\blacktriangleleft \mathbb{A}^{\rm op,cop})^{\rm op}, \quad \mathbb{A} \blacktriangleright\!\!\blacktriangleleft \mathscr{A} \equiv (\mathscr{A}^{\rm op} \bowtie \mathbb{A}^{\rm op,cop})^{\rm op}, \qquad (5.6.23)$$

where the left diagonal crossed products are made over $H^{\rm op,cop}$. Note that $\mathscr{A}^{\rm op}$ becomes an $H^{\rm op,cop}$-bimodule algebra via the actions $h \cdot_{\rm op} \varphi \cdot_{\rm op} h' = h' \cdot \varphi \cdot h$, for all $h,h' \in H$ and $\varphi \in \mathscr{A}$.

**Lemma 5.50** *Let $H$ be a quasi-Hopf algebra with bijective antipode, $\mathscr{A}$ an $H$-bimodule algebra and $\mathbb{A}$ an $H$-bicomodule algebra. Then, for all $\varphi \in \mathscr{A}$, we have*

$$\varphi \bowtie 1_{\mathbb{A}} = (1_{\mathscr{A}} \bowtie \tilde{q}^1_\rho)((\tilde{p}^1_\rho)_{[-1]}\cdot\varphi\cdot\tilde{q}^2_\rho S^{-1}(\tilde{p}^2_\rho) \bowtie (\tilde{p}^1_\rho)_{[0]})$$

*in $\mathscr{A} \bowtie \mathbb{A}$, where $\tilde{p}_\rho$ and $\tilde{q}_\rho$ are given by* (4.3.9).

*Proof* We compute:

$$(1_{\mathscr{A}} \bowtie \tilde{q}^1_\rho)((\tilde{p}^1_\rho)_{[-1]}\cdot\varphi\cdot\tilde{q}^2_\rho S^{-1}(\tilde{p}^2_\rho) \bowtie (\tilde{p}^1_\rho)_{[0]})$$
$$\overset{(5.6.15)}{=} (\tilde{q}^1_\rho)_{\langle 0\rangle_{[-1]}}(\tilde{p}^1_\rho)_{[-1]}\cdot\varphi\cdot\tilde{q}^2_\rho S^{-1}(\tilde{p}^2_\rho)S^{-1}((\tilde{q}^1_\rho)_{\langle 1\rangle}) \bowtie (\tilde{q}^1_\rho)_{\langle 0\rangle_{[0]}}(\tilde{p}^1_\rho)_{[0]}$$
$$\overset{(4.3.12)}{=} \varphi \bowtie 1_{\mathbb{A}},$$

finishing the proof. □

**Proposition 5.51** *Let $H$ be a quasi-Hopf algebra with bijective antipode, $\mathscr{A}$ an $H$-bimodule algebra and $\mathbb{A}$ an $H$-bicomodule algebra. Define the map $\Gamma : \mathscr{A} \to \mathscr{A} \bowtie \mathbb{A}$,*

$$\Gamma(\varphi) = (\tilde{p}^1_\rho)_{[-1]}\cdot\varphi\cdot S^{-1}(\tilde{p}^2_\rho) \bowtie (\tilde{p}^1_\rho)_{[0]}, \qquad (5.6.24)$$

*for all $\varphi \in \mathscr{A}$. Then $\mathscr{A} \bowtie \mathbb{A}$ is generated as an algebra by $\mathbb{A}$ and $\Gamma(\mathscr{A})$.*

*Proof* By the previous lemma it follows that $\varphi \bowtie 1_{\mathbb{A}} = (1_{\mathscr{A}} \bowtie \tilde{q}^1_\rho)\Gamma(\varphi\cdot\tilde{q}^2_\rho)$, for all $\varphi \in \mathscr{A}$, so for $\varphi \in \mathscr{A}$ and $u \in \mathbb{A}$ we can write $\varphi \bowtie u = (1_{\mathscr{A}} \bowtie \tilde{q}^1_\rho)\Gamma(\varphi\cdot\tilde{q}^2_\rho)(1_{\mathscr{A}} \bowtie u)$, finishing the proof. □

We prove now a sort of associativity property of diagonal crossed products with respect to tensoring by an arbitrary algebra.

**Proposition 5.52** *Let $H$ be a quasi-Hopf algebra with bijective antipode, $\mathscr{A}$ an $H$-bimodule algebra, $\mathbb{A}$ an $H$-bicomodule algebra and $C$ an algebra. On $\mathbb{A}\otimes C$ we have a (canonical) $H$-bicomodule algebra structure, yielding algebra isomorphisms*

$$\mathscr{A} \bowtie (\mathbb{A}\otimes C) \equiv (\mathscr{A} \bowtie \mathbb{A})\otimes C, \quad \mathscr{A} \blacktriangleright\!\!\blacktriangleleft (\mathbb{A}\otimes C) \equiv (\mathscr{A} \blacktriangleright\!\!\blacktriangleleft \mathbb{A})\otimes C,$$

*defined by the trivial identifications.*

*Proof* The $H$-bicomodule algebra structure on $\mathbb{A}\otimes C$ is given such that everything happening on $C$ is trivial, for instance the right $H$-comodule algebra structure is:

$$\rho_{\mathbb{A}\otimes C} : \mathbb{A}\otimes C \ni u\otimes c \mapsto (u_{\langle 0\rangle}\otimes c)\otimes u_{\langle 1\rangle} \in (\mathbb{A}\otimes C)\otimes H,$$
$$(\Phi_\rho)_{\mathbb{A}\otimes C} = (\tilde{X}^1_\rho \otimes 1_C)\otimes \tilde{X}^2_\rho \otimes \tilde{X}^3_\rho \in (\mathbb{A}\otimes C)\otimes H\otimes H,$$

and one can easily check that $\mathbb{A}\otimes C$ indeed becomes an $H$-bicomodule algebra. Also, it is easy to see that the elements $\Omega$ and $\omega$ for $\mathbb{A}\otimes C$ are given by

$$\Omega_{\mathbb{A}\otimes C} = \Omega^1\otimes\Omega^2\otimes(\Omega^3\otimes 1_C)\otimes\Omega^4\otimes\Omega^5,$$
$$\omega_{\mathbb{A}\otimes C} = \omega^1\otimes\omega^2\otimes(\omega^3\otimes 1_C)\otimes\omega^4\otimes\omega^5,$$

where $\Omega = \Omega^1\otimes\cdots\otimes\Omega^5$ and $\omega = \omega^1\otimes\cdots\otimes\omega^5$ are the elements for $\mathbb{A}$. By using these facts, one obtains that the multiplications in $\mathscr{A}\bowtie(\mathbb{A}\otimes C)$ and $\mathscr{A}\blacktriangleright\!\!\blacktriangleleft(\mathbb{A}\otimes C)$ coincide with those in $(\mathscr{A}\bowtie\mathbb{A})\otimes C$ and $(\mathscr{A}\blacktriangleright\!\!\blacktriangleleft\mathbb{A})\otimes C$, respectively, via the trivial identifications. □

Let $H$ be a quasi-Hopf algebra with bijective antipode and $\mathbb{A}$ an $H$-bicomodule algebra. We define two left $H\otimes H^{\rm op}$-coactions on $\mathbb{A}$, as follows:

$$\lambda_1,\lambda_2 : \mathbb{A}\to(H\otimes H^{\rm op})\otimes\mathbb{A},$$
$$\lambda_1(u) = (u_{\langle 0\rangle_{[-1]}}\otimes S^{-1}(u_{\langle 1\rangle}))\otimes u_{\langle 0\rangle_{[0]}} := u_{(-1)}\otimes u_{(0)},\quad \forall\, u\in\mathbb{A},$$
$$\lambda_2(u) = (u_{[-1]}\otimes S^{-1}(u_{[0]_{\langle 1\rangle}}))\otimes u_{[0]_{\langle 0\rangle}} := u^{(-1)}\otimes u^{(0)},\quad \forall\, u\in\mathbb{A}.$$

If we look at the element $\Omega\in H^{\otimes 2}\otimes\mathbb{A}\otimes H^{\otimes 2}$ given by (5.6.9) and consider the element $(\Omega^1\otimes\Omega^5)\otimes(\Omega^2\otimes\Omega^4)\otimes\Omega^3$, then one can check that this element is invertible in $(H\otimes H^{\rm op})\otimes(H\otimes H^{\rm op})\otimes\mathbb{A}$, its inverse being given by

$$(\Theta^1\tilde{X}^1_\lambda(\tilde{x}^1_\rho)_{[-1]_1}\otimes S^{-1}(\tilde{x}^3_\rho g^2))\otimes(\Theta^2_{[-1]}\tilde{X}^2_\lambda(\tilde{x}^1_\rho)_{[-1]_2}\otimes S^{-1}(\Theta^3\tilde{x}^2_\rho g^1))\otimes\Theta^2_{[0]}\tilde{X}^3_\lambda(\tilde{x}^1_\rho)_{[0]},$$

where $f^{-1} = g^1\otimes g^2$ is the element given by (3.2.16). We will denote this inverse by $\Phi_{\lambda_1}\in(H\otimes H^{\rm op})\otimes(H\otimes H^{\rm op})\otimes\mathbb{A}$.

Similarly, if we look at the element $\omega$ given by (5.6.10) and consider the element $(\omega^1\otimes\omega^5)\otimes(\omega^2\otimes\omega^4)\otimes\omega^3$, then one can check that this element is invertible in $(H\otimes H^{\rm op})\otimes(H\otimes H^{\rm op})\otimes\mathbb{A}$, with inverse defined by

$$(\tilde{Y}^1_\lambda\otimes S^{-1}(\theta^3\tilde{y}^3_\rho(\tilde{Y}^3_\lambda)_{\langle 1\rangle_2}g^2))\otimes(\theta^1\tilde{Y}^2_\lambda\otimes S^{-1}(\theta^2_{\langle 1\rangle}\tilde{y}^2_\rho(\tilde{Y}^3_\lambda)_{\langle 1\rangle_1}g^1))\otimes\theta^2_{\langle 0\rangle}\tilde{y}^1_\rho(\tilde{Y}^3_\lambda)_{\langle 0\rangle}.$$

We will denote this inverse by $\Phi_{\lambda_2}\in(H\otimes H^{\rm op})\otimes(H\otimes H^{\rm op})\otimes\mathbb{A}$.

**Proposition 5.53** *With notation as above, $(\mathbb{A},\lambda_1,\Phi_{\lambda_1})$ and $(\mathbb{A},\lambda_2,\Phi_{\lambda_2})$ are left $H\otimes H^{\rm op}$-comodule algebras, denoted by $\mathbb{A}_1$ and $\mathbb{A}_2$, respectively.*

*Proof* It is easy to see that $\lambda_1$ and $\lambda_2$ are algebra maps, and also that the conditions (4.3.7) and (4.3.8) in the definition of a left comodule algebra are satisfied. Then the conditions (4.3.5) and (4.3.6) for $(\mathbb{A},\lambda_1,\Phi_{\lambda_1})$ (resp. for $(\mathbb{A},\lambda_2,\Phi_{\lambda_2})$) to be a left $H\otimes H^{\rm op}$-comodule algebra are equivalent to the relations (5.6.11) and (5.6.12) fulfilled by $\Omega$ (resp. to the relations (5.6.13) and (5.6.14) fulfilled by $\omega$). □

We are now able to express the diagonal crossed products over $H$ as some generalized smash products over $H\otimes H^{\rm op}$.

**Proposition 5.54** *Let $H$ be a quasi-Hopf algebra with bijective antipode, $\mathscr{A}$ an $H$-bimodule algebra and $\mathbb{A}$ an $H$-bicomodule algebra. View $\mathscr{A}$ as a left $H\otimes H^{\rm op}$-module algebra with action $(h\otimes h')\cdot\varphi=h\cdot\varphi\cdot h'$ for all $h,h'\in H$ and $\varphi\in\mathscr{A}$, and consider the two left $H\otimes H^{\rm op}$-comodule algebras $\mathbb{A}_1$ and $\mathbb{A}_2$ obtained from $\mathbb{A}$ as above. Then we have algebra isomorphisms*

$$\mathscr{A}\bowtie\mathbb{A}\equiv\mathscr{A}\blacktriangleright\!\!<\mathbb{A}_1,\quad \mathscr{A}\blacktriangleright\!\!\blacktriangleleft\mathbb{A}\equiv\mathscr{A}\blacktriangleright\!\!<\mathbb{A}_2,$$

*defined by the trivial identifications.*

*Proof* We prove only the first isomorphism, the second being similar. The multiplication in $\mathscr{A}\blacktriangleright\!\!<\mathbb{A}_1$ looks as follows (for all $\varphi,\varphi'\in\mathscr{A}$ and $u,u'\in\mathbb{A}$):

$$\begin{aligned}
&(\varphi\blacktriangleright\!\!< u)(\varphi'\blacktriangleright\!\!< u')\\
&=((\tilde{x}^1_\lambda)_{\mathbb{A}_1}\cdot\varphi)((\tilde{x}^2_\lambda)_{\mathbb{A}_1}u_{(-1)}\cdot\varphi')\blacktriangleright\!\!<(\tilde{x}^3_\lambda)_{\mathbb{A}_1}u_{(0)}u'\\
&=((\Omega^1\otimes\Omega^5)\cdot\varphi)((\Omega^2\otimes\Omega^4)(u_{\langle 0\rangle_{[-1]}}\otimes S^{-1}(u_{\langle 1\rangle}))\cdot\varphi')\blacktriangleright\!\!<\Omega^3u_{\langle 0\rangle_{[0]}}u'\\
&=(\Omega^1\cdot\varphi\cdot\Omega^5)(\Omega^2u_{\langle 0\rangle_{[-1]}}\cdot\varphi'\cdot S^{-1}(u_{\langle 1\rangle})\Omega^4)\blacktriangleright\!\!<\Omega^3u_{\langle 0\rangle_{[0]}}u',
\end{aligned}$$

and via the trivial identification this is exactly the multiplication of $\mathscr{A}\bowtie\mathbb{A}$. □

Let us also record the fact that the two left $H\otimes H^{\rm op}$-comodule algebra structures on $H$ are defined as follows:

$$\begin{aligned}
&\lambda_1,\lambda_2:H\to(H\otimes H^{\rm op})\otimes H,\\
&\lambda_1(h)=(h_{(1,1)}\otimes S^{-1}(h_2))\otimes h_{(1,2)},\quad\forall\,h\in H,\\
&\lambda_2(h)=(h_1\otimes S^{-1}(h_{(2,2)}))\otimes h_{(2,1)},\quad\forall\,h\in H,
\end{aligned}$$

$$\begin{aligned}
&\Phi_{\lambda_1},\Phi_{\lambda_2}\in(H\otimes H^{\rm op})\otimes(H\otimes H^{\rm op})\otimes H,\\
&\Phi_{\lambda_1}=(Y^1X^1x^1_{(1,1)}\otimes S^{-1}(x^3g^2))\otimes(Y^2_1X^2x^1_{(1,2)}\otimes S^{-1}(Y^3x^2g^1))\otimes Y^2_2X^3x^1_2,\\
&\Phi_{\lambda_2}=(Y^1\otimes S^{-1}(x^3y^3Y^3_{(2,2)}g^2))\otimes(x^1Y^2\otimes S^{-1}(x^2_2y^2Y^3_{(2,1)}g^1))\otimes x^2_1y^1Y^3_1,
\end{aligned}$$

where $f^{-1}=g^1\otimes g^2$ is the element given by (3.2.16).

Again let $H$ be a quasi-Hopf algebra with bijective antipode, $\mathscr{A}$ an $H$-bimodule algebra and $\mathbb{A}$ an $H$-bicomodule algebra. We intend to prove that the two left diagonal crossed products $\mathscr{A}\bowtie\mathbb{A}$ and $\mathscr{A}\blacktriangleright\!\!\blacktriangleleft\mathbb{A}$ are isomorphic as algebras, using their description as generalized smash products.

First we need a result on generalized smash products. Namely, let $H$ be a quasi-bialgebra, $A$ a left $H$-module algebra, $\mathfrak{B}$ a left $H$-comodule algebra and $U\in H\otimes\mathfrak{B}$ an invertible element such that $(\varepsilon\otimes{\rm Id}_{\mathfrak{B}})(U)=1_{\mathfrak{B}}$. Recall from Section 4.3 that if we define a map $\lambda':\mathfrak{B}\to H\otimes\mathfrak{B}$, $\lambda'(\mathfrak{b})=U\lambda(\mathfrak{b})U^{-1}$, for all $\mathfrak{b}\in\mathfrak{B}$, then this is a new left $H$-comodule algebra structure on $\mathfrak{B}$, with

$$\Phi_{\lambda'}=(1_H\otimes U)({\rm Id}_H\otimes\lambda)(U)\Phi_\lambda(\Delta\otimes{\rm Id}_{\mathfrak{B}})(U^{-1}),$$

which is denoted by $\mathfrak{B}'$ (and we say that $\mathfrak{B}$ and $\mathfrak{B}'$ are twist equivalent). So, we can consider the generalized smash products $A \blacktriangleright\!\!< \mathfrak{B}$ and $A \blacktriangleright\!\!< \mathfrak{B}'$.

**Proposition 5.55** *The linear map*

$$f : A \blacktriangleright\!\!< \mathfrak{B} \to A \blacktriangleright\!\!< \mathfrak{B}', \quad f(a \blacktriangleright\!\!< \mathfrak{b}) = U \cdot (a \blacktriangleright\!\!< \mathfrak{b}) = U^1 \cdot a \blacktriangleright\!\!< U^2\mathfrak{b},$$

*is an algebra isomorphism, and moreover* $f(1_A \blacktriangleright\!\!< \mathfrak{b}) = 1_A \blacktriangleright\!\!< \mathfrak{b}$, *for all* $\mathfrak{b} \in \mathfrak{B}$ *(so* $A \blacktriangleright\!\!< \mathfrak{B}$ *and* $A \blacktriangleright\!\!< \mathfrak{B}'$ *are equivalent extensions of* $\mathfrak{B}$*).*

*Proof* This follows by a direct computation. □

In view of this proposition, it suffices to prove that if $\mathbb{A}$ is an $H$-bicomodule algebra, then the two left $H \otimes H^{\rm op}$-comodule algebras $\mathbb{A}_1$ and $\mathbb{A}_2$ constructed earlier are twist equivalent. To prove this, we first need a technical lemma.

**Lemma 5.56** *Let $H$ be a quasi-Hopf algebra with bijective antipode and $\mathbb{A}$ an $H$-bicomodule algebra. Consider the elements $\Omega$ and $\omega$ given by* (5.6.9) *and* (5.6.10). *Then the following relations hold:*

$$\begin{aligned}\Theta^1_1\Omega^1 \otimes \Theta^1_2\Omega^2 \otimes \Theta^2\Omega^3 \otimes \Omega^5 S^{-1}(\Theta^3)_1 \otimes \Omega^4 S^{-1}(\Theta^3)_2 &= \Theta^1_1\tilde{x}^1_\lambda \otimes \Theta^1_2\tilde{x}^2_\lambda\overline{\Theta}^1\\ \otimes \tilde{X}^1_\rho\Theta^2_{\langle 0\rangle}(\tilde{x}^3_\lambda)_{\langle 0\rangle}\overline{\Theta}^2 \otimes S^{-1}(f^2\tilde{X}^3_\rho\Theta^3) &\otimes S^{-1}(f^1\tilde{X}^2_\rho\Theta^2_{\langle 1\rangle}(\tilde{x}^3_\lambda)_{\langle 1\rangle}\overline{\Theta}^3),\end{aligned} \tag{5.6.25}$$

$$\begin{aligned}\Theta^1_1\Omega^1\theta^1 \otimes S^{-1}(\theta^3)\Omega^5 S^{-1}(\Theta^3)_1 \otimes \Theta^1_2\Omega^2\theta^2_{\langle 0\rangle_{[-1]}} \otimes S^{-1}(\theta^2_{\langle 1\rangle})\Omega^4 S^{-1}(\Theta^3)_2&\\ \otimes \Theta^2\Omega^3\theta^2_{\langle 0\rangle_{[0]}} = \omega^1 \otimes \omega^5 \otimes \omega^2\Theta^1 \otimes S^{-1}(\Theta^3)\omega^4 \otimes \omega^3\Theta^2.&\end{aligned} \tag{5.6.26}$$

*Proof* The relation (5.6.25) follows by applying (3.2.13), (4.4.3) and (4.4.2), we leave the details to the reader. We prove now (5.6.26). We compute:

$$\begin{aligned}
&\Theta^1_1\Omega^1\theta^1 \otimes S^{-1}(\theta^3)\Omega^5 S^{-1}(\Theta^3)_1 \otimes \Theta^1_2\Omega^2\theta^2_{\langle 0\rangle_{[-1]}}\\
&\qquad \otimes S^{-1}(\theta^2_{\langle 1\rangle})\Omega^4 S^{-1}(\Theta^3)_2 \otimes \Theta^2\Omega^3\theta^2_{\langle 0\rangle_{[0]}}\\
&\overset{(5.6.25)}{=} \Theta^1_1\tilde{x}^1_\lambda\theta^1 \otimes S^{-1}(f^2\tilde{X}^3_\rho\Theta^3\theta^3) \otimes \Theta^1_2\tilde{x}^2_\lambda\overline{\Theta}^1\theta^2_{\langle 0\rangle_{[-1]}}\\
&\qquad \otimes S^{-1}(f^1\tilde{X}^2_\rho\Theta^2_{\langle 1\rangle}(\tilde{x}^3_\lambda)_{\langle 1\rangle}\overline{\Theta}^3\theta^2_{\langle 1\rangle}) \otimes \tilde{X}^1_\rho\Theta^2_{\langle 0\rangle}(\tilde{x}^3_\lambda)_{\langle 0\rangle}\overline{\Theta}^2\theta^2_{\langle 0\rangle_{[0]}}\\
&\overset{(4.4.2)}{=} \tilde{x}^1_\lambda\tilde{\Theta}^1\theta^1 \otimes S^{-1}(f^2\tilde{X}^3_\rho(\tilde{x}^3_\lambda)_{\langle 1\rangle}\Theta^3\tilde{\Theta}^3\theta^3) \otimes \tilde{x}^2_\lambda\Theta^1\tilde{\Theta}^2_{[-1]}\overline{\Theta}^1\theta^2_{\langle 0\rangle_{[-1]}}\\
&\qquad \otimes S^{-1}(f^1\tilde{X}^2_\rho(\tilde{x}^3_\lambda)_{\langle 0\rangle_{\langle 1\rangle}}\Theta^2_{\langle 1\rangle}\tilde{\Theta}^2_{[0]_{\langle 1\rangle}}\overline{\Theta}^3\theta^2_{\langle 1\rangle}) \otimes \tilde{X}^1_\rho(\tilde{x}^3_\lambda)_{\langle 0\rangle_{\langle 0\rangle}}\Theta^2_{\langle 0\rangle}\tilde{\Theta}^2_{[0]_{\langle 0\rangle}}\overline{\Theta}^2\theta^2_{\langle 0\rangle_{[0]}}\\
&\overset{(4.4.1)}{=} \tilde{x}^1_\lambda \otimes S^{-1}(f^2\tilde{X}^3_\rho(\tilde{x}^3_\lambda)_{\langle 1\rangle}\Theta^3) \otimes \tilde{x}^2_\lambda\Theta^1\overline{\Theta}^1\\
&\qquad \otimes S^{-1}(f^1\tilde{X}^2_\rho(\tilde{x}^3_\lambda)_{\langle 0\rangle_{\langle 1\rangle}}\Theta^2_{\langle 1\rangle}\overline{\Theta}^3) \otimes \tilde{X}^1_\rho(\tilde{x}^3_\lambda)_{\langle 0\rangle_{\langle 0\rangle}}\Theta^2_{\langle 0\rangle}\overline{\Theta}^2\\
&\overset{(4.3.1)}{=} \tilde{x}^1_\lambda \otimes S^{-1}(f^2(\tilde{x}^3_\lambda)_{\langle 1\rangle_2}\tilde{X}^3_\rho\Theta^3) \otimes \tilde{x}^2_\lambda\Theta^1\overline{\Theta}^1\\
&\qquad \otimes S^{-1}(f^1(\tilde{x}^3_\lambda)_{\langle 1\rangle_1}\tilde{X}^2_\rho\Theta^2_{\langle 1\rangle}\overline{\Theta}^3) \otimes (\tilde{x}^3_\lambda)_{\langle 0\rangle}\tilde{X}^1_\rho\Theta^2_{\langle 0\rangle}\overline{\Theta}^2\\
&\overset{(5.6.10)}{=} \omega^1 \otimes \omega^5 \otimes \omega^2\overline{\Theta}^1 \otimes S^{-1}(\overline{\Theta}^3)\omega^4 \otimes \omega^3\overline{\Theta}^2,
\end{aligned}$$

as required. □

**Proposition 5.57** *Let $H$ be a quasi-Hopf algebra with bijective antipode and $\mathbb{A}$ an $H$-bicomodule algebra. Then the left $H \otimes H^{\mathrm{op}}$-comodule algebras $\mathbb{A}_1$ and $\mathbb{A}_2$ are twist equivalent. More precisely, for the element $U \in (H \otimes H^{\mathrm{op}}) \otimes \mathbb{A}$ given by $U = (\Theta^1 \otimes S^{-1}(\Theta^3)) \otimes \Theta^2$, we have*

$$\lambda_2(u) = U\lambda_1(u)U^{-1}, \ \ \forall\, u \in \mathbb{A},$$
$$\Phi_{\lambda_2} = (1 \otimes U)(\mathrm{Id} \otimes \lambda_1)(U)\Phi_{\lambda_1}(\Delta \otimes \mathrm{Id})(U^{-1}).$$

*Proof* The first relation follows immediately from (4.4.1), and the second is equivalent to the relation (5.6.26) proved in the previous lemma. □

As a consequence of these results and (5.6.23), we obtain:

**Corollary 5.58** *Let $H$ be a quasi-Hopf algebra with bijective antipode, $\mathscr{A}$ an $H$-bimodule algebra and $\mathbb{A}$ an $H$-bicomodule algebra. Then the two left (resp. right) diagonal crossed products $\mathscr{A} \bowtie \mathbb{A}$ and $\mathscr{A} \blacktriangleright\!\!\blacktriangleleft \mathbb{A}$ (resp. $\mathbb{A} \bowtie \mathscr{A}$ and $\mathbb{A} \blacktriangleright\!\!\blacktriangleleft \mathscr{A}$) are isomorphic as algebras, and moreover they are equivalent extensions of $\mathbb{A}$.*

**Remark 5.59** Let $H$ be a quasi-Hopf algebra with bijective antipode, $\mathscr{A}$ an $H$-bimodule algebra and $\mathbb{A}$ an $H$-bicomodule algebra with $\Phi_{\lambda,\rho} = 1_H \otimes 1_{\mathbb{A}} \otimes 1_H$. Then, by (4.4.1), it follows that $(\lambda \otimes \mathrm{Id}_H) \circ \rho = (\mathrm{Id}_H \otimes \rho) \circ \lambda$, and by (5.6.26) it follows that $\Omega = \omega$. So, in this case we have that $\mathscr{A} \bowtie \mathbb{A}$ and $\mathscr{A} \blacktriangleright\!\!\blacktriangleleft \mathbb{A}$ are not only isomorphic, but actually coincide, and $\mathbb{A}_1$ and $\mathbb{A}_2$ also coincide.

Our aim now is to show that the left generalized diagonal crossed products are isomorphic, as algebras, to the right generalized diagonal crossed products.

**Proposition 5.60** *Let $H$ be a quasi-Hopf algebra with bijective antipode, $(\delta, \Psi)$ a two-sided coaction of $H$ on an algebra $\mathbb{A}$, and $\mathscr{A}$ an $H$-bimodule algebra. Then the linear map $\vartheta : \mathscr{A} \bowtie_\delta \mathbb{A} \to \mathbb{A} \bowtie_\delta \mathscr{A}$ defined for all $\varphi \in \mathscr{A}$ and $u \in \mathbb{A}$ by*

$$\vartheta(\varphi \bowtie_\delta u) = q_\delta^2 u_{(0)} \bowtie S^{-1}(q_\delta^1 u_{(-1)}) \cdot \varphi \cdot q_\delta^3 u_{(1)}$$

*is an algebra isomorphism, where $q_\delta = q_\delta^1 \otimes q_\delta^2 \otimes q_\delta^3$ is the element defined in (4.4.16). In particular, if $\mathbb{A}$ is an $H$-bicomodule algebra then we get that all four diagonal crossed products $\mathscr{A} \bowtie \mathbb{A}$, $\mathbb{A} \bowtie \mathscr{A}$, $\mathscr{A} \blacktriangleright\!\!\blacktriangleleft \mathbb{A}$ and $\mathbb{A} \blacktriangleright\!\!\blacktriangleleft \mathscr{A}$ are isomorphic as associative unital algebras.*

*Proof* We show that $\vartheta$ is multiplicative. For any $\varphi, \varphi' \in \mathscr{A}$ and $u, u' \in \mathbb{A}$ we have (we denote by $Q_\delta^1 \otimes Q_\delta^2 \otimes Q_\delta^3$ another copy of $q_\delta$ and by $F^1 \otimes F^2$ another copy of $f$):

$$\begin{aligned}
&\vartheta((\varphi \bowtie_\delta u)(\varphi' \bowtie_\delta u')) \\
&\overset{(5.6.3),(5.6.1)}{=} \vartheta\Big((\overline{\Psi}^1 \cdot \varphi \cdot S^{-1}(f^2\overline{\Psi}^5))(\overline{\Psi}^2 u_{(-1)} \cdot \varphi' \cdot S^{-1}(f^1\overline{\Psi}^4 u_{(1)})) \bowtie_\delta \overline{\Psi}^3 u_{(0)}u'\Big) \\
&\underset{(3.2.13)}{\overset{(4.1.14)}{=}} q_\delta^2 \overline{\Psi}^3_{(0)} u_{(0,0)} u'_{(0)} \bowtie_\delta \Big(S^{-1}(F^2(q_\delta^1)_2 \overline{\Psi}^3_{(-1)_2} u_{(0,-1)_2} u'_{(-1)_2} g^2)\overline{\Psi}^1 \\
&\quad \cdot \varphi \cdot S^{-1}(f^2\overline{\Psi}^5)(q_\delta^3)_1 \overline{\Psi}^3_{(1)_1} u_{(0,1)_1} u'_{(1)_1}\Big) \\
&\quad \Big(S^{-1}(F^1(q_\delta^1)_1 \overline{\Psi}^3_{(-1)_1} u_{(0,-1)_1} u'_{(-1)_1} g^1)
\end{aligned}$$

$$
\begin{aligned}
&\overline{\Psi}^2 u_{(-1)}\cdot\varphi'\cdot S^{-1}(f^1\overline{\Psi}^4 u_{(1)})(q^3_\delta)_2\overline{\Psi}^3_{(1)_2}u_{(0,1)_2}u'_{(1)_2}\Big)\\
\overset{(4.4.22)}{=}\;& q^2_\delta(Q^2_\delta)_{(0)}\Psi^3 u_{(0,0)}u'_{(0)}\bowtie_\delta\Big(S^{-1}(q^1_\delta(Q^2_\delta)_{(-1)}\Psi^2 u_{(0,-1)_2}u'_{(-1)_2}g^2)\cdot\varphi\\
&\cdot q^3_\delta(Q^2_\delta)_{(1)}\Psi^4 u_{(0,1)_1}u'_{(1)_1}\Big)\Big(S^{-1}(Q^1_\delta\Psi^1 u_{(0,-1)_1}u'_{(-1)_1}g^1)u_{(-1)}\cdot\varphi'\\
&\cdot S^{-1}(u_{(1)})Q^3_\delta\Psi^5 u_{(0,1)_2}u'_{(1)_2}\Big)\\
\underset{(4.4.6)}{\overset{(4.4.23)}{=}}\;& q^2_\delta u_{(0)}\overline{\Psi}^3 u'_{(0)}\bowtie_\delta\Big(S^{-1}(q^1_\delta q^2_L u_{(-1)_{(2,2)}}\overline{\Psi}^2_2 u'_{(-1)_2}g^2)\\
&\cdot\varphi\cdot q^3_\delta q^1_R u_{(1)_{(1,1)}}\overline{\Psi}^4_1 u'_{(1)_1}\Big)\Big(S^{-1}(q^1_L u_{(-1)_{(2,1)}}\overline{\Psi}^2_1 u'_{(-1)_1}g^1)u_{(-1)_1}\overline{\Psi}^1\cdot\varphi'\\
&\cdot S^{-1}(u_{(1)_2}\overline{\Psi}^5)q^2_R u_{(1)_{(1,2)}}\overline{\Psi}^4_2 u'_{(1)_2}\Big)\\
\underset{(4.4.23)}{\overset{(3.2.22),(3.2.21)}{=}}\;& q^2_\delta u_{(0)}(Q^2_\delta)_{(0)}\Psi^3 u'_{(0)}\bowtie_\delta\Big(S^{-1}(q^1_\delta u_{(-1)}(Q^2_\delta)_{(-1)}\Psi^2 u'_{(-1)_2}g^2)\cdot\varphi\\
&\cdot q^3_\delta u_{(1)}(Q^2_\delta)_{(1)}\Psi^4 u'_{(1)_1}\Big)\Big(S^{-1}(Q^1_\delta\Psi^1 u'_{(-1)_1}g^1)\cdot\varphi'\cdot Q^3_\delta\Psi^5 u'_{(1)_2}\Big)\\
\underset{(5.6.2)}{\overset{(4.4.5)}{=}}\;& q^2_\delta u_{(0)}(Q^2_\delta)_{(0)}u'_{(0,0)}\Omega'^3\bowtie_\delta(\Omega'^2 S^{-1}(q^1_\delta u_{(-1)}(Q^2_\delta)_{(-1)}u'_{(0,-1)})\cdot\varphi\\
&\cdot q^3_\delta u_{(1)}(Q^2_\delta)_{(1)}u'_{(0,1)}\Omega'^4)(\Omega'^1 S^{-1}(Q^1_\delta u'_{(-1)})\cdot\varphi'\cdot Q^3_\delta u'_{(1)}\Omega'^5)\\
\overset{(5.6.4)}{=}\;& (q^2_\delta u_{(0)}\bowtie_\delta S^{-1}(q^1_\delta u_{(-1)})\cdot\varphi\cdot q^3_\delta u_{(1)})\\
&(Q^2_\delta u'_{(0)}\bowtie_\delta S^{-1}(Q^1_\delta u'_{(-1)})\cdot\varphi'\cdot Q^3_\delta u'_{(1)})\\
=\;& \vartheta(\varphi\bowtie_\delta u)\vartheta(\varphi'\bowtie_\delta u').\quad\text{QED}
\end{aligned}
$$

One can see that the unit and counit properties imply $\vartheta(1_{\mathscr{A}}\bowtie_\delta 1_{\mathbb{A}})=1_{\mathbb{A}}\bowtie_\delta 1_{\mathscr{A}}$, so it remains to show that $\vartheta$ is bijective. For this, define $\vartheta^{-1}:\mathbb{A}\bowtie_\delta\mathscr{A}\to\mathscr{A}\bowtie_\delta\mathbb{A}$ given for all $u\in\mathbb{A}$ and $\varphi\in\mathscr{A}$ by

$$\vartheta^{-1}(u\bowtie_\delta\varphi)=u_{(-1)}p^1_\delta\cdot\varphi\cdot S^{-1}(u_{(1)}p^3_\delta)\bowtie_\delta u_{(0)}p^2_\delta,$$

where $p_\delta=p^1_\delta\otimes p^2_\delta\otimes p^3_\delta$ is the element defined in (4.4.15).

We claim that $\vartheta$ and $\vartheta^{-1}$ are inverses. Indeed, $\vartheta\circ\vartheta^{-1}=\mathrm{Id}_{\mathbb{A}\bowtie_\delta\mathscr{A}}$ because of (4.4.18) and (4.4.20), and $\vartheta\circ\vartheta^{-1}=\mathrm{Id}_{\mathscr{A}\bowtie_\delta\mathbb{A}}$ because of (4.4.17) and (4.4.19) (we leave the verification of the details to the reader). □

We now present a Universal Property of the diagonal crossed product.

**Proposition 5.61** *Let $H$ be a quasi-Hopf algebra with bijective antipode, $\mathscr{A}$ an $H$-bimodule algebra, $\mathbb{A}$ an $H$-bicomodule algebra, $B$ an algebra, $\gamma:\mathbb{A}\to B$ an algebra map and $v:\mathscr{A}\to B$ a linear map such that the following conditions are satisfied:*

$$\gamma(u_{\langle 0\rangle})v(\varphi\cdot u_{\langle 1\rangle})=v(u_{[-1]}\cdot\varphi)\gamma(u_{[0]}),\tag{5.6.27}$$
$$v(\varphi\varphi')=\gamma(\tilde{X}^1_\rho)v(\theta^1\tilde{X}^1_\lambda\cdot\varphi\cdot\tilde{X}^2_\rho)\gamma(\theta^2)v(\tilde{X}^2_\lambda\cdot\varphi'\cdot\tilde{X}^3_\rho\theta^3)\gamma(\tilde{X}^3_\lambda),\tag{5.6.28}$$
$$v(1_{\mathscr{A}})=1_B,\tag{5.6.29}$$

*for all $\varphi,\varphi'\in\mathscr{A}$ and $u\in\mathbb{A}$. Consider the algebra map $j:\mathbb{A}\to\mathscr{A}\bowtie\mathbb{A}$, $j(u)=1_{\mathscr{A}}\bowtie u$, and the map $\Gamma:\mathscr{A}\to\mathscr{A}\bowtie\mathbb{A}$ defined in Proposition 5.51. Then there exists a unique algebra map $w:\mathscr{A}\bowtie\mathbb{A}\to B$ such that $w\circ\Gamma=v$ and $w\circ j=\gamma$. Moreover,* *w is given by the formula*

$$w(\varphi \bowtie u) = \gamma(\tilde{q}_\rho^1)v(\varphi \cdot \tilde{q}_\rho^2)\gamma(u), \tag{5.6.30}$$

*for all* $\varphi \in \mathscr{A}$ *and* $u \in \mathbb{A}$, *where* $\tilde{q}_\rho = \tilde{q}_\rho^1 \otimes \tilde{q}_\rho^2$ *is given by formula* (4.3.9).

*Proof* We first prove the uniqueness of $w$. From the proof of Proposition 5.51, we know that $\varphi \bowtie u = (1_{\mathscr{A}} \bowtie \tilde{q}_\rho^1)\Gamma(\varphi \cdot \tilde{q}_\rho^2)(1_{\mathscr{A}} \bowtie u)$, for all $\varphi \in \mathscr{A}$ and $u \in \mathbb{A}$, hence we can write

$$\begin{aligned} w(\varphi \bowtie u) &= w(j(\tilde{q}_\rho^1)\Gamma(\varphi \cdot \tilde{q}_\rho^2)j(u)) \\ &= w(j(\tilde{q}_\rho^1))w(\Gamma(\varphi \cdot \tilde{q}_\rho^2))w(j(u)) \\ &= \gamma(\tilde{q}_\rho^1)v(\varphi \cdot \tilde{q}_\rho^2)\gamma(u), \end{aligned}$$

showing that $w$ is unique. We prove the existence part. Define $w$ by formula (5.6.30); it is obvious that $w$ is unital and satisfies $w \circ j = \gamma$. We check that $w \circ \Gamma = v$:

$$\begin{aligned} (w \circ \Gamma)(\varphi) &= w((\tilde{p}_\rho^1)_{[-1]} \cdot \varphi \cdot S^{-1}(\tilde{p}_\rho^2) \bowtie (\tilde{p}_\rho^1)_{[0]}) \\ &= \gamma(\tilde{q}_\rho^1)v((\tilde{p}_\rho^1)_{[-1]} \cdot \varphi \cdot S^{-1}(\tilde{p}_\rho^2)\tilde{q}_\rho^2)\gamma((\tilde{p}_\rho^1)_{[0]}) \\ &\overset{(5.6.27)}{=} \gamma(\tilde{q}_\rho^1)\gamma((\tilde{p}_\rho^1)_{\langle 0\rangle})v(\varphi \cdot S^{-1}(\tilde{p}_\rho^2)\tilde{q}_\rho^2(\tilde{p}_\rho^1)_{\langle 1\rangle}) \overset{(4.3.13)}{=} v(\varphi). \end{aligned}$$

Thus, the only thing left to prove is that $w$ is multiplicative. We denote by $\tilde{Q}_\rho^1 \otimes \tilde{Q}_\rho^2$ another copy of the element $\tilde{q}_\rho$, and we record the obvious relation

$$\tilde{Q}_\rho^1\tilde{x}_\rho^1 \otimes S^{-1}(\tilde{x}_\rho^3)\tilde{Q}_\rho^2\tilde{x}_\rho^2 = 1_H \otimes S^{-1}(\alpha). \tag{5.6.31}$$

Now we compute:

$$\begin{aligned} & w((\varphi \bowtie u)(\varphi' \bowtie u')) \\ &= \gamma(\tilde{q}_\rho^1)v([\Omega^1 \cdot \varphi \cdot \Omega^5(\tilde{q}_\rho^2)_1][\Omega^2 u_{\langle 0\rangle_{[-1]}} \cdot \varphi' \cdot S^{-1}(u_{\langle 1\rangle})\Omega^4(\tilde{q}_\rho^2)_2]) \\ &\quad \gamma(\Omega^3 u_{\langle 0\rangle_{[0]}} u') \\ &\overset{(5.6.28)}{=} \gamma(\tilde{q}_\rho^1)\gamma(\tilde{X}_\rho^1)v(\theta^1\tilde{X}_\lambda^1\Omega^1 \cdot \varphi \cdot \Omega^5(\tilde{q}_\rho^2)_1\tilde{X}_\rho^2)\gamma(\theta^2) \\ &\quad v(\tilde{X}_\lambda^2\Omega^2 u_{\langle 0\rangle_{[-1]}} \cdot \varphi' \cdot S^{-1}(u_{\langle 1\rangle})\Omega^4(\tilde{q}_\rho^2)_2\tilde{X}_\rho^3\theta^3)\gamma(\tilde{X}_\lambda^3)\gamma(\Omega^3 u_{\langle 0\rangle_{[0]}} u') \\ &\overset{(5.6.9)}{=} \gamma(\tilde{q}_\rho^1\tilde{X}_\rho^1)v(\theta^1(\tilde{Y}_\rho^1)_{[-1]}\overline{\theta}^1 \cdot \varphi \cdot S^{-1}(f^2\tilde{Y}_\rho^3)(\tilde{q}_\rho^2)_1\tilde{X}_\rho^2)\gamma(\theta^2) \\ &\quad v\Big((\tilde{Y}_\rho^1)_{[0]_{[-1]}}\overline{\theta}^2_{[-1]}u_{\langle 0\rangle_{[-1]}} \cdot \varphi' \cdot S^{-1}(f^1\tilde{Y}_\rho^2\overline{\theta}^3 u_{\langle 1\rangle})(\tilde{q}_\rho^2)_2\tilde{X}_\rho^3\theta^3\Big) \\ &\quad \gamma\Big((\tilde{Y}_\rho^1)_{[0]_{[0]}}\overline{\theta}^2_{[0]}u_{\langle 0\rangle_{[0]}}\Big)\gamma(u') \\ &\overset{(5.6.27)}{=} \gamma(\tilde{q}_\rho^1\tilde{X}_\rho^1)v\Big(\theta^1(\tilde{Y}_\rho^1)_{[-1]}\overline{\theta}^1 \cdot \varphi \\ &\qquad \cdot S^{-1}(f^2\tilde{Y}_\rho^3)(\tilde{q}_\rho^2)_1\tilde{X}_\rho^2)\gamma(\theta^2(\tilde{Y}_\rho^1)_{[0]_{\langle 0\rangle}}\overline{\theta}^2_{\langle 0\rangle}u_{\langle 0\rangle_{\langle 0\rangle}}\Big) \\ &\quad v\Big(\varphi' \cdot S^{-1}(f^1\tilde{Y}_\rho^2\overline{\theta}^3 u_{\langle 1\rangle})(\tilde{q}_\rho^2)_2\tilde{X}_\rho^3\theta^3(\tilde{Y}_\rho^1)_{[0]_{\langle 1\rangle}}\overline{\theta}^2_{\langle 1\rangle}u_{\langle 0\rangle_{\langle 0\rangle}}\Big)\gamma(u') \\ &\overset{(4.4.1)}{=} \gamma(\tilde{q}_\rho^1\tilde{X}_\rho^1)v\Big((\tilde{Y}_\rho^1)_{\langle 0\rangle_{[-1]}}\theta^1\overline{\theta}^1 \cdot \varphi \end{aligned}$$

$$
\begin{aligned}
&\qquad \cdot S^{-1}(f^2\tilde{Y}^3_\rho)(\tilde{q}^2_\rho)_1\tilde{X}^2_\rho)\gamma((\tilde{Y}^1_\rho)_{\langle 0\rangle_{[0]}}\theta^2\overline{\theta}^2_{\langle 0\rangle}u_{\langle 0\rangle_{\langle 0\rangle}}\Big)\\
&v\Big(\varphi'\cdot S^{-1}(f^1\tilde{Y}^2_\rho\overline{\theta}^3u_{\langle 1\rangle})(\tilde{q}^2_\rho)_2\tilde{X}^3_\rho(\tilde{Y}^1_\rho)_{\langle 1\rangle}\theta^3\overline{\theta}^2_{\langle 1\rangle}u_{\langle 0\rangle_{\langle 1\rangle}}\Big)\gamma(u')\\
\stackrel{(4.4.3)}{=}\;& \gamma(\tilde{q}^1_\rho\tilde{X}^1_\rho)v\Big((\tilde{Y}^1_\rho)_{\langle 0\rangle_{[-1]}}(\tilde{y}^1_\rho)_{[-1]}\theta^1\cdot\varphi\cdot S^{-1}(f^2\tilde{Y}^3_\rho)(\tilde{q}^2_\rho)_1\tilde{X}^2_\rho\Big)\\
&\gamma\Big((\tilde{Y}^1_\rho)_{\langle 0\rangle_{[0]}}(\tilde{y}^1_\rho)_{[0]}\theta^2\tilde{Z}^1_\rho u_{\langle 0\rangle_{\langle 0\rangle}}\Big)\\
&v\Big(\varphi'\cdot S^{-1}(f^1\tilde{Y}^2_\rho\tilde{y}^3_\rho\theta^3_2\tilde{Z}^3_\rho u_{\langle 1\rangle})(\tilde{q}^2_\rho)_2\tilde{X}^3_\rho(\tilde{Y}^1_\rho)_{\langle 1\rangle}\tilde{y}^2_\rho\theta^3_1\tilde{Z}^2_\rho u_{\langle 0\rangle_{\langle 1\rangle}}\Big)\gamma(u')\\
\stackrel{(5.6.27)}{=}\;& \gamma(\tilde{q}^1_\rho\tilde{X}^1_\rho(\tilde{Y}^1_\rho)_{\langle 0\rangle_{\langle 0\rangle}}(\tilde{y}^1_\rho)_{\langle 0\rangle})\\
&v\Big(\theta^1\cdot\varphi\cdot S^{-1}(f^2\tilde{Y}^3_\rho)(\tilde{q}^2_\rho)_1\tilde{X}^2_\rho(\tilde{Y}^1_\rho)_{\langle 0\rangle_{\langle 1\rangle}}(\tilde{y}^1_\rho)_{\langle 1\rangle}\Big)\gamma(\theta^2\tilde{Z}^1_\rho u_{\langle 0\rangle_{\langle 0\rangle}})\\
&v\Big(\varphi'\cdot S^{-1}(f^1\tilde{Y}^2_\rho\tilde{y}^3_\rho\theta^3_2\tilde{Z}^3_\rho u_{\langle 1\rangle})(\tilde{q}^2_\rho)_2\tilde{X}^3_\rho(\tilde{Y}^1_\rho)_{\langle 1\rangle}\tilde{y}^2_\rho\theta^3_1\tilde{Z}^2_\rho u_{\langle 0\rangle_{\langle 1\rangle}}\Big)\gamma(u')\\
\underset{(4.3.2)}{\stackrel{(4.3.1)}{=}}\;& \gamma(\tilde{q}^1_\rho(\tilde{Y}^1_\rho)_{\langle 0\rangle}\tilde{y}^1_\rho\tilde{X}^1_\rho)v\Big(\theta^1\cdot\varphi\cdot S^{-1}(f^2\tilde{Y}^3_\rho)(\tilde{q}^2_\rho)_1(\tilde{Y}^1_\rho)_{\langle 1\rangle_1}(\tilde{y}^2_\rho)_1x^1\tilde{X}^2_\rho\Big)\\
&\gamma(\theta^2\tilde{Z}^1_\rho u_{\langle 0\rangle_{\langle 0\rangle}})v\Big(\varphi'\cdot S^{-1}(f^1\tilde{Y}^2_\rho\tilde{y}^3_\rho x^3(\tilde{X}^3_\rho)_2\theta^3_2\tilde{Z}^3_\rho u_{\langle 1\rangle})\\
&\qquad\qquad (\tilde{q}^2_\rho)_2(\tilde{Y}^1_\rho)_{\langle 1\rangle_2}(\tilde{y}^2_\rho)_2x^2(\tilde{X}^3_\rho)_1\theta^3_1\tilde{Z}^2_\rho u_{\langle 0\rangle_{\langle 1\rangle}}\Big)\gamma(u')\\
\underset{(4.3.15)}{\stackrel{(4.3.1)}{=}}\;& \gamma(\tilde{q}^1_\rho(\tilde{Q}^1_\rho)_{\langle 0\rangle}\tilde{x}^1_\rho\tilde{y}^1_\rho\tilde{X}^1_\rho)v\Big(\theta^1\cdot\varphi\cdot\tilde{q}^2_\rho(\tilde{Q}^1_\rho)_{\langle 1\rangle}\tilde{x}^2_\rho(\tilde{y}^2_\rho)_1x^1\tilde{X}^2_\rho\Big)\gamma(\theta^2u_{\langle 0\rangle}\tilde{Z}^1_\rho)\\
&v\Big(\varphi'\cdot S^{-1}(\tilde{y}^3_\rho x^3(\tilde{X}^3_\rho)_2\theta^3_2u_{\langle 1\rangle_2}\tilde{Z}^3_\rho)\tilde{Q}^2_\rho\tilde{x}^3_\rho(\tilde{y}^2_\rho)_2x^2(\tilde{X}^3_\rho)_1\theta^3_1u_{\langle 1\rangle_1}\tilde{Z}^2_\rho\Big)\gamma(u')\\
\stackrel{(4.3.2)}{=}\;& \gamma(\tilde{q}^1_\rho(\tilde{Q}^1_\rho)_{\langle 0\rangle}(\tilde{x}^1_\rho)_{\langle 0\rangle})v\Big(\theta^1\cdot\varphi\cdot\tilde{q}^2_\rho(\tilde{Q}^1_\rho)_{\langle 1\rangle}(\tilde{x}^1_\rho)_{\langle 1\rangle}\Big)\gamma(\theta^2u_{\langle 0\rangle}\tilde{Z}^1_\rho)\\
&v\Big(\varphi'\cdot S^{-1}(\tilde{x}^3_\rho\theta^3_2u_{\langle 1\rangle_2}\tilde{Z}^3_\rho)\tilde{Q}^2_\rho\tilde{x}^2_\rho\theta^3_1u_{\langle 1\rangle_1}\tilde{Z}^2_\rho\Big)\gamma(u')\\
\stackrel{(5.6.31)}{=}\;& \gamma(\tilde{q}^1_\rho)v(\varphi\cdot\tilde{q}^2_\rho)\gamma(u\tilde{Z}^1_\rho)v(\varphi'\cdot S^{-1}(\tilde{Z}^3_\rho)S^{-1}(\alpha)\tilde{Z}^2_\rho)\gamma(u')\\
=\;& \gamma(\tilde{q}^1_\rho)v(\varphi\cdot\tilde{q}^2_\rho)\gamma(u)\gamma(\tilde{Q}^1_\rho)v(\varphi'\cdot\tilde{Q}^2_\rho)\gamma(u')\\
=\;& w(\varphi\bowtie u)w(\varphi'\bowtie u'),
\end{aligned}
$$

finishing the proof. □

As a consequence of Proposition 5.61, we immediately obtain a new kind of Universal Property for the quasi-Hopf smash product:

**Proposition 5.62** *Let $H$ be a quasi-Hopf algebra with bijective antipode and $A$ a left $H$-module algebra. Denote by $i: A\to A\#H$, $i(a)=a\#1_H$ and $j: H\to A\#H$, $j(h)=1_A\#h$. Let $B$ be an algebra, $\gamma: H\to B$ an algebra map and $v: A\to B$ a linear map satisfying the following conditions, for all $a,a'\in A$ and $h\in H$:*

$$\gamma(h)v(a)=v(h_1\cdot a)\gamma(h_2), \tag{5.6.32}$$

$$v(aa')=v(X^1\cdot a)v(X^2\cdot a')\gamma(X^3), \tag{5.6.33}$$

$$v(1_A)=1_B. \tag{5.6.34}$$

*Then there exists a unique algebra map $w : A\#H \to B$ such that $w \circ i = v$ and $w \circ j = \gamma$. Moreover, $w$ is given by the formula $w(a\#h) = v(a)\gamma(h)$, for $a \in A$, $h \in H$.*

Proposition 5.62 may be easily extended to a Universal Property of the two-sided smash product (the proof is left to the reader):

**Proposition 5.63** *Let $H$ be a quasi-Hopf algebra with bijective antipode, $A$ a left $H$-module algebra and $B$ a right $H$-module algebra. Denote by $i_A$, $i_B$, $j$ the standard inclusions of $A$, $B$ and $H$, respectively, into $A\#H\#B$. Let $X$ be an algebra, $\gamma : H \to X$ an algebra map and $v_A : A \to X$, $v_B : B \to X$ two linear maps satisfying the conditions:*

$$\begin{aligned}
&\gamma(h)v_A(a) = v_A(h_1 \cdot a)\gamma(h_2), \quad v_A(aa') = v_A(X^1 \cdot a)v_A(X^2 \cdot a')\gamma(X^3),\\
&v_B(b)\gamma(h) = \gamma(h_1)v_B(b \cdot h_2), \quad v_B(bb') = \gamma(X^1)v_B(b \cdot X^2)v_B(b' \cdot X^3),\\
&v_A(1_A) = 1_X = v_B(1_B), \quad v_B(b)v_A(a) = v_A(x^1 \cdot a)\gamma(x^2)v_B(b \cdot x^3),
\end{aligned}$$

*for all $a, a' \in A$, $b, b' \in B$ and $h \in H$. Then there exists a unique algebra map $w : A\#H\#B \to X$ such that $w \circ i_A = v_A$, $w \circ i_B = v_B$ and $w \circ j = \gamma$. Moreover, $w$ is given by the formula $w(a\#h\#b) = v_A(a)\gamma(h)v_B(b)$, for all $a \in A$, $h \in H$, $b \in B$.*

## 5.7 L–R-smash Products

We introduce a new type of crossed product associated to a quasi-bialgebra, which in the case of a quasi-Hopf algebra with bijective antipode will turn out to be isomorphic to a diagonal crossed product.

**Proposition 5.64** *Let $H$ be a quasi-bialgebra, $\mathscr{A}$ an $H$-bimodule algebra and $\mathbb{A}$ an $H$-bicomodule algebra. Define on $\mathscr{A} \otimes \mathbb{A}$ the product*

$$(\varphi \natural u)(\psi \natural u') = (\tilde{x}^1_\lambda \cdot \varphi \cdot \theta^3 u'_{\langle 1\rangle}\tilde{x}^2_\rho)(\tilde{x}^2_\lambda u_{[-1]}\theta^1 \cdot \psi \cdot \tilde{x}^3_\rho) \natural \tilde{x}^3_\lambda u_{[0]}\theta^2 u'_{\langle 0\rangle}\tilde{x}^1_\rho, \quad (5.7.1)$$

*for $\varphi, \psi \in \mathscr{A}$ and $u, u' \in \mathbb{A}$, where $\Phi_\rho^{-1} = \tilde{x}^1_\rho \otimes \tilde{x}^2_\rho \otimes \tilde{x}^3_\rho$, $\Phi_\lambda^{-1} = \tilde{x}^1_\lambda \otimes \tilde{x}^2_\lambda \otimes \tilde{x}^3_\lambda$, $\Phi_{\lambda,\rho}^{-1} = \theta^1 \otimes \theta^2 \otimes \theta^3$, and we write $\varphi \natural u$ instead of $\varphi \otimes u$ to distinguish the new algebraic structure. Then this product defines on $\mathscr{A} \otimes \mathbb{A}$ a structure of an associative algebra with unit $1_{\mathscr{A}} \natural 1_{\mathbb{A}}$, denoted by $\mathscr{A} \natural \mathbb{A}$ and called the L–R-smash product.*

*Proof* For $\varphi, \psi, \xi \in \mathscr{A}$ and $u, u', u'' \in \mathbb{A}$ we compute:

$$\begin{aligned}
&[(\varphi \natural u)(\psi \natural u')](\xi \natural u'')\\
&\overset{(5.7.1)}{=} [(\tilde{x}^1_\lambda \cdot \varphi \cdot \theta^3 u'_{\langle 1\rangle}\tilde{x}^2_\rho)(\tilde{x}^2_\lambda u_{[-1]}\theta^1 \cdot \psi \cdot \tilde{x}^3_\rho) \natural \tilde{x}^3_\lambda u_{[0]}\theta^2 u'_{\langle 0\rangle}\tilde{x}^1_\rho](\xi \natural u'')\\
&\overset{(5.7.1)}{=} [((\tilde{y}^1_\lambda)_1\tilde{x}^1_\lambda \cdot \varphi \cdot \theta^3 u'_{\langle 1\rangle}\tilde{x}^2_\rho\overline{\theta}^3_1 u''_{\langle 1\rangle_1}(\tilde{y}^2_\rho)_1)((\tilde{y}^1_\lambda)_2\tilde{x}^2_\lambda u_{[-1]}\theta^1 \cdot \psi\\
&\qquad \cdot \tilde{x}^3_\rho\overline{\theta}^3_2 u''_{\langle 1\rangle_2}(\tilde{y}^2_\rho)_2)](\tilde{y}^2_\lambda(\tilde{x}^3_\lambda)_{[-1]}u_{[0,-1]}\theta^2_{[-1]}u'_{\langle 0\rangle_{[-1]}}(\tilde{x}^1_\rho)_{[-1]}\overline{\theta}^1 \cdot \xi \cdot \tilde{y}^3_\rho)\\
&\qquad \natural \tilde{y}^3_\lambda(\tilde{x}^3_\lambda)_{[0]}u_{[0,0]}\theta^2_{[0]}u'_{\langle 0\rangle_{[0]}}(\tilde{x}^1_\rho)_{[0]}\overline{\theta}^2 u''_{\langle 0\rangle}\tilde{y}^1_\rho\\
&\underset{(4.4.3)}{\overset{(4.3.6)}{=}} [(t^1\tilde{y}^1_\lambda \cdot \varphi \cdot \theta^3 u'_{\langle 1\rangle}\tilde{\theta}^3(\overline{\theta}^2)_{\langle 1\rangle}\tilde{x}^2_\rho u''_{\langle 1\rangle_1}(\tilde{y}^2_\rho)_1)(t^2(\tilde{y}^2_\lambda)_1\tilde{x}^1_\lambda u_{[-1]}\theta^1 \cdot \psi
\end{aligned}$$

$$
\begin{aligned}
&\cdot\overline{\theta}^3\tilde{x}^3_\rho u''_{\langle 1\rangle_2}(\tilde{y}^2_\rho)_2)](t^3(\tilde{y}^2_\lambda)_2\tilde{x}^2_\lambda u_{[0,-1]}\theta^2_{[-1]}u'_{\langle 0\rangle_{[-1]}}\tilde{\theta}^1\overline{\theta}^1\cdot\xi\cdot\tilde{y}^3_\rho)\\
&\natural\,\tilde{y}^3_\lambda\tilde{x}^3_\lambda u_{[0,0]}\theta^2_{[0]}u'_{\langle 0\rangle_{[0]}}\tilde{\theta}^2(\overline{\theta}^2)_{\langle 0\rangle}\tilde{x}^1_\rho u''_{\langle 0\rangle}\tilde{y}^1_\rho\\
\overset{(4.1.13),(4.3.1)}{\underset{(4.3.5)}{=}}\ &(\tilde{y}^1_\lambda\cdot\varphi\cdot\theta^3u'_{\langle 1\rangle}\tilde{\theta}^3(\overline{\theta}^2)_{\langle 1\rangle}u''_{\langle 0,1\rangle}\tilde{x}^2_\rho(\tilde{y}^2_\rho)_1t^1)[((\tilde{y}^2_\lambda)_1u_{[-1]_1}\tilde{x}^1_\lambda\theta^1\cdot\psi\\
&\cdot\overline{\theta}^3u''_{\langle 1\rangle}\tilde{x}^3_\rho(\tilde{y}^2_\rho)_2t^2)((\tilde{y}^2_\lambda)_2u_{[-1]_2}\tilde{x}^2_\lambda\theta^2_{[-1]}u'_{\langle 0\rangle_{[-1]}}\tilde{\theta}^1\overline{\theta}^1\cdot\xi\cdot\tilde{y}^3_\rho t^3)]\\
&\natural\,\tilde{y}^3_\lambda u_{[0]}\tilde{x}^3_\lambda\theta^2_{[0]}u'_{\langle 0\rangle_{[0]}}\tilde{\theta}^2(\overline{\theta}^2)_{\langle 0\rangle}u''_{\langle 0,0\rangle}\tilde{x}^1_\rho\tilde{y}^1_\rho\\
\overset{(4.4.1)}{=}\ &(\tilde{y}^1_\lambda\cdot\varphi\cdot\theta^3\tilde{\theta}^3u'_{[0]_{\langle 1\rangle}}(\overline{\theta}^2)_{\langle 1\rangle}u''_{\langle 0,1\rangle}\tilde{x}^2_\rho(\tilde{y}^2_\rho)_1t^1)[((\tilde{y}^2_\lambda)_1u_{[-1]_1}\tilde{x}^1_\lambda\theta^1\cdot\psi\\
&\cdot\overline{\theta}^3u''_{\langle 1\rangle}\tilde{x}^3_\rho(\tilde{y}^2_\rho)_2t^2)((\tilde{y}^2_\lambda)_2u_{[-1]_2}\tilde{x}^2_\lambda\theta^2_{[-1]}\tilde{\theta}^1u'_{[-1]}\overline{\theta}^1\cdot\xi\cdot\tilde{y}^3_\rho t^3)]\\
&\natural\,\tilde{y}^3_\lambda u_{[0]}\tilde{x}^3_\lambda\theta^2_{[0]}\tilde{\theta}^2u'_{[0]_{\langle 0\rangle}}(\overline{\theta}^2)_{\langle 0\rangle}u''_{\langle 0,0\rangle}\tilde{x}^1_\rho\tilde{y}^1_\rho\\
\overset{(4.3.2)}{\underset{(4.4.2)}{=}}\ &(\tilde{y}^1_\lambda\cdot\varphi\cdot\theta^3(\tilde{x}^3_\lambda)_{\langle 1\rangle}u'_{[0]_{\langle 1\rangle}}(\overline{\theta}^2)_{\langle 1\rangle}u''_{\langle 0,1\rangle}(\tilde{x}^1_\rho)_{\langle 1\rangle}\tilde{y}^2_\rho)[((\tilde{y}^2_\lambda)_1u_{[-1]_1}\theta^1_1\tilde{x}^1_\lambda\cdot\psi\\
&\cdot\overline{\theta}^3u''_{\langle 1\rangle}\tilde{x}^2_\rho(\tilde{y}^3_\rho)_1)((\tilde{y}^2_\lambda)_2u_{[-1]_2}\theta^1_2\tilde{x}^2_\lambda u'_{[-1]}\overline{\theta}^1\cdot\xi\cdot\tilde{x}^3_\rho(\tilde{y}^3_\rho)_2)]\\
&\natural\,\tilde{y}^3_\lambda u_{[0]}\theta^2(\tilde{x}^3_\lambda)_{\langle 0\rangle}u'_{[0]_{\langle 0\rangle}}(\overline{\theta}^2)_{\langle 0\rangle}u''_{\langle 0,0\rangle}(\tilde{x}^1_\rho)_{\langle 0\rangle}\tilde{y}^1_\rho\\
\overset{(5.7.1)}{=}\ &(\varphi\natural u)[(\tilde{x}^1_\lambda\cdot\psi\cdot\overline{\theta}^3u''_{\langle 1\rangle}\tilde{x}^2_\rho)(\tilde{x}^2_\lambda u'_{[-1]}\overline{\theta}^1\cdot\xi\cdot\tilde{x}^3_\rho)\natural\,\tilde{x}^3_\lambda u'_{[0]}\overline{\theta}^2u''_{\langle 0\rangle}\tilde{x}^1_\rho]\\
\overset{(5.7.1)}{=}\ &(\varphi\natural u)[(\psi\natural u')(\xi\natural u'')],
\end{aligned}
$$

hence the multiplication is associative. It is easy to check that $1_{\mathscr{A}}\natural 1_{\mathbb{A}}$ is the unit. □

**Remark 5.65** It is easy to see that, in $\mathscr{A}\natural\mathbb{A}$, we have $(1_{\mathscr{A}}\natural u)(1_{\mathscr{A}}\natural u')=1_{\mathscr{A}}\natural uu'$ for all $u,u'\in\mathbb{A}$, hence the map $\mathbb{A}\to\mathscr{A}\natural\mathbb{A}$, $u\mapsto 1_{\mathscr{A}}\natural u$, is an algebra map, and $(\varphi\natural 1_{\mathbb{A}})(1_{\mathscr{A}}\natural u)=\varphi\cdot u_{\langle 1\rangle}\natural u_{\langle 0\rangle}$, for all $\varphi\in\mathscr{A}$ and $u\in\mathbb{A}$.

The examples below justify the name of this construction.

**Examples 5.66** (1) Let $A$ be a left $H$-module algebra. Then $A$ becomes an $H$-bimodule algebra, with right $H$-action given via $\varepsilon$. In this case the multiplication of $A\natural\mathbb{A}$ becomes

$$(a\natural u)(a'\natural u')=(\tilde{x}^1_\lambda\cdot a)(\tilde{x}^2_\lambda u_{[-1]}\cdot a')\natural\tilde{x}^3_\lambda u_{[0]}u',$$

for all $a,a'\in A$ and $u,u'\in\mathbb{A}$, hence in this case $A\natural\mathbb{A}$ coincides with the generalized smash product $A\blacktriangleright\!\!<\mathbb{A}$.

(2) As we know, $H$ itself is an $H$-bicomodule algebra. So, in this case, the multiplication of $\mathscr{A}\natural H$ specializes to

$$(\varphi\natural h)(\psi\natural h')=(x^1\cdot\varphi\cdot t^3h'_2y^2)(x^2h_1t^1\cdot\psi\cdot y^3)\natural x^3h_2t^2h'_1y^1, \tag{5.7.2}$$

for all $\varphi,\psi\in\mathscr{A}$ and $h,h'\in H$. If the right $H$-module structure of $\mathscr{A}$ is trivial, then $\mathscr{A}\natural H$ coincides with the smash product $\mathscr{A}\#H$.

Next, we show that either a two-sided smash product or a two-sided crossed product can be identified, up to isomorphism, to certain L–R-smash products.

**Proposition 5.67** *Let $H$ be a quasi-bialgebra, $A$ a left $H$-module algebra, $B$ a right $H$-module algebra and $\mathbb{A}$ an $H$-bicomodule algebra. If we consider $A\otimes B$ as an $H$-bimodule algebra, with $H$-actions*

$$h\cdot(a\otimes b)\cdot h'=h\cdot a\otimes b\cdot h',\ \ \forall\, a\in A,\ h,\ h'\in H,\ b\in B, \tag{5.7.3}$$

*then we have an algebra isomorphism*

$$\phi:(A\otimes B)\natural\,\mathbb{A}\cong A\blacktriangleright\!\!<\mathbb{A}>\!\!\blacktriangleleft B,$$
$$\phi((a\otimes b)\natural\,u)=a\blacktriangleright\!\!<u>\!\!\blacktriangleleft b,\ \ \forall\, a\in A,\ b\in B,\ u\in\mathbb{A}.$$

*Proof* We compute:

$$\begin{aligned}
&\phi([(a\otimes b)\natural\,u)][(a'\otimes b')\natural\,u'])\\
&\quad=\phi((\tilde{x}^1_\lambda\cdot(a\otimes b)\cdot\theta^3u'_{\langle 1\rangle}\tilde{x}^2_\rho)(\tilde{x}^2_\lambda u_{[-1]}\theta^1\cdot(a'\otimes b')\cdot\tilde{x}^3_\rho)\natural\,\tilde{x}^3_\lambda u_{[0]}\theta^2u'_{\langle 0\rangle}\tilde{x}^1_\rho)\\
&\quad=\phi((\tilde{x}^1_\lambda\cdot a\otimes b\cdot\theta^3u'_{\langle 1\rangle}\tilde{x}^2_\rho)(\tilde{x}^2_\lambda u_{[-1]}\theta^1\cdot a'\otimes b'\cdot\tilde{x}^3_\rho)\natural\,\tilde{x}^3_\lambda u_{[0]}\theta^2u'_{\langle 0\rangle}\tilde{x}^1_\rho)\\
&\quad=\phi(((\tilde{x}^1_\lambda\cdot a)(\tilde{x}^2_\lambda u_{[-1]}\theta^1\cdot a')\otimes(b\cdot\theta^3u'_{\langle 1\rangle}\tilde{x}^2_\rho)(b'\cdot\tilde{x}^3_\rho))\natural\,\tilde{x}^3_\lambda u_{[0]}\theta^2u'_{\langle 0\rangle}\tilde{x}^1_\rho)\\
&\quad=(\tilde{x}^1_\lambda\cdot a)(\tilde{x}^2_\lambda u_{[-1]}\theta^1\cdot a')\blacktriangleright\!\!<\tilde{x}^3_\lambda u_{[0]}\theta^2u'_{\langle 0\rangle}\tilde{x}^1_\rho>\!\!\blacktriangleleft(b\cdot\theta^3u'_{\langle 1\rangle}\tilde{x}^2_\rho)(b'\cdot\tilde{x}^3_\rho)\\
&\quad=(a\blacktriangleright\!\!<u>\!\!\blacktriangleleft b)(a'\blacktriangleright\!\!<u'>\!\!\blacktriangleleft b')\\
&\quad=\phi((a\otimes b)\natural\,u)\phi((a'\otimes b')\natural\,u'),
\end{aligned}$$

finishing the proof. □

**Proposition 5.68** *Let $H$ be a quasi-bialgebra, $\mathfrak{A}$ a right $H$-comodule algebra, $\mathfrak{B}$ a left $H$-comodule algebra and $\mathscr{A}$ an $H$-bimodule algebra. Consider $\mathfrak{A}\otimes\mathfrak{B}$ as an $H$-bicomodule algebra, with the following structure: $\rho(\mathfrak{a}\otimes\mathfrak{b})=(\mathfrak{a}_{\langle 0\rangle}\otimes\mathfrak{b})\otimes\mathfrak{a}_{\langle 1\rangle}$, $\lambda(\mathfrak{a}\otimes\mathfrak{b})=\mathfrak{b}_{[-1]}\otimes(\mathfrak{a}\otimes\mathfrak{b}_{[0]})$, $\Phi_\rho=(\tilde{X}^1_\rho\otimes 1_{\mathfrak{B}})\otimes\tilde{X}^2_\rho\otimes\tilde{X}^3_\rho$, $\Phi_\lambda=\tilde{X}^1_\lambda\otimes\tilde{X}^2_\lambda\otimes(1_{\mathfrak{A}}\otimes\tilde{X}^3_\lambda)$, $\Phi_{\lambda,\rho}=1_H\otimes(1_{\mathfrak{A}}\otimes 1_{\mathfrak{B}})\otimes 1_H$, for all $\mathfrak{a}\in\mathfrak{A}$ and $\mathfrak{b}\in\mathfrak{B}$. Then we have an algebra isomorphism*

$$\tau:\mathscr{A}\natural\,(\mathfrak{A}\otimes\mathfrak{B})\cong\mathfrak{A}\bowtie\!\!\!\triangleleft\,\mathscr{A}\ \triangleright\!\!\!<\ \mathfrak{B},$$
$$\tau(\varphi\natural\,(\mathfrak{a}\otimes\mathfrak{b}))=\mathfrak{a}\bowtie\!\!\!\triangleleft\,\varphi\ \triangleright\!\!\!<\ \mathfrak{b},\ \ \forall\,\varphi\in\mathscr{A},\ \mathfrak{a}\in\mathfrak{A},\ \mathfrak{b}\in\mathfrak{B}.$$

*Proof* We compute:

$$\begin{aligned}
&\tau((\varphi\natural\,(\mathfrak{a}\otimes\mathfrak{b}))(\varphi'\natural\,(\mathfrak{a}'\otimes\mathfrak{b}')))\\
&\quad=\tau((\tilde{x}^1_\lambda\cdot\varphi\cdot(\mathfrak{a}'\otimes\mathfrak{b}')_{\langle 1\rangle}\tilde{x}^2_\rho)(\tilde{x}^2_\lambda(\mathfrak{a}\otimes\mathfrak{b})_{[-1]}\cdot\varphi'\cdot\tilde{x}^3_\rho)\\
&\qquad\natural\,(1_{\mathfrak{A}}\otimes\tilde{x}^3_\lambda)(\mathfrak{a}\otimes\mathfrak{b})_{[0]}(\mathfrak{a}'\otimes\mathfrak{b}')_{\langle 0\rangle}(\tilde{x}^1_\rho\otimes 1_{\mathfrak{B}}))\\
&\quad=\tau((\tilde{x}^1_\lambda\cdot\varphi\cdot\mathfrak{a}'_{\langle 1\rangle}\tilde{x}^2_\rho)(\tilde{x}^2_\lambda\mathfrak{b}_{[-1]}\cdot\varphi'\cdot\tilde{x}^3_\rho)\natural\,(1_{\mathfrak{A}}\otimes\tilde{x}^3_\lambda)(\mathfrak{a}\otimes\mathfrak{b}_{[0]})(\mathfrak{a}'_{\langle 0\rangle}\otimes\mathfrak{b}')(\tilde{x}^1_\rho\otimes 1_{\mathfrak{B}}))\\
&\quad=\tau((\tilde{x}^1_\lambda\cdot\varphi\cdot\mathfrak{a}'_{\langle 1\rangle}\tilde{x}^2_\rho)(\tilde{x}^2_\lambda\mathfrak{b}_{[-1]}\cdot\varphi'\cdot\tilde{x}^3_\rho)\natural\,(\mathfrak{a}\mathfrak{a}'_{\langle 0\rangle}\tilde{x}^1_\rho\otimes\tilde{x}^3_\lambda\mathfrak{b}_{[0]}\mathfrak{b}'))\\
&\quad=\mathfrak{a}\mathfrak{a}'_{\langle 0\rangle}\tilde{x}^1_\rho\bowtie\!\!\!\triangleleft\,(\tilde{x}^1_\lambda\cdot\varphi\cdot\mathfrak{a}'_{\langle 1\rangle}\tilde{x}^2_\rho)(\tilde{x}^2_\lambda\mathfrak{b}_{[-1]}\cdot\varphi'\cdot\tilde{x}^3_\rho)\ \triangleright\!\!\!<\ \tilde{x}^3_\lambda\mathfrak{b}_{[0]}\mathfrak{b}'\\
&\quad=(\mathfrak{a}\bowtie\!\!\!\triangleleft\,\varphi\ \triangleright\!\!\!<\ \mathfrak{b})(\mathfrak{a}'\bowtie\!\!\!\triangleleft\,\varphi'\ \triangleright\!\!\!<\ \mathfrak{b}')\\
&\quad=\tau(\varphi\natural\,(\mathfrak{a}\otimes\mathfrak{b}))\tau(\varphi'\natural\,(\mathfrak{a}'\otimes\mathfrak{b}')),
\end{aligned}$$

and the proof is finished. □

**Lemma 5.69** *Let $H$ be a quasi-Hopf algebra with bijective antipode and $\mathbb{A}$ an $H$-bicomodule algebra. Consider the element $\Omega \in H^{\otimes 2} \otimes \mathbb{A} \otimes H^{\otimes 2}$ given by (5.6.9). If we denote by $\tilde{Q}^1_\rho \otimes \tilde{Q}^2_\rho$ another copy of the element $\tilde{q}_\rho$ given by formula (4.3.9) and by $\mho^1 \otimes \mho^2 \otimes \mho^3$ another copy of $\Phi_{\lambda,\rho}$, then we have:*

$$\begin{aligned}
&\Theta^1_1\Omega^1 \otimes \Theta^1_2\Omega^2 \otimes \tilde{q}^1_\rho(\Theta^2\Omega^3)_{\langle 0\rangle} \otimes \Omega^5 S^{-1}(\Theta^3)_1(\tilde{q}^2_\rho)_1(\Theta^2\Omega^3)_{\langle 1\rangle_1}\\
&\otimes \Omega^4 S^{-1}(\Theta^3)_2(\tilde{q}^2_\rho)_2(\Theta^2\Omega^3)_{\langle 1\rangle_2}\\
&\qquad = \tilde{x}^1_\lambda\Theta^1 \otimes \tilde{x}^2_\lambda \mho^1\Theta^2_{[-1]}\overline{\Theta}^1 \otimes \tilde{x}^3_\lambda \tilde{q}^1_\rho(\mho^2\Theta^2_{[0]}\tilde{Q}^1_\rho\overline{\Theta}^2_{\langle 0\rangle})_{\langle 0\rangle}\tilde{x}^1_\rho\\
&\qquad\quad \otimes S^{-1}(\mho^3\Theta^3)\tilde{q}^2_\rho(\mho^2\Theta^2_{[0]}\tilde{Q}^1_\rho\overline{\Theta}^2_{\langle 0\rangle})_{\langle 1\rangle}\tilde{x}^2_\rho \otimes S^{-1}(\overline{\Theta}^3)\tilde{Q}^2_\rho\overline{\Theta}^2_{\langle 1\rangle}\tilde{x}^3_\rho, \qquad (5.7.4)
\end{aligned}$$

$$\begin{aligned}
&\overline{\Theta}^1 u_{\langle 0\rangle_{[-1]}} \otimes (\tilde{Q}^1_\rho\overline{\Theta}^2_{\langle 0\rangle})_{\langle 0\rangle}\tilde{x}^1_\rho u_{\langle 0\rangle_{[0]_{\langle 0\rangle}}} u'_{\langle 0\rangle} \otimes (\tilde{Q}^1_\rho\overline{\Theta}^2_{\langle 0\rangle})_{\langle 1\rangle}\tilde{x}^2_\rho u_{\langle 0\rangle_{[0]_{\langle 1\rangle_1}}} u'_{\langle 1\rangle_1}\\
&\quad \otimes S^{-1}(\overline{\Theta}^3 u_{\langle 1\rangle})\tilde{Q}^2_\rho\overline{\Theta}^2_{\langle 1\rangle}\tilde{x}^3_\rho u_{\langle 0\rangle_{[0]_{\langle 1\rangle_2}}} u'_{\langle 1\rangle_2} = u_{[-1]}\overline{\Theta}^1 \otimes (u_{[0]}\tilde{Q}^1_\rho)_{\langle 0\rangle}(\overline{\Theta}^2 u')_{\langle 0,0\rangle}\tilde{x}^1_\rho\\
&\qquad \otimes (u_{[0]}\tilde{Q}^1_\rho)_{\langle 1\rangle}(\overline{\Theta}^2 u')_{\langle 0,1\rangle}\tilde{x}^2_\rho \otimes S^{-1}(\overline{\Theta}^3)\tilde{Q}^2_\rho(\overline{\Theta}^2 u')_{\langle 1\rangle}\tilde{x}^3_\rho, \qquad (5.7.5)
\end{aligned}$$

$$\Theta^1 \otimes \tilde{q}^1_\rho\Theta^2_{\langle 0\rangle} \otimes S^{-1}(\Theta^3)\tilde{q}^2_\rho\Theta^2_{\langle 1\rangle} = (\tilde{q}^1_\rho)_{[-1]}\theta^1 \otimes (\tilde{q}^1_\rho)_{[0]}\theta^2 \otimes \tilde{q}^2_\rho\theta^3. \qquad (5.7.6)$$

*Proof* We only indicate the main steps and leave details to the reader. The relation (5.7.4) follows by using (5.6.25), (4.3.15), (4.4.2) and using (4.3.11) and (4.3.1) several times. By using (4.3.1), (4.4.1) and (4.3.11) one obtains (5.7.5). Finally, (5.7.6) follows by using (4.3.9), (4.4.3), (3.2.1) and (4.4.4). □

**Theorem 5.70** *Let $H$ be a quasi-Hopf algebra with bijective antipode, $\mathscr{A}$ an $H$-bimodule algebra and $\mathbb{A}$ an $H$-bicomodule algebra. Then the linear map*

$$\nu : \mathscr{A} \bowtie \mathbb{A} \to \mathscr{A} \natural \mathbb{A}, \quad \nu(\varphi \bowtie u) = \Theta^1 \cdot \varphi \cdot S^{-1}(\Theta^3)\tilde{q}^2_\rho\Theta^2_{\langle 1\rangle}u_{\langle 1\rangle} \natural \tilde{q}^1_\rho\Theta^2_{\langle 0\rangle}u_{\langle 0\rangle}, \quad (5.7.7)$$

*for all $\varphi \in \mathscr{A}$ and $u \in \mathbb{A}$, is an algebra isomorphism, with inverse*

$$\nu^{-1} : \mathscr{A} \natural \mathbb{A} \to \mathscr{A} \bowtie \mathbb{A}, \quad \nu^{-1}(\varphi \natural u) = \theta^1 \cdot \varphi \cdot S^{-1}(\theta^3 u_{\langle 1\rangle}\tilde{p}^2_\rho) \bowtie \theta^2 u_{\langle 0\rangle}\tilde{p}^1_\rho. \quad (5.7.8)$$

*Proof* First we establish that $\nu$ is an algebra map. We compute:

$$\begin{aligned}
&\nu((\varphi \bowtie u)(\psi \bowtie u'))\\
&\overset{(5.6.15)}{\underset{(5.7.7)}{=}} (\Theta^1_1\Omega^1 \cdot \varphi \cdot \Omega^5 S^{-1}(\Theta^3)_1(\tilde{q}^2_\rho)_1(\Theta^2\Omega^3)_{\langle 1\rangle_1}u_{\langle 0\rangle_{[0]_{\langle 1\rangle_1}}} u'_{\langle 1\rangle_1})\\
&\quad \left(\Theta^1_2\Omega^2 u_{\langle 0\rangle_{[-1]}} \cdot \psi \cdot S^{-1}(u_{\langle 1\rangle})\Omega^4 S^{-1}(\Theta^3)_2(\tilde{q}^2_\rho)_2(\Theta^2\Omega^3)_{\langle 1\rangle_2}u_{\langle 0\rangle_{[0]_{\langle 1\rangle_2}}} u'_{\langle 1\rangle_2}\right)\\
&\quad \natural\, \tilde{q}^1_\rho(\Theta^2\Omega^3)_{\langle 0\rangle}u_{\langle 0\rangle_{[0]_{\langle 0\rangle}}} u'_{\langle 0\rangle}\\
&\overset{(5.7.4)}{=} \left(\tilde{x}^1_\lambda\Theta^1 \cdot \varphi \cdot S^{-1}(\mho^3\Theta^3)\tilde{q}^2_\rho\mho^2_{\langle 1\rangle}\Theta^2_{[0]_{\langle 1\rangle}}(\tilde{Q}^1_\rho\overline{\Theta}^2_{\langle 0\rangle})_{\langle 1\rangle}\tilde{x}^2_\rho u_{\langle 0\rangle_{[0]_{\langle 1\rangle_1}}} u'_{\langle 1\rangle_1}\right)\\
&\quad \left(\tilde{x}^2_\lambda\mho^1\Theta^2_{[-1]}\overline{\Theta}^1 u_{\langle 0\rangle_{[-1]}} \cdot \psi \cdot S^{-1}(\overline{\Theta}^3 u_{\langle 1\rangle})\tilde{Q}^2_\rho\overline{\Theta}^2_{\langle 1\rangle}\tilde{x}^3_\rho u_{\langle 0\rangle_{[0]_{\langle 1\rangle_2}}} u'_{\langle 1\rangle_2}\right)\\
&\quad \natural\, \tilde{x}^3_\lambda\tilde{q}^1_\rho\mho^2_{\langle 0\rangle}\Theta^2_{[0]_{\langle 0\rangle}}(\tilde{Q}^1_\rho\overline{\Theta}^2_{\langle 0\rangle})_{\langle 0\rangle}\tilde{x}^1_\rho u_{\langle 0\rangle_{[0]_{\langle 0\rangle}}} u'_{\langle 0\rangle}\\
&\overset{(5.7.5)}{=} \left(\tilde{x}^1_\lambda\Theta^1 \cdot \varphi \cdot S^{-1}(\mho^3\Theta^3)\tilde{q}^2_\rho\mho^2_{\langle 1\rangle}\Theta^2_{[0]_{\langle 1\rangle}}(u_{[0]}\tilde{Q}^1_\rho)_{\langle 1\rangle}(\overline{\Theta}^2 u')_{\langle 0,1\rangle}\tilde{x}^2_\rho\right)
\end{aligned}$$

$$
\begin{aligned}
&\Big(\tilde{x}^2_\lambda \mho^1 \Theta^2_{[-1]} u_{[-1]} \overline{\Theta}^1 \cdot \psi \cdot S^{-1}(\overline{\Theta}^3) \tilde{Q}^2_\rho (\overline{\Theta}^2 u')_{\langle 1\rangle} \tilde{x}^3_\rho\Big)\\
&\natural\, \tilde{x}^3_\lambda \tilde{q}^1_\rho \mho^2_{\langle 0\rangle} \Theta^2_{[0]_{\langle 0\rangle}} (u_{[0]} \tilde{Q}^1_\rho)_{\langle 0\rangle} (\overline{\Theta}^2 u')_{\langle 0,0\rangle} \tilde{x}^1_\rho\\
\overset{(5.7.6)}{=}\ &\Big(\tilde{x}^1_\lambda \Theta^1 \cdot \varphi \cdot S^{-1}(\Theta^3) \tilde{q}^2_\rho \theta^3 \Theta^2_{[0]_{\langle 1\rangle}} (u_{[0]} \tilde{Q}^1_\rho)_{\langle 1\rangle} (\overline{\Theta}^2 u')_{\langle 0,1\rangle} \tilde{x}^2_\rho\Big)\\
&\Big(\tilde{x}^2_\lambda (\tilde{q}^1_\rho)_{[-1]} \theta^1 \Theta^2_{[-1]} u_{[-1]} \overline{\Theta}^1 \cdot \psi \cdot S^{-1}(\overline{\Theta}^3) \tilde{Q}^2_\rho (\overline{\Theta}^2 u')_{\langle 1\rangle} \tilde{x}^3_\rho\Big)\\
&\natural\, \tilde{x}^3_\lambda (\tilde{q}^1_\rho)_{[0]} \theta^2 \Theta^2_{[0]_{\langle 0\rangle}} (u_{[0]} \tilde{Q}^1_\rho)_{\langle 0\rangle} (\overline{\Theta}^2 u')_{\langle 0,0\rangle} \tilde{x}^1_\rho\\
\overset{(4.4.1)}{=}\ &\Big(\tilde{x}^1_\lambda \Theta^1 \cdot \varphi \cdot S^{-1}(\Theta^3) \tilde{q}^2_\rho \Theta^2_{\langle 1\rangle} u_{\langle 1\rangle} \theta^3 (\tilde{Q}^1_\rho)_{\langle 1\rangle} (\overline{\Theta}^2 u')_{\langle 0,1\rangle} \tilde{x}^2_\rho\Big)\\
&\Big(\tilde{x}^2_\lambda (\tilde{q}^1_\rho)_{[-1]} \Theta^2_{\langle 0\rangle_{[-1]}} u_{\langle 0\rangle_{[-1]}} \theta^1 \overline{\Theta}^1 \cdot \psi \cdot S^{-1}(\overline{\Theta}^3) \tilde{Q}^2_\rho (\overline{\Theta}^2 u')_{\langle 1\rangle} \tilde{x}^3_\rho\Big)\\
&\natural\, \tilde{x}^3_\lambda (\tilde{q}^1_\rho)_{[0]} \Theta^2_{\langle 0\rangle_{[0]}} u_{\langle 0\rangle_{[0]}} \theta^2 (\tilde{Q}^1_\rho)_{\langle 0\rangle} (\overline{\Theta}^2 u')_{\langle 0,0\rangle} \tilde{x}^1_\rho\\
\overset{(5.7.1)}{=}\ &\Big(\Theta^1 \cdot \varphi \cdot S^{-1}(\Theta^3) \tilde{q}^2_\rho \Theta^2_{\langle 1\rangle} u_{\langle 1\rangle} \natural\, \tilde{q}^1_\rho \Theta^2_{\langle 0\rangle} u_{\langle 0\rangle}\Big)\\
&\Big(\overline{\Theta}^1 \cdot \psi \cdot S^{-1}(\overline{\Theta}^3) \tilde{Q}^2_\rho \overline{\Theta}^2_{\langle 1\rangle} u'_{\langle 1\rangle} \natural\, \tilde{Q}^1_\rho \overline{\Theta}^2_{\langle 0\rangle} u'_{\langle 0\rangle}\Big)\\
\overset{(5.7.7)}{=}\ &\nu(\varphi \bowtie u)\nu(\psi \bowtie u'),
\end{aligned}
$$

as needed. The fact that $\nu(1_{\mathscr{A}} \bowtie 1_{\mathbb{A}}) = 1_{\mathscr{A}} \natural 1_{\mathbb{A}}$ is trivial.

We prove now that $\nu$ and $\nu^{-1}$ are inverses. Indeed, we have:

$$
\begin{aligned}
\nu\nu^{-1}(\varphi \natural u) = {}& \Theta^1 \theta^1 \cdot \varphi \cdot S^{-1}(\theta^3 u_{\langle 1\rangle} \tilde{p}^2_\rho) S^{-1}(\Theta^3) \tilde{q}^2_\rho \Theta^2_{\langle 1\rangle} \theta^2_{\langle 1\rangle} u_{\langle 0,1\rangle} (\tilde{p}^1_\rho)_{\langle 1\rangle}\\
&\natural\, \tilde{q}^1_\rho \Theta^2_{\langle 0\rangle} \theta^2_{\langle 0\rangle} u_{\langle 0,0\rangle} (\tilde{p}^1_\rho)_{\langle 0\rangle} \underset{(4.3.13)}{\overset{(4.3.11)}{=}} \varphi \natural u,
\end{aligned}
$$

and similarly

$$
\begin{aligned}
&\nu^{-1}\nu\,(\varphi \bowtie u)\\
&\quad= \quad \theta^1 \Theta^1 \cdot \varphi \cdot S^{-1}(\Theta^3) \tilde{q}^2_\rho \Theta^2_{\langle 1\rangle} u_{\langle 1\rangle} S^{-1}(\theta^3 (\tilde{q}^1_\rho)_{\langle 1\rangle} \Theta^2_{\langle 0,1\rangle} u_{\langle 0,1\rangle} \tilde{p}^2_\rho)\\
&\qquad\qquad \bowtie \theta^2 (\tilde{q}^1_\rho)_{\langle 0\rangle} \Theta^2_{\langle 0,0\rangle} u_{\langle 0,0\rangle} \tilde{p}^1_\rho\\
&\overset{(4.3.10),(4.3.12)}{=} \theta^1 \Theta^1 \cdot \varphi \cdot S^{-1}(\theta^3 \Theta^3) \bowtie \theta^2 \Theta^2 u = \varphi \bowtie u,
\end{aligned}
$$

so we are done. □

**Examples 5.71** (1) If $A$ is a left $H$-module algebra regarded as an $H$-bimodule algebra with trivial right $H$-action, then $A \bowtie \mathbb{A}$ and $A \natural \mathbb{A}$ both coincide with $A \blacktriangleright\!\!< \mathbb{A}$, and the isomorphism $\nu$ is just the identity.

(2) If $\mathbb{A} = H$ then $\nu : \mathscr{A} \bowtie H \to \mathscr{A} \natural H$ and $\nu^{-1} : \mathscr{A} \natural H \to \mathscr{A} \bowtie H$ are given by

$$
\begin{aligned}
\nu(\varphi \bowtie h) &= X^1 \cdot \varphi \cdot S^{-1}(X^3) q^2 X^2_2 h_2 \natural q^1 X^2_1 h_1,\\
\nu^{-1}(\varphi \natural h) &= x^1 \cdot \varphi \cdot S^{-1}(x^3 h_2 p^2) \bowtie x^2 h_1 p^1,
\end{aligned}
$$

for all $\varphi \in \mathscr{A}$ and $h \in H$, where $q_R = q^1 \otimes q^2$ and $p_R = p^1 \otimes p^2$ are from (3.2.19).

As consequences of Propositions 5.67, 5.68 and Theorem 5.70, we obtain:

**Corollary 5.72** *With the hypotheses of Proposition 5.67 and assuming, moreover, that $H$ is a quasi-Hopf algebra with bijective antipode, we have an algebra isomorphism*

$$\mu : (A \otimes B) \bowtie \mathbb{A} \cong A \blacktriangleright\!\!< \mathbb{A} >\!\!\blacktriangleleft B,$$
$$\mu((a \otimes b) \bowtie u) = \Theta^1 \cdot a \blacktriangleright\!\!< \tilde{q}^1_\rho \Theta^2_{\langle 0 \rangle} u_{\langle 0 \rangle} >\!\!\blacktriangleleft b \cdot S^{-1}(\Theta^3) \tilde{q}^2_\rho \Theta^2_{\langle 1 \rangle} u_{\langle 1 \rangle},$$

*for all $a \in A$, $b \in B$ and $u \in \mathbb{A}$.*

**Corollary 5.73** *With the hypotheses of Proposition 5.68 and assuming, moreover, that $H$ is a quasi-Hopf algebra with bijective antipode, we have an algebra isomorphism*

$$\eta : \mathfrak{A} \rtimes \mathscr{A} \ltimes \mathfrak{B} \cong \mathscr{A} \bowtie (\mathfrak{A} \otimes \mathfrak{B}),$$
$$\eta(\mathfrak{a} \rtimes \varphi \ltimes \mathfrak{b}) = \varphi \cdot S^{-1}(\mathfrak{a}_{\langle 1 \rangle} \tilde{p}^2_\rho) \bowtie (\mathfrak{a}_{\langle 0 \rangle} \tilde{p}^1_\rho \otimes \mathfrak{b}),$$

*for all $\mathfrak{a} \in \mathfrak{A}$, $\mathfrak{b} \in \mathfrak{B}$ and $\varphi \in \mathscr{A}$.*

As a consequence of these two results, we obtain:

**Corollary 5.74** *Let $H$ be a quasi-Hopf algebra with bijective antipode, $A$ a left $H$-module algebra, $B$ a right $H$-module algebra, $\mathfrak{A}$ a right $H$-comodule algebra and $\mathfrak{B}$ a left $H$-comodule algebra. Then we have algebra isomorphisms*

$$A \blacktriangleright\!\!< (\mathfrak{A} \otimes \mathfrak{B}) >\!\!\blacktriangleleft B \cong (A \otimes B) \bowtie (\mathfrak{A} \otimes \mathfrak{B}) \cong \mathfrak{A} \rtimes (A \otimes B) \ltimes \mathfrak{B}.$$

We now study a kind of invariance under twisting of the L–R-smash product.

Let $H$ be a quasi-bialgebra, $\mathscr{A}$ an $H$-bimodule algebra and $F \in H \otimes H$ a gauge transformation. If we introduce on $\mathscr{A}$ another multiplication, $\varphi \circ \varphi' = (G^1 \cdot \varphi \cdot F^1)(G^2 \cdot \varphi' \cdot F^2)$ for all $\varphi, \varphi' \in \mathscr{A}$, where $F^{-1} = G^1 \otimes G^2$, and denote this structure by ${}_F\mathscr{A}_{F^{-1}}$, then one can check that ${}_F\mathscr{A}_{F^{-1}}$ is an $H_F$-bimodule algebra, with the same unit and $H$-actions as for $\mathscr{A}$.

Suppose that $\mathfrak{B}$ is a left $H$-comodule algebra; then on the algebra structure of $\mathfrak{B}$ one can introduce a left $H_F$-comodule algebra structure (denoted by $\mathfrak{B}^{F^{-1}}$ in what follows) by putting $\lambda^{F^{-1}} = \lambda$ and $\Phi_\lambda^{F^{-1}} = \Phi_\lambda(F^{-1} \otimes 1_{\mathfrak{B}})$. Similarly, if $\mathfrak{A}$ is a right $H$-comodule algebra, one can introduce on the algebra structure of $\mathfrak{A}$ a right $H_F$-comodule algebra structure (denoted by ${}^F\mathfrak{A}$ in what follows) by putting ${}^F\rho = \rho$ and ${}^F\Phi_\rho = (1_{\mathfrak{A}} \otimes F)\Phi_\rho$. One checks that if $\mathbb{A}$ is an $H$-bicomodule algebra, the left and right $H_F$-comodule algebras $\mathbb{A}^{F^{-1}}$ and ${}^F\mathbb{A}$, respectively, actually define the structure of an $H_F$-bicomodule algebra on $\mathbb{A}$, denoted by ${}^F\mathbb{A}^{F^{-1}}$, which has the same $\Phi_{\lambda,\rho}$ as $\mathbb{A}$.

**Proposition 5.75** *With notation as above, we have an algebra isomorphism*

$$\mathscr{A} \natural \mathbb{A} \equiv {}_F\mathscr{A}_{F^{-1}} \natural {}^F\mathbb{A}^{F^{-1}},$$

*given by the trivial identification.*

*Proof* Let $\mathscr{F}^1\otimes\mathscr{F}^2$ and $\mathscr{G}^1\otimes\mathscr{G}^2$ be two more copies of $F$ and $F^{-1}$, respectively. We compute the multiplication in ${}_F\mathscr{A}_{F^{-1}}\natural\,{}^F\mathbb{A}^{F^{-1}}$:

$$\begin{aligned}&(\varphi\natural u)(\psi\natural u')\\&=(F^1\tilde{x}^1_\lambda\cdot\varphi\cdot\theta^3u'_{\langle 1\rangle}\tilde{x}^2_\rho G^1)\circ(F^2\tilde{x}^2_\lambda u_{[-1]}\theta^1\cdot\psi\cdot\tilde{x}^3_\rho G^2)\natural\tilde{x}^3_\lambda u_{[0]}\theta^2u'_{\langle 0\rangle}\tilde{x}^1_\rho\\&=(\mathscr{G}^1F^1\tilde{x}^1_\lambda\cdot\varphi\cdot\theta^3u'_{\langle 1\rangle}\tilde{x}^2_\rho G^1\mathscr{F}^1)(\mathscr{G}^2F^2\tilde{x}^2_\lambda u_{[-1]}\theta^1\cdot\psi\cdot\tilde{x}^3_\rho G^2\mathscr{F}^2)\natural\tilde{x}^3_\lambda u_{[0]}\theta^2u'_{\langle 0\rangle}\tilde{x}^1_\rho\\&=(\tilde{x}^1_\lambda\cdot\varphi\cdot\theta^3u'_{\langle 1\rangle}\tilde{x}^2_\rho)(\tilde{x}^2_\lambda u_{[-1]}\theta^1\cdot\psi\cdot\tilde{x}^3_\rho)\natural\tilde{x}^3_\lambda u_{[0]}\theta^2u'_{\langle 0\rangle}\tilde{x}^1_\rho,\end{aligned}$$

which is the multiplication of $\mathscr{A}\natural\mathbb{A}$. □

## 5.8 A Duality Theorem for Quasi-Hopf Algebras

We present a situation when a certain quasi-Hopf smash product is isomorphic to a usual tensor product of associative algebras. As an application, we obtain a duality theorem for quasi-Hopf algebras.

**Proposition 5.76** *Let $H$ be a quasi-Hopf algebra, $B$ an algebra and $v:H\to B$ an algebra map. Denote by $\eta$ the algebra map $\eta:H\to B\otimes H$, $\eta(h)=v(h_1)\otimes h_2$. Define the map*

$$u:B\to B\otimes H,\quad u(b)=v(x^1)bv(S(x^3_2X^3)f^1)\otimes x^2X^1\beta S(x^3_1X^2)f^2,\tag{5.8.1}$$

*where $f=f^1\otimes f^2$ is the Drinfeld twist given by* (3.2.15). *Then $u$ is a morphism of left $H$-module algebras from $B^v$ to $(B\otimes H)^\eta$.*

*Proof* The fact that $u(v(\beta))=\eta(\beta)$ follows immediately from (3.2.14). We check now that $u$ is a morphism of left $H$-modules:

$$\begin{aligned}&h\triangleright_\eta u(b)\\&=\eta(h_1)u(b)\eta(S(h_2))\\&=v(h_{(1,1)})v(x^1)bv(S(x^3_2X^3)f^1)v(S(h_2)_1)\otimes h_{(1,2)}x^2X^1\beta S(x^3_1X^2)f^2S(h_2)_2\\&\overset{(3.2.13)}{=}v(h_{(1,1)}x^1)bv(S(h_{(2,2)}x^3_2X^3)f^1)\otimes h_{(1,2)}x^2X^1\beta S(h_{(2,1)}x^3_1X^2)f^2\\&\overset{(3.1.7)}{=}v(x^1h_1)bv(S(x^3_2h_{(2,2,2)}X^3)f^1)\otimes x^2h_{(2,1)}X^1\beta S(x^3_1h_{(2,2,1)}X^2)f^2\\&\overset{(3.1.7),(3.2.1)}{=}v(x^1h_1)bv(S(x^3_2X^3h_2)f^1)\otimes x^2X^1\beta S(x^3_1X^2)f^2=u(h\triangleright_v b).\end{aligned}$$

Now we check that $u$ is multiplicative ($F=F^1\otimes F^2$ is another copy of $f$):

$$\begin{aligned}&u(b)\circ u(b')\\&=\eta(Y^1)u(b)\eta(S(y^1Y^2)\alpha y^2Y^3_1)u(b')\eta(S(y^3Y^3_2))\\&=v(Y^1_1x^1)bv(S(x^3_2X^3)f^1S(y^1Y^2)_1\alpha_1y^2_1Y^3_{(1,1)}z^1)b'\\&\quad v(S(z^3_2Z^3)F^1S(y^3Y^3_2)_1)\otimes Y^1_2x^2X^1\beta S(x^3_1X^2)f^2S(y^1Y^2)_2\alpha_2\\&\quad y^2_2Y^3_{(1,2)}z^2Z^1\beta S(z^3_1Z^2)F^2S(y^3Y^3_2)_2\end{aligned}$$

$$
\begin{aligned}
&\overset{(3.2.13)}{\underset{(3.2.14)}{=}} v(Y_1^1x^1)bv(S(T^2t_2^1y_2^1Y_2^2x_2^3X^3)\alpha T^3t^2y_1^2Y_{(1,1)}^3z^1)b'v(S(y_2^3Y_{(2,2)}^3z_2^3Z^3)f^1)\\
&\qquad\otimes Y_2^1x^2X^1\beta S(T^1t_1^1y_1^1Y_1^2x_1^3X^2)\alpha t^3y_2^2Y_{(1,2)}^3z^2Z^1\beta S(y_1^3Y_{(2,1)}^3z_1^3Z^2)f^2\\
&\overset{(3.1.7)}{\underset{(3.2.1)}{=}} v(Y_1^1x^1)bv(S(T^2t_2^1y_2^1Y_2^2x_2^3X^3)\alpha T^3t^2y_1^2z^1Y_1^3)b'v(S(y_2^3z_2^3Z^3Y_2^3)f^1)\\
&\qquad\otimes Y_2^1x^2X^1\beta S(T^1t_1^1y_1^1Y_1^2x_1^3X^2)\alpha t^3y_2^2z^2Z^1\beta S(y_1^3z_1^3Z^2)f^2\\
&\overset{(3.1.9),(3.1.7)}{\underset{(3.2.1)}{=}} v(Y_1^1x^1)bv(S(T^2t_2^1Y_2^2x_2^3X^3)\alpha T^3t^2Y_1^3)b'v(S(y_2^3Z^3t^3Y_2^3)f^1)\\
&\qquad\otimes Y_2^1x^2X^1\beta S(y^1T^1t_1^1Y_1^2x_1^3X^2)\alpha y^2Z^1\beta S(y_1^3Z^2)f^2\\
&\overset{(3.1.9)}{\underset{(3.1.7)}{=}} v(x^1Y^1)bv(S(T^2x_{(1,1,2)}^3t_2^1W_2^2Y_{(2,2)}^2X^3)\alpha T^3x_{(1,2)}^3t^2W_1^3Y_1^3)b'\\
&\qquad v(S(y_2^3Z^3x_2^3t^3W_2^3Y_2^3)f^1)\otimes x^2W^1Y_1^2X^1\beta\\
&\qquad S(y^1T^1x_{(1,1,1)}^3t_1^1W_1^2Y_{(2,1)}^2X^2)\alpha y^2Z^1\beta S(y_1^3Z^2)f^2\\
&\overset{(3.1.7)}{\underset{(3.2.1)}{=}} v(x^1Y^1)bv(S(T^2t_2^1W_2^2Y_{(2,2)}^2X^3)\alpha T^3t^2W_1^3Y_1^3)b'v(S(y_2^3Z^3x_2^3t^3W_2^3Y_2^3)f^1)\\
&\qquad\otimes x^2W^1Y_1^2X^1\beta S(y^1x_1^3T^1t_1^1W_1^2Y_{(2,1)}^2X^2)\alpha y^2Z^1\beta S(y_1^3Z^2)f^2\\
&\overset{(3.1.7)}{\underset{(3.2.1)}{=}} v(x^1Y^1)bv(S(T^2t_2^1W_2^2X^3Y^2)\alpha T^3t^2W_1^3Y_1^3)b'v(S(y_2^3Z^3x_2^3t^3W_2^3Y_2^3)f^1)\\
&\qquad\otimes x^2W^1X^1\beta S(y^1x_1^3T^1t_1^1W_1^2X^2)\alpha y^2Z^1\beta S(y_1^3Z^2)f^2\\
&\overset{(3.1.9)}{\underset{(3.2.1)}{=}} v(x^1Y^1)bv(S(T^2t_2^1z^2X_1^3Y^2)\alpha T^3t^2z_1^3X_{(2,1)}^3Y_1^3)b'\\
&\qquad v(S(y_2^3Z^3x_2^3t^3z_2^3X_{(2,2)}^3Y_2^3)f^1)\\
&\qquad\otimes x^2X^1\beta S(y^1x_1^3T^1t_1^1z^1X^2)\alpha y^2Z^1\beta S(y_1^3Z^2)f^2\\
&\overset{(3.1.9)}{\underset{(3.2.1)}{=}} v(x^1Y^1)bv(S(T^2t_2^1z^2X_1^3Y^2)\alpha T^3t^2z_1^3X_{(2,1)}^3Y_1^3)b'\\
&\qquad v(S(x_2^3t^3z_2^3X_{(2,2)}^3Y_2^3)f^1)\otimes x^2X^1\beta S(x_1^3T^1t_1^1z^1X^2)S(y^1)\alpha y^2\beta S(y^3)f^2\\
&\overset{(3.2.2)}{\underset{(3.1.9)}{=}} v(x^1Y^1)bv(S(y_1^2t^1X_1^3Y^2)\alpha y_2^2t^2X_{(2,1)}^3Y_1^3)b'v(S(x_2^3y^3t^3X_{(2,2)}^3Y_2^3)f^1)\\
&\qquad\otimes x^2X^1\beta S(x_1^3y^1X^2)f^2\\
&\overset{(3.1.7),(3.2.1)}{=} v(x^1Y^1)bv(S(t^1Y^2)\alpha t^2Y_1^3)b'v(S(x_2^3X^3t^3Y_2^3)f^1)\otimes x^2X^1\beta S(x_1^3X^2)f^2\\
&= u(v(Y^1)bv(S(t^1Y^2)\alpha t^2Y_1^3)b'v(S(t^3Y_2^3))) = u(b\circ b'),
\end{aligned}
$$

finishing the proof. □

**Corollary 5.77** *The linear map $\psi : B^v\#H \to B\otimes H$, given by*

$$\psi(b\#h) = v(X^1x_1^1)bv(S(X^2x_2^1)\alpha X^3x^2h_1)\otimes x^3h_2, \tag{5.8.2}$$

*for all $b\in B$ and $h\in H$, is an algebra map.*

*Proof* The Universal Property of the smash product (Theorem 5.12), applied to the maps $\eta$ and $u$ from Proposition 5.76, provides an algebra map $B^v\#H\to B\otimes H$, which is exactly the map $\psi$ given by (5.8.2); we leave the details to the reader. □

**Proposition 5.78** *If $H$ is a quasi-Hopf algebra, $B$ an algebra, $v : H\to B$ an algebra map, then:*

*(i) $\theta : B\to B^v\#H$, $\theta(b) = v(z^1)bv(Z^1\beta S(z^2Z^2))\#z^3Z^3$, is an algebra map;*
*(ii) $\mu : H\to B^v\#H$, $\mu(h) = v(z^1Z^1\beta S(z^2h_1Z^2))\#z^3h_2Z^3$, is an algebra map;*

*(iii) for all $h\in H$ and $b\in B$, the following relation holds:*

$$\theta(b)\mu(h)=\mu(h)\theta(b)=v(z^1)bv(Z^1\beta S(z^2h_1Z^2))\#z^3h_2Z^3;$$

*(iv) consequently, the linear map*

$$\xi: B\otimes H\to B^v\#H,\quad \xi(b\otimes h)=v(z^1)bv(Z^1\beta S(z^2h_1Z^2))\#z^3h_2Z^3, \tag{5.8.3}$$

*is an algebra map.*

*Proof* We only prove (i) and leave the rest to the reader. Clearly $\theta(1_B)=v(\beta)\#1_H$, so we only have to check that $\theta$ is multiplicative. We compute:

$$\begin{aligned}
&\theta(b)\theta(b')\\
&\quad= [x^1\triangleright_v(v(z^1)bv(Z^1\beta S(z^2Z^2)))]\circ[x^2z^3_1Z^3_1\triangleright_v(v(t^1)b'v(T^1\beta S(t^2T^2)))]\\
&\qquad\#x^3z^3_2Z^3_2t^3T^3\\
&\quad= v(X^1x^1_1z^1)bv(Z^1\beta S(y^1X^2x^1_2z^2Z^2)\alpha y^2X^3_1x^2_1z^3_{(1,1)}Z^3_{(1,1)}t^1)\\
&\qquad b'v(T^1\beta S(y^3X^3_2x^2_2z^3_{(1,2)}Z^3_{(1,2)}t^2T^2))\#x^3z^2_2Z^3_2t^3T^3\\
&\overset{(3.1.9)}{\underset{(3.2.1)}{=}} v(X^1w^1_1z^1y^1)bv(Z^1\beta S(X^2w^1_2z^2y^2_1x^1Z^2)\alpha X^3w^2z^3_1y^2_{(2,1)}x^2_1Z^3_{(1,1)}t^1)\\
&\qquad b'v(T^1\beta S(w^3z^3_2y^2_{(2,2)}x^2_2Z^3_{(1,2)}t^2T^2))\#y^3x^3Z^3_2t^3T^3\\
&\overset{(3.1.9)}{\underset{(3.2.1)}{=}} v(y^1)bv(Z^1\beta S(z^1y^2_1x^1Z^2)\alpha z^2y^2_{(2,1)}x^2_1Z^3_{(1,1)}t^1)\\
&\qquad b'v(T^1\beta S(z^3y^2_{(2,2)}x^2_2Z^3_{(1,2)}t^2T^2))\#y^3x^3Z^3_2t^3T^3\\
&\overset{(3.1.7)}{\underset{(3.2.1)}{=}} v(y^1)bv(Z^1\beta S(z^1x^1Z^2)\alpha z^2x^2_1Z^3_{(1,1)}t^1)b'v(T^1\beta S(y^2z^3x^2_2Z^3_{(1,2)}t^2T^2))\\
&\qquad\#y^3x^3Z^3_2t^3T^3\\
&\overset{(3.1.9)}{\underset{(3.2.1)}{=}} v(y^1)bv(Y^1Z^1\beta S(z^1Y^2_1Z^2)\alpha z^2Y^2_{(2,1)}Z^3_1t^1)b'v(T^1\beta S(y^2z^3Y^2_{(2,2)}Z^3_2t^2T^2))\\
&\qquad\#y^3Y^3t^3T^3\\
&\overset{(3.1.7)}{\underset{(3.2.1)}{=}} v(y^1)bv(Y^1Z^1\beta S(z^1Z^2)\alpha z^2Z^3_1t^1)b'v(T^1\beta S(y^2Y^2z^3Z^3_2t^2T^2))\\
&\qquad\#y^3Y^3t^3T^3\\
&\overset{(3.1.9),\,(3.2.1)}{=} v(y^1)bv(Y^1Z^1\beta S(Z^2)\alpha Z^3t^1)b'v(T^1\beta S(y^2Y^2t^2T^2))\#y^3Y^3t^3T^3\\
&\overset{(3.2.2)}{=} v(y^1)bb'v(T^1\beta S(y^2T^2))\#y^3T^3=\theta(bb'),
\end{aligned}$$

finishing the proof. □

**Theorem 5.79** *The maps $\psi$ and $\xi$ given by* (5.8.2) *and* (5.8.3), *respectively, are inverse to each other, providing thus an algebra isomorphism $B^v\#H\cong B\otimes H$.*

*Proof* We compute:

$$\begin{aligned}
&\xi(\psi(b\#h))\\
&\quad= v(z^1X^1x^1_1)bv(S(X^2x^1_2)\alpha X^3x^2h_1Z^1\beta S(z^2x^3_1h_{(2,1)}Z^2))\#z^3x^3_2h_{(2,2)}Z^3\\
&\overset{(3.1.7),\,(3.2.1)}{=} v(z^1X^1x^1_1)bv(S(X^2x^1_2)\alpha X^3x^2Z^1\beta S(z^2x^3_1Z^2))\#z^3x^3_2Z^3h\\
&\overset{(3.1.9),\,(3.2.1)}{=} v(z^1X^1Y^1_{(1,1)}t^1_1)bv(S(X^2Y^1_{(1,2)}t^1_2)\alpha X^3Y^1_2t^2\beta S(z^2Y^2t^3))\#z^3Y^3h
\end{aligned}$$

$$\stackrel{(3.1.7),(3.2.1)}{=} v(X^1t_1^1)bv(S(X^2t_2^1)\alpha X^3t^2\beta S(t^3))\#h$$
$$\stackrel{(3.1.9),(3.2.1)}{=} bv(S(t^1)\alpha t^2\beta S(t^3))\#h \stackrel{(3.2.2)}{=} b\#h,$$

and similarly

$$\begin{aligned}\psi(\xi(b\otimes h)) &= v(X^1x_1^1z^1)bv(Z^1\beta S(X^2x_2^1z^2h_1Z^2)\alpha X^3x^2z_1^3h_{(2,1)}Z_1^3)\otimes x^3z_2^3h_{(2,2)}Z_2^3\\ &\stackrel{(3.1.9),(3.2.1)}{=} bv(Z^1\beta S(t^1h_1Z^2)\alpha t^2h_{(2,1)}Z_1^3)\otimes t^3h_{(2,2)}Z_2^3\\ &\stackrel{(3.1.7),(3.2.1)}{=} bv(Z^1\beta S(t^1Z^2)\alpha t^2Z_1^3)\otimes ht^3Z_2^3\\ &\stackrel{(3.1.9),(3.2.1)}{=} bv(Z^1\beta S(Z^2)\alpha Z^3)\otimes h \stackrel{(3.2.2)}{=} b\otimes h,\end{aligned}$$

finishing the proof. □

**Corollary 5.80** *Let $H$ be a quasi-Hopf algebra and consider the left $H$-module algebra $H_0$ that appears in Definition 4.4. Then we have an algebra isomorphism*

$$\Psi : H_0\#H \to H\otimes H,\quad \Psi(h'\otimes h) = X^1x_1^1h'S(X^2x_2^1)\alpha X^3x^2h_1\otimes x^3h_2. \tag{5.8.4}$$

*Proof* Take $B = H$ and $v = \mathrm{Id}_H$ in Theorem 5.79. □

As an application of Theorem 5.79, we obtain a duality theorem for quasi-Hopf algebras:

**Theorem 5.81** *If $H$ is a finite-dimensional quasi-Hopf algebra with bijective antipode, then the two-sided crossed product $H \bowtie H^* \ltimes H$ is isomorphic to* $\mathrm{End}(H)\otimes H$ *as associative unital algebras.*

*Proof* By Proposition 5.34, we have the identification of algebras

$$H \bowtie H^* \ltimes H \equiv (H\overline{\#}H^*)\#H.$$

By Propositions 5.21 and 5.28, $H\overline{\#}H^*$ is isomorphic, as left $H$-module algebras, to $\mathrm{End}(H)^v$, where $\mathrm{End}(H)$ is regarded as an associative algebra in the usual way and $v : H \to \mathrm{End}(H)$ is a certain algebra map. We can thus apply Theorem 5.79 for $B = \mathrm{End}(H)$ and obtain the desired result. □

## 5.9 Notes

The content of Section 5.1 is taken from [59]. The content of Section 5.2 is taken from [46] and [60]. The results in Section 5.3 appear in [178]. The two-sided crossed and smash products in Section 5.4 appear in [107] and [60], respectively. Theorem 5.35 is from [107], but the proof is taken from [60]. The content of Section 5.5 is taken from [60]. The content of Section 5.6 is taken from [107] and [60], except for Propositions 5.61, 5.62 and 5.63, which appear in [8]. The content of Section 5.7 is taken from [179], while the duality theorem for quasi-Hopf algebras and the rest of Section 5.8 are taken from [8].

# 6

# Quasi-Hopf Bimodule Categories

We present some structure theorems for quasi-Hopf bimodules. We also show that for a quasi-Hopf algebra $H$ the category of quasi-Hopf $H$-bimodules is monoidally equivalent to the category of left $H$-representations. As an application, we prove a structure theorem for quasi-Hopf comodule algebras.

## 6.1 Quasi-Hopf Bimodules

Let $H$ be a quasi-bialgebra over a field $k$. We saw at the end of Section 4.2 that $H$ has a natural structure of a coalgebra within the monoidal category ${}_H\mathscr{M}_H$. So we can consider left and right $H$-comodules within ${}_H\mathscr{M}_H$.

**Definition 6.1** Let $H$ be a quasi-bialgebra. We call a right (resp. left) $H$-comodule within ${}_H\mathscr{M}_H$ a right (resp. left) quasi-Hopf $H$-bimodule.

More precisely, a right quasi-Hopf $H$-bimodule $M$ is an $H$-bimodule together with an $H$-bimodule map $\rho:\ M\to M\otimes H$ such that the following relations hold:

$$(\mathrm{Id}_M\otimes\varepsilon)\circ\rho=\mathrm{Id}_M, \tag{6.1.1}$$

$$\Phi\cdot(\rho\otimes\mathrm{Id}_M)(\rho(m))=(\mathrm{Id}_M\otimes\Delta)(\rho(m))\cdot\Phi,\ \ \forall\, m\in M. \tag{6.1.2}$$

A morphism between two right quasi-Hopf $H$-bimodules is an $H$-comodule morphism in ${}_H\mathscr{M}_H$, that is, an $H$-bimodule map $f:\ M\to M'$ satisfying $\rho'\circ f=(f\otimes\mathrm{Id}_H)\circ\rho$.

We denote by ${}_H\mathscr{M}_H^H$ the category of right quasi-Hopf $H$-bimodules and morphisms of right quasi-Hopf $H$-bimodules.

Likewise, a left quasi-Hopf $H$-bimodule is an $H$-bimodule $N$ together with an $H$-bimodule map $\lambda: N\to H\otimes N$ such that the following relations hold:

$$(\varepsilon\otimes\mathrm{Id}_N)\circ\lambda=\mathrm{Id}_N, \tag{6.1.3}$$

$$(\mathrm{Id}_H\otimes\lambda)(\lambda(n))\cdot\Phi=\Phi\cdot(\Delta\otimes\mathrm{Id}_N)(\lambda(n)),\ \ \forall\, n\in N. \tag{6.1.4}$$

A morphism between two left quasi-Hopf $H$-bimodules is an $H$-bimodule map that is at the same time left $H$-colinear. By ${}_H^H\mathscr{M}_H$ we denote the category of left quasi-Hopf $H$-bimodules and morphisms of left quasi-Hopf $H$-bimodules.

It can be easily checked that ${}^H_H\mathscr{M}_H \equiv {}_{H^{\rm op}}\mathscr{M}^{H^{\rm op}}_{H^{\rm op}}$ via natural identification. Hence any result proved for right quasi-Hopf $H$-bimodules is valid for left quasi-Hopf $H$-bimodules as well, and vice versa.

Definition 6.1 admits the following natural generalization.

**Definition 6.2** Let $H$ be a quasi-bialgebra, $\mathfrak{A}$ a right $H$-comodule algebra and $C$ an $H$-bimodule coalgebra. Then a right quasi-Hopf $(H,\mathfrak{A},C)$-module is an $(H,\mathfrak{A})$-bimodule $M$ together with a $k$-linear map $\rho : M \ni m \mapsto m_{(0)} \otimes m_{(1)} \in M \otimes C$ such that the following relations hold:

$$(\mathrm{Id}_M \otimes \varepsilon) \circ \rho = \mathrm{Id}_M, \tag{6.1.5}$$

$$\rho(h \cdot m \cdot \mathfrak{a}) = h_1 \cdot m_{(0)} \cdot \mathfrak{a}_{\langle 0 \rangle} \otimes h_2 \cdot m_{(1)} \cdot \mathfrak{a}_{\langle 1 \rangle}, \tag{6.1.6}$$

$$\Phi \cdot (\rho \otimes \mathrm{Id}_H)(\rho(m)) = (\mathrm{Id}_M \otimes \Delta)(\rho(m)) \cdot \Phi_\rho, \tag{6.1.7}$$

for all $h \in H$, $m \in M$ and $\mathfrak{a} \in \mathfrak{A}$. Denote by ${}_H\mathscr{M}^C_{\mathfrak{A}}$ the category of quasi-Hopf $(H,\mathfrak{A},C)$-modules and $(H,\mathfrak{A})$-bimodule maps that are at the same time right $C$-colinear maps.

In a similar manner, for $\mathfrak{B}$ a left comodule algebra and $C$ a bimodule coalgebra over a quasi-bialgebra $H$, one can introduce the category of left quasi-Hopf $(\mathfrak{B},H,C)$-modules, denoted in what follows by ${}^C_{\mathfrak{B}}\mathscr{M}_H$.

For $\mathfrak{A} = H$ (resp. $\mathfrak{B} = H$) we have that ${}_H\mathscr{M}^C_H$ (resp. ${}^C_H\mathscr{M}_H$) is the category of right (resp. left) corepresentations over $C$ within ${}_H\mathscr{M}_H$. Of course, when $C = H$, too, we reduce to the category of right (resp. left) quasi-Hopf $H$-bimodules.

Examples of right quasi-Hopf $(H,\mathfrak{A},C)$-modules can be obtained from the input datum $(H,\mathfrak{A},C)$. In what follows we call a right quasi-Hopf bimodule datum a triple $(H,\mathfrak{A},C)$ consisting of a quasi-bialgebra $H$, a right $H$-comodule algebra $\mathfrak{A}$ and an $H$-bimodule coalgebra $C$.

**Proposition 6.3** *Let $(H,\mathfrak{A},C)$ be a right quasi-Hopf bimodule datum. If $M$ is an $(H,\mathfrak{A})$-bimodule then $M \otimes C$ becomes an object in ${}_H\mathscr{M}^C_{\mathfrak{A}}$ via the structure given by*

$$h \cdot (m \otimes c) \cdot \mathfrak{a} = h_1 \cdot m \cdot \mathfrak{a}_{\langle 0 \rangle} \otimes h_2 \cdot c \cdot \mathfrak{a}_{\langle 1 \rangle}, \tag{6.1.8}$$

$$\rho : M \otimes C \ni m \otimes c \mapsto x^1 \cdot m \cdot \tilde{X}^1_\rho \otimes x^2 \cdot c_{\underline{1}} \cdot \tilde{X}^2_\rho \otimes x^3 \cdot c_{\underline{2}} \cdot \tilde{X}^3_\rho \in M \otimes C \otimes C, \tag{6.1.9}$$

*for all $h \in H$, $m \in M$, $c \in C$ and $\mathfrak{a} \in \mathfrak{A}$. Consequently, we have a functor $F : {}_H\mathscr{M}_{\mathfrak{A}} \to {}_H\mathscr{M}^C_{\mathfrak{A}}$. $F$ sends an $(H,\mathfrak{A})$-bimodule $M$ to $M \otimes C$ regarded as object in ${}_H\mathscr{M}^C_{\mathfrak{A}}$ via* (6.1.8) *and* (6.1.9)*, and an $(H,\mathfrak{A})$-bimodule morphism $f : M \to M'$ to $f \otimes \mathrm{Id}_C : M \otimes C \to M' \otimes C$.*

*Proof* It can be easily proved that $M \otimes C$ with the structure in (6.1.8) is an $(H,\mathfrak{A})$-bimodule. Now the coaction $\rho$ in (6.1.9) is an $(H,\mathfrak{A})$-bimodule map since

$$\begin{aligned}
\rho(h \cdot (m \otimes c)) &= \rho(h_1 \cdot m \otimes h_2 \cdot c) \\
&= x^1 h_1 \cdot m \cdot \tilde{X}^1_\rho \otimes x^2 h_{(2,1)} \cdot c_{\underline{1}} \cdot \tilde{X}^2_\rho \otimes x^3 h_{(2,2)} \cdot c_{\underline{2}} \cdot \tilde{X}^3_\rho \\
&\overset{(3.1.7)}{=} h_{(1,1)} x^1 \cdot m \cdot \tilde{X}^1_\rho \otimes h_{(1,2)} x^2 \cdot c_{\underline{1}} \cdot \tilde{X}^2_\rho \otimes h_2 x^3 \cdot c_{\underline{2}} \cdot \tilde{X}^3_\rho
\end{aligned}$$

$$= h_1 \cdot (m \otimes c)_{(0)} \otimes h_2 \cdot (m \otimes c)_{(1)} = h \cdot \rho(m \otimes c),$$

and similarly

$$\begin{aligned}
\rho((m \otimes c) \cdot \mathfrak{a}) &= \rho(m \cdot \mathfrak{a}_{\langle 0\rangle} \otimes c \cdot \mathfrak{a}_{\langle 1\rangle}) \\
&= x^1 \cdot m \cdot \mathfrak{a}_{\langle 0\rangle} \tilde{X}^1_\rho \otimes x^2 \cdot c_{\underline{1}} \cdot \mathfrak{a}_{\langle 1\rangle_{\underline{1}}} \tilde{X}^2_\rho \otimes x^3 \cdot c_{\underline{2}} \cdot \mathfrak{a}_{\langle 1\rangle_{\underline{2}}} \tilde{X}^3_\rho \\
&\overset{(4.3.1)}{=} x^1 \cdot m \cdot \tilde{X}^1_\rho \mathfrak{a}_{\langle 0,0\rangle} \otimes x^2 \cdot c_{\underline{1}} \cdot \mathfrak{a}_{\langle 0,1\rangle} \tilde{X}^2_\rho \otimes x^3 \cdot c_{\underline{2}} \cdot \tilde{X}^3_\rho \mathfrak{a}_{\langle 1\rangle} \\
&= (m \otimes c)_{(0)} \cdot \mathfrak{a}_{\langle 0\rangle} \otimes (m \otimes c)_{(1)} \cdot \mathfrak{a}_{\langle 1\rangle} = \rho(m \otimes c) \cdot \mathfrak{a},
\end{aligned}$$

for all $h \in H$, $m \in M$, $c \in C$ and $\mathfrak{a} \in \mathfrak{A}$. The map $\rho$ is also right $C$-colinear since

$$\begin{aligned}
&\Phi \cdot (\rho \otimes \mathrm{Id}_C)(\rho(m \otimes c)) \\
&= \Phi \cdot (\rho(x^1 \cdot m \cdot \tilde{X}^1_\rho \otimes x^2 \cdot c_{\underline{1}} \cdot \tilde{X}^2_\rho) \otimes x^3 \cdot c_{\underline{2}} \cdot \tilde{X}^3_\rho) \\
&= X^1_1 y^1 x^1 \cdot m \cdot \tilde{X}^1_\rho \tilde{Y}^1_\rho \otimes X^1_2 y^2 x^2_1 \cdot c_{(\underline{1},\underline{1})} \cdot (\tilde{X}^1_\rho)_1 \tilde{Y}^2_\rho \otimes X^2 y^3 x^2_2 \cdot c_{(\underline{1},\underline{2})} \cdot (\tilde{X}^2_\rho)_2 \tilde{Y}^3_\rho \\
&\qquad \otimes X^3 x^3 \cdot c_{\underline{2}} \cdot \tilde{X}^3_\rho \\
&\overset{(3.1.7)}{\underset{(4.3.2)}{=}} x^1 \cdot m \cdot \tilde{X}^1_\rho (\tilde{Y}^1_\rho)_{\langle 0\rangle} \otimes x^2 X^1 \cdot c_{(\underline{1},\underline{1})} \cdot y^1 \tilde{X}^2_\rho (\tilde{Y}^1_\rho)_{\langle 1\rangle} \otimes x^3_1 X^2 \cdot c_{(\underline{1},\underline{2})} \cdot y^2 (\tilde{X}^3_\rho)_1 \tilde{Y}^2_\rho \\
&\qquad \otimes x^3_2 X^3 \cdot c_{\underline{2}} \cdot y^3 (\tilde{X}^3_\rho)_2 \tilde{Y}^3_\rho \\
&\overset{(4.2.9)}{=} x^1 \cdot m \cdot \tilde{X}^1_\rho (\tilde{Y}^1_\rho)_{\langle 0\rangle} \otimes x^2 \cdot c_{\underline{1}} \cdot \tilde{X}^2_\rho (\tilde{Y}^1_\rho)_{\langle 1\rangle} \otimes x^3_1 \cdot c_{(\underline{2},\underline{1})} \cdot (\tilde{X}^3_\rho)_1 \tilde{Y}^2_\rho \\
&\qquad \otimes x^3_2 \cdot c_{(\underline{2},\underline{2})} \cdot (\tilde{X}^3_\rho)_2 \tilde{Y}^3_\rho \\
&= (m \otimes c)_{(0)} \cdot \tilde{Y}^1_\rho \otimes (m \otimes c)_{(1)_{\underline{1}}} \cdot \tilde{Y}^2_\rho \otimes (m \otimes c)_{(1)_{\underline{2}}} \cdot \tilde{Y}^3_\rho \\
&= (\mathrm{Id}_{M \otimes C} \otimes \underline{\Delta}_C)(\rho(m \otimes c)) \cdot \Phi_\rho,
\end{aligned}$$

as required. Finally, it can be easily checked that any $(H, \mathfrak{A})$-bimodule morphism $f : M \to M'$ gives rise to a morphism $f \otimes \mathrm{Id}_C : M \otimes C \to M' \otimes C$ in ${}_H\mathscr{M}^C_{\mathfrak{A}}$, and so we have a well-defined functor $F : {}_H\mathscr{M}_{\mathfrak{A}} \to {}_H\mathscr{M}^C_{\mathfrak{A}}$. This finishes the proof. □

We specialize Proposition 6.3 for various types of right quasi-Hopf bimodule data.

**Example 6.4** Let $(H, \mathfrak{A}, C)$ be a right quasi-Hopf bimodule datum. Then for any right $\mathfrak{A}$-module $M$ the $k$-vector space $M \otimes C$ is an object in ${}_H\mathscr{M}^H_{\mathfrak{A}}$ with the following structure:

- $M \otimes C$ is an $(H, \mathfrak{A})$-bimodule via

$$h \cdot (m \otimes c) \cdot \mathfrak{a} = m \cdot \mathfrak{a}_{\langle 0\rangle} \otimes h \cdot c \cdot \mathfrak{a}_{\langle 1\rangle}; \tag{6.1.10}$$

- the right coaction of $H$ on $M \otimes C$ is given by

$$\rho(m \otimes c) = m \cdot \tilde{X}^1_\rho \otimes c_{\underline{1}} \cdot \tilde{X}^2_\rho \otimes c_{\underline{2}} \cdot \tilde{X}^3_\rho, \tag{6.1.11}$$

for all $m \in M$, $h \in H$, $c \in C$ and $\mathfrak{a} \in \mathfrak{A}$.

*Proof* Consider $M$ as a left $H$-module via the trivial action $h \cdot m = \varepsilon(h)m$, for all $h \in H$ and $m \in M$. Then $M$ is an $(H, \mathfrak{A})$-bimodule and so Proposition 6.3 applies. The resulting structures on $M \otimes C$ that make it an object in ${}_H\mathscr{M}^C_{\mathfrak{A}}$ are exactly the ones stated in (6.1.10) and (6.1.11). □

**Examples 6.5** Let $H$ be a quasi-Hopf algebra with bijective antipode, $(\mathfrak{A}, \rho, \Phi_\rho)$ a right $H$-comodule algebra and $C$ an $H$-bimodule coalgebra.

(1) $\mathscr{V} := \mathfrak{A} \otimes C \in {}_H\mathscr{M}_{\mathfrak{A}}^C$. The structure maps are

$$h \succ (\mathfrak{a} \otimes c) = \mathfrak{a} \otimes h \cdot c, \quad (\mathfrak{a} \otimes c) \prec \mathfrak{a}' = \mathfrak{a}\mathfrak{a}'_{\langle 0 \rangle} \otimes c \cdot \mathfrak{a}'_{\langle 1 \rangle},$$
$$\rho_{\mathscr{V}}(\mathfrak{a} \otimes c) = \mathfrak{a}\tilde{X}^1 \otimes c_{\underline{1}} \cdot \tilde{X}^2 \otimes c_{\underline{2}} \cdot \tilde{X}^3,$$

for all $h \in H$, $\mathfrak{a}, \mathfrak{a}' \in \mathfrak{A}$ and $c \in C$.

(2) $\mathscr{U} := C \otimes \mathfrak{A} \in {}_H\mathscr{M}_{\mathfrak{A}}^C$. Now the structure maps are given by the following formulas, for all $h \in H$, $\mathfrak{a}, \mathfrak{a}' \in \mathfrak{A}$ and $c \in C$:

$$h \succ (c \otimes \mathfrak{a}) = h \cdot c \otimes \mathfrak{a}, \quad (c \otimes \mathfrak{a}) \prec \mathfrak{a}' = c \otimes \mathfrak{a}\mathfrak{a}',$$
$$\rho_{\mathscr{U}}(c \otimes \mathfrak{a}) = c_{\underline{1}} \cdot S^{-1}(q_L^2 \tilde{X}_2^3 g^2) \otimes \tilde{X}^1 \mathfrak{a}_{\langle 0 \rangle} \otimes c_{\underline{2}} \cdot S^{-1}(q_L^1 \tilde{X}_1^3 g^1)\tilde{X}^2 \mathfrak{a}_{\langle 1 \rangle}. \tag{6.1.12}$$

Here $q_L = q_L^1 \otimes q_L^2$ and $f^{-1} = g^1 \otimes g^2$ are the elements defined by the formulas (3.2.20) and (3.2.16).

*Proof* (1) This follows from Example 6.4 specialized for $M = \mathfrak{A}$, regarded as a right $\mathfrak{A}$-module via its multiplication.

(2) Consider $\theta : \mathscr{V} \to \mathscr{U}$ given by

$$\theta(\mathfrak{a} \otimes c) = c \cdot S^{-1}(\mathfrak{a}_{\langle 1 \rangle} \tilde{p}_\rho^2) \otimes \mathfrak{a}_{\langle 0 \rangle} \tilde{p}_\rho^1,$$

for all $c \in C$ and $\mathfrak{a} \in \mathfrak{A}$, where, as before, we use the notation $\tilde{p}_\rho = \tilde{p}_\rho^1 \otimes \tilde{p}_\rho^2 = \tilde{x}^1 \otimes \tilde{x}^2 \beta S(\tilde{x}^3) \in \mathfrak{A} \otimes H$. We claim that $\theta$ is bijective; its inverse $\theta^{-1} : \mathscr{U} \to \mathscr{V}$ is defined as follows:

$$\theta^{-1}(c \otimes \mathfrak{a}) = \tilde{q}_\rho^1 \mathfrak{a}_{\langle 0 \rangle} \otimes c \cdot \tilde{q}_\rho^2 \mathfrak{a}_{\langle 1 \rangle},$$

with notation $\tilde{q}_\rho = \tilde{q}_\rho^1 \otimes \tilde{q}_\rho^2 = \tilde{X}^1 \otimes S^{-1}(\alpha \tilde{X}^3)\tilde{X}^2 \in \mathfrak{A} \otimes H$.

Furthermore, $\theta$ is a morphism of quasi-Hopf $(H, \mathfrak{A})$-bimodules, and we conclude that $\mathscr{U} = C \otimes \mathfrak{A}$ and $\mathfrak{A} \otimes C = \mathscr{V}$ are isomorphic in ${}_H\mathscr{M}_{\mathfrak{A}}^C$. Indeed, by using (4.3.10) and (4.3.13) one can show easily that $\theta$ and $\theta^{-1}$ are inverses, and that $\mathscr{U}$ is an $(H, \mathfrak{A})$-bimodule via the actions $\succ$ and $\prec$. One can finally compute the right $H$-coaction on $\mathscr{U}$ carried from the coaction on $\mathscr{V}$ by using $\theta$, and then see that it coincides with (6.1.12). For, observe that (4.3.9), (4.3.2) and (4.3.4) imply

$$\tilde{X}^1_{\langle 1 \rangle} \tilde{p}_\rho^2 S(\tilde{X}^2) \otimes \tilde{X}^1_{\langle 0 \rangle} \tilde{p}_\rho^1 \otimes \tilde{X}^3 = \tilde{x}^2 S(\tilde{x}_1^3 p_L^1) \otimes \tilde{x}^1 \otimes \tilde{x}_2^3 p_L^2, \tag{6.1.13}$$

where $p_L = p_L^1 \otimes p_L^2$ is the element defined in (3.2.20). We also mention that the computation uses the formula (4.3.15); the details are left to the reader. □

We now consider the other way around.

**Example 6.6** For any left $H$-module $V$ the $k$-vector space $V \otimes H$ is a right quasi-Hopf $H$-bimodule via the structure

$$h \cdot (v \otimes x) \cdot h' = h_1 \cdot v \otimes h_2 x h' \quad \text{and} \quad \rho_{V \otimes H}(v \otimes h) = x^1 \cdot v \otimes x^2 h_1 \otimes x^3 h_2,$$

defined for all $h, x, h' \in H$ and $v \in V$.

*Proof* In Proposition 6.3 take $\mathfrak{A} = C = H$ and regard a left $H$-module $V$ as an $H$-bimodule with the trivial right $H$-action, that is, $v \cdot h = \varepsilon(h)v$, for all $h \in H$, $v \in V$. □

For $H$ a quasi-bialgebra denote by $\Gamma(H)$ the set of algebra morphisms from $H$ to $k$, which is a monoid with unit $\varepsilon$ via the multiplication $\chi * \chi'(h) = \chi(h_1)\chi'(h_2)$, for all $\chi, \chi' \in \Gamma(H)$ and $h \in H$. If, moreover, $H$ is a quasi-Hopf algebra then $\Gamma(H)$ is a group since any $\chi \in \Gamma(H)$ admits $\chi^{-1} := \chi \circ S$ as inverse; see Remark 3.15. Note that if $S$ is bijective then $\chi^{-1} = \chi \circ S^{-1}$, too, and that these assertions hold because of (3.2.1) and (3.2.2).

**Example 6.7** Let $\chi \in \Gamma(H)$ and denote by $H_\chi$ the $k$-vector space $H$ endowed with the left and right $H$-actions given by $h \cdot \hbar := \chi(h_1)h_2\hbar$ and $\hbar \cdot h = \hbar h$, for all $h, \hbar \in H$. Then $H_\chi$ is an $H$-bimodule and, moreover, a right quasi-Hopf $H$-bimodule with

$$\rho_\chi : H_\chi \ni \hbar \mapsto \chi(x^1)x^2\hbar_1 \otimes x^3\hbar_2 \in H_\chi \otimes H.$$

*Proof* Take $V = k$ in Example 6.6 and consider it as a left $H$-module via the action defined by $\hbar \cdot \kappa := \chi(\hbar)\kappa$, for all $\hbar \in H$ and $\kappa \in k$. Then the stated quasi-Hopf $H$-bimodule structure on $H_\chi$ is obtained through the identification $H_\chi \equiv k \otimes H$. □

Examples of quasi-Hopf bimodules can be also obtained by tensoring two quasi-Hopf bimodules.

**Proposition 6.8** *Let $H$ be a quasi-bialgebra and $M, N$ two objects in ${}_H\mathscr{M}_H^H$. Then $M \otimes_H N$ is a right quasi-Hopf $H$-bimodule with the structure given by*

$$h \cdot (m \otimes_H n) \cdot h' = h \cdot m \otimes_H n \cdot h', \tag{6.1.14}$$

$$\rho_{M \otimes_H N} : M \otimes_H N \ni m \otimes_H n \mapsto \left(m_{(0)} \otimes_H n_{(0)}\right) \otimes m_{(1)}n_{(1)} \in (M \otimes_H N) \otimes H, \tag{6.1.15}$$

*for all $m \in M$, $n \in N$ and $h, h' \in H$. In this way the category ${}_H\mathscr{M}_H^H$ becomes a strict monoidal category with unit object $H = H_\varepsilon$, considered in ${}_H\mathscr{M}_H^H$ as in Example 6.7.*

*Proof* It is immediate that the left and right actions in (6.1.14) are well defined, and that they endow $M \otimes_H N$ with an $H$-bimodule structure. The coaction $\rho_{M \otimes_H N}$ is also well defined since, for all $m \in M$, $h \in H$ and $n \in N$, we have

$$\begin{aligned} \rho(m \cdot h \otimes_H n) &= m_{(0)}h_1 \otimes_H n_{(0)} \otimes m_{(1)}h_2n_{(1)} \\ &= m_{(0)} \otimes_H h_1 \cdot n_{(0)} \otimes m_{(1)}h_2n_{(1)} \\ &= m_{(0)} \otimes_H (h \cdot n)_{(0)} \otimes m_{(1)}(h \cdot n)_{(1)} \\ &= \rho_{M \otimes_H N}(m \otimes_H h \cdot n). \end{aligned}$$

Furthermore, it is coassociative up to conjugation by $\Phi$ since

$$\begin{aligned} &\Phi(\rho_{M \otimes_H N} \otimes \mathrm{Id}_H)(\rho_{M \otimes_H N}(m \otimes_H n)) \\ &\quad = X^1 \cdot m_{(0,0)} \otimes_H n_{(0,0)} \otimes X^2 m_{(0,1)}n_{(0,1)} \otimes X^3 m_{(1)}n_{(1)} \\ &\quad = m_{(0)} \cdot X^1 \otimes_H n_{(0,0)} \otimes m_{(1)_1}X^2 n_{(0,1)} \otimes m_{(1)_2}X^3 n_{(1)} \\ &\quad = m_{(0)} \otimes_H X^1 \cdot n_{(0,0)} \otimes m_{(1)_1}X^2 n_{(0,1)} \otimes m_{(1)_2}X^3 n_{(1)} \end{aligned}$$

$$= m_{(0)} \otimes_H n_{(0)} \cdot X^1 \otimes m_{(1)_1} n_{(1)_1} X^2 \otimes m_{(1)_2} n_{(1)_2} X^3$$
$$= (\mathrm{Id}_{M\otimes_H N} \otimes \Delta)(\rho_{M\otimes_H N}(m \otimes_H n))\Phi,$$

for all $m \in M$ and $n \in N$. Here, and everywhere else, we adopt the notation

$$(\rho \otimes \mathrm{Id}_H)(\rho(m)) = m_{(0,0)} \otimes m_{(0,1)} \otimes m_{(1)}, \text{ etc.}$$

Now observe that for any two right quasi-Hopf $H$-bimodule morphisms $f : M \to M'$ and $g : N \to N'$ we have that $f \otimes_H g : M \otimes_H N \to M' \otimes_H N'$ is a right quasi-Hopf bimodule morphism as well. Hence $\otimes_H : {}_H\mathscr{M}_H \times {}_H\mathscr{M}_H \to {}_H\mathscr{M}_H$ induces a functor, still denoted by $\otimes_H$, from ${}_H\mathscr{M}_H^H \times {}_H\mathscr{M}_H^H$ to ${}_H\mathscr{M}_H^H$. Furthermore, it endows ${}_H\mathscr{M}_H^H$ with a strict monoidal structure since the isomorphisms $M \otimes_H H \cong M$ and $H \otimes_H M \cong M$ in ${}_H\mathscr{M}_H$ are in ${}_H\mathscr{M}_H^H$, provided that $M \in {}_H\mathscr{M}_H^H$. □

## 6.2 The Dual of a Quasi-Hopf Bimodule

The category of left or right quasi-Hopf bimodules does not have duality. However, owing to the general result in Proposition 2.43 we can associate to any finite-dimensional left (resp. right) quasi-Hopf $H$-bimodule a right (resp. left) one. Actually we will apply this general result to the case when $\mathscr{C} = {}_H\mathscr{M}_H$, the category of $H$-bimodules. This category is monoidal since it can be identified with the category of left modules over the quasi-Hopf algebra $H \otimes H^{\mathrm{op}}$. The monoidal structure on ${}_H\mathscr{M}_H$ obtained in this way was described explicitly in Lemma 4.13. Furthermore, if we restrict to the category of finite-dimensional $H$-bimodules then it is left and right rigid, provided that the antipode of $S$ is bijective (this follows from Proposition 3.33 applied to the quasi-Hopf algebra $H \otimes H^{\mathrm{op}}$). In fact, we have the following result. Recall that the linear dual $V^*$ of a right (resp. left) $H$-module $V$ is a left (resp. right) $H$-module via $(h \rightharpoonup v^*)(v) = v^*(v \cdot h)$ (resp. $(v^* \leftharpoonup h)(v) = v^*(h \cdot v)$).

**Lemma 6.9** *Let $\{v_i\}_i$ be a basis of a finite-dimensional $H$-bimodule $V$, with dual basis $\{v^i\}_i$ of its linear dual $V^*$.*

*The left dual of $V$ is $V^*$ with $H$-bimodule structure*

$$h \cdot v^* \cdot h' = (h' \otimes h) \cdot v^* = v^* \leftharpoonup (S^{-1}(h') \otimes S(h)) = S^{-1}(h') \rightharpoonup v^* \leftharpoonup S(h). \quad (6.2.1)$$

*The evaluation morphism $\mathrm{ev}_V : V^* \otimes V \to k$ and the coevaluation morphism $\mathrm{coev}_V : k \to V \otimes V^*$ are given by the formulas*

$$\mathrm{ev}_V(v^* \otimes v) = v^*((S^{-1}(\beta) \otimes \alpha) \cdot v) = v^*(\alpha \cdot v \cdot S^{-1}(\beta)), \quad (6.2.2)$$
$$\mathrm{coev}_V(1) = \sum_i (S^{-1}(\alpha) \otimes \beta) \cdot v_i \otimes v^i = \sum_i \beta \cdot v_i \cdot S^{-1}(\alpha) \otimes v^i. \quad (6.2.3)$$

*The right dual of $V$ is $V^*$, now with $H$-bimodule structure*

$$h \cdot v^* \cdot h' = (h' \otimes h) \cdot v^* = v^* \leftharpoonup (S(h') \otimes S^{-1}(h)) = S(h') \rightharpoonup v^* \leftharpoonup S^{-1}(h).$$

*The evaluation* $\mathrm{ev}'_V : V \otimes V^* \to k$ *and coevaluation* $\mathrm{coev}'_V : k \to V^* \otimes V$ *are given by*

$$\mathrm{ev}'_V(v \otimes v^*) = v^*((\beta \otimes S^{-1}(\alpha)) \cdot v) = v^*(S^{-1}(\alpha) \cdot v \cdot \beta),$$
$$\mathrm{coev}'_V(1) = \sum_i v^i \otimes (\alpha \otimes S^{-1}(\beta)) \cdot v_i = \sum_i v^i \otimes S^{-1}(\beta) \cdot v_i \cdot \alpha.$$

*Proof* The two assertions follow easily from the canonical monoidal identification ${}_H\mathscr{M}_H = {}_{H^{\mathrm{op}} \otimes H}\mathscr{M}$ and the rigid monoidal structure of the category of finite-dimensional modules over a quasi-Hopf algebra. The details are left to the reader. □

With all these structures in mind one can prove the following result.

**Proposition 6.10** *Let H be a quasi-Hopf algebra with bijective antipode, C an H-bimodule coalgebra and M a finite-dimensional H-bimodule with basis* $\{m_i\}_i$, *and let* $\{m^i\}_i$ *be the corresponding dual basis of* $M^*$.

*(i) If M is a right quasi-Hopf* $(H,H,C)$*-module then* $M^*$ *is a left quasi-Hopf* $(H,H,C)$*-module with structure*

$$h \cdot m^* \cdot h' = S^{-1}(h') \rightharpoonup m^* \leftharpoonup S(h),$$
$$\lambda_{M^*}(m^*) = \sum_i m^*(S(\tilde{p}^2) f^1 \cdot (m_i)_{(0)} \cdot S^{-1}(\tilde{q}^2 g^2)) S(\tilde{p}^1) f^2 \cdot (m_i)_{(1)} \cdot S^{-1}(\tilde{q}^1 g^1) \otimes m^i.$$

*Here* $p_L = \tilde{p}^1 \otimes \tilde{p}^2$ *and* $q_L = \tilde{q}^1 \otimes \tilde{q}^2$ *are the elements defined in* (3.2.20), $f = f^1 \otimes f^2$ *is the Drinfeld twist from* (3.2.15) *and* $f^{-1} = g^1 \otimes g^2$ *is its inverse from* (3.2.16).

*(ii) If M is a left quasi-Hopf* $(H,H,C)$*-module then* ${}^*M$ *is a right quasi-Hopf* $(H,H,C)$*-module with structure*

$$h \cdot {}^*m \cdot h' = S(h') \rightharpoonup {}^*m \leftharpoonup S^{-1}(h),$$
$$\rho_{{}^*M}({}^*m) = \sum_i {}^*m(S^{-1}(f^1 p^1) \cdot (m_i)_{[0]} \cdot g^2 S(q^1)) m^i \otimes S^{-1}(f^2 p^2) \cdot (m_i)_{[-1]} \cdot g^1 S(q^2),$$

*where* $p_R = p^1 \otimes p^2$ *and* $q_R = q^1 \otimes q^2$ *are the elements presented in* (3.2.19).

*Proof* We will prove (i), and leave (ii) to the reader. Actually, we will show that the structure on $M^*$ as stated in (i) is precisely the structure that we obtain after applying part (ii) of Proposition 2.43 in the case when we have a coalgebra $C$ in $\mathscr{C} = {}_H\mathscr{M}_H$ and $V = M$. Consider $H$ as a bimodule via multiplication.

An object $M \in {}_H\mathscr{M}_H^C$ is actually a right $C$-comodule in ${}_H\mathscr{M}_H$. Since $M$ is finite dimensional we get that $M^*$, the left dual of $M$ in ${}_H\mathscr{M}_H$, is a left $H$-comodule in $\mathscr{C}$, and therefore an object of ${}_H^C\mathscr{M}_H$. By (6.2.1) we deduce that the $H$-bimodule structure of $H^*$ is the one mentioned in part (i) of the statement. In order to find the left $C$-coaction on $M^*$ we specialize Proposition 2.43 (ii) for the monoidal structure of ${}_H\mathscr{M}_H$, and use (6.2.2) and (6.2.3) to compute $\lambda_{M^*}$ as the following composition:

$$m^* \overset{\mathrm{Id}_{M^*} \otimes \mathrm{coev}_M}{\longmapsto} \sum_i m^* \otimes (\beta \cdot m_i \cdot S^{-1}(\alpha) \otimes m^i) = \sum_{i,j} m^j(\beta \cdot m_i \cdot S^{-1}(\alpha)) m^* \otimes (m_j \otimes m^i)$$

$$\overset{\mathrm{Id}_{M^*} \otimes (\rho_M \otimes \mathrm{Id}_{M^*})}{\longmapsto} \sum_{i,j} m^j(\beta \cdot m_i \cdot S^{-1}(\alpha)) m^* \otimes (((m_j)_{(0)} \otimes (m_j)_{(1)}) \otimes m^i)$$

$$\overset{\mathrm{Id}_{M^*}\otimes a_{M,C,M^*}}{\longmapsto} \sum_{i,j} m^j(\beta S(X^3)\cdot m_i\cdot S^{-1}(\alpha x^3))$$
$$m^*\otimes(X^1\cdot(m_j)_{(0)}x^1\otimes(X^2\cdot(m_j)_{(1)}x^2\otimes m^i))$$
$$\overset{a^{-1}_{M^*,M,C\otimes M^*}}{\longmapsto} \sum_{i,j} m^j(\beta S(y_2^3X^3)\cdot m_i\cdot S^{-1}(\alpha x^3Y_2^3))\,(S^{-1}(Y^1)\rightharpoonup m^*\leftharpoonup S(y^1)$$
$$\otimes y^2X^1\cdot(m_j)_{(0)}\cdot x^1Y^2)\otimes(y_1^3X^2\cdot(m_j)_{(1)}\cdot x^2Y_1^3\otimes m^i)$$
$$\overset{\mathrm{ev}_M\otimes\mathrm{Id}_{C\otimes M^*}}{\longmapsto} \sum_j m^*(S(y^1)\alpha y^2X^1\cdot(m_j)_{(0)}\cdot x^1Y^2S^{-1}(Y^1\beta))$$
$$y_1^3X^2\cdot(m_j)_{(1)}\cdot x^2Y_1^3\otimes S^{-1}(\alpha x^3Y_2^3)\rightharpoonup m^j\leftharpoonup\beta S(y_2^3X^3).$$

We then have, for all $m^*\in M^*$, that

$$\begin{aligned}
\lambda_{M^*}(m^*) &\overset{(3.2.20)}{=} \sum_j\langle m^*,\tilde{q}^1X^1\cdot(m_j)_{(0)}\cdot x^1\tilde{p}^1\rangle\\
&\qquad \tilde{q}_1^2X^2\cdot(m_j)_{(1)}\cdot x^2\tilde{p}_1^2\otimes S^{-1}(\alpha x^3\tilde{p}_2^2)\rightharpoonup m^j\leftharpoonup\beta S(\tilde{q}_2^2X^3)\\
&= \sum_i\langle m^*,\tilde{q}^1X^1\beta_1S(\tilde{q}_2^2X^3)_1\cdot(m_i)_{(0)}\cdot S^{-1}(\alpha x^3\tilde{p}_2^2)_1x^1\tilde{p}^1\rangle\\
&\qquad \tilde{q}_1^2X^2\beta_2S(\tilde{q}_2^2X^3)_2\cdot(m_i)_{(1)}\cdot S^{-1}(\alpha x^3\tilde{p}_2^2)_2x^2\tilde{p}_1^2\otimes m^i\\
&\underset{(3.2.13)}{\overset{(3.2.14)}{=}} \langle m^*,\tilde{q}^1X^1\delta^1S(\tilde{q}_{(2,2)}^2X_2^3)f^1\cdot(m_i)_{(0)}\cdot S^{-1}(\gamma^2x_2^3\tilde{p}_{(2,2)}^2g^2)x^1\tilde{p}^1\rangle\\
&\qquad \tilde{q}_1^2X^2\delta^2S(\tilde{q}_{(2,1)}^2X_1^3)f^2\cdot(m_i)_{(1)}\cdot S^{-1}(\gamma^1x_1^3\tilde{p}_{(2,1)}^2g^1)x^2\tilde{p}_1^2\otimes m^i\\
&\underset{(3.2.6)}{\overset{(3.2.5)}{=}} \sum_i\langle m^*,\tilde{q}^1\beta S(\tilde{q}_{(2,2)}^2X^3)f^1\cdot(m_i)_{(0)}\cdot S^{-1}(\alpha x^3\tilde{p}_{(2,2)}^2g^2)\tilde{p}^1\rangle\\
&\qquad \tilde{q}_1^2X^1\beta S(\tilde{q}_{(2,1)}^2X^2)f^2\cdot(m_i)_{(1)}\cdot S^{-1}(\alpha x^2\tilde{p}_{(2,1)}^2g^1)x^1\tilde{p}_1^2\otimes m^i\\
&\underset{(3.2.1)}{\overset{(3.1.7)}{=}} \sum_i\langle m^*,\tilde{q}^1\beta S(X^3\tilde{q}^2)f^1\cdot(m_i)_{(0)}\cdot S^{-1}(\alpha\tilde{p}^2x^3g^2)\tilde{p}^1\rangle\\
&\qquad X^1\beta S(X^2)f^2\cdot(m_i)_{(1)}\cdot S^{-1}(\alpha x^2g^1)x^1\otimes m^i\\
&\underset{(3.2.2)}{\overset{(3.2.20)}{=}} \sum_i\langle m^*,S(X^3)f^1\cdot(m_i)_{(0)}\cdot S^{-1}(x^3g^2)\rangle\\
&\qquad X^1\beta S(X^2)f^2\cdot(m_i)_{(1)}\cdot S^{-1}(\alpha x^2g^1)x^1\otimes m^i\\
&\overset{(3.2.20)}{=} \sum_i\langle m^*,S(\tilde{p}^2)f^1\cdot(m_i)_{(0)}\cdot S^{-1}(\tilde{q}^2g^2)\rangle\\
&\qquad S(\tilde{p}^1)f^2\cdot(m_i)_{(1)}\cdot S^{-1}(\tilde{q}^1g^1)\otimes m^i,
\end{aligned}$$

as stated. □

We now describe the morphisms of quasi-Hopf $(H,H,C)$-bimodules between $C$ and one of its dual objects. We start with a more general result.

**Lemma 6.11** *Let $C$ be a coalgebra in a monoidal category $\mathscr{C}$ and $V,W$ objects of $\mathscr{C}$. Then the following assertions hold.*

*(i) If $V$ has a right dual object ${}^*V$ in $\mathscr{C}$ then* $\mathrm{Hom}_{\mathscr{C}}(W,{}^*V)\cong\mathrm{Hom}_{\mathscr{C}}(V\otimes W,\underline{1})$.

*Furthermore, if $V$ is a left $C$-comodule and $W$ a right $C$-comodule in $\mathscr{C}$ then giving a right $C$-colinear morphism from $W$ to ${}^*V$ is equivalent to giving a morphism $\Sigma : V \otimes W \to \underline{1}$ that satisfies*

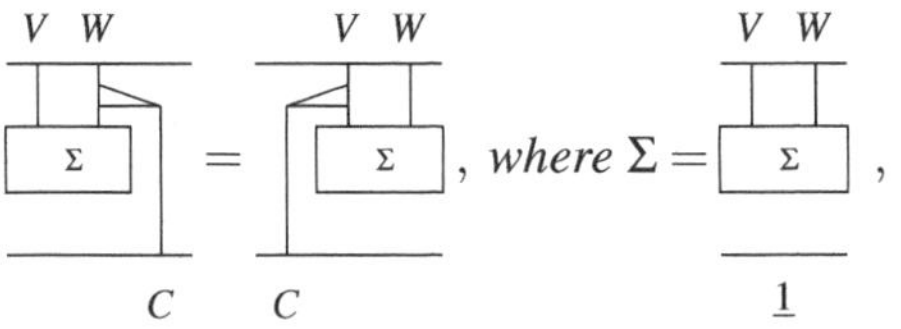
(6.2.4)

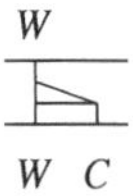 *is the right $C$-coaction on $W$ and* 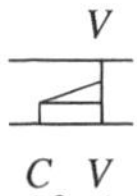 *is the left $C$-coaction on $V$.*

*(ii) If $W$ has a left dual object $W^*$ in $\mathscr{C}$ then $\mathrm{Hom}_{\mathscr{C}}(V, W^*) \cong \mathrm{Hom}_{\mathscr{C}}(V \otimes W, \underline{1})$. If, moreover, $V$ is a left $C$-comodule and $W$ a right $C$-comodule in $\mathscr{C}$ then giving a left $C$-colinear morphism from $V$ to $W^*$ in $\mathscr{C}$ is equivalent to giving a morphism $\Sigma : V \otimes W \to \underline{1}$ obeying* (6.2.4).

*Proof* The assertions in (ii) follow from those in (i) by replacing $\mathscr{C}$ with $\overline{\mathscr{C}}$.

The first assertion in (i) is a particular case of Proposition 1.81. To be concise, the two correspondences below

$\mathrm{Hom}_{\mathscr{C}}(W, {}^*V) \ni f \mapsto \Sigma_f :=$ 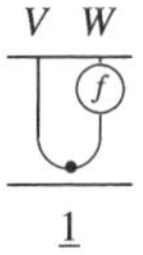 $\in \mathrm{Hom}_{\mathscr{C}}(V \otimes W, \underline{1})$ ,

$\mathrm{Hom}_{\mathscr{C}}(V \otimes W, \underline{1}) \ni \Sigma \mapsto f_\Sigma :=$ 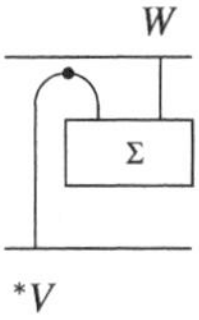 $\in \mathrm{Hom}_{\mathscr{C}}(W, {}^*V)$

are inverse to each other. Thus, if $V$ is a left and $W$ is a right comodule over $C$ then from Proposition 2.43 we have that $f : W \to {}^*V$ is right $C$-colinear if and only if

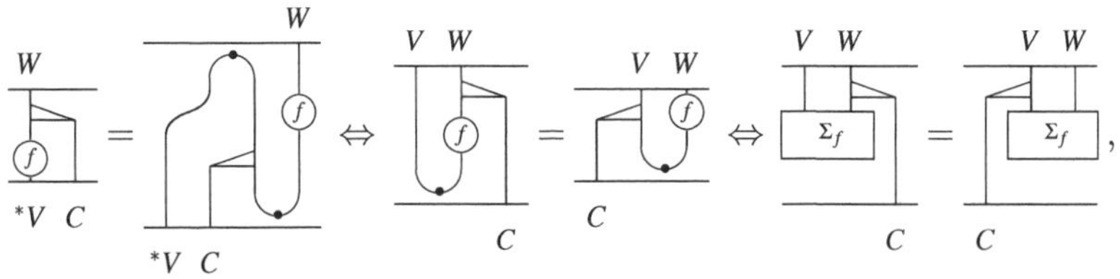

as stated. This finishes the proof of the lemma. □

Consider again the situation when $C$ is a coalgebra in the category of bimodules over a quasi-Hopf algebra.

**Proposition 6.12** *Let $H$ be a quasi-Hopf algebra with bijective antipode, $C$ an $H$-bimodule coalgebra, and take $N \in {}^C_H\mathscr{M}_H$ and $M \in {}_H\mathscr{M}^C_H$. Then:*

*(i) If N is finite dimensional then the assignment* $f \mapsto \Sigma_f$, *where*

$$\Sigma_f(n \otimes m) = f(m)(S^{-1}(\alpha) \cdot n \cdot \beta),$$

*for all* $m \in M$ *and* $n \in N$, *provides a one-to-one correspondence between H-bilinear maps* $f : M \to {}^*N$ *and H-bilinear forms* $\Sigma_f : N \otimes M \to k$. *Its inverse sends* $\Sigma : N \otimes M \to k$ *in* ${}_H\mathscr{M}_H$ *to*

$$f_\Sigma : M \ni m \mapsto \Sigma(p_L^1 \cdot n_i \cdot q_L^1 \otimes p_L^2 \cdot m \cdot q_L^2)n^i \in {}^*N,$$

*an H-bimodule map, where* $\{n_i, n^i\}_i$ *are dual bases in N and* $N^*$. *Consequently, giving a morphism* $f : M \to {}^*N$ *in* ${}_H\mathscr{M}_H^C$ *is equivalent to giving an H-bilinear form* $\Sigma_f : N \otimes M \to k$ *satisfying, for all* $m \in M$ *and* $n \in N$,

$$\begin{aligned} \Sigma_f(x^1 \cdot n \cdot X^1 \otimes x^2 \cdot m_{(0)} \cdot X^2)x^3 \cdot m_{(1)} \cdot X^3 \\ = \Sigma_f(X^2 \cdot n_{[0]} \cdot x^2 \otimes X^3 \cdot m \cdot x^3)X^1 \cdot n_{[-1]} \cdot x^1. \end{aligned} \tag{6.2.5}$$

*(ii) Likewise, if M is finite dimensional then the assignment* $g \mapsto \Sigma_g$, *where*

$$\Sigma_g(n \otimes m) = g(n)(\alpha \cdot m \cdot S^{-1}(\beta)),$$

*for all* $m \in M$ *and* $n \in N$, *provides a one-to-one correspondence between H-bilinear morphisms* $g : N \to M^*$ *and H-bilinear forms* $\Sigma_g : N \otimes M \to k$. *Its inverse associates to an H-bilinear map* $\Sigma : N \otimes M \to k$ *the H-bilinear morphism*

$$g_\Sigma : N \ni n \mapsto \Sigma(p_R^1 \cdot n \cdot q_R^1 \otimes p_R^2 \cdot m_j \cdot q_R^2)m^j \in M^*,$$

*where this time* $\{m_j, m^j\}_j$ *are dual bases in M and* $M^*$. *Consequently, giving a morphism* $g : N \to M^*$ *in* ${}_H\mathscr{M}_H^C$ *is equivalent to giving an H-bilinear form* $\Sigma_g : N \otimes M \to k$ *obeying* (6.2.5), *for all* $m \in M$ *and* $n \in N$.

*Proof* This follows by specializing Lemma 6.11 for $\mathscr{C} = {}_H\mathscr{M}_H$. The computations are similar to those in the proof of Proposition 6.10, and this is why we leave all the details to the reader. For instance to a morphism $f : M \to {}^*N$ in ${}_H\mathscr{M}_H$ we associate

$$\Sigma_f(n \otimes m) = \mathrm{ev}'_N(n \otimes f(m)) = f(m)(S^{-1}(\alpha) \cdot n \cdot \beta),$$

an $H$-bilinear form from $N \otimes M$ to $k$, as stated. □

**Corollary 6.13** *Let H be a quasi-Hopf algebra with bijective antipode, C a finite-dimensional H-bimodule coalgebra. There are bijective correspondences between*

*(i) morphisms* $f : C \to {}^*C$ *in* ${}_H\mathscr{M}_H^C$;

*(ii) H-bimodule maps* $\Sigma_f : C \otimes C \to k$ *satisfying, for all* $c, c' \in C$,

$$\Sigma(x^1 \cdot c \cdot X^1 \otimes x^2 \cdot c'_{\underline{1}} \cdot X^2)x^3 \cdot c'_{\underline{2}} \cdot X^3 = \Sigma_f(X^2 \cdot c_{\underline{2}} \cdot x^2 \otimes X^3 \cdot c' \cdot x^3)X^1 \cdot c_{\underline{1}} \cdot x^1;$$

*(iii) morphisms* $g : C \to C^*$ *in* ${}_H^C\mathscr{M}_H$.

*Proof* Consider $M = N = C$ in Proposition 6.12 and view it alternatively as a left or right quasi-Hopf $(H, H, C)$-module. □

## 6.3 Structure Theorems for Quasi-Hopf Bimodules

If $H$ is a quasi-Hopf algebra and $\mathfrak{A}$ is a right $H$-comodule algebra we will show that ${}_H\mathscr{M}_{\mathfrak{A}}^H$ is equivalent to the category of right $\mathfrak{A}$-modules, $\mathscr{M}_{\mathfrak{A}}$. The essential role in the proof is played by the definition of the set of coinvariants of an object in ${}_H\mathscr{M}_{\mathfrak{A}}^H$, and then by a certain structure theorem for quasi-Hopf $(H, \mathfrak{A}, H)$-modules.

Of course we can formulate and prove a left-handed version for the result stated above: if $\mathfrak{A}$ is a right $H$-comodule algebra then the categories ${}_{\mathfrak{A}}\mathscr{M}_H^H$ and ${}_{\mathfrak{A}}\mathscr{M}$ are equivalent. As we shall see, this requires a different definition for the set of coinvariants of an object $N \in {}_{\mathfrak{A}}\mathscr{M}_H^H$. Thus when $\mathfrak{A} = H$ we get two different concepts of coinvariants for a quasi-Hopf $H$-bimodule. For consistency, we will call them the set of coinvariants of the first type and the set of alternative coinvariants, respectively. More details will be presented in what follows.

**Definition 6.14** Let $H$ be a quasi-Hopf algebra with bijective antipode, $\mathfrak{A}$ a right $H$-comodule algebra and $M \in {}_H\mathscr{M}_{\mathfrak{A}}^H$. We call $m \in M$ a coinvariant element of $M$ of the first type if

$$\rho_M(m) = S^{-1}(\mathfrak{q}^2(\tilde{X}^3_\rho)_2 g^2) \cdot m \cdot \tilde{X}^1_\rho \otimes S^{-1}(\mathfrak{q}^1(\tilde{X}^3_\rho)_1 g^1)\tilde{X}^2_\rho, \tag{6.3.1}$$

where $q_L := \mathfrak{q}^1 \otimes \mathfrak{q}^2$ is the element defined by (3.2.20) and $f^{-1} = g^1 \otimes g^2$ is the element defined by (3.2.16).

We will denote by $M^{\underline{\mathrm{co}(H)}}$ the set of coinvariants of $M$ of the first type.

**Lemma 6.15** *Let $M \in {}_H\mathscr{M}_{\mathfrak{A}}^H$ and let $M^{\underline{\mathrm{co}(H)}}$ be its set of coinvariants of the first type. Then $M^{\underline{\mathrm{co}(H)}}$ becomes a right $\mathfrak{A}$-module via*

$$m \leftarrow \mathfrak{a} := S^{-1}(\mathfrak{a}_{\langle 1\rangle}) \cdot m \cdot \mathfrak{a}_{\langle 0\rangle}, \tag{6.3.2}$$

*for all $m \in M^{\underline{\mathrm{co}(H)}}$ and $\mathfrak{a} \in \mathfrak{A}$. Consequently, we have a functor $G : {}_H\mathscr{M}_{\mathfrak{A}}^H \to \mathscr{M}_{\mathfrak{A}}$. $G$ sends $M \in {}_H\mathscr{M}_{\mathfrak{A}}^H$ to $M^{\underline{\mathrm{co}(H)}}$ and a morphism to its restriction to $M^{\underline{\mathrm{co}(H)}}$.*

*Proof* The difficult part is to show that $\leftarrow$ is well defined. To this end we compute, for all $m \in M^{\underline{\mathrm{co}(H)}}$ and $\mathfrak{a} \in \mathfrak{A}$, that

$$\begin{aligned}
&\rho(m \leftarrow \mathfrak{a})\\
&\quad = S^{-1}(\mathfrak{a}_{\langle 1\rangle})_1 \cdot m_{(0)} \cdot \mathfrak{a}_{\langle 0,0\rangle} \otimes S^{-1}(\mathfrak{a}_{\langle 1\rangle})_2 m_{(1)} \mathfrak{a}_{\langle 0,1\rangle}\\
&\overset{(3.2.13)}{=} S^{-1}(\mathfrak{q}^2(\tilde{X}^3_\rho)_2 \mathfrak{a}_{\langle 1\rangle_2} g^2) \cdot m \cdot \tilde{X}^1_\rho \mathfrak{a}_{\langle 0,0\rangle} \otimes S^{-1}(\mathfrak{q}^1(\tilde{X}^3_\rho)_1 \mathfrak{a}_{\langle 1\rangle_1} g^1)\tilde{X}^2_\rho \mathfrak{a}_{\langle 0,1\rangle}\\
&\overset{(4.3.1)}{=} S^{-1}(\mathfrak{q}^2 \mathfrak{a}_{\langle 1\rangle_{(2,2)}}(\tilde{X}^3_\rho)_2 g^2) \cdot m \cdot \mathfrak{a}_{\langle 0\rangle}\tilde{X}^1_\rho \otimes S^{-1}(\mathfrak{q}^1 \mathfrak{a}_{\langle 1\rangle_{(2,1)}}(\tilde{X}^3_\rho)_1 g^1)\mathfrak{a}_{\langle 1\rangle_1}\tilde{X}^2_\rho\\
&\overset{(3.2.22)}{=} S^{-1}(\mathfrak{q}^2(\tilde{X}^2_\rho)_2 g^2) \cdot (m \leftarrow \mathfrak{a}) \cdot \tilde{X}^1_\rho \otimes S^{-1}(\mathfrak{q}^1(\tilde{X}^3_\rho)_1 g^1)\tilde{X}^2_\rho,
\end{aligned}$$

as required. The fact that $\leftarrow$ defines on $M^{\underline{\mathrm{co}(H)}}$ a right $\mathfrak{A}$-module structure is immediate, thus $G$ in the statement is a well-defined functor. □

We can now prove the desired equivalence of categories.

**Theorem 6.16** *If $H$ is a quasi-Hopf algebra with bijective antipode and $\mathfrak{A}$ is a right $H$-comodule algebra then the category of quasi-Hopf $(H,\mathfrak{A},H)$-modules is equivalent to the category of right $\mathfrak{A}$-modules, $\mathscr{M}_{\mathfrak{A}}$.*

*Proof* From Example 6.4 we have a functor $F=\bullet\otimes H:\mathscr{M}_{\mathfrak{A}}\to {}_H\mathscr{M}_{\mathfrak{A}}^H$. We recall that for any $M\in\mathscr{M}_{\mathfrak{A}}$ we can endow $M\otimes H$ with a structure of an object in ${}_H\mathscr{M}_{\mathfrak{A}}^H$ as follows (for all $m\in M$, $h,h'\in H$ and $\mathfrak{a}\in\mathfrak{A}$):

$$h\cdot(m\otimes h')\cdot\mathfrak{a}=m\cdot\mathfrak{a}_{\langle 0\rangle}\otimes hh'\mathfrak{a}_{\langle 1\rangle}, \tag{6.3.3}$$

$$\rho(m\otimes h)=m\cdot\tilde{X}^1_\rho\otimes h_1\tilde{X}^2_\rho\otimes h_2\tilde{X}^3_\rho. \tag{6.3.4}$$

We claim that $F$ and $G$ define inverse equivalences, where $G$ is the functor defined in Lemma 6.15. Indeed, let $M\in\mathscr{M}_{\mathfrak{A}}$ and consider $M\otimes H$ as an object in ${}_H\mathscr{M}_{\mathfrak{A}}^H$ with the structure defined above. We then have that

$$(M\otimes H)^{\underline{\mathrm{co}(H)}}=\{m\cdot\tilde{q}^1_\rho\otimes\tilde{q}^2_\rho\mid m\in M\},$$

where $\tilde{q}_\rho=\tilde{q}^1_\rho\otimes\tilde{q}^2_\rho$ is the element defined in (4.3.9). Actually, by using the above definitions we have that $m\otimes h\in(M\otimes H)^{\underline{\mathrm{co}(H)}}$ if and only if

$$\begin{aligned}&m\cdot\tilde{X}^1_\rho\otimes h_1\tilde{X}^2_\rho\otimes h_2\tilde{X}^3_\rho\\&\quad=m\cdot(\tilde{X}^1_\rho)_{\langle 0\rangle}\otimes S^{-1}(\mathfrak{q}^2(\tilde{X}^3_\rho)_2g^2)h(\tilde{X}^1_\rho)_{\langle 1\rangle}\otimes S^{-1}(\mathfrak{q}^1(\tilde{X}^3_\rho)_1g^1)\tilde{X}^2_\rho,\end{aligned} \tag{6.3.5}$$

where, as usual, $q_L=\mathfrak{q}^1\otimes\mathfrak{q}^2$ is the element defined in (3.2.20). Therefore, if $m\otimes h\in(M\otimes H)^{\underline{\mathrm{co}(H)}}$ then by applying $\mathrm{Id}_M\otimes\varepsilon\otimes\mathrm{Id}_H$ to the equality (6.3.5) we obtain

$$m\otimes h=\varepsilon(h)m\cdot\tilde{X}^1_\rho\otimes S^{-1}(\alpha\tilde{X}^3_\rho)\tilde{X}^2_\rho=\varepsilon(h)m\cdot\tilde{q}^1_\rho\otimes\tilde{q}^2_\rho,$$

as wanted. For the converse inclusion we need the formula

$$\tilde{Q}^1_\rho(\tilde{x}^1_\rho)_{\langle 0\rangle}\otimes S^{-1}(\tilde{x}^2_\rho)\tilde{Q}^2_\rho(\tilde{x}^1_\rho)_{\langle 1\rangle}\otimes\tilde{x}^3_\rho=\tilde{X}^1_\rho\otimes S^{-1}(\mathfrak{q}^1(\tilde{X}^3_\rho)_1)\tilde{X}^2_\rho\otimes\mathfrak{q}^2(\tilde{X}^3_\rho)_2, \tag{6.3.6}$$

which follows easily from the definitions of $\tilde{q}_\rho=\tilde{Q}^1_\rho\otimes\tilde{Q}^2_\rho$ and $q_L$ and the equation (4.3.1). Then we compute, for all $m\in M$:

$$\begin{aligned}&\rho(m\cdot\tilde{q}^1_\rho\otimes\tilde{q}^2_\rho)\\&\quad= m\cdot\tilde{q}^1_\rho\tilde{X}^1_\rho\otimes(\tilde{q}^2_\rho)_1\tilde{X}^2_\rho\otimes(\tilde{q}^2_\rho)_2\tilde{X}^3_\rho\\&\overset{(4.3.15)}{=} m\cdot\tilde{q}^1_\rho(\tilde{Q}^1_\rho(\tilde{x}^1_\rho)_{\langle 0\rangle})_{\langle 0\rangle}\otimes S^{-1}(\tilde{x}^3_\rho g^2)\tilde{q}^2_\rho(\tilde{Q}^1_\rho(\tilde{x}^1_\rho)_{\langle 0\rangle})_{\langle 1\rangle}\otimes S^{-1}(\tilde{x}^2_\rho g^1)\tilde{Q}^2_\rho(\tilde{x}^1_\rho)_{\langle 1\rangle}\\&\overset{(6.3.6)}{=} m\cdot\tilde{q}^1_\rho(\tilde{X}^1_\rho)_{\langle 0\rangle}\otimes S^{-1}(\mathfrak{q}^2(\tilde{X}^3_\rho)_2g^2)\tilde{q}^2_\rho(\tilde{X}^1_\rho)_{\langle 1\rangle}\otimes S^{-1}(\mathfrak{q}^1(\tilde{X}^3_\rho)_1g^1)\tilde{X}^2_\rho\\&\quad= S^{-1}(\mathfrak{q}^2(\tilde{X}^3_\rho)_2g^2)\cdot(m\cdot\tilde{q}^1_\rho\otimes\tilde{q}^2_\rho)\cdot\tilde{X}^1_\rho\otimes S^{-1}(\mathfrak{q}^1(\tilde{X}^3_\rho)_1g^1)\tilde{X}^2_\rho,\end{aligned}$$

as required. Now, it is not hard to see that

$$\xi_M:GF(M)=\{m\cdot\tilde{q}^1_\rho\otimes\tilde{q}^2_\rho\mid m\in M\}\to M,$$
$$\xi_M(m\cdot\tilde{q}^1_\rho\otimes\tilde{q}^2_\rho)=m$$

is a well-defined isomorphism in $\mathscr{M}_{\mathfrak{A}}$ with $\xi_M^{-1}(m)=m\cdot\tilde{q}^1_\rho\otimes\tilde{q}^2_\rho$, for any $m\in M$.

Conversely, take $M \in {}_H\mathscr{M}_{\mathfrak{A}}^H$ and define $\underline{E}: M \to M$ by

$$\underline{E}(m) = S^{-1}(\alpha m_{(1)}) \cdot m_{(0)}, \quad \forall\, m \in M.$$

We have $\mathrm{Im}(\underline{E}) \subseteq M^{\underline{\mathrm{co}(H)}}$ since

$$\begin{aligned}
\rho(\underline{E}(m)) \quad &= \quad S^{-1}(\alpha m_{(1)})_1 \cdot m_{(0,0)} \otimes S^{-1}(\alpha m_{(1)})_2 \cdot m_{(0,1)} \\
&\overset{(3.2.13),(3.2.14)}{=} S^{-1}(\gamma^2 m_{(1)_2} g^2) \cdot m_{(0,0)} \otimes S^{-1}(\gamma^1 m_{(1)_1} g^1) m_{(0,1)} \\
&\overset{(3.2.5)}{=} S^{-1}(\alpha x^3 X_2^3 m_{(1)_2} g^2) X^1 \cdot m_{(0,0)} \\
&\qquad \otimes S^{-1}(\alpha x^2 X_1^3 m_{(1)_1} g^1) x^1 X^2 m_{(0,0)} \\
&\overset{(6.1.5)}{=} S^{-1}(\alpha x^3 m_{(1)_{(2,2)}} (\tilde{X}_\rho^3)_2 g^2) m_{(0)} \cdot \tilde{X}_\rho^1 \\
&\qquad \otimes S^{-1}(\alpha x^2 m_{(1)_{(2,1)}} (\tilde{X}_\rho^3)_1 g^1) x^1 m_{(1)_1} \tilde{X}_\rho^2 \\
&\overset{(3.1.7),(3.2.1)}{=} S^{-1}(\alpha m_{(1)} x^3 (\tilde{X}_\rho^3)_2 g^2) m_{(0)} \cdot \tilde{X}_\rho^1 \otimes S^{-1}(\alpha x^2 (\tilde{X}_\rho^3)_1 g^1) x^1 \tilde{X}_\rho^2 \\
&\overset{(3.2.20)}{=} S^{-1}(\mathfrak{q}^2 (\tilde{X}_\rho^3)_2 g^2) \cdot \underline{E}(m) \cdot \tilde{X}_\rho^1 \otimes S^{-1}(\mathfrak{q}^1 (\tilde{X}_\rho^3)_1 g^1),
\end{aligned}$$

for all $m \in M$. Therefore, the map

$$\zeta_M : M \to M^{\underline{\mathrm{co}(H)}} \otimes H = \mathfrak{F}(\mathfrak{G}(M)), \quad \zeta_M(m) = \underline{E}(m_{(0)}) \otimes m_{(1)}, \ \forall\, m \in M$$

is well defined. We show that $\zeta_M$ is an isomorphism in ${}_H\mathscr{M}_{\mathfrak{A}}^H$ with inverse

$$\zeta_M^{-1}(m \otimes h) = h S^{-1}(\tilde{p}_\rho^2) \cdot m \cdot \tilde{p}_\rho^1,$$

for all $m \in M^{\underline{\mathrm{co}(H)}}$ and $h \in H$. Here $\tilde{p}_\rho = \tilde{p}_\rho^1 \otimes \tilde{p}_\rho^2$ is the element defined in (4.3.9).

Indeed, the fact that $\zeta_M$ is left $H$-linear is a consequence of the fact that

$$\underline{E}(h \cdot m) = S^{-1}(h_2 m_{(1)}) h_1 \cdot m_{(0)} = \varepsilon(h) \underline{E}(m), \ \forall\, h \in H, \ m \in M.$$

Likewise, the fact that

$$\underline{E}(m \cdot \mathfrak{a}) = S^{-1}(\alpha m_{(1)} \mathfrak{a}_{\langle 1 \rangle}) \cdot m_{(0)} \cdot \mathfrak{a}_{\langle 0 \rangle} = \underline{E}(m) \leftarrow \mathfrak{a}, \ \forall\, m \in M, \ \mathfrak{a} \in \mathfrak{A}$$

implies that $\zeta_M$ is right $\mathfrak{A}$-linear, and so an $(H, \mathfrak{A})$-bimodule morphism. It is right $H$-colinear as well since, for all $m \in M$, we have

$$\begin{aligned}
\underline{E}(m_{(0)}) \leftarrow \tilde{X}_\rho^1 \otimes m_{(1)_1} \tilde{X}_\rho^2 \otimes m_{(1)_2} \tilde{X}_\rho^3 \quad &= \quad \underline{E}(m_{(0)} \cdot \tilde{X}_\rho^1) \otimes m_{(1)_1} \tilde{X}_\rho^2 \otimes m_{(1)_2} \tilde{X}_\rho^3 \\
&\overset{(6.1.5)}{=} \underline{E}(X^1 \cdot m_{(0,0)}) \otimes X^2 m_{(0,1)} \otimes X^3 m_{(1)} \\
&= \quad \underline{E}(m_{(0,0)}) \otimes m_{(0,1)} \otimes m_{(1)}.
\end{aligned}$$

We now check that $\zeta_M$ and $\zeta_M^{-1}$ are inverses. On the one hand we have

$$\begin{aligned}
\zeta_M^{-1} \zeta_M(m) \quad &= \quad m_{(1)} S^{-1}(\tilde{p}_\rho^2) \cdot \underline{E}(m_{(0)}) \cdot \tilde{p}_\rho^1 \\
&= \quad m_{(1)} S^{-1}(\alpha m_{(0,1)} \tilde{p}_\rho^2) \cdot m_{(0,0)} \cdot \tilde{p}_\rho^1 \\
&= \quad m_{(1)} \tilde{x}_\rho^3 S^{-1}(\alpha m_{(0,1)} \tilde{x}_\rho^2 \beta) \cdot m_{(0,0)} \cdot \tilde{x}_\rho^1 \\
&\overset{(6.1.5)}{=} x^3 m_{(1)_2} S^{-1}(\alpha x^2 m_{(1)_1} \beta) x^1 \cdot m_{(0)}
\end{aligned}$$

$$= x^3S^{-1}(\alpha x^2\beta)x^1\cdot m \overset{(3.2.2)}{=} m,$$

for all $m\in M$. On the other hand, for all $m\in M^{\underline{\mathrm{co}(H)}}$ and $h\in H$ we have

$$\begin{aligned}
&\zeta_M\zeta_M^{-1}(m\otimes h)\\
&\quad =\underline{E}(h_1S^{-1}(\tilde{p}_\rho^2)_1\cdot m_{(0)}\cdot(\tilde{p}_\rho^1)_{\langle 0\rangle}\otimes h_2S^{-1}(\tilde{p}_\rho^2)_2m_{(1)}(\tilde{p}_\rho^2)_{\langle 1\rangle}\\
&\quad =\underline{E}(m_{(0)})\leftarrow(\tilde{p}_\rho^1)_{\langle 0\rangle}\otimes hS^{-1}(\tilde{p}_\rho^2)m_{(1)}(\tilde{p}_\rho^1)_{\langle 1\rangle}\\
&\quad =\underline{E}(S^{-1}(\mathfrak{q}^2(\tilde{X}_\rho^3)_2g^2)\cdot m\cdot\tilde{X}_\rho^1)\leftarrow(\tilde{p}_\rho^1)_{\langle 0\rangle}\otimes hS^{-1}(\mathfrak{q}^1(\tilde{X}_\rho^3)_1g^1\tilde{p}_\rho^2)\tilde{X}_\rho^2(\tilde{p}_\rho^1)_{\langle 1\rangle}\\
&\quad =\underline{E}(m)\leftarrow\tilde{X}_\rho^1(\tilde{p}_\rho^1)_{\langle 0\rangle}\otimes S^{-1}(\alpha\tilde{X}_\rho^3\tilde{p}_\rho^2)\tilde{X}_\rho^2(\tilde{p}_\rho^1)_{\langle 1\rangle}\\
&\quad =\underline{E}(m)\leftarrow\tilde{q}_\rho^1(\tilde{p}_\rho^1)_{\langle 0\rangle}\otimes hS^{-1}(\tilde{p}_\rho^2)\tilde{q}_\rho^2(\tilde{p}_\rho^1)_{\langle 1\rangle}\overset{(4.3.13)}{=}m\otimes h,
\end{aligned}$$

where in the last equality we also use the fact that $m\in M^{\underline{\mathrm{co}(H)}}$ implies

$$\underline{E}(m)=S^{-1}(\alpha m_{(1)})\cdot m_{(0)}=S^{-1}(\mathfrak{q}^2(\tilde{X}_\rho^3)_2g^2\alpha S^{-1}(\mathfrak{q}^1(\tilde{X}_\rho^3)_1g^1)\tilde{X}_\rho^2)\cdot m\cdot\tilde{X}_\rho^1=m.$$

Finally, it is straightforward to check that the maps $\xi_M$ and $\zeta_M$ define natural transformations, so the proof is finished. □

**Remark 6.17** By the above proposition it follows that the category ${}_H\mathscr{M}_H^H$ is equivalent to the category of right $H$-modules. It is equivalent to the category of left $H$-modules, too. To see this we use the equivalence between the categories ${}_{\mathfrak{A}}\mathscr{M}_H^H(\equiv {}_{H^{\mathrm{op}}}\mathscr{M}_{\mathfrak{A}^{\mathrm{op}}}^{H^{\mathrm{op}}})$ and ${}_{\mathfrak{A}}\mathscr{M}(\equiv\mathscr{M}_{\mathfrak{A}^{\mathrm{op}}})$. More precisely, the equivalence is produced by the functors

$${}_{\mathfrak{A}}\mathscr{M}\ \underset{\mathbb{G}}{\overset{\mathbb{F}}{\rightleftarrows}}\ {}_{\mathfrak{A}}\mathscr{M}_H^H.$$

If $M\in{}_{\mathfrak{A}}\mathscr{M}$ then $\mathbb{F}(M)=M\otimes H$ regarded as an object in ${}_{\mathfrak{A}}\mathscr{M}_H^H$ via

$$\begin{aligned}
\mathfrak{a}(m\otimes h)h' &= \mathfrak{a}_{\langle 0\rangle}\cdot m\otimes\mathfrak{a}_{\langle 1\rangle}hh',\\
\rho(m\otimes h) &= \tilde{x}_\rho^1\cdot m\otimes\tilde{x}_\rho^2h_1\otimes\tilde{x}_\rho^3h_2,
\end{aligned}$$

for all $\mathfrak{a}\in\mathfrak{A}$, $m\in M$ and $h,h'\in H$. If $M\in{}_{\mathfrak{A}}\mathscr{M}_H^H$ then $\mathbb{G}(M)=M^{\overline{\mathrm{co}(H)}}$, where

$$M^{\overline{\mathrm{co}(H)}}:=\{m\in M\mid\rho(m)=\tilde{x}_\rho^1\cdot m\cdot S((\tilde{x}_\rho^3)_2X^3)f^1\otimes\tilde{x}_\rho^2X^1\beta S((\tilde{x}_\rho^3)_1X^2)f^2\}.$$

Note that $M^{\overline{\mathrm{co}(H)}}=\{\overline{E}(m)\mid m\in M\}$, where $\overline{E}:\ M\to M$ is given by

$$\overline{E}(m):=m_{(0)}\cdot\beta S(m_{(1)}),\ \ \forall\, m\in M, \tag{6.3.7}$$

and that $M^{\overline{\mathrm{co}(H)}}$ can be regarded as a left $\mathfrak{A}$-module via the action

$$\mathfrak{a}\to m:=\mathfrak{a}_{\langle 0\rangle}\cdot m\cdot S(\mathfrak{a}_{\langle 1\rangle}),\ \ \forall\,\mathfrak{a}\in\mathfrak{A},\ m\in M. \tag{6.3.8}$$

Then $M\ni m\mapsto\overline{E}(m_{(0)})\otimes m_{(1)}\in M^{\overline{\mathrm{co}(H)}}\otimes H$ is an isomorphism in ${}_{\mathfrak{A}}\mathscr{M}_H^H$ with inverse given by $M^{\overline{\mathrm{co}(H)}}\otimes H\ni m\otimes h\mapsto\tilde{q}_\rho^1\cdot m\cdot S(\tilde{q}_\rho^2)h\in M$. By taking $\mathfrak{A}=H$ this gives a second structure theorem for right quasi-Hopf $H$-bimodules.

## 6.4 The Categories ${}_H\mathscr{M}_H^H$ and ${}_H\mathscr{M}$

For $H$ a quasi-Hopf algebra, we have proved so far two structure theorems for quasi-Hopf bimodules. A third one will be presented in this section. As we shall see, it will provide a monoidal equivalence between the categories ${}_H\mathscr{M}_H^H$ and ${}_H\mathscr{M}$. Once more this third structure theorem is possible due to the choice of the set of coinvariants for a right quasi-Hopf $H$-bimodule.

**Definition 6.18** Let $H$ be a quasi-Hopf algebra with bijective antipode and $M$ a right quasi-Hopf $H$-bimodule. We define $E: M \to M$ by

$$E(m) = X^1 \cdot m_{(0)} \cdot \beta S(X^2 m_{(1)})\alpha X^3 = q^1 \cdot m_{(0)} \cdot \beta S(q^2 m_{(1)}), \tag{6.4.1}$$

for all $m \in M$, where $M \ni m \mapsto \rho_M(m) := m_{(0)} \otimes m_{(1)} \in M \otimes H$ denotes the right coaction of $H$ on $M$ and $q_R = q^1 \otimes q^2$ is the element defined in (3.2.19). The space

$$M^{\mathrm{co}(H)} = \{n \in M \mid E(n) = n\}$$

is called the space of coinvariants of $M$ of the second type.

Let $M \in {}_H\mathscr{M}_H^H$. To avoid confusion, we call the elements of $M^{\overline{\mathrm{co}(H)}}$ (see Remark 6.17) alternative coinvariants for the right quasi-Hopf $H$-bimodule $M$. As we see next, $M^{\overline{\mathrm{co}(H)}}$ and $M^{\mathrm{co}(H)}$ are isomorphic as left $H$-modules.

**Lemma 6.19** *For $h \in H$ and $m \in M$ define $h \neg m = E(h \cdot m)$. Then, for all $m \in M$ and $h, h' \in H$, the following relations hold:*

$$E(m \cdot h) = \varepsilon(h)E(m), \quad E(h \cdot E(m)) = E(h \cdot m), \tag{6.4.2}$$
$$h \cdot E(m) = E(h_1 \cdot m) \cdot h_2 = [h_1 \neg m] \cdot h_2, \tag{6.4.3}$$
$$E^2 = E,\ E(m_{(0)}) \cdot m_{(1)} = m \text{ and } E(E(m)_{(0)}) \otimes E(m)_{(1)} = E(m) \otimes 1_H, \tag{6.4.4}$$
$$(hh') \neg m = h \neg (h' \neg m). \tag{6.4.5}$$

*Proof* For all $m \in M$ and $h \in H$ we have

$$E(m \cdot h) = q^1 \cdot m_{(0)} \cdot h_1 \beta S(q^2 m_{(1)} h_2) \overset{(3.2.1)}{=} \varepsilon(h)E(m),$$

and this implies that

$$E(h \cdot E(m)) = E(hq^1 \cdot m_{(0)} \cdot \beta S(q^2 m_{(1)})) = E(h \cdot m).$$

Next we compute:

$$\begin{aligned} [h_1 \neg m] \cdot h_2 &= E(h_1 \cdot m) \cdot h_2 \\ &= q^1 h_{(1,1)} \cdot m_{(0)} \cdot \beta S(q^2 h_{(1,2)} m_{(1)}) h_2 \\ &\overset{(3.2.21)}{=} h \cdot E(m). \end{aligned}$$

Also, by using the first relation in (6.4.2) we have

$$\begin{aligned} E^2(m) &= E(q^1 \cdot m_{(0)} \cdot \beta S(q^2 m_{(1)})) \\ &= \varepsilon(\beta S(q^2 m_{(1)}))E(q^1 \cdot m_{(0)}) = E(m), \end{aligned}$$

for all $m \in M$. Finally, for $m \in M$, we have

$$\begin{aligned} E(m_{(0)}) \cdot m_{(1)} &= X^1 \cdot m_{(0,0)} \cdot \beta S(X^2 m_{(0,1)}) \alpha X^3 m_{(1)} \\ &\overset{(6.1.2)}{=} m_{(0)} \cdot X^1 \beta S(m_{(1)_1} X^2) \alpha m_{(1)_2} X^3 \\ &\overset{(3.2.1)}{=} m \cdot X^1 \beta S(X^2) \alpha X^3 \overset{(3.2.2)}{=} m, \end{aligned}$$

and similarly

$$\begin{aligned} &E(E(m)_{(0)}) \otimes E(m)_{(1)} \\ &= E(q_1^1 \cdot m_{(0,0)} \cdot \beta_1 S(q^2 m_{(1)})_1) \otimes q_2^1 m_{(0,1)} \beta_2 S(q^2 m_{(1)})_2 \\ &\overset{(6.4.2)}{=} E(q_1^1 \cdot m_{(0,0)}) \otimes q_2^1 m_{(0,1)} \beta S(q^2 m_{(1)}) \\ &\overset{(6.1.2)}{=} E(q_1^1 x^1 \cdot m_{(0)} \cdot X^1) \otimes q_2^1 x^2 m_{(1)_1} X^2 \beta S(q^2 x^3 m_{(1)_2} X^3) \\ &\overset{(6.4.2),(3.2.1)}{=} E(q_1^1 x^1 \cdot m) \otimes q_2^1 x^2 \beta S(q^2 x^3) \\ &\overset{(3.2.19),(3.2.23)}{=} E(m) \otimes 1_H. \end{aligned}$$

The relation (6.4.5) follows immediately from (6.4.2). □

It is clear that $\neg$ defines a left $H$-module structure on $M^{\mathrm{co}(H)}$. Also, it follows that

$$\overline{E}(m) = E(p^1 \cdot m) \cdot p^2, \quad E(m) = X^1 \cdot \overline{E}(m) \cdot S(X^2) \alpha X^3, \ \forall\, m \in M, \tag{6.4.6}$$

where $\overline{E}$ is the projection onto $M^{\overline{\mathrm{co}(H)}}$ defined in (6.3.7). By using (6.4.2) we get that the maps

$$\overline{E} : M^{\mathrm{co}(H)} \to M^{\overline{\mathrm{co}(H)}} \text{ and } E : M^{\overline{\mathrm{co}(H)}} \to M^{\mathrm{co}(H)} \tag{6.4.7}$$

are inverse to each other. Note that in the case of a Hopf algebra, the maps $E$ and $\overline{E}$ are equal to the identity on $M^{\mathrm{co}(H)} = M^{\overline{\mathrm{co}(H)}}$. Coming back to the quasi-Hopf case we have that $\overline{E} : M^{\mathrm{co}(H)} \to M^{\overline{\mathrm{co}(H)}}$ is a left $H$-linear isomorphism since

$$\begin{aligned} \overline{E}(h \neg n) &= \overline{E}(E(h \cdot n)) \\ &\overset{(6.4.6),(6.4.2)}{=} E(p^1 h \cdot n) \cdot p^2 \\ &\overset{(3.2.21),(6.4.3)}{=} h_1 \cdot E(p^1 \cdot n) \cdot p^2 S(h_2) \\ &\overset{(6.4.6),(6.3.8)}{=} h \to \overline{E}(n), \end{aligned}$$

for all $h \in H$ and $n \in M^{\mathrm{co}(H)}$. We then have the following structure theorem for right quasi-Hopf $H$-bimodules.

**Theorem 6.20** *The linear map*

$$\nu_M : M^{\mathrm{co}(H)} \otimes H \to M, \quad \nu_M(n \otimes h) = n \cdot h, \ \forall\, n \in M^{\mathrm{co}(H)},\ h \in H, \tag{6.4.8}$$

*is an isomorphism of right quasi-Hopf $H$-bimodules. Here $M^{\mathrm{co}(H)} \otimes H$ is a right quasi-Hopf $H$-bimodule with structures*

$$a \cdot (n \otimes h) \cdot b = E(a_1 \cdot n) \otimes a_2 h b \quad \text{and} \quad \rho(n \otimes h) = E(x^1 \cdot n) \otimes x^2 h_1 \otimes x^3 h_2,$$

*for all $n\in N$, $a,h,b\in H$. The inverse of $\nu$ is given by*

$$\nu_M^{-1}(m)=E(m_{(0)})\otimes m_{(1)},\ \ \forall\, m\in M. \tag{6.4.9}$$

*Proof* We have seen that $\overline{M^{\mathrm{co}(H)}}\cong M^{\mathrm{co}(H)}$ are isomorphic as left $H$-modules, hence $\overline{M^{\mathrm{co}(H)}}\otimes H\cong M^{\mathrm{co}(H)}\otimes H$ as quasi-Hopf $H$-bimodules (in both cases, the structure is determined as in Example 6.6). From the structure theorem in Remark 6.17 it follows that $M\cong\overline{M^{\mathrm{co}(H)}}\otimes H$ as quasi-Hopf $H$-bimodules. Thus we find that $M^{\mathrm{co}(H)}\otimes H\cong M$ as quasi-Hopf $H$-bimodules, and it is straightforward to verify that the connecting isomorphism is

$$M^{\mathrm{co}(H)}\otimes H\ni m\otimes h\mapsto \overline{E}(m)\otimes h\mapsto q^1\cdot\overline{E}(m)\cdot S(q^2)h\in M,$$

which is exactly $\nu_M$ since

$$\begin{aligned} q^1\cdot\overline{E}(m)\cdot S(q^2)h &\overset{(6.4.6)}{=} q^1\cdot E(p^1\cdot m)\cdot p^2S(q^2)h\\ &\overset{(6.4.3)}{=} E(q_1^1p^1\cdot m)\cdot q_2^1p^2S(q^2)h\overset{(3.2.23)}{=}E(m)\cdot h=m\cdot h. \end{aligned}$$

This finishes our proof. □

We give more descriptions for the set of coinvariants of the second type.

**Proposition 6.21** *We also have*

$$\begin{aligned} M^{\mathrm{co}(H)} &= \{n\in M\mid E(n_{(0)})\otimes n_{(1)}=E(n)\otimes 1_H\} &(6.4.10)\\ &= \{n\in M\mid \rho(n)=E(x^1\cdot n)\cdot x^2\otimes x^3\}. &(6.4.11) \end{aligned}$$

*Proof* If $n\in M^{\mathrm{co}(H)}$ then $E(n)=n$, and therefore

$$E(n_{(0)})\otimes n_{(1)}=E(E(n)_{(0)})\otimes E(n)_{(1)}\overset{(6.4.4)}{=}E(n)\otimes 1_H.$$

Conversely, if $E(n_{(0)})\otimes n_{(1)}=E(n)\otimes 1_H$ then $n\overset{(6.4.4)}{=}E(n_{(0)})\cdot n_{(1)}=E(n)\cdot 1_H=E(n)$. Hence we have proved the equality in (6.4.10).

We now prove (6.4.11). If $n\in M$ such that $\rho(n)=E(x^1\cdot n)\cdot x^2\otimes x^3$ then

$$E(n)\overset{(6.4.2)}{=}E(E(x^1\cdot n)\cdot x^2)\cdot x^3=E(n_{(0)})\cdot n_{(1)}\overset{(6.4.4)}{=}n,$$

and so $n\in M^{\mathrm{co}(H)}$. Conversely, if $n\in M^{\mathrm{co}(H)}$ then

$$\begin{aligned} \rho(n) &= (\nu\otimes \mathrm{Id}_H)\rho_{M^{\mathrm{co}(H)}\otimes H}(\nu^{-1}(n))=(\nu\otimes\mathrm{Id}_H)\rho_{M^{\mathrm{co}(H)}\otimes H}(n\otimes 1_H)\\ &= E(x^1\cdot n)\cdot x^2\otimes x^3, \end{aligned}$$

finishing the proof. □

The equivalence described in Remark 6.17 is a monoidal equivalence if we specialize and reconsider it as in Theorem 6.20. So the equivalence of categories described in Remark 6.17, specialized for $\mathfrak{A}=H$, is a monoidal equivalence.

In what follows we denote by $E_M,\overline{E}_M:M\to M$ the projection associated to a right quasi-Hopf $H$-bimodule $M$ as in Definition 6.18 and (6.3.7), respectively.

**Proposition 6.22** *Let $H$ be a quasi-Hopf algebra with bijective antipode and $M,N \in {}_H\mathscr{M}_H^H$. Define the linear maps $M^{\mathrm{co}(H)} \otimes N^{\mathrm{co}(H)} \underset{j_{M,N}}{\overset{i_{M,N}}{\rightleftarrows}} M \otimes_H N$,*

$$i_{M,N}(m \otimes n) = E_M(X^1 \cdot m) \otimes_H E_N(X^2 \cdot n) \cdot X^3,$$
$$j_{M,N}(m \otimes_H n) = E_M(m_{(0)}) \otimes E_N(m_{(1)} \cdot n),$$

*for all $m \in M$ and $n \in N$. Then we have $j_{M,N} i_{M,N} = \mathrm{Id}_{M^{\mathrm{co}(H)} \otimes N^{\mathrm{co}(H)}}$ and $i_{M,N} j_{M,N} = E_{M \otimes_H N}$, where $M \otimes_H N$ is considered as object in ${}_H\mathscr{M}_H^H$ with the structure from Proposition 6.8.*

*Consequently, the image of $i_{M,N}$ is $(M \otimes_H N)^{\mathrm{co}(H)}$, and so $i_{M,N}$ induces a left $H$-module isomorphism between $M^{\mathrm{co}(H)} \otimes N^{\mathrm{co}(H)}$ and $(M \otimes_H N)^{\mathrm{co}(H)}$.*

*Proof* First of all, $j_{M,N}$ is well defined since

$$\begin{aligned} j_{M,N}(m \cdot h \otimes_H n) &= E_M(m_{(0)} \cdot h_1) \otimes_H E_N(m_{(1)} h_2 \cdot n) \\ &\overset{(6.4.2)}{=} E_M(m_{(0)}) \otimes E_N(m_{(1)} \cdot (h \cdot n)) = j_{M,N}(m \otimes_H h \cdot n), \end{aligned}$$

for all $m \in M$, $h \in H$ and $n \in N$. Furthermore, if $m \in M^{\mathrm{co}(H)}$ and $n \in N$ then

$$j_{M,N}(m \otimes_H n) = E_M(m_{(0)}) \otimes E_N(m_{(1)} \cdot n) \overset{(6.4.10)}{=} E_M(m) \otimes E_N(1_H \cdot n) = m \otimes E_N(n), \tag{6.4.12}$$

and this fact allows to compute, for all $m \in M^{\mathrm{co}(H)}$ and $n \in N^{\mathrm{co}(H)}$:

$$\begin{aligned} j_{M,N} i_{M,N}(m \otimes n) &= j_{M,N}(E_M(X^1 \cdot m) \otimes_H E_N(X^2 \cdot n) \cdot X^3) \\ &= E_M(X^1 \cdot m) \otimes E_N(E_N(X^2 \cdot n) \cdot X^3) \\ &\overset{(6.4.2)}{=} E_M(m) \otimes E_N^2(n) \overset{(6.4.4)}{=} E_M(m) \otimes E_N(n) = m \otimes n, \end{aligned}$$

as stated. To prove the second equality in the statement, for all $m \in M^{\mathrm{co}(H)}$ and $n \in N$ we compute:

$$\begin{aligned} E_{M \otimes_H N}(m \otimes_H n) &= q^1 \cdot m_{(0)} \otimes_H n_{(0)} \cdot \beta S(q^2 m_{(1)} n_{(1)}) \\ &\overset{(6.4.11)}{=} q^1 \cdot E_M(x^1 \cdot m) \cdot x^2 \otimes_H n_{(0)} \cdot \beta S(q^2 x^3 n_{(1)}) \\ &\overset{(6.4.3)}{=} E_M(q^1_1 x^1 \cdot m) \otimes_H q^1_2 x^2 \cdot n_{(0)} \cdot \beta S(q^2 x^3 n_{(1)}) \\ &\overset{(5.5.17)}{=} E_M(X^1 \cdot m) \otimes_H q^1 X^2_1 \cdot n_{(0)} \cdot \beta S(q^2 X^2_2 n_{(1)}) X^3 \\ &= E_M(X^1 \cdot m) \otimes_H E_N(X^2 \cdot n) \cdot X^3. \end{aligned}$$

So we have shown that

$$E_{M \otimes_H N}(m \otimes_H n) = E_M(X^1 \cdot m) \otimes_H E_N(X^2 \cdot n) \cdot X^3, \quad \forall\, m \in M^{\mathrm{co}(H)},\ n \in N. \tag{6.4.13}$$

Therefore, for arbitrary $m \in M$ and $n \in N$ we have

$$\begin{aligned} E_{M \otimes_H N}(m \otimes_H n) &\overset{(6.4.4)}{=} E_{M \otimes_H N}(E_M(m_{(0)}) \cdot m_{(1)} \otimes_H n) \\ &= E_{M \otimes_H N}(E_M(m_{(0)}) \otimes_H m_{(1)} \cdot n) \\ &= E_M(X^1 \cdot E_M(m_{(0)})) \otimes_H E_N(X^2 m_{(1)} \cdot n) \cdot X^3 \end{aligned}$$

$$\overset{(6.4.2)}{=} E_M(X^1\cdot E_M(m_{(0)}))\otimes_H E_N(X^2\cdot E_N(m_{(1)}\cdot n))\cdot X^3$$
$$= i_{M,N}j_{M,N}(m\otimes_H n),$$

as desired. Now, from $j_{M,N}i_{M,N}=\mathrm{Id}_{M^{\mathrm{co}(H)}\otimes N^{\mathrm{co}(H)}}$ we get that $j_{M,N}$ is surjective, and therefore that $(M\otimes_H N)^{\mathrm{co}(H)}=\mathrm{Im}(E_{M\otimes_H N})=\mathrm{Im}(i_{M,N}j_{M,N})=\mathrm{Im}(i_{M,N})$. Since $i_{M,N}$ is injective it follows that its corestriction to $(M\otimes_H N)^{\mathrm{co}(H)}$ is a bijection. We will denote it by $\phi_{2,M,N}: M^{\mathrm{co}(H)}\otimes N^{\mathrm{co}H}\to (M\otimes_H N)^{\mathrm{co}(H)}$.

If $\iota:(M\otimes_H N)^{\mathrm{co}(H)}\hookrightarrow M\otimes_H N$ is the inclusion map then $E_{M\otimes_H N}\iota=\iota$ and $i_{M,N}=\iota\phi_{2,M,N}$. Thus if $\phi_{2,M,N}^{-1}:=j_{M,N}\iota:(M\otimes_H N)^{\mathrm{co}(H)}\to M^{\mathrm{co}(H)}\otimes N^{\mathrm{co}(H)}$ then

$$\iota\phi_{2,M,N}\phi_{2,M,N}^{-1}=i_{M,N}j_{M,N}\iota=E_{M\otimes_H N}\iota=\iota,$$

and so $\phi_{2,M,N}\phi_{2,M,N}^{-1}=\mathrm{Id}_{(M\otimes_H N)^{\mathrm{co}(H)}}$. Similarly,

$$\phi_{2,M,N}^{-1}\phi_{2,M,N}=j_{M,N}\iota\phi_{2,M,N}=j_{M,N}i_{M,N}=\mathrm{Id}_{M^{\mathrm{co}(H)}\otimes N^{\mathrm{co}(H)}}.$$

We conclude that $\phi_{2,M,N}$ is bijective and $\phi_{2,M,N}^{-1}$ is its inverse. Finally, observe that

$$\begin{aligned} h\cdot i_{M,N}(m\otimes n) &= h\cdot E_M(X^1\cdot m)\otimes_H E_N(X^2\cdot n)\cdot X^3\\ &\overset{(6.4.3)}{=} E_M(h_1X^1\cdot m)\cdot h_2\otimes_H E_N(X^2\cdot n)\cdot X^3\\ &\overset{(6.4.3)}{=} E_M(h_1X^1\cdot m)\otimes_H E_N(h_{(2,1)}X^2\cdot n)\cdot h_{(2,2)}X^3\\ &\overset{(3.1.7),(6.4.2)}{=} E_M(X^1\cdot E_M(h_{(1,1)}\cdot m))\otimes_H E_N(X^2\cdot E_N(h_{(1,2)}\cdot n))\cdot X^3h_2\\ &= i_{M,N}(h_{(1,1)}\neg m\otimes h_{(1,2)}\neg n)\cdot h_2, \end{aligned} \tag{6.4.14}$$

for all $h\in H$, $m\in M^{\mathrm{co}(H)}$ and $n\in N^{\mathrm{co}(H)}$. Thus $\phi_{2,M,N}$ is left $H$-linear since

$$\begin{aligned} \phi_{2,M,N}(h\neg(m\otimes n)) &= i_{M,N}(h_1\neg m\otimes h_2\neg n)\\ &= i_{M,N}j_{M,N}i_{M,N}(h_1\neg m\otimes h_2\neg m)\\ &= E_{M\otimes_H N}i_{M,N}(h_1\neg m\otimes h_2\neg m)\\ &\overset{(6.4.2)}{=} E_{M\otimes_H N}(i_{M,N}(h_{(1,1)}\neg m\otimes h_{(1,2)}\neg n)\cdot h_2)\\ &\overset{(6.4.14)}{=} E_{M\otimes_H N}(h\cdot i_{M,N}(m\otimes n))\\ &= h\neg i_{M,N}(m\otimes n)\\ &= h\neg \phi_{2,M,N}(m\otimes n), \end{aligned}$$

for all $h\in H$, $m\in M^{\mathrm{co}(H)}$ and $n\in N^{\mathrm{co}(H)}$, as required. □

We now state and prove the main result of this section.

**Theorem 6.23** *If $H$ is a quasi-Hopf algebra with bijective antipode then the categories ${}_H\mathcal{M}_H^H$ and ${}_H\mathcal{M}$ are strong monoidally equivalent.*

*Proof* We show that the functor $G:{}_H\mathcal{M}_H^H\to{}_H\mathcal{M}$ defined by $G(M)=M^{\mathrm{co}(H)}$ and $G(f)=f\mid_{M^{\mathrm{co}(H)}}$ produces the desired monoidal equivalence. Towards this end, we first show that $G$ is a strong monoidal functor. If $\phi_{2,M,N}$ are as in Proposition 6.22, we prove that the first corresponding diagram for $\phi$ in Definition 1.22 is commutative. We compute, for all $M,N,P \in {}_H\mathscr{M}_H^H$:

$$\begin{aligned}
&\phi_{2,M\otimes_H N,P}\circ(\phi_{2,M,N}\otimes \mathrm{Id}_{P^{\mathrm{co}(H)}})((m\otimes n)\otimes p)\\
&= \phi_{2,M\otimes_H N,P}((E_M(X^1\cdot m)\otimes_H E_N(X^2\cdot n)\cdot X^3)\otimes p)\\
&= E_{M\otimes_H N}(Y^1\cdot E_M(X^1\cdot m)\otimes_H E_N(X^2\cdot n)\cdot X^3)\otimes_H E_P(Y^2\cdot p)\cdot Y^3\\
&\overset{(6.4.3)}{=} E_{M\otimes_H N}\left((E_M(Y_1^1X^1\cdot m)\otimes_H E_N(Y^1_{(2,1)}X^2\cdot n))\cdot Y^1_{(2,2)}X^3\right)\\
&\qquad \otimes_H E_P(Y^2\cdot p)\cdot Y^3\\
&\overset{(6.4.2)}{=} E_{M\otimes_H N}\left(E_M(Y_1^1\cdot m)\otimes_H E_N(Y_2^1\cdot n)\right)\otimes_H E_P(Y^2\cdot p)\cdot Y^3\\
&\overset{(6.4.13),(6.4.2)}{=} E_M(X^1Y_1^1\cdot m)\otimes_H E_N(X^2Y_2^1\cdot n)\cdot X^3\otimes_H E_P(Y^2\cdot p)\cdot Y^3\\
&\overset{(6.4.3)}{=} E_M(X^1Y_1^1\cdot m)\otimes_H E_N(X^2Y_2^1\cdot n)\otimes_H E_P(X_1^3Y^2\cdot p)\cdot X_2^3Y^3,
\end{aligned}$$

for all $m\in M^{\mathrm{co}(H)}$, $n\in N^{\mathrm{co}(H)}$ and $p\in P^{\mathrm{co}(H)}$, and on the other hand,

$$\begin{aligned}
&\phi_{2,M,N\otimes_H P}\circ(\mathrm{Id}_{M^{\mathrm{co}(H)}}\otimes\phi_{2,N,P})\circ a_{M^{\mathrm{co}(H)},N^{\mathrm{co}(H)},P^{\mathrm{co}(H)}}((m\otimes n)\otimes p)\\
&= \phi_{2,M,N\otimes_H P}\circ(\mathrm{Id}_{M^{\mathrm{co}(H)}}\otimes\phi_{2,N,P})(E_M(X^1\cdot m)\otimes(E_N(X^2\cdot n)\otimes E_P(X^3\cdot p)))\\
&= \varphi_{2,M,N\otimes_H P}\left(E_M(X^1\cdot m)\otimes[E_N(Y^1\cdot E_N(X^2\cdot n))\right.\\
&\qquad\qquad \left.\otimes_H E_P(Y^2\cdot E_P(X^3\cdot p))\cdot Y^3]\right)\\
&\overset{(6.4.2)}{=} E_M(Z^1X^1\cdot m)\otimes_H E_{N\otimes_H P}(Z^2\cdot E_N(Y^1X^2\cdot n)\otimes_H E_P(Y^2X^3\cdot p)\cdot Y^3)\cdot Z^3\\
&\overset{(6.4.3)}{=} E_M(Z^1X^1\cdot m)\\
&\qquad \otimes_H E_{N\otimes_H P}\left((E_N(Z_1^2Y^1X^2\cdot n)\otimes_H E_P(Z^2_{(2,1)}Y^2X^3\cdot p))\cdot Z^2_{(2,2)}Y^3\right)\cdot Z^3\\
&\overset{(6.4.2)}{=} E_M(Z^1X^1\cdot m)\otimes_H E_{N\otimes_H P}\left(E_N(Z_1^2X^2\cdot n)\otimes_H E_P(Z_2^2X^3\cdot p)\right)\cdot Z^3\\
&\overset{(6.4.13),(6.4.2)}{=} E_M(Z^1X^1\cdot m)\otimes_H E_N(Y^1Z_1^2X^2\cdot n)\otimes_H E_P(Y^2Z_2^2X^3\cdot p)\cdot Y^3Z^3\\
&\overset{(3.1.9)}{=} E_M(X^1Y_1^1\cdot m)\otimes_H E_N(X^2Y_2^1\cdot n)\otimes_H E_P(X_1^3Y^2\cdot p)\cdot X_2^3Y^3,
\end{aligned}$$

as claimed. Secondly, the unit object of ${}_H\mathscr{M}_H^H$ is $H$, with the structure given by its multiplication and comultiplication. We have $H^{\mathrm{co}(H)}=k1_H$ since

$$E_H(h)=q^1h_1\beta S(q^2h_2)=\varepsilon(h)X^1\beta S(X^2)\alpha X^3=\varepsilon(h)1_H,\ \ \forall\, h\in H.$$

It follows that $\phi_0 : k\to G(H)=H^{\mathrm{co}(H)}$, $\phi_0(\kappa)=\kappa 1_H$, for all $\kappa\in k$, is a left $H$-module isomorphism and closes commutatively the two square diagrams in (1.3.2).

So far we have shown that $(G,\phi_0,\phi_2)$ is a strong monoidal functor which at the same time defines an equivalence between the categories ${}_H\mathscr{M}_H^H$ and ${}_H\mathscr{M}$. By Proposition 1.31 it follows that ${}_H\mathscr{M}_H^H$ and ${}_H\mathscr{M}$ are strong monoidally equivalent. □

**Remark 6.24** The equivalence inverse of the functor $G$ from the proof of Theorem 6.23 is the functor $F=\bullet\otimes H : {}_H\mathscr{M}\to {}_H\mathscr{M}_H^H$, which coincides with the functor $\mathbb{F}$ defined in Remark 6.17 in the case when $\mathfrak{A}=H$. The functor $F$ is strong monoidal via the structure given by $\varphi_2$ determined, for all $M,N\in {}_H\mathscr{M}$, by the following composition of isomorphisms:

$$F(M)\otimes_H F(N) = (M\otimes H)\otimes_H (N\otimes H) \cong M\otimes (N\otimes H) \cong (M\otimes N)\otimes H = F(M\otimes N),$$

and $\varphi_0 = \mathrm{Id}_H : H \to F(k) = k\otimes H \cong H$.

Explicitly, we have that

$$\varphi_{2,M,N}((m\otimes h)\otimes_H (n\otimes h')) = (x^1\cdot m\otimes x^2 h_1\cdot n)\otimes x^3 h_2 h', \tag{6.4.15}$$

for all $m\in M$, $h, h'\in H$ and $n\in N$.

By working with alternative coinvariants instead of coinvariants of the second type we get a second strong monoidal equivalence between the categories ${}_H\mathcal{M}_H^H$ and ${}_H\mathcal{M}$ as follows.

**Corollary 6.25** *Let $H$ be a quasi-Hopf algebra with bijective antipode and consider ${}_H\mathcal{M} \underset{\mathbb{G}}{\overset{\mathbb{F}}{\rightleftarrows}} {}_H\mathcal{M}_H^H$ the pair of functors defined in Remark 6.17, specialized for $\mathfrak{A} = H$. Then $\mathbb{F}$, $\mathbb{G}$ are strong monoidal functors and they induce a strong monoidal equivalence between ${}_H\mathcal{M}_H^H$ and ${}_H\mathcal{M}$.*

*Proof* The functor $\mathbb{G}$ is naturally isomorphic to the functor $G$ defined in the proof of Theorem 6.23. The natural isomorphism between them is given by the natural transformation

$$\overline{E} = \left(\overline{E}_M : G(M) = M^{\mathrm{co}(H)} \to \mathbb{G}(M) = M^{\overline{\mathrm{co}(H)}}\right)_{M\in {}_H\mathcal{M}_H^H}.$$

Note that the inverse natural transformation of $\overline{E}$ is $E$, and that $F = \mathbb{F}$. Thus, by Theorem 6.23 it follows that $\mathbb{F}$, $\mathbb{G}$ are strong monoidal functors, and that they provide a strong monoidal equivalence of categories.

We end by pointing out that the strong monoidal structure of $\mathbb{G}$ is given by $\overline{\phi}_{2,M,N}$ : $M^{\overline{\mathrm{co}(H)}}\otimes N^{\overline{\mathrm{co}(H)}} \to (M\otimes_H N)^{\overline{\mathrm{co}(H)}}$ determined by the composition

$$M^{\overline{\mathrm{co}(H)}}\otimes N^{\overline{\mathrm{co}(H)}} \overset{E_M\otimes E_N}{\longrightarrow} M^{\mathrm{co}(H)}\otimes N^{\mathrm{co}(H)} \overset{\phi_{2,M,N}}{\longrightarrow} (M\otimes_H N)^{\mathrm{co}(H)} \overset{\overline{E}_{M\otimes_H N}}{\longrightarrow} (M\otimes_H N)^{\overline{\mathrm{co}(H)}},$$

and $\overline{\phi}_0 : k\to \mathbb{G}(H) = k\beta$ defined by $\overline{\phi}_0(\kappa) = \kappa\beta$, for all $\kappa\in k$. Explicitly, we have

$$\overline{\phi}_{2,M,N}(m\otimes n) = q^1 x_1^1\cdot m\cdot S(q^2 x_2^1)x^2\otimes_H n\cdot S(x^3), \tag{6.4.16}$$

for all $M, N\in {}_H\mathcal{M}_H^H$ and $m\in M^{\overline{\mathrm{co}(H)}}$, $n\in N^{\overline{\mathrm{co}(H)}}$. Indeed, by taking into account the above definitions and structures we have that

$$\begin{aligned}
&\overline{\phi}_{2,M,N}(m\otimes n)\\
&\quad = \overline{E}_{M\otimes_H N}\phi_{2,M,N}(E_M(m)\otimes E_N(n))\\
&\quad = \overline{E}_{M\otimes_H N}(E_M(X^1\cdot E_M(m))\otimes_H E_N(X^2\cdot E_N(n))\cdot X^3\\
&\quad \overset{(6.4.2)}{=} \overline{E}_{M\otimes_H N}(E_M(X^1\cdot m)\otimes_H E_N(X^2\cdot n)\cdot X^3\\
&\quad = E_M(X^1\cdot m)_{(0)}\otimes_H E_N(X^2\cdot n)_{(0)}\cdot X_1^3\beta S(E_M(X^1\cdot m)_{(1)}E_N(X^2\cdot n)_{(1)}X_2^3)\\
&\quad \overset{(3.2.1)}{=} E_M(m)_{(0)}\otimes_H \overline{E}E(n)\cdot S(E_M(m)_{(1)})
\end{aligned}$$

$$\overset{(6.4.11)}{=} E_M(x^1 \cdot E_M(m)) \cdot x^2 \otimes_H n \cdot S(x^3)$$
$$\overset{(6.4.2)}{=} E_M(x^1 \cdot m) \cdot x^2 \otimes_H n \cdot S(x^3)$$
$$= q^1 x^1_1 \cdot m_{(0)} \cdot \beta S(q^2 x^1_2 m_{(1)}) x^2 \otimes_H n \cdot S(x^3)$$
$$= q^1 x^1_1 \cdot \overline{E}_M(m) \cdot S(q^2 x^1_2) x^2 \otimes_H n \cdot S(x^3)$$
$$= q^1 x^1_1 \cdot m \cdot S(q^2 x^1_2) x^2 \otimes_H n \cdot S(x^3),$$

for all $m \in \overline{M^{\mathrm{co}(H)}}$ and $n \in \overline{N^{\mathrm{co}(H)}}$, as stated. □

## 6.5 A Structure Theorem for Comodule Algebras

We shall see that the structure theorem for quasi-Hopf bimodules provides a structure theorem for algebras within categories of quasi-Hopf bimodules.

We begin with a lemma of independent interest.

**Lemma 6.26** *Let $H$ be a quasi-bialgebra and ${}_H\mathscr{M}_H^H$ the category of right quasi-Hopf bimodules over $H$ equipped with the monoidal structure presented in Proposition 6.8. Then giving an algebra $\mathfrak{A}$ in ${}_H\mathscr{M}_H^H$ is equivalent to giving a triple $(\mathfrak{A}, \rho, i)$ consisting of a $k$-algebra $\mathfrak{A}$, a $k$-linear map $\rho : \mathfrak{A} \to \mathfrak{A} \otimes H$ and a $k$-algebra morphism $i : H \to \mathfrak{A}$ such that $(\mathfrak{A}, \rho, \Phi_\rho := i(X^1) \otimes X^2 \otimes X^3)$ is a right $H$-comodule algebra and $i$ is a right $H$-comodule morphism, that is, in addition*

$$\rho(i(h)) = i(h_1) \otimes h_2, \quad \forall\, h \in H.$$

*Proof* Assume that $(\mathfrak{A}, \underline{m} : \mathfrak{A} \otimes_H \mathfrak{A} \to \mathfrak{A}, i : H \to \mathfrak{A})$ is an algebra in ${}_H\mathscr{M}_H^H$. Since the forgetful functor from $({}_H\mathscr{M}_H^H, \otimes_H, H)$ to $({}_H\mathscr{M}_H, \otimes_H, H)$ is strong monoidal we get that $(\mathfrak{A}, \underline{m}, i)$ is an algebra in $({}_H\mathscr{M}_H, \otimes_H, H)$, too (see Proposition 2.3). Otherwise stated, $(\mathfrak{A}, \underline{m}, i)$ is an $H$-ring. Thus, by Example 2.2(5) we obtain that $\mathfrak{A}$ is a $k$-algebra with multiplication $m = q^H_{\mathfrak{A},\mathfrak{A}}\underline{m}$ and unit $1_{\mathfrak{A}} = i(1_H)$. Furthermore, the input $H$-bimodule structure of $\mathfrak{A}$ is completely determined by $h \cdot \mathfrak{a} \cdot h' = i(h)\mathfrak{a} i(h')$, for all $h, h' \in H$ and $\mathfrak{a} \in \mathfrak{A}$, and $i : H \to \mathfrak{A}$ becomes a $k$-algebra morphism.

Now, since $\mathfrak{A}$ is an object in ${}_H\mathscr{M}_H^H$ we have a $k$-linear map $\rho : \mathfrak{A} \ni \mathfrak{a} \mapsto \mathfrak{a}_{\langle 0\rangle} \otimes \mathfrak{a}_{\langle 1\rangle} \in \mathfrak{A} \otimes H$ such that $\varepsilon(\mathfrak{a}_{\langle 1\rangle})\mathfrak{a}_{\langle 0\rangle} = \mathfrak{a}$ and

$$i(X^1)\mathfrak{a}_{\langle 0,0\rangle} \otimes X^2 \mathfrak{a}_{\langle 0,1\rangle} \otimes X^3 \mathfrak{a}_{\langle 1\rangle} = \mathfrak{a}_{\langle 0\rangle} i(X^1) \otimes \mathfrak{a}_{\langle 1\rangle_1} X^2 \otimes \mathfrak{a}_{\langle 1\rangle_2} X^3, \quad \forall\, \mathfrak{a} \in \mathfrak{A},$$

that is, (4.3.3) and (4.3.1) hold. Furthermore, $\rho$ is an $H$-bimodule morphism, and so

$$\rho(i(h)\mathfrak{a} i(h')) = i(h_1)\mathfrak{a}_{\langle 0\rangle} i(h'_1) \otimes h_2 \mathfrak{a}_{\langle 1\rangle} h'_2, \quad \forall\, \mathfrak{a} \in \mathfrak{A} \text{ and } h,\ h' \in H.$$

Clearly, this implies $\rho(i(h)) = i(h_1) \otimes h_2$, for all $h \in H$. The latter equality allows us to show that $\Phi_\rho := i(X^1) \otimes X^2 \otimes X^3$ satisfies (4.3.2). (4.3.4) is automatic.

It remains to prove that $\rho$ is a $k$-algebra morphism. This follows easily from the fact that $\underline{m}$ and $i$ are right $H$-colinear morphisms; we leave the verification of this detail to the reader. So we have shown that $(\mathfrak{A}, \rho, \Phi_\rho)$ is a right $H$-comodule algebra and $i : H \to \mathfrak{A}$ is a right $H$-comodule algebra morphism.

For the converse, assume that we have a datum $(\mathfrak{A},\rho,i)$ as in the statement. First, $\mathfrak{A}$ becomes an $H$-bimodule via $i$, that is, $h\cdot\mathfrak{a}\cdot h' = i(h)\mathfrak{a}i(h')$, for all $h,h'\in H$ and $\mathfrak{a}\in\mathfrak{A}$. Together with $\rho$ this makes $\mathfrak{A}$ an object in ${}_H\mathcal{M}_H^H$.

Indeed, we now prove that $\rho:\mathfrak{A}\to\mathfrak{A}\otimes H$ is an $H$-bimodule map. We compute:

$$\begin{aligned}
\rho(h\cdot\mathfrak{a}\cdot h') &= \rho(i(h)\mathfrak{a}i(h'))\\
&= \rho(i(h))\rho(\mathfrak{a})\rho(i(h'))\\
&= (i(h_1)\otimes h_2)(\mathfrak{a}_{\langle 0\rangle}\otimes\mathfrak{a}_{\langle 1\rangle})(i(h'_1)\otimes h'_2)\\
&= i(h_1)\mathfrak{a}_{\langle 0\rangle}i(h'_1)\otimes h_2\mathfrak{a}_{\langle 1\rangle}h'_2\\
&= h_1\cdot\mathfrak{a}_{\langle 0\rangle}\cdot h'_1\otimes h_2\mathfrak{a}_{\langle 1\rangle}h'_2\\
&= h\cdot\rho(\mathfrak{a})\cdot h',
\end{aligned}$$

as required. Obviously we have $(\mathrm{Id}_{\mathfrak{A}}\otimes\varepsilon)\circ\rho = \mathrm{Id}_{\mathfrak{A}}$. Finally, it is easy to see that

$$\Phi\cdot(\rho\otimes\mathrm{Id}_{\mathfrak{A}})(\rho(\mathfrak{a})) = (\mathrm{Id}_{\mathfrak{A}}\otimes\Delta)(\rho(\mathfrak{a}))\cdot\Phi,$$

because this is the condition $\Phi_\rho(\rho\otimes\mathrm{Id}_{\mathfrak{A}})(\rho(\mathfrak{a})) = (\mathrm{Id}_{\mathfrak{A}}\otimes\Delta)(\rho(\mathfrak{a}))\Phi_\rho$ from the definition of a right $H$-comodule algebra, due to the fact that $\Phi_\rho = i(X^1)\otimes X^2\otimes X^3$. Hence with the above structure $\mathfrak{A}$ is indeed a right quasi-Hopf $H$-bimodule.

Since $\mathfrak{A}$ is an associative unital $k$-algebra and $i:H\to\mathfrak{A}$ is a $k$-algebra morphism we get from Example 2.2(5) that $(\mathfrak{A},\underline{m},i)$ with $\underline{m}:\mathfrak{A}\otimes_H\mathfrak{A}\to\mathfrak{A}$, $\underline{m}(\mathfrak{a}\otimes_H\mathfrak{a}') = \mathfrak{a}\mathfrak{a}'$, for all $\mathfrak{a},\mathfrak{a}'\in\mathfrak{A}$, is an algebra in $({}_H\mathcal{M}_H,\otimes_H,H)$. A simple inspection shows that $(\mathfrak{A},\underline{m},i)$ is, moreover, an algebra in ${}_H\mathcal{M}_H^H$, where ${}_H\mathcal{M}_H^H$ has the monoidal structure from Proposition 6.8. □

**Corollary 6.27** *Let $H$ be a quasi-bialgebra and $A$ a left $H$-module algebra. Then $A\#H$, the smash product between $A$ and $H$, is an algebra in ${}_H\mathcal{M}_H^H$.*

*Proof* By Proposition 5.1, $A\#H$ is a $k$-algebra and $j:H\ni h\mapsto 1_A\#h\in A\#H$ is a $k$-algebra map. Furthermore, with the structure as in Proposition 5.9, $A\#H$ becomes a right $H$-comodule algebra; note that its reassociator is just $j(X^1)\otimes X^2\otimes X^3$. Since

$$\rho(j(h)) = \rho(1_A\#h) = x^1\cdot 1_A\#x^2h_1\otimes x^3h_2 = 1_A\#h_1\otimes h_2 = j(h_1)\otimes h_2,$$

for all $h\in H$, by Lemma 6.26 we conclude that $A\#H$ is an algebra in ${}_H\mathcal{M}_H^H$. □

We next show that, in the case when $H$ is a quasi-Hopf algebra, any algebra $\mathfrak{A}$ in ${}_H\mathcal{M}_H^H$ is of the form presented in Corollary 6.27, for a certain algebra in ${}_H\mathcal{M}$.

Unless otherwise specified, from now on $H$ is a quasi-Hopf algebra and $\mathfrak{A}$ is an algebra within ${}_H\mathcal{M}_H^H$ with structure $(\rho,i)$ as in Lemma 6.26. For the $H$-comodule algebra $(\mathfrak{A},\rho,\Phi_\rho)$ we keep the same type of notation as in Section 4.3.

**Theorem 6.28** *Let $H$ be a quasi-Hopf algebra with bijective antipode and $\mathfrak{A}$ an algebra in ${}_H\mathcal{M}_H^H$. Then there exists a left $H$-module algebra $A$ such that $\mathfrak{A}\cong A\#H$, as algebras in ${}_H\mathcal{M}_H^H$.*

*Proof* We know from Corollary 6.25 that we have a strong monoidal functor $\mathbb{G}: {}_H\mathscr{M}_H^H \to {}_H\mathscr{M}$. So to the algebra $\mathfrak{A}$ in ${}_H\mathscr{M}_H^H$ there corresponds an algebra $\mathbb{G}(\mathfrak{A}) = \overline{\mathfrak{A}^{\mathrm{co}(H)}}$ in ${}_H\mathscr{M}$, which will be denoted by $A$ in what follows.

By the definition of $\mathbb{G}$ in Remark 6.17 we get that $A$ is a left $H$-module via the action given by

$$h \rightarrow a = h_1 \cdot a \cdot S(h_2) = i(h_1)ai(S(h_2)) := h \triangleright_i a,$$

for all $h \in H$ and $a \in A \subseteq \mathfrak{A}$. Keeping in mind the strong monoidal structure of $\mathbb{G}$ obtained in (6.4.16) and the proof of Proposition 2.3, we deduce that the multiplication of $A$ in ${}_H\mathscr{M}$ is

$$\begin{aligned} a * a' &= \mathbb{G}(\underline{m})\overline{\phi}_{2,\mathfrak{A},\mathfrak{A}}(a \otimes a') \\ &= (q^1x_1^1 \cdot a \cdot S(q^2x_2^1)x^2)(a' \cdot S(x^3)) \\ &= i(q^1x_1^1)ai(S(q^2x_2^1)x^2)a'i(S(x^3)), \end{aligned}$$

for all $a, a' \in A$, while its unit is given by $\mathbb{G}(i)\overline{\phi}_0(1_k) = \mathbb{G}(i)(\beta) = i(\beta)$. But, using the definitions of $q_R$, $q_L$, (3.1.9) and (3.2.1) we deduce easily that

$$X^1 \otimes S(X^2)\tilde{q}^1X_1^3 \otimes \tilde{q}^2X_2^3 = q^1x_1^1 \otimes S(q^2x_2^1)x^2 \otimes x^3, \tag{6.5.1}$$

and so we get that, for all $a, a' \in H$,

$$a * a' = i(X^1)ai(S(X^2)\tilde{q}^1X_1^3)a'i(S(\tilde{q}^2X_2^3) := a \circ a'. \tag{6.5.2}$$

The notation $\triangleright_i$ and $\circ$ is imposed by analogy with the structure in Proposition 4.3. Hence, summing up, by (5.1.1) the multiplication in $A\#H$ is given by

$$(a\#h)(a'\#h') = i(X^1x_1^1)bi(S(y^1X^2x_2^1)\alpha y^2X_1^3x_1^2h_{(1,1)})b'i(S(y^3X_2^3x_2^2h_{(1,2)}))\#x^3h_2h', \tag{6.5.3}$$

for all $a, a' \in A$ and $h, h' \in H$.

On the other hand, by the structure theorem in Remark 6.17 we get that $\chi: A \otimes H \to \mathfrak{A}$ given by $\chi(a \otimes h) = q^1 \cdot a \cdot S(q^2)h = i(q^1)ai(S(q^2)h)$, for all $a \in A$ and $h \in H$, is an isomorphism in ${}_H\mathscr{M}_H^H$ with inverse $\chi^{-1}: \mathfrak{A} \to A \otimes H$ defined by

$$\begin{aligned} \chi^{-1}(\mathfrak{a}) &= \overline{E}(\mathfrak{a}_{\langle 0\rangle}) \otimes \mathfrak{a}_{\langle 1\rangle} \\ &= \mathfrak{a}_{\langle 0,0\rangle} \cdot \beta S(\mathfrak{a}_{\langle 0,1\rangle}) \otimes \mathfrak{a}_{\langle 1\rangle} \\ &= \mathfrak{a}_{\langle 0,0\rangle} i(\beta S(\mathfrak{a}_{\langle 0,1\rangle})) \otimes \mathfrak{a}_{\langle 1\rangle}, \end{aligned}$$

for all $\mathfrak{a} \in \mathfrak{A}$. So to end the proof it suffices to show that $\chi$ is an algebra morphism in ${}_H\mathscr{M}_H^H$, if it is considered as a morphism between $A\#H$ and $\mathfrak{A}$. To this end, we compute:

$$\begin{aligned} &\chi((a\#h)(a'\#h')) \\ &\quad = i(Z^1X^1x_1^1)ai(S(y^1X^2x_2^1)\alpha y^2X_1^3x_1^2h_{(1,1)})a' \\ &\qquad\; i(S(Z^2y^3X_2^3x_2^2h_{(1,2)})\alpha Z^3x^3h_2h') \\ &\overset{(3.1.9),(3.1.7)}{\underset{(3.2.1)}{=}} i(X^1)ai(S(y^1z^1X^2)\alpha y^2z_1^2X_{(1,1)}^3h_{(1,1)})a' \end{aligned}$$

$$
\begin{aligned}
& i(S(y^3z_2^2X^3_{(1,2)}h_{(1,2)})\alpha z^3X_2^3h_2h') \\
\overset{(3.1.9),(3.2.1)}{=} & i(X^1)ai(S(X^2)\alpha Y^1(X^3h)_{(1,1)})a'i(S(Y^2(X^3h)_{(1,2)})\alpha Y^3(X^3h)_2h') \\
\overset{(3.1.7),(3.2.1)}{=} & i(X^1)ai(S(X^2)\alpha X^3h)i(Y^1)a'i(S(Y^2)\alpha Y^3h') \\
= \quad & \chi(a\#h)\chi(a'\#h'),
\end{aligned}
$$

for all $a,a' \in A$ and $h,h' \in H$. It follows that $\chi(i(\beta)\#1_H) = 1_{\mathfrak{A}}$, hence $\chi$ is an algebra morphism, as stated. Observe also that

$$
\begin{aligned}
\chi(j(X^1))\otimes X^2\otimes X^3 &= \chi(1_A\#X^1)\otimes X^2\otimes X^3 \\
&= i(q^1)i(\beta)i(S(q^2)X^1)\otimes X^2\otimes X^3 \\
&\overset{(3.2.2)}{=} i(X^1)\otimes X^2\otimes X^3,
\end{aligned}
$$

and this completes our proof. □

**Remarks 6.29** (1) In the hypothesis of Theorem 6.28, since $i : H \to \mathfrak{A}$ is an algebra morphism, we have an algebra structure on $\mathfrak{A}$ within ${}_H\mathscr{M}$; see Proposition 4.3. It is easy to see that the algebra $A$ in ${}_H\mathscr{M}$ constructed in the proof of the cited theorem is actually a subalgebra of $\mathfrak{A}^i$ in ${}_H\mathscr{M}$.

(2) Let $H$ be a quasi-Hopf algebra with bijective antipode, $A$ a left $H$-module algebra and $\mathfrak{A} := A\#H$; then, by Corollary 6.27, we have that $\mathfrak{A}$ is an algebra in ${}_H\mathscr{M}_H^H$. In addition,

$$
\mathfrak{A}^{\overline{\mathrm{co}(H)}} = \mathbb{GF}(A) = \{p^1\cdot a\otimes p^2 \mid a\in A\} \cong A,
$$

an isomorphism of algebras in ${}_H\mathscr{M}$. Hence, the structure theorem allows us to recover the structure of $A$ from the one of $A\#H$; the details are left to the reader.

## 6.6 Coalgebras in ${}_H\mathscr{M}_H^H$

We now move to the coalgebra case. As we pointed out several times, the quasi-Hopf algebra notion is not selfdual, thus the results of this section cannot be viewed as the formal dual of those proved for quasi-Hopf comodule algebras. But we should stress the fact that in both situations the key role is played by the monoidal equivalence between ${}_H\mathscr{M}_H^H$ and ${}_H\mathscr{M}$.

Recall from Example 2.14(5) that if $A$ is a $k$-algebra then a coalgebra in ${}_A\mathscr{M}_A$ is called an $A$-coring. As we shall see next, any left module coalgebra $C$ over a quasi-bialgebra $H$ defines an $H$-coring structure on the $k$-vector space $C\otimes H$.

**Proposition 6.30** *Let $H$ be a quasi-bialgebra and $C$ a left $H$-module coalgebra. Then $C\otimes H$ is an $H$-coring, with structure given by*

$$
h\cdot(c\otimes h')\cdot h'' = h_1\cdot c\otimes h_2h'h'', \tag{6.6.1}
$$

$$
\underline{\Delta}(c\otimes h) = (X^1\cdot c_{\underline{1}}\otimes 1_H)\otimes_H(X^2\cdot c_{\underline{2}}\otimes X^3h) \quad \textit{and} \quad \underline{\varepsilon}(c\otimes h) = \varepsilon_C(c)h, \tag{6.6.2}
$$

*for all $c \in C$ and $h, h', h'' \in H$, where $\Delta_C(c) := c_{\underline{1}} \otimes c_{\underline{2}}$, for all $c \in C$, is the comultiplication of the coalgebra $C$ in ${}_H\mathcal{M}$, and $\cdot$ is the left action of $H$ on $C$.*

*Proof* Clearly $C \otimes H$ is an $H$-bimodule with structure as in (6.6.1). Secondly, $\underline{\Delta}$ is $H$-bilinear since

$$
\begin{aligned}
\underline{\Delta}(h_1 \cdot c \otimes h_2 h' h'') &= (X^1 h_{(1,1)} \cdot c_{\underline{1}} \otimes 1_H) \otimes_H (X^2 h_{(1,2)} \cdot c_{\underline{2}} \otimes X^3 h_2 h' h'') \\
&\overset{(3.1.7)}{=} (h_1 X^1 \cdot c_{\underline{1}} \otimes 1_H) \otimes_H (h_{(2,1)} X^2 \cdot c_{\underline{2}} \otimes h_{(2,2)} X^3 h' h'') \\
&= (h_1 X^1 \cdot c_{\underline{1}} \otimes 1_H) \otimes_H h_2 \cdot (X^2 \cdot c_{\underline{2}} \otimes X^3 h' h'') \\
&= (h_1 X^1 \cdot c_{\underline{1}} \otimes h_2) \otimes_H (X^2 \cdot c_{\underline{2}} \otimes X^3 h' h'') \\
&= h \cdot \underline{\Delta}(c \otimes h') \cdot h'',
\end{aligned}
$$

for all $c \in C$ and $h, h', h'' \in H$. It can be easily checked that $\underline{\varepsilon}$ is $H$-bilinear, too, and that $\underline{\varepsilon}$ is a counit for $\underline{\Delta}$. Also, the fact that $\underline{\Delta}$ is coassociative follows from the coassociativity of $\Delta_C$ in ${}_H\mathcal{M}$ and the 3-cocycle property of the reassociator $\Phi$ of $H$. We leave the verification of these details to the reader. □

We call the structure in Proposition 6.30 an $H$-coring structure defined by a left $H$-module coalgebra.

The next result describes the structure of a coalgebra within ${}_H\mathcal{M}_H^H$, in the case when $H$ is a quasi-Hopf algebra with bijective antipode.

**Theorem 6.31** *Let $H$ be a quasi-Hopf algebra with bijective antipode. Then there exists a one-to-one correspondence between*

*(i) coalgebra structures in ${}_H\mathcal{M}_H^H$;*

*(ii) coalgebra structures in ${}_H\mathcal{M}$;*

*(iii) $H$-coring structures defined by left $H$-module coalgebras.*

*Proof* The one-to-one correspondence between (i) and (ii) is established by the monoidal category equivalence between ${}_H\mathcal{M}_H^H$ and ${}_H\mathcal{M}$.

Up to an isomorphism, any coalgebra $\mathbf{C}$ in ${}_H\mathcal{M}_H^H$ is of the form $C \otimes H$ for a suitable coalgebra $C$ in ${}_H\mathcal{M}$. Once more, note that $C \otimes H$ is an object in ${}_H\mathcal{M}_H^H$ via the structure determined by

$$
\begin{aligned}
h \cdot (c \otimes h') \cdot h'' &= h_1 \cdot c \otimes h_2 h' h'', \\
\rho_{C \otimes H}(c \otimes h) &= (x^1 \cdot c \otimes x^2 h_1) \otimes x^3 h_2,
\end{aligned}
$$

for all $c \in H$ and $h, h', h'' \in H$. Furthermore, by the strong monoidal structure of the functor $F$ defined in Remark 6.24, we deduce that $C \otimes H$ is a coalgebra in ${}_H\mathcal{M}_H^H$ with comultiplication and counit given by

$$
\underline{\Delta} : C \otimes H = F(C) \overset{F(\Delta_C)}{\longrightarrow} F(C \otimes C) \overset{\varphi_{2,C,C}^{-1}}{\longrightarrow} F(C) \otimes_H F(C) = (C \otimes H) \otimes_H (C \otimes H)
$$

$$
\text{and } \underline{\varepsilon} : C \otimes H = F(C) \overset{F(\varepsilon_C)}{\longrightarrow} F(k) \overset{\varphi_0^{-1}}{\longrightarrow} H,
$$

where, as before, $\Delta_C$ and $\varepsilon_C$ are the comultiplication and the counit of the coalgebra $C$ in ${}_H\mathcal{M}$. According to Theorem 6.23, we find that $\underline{\Delta}$ and $\underline{\varepsilon}$ are given by (6.6.2).

On the other hand, since the forgetful functor from ${}_H\mathcal{M}_H^H$ to ${}_H\mathcal{M}_H$ is strong monoidal, it follows that a coalgebra in ${}_H\mathcal{M}_H^H$ is nothing but an $H$-coring $(\mathbf{C}, \underline{\Delta}_{\mathbf{C}}, \underline{\varepsilon}_{\mathbf{C}})$ for which the comultiplication $\underline{\Delta}_{\mathbf{C}} : \mathbf{C} \to \mathbf{C} \otimes_H \mathbf{C}$ and the counit $\underline{\varepsilon}_{\mathbf{C}}$ are right $H$-colinear maps. Since $\mathbf{C} \equiv C \otimes H$ in ${}_H\mathcal{M}_H^H$, with $C = \overline{\mathbf{C}^{\mathrm{co}(H)}}$, a left $H$-module coalgebra, from the arguments presented above we conclude that $\underline{\Delta}_{\mathbf{C}}$ and $\underline{\varepsilon}_{\mathbf{C}}$ are as in (6.6.2), and therefore right $H$-colinear maps. Thus the one-to-one correspondence between (i) and (iii) is established, too. □

## 6.7 Notes

Quasi-Hopf $H$-bimodules were introduced by Hausser and Nill in [109]. Afterwards, this concept was generalized in [63]. The structure theorems for generalized quasi-Hopf bimodules were proved in [63] as well, inspired at that time by the alternative structure theorem given in [47]. But the first structure theorem for quasi-Hopf bimodules is due to Hausser and Nill [109], as well as the monoidal equivalence between the category of quasi-Hopf $H$-bimodules and the category of $H$-modules. The duality theory for quasi-Hopf bimodules is based on the results published in [49], which gave a categorical flavour to some results obtained previously in [109]. Schauenburg noticed in [200] that $H^*$ being a quasi-Hopf $H$-bimodule is a consequence of a categorical monoidal result proved by Pareigis in [182]. The content of Section 6.5 is taken from [44, 180], while that of Section 6.6 is taken from [44].

# 7
# Finite-Dimensional Quasi-Hopf Algebras

The main goal of this chapter is to show that for a finite-dimensional quasi-Hopf algebra $H$ the space of integrals in $H$, and the space of cointegrals on $H$, has dimension 1. We characterize semisimple and symmetric quasi-Hopf algebras with the help of integrals, and prove a formula for the fourth power of the antipode in terms of the modular elements by using the machinery provided by Frobenius algebras. The chapter ends with a freeness theorem stating that any finite-dimenisonal quasi-Hopf algebra is free over any quasi-Hopf subalgebra.

## 7.1 Frobenius Algebras

Throughout this section $A$ is a finite-dimensional algebra with unit $1_A$ over a field $k$. By $A^*$ we denote the $k$-linear dual of $A$, that is $A^* = \mathrm{Hom}_k(A,k)$, the set of $k$-linear maps from $A$ to $k$. It can be easily checked that $A^*$ is an $A$-bimodule via the $A$-actions defined by

$$\langle a \rightharpoonup a^*, a' \rangle = \langle a^*, a'a \rangle \quad \text{and} \quad \langle a^* \leftharpoonup a, a' \rangle = \langle a^*, aa' \rangle,$$

for all $a^* \in A^*$ and $a, a' \in A$. Also, the multiplication of $A$ induces on $A$ an $A$-bimodule structure as well.

**Proposition 7.1** *For a finite-dimensional k-algebra A the following assertions are equivalent:*

*(i) There exists a pair $(\phi, e)$ consisting of a k-linear map $\phi : A \to k$ and an element $e = e^1 \otimes e^2 \in A \otimes A$ (formal notation, summation implicitly understood) such that*

$$ae^1 \otimes e^2 = e^1 \otimes e^2 a\,, \quad \forall\, a \in A, \text{and } \phi(e^1)e^2 = \phi(e^2)e^1 = 1_A;$$

*(ii) A is isomorphic to $A^*$ as a right A-module;*

*(iii) A is isomorphic to $A^*$ as a left A-module;*

*(iv) A has a k-coalgebra structure $(\Delta, \varepsilon)$ such that $\Delta : A \to A \otimes A$ is an A-bimodule morphism, where $A \otimes A$ is considered as an A-bimodule via the multiplication of A;*

*(v) there exists a bilinear map $B_r : A \times A \to k$ which is right non-degenerate and associative, that is, $B_r(x, a) = 0$, for all $a \in A$, implies $x = 0$, and $B_r(ab, c) = B_r(a, bc)$, for all $a, b, c \in A$;*

*(vi) there exists a bilinear map $B_l : A \times A \to k$ which is left non-degenerate (i.e. $B_l(a,x) = 0$ for all $a \in A$ implies $x = 0$) and associative;*

*(vii) there exists a hyperplane (linear subspace of codimension 1) in A that does not contain either left or right non-zero ideals.*

*Proof* We sketch the proof, leaving the verification of the details to the reader.

$(i) \Rightarrow (ii)$. It can be easily seen that $f : A \to A^*$ given by $f(a) = \phi \leftharpoonup a$, for all $a \in A$, is an isomorphism of right $A$-modules. Its inverse is $f^{-1} : A^* \to A$ defined by $f^{-1}(a^*) = a^*(e^1)e^2$, for all $a^* \in A^*$.

$(ii) \Rightarrow (i)$. Let $\{a_i, a^i\}_i$ be dual bases in $A$ and $A^*$. If $f : A \to A^*$ is a right $A$-linear isomorphism then the pair $(f(1_A), \sum_i a_i \otimes f^{-1}(a^i))$ obeys the required conditions.

$(ii) \Rightarrow (iii)$. If $f : A \to A^*$ is an isomorphism of right $A$-modules then

$$g : A \overset{\theta_A}{\to} A^{**} \overset{f^*}{\to} A^*$$

is an isomorphism of left $A$-modules, where $\theta_A$ is the canonical isomorphism (see also Corollary 1.76) and $f^*$ is the transpose of $f$. The implication $(iii) \Rightarrow (ii)$ can be proved in a similar manner.

$(ii) \Rightarrow (iv)$. Let $f : A \to A^*$ be a right $A$-linear isomorphism. As, for instance, the left dual functor $(-)^*$ is monoidal, it follows that $A^*$ admits a coalgebra structure within the reverse monoidal category associated to ${}_k\mathcal{M}$, and therefore in ${}_k\mathcal{M}$, too. If we carry on $A$ the coalgebra structure on $A^*$ through the isomorphism $f$ we get that $A$ is a $k$-coalgebra via the structure determined by

$$\Delta(a) = \sum_{i,j} \langle f(a), a_i a_j \rangle f^{-1}(a^j) \otimes f^{-1}(a^i) \quad \text{and} \quad \varepsilon(a) = \langle f(a), 1_A \rangle,$$

for all $a \in A$. From the right $A$-linearity of $f$ we deduce that the above morphism $\Delta$ is an $A$-bimodule morphism, as required.

$(iv) \Rightarrow (i)$. If $(A, \Delta, \varepsilon)$ is a $k$-coalgebra with $\Delta$ an $A$-bimodule morphism, then the pair $(\varepsilon, \Delta(1_A))$ satisfies the conditions in (i).

$(ii) \Leftrightarrow (v)$. For $f : A \to A^*$ a right $A$-module isomorphism define $B_r : A \times A \to k$, $B_r(a,b) = f(a)(b)$, for all $a, b \in A$. Then $B_r$ is an associative and right non-degenerate bilinear form. Conversely, given $B_r$ we have that $f : A \to A^*$, $f(a)(b) = B_r(a,b)$, for all $a, b \in A$, is a right $A$-linear isomorphism.

$(iii) \Leftrightarrow (vi)$. Similar to $(ii) \Leftrightarrow (v)$: given $g$ we define $B_l : A \times A \to k$ by $B_l(a,b) = g(b)(a)$, for all $a, b \in A$, and vice versa.

$(v) \Rightarrow (vii)$. Let $B_r$ be a right non-degenerate and associative bilinear map on $A$. Then $A \ni a \mapsto B_r(1_A, a) \in k$ is non-zero, and therefore its kernel $H := \{a \in A \mid B_r(1_A, a) = 0\}$ is a hyperplane in $A$. If $I$ is a right ideal in $A$ contained in $H$ then, for any $x \in I$, we have $0 = B_r(1_A, xa) = B_r(x, a)$, for all $a \in A$. Now using that $B_r$ is right non-degenerate we get $x = 0$, and so $I = 0$.

$(vii) \Rightarrow (ii)$. Let $H$ be a hyperplane in $A$. As $k$ is a field one can find $\phi \in A^*$ such that its kernel is $H$.

Assume that $H$ does not contain non-zero right ideals. If $f : A \to A^*$ is defined by $f(a)(b) = \phi(ab)$, for all $a, b \in A$, then

$$(f(a) \leftharpoonup b)(c) = f(a)(bc) = \phi(a(bc)) = \phi((ab)c) = f(ab)(c),$$

for all $a, b, c \in A$, hence $f$ is right $A$-linear. In addition, if $x \in A$ is such that $f(x) = 0$ then $\phi(xa) = 0$, for all $a \in A$, thus $xA \subseteq H$. Consequently, $xA = 0$, and so $x = 0$. Therefore $f$ is injective, and so an isomorphism because $A$ and $A^*$ have the same dimension over $k$.

If $A$ does not contain non-zero left ideals then in a manner similar to the one above one can show that $f' : A \to A^*$ given by $f'(a)(b) = \phi(ba)$, for all $a, b \in A$, is left $A$-linear and an isomorphism. So we are done. □

**Definition 7.2** If a finite-dimensional $k$-algebra $A$ satisfies one, and therefore all, of the equivalent conditions in Proposition 7.1 we say that $A$ is a Frobenius algebra. Furthermore, if $(\phi, e)$ is a Frobenius system for a Frobenius algebra $A$ (i.e. a pair as in Proposition 7.1 (i)) we refer to $\phi$ as a Frobenius morphism for $A$, and call $e$ a Frobenius element for $A$.

**Remark 7.3** Let $f : A \to A^*$ be an isomorphism of left/right $A$-modules and $g : A \to A^*$ the isomorphism of right/left $A$-modules corresponding to $f$ as in the above proposition. Then the Frobenius systems defined by $f$ and $g$ coincide.

Indeed, by the proof of Proposition 7.1 we have that $g(a)(b) = (f^* \circ \theta_A)(a)(b) = \theta_A(a)(f(b)) = f(b)(a)$, for all $a, b \in A$. Thus, if $f$ is left (resp. right) $A$-linear and $\phi = f(1_A)$ then $f(b) = b \rightharpoonup \phi$ (resp. $f(b) = \phi \leftharpoonup b$), for all $b \in A$, and so $g(a)(b) = \phi(ab) = (\phi \leftharpoonup a)(b)$ (resp. $g(a)(b) = \phi(ba) = (a \rightharpoonup \phi)(b)$), for all $a, b \in A$. We deduce that $g(a) = \phi \leftharpoonup a$ (resp. $g(a) = a \rightharpoonup \phi$), for all $a \in A$, thus $g(1_A) = \phi$ in both situations. This says that the Frobenius morphism associated to $g$ is $\phi$, as required.

Now, if $f$ is left $A$-linear then $(\phi, \sum_i f^{-1}(a^i) \otimes a_i)$ is the Frobenius system associated to $f$, while $(\phi, \sum_i a_i \otimes g^{-1}(a^i))$ is the Frobenius system associated to $g$. Since

$$g^{-1}(a^*) = \theta_A^{-1}(a^* \circ f^{-1}) = \sum_i a^*(f^{-1}(a^i))a_i,$$

it follows that

$$\sum_i a_i \otimes g^{-1}(a^i) = \sum_{i,j} a^i(f^{-1}(a^j))a_i \otimes a_j = \sum_j f^{-1}(a^j) \otimes a_j,$$

as required. Likewise, if $f$ is right $A$-linear then the Frobenius system induced by it is $(\phi, \sum_i a_i \otimes f^{-1}(a^i))$, and the Frobenius system corresponding to $g$ is $(\phi, \sum_i g^{-1}(a^i) \otimes a_i)$. According to the last computation they coincide, so the proof is complete.

**Examples 7.4** (1) If $A$, $A'$ are Frobenius algebras then so are $A \times A'$ and $A \otimes A'$. More precisely, assume that $B_l$ and $B'_l$ are left non-degenerate, associative bilinear maps on $A$ and $A'$, respectively. One can easily see that $\mathbf{B}_l : (A \times A') \times (A \times A') \to k$

given by

$$\mathbf{B}_l((a,a'),(b,b')) = B_l(a,b) + B'_l(a',b'), \ \ \forall\, a,b \in A \ \text{ and } \ a',b' \in A',$$

is a left non-degenerate, associative bilinear map on $A \times A'$. Thus $A \times A'$ is a Frobenius algebra as well.

Similarly, if we define $\widetilde{\mathbf{B}}_l : (A \otimes A') \times (A \otimes A') \to k$ by

$$\widetilde{\mathbf{B}}_l(a \otimes a', b \otimes b') = B_l(a,b)B'_l(a',b'), \ \ \forall\, a,b \in A \ \text{ and } \ a',b' \in A',$$

then a simple verification guarantees that $\widetilde{\mathbf{B}}_l$ is a left non-degenerate, associative bilinear map on $A \otimes A'$. Therefore $A \otimes A'$ is a Frobenius algebra, as stated.

(2) For any non-zero natural number $n$ the $n \times n$-matrix algebra $M_n(k)$ is a Frobenius algebra. To see this, consider the pair $(\phi = \mathrm{Tr}, e = \sum_{i,j} E_{ij} \otimes E_{ji})$, where $\mathrm{Tr} : M_n(k) \to k$ is the trace morphism and $\{E_{ij} \mid 1 \le i,j \le n\}$ is the canonical basis of $M_n(k)$. We have that Tr is a Frobenius morphism for $M_n(k)$ since

$$\sum_{i,j} \mathrm{Tr}(E_{ij})E_{ji} = \sum_{i,j} \delta_{i,j}E_{ji} = \sum_i E_{ii} = I_n,$$

and similarly $\sum_{i,j} \mathrm{Tr}(E_{ji})E_{ij} = I_n$. Here $\delta_{i,j}$ is the Kronecker delta and $I_n$ is the unit matrix of $M_n(k)$.

Now, $e$ is a Frobenius element for $M_n(k)$ since, for any $X = (x_{ij})_{1 \le i,j \le n} \in M_n(k)$,

$$\begin{aligned}\sum_{i,j} XE_{ij} \otimes E_{ji} &= \sum_{i,j,u,v} x_{uv}E_{uv}E_{ij} \otimes E_{ji}\\ &= \sum_{i,j,u} x_{ui}E_{uj} \otimes E_{ji} = \sum_{i,j,u,v} E_{uj} \otimes x_{vi}E_{ju}E_{vi} = \sum_{j,u} E_{uj} \otimes E_{ju}X,\end{aligned}$$

as required.

(3) For any non-zero natural number $n$, $A = \frac{k[X]}{(X^n)}$ is a Frobenius algebra. To prove this, denote by $x$ the class of $X$ modulo $(X^n)$ and take $\{1, x, \ldots, x^{n-1}\}$ the canonical basis of $A$. If we define $\phi : A \to k$ by

$$\phi\left(\sum_{j=1}^{n} a_j x^{j-1}\right) = a_n \text{ and } e = \sum_{i=1}^{n} x^{i-1} \otimes x^{n-i}$$

we claim that $(\phi, e)$ is a Frobenius system for $A$. Indeed, one can easily see that

$$\sum_{i=1}^{n} \phi(x^{i-1})x^{n-i} = \sum_{i=1}^{n} \phi(x^{n-i})x^{i-1} = \phi(x^{n-1}) = 1.$$

Also, for any $1 \le j \le n-1$, we have

$$\begin{aligned}x^j e &= \sum_{i=1}^{n} x^{j+i-1} \otimes x^{n-i}\\ &= \sum_{t=0}^{j-1} x^t \otimes x^{j-t-1} + \sum_{t=0}^{n-j-1} x^{j+t} \otimes x^{n-t-1}\end{aligned}$$

$$= \sum_{i=1}^{n} x^{i-1} \otimes x^{n+j-i} = ex^j,$$

and so $ae = ea$, for all $a \in A$, because $\{1, x, \ldots, x^{n-1}\}$ is a basis for $A$.

Let $A$ be a Frobenius algebra and $(\phi, e)$ a Frobenius system for $A$. We have seen that $f : A \to A^*$ defined by $f(a) = \phi \leftharpoonup a$, for all $a \in A$, is a right $A$-linear isomorphism. So for any $a \in A$ there exists a unique element $\chi(a) \in A$ such that $a \rightharpoonup \phi = \phi \leftharpoonup \chi(a)$. In this way we have a well-defined map $\chi : A \to A$.

**Proposition 7.5** *The map $\chi$ is an algebra automorphism of $A$. Furthermore, it can be explicitly computed in terms of the Frobenius system $(\phi, e = e^1 \otimes e^2)$ of $A$ as*

$$\chi(a) = \phi(e^1 a)e^2, \ \ \forall\, a \in A. \tag{7.1.1}$$

*Proof* On the one hand, for all $a, b \in A$ we have

$$\begin{aligned} a \rightharpoonup (b \rightharpoonup \phi) &= a \rightharpoonup (\phi \leftharpoonup \chi(b)) \\ &= (a \rightharpoonup \phi) \leftharpoonup \chi(b) \\ &= (\phi \leftharpoonup \chi(a)) \leftharpoonup \chi(b) \\ &= \phi \leftharpoonup \chi(a)\chi(b), \end{aligned}$$

from which we deduce that $a \rightharpoonup (b \rightharpoonup \phi) = f(\chi(a)\chi(b))$.

On the other hand, $a \rightharpoonup (b \rightharpoonup \phi) = ab \rightharpoonup \phi = \phi \leftharpoonup \chi(ab) = f(\chi(ab))$. By the injectivity of $f$ it follows that $\chi$ is multiplicative. Clearly $1_A \rightharpoonup \phi = \phi = \phi \leftharpoonup 1_A$, so $\chi(1_A) = 1_A$. We can now conclude that $\chi$ is an algebra endomorphism of $A$.

It has also been proved that $f' : A \to A^*$ given by $f'(a) = a \rightharpoonup \phi$ is an isomorphism of left $A$-modules. Thus for any $a \in A$ there exists a unique $\chi^{-1}(a) \in A$ such that $\phi \leftharpoonup a = \chi^{-1}(a) \rightharpoonup \phi$, and so we have a well-defined map $\chi^{-1} : A \to A$. By the two definitions of $\chi$ and $\chi^{-1}$ we have

$$\begin{aligned} f'(a) &= a \rightharpoonup \phi = \phi \leftharpoonup \chi(a) = \chi^{-1}(\chi(a)) \rightharpoonup \phi = f'(\chi^{-1}(\chi(a))), \\ f(a) &= \phi \leftharpoonup a = \chi^{-1}(a) \rightharpoonup \phi = \phi \leftharpoonup \chi(\chi^{-1}(a)) = f(\chi(\chi^{-1}(a))), \end{aligned}$$

for all $a \in A$. Since $f, f'$ are injective maps it follows that $\chi$ and $\chi^{-1}$ are bijective inverses. Hence $\chi$ is an algebra automorphism of $A$, as desired.

In addition, for all $a, b \in A$, we compute

$$\begin{aligned} \langle \phi \leftharpoonup \phi(e^1 a)e^2, b \rangle &= \phi(e^1 a)\phi(e^2 b) \\ &= \phi(be^1 a)\phi(e^2) \\ &= \phi(b\phi(e^2)e^1 a) \\ &= \phi(ba) = \langle a \rightharpoonup \phi, b \rangle, \end{aligned}$$

and therefore $\phi \leftharpoonup \phi(e^1 a)e^2 = a \rightharpoonup \phi$, for all $a \in A$, proving (7.1.1). □

**Definition 7.6** The Nakayama automorphism of a Frobenius algebra $A$ corresponding to a Frobenius system $(\phi, e)$ is the algebra automorphism $\chi$ of $A$ defined by $a \rightharpoonup \phi = \phi \leftharpoonup \chi(a)$, for all $a \in A$.

**Remark 7.7** The inverse of the Nakayama automorphism $\chi$ associated to a Frobenius algebra $A$ with Frobenius system $(\phi, e)$ is given by

$$\chi^{-1}(a) = \phi(ae^2)e^1, \ \forall\, a \in A. \tag{7.1.2}$$

Indeed, we compute

$$\begin{aligned}\chi\chi^{-1}(a) &\overset{(7.1.2)}{=} \phi(ae^2)\chi(e^1)\\ &\overset{(7.1.1)}{=} \phi(ae^2)\phi(\mathfrak{e}^1e^1)\mathfrak{e}^2\\ &= \phi(ae^2\mathfrak{e}^1)\phi(e^1)\mathfrak{e}^2\\ &= \phi(a\mathfrak{e}^1)\mathfrak{e}^2 = \phi(\mathfrak{e}^1)\mathfrak{e}^2a = a,\end{aligned}$$

where $\mathfrak{e}^1 \otimes \mathfrak{e}^2$ is another copy of $e$, and

$$\begin{aligned}\chi^{-1}\chi(a) &\overset{(7.1.1)}{=} \phi(e^1a)\chi^{-1}(e^2)\\ &\overset{(7.1.2)}{=} \phi(e^1a)\phi(e^2\mathfrak{e}^2)\mathfrak{e}^1\\ &= \phi(\mathfrak{e}^2e^1a)\phi(e^2)\mathfrak{e}^1\\ &= \phi(\mathfrak{e}^2a)\mathfrak{e}^1 = \phi(\mathfrak{e}^2)a\mathfrak{e}^1 = a,\end{aligned}$$

for all $a \in A$, as needed.

**Examples 7.8** (1) For $M_n(k)$ with Frobenius system $(\mathrm{Tr}, \sum_{i,j} E_{ij} \otimes E_{ji})$ as in Example 7.4(2) the Nakayama automorphism is the identity morphism of $M_n(k)$. To see this we use (7.1.1) to compute

$$\chi(X) = \sum_{i,j}\mathrm{Tr}(E_{ij}X)E_{ji} = \sum_{i,j,u,v}\mathrm{Tr}(x_{uv}E_{ij}E_{uv})E_{ji} = \sum_{i,j,v}\mathrm{Tr}(x_{jv}E_{iv})E_{ji} = \sum_{i,j}x_{ji}E_{ji} = X,$$

for all $X = (x_{uv})_{1 \le u,v \le n}$, as stated.

(2) For $A = \frac{k[X]}{(X^n)}$ with the Frobenius system as in Example 7.4(3), the Nakayama automorphism is the identity map. This follows from

$$\chi(x^j) = \sum_{i=1}^{n}\phi(x^{j+i-1})x^{n-i} = \sum_{i=1}^{n}\delta_{i,n-j}x^{n-i} = x^j,$$

for all $0 \le j \le n-1$, and the fact that $\{1, x, \ldots, x^{n-1}\}$ is a basis for $A$.

We next show that a Frobenius system is unique up to an invertible element, in the following sense.

**Proposition 7.9** *Let $A$ be a Frobenius algebra with a Frobenius system $(\phi, e = e^1 \otimes e^2)$. Then any other Frobenius system for $A$ is of the form*

$$(\phi \leftharpoonup d, e^1 \otimes d^{-1}e^2),$$

*for some invertible element $d \in A$ with inverse $d^{-1}$, or equivalently of the form*

$$(d' \rightharpoonup \phi, e^1d'^{-1} \otimes e^2),$$

*for some invertible element $d' \in A$ with inverse $d'^{-1}$.*

*Proof* If $d \in A$ is an invertible element with inverse $d^{-1}$ then one can easily check that both $(\phi \leftharpoondown d, e^1 \otimes d^{-1}e^2)$ and $(d \rightharpoonup \phi, e^1 d^{-1} \otimes e^2)$ are Frobenius systems for $A$.

Conversely, let $(\psi, b = b^1 \otimes b^2)$ be another Frobenius system for $A$. Then $f, g : A \to A^*$ given by $f(a) = \phi \leftharpoondown a$, and by $g(a) = \psi \leftharpoondown a$, respectively, for all $a \in A$, are right $A$-linear isomorphisms. Furthermore, $f^{-1}(a^*) = a^*(e^1)e^2$ and $g^{-1}(a^*) = a^*(b^1)b^2$, for all $a^* \in A^*$.

For $\psi \in A^*$ there exists a unique $d \in A$ such that $f(d) = \psi$, that is, $\phi \leftharpoondown d = \psi$. Similarly, for $\phi \in A^*$ there exists a unique $d^{-1} \in A$ such that $g(d^{-1}) = \phi$, that is, $\psi \leftharpoondown d^{-1} = \phi$. We get

$$g(1_A) = \psi = \phi \leftharpoondown d = \psi \leftharpoondown d^{-1}d = g(d^{-1}d),$$
$$f(1_A) = \phi = \psi \leftharpoondown d^{-1} = \phi \leftharpoondown dd^{-1} = f(dd^{-1}),$$

and since $f$ and $g$ are injective maps it follows that $d$ is invertible in $A$ with inverse $d^{-1}$. For further use note that $d$ and $d^{-1}$ can be computed explicitly as

$$d = \psi(e^1)e^2 \quad \text{and } d^{-1} = \phi(b^1)b^2, \tag{7.1.3}$$

since $(\phi \leftharpoondown \psi(e^1)e^2)(a) = \psi(e^1)\phi(e^2a) = \psi(ae^1\phi(e^2)) = \psi(a)$, for all $a \in A$, and similarly $\psi \leftharpoondown \phi(b^1)b^2 = \phi$.

Note now that

$$f(a) = \phi \leftharpoondown a = \psi \leftharpoondown d^{-1}a = g(d^{-1}a), \ \forall\, a \in A,$$

from which we obtain that $f = g \leftharpoondown d^{-1}$ or, equivalently, that $g = f \leftharpoondown d$. It is clear at this point that $g^{-1}(a^*) = d^{-1}f^{-1}(a^*)$, for all $a^* \in A^*$. Thus $a^*(b^1)b^2 = a^*(e^1)d^{-1}e^2$, for all $a^* \in A^*$, and this is equivalent to $b^1 \otimes b^2 = e^1 \otimes d^{-1}e^2$, as required.

To land at the second possibility for $(\psi, b)$ we use the Nakayama automorphism $\chi$ associated to the Frobenius system $(\phi, e)$. We obtain that $\psi = \phi \leftharpoondown d = \chi^{-1}(d) \rightharpoonup \phi$, and therefore we set $d' = \chi^{-1}(d)$. If follows that $d'$ is invertible with inverse $d'^{-1} = \chi^{-1}(d^{-1})$, and so

$$\begin{aligned} e^1 d'^{-1} \otimes e^2 &= e^1 \chi^{-1}(d^{-1}) \otimes e^2 \\ &\overset{(7.1.2)}{=} \phi(d^{-1}\mathfrak{e}^2)e^1\mathfrak{e}^1 \otimes e^2 \\ &= \phi(d^{-1}\mathfrak{e}^2 e^1)\mathfrak{e}^1 \otimes e^2 \\ &= \phi(e^1)\mathfrak{e}^1 \otimes e^2 d^{-1}\mathfrak{e}^2 \\ &= \mathfrak{e}^1 \otimes d^{-1}\mathfrak{e}^2 = b^1 \otimes b^2, \end{aligned}$$

where $\mathfrak{e}^1 \otimes \mathfrak{e}^2$ denotes a second copy of $e$. So our proof is complete. □

**Corollary 7.10** *Let $A$ be a Frobenius algebra, $\chi$ the Nakayama automorphism associated to the Frobenius system $(\phi, e)$ of $A$ and $\eta$ the Nakayama automorphism corresponding to a second Frobenius system $(\psi, b)$ for $A$. Then there exists an invertible element $d$ in $A$ such that $\chi(a) = d\eta(a)d^{-1}$, for all $a \in A$.*

*Proof* By the previous result there exists an invertible element $d \in A$ such that $\psi = \phi \leftharpoonup d$ and $b^1 \otimes b^2 = e^1 \otimes d^{-1}e^2$. We can compute:

$$\begin{aligned} \eta(a) &\overset{(7.1.1)}{=} \psi(b^1 a)b^2 \\ &= (\phi \leftharpoonup d)(e^1 a)d^{-1}e^2 \\ &= (a \rightharpoonup \phi)(de^1)d^{-1}e^2 \\ &= (\phi \leftharpoonup \chi(a))(de^1)d^{-1}e^2 \\ &= \phi(\chi(a)de^1)d^{-1}e^2 \\ &= \phi(e^1)d^{-1}e^2\chi(a)d = d^{-1}\chi(a)d, \end{aligned}$$

for all $a \in A$, which is clearly equivalent to $\chi(a) = d\eta(a)d^{-1}$, for all $a \in A$. □

We end this section by showing that any Frobenius algebra is a self injective module. This result will be used later on in the proof of the Nichols–Zoeller type theorem for quasi-Hopf algebras. First we need some preparatory work.

**Lemma 7.11** *Let $R$ and $S$ be two rings and $P$ an $(R,S)$-bimodule which is flat as a left $R$-module. If $M$ is an injective right $S$-module then $\widehat{M} := \mathrm{Hom}_S(P,M)$ is an injective right $R$-module, where $\widehat{M}$ is considered as a right $R$-module via the action $(f \cdot r)(p) = f(r \cdot p)$, for all $f \in \widehat{M}$, $r \in R$ and $p \in P$.*

*Proof* We show that the contravariant functor $\mathrm{Hom}_R(-,\widehat{M})$ is exact, and this will imply that $\widehat{M}$ is injective as a right $R$-module.

Consider $0 \to X \to Y \to Z \to 0$, a short exact sequence of right $R$-modules. Since $P$ is flat as a left $R$-module we get that

$$0 \to X \otimes_R P \to Y \otimes_R P \to Z \otimes_R P \to 0$$

is a short exact sequence of right $S$-modules. Now using that $M$ is an injective right $S$-module we obtain that

$$0 \to \mathrm{Hom}_S(Z \otimes_R P, M) \to \mathrm{Hom}_S(Y \otimes_R P, M) \to \mathrm{Hom}_S(X \otimes_R P, M) \to 0$$

is a short exact sequence of abelian groups. For any right $R$-module $U$ we have $\mathrm{Hom}_R(U,\widehat{M}) = \mathrm{Hom}_R(U,\mathrm{Hom}_S(P,M)) \cong \mathrm{Hom}_S(U \otimes_R P, M)$, the isomorphism being produced by

$$\mathrm{Hom}_R(U,\mathrm{Hom}_S(P,M)) \ni \lambda \mapsto (x \otimes_R p \mapsto \lambda(x)(p)) \in \mathrm{Hom}_S(U \otimes_R P, M).$$

Its inverse is

$$\mathrm{Hom}_S(U \otimes_R P, M) \ni \mu \mapsto (x \mapsto (p \mapsto \mu(x \otimes_R p))) \in \mathrm{Hom}_R(U,\mathrm{Hom}_S(P,M)).$$

It now follows that $0 \to \mathrm{Hom}_R(Z,\widehat{M}) \to \mathrm{Hom}_R(Y,\widehat{M}) \to \mathrm{Hom}_R(X,\widehat{M}) \to 0$ is a short exact sequence of abelian groups, so our proof is finished. □

**Corollary 7.12** *Let $R$ be a $k$-algebra, $P$ a projective left $R$-module and consider $P$ as an $(R,k)$-bimodule in the obvious way. Then for any $k$-vector space $M$ the right $R$-module $\mathrm{Hom}_k(P,M)$ is injective. Consequently, $P^* := \mathrm{Hom}_k(P,k)$ is injective as a* *right R-module, where the R-action on $P^*$ is given by $(p^* \cdot r)(p) = p^*(r \cdot p)$, for all $p^* \in P^*$, $r \in R$ and $p \in P$.*

*Proof* Any projective module is flat, hence $P$ is flat as a left $R$-module. Moreover, since $k$ is a field any $k$-vector space is an injective $k$-module, so by the previous result we get that $\mathrm{Hom}_k(P,M)$ is an injective right $R$-module. The second assertion in the statement follows by taking $M = k$. □

One can now prove the result announced above.

**Proposition 7.13** *If A is a Frobenius algebra then A is injective as a left and right A-module.*

*Proof* Since $A$ is a Frobenius algebra we have that $A \cong A^*$ both as left and right $A$-modules. As $A$ is a free left $A$-module, it is projective as a left $A$-module. By Corollary 7.12 we deduce that $A^*$ is injective as a right $A$-module, and therefore $A$ is injective as a right $A$-module. By working now with the right-handed version we get that $A$ is injective as a left $A$-module as well, so the proof is finished. □

## 7.2 Integral Theory

Unless otherwise specified, throughout this section $H$ is a finite-dimensional quasi-Hopf algebra with an antipode $S$, $\{e_i\}_i$ is a basis of $H$ and $\{e^i\}_i$ is the corresponding dual basis of $H^*$.

In the first part of this section we show that the spaces of left and right integrals in $H$ are one-dimensional. We then prove that $S$ is bijective and that $H$ is a Frobenius algebra. In the end we will see that the space of integrals of an infinite-dimensional quasi-Hopf algebra with bijective antipode is zero.

**Definition 7.14** Let $H$ be a quasi-bialgebra. An element $t \in H$ is called a left (resp. right) integral in $H$ if $ht = \varepsilon(h)t$ (resp. $th = \varepsilon(h)t$), for all $h \in H$. We denote by $\int_l^H$ (resp. $\int_r^H$) the space of left (resp. right) integrals in $H$. If there exists a non-zero left integral in $H$ which is at the same time a right integral, then $H$ is called unimodular.

**Examples 7.15** (1) For the two-dimensional quasi-Hopf algebra $H(2)$ constructed in Example 3.26 it can be easily checked that $t = 1 + g$ is both a non-zero left and right integral in $H(2)$. Thus $H(2)$ is a unimodular quasi-Hopf algebra.

(2) Let $H_\pm(8)$ be the two eight-dimensional quasi-Hopf algebras considered in Example 3.30. If $t = (1+g)x^3$ then

$$gt = g(1+g)x^3 = (g+1)x^3 = t = \varepsilon(g)t,$$
$$xt = x(1+g)x^3 = (x+xg)x^3 = (x-gx)x^3 = (1-g)x^4 = 0 = \varepsilon(x)t,$$

and so $t$ is a non-zero left integral in $H_\pm(8)$, because $g$ and $x$ generate $H_\pm(8)$ as an algebra. In a similar manner one can show that $r = (1-g)x^3$ is a non-zero right integral in $H_\pm(8)$. Since the characteristic of $k$ is not 2 it follows that $H_\pm(8)$ are not unimodular quasi-Hopf algebras.

(3) By computations similar to the ones performed in the previous example one can show that $t = (1+g)x^3y^3$ is both a non-zero left and right integral in the 32-dimensional quasi-Hopf algebra described in Example 3.32. Therefore $H(32)$ is a unimodular quasi-Hopf algebra.

One can construct a projection onto the space of left integrals as follows.

**Proposition 7.16** *Let $H$ be a finite-dimensional quasi-Hopf algebra with antipode $S$ and for any $h \in H$ define*

$$P(h) = \sum_i \langle e^i, \beta S(S(X^2(e_i)_2)\alpha X^3)h\rangle X^1(e_i)_1. \tag{7.2.1}$$

*Then $P(h) \in \int_l^H$, for all $h \in H$, and $\sum_i \langle e^i, S(P(e_i)\beta)\rangle = 1$. Consequently, at least one of the elements $P(e_i)$ is nonzero, and therefore $\int_l^H \neq 0$.*

*Proof* We check that $P(h)$ is a left integral in $H$, for all $h \in H$:

$$\begin{aligned}
aP(h) &= \sum_i \langle e^i, \beta S(S(X^2(e_i)_2)\alpha X^3)h\rangle aX^1(e_i)_1 \\
&\overset{(3.2.1)}{=} \sum_i \langle e^i, \beta S(S(a_{(2,1)}X^2(e_i)_2)\alpha a_{(2,2)}X^3)h\rangle a_1X^1(e_i)_1 \\
&\overset{(3.1.7)}{=} \sum_i \langle e^i, \beta S(S(X^2(a_1e_i)_2)\alpha X^3a_2)h\rangle X^1(a_1e_i)_1 \\
&= \sum_{i,j} \langle e^j, a_1e_i\rangle\langle e^i, \beta S(S(X^2(e_j)_2)\alpha X^3a_2)h\rangle X^1(e_j)_1 \\
&= \sum_j \langle e^j, a_1\beta S(S(X^2(e_j)_2)\alpha X^3a_2)h\rangle X^1(e_j)_1 \\
&\overset{(3.2.1)}{=} \varepsilon(a)\sum_i \langle e^i, \beta S(S(X^2(e_i)_2)\alpha X^3)h\rangle X^1(e_i)_1 \overset{(7.2.1)}{=} \varepsilon(a)P(h),
\end{aligned}$$

for all $a \in H$, as needed. We next compute that

$$\begin{aligned}
\sum_j \langle e^j, S(P(e_j)\beta)\rangle &\overset{(7.2.1)}{=} \sum_{i,j} \langle e^i, \beta S(S(X^2(e_i)_2)\alpha X^3)e_j\rangle\langle e^j, S(X^1(e_i)_1\beta)\rangle \\
&= \sum_i \langle e^i, \beta S(\alpha X^3)S(X^1(e_i)_1\beta S((e_i)_2)S(X^2))\rangle \\
&\overset{(3.2.1)}{=} \sum_i \varepsilon(e_i)\langle e^i, \beta S(X^1\beta S(X^2)\alpha X^3)\rangle \\
&\overset{(3.2.2)}{=} \sum_i \varepsilon(e_i)\langle e^i, \beta\rangle = \varepsilon(\beta) = 1,
\end{aligned}$$

as stated. □

The proof of the above proposition provides a $k$-linear map $P : H \to \int_l^H$. If we denote the inclusion of $\int_l^H$ into $H$ by $i : \int_l^H \to H$, then a simple calculation shows that $P \circ i = \mathrm{Id}$. Moreover, by repeating the above arguments for the quasi-Hopf algebra $H^{\mathrm{op,cop}}$ instead of $H$ we get that $\int_r^H \neq 0$, too, and this follows by proving the existence of a projection of $H$ onto the space of right integrals in $H$ that covers the natural inclusion.

In order to prove the uniqueness of integrals for finite-dimensional quasi-Hopf algebras we need the following lemma.

**Lemma 7.17** *Let $t$ be a left integral in a quasi-Hopf algebra $H$. Then, for all $h \in H$, we have:*

$$hX^1t_1 \otimes S(X^2t_2)\alpha X^3 = X^1t_1 \otimes S(X^2t_2)\alpha X^3h, \tag{7.2.2}$$

$$t_1 \otimes S(t_2) = X^1t_1 \otimes S(X^2t_2)\alpha X^3\beta = \beta X^1t_1 \otimes S(X^2t_2)\alpha X^3. \tag{7.2.3}$$

*Proof* For all $h \in H$ we calculate, by using (3.2.1), (3.1.7) and $t \in \int_l^H$:

$$\begin{aligned} hX^1t_1 \otimes S(X^2t_2)\alpha X^3 &= h_1X^1t_1 \otimes S(h_{(2,1)}X^2t_2)\alpha h_{(2,2)}X^3 \\ &= X^1(h_1t)_1 \otimes S(X^2(h_1t)_2)\alpha X^3h_2 = X^1t_1 \otimes S(X^2t_2)\alpha X^3h. \end{aligned}$$

To prove the first equality in (7.2.3), we take $X^1t_1 \otimes S(X^2t_2)\alpha X^3\beta$. First we apply the 3-cocycle condition

$$\Phi \otimes 1_H = (\mathrm{Id}_H \otimes \Delta \otimes \mathrm{Id}_H)(\Phi^{-1})(1_H \otimes \Phi^{-1})(\mathrm{Id}_H \otimes \mathrm{Id}_H \otimes \Delta)(\Phi)(\Delta \otimes \mathrm{Id}_H \otimes \mathrm{Id}_H)(\Phi)$$

and then, successively using the fact that $t \in \int_l^H$, (3.2.1), (3.1.10), (3.1.11) and (3.2.2), we find the left-hand side of (7.2.3). The second equality in (7.2.3) follows from (7.2.2). □

We also need the dual structure of $H$. Recall that $H^*$ has a natural multiplication $\langle h^*g^*, h\rangle = h^*(h_1)g^*(h_2)$, where $h^*, g^* \in H^*$ and $h \in H$. Since $H$ is finite dimensional, $H^*$ is also equipped with a natural coassociative coalgebra structure $(\tilde{\Delta}, \tilde{\varepsilon})$ given by $\langle \tilde{\Delta}(h^*), h \otimes h'\rangle = \langle h^*, hh'\rangle$ and $\tilde{\varepsilon}(h^*) = h^*(1_H)$, where $h^* \in H^*$, $h, h' \in H$ and $\langle , \rangle : H^* \otimes H \to k$ denotes the dual pairing. As $H$ is an algebra, on $H^*$ we have the natural left and right $H$-actions

$$\langle h \rightharpoonup h^*, h'\rangle = \langle h^*, h'h\rangle, \qquad \langle h^* \leftharpoonup h, h'\rangle = \langle h^*, hh'\rangle,$$

where $h, h' \in H$ and $h^* \in H^*$. This makes $H^*$ into an $H$-bimodule.

We also introduce $\overline{S} : H^* \to H^*$ as the anti-coalgebra homomorphism dual to $S$, that is, $\langle \overline{S}(h^*), h\rangle = \langle h^*, S(h)\rangle$, for all $h^* \in H^*$, $h \in H$. As we have seen before, all these endow $H^*$ with a dual quasi-Hopf algebra structure.

**Theorem 7.18** *Let $H$ be a finite-dimensional quasi-Hopf algebra, $\{e_i\}_i$ a basis of $H$ with dual basis $\{e^i\}_i$ of $H^*$, and define the linear map $\theta : \int_l^H \otimes H^* \to H$ by*

$$\theta(t \otimes h^*) = h^*(S(X^2t_2p^2)\alpha X^3)X^1t_1p^1, \ \forall\, t \in \int_l^H, \ \ h^* \in H^*, \tag{7.2.4}$$

*where $p_R = p^1 \otimes p^2$ is defined in* (3.2.19). *Then the following assertions hold:*

*(i) $\theta$ is an isomorphism of left $H$-modules, where $\int_l^H \otimes H^*$ is a left $H$-module via*

*$h\cdot(t\otimes h^*)=t\otimes h\rightharpoonup h^*$, for all $h\in H$, $t\in\int_l^H$, $h^*\in H^*$, and $H$ is a left $H$-module via left multiplication. Consequently,* $\dim_k\int_l^H=1$. *The inverse of $\theta$ is given by*

$$\theta^{-1}(h)=\sum_i P(e_ih)\otimes e^i,\ \ \forall\, h\in H, \tag{7.2.5}$$

*where $P$ is the projection onto the space of left integrals defined in* (7.2.1).

*(ii) The antipode $S$ is bijective.*

*(iii)* $S(\int_l^H)=\int_r^H$, $S(\int_r^H)=\int_l^H$ *and* $\dim_k\int_r^H=1$.

*Proof* (i) First we show that $\theta$ and $\theta^{-1}$ are bijective inverses. For all $h\in H$ we have:

$$\begin{aligned}
&\theta(\theta^{-1}(h))\\
&\overset{(7.2.4),(7.2.1)}{=}\sum_{i,j}\langle e^j,\beta S(S(X^2(e_j)_2)\alpha X^3)e_ih\rangle\langle e^i,S(Y^2X_2^1(e_j)_{(1,2)}p^2)\alpha Y^3\rangle\\
&\qquad\qquad Y^1X_1^1(e_j)_{(1,1)}p^1\\
&\overset{(3.2.21)}{=}\sum_j\langle e^j,\beta S(Y^2X_2^1p^2S(X^2)\alpha X^3)\alpha Y^3h\rangle Y^1X_1^1p^1e_j\\
&\overset{(3.2.19),(3.1.9)}{=}Y^1x^1Z^1\beta S(Y^2x^2X^1Z_1^2\beta S(x_1^3X^2Z_2^2)\alpha x_2^3X^3Z^3)\alpha Y^3h\\
&\overset{\substack{(3.2.1),(3.1.10)\\(3.1.11),(3.2.2)}}{=}Y^1\beta S(Y^2X^1\beta S(X^2)\alpha X^3)\alpha Y^3h=h.
\end{aligned}$$

For all $t\in\int_l^H$ and $h^*\in H^*$ we compute

$$\begin{aligned}
&\theta^{-1}(\theta(t\otimes h^*))\\
&=\sum_i h^*(S(X^2t_2p^2)\alpha X^3)P(e_iX^1t_1p^1)\otimes e^i\\
&\overset{(7.2.1),(7.2.2)}{=}\sum_{i,j}h^*(S(X^2t_2p^2)\alpha X^3\beta S(S(Y^2(e_j)_2)\alpha Y^3)e_i)\\
&\qquad\qquad\langle e^j,X^1t_1p^1\rangle Y^1(e_j)_1\otimes e^i\\
&\overset{(7.2.3)}{=}\sum_i h^*(S(S(Y^2t_{(1,2)}p_2^1)\alpha Y^3t_2p^2)e_i)Y^1t_{(1,1)}p_1^1\otimes e^i\\
&\overset{(3.1.7),(3.2.1)}{=}\sum_i h^*(S(S(Y^2p_2^1)\alpha Y^3p^2)e_i)tY^1p_1^1\otimes e^i\\
&\overset{(3.2.19),(3.1.9)}{=}\sum_i h^*(S(S(y_1^2x^1X^2)\alpha y_2^2x^2X_1^3\beta S(y^3x^3X_2^3))e_i)ty^1X^1\otimes e^i\\
&\overset{(3.2.1),(3.1.10),(3.1.11)}{=}\sum_i h^*(S(S(x^1)\alpha x^2\beta S(x^3))e_i)t\otimes e^i\\
&\overset{(3.2.2)}{=}\sum_i h^*(e_i)t\otimes e^i=t\otimes h^*.
\end{aligned}$$

Since $\theta$ is a bijection and $\dim_kH=\dim_kH^*$ is finite, it follows that $\dim_k\int_l^H=1$. We are left to show that $\theta$ is $H$-linear. For all $h\in H$, $t\in\int_l^H$ and $h^*\in H^*$ we have:

$$\begin{aligned}
h\theta(t\otimes h^*)&=h^*(S(X^2t_2p^2)\alpha X^3)hX^1t_1p^1\\
&\overset{(7.2.2)}{=}\langle h\rightharpoonup h^*,S(X^2t_2p^2)\alpha X^3\rangle X^1t_1p^1\\
&=\theta(t\otimes h\rightharpoonup h^*).
\end{aligned}$$

(ii) First we prove that $\overline{S}$ is bijective. $H^*$ is finite dimensional, so it suffices to show that $\overline{S}$ is injective. Let $h^* \in H^*$ be such that $\overline{S}(h^*) = 0$, and take $0 \neq t \in \int_l^H$. For all $h \in H$ we have

$$\begin{aligned}\theta(t \otimes \beta S(h) \rightharpoonup h^*) &= \langle \beta S(h) \rightharpoonup h^*, S(X^2 t_2 p^2)\alpha X^3\rangle X^1 t_1 p^1 \\ &= \langle h^*, S(X^2 t_2 p^2)\alpha X^3 \beta S(h)\rangle X^1 t_1 p^1 \\ &\overset{(7.2.3)}{=} \langle h^*, S(h t_2 p^2)\rangle t_1 p^1 \\ &= \langle \overline{S}(h^*), h t_2 p^2\rangle t_1 p^1 = 0.\end{aligned}$$

Since $\theta$ is bijective we obtain that $t \otimes \beta S(h) \rightharpoonup h^* = 0$. Now, because $t \neq 0$ and $\dim_k \int_l^H = 1$, it follows that $\beta S(h) \rightharpoonup h^* = 0$, for all $h \in H$. Therefore, by (3.2.2), for all $h' \in H$ we have

$$h^*(h') = \langle h^*, h' S(x^1)\alpha x^2 \beta S(x^3)\rangle = \langle \beta S(x^3) \rightharpoonup h^*, h' S(x^1)\alpha x^2\rangle = 0.$$

It is not hard to see that $\overline{S}^* : H^{**} \to H^{**}$, $\overline{S}^*(h^{**}) = h^{**} \circ \overline{S}$, for all $h^{**} \in H^{**}$, is a bijective map. If we define $\xi : H \to H^{**}$ by $\xi(h)(h^*) = h^*(h)$, for all $h \in H$, $h^* \in H^*$, then one can easily show that $\{\xi(e_i)\}_{i=\overline{1,n}}$ is a basis of $H^{**}$ dual to the basis $\{e^i\}_{i=\overline{1,n}}$ of $H^*$, and it follows that $\xi$ is bijective. Moreover, $\xi^{-1}$ is given by $\xi^{-1}(h^{**}) = \sum_i h^{**}(e^i)e_i$, for all $h^{**} \in H^{**}$. In addition, $\xi^{-1} \circ \overline{S}^* \circ \xi = S$, so $S$ is bijective.

(iii) We have seen that $S$ is an anti-algebra automorphism of $H$ and $\dim_k \int_l^H = 1$. This guarantees that $S(\int_l^H) = \int_r^H$, $S(\int_r^H) = \int_l^H$ and $\dim_k \int_r^H = 1$, as required. □

Let $H$ be a quasi-Hopf algebra and $t \in \int_l^H$. Since $H$ is an associative algebra, $th$ is also a left integral in $H$, for all $h \in H$, hence the space of left (right) integrals in $H$ is a two-sided ideal. Moreover, if $H$ is finite dimensional, then it follows from the uniqueness of the integrals in $H$ that there exists $\mu \in H^*$ such that

$$th = \mu(h)t, \ \ \forall\, t \in \int_l^H \ \text{ and } \ h \in H. \tag{7.2.6}$$

One can easily see that $\mu : H \to k$ is an algebra map, and so its inverse in the group $\mathrm{Alg}(H,k)$ defined in Remark 3.15 is $\mu^{-1} = \mu \circ S$, which is also equal to $\mu \circ S^{-1}$. We call $\mu$ the modular element of $H^*$, or the distinguished grouplike element of $H^*$.

Observe that $\mu = \varepsilon$ if and only if $H$ is unimodular. Also, from the bijectivity of the antipode we get that

$$hr = \mu^{-1}(h)r = \mu(S(h))r, \ \ \forall\, r \in \int_r^H \ \text{ and } \ h \in H. \tag{7.2.7}$$

Furthermore, by using the fact that the inverse of $\mu$ in the group $\mathrm{Alg}(H,k)$ is given either by $\mu^{-1} = \mu \circ S$ or by $\mu^{-1} = \mu \circ S^{-1}$, one can see that

$$\mu(\alpha\beta)\mu^{-1}(\alpha\beta) = \mu(X^1\beta S(X^2)\alpha X^3)\mu^{-1}(S(x^1)\alpha x^2 \beta S(x^3)) \overset{(3.2.2)}{=} 1. \tag{7.2.8}$$

**Examples 7.19** By using the non-zero integrals found in Examples 7.15 we get that for $H(2)$ and $H(32)$ the modular element is just the counit, while for $H_\pm(8)$ it is given by $\mu(1) = 1$, $\mu(g) = -1$ and $\mu(x) = 0$, extended to a $k$-algebra morphism.

Since a finite-dimensional quasi-Hopf algebra has bijective antipode we can make use of the elements $q_R$, $p_R$ defined in (3.2.19). If for $h^* \in H^*$ and $h \in H$ we define $h^* \rightarrow h := h^*(q^2h_2p^2)q^1h_1p^1$ then the bijective map $\theta$ defined in (7.2.4) takes the form $\theta(t \otimes h^*) = h^* \circ S \rightarrow t$.

**Proposition 7.20** *Any finite-dimensional quasi-Hopf algebra $H$ is a Frobenius algebra. A Frobenius system for $H$ is $(\phi, q^1t_1p^1 \otimes S(q^2t_2p^2))$, where $\phi \in H^*$ is the unique map that satisfies*

$$\phi(q^1t_1p^1)q^2t_2p^2 = 1_H \quad \text{or, equivalently,} \quad \phi(S(q^2t_2p^2))q^1t_1p^1 = 1_H, \tag{7.2.9}$$

*and where $t$ is a non-zero left integral in $H$. Furthermore, the Nakayama automorphism corresponding to this Frobenius system is given by*

$$\chi(h) = \mu(h_1)S^2(h_2), \quad \forall\, h \in H, \tag{7.2.10}$$

*where $\mu$ is the modular element of $H^*$.*

*Proof* Theorem 7.18 implies that the map

$$\xi : H^* \rightarrow H, \ \xi(h^*) = (h^* \circ S) \rightarrow t := h^*(S(q^2t_2p^2))q^1t_1p^1, \ \forall\, h^* \in H^*, \tag{7.2.11}$$

is bijective. Moreover, $\xi$ respects the natural left $H$-module structures of $H$ and $H^*$, and so $H$ and $H^*$ are isomorphic as left $H$-modules. Thus $H$ is a Frobenius algebra.

A right $H$-module isomorphism between $H^*$ and $H$ can be obtained from $\xi$ and the natural isomorphism $\theta_H^{-1} : H^{**} \rightarrow H$ as in the proof of $(iii) \Rightarrow (ii)$ of Proposition 7.1. Namely, there we have shown that

$$\xi' : H^* \xrightarrow{\ \xi^*\ } H^{**} \xrightarrow{\ \theta_H^{-1}\ } H$$

is a right $H$-module isomorphism. Explicitly,

$$\begin{aligned}\xi'(h^*) = \theta_H^{-1}(\xi^*(h^*)) &= \sum_i h^*(\xi(e^i))e_i = \sum_i \langle h^*, e^i \circ S \rightarrow t\rangle e_i \\ &= \sum_i \langle h^*, e^i(S(q^2t_2p^2))q^1t_1p^1\rangle e_i = h^*(q^1t_1p^1)S(q^2t_2p^2),\end{aligned}$$

for all $h^* \in H^*$. Now from Remark 7.3 we obtain that $(\phi, e)$ is a Frobenius system for $H$, where $\phi \in H^*$ is the unique element in $H^*$ satisfying $\xi'(\phi) = 1_H$ or, equivalently, $\xi(\phi) = 1_H$ and $e = \sum_i e_i \otimes \xi'(e^i)$. Since $S$ is bijective it follows that a Frobenius morphism for $H$ is the unique $\phi \in H^*$ that obeys one of the (equivalent) conditions in (7.2.9). Also, in this case the Frobenius element comes out as

$$e = \sum_i e^i(q^1t_1p^1)e_i \otimes S(q^2t_2p^2) = q^1t_1p^1 \otimes S(q^2t_2p^2),$$

as stated. Finally, to compute $\chi$ we use the formula in (7.1.1) to get

$$\begin{aligned}\chi(h) &= \phi(q^1t_1p^1h)S(q^2t_2p^2) \overset{(3.2.21)}{=} \phi(q^1(th_1)_1p^1)S(q^2(th_1)_2p^2S(h_2)) \\ &\overset{(7.2.6)}{=} \mu(h_1)\phi(q^1t_1p^1)S(q^2t_2p^2S(h_2)) \overset{(7.2.9)}{=} \mu(h_1)S^2(h_2),\end{aligned}$$

for all $h \in H$, as desired. □

We end this section by showing that for an infinite-dimensional quasi-Hopf algebra with bijective antipode the space of integrals is zero.

**Lemma 7.21** *Let $H$ be a quasi-Hopf algebra with bijective antipode and define*

$$\underline{\Delta}: H \to H \otimes H,\ \ \underline{\Delta}(h) = h_{\underline{1}} \otimes h_{\underline{2}} := q^1 h_1 p^1 \otimes q^2 h_2 p^2,\ \ \forall\, h \in H, \tag{7.2.12}$$

*where $p_R = p^1 \otimes p^2$ and $q_R = q^1 \otimes q^2$ are defined by* (3.2.19). *If $J$ is a non-zero two-sided ideal of $H$ such that $\underline{\Delta}(J) \subseteq J \otimes H$, then $J = H$.*

*Proof* From (3.2.23), we easily deduce that

$$(1 \otimes S^{-1}(p^2))\underline{\Delta}(p^1 h q^1)(1 \otimes S(q^2)) = \Delta(h),\ \ \forall\, h \in H.$$

This implies $\Delta(J) \subseteq J \otimes H$, since $J$ is a two-sided ideal of $H$ and $\underline{\Delta}(J) \subseteq J \otimes H$. If $\varepsilon(J) = 0$, then for any $h \in H$ we have $h = \varepsilon(h_1)h_2 \in \varepsilon(J)H = 0$, so $J = 0$, a contradiction. Thus $\varepsilon(J) \neq 0$, and there exists $a \in J$ with $\varepsilon(a) = 1$. By using (3.2.1), we obtain $\beta = \varepsilon(a)\beta = a_1\beta S(a_2) \in JH \subseteq J$, so $\beta \in J$. By using (3.2.2) and the fact that $J$ is a two-sided ideal of $H$, we find $1_H = X^1\beta S(X^2)\alpha X^3 \in J$, and $J = H$. □

For $h \in H$ and $h^* \in H^*$, we have defined $h^* \to h = h^*(h_{\underline{2}})h_{\underline{1}}$. For a two-sided ideal $I$ of $H$, we let $H^* \to I$ be the subspace of $H$ generated by all the elements of the form $h^* \to a$, with $h^* \in H^*$ and $a \in I$.

**Lemma 7.22** *Let $H$ be a quasi-Hopf algebra with bijective antipode and $I$ a non-zero two-sided ideal of $H$. Then $J := H^* \to I = H$.*

*Proof* The statement follows from Lemma 7.21 if we can show that $J$ is a non-zero two-sided ideal of $H$ such that $\underline{\Delta}(J) \subseteq J \otimes H$.

Obviously $\varepsilon \to h = h$, therefore $I \subseteq J$. For all $h \in H$, $h^* \in H^*$, $a \in I$ we have

$$\begin{aligned}(h^* \to a)h &= h^*(q^2 a_2 p^2)q^1 a_1 p^1 h\\ &\overset{(3.2.21)}{=} h^*(q^2(ah_1)_2 p^2 S(h_2))q^1(ah_1)_1 p^1\\ &= (S(h_2) \rightharpoonup h^*) \to (ah_1),\end{aligned}$$

and so $J$ is a right ideal. $J$ is also a left ideal, since

$$\begin{aligned}h(h^* \to a) &= h^*(q^2 a_2 p^2)hq^1 a_1 p^1\\ &\overset{(3.2.21)}{=} h^*(S^{-1}(h_2)q^2(h_1a)_2 p^2)q^1(h_1a)_1 p^1\\ &= (h^* \leftharpoonup S^{-1}(h_2)) \to (h_1 a).\end{aligned}$$

Let $f = f^1 \otimes f^2$ be the Drinfeld element as in (3.2.15). By using (3.2.26), (3.1.7) and (3.2.25) one can show that

$$\begin{aligned}&h^* \to (g^* \to h)\\ &= [(g^1S(x^3) \rightharpoonup h^* \leftharpoonup S^{-1}(f^2X^3))(g^2S(x^2) \rightharpoonup g^* \leftharpoonup S^{-1}(f^1X^2))] \to (X^1ax^1),\end{aligned}$$

for all $h^*, g^* \in H^*$ and $h \in H$. $I$ is a two-sided ideal of $H$, so the above equality shows that $H^* \to J \subseteq J$. To prove that $\underline{\Delta}(J) \subseteq J \otimes H$, we proceed as follows. Take

$a \in J$, and write $\underline{\Delta}(a) = \sum_i a_i \otimes a'_i$, where $a_1, \ldots, a_m \in J$ and $a_{m+1}, \ldots, a_n$ are linearly independent modulo $J$. For any $h^* \in H^*$, $h^* \rightarrow a = \sum_i h^*(a'_i)a_i \in J$. The linear independence of $a_{m+1}, \ldots, a_n$ modulo $J$ implies that $h^*(a'_i) = 0$, and therefore $a'_i = 0$ ($h^*$ is arbitrary), for all $i > m$. We find that $\underline{\Delta}(a) \in J \otimes H$, as required. □

As we next see, Theorem 7.18 has a converse.

**Theorem 7.23** *If $H$ is a quasi-Hopf algebra with an antipode $S$, then $H$ is finite dimensional if and only if $S$ is bijective and $\int_l^H \neq 0$.*

*Proof* One implication follows from Theorem 7.18. Conversely, assume that $S$ is bijective and $I = \int_l^H \neq 0$. Then $I$ is a non-zero two-sided ideal of $H$ and Lemma 7.22 tells us that $H^* \rightarrow I = H$. Thus there exist $\{h_i^*\}_{i=\overline{1,n}} \subseteq H^*$ and $\{t_i\}_{i=\overline{1,n}} \subseteq \int_l^H$ such that $1_H = \sum_{i=1}^n h_i^* \rightarrow t_i$. For any $i$ we have $\underline{\Delta}(t_i) = \sum_{j=1}^{n_i} a^i_j \otimes b^i_j$, for some $\{a^i_j\}_{j=\overline{1,n_i}} \subseteq H$ and $\{b^i_j\}_{j=\overline{1,n_i}} \subseteq H$. Therefore, for any $h^* \in H^*$ and $i = \overline{1,n}$ we have $h^* \rightarrow t_i = \sum_{j=1}^{n_i} h^*(b^i_j)a^i_j$. For all $h \in H$ we obtain that

$$
\begin{aligned}
h &= \sum_{i=1}^{n} h(h_i^* \rightarrow t_i) \\
&= \sum_{i=1}^{n} (h^* \leftharpoonup S^{-1}(h_2)) \rightarrow h_1 t_i \\
&= \sum_{i=1}^{n} (h^* \leftharpoonup S^{-1}(h)) \rightarrow t_i \quad (\text{since } t_i \in \textstyle\int_l^H,\ \forall\, i = \overline{1,n}) \\
&= \sum_{i=1}^{n} \sum_{j=1}^{n_i} h^*(S^{-1}(h) b^i_j) a^i_j.
\end{aligned}
$$

Thus $H$ is a subspace of the span of $\{a^i_j \mid i = \overline{1,n}, j = \overline{1,n_i}\}$, and therefore it is finite dimensional. □

**Corollary 7.24** *If $H$ is an infinite-dimensional quasi-Hopf algebra with bijective antipode then $\int_l^H = \int_r^H = 0$.*

*Proof* The bijectivity of $S$ implies $S(\int_l^H) = \int_r^H$, and so $\dim_k \int_l^H = \dim_k \int_r^H$. By Theorem 7.23, if $H$ is infinite dimensional then $\int_l^H = \int_r^H = 0$. □

## 7.3 Semisimple Quasi-Hopf Algebras

We say that a quasi-Hopf algebra $H$ is semisimple if it is semisimple as an algebra. Similarly, we say that $H$ is separable if it is separable as an algebra. One of the characterizations of a separable algebra $A$ is the following: there exists an element $e = e^1 \otimes e^2 \in A \otimes A$, called a separability element, satisfying $e^1 e^2 = 1_A$ and $ae^1 \otimes e^2 = e^1 \otimes e^2 a$, for all $a \in A$, where we again have suppressed summation and indices. The well-known example of separable algebra is given by the matrix algebra: for any $i \in \{1,\dots,n\}$ the element $e_i = \sum_{j=1}^{n} E_{ji} \otimes E_{ij}$ is a separability element for $M_n(k)$.

The result below measures how far a Frobenius algebra is from being separable.

**Proposition 7.25** *Let A be a Frobenius algebra with Frobenius system $(\phi, e^1 \otimes e^2)$. Then A is separable if and only if there exists $a \in A$ such that $e^1 a e^2 = 1_A$.*

*Proof* If there exists $a \in A$ such that $e^1 a e^2 = 1_A$ it is clear that either $e^1 a \otimes e^2$ or $e^1 \otimes a e^2$ is a separability element for $A$.

Conversely, if $A$ is separable we consider $f^1 \otimes f^2 \in A \otimes A$ a separability element for it and define $a = \phi(f^2) f^1$. We have

$$e^1 a \otimes e^2 = \phi(f^2) e^1 f^1 \otimes e^2 = \phi(f^2 e^1) f^1 \otimes e^2 = \phi(e^1) f^1 \otimes e^2 f^2 = f^1 \otimes f^2,$$

where we used that both $e^1 \otimes e^2$ and $f^1 \otimes f^2$ satisfy the property of a Frobenius element. We then get $e^1 a e^2 = f^1 f^2 = 1_A$, as needed.

Note that, if we define $a' = \phi(f^1) f^2$, computations similar to those above prove that $e^1 \otimes a' e^2 = f^1 \otimes f^2$, and so $e^1 a' e^2 = 1_A$ as well. □

By using integral theory we show that for finite-dimensional quasi-Hopf algebras the concepts of separable and semisimple are equivalent. Actually, we characterize both of them in terms of normalized integrals, providing in this way a Maschke type theorem for quasi-Hopf algebras.

Let us start with the following general result.

**Proposition 7.26** *Any separable algebra A over a field k is semisimple.*

*Proof* Let $e = e^1 \otimes e^2 \in A \otimes A$ be a separability element for $A$. Take $M$ a left $A$-module and $N$ an $A$-submodule of it. Since $k$ is a field there exists a $k$-linear map $f : M \to N$ that covers the natural inclusion $i : N \to M$, that is, $f(n) = n$, for all $n \in N$. If we define

$$\tilde{f} : M \to N, \quad \tilde{f}(m) = e^1 \cdot f(e^2 \cdot m), \quad \forall\, m \in M,$$

then by $ae^1 \otimes e^2 = e^1 \otimes e^2 a$, for all $a \in A$, it follows that $\tilde{f}$ is left $A$-linear. Also, the fact that $e^1 e^2 = 1_A$ implies $\tilde{f}(n) = n$, for all $n \in N$. Hence $N$ is an $A$-direct summand of $M$ and this proves that $M$ is completely reducible. Since $M$ was arbitrary we obtain that $A$ is semisimple, as desired. □

**Definition 7.27** We call a left (right) integral $t$ in $H$ normalized if $\varepsilon(t) = 1$. We call a Haar integral in $H$ a normalized left integral which is at the same time a right integral.

The equivalence between $(i)$ and $(ii)$ below is known as the Maschke theorem for quasi-Hopf algebras.

**Theorem 7.28** *For a finite-dimensional quasi-Hopf algebra H the following assertions are equivalent:*

*(i) $H$ is semisimple;*
*(ii) $H$ has a normalized left or right integral;*
*(iii) $k$ is a projective left or right $H$-module via the $H$-action defined by $\varepsilon$;*
*(iv) $H$ has a Haar integral;*
*(v) $H$ is a separable algebra.*

*Proof* $(i) \Rightarrow (ii)$. Since $H$ is semisimple there exists a left ideal $I$ in $H$ such that $H = I \oplus \mathrm{Ker}(\varepsilon)$, where $\mathrm{Ker}(\varepsilon)$ is the kernel of $\varepsilon \in H^*$. For $x \in \mathrm{Ker}(\varepsilon)$ and $y \in I$, we have $xy \in \mathrm{Ker}(\varepsilon) \cap I$, hence $xy = 0 = \varepsilon(x)y$. Then, for any $h \in H$, $h = (h - \varepsilon(h)1_H) + \varepsilon(h)1_H$, and since $(h - \varepsilon(h)1_H) \in \mathrm{Ker}(\varepsilon)$ we get that $hy = \varepsilon(h)y$. So we have proved that $I \subseteq \int_l^H$. As $\mathrm{Ker}(\varepsilon) \cap I = 0$ it follows that $\varepsilon(\int_l^H) \neq 0$, for otherwise $H = \mathrm{Ker}(\varepsilon)$, a contradiction. This guarantees the existence of a left integral $t$ in $H$ such that $\varepsilon(t) = 1$, as needed.

$(ii) \Rightarrow (iii)$. Let $t \in H$ be a left (resp. right) normalized integral and define $f : k \to H$ by $f(\kappa) = \kappa t$, for all $\kappa \in k$. Since $k$ is viewed as a left (resp. right) $H$-module via $\varepsilon$ it follows that $f$ is a left (resp. right) $H$-module morphism obeying $\varepsilon \circ f = \mathrm{Id}_k$. Thus $k$ is isomorphic to a direct summand of the free $H$-module $H$, and therefore it is a projective left (resp. right) $H$-module.

$(iii) \Rightarrow (iv)$. Assume $k$ is a projective left (resp. right) $H$-module. Clearly $\varepsilon$ is left (resp. right) $H$-linear and surjective and from here we get that there exists a left (resp. right) $H$-linear morphism $\vartheta : k \to H$ such that $\varepsilon \circ \vartheta = \mathrm{Id}_k$. Then $t := \vartheta(1_k)$ is a left (resp. right) integral since, for instance,

$$ht = h\vartheta(1_k) = \vartheta(h \cdot 1_k) = \varepsilon(h)\vartheta(1_k) = \varepsilon(h)t,$$

for all $h \in H$. Moreover, $\varepsilon(t) = \varepsilon(\vartheta(1_k)) = 1_k$, hence $t$ is a normalized left (resp. right) integral in $H$. If $\mu$ is the modular element of $H^*$ then by applying $\varepsilon$ to the both sides of (7.2.6) (resp. (7.2.7)) we obtain that $\mu = \varepsilon$ (resp. $\mu \circ S = \varepsilon$). But $S$ is bijective and $\varepsilon \circ S = \varepsilon$, so in both cases we obtain that $\mu = \varepsilon$. This shows that $t$ is a Haar integral in $H$, so the implication is proved.

$(iv) \Rightarrow (v)$. If $t$ is a normalized left integral in $H$ then both

$$e_1 = q^1 t_1 \beta \otimes S(q^2 t_2) \quad \text{and} \quad e_2 = \tilde{q}^1 t_1 \beta \otimes S(\tilde{q}^2 t_2)$$

are separability elements for $H$, where $q_R = q^1 \otimes q^2$ and $q_L = \tilde{q}^1 \otimes \tilde{q}^2$ are the elements defined in (3.2.19) and (3.2.20), respectively. For instance, $e_1$ is a separability element because of (7.2.2) and since

$$q^1 t_1 \beta S(q^2 t_2) \overset{(3.2.1)}{=} \varepsilon(t) q^1 \beta S(q^2) = X^1 \beta S(X^2) \alpha X^3 \overset{(3.2.2)}{=} 1_H.$$

Likewise, if $r$ is a normalized right integral in $H$ then either $e_1' = S(r_1 p^1) \otimes \alpha r_2 p^2$ or $e_2' = S(r_1 \tilde{p}^1) \otimes \alpha r_2 \tilde{p}^2$ is a separability element for $H$, where $p_R = p^1 \otimes p^2$ and $p_L = \tilde{p}^1 \otimes \tilde{p}^2$ are the elements defined in (3.2.19) and (3.2.20), respectively. We leave the verification of all these details to the reader.

$(v) \Rightarrow (i)$. This follows from Proposition 7.26. □

**Remark 7.29** If $H$ is a semisimple quasi-Hopf algebra with bijective antipode then it is finite dimensional. Indeed, $\mathrm{Ker}(\varepsilon)$ is an ideal of $H$, and so there exists a left ideal $I$ of $H$ such that $H = I \oplus \mathrm{Ker}(\varepsilon)$. Then as in the proof of $(i) \Rightarrow (ii)$ above one can see that $I \subseteq \int_l^H$, and since $\mathrm{Ker}(\varepsilon)$ has codimension 1 in $H$ it follows that $\int_l^H \neq 0$. Combined with $S$ bijective this implies that $H$ is finite dimensional; see Theorem 7.23.

**Remark 7.30** Obviously, by (iv) in Theorem 7.28, a finite-dimensional semisimple quasi-Hopf algebra is unimodular.

**Examples 7.31** (1) The quasi-Hopf algebra $H(2)$ from Example 3.26 is semisimple since $l = \frac{1}{2}(1+g)$ is a Haar integral for it. In this case the four separability elements considered in the proof of $(iv) \Rightarrow (v)$ above are all equal to $\frac{1}{2}(1\otimes 1 + g\otimes g)$. To see this write $\Phi = 1 - 2p_- \otimes p_- \otimes p_-$ under the form

$$\Phi = \frac{3}{4}1\otimes 1\otimes 1 + \frac{1}{4}(1\otimes 1\otimes g + 1\otimes g\otimes 1 + g\otimes 1\otimes 1)$$
$$-\frac{1}{4}(1\otimes g\otimes g + g\otimes 1\otimes g + g\otimes g\otimes 1) + \frac{1}{4}g\otimes g\otimes g,$$

to compute that

$$q_R = X^1 \otimes S^{-1}(\alpha X^3)X^2 = X^1 \otimes X^2 g X^3 = \frac{1}{2}(1\otimes g + g\otimes g + 1\otimes 1 - g\otimes 1)$$

and

$$p_R = x^1 \otimes x^2 \beta S(x^3) = x^1 \otimes x^2 x^3 = \frac{1}{2}(1\otimes 1 + 1\otimes g + g\otimes 1 - g\otimes g),$$

since, as can be easily checked, $\Phi^{-1} = \Phi$. Therefore

$$e_1 = q_R \Delta(l) = \frac{1}{2} q_R (1\otimes 1 + g\otimes g) = \frac{1}{2}(1\otimes 1 + g\otimes g),$$

and, likewise,

$$e_1' = (1\otimes g)\Delta(l) p_R = \frac{1}{2}(1\otimes g + g\otimes 1) p_R = \frac{1}{2}(1\otimes 1 + g\otimes g).$$

In a similar manner we compute that $e_2 = e_2' = \frac{1}{2}(1\otimes 1 + g\otimes g)$; the details are left to the reader.

(2) The quasi-Hopf algebras presented in Example 3.30 are not semisimple since, as we have seen, they are not unimodular.

(3) The quasi-Hopf algebra $H(32)$ from Example 3.32 is unimodular but not semisimple because, according to Example 7.15(3), $t = (1+g)x^3y^3$ is a left and right non-zero integral in $H(32)$ but $\varepsilon(t) = 0$.

We next show that we always have $e_1 = e_2$ and $e_1' = e_2'$, and so for a semisimple quasi-Hopf algebra we can only construct one separability element from a fixed left (or right) normalized integral of it. To this end we need several formulas that will be intensively used from now on.

**Lemma 7.32** *For a quasi-Hopf algebra $H$ with bijective antipode define the elements $U = U^1 \otimes U^2$ and $V = V^1 \otimes V^2$ in $H \otimes H$ by*

$$U := g^1 S(q^2) \otimes g^2 S(q^1) \quad \text{and} \quad V := S^{-1}(f^2 p^2) \otimes S^{-1}(f^1 p^1), \tag{7.3.1}$$

*where $p_R = p^1 \otimes p^2$ and $q_R = q^1 \otimes q^2$ are the elements defined in* (3.2.19), *and $f = f^1 \otimes f^2$ is the Drinfeld twist considered in* (3.2.15) *with its inverse $f^{-1} = g^1 \otimes g^2$ as in* (3.2.16). *Then the following relations hold:*

$$U[1_H \otimes S(h)] = \Delta(S(h_1))U[h_2 \otimes 1_H], \ \ \forall\, h \in H, \tag{7.3.2}$$
$$[1_H \otimes S^{-1}(h)]V = [h_2 \otimes 1_H]V\Delta(S^{-1}(h_1)), \ \ \forall\, h \in H, \tag{7.3.3}$$
$$q_R = [\tilde{q}^2 \otimes 1_H]V\Delta(S^{-1}(\tilde{q}^1)), \tag{7.3.4}$$
$$p_R = \Delta(S(\tilde{p}^1))U[\tilde{p}^2 \otimes 1_H], \tag{7.3.5}$$

*where $q_L = \tilde{q}^1 \otimes \tilde{q}^2$ and $p_L = \tilde{p}^1 \otimes \tilde{p}^2$ are the elements in $H \otimes H$ defined in* (3.2.20).

*Proof* We prove only (7.3.2) and (7.3.4); the other two can be obtained from these if we think of them in $H^{\rm op}$ instead of $H$. Note only that when we pass from $H$ to $H^{\rm op}$ the roles of $U$ and $V$, as well as of $q_R$ and $p_R$ and of $q_L$ and $p_L$, interchange.

To prove (7.3.2) we compute:

$$\begin{aligned}
\Delta(S(h_1))U[h_2 \otimes 1_H] &= S(h_1)_1 g^1 S(q^2) h_2 \otimes S(h_1)_2 g^2 S(q^1) \\
&\overset{(3.2.13)}{=} g^1 S(q^2 h_{(1,2)}) h_2 \otimes g^2 S(q^1 h_{(1,1)}) \\
&\overset{(3.2.21)}{=} g^1 S(q^2) \otimes g^2 S(hq^1) = U[1_H \otimes S(h)],
\end{aligned}$$

for all $h \in H$, as required. (7.3.4) follows from:

$$\begin{aligned}
&[\tilde{q}^2 \otimes 1_H]V\Delta(S^{-1}(\tilde{q}^1)) \\
&\quad = \tilde{q}^2 S^{-1}(f^2 p^2) S^{-1}(\tilde{q}^1)_1 \otimes S^{-1}(f^1 p^1) S^{-1}(\tilde{q}^1)_2 \\
&\quad \overset{(3.2.13)}{=} \tilde{q}^2 S^{-1}(f^2 \tilde{q}^1_2 p^2) \otimes S^{-1}(f^1 \tilde{q}^1_1 p^1) \\
&\quad \overset{(3.2.20),(3.2.13)}{=} x^3 S^{-1}(S(x^1_1) f^2 \alpha_2 x^2_2 p^2) \otimes S^{-1}(S(x^1_2) f^1 \alpha_1 x^2_1 p^1) \\
&\quad \overset{(3.2.14)}{=} x^3 S^{-1}(S(x^1_1) \gamma^2 x^2_2 p^2) \otimes S^{-1}(S(x^1_2) \gamma^1 x^2_1 p^1) \\
&\quad \overset{(3.2.5)}{=} S^{-1}(S(X^1 y^1_1 x^1_1) \alpha y^3 x^2_2 p^2 S(x^3)) \otimes S^{-1}(S(X^2 y^1_2 x^1_2) \alpha X^3 y^2 x^2_1 p^1) \\
&\quad \overset{(3.2.19),(3.1.9)}{=} S^{-1}(S(X^1 y^1_{(1,1)} x^1_1) \alpha y^2 x^3_1 \beta S(y^3 x^3_2)) \otimes S^{-1}(S(X^2 y^1_{(1,2)} x^1_2) \alpha X^3 y^1_2 x^2) \\
&\quad \overset{(3.1.7),(3.2.1)}{=} S^{-1}(S(y^1 X^1) \alpha y^2 \beta S(y^3)) \otimes S^{-1}(S(X^2) \alpha X^3) \\
&\quad \overset{(3.2.2)}{=} X^1 \otimes S^{-1}(\alpha X^3) X^2 \overset{(3.2.19)}{=} q_R.
\end{aligned}$$

So the proof is complete. □

**Proposition 7.33** *Let $H$ be a finite-dimensional quasi-Hopf algebra, so its antipode $S$ is bijective. If $p_R, q_R$ and $p_L, q_L$ are as in* (3.2.19) *and* (3.2.20), *respectively, then*

$$q^1 t_1 \otimes q^2 t_2 = \tilde{q}^1 t_1 \otimes \tilde{q}^2 t_2 \quad \text{and} \quad r_1 p^1 \otimes r_2 p^2 = r_1 \tilde{p}^1 \otimes r_2 \tilde{p}^2, \tag{7.3.6}$$

*for any left integral $t$ and right integral $r$ in $H$. Consequently, when $H$ is semisimple* *we have $e_1 = e_2$ and $e_1' = e_2'$, where $e_1$, $e_2$ and $e_1'$, $e_2'$ are the separability elements defined in the proof of Theorem 7.28.*

*Proof* We only prove the first relation, the second one is the first one regarded in $H^{\rm op}$ instead of $H$.

The fact that $t \in \int_l^H$ and (7.3.4) imply

$$q^1 t_1 \otimes q^2 t_2 = V^1 t_1 \otimes V^2 t_2. \tag{7.3.7}$$

We use now the quasi-Hopf algebra structure of $H^{\rm cop}$ to see that in $H^{\rm cop}$ we have $(q_R)_{\rm cop} = \tilde{q}^2 \otimes \tilde{q}^1$, $(p_R)_{\rm cop} = \tilde{p}^2 \otimes \tilde{p}^1$ and $f_{\rm cop} = (S^{-1} \otimes S^{-1})(f)$, and so $V_{\rm cop} = S(\tilde{p}^1) f^2 \otimes S(\tilde{p}^2) f^1$. Thus, if we consider (7.3.7) in $H^{\rm cop}$ instead of $H$, we obtain

$$\tilde{q}^1 t_1 \otimes \tilde{q}^2 t_2 = S(\tilde{p}^2) f^1 t_1 \otimes S(\tilde{p}^1) f^2 t_2.$$

On the other hand, one can see that (3.2.20), (3.1.5), (3.2.17) and $S^{-1}(f^2)\beta f^1 = S^{-1}(\alpha)$, which has been proved before, imply

$$S(\tilde{p}^2) f^1 \otimes S(\tilde{p}^1) f^2 = q^1 g_1^1 \otimes S^{-1}(g^2) q^2 g_2^1, \tag{7.3.8}$$

where, as usual, we denote $f^{-1} = g^1 \otimes g^2$. From the above, we conclude that

$$\tilde{q}^1 t_1 \otimes \tilde{q}^2 t_2 = q^1 g_1^1 t_1 \otimes S^{-1}(g^2) q^2 g_2^1 t_2 \overset{t \in \int_l^H}{=} q^1 t_1 \otimes q^2 t_2.$$

The consequence now follows directly from the definitions of $e_1, e_2, e_1'$ and $e_2'$. □

## 7.4 Symmetric Quasi-Hopf Algebras

In this section we characterize symmetric quasi-Hopf algebras by making use of the integral theory that we have developed so far. To this end we first check when a Frobenius algebra is symmetric and we develop an integral theory for augmented Frobenius algebras. These allow us to provide elegant proofs when we come back to the quasi-Hopf algebra setting.

**Definition 7.34** A symmetric algebra is a finite-dimensional $k$-algebra $A$ that is isomorphic to $A^*$ as an $A$-bimodule.

If follows that any symmetric algebra is a Frobenius algebra, and that the two notions coincide in the commutative case. To give some non-trivial examples we need first the following list of characterizations. The last two say how far a Frobenius algebra is from being symmetric.

**Proposition 7.35** *Let $A$ be a finite-dimensional $k$-algebra. Then the following are equivalent:*

*(i) $A$ is a symmetric algebra;*

*(ii) there exists a bilinear map $B : A \times A \to k$ which is non-degenerate (i.e. $B$ is at the same time left and right non-degenerate), associative and symmetric (i.e. $B(a,b) = B(b,a)$, for all $a, b \in A$);*

*(iii) there exists a k-linear map $\phi : A \to k$ that is a trace, that is, $\phi(ab) = \phi(ba)$, for all $a, b \in A$, and such that* $\mathrm{Ker}(\phi)$ *does not contain non-zero left or right ideals;*

*(iv) there exists a Frobenius system for A such that $\phi$ is a trace and e is symmetric, in the sense that $e^1 \otimes e^2 = e^2 \otimes e^1$ in $A \otimes A$;*

*(v) A is a Frobenius algebra for which the Nakayama automorphism is inner.*

*Proof* $(i) \Rightarrow (ii)$. Let $f : A \to A^*$ be an isomorphism of $A$-bimodules. Since $f$ is right $A$-linear, as in the proof of $(ii) \Rightarrow (v)$ in Proposition 7.1 we get that $B : A \times A \to k$ given by $B(a,b) = f(a)(b)$, for all $a, b \in A$, is associative and right non-degenerate. Using now the left $A$-linearity of $f$, which comes out as $f(ab) = a \rightharpoonup f(b)$, for all $a, b \in A$, we compute that

$$B(ab,c) = f(ab)(c) = (a \rightharpoonup f(b))(c) = f(b)(ca) = B(b,ca), \ \ \forall\, a,b,c \in A.$$

By taking $c = 1_A$ and using the associativity of $B$ we obtain that $B$ is symmetric, and so left-non-degenerate as well.

$(ii) \Rightarrow (i)$. Once we have $B$ as in the statement define $f : A \to A^*$ by $f(a)(b) = B(a,b)$, for all $a, b \in A$. The proof of $(v) \Rightarrow (ii)$ in Proposition 7.1 ensures that $f$ is an isomorphism of right $A$-modules. It is also left $A$-linear since

$$f(ba)(c) = B(ba,c) = B(c,ba) = B(cb,a) = B(a,cb) = f(a)(cb) = (b \rightharpoonup f(a))(c),$$

for all $a, b, c \in A$, as needed.

$(ii) \Rightarrow (iii)$. For $B$ as in (ii) define $\phi \in A^*$ by $\phi(a) = B(1_A, a) = B(a, 1_A)$, for all $a \in A$. We have

$$\phi(ab) = B(ab, 1_A) = B(a,b) = B(b,a) = B(1_A, ba) = \phi(ba),$$

for all $a, b \in A$, so $\phi$ is a trace.

Now let $I \subseteq \mathrm{Ker}(\phi)$ be a right ideal. Then $\phi(I) = 0$, so $B(I, 1_A) = 0$. As $I = IA$ we obtain $B(I, A) = 0$, and therefore $I = 0$, because $B$ is non-degenerate. Similarly, if $I \subseteq \mathrm{Ker}(\phi)$ is a left ideal then $I = 0$.

$(iii) \Rightarrow (ii)$. If $\phi$ is as in (iii) we define $B : A \times A \to k$ by $B(a,b) = \phi(ab)$, for all $a, b \in A$. Then

$$B(ab,c) = \phi((ab)c) = \phi(a(bc)) = B(a,bc),$$

and this shows that $B$ is associative. It is also symmetric since $\phi$ is a trace. Moreover, if $a \in A$ is such that $B(a, A) = 0$ then $\phi(aA) = 0$, hence $a = 0$ because $aA$ is a right ideal in $A$. Thus $B$ is non-degenerate, too.

$(i) \Rightarrow (iv)$. Let $f : A \to A^*$ be an isomorphism of $A$-bimodules, and $\{a_i\}_i$ a basis in $A$ with corresponding dual basis $\{a^i\}_i$ in $A^*$. From the proof of $(ii) \Rightarrow (i)$ in Proposition 7.1 we know that $(\phi := f(1_A), e = \sum_i a_i \otimes f^{-1}(a^i))$ is a Frobenius system for $A$. The fact that $f$ is left and right $A$-linear implies

$$a \rightharpoonup \phi = a \rightharpoonup f(1_A) = f(a) = f(1_A) \leftharpoonup a = \phi \leftharpoonup a, \ \ \forall\, a \in A,$$

and so $\phi(ab) = \phi(ba)$, for all $a, b \in A$, which means that $\phi$ is a trace. Furthermore, if $\mathfrak{e}^1 \otimes \mathfrak{e}^2$ is another copy of $e$ then, for all $a \in A$,

$$\begin{aligned} ae^1 \otimes e^2 &= \phi(\mathfrak{e}^1)\mathfrak{e}^2 ae^1 \otimes e^2 \\ &= \phi(ae^1\mathfrak{e}^1)\mathfrak{e}^2 \otimes e^2 \\ &= \phi(\mathfrak{e}^1 ae^1)\mathfrak{e}^2 \otimes e^2 \\ &= \phi(e^1)\mathfrak{e}^2 \otimes e^2\mathfrak{e}^1 a \\ &= \mathfrak{e}^2 \otimes \phi(e^1)e^2\mathfrak{e}^1 a = \mathfrak{e}^2 \otimes \mathfrak{e}^1 a, \end{aligned}$$

where we used that $\phi$ is a trace. So $e$ is a symmetric element.

$(iv) \Rightarrow (i)$. From the proof of $(i) \Rightarrow (ii)$ in Proposition 7.1 we know that $f : A \to A^*$ given by $f(a) = \phi \leftharpoonup a$, for all $a \in A$, is a right $A$-module isomorphism. The fact that $\phi$ is a trace is equivalent to $a \rightharpoonup \phi = \phi \leftharpoonup a$, for all $a \in A$, and this allows us to show that $f$ is also left $A$-linear, because for all $a, b \in A$ we have

$$f(ba) = \phi \leftharpoonup ba = (\phi \leftharpoonup b) \leftharpoonup a = (b \rightharpoonup \phi) \leftharpoonup a = b \rightharpoonup (\phi \leftharpoonup a) = b \rightharpoonup f(a).$$

$(i) \Rightarrow (v)$. Assume again that $f : A \to A^*$ is an $A$-bimodule isomorphism, so in particular $A$ is a Frobenius algebra. Then we have seen that $A$ admits a Frobenius system $(\phi, e)$ with $e$ symmetric. If $\chi$ is its corresponding Nakayama automorphism then by (7.1.1) we deduce that $\chi(a) = \phi(e^1 a)e^2 = \phi(e^2 a)e^1 = \phi(e^2)ae^1 = a$, for all $a \in A$, thus $\chi$ is inner.

$(v) \Rightarrow (i)$. Let $(\phi, e)$ be a Frobenius system such that the Nakayama automorphism defined by it is inner via, say, $u \in A$. In other words, $u$ is invertible in $A$ and $\chi(a) = \phi(e^1 a)e^2 = uau^{-1}$, for all $a \in A$.

Now define $g : A \to A^*$ by $g(a) = \phi \leftharpoonup ua$, for all $a \in A$. Since $(\phi \leftharpoonup u, e^1 \otimes u^{-1}e^2)$ is another Frobenius system for $A$ (see Proposition 7.9) it follows that $g$ is a right $A$-module isomorphism. In addition,

$$\begin{aligned} a \rightharpoonup g(b) &= a \rightharpoonup (\phi \leftharpoonup ub) \\ &= (a \rightharpoonup \phi) \leftharpoonup ub \\ &= (\phi \leftharpoonup \chi(a)) \leftharpoonup ub \\ &= \phi \leftharpoonup \chi(a)ub \\ &= \phi \leftharpoonup uab = g(ab), \end{aligned}$$

for all $a, b \in A$, and so $g$ is an isomorphism of $A$-bimodules. □

**Remark 7.36** If $A$ is a Frobenius algebra with Frobenius system $(\phi, e)$, then $\phi$ is a trace if and only if $e$ is a symmetric element.

Indeed, if $\phi$ is a trace then as in the proof of $(i) \Rightarrow (iv)$ above we obtain that $e$ is symmetric. Conversely, if $e$ is symmetric then $\phi$ is a trace since

$$\phi(ab) = \phi(e^1)\phi(ae^2 b) = \phi(be^1)\phi(ae^2) = \phi(be^2)\phi(ae^1) = \phi(be^2 a)\phi(e^1) = \phi(ba),$$

for all $a, b \in A$, as required.

We use the above characterizations to provide examples of symmetric algebras.

**Examples 7.37** (1) If $A$ and $A'$ are symmetric algebras then so are $A\times A'$ and $A\otimes A'$. More precisely, if $B$ and $B'$ are non-degenerate, associative, symmetric bilinear maps on $A$ and $A'$, respectively, then $\mathbf{B}$ and $\widetilde{\mathbf{B}}$ defined in exactly the same way as $\mathbf{B}_l$ and $\widetilde{\mathbf{B}}_l$ were defined in Example 7.4(1), are non-degenerate, associative, symmetric bilinear maps on $A\times A'$ and $A\otimes A'$, respectively.

(2) For any finite group $G$ the group algebra $k[G]$ is symmetric. To see this, consider $\phi(\sum_g \alpha_g g)=\alpha_e$, for all $\alpha=\sum_g \alpha_g g\in k[G]$, where $e$ is the neutral element of $G$. Then $\phi$ is a trace since, for any other element $\beta=\sum_g \beta_g g\in k[G]$, we have

$$\phi(\alpha\beta)=\sum_g \alpha_g\beta_{g^{-1}}=\sum_h \beta_h\alpha_{h^{-1}}=\phi(\beta\alpha).$$

Moreover, if $I\subseteq \mathrm{Ker}(\phi)$ is a non-zero right ideal take $0\neq\alpha=\sum_g \alpha_g g\in I$, so there exists $g\in G$ such that $\alpha_g\neq 0$. We have

$$0\neq \alpha g^{-1}=\alpha_g e+\sum_{h\in G\setminus\{g\}}\alpha_h hg^{-1}\in I\subseteq \mathrm{Ker}(\phi),$$

thus $\phi(\alpha g^{-1})=0$. But the definition of $\phi$ says that $\phi(\alpha g^{-1})=\alpha_g\neq 0$, a contradiction. Hence $I=0$. In a similar manner one can show that $\mathrm{Ker}(\phi)$ does not contain non-zero left ideals; we leave the details to the reader.

(3) For any field $k$ and non-zero natural number $n$ the algebra $M_n(k)$ is symmetric. The trace in this case is given by the usual trace of a matrix, that is, $\phi(A)=\mathrm{Tr}(A):=\sum_i a_{ii}$, for any $A=(a_{ij})_{1\leq i,j\leq n}\in M_n(k)$. It is well-known that $\mathrm{Tr}(AB)=\mathrm{Tr}(BA)$, for any $A,B\in M_n(k)$, and this justifies, in general, the definition of a trace.

If $I\subseteq \mathrm{Ker}(\phi)$ is a right ideal then $\mathrm{Tr}(AE_{ij})=0$, for all $A\in I$ and $1\leq i,j\leq n$, where $\{E_{ij}\mid 1\leq i,j\leq n\}$ is the canonical basis of $M_n(k)$. As $AE_{ij}=\sum_u a_{ui}E_{uj}$ we get that $0=\mathrm{Tr}(AE_{ij})=a_{ji}$, for all $1\leq i,j\leq n$, hence $A=0$. Similarly, one can shown that $\mathrm{Ker}(\phi)$ does not contain non-zero left ideals, therefore $M_n(k)$ is a symmetric algebra.

Our next aim is to find a necessary and sufficient condition for a finite-dimensional quasi-Hopf algebra $H$ to be symmetric as an algebra. If $H$ is symmetric then we will show that $H$ is unimodular. Due to the "Frobenius" flavor of the proof, we develop first an integral theory for augmented Frobenius algebras.

**Definition 7.38** We say that a $k$-algebra is an augmented algebra if there exists an algebra morphism $\varepsilon: A\to k$, called an augmentation morphism.

If $A$ is an augmented algebra with augmentation morphism $\varepsilon$ then a left (resp. right) integral in $A$ is an element $t$ (resp. $r$) in $A$ satisfying $at=\varepsilon(a)t$ (resp. $ra=\varepsilon(a)r$), for all $a\in A$. By $\int_l^A$ (resp. $\int_r^A$) we denote the space of left (resp. right) integrals in $A$, and call $A$ unimodular if $\int_l^A\cap\int_r^A\neq 0$.

**Proposition 7.39** *If $A$ is a Frobenius augmented algebra then*

$$\dim_k \int_l^A = \dim_k \int_r^A = 1.$$

*Proof* Let $(\phi, e)$ be a Frobenius system for $A$ and $f : A \to A^*$ the right $A$-module isomorphism induced by it, that is, $f(a) = \phi \leftharpoonup a$, for all $a \in A$.

If $\varepsilon \in A^*$ is the augmentation morphism of $A$, there exists a unique element $r \in A$ such that $\phi \leftharpoonup r = \varepsilon$. As $f(r) = \varepsilon \neq 0$ we get $r \neq 0$. Furthermore, for all $a \in A$,

$$f(ra) = \phi \leftharpoonup ra = (\phi \leftharpoonup r) \leftharpoonup a = \varepsilon \leftharpoonup a = \varepsilon(a)\varepsilon = \phi \leftharpoonup \varepsilon(a)r = f(\varepsilon(a)r),$$

and by using the fact that $f$ is injective we conclude that $0 \neq r \in \int_r^A$. To prove the uniqueness of right integrals in $A$ we take $r'$ an arbitrary right integral in $A$ and compute:

$$(\phi \leftharpoonup r')(a) = \phi(r'a) = \phi(r')\varepsilon(a) = (\phi \leftharpoonup \phi(r')r)(a), \ \forall\, a \in A.$$

Thus $f(r') = f(\phi(r')r)$, and so $r' = \phi(r')r$. We have proved that $\int_r^A = kr$ with $r \neq 0$, therefore $\dim_k \int_r^A = 1$.

To see that $\dim_k \int_l^A = 1$ we have to construct from $(\phi, e)$ a left $A$-module isomorphism $f' : A \to A^*$. As we have pointed out several times this can be done by taking $f' : A \xrightarrow{\theta_A} A^{**} \xrightarrow{f^*} A^*$, where $\theta_A$ is the canonical isomorphism. It comes out that $f'(a) = a \rightharpoonup \phi$, for all $a \in A$, with inverse $f'^{-1}(a^*) = a^*(e^2)e^1$, for all $a^* \in A^*$. Now we have to repeat the above arguments for $f'$ instead of $f$ to conclude that $\dim_k \int_l^A = 1$, as required. □

As in the quasi-Hopf algebra case, one can show that for any Frobenius augmented algebra $A$ the spaces $\int_l^A$ and $\int_r^A$ are ideals in $A$. So, by the uniqueness of integrals in $A$ it follows that there exists $\mu' \in A^*$ such that $ar = \mu'(a)r$, for all $r \in \int_r^A$ and $a \in A$. Note that, when $A$ is a finite-dimensional quasi-Hopf algebra, $\mu'$ is nothing else than the convolution inverse of the modular element $\mu$ defined by (7.2.6).

**Proposition 7.40** *Let $A$ be a Frobenius augmented algebra with augmentation morphism $\varepsilon$ and Nakayama automorphism $\chi$ associated to a Frobenius system $(\phi, e)$. If $\mu'$ is the map defined above then $\mu'$ is an algebra morphism and $\mu' \circ \chi = \varepsilon$. Consequently, any symmetric augmented algebra is unimodular.*

*Proof* Let $r$ be the non-zero right integral in $A$ defined by $\phi \leftharpoonup r = \varepsilon$.

Clearly $\mu'(1_A) = 1$, and $\mu'$ is multiplicative since

$$\mu'(ab)r = (ab)r = a(br) = \mu'(b)ar = \mu'(a)\mu'(b)r$$

and $r$ is non-zero. Hence $\mu'$ is another augmentation morphism for $A$.

To prove the second assertion observe first that

$$(r \rightharpoonup \phi)(a) = \phi(ar) = \mu'(a)\phi(r) = \mu'(a)(\phi \leftharpoonup r)(1_A) = \mu'(a)\varepsilon(1_A) = \mu'(a),$$

for all $a \in A$, and therefore $\mu' = r \rightharpoonup \phi$. Then we compute:

$$\begin{aligned}\mu'\chi(a) &= \langle r \rightharpoonup \phi, \chi(a)\rangle = \langle \phi \leftharpoonup \chi(r), \chi(a)\rangle \\ &= \langle \phi, \chi(r)\chi(a)\rangle = \langle \phi, \chi(ra)\rangle \\ &= \varepsilon(a)\langle \phi, \chi(r)\rangle = \varepsilon(a)\langle \phi \leftharpoonup \chi(r), 1_A\rangle \\ &= \varepsilon(a)\langle r \rightharpoonup \phi, 1_A\rangle = \varepsilon(a)\phi(r) \\ &= \varepsilon(a)\langle \phi \leftharpoonup r, 1_A\rangle = \varepsilon(a)\varepsilon(1_A) = \varepsilon(a),\end{aligned}$$

for all $a \in A$, where we have used that $\chi$ is multiplicative and the fact that $r$ is a right integral in $A$. Thus $\mu' \circ \chi = \varepsilon$.

If $A$ is, moreover, symmetric then $\chi$ is inner; see Proposition 7.35 (v). If $u$ is invertible in $A$ and such that $\chi(a) = uau^{-1}$, for all $a \in A$, then

$$(\mu' \circ \chi)(a) = \mu'(uau^{-1}) = \mu'(u)\mu'(a)\mu'(u^{-1}) = \mu'(a),$$

for all $a \in A$. This means that $\varepsilon = \mu' \circ \chi = \mu'$. In other words, $r$ is a non-zero left integral in $A$ as well, and so $A$ is unimodular. □

We are now in the position to apply all the above results to quasi-Hopf algebras.

**Theorem 7.41** *A finite-dimensional quasi-Hopf algebra with antipode $S$ is a symmetric algebra if and only if it is unimodular and $S^2$ is an inner automorphism.*

*Proof* Let $H$ be a finite-dimensional quasi-Hopf algebra with antipode $S$. From Proposition 7.20 we know that $H$ is a Frobenius algebra with Nakayama automorphism $\chi$ given by $\chi(h) = \mu(h_1)S^2(h_2)$, for all $h \in H$, where $\mu$ is the modular element of $H^*$.

Assume $H$ is symmetric. As it is an augmented algebra as well, from the previous proposition it follows that $H$ is unimodular. Hence $\mu = \varepsilon$, and this implies $\chi = S^2$. Now Proposition 7.35 (v) ensures that $\chi$, and therefore $S^2$, is an inner automorphism of $H$, as required.

Conversely, suppose that $H$ is unimodular and $S^2$ is inner. Then $H$ is a Frobenius algebra with the Nakayama automorphism $S^2$, an inner automorphism. Thus $H$ is symmetric; see Proposition 7.35. □

As we shall see later on, classes of symmetric quasi-Hopf algebras are given by involutory quasi-Hopf algebras, factorizable quasi-Hopf algebras and quasi-Hopf algebras of the type $D^{\omega}(H)$. For the moment we restrict ourselves to the following:

**Example 7.42** The two-dimensional quasi-Hopf algebra $H(2)$ is symmetric since it is unimodular and has the antipode defined by the identity morphism.

More generally, we have the following result due to Eilenberg and Nakayama. In particular, it implies that any finite-dimensional semisimple (or, equivalently, separable) quasi-Hopf algebra is symmetric.

**Theorem 7.43** *Any finite-dimensional semisimple algebra is symmetric.*

*Proof* By the Wedderburn theorem, $A$ is isomorphic to $M_{n_1}(D_1) \times \cdots \times M_{n_t}(D_t)$, where each $D_j$ is a finite-dimensional division algebra over $k$. Moreover, $M_{n_i}(D_i) \cong M_{n_i}(k) \otimes D_i$, for all $1 \le i \le t$, thus, according to Examples 7.37 (1) and (3) it suffices to show that any finite-dimensional division $k$-algebra is symmetric.

Let $D$ be a finite-dimensional division $k$-algebra and $[D,D]$ the commutator space of $D$. We show that $[D,D] \neq D$, and this will imply that any non-zero $k$-linear map $\phi : D \to k$ whose kernel contains $[D,D]$ satisfies the condition (iii) in Proposition 7.35. Indeed, $\phi$ is a trace since $[D,D] \subseteq \mathrm{Ker}(\phi)$, and $\mathrm{Ker}(\phi)$ does not contain non-zero left or right ideals since $D$ is a division algebra over $k$.

Take $\overline{k}$ the algebraic closure of $k$ and $D^{\overline{k}} := D \otimes_k \overline{k}$. Then $D^{\overline{k}} \cong M_m(\overline{k})$, for a certain non-zero natural number $m$. The trace morphism $\mathrm{Tr} : M_m(\overline{k}) \to \overline{k}$ is non-zero and $[M_m(\overline{k}), M_m(\overline{k})] \subseteq \mathrm{Ker}(\mathrm{Tr})$, and therefore $[D^{\overline{k}}, D^{\overline{k}}] \neq D^{\overline{k}}$. This together with $[D,D]^{\overline{k}} := [D,D] \otimes_k \overline{k} \subseteq [D^{\overline{k}}, D^{\overline{k}}]$ implies $[D,D] \neq D$, as needed. □

**Corollary 7.44** *If $H$ is a finite-dimensional semisimple quasi-Hopf algebra with antipode $S$ then $S^2$ is an inner automorphism of $H$.*

## 7.5 Cointegral Theory

Unless otherwise specified, throughout this section $H$ is a finite-dimensional quasi-Hopf algebra with antipode $S$, and $\{e_i\}_i$ is a basis in $H$ with dual basis $\{e^i\}_i$ in $H^*$. We know that $S$ is bijective, thus we will make use of the elements $p_R, q_R$ from (3.2.19) or $p_L, q_L$ from (3.2.20) without any restriction. Also, we will freely use the dual quasi-Hopf algebra structure of $H^*$, the linear dual of $H$.

We next show that $H^*$ is a right quasi-Hopf $H$-bimodule, and that the corresponding space of coinvariants has dimension 1. As we shall see, these coinvariants of $H^*$ play an important role in the structure of both $H^*$ (by generating it as a left $H$-module) and $H$ (by defining a Frobenius morphism for it). We end by defining the set of alternative cointegrals on $H$, elements in $H^*$ that can be defined even in the infinite-dimensional case.

We start with the following technical result.

**Lemma 7.45** *Let $H$ be an arbitrary quasi-Hopf algebra with bijective antipode $S$ and let $U, V \in H \otimes H$ be the elements defined in* (7.3.1). *Then we have:*

$$\Phi^{-1}(\mathrm{Id}_H \otimes \Delta)(U)(1_H \otimes U) = (\Delta \otimes \mathrm{Id}_H)(\Delta(S(X^1))U)(X^2 \otimes X^3 \otimes 1_H), \tag{7.5.1}$$

$$(\Delta \otimes \mathrm{Id}_H)(V)\Phi^{-1} = (X^2 \otimes X^3 \otimes 1_H)(1_H \otimes V)(\mathrm{Id}_H \otimes \Delta)(V\Delta(S^{-1}(X^1))). \tag{7.5.2}$$

*Proof* In $H^{\mathrm{op}}$ the roles of $U$ and $V$ interchange, and (7.5.2) reduces to (7.5.1) in $H^{\mathrm{op}}$, so it suffices to prove only (7.5.1). To this end denote by $Q^1 \otimes Q^2$ another copy of $q_R$, by $G^1 \otimes G^2$ another copy of $f^{-1}$, and then compute that

$$\begin{aligned} &\Phi^{-1}(\mathrm{Id}_H \otimes \Delta)(U)(1_H \otimes U) \\ &\overset{(7.3.1)}{=} \quad x^1 g^1 S(q^2) \otimes x^2 g^2_1 S(q^1)_1 U^1 \otimes x^3 g^2_2 S(q^1)_2 U^2 \end{aligned}$$

$$\begin{aligned}
&\overset{(7.3.1),(3.2.13)}{=} x^1g^1S(q^2)\otimes x^2g_1^2G^1S(Q^2q_2^1)\otimes x^3g_2^2G^2S(Q^1q_1^1)\\
&\overset{(3.2.26)}{=} x^1g^1S(q_2^2X^1_{(2,2)}Y^3)f^1X^2\otimes x^2g_1^2G^1S(q_1^2X^1_{(2,1)}Y^2)f^2X^3\\
&\qquad \otimes x^3g_2^2G^2S(q^1X_1^1Y^1)\\
&\overset{(3.1.5),(3.2.17)}{=} g_1^1G^1S(q_2^2X^1_{(2,2)})f^1X^2\otimes g_2^1G^2S(q_1^2X^1_{(2,1)})f^2X^3\otimes g^2S(q^1X_1^1)\\
&\overset{(3.2.13)}{=} (\Delta(g^1S(q^2X_2^1))\otimes g^2S(q^1X_1^1))(X^2\otimes X^3\otimes 1_H)\\
&\overset{(3.2.13),(7.3.1)}{=} (\Delta(S(X^1)_1U^1)\otimes S(X^1)_2U^2)(X^2\otimes X^3\otimes 1_H)\\
&= (\Delta\otimes \mathrm{Id}_H)(\Delta(S(X^1))U)(X^2\otimes X^3\otimes 1_H),
\end{aligned}$$

as required. □

By $*$ we denote the multiplication $*:\ H^*\otimes H^*\to H^*$ given by

$$\langle h^* * g^*, h\rangle = h^*(V^1h_1U^1)g^*(V^2h_2U^2),\ \ \forall\, h^*,\ g^*\in H^*,\ h\in H. \tag{7.5.3}$$

**Proposition 7.46** *If $H$ is a finite-dimensional quasi-Hopf algebra then $H^*\in {}_H\mathscr{M}_H^H$ with the following structure:*

- *$H^*$ is an $H$-bimodule via the $H$-actions induced by $S$, that is,*

$$h\cdot h^*\cdot h' = S(h')\rightharpoonup h^*\leftharpoonup S^{-1}(h),\ \ \forall\, h,h'\in H,\ \ h^*\in H^*; \tag{7.5.4}$$

- *$H$ coacts from the right on $H^*$ by*

$$\rho(h^*) = \sum_{i=1}^{n} e^i * h^*\otimes e_i,\ \ \forall\, h^*\in H^*, \tag{7.5.5}$$

*where $*$ is the multiplication on $H^*$ defined above.*

*Proof* This follows from Proposition 6.10, specialized for $C=H$. Note that the structures above come from the right dual structure ${}^*H$ of the coalgebra $H$ in ${}_H\mathscr{M}_H$, viewed in a canonical way as an object in ${}_H^H\mathscr{M}_H$. □

**Definition 7.47** The coinvariants (of the second type) $\lambda\in H^{*\mathrm{co}(H)}$ are called left cointegrals on $H$ and the space of left cointegrals is denoted by $\mathscr{L}$.

For an algebra $A$, we call an element of $A^*$ non-degenerate if its kernel does not contain non-zero left or right ideals.

**Theorem 7.48** *Let $H$ be a finite-dimensional quasi-Hopf algebra. Then* $\dim_k\mathscr{L}=1$ *and all non-zero left cointegrals on $H$ are non-degenerate.*

*Proof* Theorem 6.20 implies that

$$\mathscr{L}\otimes H\ni\lambda\otimes h\mapsto \lambda\cdot h = S(h)\rightharpoonup\lambda\in H^* \tag{7.5.6}$$

is an isomorphism of right quasi-Hopf $H$-bimodules. As $\dim_k H=\dim_k H^*$ is finite it follows that $\dim_k\mathscr{L}=1$.

Let $\lambda\in\mathscr{L}$ be non-zero. If $I$ is a left ideal in $H$ such that $\lambda(I)=0$ then $\lambda(HI)=0$, and so $I\rightharpoonup\lambda=0$, from which we conclude that $I=0$. Likewise, if $I$ is a right ideal in $H$ such that $\lambda(I)=0$ then $(H \rightharpoonup \lambda)(I)=0$, thus $h^*(I)=0$, for all $h^*\in H^*$ because $S$ is bijective. Therefore $I=0$, hence $\lambda$ is non-degenerate. □

According to Proposition 6.21 and (7.5.4), an element $\lambda \in H^*$ is a left cointegral on $H$ if and only if

$$\sum_i e^i * \lambda \otimes e_i = S(x^2) \rightharpoonup E(\lambda \leftharpoonup S^{-1}(x^1)) \otimes x^3,$$

which is equivalent to

$$\lambda(V^2 h_2 U^2)V^1 h_1 U^1 = \langle E(\lambda \leftharpoonup S^{-1}(x^1)), hS(x^2)\rangle x^3, \ \ \forall\, h \in H. \tag{7.5.7}$$

Here $E : H^* \to \mathscr{L}$ is the projection from Definition 6.18, specialized for our context. More precisely,

$$\langle E(h^*), h\rangle = \sum_i \langle e^i * h^*, S^{-1}(q^1) h S(\beta S(q^2 e_i))\rangle, \tag{7.5.8}$$

for all $h^* \in H^*$ and $h \in H$. It can be expressed in terms of the projection $P$ onto the space of left integrals defined in (7.2.1) as follows.

**Lemma 7.49** *Let $H$ be a finite-dimensional quasi-Hopf algebra and $E, P$ the two projections mentioned above. Then*

$$\langle E(h^*), h\rangle = \langle h^*, S^{-1}(P(S(h)))\rangle, \tag{7.5.9}$$

*for all $h^* \in H^*$ and $h \in H$. In particular,*

$$E(h^* \leftharpoonup S^{-1}(h)) = \mu(h)E(h^*), \ \ \forall\, h^* \in H^* \ \text{ and } \ h \in H, \tag{7.5.10}$$

*which implies*

$$\lambda(S^{-1}(h)h') = \mu(h_1)\lambda(h'S(h_2)), \ \ \forall\, \lambda \in \mathscr{L} \ \text{ and } \ h, h' \in H. \tag{7.5.11}$$

*As usual, $\mu \in H^*$ is the modular element of $H^*$.*

*Proof* Take $f = f^1 \otimes f^2 = F^1 \otimes F^2$ and $f^{-1} = g^1 \otimes g^2 = G^1 \otimes G^2$ as in (3.2.15) and (3.2.16). As we have seen before, we have

$$g^1 S(g^2 \alpha) = \beta, \ S(\beta f^1) f^2 = \alpha, \quad f^1 \beta S(f^2) = S(\alpha), \tag{7.5.12}$$

and so we compute, for $h^* \in H^*$ and $h \in H$:

$$\begin{aligned}
&\langle E(h^*), h\rangle \\
&\quad = \quad \sum_i \langle e^i * h^*, S^{-1}(q^1) h S^2(q^2 e_i) S(\beta)\rangle \\
&\overset{(7.3.1),(3.2.13)}{=} \sum_j \langle e^j, hS^2(q^2 S^{-1}(f^2 q^1_2 p^2)(e_j)_1 U^1) S(\beta)\rangle \langle h^*, S^{-1}(f^1 q^1_1 p^1)(e_j)_2 U^2\rangle \\
&\underset{(3.2.19)}{\overset{(3.2.23),(7.3.1)}{=}} \sum_j \langle e^j, hS^2(S^{-1}(f^2)(e_j)_1 g^1 S(X^2)\alpha X^3 S^{-1}(\beta))\rangle \\
&\qquad\qquad \langle h^*, S^{-1}(f^1)(e_j)_2 g^2 S(X^1)\rangle
\end{aligned}$$

$$\overset{(3.1.5)}{\underset{(3.2.17)}{=}} \sum_j \langle e^j, hS^2(S^{-1}(f^2)(e_j)_1 F_1^2 X^2 g_2^1 G^2 \alpha S^{-1}(\beta F^1 X^1 g_1^1 G^1))\rangle$$

$$\langle h^*, S^{-1}(f^1)(e_j)_2 F_2^2 X^3 g^2\rangle$$

$$\overset{(7.5.12),(3.2.1)}{=} \sum_i \langle e^i, hS^2(S^{-1}(f^2)(e_i)_1 X^2)S(\beta F^1 X^1 \beta)F^2\rangle\langle h^*, S^{-1}(f^1)(e_i)_2 X^3\rangle$$

$$\overset{(7.5.12),(3.2.13)}{=} = \sum_i \langle e^i, hS^2(S^{-1}(S(e_i)_2)S^{-1}(f^2)X^2 S^{-1}(S^{-1}(\alpha)X^1\beta))\rangle$$

$$\langle h^*, S^{-1}(S(e_i)_1)S^{-1}(f^1)X^3\rangle$$

$$\overset{(3.1.5)}{\underset{(3.2.17)}{=}} \sum_i \langle e^i, hS(S(e_i)_2)S(\beta F^1 f_1^2 X^2 g_2^1)F^2 f_2^2 X^3 g^2 \alpha\rangle$$

$$\langle h^*, S^{-1}(S(e_i)_1)S^{-1}(f^1 X^1 g_1^1)\rangle$$

$$\overset{(7.5.12),(3.2.1)}{=} \sum_j \langle e^j, g^1 S(hS((e_j)_2)S(X^2)\alpha X^3 g^2 \alpha)\rangle\langle h^*, S^{-1}(X^1(e_j)_1)\rangle$$

$$\overset{(7.5.12),(3.2.19)}{\underset{(7.2.1)}{=}} \langle h^*, S^{-1}(P(S(h)))\rangle,$$

as desired. Now, the relation (7.5.10) follows easily from (7.5.9) and the fact that $P(h) \in \int_l^H$, for all $h \in H$. To prove (7.5.11), let $q_R$ be the element defined by (3.2.19). For all $h, h' \in H$ and $\lambda \in \mathscr{L}$ we have:

$$\begin{aligned}\lambda(S^{-1}(h)h') &= \langle E(\lambda), S^{-1}(h)h'\rangle\\ &\overset{(7.5.9),(7.2.1)}{=} \sum_i \langle e^i, \beta S^2(q^2(e_i)_2)S(h')h\rangle\langle \lambda, S^{-1}(q^1(e_i)_1)\rangle\\ &= \sum_j \langle e^j, \beta S^2(q^2(e_j)_2 h_2)S(h')\rangle\langle \lambda, S^{-1}(q^1(e_j)_1 h_1)\rangle\\ &\overset{(7.2.1),(7.5.9)}{=} \langle E(\lambda \leftharpoonup S^{-1}(h_1)), h'S(h_2)\rangle \overset{(7.5.10)}{=} \mu(h_1)\lambda(h'S(h_2)),\end{aligned}$$

as claimed, so the proof is finished. □

Lemma 7.49 and (7.5.7) say that $\lambda \in H^*$ is a left cointegral on $H$ if and only if

$$\lambda(V^2 h_2 U^2)V^1 h_1 U^1 = \mu(x^1)\lambda(hS(x^2))x^3, \ \forall\, h \in H. \tag{7.5.13}$$

In what follows we will give other characterizations for a left cointegral on $H$. We first need the following result.

**Lemma 7.50** *If $\phi$ is the Frobenius morphism for $H$ defined in Proposition 7.20 then $\lambda := \phi \circ S$ is a non-zero left cointegral on $H$.*

*Proof* $\lambda$ is non-zero since $\xi(\lambda \circ S^{-1}) = \xi(\phi) = 1_H$, where $\xi$ is the isomorphism considered in the proof of Proposition 7.20. We next show that

$$\lambda_0(q^2 t_2 p^2)q^1 t_1 p^1 = \mu(\beta)\lambda_0(t)1_H, \ \forall\, \lambda_0 \in \mathscr{L} \text{ and } t \in \int_l^H, \tag{7.5.14}$$

where $\mu$ is the modular element of $H^*$ defined in (7.2.6). We have

$$\lambda_0(q^2 h_2 p^2)q^1 h_1 p^1$$
$$\overset{(7.3.4),(7.3.5)}{=} \lambda_0(V^2[S^{-1}(\tilde{q}^1)hS(\tilde{p}^1)]_2 U^2)\tilde{q}^2 V^1[S^{-1}(\tilde{q}^1)hS(\tilde{p}^1)]_1 U^1 \tilde{p}^2$$

$$\overset{(7.5.13)}{=} \mu(x^1)\lambda_0(S^{-1}(\tilde{q}^1)hS(x^2\tilde{p}^1))\tilde{q}^2x^3\tilde{p}^2,$$

for all $h \in H$. Thus we have shown that

$$\lambda_0(q^2h_2p^2)q^1h_1p^1 = \mu(x^1)\lambda_0(S^{-1}(\tilde{q}^1)hS(x^2\tilde{p}^1))\tilde{q}^2x^3\tilde{p}^2, \ \forall\, h \in H. \tag{7.5.15}$$

Specializing the above formula for $h = t$, a left integral in $H$, we obtain that

$$\begin{aligned} \lambda_0(q^2t_2p^2)q^1t_1p^1 &\overset{(3.2.20)}{=} \mu(x^1)\lambda_0(tS(x^2\tilde{p}^1))x^3\tilde{p}^2 \\ &\overset{(7.2.6),(3.2.20)}{=} \mu(x^1)\mu(X^1\beta S(X^2))\mu(S(x^2))\lambda_0(t)x^3X^3 \\ &= \mu(\beta)\lambda_0(t)1_H, \end{aligned}$$

as desired. Observe now that (7.5.14) implies

$$\xi(\lambda_0 \circ S^{-1}) = \mu(\beta)\lambda_0(t)1_H = \xi(\mu(\beta)\lambda_0(t)\phi),$$

and so $\lambda_0 = \mu(\beta)\lambda_0(t)\phi \circ S = \mu(\beta)\lambda_0(t)\lambda$, for all $\lambda_0 \in \mathscr{L}$. Taking a non-zero $\lambda_0$ in $\mathscr{L}$ and using the uniqueness of left cointegrals on $H$ we conclude that $\lambda$ is a non-zero left cointegral on $H$, as needed. □

**Theorem 7.51** *For a finite-dimensional quasi-Hopf algebra $H$ and $\lambda \in H^*$, the following assertions are equivalent:*

*(i) $\lambda$ is a non-zero left cointegral on $H$;*

*(ii) $\lambda(t) \neq 0$ and $\lambda(q^2t_2p^2)q^1t_1p^1 = \mu(\beta)\lambda(t)1_H$, for any left integral $t \neq 0$ in $H$;*

*(iii) $\lambda(t) \neq 0$ and $\lambda(ht_2p^2)t_1p^1 = \mu(\beta)\lambda(t)\beta S(h)$, for all $t \in \int_l^H \backslash \{0\}$ and $h \in H$.*

*Furthermore, if $\beta$ has a left inverse in $H$ then (i)–(iii) above are also equivalent to*

*(iv) $\lambda(t) \neq 0$ and $\lambda(t_2p^2)t_1p^1 = \mu(\beta)\lambda(t)\beta$, for all $t \in \int_l^H \backslash \{0\}$.*

*Here $p_R$ and $q_R$ are the elements defined in* (3.2.19).

*Proof* $(i) \Rightarrow (ii)$. Take $t$ a non-zero left integral in $H$. If $\lambda(t) = 0$ then $\lambda(Ht) = 0$, and so $\mathrm{Ker}(\lambda)$ contains a non-zero left ideal, a contradiction (see Theorem 7.48). Thus $\lambda(t) \neq 0$. The rest of the proof follows from (7.5.14).

$(ii) \Rightarrow (i)$. Take $t$ a non-zero left integral in $H$. The fact that $\lambda(t) \neq 0$ implies that $\lambda$ is non-zero. We know from Remark 3.15 that $\mu(\beta) \neq 0$, and so we can write

$$(\mu(\beta)\lambda(t))^{-1}\lambda(q^2t_2p^2)q^1t_1p^1 = 1_H,$$

which means that $\xi((\mu(\beta)\lambda(t))^{-1}\lambda \circ S^{-1}) = 1_H = \xi(\phi)$. As $\xi$ is an injective map, Lemma 7.50 implies that $(\mu(\beta)\lambda(t))^{-1}\lambda$, and so $\lambda$ as well, is a left cointegral on $H$.

$(i) \Rightarrow (iii)$. By (3.2.21) and (7.2.6) it follows that

$$t_1p^1h \otimes t_2p^2 = \mu(h_1)t_1p^1 \otimes t_2p^2S(h_2), \ \forall\, h \in H. \tag{7.5.16}$$

Also, as an immediate consequence of (7.2.3), we obtain

$$t_1 \otimes t_2 = \beta q^1t_1 \otimes q^2t_2 = q^1t_1 \otimes S^{-1}(\beta)q^2t_2. \tag{7.5.17}$$

These formulas allow us to compute:

$$\lambda(ht_2p^2)t_1p^1 \ = \ \langle \lambda, S^{-1}(S(h))t_2p^2\rangle t_1p^1$$

$$\stackrel{(7.5.11)}{=} \mu(S(h)_1)\langle\lambda, t_2p^2S(S(h)_2)\rangle t_1p^1$$
$$\stackrel{(7.5.16)}{=} \lambda(t_2p^2)t_1p^1S(h) \stackrel{(7.5.17),(7.5.14)}{=} \mu(\beta)\lambda(t)\beta S(h),$$

for all $h \in H$, as required.

$(iii) \Rightarrow (ii)$. We use (iii) to compute

$$\begin{aligned}\lambda(q^2t_2p^2)q^1t_1p^1 &= \mu(\beta)\lambda(t)q^1\beta S(q^2)\\ &\stackrel{(3.2.19)}{=} \mu(\beta)\lambda(t)X^1\beta S(X^2)\alpha X^3\\ &\stackrel{(3.2.2)}{=} \mu(\beta)\lambda(t)1_H,\end{aligned}$$

as desired.

It is clear that (iii) implies (iv) in general, without the extra hypothesis on $\beta$. For the converse, we need $\beta$ to be left invertible in $H$ since

$$\lambda(q^2t_2p^2)\beta q^1t_1p^1 \stackrel{(7.5.17)}{=} \lambda(t_2p^2)t_1p^1 \stackrel{(iv)}{=} \mu(\beta)\lambda(t)\beta.$$

Thus the relation in (ii) holds, and so the one in (iii) also holds. □

**Proposition 7.52** *A cointegral $\lambda$ on $H$ is non-zero if and only if $\lambda(t) \neq 0$ for some non-zero left integral $t$ in $H$.*

*Proof* We prove the opposite statement. Namely, $\lambda = 0$ if and only if $\lambda(t) = 0$ for some non-zero left integral $t$ in $H$.

The direct implication is obvious. For the converse, if $\lambda(t) = 0$ then $\lambda$ is zero on a non-zero left ideal of $H$, which implies $\lambda = 0$ by Theorem 7.48. □

The above characterizations show how to find left cointegrals on a finite-dimensional quasi-Hopf algebra $H$. First we have to find a non-zero left integral in $H$. Secondly, working maybe with dual bases, we have to determine the element $\lambda \in H^*$ that satisfies, for instance, (ii) of the above theorem. The simplest condition that we have is (iv) but it can be applied only in the case when $\beta$ is invertible. Nevertheless, this situation occurs in all the forthcoming examples in this section. Notice that when $H$ is unimodular and $\alpha, \beta$ are invertible, (iv) simplifies as follows: $\lambda(t_2)t_1 = \lambda(t)\beta S^{-1}(\alpha)$, and this is because $\mu = \varepsilon$, and $t$, which this time is a right integral as well, obeys

$$t_1 \otimes t_2 \stackrel{(3.2.23)}{=} t_1q_1^1p^1 \otimes t_2q_2^1p^2S(q^2) = t_1p^1 \otimes t_2p^2\alpha \stackrel{(7.5.16)}{=} t_1p^1S^{-1}(\alpha) \otimes t_2p^2.$$

We next apply these observations to $H(2)$ and $H_\pm(8)$, respectively.

**Example 7.53** For the quasi-Hopf algebra $H(2)$ defined in Example 3.26, denote by $\{P_1, P_g\}$ the dual basis in $H(2)^*$ corresponding to the basis $\{1, g\}$ of $H(2)$. Then $P_g$ is a (non-zero) left cointegral on $H(2)$.

*Proof* For $H(2)$ we have $\beta = 1$, $\alpha = g$ is invertible and $H(2)$ is unimodular (see Example 7.15(1)), a left and right non-zero integral being given by $t = 1 + g$. Thus, finding all the left cointegrals on $H(2)$ is equivalent to finding all those elements

$\lambda \in H(2)^*$ satisfying $\lambda(1)1 + \lambda(g)g = \lambda(1+g)g$. It is clear that we must have $\lambda(1) = 0$, and so $\mathscr{L} = kP_g$. Note that $P_g(t) = 1$, so $P_g$ is non-zero, as needed. □

**Example 7.54** The spaces of left cointegrals for $H_\pm(8)$, the quasi-Hopf algebras from Example 3.30, are both equal to $kP_{x^3}$. Here we have denoted by $\{P_{g^ix^j} \mid 0 \le i \le 1,\ 0 \le j \le 3\}$ the dual basis corresponding to the canonical basis $\{g^ix^j \mid 0 \le i \le 1,\ 0 \le j \le 3\}$ of $H_\pm(8)$.

*Proof* This time the computations are much more complicated. Recall first that $t = (1+g)x^3 = x^3(1-g)$ is a non-zero left integral in $H_\pm(8)$ which is not a right integral; see Example 7.15(2).

Denote $\omega := \frac{1}{2}(1 \pm i)$ and let $\overline{\omega} = \frac{1}{2}(1 \mp i)$ be its conjugate. In order to compute $\Delta(t)$ we rewrite $\Delta(x)$ as

$$\Delta(x) = \omega x \otimes 1 + \overline{\omega} x \otimes g + p_+ \otimes x + p_- \otimes gx,$$

to compute that

$$\Delta(x^2) = x^2 \otimes g + g \otimes x^2 + (p_+ \pm ip_-)x \otimes x + (p_- \pm ip_+)x \otimes gx,$$

from which we obtain

$$\begin{aligned}\Delta(x^3) = \overline{\omega}x^3 \otimes 1 + \omega x^3 \otimes g \pm ip_-x^2 \otimes x + \overline{\omega} gx \otimes x^2 + p_+ \otimes x^3 \\ \pm ip_+x^2 \otimes gx - \omega gx \otimes gx^2 - p_- \otimes gx^3.\end{aligned}$$

We have $\Delta(t)p_R = \Delta(x^3)\Delta(1-g)p_R$. By taking the formula of $p_R$ from Example 7.31(1) and using $\Delta(1-g)p_R = 1 \otimes 1 - g \otimes g$ we conclude that

$$\begin{aligned}\Delta(t)p_R = (\overline{\omega} + \omega g)x^3 \otimes 1 + (\omega + \overline{\omega} g)x^3 \otimes g \pm ix^2 \otimes x + (\overline{\omega} g - \omega)x \\ \otimes x^2 + 1 \otimes x^3 \pm igx^2 \otimes gx - (\omega g - \overline{\omega})x \otimes gx^2 + g \otimes gx^3.\end{aligned}$$

Now take $\lambda = \sum_{i,j} c_{ij} P_{g^ix^j}$ be an element of $H^*$. It follows that $\lambda$ satisfies (iv) in the statement of Theorem 7.51 if and only if $c_{01} = c_{11} = c_{13} = 0$ and the following relations hold:

$$\overline{\omega}c_{12} - \omega c_{02} = 0, \quad \overline{\omega}c_{00} + \omega c_{10} = 0, \quad \overline{\omega}c_{02} - \omega c_{12} = 0 \quad \text{and} \quad \omega c_{00} + \overline{\omega}c_{10} = 0.$$

We get $c_{00} = c_{02} = c_{10} = c_{12} = 0$ as well, and so $\lambda = c_{03}P_{x^3}$. We thus have $\mathscr{L} = kP_{x^3}$, since $P_{x^3}(t) = 1$. □

Other characterizations for left cointegrals are included in the result below. This time they involve all the elements and not only the "generator," that is, an integral.

**Proposition 7.55** *Let $H$ be a finite-dimensional quasi-Hopf algebra and $\mu$ the modular element of $H^*$. For $\lambda \in H^*$, the following statements are equivalent:*

*(i) $\lambda$ is a left cointegral on $H$;*

*(ii) for all $h \in H$, we have*

$$\begin{aligned}&\lambda(S^{-1}(f^1)h_2S^{-1}(q^1g^1))S^{-1}(f^2)h_1S^{-1}(q^2g^2) \\ &\qquad = \mu(q_1^1x^1)\langle \lambda, hS^{-1}(f^1)g^2S(q_2^1x^2)\rangle q^2x^3S^{-1}(S^{-1}(f^2)g^1); \qquad (7.5.18)\end{aligned}$$

*(iii) for all $h \in H$, we have*

$$\lambda(S^{-1}(f^1)h_2U^2)S^{-1}(f^2)h_1U^1 = \mu(q_1^1x^1)\langle\lambda, hS(q_2^1x^2)\rangle q^2x^3.$$

*Here $f = f^1 \otimes f^2$, $q_R = q^1 \otimes q^2$ and $U = U^1 \otimes U^2$ are defined by* (3.2.15), (3.2.19) *and* (7.3.1), *respectively, and $f^{-1} = g^1 \otimes g^2$ is as in* (3.2.16).

*Proof* $(i) \Rightarrow (ii)$. Suppose that $\lambda$ is a left cointegral. As before, we write $f = f^1 \otimes f^2 = F^1 \otimes F^2 = \mathbf{F}^1 \otimes \mathbf{F}^2$, $f^{-1} = g^1 \otimes g^2 = G^1 \otimes G^2 = \mathbf{G}^1 \otimes \mathbf{G}^2$, $q_R = q^1 \otimes q^2$ and $p_R = p^1 \otimes p^2$. We compute:

$$\begin{aligned}
&\mu(q_1^1x^1)\langle\lambda, hS^{-1}(f^1)g^2S(q_2^1x^2)\rangle q^2x^3S^{-1}(S^{-1}(f^2)g^1)\\
&\overset{(7.5.10)}{\underset{(7.5.13)}{=}} \langle\lambda, V^2[S^{-1}(q^1)hS^{-1}(f^1)g^2]_2U^2\rangle q^2V^1[S^{-1}(q^1)hS^{-1}(f^1)g^2]_1U^1\\
&\qquad S^{-1}(S^{-1}(f^2)g^1)\\
&\overset{(7.3.1)}{\underset{(3.2.13)}{=}} \langle\lambda, S^{-1}(\mathbf{F}^1q_1^1p^1)h_2S^{-1}(F^1f_1^1\mathbf{G}^1)g_2^2U^2\rangle q^2S^{-1}(\mathbf{F}^2q_2^1p^2)h_1\\
&\qquad S^{-1}(F^2f_2^1\mathbf{G}^2)g_1^2U^1S^{-1}(S^{-1}(f^2)g^1)\\
&\overset{(3.2.23)}{\underset{(3.2.13)}{=}} \langle\lambda, S^{-1}(S(h)_1)S^{-1}(F^1f_1^1)g_2^2G^2S(X^1)\rangle S^{-1}(S(h)_2)S^{-1}(F^2f_2^1)\\
&\qquad g_1^2G^1S(X^2)\alpha S^{-1}(S^{-1}(f^2)g^1S(X^3))\\
&\overset{(3.1.5)}{\underset{(3.2.17)}{=}} \langle\lambda, S^{-1}(S(h)_1)S^{-1}(S(X^3)F^1f_1^1)g^2\rangle S^{-1}(S(h)_2)S^{-1}(S(X^2)F^2f_2^1)\\
&\qquad g_2^1G^2\alpha S^{-1}(S^{-1}(S(X^1)f^2)g_1^1G^1)\\
&\overset{(7.5.12)}{\underset{(3.2.1)}{=}} \langle\lambda, S^{-1}(S(h)_1)S^{-1}(S(X^3)F^1f_1^1)\rangle S^{-1}(S(h)_2)\\
&\qquad S^{-1}(S^{-1}(S(X^1)f^2)\beta S(X^2)F^2f_2^1)\\
&\overset{(3.1.5),(3.2.17)}{=} \langle\lambda, S^{-1}(S(h)_1)S^{-1}(f^1X^1)\rangle S^{-1}(S(h)_2)S^{-1}(S^{-1}(F^2f_2^2X^3)\beta F^1f_1^2X^2)\\
&\overset{(7.5.12),(3.2.1)}{=} \langle\lambda, S^{-1}(X^1S(h)_1)\rangle S^{-1}(S^{-1}(\alpha X^3)X^2S(h)_2)\\
&\overset{(3.2.19),(3.2.13)}{=} \langle\lambda, S^{-1}(f^1)h_2S^{-1}(q^1g^1)\rangle S^{-1}(f^2)h_1S^{-1}(q^2g^2).
\end{aligned}$$

$(ii) \Rightarrow (i)$. Assume that $\lambda \in H^*$ satisfies (7.5.18). It follows from (3.2.13) that

$$\lambda(S^{-1}(q^1h_1))q^2h_2 = \mu(q_1^1x^1)\langle\lambda, S^{-1}(f^1h)g^2S(q_2^1x^2)\rangle S^{-1}(f^2)g^1S(q^2x^3), \quad (7.5.19)$$

for all $h \in H$, and

$$\begin{aligned}
&\langle\lambda, S^{-1}(P(S(h)))\rangle\\
&\overset{(7.2.1),(7.5.19)}{=} \mu(q_1^1x^1)\langle\lambda, S^{-1}(f^1\beta S(f^2)S^2(g^1)S^3(q^2x^3)S(h))g^2S(q_2^1x^2)\rangle\\
&\overset{(7.5.12)}{=} \mu(q_1^1x^1)\langle\lambda, hS^2(q^2x^3)S(g^1)\alpha g^2S(q_2^1x^2)\rangle\\
&\overset{(7.5.12),(3.2.19),(3.2.23)}{=} \lambda(h).
\end{aligned}$$

It follows from Lemma 7.49 that $E(\lambda) = \lambda$, so $\lambda \in \mathscr{L}$.

$(iii) \Rightarrow (i)$. Repeating the computations made in the first part of the proof of Lemma 7.49, we find that the projection $E$ (see (7.5.8)) is given by

$$\langle E(h^*), h\rangle = \sum_{i=1}^{n}\langle e^i, hS(f^2)S^2((e_i)_1U^1)S(\beta)\rangle\langle h^*, S^{-1}(f^1)(e_i)_2U^2\rangle, \qquad (7.5.20)$$

for all $h^* \in H$, $h \in H$. Using (7.5.20), we can compute that $E(\lambda) = \lambda$, so $\lambda \in \mathscr{L}$.

$(i) \Rightarrow (iii)$. Assume that $\lambda \in \mathscr{L}$. We calculate:

$$\mu(q_1^1x^1)\langle\lambda, hS(q_2^1x^2)\rangle q^2x^3 \overset{(7.5.11),(7.5.13)}{=} \langle\lambda, V^2[S^{-1}(q^1)h]_2U^2\rangle q^2V^1[S^{-1}(q^1)h]_1U^1 \overset{(7.3.1),(3.2.13),(3.2.23)}{=} \langle\lambda, S^{-1}(f^1)h_2U^2\rangle S^{-1}(f^2)h_1U^1$$

and the proof is complete. □

The notion of right cointegral on a finite-dimensional quasi-Hopf algebra $H$ can be introduced, too. In case of integrals we have that $r$ is a right integral in $H$ if and only if it is a left integral in $H^{\mathrm{op}}$. By analogy, we say that $\Lambda \in H^*$ is a right cointegral on $H$ if and only if it is a left cointegral on $H^{\mathrm{cop}}$. Since $H^{\mathrm{cop}}$ is also a finite-dimensional quasi-Hopf algebra all the results obtained for left cointegrals have counterparts for right cointegrals: all we have to do is to transfer them through the "cop" machinery. For instance, in $H^{\mathrm{cop}}$ we have $U_{\mathrm{cop}} = S^{-1}(\tilde{q}^1g^1) \otimes S^{-1}(\tilde{q}^2g^2)$, $V_{\mathrm{cop}} = S(\tilde{p}^1)f^2 \otimes S(\tilde{p}^2)f^1$ and $\mu_{\mathrm{cop}} = \mu$. Thus $\Lambda$ is a right cointegral on $H$ if and only if

$$\Lambda(S(\tilde{p}^2)f^1h_1S^{-1}(\tilde{q}^2g^2))S(\tilde{p}^1)f^2h_2S^{-1}(\tilde{q}^1g^1) = \mu(X^3)\Lambda(hS^{-1}(X^2))X^1, \quad \forall\, h \in H.$$

The $H^{\mathrm{cop}}$-version of Theorem 7.51 is the following:

**Theorem 7.56** *For a finite-dimensional quasi-Hopf algebra $H$ and $\Lambda \in H^*$, the following assertions are equivalent:*

*(i) $\Lambda$ is a non-zero right cointegral on $H$;*

*(ii) $\Lambda(t) \neq 0$ and $\Lambda(\tilde{q}^1t_1\tilde{p}^1)\tilde{q}^2t_2\tilde{p}^2 = \mu^{-1}(\beta)\Lambda(t)1_H$, for all $0 \neq t \in \int_l^H$;*

*(iii) $\Lambda(t) \neq 0$ and $\Lambda(ht_1\tilde{p}^1)t_2\tilde{p}^2 = \mu^{-1}(\beta)\Lambda(t)S^{-1}(h\beta)$, for all $0 \neq t \in \int_l^H$ and $h \in H$.*

*Furthermore, if $\beta$ is invertible then (i)–(iii) above are also equivalent to*

*(iv) $\Lambda(t) \neq 0$ and $\Lambda(t_1\tilde{p}^1)t_2\tilde{p}^2 = \mu^{-1}(\beta)\Lambda(t)S^{-1}(\beta)$, for all $0 \neq t \in \int_l^H$.*

Also, for a right cointegral $\Lambda$ on $H$ we have

$$\Lambda(S(h)h') = \mu(h_2)\Lambda(h'S^{-1}(h_1)), \quad \forall\, h, h' \in H.$$

The space of right cointegrals on $H$ will be denoted by $\mathscr{R}$. Then

$$H \otimes \mathscr{R} \ni h \otimes \Lambda \mapsto S^{-1}(h) \rightharpoonup \Lambda \in H^*$$

is an isomorphism of left quasi-Hopf $H$-bimodules, and so any non-zero right cointegral is non-degenerate and $\dim_k \mathscr{R} = 1$, according to the "cop" version of Theorem 7.48. We do not mention here the structure of $H^*$ in ${}_H^H\mathscr{M}_H$ but we will explain in the notes of this chapter a natural monoidal way in which one can obtain it. Also, in the next section we will see connections between left and right cointegrals given by the antipode.

We end with few comments about the infinite-dimensional case. As we have seen, the concept of cointegral only makes sense in the case where $H$ is finite dimensional: indeed, we need dual bases in $H$ and $H^*$ in order to make $H^*$ into a right quasi-Hopf bimodule. Also, the equivalent characterizations from Theorem 7.51 and Proposition 7.55 make no sense in the infinite-dimensional case, as they involve the modular element and/or left integrals, which can only be defined in the finite-dimensional case. To avoid this we can make use of the second structure theorem for quasi-Hopf bimodules presented in Chapter 6. Suppose first that $H$ is finite dimensional and consider $H^* \in {}_H\mathscr{M}_H^H$ exactly as in the statement of Proposition 7.46. The coinvariants $\overline{\lambda} \in \overline{H^{*\mathrm{co}(H)}}$ are called left alternative cointegrals on $H$, and the space of left alternative cointegrals is denoted by $\overline{\mathscr{L}} = \overline{H^{*\mathrm{co}(H)}}$. From the second structure theorem for quasi-Hopf bimodules we obtain immediately that $\dim_k \overline{\mathscr{L}} = 1$, assuming $H$ is finite dimensional. Furthermore, from the results obtained in Chapter 6 we have $\mathscr{L} \cong \overline{\mathscr{L}}$ as left $H$-modules.

By applying (3.1.9) and (3.2.1), we find that $\overline{\lambda} \in H^*$ is an alternative left cointegral if and only if

$$\overline{\lambda}(V^2 h_2 U^2)V^1 h_1 U^1 = \overline{\lambda}(S^{-1}(X_1^1 p^1)hS(S(X^3)f^1))X_2^1 p^2 S(X^2)f^2 \qquad (7.5.21)$$

for all $h \in H$. Then (7.5.21) can be used to extend the definition of left alternative cointegrals to infinite-dimensional quasi-Hopf algebras with bijective antipode. The same can be done in the case of right cointegrals.

## 7.6 Integrals, Cointegrals and the Fourth Power of the Antipode

We will present various Frobenius systems for a finite-dimensional quasi-Hopf algebra $H$ in terms of integrals and cointegrals in and on $H$, respectively. Combined with the uniqueness of a Frobenius system this yields a formula for the fourth power of the antipode of $H$, known as the Radford $S^4$ formula.

**Lemma 7.57** *Let $A$ be a Frobenius algebra and $S : A \to A$ an anti-algebra automorphism. If $(\phi, e = e^1 \otimes e^2)$ is a Frobenius system for $A$ then $(\phi \circ S, S^{-1}(e^2) \otimes S^{-1}(e^1))$ is a Frobenius system for $A$, too. In addition, if $\chi$ is the Nakayama automorphism associated to $(\phi, e)$ then $S^{-1} \circ \chi^{-1} \circ S$ is the Nakayama automorphism corresponding to $(\phi \circ S, S^{-1}(e^2) \otimes S^{-1}(e^1))$.*

*Proof* To see that $(\phi \circ S, S^{-1}(e^2) \otimes S^{-1}(e^1))$ is a Frobenius system for $A$ is straightforward, so we leave the verification of the details to the reader. Now, if we denote by $\eta$ the Nakayama automorphism defined by this new Frobenius system then

$$\eta(a) \overset{(7.1.1)}{=} (\phi \circ S)(S^{-1}(e^2)a)S^{-1}(e^1) = \phi(S(a)e^2)S^{-1}(e^1) \overset{(7.1.2)}{=} S^{-1}(\chi^{-1}(S(a))),$$

for all $a \in A$, as needed. □

From now on $H$ is a finite-dimensional quasi-Hopf algebra. From Proposition 7.20 we know that $(\phi, q^1 t_1 p^1 \otimes S(q^2 t_2 p^2))$ is a Frobenius system for $H$, where $t$ is a non-zero left integral in $H$ and $\phi \in H^*$ is the unique map satisfying one of the equivalent conditions in (7.2.9). By Lemma 7.50, $\lambda := \phi \circ S$ is a non-zero left cointegral on $H$, and so this Frobenius system can be rewritten as $(\lambda \circ S^{-1}, q^1 t_1 p^1 \otimes S(q^2 t_2 p^2))$, with $0 \neq t \in \int_l^H$ and $\lambda \in \mathscr{L}$ the unique map satisfying $\lambda(S^{-1}(q^1 t_1 p^1))q^2 t_2 p^2 = 1_H$ or, equivalently, $\lambda(q^2 t_2 p^2)q^1 t_1 p^1 = 1_H$. By Theorem 7.51 this reduces to $\mu(\beta)\lambda(t) = 1$, and we shall see that this is equivalent to $\lambda(S^{-1}(t)) = 1$.

We apply the previous lemma to the above Frobenius system for $H$. In this way we get that $(\lambda, q^2 t_2 p^2 \otimes S^{-1}(q^1 t_1 p^1))$ is another Frobenius system for $H$. From Proposition 7.9 there exists an invertible element $\underline{g}$ in $H$ such that

$$\lambda \circ S^{-1} = \lambda \leftharpoonup \underline{g} \text{ and } q^1 t_1 p^1 \otimes S(q^2 t_2 p^2) = q^2 t_2 p^2 \otimes \underline{g}^{-1} S^{-1}(q^1 t_1 p^1). \tag{7.6.1}$$

Furthermore, by (7.1.3) we have

$$\underline{g} = \lambda(S^{-1}(q^2 t_2 p^2))S^{-1}(q^1 t_1 p^1) \quad \text{and } \underline{g}^{-1} = \lambda(q^1 t_1 p^1)S(q^2 t_2 p^2). \tag{7.6.2}$$

**Definition 7.58** The invertible element $\underline{g}$ of $H$ defined by (7.6.2) is called the modular element of $H$.

Clearly, $\underline{g}$ is independent of the choice of the pair $(\lambda, t)$ with the above properties.

**Proposition 7.59** *Let $H$ be a finite-dimensional quasi-Hopf algebra and $(\lambda, t) \in \mathscr{L} \times \int_l^H$ satisfying $\lambda(S^{-1}(t)) = 1$. Then $(\lambda \circ S^{-1}, q^1 t_1 p^1 \otimes S(q^2 t_2 p^2))$ is a Frobenius system for $H$.*

*Proof* All we have to prove is that giving a pair $(\lambda, t) \in \mathscr{L} \times \int_l^H$ such that, for instance, $\lambda(S^{-1}(q^1 t_1 p^1))q^2 t_2 p^2 = 1_H$ is equivalent to giving a pair $(\lambda, t) \in \mathscr{L} \times \int_l^H$ such that $\lambda(S^{-1}(t)) = 1$. Then everything follows because of Proposition 7.20.

Indeed, if $\lambda(S^{-1}(q^1 t_1 p^1))q^2 t_2 p^2 = 1_H$ then by applying $\varepsilon$ to both sides we get $\lambda(S^{-1}(t)) = 1$. Conversely, if $\lambda(S^{-1}(t)) = 1$ then

$$1 = \lambda(S^{-1}(t)) = (\lambda \leftharpoonup \underline{g})(t) = \lambda(\underline{g}t) = \varepsilon(\underline{g})\lambda(t) = \mu(\beta)\lambda(S^{-1}(t))\lambda(t) = \mu(\beta)\lambda(t).$$

Then Theorem 7.51 says that $\lambda(q^2 t_2 p^2)q^1 t_1 p^1 = 1_H$ and as we mentioned several times this is equivalent to $\lambda(S^{-1}(q^1 t_1 p^1))q^2 t_2 p^2 = 1_H$, as required. □

**Remark 7.60** If $(\lambda, t) \in \mathscr{L} \times \int_l^H$ is such that $\lambda(S^{-1}(t)) = 1$ then the inverse of $\xi : H^* \to H$ considered in the proof of Proposition 7.20 is given by $\xi^{-1}(h) = h \rightharpoonup \lambda \circ S^{-1}$, for all $h \in H$. This is because $\xi$ is left $H$-linear and so $\xi(h \rightharpoonup \lambda \circ S^{-1}) = h\xi(\lambda \circ S^{-1}) = h\xi(\phi) = h$, for all $h \in H$. Thus the pair $(\lambda, t)$ also has the property that $t \rightharpoonup \lambda \circ S^{-1} = \varepsilon$ (since $\xi(\varepsilon) = t$).

Note also that $\lambda \circ S^{-1} = \lambda \leftharpoonup \underline{g}$ implies

$$\lambda \circ S = (\lambda \circ S^{-1} \leftharpoonup \underline{g}^{-1}) \circ S = S^{-1}(\underline{g}^{-1}) \rightharpoonup \lambda. \tag{7.6.3}$$

Another Frobenius system for $H$ can be obtained by working with $H^{\mathrm{cop}}$ instead of $H$. This allows to find a bijection between the space of left and right cointegrals.

**Proposition 7.61** *Let $H$ be a finite-dimensional quasi-Hopf algebra, $t$ a non-zero left integral in $H$ and $\lambda \in \mathscr{L}$ and $\Lambda \in \mathscr{R}$ such that $\lambda(S^{-1}(t)) = 1$ and $\Lambda(S(t)) = 1$.*

*Then* $u := \mu(V^1)S^2(V^2)$ *is invertible in* $H$ *and* $\lambda\circ S^{-1} = \Lambda \leftharpoonup u$*, where the element* $V = V^1\otimes V^2$ *is given by* (7.3.1)*. Consequently,*

$$\mathscr{L}\ni\lambda\mapsto\lambda\circ S^{-1}\leftharpoonup u^{-1}\in\mathscr{R}$$

*is a well-defined bijection.*

*Proof* The Frobenius system $(\lambda\circ S^{-1}, q^1t_1p^1\otimes S(q^2t_2p^2))$ of $H$ turns into $(\Lambda\circ S, \tilde{q}^2t_2\tilde{p}^2\otimes S^{-1}(\tilde{q}^1t_1\tilde{p}^1))$ for $H^{\rm cop}$, where $\Lambda$ is the unique right cointegral on $H$ such that $\Lambda(S(t)) = 1$. By Lemma 7.57 we have that $(\Lambda, \tilde{q}^1t_1\tilde{p}^1\otimes S(\tilde{q}^1t_2\tilde{p}^1))$ is also a Frobenius system for $H^{\rm cop}$, and therefore for $H$ as well. Now, according to Proposition 7.9, the Frobenius systems $(\lambda\circ S^{-1}, q^1t_1p^1\otimes S(q^2t_2p^2))$ and $(\Lambda, \tilde{q}^1t_1\tilde{p}^1\otimes S(\tilde{q}^2t_2\tilde{p}^2))$ are related through an invertible element $u\in H$ satisfying

$$\lambda\circ S^{-1} = \Lambda\leftharpoonup u,\quad q^1t_1p^1\otimes S(q^2t_2p^2) = \tilde{q}^1t_1\tilde{p}^1\otimes u^{-1}S(\tilde{q}^2t_2\tilde{p}^2). \tag{7.6.4}$$

Thus to prove the first assertion it suffices to show that $u = \mu(V^1)S^2(V^2)$. To this end, we first use (7.1.3) to find

$$u = \lambda(S^{-1}(\tilde{q}^1t_1\tilde{p}^1))S(\tilde{q}^2t_2\tilde{p}^2).$$

We have

$$\begin{aligned}
f^1\tilde{p}^1\otimes f^2\tilde{p}^2 &\overset{(3.2.20)}{=} f^1X^2S^{-1}(X^1\beta)\otimes f^2X^3\\
&\overset{(7.5.12),(3.2.1)}{=} f^1X^1g_1^2G^2\alpha S^{-1}(X^1g_1^1G^1)\otimes f^2X^3g^2\\
&\overset{(3.1.5),(3.2.17)}{=} f^1g_1^2G^1S(X^2)\alpha X^3S^{-1}(g^1)\otimes f^2g_2^2G^2S(X^1)\\
&\overset{(3.2.13),(3.2.19)}{=} S(q^2S^{-1}(g^2)_2)S^{-1}(g^1)\otimes S(q^1S^{-1}(g^2)_1),
\end{aligned}$$

where $f = f^1\otimes f^2 = F^1\otimes F^2$ and $f^{-1} = g^1\otimes g^2 = G^1\otimes G^2$, respectively. Also, by (3.2.21) it follows that any right integral $r$ in $H$ satisfies

$$r_1p^1h\otimes r_2p^2 = r_1p^1\otimes r_2p^2S(h),\ \ \forall\, h\in H. \tag{7.6.5}$$

In particular, if we set $r := S^{-1}(t)\in\int_r^H$ then

$$\begin{aligned}
t_1\tilde{p}^1\otimes t_2\tilde{p}^2 &\overset{(3.2.13)}{=} g^1S(r_2)f^1\tilde{p}^1\otimes g^2S(r_1)f^2\tilde{p}^2\\
&= g^1S(q^2S^{-1}(G^2)_2r^2)S^{-1}(G^1)\otimes g^2S(q^1S^{-1}(G^2)_1r_1)\\
&\overset{(7.2.7),(3.2.16)}{=} \mu(G^2)\alpha_1\delta^1S(q^2r_2)S^{-1}(G^1)\otimes\alpha_2\delta^2S(q^1r_1)\\
&\overset{(3.2.6)}{=} \mu(G^2)\alpha_1\beta S(q^2r_2X^3)S^{-1}(G^1)\otimes\alpha_2X^1\beta S(q^1r_1X^2)\\
&\overset{(3.2.20),(7.3.6)}{=} \mu(G^2)\alpha_1\beta S(q^2r_2p^2)S^{-1}(G^1)\otimes\alpha_2S(q^1r_1p^1)\\
&\overset{(7.6.5)}{=} \mu(G^2)\alpha_1\beta S(q^2r_2p^2\alpha_2)S^{-1}(G^1)\otimes S(q^1r_1p^1)\\
&\overset{(3.2.1)}{=} \mu(G^2)\beta S(q^2r_2p^2)S^{-1}(G^1)\otimes S(q^1r_1p^1),
\end{aligned}$$

and therefore

$$\begin{aligned}
\tilde{q}^1t_1\tilde{p}^1\otimes\tilde{q}^2t_2\tilde{p}^2 &= \mu(G^2)\tilde{q}^1\beta S(q^2r_2p^2)S^{-1}(G^1)\otimes\tilde{q}^2S(q^1r_1p^1)\\
&\overset{(7.6.5)}{=} \mu(G^2)\tilde{q}^1\beta S(q^2r_2p^2\tilde{q}^2)S^{-1}(G^1)\otimes S(q^1r_1p^1)
\end{aligned}$$

$$\overset{(3.2.20),(3.2.2)}{=} \mu(G^2)S(q^2r_2p^2)S^{-1}(G^1)\otimes S(q^1r_1p^1).$$

From here we compute

$$\begin{aligned} u &= \mu(G^2)\lambda(S^{-2}(G^1)q^2r_2p^2)S^2(q^1r_1p^1) \\ &\overset{(7.5.11)}{=} \mu(g^2)\mu(S^{-1}(g^1)_1)\lambda(q^2r_2p^2S(S^{-1}(g^1)_2))S^2(q^1r_1p^1) \\ &\overset{(7.6.5)}{=} \mu(g^2)\mu(S^{-1}(g^1)_1)\lambda(q^2r_2p^2)S^2(q^1r_1p^1S^{-1}(g^1)_2). \end{aligned}$$

Now, since

$$\begin{aligned} \lambda(q^2r_2p^2)q^1r_1p^1 &\overset{(7.3.4),(7.3.5)}{=} \lambda(V^2[S^{-1}(\tilde{q}^1)rS(\tilde{p}^1)]_2U^2)\tilde{q}^2V^1[S^{-1}(\tilde{q}^1)rS(\tilde{p}^1)]_1U^1\tilde{p}^2 \\ (r\in \textstyle\int_r^H, (7.2.7)) &= \mu(\tilde{q}^1)\lambda(V^2r_2U^2)\tilde{q}^2V^1r_1U^1 \\ &\overset{(7.5.13)}{=} \mu(\tilde{q}^1)\mu(x^1)\lambda(rS(x^2))\tilde{q}^2x^3 \\ &= \mu(\tilde{q}^1)\lambda(S^{-1}(t))\tilde{q}^2 \\ &= \mu(\tilde{q}^1)\tilde{q}^2, \end{aligned}$$

we get that

$$\begin{aligned} u &= \mu(g^2\tilde{q}^1)\mu(S^{-1}(g^1)_1)S^2(\tilde{q}^2S^{-1}(g^1)_2) \\ &\overset{(3.2.20),(3.2.13)}{=} \mu(g^2S(x^1)\alpha S^{-1}(f^2g_2^1G^2S(x^2)))S(f^1g_1^1G^1S(x^3)) \\ &\overset{(3.1.5),(3.2.17)}{=} \mu(x^3S^{-1}(f^2x^2\beta))S(f^1x^1) \overset{(3.2.19)}{=} \mu^{-1}(f^2p^2)S(f^1p^1) \\ &\overset{(7.3.1)}{=} \mu(V^1)S^2(V^2), \end{aligned}$$

as stated. Using (7.1.3) or a simple observation guarantees that

$$u^{-1} = \mu^{-1}(q_2^1g^2S(q^2))q_1^1g^1. \tag{7.6.6}$$

From the uniqueness of left and right cointegrals on $H$ it is clear at this point that $\mathscr{L}\ni\lambda\mapsto\lambda\circ S^{-1}\leftharpoonup u^{-1}\in\mathscr{R}$ is well defined. It is also bijective with the inverse defined by

$$\mathscr{R}\ni\Lambda\mapsto(\Lambda\leftharpoonup u)\circ S\in\mathscr{L}.$$

So our proof is complete. □

**Corollary 7.62** *If $H$ is a finite-dimensional unimodular quasi-Hopf algebra and $\overline{S}$ is the antipode of $H^*$ then $\overline{S}^{-1}(\mathscr{L})=\mathscr{R}$.*

*Proof* $\mu=\varepsilon$ implies $u=1$, and so $\lambda\circ S^{-1}=\Lambda$. Everything follows now from the uniqueness of left and right cointegrals on $H$. □

**Corollary 7.63** *Consider $(\Lambda,t)\in\mathscr{R}\times\int_l^H$ such that $\Lambda(S(t))=1$. Then $\Lambda\circ S=\Lambda\leftharpoonup uS^2(S^{-1}(u^{-1})\leftharpoonup\mu)\underline{g}^{-1}$, where $\underline{g}$ is the modular element of $H$, and where we define $h\leftharpoonup h^*:=h^*(h_1)h_2$, for all $h^*\in H^*$ and $h\in H$.*

*Proof* If $(\lambda,t)\in\mathscr{L}\times\int_l^H$ satisfies $\lambda(S^{-1}(t))=1$ we know that $\lambda\circ S^{-1}=\lambda\leftharpoonup\underline{g}$. If we think of this relation in $H^{\rm cop}$ then for any pair $(\Lambda,t)\in\mathscr{R}\times\int_l^H$ such that

$\Lambda(S(t)) = 1$ we have $\Lambda \circ S = \Lambda \leftharpoonup g_{\underline{\mathrm{cop}}}$ with $g_{\underline{\mathrm{cop}}} = \Lambda(S(\tilde{q}^1 t_1 \tilde{p}^1))S(\tilde{q}^2 t_2 \tilde{p}^2)$. We next show that $g_{\underline{\mathrm{cop}}} = uS^2(S^{-1}(u^{-1}) \leftharpoonup \mu)\underline{g}^{-1}$ and this would finish the proof.

By $\lambda \circ S^{-1} = \Lambda \leftharpoonup u$ we get $\Lambda \circ S = (\lambda \circ S^{-1} \leftharpoonup u^{-1}) \circ S = S^{-1}(u^{-1}) \rightharpoonup \lambda$, so

$$\begin{aligned} g_{\underline{\mathrm{cop}}} &= \lambda(\tilde{q}^1 t_1 \tilde{p}^1 S^{-1}(u^{-1}))S(\tilde{q}^2 t_2 \tilde{p}^2) \\ &\overset{(7.6.4)}{=} \lambda(q^1 t_1 p^1 S^{-1}(u^{-1}))uS(q^2 t_2 p^2) \\ &\overset{(7.5.16)}{=} \mu(S^{-1}(u^{-1})_1)\lambda(q^1 t_1 p^1)uS^2(S^{-1}(u^{-1})_2)S(q^2 t_2 p^2) \\ &\overset{(7.6.2)}{=} uS^2(S^{-1}(u^{-1}) \leftharpoonup \mu)\underline{g}^{-1}, \end{aligned}$$

as desired. Notice that $g_{\underline{\mathrm{cop}}} = \underline{g}^{-1}$ in the case when $H$ is unimodular. □

**Corollary 7.64** *For a finite-dimensional quasi-Hopf algebra $H$ assume that $\mathscr{L} = \mathscr{R}$. Then $\underline{g} = \mu(\beta)\mu^{-1}(\beta)^{-1}u$. Consequently, if $H$ is unimodular and admits a non-zero left cointegral that is at the same time right cointegral then $\underline{g} = 1_H$.*

*Proof* Since $\dim_k \mathscr{L} = \dim_k \mathscr{R} = 1$ it follows that $\mathscr{L} = \mathscr{R}$ if and only if $\mathscr{L} \cap \mathscr{R} \neq 0$.

Let $0 \neq T \in \mathscr{L} \cap \mathscr{R}$ and $t \in \int_l^H$ such that $T(S^{-1}(t)) = 1$. Then $T \circ S^{-1} = \Lambda \leftharpoonup u$, for some non-zero $\Lambda \in \mathscr{R}$. But $\Lambda = cT$ for a certain $c \in k$, so $T \circ S^{-1} = cT \leftharpoonup u$. We have $\mu(\beta)T(t) = 1$, hence

$$1 = T(S^{-1}(t)) = cT(ut) = c\varepsilon(u)T(t) = c\mu^{-1}(\beta)\mu(\beta)^{-1}.$$

We get $c = \mu(\beta)\mu^{-1}(\beta)^{-1}$ and therefore

$$\begin{aligned} \underline{g} &= T(S^{-1}(q^2 t_2 p^2))S^{-1}(q^1 t_1 p^1) = cT(uq^2 t_2 p^2)S^{-1}(q^1 t_1 p^1) \\ &\overset{(7.2.2)}{=} cT(q^2 t_2 p^2)S^{-1}(q^1 t_1 p^1)u = \mu(\beta)\mu^{-1}(\beta)^{-1}u, \end{aligned}$$

as stated. □

**Corollary 7.65** *The inverse of the Nakayama isomorphism $\chi$ introduced in Proposition 7.20 is $\chi^{-1}(h) = \mu(S^{-1}(uhu^{-1})_2)S^{-1}(S^{-1}(uhu^{-1})_1)$, for all $h \in H$, where $u$ is the element of $H$ introduced in Proposition 7.61. Consequently,*

$$\mu^{-1}(\tilde{q}^1 h_1 \tilde{p}^1)\lambda \leftharpoonup S^{-1}(\tilde{q}^2 h_2 \tilde{p}^2) = \mu^{-1}(\alpha)\mu(\beta)S(h) \rightharpoonup \lambda, \tag{7.6.7}$$

*for all $\lambda \in \mathscr{L}$ and $h \in H$.*

*Proof* It follows from Proposition 7.61 that

$$\begin{aligned} \lambda(\tilde{q}^2 t_2 \tilde{p}^2)\tilde{q}^1 t_1 \tilde{p}^1 &= \Lambda(uS(\tilde{q}^2 t_2 \tilde{p}^2))\tilde{q}^1 t_1 \tilde{p}^1 \\ &\overset{(3.2.22)}{=} \mu(S^{-1}(u)_2)\Lambda(S(\tilde{q}^2 t_2 \tilde{p}^2))\tilde{q}^1 t_1 \tilde{p}^1 S^{-1}(S^{-1}(u)_1). \end{aligned}$$

Hence, by (7.1.1) we obtain that

$$\begin{aligned} \chi^{-1}(h) &= \phi(hS(q^2 t_2 p^2))q^1 t_1 p^1 = \lambda(q^2 t_2 p^2 S^{-1}(h))q^1 t_1 p^1 \\ &= \lambda(\tilde{q}^2 t_2 \tilde{p}^2 S^{-1}(hu^{-1}))\tilde{q}^1 t_1 \tilde{p}^1 \\ &\overset{(3.2.22)}{=} \mu(S^{-1}(hu^{-1})_2)\lambda(\tilde{q}^2 t_2 \tilde{p}^2)\tilde{q}^1 t_1 \tilde{p}^1 S^{-1}(S^{-1}(hu^{-1})_1) \\ &= \mu(S^{-1}(uhu^{-1})_2)S^{-1}(S^{-1}(uau^{-1})_1). \text{ QED} \end{aligned}$$

The above formula allows us to compute:

$$\begin{aligned}
&S^{-1}\chi^{-1}S^2(h)\\
&\quad = \quad \mu(S^{-1}(u^{-1})_2 S(h)_2 S^{-1}(u)_2)S^{-2}(S^{-1}(u^{-1})_1 S(h)_1 S^{-1}(u)_1)\\
&\quad = \quad \mu^{-1}(q^1_2 g^2 S(q^2))\mu(V^1)\mu(q^1_{(1,2)} g^1_2 S(V^2 h)_2)S^{-2}(q^1_{(1,1)} g^1_1 S(V^2 h)_1)\\
&\overset{(3.2.13),(3.2.17)}{\underset{(3.1.7)}{=}} \mu^{-1}(x^3 G^2 S(q^2 X^1))\mu(V^1)\mu(x^2 G^1 S(V^2_1 h_1 X^2) f^2)\\
&\qquad\qquad S^{-2}(x^1 q^1 S(V^2_2 h_2 X^3) f^1)\\
&\overset{(7.3.1),(3.2.13)}{\underset{(3.2.17)}{=}} \mu^{-1}(x^3 G^2 S(q^2 X^1))\mu(S^{-1}(F^2 Y^3 p^2) y^1)\mu(x^2 G^1 S(y^2 h_1 X^2) F^1 Y^2 p^1_2)\\
&\qquad\qquad S^{-2}(x^1 q^1 S(y^3 h_2 X^3) Y^1 p^1_1)\\
&\overset{(3.2.19),(5.5.16)}{=} \mu(X^1\beta)\mu(S^{-1}(F^2 p^2) y^1)\mu(S(y^2 h_1 X^2) F^1 p^1) S^{-1}(y^3 h_2 X^3)\\
&\overset{(3.2.20),(7.5.12)}{=} (\mu^{-1}(\alpha)\mu(\beta))^{-1}\mu^{-1}(\tilde{q}^1 h_1 \tilde{p}^1) S^{-1}(\tilde{q}^2 h_2 \tilde{p}^2),
\end{aligned}$$

for all $h \in H$. Consequently,

$$\begin{aligned}
S(h) \rightharpoonup \lambda &= S(h) \rightharpoonup \phi \circ S = (\phi \leftharpoonup S^2(h)) \circ S = (\chi^{-1}S^2(h) \rightharpoonup \phi) \circ S\\
&= (\chi^{-1}S^2(h) \rightharpoonup \lambda \circ S^{-1}) \circ S = \lambda \leftharpoonup S^{-1}\chi^{-1}S^2(h)\\
&= (\mu^{-1}(\alpha)\mu(\beta))^{-1}\mu^{-1}(\tilde{q}^1 h_1 \tilde{p}^1)\lambda \leftharpoonup S^{-1}(\tilde{q}^2 h_2 \tilde{p}^2),
\end{aligned}$$

finishing our proof. □

The formulas that we have just proved indicate how to find right cointegrals when we know a left cointegral, and vice versa.

**Example 7.66** For $H(2)$, $P_g$ is both a left and right cointegral and $\underline{g} = 1$.

*Proof* $H(2)$ is unimodular and has the antipode defined by the identity map. Thus in this particular case the formula $\lambda \circ S^{-1} = \Lambda \leftharpoonup u$ reduces to $\lambda = \Lambda$, and so $\mathscr{L} = \mathscr{R}$. From Example 7.53 we deduce that $P_g$ is a left and right non-zero cointegral on $H(2)$, and from Corollary 7.64 we get $\underline{g} = 1$. □

**Example 7.67** For $H_\pm(8)$ we have $\mathscr{R} = k(\omega P_{x^3} + \overline{\omega} P_{gx^3})$, $\underline{g} = \omega 1 + \overline{\omega} g$ and $\underline{g}^{-1} = \overline{\omega} 1 + \omega g$.

*Proof* To find a right cointegral on $H_\pm(8)$ we compute $\lambda \circ S^{-1}$ and the element $u$. Then $\lambda \circ S^{-1} \leftharpoonup u^{-1}$ will be a non-zero right cointegral on $H_\pm(8)$.

Consider the left integral $t = (1+g)x^3$ and take $\lambda = cP_{x^3}$ with $c \in k$ that has to be determined such that $\lambda(S^{-1}(t)) = 1$. Actually, since $\beta = 1$ we need to find that unique $c \in k$ such that $\lambda(t) = 1$ and it then follows that we should have $c = 1$, and thus $\lambda = P_{x^3}$.

We use now $(p_+ \pm ip_-)(p_+ \mp ip_-) = 1$ to see that $S^{-1}(x) = -(p_+ \mp ip_-)x$, and

$$\begin{aligned}
&S^{-1}(x^2) = \mp ix^2, \quad S^{-1}(x^3) = \pm i(p_+ \mp ip_-)x^3, \quad S^{-1}(gx) = (p_+ \pm ip_-)x,\\
&\qquad S^{-1}(gx^2) = \mp igx^2, \quad S^{-1}(gx^3) = \mp i(p_+ \pm ip_-)x^3.
\end{aligned}$$

In particular, we get $\lambda(S^{-1}(g^ix^j)) = 0$ except in the following two cases where

$$\lambda(S^{-1}(x^3)) = \pm i\lambda(p_+x^3) + \lambda(p_-x^3) = \frac{1}{2}(1 \pm i) = \omega,$$
$$\lambda(S^{-1}(gx^3)) = \mp i\lambda(p_+x^3) + \lambda(p_-x^3) = \frac{1}{2}(\mp i + 1) = \overline{\omega}.$$

In other words, we have $\lambda \circ S^{-1} = \omega P_{x^3} + \overline{\omega} P_{gx^3}$. It can be easily checked that $f = f^{-1} = p_R$ in the case when $H = H_{\pm}(8)$. We conclude that $u = 1$, even if $H$ is not unimodular. Thus $\omega P_{x^3} + \overline{\omega} P_{gx^3}$ is a right non-zero cointegral on $H_{\pm}(8)$.

We end by computing $\underline{g}$. Since $\beta = 1$, formula (7.2.3) implies $q^2 t_2 p^2 \otimes q^1 t_1 p^1 = t_1 p^1 \otimes t_2 p^2$, hence $\underline{g} = \lambda(S^{-1}(t_2 p^2))S^{-1}(t_1 p^1)$. By the expression of $\Delta(t)p_R$ found in Example 7.54 we then obtain

$$\underline{g} = \lambda(S^{-1}(x^3))1 + \lambda(S^{-1}(gx^3))g = \omega 1 + \overline{\omega} g,$$

as desired. A simple inspection shows that $\underline{g}^{-1} = \overline{\omega} 1 + \omega g$. □

We next indicate a connection between $\lambda \circ S$ and $\Lambda$.

**Proposition 7.68** *Let $t$ be a left integral in $H$ and $\lambda \in \mathscr{L}$ and $\Lambda \in \mathscr{R}$ such that $\lambda(S^{-1}(t)) = 1$ and $\Lambda(S(t)) = 1$. Then*

$$v := (\mu^{-1}(\underline{g})\mu(\beta))^{-1}\mu(S(p^2)f^1)S(p^1)f^2$$

*is invertible in $H$ and $\lambda \circ S = \Lambda \leftharpoonup v$. Consequently, $\mathscr{L} \ni \lambda \mapsto \lambda \circ S \leftharpoonup v^{-1} \in \mathscr{R}$ is a well-defined bijective map, too.*

*Proof* By the "op" version of Proposition 7.20 we have that $(\phi_{\rm op}, S^{-1}(q^2 r_2 p^2) \otimes q^1 r_1 p^1)$ is a Frobenius system for $H$, where $r$ is a non-zero right integral in $H$ and $\phi_{\rm op}$ is the unique element of $H^*$ satisfying $\phi_{\rm op}(S^{-1}(q^2 r_2 p^2))q^1 r^1 p^1 = 1$ or, equivalently, $\phi_{\rm op}(q^1 r_1 p^1)q^2 r_2 p^2 = 1$.

We set now $t = S(r) \in \int_l^H$ and take $\lambda \in \mathscr{L}$ and $\Lambda \in \mathscr{R}$ as in the statement, that is, such that $\lambda(S^{-1}(t)) = 1$ and $\Lambda(S(t)) = 1$. We will prove that $\phi_{\rm op} = \mu^{-1}(\tilde{p}^1)S^{-2}(\tilde{p}^2) \rightharpoonup \lambda \circ S$. We begin by showing that

$$V^1 r_1 U^1 \otimes V^2 r_2 U^2 = S^{-1}(q^2 t_2 p^2) \otimes S^{-1}(q^1 t_1 p^1). \tag{7.6.8}$$

Indeed, by (7.3.4) and (7.3.5) we have

$$\begin{aligned}
\tilde{q}^1_1 p^1 \otimes \tilde{q}^1_2 p^2 S(\tilde{q}^2) &= [\tilde{q}^1 S(\tilde{p}^1)]_1 U^1 \tilde{p}^2 \otimes [\tilde{q}^1 S(\tilde{p}^1)]_2 U^2 S(\tilde{q}^2) \\
&\overset{(7.3.2)}{=} [\tilde{q}^1 S(\tilde{q}^2_1 \tilde{p}^1)]_1 U^1 \tilde{q}^2_2 \tilde{p}^2 \otimes [\tilde{q}^1 S(\tilde{q}^2_1 \tilde{p}^1)]_2 U^2 \\
&\overset{(3.2.24)}{=} U^1 \otimes U^2.
\end{aligned}$$

So we have shown that

$$U = \tilde{q}^1_1 p^1 \otimes \tilde{q}^1_2 p^2 S(\tilde{q}^2). \tag{7.6.9}$$

Notice that the "op" version of the above relation is

$$V^1 \otimes V^2 = q^1 \tilde{p}^1_1 \otimes S^{-1}(\tilde{p}^2) q^2 \tilde{p}^1_2. \tag{7.6.10}$$

We next compute

$$
\begin{aligned}
V^1r_1U^1\otimes V^2r_2U^2 &= V^1r_1\tilde{q}_1^1p^1\otimes V^2r_2\tilde{q}_2^1p^2S(\tilde{q}^2)\\
&\overset{(7.3.1)}{=} S^{-1}(f^2P^2)S^{-1}(t)_1p^1\otimes S^{-1}(f^1P^1)S^{-1}(t)_2p^2\\
&\overset{(3.2.13),(3.2.19)}{=} S^{-1}(S(x^1)f^2t_2P^2)\otimes S^{-1}(S(x^2)f^1t_1P^1)\beta S(x^3)\\
&\overset{(3.1.5),(3.2.17)}{=} S^{-1}(F^2x^3g_2^2t_2P^2)\otimes S^{-1}(f^2F_2^1x^2g_1^2t_1P^1)\beta f^1F_1^1x^1\\
&\overset{(7.5.12),(3.2.1)}{=} S^{-1}(x^3t_2P^2)\otimes S^{-1}(S(x^1)\alpha x^2t_1P^1)\\
&\overset{(3.2.20)}{=} S^{-1}(\tilde{q}^2t_2P^2)\otimes S^{-1}(\tilde{q}^1t_1P^1)\\
&\overset{(7.3.6)}{=} S^{-1}(q^2t_2P^2)\otimes S^{-1}(q^1t_1P^1),
\end{aligned}
$$

as desired. On the other hand,

$$
\begin{aligned}
V^1r_1U^1\otimes V^2r_2U^2 &= q^1\tilde{p}_1^1r_1\tilde{q}_1^1p^1\otimes S^{-1}(\tilde{p}^2)q^2\tilde{p}_2^1r_2\tilde{q}_2^1p^2S^{-1}(\tilde{q}^2)\\
&= \mu^{-1}(\tilde{p}^1)q^1r_1p^1\otimes S^{-1}(\tilde{p}^2)q^2r_2p^2,
\end{aligned}
$$

from which we conclude that

$$\mu^{-1}(\tilde{p}^1)q^1r_1p^1\otimes S^{-1}(\tilde{p}^2)q^2r_2p^2 = S^{-1}(q^2t_2p^2)\otimes S^{-1}(q^1t_1p^1).$$

Note that

$$hq^1r_1\otimes q^2r_2 \overset{(3.2.21)}{=} q^1h_{(1,1)}r_1\otimes S^{-1}(h_2)q^2h_{(1,2)}r_2 \overset{(7.2.7)}{=} \mu^{-1}(h_1)q^1r_1\otimes S^{-1}(h_2)q^2r_2,$$

for all $h\in H$. Since

$$\mu(\tilde{q}^1)\mu^{-1}(\tilde{p}^1)\tilde{q}^2q^1r_1p^1\otimes S^{-1}(\tilde{p}^2)q^2r_2p^2 = \mu(\tilde{q}^1)S^{-1}(q^2t_2p^2S(\tilde{q}^2))\otimes S^{-1}(q^1t_1p^1),$$

the latest relation implies

$$q^1r_1p^1\otimes q^2r_2p^2 = \mu(\tilde{q}^1)\tilde{q}^2S^{-1}(q^2t_2p^2)\otimes S^{-1}(q^1t_1p^1), \tag{7.6.11}$$

where we also made use of (3.2.24).

From the proof of the previous proposition, $\lambda(q^2r_2p^2)q^1r_1p^1=\mu(\tilde{q}^1)\tilde{q}^2$. Hence

$$
\begin{aligned}
1_H &\overset{(3.2.24)}{=} \mu(S(\tilde{p}^1)\tilde{q}^1\tilde{p}_1^2)\tilde{q}^2\tilde{p}_2^2\\
&= \mu^{-1}(\tilde{p}^1)\mu(\tilde{p}_1^2)\lambda(q^2r_2p^2)q^1r_1p^1\tilde{p}_2^2\\
&\overset{(7.6.5)}{=} \mu^{-1}(\tilde{p}^1)\mu(\tilde{p}_1^2)\lambda(q^2r_2p^2S(\tilde{p}_2^2))q^1r_1p^1\\
&\overset{(7.5.11)}{=} \mu^{-1}(\tilde{p}^1)\lambda(S^{-1}(\tilde{p}^2)q^2r_2p^2)q^1r_1p^1\\
&= \langle\lambda\leftharpoonup\mu^{-1}(\tilde{p}^1)S^{-1}(\tilde{p}^2),q^2r_2p^2\rangle q^1r_1p^1.
\end{aligned}
$$

From the uniqueness of the map $\phi_{\rm op}$ it follows that

$$\phi_{\rm op}=(\lambda\leftharpoonup\mu^{-1}(\tilde{p}^1)S^{-1}(\tilde{p}^2))\circ S=\mu^{-1}(\tilde{p}^1)S^{-2}(\tilde{p}^2)\rightharpoonup\lambda\circ S,$$

as claimed. By the definition of $p_L$ it is immediate that $d:=\mu^{-1}(\tilde{p}^1)S^{-2}(\tilde{p}^2)$ is invertible in $H$, therefore $(\lambda\circ S,S^{-1}(q^2r_2p^2)d\otimes q^1r_1p^1)$ is a Frobenius system for $H$ (see Proposition 7.9) whenever $(\lambda,r)\in\mathscr{L}\times\int_r^H$ is such that $\lambda(r)=1$. Comparing it with $(\Lambda, \tilde{q}^1 t_1 \tilde{p}^1 \otimes S(\tilde{q}^2 t_2 \tilde{p}^2))$ we conclude that there is an invertible element $v \in H$ such that $\lambda \circ S = \Lambda \leftharpoonup v$. According to (7.1.3) we have

$$\begin{aligned} v &= \lambda(S(\tilde{q}^1 t_1 \tilde{p}^1))S(\tilde{q}^2 t_2 \tilde{p}^2) \\ &\overset{(7.3.6),(3.2.13)}{=} \langle \lambda, S(\tilde{p}^1) f^2 S(t)_2 g^2 S(q^1)\rangle S(\tilde{p}^2) f^1 S(t)_1 g^1 S(q^2) \\ &\overset{(7.3.8),(7.3.1)}{=} \langle \lambda, S^{-1}(g^2) q^2 (g^1 S(t))_2 U^2\rangle q^1 (g^1 S(t))_1 U^1 \\ &\overset{(7.5.11),(7.2.7)}{=} \mu(g^2_1)\mu^{-1}(g^1)\langle \lambda, q^2 S(t)_2 U^2 S(g^2_2)\rangle q^1 S(t)_1 U^1 \\ ((7.3.2),\ S(t) \in \textstyle\int_r^H) \quad &= \mu(g^2_1)\mu^{-1}(g^1)\langle \lambda, q^2 S(t)_2 U^2\rangle q^1 S(t)_1 U^1 g^2_2 \\ &\overset{(7.3.4),(7.2.7)}{=} \mu^{-1}(g^1)\mu(\tilde{q}^1 g^2_1)\langle \lambda, V^2 S(t)_2 U^2\rangle \tilde{q}^2 V^1 S(t)_1 U^1 g^2_2 \\ ((7.5.13),\ S(t) \in \textstyle\int_r^H) \quad &= \mu(S(g^1)\tilde{q}^1 g^2_1)\lambda(S(t))\tilde{q}^2 g^2_2. \end{aligned}$$

By (3.2.20), (3.1.5) and (3.2.17) we compute

$$\begin{aligned} &S(g^1)\tilde{q}^1 g^2_1 \otimes \tilde{q}^2 g^2_2 \\ &\quad = S(g^1_1 G^1 S(x^3))\alpha g^1_2 G^2 S(x^2) f^1 \otimes g^1 S(x^1) f^2 \\ &\quad \overset{(3.2.1),(7.5.12)}{=} S(x^2 \beta S(x^3)) f^1 \otimes S(x^1) f^2 = S(p^2) f^1 \otimes S(p^1) f^2. \qquad (7.6.12) \end{aligned}$$

We also have $\lambda(S(t)) = (S^{-1}(\underline{g}^{-1}) \rightharpoonup \lambda)(t) = \mu^{-1}(\underline{g}^{-1})\lambda(t) = (\mu^{-1}(\underline{g})\mu(\beta))^{-1}$. Thus $v = (\mu^{-1}(\underline{g})\mu(\beta))^{-1}\mu(S(p^2)f^1)S(p^1)f^2$, as claimed. It is easy to see that

$$v^{-1} = \mu^{-1}(\underline{g})\mu(\beta q^2 g^1 S(q^1_2))g^2 S(q^1_1),$$

and so the proof is finished. □

**Corollary 7.69** *Let $H$ be a finite-dimensional unimodular quasi-Hopf algebra and denote by $\overline{S}$ the antipode of $H^*$. Then $\overline{S}(\mathscr{L}) = \mathscr{R}$.*

*Proof* In this case we have $\mu = \varepsilon$ and therefore $v = 1_H$. □

We can now prove the formula for the fourth power of the antipode.

**Theorem 7.70** *Let $H$ be a finite-dimensional quasi-Hopf algebra with modular elements $\underline{g} \in H$ and $\mu \in H^*$. Then, for all $h \in H$, we have*

$$S^4(\mu^{-1} \rightharpoonup (h \leftharpoonup \mu)) = S^3(f_\mu^{-1})S(\underline{g})hS(\underline{g}^{-1})S^3(f_\mu),$$

*where $f_\mu := \mu(f^1)f^2$ and, for all $h^* \in H^*$ and $h \in H$,*

$$h^* \rightharpoonup h := h^*(h_2)h_1, \quad h \leftharpoonup h^* = h^*(h_1)h_2.$$

*Proof* As before, consider $\lambda \in \mathscr{L}$ and $t \in \int_l^H$ such that $\lambda(S^{-1}(t)) = 1$. As we have seen, $(\lambda, t)$ defines two Frobenius systems on $H$, namely $(\lambda \circ S^{-1}, q^1 t_1 p^1 \otimes S(q^2 t_2 p^2))$ and $(\lambda, q^2 t_2 p^2 \otimes S^{-1}(q^1 t_1 p^1))$. Moreover, they are related through the modular element $\underline{g}$ of $H$, in the sense of relations (7.6.1). Denote by $\chi$ and $\eta$, respectively, the Nakayama automorphisms corresponding to them.

On the one hand, by Corollary 7.10, we have $\chi(h) = \underline{g}^{-1}\eta(h)\underline{g}$, for all $h \in H$. On the other hand, by Lemma 7.57, we know that $\eta = S^{-1} \circ \chi^{-1} \circ S$. From these two expressions of $\eta$ we deduce that $\chi(h) = \underline{g}^{-1}S^{-1}(\chi^{-1}(S(h)))\underline{g}$, for all $h \in H$, which is clearly equivalent to $\chi \circ S^{-1} \circ \chi \circ S = \mathrm{Inn}_{\underline{g}^{-1}}$, where by $\mathrm{Inn}_{\underline{g}^{-1}}$ we denote the inner automorphism of $H$ produced by $\underline{g}^{-1}$, that is, $\mathrm{Inn}_{\underline{g}^{-1}}(h) = \underline{g}^{-1}h\underline{g}$, for all $h \in H$.

Now, we use (7.2.10) to get $(\chi \circ S)(h) = \mu(S(h)_1)S^2(S(h)_2)$, for all $h \in H$, and so $S^{-1} \circ \chi \circ S = S(S(h) \leftharpoonup \mu)$, for all $h \in H$. If we define $S_\mu(h) := S(h) \leftharpoonup \mu$, for all $h \in H$, then the last relation reads $S^{-1} \circ \chi \circ S = S \circ S_\mu$. Therefore

$$\mathrm{Inn}_{\underline{g}^{-1}} = \chi \circ S^{-1} \circ \chi \circ S = \chi \circ S \circ S_\mu = S^2 \circ S_\mu^2.$$

Observe next that

$$S_\mu(h) \overset{(3.2.13)}{=} \mu(g^1 S(h_2) f^1) g^2 S(h_1) f^2 = f_\mu^{-1} S(\mu^{-1} \rightharpoonup h) f_\mu, \ \forall\, h \in H.$$

Thus $S_\mu^2(h) = f_\mu^{-1} S(\mu^{-1} \rightharpoonup S_\mu(h)) f_\mu = f_\mu^{-1} S(\mu^{-1} \rightharpoonup (S(h) \leftharpoonup \mu)) f_\mu$, for all $h \in H$, and since $S$ is bijective we conclude that

$$\mathrm{Inn}_{\underline{g}^{-1}}(S^{-1}(h)) = S^2(f_\mu^{-1}) S^3(\mu^{-1} \rightharpoonup (h \leftharpoonup \mu)) S^2(f_\mu), \ \forall\, h \in H.$$

By applying $S$ to both sides we get the expression for $S^4$ stated above. □

Observe that the formula for the fourth power of the antipode in the statement of Theorem 7.70 can be restated, for all $h \in H$, as

$$\mu(f^1) S^{-2}(h) S^{-1}(\underline{g}^{-1}) S(f^2) = \mu(h_1 f^1) \mu^{-1}(h_{(2,2)}) S^{-1}(\underline{g}^{-1}) S(S(h_{(2,1)}) f^2). \tag{7.6.13}$$

**Corollary 7.71** *If $H$ is a finite-dimensional unimodular quasi-Hopf algebra then $S^4 = \mathrm{Inn}_{S(\underline{g})}$, where $\underline{g}$ is the modular element of $H$. Furthermore, if $\mathscr{L} = \mathscr{R}$ then $S^4$ is the identity morphism of $H$.*

*Proof* In the unimodular case we have $\mu = \varepsilon$ and $f_\mu = 1$ and if, moreover, $\mathscr{L} = \mathscr{R}$ then $\underline{g} = 1$; see Corollary 7.64. □

We know how the antipode of $H^*$, or its inverse, carries a left or right cointegral on $H$. In the result below we will see how the antipode of $H$ itself carries left or right integrals in $H$.

**Proposition 7.72** *Let $t, r$ be a non-zero left, right integral, respectively, in $H$. Then*

$$\begin{aligned} S(t) &= \mu(\beta)^{-1} \mu(q^2 t_2 p^2) q^1 t_1 p^1, \\ S^{-1}(t) &= \mu^{-1}(\underline{g}) \mu(q^2 t_2 p^2) q^1 t_1 p^1, \\ S(r) &= (\mu^{-1}(\underline{g}) \mu(\alpha\beta))^{-1} \mu^{-1}(q^2 r_2 p^2) q^1 r_1 p^1, \\ S^{-1}(r) &= \mu(\alpha)^{-1} \mu^{-1}(q^2 r_2 p^2) q^1 r_1 p^1. \end{aligned}$$

*Proof* Consider $\lambda \in \mathscr{L}$ obeying $\lambda(S^{-1}(t)) = 1$, so that $\mu(\beta)\lambda(t) = 1$. If we define $r' = \mu(t_2 p^2) t_1 p^1$ then

$$\begin{aligned} r'h &= \mu(t_2 p^2) t_1 p^1 h \\ &\overset{(3.2.21)}{=} \mu(t_2 h_{(1,2)} p^2 S(h_2)) t_1 h_{(1,1)} p^1 \end{aligned}$$

$$\overset{(7.2.6)}{=} \mu(h_1)\mu(t_2p^2)\mu^{-1}(h_2)t_1p^1$$
$$= \varepsilon(h)r',$$

for all $h \in H$. Thus $r'$ is a right integral in $H$. As $\dim_k \int_r^H = 1$ there exist $c, c' \in k$ such that $S(t) = cr'$ and $S^{-1}(t) = c'r'$. We have

$$\lambda(S(t)) \overset{(7.6.3)}{=} (S^{-1}(\underline{g}^{-1}) \rightharpoonup \lambda)(t)$$
$$\overset{(7.2.6)}{=} \mu^{-1}(\underline{g}^{-1})\lambda(t) = (\mu^{-1}(\underline{g})\mu(\beta))^{-1},$$

$$\lambda(r') = \mu(t_2p^2)\lambda(t_1p^1) \overset{(7.2.3)}{=} \mu(S^{-1}(\beta)q^2t_2p^2)\lambda(q^1t_1p^1) \overset{(7.6.2)}{=} \mu^{-1}(\beta)\mu^{-1}(\underline{g}^{-1}).$$

Thus $c = (\mu(\beta)\mu^{-1}(\beta))^{-1}$ and $c' = \mu^{-1}(\beta\underline{g}^{-1})^{-1}$. These together with (7.2.3) prove the relations stated for $S(t)$ and $S^{-1}(t)$, respectively.

If $r$ is a right integral then denote $t = S(r)$ and consider $\lambda \in \mathscr{L}$ such that $\lambda(r) = \lambda(S^{-1}(t)) = 1$. As in the left-handed case one can show that $t' := \mu^{-1}(q^2r_2)q^1r_1$ is a left integral, hence there exist $b, b' \in k$ such that $S(r) = bt'$ and $S^{-1}(r) = b't'$. The formula for $S^{-1}(r)$ can be obtained from that for $S(t)$ by replacing $H$ with $H^{\rm op}$. As the formula for $S^{-1}(t)$ contains $\underline{g}$ and we do not have an analogue of $\underline{g}$ in $H^{\rm op}$ we cannot derive the formula for $S(r)$ from that for $S^{-1}(t)$. Nevertheless, we can obtain it by computing $b$ as follows.

We have $\lambda(S(r)) = (S^{-1}(\underline{g}^{-1}) \rightharpoonup \lambda)(r) = \varepsilon(\underline{g}^{-1})\lambda(r) = \lambda(t) = \mu(\beta)^{-1}$ and

$$\begin{aligned}
\lambda(t') &= \lambda(q^1r_1)\mu^{-1}(q^2r_2)\\
&= \lambda(q^1r_1p^1)\mu^{-1}(q^2r_2p^2\alpha)\\
&\overset{(7.6.11)}{=} \mu(\tilde{q}^1)\lambda(\tilde{q}^2S^{-1}(q^2t_2p^2))\mu^{-1}(S^{-1}(q^1t_1p^1)\alpha)\\
&= \mu(\tilde{q}^1)\phi(q^2t_2p^2S(\tilde{q}^2))\mu(q^1t_1p^1)\mu^{-1}(\alpha),
\end{aligned}$$

where in the second equality we applied the "op" version of (7.2.3), and where $\phi = \lambda \circ S^{-1}$ is the notation used for the Frobenius morphism of $H$. We now make use of the Nakayama automorphism $\chi$ of $H$ associated to $\phi$ to continue to compute

$$\begin{aligned}
\lambda(t') &= \mu(\tilde{q}^1)\phi(\chi(S(\tilde{q}^2))q^2t_2p^2)\mu(q^1t_1p^1)\mu^{-1}(\alpha)\\
&\overset{(7.2.2)}{=} \mu(\tilde{q}^1)\phi(q^2t_2p^2)\mu(S(\chi(S(\tilde{q}^2)))q^1t_1p^1)\mu^{-1}(\alpha)\\
&= \mu(\tilde{q}^1)\lambda(S^{-1}(q^2t_2p^2))\mu^{-1}(S^{-1}(q^1t_1p^1))\mu^{-1}(\chi(S(\tilde{q}^2)))\mu^{-1}(\alpha)\\
&\overset{(7.6.2)}{=} \mu(\tilde{q}^1)\mu^{-1}(\underline{g})\mu^{-1}(S_\mu(\tilde{q}^2))\mu^{-1}(\alpha),
\end{aligned}$$

where the last equality follows because, as we have seen in the proof of the last theorem, $\chi \circ S = S^2 \circ S_\mu$. Recall that $S_\mu(h) = S(h) \leftharpoonup \mu$, so

$$\mu^{-1}(S_\mu(h)) = \mu(S(h)_1)\mu^{-1}(S(h)_2) = \varepsilon(S(h)) = \varepsilon(h),$$

for all $h \in H$. From here we deduce that $\lambda(t') = \mu(\alpha)\mu^{-1}(\underline{g}\alpha)$, and thus $\mu(\beta)^{-1} = b\mu(\alpha)\mu^{-1}(\underline{g}\alpha)$. Hence

$$\begin{aligned}
S(r) &= \mu(\alpha\beta)^{-1}\mu^{-1}(\underline{g}\alpha)^{-1}\mu^{-1}(q^2r_2)q^1r_1\\
&= (\mu^{-1}(\underline{g})\mu(\alpha\beta))^{-1}\mu^{-1}(q^2r_2p^2)q^1r_1p^1,
\end{aligned}$$

and the proof is complete. □

By applying Proposition 7.72 and (7.2.8), we obtain the following result.

**Corollary 7.73** *Let $H$ be a finite-dimensional quasi-Hopf algebra. For all $t \in \int_l^H$ and $r \in \int_r^H$, we have that*

$$S^2(t) = (\mu^{-1}(\underline{g})\mu(\beta))^{-1}t \quad \text{and} \quad S^2(r) = (\mu^{-1}(\underline{g})\mu(\beta))^{-1}r.$$

## 7.7 A Freeness Theorem for Quasi-Hopf Algebras

The aim of this section is to show that a finite-dimensional quasi-Hopf algebra $K$ is a free module over any quasi-Hopf subalgebra $H$. This is by definition a subalgebra of $K$ such that $\Delta(H) \subseteq H \otimes H$ and $H \otimes H \otimes H$ contains the reassociator $\Phi$ of $K$. In addition, the distinguished elements $\alpha, \beta$ that define the antipode $S$ of $K$ belong to $H$, and $S(H) \subseteq H$. In other words, the quasi-Hopf algebra structure of $K$ restricts to a quasi-Hopf algebra structure on $H$.

As we shall see, the proof of this freeness theorem has a Frobenius-module theory flavour. This is why we have to start by recalling some concepts and well-known results from this area.

The result below is due to Krull and Schmidt and a proof for it can be found in any module theory book; we will indicate several such books in the notes of this chapter. Unless otherwise specified, "module" means either a left or a right module.

**Theorem 7.74** *Every module of finite length over a ring $R$ admits a finite indecomposable decomposition $M = \bigoplus_{i=1}^t M_i$, with each $M_i$ an indecomposable $R$-module. It is unique in the sense that if $M = \bigoplus_{j=1}^s N_j$ is another indecomposable decomposition of $M$ then $t = s$ and there exists a bijection $\sigma : \{1, \ldots, t\} \to \{1, \ldots, s\}$ such that $M_i \cong N_{\sigma(i)}$ as $R$-modules, for any $1 \leq i \leq t$.*

If $R$ is a finite-dimensional $k$-algebra then $R$ admits a decomposition of the form

$$R = P_1^{(n_1)} \oplus \cdots P_t^{(n_t)},$$

where $n_i$ are non-zero natural numbers and $P_i$ are non-isomorphic indecomposable $R$-modules; here $P^{(t)}$ denotes the direct sum of $t$ copies of $P$. In what follows we call $P_1, \ldots, P_t$ the principal indecomposable $R$-modules of $R$.

We say that a left $R$-module $M$ is faithful if $\mathrm{Ann}_R(M)$, its annihilator, is the null space. That is, $\{r \in R \mid r \cdot m = 0, \ \forall\, m \in M\} = (0)$. We have a similar definition for the case when $M$ is a right $R$-module.

**Lemma 7.75** *Let $R$ be a finite-dimensional $k$-algebra which is left self injective and $M$ a left $R$-module which is finite dimensional as a $k$-vector space. Then $M$ is faithful if and only if for any $1 \leq i \leq t$ the left principal indecomposable $R$-module $P_i$ of $R$ is isomorphic to a direct summand of $M$. A right-handed version holds as well.*

*Proof* Consider $\{m_1,\ldots,m_n\}$ a basis of $M$ over $k$.

If $M$ is $R$-faithful then $f: R \to M^n$ given by $f(r) = (r\cdot m_1,\ldots,r\cdot m_n)$ is $R$-linear and injective. As $R$ is injective as a left $R$-module there exists a left $R$-module $X$ such that $M^n \cong R\oplus X$. Writing $M$, $R$ and $X$ as direct sums of indecomposable $R$-modules, by the uniqueness of such a decomposition it follows that each $P_i$ is isomorphic (as an $R$-module) to a direct summand of $M$.

Conversely, assume each $P_i$ is isomorphic to a direct summand of $M$, say $M_i$. Then

$$R = P_1^{(n_1)} \oplus\cdots\oplus P_t^{(n_t)} \cong M_1^{(n_1)} \oplus\cdots\oplus M_t^{(n_t)},$$

as left $R$-modules. So if we take $q$ to be the maximum of $n_1$, …, $n_t$ then $R$ embeds into $M^{(q)}$ as a left $R$-module. Hence, if $r\in R$ such that $r\cdot m = 0$, for all $m\in M$, then $rM^{(q)} = 0$, and consequently $rR = 0$. We conclude that $r = 0$, and therefore $M$ is a faithful left $R$-module. The right-handed case can be proved in a similar manner, so we are done. □

**Proposition 7.76** *Let $R$ be a finite-dimensional $k$-algebra that is left self injective, and consider $M$ a left $R$-module which has finite dimension as a $k$-vector space. Then $M$ is faithful if and only if there exist a positive integer $s$, a free left $R$-module $F$ and a non faithful left $R$-module $E$ such that $M^{(s)} \cong F\oplus E$ as left $R$-modules. The result is valid for right modules, too.*

*Proof* Assume first that $M$ is $R$-faithful. By the above lemma we decompose $M$ as

$$M = P_1^{(m_1)} \oplus\cdots\oplus P_t^{(m_t)} \oplus Q,$$

where $Q$ is an $R$-submodule of $M$ that does not contain submodules isomorphic to one of the $P_i$s. If we take $s$ to be the smallest common multiple of $n_1$, …, $n_t$ with $n_1$, …, $n_t$ as in the indecomposable decomposition of $R$, then

$$M^{(s)} \cong P_1^{(sm_1)} \oplus\cdots\oplus P_t^{(sm_t)} \oplus Q^{(s)},$$

and so if $p = \frac{sm_i}{n_i}$ is the minimum of $\frac{sm_1}{n_1}$, …, $\frac{sm_t}{n_t}$ we have

$$\begin{aligned}M^{(s)} &\cong P_1^{(pn_1)} \oplus\cdots\oplus P_t^{(pn_t)} \oplus \bigoplus_{1\leq j\neq i\leq t} P_j^{(sm_j-pn_j)} \oplus Q^{(s)}\\ &\cong R^{(p)} \oplus \bigoplus_{1\leq j\neq i\leq t} P_j^{(sm_j-pn_j)} \oplus Q^{(s)}.\end{aligned}$$

Set $F = R^{(p)}$ and $E = \bigoplus_{1\leq j\neq i\leq t} P_j^{(sm_j-pn_j)} \oplus Q^{(s)}$, so that $M^{(s)} \cong E\oplus F$ as left $R$-modules. Clearly $F$ is a free $R$-module of rank $p$. By the way of contradiction, if $E$ is faithful then the previous lemma says that $P_i$ is isomorphic to a (indecomposable) direct summand of $Q^{(s)}$, and so of $Q$. This contradicts the choice of $Q$, and therefore $E$ is not faithful as a left $R$-module.

For the converse, if there is a positive integer $s$ such that $M^{(s)} \cong E\oplus F$ with $E,F$ as in the statement then $rM = 0$ implies $rM^{(s)} = 0$, and thus $rF = 0$. Since $F$ is a free left $R$-module it follows that $r = 0$. We conclude that $M$ is $R$-faithful, as needed. □

**Corollary 7.77** *Let $R$ be a finite-dimensional $k$-algebra that is left self injective, and $M$ a left $R$-module of finite dimension over $k$. Then there exist a positive integer $s$, a free left $R$-module $F$ and a non faithful left $R$-module $E$ such that $M^{(s)} \cong F \oplus E$ as left $R$-modules. We have a similar statement for right modules as well.*

*Proof* If $M$ is not faithful we take $s = 1$, $F = 0$ and $E = M$. Otherwise, the last proved result applies. □

Another important result we need is the following. Recall that if $H$ is a quasi-Hopf algebra and $X$, $Y$ are left or right $H$-modules then so is $X \otimes Y$ via the action defined by the comultiplication of $H$.

**Proposition 7.78** *Let $H$ be a finite-dimensional quasi-Hopf algebra and $M$ a left $H$-module that is finite dimensional over $k$. If there exists a finite-dimensional faithful left $H$-module $V$ such that $V \otimes M \cong M^{\dim_k(V)}$ as left $H$-modules, then $M$ is free as a left $H$-module.*

*Proof* $H$ is a Frobenius algebra and thus left and right self injective; see Proposition 7.13. So Corollary 7.77 applies and we find that $V^{(l)} \cong F' \oplus E'$ and $M^{(s)} \cong F \oplus E$ as left $H$-modules, for some positive integers $l, s$, free left $H$-modules $F, F'$ and non-faithful left $H$-modules $E, E'$. We prove first that $N := M^{(s)}$ is a free left $H$-module. If we denote $W := V^{(l)}$ we have

$$W \otimes N \cong (V \otimes M)^{(ls)} \cong M^{(ls\dim_k V)} \cong N^{(\dim_k W)}$$

as left $H$-modules. Furthermore, $W \cong F' \oplus E'$ and $N \cong F \oplus E$ as left $H$-modules, so

$$(W \otimes F) \oplus (W \otimes E) \cong F^{(\dim_k W)} \oplus E^{(\dim_k W)}$$

as left $H$-modules. We now show that $W \otimes F \cong F^{(\dim_k W)}$ as left $H$-modules and this would imply that $W \otimes E \cong E^{(\dim_k W)}$ in ${}_H\mathcal{M}$; see the Krull–Schmidt theorem. Indeed, if $Q$ is an arbitrary left $H$-module then $.Q \otimes .H \cong Q \otimes .H$ as left $H$-modules, where the dots indicate that $.Q \otimes .H$ is endowed with the left $H$-module structure defined by $\Delta$, while $Q \otimes .H$ has the left $H$-module structure given by the multiplication of $H$. An isomorphism is produced by

$$\varsigma : Q \otimes .H \to .Q \otimes .H, \quad \varsigma(x \otimes h) = h_1 \tilde{p}^1 \cdot x \otimes h_2 \tilde{p}^2$$

for all $x \in Q$ and $h \in H$. Actually, if $\varsigma^{-1} : Q \otimes .H \to .Q \otimes .H$ is given by $\varsigma^{-1}(x \otimes h) = S^{-1}(\tilde{q}^1 h_1) \cdot x \otimes \tilde{q}^2 h_2$, for all $x \in Q$ and $h \in H$, then

$$\begin{aligned} \varsigma(\varsigma^{-1}(x \otimes h)) &= \tilde{q}_1^2 h_{(2,1)} \tilde{p}^1 S^{-1}(\tilde{q}^1 h_1) \cdot x \otimes \tilde{q}_2^2 h_{(2,2)} \tilde{p}^2 \\ &\overset{(3.2.22)}{=} \tilde{q}_1^2 \tilde{p}^1 S^{-1}(\tilde{q}^1) \cdot x \otimes \tilde{q}_2^2 \tilde{p}^2 h \overset{(3.2.24)}{=} x \otimes h. \end{aligned}$$

In a similar manner it can be checked that $\varsigma^{-1}(\varsigma(x \otimes h)) = x \otimes h$, for all $x \in Q$ and $h \in H$, hence $\varsigma$ is bijective. It is also left $H$-linear since $\Delta$ is multiplicative. This $H$-linear isomorphism shows in particular that $.Q \otimes .H \cong H^{(\dim_k Q)}$ as left $H$-modules, for any finite-dimensional left $H$-module $Q$.

In our context, since $F$ is $H$-free, say $F \cong H^{(q)}$ for some positive integer $q$, we get

$$W \otimes F \cong W \otimes H^{(q)} \cong (W \otimes H)^{(q)} \cong H^{(q\dim_k W)} \cong F^{(\dim_k W)}$$

as left $H$-modules, as claimed. We get $W \otimes E \cong E^{(\dim_k W)}$, which implies

$$E^{(\dim_k W)} \cong W \otimes E \cong (F' \otimes E) \oplus (E' \otimes E) \tag{7.7.1}$$

as left $H$-modules. Note that $F'$ is not zero, for otherwise $W = E'$ is not $H$-faithful, and this cannot happen because $V$ is $H$-faithful and so is $V^{(l)} = W$. Now using that $F'$ is $H$-free, by similar arguments to the ones above we obtain that $F' \otimes E$ is also free as a left $H$-module. More precisely, it can be proved that for any left $H$-module $Q$ the map $\varsigma' : .H \otimes .Q \to Q \otimes .H$ defined by $\varsigma'(h \otimes x) = S(q^2 h_2) \cdot x \otimes q^1 h_1$, for all $h \in H$ and $x \in Q$, is left $H$-linear and an isomorphism with inverse given by $\varsigma'^{-1}(x \otimes h) = h_1 p^1 \otimes h_2 p^2 \cdot x$, for all $h \in H$ and $x \in Q$, where the dots have the meaning explained a few lines above. Notice that this assertion can be proved by using equations (3.2.21) and (3.2.23); we leave it to the reader to check out all the details. Consequently, since $Q \otimes .H$ is left $H$-free, $.H \otimes .Q$ is also left $H$-free, and from here we get that $U \otimes Q$ is $H$-free for any finite-dimensional left $H$-free module $U$.

Therefore, if $E$ is not zero then $F' \otimes E$ is a non-zero free $H$-module, so a faithful $H$-module. Then the isomorphism (7.7.1) implies that $E^{(\dim_k W)}$ is $H$-faithful, and thus $E$ is $H$-faithful, a contradiction. Thus we must have $E = 0$, and consequently $M^{(s)} = N = F$, a free left $H$-module.

Finally, we are in the position to prove that $M$ itself is $H$-free. Toward this end consider a positive integer $n$ such that $M^{(s)} \cong H^{(n)}$ as left $H$-modules. Now view $k$ as a right $H$-module via the counit $\varepsilon$ of $H$, that is, $\kappa \cdot h = \varepsilon(h)\kappa$, for all $\kappa \in k$, $h \in H$. Then

$$(k \otimes_H M)^{(s)} \cong k \otimes_H M^{(s)} \cong k \otimes_H H^{(n)} \cong (k \otimes_H H)^{(n)} \cong k^{(n)}.$$

If $d$ is the dimension of $k \otimes_H M$ we obtain $ds = n$, hence $M^{(s)} \cong H^{(ds)}$ as left $H$-modules. By the Krull–Schmidt theorem we conclude that $M \cong H^{(d)}$, a free left $H$-module. The proof is finished. □

We can now prove the announced result. For reasons that will be explained in the notes of this chapter we will refer to it as the quasi-Hopf version of the Nichols–Zoeller theorem.

**Theorem 7.79** *Let $K$ be a finite-dimensional quasi-Hopf algebra and $H \subseteq K$ a quasi-Hopf subalgebra. Then $K$ is a free left and right $H$-module.*

*Proof* In view of the above proposition it suffices to show that for the finite-dimensional left $H$-module $K$ there exists a finite-dimensional left $H$-faithful module $V$ such that $V \otimes K \cong K^{(\dim_k V)}$ as left $H$-modules. We can take $V = K$, which clearly has the desired properties. The isomorphisms $\varsigma$ defined in the proof of Proposition 7.78, considered for $K$ in place of $H$, tells us that $Q \otimes K \cong K^{(\dim_k Q)}$ as left $K$-modules, for any left $K$-module $Q$. We deduce in particular that $K \otimes K \cong K^{(\dim_k K)}$ as left $K$-modules, and therefore as left $H$-modules as well. Thus $K$ is left $H$-free.

Working with $K^{\mathrm{op}}$ and $H^{\mathrm{op}}$ instead of $K$ and $H$, respectively, we deduce that $K$ is also right $H$-free, and this finishes the proof. □

We present some applications of the freeness theorem. The next result can be seen as an extension of the Lagrange theorem for groups to quasi-Hopf algebras.

**Corollary 7.80** *If $K$ is a finite-dimensional quasi-Hopf algebra and $H \subseteq K$ is a quasi-Hopf subalgebra then* $\dim_k H \mid \dim_k K$.

*Proof* As we have seen, $K$ is left $H$-free, so there is a positive integer $s$ such that $K \cong H^{(s)}$ as left $H$-modules. Thus $\dim_k K = s\dim_k H$ and this proves that $\dim_k H$ is a divisor of $\dim_k K$. □

**Corollary 7.81** *A quasi-Hopf algebra of prime dimension does not admit proper quasi-Hopf subalgebras.*

**Corollary 7.82** *If $K$ is a finite-dimensional semisimple quasi-Hopf algebra then every quasi-Hopf subalgebra $H$ of it is also semisimple.*

*Proof* By Theorem 7.28 $K$ contains a left integral $t$ satisfying $\varepsilon(t) = 1$. Since $K$ is a left $H$-free module there exist $h_1, \cdots, h_n \in H$ such that $t = \sum_i h_i k_i$, where $\{k_i\}_i$ is a fixed basis of $K$ over $H$. Then for all $x \in H$ we have

$$\sum_i (xh_i)k_i = x\sum_i h_i k_i = xt = \varepsilon(x)t = \varepsilon(x)\sum_i h_i k_i,$$

and therefore, for each $i$, $xh_i = \varepsilon(x)h_i$, for all $x \in H$. Thus each $h_i$ is a left integral in $H$ and since $1 = \varepsilon(t) = \sum_i \varepsilon(h_i)\varepsilon(k_i)$ we obtain that at least one of the $\varepsilon(h_i)$s is non-zero. By applying Theorem 7.28 again we conclude that $H$ is semisimple, as needed. □

## 7.8 Notes

Frobenius algebras started to be studied in the 1930s by Brauer and Nesbitt, who named these algebras after Frobenius; they also introduced the concept of symmetric algebra in [37]. Nakayama discovered the beginnings of the duality property of a Frobenius algebra in [161, 162], and Dieudonné used this to characterize Frobenius algebras in [76]. Nakayama also studied symmetric algebras in [160]; the automorphism that carries his name was defined in [162]. Recently, interest in these algebras has been renewed due to connections to monoidal categories and topological quantum field theory; see [208]. Our main sources of inspiration in the presentation of these concepts were the books of Lam [131] and Kadison [123], and the papers [124, 125, 181].

The existence of integrals in finite-dimensional quasi-Hopf algebras was proved in [176], where a short argument of Van Daele [214] that guarantees the existence (and uniqueness) of integrals for finite-dimensional Hopf algebras was generalized to the quasi-Hopf algebra setting. For Hopf algebras this result has been known since the late 1960s. Actually, Larson and Sweedler [135] used the fundamental theorem for Hopf modules to prove the existence and uniqueness of integrals in a Hopf algebra, and then the bijectivity of the antipode follows. As we have seen, this fundamental theorem was adapted in [109] for quasi-Hopf bimodules but the proof there relies on the fact that the antipode is bijective. The bijectivity of the antipode has been proved since then in [47], a second proof being given one year later in [199]. The integral theory that was presented here is taken from [47], where it has been proved also that there are no non-zero integrals in the case of an infinite-dimensional quasi-Hopf algebra with bijective antipode. The Maschke theorem is essentially from [172], and the other characterizations for semisimple quasi-Hopf algebras are from [109]. We have used also [109, 124] in the presentation of symmetric algebras and symmetric quasi-Hopf algebras. We reproduced the general theory on cointegrals from [109] but the two sets of characterizations, the examples and the connections between left and right cointegrals are from [47, 49]. As $H$ is a coalgebra in the monoidal category of $H$-bimodules, and so has a natural left and right $H$-comodule structure within this category, we get for free that $H^*$ is a quasi-Hopf $H$-bimodule with the structure as in [109]. But we should point out that (apart from the Hopf algebra case) we have multiple choices for the set of coinvariants, and this gives us alternative definitions for cointegrals.

The formula for the fourth power of the antipode was proved for unimodular Hopf algebras by Larson [132], and was extended by Radford [183] to any finite-dimensional Hopf algebra. Recently, it was generalized for co-Frobenius Hopf algebras in [27]. For quasi-Hopf algebras a first proof appears in [109] but, owing to the Frobenius arguments, we preferred here to present the one given by Kadison in [122].

For module theoretical aspects related to the freeness theorem for quasi-Hopf algebras we refer to the books [131, 10, 70]. It was conjectured by Kaplansky [126] that any Hopf algebra is free over any Hopf subalgebra. In the finite-dimensional case a positive answer was given by Nichols and Zoeller in [169], and this is why the stated freeness theorem is known as the Nichols–Zoeller theorem. In the infinite-dimensional case it has been known for a long time that the conjecture is false, examples of Hopf algebras which are not free over a Hopf subalgebra having been given by Oberst and Schneider [170] in 1974. Afterwards, in [184, 185], Radford proved that this conjecture of Kaplansky is also true in some special infinite-dimensional cases. For finite-dimensional quasi-Hopf algebras the freeness theorem has been proved by Schauenburg in [200], and our presentation is taken mostly from there. We also used the books [157, 73], and a simplified argument from [202].

# 8

# Yetter–Drinfeld Module Categories

We introduce the categories of Yetter–Drinfeld modules by computing the left and right centers of a category of modules over a quasi-bialgebra $H$. We then show that all four categories of Yetter–Drinfeld modules are braided isomorphic. We also introduce the quasi-Hopf algebra structure of the quantum double of a finite-dimensional quasi-Hopf algebra.

## 8.1 The Left and Right Center Constructions

The center construction associates to a monoidal category a braided monoidal one.

**Definition 8.1** Let $\mathscr{C}$ be a monoidal category.

Denote by $\mathscr{W}_r(\mathscr{C})$ the category whose objects are pairs $(V, c_{-,V})$, with $V$ an object of $\mathscr{C}$ and $c_{-,V} = (c_{X,V} : X \otimes V \to V \otimes X)_{X \in \mathscr{C}}$ a natural transformation, satisfying $c_{\underline{1},V} = r_V^{-1} l_V$ and, for all $X, Y \in \mathscr{C}$,

$c_{X\otimes Y,V} =$ 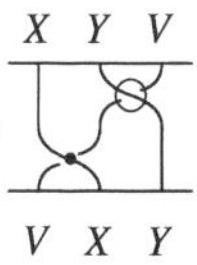

where $c_{X,V} =$ 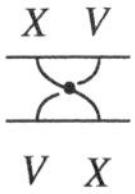

and $c_{Y,V} =$ 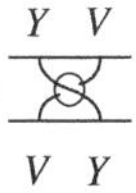

, (8.1.1)

and where we assume, for simplicity, that $\mathscr{C}$ is strict monoidal.

A morphism between $(V, c_{-,V})$ and $(W, c_{-,W})$ in $\mathscr{W}_r(\mathscr{C})$ is a morphism $f : V \to W$ in $\mathscr{C}$ obeying 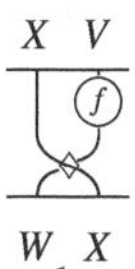

$=$ 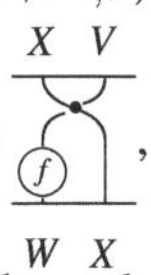

, where $c_{X,W} =$ 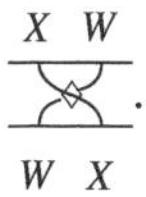

. It is clear that $\mathscr{W}_r(\mathscr{C})$ is a category, called in what follows the right weak center of $\mathscr{C}$. Note that $\mathrm{Id}_{(V,c_{-,V})} = \mathrm{Id}_V$.

We will show that $\mathscr{W}_r(\mathscr{C})$ is a pre-braided monoidal category via the following structure:

**Proposition 8.2** *The right weak center $\mathscr{W}_r(\mathscr{C})$ of a monoidal category $\mathscr{C}$ is* *monoidal with tensor product*

$(V, c_{-,V}) \otimes (W, c_{-,W}) = (V \otimes W, c_{-,V\otimes W})$, where $c_{X,V\otimes W} =$ 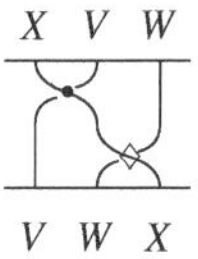  , $\forall X \in \mathscr{C}$. (8.1.2)

*The unit object of* $\mathscr{W}_r(\mathscr{C})$ *is* $(\underline{1}, c_{-,\underline{1}} := (l_X^{-1} r_X : X \otimes \underline{1} \to \underline{1} \otimes X)_{X\in\mathscr{C}})$, *and the left and right unit constraints are the same as those of* $\mathscr{C}$. *The monoidal category* $\mathscr{W}_r(\mathscr{C})$ *is, moreover, pre-braided with pre-braiding*

$$c_{(V,c_{-,V}),(W,c_{-,W})} = c_{V,W} : (V, c_{-,V}) \otimes (W, c_{-,W}) \to (W, c_{-,W}) \otimes (V, c_{-,V}), \quad (8.1.3)$$

*for all* $(V, c_{-,V})$, $(W, c_{-,W})$ *in* $\mathscr{W}_r(\mathscr{C})$.

*Proof* We first prove that the tensor product of $\mathscr{W}_r(\mathscr{C})$ is well defined. Actually, $(V \otimes W, c_{-,V\otimes W})$ is an object of the right weak center of $\mathscr{C}$ since, for all $X, Y \in \mathscr{C}$,

$c_{X\otimes Y,V\otimes W} =$ 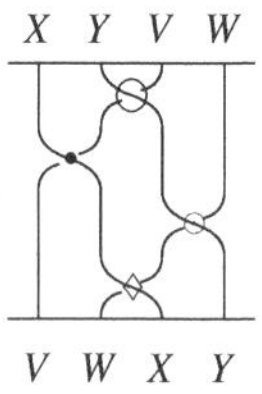  $=$ 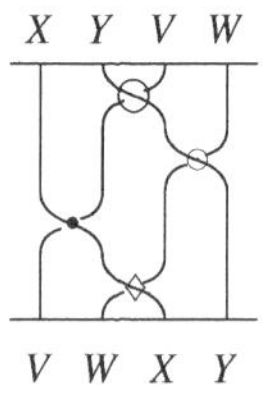  ,

where

$c_{X,V} =$ 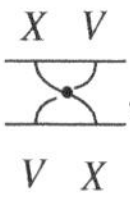  , $c_{Y,V} =$ 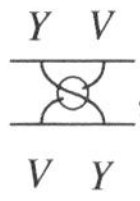  , $c_{X,W} =$ 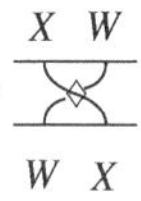  and $c_{Y,W} =$ 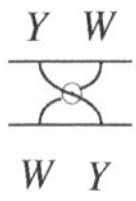  .

According to the definition we have

$$\begin{aligned} c_{\underline{1},V\otimes W} &= a^{-1}_{V,W,\underline{1}}(\mathrm{Id}_V \otimes r_W^{-1} l_W) a_{V,\underline{1},W}(r_V^{-1} l_V \otimes \mathrm{Id}_W) a^{-1}_{\underline{1},V,W} \\ &= r^{-1}_{V\otimes W}(\mathrm{Id}_V \otimes l_W) a_{V,\underline{1},W}(r_V^{-1} \otimes \mathrm{Id}_W) l_{V\otimes W} \\ &= r^{-1}_{V\otimes W} l_{V\otimes W}, \end{aligned}$$

as needed, where in the second equality we used Proposition 1.5 and in the last one (1.1.2). Thus $(V \otimes W, c_{-,V\otimes W})$ is indeed an object of $\mathscr{W}_r(\mathscr{C})$, as claimed.

We prove that the tensor product in $\mathscr{C}$ of two morphisms $(V, c_{-,V}) \xrightarrow{f} (W, c_{-,W})$ and $(V', c_{-,V'}) \xrightarrow{f'} (W', c_{-,W'})$ in $\mathscr{W}_r(\mathscr{C})$ is a morphism in $\mathscr{W}_r(\mathscr{C})$. We compute

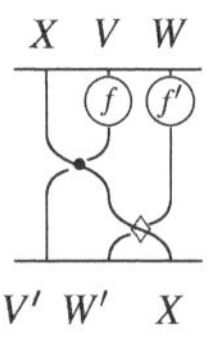  $=$ 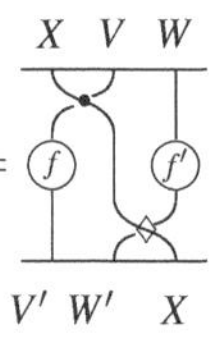  $=$ 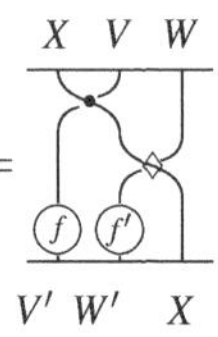  ,

where the diagrammatic notation for $c_{X,V'}$ and $c_{X,W'}$ is the same as the one used for $c_{X,V}$ and $c_{X,W}$, respectively.

Clearly $(\underline{1}, l_-^{-1} r_-)$ together with $l$ and $r$ define a unit object for $\mathscr{W}_r(\mathscr{C})$, and so $\mathscr{W}_r(\mathscr{C})$ is a monoidal category, as desired.

It remains to show that $c$ defined in the statement endows $\mathscr{W}_r(\mathscr{C})$ with a pre-braided structure. First, $c_{V,W}$ is a morphism in $\mathscr{W}_r(\mathscr{C})$ if and only if

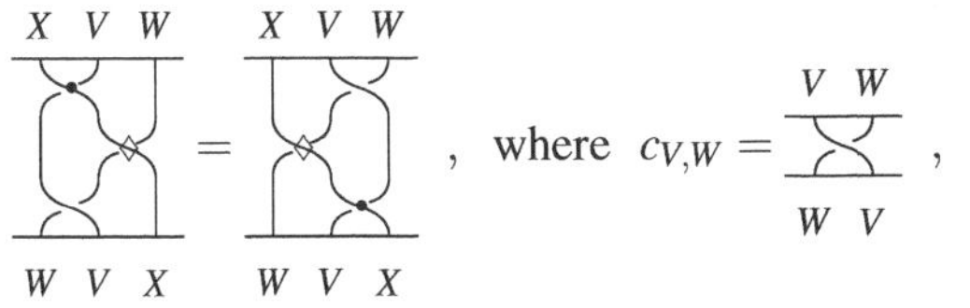


and this is true since $c_{-,W}$ is natural and satisfies (8.1.1) (of course, with $V$ replaced by $W$), and since $c_{X,V} : X \otimes V \to V \otimes X$ is a morphism in $\mathscr{C}$.

Now, $c$ obeys (1.5.2) because of its definition and (8.1.1). (1.5.1) in our situation becomes $c_{\underline{U},\underline{V}\otimes\underline{W}} = c_{U,V\otimes W} = (\mathrm{Id}_{\underline{V}} \otimes c_{\underline{W},\underline{U}})(c_{\underline{U},\underline{V}} \otimes \mathrm{Id}_{\underline{W}})$, where $\underline{V} = (V, c_{-,V})$, etc. and follows from the definition of the tensor product in $\mathscr{W}_r(\mathscr{C})$. □

In the definition of the objects of the right weak center one can require $c_{-,V}$ to be a natural isomorphism. The new category obtained in this way is denoted by $\mathscr{Z}_r(\mathscr{C})$, and called the right center of $\mathscr{C}$. It follows that $\mathscr{Z}_r(\mathscr{C})$ is a braided category.

Let $\Pi : \mathscr{Z}_r(\mathscr{C}) \to \mathscr{C}$ be the forgetful functor, that is, $\Pi$ acts as identity on morphisms and $\Pi((V, c_{-,V})) = V$, for all $(V, c_{-,V}) \in \mathscr{Z}_r(\mathscr{C})$. We have that $\Pi$ is a strict monoidal functor, and is bijective on objects. These observations provide a Universal Property for the center construction that we have just described.

**Proposition 8.3** *Let $\mathscr{C}'$ be a monoidal category with right center $\mathscr{Z}_r(\mathscr{C}')$ and canonical monoidal functor $\Pi' : \mathscr{Z}_r(\mathscr{C}') \to \mathscr{C}'$. Then for any braided category $\mathscr{C}$ and any monoidal functor $F : \mathscr{C} \to \mathscr{C}'$ which is bijective on objects and surjective on morphisms, there exists a unique braided monoidal functor $\mathscr{Z}_r(F) : \mathscr{C} \to \mathscr{Z}_r(\mathscr{C}')$ such that $\Pi' \circ \mathscr{Z}_r(F) = F$.*

*Proof* As usual, assume that $\mathscr{C}, \mathscr{C}'$ are strict monoidal categories, and that $F$ is a strict monoidal functor.

If $V$ is an object of $\mathscr{C}$ we define $\mathscr{Z}_r(F)(V) := (F(V), c_{-,F(V)})$, with $c_{X',F(V)} = F(c_{F^{-1}(X'),V})$, for all $X' \in \mathscr{C}'$, where $F^{-1}(X')$ is the unique object of $\mathscr{C}$ that is mapped by $F$ to $X'$. Note that, if $(F, \varphi_0, \varphi_2)$ is an arbitrary monoidal functor, we have

$$c_{X',F(V)} : X' \otimes F(V) = FF^{-1}(X') \otimes F(V) \xrightarrow{\varphi_{2,F^{-1}(X'),V}} F(F^{-1}(X') \otimes V)$$
$$\xrightarrow{F(c_{F^{-1}(X'),V})} F(V \otimes F^{-1}(X')) \xrightarrow{\varphi^{-1}_{2,V,F^{-1}(X')}} F(V) \otimes FF^{-1}(X') = F(V) \otimes X'.$$

Clearly $c_{-,\underline{1}'}$ is defined by identity morphisms of $\mathscr{C}'$ when $\mathscr{C}'$ is strict monoidal, otherwise it is defined by the left and right unit constraints of $\mathscr{C}'$. We next compute

$$\begin{aligned} c_{X'\otimes Y',F(V)} &= F(c_{F^{-1}(X'\otimes Y'),V}) \\ &= F(c_{F^{-1}(X')\otimes F^{-1}(Y'),V}) \end{aligned}$$

$$
\begin{aligned}
&= F(c_{F^{-1}(X'),V} \otimes \mathrm{Id}_{F^{-1}(Y')})F(\mathrm{Id}_{F^{-1}(X')} \otimes c_{F^{-1}(Y'),V}) \\
&= (F(c_{F^{-1}(X'),V}) \otimes \mathrm{Id}_{Y'})(\mathrm{Id}_{X'} \otimes F(c_{F^{-1}(Y'),V})) \\
&= c_{X',F(V)} \otimes c_{Y',F(V)},
\end{aligned}
$$

and so $\mathscr{Z}_r(F)$ is well defined on objects. It is well defined on morphisms as well since for all $V \xrightarrow{f} W$ in $\mathscr{C}$ one has $\mathscr{Z}_r(F)(f) := F(f) : F(V) \to F(W)$ and it satisfies the required property to be a morphism in $\mathscr{Z}_r(\mathscr{C}')$ since

$$
\begin{aligned}
(F(f) \otimes \mathrm{Id}_{X'})c_{X',F(V)} &= (F(f) \otimes \mathrm{Id}_{X'})F(c_{F^{-1}(X'),V}) \\
&= F((f \otimes \mathrm{Id}_{F^{-1}(X')})c_{F^{-1}(X'),V}) \\
&= F(c_{F^{-1}(X'),W}(\mathrm{Id}_{F^{-1}(X')} \otimes f)) \\
&= F(c_{F^{-1}(X'),W})(\mathrm{Id}_{X'} \otimes F(f)) \\
&= c_{X',F(W)}(\mathrm{Id}_{X'} \otimes F(f)).
\end{aligned}
$$

We also have $\Pi' \circ \mathscr{Z}_r(F)(V) = \Pi'((F(V), c_{-,F(V)})) = F(V)$ and $\Pi' \circ \mathscr{Z}_r(F)(f) = \Pi'(F(f)) = F(f)$. Thus $\mathscr{Z}_r(F)$ is a well-defined functor that satisfies $\Pi' \circ \mathscr{Z}_r(F) = F$. To complete the proof of the existence we have to check that $\mathscr{Z}_r(F)$ is a braided monoidal functor. In general, the monoidal structure of $F$ induces a monoidal structure on $\mathscr{Z}_r(F)$ since, for instance, $\varphi_{2,V,W} : F(V) \otimes F(W) \to F(V \otimes W)$ can be viewed also as a morphism in $\mathscr{Z}_r(\mathscr{C})$. Namely,

$$
\varphi_{2,V,W} : (F(V) \otimes F(W), c_{-,F(V)\otimes F(W)}) \to (F(V \otimes W), c_{-,F(V\otimes W)}),
$$

or, equivalently,

$$
\varphi_{2,V,W} : \mathscr{Z}_r(F)(V) \otimes \mathscr{Z}_r(F)(W) \to \mathscr{Z}_r(F)(V \otimes W).
$$

When $F$ is strict monoidal this fact follows from

$$
\begin{aligned}
F(c_{F^{-1}(X'),V\otimes W}) &= F((\mathrm{Id}_V \otimes c_{F^{-1}(X'),W})(c_{F^{-1}(X'),V} \otimes \mathrm{Id}_W)) \\
&= (\mathrm{Id}_{F(V)} \otimes F(c_{F^{-1}(X'),W}))(F(c_{F^{-1}(X'),V}) \otimes \mathrm{Id}_W) \\
&= (\mathrm{Id}_{F(V)} \otimes c_{X',F(W)})(c_{X',F(V)} \otimes \mathrm{Id}_W) \\
&= c_{X',F(V)\otimes F(W)}.
\end{aligned}
$$

The required property for $\mathscr{Z}_r(F)$ to be braided monoidal reduces to the definition of this functor: $\mathscr{Z}_r(F)(c_{V,W}) = F(c_{V,W})$, and so the first part of the proof is done.

Suppose now that $F' : \mathscr{C} \to \mathscr{Z}_r(\mathscr{C}')$ is a braided monoidal functor satisfying $\Pi' \circ F' = F$. If we denote $F'(X) = (X', c_{-,X'})$, then we have:

- on objects: $\Pi' \circ F'(X) = F(X)$, which is equivalent to $X' = F(X)$, and so $F'(X) = (F(X), c_{-,F(X)})$, for any object $X$ of $\mathscr{C}$;
- on morphisms: $\Pi' \circ F'(f) = F(f)$, thus $F'(f) = F(f)$, for any morphism $f$ in $\mathscr{C}$.

Since $F'$ is braided monoidal, we have $\varphi_{2,Y,X} c_{F(X),F(Y)} = F(c_{X,Y})\varphi_{2,X,Y}$. Thus

$$
c_{X',F(Y)} = \varphi^{-1}_{2,Y,F^{-1}(X')} F(c_{F^{-1}(X'),Y})\varphi_{2,F^{-1}(X'),Y},
$$

for all $X' \in \mathscr{C}'$ and $Y \in \mathscr{C}$, implying that $F' = \mathscr{Z}_r(F)$ and finishing the proof. □

**Corollary 8.4** *If $\mathscr{C}$ is a braided category, there exists a unique braided monoidal functor $\mathscr{Z}_r : \mathscr{C} \to \mathscr{Z}_r(\mathscr{C})$ such that $\Pi \circ \mathscr{Z}_r = \mathrm{Id}_{\mathscr{C}}$.*

*Proof* In Proposition 8.3 take $\mathscr{C}' = \mathscr{C}$ and $F = \mathrm{Id}_{\mathscr{C}}$. Then $\mathscr{Z}_r = \mathscr{Z}_r(\mathrm{Id}_{\mathscr{C}})$. □

The above result says that, in general, the center of a braided category $\mathscr{C}$ is not isomorphic to $\mathscr{C}$ itself. All we can say is that $\mathscr{C}$ can be identified with a braided subcategory of its right center. This fact will be illustrated when we compute the center of the monoidal category of representations of a quasi-Hopf algebra; see the next sections.

In a similar manner we can introduce the left (weak) center of a monoidal category. More precisely, let $\mathscr{C}$ be a monoidal category.

By $\mathscr{W}_l(\mathscr{C})$ we denote the category whose objects are pairs $(V, d_{V,-})$ consisting of an object $V$ of $\mathscr{C}$ and a natural transformation $d_{V,-} = (d_{V,X} : V \otimes X \to X \otimes V)_{X \in \mathscr{C}}$, such that, for all $X, Y \in \mathscr{C}$,

$d_{V,X\otimes Y} =$ 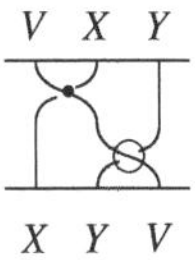

, where $d_{V,X} =$ 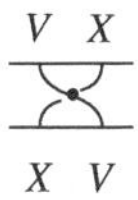

and $d_{V,Y} =$ 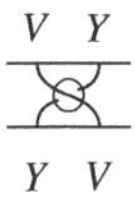

, (8.1.4)

and where, once more, we assumed that $\mathscr{C}$ is strict monoidal.

A morphism $f : (V, d_{V,-}) \to (W, d_{W,-})$ in $\mathscr{W}_l(\mathscr{C})$ is a morphism $f : V \to W$ in $\mathscr{C}$ such that $d_{X,W}(f \otimes \mathrm{Id}_X) = (\mathrm{Id}_X \otimes f)d_{V,X}$, for all $X \in \mathscr{C}$.

We call $\mathscr{W}_l(\mathscr{C})$ the left weak center of $\mathscr{C}$. It is a monoidal category with tensor product

$(V, d_{V,-}) \otimes (W, d_{W,-}) = (V \otimes W, d_{V\otimes W,-})$, with $d_{V\otimes W,X} =$ 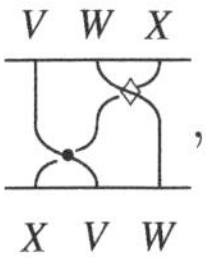

, (8.1.5)

where the notation is similar to the one used in the right-handed case. The unit object of $\mathscr{W}_l(\mathscr{C})$ is $(\underline{1}, d_{\underline{1},-} := (r_X^{-1} l_X)_{X \in \mathscr{C}})$, and the left and right unit constraints coincide with those of $\mathscr{C}$. $\mathscr{W}_l(\mathscr{C})$ is, moreover, pre-braided via the pre-braiding given by

$$d_{V,W} : (V, d_{V,-}) \otimes (W, d_{W,-}) \to (W, d_{W,-}) \otimes (V, d_{V,-}), \tag{8.1.6}$$

for all $(V, d_{V,-})$, $(W, d_{W,-})$ objects of $\mathscr{W}_l(\mathscr{C})$.

If we require $d_{V,-}$ to be a natural isomorphism we obtain the notion of left center, denoted by $\mathscr{Z}_l(\mathscr{C})$. As in the right-handed case, $\mathscr{Z}_l(\mathscr{C})$ is a braided category. The connection with $\mathscr{Z}_r(\mathscr{C})$ is presented in the following result.

**Proposition 8.5** *Let $\mathscr{C}$ be a monoidal category, and $\mathscr{Z}_l(\mathscr{C})$ and $\mathscr{Z}_r(\mathscr{C})$ the left and right centers associated to it. Then $F : \mathscr{Z}_r(\mathscr{C}) \to \mathscr{Z}_l(\mathscr{C})^{\mathrm{in}}$ defined by $F((V, c_{-,V})) = (V, c_{-,V}^{-1})$, for all $(V, c_{-,V}) \in \mathscr{Z}_r(\mathscr{C})$, provides a braided monoidal isomorphism. $F$ acts as identity on morphisms.*

*Proof* If $c_{X\otimes Y,V} =$ 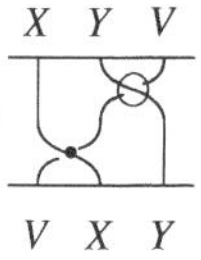

then $c^{-1}_{X\otimes Y,V} =$ 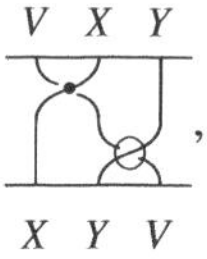

, where, if $c_{Y,V} =$ 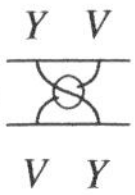

then $c^{-1}_{Y,V} =$ 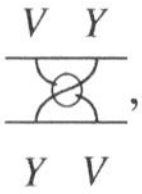

, etc. This shows that $(V, c^{-1}_{-,V}) \in \mathscr{Z}_l(\mathscr{C})$, and so $F$ is well defined on objects. It can be easily checked that $F$ maps a morphism in $\mathscr{Z}_r(\mathscr{C})$ to a morphism in $\mathscr{Z}_l(\mathscr{C})$, thus $F$ is well defined on morphisms, too. By the definition of $(V, c_{-,V}) \otimes (W, c_{-,W}) = (V \otimes W, c_{-,V\otimes W})$ it follows that $(V, c^{-1}_{-,V}) \otimes (W, c^{-1}_{-,W}) = (V \otimes W, c^{-1}_{-,V\otimes W})$, so $F$ is a strict monoidal functor. Viewed as a functor from $\mathscr{Z}_r(\mathscr{C})$ to $\mathscr{Z}_l(\mathscr{C})^{\mathrm{in}}$ it is, moreover, braided monoidal since the commutativity of the diagram

$$\begin{array}{ccc} F(\underline{V}) \otimes F(\underline{W}) & \xrightarrow{c^{\mathrm{in}}_{\underline{V},\underline{W}} = c_{V,W}} & F(\underline{W}) \otimes F(\underline{V}) \\ \| & & \| \\ F(\underline{V} \otimes \underline{W}) & \xrightarrow{F(c_{\underline{V},\underline{W}}) = c_{\underline{V},\underline{W}}} & F(\underline{W} \otimes \underline{V}) \end{array}$$

follows from the definition of the braiding $c$ of $\mathscr{Z}_r(\mathscr{C})$. Here we have denoted $\underline{V} := (V, c_{-,V})$ and $\underline{W} := (W, c_{-,W})$. Hence the proof is finished. □

The following result is completely obvious; we leave the details to the reader.

**Proposition 8.6** *Let $\mathscr{C}$ be a monoidal category. Then*

$$\overline{\mathscr{W}_l(\mathscr{C})} \simeq \mathscr{W}_r(\overline{\mathscr{C}}) \ \ \textit{and} \ \ \overline{\mathscr{W}_r(\mathscr{C})} \simeq \mathscr{W}_l(\overline{\mathscr{C}})$$

*as pre-braided categories, and*

$$\overline{\mathscr{Z}_l(\mathscr{C})} \simeq \mathscr{Z}_r(\overline{\mathscr{C}}) \ \ \textit{and} \ \ \overline{\mathscr{Z}_r(\mathscr{C})} \simeq \mathscr{Z}_l(\overline{\mathscr{C}})$$

*as braided categories, where $\overline{\mathscr{C}}$ is the reverse monoidal category associated to $\mathscr{C}$.*

## 8.2 Yetter–Drinfeld Modules over Quasi-bialgebras

Throughout this section $H$ is a quasi-bialgebra or a quasi-Hopf algebra (sometimes with bijective antipode) over a field $k$. We introduce several categories of Yetter–Drinfeld modules over $H$ and show that they are isomorphic to the left or right weak centers of ${}_H\mathscr{M}$ or $\mathscr{M}_H$.

**Definition 8.7** Let $H$ be a quasi-bialgebra.

(1) A left Yetter–Drinfeld module over $H$ is a left $H$-module $M$ together with a $k$-linear map (called the left $H$-coaction)

$$\lambda_M : M \to H \otimes M, \quad \lambda_M(m) = m_{(-1)} \otimes m_{(0)}$$

such that the following conditions hold, for all $h \in H$ and $m \in M$:

$$X^1 m_{(-1)} \otimes (X^2 \cdot m_{(0)})_{(-1)} X^3 \otimes (X^2 \cdot m_{(0)})_{(0)}$$
$$= X^1 (Y^1 \cdot m)_{(-1)_1} Y^2 \otimes X^2 (Y^1 \cdot m)_{(-1)_2} Y^3 \otimes X^3 \cdot (Y^1 \cdot m)_{(0)}, \quad (8.2.1)$$
$$\varepsilon(m_{(-1)}) m_{(0)} = m, \quad (8.2.2)$$
$$h_1 m_{(-1)} \otimes h_2 \cdot m_{(0)} = (h_1 \cdot m)_{(-1)} h_2 \otimes (h_1 \cdot m)_{(0)}. \quad (8.2.3)$$

The category of left Yetter–Drinfeld modules and $k$-linear maps that preserve the $H$-action and $H$-coaction is denoted by ${}_H^H\mathcal{YD}$.

(2) A left–right Yetter–Drinfeld module over $H$ is a left $H$-module $M$ together with a $k$-linear map (called the right $H$-coaction)

$$\rho_M : M \to M \otimes H, \quad \rho_M(m) = m_{(0)} \otimes m_{(1)}$$

such that the following conditions hold, for all $h \in H$ and $m \in M$:

$$(x^2 \cdot m_{(0)})_{(0)} \otimes (x^2 \cdot m_{(0)})_{(1)} x^1 \otimes x^3 m_{(1)}$$
$$= x^1 \cdot (y^3 \cdot m)_{(0)} \otimes x^2 (y^3 \cdot m)_{(1)_1} y^1 \otimes x^3 (y^3 \cdot m)_{(1)_2} y^2, \quad (8.2.4)$$
$$\varepsilon(m_{(1)}) m_{(0)} = m, \quad (8.2.5)$$
$$h_1 \cdot m_{(0)} \otimes h_2 m_{(1)} = (h_2 \cdot m)_{(0)} \otimes (h_2 \cdot m)_{(1)} h_1. \quad (8.2.6)$$

The category of left–right Yetter–Drinfeld modules and $k$-linear maps that preserve the $H$-action and $H$-coaction is denoted by ${}_H\mathcal{YD}^H$.

(3) A right–left Yetter–Drinfeld module over $H$ is a right $H$-module $M$ together with a $k$-linear map (called the left $H$-coaction)

$$\lambda_M : M \to H \otimes M, \quad \lambda_M(m) = m_{(-1)} \otimes m_{(0)}$$

such that the following conditions hold, for all $h \in H$ and $m \in M$:

$$m_{(-1)} x^1 \otimes x^3 (m_{(0)} \cdot x^2)_{(-1)} \otimes (m_{(0)} \cdot x^2)_{(0)}$$
$$= y^2 (m \cdot y^1)_{(-1)_1} x^1 \otimes y^3 (m \cdot y^1)_{(-1)_2} x^2 \otimes (m \cdot y^1)_{(0)} \cdot x^3, \quad (8.2.7)$$
$$\varepsilon(m_{(-1)}) m_{(0)} = m, \quad (8.2.8)$$
$$m_{(-1)} h_1 \otimes m_{(0)} \cdot h_2 = h_2 (m \cdot h_1)_{(-1)} \otimes (m \cdot h_1)_{(0)}. \quad (8.2.9)$$

The category of right–left Yetter–Drinfeld modules and $k$-linear maps that preserve the $H$-action and $H$-coaction is denoted by ${}^H\mathcal{YD}_H$.

(4) A right Yetter–Drinfeld module over $H$ is a right $H$-module $M$ together with a $k$-linear map (called the right $H$-coaction)

$$\rho_M : M \to M \otimes H, \quad \rho_M(m) = m_{(0)} \otimes m_{(1)}$$

such that the following conditions hold, for all $h \in H$ and $m \in M$:

$$(m_{(0)} \cdot X^2)_{(0)} \otimes X^1 (m_{(0)} \cdot X^2)_{(1)} \otimes m_{(1)} X^3$$
$$= (m \cdot Y^3)_{(0)} \cdot X^1 \otimes Y^1 (m \cdot Y^3)_{(1)_1} X^2 \otimes Y^2 (m \cdot Y^3)_{(1)_2} X^3, \quad (8.2.10)$$

$$\varepsilon(m_{(1)})m_{(0)} = m, \tag{8.2.11}$$

$$m_{(0)} \cdot h_1 \otimes m_{(1)}h_2 = (m \cdot h_2)_{(0)} \otimes h_1(m \cdot h_2)_{(1)}. \tag{8.2.12}$$

The category of right Yetter–Drinfeld modules and $k$-linear maps that preserve the $H$-action and $H$-coaction is denoted by $\mathscr{YD}_H^H$.

The relation between Yetter–Drinfeld modules and the center construction is the following.

**Theorem 8.8** *Let $H$ be a quasi-bialgebra. Then we have the following isomorphisms of categories:*

$$\mathscr{W}_l({}_H\mathscr{M}) \cong {}_H^H\mathscr{YD}\ ; \quad \mathscr{W}_r({}_H\mathscr{M}) \cong {}_H\mathscr{YD}^H;$$

$$\mathscr{W}_r(\mathscr{M}_H) \cong \mathscr{YD}_H^H\ ; \quad \mathscr{W}_l(\mathscr{M}_H) \cong {}^H\mathscr{YD}_H.$$

*Proof* Take $(M, s_{M,-}) \in \mathscr{W}_l({}_H\mathscr{M})$, and consider $\lambda_M = s_{M,H} \circ (\mathrm{Id}_M \otimes \eta_H) :\ M \to H \otimes M$, where we denote by $\eta_H : k \to H$, $\eta_H(1) = 1_H$. We use the notation $\lambda_M(m) = m_{(-1)} \otimes m_{(0)} = s_{M,H}(m \otimes 1_H)$. The map $\lambda_M$ determines $s$ completely; indeed, if $x \in X \in {}_H\mathscr{M}$ and we consider $f :\ H \to X,\ f(h) = h \cdot x$, a left $H$-linear map, the naturality of $s$ entails that $(f \otimes \mathrm{Id}_M) \circ s_{M,H} = s_{M,X}(\mathrm{Id}_M \otimes f)$. Hence, for all $m \in M$, we have

$$\begin{aligned} s_{M,X}(m \otimes x) &= s_{M,X}((\mathrm{Id}_M \otimes f)(m \otimes 1_H)) \\ &= f(m_{(-1)}) \otimes m_{(0)} = m_{(-1)} \cdot x \otimes m_{(0)}. \end{aligned} \tag{8.2.13}$$

In particular, $m = s_{M,k}(m \otimes 1) = m_{(-1)} \cdot 1 \otimes m_{(0)} = \varepsilon(m_{(-1)})m_{(0)}$, so (8.2.2) holds. If we evaluate (8.1.4), with $X = Y = H$, at $m \otimes 1_H \otimes 1_H$, we find (8.2.1). Finally,

$$h \cdot s_{M,H}(m \otimes 1_H) = s_{M,H}(h_1 \cdot m \otimes h_2) = (h_1 \cdot m)_{(-1)}h_2 \otimes (h_1 \cdot m)_{(0)},$$
$$h \cdot s_{M,H}(m \otimes 1_H) = h \cdot (m_{(-1)} \otimes m_{(0)}) = h_1 m_{(-1)} \otimes h_2 \cdot m_{(0)},$$

proving (8.2.3). So we have shown that $(M, \lambda_M)$ is a left Yetter–Drinfeld module.

Conversely, if $(M, \lambda_M)$ is a left Yetter–Drinfeld module, then $(M, s_{M,-})$, with $s$ given by (8.2.13), is an object of $\mathscr{W}_l({}_H\mathscr{M})$; the details are left to the reader.

The proof of the other three isomorphisms can be done in a similar way. □

**Corollary 8.9** ${}_H^H\mathscr{YD}$, ${}_H\mathscr{YD}^H$, ${}^H\mathscr{YD}_H$ *and* $\mathscr{YD}_H^H$ *are pre-braided categories.*

*Proof* The pre-braided structure of $\mathscr{W}_l({}_H\mathscr{M})$ induces a pre-braided structure on ${}_H^H\mathscr{YD}$. This structure is such that the forgetful functor ${}_H^H\mathscr{YD} \to {}_H\mathscr{M}$ is strict monoidal. By using (3.1.2) and (8.1.5), we find that the $H$-coaction on the tensor product $M \otimes N$ of two left Yetter–Drinfeld modules $M$ and $N$ is given by

$$\begin{aligned} \lambda_{M\otimes N}(m \otimes n) = {} & X^1(x^1Y^1 \cdot m)_{(-1)}x^2(Y^2 \cdot n)_{(-1)}Y^3 \\ & \otimes X^2 \cdot (x^1Y^1 \cdot m)_{(0)} \otimes X^3x^3 \cdot (Y^2 \cdot n)_{(0)}. \end{aligned} \tag{8.2.14}$$

By using (8.1.6), we also find the pre-braiding:

$$c_{M,N}(m \otimes n) = m_{(-1)} \cdot n \otimes m_{(0)}. \tag{8.2.15}$$

By applying the same procedure, we can make ${}_H\mathcal{YD}^H$, ${}^H\mathcal{YD}_H$ and $\mathcal{YD}^H_H$ into pre-braided categories.

On ${}_H\mathcal{YD}^H$, we find the following structure. The right $H$-coaction on the tensor product $M\otimes N$ of $M,N\in {}_H\mathcal{YD}^H$ can be computed by using (8.1.2). Explicitly, it is given, for all $m\in M$, $n\in N$, by

$$\rho_{M\otimes N}(m\otimes n) = x^1X^1\cdot(y^2\cdot m)_{(0)}\otimes x^2\cdot(X^3y^3\cdot n)_{(0)} \otimes x^3(X^3y^3\cdot n)_{(1)}X^2(y^2\cdot m)_{(1)}y^1. \tag{8.2.16}$$

We have seen that the functor forgetting the coaction is strict monoidal, so

$$h\cdot(m\otimes n) = h_1\cdot m\otimes h_2\cdot n. \tag{8.2.17}$$

The pre-braiding $\mathfrak{c}$ can be deduced from (8.1.3): $\mathfrak{c}_{M,N}: M\otimes N\to N\otimes M$ is given by

$$\mathfrak{c}_{M,N}(m\otimes n) = n_{(0)}\otimes n_{(1)}\cdot m, \tag{8.2.18}$$

for $m\in M$ and $n\in N$.

For completeness' sake, let us also describe the pre-braided structure of $\mathcal{YD}^H_H$ and ${}^H\mathcal{YD}_H$. For $M,N\in\mathcal{YD}^H_H$, the coaction on $M\otimes N$ is given by the formula

$$\rho(m\otimes n) = (m\cdot X^2)_{(0)}\cdot x^1Y^1\otimes(n\cdot X^3x^3)_{(0)}\cdot Y^2 \otimes X^1(m\cdot X^2)_{(1)}x^2(n\cdot X^3x^3)_{(0)}Y^3.$$

The pre-braiding $d$ is the following:

$$d_{M,N}(m\otimes n) = n_{(0)}\otimes m\cdot n_{(1)}. \tag{8.2.19}$$

Now take $M,N\in {}^H\mathcal{YD}_H$. The coaction on $M\otimes N$ is the following:

$$\lambda(m\otimes n) = x^3(n\cdot x^2)_{(-1)}X^2(m\cdot x^1X^1)_{(-1)}y^1 \otimes(m\cdot x^1X^1)_{(0)}\cdot y^2\otimes(n\cdot x^2)_{(0)}\cdot X^3y^3.$$

The pre-braiding $\mathfrak{d}$ is given by

$$\mathfrak{d}_{M,N}(m\otimes n) = n\cdot m_{(-1)}\otimes m_{(0)}. \tag{8.2.20}$$

The verification of all these facts is left to the reader. □

There exist connections between the four Yetter–Drinfeld categories defined above. Recall that if $H$ is a quasi-bialgebra then by $H^{\mathrm{op,cop}}$ we have denoted the $k$-vector space $H$ endowed with the opposite multiplication and comultiplication of $H$. It is also a quasi-bialgebra with reassociator $\Phi^{321}$, where $\Phi$ is the reassociator of $H$.

**Proposition 8.10** *We have an isomorphism of monoidal categories*

$$F: \overline{{}_{H^{\mathrm{op,cop}}}\mathcal{M}}\to\mathcal{M}_H.$$

*$F(M) = M$ as a k-vector space, with right $H$-action $mh = h\cdot m$.*

*Proof* It is obvious that $F$ is an isomorphism of categories. So we only need to show that it preserves the monoidal structure. Let us first describe the monoidal structure of ${}_{H^{\mathrm{op,cop}}}\mathscr{M}$. The left $H^{\mathrm{op,cop}}$-action on $N\otimes M$ is

$$h\cdot(n\otimes m)=h_2\cdot n\otimes h_1\cdot m.$$

The associativity constraint $a_{P,N,M}:\ (P\otimes N)\otimes M\to P\otimes(N\otimes M)$ is

$$a_{P,N,M}((p\otimes n)\otimes m)=X^3\cdot p\otimes(X^2\cdot n\otimes X^1\cdot m).$$

Now we describe the monoidal structure of $\overline{{}_{H^{\mathrm{op,cop}}}\mathscr{M}}$. We have $M\overline{\otimes}N=N\otimes M$. For $m\in M$, $n\in N$, we write $m\overline{\otimes}n=n\otimes m\in M\overline{\otimes}N=N\otimes M$. Then

$$h\cdot(m\overline{\otimes}n)=h_2\cdot m\overline{\otimes}h_1\cdot n. \tag{8.2.21}$$

The associativity constraint

$$\overline{a}_{M,N,P}=a^{-1}_{P,N,M}:\ (M\overline{\otimes}N)\overline{\otimes}P=P\otimes(N\otimes M)\to M\overline{\otimes}(N\overline{\otimes}P)=(P\otimes N)\otimes M$$

is given by

$$\overline{a}_{M,N,P}((m\overline{\otimes}n)\overline{\otimes}p)=x^1\cdot m\overline{\otimes}(x^2\cdot n\overline{\otimes}x^3\cdot p). \tag{8.2.22}$$

It is then clear from (8.2.21–8.2.22) that $F$ behaves well with respect to the monoidal structures of $\overline{{}_{H^{\mathrm{op,cop}}}\mathscr{M}}$ and $\mathscr{M}_H$. More precisely, $F$ is strong monoidal with the structure given by

$$\varphi_2=(\varphi_{2,M,N}:F(M)\otimes F(N)\ni m\otimes n\mapsto n\otimes m\in F(M\overline{\otimes}N))_{M,N\in\overline{{}_{H^{\mathrm{op,cop}}}\mathscr{M}}}$$

and $\varphi_0=\mathrm{Id}_k$. So the proof is finished. □

An immediate consequence of Proposition 8.6, Theorem 8.8 and Proposition 8.10 is then the following.

**Proposition 8.11** *Let $H$ be a quasi-bialgebra. Then we have the following isomorphisms of pre-braided categories, induced by the functor $F$ from Proposition 8.10:*

$${}_H\mathscr{YD}^H\cong\overline{{}^{H^{\mathrm{op,cop}}}_{H^{\mathrm{op,cop}}}\mathscr{YD}}\ \text{ and }\ {}^H\mathscr{YD}_H\cong\overline{{}_{H^{\mathrm{op,cop}}}\mathscr{YD}^{H^{\mathrm{op,cop}}}}.$$

*Proof* We have the following isomorphisms of pre-braided categories:

$$\mathscr{YD}^H_H\cong\mathscr{W}_r(\mathscr{M}_H)\cong\mathscr{W}_r(\overline{{}_{H^{\mathrm{op,cop}}}\mathscr{M}})\cong\overline{\mathscr{W}_l({}_{H^{\mathrm{op,cop}}}\mathscr{M})}\cong\overline{{}^{H^{\mathrm{op,cop}}}_{H^{\mathrm{op,cop}}}\mathscr{YD}}.$$

Similarly,

$${}^H\mathscr{YD}_H\cong\mathscr{W}_l(\mathscr{M}_H)\cong\mathscr{W}_l(\overline{{}_{H^{\mathrm{op,cop}}}\mathscr{M}})\cong\overline{\mathscr{W}_r({}_{H^{\mathrm{op,cop}}}\mathscr{M})}\cong\overline{{}_{H^{\mathrm{op,cop}}}\mathscr{YD}^{H^{\mathrm{op,cop}}}},$$

as pre-braided categories. This finishes the proof. □

If $H$ is a quasi-Hopf algebra with bijective antipode then the four weak centers in the statement of Theorem 8.8 are equal to the centers.

**Proposition 8.12** *If the antipode of $H$ is bijective then $\mathscr{W}_r({}_H\mathscr{M}) = \mathscr{Z}_r({}_H\mathscr{M})$, $\mathscr{W}_l(\mathscr{M}_H) = \mathscr{Z}_l(\mathscr{M}_H)$, $\mathscr{W}_l({}_H\mathscr{M}) = \mathscr{Z}_l({}_H\mathscr{M})$ and $\mathscr{W}_r(\mathscr{M}_H) = \mathscr{Z}_r(\mathscr{M}_H)$.*

*Consequently, ${}_H\mathscr{YD}^H$, ${}^H\mathscr{YD}_H$, ${}^H_H\mathscr{YD}$ and $\mathscr{YD}^H_H$ are braided categories.*

*Proof* If $H$ is a quasi-Hopf algebra with bijective antipode then the pre-braiding $c$ defined in (8.2.15) is a natural isomorphism. For any $M,N \in {}^H_H\mathscr{YD}$, the inverse of $c_{M,N}$ is given by

$$c^{-1}_{M,N}(n\otimes m) = \tilde{q}^2_1X^2\cdot(p^1\cdot m)_{(0)}\otimes S^{-1}(\tilde{q}^1X^1(p^1\cdot m)_{(-1)}p^2S(\tilde{q}^2_2X^3))\cdot n, \quad (8.2.23)$$

where $p_R = p^1\otimes p^2$ and $q_L = \tilde{q}^1\otimes\tilde{q}^2$ are the elements defined by (3.2.19) and (3.2.20), respectively.

Indeed, on the one hand we have

$$\begin{aligned}
&c_{M,N}\circ c^{-1}_{M,N}(n\otimes m)\\
&= \left(\tilde{q}^2_1X^2\cdot(p^1\cdot m)_{(0)}\right)_{(-1)}\tilde{q}^2_2X^3S^{-1}\left(\tilde{q}^1X^1(p^1\cdot m)_{(-1)}p^2\right)\cdot n\\
&\quad\otimes\left(\tilde{q}^2_1X^2\cdot(p^1\cdot m)_{(0)}\right)_{(0)}\\
&\overset{(8.2.3)}{=} \tilde{q}^2_1(X^2\cdot(p^1\cdot m)_{(0)})_{(-1)}X^3S^{-1}(\tilde{q}^1X^1(p^1\cdot m)_{(-1)}p^2)\cdot n\\
&\quad\otimes\tilde{q}^2_2\cdot(X^2\cdot(p^1\cdot m)_{(0)})_{(0)}\\
&\overset{(8.2.1),(3.2.19)}{=} \tilde{q}^2_1X^2m_{(-1)_2}S^{-1}(\tilde{q}^1X^1m_{(-1)_1}\beta)\cdot n\otimes\tilde{q}^2_2X^3\cdot m_{(0)}\\
&= \tilde{q}^2_1\tilde{p}^1S^{-1}(\tilde{q}^1)\cdot n\otimes\tilde{q}^2_2\tilde{p}^2\cdot m\overset{(3.2.24)}{=}n\otimes m,
\end{aligned}$$

for all $n\in N$ and $m\in M$. On the other hand, we have:

$$\begin{aligned}
&c^{-1}_{M,N}\circ c_{M,N}(m\otimes n)\\
&= \tilde{q}^2_1X^2\cdot(p^1\cdot m_{(0)})_{(0)}\otimes S^{-1}(\tilde{q}^1X^1(p^1\cdot m_{(0)})_{(-1)}p^2S(\tilde{q}^2_2X^3))m_{(-1)}\cdot n\\
&\overset{(5.2.7)}{=} z^3y^2_2\cdot(p^1\cdot m_{(0)})_{(0)}\otimes S^{-1}(\alpha z^2y^2_1(p^1\cdot m_{(0)})_{(-1)}p^2S(y^3))z^1y^1m_{(-1)}\cdot n\\
&\overset{(8.2.3)}{=} z^3\cdot(y^2_1p^1\cdot m_{(0)})_{(0)}\otimes S^{-1}(\alpha z^2(y^2_1p^1\cdot m_{(0)})_{(-1)}y^2_2p^2S(y^3))z^1y^1m_{(-1)}\cdot n\\
&\underset{(8.2.3)}{\overset{(5.5.16)}{=}} z^3\cdot(Y^2\cdot(p^1_1\cdot m)_{(0)})_{(0)}\\
&\quad\otimes S^{-1}(\alpha z^2(Y^2\cdot(p^1_1\cdot m)_{(0)})_{(-1)}Y^3p^2)z^1Y^1(p^1_1\cdot m)_{(-1)}p^1_2\cdot n\\
&\overset{(8.2.1)}{=} (Y^1p^1_1\cdot m)_{(0)}\otimes S^{-1}(\alpha(Y^1p^1_1\cdot m)_{(-1)_2}Y^3p^2)(Y^1p^1_1\cdot m)_{(-1)_1}Y^2p^1_2\cdot n\\
&= q^1p^1_1\cdot m\otimes S^{-1}(p^2)q^2p^1_2\cdot n\overset{(3.2.23)}{=}m\otimes n,
\end{aligned}$$

for all $m\in M$ and $n\in N$, as needed.

Since $\mathscr{YD}^H_H$ and ${}^{H^{\rm op,cop}}_{H^{\rm op,cop}}\mathscr{YD}$ identify as ordinary categories, it follows that $\mathscr{YD}^H_H$ is a braided category, too. We have that the inverse of the (pre-)braiding $d$ defined in (8.2.19) is given by

$$d^{-1}_{M,N}(n\otimes m) = (m\cdot\tilde{q}^2)_{(0)}\cdot X^2p^1_2\otimes n\cdot S(S^{-1}(X^1p^1_1)\tilde{q}^1(m\cdot\tilde{q}^2)_{(1)}X^3p^2), \quad (8.2.24)$$

for all $n\in N\in\mathscr{YD}^H_H$ and $m\in M\in\mathscr{YD}^H_H$.

Similarly, for $H$ a quasi-bialgebra we have that ${}_H\mathscr{YD}^H$ can be identified as an ordinary category with ${}^{H^{\rm cop}}_{H^{\rm cop}}\mathscr{YD}$. Thus if $H$ is a quasi-Hopf algebra then ${}_H\mathscr{YD}^H$ is braided; the inverse of $\mathfrak{c}$ defined in (8.2.18) is given by

$$\begin{aligned}\mathfrak{c}^{-1}_{M,N}(n\otimes m) &= q^1_1x^1S(q^2x^3(\tilde{p}^2\cdot n)_{(1)}\tilde{p}^1)\cdot m\otimes q^1_2x^2\cdot(\tilde{p}^2\cdot n)_{(0)}\\ &= X^1_1x^1S(X^2x^3(\tilde{p}^2\cdot n)_{(1)}\tilde{p}^1)\alpha X^3\cdot m\otimes X^1_2x^2\cdot(\tilde{p}^2\cdot n)_{(0)},\end{aligned}$$

for all $n\in N\in {}_H\mathscr{YD}^H$ and $m\in M\in {}_H\mathscr{YD}^H$. Finally, ${}^H\mathscr{YD}_H\equiv {}_{H^{\rm op,cop}}\mathscr{YD}^{H^{\rm op,cop}}$ as ordinary categories, hence when $H$ is a quasi-Hopf algebra ${}^H\mathscr{YD}_H$ is a braided category. The inverse of the (pre-)braiding $\mathfrak{d}$ in (8.2.20) is given by

$$\begin{aligned}\mathfrak{d}^{-1}_{M,N}(n\otimes m) &= m\cdot S(q^2(n\cdot q^1)_{(-1)}x^1\tilde{p}^1)x^3\tilde{p}^2_2\otimes(n\cdot q^1)_{(0)}\cdot x^2\tilde{p}^2_1\\ &= m\cdot S(X^2(n\cdot X^1)_{(-1)}x^1\tilde{p}^1)\alpha X^3x^3\tilde{p}^2_2\otimes(n\cdot X^1)_{(0)}\cdot x^2\tilde{p}^2_1,\end{aligned}$$

for all $n\in N\in {}^H\mathscr{YD}_H$ and $m\in M\in {}^H\mathscr{YD}_H$. □

**Remark 8.13** By using the definitions of $q_R$ and $p_L$ one can rewrite the formulas for $\mathfrak{c}^{-1}$ and $\mathfrak{d}^{-1}$ in such a way that the bijectivity of the antipode becomes unnecessary for defining them. Therefore, the categories of left–right and right–left Yetter–Drinfeld modules are braided over an arbitrary quasi-Hopf algebra.

By combining Proposition 8.5 and Theorem 8.8, we find the following result.

**Theorem 8.14** *Let $H$ be a quasi-Hopf algebra with bijective antipode. Then we have an isomorphism of braided categories $F:\ {}_H\mathscr{YD}^{H\,{\rm in}}\to {}^H_H\mathscr{YD}$, defined as follows. For $M\in {}_H\mathscr{YD}^H$, $F(M)=M$ as a left $H$-module; the left $H$-coaction is*

$$\lambda_M(m)=m_{(-1)}\otimes m_{(0)}=q^1_1x^1S(q^2x^3(\tilde{p}^2\cdot m)_{(1)}\tilde{p}^1)\otimes q^1_2x^2\cdot(\tilde{p}^2\cdot m)_{(0)},\qquad(8.2.25)$$

*for all $m\in M$, where $q_R=q^1\otimes q^2$ and $p_L=\tilde{p}^1\otimes\tilde{p}^2$ are the elements defined by* (3.2.19) *and* (3.2.20), *and $\rho_M(m)=m_{(0)}\otimes m_{(1)}$ is the right coaction of $H$ on $M$. The functor $F$ sends a morphism to itself.*

*Proof* $F$ is nothing else than the composition of the isomorphisms

$${}_H\mathscr{YD}^{H\,{\rm in}}\to\mathscr{Z}_r({}_H\mathscr{M})^{\rm in}\to\mathscr{Z}_l({}_H\mathscr{M})\to{}^H_H\mathscr{YD}.$$

For $M\in {}_H\mathscr{YD}^H$, we compute that the corresponding left Yetter–Drinfeld module structure $\lambda_M$ on $M$ is the following

$$\begin{aligned}\lambda_M(m) &= \mathfrak{c}^{-1}_{H,M}(m\otimes 1_H)\\ &= m_{(-1)}\otimes m_{(0)}=q^1_1x^1S(q^2x^3(\tilde{p}^2\cdot m)_{(1)}\tilde{p}^1)\otimes q^1_2x^2\cdot(\tilde{p}^2\cdot m)_{(0)},\end{aligned}$$

as needed. Note that the inverse of $F$ is the functor $G:{}^H_H\mathscr{YD}\to {}_H\mathscr{YD}^{H\,{\rm in}}$ defined as follows. If $M\in {}^H_H\mathscr{YD}$ then $G(M)=M$ as a left $H$-module, with right $H$-coaction

$$\begin{aligned}\rho_M(m) &= m_{(0)}\otimes m_{(1)}\\ &= \tilde{q}^2_1X^2\cdot(p^1\cdot m)_{(0)}\otimes\tilde{q}^2_2X^3S^{-1}(\tilde{q}^1X^1(p^1\cdot m)_{(-1)}p^2),\qquad(8.2.26)\end{aligned}$$

for all $m\in M$, where $q_L=\tilde{q}^1\otimes\tilde{q}^2$ and $p_R=p^1\otimes p^2$ are the elements defined by (3.2.20) and (3.2.19), and $\lambda_M(m)=m_{(-1)}\otimes m_{(0)}$ is the left coaction of $H$ on $M$. □

Similarly, we have the following result.

**Theorem 8.15** *Let $H$ be a quasi-Hopf algebra with bijective antipode. Then the categories $\mathscr{YD}_H^H$ and ${}^H\mathscr{YD}_H^{\rm in}$ are isomorphic as braided categories.*

Let $H$ be a quasi-bialgebra and $\mathfrak{F} = \mathfrak{F}^1 \otimes \mathfrak{F}^2 \in H \otimes H$ a twist with inverse $\mathfrak{F}^{-1} = \mathfrak{G}^1 \otimes \mathfrak{G}^2$. Then, by Proposition 3.5 we have an isomorphism of monoidal categories $\Pi : {}_H\mathscr{M} \to {}_{H_\mathfrak{F}}\mathscr{M}$, where $\Pi(M) = M$, with the same left $H$-action. If $H$ is a quasi-Hopf algebra with bijective antipode, then we can consider the Drinfeld twist $f$ defined in (3.2.15). The antipode $S : H^{\rm op,cop} \to H_f$ is a quasi-Hopf algebra isomorphism (see Proposition 3.23) and therefore the monoidal categories ${}_{H^{\rm op,cop}}\mathscr{M}$ and ${}_{H_f}\mathscr{M}$ are isomorphic. We have seen in Proposition 8.10 that ${}_{H^{\rm op,cop}}\mathscr{M}$ is isomorphic to $\overline{\mathscr{M}_H}$ as a monoidal category. Thus we conclude that the monoidal categories ${}_H\mathscr{M}$ and $\overline{\mathscr{M}_H}$ are isomorphic. By using Proposition 8.6 and Theorem 8.8, we find braided monoidal isomorphisms

$$
{}_H^H\mathscr{YD} \cong \mathscr{Z}_l({}_H\mathscr{M}) \cong \mathscr{Z}_l(\overline{\mathscr{M}_H}) \cong \overline{\mathscr{Z}_r(\mathscr{M}_H)} \cong \overline{\mathscr{YD}_H^H},
$$
$$
{}_H\mathscr{YD}^H \cong \mathscr{Z}_r({}_H\mathscr{M}) \cong \mathscr{Z}_r(\overline{\mathscr{M}_H}) \cong \overline{\mathscr{Z}_l(\mathscr{M}_H)} \cong \overline{{}^H\mathscr{YD}_H}.
$$

We summarize our results as follows:

**Theorem 8.16** *Let $H$ be a quasi-Hopf algebra with bijective antipode. Then we have the following isomorphisms of braided categories:*

$$
{}_H^H\mathscr{YD} \cong {}_H\mathscr{YD}^{H\,{\rm in}} \cong \overline{{}^H\mathscr{YD}_H}^{\rm in} \cong \overline{\mathscr{YD}_H^H}.
$$

The braided monoidal isomorphisms ${}_H^H\mathscr{YD} \cong \overline{\mathscr{YD}_H^H}$ and ${}_H\mathscr{YD}^H \cong \overline{{}^H\mathscr{YD}_H}$ can be described explicitly. Let us compute the functor ${}_H^H\mathscr{YD} \to \overline{\mathscr{YD}_H^H}$.

We have a monoidal isomorphism $\Pi : {}_H\mathscr{M} \to {}_{H_f}\mathscr{M}$, where $\Pi(M) = M$ with the same left $H$-action. We will denote the tensor product on ${}_{H_f}\mathscr{M}$ by $\otimes^f$. For $M, N \in {}_H\mathscr{M}$, the isomorphism $\psi : \Pi(M \otimes N) \to \Pi(M) \otimes^f \Pi(N)$ is given by the formula $\psi(m \otimes n) = f^1 \cdot m \otimes^f f^2 \cdot n$. This isomorphism induces an isomorphism between the two left centers, hence between the categories ${}_H^H\mathscr{YD}$ and ${}_{H_f}^{H_f}\mathscr{YD}$. Take $(M, s_{M,-}) \in \mathscr{Z}_l({}_H\mathscr{M})$ and $(M, s^f_{M,-})$ the corresponding object in $\mathscr{Z}_l({}_{H_f}\mathscr{M})$, and let $\lambda$ and $\lambda^f$ be the associated coactions. Then we have $\psi s_{M,H} = s^f_{M,H}\psi$, and so we compute

$$
\lambda^f(m) = s^f_{M,H}(m \otimes 1_H) = \psi((s_{M,H}(g^1 \cdot m \otimes g^2)) = f^1(g^1 \cdot m)_{(-1)}g^2 \otimes f^2(g^1 \cdot m)_{(0)}.
$$

The quasi-Hopf algebra isomorphism $S^{-1} : H_f \to H^{\rm op,cop}$ induces an isomorphism of monoidal categories ${}_{H_f}\mathscr{M} \to {}_{H^{\rm op,cop}}\mathscr{M} \cong \overline{\mathscr{M}_H}$. The image of $M$ is $M$ as a $k$-vector space, with right $H$-action given by $m \cdot h = S(h) \cdot m$. Take $(M, s^f_{M,-}) \in \mathscr{Z}_l({}_{H_f}\mathscr{M})$ and the corresponding object $(M, t_{-,M}) \in \overline{\mathscr{Z}_r(\mathscr{M}_H)}$. Then the diagram

$$\begin{array}{ccc}
M \otimes^f H & \xrightarrow{\;s^f_{M,H}\;} & H \otimes^f M \\
{\scriptstyle \mathrm{Id}_M \otimes S^{-1}}\downarrow & & \downarrow{\scriptstyle S^{-1} \otimes \mathrm{Id}_M} \\
H \overline{\otimes} M = M \otimes H & \xrightarrow{\;t_{H,M}\;} & M \overline{\otimes} H = H \otimes M
\end{array}$$

is commutative, and so we compute the right $H$-coaction $\rho_M$ on $M$ as follows:

$$\rho_M(m) = t_{H,M}(1_H \overline{\otimes} m) = f^2 \cdot (g^1 \cdot m)_{(0)} \otimes S^{-1}(f^1 (g^1 \cdot m)_{(-1)} g^2).$$

We conclude that the braided monoidal isomorphism $K : {}_H^H\mathscr{YD} \to \overline{\mathscr{YD}_H^H}$ is defined as follows: $K(M) = M$, with

$$m \cdot h = S(h) \cdot m,$$
$$\rho_M(m) = f^2 \cdot (g^1 \cdot m)_{(0)} \otimes S^{-1}(f^1 (g^1 \cdot m)_{(-1)} g^2).$$

The inverse functor $K^{-1}$ can be computed in a similar way: $K^{-1}(M) = M$ with

$$h \cdot m = m \cdot S^{-1}(h),$$
$$\lambda_M(m) = g^1 S((m \cdot S^{-1}(f^1))_{(1)}) f^2 \otimes (m \cdot S^{-1}(f^1))_{(0)} \cdot S^{-1}(g^2).$$

For completeness' sake, we also give the formulas for the braided monoidal isomorphism $G : \overline{{}^H\mathscr{YD}_H} \to {}_H\mathscr{YD}^H$. We have $G(M) = M$ with

$$h \cdot m = m \cdot S^{-1}(h),$$
$$\rho_M(m) = g^1 \cdot (f^2 \cdot m)_{(0)} \otimes g^2 S((f^2 \cdot m)_{(-1)}) f^1.$$

Conversely, $G^{-1}(M) = M$ with

$$m \cdot h = S(h) \cdot m,$$
$$\lambda_M(m) = S^{-1}(f^2 (g^2 \cdot m)_{(1)} g^1) \otimes f^1 \cdot (g^2 \cdot m)_{(0)}.$$

## 8.3 The Rigid Braided Category ${}_H^H\mathscr{YD}^{\mathrm{fd}}$

In this section we investigate when the center of a monoidal category is left or right rigid. We then apply the results to the category of left Yetter–Drinfeld modules over a quasi-Hopf algebra with bijective antipode.

Recall that the "mate" notion was defined in the last part of Section 1.8.

**Theorem 8.17** *Let $\mathscr{C}$ be a monoidal category and $\mathscr{Z}_l(\mathscr{C})$ its left center. An object $(V, c_{V,-})$ of the left center admits a left dual if and only $V$ has a left dual in $\mathscr{C}$ and the mate of $c_{V,X}^{-1}$ is an isomorphism, for any object $X \in \mathscr{C}$. If this is the case then the left dual of $(V, c_{V,-})$ is $(V^*, s_{V^*,-} := (c_{V,-}^{-1})^\flat)$.*

*Similarly, an object $(V, c_{V,-})$ of $\mathscr{Z}_l(\mathscr{C})$ admits a right dual if and only if $V$ has a right dual in $\mathscr{C}$ and ${}^\flat c_{V,X}$ is an isomorphism, for any object $X \in \mathscr{C}$. If this is the case then $({}^*V, t_{{}^*V,-} := ({}^\flat c_{V,-})^{-1})$ is a right dual for $(V, c_{V,-})$.*

*Proof* We prove only the first assertion; the second follows in a similar manner.

Assume first that $(V, c_{V,-})$ has a left dual $(V^*, s_{V^*,-})$ in $\mathscr{Z}_l(\mathscr{C})$, and let $\varepsilon, \eta$ be the associated evaluation and coevaluation morphisms in $\mathscr{Z}_l(\mathscr{C})$. Since the forgetful functor $\Pi : \mathscr{Z}_l(\mathscr{C}) \to \mathscr{C}$, defined by $\Pi(W, c_{W,-}) = W$ and $\Pi(f) = f$, is a strict monoidal functor it follows from Proposition 1.67 that $V^*$ is a left dual object of $V$ in $\mathscr{C}$ via the same morphisms $\varepsilon$, $\eta$, viewed now as morphisms in $\mathscr{C}$. Furthermore, the fact that $\varepsilon =$ 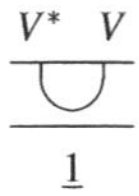

is a morphism in $\mathscr{Z}_l(\mathscr{C})$ means

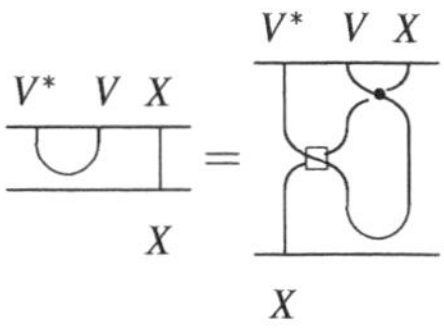

, $\forall X \in \mathscr{C}$, where $c_{V,X} =$ 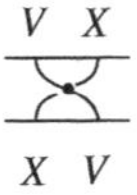

and $s_{V^*,X} =$ 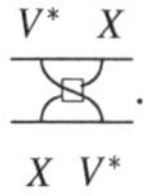

.

From here we deduce that

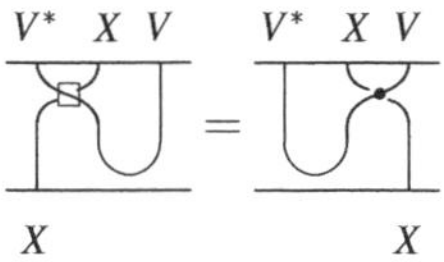

, so $s_{V^*,X} =$ 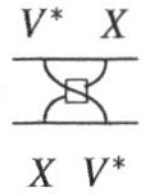

$=$ 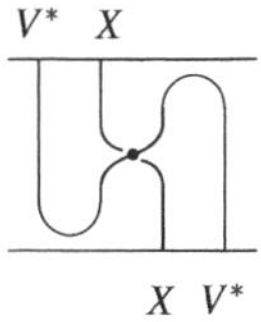

$= (c_{V,X}^{-1})^\flat$ ; see (1.6.6).

We conclude that $(c_{V,X}^{-1})^\flat$ is an isomorphism, for all $X$ in $\mathscr{C}$, so the direct implication is proved.

Conversely, if $(c_{V,X}^{-1})^\flat$ is an isomorphism for any $X \in \mathscr{C}$ and $V$ admits a left dual $V^*$ we show that $(V^*, s_{V^*,-} := (c_{V,-}^{-1})^\flat)$ is a left dual of $(V, c_{V,-})$ in $\mathscr{Z}_l(\mathscr{C})$. To this end we first need to show that $(V^*, s_{V^*,-} := (c_{V,-}^{-1})^\flat)$ is an object in $\mathscr{Z}_l(\mathscr{C})$. Indeed, since $(V, c_{V,-})$ belongs to the left center of $\mathscr{C}$ we have that

$c_{V,X\otimes Y} =$ 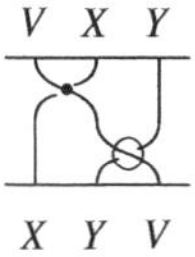

, so $c_{V,X\otimes Y}^{-1} =$ 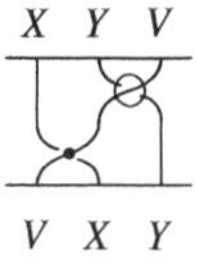

and $s_{V^*,X\otimes Y} =$ 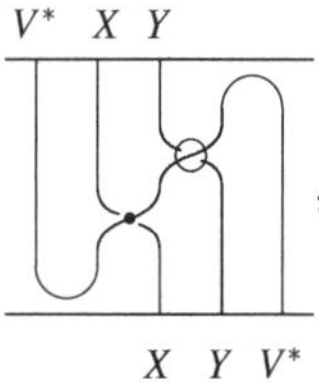

,

for any objects $X$, $Y$ of $\mathscr{C}$. The latest equality is equivalent to

$$s_{V^*,X\otimes Y} = (\mathrm{Id}_{V^*} \otimes s_{V^*,Y})(s_{V^*,X} \otimes \mathrm{Id}_Y);$$

see (1.6.6). It is clear that $s_{V^*,\underline{1}}$ identifies with $\mathrm{Id}_{V^*}$, therefore $(V^*, s_{V^*,-})$ is an object in the left center of $\mathscr{C}$, as claimed.

Consider now $\mathrm{ev}_V$ and $\mathrm{coev}_V$ the evaluation and coevaluation morphisms attached to the left dual $V^*$ of $V$. We prove that they are morphisms in $\mathscr{Z}_l(\mathscr{C})$ and this would finish the proof, since the equations required for $\mathrm{ev}_V$ and $\mathrm{coev}_V$ for an adjunction are the same in $\mathscr{Z}_l(\mathscr{C})$ as in $\mathscr{C}$. So we compute

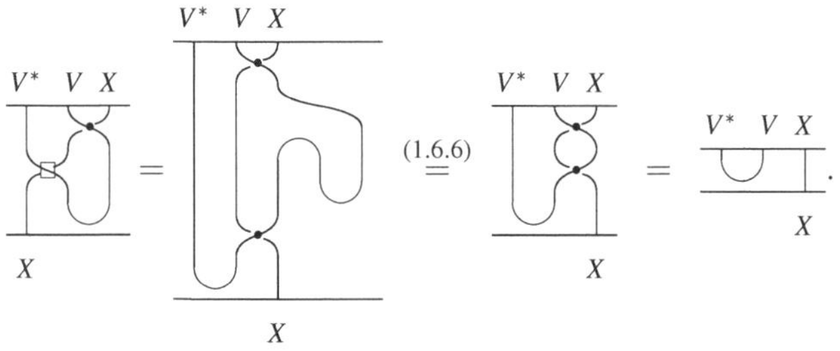


This shows that $\mathrm{ev}_V$ is a morphism in $\mathscr{Z}_l(\mathscr{C})$. In a similar manner it can be proved that $\mathrm{coev}_V$ is a morphism in $\mathscr{Z}_l(\mathscr{C})$, so the proof is finished. □

We next supply, without proof, the right-handed version of Theorem 8.17.

**Theorem 8.18** *Let $\mathscr{C}$ be a monoidal category and $\mathscr{Z}_r(\mathscr{C})$ its right center. An object $(V, c_{-,V})$ of the right center admits a left dual if and only $V$ has a left dual in $\mathscr{C}$ and the mate of $c_{X,V}$ is an isomorphism, for any object $X \in \mathscr{C}$. If this is the case then the left dual of $(V, c_{-,V})$ is $(V^*, s_{-,V^*} := (c^\flat_{-,V})^{-1})$.*

*Likewise, an object $(V, c_{-,V})$ of $\mathscr{Z}_r(\mathscr{C})$ admits a right dual if and only if $V$ has a right dual in $\mathscr{C}$ and ${}^\flat(c^{-1}_{X,V})$ is an isomorphism, for any object $X \in \mathscr{C}$. If this is the case then $({}^*V, t_{-,{}^*V} := {}^\flat(c^{-1}_{-,V}))$ is a right dual for $(V, c_{-,V})$.*

**Remark 8.19** Observe that, by the first part of Theorem 8.17, when $\mathscr{C}$ is a left rigid monoidal category then any object $(V, c_{V,-})$ of $\mathscr{Z}_l(\mathscr{C})$ has $(V^*, s_{V^*,-} = (c^{-1}_{V,-})^\flat)$ as a left dual object in $\mathscr{W}_l(\mathscr{C})$. Similarly, by the second part of Theorem 8.18 it follows that when $\mathscr{C}$ is a right rigid monoidal category then any object $(V, c_{-,V})$ of $\mathscr{Z}_r(\mathscr{C})$ admits $({}^*V, t_{-,{}^*V} := {}^\flat(c^{-1}_{-,V}))$ as a right dual object in $\mathscr{W}_r(\mathscr{C})$.

We next see that the weak center and the center coincide if $\mathscr{C}$ is a rigid monoidal category.

**Theorem 8.20** *Let $\mathscr{C}$ be a monoidal category.*

*(i) If $\mathscr{C}$ is left rigid then $\mathscr{W}_r(\mathscr{C}) = \mathscr{Z}_r(\mathscr{C})$ and, similarly, if $\mathscr{C}$ is right rigid then $\mathscr{W}_l(\mathscr{C}) = \mathscr{Z}_l(\mathscr{C})$.*

*(ii) If $\mathscr{C}$ is a rigid monoidal category then $\mathscr{W}_l(\mathscr{C}) = \mathscr{Z}_l(\mathscr{C})$ and $\mathscr{W}_r(\mathscr{C}) = \mathscr{Z}_r(\mathscr{C})$ are rigid monoidal categories.*

*Proof* (i) We prove only the first assertion, the second one can be proved in a similar manner.

Suppose that $\mathscr{C}$ is a left rigid monoidal category and let $(V, c_{-,V})$ be an object of $\mathscr{W}_r(\mathscr{C})$. We state that, for any object $X$ of $\mathscr{C}$, $c_{X,V}$ is an isomorphism in $\mathscr{C}$ with $c^{-1}_{X,V} = (c_{X^*,V})^\sharp$ (see the proof of Proposition 1.81 for the concrete definition of $(c_{X^*,V})^\sharp$).

Indeed, since $c_{-,V}$ is natural and $\mathrm{coev}_X : \underline{1} \to X \otimes X^*$ is a morphism in $\mathscr{C}$ we get

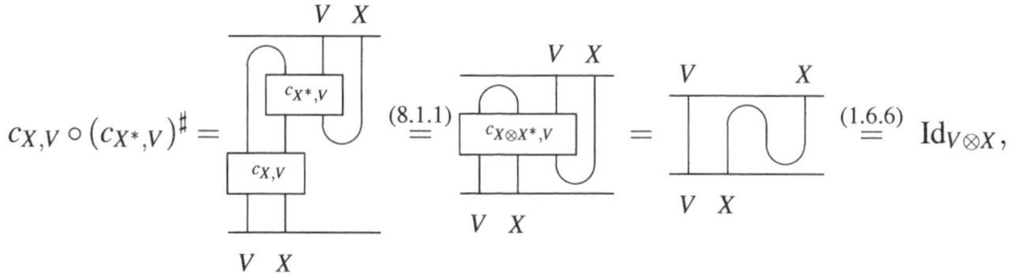
$c_{X,V} \circ (c_{X^*,V})^\sharp =$
V X
$c_{X^*,V}$
$c_{X,V}$
V X
(8.1.1)
V X
$c_{X\otimes X^*,V}$
V X
=
V X
V X
(1.6.6)
$\mathrm{Id}_{V\otimes X}$,

where in the last but one equality, apart from the naturality of $c_{-,V}$, we also used the fact that $c_{\underline{1},V} = \mathrm{Id}_V$. Similarly, by the naturality of $c_{-,V}$ and the fact that $\mathrm{ev}_X : X^* \otimes X \to \underline{1}$ is a morphism in $\mathscr{C}$, we have that

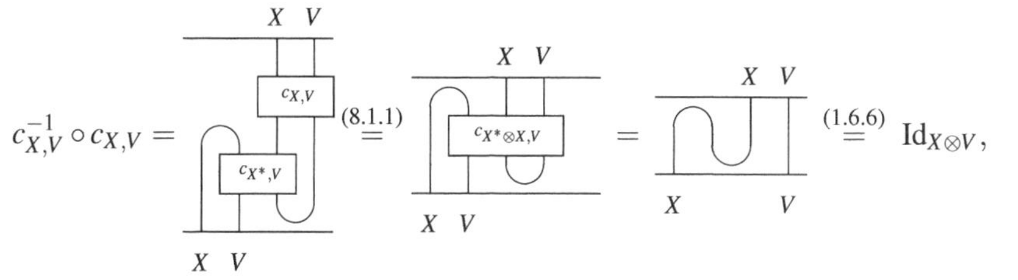
$c^{-1}_{X,V} \circ c_{X,V} =$
X V
$c_{X,V}$
$c_{X^*,V}$
X V
(8.1.1)
X V
$c_{X^*\otimes X,V}$
X V
=
X V
X V
(1.6.6)
$\mathrm{Id}_{X\otimes V}$,

and this finishes the proof of (i).

(ii) Let $\mathscr{C}$ be a rigid monoidal category. By part (i) we have $\mathscr{W}_r(\mathscr{C}) = \mathscr{Z}_r(\mathscr{C})$ and $\mathscr{W}_l(\mathscr{C}) = \mathscr{Z}_l(\mathscr{C})$. Then by Remark 8.19 it follows that $\mathscr{W}_l(\mathscr{C}) = \mathscr{Z}_l(\mathscr{C})$ is left rigid, while $\mathscr{W}_r(\mathscr{C}) = \mathscr{Z}_r(\mathscr{C})$ is right rigid. Since $\mathscr{W}_l(\mathscr{C}) = \mathscr{Z}_l(\mathscr{C})$ and $\mathscr{W}_r(\mathscr{C}) = \mathscr{Z}_r(\mathscr{C})$ are braided categories, by Proposition 1.74 we get that $\mathscr{W}_l(\mathscr{C}) = \mathscr{Z}_l(\mathscr{C})$ and $\mathscr{W}_r(\mathscr{C}) = \mathscr{Z}_r(\mathscr{C})$ are braided rigid categories, as stated. □

From now on, throughout this section $H$ is a quasi-Hopf algebra with bijective antipode, so ${}_H\mathscr{M}^{\rm fd}$ is rigid monoidal and ${}^H_H\mathcal{YD}$ is a braided category. We will see that the category of finite-dimensional left Yetter–Drinfeld modules ${}^H_H\mathcal{YD}^{\rm fd}$ is rigid and supply the explicit structures for the left and right dual objects. We first need a lemma.

**Lemma 8.21** *Let $H$ be a quasi-Hopf algebra and $p_R = p^1 \otimes p^2$, $q_L = \tilde{q}^1 \otimes \tilde{q}^2 = \tilde{Q}^1 \otimes \tilde{Q}^2$ and $f = f^1 \otimes f^2$ the elements defined by* (3.2.19), (3.2.20) *and* (3.2.15), *respectively. Then the following relation holds:*

$$S(p^1)\tilde{q}^1 p^2_1 S(\tilde{Q}^2)_1 \otimes \tilde{Q}^1 \tilde{q}^2 p^2_2 S(\tilde{Q}^2)_2 = f. \tag{8.3.1}$$

*Proof* In order to prove (8.3.1), we denote by $\delta = \delta^1 \otimes \delta^2$ the element defined in (3.2.6), and then compute that

$$\begin{aligned} & S(p^1)\tilde{q}^1 p^2_1 S(\tilde{Q}^2)_1 \otimes \tilde{Q}^1 \tilde{q}^2 p^2_2 S(\tilde{Q}^2)_2 \\ &\overset{(3.2.19)}{=} S(x^1)\tilde{q}^1 x^2_1 \beta_1 S(\tilde{Q}^2 x^3)_1 \otimes \tilde{Q}^1 \tilde{q}^2 x^2_2 \beta_2 S(\tilde{Q}^2 x^3)_2 \\ &\overset{(3.2.13),(3.2.14)}{=} S(x^1)\tilde{q}^1 x^2_1 \delta^1 S(\tilde{Q}^2_2 x^3_2) f^1 \otimes \tilde{Q}^1 \tilde{q}^2 x^2_2 \delta^2 S(\tilde{Q}^2_1 x^3_1) f^2 \\ &\overset{(3.2.6),(3.1.9),(3.2.1)}{=} S(z^1 x^1)\alpha z^2 x^2_1 y^1 \beta S(\tilde{Q}^2_2 x^3_2 y^3_2 X^3) f^1 \end{aligned}$$

$$
\begin{aligned}
&\otimes \tilde{Q}^1 z^3 x_2^2 y^2 X^1 \beta S(\tilde{Q}_1^2 x_1^3 y_1^3 X^2) f^2 \\
\overset{(3.1.9),(3.2.1)}{=}\ & S(x^1)\alpha x^2 \beta S(\tilde{Q}_2^2 X^3 x^3) f^1 \otimes \tilde{Q}^1 X^1 \beta S(\tilde{Q}_1^2 X^2) f^2 \\
\overset{(3.2.2),(3.2.20)}{=}\ & S(\tilde{Q}_2^2 \tilde{p}^2) f^1 \otimes \tilde{Q}^1 S(\tilde{Q}_1^2 \tilde{p}^1) f^1 \\
\overset{(3.2.24)}{=}\ & f^1 \otimes f^2 = f,
\end{aligned}
$$

as needed, and this finishes the proof. □

**Theorem 8.22** *Let $H$ be a quasi-Hopf algebra with bijective antipode. Then ${}_H^H\mathcal{YD}^{\rm fd}$ is a rigid braided category. For a finite-dimensional left Yetter–Drinfeld module $M$ with basis $({}_im)_{i=\overline{1,n}}$ and corresponding dual basis $({}^im)_{i=\overline{1,n}}$, the left and right duals $M^*$ and ${}^*M$ are equal to* Hom$(M,k)$ *as a vector space, with the following $H$-action and $H$-coaction:*

*For $M^*$:*

$$(h\cdot m^*)(m) = m^*(S(h)\cdot m); \tag{8.3.2}$$

$$
\begin{aligned}
\lambda_{M^*}(m^*) &= m^*_{(-1)} \otimes m^*_{(0)} \\
&= \sum_{i=1}^{n} \langle m^*, f^2\cdot(g^1\cdot {}_im)_{(0)}\rangle S^{-1}(f^1(g^1\cdot {}_im)_{(-1)}g^2) \otimes {}^im.
\end{aligned} \tag{8.3.3}
$$

*For ${}^*M$:*

$$(h\cdot {}^*m)(m) := {}^*m(S^{-1}(h)\cdot m); \tag{8.3.4}$$

$$
\begin{aligned}
\lambda_{{}^*M}({}^*m) &= {}^*m_{(-1)} \otimes {}^*m_{(0)} \\
&= \sum_{i=1}^{n} \langle {}^*m, S^{-1}(f^1)\cdot(S^{-1}(g^2)\cdot {}_im)_{(0)}\rangle g^1 S((S^{-1}(g^2)\cdot {}_im)_{(-1)}) f^2 \otimes {}^im,
\end{aligned} \tag{8.3.5}
$$

*for all $h\in H$, $m^*\in M^*$, ${}^*m\in {}^*M$ and $m\in M$. Here $f = f^1\otimes f^2$ is the twist defined by* (3.2.15), *with inverse $f^{-1} = g^1\otimes g^2$.*

*Proof* The left $H$-action on $M^*$ viewed as an object of ${}_H^H\mathcal{YD}$ is the same as the left $H$-action on $M^*$ viewed as an object of ${}_H\mathcal{M}$. We compute the left $H$-coaction, by using Theorem 8.17. By (8.2.23) in $\mathscr{Z}_l({}_H\mathcal{M}) \cong {}_H^H\mathcal{YD}$ we have

$$s^{-1}_{V,X}(v\otimes x) = \tilde{q}_1^2 X^2\cdot(p^1\cdot x)_{(0)} \otimes S^{-1}(\tilde{q}^1 X^1 (p^1\cdot x)_{(-1)} p^2 S(\tilde{q}_2^2 X^3))\cdot v$$

for all $V,X\in {}_H\mathcal{M}$, $x\in X$ and $v\in V$. Now, if we denote by $P^1\otimes P^2$ another copy of $p_R$ and by $\tilde{Q}^1\otimes\tilde{Q}^2$ another copy of $q_L$, we can compute:

$$
\begin{aligned}
\lambda(m^*) &= s_{M^*,H}(m^*\otimes 1_H) \\
&= \langle x^1 Z^1\cdot m^*, \alpha x^2 Y^1 \tilde{q}_1^2 X^2\cdot(p^1 y^2 Z_1^3 \beta\cdot {}_im)_{(0)}\rangle \\
&\quad x_1^3 Y^2 \tilde{q}_2^2 X^3 S^{-1}\Big(\tilde{q}^1 X^1 (p^1 y^2 Z_1^3\beta\cdot {}_im)_{(-1)} p^2\Big) y^1 Z^2 \otimes x_2^3 Y^3 y^3 Z_2^3\cdot {}^im \\
&\overset{(3.2.1)}{\underset{(3.2.19)}{=}} \langle m^*, \tilde{Q}^1 Y^1 \tilde{q}_1^2 X^2\cdot(p^1 P^2 S(\tilde{Q}_2^2 Y^3)\cdot {}_im)_{(0)}\rangle \\
&\quad \tilde{Q}_1^2 Y^2 \tilde{q}_2^2 X^3 S^{-1}\Big(\tilde{q}^1 X^1 (p^1 P^2 S(\tilde{Q}_2^2 Y^3)\cdot {}_im)_{(-1)} p^2\Big) P^1 \otimes {}^im
\end{aligned}
$$

$$\overset{(5.2.7)}{\underset{(8.2.3)}{=}} \langle m^*, \tilde{Q}^1Y^1\tilde{q}^2\cdot\left(y_1^2p^1P^2S(\tilde{Q}_2^2Y^3)\cdot {}_im\right)_{(0)}\rangle$$

$$\tilde{Q}_1^2Y^2y^3S^{-1}\left(\tilde{q}^1\left(y_1^2p^1P^2S(\tilde{Q}_2^2Y^3)\cdot {}_im\right)_{(-1)}y_2^2p^2\right)y^1P^1\otimes {}^im$$

$$\overset{(3.2.25)}{\underset{(8.2.3)}{=}} \langle m^*, \tilde{Q}^1Y^1\tilde{q}^2x^1_{(2,2)}p_2^2\cdot\left(g^1S(\tilde{Q}_2^2Y^3x^3)\cdot {}_im\right)_{(0)}\rangle$$

$$\tilde{Q}_1^2Y^2S^{-1}\left(\tilde{q}^1x^1_{(2,1)}p_1^2\cdot\left(g^1S(\tilde{Q}_2^2Y^3x^3)\cdot {}_im\right)_{(-1)}g^2S(x^2)\right)x_1^1p^1\otimes {}^im$$

$$\overset{(3.2.22)}{\underset{(3.2.13)}{=}} \langle m^*, \tilde{Q}^1\tilde{q}^2p_2^2\cdot\left(S(\tilde{Q}^2)_1g^1\cdot {}_im\right)_{(0)}\rangle$$

$$S^{-1}\left(\tilde{q}^1p_1^2\cdot\left(S(\tilde{Q}^2)_1g^1\cdot {}_im\right)_{(-1)}S(\tilde{Q}^2)_2g^2\right)p^1\otimes {}^im$$

$$\overset{(8.2.3)}{\underset{(8.3.1)}{=}} \langle m^*, f^2\cdot\left(g^1\cdot {}_im\right)_{(0)}\rangle S^{-1}\left(f^1\cdot\left(g^1\cdot {}_im\right)_{(-1)}g^2\right)\otimes {}^im,$$

as claimed. The structure on ${}^*M$ can be computed in a similar way, we leave the details to the reader. □

Since all four categories of Yetter–Drinfeld modules over a quasi-Hopf algebra with bijective antipode are somehow braided isomorphic it follows that the rigidity property of ${}^H_H\mathscr{YD}^{\rm fd}$ transfers to the other three categories. For instance, for later use we record without further details the following:

**Proposition 8.23** *If $H$ is a quasi-Hopf algebra with bijective antipode then the braided category ${}_H\mathscr{YD}^{H^{\rm fd}}$ is rigid. For a finite-dimensional left–right Yetter–Drinfeld module $M$ with basis $({}_im)_{i=\overline{1,n}}$ and corresponding dual basis $({}^im)_{i=\overline{1,n}}$, the left and right duals $M^*$ and ${}^*M$ are equal to* Hom$(M,k)$ *as a vector space, with the following $H$-action and $H$-coaction:*

*For $M^*$:*

$$(h\cdot m^*)(m) = m^*(S(h)\cdot m); \tag{8.3.6}$$

$$\begin{aligned}\rho_{M^*}(m^*) &= m^*_{(0)}\otimes m^*_{(1)}\\ &= \sum_{i=1}^n \langle m^*, f^1\cdot(g^2\cdot {}_im)_{(0)}\rangle {}^im\otimes S^{-1}(f^2(g^2\cdot {}_im)_{(1)}g^1).\end{aligned} \tag{8.3.7}$$

*For ${}^*M$:*

$$(h\cdot {}^*m)(m) := {}^*m(S^{-1}(h)\cdot m); \tag{8.3.8}$$

$$\begin{aligned}\rho_{{}^*M}({}^*m) &= {}^*m_{(0)}\otimes {}^*m_{(1)}\\ &= \sum_{i=1}^n \langle {}^*m, S^{-1}(f^2)\cdot(S^{-1}(g^1)\cdot {}_im)_{(0)}\rangle {}^im\otimes g^2S((S^{-1}(g^1)\cdot {}_im)_{(1)})f^1,\end{aligned} \tag{8.3.9}$$

*for all $h\in H$, $m^*\in M^*$, ${}^*m\in {}^*M$ and $m\in M$. Here, as everywhere else, $f=f^1\otimes f^2$ is the twist defined by* (3.2.15), *with inverse $f^{-1}=g^1\otimes g^2$.*

**Remark 8.24** The left rigid monoidal structure of ${}_H\mathscr{YD}^{H^{\rm fd}}$ presented in Proposition 8.23 is designed in such a way that the forgetful functor $F: {}_H\mathscr{YD}^{H^{\rm fd}} \to {}_k\mathscr{M}^{\rm fd}$ is a left rigid quasi-monoidal functor, in the sense of Definition 3.36. This follows mostly from the fact that the left dual of an object $M$ in ${}_H\mathscr{YD}^{H\mathrm{fd}}$ is built on the left dual object of $M$ regarded now in ${}_k\mathscr{M}^{\mathrm{fd}}$, of course with a different evaluation and coevaluation morphisms, and the fact that the left transpose of a morphism $f$ in ${}_H\mathscr{YD}^{H\mathrm{fd}}$ coincides with the usual left transpose morphism of $f$ in ${}_k\mathscr{M}^{\mathrm{fd}}$. The latter follows by Proposition 3.34 since the evaluation and coevaluation morphisms of a left dual of an object $M$ in ${}_H\mathscr{YD}^{H\mathrm{fd}}$ coincide with those of the left dual of $M$ regarded in ${}_H\mathscr{M}^{\mathrm{fd}}$, and since ${}_H\mathscr{YD}^{H\mathrm{fd}}$ has the associativity constraint defined in the same manner as that of ${}_H\mathscr{M}^{\mathrm{fd}}$.

We present a sufficient condition for ${}_H\mathscr{YD}^{H\mathrm{fd}}$ to be a sovereign category.

**Theorem 8.25** *If $H$ is a sovereign quasi-Hopf algebra then ${}_H\mathscr{YD}^{H\mathrm{fd}}$ is a sovereign category.*

*Proof* A left dual of $V$ in $\mathscr{C} := {}_H\mathscr{YD}^{H\mathrm{fd}}$ is nothing but the left dual of $V$ in ${}_H\mathscr{M}^{\mathrm{fd}}$, the evaluation and coevaluations morphisms being the same in $\mathscr{C}$ and ${}_H\mathscr{M}^{\mathrm{fd}}$; the difference is made only by the $H$-coaction that turns $V^*$ into a left–right Yetter–Drinfeld module, and for this we need $V \in \mathscr{C}$. As the $H$-coactions do not contribute to the definition of the isomorphisms $\lambda$ in (1.7.1), it follows that they are defined by the same formula in both $\mathscr{C}$ and ${}_H\mathscr{M}^{\mathrm{fd}}$. For the last assertion, note also that the monoidal structure of ${}_H\mathscr{YD}^{H\mathrm{fd}}$ was defined in such a way that the forgetful functor to ${}_H\mathscr{M}^{\mathrm{fd}}$ is a strict monoidal functor.

Assume that $H$ is sovereign. By Theorem 3.40, we have a natural monoidal isomorphism $i : \mathrm{Id}_{{}_H\mathscr{M}^{\mathrm{fd}}} \to (-)^{**}$. In other words, for any $V \in {}_H\mathscr{M}^{\mathrm{fd}}$ we have an isomorphism $i_V : V \to V^{**}$ in ${}_H\mathscr{M}^{\mathrm{fd}}$, natural in $V$, such that the conditions in (3.6.1) are satisfied. From the above comments, to prove that $\mathscr{C}$ is sovereign it suffices to show that, for any object $W$ of $\mathscr{C}$, a subcategory of ${}_H\mathscr{M}^{\mathrm{fd}}$, the map $i_W : W \ni w \mapsto \mathfrak{g}^{-1} \cdot w \in \underline{W}$ defined in the proof of Proposition 3.41 is an isomorphism in $\mathscr{C}$ rather than in ${}_H\mathscr{M}^{\mathrm{fd}}$. Here, to avoid any confusion, we denoted by $\underline{W}$ the $k$-vector space $W$ endowed with the structure of a left-right Yetter–Drinfeld module obtained from the natural identification $W \cong W^{**}$. Let us compute this structure explicitly.

It is clear that the left $H$-module structure on $\underline{W}$ given by the identification $W \cong W^{**}$ is $h \to w = S^2(h) \cdot w$, for all $h \in H$, $w \in W$. We next compute the right $H$-coaction on $\underline{W}$, and to this end we need dual bases $\{w_i, w^i\}_i$ in $W$ and $W^*$, and $\{w^i, w^{*i}\}_i$ in $W^*$ and $W^{**}$, respectively. We have (summations implicitly understood):

$$
\begin{aligned}
w &\mapsto w^i(w)w^{*i}\\
&\mapsto w^i(w)\langle w^{*i}, f^1 \cdot (g^2 \cdot w^j)_{(0)}\rangle w^{*j} \otimes S^{-1}(f^2(g^2 \cdot w^j)_{(1)}g^1)\\
&= w^i(w)\langle w^{*i}, f^1 \cdot w^s\rangle \langle g^2 \cdot w^j, F^1 \cdot (G^2 \cdot w_s)_{(0)}\rangle\\
&\qquad\qquad w^{*j} \otimes S^{-1}(f^2 S^{-1}(F^2(G^2 \cdot w_s)_{(1)}G^1)g^1)\\
&\mapsto \langle f^1 \cdot w^s, w\rangle \langle g^2 \cdot v^j, F^1 \cdot (G^2 \cdot w_s)_{(0)}\rangle w_j \otimes S^{-1}(f^2 S^{-1}(F^2(G^2 \cdot w_s)_{(1)}G^1)g^1)\\
&= S(g^2)F^1 \cdot (G^2 S(f^1) \cdot w)_{(0)} \otimes S^{-1}(f^2 S^{-1}(F^2(G^2 S(f^1) \cdot w)_{(1)}G^1)g^1).
\end{aligned}
$$

Here we used the right $H$-coactions on $W^*$ and $W^{**} = (W^*)^*$ defined by (8.3.7) and the canonical isomorphism $W \cong W^{**}$ and its inverse, respectively.

We can see now that $i_W$ is right $H$-colinear:

$$\begin{aligned}
&\lambda_{\underline{W}} i_W(w) \\
&\quad = S(g^2)F^1 \cdot (G^2S(f^1)\mathfrak{g}^{-1} \cdot w)_{(0)} \otimes S^{-1}(f^2S^{-1}(F^2(G^2S(f^1)\mathfrak{g}^{-1} \cdot w)_{(1)}G^1)g^1) \\
&\overset{(3.6.4)}{=} S(g^2)F^1 \cdot (\mathfrak{g}_2^{-1} \cdot w)_{(0)} \otimes S^{-2}(S(g^1)F^2(\mathfrak{g}_2^{-1} \cdot w)_{(1)}\mathfrak{g}_1^{-1}\mathfrak{g}) \\
&\overset{(8.2.6)}{=} S(g^2)F^1\mathfrak{g}_1^{-1} \cdot w_{(0)} \otimes S^{-2}(S(g^1)F^2\mathfrak{g}_2^{-1}w_{(1)}\mathfrak{g}) \\
&\overset{(3.6.4)}{=} \mathfrak{g}^{-1} \cdot w_{(0)} \otimes S^{-2}(\mathfrak{g}^{-1}w_{(1)}\mathfrak{g}) \\
&\overset{(3.6.3)}{=} \mathfrak{g}^{-1} \cdot w_{(0)} \otimes w_{(1)} = (i_W \otimes \mathrm{Id}_H)\lambda_W(w),
\end{aligned}$$

for all $w \in W$. This finishes the proof. □

## 8.4 Yetter–Drinfeld Modules as Modules over an Algebra

The goal of this section is to prove that for a finite-dimensional quasi-Hopf algebra $H$ the category of Yetter–Drinfeld modules over $H$ is isomorphic to a certain category of representations. As we shall see, this can be done in a more general framework.

Our next definition extends the definition of Yetter–Drinfeld modules from the previous sections.

**Definition 8.26** Let $H$ be a quasi-bialgebra, $C$ an $H$-bimodule coalgebra and $\mathbb{A}$ an $H$-bicomodule algebra. A left–right Yetter–Drinfeld module is a $k$-vector space $M$ with the following additional structure:

- $M$ is a left $\mathbb{A}$-module; we write $\cdot$ for the left $\mathbb{A}$-action;
- we have a $k$-linear map $\rho_M : M \to M \otimes C$, $\rho_M(m) = m_{(0)} \otimes m_{(1)}$, called the right $C$-coaction on $M$, such that for all $m \in M$, $\underline{\varepsilon}(m_{(1)})m_{(0)} = m$ and
$$\begin{aligned}
&(\theta^2 \cdot m_{(0)})_{(0)} \otimes (\theta^2 \cdot m_{(0)})_{(1)} \cdot \theta^1 \otimes \theta^3 \cdot m_{(1)} \\
&\qquad = \tilde{x}_\rho^1 \cdot (\tilde{x}_\lambda^3 \cdot m)_{(0)} \otimes \tilde{x}_\rho^2 \cdot (\tilde{x}_\lambda^3 \cdot m)_{(1)_{\underline{1}}} \cdot \tilde{x}_\lambda^1 \otimes \tilde{x}_\rho^3 \cdot (\tilde{x}_\lambda^3 \cdot m)_{(1)_{\underline{2}}} \cdot \tilde{x}_\lambda^2; \quad (8.4.1)
\end{aligned}$$
- the following compatibility relation holds:
$$u_{\langle 0\rangle} \cdot m_{(0)} \otimes u_{\langle 1\rangle} \cdot m_{(1)} = (u_{[0]} \cdot m)_{(0)} \otimes (u_{[0]} \cdot m)_{(1)} \cdot u_{[-1]}, \quad (8.4.2)$$
for all $u \in \mathbb{A}$, $m \in M$. ${}_{\mathbb{A}}\mathcal{YD}(H)^C$ will be the category of left–right Yetter–Drinfeld modules and maps preserving the actions by $\mathbb{A}$ and the coactions by $C$.

Let $H$ be a quasi-bialgebra, $\mathbb{A}$ an $H$-bicomodule algebra and $C$ an $H$-bimodule coalgebra. Let us call the 3-tuple $(H, \mathbb{A}, C)$ a Yetter–Drinfeld datum. We note that, for an arbitrary $H$-bimodule coalgebra $C$, the linear dual space of $C$, $C^*$, is an $H$-bimodule algebra. The multiplication of $C^*$ is the convolution, that is, $(c^*d^*)(c) = c^*(c_{\underline{1}})d^*(c_{\underline{2}})$, the unit is $\underline{\varepsilon}$ and the left and right $H$-module structures are given by $(h \rightharpoonup c^* \leftharpoonup h')(c) = c^*(h' \cdot c \cdot h)$, for all $h, h' \in H$, $c^*, d^* \in C^*$, $c \in C$.

In the rest of this section we establish that if $H$ is a quasi-Hopf algebra and $C$ is finite dimensional then the category ${}_{\mathbb{A}}\mathcal{YD}(H)^C$ is isomorphic to the category of left $C^*\bowtie\mathbb{A}$-modules, ${}_{C^*\bowtie\mathbb{A}}\mathcal{M}$, where $C^*\bowtie\mathbb{A}$ is the diagonal crossed product algebra between the $H$-bimodule algebra $C^*$ and the $H$-bicomodule algebra $\mathbb{A}$ as in Section 5.6. First we need some lemmas.

**Lemma 8.27** *Let $H$ be a quasi-Hopf algebra with bijective antipode and $(H,\mathbb{A},C)$ a Yetter–Drinfeld datum. We have a functor $F : {}_{\mathbb{A}}\mathcal{YD}(H)^C \to {}_{C^*\bowtie\mathbb{A}}\mathcal{M}$, given by $F(M)=M$ as k-vector space, with the $C^*\bowtie\mathbb{A}$-module structure defined by*

$$(c^*\bowtie u)m := \langle c^*, \tilde{q}^2_\rho\cdot(u\cdot m)_{(1)}\rangle \tilde{q}^1_\rho\cdot(u\cdot m)_{(0)}, \tag{8.4.3}$$

*for all $c^*\in C^*$, $u\in\mathbb{A}$ and $m\in M$, where $\tilde{q}_\rho=\tilde{q}^1_\rho\otimes\tilde{q}^2_\rho$ is the element defined in (4.3.9). $F$ maps a morphism to itself.*

*Proof* Let $\tilde{Q}^1_\rho\otimes\tilde{Q}^2_\rho$ be another copy of $\tilde{q}_\rho$. For all $c^*,d^*\in C^*$, $u,u'\in\mathbb{A}$ and $m\in M$ we compute:

$$\begin{aligned}
&[(c^*\bowtie u)(d^*\bowtie u')]m\\
&\stackrel{(5.6.15)}{=} [(\Omega^1\rightharpoonup c^*\leftharpoonup\Omega^5)(\Omega^2 u_{\langle 0\rangle_{[-1]}}\rightharpoonup d^*\leftharpoonup S^{-1}(u_{\langle 1\rangle})\Omega^4)\bowtie\Omega^3 u_{\langle 0\rangle_{[0]}}u']m\\
&= \langle d^*, S^{-1}(u_{\langle 1\rangle})\Omega^4(\tilde{q}^2_\rho)_2\cdot(\Omega^3 u_{\langle 0\rangle_{[0]}}u'\cdot m)_{(1)_{\underline{2}}}\cdot\Omega^2 u_{\langle 0\rangle_{[-1]}}\rangle\\
&\quad \langle c^*, \Omega^5(\tilde{q}^2_\rho)_1\cdot(\Omega^3 u_{\langle 0\rangle_{[0]}}u'\cdot m)_{(1)_{\underline{1}}}\cdot\Omega^1\rangle\tilde{q}^1_\rho\cdot(\Omega^3 u_{\langle 0\rangle_{[0]}}u'\cdot m)_{(0)}\\
&\stackrel{(5.6.9)}{=} \langle d^*, S^{-1}(f^1\tilde{X}^2_\rho\theta^3 u_{\langle 1\rangle})(\tilde{q}^2_\rho)_2\cdot((\tilde{X}^1_\rho)_{[0]}\tilde{x}^3_\lambda\theta^2_{[0]}u_{\langle 0\rangle_{[0]}}u'\cdot m)_{(1)_{\underline{2}}}\cdot(\tilde{X}^1_\rho)_{[-1]_{\underline{2}}}\\
&\quad \tilde{x}^2_\lambda\theta^2_{[-1]}u_{\langle 0\rangle_{[-1]}}\rangle\langle c^*, S^{-1}(f^2\tilde{X}^3_\rho)(\tilde{q}_\rho)_1\cdot((\tilde{X}^1_\rho)_{[0]}\tilde{x}^3_\lambda\theta^2_{[0]}u_{\langle 0\rangle_{[0]}}u'\cdot m)_{(1)_{\underline{1}}}\\
&\qquad\qquad \cdot\tilde{x}^1_\lambda\theta^1\rangle\tilde{q}^1_\rho\cdot((\tilde{X}^1_\rho)_{[0]}\tilde{x}^3_\lambda\theta^2_{[0]}u_{\langle 0\rangle_{[0]}}u'\cdot m)_{(0)}\\
&\stackrel{(8.4.2)}{\underset{(4.3.15)}{=}} \langle d^*, S^{-1}(\theta^3 u_{\langle 1\rangle})\tilde{Q}^2_\rho\tilde{x}^3_\rho\cdot(\tilde{x}^3_\lambda\theta^2_{[0]}u_{\langle 0\rangle_{[0]}}u'\cdot m)_{(1)_{\underline{2}}}\cdot\tilde{x}^2_\lambda\theta^2_{[-1]}u_{\langle 0\rangle_{[-1]}}\rangle\\
&\quad \langle c^*, \tilde{q}^2_\rho(\tilde{Q}^1_\rho)_{\langle 1\rangle}\tilde{x}^2_\rho\cdot(\tilde{x}^3_\lambda\theta^2_{[0]}u_{\langle 0\rangle_{[0]}}u'\cdot m)_{(1)_{\underline{1}}}\cdot\tilde{x}^1_\lambda\theta^1\rangle\\
&\quad \tilde{q}^1_\rho(\tilde{Q}^1_\rho)_{\langle 0\rangle}\tilde{x}^1_\rho\cdot(\tilde{x}^3_\lambda\theta^2_{[0]}u_{\langle 0\rangle_{[0]}}u'\cdot m)_{(0)}\\
&\stackrel{(8.4.1)}{=} \langle d^*, S^{-1}(\theta^3 u_{\langle 1\rangle})\tilde{Q}^2_\rho\overline{\theta}^3\cdot(\theta^2_{[0]}u_{\langle 0\rangle_{[0]}}u'\cdot m)_{(1)}\cdot\theta^2_{[-1]}u_{\langle 0\rangle_{[-1]}}\rangle\\
&\quad \langle c^*, \tilde{q}^2_\rho(\tilde{Q}^1_\rho)_{\langle 1\rangle}\cdot[\overline{\theta}^2\cdot(\theta^2_{[0]}u_{\langle 0\rangle_{[0]}}u'\cdot m)_{(0)}]_{(1)}\cdot\overline{\theta}^1\theta^1\rangle\\
&\quad \tilde{q}^1_\rho(\tilde{Q}^1_\rho)_{\langle 0\rangle}\cdot[\overline{\theta}^2\cdot(\theta^2_{[0]}u_{\langle 0\rangle_{[0]}}u'\cdot m)_{(0)}]_{(0)}\\
&\stackrel{(8.4.2)}{\underset{(4.4.3)}{=}} \langle d^*, S^{-1}(\alpha\tilde{X}^3_\rho\theta^3 u_{\langle 1\rangle})\tilde{X}^2_\rho\overline{\theta}^3\theta^2_{\langle 1\rangle}u_{\langle 0,1\rangle}\cdot(u'\cdot m)_{(1)}\rangle\\
&\quad \langle c^*, \tilde{q}^2_\rho\cdot[(\tilde{X}^1_\rho)_{[0]}\overline{\theta}^2\theta^2_{\langle 0\rangle}u_{\langle 0,0\rangle}\cdot(u'\cdot m)_{(0)}]_{(1)}\cdot(\tilde{X}^1_\rho)_{[-1]}\overline{\theta}^1\theta^1\rangle\\
&\quad \tilde{q}^1_\rho\cdot[(\tilde{X}^1_\rho)_{[0]}\overline{\theta}^2\theta^2_{\langle 0\rangle}u_{\langle 0,0\rangle}\cdot(u'\cdot m)_{(0)}]_{(0)}\\
&\stackrel{(4.4.3)}{\underset{(4.3.1)}{=}} \langle d^*, S^{-1}(\alpha\theta^3_2 u_{\langle 1\rangle_2}\tilde{X}^3_\rho)\theta^3_1 u_{\langle 1\rangle_1}\tilde{X}^2_\rho\cdot(u'\cdot m)_{(1)}\rangle\\
&\quad \langle c^*, \tilde{q}^2_\rho\cdot[\theta^2 u_{\langle 0\rangle}\tilde{X}^1_\rho\cdot(u'\cdot m)_{(0)}]_{(1)}\cdot\theta^1\rangle\tilde{q}^1_\rho\cdot[\theta^2 u_{\langle 0\rangle}\tilde{X}^1_\rho\cdot(u'\cdot m)_{(0)}]_{(0)}\\
&\stackrel{(3.2.1)}{\underset{(4.3.9)}{=}} \langle c^*, \tilde{q}^2_\rho\cdot[u\tilde{Q}^1_\rho\cdot(u'\cdot m)_{(0)}]_{(1)}\rangle\langle d^*, \tilde{Q}^2_\rho\cdot(u'\cdot m)_{(1)}\rangle
\end{aligned}$$

$$
\begin{aligned}
&\tilde{q}^1_\rho \cdot [u\tilde{Q}^1_\rho \cdot (u'\cdot m)_{(0)}]_{(0)} \\
&\overset{(8.4.3)}{=} \langle d^*, \tilde{Q}^2_\rho \cdot (u'\cdot m)_{(1)}\rangle (c^* \bowtie u)[\tilde{Q}^1_\rho \cdot (u'\cdot m)_{(0)}] = (c^* \bowtie u)[(d^* \bowtie u')m],
\end{aligned}
$$

as needed. It is not hard to see that $(\underline{\varepsilon} \bowtie 1_{\mathbb{A}})m = m$ for all $m \in M$, so $M$ is a left $C^* \bowtie \mathbb{A}$-module. The fact that a morphism in ${}_{\mathbb{A}}\mathscr{YD}(H)^C$ becomes a morphism in ${}_{C^*\bowtie\mathbb{A}}\mathscr{M}$ can be proved more easily; we leave the details to the reader. $\square$

We construct now a functor in the opposite direction.

**Lemma 8.28** *Let $H$ be a quasi-Hopf algebra with bijective antipode, $(H,\mathbb{A},C)$ a Yetter–Drinfeld datum, and assume that $C$ is finite dimensional. We have a functor $G : {}_{C^*\bowtie\mathbb{A}}\mathscr{M} \to {}_{\mathbb{A}}\mathscr{YD}(H)^C$, given by $G(M) = M$ as $k$-vector space, with structure defined, for $m \in M$ and $u \in \mathbb{A}$, by*

$$u \cdot m = (\underline{\varepsilon} \bowtie u)m, \tag{8.4.4}$$

$$\rho_M : M \to M \otimes C,\ \ \rho_M(m) = \sum_{i=1}^{n} (c^i \bowtie (\tilde{p}^1_\rho)_{[0]})m \otimes S^{-1}(\tilde{p}^2_\rho)\cdot c_i \cdot (\tilde{p}^1_\rho)_{[-1]}. \tag{8.4.5}$$

*Here $\tilde{p}_\rho = \tilde{p}^1_\rho \otimes \tilde{p}^2_\rho$ is the element defined in (4.3.9), $\{c_i\}_{i=\overline{1,n}}$ is a basis of $C$ and $\{c^i\}_{i=\overline{1,n}}$ is the corresponding dual basis of $C^*$. $G$ maps a morphism to itself.*

*Proof* The most difficult part of the proof is to show that $G(M)$ satisfies the relations (8.4.1) and (8.4.2). It is then straightforward to show that a morphism in ${}_{C^*\bowtie\mathbb{A}}\mathscr{M}$ is also a morphism in ${}_{\mathbb{A}}\mathscr{YD}(H)^C$, and that $G$ is a functor.

It is not hard to see that (4.4.3), (3.2.1) and (4.4.4) imply

$$\overline{\theta}^1\theta^1 \otimes \overline{\theta}^2\theta^2_{\langle 0\rangle}\tilde{p}^1_\rho \otimes \overline{\theta}^3\theta^2_{\langle 1\rangle}\tilde{p}^2_\rho S(\theta^3) = (\tilde{p}^1_\rho)_{[-1]} \otimes (\tilde{p}^1_\rho)_{[0]} \otimes \tilde{p}^2_\rho. \tag{8.4.6}$$

Write $\tilde{p}_\rho = \tilde{p}^1_\rho \otimes \tilde{p}^2_\rho = \tilde{P}^1_\rho \otimes \tilde{P}^2_\rho$. For all $m \in M$ we compute:

$$
\begin{aligned}
&(\theta^2\cdot m_{(0)})_{(0)} \otimes (\theta^2\cdot m_{(0)})_{(1)}\cdot\theta^1 \otimes \theta^3\cdot m_{(1)} \\
&= \sum_{i=1}^{n} ((\underline{\varepsilon}\bowtie\theta^2)(c^i \bowtie (\tilde{p}^1_\rho)_{[0]})m)_{(0)} \otimes ((\underline{\varepsilon}\bowtie\theta^2)(c^i \bowtie (\tilde{p}^1_\rho)_{[0]})m)_{(1)}\cdot\theta^1 \\
&\qquad \otimes \theta^3 S^{-1}(\tilde{p}^2_\rho)\cdot c_i \cdot (\tilde{p}^1_\rho)_{[-1]} \\
&\overset{(5.6.15)}{\underset{(8.4.5)}{=}} \sum_{i,j=1}^{n} (c^j \bowtie (\tilde{P}^1_\rho)_{[0]})(c^i \bowtie (\theta^2_{\langle 0\rangle}\tilde{p}^1_\rho)_{[0]})m \otimes S^{-1}(\tilde{P}^2_\rho)\cdot c_j \cdot (\tilde{P}^1_\rho)_{[-1]}\theta^1 \\
&\qquad \otimes \theta^3 S^{-1}(\theta^2_{\langle 1\rangle}\tilde{p}^2_\rho)\cdot c_i \cdot (\theta^2_{\langle 0\rangle}\tilde{p}^1_\rho)_{[-1]} \\
&\overset{(5.6.15)}{\underset{(5.6.9)}{=}} \sum_{i,j=1}^{n} [c^jc^i \bowtie (\tilde{X}^1_\rho)_{[0]}\tilde{x}^3_\lambda(\overline{\theta}^2(\tilde{P}^1_\rho)_{[0]_{\langle 0\rangle}}\theta^2_{\langle 0\rangle}\tilde{p}^1_\rho)_{[0]}]m \\
&\qquad \otimes S^{-1}(f^2\tilde{X}^3_\rho\tilde{P}^2_\rho)\cdot c_j \cdot (\tilde{X}^1_\rho)_{[-1]_1}\tilde{x}^1_\lambda\overline{\theta}^1(\tilde{P}^1_\rho)_{[-1]}\theta^1 \\
&\qquad \otimes \theta^3 S^{-1}(f^1\tilde{X}^2_\rho\overline{\theta}^3(\tilde{P}^1_\rho)_{[0]_{\langle 1\rangle}}\theta^2_{\langle 1\rangle}\tilde{p}^2_\rho)\cdot c_i \\
&\qquad \cdot (\tilde{X}^1_\rho)_{[-1]_2}\tilde{x}^2_\lambda(\overline{\theta}^2(\tilde{P}^1_\rho)_{[0]_{\langle 0\rangle}}\theta^2_{\langle 0\rangle}\tilde{p}^1_\rho)_{[-1]}
\end{aligned}
$$

$$\overset{(4.4.1),(8.4.6)}{\underset{(4.3.5)}{=}} \sum_{i,j=1}^{n} [c^j c^i \bowtie (\tilde{X}^1_\rho (\tilde{P}^1_\rho)_{\langle 0\rangle} \tilde{p}^1_\rho)_{[0]} \tilde{x}^3_\lambda] m$$
$$\otimes S^{-1}(f^2 \tilde{X}^3_\rho \tilde{P}^2_\rho) \cdot c_j \cdot (\tilde{X}^1_\rho (\tilde{P}^1_\rho)_{\langle 0\rangle} \tilde{p}^1_\rho)_{[-1]_1} \tilde{x}^1_\lambda$$
$$\otimes S^{-1}(f^1 \tilde{X}^2_\rho (\tilde{P}^1_\rho)_{\langle 1\rangle} \tilde{p}^2_\rho) \cdot c_i \cdot (\tilde{X}^1_\rho (\tilde{P}^1_\rho)_{\langle 0\rangle} \tilde{p}^1_\rho)_{[-1]_2} \tilde{x}^2_\lambda$$
$$\overset{(4.3.14)}{=} \sum_{i,j=1}^{n} [c^j c^i \bowtie ((\tilde{x}^1_\rho)_{\langle 0\rangle} \tilde{p}^1_\rho)_{[0]} \tilde{x}^3_\lambda] m$$
$$\otimes \tilde{x}^2_\rho S^{-1}(f^2 ((\tilde{x}^1_\rho)_{\langle 1\rangle} \tilde{p}^2_\rho)_2 g^2) \cdot c_j \cdot ((\tilde{x}^1_\rho)_{\langle 0\rangle} \tilde{p}^1_\rho)_{[-1]_1} \tilde{x}^1_\lambda$$
$$\otimes \tilde{x}^3_\rho S^{-1}(f^1 ((\tilde{x}^1_\rho)_{\langle 1\rangle} \tilde{p}^2_\rho)_1 g^1) \cdot c_i \cdot ((\tilde{x}^1_\rho)_{\langle 0\rangle} \tilde{p}^1_\rho)_{[-1]_2} \tilde{x}^2_\lambda$$
$$\overset{(3.2.13)}{\underset{(4.2.10)}{=}} \sum_{i=1}^{n} [c^i \bowtie ((\tilde{x}^1_\rho)_{\langle 0\rangle} \tilde{p}^1_\rho)_{[0]} \tilde{x}^3_\lambda] m$$
$$\otimes \tilde{x}^2_\rho \cdot (S^{-1}((\tilde{x}^1_\rho)_{\langle 1\rangle} \tilde{p}^2_\rho) \cdot c_i \cdot ((\tilde{x}^1_\rho)_{\langle 0\rangle} \tilde{p}^1_\rho)_{[-1]})_{\underline{1}} \cdot \tilde{x}^1_\lambda$$
$$\otimes \tilde{x}^3_\rho \cdot (S^{-1}((\tilde{x}^1_\rho)_{\langle 1\rangle} \tilde{p}^2_\rho) \cdot c_i \cdot ((\tilde{x}^1_\rho)_{\langle 0\rangle} \tilde{p}^1_\rho)_{[-1]})_{\underline{2}} \cdot \tilde{x}^2_\lambda$$
$$= \sum_{i=1}^{n} [(\tilde{x}^1_\rho)_{\langle 0\rangle_{[-1]}} \rightharpoonup c^i \leftharpoonup S^{-1}((\tilde{x}^1_\rho)_{\langle 1\rangle}) \bowtie ((\tilde{x}^1_\rho)_{\langle 0\rangle} \tilde{p}^1_\rho)_{[0]} \tilde{x}^3_\lambda] m$$
$$\otimes \tilde{x}^2_\rho \cdot (S^{-1}(\tilde{p}^2_\rho) \cdot c_i \cdot (\tilde{p}^1_\rho)_{[-1]})_{\underline{1}} \cdot \tilde{x}^1_\lambda$$
$$\otimes \tilde{x}^3_\rho \cdot (S^{-1}(\tilde{p}^2_\rho) \cdot c_i \cdot (\tilde{p}^1_\rho)_{[-1]})_{\underline{2}} \cdot \tilde{x}^2_\lambda$$
$$\overset{(5.6.15)}{=} \sum_{i=1}^{n} [(\underline{\varepsilon} \bowtie \tilde{x}^1_\rho)(c^i \bowtie (\tilde{p}^1_\rho)_{[0]})(\underline{\varepsilon} \bowtie \tilde{x}^3_\lambda)] m$$
$$\otimes \tilde{x}^2_\rho \cdot (S^{-1}(\tilde{p}^2_\rho) \cdot c_i \cdot (\tilde{p}^1_\rho)_{[-1]})_{\underline{1}} \cdot \tilde{x}^1_\lambda$$
$$\otimes \tilde{x}^3_\rho \cdot (S^{-1}(\tilde{p}^2_\rho) \cdot c_i \cdot (\tilde{p}^1_\rho)_{[-1]})_{\underline{2}} \cdot \tilde{x}^2_\lambda$$
$$\overset{(8.4.4)}{\underset{(8.4.5)}{=}} \tilde{x}^1_\rho \cdot (\tilde{x}^3_\lambda \cdot m)_{(0)} \otimes \tilde{x}^2_\rho \cdot (\tilde{x}^3_\lambda \cdot m)_{(1)_{\underline{1}}} \cdot \tilde{x}^1_\lambda \otimes \tilde{x}^3_\rho \cdot (\tilde{x}^3_\lambda \cdot m)_{(1)_{\underline{2}}} \cdot \tilde{x}^2_\lambda .$$

Similarly, we compute:

$$u_{\langle 0\rangle} \cdot m_{(0)} \otimes u_{\langle 1\rangle} \cdot m_{(1)}$$
$$= \sum_{i=1}^{n} (\underline{\varepsilon} \bowtie u_{\langle 0\rangle})(c^i \bowtie (\tilde{p}^1_\rho)_{[0]}) m \otimes u_{\langle 1\rangle} S^{-1}(\tilde{p}^2_\rho) \cdot c_i \cdot (\tilde{p}^1_\rho)_{[-1]}$$
$$\overset{(5.6.15)}{=} \sum_{i=1}^{n} (u_{\langle 0,0\rangle_{[-1]}} \rightharpoonup c^i \leftharpoonup S^{-1}(u_{\langle 0,1\rangle}) \bowtie u_{\langle 0,0\rangle_{[0]}} (\tilde{p}^1_\rho)_{[0]}) m$$
$$\otimes u_{\langle 1\rangle} S^{-1}(\tilde{p}^2_\rho) \cdot c_i \cdot (\tilde{p}^1_\rho)_{[-1]}$$
$$= \sum_{i=1}^{n} (c^i \bowtie (u_{\langle 0,0\rangle} \tilde{p}^1_\rho)_{[0]}) m \otimes u_{\langle 1\rangle} S^{-1}(u_{\langle 0,1\rangle} \tilde{p}^2_\rho) \cdot c_i \cdot (u_{\langle 0,0\rangle} \tilde{p}^1_\rho)_{[-1]}$$
$$\overset{(4.3.10)}{=} \sum_{i=1}^{n} (c^i \bowtie (\tilde{p}^1_\rho u)_{[0]}) m \otimes S^{-1}(\tilde{p}^2_\rho) \cdot c_i \cdot (\tilde{p}^1_\rho u)_{[-1]}$$
$$\overset{(5.6.15)}{=} \sum_{i=1}^{n} (c^i \bowtie (\tilde{p}^1_\rho)_{[0]})(\underline{\varepsilon} \bowtie u_{[0]}) m \otimes S^{-1}(\tilde{p}^2_\rho) \cdot c_i \cdot (\tilde{p}^1_\rho)_{[-1]} u_{[-1]}$$

$$\overset{(8.4.5)}{=} (u_{[0]}\cdot m)_{(0)}\otimes (u_{[0]}\cdot m)_{(1)}\cdot u_{[-1]},$$

for all $u\in\mathbb{A}$ and $m\in M$, and this finishes the proof. □

One can now prove the desired isomorphism of categories.

**Theorem 8.29** *Let $H$ be a quasi-Hopf algebra with bijective antipode and $(H,\mathbb{A},C)$ a Yetter–Drinfeld datum, assuming $C$ to be finite dimensional. Then the categories ${}_{\mathbb{A}}\mathscr{YD}(H)^C$ and ${}_{C^*\bowtie\mathbb{A}}\mathscr{M}$ are isomorphic.*

*Proof* We have to verify that the functors $F$ and $G$ defined in Lemmas 8.27 and 8.28 are inverse to each other. Let $M\in{}_{\mathbb{A}}\mathscr{YD}(H)^C$. The structures on $G(F(M))$ (using first Lemma 8.27 and then Lemma 8.28) are denoted by $\cdot'$ and $\rho'_M$. For any $u\in\mathbb{A}$ and $m\in M$ we have that

$$u\cdot' m=(\underline{\varepsilon}\bowtie u)m=\langle\underline{\varepsilon},\tilde{q}^2_\rho\cdot(u\cdot m)_{(1)}\rangle\tilde{q}^1_\rho\cdot(u\cdot m)_{(0)}=u\cdot m$$

because $\underline{\varepsilon}(h\cdot c)=\varepsilon(h)\underline{\varepsilon}(c)$ and $\underline{\varepsilon}(m_{(1)})m_{(0)}=m$ for all $h\in H$, $c\in C$, $m\in M$. We now compute for $m\in M$ that

$$\begin{aligned}
\rho'_M(m) &= \sum_{i=1}^n (c^i\bowtie(\tilde{p}^1_\rho)_{[0]})m\otimes S^{-1}(\tilde{p}^2_\rho)\cdot c_i\cdot(\tilde{p}^1_\rho)_{[-1]}\\
&\overset{(8.4.3)}{=} \sum_{i=1}^n \langle c^i,\tilde{q}^2_\rho\cdot((\tilde{p}^1_\rho)_{[0]}\cdot m)_{(1)}\rangle\tilde{q}^1_\rho\cdot((\tilde{p}^1_\rho)_{[0]}\cdot m)_{(0)}\otimes S^{-1}(\tilde{p}^2_\rho)\cdot c_i\cdot(\tilde{p}^1_\rho)_{[-1]}\\
&\overset{(8.4.2)}{=} \tilde{q}^1_\rho(\tilde{p}^1_\rho)_{\langle 0\rangle}\cdot m_{(0)}\otimes S^{-1}(\tilde{p}^2_\rho)\tilde{q}^2_\rho(\tilde{p}^1_\rho)_{\langle 1\rangle}\cdot m_{(1)}\\
&\overset{(4.3.13)}{=} m_{(0)}\otimes m_{(1)}=\rho_M(m).
\end{aligned}$$

Conversely, take $M\in{}_{C^*\bowtie\mathbb{A}}\mathscr{M}$. We want to show that $F(G(M))=M$. If we denote the left $C^*\bowtie\mathbb{A}$-action on $F(G(M))$ by $\mapsto$ then by using Lemmas 8.27 and 8.28 we find, for all $c^*\in C^*$, $u\in\mathbb{A}$ and $m\in M$:

$$\begin{aligned}
&(c^*\bowtie u)\mapsto m\\
&= \langle c^*,\tilde{q}^2_\rho\cdot(u\cdot m)_{(1)}\rangle\tilde{q}^1_\rho\cdot(u\cdot m)_{(0)}\\
&= \sum_{i=1}^n\langle c^*,\tilde{q}^2_\rho S^{-1}(\tilde{p}^2_\rho)\cdot c_i\cdot(\tilde{p}^1_\rho)_{[-1]}\rangle(\underline{\varepsilon}\bowtie\tilde{q}^1_\rho)(c^i\bowtie(\tilde{p}^1_\rho)_{[0]})(\underline{\varepsilon}\bowtie u)m\\
&\overset{(5.6.15)}{=} \sum_{i=1}^n\langle c^*,\tilde{q}^2_\rho S^{-1}((\tilde{q}^1_\rho)_{\langle 1\rangle}\tilde{p}^2_\rho)\cdot c_i\cdot((\tilde{q}^1_\rho)_{\langle 0\rangle}\tilde{p}^1_\rho)_{[-1]}\rangle\\
&\qquad (c^i\bowtie((\tilde{q}^1_\rho)_{\langle 0\rangle}\tilde{p}^1_\rho)_{[0]})(\underline{\varepsilon}\bowtie u)m\\
&\overset{(4.3.12),(5.6.15)}{=} (c^*\bowtie 1_{\mathbb{A}})(\underline{\varepsilon}\bowtie u)m=(c^*\bowtie u)m,
\end{aligned}$$

and this finishes our proof. □

There is a relation between the functor $F$ from Lemma 8.27 and the map $\Gamma$ as in Proposition 5.51.

**Proposition 8.30** *Let $H$ be a quasi-Hopf algebra with bijective antipode, $(H,\mathbb{A},C)$ a Yetter–Drinfeld datum and $M$ an object in ${}_{\mathbb{A}}\mathscr{YD}(H)^C$; consider the map $\Gamma: C^*\to$*

*$C^* \bowtie \mathbb{A}$ as in Proposition 5.51. Then the left $C^* \bowtie \mathbb{A}$-module structure on $M$ given in Lemma 8.27 and the map $\Gamma$ are related by the formula:*

$$\Gamma(c^*)m = \langle c^*, m_{(1)}\rangle m_{(0)}, \quad \forall\, c^* \in C^*,\ m \in M.$$

*Proof* We compute:

$$\begin{aligned}
\Gamma(c^*)m &= ((\tilde{p}_\rho^1)_{[-1]} \rightharpoonup c^* \leftharpoonup S^{-1}(\tilde{p}_\rho^2) \bowtie (\tilde{p}_\rho^1)_{[0]})m \\
&= \langle(\tilde{p}_\rho^1)_{[-1]} \rightharpoonup c^* \leftharpoonup S^{-1}(\tilde{p}_\rho^2), \tilde{q}_\rho^2 \cdot ((\tilde{p}_\rho^1)_{[0]} \cdot m)_{(1)}\rangle \tilde{q}_\rho^1 \cdot ((\tilde{p}_\rho^1)_{[0]} \cdot m)_{(0)} \\
&= \langle c^*, S^{-1}(\tilde{p}_\rho^2)\tilde{q}_\rho^2 \cdot ((\tilde{p}_\rho^1)_{[0]} \cdot m)_{(1)} \cdot (\tilde{p}_\rho^1)_{[-1]}\rangle \tilde{q}_\rho^1 \cdot ((\tilde{p}_\rho^1)_{[0]} \cdot m)_{(0)} \\
&\overset{(8.4.2)}{=} \langle c^*, S^{-1}(\tilde{p}_\rho^2)\tilde{q}_\rho^2 (\tilde{p}_\rho^1)_{\langle 1\rangle} \cdot m_{(1)}\rangle \tilde{q}_\rho^1 (\tilde{p}_\rho^1)_{\langle 0\rangle} \cdot m_{(0)} \\
&\overset{(4.3.13)}{=} \langle c^*, m_{(1)}\rangle m_{(0)},
\end{aligned}$$

finishing the proof. □

By taking $C = \mathbb{A} = H$ in Theorem 8.29 we get the following:

**Corollary 8.31** *Let $H$ be a finite-dimensional quasi-Hopf algebra. Then the categories ${}_H\mathscr{YD}^H$ and ${}_{H^*\bowtie H}\mathscr{M}$ are isomorphic, where $H^* \bowtie H$ is the diagonal crossed product algebra between the $H$-bimodule algebra $H^*$ and the $H$-bicomodule algebra $H$ as in Examples 5.49 (i), specialized for $\mathbb{A} = H$.*

## 8.5 The Quantum Double of a Quasi-Hopf Algebra

Throughout this section, $H$ is a finite-dimensional quasi-Hopf algebra, so has bijective antipode (see Theorem 7.18). The category of finite-dimensional left–right Yetter–Drinfeld modules, ${}_H\mathscr{YD}^{H\,\mathrm{fd}}$, is a left rigid monoidal category; see Proposition 8.23. It is, moreover, a braided category, but to define the quantum double of $H$ as a quasi-Hopf algebra we will not need this extra property of ${}_H\mathscr{YD}^{H\,\mathrm{fd}}$. Actually, we will exploit it in Section 10.4, where we will show that the quantum double of $H$ is, moreover, a quasitriangular quasi-Hopf algebra.

By Remark 8.24, the forgetful functor $F : {}_H\mathscr{YD}^{H\,\mathrm{fd}} \to {}_k\mathscr{M}^{\mathrm{fd}}$ is a left rigid quasi-monoidal functor. By Corollary 8.31 it follows that the categories ${}_H\mathscr{YD}^{H\,\mathrm{fd}}$ and ${}_{H^*\bowtie H}\mathscr{M}^{\mathrm{fd}}$ are isomorphic, where $H^* \bowtie H$ is the diagonal crossed product algebra between the $H$-bimodule algebra $H^*$ and the $H$-bicomodule algebra $H$. From now on we will denote $H^* \bowtie H$ by $D(H)$ and call it the quantum double of $H$. Recall from Section 5.6 that if $\Omega \in H^{\otimes 5}$ is given by

$$\begin{aligned}
\Omega &= \Omega^1 \otimes \Omega^2 \otimes \Omega^3 \otimes \Omega^4 \otimes \Omega^5 \\
&= X^1_{(1,1)}y^1x^1 \otimes X^1_{(1,2)}y^2x^2_1 \otimes X^1_2 y^3 x^2_2 \otimes S^{-1}(f^1X^2x^3) \otimes S^{-1}(f^2X^3), \quad (8.5.1)
\end{aligned}$$

where $f \in H \otimes H$ is the twist defined in (3.2.15), then the quantum double $D(H) = H^* \bowtie H$ is the $k$-vector space $H^* \otimes H$ endowed with the associative unital $k$-algebra structure given by

$$
\begin{aligned}
&(\varphi \bowtie h)(\psi \bowtie h') \\
&= [(\Omega^1 \rightharpoonup \varphi \leftharpoonup \Omega^5)(\Omega^2 h_{(1,1)} \rightharpoonup \psi \leftharpoonup S^{-1}(h_2)\Omega^4)] \bowtie \Omega^3 h_{(1,2)} h' \qquad (8.5.2)\\
&= [(\Omega^1 \rightharpoonup \varphi \leftharpoonup \Omega^5)(\Omega^2 \rightharpoonup \psi_2 \leftharpoonup \Omega^4)] \bowtie \Omega^3[(\overline{S}^{-1}(\psi_1) \rightharpoonup h) \leftharpoonup \psi_3]h'
\end{aligned}
$$

and $1_{D(H)} = \varepsilon \bowtie 1_H$. Note that in the above relations we used the dual quasi-Hopf algebra structure of $H^*$ and the $H$-bimodule structure of $H^*$ from Example 4.15.

It is easy to see that

$$
(\varepsilon \bowtie h)(\varphi \bowtie h') = h_{(1,1)} \rightharpoonup \varphi \leftharpoonup S^{-1}(h_2) \bowtie h_{(1,2)} h' \qquad (8.5.3)
$$

and $(\varphi \bowtie h)(\varepsilon \bowtie h') = \varphi \bowtie hh'$ for all $\varphi \in H^*$ and $h, h' \in H$. Thus $D(H)$ contains $H$ as a $k$-subalgebra, modulo the identification $h \equiv \varepsilon \bowtie h$, $h \in H$.

Summing up, we have a finite-dimensional $k$-algebra, namely $D(H)$, with the property that ${}_{D(H)}\mathscr{M}^{\rm fd}$ is a left rigid monoidal category (as it identifies with ${}_H\mathscr{YD}^{H\rm fd}$ as an ordinary category) such that the forgetful functor $U : {}_{D(H)}\mathscr{M} \to {}_k\mathscr{M}^{\rm fd}$ is a left rigid quasi-monoidal functor. By Theorem 3.38 it follows that $D(H)$ admits a quasi-Hopf algebra structure. The purpose of this section is to describe this structure of $D(H)$ explicitly. Since we will very often use the categorical isomorphism between ${}_{D(H)}\mathscr{M}^{\rm fd}$ and ${}_H\mathscr{YD}^{H\rm fd}$, for the convenience of the reader, note that, by Lemma 8.28, $D(H)$ is a left–right Yetter–Drinfeld module over $H$ via the structure given by:

$$
h \cdot (\varphi \bowtie h') = (\varepsilon \bowtie h)(\varphi \bowtie h') = h_{(1,1)} \rightharpoonup \varphi \leftharpoonup S^{-1}(h_2) \bowtie h_{(1,2)} h', \qquad (8.5.4)
$$

$$
\begin{aligned}
\varphi \bowtie h \mapsto & \sum_i (e^i \bowtie p_2^1)(\varphi \bowtie h) \otimes S^{-1}(p^2) e_i p_1^1 \\
&= \sum_i e^i (\Omega^2 (p_2^1)_{(1,1)} \rightharpoonup \varphi \leftharpoonup S^{-1}((p_2^1)_2)\Omega^4) \\
&\quad \bowtie \Omega^3 (p_2^1)_{(1,2)} h \otimes S^{-1}(p^2)\Omega^5 e_i \Omega^1 p_1^1, \qquad (8.5.5)
\end{aligned}
$$

for all $\varphi \in H^*$ and $h, h' \in H$, where $\{e_i, e^i\}_i$ are dual bases in $H$ and $H^*$.

Similarly, by Lemma 8.27, any left–right Yetter–Drinfeld module $M$ is a left $D(H)$-module via the left $D(H)$-action given by

$$
(\varphi \bowtie h)m = \langle \varphi, q^2 (h \cdot m)_{(1)} \rangle q^1 \cdot (h \cdot m)_{(0)}, \quad \forall\, \varphi \in H^*,\ h \in H,\ m \in M. \qquad (8.5.6)
$$

Consequently, if $M, N \in {}_H\mathscr{YD}^H$ then $M \otimes N \in {}_H\mathscr{YD}^H$ with structure as in (8.2.17) and (8.2.16), and therefore $M \otimes N$ is a left $D(H)$-module with structure given by

$$
\begin{aligned}
(\varphi \bowtie h)(m \otimes n) &= \langle \varphi, q^2 (h \cdot (m \otimes n))_{(1)} \rangle q^1 \cdot (h \cdot (m \otimes n))_{(0)} \\
&= \langle \varphi, q^2 (h_1 \cdot m \otimes h_2 \cdot n)_{(1)} \rangle q^1 \cdot (h_1 \cdot m \otimes h_2 \cdot n)_{(0)} \\
&= \langle \varphi, q^2 x^3 (X^3 y^3 h_2 \cdot n)_{(1)} X^2 (y^2 h_1 \cdot m)_{(1)} y^1 \rangle \\
&\quad q_1^1 x^1 X^1 \cdot (y^2 h_1 \cdot m)_{(0)} \otimes q_2^1 x^2 \cdot (X^3 y^3 h_2 \cdot n)_{(0)}. \qquad (8.5.7)
\end{aligned}
$$

By using the reconstruction theorem for quasi-bialgebras (Proposition 3.3) we find the quasi-bialgebra structure of $D(H)$.

**Proposition 8.32** *Let $H$ be a finite-dimensional quasi-Hopf algebra $H$. Then $D(H)$ has a unique quasi-bialgebra structure with respect to which the isomorphism of categories in Corollary 8.31 becomes a monoidal isomorphism.*

*Proof* By the proof of Proposition 3.3, the comultiplication of $D(H)$, denoted in what follows by $\Delta_D$, is given by

$$\begin{aligned}
\Delta_D(\varphi\bowtie h) &= (\varphi\bowtie h)(1_{D(H)}\otimes 1_{D(H)})\\
&= (\varphi\bowtie h)((\varepsilon\bowtie 1_H)\otimes(\varepsilon\bowtie 1_H))\\
&\overset{(8.5.7)}{=} \langle\varphi, q^2x^3(X^3y^3h_2\cdot(\varepsilon\bowtie 1_H))_{(1)}X^2(y^2h_1\cdot(\varepsilon\bowtie 1_H))_{(1)}y^1\rangle\\
&\qquad q^1_1x^1X^1\cdot(y^2h_1\cdot(\varepsilon\bowtie 1_H))_{(0)}\otimes q^1_2x^2\cdot(X^3y^3h_2\cdot(\varepsilon\bowtie 1_H))_{(0)}\\
&\overset{(8.5.4)}{=} \langle\varphi, q^2x^3(\varepsilon\bowtie X^3y^3h_2)_{(1)}X^2(\varepsilon\bowtie y^2h_1)_{(1)}y^1\rangle\\
&\qquad q^1_1x^1X^1\cdot(\varepsilon\bowtie y^2h_1)_{(0)}\otimes q^1_2x^2\cdot(\varepsilon\bowtie X^3y^3h_2)_{(0)},
\end{aligned}$$

for all $\varphi\in H^*$ and $h\in H$. If we take $\varphi=\varepsilon$ in (8.5.5) we get that

$$(\varepsilon\bowtie h)_{(0)}\otimes(\varepsilon\bowtie h)_{(1)}=\sum_i e^i\bowtie p^1_2h\otimes S^{-1}(p^2)e_ip^1_1, \tag{8.5.8}$$

for all $h\in H$. Thus, if we denote by $P^1\otimes P^2$ a second copy of $p_R$, we can compute:

$$\begin{aligned}
&\Delta_D(\varphi\bowtie h)\\
&= \sum_{i,j}\langle\varphi, q^2x^3S^{-1}(p^2)e_ip^1_1X^2S^{-1}(P^2)e_jP^1_1y^1\rangle\\
&\qquad q^1_1x^1X^1\cdot(e^j\bowtie P^1_2y^2h_1)\otimes q^1_2x^2\cdot(e^i\bowtie p^1_2X^3y^3h_2)\\
&\overset{(8.5.4)}{=} \sum_{i,j}\langle\varphi, q^2x^3S^{-1}(p^2)e_ip^1_1X^2S^{-1}(P^2)e_jP^1_1y^1\rangle\\
&\qquad (q^1_1x^1X^1)_{(1,1)}\rightharpoonup e^j\leftharpoonup S^{-1}((q^1_1x^1X^1)_2)\bowtie(q^1_1x^1X^1)_{(1,2)}P^1_2y^2h_1\\
&\qquad \otimes(q^1_2x^2)_{(1,1)}\rightharpoonup e^i\leftharpoonup S^{-1}((q^1_2x^2)_2)\bowtie(q^1_2x^2)_{(1,2)}p^1_2X^3y^3h_2\\
&\overset{(5.5.17)}{=} \sum_{i,j}\langle\varphi, S^{-1}(Y^3)q^2Y^2_2S^{-1}((q^1Y^2_1)_2p^2)e_i(q^1Y^2_1)_{(1,1)}p^1_1X^2\\
&\qquad S^{-1}((Y^1X^1)_2P^2)e_j(Y^1X^1)_{(1,1)}P^1_1y^1\rangle e^j\bowtie(Y^1X^1)_{(1,2)}P^1_2y^2h_1\\
&\qquad \otimes e^i\bowtie(q^1Y^2_1)_{(1,2)}p^1_2X^3y^3h_2\\
&\overset{(3.2.21)}{=} \sum_{i,j}\langle\varphi, S^{-1}(Y^3)q^2S^{-1}(q^1_2p^2)e_i(q^1_1p^1)_1Y^2_1X^2S^{-1}((Y^1X^1)_2P^2)e_j\\
&\qquad (Y^1X^1)_{(1,1)}P^1_1y^1\rangle e^j\bowtie(Y^1X^1)_{(1,2)}P^1_2y^2h_1\otimes e^i\bowtie(q^1_1p^1)_2Y^2_2X^3y^3h_2\\
&\overset{(3.2.23)}{=} (Y^1X^1)_{(1,1)}P^1_1y^1\rightharpoonup\varphi_2\leftharpoonup X^2S^{-1}((Y^1X^1)_2P^2)\bowtie(Y^1X^1)_{(1,2)}P^1_2y^2h_1\\
&\qquad \otimes Y^2_1\rightharpoonup\varphi_1\leftharpoonup S^{-1}(Y^3)\bowtie Y^2_2X^3y^3h_2.
\end{aligned}$$

Combined with (8.5.3), this computation shows that, for all $\varphi\in H^*$, $h\in H$:

$$\begin{aligned}
\Delta_D(\varphi\bowtie h) = {}&(\varepsilon\bowtie X^1Y^1)(p^1_1x^1\rightharpoonup\varphi_2\leftharpoonup Y^2S^{-1}(p^2)\bowtie p^1_2x^2h_1)\\
&\otimes(X^2_1\rightharpoonup\varphi_1\leftharpoonup S^{-1}(X^3)\bowtie X^2_2Y^3x^3h_2).
\end{aligned} \tag{8.5.9}$$

Similar computations yield explicit formulas for the counit $\varepsilon_D$ and the reassociator $\Phi_D$ of $D(H)$. Namely, by the proof of Proposition 3.3 we have

$$\begin{aligned}\varepsilon_D(\varphi \bowtie h) &= (\varphi \bowtie h)\cdot 1_k\\ &\overset{(8.5.6)}{=} \langle \varphi, q^2(h\cdot 1_k)_{(1)}\rangle q^1\cdot (h\cdot 1_k)_{(0)}\\ &= \varepsilon(h)\langle \varphi, q^2\rangle q^1\cdot 1_k = \varepsilon(h)\varepsilon(q^1)\varphi(q^2) = \varepsilon(h)\varphi(S^{-1}(\alpha)),\end{aligned}$$

for all $\varphi \in H^*$ and $h \in H$, while

$$\begin{aligned}\Phi_D &= a_{D(H),D(H),D(H)}(1_{D(H)}\otimes 1_{D(H)}\otimes 1_{D(H)})\\ &= X^1\cdot(\varepsilon \bowtie 1_H)\otimes X^2\cdot(\varepsilon \bowtie 1_H)\otimes X^3\cdot(\varepsilon \bowtie 1_H)\\ &\overset{(8.5.4)}{=} (\varepsilon \bowtie X^1)\otimes(\varepsilon \bowtie X^2)\otimes(\varepsilon \bowtie X^3).\end{aligned}$$

Concluding, $D(H)$ is a quasi-bialgebra with multiplication given in (8.5.2), unit $1_D = \varepsilon \bowtie 1_H$, comultiplication $\Delta_D$ as in (8.5.9), and counit and reassociator given by

$$\varepsilon_D(\varphi \bowtie h) = \varepsilon(h)\varphi(S^{-1}(\alpha)),\ \ \forall\ \varphi \bowtie h \in D(H), \tag{8.5.10}$$

$$\Phi_D = (\varepsilon \bowtie X^1)\otimes(\varepsilon \bowtie X^2)\otimes(\varepsilon \bowtie X^3), \tag{8.5.11}$$

and this is the unique quasi-bialgebra structure on the diagonal crossed product algebra $D(H)$ that turns the isomorphism in Corollary 8.31 into an isomorphism of monoidal categories. □

We end this section by computing the quasi-Hopf algebra structure of $D(H)$. For this, we will make use of the reconstruction theorem for quasi-Hopf algebras as it was stated in Theorem 3.38.

**Theorem 8.33** *Let $H$ be a finite-dimensional quasi-Hopf algebra. Then $D(H)$, the quantum double of $H$, is a quasi-Hopf algebra that contains $H$ as a quasi-Hopf subalgebra.*

*Proof* Let $\{e_i, e^i\}_i$ be dual bases in $H$ and $H^*$ and denote by $\{\theta_{ij}\}_{i,j}$ the basis of $D(H)^*$ dual to the basis $\{e^i \bowtie e_j\}_{i,j}$ of $D(H)$, that is, $\theta_{ij}(e^s \bowtie e_t) = \delta_{i,s}\delta_{j,t}$, where $\delta_{i,j}$ is the Kronecker delta.

By (3.5.1), specialized to our context, the antipode $S_D$ of $D(H)$ is given by

$$\begin{aligned}S_D(\varphi \bowtie h) &= \sum_{i,j}\langle(\varphi \bowtie h)\cdot\theta_{i,j}, 1_D\rangle e^i \bowtie e_j\\ &\overset{(8.5.6)}{=} \sum_{i,j}\langle \varphi, q^2(h\cdot\theta_{i,j})_{(1)}\rangle\langle q^1\cdot(h\cdot\theta_{i,j})_{(0)}, \varepsilon \bowtie 1_H\rangle e^i \bowtie e_j\\ &\overset{(8.3.6),(8.5.4)}{=} \sum_{i,j}\langle \varphi, q^2(h\cdot\theta_{i,j})_{(1)}\rangle\langle(h\cdot\theta_{i,j})_{(0)}, \varepsilon \bowtie S(q^1)\rangle e^i \bowtie e_j,\end{aligned}$$

for all $\varphi \in H^*$ and $h \in H$. By Proposition 8.23 we have that $D(H)^*$ is a left–right Yetter–Drinfeld module over $H$ with the left $H$-action defined by

$$h\cdot\theta_{i,j} = \sum_{s,t}(h\cdot\theta_{i,j})(e^s \bowtie e_t)\theta_{s,t}$$

$$
\begin{aligned}
&\overset{(8.3.6)}{=} \sum_{s,t} \theta_{i,j}(S(h)\cdot(e^s \bowtie e_t))\theta_{s,t}\\
&\overset{(8.5.4)}{=} \sum_{s,t} \theta_{i,j}(S(h)_{(1,1)} \rightharpoonup e^s \leftharpoonup S^{-1}(S(h)_2) \bowtie S(h)_{(1,2)}e_t)\theta_{s,t}\\
&= \sum_{s,t} e^s(S^{-1}(S(h)_2)e_iS(h)_{(1,1)})e^j(S(h)_{(1,2)}e_t)\theta_{s,t}, \qquad (8.5.12)
\end{aligned}
$$

and the right $H$-action, see (8.3.7), given by

$$
\theta_{i,j} \mapsto \sum_{s,t} \langle \theta_{i,j}, f^1 \cdot (g^2 \cdot (e^s \bowtie e_t))_{(0)} \rangle \theta_{s,t} \otimes S^{-1}(f^2(g^2 \cdot (e^s \bowtie e_t))_{(1)}g^1),
$$

for all $i, j$, extended by linearity. We conclude that

$$
\begin{aligned}
&S_D(\varphi \bowtie h)\\
&\overset{(8.5.4)}{=} \sum_{s,t} \langle \theta_{s,t}, f^1 \cdot (\varepsilon \bowtie g^2 S(q^1))_{(0)} \rangle \langle \varphi, q^2 S^{-1}(f^2(\varepsilon \bowtie g^2 S(q^1))_{(1)} g^1) \rangle\\
&\qquad S(h)_{(1,1)} \rightharpoonup e^s \leftharpoonup S^{-1}(S(h)_2) \bowtie S(h)_{(1,2)} e_t\\
&\underset{(8.5.2)}{\overset{(8.5.8)}{=}} \sum_{i,s,t} \langle \theta_{st}, f^1 \cdot (e^i \bowtie p_2^1 g^2 S(q^1)) \rangle \langle \varphi, q^2 S^{-1}(f^2 S^{-1}(p^2) e_i p_1^1 g^1) \rangle\\
&\qquad (\varepsilon \bowtie S(h))(e^s \bowtie e_t)\\
&\overset{(8.5.4)}{=} \sum_{s} \langle \varphi, q^2 S^{-1}(f^2 S^{-1}(f_2^1 p^2) e_s f^1_{(1,1)} p_1^1 g^1) \rangle (\varepsilon \bowtie S(h))(e^s \bowtie f^1_{(1,2)} p_2^1 g^2 S(q^1))\\
&\underset{(8.5.2)}{\overset{(7.3.1)}{=}} (\varepsilon \bowtie S(h) f^1)(p_1^1 U^1 \rightharpoonup \varphi \circ S^{-1} \leftharpoonup f^2 S^{-1}(p^2) \bowtie p_2^1 U^2),
\end{aligned}
$$

for all $\varphi \in H^*$ and $h \in H$. Similarly, the general formulas in (3.5.3) give the elements $\alpha_D$ and $\beta_D$ that together with $S_D$ define the antipode of $D(H)$. More precisely,

$$
\alpha_D = \sum_{i,j} \mathrm{ev}_D(\theta_{i,j} \otimes 1_D) e^i \bowtie e_j \overset{(8.5.4)}{=} \sum_{i,j} \theta_{i,j}(\varepsilon \bowtie \alpha) e^i \bowtie e_j = \varepsilon \bowtie \alpha,
$$

and, similarly, since $\mathrm{coev}_D(1_k) = \sum_{i,j} \beta \cdot (e^i \bowtie e_j) \otimes \theta_{i,j}$ we get that

$$
\beta_D = \sum_{i,j} \theta_{i,j}(1_D)\beta \cdot (e^i \bowtie e_j) = \beta \cdot (\varepsilon \bowtie 1) = \varepsilon \bowtie \beta.
$$

Here $\mathrm{ev}_D$ and $\mathrm{coev}_D$ are the evaluation and coevaluation morphisms of the left dual object of $D(H)$ in ${}_H\mathscr{YD}^{H^{\mathrm{fd}}}$ as in Proposition 8.23.

Hence, the formulas

$$
\alpha_D = \varepsilon \bowtie \alpha, \quad \beta_D = \varepsilon \bowtie \beta, \qquad (8.5.13)
$$

$$
S_D(\varphi \bowtie h) = (\varepsilon \bowtie S(h) f^1)(p_1^1 U^1 \rightharpoonup \overline{S}^{-1}(\varphi) \leftharpoonup f^2 S^{-1}(p^2) \bowtie p_2^1 U^2), \qquad (8.5.14)
$$

together with the ones found in the proof of Proposition 8.32 define on $D(H)$ a quasi-Hopf algebra structure.

It is immediate that $i_D : H \ni h \mapsto \varepsilon \bowtie h \in D(H)$ is an injective quasi-Hopf algebra morphism, so $H$ can be regarded as a quasi-Hopf subalgebra of $D(H)$. □

## 8.6 The Quasi-Hopf Algebras $D^{\omega}(H)$ and $D^{\omega}(G)$

Let $H$ be a cocommutative Hopf algebra with antipode S over a base field $k$ (in particular, the cocommutativity implies $S^2 = \mathrm{Id}_H$). Since $H$ is cocommutative, we can introduce an even more simplified version of Sweedler's sigma notation: for $h \in H$, we denote

$$\Delta(h) = h \otimes h, \quad (\mathrm{Id}_H \otimes \Delta)(\Delta(h)) = (\Delta \otimes \mathrm{Id}_H)(\Delta(h)) = h \otimes h \otimes h,$$

and so on. With this notation, the antipode and counit axioms read:

$$S(h)h = hS(h) = \varepsilon(h)1_H, \quad \varepsilon(h)h = h\varepsilon(h) = h.$$

We now recall some facts concerning Hopf crossed products and cohomology. Let $H$ be a cocommutative Hopf algebra and $A$ a commutative left $H$-module algebra, with $H$-action denoted by $H \otimes A \to A$, $h \otimes a \mapsto h \cdot a$. Assume that we are given a linear map $\sigma : H \otimes H \to A$, which is normalized (i.e. $\sigma(1_H, h) = \sigma(h, 1_H) = \varepsilon(h)1_A$ for all $h \in H$) and convolution invertible. Suppose that, moreover, $\sigma$ satisfies the 2-cocycle condition:

$$\sigma(x,y)\sigma(xy,z) = [x \cdot \sigma(y,z)]\sigma(x,yz), \quad \forall\, x,y,z \in H.$$

Then, if we define a multiplication on $A \otimes H$ by

$$(a\#h)(b\#g) = a(h \cdot b)\sigma(h,g)\#hg,$$

(we denoted $a\#h := a \otimes h$, for $a \in A$, $h \in H$), this multiplication is associative and $1_A \otimes 1_H$ is a unit, hence $A \otimes H$ becomes an algebra, which will be denoted by $A\#_{\sigma}H$ and will be called the Hopf crossed product of $A$ and $H$.

Suppose again that $H$ is a cocommutative Hopf algebra and $A$ is a commutative left $H$-module algebra, and denote by $\Psi : H \otimes A \to A$ the $H$-module structure of $A$. We denote by $Reg_+^q(H,A)$ the set of $k$-linear maps $g : H^{\otimes q} \to A$ which are normalized (i.e. $g(h_1 \otimes \cdots \otimes h_q) = \varepsilon(h_1)\cdots\varepsilon(h_q)1_A$ whenever at least one of the $h_i$s equals $1_H$) and convolution invertible. We denote by $Z^q(H,A)$, $B^q(H,A)$ and $H^q(H,A)$ the $q$-cocycles, $q$-coboundaries and $q$-cohomology group of the complex determined by $Reg_+^q(H,A)$ and the maps $D^q : Reg_+^q(H,A) \to Reg_+^{q+1}(H,A)$, given by

$$\begin{aligned} D^q(u) = {} & [\Psi(\mathrm{Id} \otimes u)] * [u^{-1}(m \otimes \mathrm{Id} \otimes \cdots \otimes \mathrm{Id})] * [u(\mathrm{Id} \otimes m \otimes \mathrm{Id} \otimes \cdots \otimes \mathrm{Id})] * \\ & \cdots * [u^{\pm 1}(\mathrm{Id} \otimes \cdots \otimes m)] * [u^{\mp 1} \otimes \varepsilon]. \end{aligned}$$

Here $m$ denotes multiplication on $H$ and $u^{-1}$ the convolution inverse of $u$. This cohomology is called the Sweedler cohomology of $H$ with coefficients in $A$.

From now on, for the remainder of this section, we assume that $H$ is a finite-dimensional cocommutative Hopf algebra. Thus, $H^*$ is a commutative Hopf algebra, with unit $\varepsilon$, counit $\varepsilon(\varphi) = \varphi(1_H)$, for all $\varphi \in H^*$, multiplication $(\varphi\psi)(h) = \varphi(h)\psi(h)$, for all $\varphi, \psi \in H^*$ and $h \in H$, comultiplication $\Delta(\varphi) = \varphi_1 \otimes \varphi_2$ if and only if $\varphi(hg) = \varphi_1(h)\varphi_2(g)$, for all $h, g \in H$, and antipode $\overline{S}(\varphi) = \varphi \circ S$, for all $\varphi \in H^*$.

Assume that we are given a $k$-linear map $\omega : H\otimes H\otimes H \to k$ that is convolution invertible and satisfies the conditions:

$$\omega(x,y,zt)\omega(xy,z,t) = \omega(y,z,t)\omega(x,yz,t)\omega(x,y,z),\ \ \forall\, x,y,z,t \in H, \quad (8.6.1)$$
$$\omega(1_H,x,y) = \omega(x,1_H,y) = \omega(x,y,1_H) = \varepsilon(x)\varepsilon(y),\ \ \forall\, x,y \in H. \quad (8.6.2)$$

Such a map $\omega$ is exactly a 3-cocycle in the Sweedler cohomology defined above.

Since $H$ is finite dimensional, we can identify $(H\otimes H\otimes H)^*$ with $H^*\otimes H^*\otimes H^*$, so we can consider $\omega \in H^*\otimes H^*\otimes H^*$; we denote $\omega = \omega_1\otimes\omega_2\otimes\omega_3$ and its convolution inverse $\omega^{-1} = \overline{\omega}_1\otimes\overline{\omega}_2\otimes\overline{\omega}_3$.

We define the element $\Phi \in H^*\otimes H^*\otimes H^*$ by $\Phi := \omega^{-1} = \overline{\omega}_1\otimes\overline{\omega}_2\otimes\overline{\omega}_3$. Since $H^*$ is a commutative algebra, obviously $(H^*,\Delta,\varepsilon,\Phi)$ is a quasi-bialgebra, where $\Delta$ and $\varepsilon$ are the ones that give the usual coalgebra structure of $H^*$ (dual to the algebra structure of $H$). Moreover, if we define $\beta \in H^*$ by the formula $\beta(h) = \omega(h,S(h),h)$, then it is easy to see that $(H^*,\Delta,\varepsilon,\Phi,\overline{S},\alpha = \varepsilon,\beta)$ is a quasi-Hopf algebra, which will be denoted by $H^*_\omega$.

We can consider the diagonal crossed product $(H^*_\omega)^* \bowtie H^*_\omega$ as in Section 5.6. On the other hand, we will construct a certain Hopf crossed product $H^*\#_\sigma H$, as follows.

We introduce first the following notation: $g\triangleleft x = S(x)gx$, for all $g,x \in H$. Next, we define the linear map $\theta : H\otimes H\otimes H \to k$, by

$$\theta(g;x,y) = \omega(g,x,y)\omega(x,y,g\triangleleft(xy))\omega^{-1}(x,g\triangleleft x,y), \quad (8.6.3)$$

for all $g,x,y \in H$, where $\omega^{-1}$ is the convolution inverse of $\omega$.

It is easy to see that $\theta$ is also normalized and convolution invertible. By using the 3-cocycle condition for $\omega$ several times, one can get the following relation:

$$\theta(g;x,y)\theta(g;xy,z) = \theta(g\triangleleft x;y,z)\theta(g;x,yz), \quad (8.6.4)$$

for all $g,x,y,z \in H$.

Since $H$ is cocommutative, $H^*$ becomes a commutative left $H$-module algebra, with action $H\otimes H^* \to H^*$, $h\otimes\varphi \mapsto h\bullet\varphi$, where $h\bullet\varphi = h\rightharpoonup\varphi\leftharpoonup S(h)$, and as before $\rightharpoonup$ and $\leftharpoonup$ denote the left and right regular actions of $H$ on $H^*$ given by $(h\rightharpoonup\varphi)(a) = \varphi(ah)$ and $(\varphi\leftharpoonup h)(a) = \varphi(ha)$ for all $h,a \in H$ and $\varphi \in H^*$. Hence, $(h\bullet\varphi)(a) = \varphi(a\triangleleft h)$ for all $h,a \in H$ and $\varphi \in H^*$.

Now define the linear map $\sigma : H\otimes H \to H^*$ by $\sigma(x,y)(g) = \theta(g;x,y)$. Since $\theta$ is normalized and convolution invertible, $\sigma$ is also normalized and convolution invertible; one can easily see that the relation (8.6.4) is equivalent to the fact that $\sigma$ is a 2-cocycle, that is:

$$\sigma(x,y)\sigma(xy,z) = [x\bullet\sigma(y,z)]\sigma(x,yz), \quad (8.6.5)$$

for all $x,y,z \in H$. Hence, we can consider the Hopf crossed product $H^*\#_\sigma H$, which will be denoted by $D^\omega(H)$, and which is an associative algebra with unit $\varepsilon\#1_H$. Its multiplication is given, for all $\varphi,\varphi' \in H^*$, $h,h' \in H$, by

$$(\varphi\otimes h)(\varphi'\otimes h') = \varphi(h\rightharpoonup\varphi'\leftharpoonup S(h))\sigma(h,h')\otimes hh'. \quad (8.6.6)$$

**Theorem 8.34** *The linear map* $w : (H^*_\omega)^* \bowtie H^*_\omega \to H^*\#_\sigma H$ *defined by*

$$w(h \bowtie \varphi) = \overline{\omega}_2(h)\overline{\omega}_3(S(h))\overline{\omega}_1(h \rightharpoonup \varphi \leftharpoonup S(h))\#h, \ \forall\, h \in H,\ \varphi \in H^*, \tag{8.6.7}$$

*is an algebra isomorphism, with inverse* $W : H^*\#_\sigma H \to (H^*_\omega)^* \bowtie H^*_\omega$ *given by*

$$W(\varphi\#h) = p^1_1(h)p^2(S(h))(\varphi_1 \rightharpoonup h \leftharpoonup S(\varphi_3)) \bowtie p^1_2\varphi_2, \ \forall\, \varphi \in H^*,\ h \in H, \tag{8.6.8}$$

*where we denote by* $p^1 \otimes p^2 = x^1 \otimes x^2\beta S(x^3)$ *the element for* $H^*_\omega$ *given by* (3.2.19) *and by* $\rightharpoonup$ *and* $\leftharpoonup$ *the regular actions of* $H$ *on* $H^*$ *and of* $H^*$ *on* $H$.

*Proof* We will construct the map $w$ by using the Universal Property of the diagonal crossed product (Proposition 5.61). We define the linear maps

$$\gamma : H^*_\omega \to H^*\#_\sigma H, \quad \gamma(\varphi) = \varphi\#1_H,$$
$$v : H = (H^*_\omega)^* \to H^*\#_\sigma H, \quad v(h) = \varepsilon\#h.$$

One can easily see that $\gamma$ is an algebra map and the relations (5.6.27) and (5.6.29) are satisfied, that is, we have

$$\gamma(\varphi_1)v(h \leftharpoonup \varphi_2) = v(\varphi_1 \rightharpoonup h)\gamma(\varphi_2), \quad v(1_H) = \varepsilon\#1_H,$$

for all $\varphi \in H^*$ and $h \in H$. So the only thing left to prove is the relation (5.6.28), that is,

$$\varepsilon\#hh' = (\overline{\overline{\omega}}_1\#1_H)(\varepsilon\#\omega_1\overline{\omega}_1 \rightharpoonup h \leftharpoonup \overline{\overline{\omega}}_2)(\omega_2\#1_H)(\varepsilon\#\overline{\omega}_2 \rightharpoonup h' \leftharpoonup \overline{\overline{\omega}}_3\omega_3)(\overline{\omega}_3\#1_H),$$

where we denote by $\omega^{-1} = \overline{\overline{\omega}}_1 \otimes \overline{\overline{\omega}}_2 \otimes \overline{\overline{\omega}}_3$ another copy of $\omega^{-1}$. We compute:

$$\begin{aligned}
&(\overline{\overline{\omega}}_1\#1_H)(\varepsilon\#\omega_1\overline{\omega}_1 \rightharpoonup h \leftharpoonup \overline{\overline{\omega}}_2)(\omega_2\#1_H)(\varepsilon\#\overline{\omega}_2 \rightharpoonup h' \leftharpoonup \overline{\overline{\omega}}_3\omega_3)(\overline{\omega}_3\#1_H)\\
&\quad = (\overline{\overline{\omega}}_1\#\omega_1\overline{\omega}_1 \rightharpoonup h \leftharpoonup \overline{\overline{\omega}}_2)(\omega_2\#\overline{\omega}_2 \rightharpoonup h' \leftharpoonup \overline{\overline{\omega}}_3\omega_3)(\overline{\omega}_3\#1_H)\\
&\quad = \omega_1(h)\overline{\omega}_1(h)\overline{\overline{\omega}}_2(h)\overline{\omega}_2(h')\overline{\overline{\omega}}_3(h')\omega_3(h')(\overline{\overline{\omega}}_1\#h)(\omega_2(h' \rightharpoonup \overline{\omega}_3 \leftharpoonup S(h'))\#h')\\
&\quad = \omega_1(h)\overline{\omega}_1(h)\overline{\overline{\omega}}_2(h)\overline{\omega}_2(h')\overline{\overline{\omega}}_3(h')\omega_3(h')\\
&\qquad (\overline{\overline{\omega}}_1(h \rightharpoonup \omega_2 \leftharpoonup S(h))(hh' \rightharpoonup \overline{\omega}_3 \leftharpoonup S(hh'))\sigma(h,h')\#hh').
\end{aligned}$$

When we evaluate this in $g \otimes \varphi \in H \otimes H^*$ we obtain:

$$\begin{aligned}
&\omega(h,S(h)gh,h')\omega^{-1}(h,h',S(hh')ghh')\omega^{-1}(g,h,h')\theta(g;h,h')\varphi(hh')\\
&\quad = \omega(h,S(h)gh,h')\omega^{-1}(h,h',S(hh')ghh')\omega^{-1}(g,h,h')\\
&\qquad \omega(g,h,h')\omega(h,h',S(hh')ghh')\omega^{-1}(h,S(h)gh,h')\varphi(hh')\\
&\quad = \varepsilon(g)\varphi(hh') = (\varepsilon\#hh')(g \otimes \varphi).
\end{aligned}$$

Thus, Proposition 5.61 yields an algebra map $w : (H^*_\omega)^* \bowtie H^*_\omega \to H^*\#_\sigma H$, defined by the formula

$$w(h \bowtie \varphi) = \gamma(q^1)v(h \leftharpoonup q^2)\gamma(\varphi),$$

where $q^1 \otimes q^2 = X^1 \otimes S^{-1}(\alpha X^3)X^2$ is the element for $H^*_\omega$ given by (3.2.19). An easy computation shows that this map is identical to the one given by (8.6.7).

To prove that $w$ is bijective with inverse $W$, since the underlying vector spaces have the same (finite) dimension, it is enough to prove that $w \circ W = \mathrm{Id}$. We compute:

$$\begin{aligned} w(W(\varphi\#h)) &= w(p_1^1(h)p^2(S(h))\varphi_1(h)\varphi_3(S(h))h \bowtie p_2^1\varphi_2) \\ &= p_1^1(h)p^2(S(h))\varphi_1(h)\varphi_3(S(h))\overline{\omega}_2(h)\overline{\omega}_3(S(h)) \\ &\quad \overline{\omega}_1(h \rightharpoonup p_2^1\varphi_2 \leftharpoonup S(h))\#h. \end{aligned}$$

When we evaluate this in $g \otimes \psi \in H \otimes H^*$ we obtain:

$$\begin{aligned} & p_1^1(h)p^2(S(h))\varphi_1(h)\varphi_3(S(h))\overline{\omega}_2(h)\overline{\omega}_3(S(h))p_2^1(S(h)gh)\varphi_2(S(h)gh)\overline{\omega}_1(g)\psi(h) \\ &= p^1(gh)p^2(S(h))\varphi(g)\omega^{-1}(g,h,S(h))\psi(h) \\ &= \omega_1(gh)\omega_2(S(h))\beta(S(h))\omega_3(h)\omega^{-1}(g,h,S(h))\varphi(g)\psi(h) \\ &= \omega(gh,S(h),h)\omega(S(h),h,S(h))\omega^{-1}(g,h,S(h))\varphi(g)\psi(h). \end{aligned}$$

To finish the proof it will be enough to prove that

$$\omega(gh,S(h),h)\omega(S(h),h,S(h))\omega^{-1}(g,h,S(h)) = \varepsilon(g)\varepsilon(h).$$

Note first that the 3-cocycle condition for $\omega$ applied to the elements $x = h$, $y = S(h)$, $z = h$, $t = S(h)$ yields

$$\omega(S(h),h,S(h)) = \omega^{-1}(h,S(h),h). \tag{8.6.9}$$

So it is enough to prove that

$$\omega(gh,S(h),h)\omega^{-1}(h,S(h),h)\omega^{-1}(g,h,S(h)) = \varepsilon(g)\varepsilon(h).$$

But this relation follows immediately by applying the 3-cocycle condition for $\omega$ to the elements $x = g$, $y = h$, $z = S(h)$, $t = h$. □

The quantum double $D(H^*_\omega)$ of the quasi-Hopf algebra $H^*_\omega$ has as underlying algebra structure the diagonal crossed product $(H^*_\omega)^* \bowtie H^*_\omega$, so Theorem 8.34 implies:

**Theorem 8.35** *$D^\omega(H) = H^*\#_\sigma H$ is a quasi-Hopf algebra.*

*Proof* Most of the structure of $D^\omega(H)$ may be obtained, by a straightforward computation, by transferring the structure from $D(H^*_\omega)$ via the isomorphism (8.6.7). We write down the structures obtained in this way:

- the counit:

$$\varepsilon : D^\omega(H) \to k, \quad \varepsilon(\varphi\#h) = \varphi(1_H)\varepsilon(h), \ \forall\ \varphi \in H^*,\ h \in H;$$

- the reassociator:

$$\Phi = (\overline{\omega}_1\#1_H) \otimes (\overline{\omega}_2\#1_H) \otimes (\overline{\omega}_3\#1_H) \in D^\omega(H) \otimes D^\omega(H) \otimes D^\omega(H);$$

- $\alpha_{D^\omega(H)} = \varepsilon\#1_H$, $\beta_{D^\omega(H)} = \beta\#1_H$;
- the comultiplication: define the linear map $\gamma : H \otimes H \otimes H \to k$ by

$$\gamma(g,h;x) = \omega(g,h,x)\omega(x,g \triangleleft x,h \triangleleft x)\omega^{-1}(g,x,h \triangleleft x), \tag{8.6.10}$$

for all $g,h,x \in H$. Then define the linear map $\nu : H \to (H \otimes H)^*$, $\nu(h)(x \otimes y) = \gamma(x,y;h)$. Identifying $(H \otimes H)^*$ with $H^* \otimes H^*$, we will write, for any $h \in H$, $\nu(h) = \nu_1(h) \otimes \nu_2(h) \in H^* \otimes H^*$. Then the comultiplication of $D^{\omega}(H)$ is defined, for all $\varphi \in H^*$, $h \in H$, by

$$\Delta : D^{\omega}(H) \to D^{\omega}(H) \otimes D^{\omega}(H), \quad \Delta(\varphi\#h) = (\nu_1(h)\varphi_1\#h) \otimes (\nu_2(h)\varphi_2\#h). \tag{8.6.11}$$

The only part of the structure of $D^{\omega}(H)$ that is difficult to obtain by transferring from $D(H^*_{\omega})$ (the computations become very unpleasant) is the antipode, so we need to give a direct proof of the fact that the linear map $s : D^{\omega}(H) \to D^{\omega}(H)$,

$$s(\varphi\#h) = [\varepsilon\#S(h)][\sigma^{-1}(h,S(h))S(\varphi\nu_1^{-1}(h))\nu_2^{-1}(h)\#1_H], \tag{8.6.12}$$

for all $\varphi \in H^*$, $h \in H$, is the antipode, where we denote by $\nu^{-1}$ the convolution inverse of $\nu$, with notation $\nu^{-1}(h) = \nu_1^{-1}(h) \otimes \nu_2^{-1}(h) \in H^* \otimes H^*$.

By using the relation (8.6.4), it follows that

$$\theta(x;S(h),h) = \theta(hxS(h);h,S(h)), \tag{8.6.13}$$

for all $x,h \in H$, and by using this relation we obtain

$$s((\varphi\#h)_1)(\varphi\#h)_2 = \varepsilon(\varphi\#h)(\varepsilon\#1_H), \tag{8.6.14}$$

for all $\varphi \in H^*$, $h \in H$, where we denote, as usual, $\Delta(\varphi\#h) = (\varphi\#h)_1 \otimes (\varphi\#h)_2$.

By using the 3-cocycle relation for $\omega$, one obtains the following identity:

$$\gamma(x,y;h)\gamma(xy,z;h)\omega(x \triangleleft h, y \triangleleft h, z \triangleleft h) = \gamma(x,yz;h)\gamma(y,z;h)\omega(x,y,z), \tag{8.6.15}$$

for all $h,x,y,z \in H$. By using this relation we obtain

$$\gamma(x,S(x);h)\beta(x \triangleleft h) = \gamma(S(x),x;h)\beta(x), \tag{8.6.16}$$

from which we get that

$$(\varphi\#h)_1\beta s((\varphi\#h)_2) = \varepsilon(\varphi\#h)\beta, \tag{8.6.17}$$

for all $\varphi \in H^*$, $h \in H$. The relation

$$(\overline{\omega}_1\#1_H)\beta s(\overline{\omega}_2\#1_H)(\overline{\omega}_3\#1_H) = \varepsilon\#1_H, \tag{8.6.18}$$

is immediate, while the relation

$$s(\omega_1\#1_H)(\omega_2\#1_H)\beta s(\omega_3\#1_H) = \varepsilon\#1_H \tag{8.6.19}$$

follows immediately by using (8.6.9).

By applying the 3-cocycle relation for $\omega$ repeatedly, we get:

$$\gamma(x,y;h)\gamma(x \triangleleft h, y \triangleleft h;h')\theta(x;h,h')\theta(y;h,h') = \gamma(x,y;hh')\theta(xy;h,h'), \tag{8.6.20}$$

for all $h,h',x,y \in H$, and by using this relation and (8.6.4) we obtain

$$s((\varphi\#h)(\varphi'\#h')) = s(\varphi'\#h')s(\varphi\#h), \tag{8.6.21}$$

for all $\varphi, \varphi' \in H^*$ and $h, h' \in H$, and since obviously $s(\varepsilon\#1_H) = \varepsilon\#1_H$, it follows that $s$ is an anti-algebra homomorphism.

Thus, $s$ is indeed the antipode of $D^\omega(H)$. Moreover, by (8.6.9) we get that $\beta$ is convolution invertible with inverse $\beta^{-1} = \beta \circ S$. By using (8.6.20), (8.6.13) and (8.6.16), $s^2(\varphi\#h) = (\beta^{-1}\#1_H)(\varphi\#h)(\beta\#1_H)$, for all $\varphi \in H^*$, $h \in H$. □

We will see how $D^\omega(H)$ depends on the cohomology class of $\omega$. Suppose that $\omega'$ is another normalized 3-cocycle on $H$, which lies in the same cohomology class as $\omega$; that is, there exists a $k$-linear map $t : H \otimes H \to k$, normalized and convolution invertible, such that $\omega' = \omega d(t)$, where $d(t) : H \otimes H \otimes H \to k$ is given by

$$d(t)(a \otimes b \otimes c) = t(b,c)t^{-1}(ab,c)t(a,bc)t^{-1}(a,b),$$

for all $a,b,c \in H$, and $t^{-1}$ is the convolution inverse of $t$.

We will denote by $\theta, \theta', \theta_{d(t)}$ and $\sigma, \sigma', \sigma_{d(t)}$ the maps associated (as above) to $\omega, \omega', d(t)$, respectively.

Now define the map $\tau = \tau_t : H \to H^*$ by $\tau(x)(g) = t(x, g \triangleleft x)t^{-1}(g,x)$, for all $x, g \in H$. Obviously $\tau(1_H) = \varepsilon$ and $\tau$ is convolution invertible. Since $\omega' = \omega d(t)$, it follows that $\theta' = \theta\theta_{d(t)}$, so we have

$$\theta'(g;x,y)\theta^{-1}(g;x,y) = \theta_{d(t)}(g;x,y), \quad \forall\, g,x,y \in H.$$

A straightforward computation yields:

$$\theta_{d(t)}(g;x,y) = \tau(x)(g)\tau(y)(g \triangleleft x)\tau^{-1}(xy)(g), \quad \forall\, g,x,y \in H.$$

Now, for the cohomology $H^n(H,H^*)$, since $\tau \in Reg^1_+(H,H^*)$, we can consider $D^1(\tau) \in B^2(H,H^*)$, and it is easy to see that

$$D^1(\tau)(x,y)(g) = \tau(x)(g)\tau(y)(g \triangleleft x)\tau^{-1}(xy)(g), \quad \forall\, g,x,y \in H,$$

hence we have $\theta_{d(t)}(g;x,y) = D^1(\tau)(x,y)(g)$, that is $\sigma_{d(t)} = D^1(\tau)$.

By identifying $(H \otimes H)^* = H^* \otimes H^*$, we will write $t = t_1 \otimes t_2 \in H^* \otimes H^*$. Now define $F \in D^\omega(H) \otimes D^\omega(H)$ by $F = (t_1 \otimes 1_H) \otimes (t_2 \otimes 1_H)$. Since $t$ is normalized and convolution invertible (let $t^{-1} = l_1 \otimes l_2$ be the convolution inverse of $t$), it follows immediately that $F$ is invertible with inverse $F^{-1} = (l_1 \otimes 1_H) \otimes (l_2 \otimes 1_H)$, and satisfies the relation $(\varepsilon \otimes \mathrm{Id})(F) = (\mathrm{Id} \otimes \varepsilon)(F) = 1$, that is, $F$ is a gauge transformation on $D^\omega(H)$.

Define the linear map $T : D^{\omega'}(H) \to D^\omega(H)$ by $T(\varphi \otimes h) = \varphi\tau(h) \otimes h$, for all $\varphi \in H^*$ and $h \in H$. By a direct computation, using the facts that $\sigma' = \sigma\sigma_{d(t)}$ and $\sigma_{d(t)} = D^1(\tau)$, it follows that $T$ is an algebra isomorphism.

Consider now $D^\omega(H)_{F^{-1}}$, that is, $D^\omega(H)$ with its algebra structure, but with comultiplication given by $\Delta_{F^{-1}}(a) = F^{-1}\Delta(a)F$, for all $a \in D^\omega(H)$. By using the fact that $\gamma' = \gamma\gamma_{d(t)}$, it follows that $(T \otimes T) \circ \Delta' = \Delta_{F^{-1}} \circ T$. By using $\omega' = \omega d(t)$, it follows that $(T \otimes T \otimes T)(\Phi') = \Phi_{F^{-1}}$, where $\Phi_{F^{-1}} = F_{23}^{-1}(\mathrm{Id} \otimes \Delta)(F^{-1})\Phi(\Delta \otimes \mathrm{Id})(F)F_{12}$. Obviously, we have $\varepsilon \circ T = \varepsilon$. Summing up, we obtain that $T$ is an isomorphism of quasi-bialgebras.

In conclusion, we have obtained the following result:

**Proposition 8.36** *The above map $T$ is an isomorphism of quasi-bialgebras between $D^{\omega'}(H)$ and $D^{\omega}(H)_{F^{-1}}$. Consequently, $D^{\omega'}(H)$ and $D^{\omega}(H)$ are twist equivalent quasi-bialgebras.*

Now let $G$ be a finite group, with multiplication denoted by juxtaposition and unit denoted by $e$. Let $\omega$ be a normalized 3-cocycle on $G$, that is, $\omega : G\times G\times G \to k^*$ is a map such that $\omega(x,y,z)\omega(tx,y,z)^{-1}\omega(t,xy,z)\omega(t,x,yz)^{-1}\omega(t,x,y) = 1$ for all $t,x,y,z\in G$, and $\omega(x,y,z) = 1$ whenever $x$, $y$ or $z$ is equal to 1. We can take $H = k[G]$, the group algebra of $G$, which is a finite-dimensional cocommutative Hopf algebra, and extend $\omega$ by linearity to a map $\omega : H\otimes H\otimes H \to k$, which turns out to be a Sweedler 3-cocycle on $H$. So, we can consider the quasi-Hopf algebra $D^{\omega}(H)$, which will be denoted by $D^{\omega}(G)$. As a linear space, $D^{\omega}(G) = k[G]^*\otimes k[G]$, which has the basis $\{p_g\otimes x\}$, with $g,x\in G$, where $p_g(h) = \delta_{g,h}$ for any $g,h\in G$ ($\delta$ is the Kronecker delta). In view of the formulas presented above, $D^{\omega}(G)$ becomes a quasi-Hopf algebra with the following structure:

- the multiplication: $(p_g\otimes x)(p_h\otimes y) = \delta_{g,xhx^{-1}}\theta(g,x,y)(p_g\otimes xy)$, where

$$\theta(g,x,y) = \omega(g,x,y)\omega(x,y,(xy)^{-1}gxy)\omega(x,x^{-1}gx,y)^{-1};$$

- the unit: $1 = \sum_{g\in G} p_g\otimes e$;
- the comultiplication: $\Delta : D^{\omega}(G)\to D^{\omega}(G)\otimes D^{\omega}(G)$,

$$\Delta(p_g\otimes x) = \sum_{uv=g}\gamma(x,u,v)(p_u\otimes x)\otimes(p_v\otimes x),\ \text{where}$$

$$\gamma(x,u,v) = \omega(u,v,x)\omega(x,x^{-1}ux,x^{-1}vx)\omega(u,x,x^{-1}vx)^{-1};$$

- the counit: $\varepsilon : D^{\omega}(G)\to k$, $\varepsilon(p_g\otimes x) = \delta_{g,e}$;
- the reassociator: $\Phi = \sum_{x,y,z\in G}\omega(x,y,z)^{-1}(p_x\otimes e)\otimes(p_y\otimes e)\otimes(p_z\otimes e)$;
- $\alpha = 1$, $\beta = \sum_{g\in G}\omega(g,g^{-1},g)(p_g\otimes e)$;
- the antipode: $s : D^{\omega}(G)\to D^{\omega}(G)$,

$$s(p_g\otimes x) = \theta(g^{-1},x,x^{-1})^{-1}\gamma(x,g,g^{-1})^{-1}(p_{x^{-1}g^{-1}x}\otimes x^{-1}).$$

Now, consider the element $\lambda = \sum_{y\in G} p_e\otimes y\in D^{\omega}(G)$. We check that $\lambda$ is a left integral in $D^{\omega}(G)$. If $g,x\in G$, we have:

$$(p_g\otimes x)\lambda = \sum_{y\in G}\delta_{g,e}\theta(g,x,y)(p_g\otimes xy).$$

If $g = e$, then $(p_e\otimes x)\lambda = \sum_{y\in G} p_e\otimes xy = \lambda = \varepsilon(p_e\otimes x)\lambda$. If $g\neq e$, then $(p_g\otimes x)\lambda = 0 = \varepsilon(p_g\otimes x)\lambda$, since $\varepsilon(p_g\otimes x) = \delta_{g,e}$.

This shows that $\lambda$ is indeed a left integral in $D^{\omega}(G)$. Similarly, one can prove that $\lambda$ is a right integral as well, so $D^{\omega}(G)$ is unimodular.

Since $\varepsilon(\lambda) = \sum_{y \in G} \delta_{e,e} = |G|$, by applying Theorem 7.28 we obtain that $D^\omega(G)$ is semisimple if and only if the characteristic of $k$ does not divide $|G|$, the order of $G$.

## 8.7 Algebras within Categories of Yetter–Drinfeld Modules

We will show that the algebra $H_0$ within the monoidal category ${}_H\mathscr{M}$ considered in Definition 4.4 actually has a commutative algebra structure within ${}_H^H\mathscr{YD}$.

We start by presenting the Yetter–Drinfeld module structure of $H_0$. Recall that $H_0 = H$ as $k$-vector spaces.

**Lemma 8.37** *Let $H$ be a quasi-Hopf algebra with bijective antipode. Then $H$ is a left Yetter–Drinfeld module with the following structure:*

$$h \triangleright h' = h_1 h' S(h_2), \tag{8.7.1}$$
$$\lambda_H(h) = h_{(-1)} \otimes h_{(0)} := X^1 Y_1^1 h_1 g^1 S(q^2 Y_2^2) Y^3 \otimes X^2 Y_2^1 h_2 g^2 S(X^3 q^1 Y_1^2), \tag{8.7.2}$$

*for all $h, h' \in H$, where $f^{-1} = g^1 \otimes g^2$ and $q_R = q^1 \otimes q^2$ are the elements defined by* (3.2.16) *and* (3.2.19)*, respectively.*

*Proof* We know that $H$ is a left $H$-module via $\triangleright$. Now we will prove that the relations (8.2.1)–(8.2.3) hold for our structure. For this, observe first that (3.2.13) implies

$$\lambda_H(h) = (X^1 \otimes X^2)\Delta(Y^1 h S(Y^2))U[Y^3 \otimes S(X^3)], \ \forall\, h \in H, \tag{8.7.3}$$

where $U = U^1 \otimes U^2 = \mathbf{U}^1 \otimes \mathbf{U}^2$ is the element defined in (7.3.1). Therefore, for any $h \in H$ we have:

$$\begin{aligned}
&Z^1(T^1 \triangleright h)_{(-1)_1} T^2 \otimes Z^2(T^1 \triangleright h)_{(-1)_2} T^3 \otimes Z^3 \triangleright (T^1 \triangleright h)_{(0)} \\
&\overset{(8.7.3)}{=} (Z^1 X_1^1 \otimes Z^2 X_2^1 \otimes Z_1^3 X^2)(\Delta \otimes \mathrm{Id})(\Delta(Y^1(T^1 \triangleright h)S(Y^2))U) \\
&\qquad [Y_1^3 T^2 \otimes Y_2^3 T^3 \otimes S(Z_2^3 X^3)] \\
&\overset{(3.1.9),(3.1.7)}{=} (Z^1 X_1^1 \otimes Z^2 X_2^1 \otimes Z_1^3 X^2)\Phi^{-1}(\mathrm{Id} \otimes \Delta)(\Delta(Y^1 T^1 h S(Y_1^2 T^2)))\Phi \\
&\qquad (\Delta \otimes \mathrm{Id})(\Delta(S(W^1)U))(W^2 \otimes W^3 \otimes 1_H)[Y_2^2 T^3 \otimes Y^3 \otimes S(Z_2^3 X^3)] \\
&\overset{(7.5.1),(3.1.9)}{=} (Z^1 \otimes X^1 Z_1^2 \otimes X^2 Z_2^2)(\mathrm{Id} \otimes \Delta)(\Delta(Y^1 T^1 h S(Y_1^2 T^2))U)(1 \otimes U) \\
&\qquad [Y_2^2 T^3 \otimes Y^3 \otimes S(X^3 Z^3)] \\
&\overset{(7.3.2)\times 2}{=} Z^1 Y_1^1 (T^1 h S(T^2))_1 \mathbf{U}^1 T^3 \otimes \{[(X^1 \otimes X^2)\Delta(Z^2 Y_2^1 (T^1 h S(T^2))_2 \mathbf{U}^2 S(Z_1^3 Y^2)] \\
&\qquad U[Z_2^3 Y^3 \otimes S(X^3)]\} \\
&\overset{(3.1.9),(8.7.3)}{=} Z^1 h_{(-1)} \otimes \{(X^1 \otimes X^2)\Delta(Y^1 Z_1^2 h_{(0)} S(Y^2 Z_2^2))U[Y^3 Z^3 \otimes S(X^3)]\} \\
&\overset{(8.7.3)}{=} Z^1 h_{(-1)} \otimes (Z^2 \triangleright h_{(0)})_{(-1)} Z^3 \otimes (Z^2 \triangleright h_{(0)})_{(0)}.
\end{aligned}$$

By $\varepsilon(\alpha) = \varepsilon(\beta) = 1$, $\varepsilon(g^1)g^2 = 1$, $\varepsilon(q^2)q^1 = 1$, (3.1.8) and (3.1.11) we deduce that $\varepsilon(h_{(-1)})h_{(0)} = h$, for all $h \in H$, thus we only have to show the compatibility relation (8.2.3). For all $h, h' \in H$ we calculate:

$$(h_1 \triangleright h')_{(-1)} h_2 \otimes (h_1 \triangleright h')_{(0)}$$

$$
\begin{aligned}
&= X^1(Y^1h_{(1,1)}h'S(Y^2h_{(1,2)}))_1\mathbf{U}^1Y^3h_2\\
&\quad \otimes X^2(Y^1h_{(1,1)}h'S(Y^2h_{(1,2)}))_2\mathbf{U}^2S(X^3)\\
&\overset{(3.1.7)}{=} X^1(h_1Y^1h'S(Y^2))_1S(h_{(2,1)})_1\mathbf{U}^1h_{(2,2)}Y^3\\
&\quad \otimes X^2(h_1Y^1h'S(Y^2))_2S(h_{(2,1)})_2\mathbf{U}^2S(X^3)\\
&\overset{(7.3.2),(8.7.3)}{=} X^1h_{(1,1)}(Y^1h'S(Y^2))_1\mathbf{U}^1Y^3\otimes X^2h_{(1,2)}(Y^1h'S(Y^2))_2\mathbf{U}^2S(X^3h_2)\\
&\overset{(3.1.7),(8.7.3)}{=} h_1h'_{(-1)}\otimes h_2\triangleright h'_{(0)},
\end{aligned}
$$

as needed. This finishes the proof. □

Let $H$ be a quasi-Hopf algebra with bijective antipode and $H_0$ the algebra within ${}_H\mathcal{M}$ associated to $H$, as in Definition 4.4. By the above lemma, it follows that $H_0$ is also an object in ${}_H^H\mathcal{YD}$ with the same action and coaction as $H$. Furthermore, we next show that $H_0$ is an algebra within ${}_H^H\mathcal{YD}$.

In general, an object $A\in {}_H^H\mathcal{YD}$ has an algebra structure within ${}_H^H\mathcal{YD}$ if and only if there exist morphisms $\underline{m}_A : A\otimes A\to A$ and $\underline{\eta}_A : k\to A$ in ${}_H^H\mathcal{YD}$ such that $\underline{m}_A$ is associative up to the associativity constraint of ${}_H^H\mathcal{YD}$ and $\underline{\eta}_A$ is a unit for $\underline{m}_A$. In other words, $A$ admits an algebra structure $(A,\underline{m}_A,\underline{\eta}_A)$ within ${}_H\mathcal{M}$ such that

$$
\begin{aligned}
\lambda_A(aa') &= X^1(x^1Y^1\cdot a)_{(-1)}x^2(Y^2\cdot a')_{(-1)}Y^3\\
&\qquad \otimes [X^2\cdot (x^1Y^1\cdot a)_{(0)}][X^3x^3\cdot (Y^2\cdot a')_{(0)}], &\quad (8.7.4)\\
\lambda_A(1_A) &= 1_H\otimes 1_A, &\quad (8.7.5)
\end{aligned}
$$

for all $a,a'\in A$, where we denote by $\cdot$ the left action of $H$ on $A$, by $\lambda_A$ the left $H$-coaction on $A$ and by $1_A$ the unit of $A$. Note that the two equalities above express the fact that $\underline{m}_A$ and $\underline{\eta}_A$ are left $H$-colinear morphisms.

**Proposition 8.38** *Let $H$ be a quasi-Hopf algebra with bijective antipode and $H_0$ the $H$-module algebra from Definition 4.4. Then $H_0$ is an algebra in the monoidal category ${}_H^H\mathcal{YD}$.*

*Proof* We only have to show that the relations (8.7.4) and (8.7.5) hold. To prove (8.7.4) we set $q_R = q^1\otimes q^2 = Q^1\otimes Q^2$, $f^{-1} = g^1\otimes g^2 = G^1\otimes G^2$ and $\gamma=\gamma^1\otimes\gamma^2$, where $q_R$, $f^{-1}$ and $\gamma$ are the elements defined by (3.2.19), (3.2.16) and (3.2.5), respectively, and for all $h,h'\in H$ we calculate:

$$
\begin{aligned}
&X^1(x^1Y^1\triangleright h)_{(-1)}x^2(Y^2\triangleright h')_{(-1)}Y^3\otimes [X^2\triangleright (x^1Y^1\triangleright h)_{(0)}]\circ [X^3x^3\triangleright (Y^2\triangleright h')_{(0)}]\\
&\overset{(8.7.1),(8.7.2)}{\underset{(3.2.13)}{=}} X^1Z^1(T^1x^1_1Y^1_1hS(T^2x^1_2Y^1_2))_1g^1S(q^2)T^3x^2U^1(V^1Y^2_1h'S(V^2Y^2_2))_1G^1\\
&\quad S(Q^2)V^3Y^3\otimes [X^2_1Z^2(T^1x^1_1Y^1_1hS(T^2x^1_2Y^1_2))_2g^2S(X^2_2Z^3q^1)]\\
&\quad \circ [X^3_1x^3_1U^2(V^1Y^2_1h'S(V^2Y^2_2))_2G^2S(X^3_2x^3_2U^3Q^1)]\\
&\overset{(3.1.9),(3.2.13)}{\underset{(3.2.21),(3.1.7)}{=}} X^1Z^1y^1_1(T^1Y^1_1hS(x^1T^2Y^1_2))_1g^1S(q^2)x^2U^1(T^3_1V^1Y^2_1h'S(V^2Y^2_2))_1G^1\\
&\quad S(Q^2)V^3Y^3\otimes [X^2_1Z^2y^1_2(T^1Y^1_1hS(x^1T^2Y^1_2))_2g^2S(X^2_2Z^3y^2q^1)]\\
&\quad \circ [(X^3y^3x^3)_1U^2(T^3_1V^1Y^2_1h'S(V^2Y^2_2))_2G^2S((X^3y^3x^3)_2U^3T^3_2Q^1)]
\end{aligned}
$$

$$
\begin{aligned}
&\overset{(3.1.9),(4.1.9)}{\underset{(3.1.7),(3.2.1)}{=}} X^1(T^1Y_1^1hS(x^1T^2Y_2^1))_1g^1S(q^2)x^2U^1(T_1^3V^1Y_1^2h'S(V^2Y_2^2))_1G^1S(Q^2)\\
&\qquad V^3Y^3\otimes X^2(T^1Y_1^1hS(x^1T^2Y_2^1))_2g^2S(z^1q^1)\alpha z^2x_1^3U^2\\
&\qquad (T_1^3V^1Y_1^2h'S(V^2Y_2^2))_2G^2S(X^3z^3x_2^3U^3T_2^3Q^1)\\
&\overset{(3.2.21),(3.2.13)}{\underset{(3.1.9)}{=}} X^1(T^1Y_1^1hS(z^1x^1T^2Y_2^1))_1g^1S(q^2)z^2(x^2T_1^3V^1Y_1^2h'S(V^2Y_2^2))_1G^1\\
&\qquad S(Q^2)V^3Y^3\otimes X^2(T^1Y_1^1hS(z^1x^1T^2Y_2^1))_2g^2S(q^1)\alpha z^3\\
&\qquad (x^2T_1^3V^1Y_1^2h'S(V^2Y_2^2))_2G^2S(X^3x^3T_2^3Q^1)\\
&\overset{(3.1.9),(3.2.13)}{\underset{(3.2.21),(3.1.9)}{=}} X^1(T^1V^1(Y_1^1y^1)_1hS(z^1T_1^2V^2(Y_1^1y^1)_2))_1g^1S(q^2)z^2T_{(2,1)}^2(V^3Y_2^1y^2h')_1\\
&\qquad S(Y^2y^3)_1G^1S(Q^2)Y^3\otimes X^2(T^1V^1(Y_1^1y^1)_1hS(z^1T_1^2V^2(Y_1^1y^1)_2))_2g^2\\
&\qquad S(q^1)\alpha z^3T_{(2,2)}^2(V^3Y_2^1y^2h')_2S(Y^2y^3)_2G^2S(X^3T^3Q^1)\\
&\overset{(3.1.7),(3.2.13)}{\underset{(3.2.21),(3.2.1)}{=}} X^1(V^1Y_{(1,1)}^1y_1^1hS(z^1V^2Y_{(1,2)}^1y_2^1))_1g^1S(q^2)z^2(V^3Y_2^1y^2h'S(Y^2y^3))_1\\
&\qquad G^1S(Q^2)Y^3\otimes X^2(V^1Y_{(1,1)}^1y_1^1hS(z^1V^2Y_{(1,2)}^1y_2^1))_2g^2\\
&\qquad S(q^1)\alpha z^3(V^3Y_2^1y^2h'S(Y^2y^3))_2G^2S(X^3Q^1)\\
&\overset{(3.1.7)\times 2,(3.2.21)}{\underset{(3.2.1),(3.2.13)\times 2}{=}} X^1Y_1^1(V^1y_1^1hS(z^1V^2y_2^1))_1g^1S(q^2)z^2(V^3y^2h'S(y^3))_1G^1S(Q^2Y_2^2)Y^3\\
&\qquad \otimes X^2Y_2^1(V^1y_1^1hS(z^1V^2y_2^1))_2g^2S(q^1)\alpha z^3(V^3y^2h'S(y^3))_2\\
&\qquad G^2S(X^3Q^1Y_1^2)\\
&\overset{(3.2.19),(3.2.13)}{\underset{(3.2.5)}{=}} X^1Y_1^1(V^1y_1^1hS(V^2y_2^1))_1g^1\gamma^1(V^3y^2h'S(y^3))_1G^1S(Q^2Y_2^2)Y^3\\
&\qquad \otimes X^2Y_2^1(V^1y_1^1hS(V^2y_2^1))_2g^2\gamma^2(V^3y^2h'S(y^3))_2G^2S(X^3Q^1Y_1^2)\\
&\overset{(3.2.14)}{=} \lambda_{H_0}(V^1y_1^1hS(V^2y_2^1)\alpha V^3y^2h'S(y^3))\overset{(3.1.9),(3.2.1)}{=}\lambda_{H_0}(h\circ h').
\end{aligned}
$$

Now we prove the relation in (8.7.5). For this, recall that the unit for $H_0$ is $\beta$, and therefore, by (8.7.3), we have:

$$
\begin{aligned}
\lambda_{H_0}(1_{H_0}) &= (X^1\otimes X^2)\Delta(Y^1\beta S(Y^2))U[Y^3\otimes S(X^3)]\\
&\overset{(7.3.5)}{=} (X^1\otimes X^2)p_R[1\otimes S(X^3)]\overset{(3.2.19)}{=} 1_H\otimes 1_{H_0},
\end{aligned}
$$

as desired. □

We end this section by showing that $H_0$ is commutative as an algebra within ${}_H^H\mathscr{YD}$. We first prove some formulas; some of them are of independent interest.

**Lemma 8.39** *Let $H$ be a quasi-Hopf algebra with bijective antipode. Then we have*

$$\Delta(S(p^1))U(p^2\otimes 1_H)=f^{-1}, \tag{8.7.6}$$

$$S(g^1)\alpha g^2=S(\beta),\qquad f^1\beta S(f^2)=S(\alpha), \tag{8.7.7}$$

$$S(q_2^2X^3)f^1\otimes S(q^1X^1\beta S(q_1^2X^2)f^2)=(\mathrm{Id}_H\otimes S)(q_L), \tag{8.7.8}$$

*where $U$, $q_R=q^1\otimes q^2$ and $p_R=p^1\otimes p^2$, $q_L$, $f=f^1\otimes f^2$ and $f^{-1}=g^1\otimes g^2$ are the elements defined by* (7.3.1), (3.2.19), (3.2.20), (3.2.15) *and* (3.2.16), *respectively.*

*Proof* The relation (8.7.6) is an immediate consequence of (3.2.13) and (3.2.23), and the equalities in (8.7.7) were checked in the proof of Proposition 3.23; for the convenience of the reader we have recorded them one more time here.

It remains to prove that the relation (8.7.8) holds. By (3.2.26) we obtain:

$$(\mathrm{Id}_H \otimes \Delta)(q_R) = (1_H \otimes S^{-1}(x^3g^2) \otimes S^{-1}(x^2g^1))(q_R \otimes 1_H)$$
$$(\Delta \otimes \mathrm{Id}_H)(q_R)\Phi^{-1}(\mathrm{Id}_H \otimes \Delta)(\Delta(x^1)).$$

By using the formula in (5.5.17) we get that

$$(\mathrm{Id}_H \otimes \Delta)(q_R) = Q^1Y^1x^1_1 \otimes S^{-1}(x^3g^2)Q^2q^1Y^2_1x^1_{(2,1)} \otimes S^{-1}(Y^3x^2g^1)q^2Y^2_2x^1_{(2,2)}, \tag{8.7.9}$$

where $q_R = q^1 \otimes q^2 = Q^1 \otimes Q^2$, so we can now compute:

$$\begin{aligned}
&S(q^2_2X^3)f^1 \otimes S(q^1X^1\beta S(q^2_1X^2)f^2)\\
&\overset{(8.7.9)}{=} S(q^2Y^2_2x^1_{(2,2)}X^3)Y^3x^2 \otimes S(Q^1Y^1x^1_1X^1\beta S(Q^2q^1Y^2_1x^1_{(2,1)}X^2)x^3)\\
&\overset{(3.1.7)}{=} S(q^2Y^2_2X^3x^1_2)Y^3x^2 \otimes S(Q^1Y^1X^1x^1_{(1,1)}\beta S(Q^2q^1Y^2_1X^2x^1_{(1,2)})x^3)\\
&\overset{(3.2.1)}{=} S(q^2Y^2_2X^3x^1)Y^3x^2 \otimes S(Q^1Y^1X^1\beta S(Q^2q^1Y^2_1X^2)x^3)\\
&\overset{(3.1.9)}{=} S(q^2y^2X^3_1Y^2x^1)y^3X^3_2Y^3x^2 \otimes S(Q^1X^1Y^1_1\beta S(Q^2q^1y^1X^2Y^1_2)x^3)\\
&\overset{(3.2.1),(3.2.19)}{=} S(X^3_1x^1)\alpha X^3_2x^2 \otimes S(Q^1X^1\beta S(Q^2X^2)x^3)\\
&\overset{(3.2.1),(3.1.11)}{=} S(x^1)\alpha x^2 \otimes S(Q^1\beta S(Q^2)x^3)\\
&\overset{(3.2.19),(3.2.2)}{=} S(x^1)\alpha x^2 \otimes S(x^3),
\end{aligned}$$

as needed. □

We next show the commutativity of the algebra $H_0$ within the braided category ${}^H_H\mathscr{YD}$.

**Proposition 8.40** *Let $H$ be a quasi-Hopf algebra with bijective antipode. Then $H_0$ is commutative as an algebra in ${}^H_H\mathscr{YD}$, that is, for all $h, h' \in H$:*

$$h \circ h' = (h_{(-1)} \triangleright h') \circ h_{(0)}.$$

*Proof* For all $h, h' \in H$ we have:

$$\begin{aligned}
&(h_{(-1)} \triangleright h') \circ h_{(0)}\\
&\overset{(8.7.2)}{=} (X^1Y^1_1h_1g^1S(q^2Y^2_2)Y^3 \triangleright h') \circ X^2Y^1_2h_2g^2S(X^3q^1Y^2_1)\\
&\overset{(8.7.1),(4.1.9)}{=} Z^1X^1_1Y^1_{(1,1)}h_{(1,1)}g^1_1S(q^2Y^2_2)_1Y^3_1h'S(x^1Z^2X^1_2Y^1_{(1,2)}h_{(1,2)}g^1_2S(q^2Y^2_2)_2Y^3_2)\\
&\qquad \alpha x^2Z^3_1X^2Y^1_2h_2g^2S(x^3Z^3_2X^3q^1Y^2_1)\\
&\overset{(3.1.9),(3.2.1)}{=} Z^1Y^1_{(1,1)}h_{(1,1)}g^1_1S(q^2Y^2_2)_1Y^3_1h'S(Z^2Y^1_{(1,2)}h_{(1,2)}g^1_2S(q^2Y^2_2)_2Y^3_2)\\
&\qquad \alpha Z^3Y^1_2h_2g^2S(q^1Y^2_1)\\
&\overset{(3.2.13)}{=} Z^1[Y^1hS(Y^2)]_{(1,1)}g^1_1S(q^2)_1Y^3_1h'S(Z^2[Y^1hS(Y^2)]_{(1,2)}g^1_2S(q^2)_2Y^3_2)\\
&\qquad \alpha Z^3[Y^1hS(Y^2)]_2g^2S(q^1)\\
&\overset{(3.1.7),(3.2.1)}{=} Y^1hS(Y^2)Z^1g^1_1S(q^2)_1Y^3_1h'S(Z^2g^1_2S(q^2)_2Y^3_2)\alpha Z^3g^2S(q^1)\\
&\overset{(3.2.17)}{=} Y^1hS(Y^2)g^1S(X^3)f^1S(q^2)_1Y^3_1h'\\
&\qquad S(g^2_1G^1S(X^2)f^2S(q^2)_2Y^3_2)\alpha g^2_2G^2S(q^1X^1)
\end{aligned}$$

$$\begin{aligned}
&\overset{(3.2.1),(8.7.7)}{=} Y^1hS(X^3Y^2)f^1S(q^2)_1Y_1^3h'S(q^1X^1\beta S(X^2)f^2S(q^2)_2Y_2^3)\\
&\overset{(3.2.13)}{=} Y^1hS(q_2^2X^3Y^2)f^1Y_1^3h'S(q^1X^1\beta S(q_1^2X^2)f^2Y_2^3)\\
&\overset{(8.7.8)}{=} Y^1hS(x^1Y^2)\alpha x^2Y_1^3h'S(x^3Y_2^3)\overset{(4.1.9)}{=} h\circ h',
\end{aligned}$$

as required. □

**Remark 8.41** In general, $H_0$ does not have a coalgebra, bialgebra or Hopf algebra structure within ${}_H^H\mathscr{YD}$. Later on we shall see that it is possible to endow $H_0$ with such structures in the case when $H$ is a quasitriangular quasi-Hopf algebra.

If $H$ is a finite-dimensional quasi-Hopf algebra then $H_0$ becomes a left $D(H)$-module algebra, where $D(H)$ is the quantum double of $H$. This follows from the following three results, which we have already proved:

- $H_0$ is an algebra within the monoidal category of left Yetter–Drinfeld modules ${}_H^H\mathscr{YD}$ (see Proposition 8.38);
- there is a braided monoidal isomorphism between ${}_H^H\mathscr{YD}$ and ${}_H\mathscr{YD}^{H^{\rm in}}$ (see Theorem 8.16);
- the category ${}_H\mathscr{YD}^H$ is monoidally isomorphic to ${}_{D(H)}\mathscr{M}$ (see Proposition 8.32).

By using these isomorphisms we can transfer the algebra structure of $H_0$ from ${}_H^H\mathscr{YD}$ to ${}_{D(H)}\mathscr{M}$. In this way we associate to any finite-dimensional quasi-Hopf algebra $H$ a left $D(H)$-module algebra, called the Schrödinger representation of $H$.

Our next goal is to find the explicit structure of $H_0$ as a left $D(H)$-module algebra. For this, first we compute the algebra structure of $H_0$ in ${}_H\mathscr{YD}^H$, and second the left $D(H)$-module algebra structure of $H_0$, as desired.

**Proposition 8.42** *Let $H$ be a quasi-Hopf algebra with bijective antipode. Then $H_0$ is an algebra in the monoidal category ${}_H\mathscr{YD}^H$ with the left $H$-module structure defined in* (8.7.1) *and with the right $H$-coaction $\rho_{H_0}: H_0 \to H_0 \otimes H$ given for all $h \in H$ by*

$$\rho_{H_0}(h) = h_{(0)} \otimes h_{(1)} = x^1\tilde{q}^2y_2^2h_2g^2S(x^2y_1^3) \otimes x^3y_2^3S^{-1}(\tilde{q}^1y_1^2h_1g^1)y^1, \tag{8.7.10}$$

*where $q_L = \tilde{q}^1 \otimes \tilde{q}^2$ and $f^{-1} = g^1 \otimes g^2$ are the elements defined in* (3.2.20) *and* (3.2.16), *respectively. Moreover, if $H$ is finite dimensional, then $H_0$ is a left $D(H)$-module algebra via the action*

$$\begin{aligned}(\varphi \bowtie h) \rightharpoonup h' = \langle \varphi, q^2x^3y_2^3S^{-1}(\tilde{q}^1y_1^2(h \triangleright h')_1g^1)y^1\rangle&\\ q_1^1x^1\tilde{q}^2y_2^2(h \triangleright h')_2g^2S(q_2^1x^2y_1^3),&\end{aligned} \tag{8.7.11}$$

*for all $\varphi \in H^*$ and $h, h' \in H$, where $q_R = q^1 \otimes q^2$ is the element defined in* (3.2.19).

*Proof* The functor $G$ described in the proof of Theorem 8.14 is strong monoidal, so it carries algebras to algebras. Moreover, $G$ is a strict monoidal functor, so if $A$ is an algebra in ${}_H^H\mathscr{YD}$ then $G(A)$ is an algebra in ${}_H\mathscr{Y}D^H$ with the same multiplication and unit. Now, $G$ acts as identity on objects at the level of actions. Thus $G(H_0) = H_0$

as left $H$-module algebras, so we only have to show that that corresponding right $H$-action on $H_0$ through the functor $G$ is the one claimed in (8.7.10).

By (3.2.13) and (7.3.1) we get a second formula for the left $H$-coaction on $H_0$ defined in (8.7.2), namely

$$\lambda_{H_0}(h) = h_{(-1)} \otimes h_{(0)} = (X^1 \otimes X^2)\Delta(Y^1 hS(Y^2))U(Y^3 \otimes S(X^3)).$$

Now, for any $h \in H$ we calculate:

$$\begin{aligned}
\rho_{H_0}(h) &\overset{(8.2.26)}{=} \tilde{q}_1^2 Z^2 \triangleright (p^1 \triangleright h)_{\langle 0\rangle} \otimes \tilde{q}_2^2 Z^3 S^{-1}(\tilde{q}^1 Z^1 (p^1 \triangleright h)_{\langle -1\rangle} p^2) \\
&= \tilde{q}_1^2 Z^2 \triangleright [X^2 \left(Y^1 (p^1 \triangleright h) S(Y^2)\right)_2 U^2 S(X^3)] \\
&\qquad \otimes \tilde{q}_2^2 Z^3 S^{-1}(\tilde{q}^1 Z^1 X^1 \left(Y^1 (p^1 \triangleright h) S(Y^2)\right)_1 U^1 Y^3 p^2) \\
&\overset{(8.7.1),(5.5.16)}{=} \tilde{q}_1^2 Z^2 \triangleright [X^2 x_2^1 h_2 S(x_1^2 p^1)_2 U^2 S(X^3)] \\
&\qquad \otimes \tilde{q}_2^2 Z^3 x^3 S^{-1}(\tilde{q}^1 Z^1 X^1 x_1^1 h_1 S(x_1^2 p^1)_1 U^1 x_2^2 p^2) \\
&\overset{(8.7.1),(7.3.2)}{=} \tilde{q}_1^2 \triangleright [Z_1^2 X^2 x_2^1 h_2 S(p^1)_2 U^2 S(Z_2^2 X^3 x^2)] \\
&\qquad \otimes \tilde{q}_2^2 Z^3 x^3 S^{-1}(\tilde{q}^1 Z^1 X^1 x_1^1 h_1 S(p^1)_1 U^1 p^2) \\
&\overset{(3.1.9),(8.7.6),(8.7.1)}{=} \tilde{q}_{(1,1)}^2 x^1 X^2 h_2 g^2 S(\tilde{q}_{(1,2)}^2 x^2 X_1^3) \otimes \tilde{q}_2^2 x^3 X_2^3 S^{-1}(\tilde{q}^1 X^1 h_1 g^1) \\
&\overset{(3.1.7),(5.2.7)}{=} x^1 \tilde{q}^2 y_2^2 h_2 g^2 S(x^2 y_1^3) \otimes x^3 y_2^3 S^{-1}(\tilde{q}^1 y_1^2 h_1 g^1) y^1,
\end{aligned}$$

as needed. The last assertion is a consequence of (8.5.6) and (8.7.10), the details are left to the reader. □

## 8.8 Cross Products of Algebras in ${}_H\mathcal{M}$, ${}_H\mathcal{M}_H$, ${}_H^H\mathcal{YD}$

We recall from Section 2.1 that, if $\mathscr{C}$ is a monoidal category with associativity constraint $a_{U,V,W} : (U \otimes V) \otimes W \to U \otimes (V \otimes W)$ and unit $\underline{1}$, and $(A, \underline{m}_A, \underline{\eta}_A)$, $(B, \underline{m}_B, \underline{\eta}_B)$ are two algebras in $\mathscr{C}$, then a morphism $R : B \otimes A \to A \otimes B$ in $\mathscr{C}$ is called a twisting morphism between $A$ and $B$ if the following conditions hold:

$$R \circ (\mathrm{Id}_B \otimes \underline{\eta}_A) = \underline{\eta}_A \otimes \mathrm{Id}_B, \quad R \circ (\underline{\eta}_B \otimes \mathrm{Id}_A) = \mathrm{Id}_A \otimes \underline{\eta}_B, \tag{8.8.1}$$

$$\begin{aligned}
R \circ (\underline{m}_B \otimes \mathrm{Id}_A) &= (\mathrm{Id}_A \otimes \underline{m}_B) \circ a_{A,B,B} \circ (R \otimes \mathrm{Id}_B) \circ a_{B,A,B}^{-1} \\
&\quad \circ (\mathrm{Id}_B \otimes R) \circ a_{B,B,A},
\end{aligned} \tag{8.8.2}$$

$$\begin{aligned}
R \circ (\mathrm{Id}_B \otimes \underline{m}_A) &= (\underline{m}_A \otimes \mathrm{Id}_B) \circ a_{A,A,B}^{-1} \circ (\mathrm{Id}_A \otimes R) \circ a_{A,B,A} \\
&\quad \circ (R \otimes \mathrm{Id}_A) \circ a_{B,A,A}^{-1}.
\end{aligned} \tag{8.8.3}$$

Given such a twisting morphism, $A \otimes B$ becomes an algebra in $\mathscr{C}$, with multiplication

$$\begin{aligned}
\mu = &(\underline{m}_A \otimes \underline{m}_B) \circ a_{A,A,B\otimes B}^{-1} \circ (\mathrm{Id}_A \otimes a_{A,B,B}) \circ (\mathrm{Id}_A \otimes R \otimes \mathrm{Id}_B) \\
&\circ (\mathrm{Id}_A \otimes a_{B,A,B}^{-1}) \circ a_{A,B,A\otimes B}
\end{aligned} \tag{8.8.4}$$

and unit $\underline{\eta} = \underline{\eta}_A \otimes \underline{\eta}_B$. This algebra structure on $A \otimes B$ is denoted by $A \#_R B$ and is called the cross product of $A$ and $B$ afforded by the twisting morphism $R$. It has, moreover, the property that the morphisms $i_A := \mathrm{Id}_A \otimes \underline{\eta}_B : A \to A \#_R B$ and $i_B := \underline{\eta}_A \otimes \mathrm{Id}_B : B \to A \#_R B$ are morphisms of algebras in $\mathscr{C}$.

Also from Section 2.1 we recall that, if $c_{U,V}: U\otimes V\to V\otimes U$ is a braiding on $\mathscr{C}$, then for any two algebras $A$ and $B$ in $\mathscr{C}$, the morphism $R=c_{B,A}: B\otimes A\to A\otimes B$ is a twisting morphism, and in this case $A\#_R B$ is denoted by $A\otimes_+ B$ and is called the $c$-tensor product of $A$ and $B$.

**Proposition 8.43** *Let $H$ be a quasi-bialgebra, $\mathscr{A}$ an $H$-bimodule algebra and $A$ an algebra in ${}_H^H\mathscr{YD}$. We regard $A$ as an $H$-bimodule algebra with trivial right $H$-action. Define the linear map*

$$R: A\otimes\mathscr{A}\to\mathscr{A}\otimes A,\quad R(a\otimes\varphi)=a_{(-1)}\cdot\varphi\otimes a_{(0)},\ \forall\, a\in A,\ \varphi\in\mathscr{A}. \tag{8.8.5}$$

*Then $R$ is a twisting morphism between $\mathscr{A}$ and $A$ in the monoidal category ${}_H\mathscr{M}_H$. We will denote by $\mathscr{A}\odot A$ the $H$-bimodule algebra $\mathscr{A}\#_R A$, the cross product of $\mathscr{A}$ and $A$ in ${}_H\mathscr{M}_H$.*

*Proof* The fact that $R$ is right $H$-linear is obvious, and the fact that it is left $H$-linear follows immediately from (8.2.3), so $R$ is indeed a morphism in ${}_H\mathscr{M}_H$. The relations (8.8.1) follow immediately from (8.2.2) and (8.7.5). The relation (8.8.2) reduces to (8.7.4), while the relation (8.8.3) reduces to (8.2.1). □

**Remark 8.44** The explicit structure of $\mathscr{A}\odot A$ is the following: the unit is $1_{\mathscr{A}}\otimes 1_A$, the left $H$-action is $h\cdot(\varphi\otimes a)=h_1\cdot\varphi\otimes h_2\cdot a$, the right $H$-action is $(\varphi\otimes a)\cdot h=\varphi\cdot h\otimes a$, and the multiplication is (by (8.8.4)):

$$\begin{aligned}(\varphi\otimes a)(\varphi'\otimes a')&=(y^1X^1\cdot\varphi)(y^2Y^1(x^1X^2\cdot a)_{(-1)}x^2X^3_1\cdot\varphi')\\&\quad\otimes(y^3_1Y^2\cdot(x^1X^2\cdot a)_{(0)})(y^3_2Y^3x^3X^3_2\cdot a').\end{aligned}$$

We are mainly interested in the following particular case of Proposition 8.43.

**Corollary 8.45** *Let $H$ be a quasi-Hopf algebra with bijective antipode, $A$ a left $H$-module algebra and $H_0$ the algebra in ${}_H^H\mathscr{YD}$ as in Section 8.7. Define the map $R: H_0\otimes A\to A\otimes H_0$,*

$$R(h\otimes a)=h_{(-1)}\cdot a\otimes h_{(0)}=X^1Y^1_1h_1g^1S(q^2Y^2_2)Y^3\cdot a\otimes X^2Y^1_2h_2g^2S(X^3q^1Y^2_1). \tag{8.8.6}$$

*Then $R$ is a twisting morphism between $A$ and $H_0$ in the monoidal category ${}_H\mathscr{M}$. We will denote by $A\diamond H_0$ the left $H$-module algebra $A\#_R H_0$, the cross product of $A$ and $H_0$ in ${}_H\mathscr{M}$. Its unit is $1_A\otimes\beta$, the $H$-action is $h\cdot(a\otimes h')=h_1\cdot a\otimes h_2\triangleright h'$, and the multiplication is*

$$\begin{aligned}(a\otimes h)(a'\otimes h')&=(y^1X^1\cdot a)(y^2Y^1(x^1X^2\triangleright h)_{(-1)}x^2X^3_1\cdot a')\\&\quad\otimes(y^3_1Y^2\triangleright(x^1X^2\triangleright h)_{(0)})\circ(y^3_2Y^3x^3X^3_2\triangleright h'),\end{aligned}$$

*where $\circ$ is the multiplication of $H_0$.*

More generally, if $C$ is a left $H$-module algebra and $A$ is an algebra in ${}_H^H\mathscr{YD}$, then $C\odot A$ is a left $H$-module algebra, which will be denoted by $C\diamond A$.

Since the braiding $c$ in ${}_H^H\mathscr{YD}$ is given by $m\otimes n\mapsto m_{(-1)}\cdot n\otimes m_{(0)}$, we obtain the following:

**Corollary 8.46** *Let $H$ be a quasi-Hopf algebra with bijective antipode and $A$ an algebra in ${}_H^H\mathcal{YD}$. Then the left $H$-module algebra $A \diamond H_0$ is actually an algebra in ${}_H^H\mathcal{YD}$ and it coincides with the c-tensor product algebra $A \otimes_+ H_0$ in ${}_H^H\mathcal{YD}$.*

**Lemma 8.47** *Let $H$ be a quasi-Hopf algebra and $A$ a left $H$-module algebra. Consider the map $j : H \to A\#H$, $j(h) = 1_A\#h$, as well as the map $i_0 : A \to A\#H$ defined by (5.1.5). Then the following relation holds, for all $h \in H$ and $a \in A$:*

$$j(h) \circ i_0(a) = i_0(Y^1X_1^1h_1g^1S(T^2X_2^2)\alpha T^3X^3 \cdot a) \circ j(Y^2X_2^1h_2g^2S(Y^3T^1X_1^2)), \quad (8.8.7)$$

*where $f^{-1} = g^1 \otimes g^2$ is given by (3.2.16) and $\circ$ is the multiplication in $(A\#H)^j$.*

*Proof* We compute:

$$\begin{aligned}
& j(h) \circ i_0(a) \\
&= (1_A\#h) \circ (x^1 \cdot a\#x^2\beta S(x^3)) \\
&= (1_A\#X^1h)(1_A\#S(y^1X^2)\alpha y^2X_1^3)(x^1 \cdot a\#x^2\beta S(x^3))(1_A\#S(y^3X_2^3)) \\
&= X_1^1h_1S(y^1X^2)_1\alpha_1y_1^2X_{(1,1)}^3x^1 \cdot a \\
&\quad \#X_2^1h_2S(y^1X^2)_2\alpha_2y_2^2X_{(1,2)}^3x^2\beta S(y^3X_2^3x^3) \\
&\overset{(3.1.7),(3.2.14)}{\underset{(3.2.13)}{=}} X_1^1h_1g^1S(y_2^1X_2^2)\gamma^1y_1^2x^1X_1^3 \cdot a \\
&\quad \#X_2^1h_2g^2S(y_1^1X_1^2)\gamma^2y_2^2x^2X_{(2,1)}^3\beta S(X_{(2,2)}^3)S(y^3x^3) \\
&\overset{(3.2.1),(3.2.5)}{=} X_1^1h_1g^1S(T^2t_2^1y_2^1X_2^2)\alpha T^3t^2y_1^2x^1X^3 \cdot a \\
&\quad \#X_2^1h_2g^2S(T^1t_1^1y_1^1X_1^2)\alpha t^3y_2^2x^2\beta S(y^3x^3) \\
&\overset{(3.1.9),(3.1.7)}{\underset{(3.2.1)}{=}} X_1^1h_1g^1S(T^2X_2^2)\alpha T^3X^3 \cdot a \\
&\quad \#X_2^1h_2g^2S(T^1X_1^2)S(y^1)\alpha y^2\beta S(y^3) \\
&\overset{(3.2.2)}{=} X_1^1h_1g^1S(T^2X_2^2)\alpha T^3X^3 \cdot a\#X_2^1h_2g^2S(T^1X_1^2) \\
&\overset{(5.1.6)}{=} i_0(Y^1X_1^1h_1g^1S(T^2X_2^2)\alpha T^3X^3 \cdot a) \circ j(Y^2X_2^1h_2g^2S(Y^3T^1X_1^2)),
\end{aligned}$$

finishing the proof. □

**Proposition 8.48** *Let $H$ be a quasi-Hopf algebra with bijective antipode and $A$ a left $H$-module algebra. Then the linear map*

$$\Pi : A \diamond H_0 \to (A\#H)^j, \quad \Pi(a \otimes h) = x^1 \cdot a\#x^2hS(x^3), \ \forall\, a \in A,\ h \in H,$$

*is an isomorphism of left $H$-module algebras.*

*Proof* Note first that $\Pi$ is bijective, with inverse given by $\Pi^{-1}(a\#h) = X^1 \cdot a \otimes X^2hS(X^3)$. For proving that $\Pi$ is a morphism of left $H$-module algebras, we will use the Universal Property of $A \diamond H_0$ as a cross product algebra in ${}_H\mathcal{M}$ (see Proposition 2.9). We know that $i_0 : A \to (A\#H)^j$ is a morphism of left $H$-module algebras, and it is easy to see that $j : H_0 \to (A\#H)^j$ is also a morphism of left $H$-module algebras. Moreover, the relation (2.1.5) reduces in this case exactly to (8.8.7). We can thus use the Universal Property of $A \diamond H_0$, which provides a morphism $w : A \diamond H_0 \to (A\#H)^j$ of left $H$-module algebras, which is moreover given by $w(a \otimes h) = i_0(a) \circ j(h)$. Relation (5.1.6) shows that actually we have $w = \Pi$. □

If we assume, moreover, that $A$ is an algebra in ${}_H^H\mathcal{YD}$, in which case $A \diamond H_0$ becomes the $c$-tensor product algebra $A \otimes_+ H_0$ in ${}_H^H\mathcal{YD}$, then we intend to show that $(A\#H)^j$ also becomes an algebra in ${}_H^H\mathcal{YD}$ in a natural way and that $\Pi$ becomes an isomorphism of algebras in ${}_H^H\mathcal{YD}$.

The next result is a generalization of the fact that $H_0$ is an algebra in ${}_H^H\mathcal{YD}$; the proof is similar to the one given for $H_0$ and will be omitted.

**Proposition 8.49** *If $H$ is a quasi-Hopf algebra with bijective antipode, $(\mathfrak{B}, \lambda, \Phi_\lambda)$ a left $H$-comodule algebra and $v : H \to \mathfrak{B}$ a morphism of left $H$-comodule algebras, then $\mathfrak{B}^v$ becomes an algebra in ${}_H^H\mathcal{YD}$ with coaction*

$$\mathfrak{B}^v \to H \otimes \mathfrak{B}^v, \quad \mathfrak{b} \mapsto X^1 Y_1^1 \mathfrak{b}_{[-1]} g^1 S(q^2 Y_2^2) Y^3 \otimes v(X^2 Y_2^1) \mathfrak{b}_{[0]} v(g^2 S(X^3 q^1 Y_1^2)),$$

*where $f^{-1} = g^1 \otimes g^2$ is given by* (3.2.16) *and $q_R = q^1 \otimes q^2 = Z^1 \otimes S^{-1}(\alpha Z^3) Z^2$. Moreover, the map $v : H_0 \to \mathfrak{B}^v$ is a morphism of algebras in ${}_H^H\mathcal{YD}$.*

The next result is a part of Proposition 9.13 from Chapter 9, so its proof will be given there.

**Proposition 8.50** *Let $H$ be a quasi-bialgebra and $A$ an algebra in ${}_H^H\mathcal{YD}$, with coaction denoted by $A \to H \otimes A$, $a \mapsto a_{(-1)} \otimes a_{(0)}$. Then $(A\#H, \lambda, \Phi_\lambda)$ is a left $H$-comodule algebra, with structures:*

$$\lambda : A\#H \to H \otimes (A\#H), \quad \lambda(a\#h) = T^1 (t^1 \cdot a)_{(-1)} t^2 h_1 \otimes (T^2 \cdot (t^1 \cdot a)_{(0)} \# T^3 t^3 h_2),$$
$$\Phi_\lambda = X^1 \otimes X^2 \otimes (1_A \# X^3) \in H \otimes H \otimes (A\#H),$$

*for all $a \in A$ and $h \in H$.*

As a consequence of these results, we obtain:

**Corollary 8.51** *Let $H$ be a quasi-Hopf algebra with bijective antipode and $A$ an algebra in ${}_H^H\mathcal{YD}$. Then $(A\#H)^j$ becomes an algebra in ${}_H^H\mathcal{YD}$, via the left $H$-coaction $\lambda_{(A\#H)^j} : (A\#H)^j \to H \otimes (A\#H)^j$, given by*

$$\lambda_{(A\#H)^j}(a\#h) = X^1 Y_1^1 T^1 (t^1 \cdot a)_{(-1)} t^2 h_1 g^1 S(q^2 Y_2^2) Y^3 \otimes [X_1^2 Y_{(2,1)}^1 T^2 \cdot (t^1 \cdot a)_{(0)}$$
$$\# X_2^2 Y_{(2,2)}^1 T^3 t^3 h_2 g^2 S(X^3 q^1 Y_1^2)],$$

*and the map $j : H_0 \to (A\#H)^j$ is a morphism of algebras in ${}_H^H\mathcal{YD}$.*

**Lemma 8.52** *Let $H$ be a quasi-Hopf algebra with bijective antipode and $A$ an algebra in ${}_H^H\mathcal{YD}$. Then $i_0$ is a morphism of algebras in ${}_H^H\mathcal{YD}$ from $A$ to $(A\#H)^j$.*

*Proof* We already know that $i_0$ is a morphism of left $H$-module algebras from $A$ to $(A\#H)^j$, so the only thing left to prove is that $\lambda_{(A\#H)^j} \circ i_0 = (\mathrm{Id}_H \otimes i_0) \circ \lambda_A$.

Now, by denoting $p_R = P^1 \otimes P^2$ another copy of $p_R$, we see that

$$
\begin{aligned}
&(\lambda_{(A\#H)^j} \circ i_0)(a)\\
&\quad = \lambda_{(A\#H)^j}(p^1 \cdot a\#p^2)\\
&\quad = X^1 Y_1^1 T^1 (t^1 p^1 \cdot a)_{(-1)} t^2 p_1^2 g^1 S(q^2 Y_2^2) Y^3\\
&\qquad \otimes [X_1^2 Y_{(2,1)}^1 T^2 \cdot (t^1 p^1 \cdot a)_{(0)} \# X_2^2 Y_{(2,2)}^1 T^3 t^3 p_2^2 g^2 S(X^3 q^1 Y_1^2)]\\
&\quad \overset{(3.2.25)}{=} X^1 Y_1^1 T^1 (Z_{(1,1)}^1 p_1^1 P^1 \cdot a)_{(-1)} Z_{(1,2)}^1 p_2^1 P^2 S(q^2 Y_2^2 Z^3) Y^3\\
&\qquad \otimes [X_1^2 Y_{(2,1)}^1 T^2 \cdot (Z_{(1,1)}^1 p_1^1 P^1 \cdot a)_{(0)} \# X_2^2 Y_{(2,2)}^1 T^3 Z_2^1 p^2 S(X^3 q^1 Y_1^2 Z^2)]\\
&\quad \overset{(8.2.3),(3.1.7)}{=} X^1 T^1 Y_{(1,1)}^1 Z_{(1,1)}^1 p_1^1 (P^1 \cdot a)_{(-1)} P^2 S(q^2 Y_2^2 Z^3) Y^3\\
&\qquad \otimes [X_1^2 T^2 Y_{(1,2)}^1 Z_{(1,2)}^1 p_2^1 \cdot (P^1 \cdot a)_{(0)} \# X_2^2 T^3 Y_2^1 Z_2^1 p^2 S(X^3 q^1 Y_1^2 Z^2)]\\
&\quad \overset{(5.5.17)}{=} X^1 T^1 q_{(1,1,1)}^1 p_1^1 (P^1 \cdot a)_{(-1)} P^2 S(q^2)\\
&\qquad \otimes [X_1^2 T^2 q_{(1,1,2)}^1 p_2^1 \cdot (P^1 \cdot a)_{(0)} \# X_2^2 T^3 q_{(1,2)}^1 p^2 S(q_2^1) S(X^3)]\\
&\quad \overset{(3.2.21)}{=} X^1 T^1 p_1^1 q_1^1 (P^1 \cdot a)_{(-1)} P^2 S(q^2)\\
&\qquad \otimes [X_1^2 T^2 p_2^1 q_2^1 \cdot (P^1 \cdot a)_{(0)} \# X_2^2 T^3 p^2 S(X^3)]\\
&\quad \overset{(8.2.3),(5.5.16)}{=} X^1 y^1 (q_1^1 P^1 \cdot a)_{(-1)} q_2^1 P^2 S(q^2)\\
&\qquad \otimes [X_1^2 y_1^2 p^1 \cdot (q_1^1 P^1 \cdot a)_{(0)} \# X_2^2 y_2^2 p^2 S(X^3 y^3)]\\
&\quad \overset{(3.2.23)}{=} a_{(-1)} \otimes (p^1 \cdot a_{(0)} \# p^2) = ((\mathrm{Id}_H \otimes i_0) \circ \lambda_A)(a),
\end{aligned}
$$

finishing the proof. □

**Theorem 8.53** *Let $H$ be a quasi-Hopf algebra with bijective antipode and $A$ an algebra in ${}_H^H\mathcal{YD}$. Then the map $\Pi : A \otimes_+ H_0 \to (A\#H)^j$, $\Pi(a \otimes h) = x^1 \cdot a\#x^2 h S(x^3)$, is an isomorphism of algebras in ${}_H^H\mathcal{YD}$.*

*Proof* We proved that $j : H_0 \to (A\#H)^j$ and $i_0 : A \to (A\#H)^j$ are morphisms of algebras in ${}_H^H\mathcal{YD}$, and together with the commutation relation (8.8.7) this allows us to apply the Universal Property of the cross product algebra $A \otimes_+ H_0 = A\#_R H_0$ in the category ${}_H^H\mathcal{YD}$, obtaining thus a morphism of algebras in ${}_H^H\mathcal{YD}$ between $A \otimes_+ H_0$ and $(A\#H)^j$, say $\omega$, which has to be given by $\omega(a \otimes h) = i_0(a) \circ j(h)$. It turns out that $\omega$ coincides with the map $\Pi$, finishing the proof. □

## 8.9 Notes

The center construction is due to Drinfeld (unpublished), Joyal and Street [120] and Majid [145]. The connection between the left and right center constructions has been observed in [51]. The category of Yetter–Drinfeld modules over a quasi-bialgebra $H$ was introduced by Majid in [147] by computing the center of the monoidal category ${}_H\mathcal{M}$. The aim in [147] was to define the quantum double $D(H)$ of a quasi-Hopf algebra $H$ by an implicit Tannaka–Krein reconstruction procedure, in such a way that ${}_{D(H)}\mathcal{M}$ identifies as braided monoidal category with the category of Yetter–Drinfeld modules over $H$. This goal was achieved afterwards in [107, 108, 195, 64]; an explicit construction of the quantum double (as a diagonal crossed product) was given for the first time by Hausser and Nill in [107, 108], an isomorphic copy of it being constructed afterwards by Schauenburg in [195]. That in the finite-dimensional case Yetter–Drinfeld modules are modules over the quantum double algebra $D(H)$ is from [108, 60].

The rigidity of the center is taken from [207, 195]; the rigid monoidal structure of the category of Yetter–Drinfeld modules over a quasi-Hopf algebra is taken from [51] as well as the suitable braided monoidal isomorphisms between the four types of categories of Yetter–Drinfeld modules. The content of Section 8.7 is taken from [54, 50], and that of Section 8.8 is taken from [8].

The quasi-Hopf algebra $D^{\omega}(G)$ was introduced in [77] for physical reasons, in relation to the work of Dijkgraaf and Witten on topological field theories. In [9], Altschuler and Coste proved that $D^{\omega}(G)$ is, moreover, ribbon and used it to construct invariants for knots and links. The quasi-Hopf algebra $D^{\omega}(H)$ was introduced in [56] as a generalization of $D^{\omega}(G)$. The quasi-Hopf algebra $H^*_{\omega}$ was introduced in [177], where it was also proved that the center of ${}_{H^*_{\omega}}\mathcal{M}$ is braided equivalent to ${}_{D^{\omega}(H)}\mathcal{M}$, implying that $D^{\omega}(H)$ should be the quantum double of $H^*_{\omega}$. The explicit isomorphism between $D^{\omega}(H)$ and $D(H^*_{\omega})$ constructed in Theorem 8.34 is from [57].

# 9

# Two-sided Two-cosided Hopf Modules

We introduce the category of two-sided two-cosided Hopf modules over a quasi-bialgebra $H$ and show that it is braided monoidally equivalent to the category of Yetter–Drinfeld modules over $H$, provided that $H$ is a quasi-Hopf algebra. We use this equivalence to obtain certain structure theorems for bicomodule algebras and bimodule coalgebras over $H$. Finally, we show that a Hopf algebra within this braided monoidal category identifies with a quasi-Hopf algebra with projection.

## 9.1 Two-sided Two-cosided Hopf Modules

Throughout this section, $H$ is a quasi-bialgebra or a quasi-Hopf algebra, $\mathbb{A}$ is an $H$-bicomodule algebra and $C$ is an $H$-bimodule coalgebra. Recall that ${}_H\mathscr{M}_H$ is a monoidal category and that the underlying quasi-coalgebra structure of $H$ provides a monoidal coalgebra structure for $H$ in ${}_H\mathscr{M}_H$. In Section 6.1 we defined the category ${}_H\mathscr{M}_H^H$ as the category of right $H$-corepresentations in ${}_H\mathscr{M}_H$. Similarly, we can define ${}_H^H\mathscr{M}_H^H$ as the category of $H$-bicomodules within ${}_H\mathscr{M}_H$, and this leads to the notion of two-sided two-cosided Hopf module over $H$. We can generalize this last concept as follows.

**Definition 9.1** Let $H$ be a quasi-bialgebra, $(\mathbb{A},\lambda,\rho,\Phi_\lambda,\Phi_\rho,\Phi_{\lambda,\rho})$ an $H$-bicomodule algebra and $C$ an $H$-bimodule coalgebra. A two-sided two-cosided $(H,\mathbb{A},C)$-Hopf module is a $k$-vector space $M$ with the following additional structure:

- $M$ is an $(H,\mathbb{A})$-two-sided Hopf module, that is, $M \in {}_H\mathscr{M}_{\mathbb{A}}^H$; as usual, we denote the left $H$-action and the right $\mathbb{A}$-action by $\cdot$, and we write $\rho_M^H(m) = m_{(0)} \otimes m_{(1)}$ for the right $H$-coaction on $m \in M$;
- we have a $k$-linear map $\lambda_M^C : M \to C \otimes M$, $\lambda_M^C(m) = m_{\{-1\}} \otimes m_{\{0\}}$, called the left $C$-coaction on $M$, such that, for all $m \in M$, $\underline{\varepsilon}(m_{\{-1\}})m_{\{0\}} = m$ and

$$\Phi \cdot (\underline{\Delta} \otimes \mathrm{Id}_M)(\lambda_M^C(m)) = (\mathrm{Id}_C \otimes \lambda_M^C)(\lambda_M^C(m)) \cdot \Phi_\lambda; \tag{9.1.1}$$

- $M$ is a $(C,H)$-"bicomodule," in the sense that, for all $m \in M$,

$$\Phi \cdot (\lambda_M^C \otimes \mathrm{Id}_H)(\rho_M^H(m)) = (\mathrm{Id}_C \otimes \rho_M^H)(\lambda_M^C(m)) \cdot \Phi_{\lambda,\rho}; \tag{9.1.2}$$

- the following compatibility relations hold:

$$\lambda_M^C(h\cdot m) = h_1\cdot m_{\{-1\}}\otimes h_2\cdot m_{\{0\}}, \tag{9.1.3}$$
$$\lambda_M^C(m\cdot u) = m_{\{-1\}}\cdot u_{[-1]}\otimes m_{\{0\}}\cdot u_{[0]}, \tag{9.1.4}$$

for all $h\in H$, $m\in M$ and $u\in \mathbb{A}$.

${}_H^C\mathscr{M}_{\mathbb{A}}^H$ will be the category of two-sided two-cosided Hopf modules and maps preserving the actions by $H$ and $\mathbb{A}$ and the coactions by $H$ and $C$.

When $C=\mathbb{A}=H$ we have that ${}_H^H\mathscr{M}_H^H$ is just the category of $H$-bicomodules within ${}_H\mathscr{M}_H$, as we have already explained.

We conclude this section by constructing a functor between the categories ${}_H\mathscr{M}_{\mathbb{A}}^H$ and ${}_H^C\mathscr{M}_{\mathbb{A}}^H$. This will allow us to construct two-sided two-cosided Hopf modules from right quasi-Hopf $(H,\mathbb{A},C)$-modules.

**Proposition 9.2** *Let $H$ be a quasi-bialgebra, $\mathbb{A}$ an $H$-bicomodule algebra and $C$ an $H$-bimodule coalgebra. Then for any right quasi-Hopf $(H,\mathbb{A},C)$-Hopf module $M$ we can define on the k-vector space $C\otimes M$ a two-sided two-cosided $(H,\mathbb{A},C)$-Hopf module structure as follows:*

- *$C\otimes M$ is an $(H,\mathbb{A})$-bimodule via the actions given for all $h\in H$, $c\in C$, $m\in M$ and $u\in\mathbb{A}$ by*

$$h\cdot(c\otimes m)\cdot u := h_1\cdot c\cdot u_{[-1]}\otimes h_2\cdot m\cdot u_{[0]};$$

- *the right $H$-coaction on $M$ is defined for all $c\in C$ and $m\in M$ by*

$$\rho_{C\otimes M}^H(c\otimes m) := (x^1\cdot c\cdot\Theta^1\otimes x^2\cdot m_{(0)}\cdot\Theta^2)\otimes x^3 m_{(1)}\Theta^3;$$

- *the left $C$-coaction on $M$ is defined for all $c\in C$ and $m\in M$ by*

$$\lambda_{C\otimes M}^C(c\otimes m) := X^1\cdot c_{\underline{1}}\cdot\tilde{x}_\lambda^1\otimes(X^2\cdot c_{\underline{2}}\cdot\tilde{x}_\lambda^2\otimes X^3\cdot m\cdot\tilde{x}_\lambda^3).$$

*In this way we have a well-defined functor $C\otimes - :\ {}_H\mathscr{M}_{\mathbb{A}}^H\to{}_H^C\mathscr{M}_{\mathbb{A}}^H$. This functor sends a morphism $\vartheta$ to $\mathrm{Id}_C\otimes\vartheta$.*

*Proof* This is a straightforward computation. □

**Example 9.3** Let $H$ be a quasi-Hopf algebra, $\mathbb{A}$ an $H$-bicomodule algebra and $C$ an $H$-bimodule coalgebra. Then $C\otimes\mathscr{V} = C\otimes\mathbb{A}\otimes C$ and $C\otimes\mathscr{U} = C\otimes C\otimes\mathbb{A}$ are isomorphic two-sided two-cosided $(H,\mathbb{A},C)$-Hopf modules, where $\mathscr{V}$ and $\mathscr{U}$ are the isomorphic quasi-Hopf $(H,\mathbb{A})$-bimodules considered in Examples 6.5.

The monoidal structure of ${}_H\mathscr{M}_H^H$ in Proposition 6.8 gives rise to a monoidal structure on ${}_H^H\mathscr{M}_H^H$. The proof of the next result is immediate.

**Proposition 9.4** *Let $H$ be a quasi-bialgebra and $M,N$ two objects in ${}_H^H\mathscr{M}_H^H$. Then $M\otimes_H N$ is a two-sided two-cosided Hopf module over $H$ with the structure given by* (6.1.14), (6.1.15) *and*

$$\lambda_{M\otimes_H N} : M\otimes_H N\ni m\otimes_H n\mapsto m_{\{-1\}}n_{\{-1\}}\otimes m_{\{0\}}\otimes_H n_{\{0\}}\in H\otimes M\otimes_H N, \tag{9.1.5}$$

*for all $m \in M$, $n \in N$ and $h, h' \in H$. In this way the category ${}_H^H\mathscr{M}_H^H$ becomes a strict monoidal category with unit object $H = H_\varepsilon$, considered in ${}_H^H\mathscr{M}_H^H$ with the structure as in Example 6.7, completed with $\lambda_H = \Delta$.*

## 9.2 Two-sided Two-cosided Hopf Modules versus Yetter–Drinfeld Modules

Our goal is to prove that the functors defined in Theorem 6.16 restrict to a category equivalence between two-sided two-cosided Hopf modules and Yetter–Drinfeld modules.

Roughly speaking, we can say that ${}_H^C\mathscr{M}_{\mathbb{A}}^H$ is obtained from ${}_H\mathscr{M}_{\mathbb{A}}^H$ by "adding" a left $C$-coaction plus compatibilities. Thus the equivalence between the categories ${}_H\mathscr{M}_{\mathbb{A}}^H$ and $\mathscr{M}_{\mathbb{A}}$ should extend "somehow" to an equivalence between the category ${}_H^C\mathscr{M}_{\mathbb{A}}^H$ and a category whose objects are right $\mathbb{A}$-modules, a sort of left $C$-comodules plus a compatibility relation between these two structures. It turns out that the category we are looking for is just ${}^C\mathscr{YD}(H)_{\mathbb{A}}$, the category of generalized right-left Yetter–Drinfeld modules associated to the datum $(H, \mathbb{A}, C)$. It can be introduced as ${}_{\mathbb{A}^{\mathrm{op}}}\mathscr{YD}(H^{\mathrm{op,cop}})^{C^{\mathrm{cop}}}$. More explicitly:

**Definition 9.5** Let $H$ be a quasi-bialgebra, $C$ an $H$-bimodule coalgebra and $\mathbb{A}$ an $H$-bicomodule algebra. A right-left $(H, \mathbb{A}, C)$-Yetter–Drinfeld module is a right $\mathbb{A}$-module together with a $k$-linear map $\lambda_M : M \to C \otimes M$ (we denote $\lambda_M(m) = m_{\{-1\}} \otimes m_{\{0\}}$ for all $m \in M$) such that the following relations hold:

$$\begin{aligned} & m_{\{-1\}} \cdot \theta^1 \otimes \theta^3 \cdot (m_{\{0\}} \cdot \theta^2)_{\{-1\}} \otimes (m_{\{0\}} \cdot \theta^2)_{\{0\}} \\ & \qquad = \tilde{x}_\rho^2 \cdot (m \cdot \tilde{x}_\rho^1)_{\{-1\}_{\underline{1}}} \cdot \tilde{x}_\lambda^1 \otimes \tilde{x}_\rho^3 \cdot (m \cdot \tilde{x}_\rho^1)_{\{-1\}_{\underline{2}}} \cdot \tilde{x}_\lambda^2 \otimes (m \cdot \tilde{x}_\rho^1)_{\{0\}} \cdot \tilde{x}_\lambda^3, \end{aligned} \tag{9.2.1}$$

$$\underline{\varepsilon}(m_{\{-1\}}) m_{\{0\}} = m, \tag{9.2.2}$$

$$m_{\{-1\}} \cdot u_{[-1]} \otimes m_{\{0\}} \cdot u_{[0]} = u_{\langle 1\rangle} \cdot (m \cdot u_{\langle 0\rangle})_{\{-1\}} \otimes (m \cdot u_{\langle 0\rangle})_{\{0\}}, \tag{9.2.3}$$

for all $m \in M$ and $u \in \mathbb{A}$. ${}^C\mathscr{YD}(H)_{\mathbb{A}}$ will be the category of right-left Yetter–Drinfeld modules and maps preserving the actions by $\mathbb{A}$ and coactions by $C$.

In order to prove the claimed equivalence of categories we first need some lemmas.

**Lemma 9.6** *Let $H$ be a quasi-bialgebra, $C$ an $H$-bimodule coalgebra and $\mathbb{A}$ an $H$-bicomodule algebra. We have a functor $\mathbf{F} : {}^C\mathscr{YD}(H)_{\mathbb{A}} \to {}_H^C\mathscr{M}_{\mathbb{A}}^H$, defined for any $M \in {}^C\mathscr{YD}(H)_{\mathbb{A}}$ by $\mathbf{F}(M) = M \otimes H$, where the structure of $M \otimes H$ in ${}_H^C\mathscr{M}_{\mathbb{A}}^H$ is:*

- *$M \otimes H$ is an $(H, \mathbb{A})$-bimodule via (for all $h, h' \in H$, $m \in M$ and $u \in \mathbb{A}$)*

$$h \cdot (m \otimes h') \cdot u = m \cdot u_{\langle 0\rangle} \otimes hh' u_{\langle 1\rangle}; \tag{9.2.4}$$

- *for all $m \in M$ and $h \in H$, the left $C$-coaction and the right $H$-coaction on $M \otimes H$ are given by*

$$\lambda_{M\otimes H}^C(m \otimes h) = h_1 \tilde{X}_\rho^2 \cdot (m \cdot \tilde{X}_\rho^1)_{\{-1\}} \cdot \theta^1 \otimes (m \cdot \tilde{X}_\rho^1)_{\{0\}} \cdot \theta^2 \otimes h_2 \tilde{X}_\rho^3 \theta^3, \tag{9.2.5}$$

$$\rho_{M\otimes H}^H(m \otimes h) = m \cdot \tilde{X}_\rho^1 \otimes h_1 \tilde{X}_\rho^2 \otimes h_2 \tilde{X}_\rho^3. \tag{9.2.6}$$

*If* $\chi : M \to N$ *is a morphism in* ${}^C\mathscr{YD}(H)_{\mathbb{A}}$ *then* $\mathbf{F}(\chi) = \chi \otimes \mathrm{Id}_H$.

*Proof* The functor $\mathbf{F}$ in the statement is the functor $F$ defined in the proof of Theorem 6.16, restricted to the category of right-left Yetter–Drinfeld modules. To see this, compare (6.3.3), (6.3.4) to (9.2.4), (9.2.6). Now, we will check that (9.1.1) holds for our structures. For all $m \in M$ and $h \in H$ we have

$$\begin{aligned}
&\Phi \cdot (\underline{\Delta} \otimes \mathrm{Id}_{M\otimes H})(\lambda^C_{M\otimes H}(m \otimes h)) \cdot \Phi_\lambda^{-1} \\
&\overset{(9.2.5)}{\underset{(9.2.4)}{=}} X^1 h_{(1,1)}(\tilde{X}^2_\rho)_1 \cdot (m \cdot \tilde{X}^1_\rho)_{\{-1\}_{\underline{1}}} \cdot \theta^1_1 \tilde{x}^1_\lambda \otimes X^2 h_{(1,2)}(\tilde{X}^2_\rho)_2 \\
&\qquad \cdot (m \cdot \tilde{X}^1_\rho)_{\{-1\}_{\underline{2}}} \cdot \theta^1_2 \tilde{x}^2_\lambda \otimes (m \cdot \tilde{X}^1_\rho)_{\{0\}} \cdot \theta^2 (\tilde{x}^3_\lambda)_{\langle 0\rangle} \otimes X^3 h_2 \tilde{X}^3_\rho \theta^3 (\tilde{x}^3_\lambda)_{\langle 1\rangle} \\
&\overset{(3.1.7),(4.3.2)}{\underset{(4.4.2)}{=}} h_1 \tilde{X}^2_\rho (\tilde{Y}^1_\rho)_{\langle 1\rangle} \tilde{x}^2_\rho \cdot (m \cdot \tilde{x}^1_\rho (\tilde{Y}^1_\rho)_{\langle 0\rangle} \tilde{x}^1_\rho)_{\{-1\}_{\underline{1}}} \cdot \tilde{x}^1_\lambda \theta^1 \\
&\qquad \otimes h_{(2,1)}(\tilde{X}^3_\rho)_1 \tilde{Y}^2_\rho \tilde{x}^3_\rho \cdot (m \cdot \tilde{X}^1_\rho (\tilde{Y}^1_\rho)_{\langle 0\rangle} \tilde{x}^1_\rho)_{\{-1\}_{\underline{2}}} \cdot \tilde{x}^2_\lambda \theta^2_{[-1]} \overline{\theta}^1 \\
&\qquad \otimes (m \cdot \tilde{X}^1_\rho (\tilde{Y}^1_\rho)_{\langle 0\rangle} \tilde{x}^1_\rho)_{\{0\}} \cdot \tilde{x}^3_\lambda \theta^2_{[0]} \overline{\theta}^2 \otimes h_{(2,2)} X^3 (\tilde{X}^3_\rho)_2 \tilde{Y}^3_\rho \theta^3 \overline{\theta}^3 \\
&\overset{(9.2.1)}{\underset{(9.2.3)}{=}} h_1 \tilde{X}^2_\rho \cdot (m \cdot \tilde{X}^1_\rho)_{\{-1\}} \cdot (\tilde{Y}^1_\rho)_{[-1]} \overline{\overline{\theta}}^1 \theta^1 \\
&\qquad \otimes h_{(2,1)}(\tilde{X}^3_\rho)_1 \tilde{Y}^2_\rho \overline{\overline{\theta}}^3 \cdot ((m \cdot \tilde{X}^1_\rho)_{\{0\}} \cdot (\tilde{Y}^1_\rho)_{[0]} \overline{\overline{\theta}}^2)_{\{-1\}} \cdot \theta^2_{[-1]} \overline{\theta}^1 \\
&\qquad \otimes ((m \cdot \tilde{X}^1_\rho)_{\{0\}} \cdot (\tilde{Y}^1_\rho)_{[0]} \overline{\overline{\theta}}^2)_{\{0\}} \cdot \theta^2_{[0]} \overline{\theta}^2 \otimes h_{(2,2)} X^3 (\tilde{X}^3_\rho)_2 \tilde{Y}^3_\rho \theta^3 \overline{\theta}^3 \\
&\overset{(9.2.3),(4.4.3)}{\underset{(9.2.5)}{=}} (\mathrm{Id}_C \otimes \lambda^C_{M\otimes H})(\lambda^C_{M\otimes H}(m \otimes h)),
\end{aligned}$$

as required. The relation (9.1.2) can be proved in a similar way. The relation (9.1.3) follows directly from the definitions, and the relation (9.1.4) follows from (4.3.1), (9.2.3), (4.4.1) and (9.2.4). □

The converse of the above result is also true if $H$ is a quasi-Hopf algebra.

**Lemma 9.7** *Let $H$ be a quasi-Hopf algebra with bijective antipode, $C$ an $H$-bimodule coalgebra and $\mathbb{A}$ an $H$-bicomodule algebra. Then we have a functor* $\mathbf{G} : {}^C_H\mathscr{M}^H_{\mathbb{A}} \to {}^C\mathscr{YD}(H)_{\mathbb{A}}$. *If $M$ is a two-sided two-cosided Hopf module then* $G(M) = M^{\underline{\mathrm{co}(H)}}$, *where* $M^{\underline{\mathrm{co}(H)}}$ *is viewed as a Yetter–Drinfeld module via the structure*

$$m \leftarrow u = S^{-1}(u_{\langle 1\rangle}) \cdot m \cdot u_{\langle 0\rangle}, \tag{9.2.7}$$

$$\begin{aligned}
\lambda^C_{\mathbf{G}(M)}(m) := m^{\{-1\}} \otimes m^{\{0\}} = \tilde{x}^3_\rho S^{-1}(f^2(\tilde{x}^2_\rho)_2 p^2) \cdot m_{\{-1\}} \cdot (\tilde{x}^1_\rho)_{[-1]} \theta^1 \\
\otimes S^{-1}(f^1(\tilde{x}^2_\rho)_1 p^1 \theta^3) \cdot m_{\{0\}} \cdot (\tilde{x}^1_\rho)_{[0]} \theta^2,
\end{aligned} \tag{9.2.8}$$

*where* $p_R = p^1 \otimes p^2$, $f = f^1 \otimes f^2$ *are the elements defined in* (3.2.19), (3.2.15).

*Proof* If $M \in {}^C_H\mathscr{M}^H_{\mathbb{A}}$ then $M$ is a two-sided $(H, \mathbb{A})$-Hopf module, so it makes sense to consider the set of coinvariants (of the first type) of $M$, $M^{\underline{\mathrm{co}(H)}}$. From the proof of Theorem 6.16 we know that $M^{\underline{\mathrm{co}(H)}}$ is a right $\mathbb{A}$-module via the action defined by (9.2.7). The most difficult part of the proof is to show that $\lambda^C$ is well defined, that is, $\lambda^C(M^{\underline{\mathrm{co}(H)}}) \subseteq C \otimes M^{\underline{\mathrm{co}(H)}}$, and that (9.2.1) and (9.2.3) hold for our context.

Write $f = f^1 \otimes f^2 = F^1 \otimes F^2$ and $f^{-1} = g^1 \otimes g^2$. For $m \in M^{\underline{\mathrm{co}(H)}}$ we compute:

$$\begin{aligned}
&\tilde{x}^3_\rho S^{-1}(f^2(\tilde{x}^2_\rho)_2 p^2)\cdot m_{\{-1\}}\cdot(\tilde{x}^1_\rho)_{[-1]}\theta^1 \otimes \left(S^{-1}(f^1(\tilde{x}^2_\rho)_1 p^1\theta^3)\cdot m_{\{0\}}\cdot(\tilde{x}^1_\rho)_{[0]}\theta^2\right)_{(0)}\\
&\qquad \otimes \left(S^{-1}(f^1(\tilde{x}^2_\rho)_1 p^1\theta^3)\cdot m_{\{0\}}\cdot(\tilde{x}^1_\rho)_{[0]}\theta^2\right)_{(1)}\\
\overset{(6.1.6)}{\underset{(3.2.13)}{=}}\ &\tilde{x}^3_\rho S^{-1}(f^2(\tilde{x}^2_\rho)_2 p^2)\cdot m_{\{-1\}}\cdot(\tilde{x}^1_\rho)_{[-1]}\theta^1\\
&\otimes S^{-1}(F^2 f^1_2(\tilde{x}^2_\rho)_{(1,2)}p^1_2\theta^3_2 g^2)\cdot m_{\{0\}_{(0)}}\cdot(\tilde{x}^1_\rho)_{[0]_{\langle 0\rangle}}\theta^2_{\langle 0\rangle}\\
&\otimes S^{-1}(F^1 f^1_1(\tilde{x}^2_\rho)_{(1,1)}p^1_1\theta^3_1 g^1)\cdot m_{\{0\}_{(1)}}\cdot(\tilde{x}^1_\rho)_{[0]_{\langle 1\rangle}}\theta^2_{\langle 1\rangle}\\
\overset{(9.1.2),(3.2.17)}{\underset{(4.4.1)}{=}}\ &\tilde{x}^3_\rho S^{-1}(F^2 f^2_2 X^3(\tilde{x}^2_\rho)_2 p^2)\cdot m_{(0)_{\{-1\}}}\cdot(\tilde{x}^1_\rho)_{\langle 0\rangle_{[-1]}}\overline{\theta}^1\theta^1\\
&\otimes S^{-1}(F^1 f^2_1 X^2(\tilde{x}^2_\rho)_{(1,2)}p^1_2\theta^3_2 g^2)\cdot m_{(0)_{\{0\}}}\cdot(\tilde{x}^1_\rho)_{\langle 0\rangle_{[0]}}\overline{\theta}^2\theta^2_{\langle 0\rangle}\\
&\otimes S^{-1}(f^1 X^1(\tilde{x}^2_\rho)_{(1,1)}p^1_1\theta^3_1 g^1)\cdot m_{(1)}\cdot(\tilde{x}^1_\rho)_{\langle 1\rangle}\overline{\theta}^3\theta^2_{\langle 1\rangle}\\
\overset{(6.3.1),(9.1.3)}{\underset{(9.1.4),(3.2.13)}{=}}\ &\tilde{x}^3_\rho S^{-1}(f^2\mathfrak{q}^2_2(\tilde{X}^3_\rho)_{(2,2)}X^3(\tilde{x}^2_\rho)_2 p^2)\cdot m_{\{-1\}}\cdot(\tilde{X}^1_\rho(\tilde{x}^1_\rho)_{\langle 0\rangle})_{[-1]}\overline{\theta}^1\theta^1\\
&\otimes S^{-1}(f^1\mathfrak{q}^2_1(\tilde{X}^3_\rho)_{(2,1)}X^2(\tilde{x}^2_\rho)_{(1,2)}p^1_2\theta^3_2 g^2)\cdot m_{\{0\}}\cdot(\tilde{X}^1_\rho(\tilde{x}^1_\rho)_{\langle 0\rangle})_{[0]}\overline{\theta}^2\theta^2_{\langle 0\rangle}\\
&\otimes S^{-1}(\mathfrak{q}^1(\tilde{X}^3_\rho)_1 X^1(\tilde{x}^2_\rho)_{(1,1)}p^1_1\theta^3_1 g^1)\tilde{X}^2_\rho(\tilde{x}^1_\rho)_{\langle 1\rangle}\overline{\theta}^3\theta^2_{\langle 1\rangle}\\
\overset{(4.4.3),(4.3.10)}{\underset{(3.1.7),(4.3.2)}{=}}\ &\tilde{x}^3_\rho y^3 S^{-1}(f^2\mathfrak{q}^2_2((\tilde{x}^2_\rho)_2 y^2)_{(2,2)}X^3 p^2)\cdot m_{\{-1\}}\cdot(\tilde{x}^1_\rho)_{[-1]}\theta^1\\
&\otimes S^{-1}(f^1\mathfrak{q}^2_1((\tilde{x}^2_\rho)_2 y^2)_{(2,1)}X^2 p^1_2\theta^3_{(2,2)}(\tilde{Y}^3_\rho)_2 g^2)\cdot m_{\{0\}}\cdot(\tilde{x}^1_\rho)_{[0]}\theta^2\tilde{Y}^1_\rho\\
&\otimes S^{-1}(\mathfrak{q}^1((\tilde{x}^2_\rho)_2 y^2)_1 X^1 p^1_1\theta^3_{(2,1)}(\tilde{Y}^3_\rho)_1 g^1)(\tilde{x}^2_\rho)_1 y^1\theta^3_1\tilde{Y}^2_\rho\\
\overset{(4.3.11),(3.1.7)}{\underset{(5.5.16),(3.1.9)}{=}}\ &\tilde{x}^3_\rho S^{-1}(f^2(\tilde{x}^2_\rho)_2 Y^3 p^2)\cdot m_{\{-1\}}\cdot(\tilde{x}^1_\rho)_{[-1]}\theta^1\\
&\otimes S^{-1}(f^1(\tilde{x}^2_\rho)_1 Y^2 y^3(p^1\theta^3)_{(2,2)}(\tilde{Y}^3_\rho)_2 g^2)\cdot m_{\{0\}}\cdot(\tilde{x}^1_\rho)_{[0]}\theta^2\tilde{Y}^1_\rho\\
&\otimes S^{-1}(\alpha Y^1_2 y^2(p^1\theta^3)_{(2,1)}(\tilde{Y}^3_\rho)_1 g^1)Y^1_1 y^1(p^1\theta^3)_1\tilde{Y}^2_\rho\\
\overset{(3.1.7)}{\underset{(3.2.1)}{=}}\ &\tilde{x}^3_\rho S^{-1}(f^2(\tilde{x}^2_\rho)_2 p^2)\cdot m_{\{-1\}}\cdot(\tilde{x}^1_\rho)_{[-1]}\theta^1\otimes S^{-1}(\mathfrak{q}^2(\tilde{Y}^3_\rho)_2 g^2)\\
&\cdot\left(S^{-1}(f^1(\tilde{x}^2_\rho)_1 p^1\theta^3)\cdot m_{\{0\}}\cdot(\tilde{x}^1_\rho)_{[0]}\theta^2\right)\cdot\tilde{Y}^1_\rho\otimes S^{-1}(\mathfrak{q}^1(\tilde{Y}^3_\rho)_1 g^1)\tilde{Y}^2_\rho,
\end{aligned}$$

which shows that $\lambda^C_{\mathbf{G}(M)}$ is well defined.

The relation in (9.2.1) follows from (9.2.8), (9.2.7) and the axioms of a two-sided Hopf module. Note that the desired relation comes out as

$$\begin{aligned}
&\tilde{x}^2_\rho\cdot\left((m\leftarrow\tilde{x}^1_\rho)^{\{-1\}}\right)_{\underline{1}}\cdot\tilde{x}^1_\lambda\otimes\tilde{x}^3_\rho\cdot\left((m\leftarrow\tilde{x}^1_\rho)^{\{-1\}}\right)_{\underline{2}}\cdot\tilde{x}^2_\lambda\otimes(m\leftarrow\tilde{x}^1_\rho)^{\{0\}}\leftarrow\tilde{x}^3_\lambda\\
&\qquad = m^{\{-1\}}\cdot\theta^1\otimes\theta^3\cdot(m^{\{-1\}}\leftarrow\theta^2)^{\{-1\}}\otimes(m^{\{-1\}}\leftarrow\theta^2)^{\{0\}}.
\end{aligned}$$

Finally, (9.2.3) follows from similar computations, left to the reader. □

We can prove now the main result of this section.

**Theorem 9.8** *Let $H$ be a quasi-Hopf algebra with bijective antipode, $C$ an $H$-bimodule coalgebra and $\mathbb{A}$ an $H$-bicomodule algebra. Then the categories ${}^C\mathscr{YD}(H)_{\mathbb{A}}$ and ${}^C_H\mathscr{M}^H_{\mathbb{A}}$ are equivalent.*

*Proof* As we have already discussed, the functors $\mathbf{F}$ and $\mathbf{G}$ from Lemma 9.6 and Lemma 9.7 provide the inverse equivalences between the categories of two-sided $(H,\mathbb{A})$-Hopf modules and right $\mathbb{A}$-modules presented in Theorem 6.16. So it suffices to check that for any $M \in {}^{C}\mathscr{YD}(H)_{\mathbb{A}}$ the right $\mathbb{A}$-module isomorphism

$$\xi_M : \mathbf{G}(\mathbf{F}(M)) = \{m \cdot \tilde{q}^1_\rho \otimes \tilde{q}^2_\rho \mid m \in M\} \to M,\ \ \xi_M(m \cdot \tilde{q}^1_\rho \otimes \tilde{q}^2_\rho) = m,\ \ \forall\, m \in M,$$

from the proof of Theorem 6.16 is left $C$-colinear, and that for all $M \in {}^{C}_{H}\mathscr{M}^{H}_{\mathbb{A}}$ the isomorphism

$$\zeta_M : M \to M^{\underline{\mathrm{co}(H)}} \otimes H = \mathbf{F}(\mathbf{G}(M)),\ \ \zeta_M(m) = \underline{E}(m_{(0)}) \otimes m_{(1)},\ \ \forall\, m \in M$$

from the same proof is also left $C$-colinear.

We check first that $\xi_M$ is left $C$-colinear. Note that (4.3.2) and (3.2.1) imply

$$\tilde{x}^1_\rho \otimes (\tilde{x}^2_\rho)_1 p^1 \otimes \tilde{x}^3_\rho S^{-1}((\tilde{x}^2_\rho)_2 p^2) = \tilde{X}^1_\rho (\tilde{p}^1_\rho)_{\langle 0\rangle} \otimes \tilde{X}^2_\rho (\tilde{p}^1_\rho)_{\langle 1\rangle} \otimes S^{-1}(\tilde{X}^3_\rho \tilde{p}^2_\rho), \quad (9.2.9)$$

where $p_R = p^1 \otimes p^2$ is as in (3.2.19). Now, if we denote by $\tilde{Q}^1_\rho \otimes \tilde{Q}^2_\rho$ another copy of $\tilde{q}_\rho$ then from (9.2.8), (9.2.5) and (9.2.4) we have

$$\begin{aligned}
&\lambda^C_{\mathbf{G}(\mathbf{F}(M))}(m \cdot \tilde{q}^1_\rho \otimes \tilde{q}^2_\rho)\\
&= \tilde{x}^3_\rho S^{-1}(f^2(\tilde{x}^2_\rho)_2 p^2)(\tilde{q}^2_\rho)_1 \tilde{X}^2_\rho \cdot (m \cdot \tilde{q}^1_\rho \tilde{X}^1_\rho)_{\{-1\}} \cdot \overline{\theta}^1 (\tilde{x}^1_\rho)_{[-1]} \theta^1\\
&\quad \otimes (m \cdot \tilde{q}^1_\rho \tilde{X}^1_\rho)_{\{0\}} \cdot \overline{\theta}^2 (\tilde{x}^1_\rho)_{[0]_{\langle 0\rangle}} \theta^2_{\langle 0\rangle} \otimes S^{-1}(f^1(\tilde{x}^2_\rho)_1 p^1 \theta^3)(\tilde{q}^2_\rho)_2 \tilde{X}^3_\rho \overline{\theta}^3 (\tilde{x}^1_\rho)_{[0]_{\langle 1\rangle}} \theta^3_{\langle 1\rangle}\\
&\overset{(4.4.1),(9.2.3)}{\underset{(9.2.9)}{=}} S^{-1}(f^2 \tilde{Y}^3_\rho \tilde{p}^2_\rho)(\tilde{q}^2_\rho)_1 \tilde{X}^2_\rho (\tilde{Y}^1_\rho (\tilde{p}^1_\rho)_{\langle 0\rangle})_{\langle 0,1\rangle}\\
&\quad \cdot (m \cdot \tilde{q}^1_\rho \tilde{X}^1_\rho (\tilde{Y}^1_\rho (\tilde{p}^1_\rho)_{\langle 0\rangle})_{\langle 0,0\rangle})_{\{-1\}} \cdot \overline{\theta}^1 \theta^1 \otimes (m \cdot \tilde{q}^1_\rho \tilde{X}^1_\rho (\tilde{Y}^1_\rho (\tilde{p}^1_\rho)_{\langle 0\rangle})_{\langle 0,0\rangle})_{\{0\}}\\
&\quad \cdot \overline{\theta}^2 \theta^2_{\langle 0\rangle} \otimes S^{-1}(f^1 \tilde{Y}^2_\rho (\tilde{p}^1_\rho)_{\langle 1\rangle} \theta^3)(\tilde{q}^2_\rho)_2 \tilde{X}^3_\rho (\tilde{Y}^1_\rho (\tilde{p}^1_\rho)_{\langle 0\rangle})_{\langle 1\rangle} \overline{\theta}^3 \theta^3_{\langle 1\rangle}\\
&\overset{(4.3.1)}{\underset{(4.3.15)}{=}} S^{-1}(\tilde{p}^2_\rho)\tilde{q}^2_\rho (\tilde{Q}^1_\rho (\tilde{p}^1_\rho)_{\langle 0,0\rangle})_{\langle 1\rangle} \cdot (m \cdot \tilde{q}^1_\rho (\tilde{Q}^1_\rho (\tilde{p}^1_\rho)_{\langle 0,0\rangle})_{\langle 0\rangle})_{\{-1\}} \cdot \overline{\theta}^1 \theta^1\\
&\quad \otimes (m \cdot \tilde{q}^1_\rho (\tilde{Q}^1_\rho (\tilde{p}^1_\rho)_{\langle 0,0\rangle})_{\langle 0\rangle})_{\{0\}} \cdot \overline{\theta}^2 \theta^2_{\langle 0\rangle} \otimes S^{-1}((\tilde{p}^1_\rho)_{\langle 1\rangle} \theta^3)\tilde{Q}^2_\rho (\tilde{p}^1_\rho)_{\langle 0,1\rangle} \overline{\theta}^3 \theta^3_{\langle 1\rangle}\\
&\overset{(4.3.11)}{\underset{(4.3.13)}{=}} (\tilde{Q}^1_\rho)_{\langle 1\rangle} \cdot (m \cdot (\tilde{Q}^1_\rho)_{\langle 0\rangle})_{\{-1\}} \cdot \overline{\theta}^1 \theta^1 \otimes (m \cdot (\tilde{Q}^1_\rho)_{\langle 0\rangle})_{\{0\}} \cdot \overline{\theta}^2 \theta^2_{\langle 0\rangle}\\
&\quad \otimes S^{-1}(\theta^3)\tilde{Q}^2_\rho \overline{\theta}^3 \theta^3_{\langle 1\rangle}\\
&\overset{(9.2.3),(4.3.9)}{=} m_{\{-1\}} \cdot (\tilde{X}^1_\rho)_{[-1]} \overline{\theta}^1 \theta^1 \otimes m_{\{0\}} \cdot (\tilde{X}^1_\rho)_{[0]} \overline{\theta}^2 \theta^2_{\langle 0\rangle} \otimes S^{-1}(\alpha \tilde{X}^3_\rho \theta^3)\tilde{X}^2_\rho \overline{\theta}^3 \theta^3_{\langle 1\rangle}\\
&\overset{(4.4.3),(3.2.1)}{\underset{(4.3.9)}{=}} m_{\{-1\}} \otimes m_{\{0\}} \cdot \tilde{q}^1_\rho \otimes \tilde{q}^2_\rho,
\end{aligned}$$

for all $m \in M$. Having this explicit formula for $\lambda^C_{\mathbf{G}(\mathbf{F}(M))}$ one can easily see that $\lambda^C_M \circ \xi_M = (\mathrm{Id}_C \otimes \xi_M) \circ \lambda^C_{\mathbf{G}(\mathbf{F}(M))}$, and this means that $\lambda^C_{\mathbf{G}(\mathbf{F}(M))}$ is left $C$-colinear.

Now let $M$ be an object of ${}^{C}_{H}\mathscr{M}^{H}_{\mathbb{A}}$. For simplicity we prove that $\zeta_M^{-1}$ is left $C$-colinear. From (9.2.5), (9.2.8) and (9.2.7) we obtain that $\lambda^C_{\mathbf{F}(\mathbf{G}(M))}$ is given for all $m \in M^{\underline{\mathrm{co}(H)}}$ and $h \in H$ by

$$\begin{aligned}
&\lambda^C_{\mathbf{F}(\mathbf{G}(M))}(m \otimes h)\\
&= h_1 \tilde{X}^2_\rho \tilde{x}^3_\rho S^{-1}(f^2(\tilde{X}^1_\rho)_{\langle 1\rangle_2}(\tilde{x}^2_\rho)_2 p^2) \cdot m_{\{-1\}} \cdot (\tilde{X}^1_\rho)_{\langle 0\rangle_{[-1]}} (\tilde{x}^1_\rho)_{[-1]} \overline{\theta}^1 \theta^1
\end{aligned}$$

$$\otimes S^{-1}(f^1(\tilde{X}^1_\rho)_{\langle 1\rangle_1}(\tilde{x}^2_\rho)_1 p^1 \overline{\theta}^3 \theta^2_{\langle 1\rangle}) \cdot m_{\{0\}} \cdot (\tilde{X}^1_\rho)_{\langle 0\rangle_{[0]}}(\tilde{x}^1_\rho)_{[0]} \overline{\theta}^2 \theta^2_{\langle 0\rangle} \otimes h_2 \tilde{X}^3_\rho \theta^3,$$

and since $\zeta_M^{-1}(m\otimes h) = hS^{-1}(\tilde{p}^2_\rho)\cdot m\cdot \tilde{p}^1_\rho$, after a straightforward computation we obtain for all $m \in M^{\underline{\mathrm{co}(H)}}$ and $h \in H$:

$$(\mathrm{Id}_C \otimes \zeta_M^{-1}) \circ \lambda^C_{\mathbf{F}(\mathbf{G}(M))}(m\otimes h) = \lambda^C_M \circ \zeta_M^{-1}(m\otimes h).$$

The above equality shows that $\lambda^C_{\mathbf{F}(\mathbf{G}(M))}$ is left $C$-colinear. □

Remark 6.17 suggests another equivalence between a category of two-sided two-cosided Hopf modules and a category of Yetter–Drinfeld modules. As before, let $H$ be a quasi-Hopf algebra with bijective antipode, $\mathbb{A}$ an $H$-bicomodule algebra and $C$ an $H$-bimodule coalgebra. As we have already seen, $\mathbb{A}^{\mathrm{op}}$ is an $H^{\mathrm{op}}$-bicomodule algebra and $C$ is also an $H^{\mathrm{op}}$-bimodule coalgebra. As in Remark 6.17, we can define the category of two-sided $(\mathbb{A}, H)$-Hopf modules as being ${}_{\mathbb{A}}\mathscr{M}_H^H := {}_{H^{\mathrm{op}}}\mathscr{M}_{\mathbb{A}^{\mathrm{op}}}^{H^{\mathrm{op}}}$, and then the category ${}^C_{\mathbb{A}}\mathscr{M}_H^H := {}^C_{H^{\mathrm{op}}}\mathscr{M}_{\mathbb{A}^{\mathrm{op}}}^{H^{\mathrm{op}}}$. By Theorem 9.8 we have an equivalence between ${}^C_{\mathbb{A}}\mathscr{M}_H^H$ and ${}^C\mathscr{YD}(H^{\mathrm{op}})_{\mathbb{A}^{\mathrm{op}}}$. The latter category will be denoted in what follows by ${}^C_{\mathbb{A}}\mathscr{YD}(H)$.

**Proposition 9.9** *Consider the functors*

$$ {}^C_{\mathbb{A}}\mathscr{YD}(H) \underset{\mathbb{G}}{\overset{\mathbb{F}}{\rightleftarrows}} {}^C_{\mathbb{A}}\mathscr{M}_H^H, $$

*defined as follows:*

- *For $M \in {}^C_{\mathbb{A}}\mathscr{YD}(H)$ we have $\mathbb{F}(M) = M\otimes H \in {}^C_{\mathbb{A}}\mathscr{M}_H^H$ with structure*

$$u\cdot(m\otimes h)\cdot h' = u_{\langle 0\rangle}\cdot m\otimes u_{\langle 1\rangle}hh', \tag{9.2.10}$$

$$\lambda^C_{M\otimes H}(m\otimes h) = \Theta^1\cdot(\tilde{x}^1_\rho\cdot m)_{\{-1\}}\cdot \tilde{x}^2_\rho h_1 \otimes \left(\Theta^2\cdot(\tilde{x}^1_\rho\cdot m)_{\{0\}}\otimes \Theta^3\tilde{x}^3_\rho h_2\right), \tag{9.2.11}$$

$$\rho^H_{M\otimes H}(m\otimes h) = (\tilde{x}^1_\rho\cdot m\otimes \tilde{x}^2_\rho h_1)\otimes \tilde{x}^3_\rho h_2, \tag{9.2.12}$$

  *for all $u \in \mathbb{A}$, $h, h' \in H$ and $m \in M$. If $f : M \to N$ is a morphisms in ${}^C_{\mathbb{A}}\mathscr{YD}(H)$ then $\mathbb{F}(f) = f\otimes \mathrm{Id}_H$.*
- *If $M \in {}^C_{\mathbb{A}}\mathscr{M}_H^H$ then $\mathbb{G}(M) = M^{\overline{\mathrm{co}(H)}}$, the set of alternative coinvariants of $M$, which is an object of ${}^C_{\mathbb{A}}\mathscr{YD}(H)$ via the structure*

$$u \to m = u_{\langle 0\rangle}\cdot m\cdot S(u_{\langle 1\rangle}), \tag{9.2.13}$$

$$\begin{aligned}\lambda^C_{M^{\overline{\mathrm{co}(H)}}}(m) = {}&\Theta^1(\tilde{X}^1_\rho)_{[-1]}\cdot m_{\{-1\}}\cdot g^1S(q^2(\tilde{X}^2_\rho)_2)\tilde{X}^3_\rho\\ &\otimes \Theta^2(\tilde{X}^1_\rho)_{[0]}\cdot m_{\{0\}}\cdot g^2S(\Theta^3q^1(\tilde{X}^2_\rho)_1),\end{aligned} \tag{9.2.14}$$

  *for all $u \in \mathbb{A}$ and $m \in M^{\overline{\mathrm{co}(H)}}$. On morphisms we have that $\mathbb{G}(f) = f|_{M^{\overline{\mathrm{co}(H)}}}$, for any morphism $f : M \to N$ in ${}^C_{\mathbb{A}}\mathscr{M}_H^H$.*

*Then $\mathbb{F}$ and $\mathbb{G}$ are inverse equivalence functors.*

## 9.3 The Categories ${}_H^H\mathcal{M}_H^H$ and ${}_H^H\mathcal{YD}$

The goal of this section is to show that in the situation when $C = \mathbb{A} = H$ the pair $(\mathbb{F}, \mathbb{G})$ in Proposition 9.9 provides a monoidal equivalence between ${}_H^H\mathcal{M}_H^H$ and ${}_H^H\mathcal{YD}$. In other words we want to show that the monoidal equivalence between ${}_H\mathcal{M}_H^H$ and ${}_H\mathcal{M}$ in Section 6.4 yields a monoidal equivalence between ${}_H^H\mathcal{M}_H^H$ and ${}_H^H\mathcal{YD}$. Towards this end we first need some preparatory work.

**Lemma 9.10** *In any quasi-Hopf algebra $H$ with bijective antipode we have:*

$$q^1Q_1^1z^1y_1^1 \otimes S(Q^2z^3y^1_{(2,2)})y^2 \otimes S(q^2Q_2^1z^2y^1_{(2,1)})y^3$$
$$= X^1 \otimes S(q^2x_2^1X_2^2)x^2X_1^3 \otimes S(q^1x_1^1X_1^2)\alpha x^3X_2^3, \qquad (9.3.1)$$

*where $q^1 \otimes q^2 = Q^1 \otimes Q^2$ are two copies of the element $q_R$ defined in* (3.2.19).

*Proof* The formula (9.3.1) is a consequence of the following computation:

$$q^1Q_1^1z^1y_1^1 \otimes S(Q^2z^3y^1_{(2,2)})y^2 \otimes S(q^2Q_2^1z^2y^1_{(2,1)})y^3$$
$$\overset{(3.1.7)}{=} q^1(Q^1y_1^1)_1z^1 \otimes S(Q^2y_2^1z^3)y^2 \otimes S(q^2(Q^1y_1^1)_2z^2)y^3$$
$$\overset{(6.5.1)}{=} q^1X_1^1z^1 \otimes S(X^2z^3)\tilde{q}^1X_1^3 \otimes S(q^2X_2^1z^2)\tilde{q}^2X_2^3$$
$$\overset{(3.2.19),(3.2.22)}{=} Y^1X_1^1z^1 \otimes S(Y_1^3X^2z^3)\tilde{q}^1(Y_2^3X^3)_1 \otimes S(Y^2X_2^1z^2)\alpha\tilde{q}^2(Y_2^3X^3)_2$$
$$\overset{(3.1.9)}{=} Y^1 \otimes S(X^2Y_2^2)\tilde{q}^1X_1^3Y_1^3 \otimes S(X^1Y_1^2)\alpha\tilde{q}^2X_2^3Y_2^3$$
$$\overset{(6.5.1)}{=} Y^1 \otimes S(q^2x_2^1Y_2^2)x^2Y_1^3 \otimes S(q^1x_1^1Y_1^2)\alpha x^3Y_2^3,$$

as we stated. □

**Theorem 9.11** *If $H$ is a quasi-Hopf algebra with bijective antipode then the categories ${}_H^H\mathcal{M}_H^H$ and ${}_H^H\mathcal{YD}$ are strong monoidally equivalent.*

*Proof* By Corollary 6.25, we have that the functors $\mathbb{F}$ and $\mathbb{G}$ that provide the equivalence between ${}_H^H\mathcal{M}_H^H$ and ${}_H^H\mathcal{YD}$ in Proposition 9.9 yield a monoidal equivalence between ${}_H\mathcal{M}_H^H$ and ${}_H\mathcal{M}$. So, according to Proposition 1.31, it is enough to prove that the strong monoidal structure of $\mathbb{G}$ (considered as functor from ${}_H\mathcal{M}_H^H$ to ${}_H\mathcal{M}$) extends to a strong monoidal structure of $\mathbb{G}$ when it is considered as functor between ${}_H^H\mathcal{M}_H^H$ and ${}_H^H\mathcal{YD}$. In other words, it suffices to prove that $\overline{\phi}_{2,M,N}$ defined in (6.4.16) is left $H$-colinear, for all $M, N \in {}_H^H\mathcal{M}_H^H$.

As before, for $M \in {}_H^H\mathcal{M}_H^H$ we denote by $M \ni m \mapsto \lambda_M(m) = m_{\{-1\}} \otimes m_{\{0\}} \in H \otimes M$ its left $H$-coaction and by $M \ni m \mapsto \rho_M(m) = m_{(0)} \otimes m_{(1)} \in M \otimes H$ its right $H$-coaction.

With this notation, we have that, for all $m \in M^{\overline{\mathrm{co}(H)}}$ and $n \in N^{\overline{\mathrm{co}(H)}}$,

$$\lambda_{(M\otimes_H N)^{\overline{\mathrm{co}(H)}}}\overline{\phi}_{2,M,N}(m \otimes_H n)$$
$$\overset{(6.4.16)}{=} \lambda_{(M\otimes_H N)^{\overline{\mathrm{co}(H)}}}(q^1x_1^1 \cdot m \cdot S(q^2x_2^1)x^2 \otimes_H n \cdot S(x^3))$$
$$\overset{(9.2.14)}{=} X^1Y_1^1(q^1x_1^1 \cdot m \cdot S(q^2x_2^1)x^2)_{\{-1\}}(n \cdot S(x^3))_{\{-1\}}g^1S(Q^2Y_2^2)Y^3$$
$$\otimes X^2Y_2^1 \cdot (q^1x_1^1 \cdot m \cdot S(q^2x_2^1)x^2)_{\{0\}} \otimes_H (n \cdot S(x^3))_{\{0\}} \cdot g^2S(X^3Q^1Y_1^2)$$

$$
\begin{aligned}
&\overset{(3.2.13)}{\underset{(3.2.26)}{=}} X^1Y_1^1(q^1\mathfrak{Q}_1^1z^1y_1^1x_1^1)_1m_{\{-1\}}G^1S(\mathfrak{Q}^2z^3y^1_{(2,2)}x^1_{(2,2)})y^2x_1^2n_{\{-1\}}g^1S(Q_2^2Y_2^2x_2^3)Y^3\\
&\qquad \otimes X^2Y_2^1(q^1\mathfrak{Q}_1^1z^1y_1^1x_1^1)_2\cdot m_{\{0\}}\cdot G^2S(q^2\mathfrak{Q}_2^1z^2y^1_{(2,1)}x^1_{(2,1)})y^3x_2^2\\
&\qquad \otimes_H n_{\{0\}}\cdot g^2S(X^3Q^1Y_1^2x_1^3)\\
&\overset{(9.3.1)}{=} X^1Y_1^1Z_1^1x^1_{(1,1)}m_{\{-1\}}G^1S(q^2y_2^1Z_2^2x^1_{(2,2)})y^2Z_1^3x_1^2n_{\{-1\}}g^1S(Q^2Y_2^2x_2^3)Y^3\\
&\qquad \otimes X^2Y_2^1Z_2^1x^1_{(1,2)}\cdot m_{\{0\}}\cdot G^2S(q^1y_1^1Z_1^2x^1_{(2,1)})\alpha y^3Z_2^3x_2^2\\
&\qquad \otimes_H n_{\{0\}}\cdot g^2S(X^3Q^1Y_1^2x_1^3).
\end{aligned}
$$

On the other hand, we compute that

$$
\begin{aligned}
&(\mathrm{Id}_H\otimes\overline{\phi}_{2,M,N})\lambda_{M^{\overline{\mathrm{co}(H)}}\otimes N^{\overline{\mathrm{co}(H)}}}(m\otimes n)\\
&\overset{(8.2.14)}{=} X^1(x^1Y^1\to m)_{[-1]}x^2(Y^2\to n)_{[-1]}Y^3\\
&\qquad \otimes\overline{\phi}_{2,M,N}(X^2\to(x^1Y^1\to m)_{[0]}\otimes X^3x^3\to(Y^2\to n)_{[0]})\\
&\overset{(9.2.14)}{=} X^1U^1Z_1^1(x^1Y^1\to m)_{\{-1\}}g^1S(q^2Z_2^2)Z^3x^2T^1V_1^1(Y^2\to n)_{\{-1\}}G^1S(Q^2V_2^2)\\
&\qquad V^3Y^3\otimes\overline{\phi}_{2,M,N}\big(X^2\to(U^2Z_2^1\cdot(x^1Y^1\to m)_{\{0\}}\cdot g^2S(U^3q^1Z_1^2))\\
&\qquad \otimes X^3x^3\to(T^2V_2^1\cdot(Y^2\to n)_{\{0\}}\cdot G^2S(T^3Q^1V_1^2))\big)\\
&\overset{(9.2.13)}{\underset{(9.3.1)}{=}} X^1U^1Z_1^1(x^1Y^1)_{(1,1)}m_{\{-1\}}g^1S(q^2Z_2^2(x^1Y^1)_{(2,2)})Z^3x^2T^1V_1^1Y^2_{(1,1)}n_{\{-1\}}G^1\\
&\qquad S(Q^2V_2^2Y^2_{(2,2)})V^3Y^3\otimes W^1X_1^2U^2Z_2^1(x^1Y^1)_{(1,2)}\cdot m_{\{0\}}\cdot\\
&\qquad g^2S(t^1W^2X_2^2U^3q^1Z_1^2(x^1Y^1)_{(2,1)})\alpha t^2W_1^3X_1^3T^2V_2^1Y^2_{(1,2)}\\
&\qquad \otimes_H n_{\{0\}}\cdot G^2S(t^3W_2^3X_2^3x_2^3T^3Q^1V_1^2Y^2_{(2,1)})\\
&\overset{(3.1.9)}{=} X^1U_1^1Z_1^1(x^1Y^1)_{(1,1)}m_{\{-1\}}g^1S(q^2Z_2^2(x^1Y^1)_{(2,2)})Z^3x^2T^1V_1^1Y^2_{(1,1)}n_{\{-1\}}\\
&\qquad G^1S(Q^2V_2^2Y^2_{(2,2)})V^3Y^3\otimes X^2U_2^1Z_2^1(x^1Y^1)_{(1,2)}\cdot m_{\{0\}}\\
&\qquad \cdot g^2S(t^1X_1^3U^2q^1Z_1^2(x^1Y^1)_{(2,1)})\alpha t^2(X_2^3U^3)_1x_1^3T^2V_2^1Y^2_{(1,2)}\\
&\qquad \otimes_H n_{\{0\}}\cdot G^2S(t^3(X_2^3U^3)_2x_2^3T^3Q^1V_1^2Y^2_{(2,1)})\\
&\overset{(3.2.21)}{\underset{(3.1.9)}{=}} X^1Z_1^1Y^1_{(1,1)}m_{\{-1\}}g^1S(q^2(x^1Z^2Y_2^1)_2)x^2Z_1^3T^1(V_1^1Y_1^2)_1n_{\{-1\}}G^1\\
&\qquad S(Q^2(V^2Y_2^2)_2)V^3Y^3\otimes X^2Z_2^1Y^1_{(1,2)}\cdot m_{\{0\}}\cdot g^2S(t^1q^1(x^1Z^2Y_2^1)_1)\alpha t^2\\
&\qquad (x^3Z_2^3)_1T^2(V^1Y_1^2)_2\otimes_H n_{\{0\}}\cdot G^2S(X^3t^3(x^3Z_2^3)_2T^3Q^1(V^2Y_2^2)_1)\\
&\overset{(3.2.21),(3.1.7)}{\underset{(3.1.9)}{=}} X^1(Z^1Y_1^1)_1m_{\{-1\}}g^1S(q^2(t^1x^1Z^2Y_2^1)_2)t^2(x^2Z_1^3V^1Y_1^2)_1n_{\{-1\}}G^1\\
&\qquad S(Q^2(V^2Y_2^2)_2)V^3Y^3\otimes X^2(Z^1Y_1^1)_2\cdot m_{\{0\}}\cdot g^2S(q^1(t^1x^1Z^2Y_2^1)_1)\alpha t^3\\
&\qquad (x^2Z_1^3V^1Y_1^2)_2\otimes_H n_{\{0\}}\cdot G^2S(X^3x^3Z_2^3Q^1(V^2Y_2^2)_1)\\
&\overset{(3.1.9),(3.2.21)}{\underset{(3.1.9)}{=}} X^1(Z^1V^1(Y_1^1x^1)_1)_1m_{\{-1\}}g^1S(q^2(t^1Z_1^2V^2(Y_1^1x^1)_2)_2)t^2(Z_2^2V^3)_1(Y_2^1x^2)_1\\
&\qquad n_{\{-1\}}G^1S(Q^2(Y^2x^3)_2)Y^3\otimes X^2(Z^1V^1(Y_1^1x^1)_1)_2\cdot m_{\{0\}}\\
&\qquad \cdot g^2S(q^1(t^1Z_1^2V^2(Y_1^1x^1)_2)_1)\alpha t^3(Z_2^3V^3)_2(Y_2^1x^2)_2\\
&\qquad \otimes_H n_{\{0\}}\cdot G^2S(X^3Z^3Q^1(Y^2x^3)_1)\\
&\overset{(3.1.7)}{\underset{(3.2.21),(3.2.1)}{=}} X^1Y_1^1V_1^1x^1_{(1,1)}m_{\{-1\}}g^1S(q^2(t^1V^2x_2^1)_2)t^2V_1^3x_1^2n_{\{-1\}}G^1S(Q^2Y_2^2x_2^3)Y^3
\end{aligned}
$$

$$\otimes X^2Y_2^1V_2^1x^1_{(1,2)}\cdot m_{\{0\}}\cdot g^2S(q^1(t^1V^2x_2^1)_1)\alpha t^3V_2^3x_2^2$$
$$\otimes_H n_{\{0\}}\cdot G^2S(X^3Q^1Y_1^2x_1^3).$$

Hence, by comparing the two computations performed above we get that $\overline{\phi}_{2,M,N}$ is left $H$-colinear, as needed. □

## 9.4 A Structure Theorem for Bicomodule Algebras

We will make use of the ideas in Section 6.5 in order to give a structure theorem for algebras within the strict monoidal category $({}^H_H\mathscr{M}^H_H,\otimes_H,H)$.

We start with the description of an algebra in ${}^H_H\mathscr{M}^H_H$.

**Lemma 9.12** *For $H$ a quasi-bialgebra, an algebra in ${}^H_H\mathscr{M}^H_H$ is a 4-tuple $(\mathbb{A},\lambda,\rho,i)$ consisting of a $k$-algebra $\mathbb{A}$, a $k$-algebra morphism $i : H\to\mathbb{A}$, and $k$-linear maps $\lambda:\mathbb{A}\to H\otimes\mathbb{A}$ and $\rho:\mathbb{A}\to\mathbb{A}\otimes H$ such that the following conditions hold:*

- *$\lambda(i(h))=h_1\otimes i(h_2)$ and $\rho(i(h))=i(h_1)\otimes h_2$, for all $h\in H$;*
- *$(\mathbb{A},\lambda,\rho,\Phi_\lambda:=X^1\otimes X^2\otimes i(X^3),\Phi_\rho:=i(X^1)\otimes X^2\otimes X^3,\Phi_{\lambda,\rho}:=X^1\otimes i(X^2)\otimes X^3)$ is an $H$-bicomodule algebra, where $\Phi=X^1\otimes X^2\otimes X^3$ is the reassociator of $H$.*

*Proof* Let $\mathbb{A}$ be an algebra in ${}^H_H\mathscr{M}^H_H$. Since the forgetful functors ${}^H_H\mathscr{M}^H_H\to{}_H\mathscr{M}^H_H$ and ${}^H_H\mathscr{M}^H_H\to{}^H_H\mathscr{M}_H$ are strong monoidal we get that $\mathbb{A}$, with the same $H$-bimodule structure, is both an algebra in ${}_H\mathscr{M}^H_H$ and ${}^H_H\mathscr{M}_H$. Hence, by Lemma 6.26 and its left-handed version, we deduce that $\mathbb{A}$ is a $k$-algebra and that there exist $i:H\to\mathbb{A}$ an algebra morphism, and $\lambda:\mathbb{A}\to H\otimes\mathbb{A}$ and $\rho:\mathbb{A}\to\mathbb{A}\otimes H$ linear maps, such that

- $\lambda(i(h)=h_1\otimes i(h_2)$ and $\rho(i(h))=i(h_1)\otimes h_2$, for all $h\in H$;
- $(\mathbb{A},\lambda,\Phi_\lambda:=X^1\otimes X^2\otimes i(X^3))$ is a left $H$-comodule algebra and $(\mathbb{A},\rho,\Phi_\rho:=i(X^1)\otimes X^2\otimes X^3)$ is a right $H$-comodule algebra.

There is only one property of $\mathbb{A}$ that we have not yet explored: namely, the compatibility between the left and right $H$-coactions on $\mathbb{A}$. More precisely, by (9.1.2) we have, for all $u\in\mathbb{A}$, that

$$X^1u_{\langle 0\rangle_{[-1]}}\otimes i(X^2)u_{\langle 0\rangle_{[0]}}\otimes X^3u_{\langle 1\rangle}=u_{[-1]}X^1\otimes u_{[0]_{\langle 0\rangle}}i(X^2)\otimes u_{[0]_{\langle 1\rangle}}X^3,$$

and this means that $(\mathbb{A},\lambda,\rho,\Phi_\lambda:=X^1\otimes X^2\otimes i(X^3),\Phi_\rho:=i(X^1)\otimes X^2\otimes X^3,\Phi_{\lambda,\rho}:=X^1\otimes i(X^2)\otimes X^3)$ is an $H$-bicomodule algebra.

The converse follows from Lemma 6.26 and its left-handed version. □

Examples of algebras in ${}^H_H\mathscr{M}^H_H$ are given by some smash product algebras.

**Proposition 9.13** *Let $H$ be a quasi-bialgebra and $A$ an algebra in ${}^H_H\mathscr{YD}$, with coaction denoted by $A\to H\otimes A$, $a\mapsto a_{[-1]}\otimes a_{[0]}$. Then $(A\#H,\lambda,\rho,j)$ is an algebra in ${}^H_H\mathscr{M}^H_H$, where $j:H\to A\#H$ is the canonical inclusion map and*

$$\lambda:A\#H\to H\otimes(A\#H),\ \lambda(a\#h)=T^1(t^1\cdot a)_{[-1]}t^2h_1\otimes(T^2\cdot(t^1\cdot a)_{[0]}\#T^3t^3h_2),$$
$$\rho:A\#H\to(A\#H)\otimes H,\ \rho(a\#h)=(x^1\cdot a\#x^2h_1)\otimes x^3h_2,$$

*for all $a \in A$ and $h \in H$.*

*Proof* By Corollary 6.27 we know that $(\mathbb{A}, \rho, j)$ is a right $H$-comodule algebra. Also, since $A$ is an algebra in ${}^H_H\mathscr{YD}$ and the functor $\mathbb{F}$ from the proof of Theorem 9.11 is strong monoidal we deduce that $\mathbb{F}(A) = A \otimes H$ is an algebra in ${}^H_H\mathscr{M}^H_H$. Firstly, by (9.2.10), (9.2.11) and (9.2.12) we deduce that $A \otimes H$ is an object in ${}^H_H\mathscr{M}^H_H$ with structure given by

$$h' \cdot (a \otimes h) \cdot h'' = h_1 \cdot a \otimes h'_2 h h'',$$
$$\lambda(a \otimes h) = X^1(x^1 \cdot a)_{[-1]} x^2 h_1 \otimes (X^2 \cdot (x^1 \cdot a)_{[0]} \otimes X^3 x^3 h_2),$$
$$\rho(a \otimes h) = (x^1 \cdot a \otimes x^2 h_1) \otimes x^3 h_2,$$

for all $a \in A$ and $h, h', h'' \in H$. Secondly, the multiplication on $\mathbb{F}(A)$ is given by

$$(a \otimes h)(b \otimes h') = \varphi_{2,A,A}((a \otimes h) \otimes_H (a' \otimes h')) \overset{(6.4.15)}{=} (x^1 \cdot a)(x^2 h_1 \cdot a') \otimes x^3 h_2 h',$$

for all $a, a' \in A$ and $h, h' \in H$, and the unit is $(\mathrm{Id}_H \otimes 1_A)\varphi_0(1_H) = 1_H \otimes 1_A$. In other words, $\mathbb{F}(A) = A\#H$ with the algebra structure in ${}^H_H\mathscr{M}^H_H$ provided by $j : H \ni h \mapsto 1_A\#h \in A\#H$ and $\lambda, \rho$ as in the statement. □

The result below is a two-sided two-cosided version of Theorem 6.28.

**Theorem 9.14** *For $H$ a quasi-Hopf algebra with bijective antipode and $(\mathbb{A}, \rho, \lambda, i)$ an algebra in ${}^H_H\mathscr{M}^H_H$, there exists an algebra $A$ in ${}^H_H\mathscr{YD}$ such that $\mathbb{A}$ is isomorphic to $A\#H$ as algebras in ${}^H_H\mathscr{M}^H_H$.*

*Proof* $(\mathbb{A}, \rho, i)$ is an algebra in ${}_H\mathscr{M}^H_H$. If $A := \mathbb{A}^{\overline{\mathrm{co}(H)}}$ then by Theorem 6.28 we know that $A$ is a left $H$-module algebra and $\mathbb{A} \cong A\#H$, as algebras in ${}_H\mathscr{M}^H_H$. Furthermore, since $\mathbb{A}$ is actually an algebra in ${}^H_H\mathscr{M}^H_H$ and from the proof of Theorem 9.11 $\mathbb{G}$ is a strong monoidal functor, we get that $A$ is an algebra in ${}^H_H\mathscr{YD}$. Consequently, $A\#H$ is an algebra in ${}^H_H\mathscr{M}^H_H$ via the structure described in Proposition 9.13.

Thus, the only thing that is left to check is the fact that the isomorphism $\chi$ in the proof of Theorem 6.28 intertwines the left $H$-coactions of $\mathbb{A}$ and $A\#H$. But this follows from Proposition 9.9, specialized for $\mathbb{A} = C = H$, since any $M \in {}^H_H\mathscr{M}^H_H$ decomposes as $M^{\overline{\mathrm{co}H}} \otimes H$ in ${}^H_H\mathscr{M}^H_M$ via an isomorphism, say $\chi_M$, and $\chi = \chi_{\mathbb{A}}$. □

## 9.5 The Structure of a Coalgebra in ${}^H_H\mathscr{M}^H_H$

Theorem 6.31 does not say that, up to an isomorphism, a coalgebra in ${}_H\mathscr{M}^H_H$ is some sort of smash product coalgebra. In other words, in the coalgebra case we cannot produce a sort of dual result for Theorem 6.28. Nevertheless, when we pass to ${}^H_H\mathscr{M}^H_H$, because of the extra corner that we have in this case, this time we can characterize coalgebras in ${}^H_H\mathscr{M}^H_H$ as some sort of smash product coalgebras. To achieve this, we start by proving the following key result.

In order to avoid any confusion, we denote by ${}_H\overline{\mathscr{M}}_H$ the category of bimodules over a quasi-bialgebra $H$, endowed with the monoidal structure defined by the structure of $H$ as in Lemma 4.13.

**Lemma 9.15** *Let $H$ be a quasi-bialgebra. Then the forgetful functor*

$$\mathscr{U} : {}_H^H\mathscr{M}_H^H = ({}_H^H\mathscr{M}_H^H, \otimes_H, H) \to {}_H\overline{\mathscr{M}}_H = ({}_H\mathscr{M}_H, \otimes, k, a, l, r)$$

*is opmonoidal under the structure given, for all $M, N \in {}_H^H\mathscr{M}_H^H$, by*

$$\psi_{2,M,N} : \mathscr{U}(M \otimes_H N) \ni m \otimes_H n \mapsto m_{(0)} \cdot n_{\{-1\}} \otimes m_{(1)} \cdot n_{(0)} \in \mathscr{U}(M) \otimes \mathscr{U}(N)$$

*and $\psi_0 = \varepsilon : \mathscr{U}(H) = H \to k$.*

*Proof* $\psi_{2,M,N}$ is well defined since, for all $m \in M$, $h \in H$ and $n \in N$ we have that

$$\begin{aligned}
\psi_{2,M,N}(m \cdot h \otimes_H n) &= (m \cdot h)_{(0)} \cdot n_{\{-1\}} \otimes (m \cdot h)_{(1)} \cdot n_{\{0\}} \\
&= m_{(0)} \cdot h_1 n_{\{-1\}} \otimes m_{(1)} h_2 \cdot n_{\{0\}} \\
&= m_{(0)} \cdot (h \cdot n)_{\{-1\}} \otimes m_{(1)} \cdot (h \cdot n)_{\{0\}} \\
&= \psi_{2,M,N}(m \otimes_H h \cdot n).
\end{aligned}$$

Also, it can be easily checked that $\psi_{2,M,N}$ is an $H$-bilinear map.

We next show that $\psi_2$ fulfills the corresponding relations in (1.3.2). Indeed, for any $M, N, P \in {}_H^H\mathscr{M}_H^H$ we have

$$\begin{aligned}
&a_{\mathscr{U}(M),\mathscr{U}(N),\mathscr{U}(P)}(\psi_{2,M,N} \otimes \mathrm{Id}_{\mathscr{U}(P)})\psi_{2,M \otimes_H N,P}(m \otimes_H n \otimes_h p) \\
&= a_{\mathscr{U}(M),\mathscr{U}(N),\mathscr{U}(P)}(\psi_{2,M,N}((m \otimes_H n)_{(0)} \cdot p_{\{-1\}}) \otimes (m \otimes_H n)_{(1)} \cdot p_{\{0\}}) \\
&= a_{\mathscr{U}(M),\mathscr{U}(N),\mathscr{U}(P)}(\psi_{2,M,N}(m_{(0)} \otimes_H n_{(0)} \cdot p_{\{-1\}}) \otimes m_{(1)} n_{(1)} \cdot p_{\{0\}}) \\
&= X^1 \cdot m_{(0,0)} \cdot n_{(0)_{\{-1\}}} p_{\{-1\}_1} x^1 \otimes (X^2 m_{(0,1)} \cdot n_{(0)_{\{0\}}} p_{\{-1\}_2} x^2 \\
&\qquad \otimes X^3 m_{(1)} n_{(1)} \cdot p_{\{0\}} \cdot x^3) \\
&\overset{(6.1.2)}{\underset{(9.1.2)}{=}} m_{(0)} \cdot n_{\{-1\}} X^1 p_{\{-1\}_1} x^1 \otimes (m_{(1)_1} \cdot n_{\{0\}_{(0)}} \cdot X^2 p_{\{-1\}_2} x^2 \\
&\qquad \otimes m_{(1)_2} n_{\{0\}_{(1)}} X^3 \cdot p_{\{0\}} \cdot x^3) \\
&\overset{(9.1.1)}{=} m_{(0)} \cdot n_{\{-1\}} p_{\{-1\}} \otimes (m_{(1)_1} \cdot n_{\{0\}_{(0)}} \cdot p_{\{0,-1\}} \otimes m_{(1)_2} n_{\{0\}_{(1)}} \cdot p_{\{0,0\}}) \\
&= m_{(0)} \cdot n_{\{-1\}} p_{\{-1\}} \otimes ((m_{(1)} \cdot n_{\{0\}})_{(0)} \cdot p_{\{0,-1\}} \otimes (m_{(1)} \cdot n_{\{0\}})_{(1)} \cdot p_{\{0,0\}}) \\
&= m_{(0)} \cdot (n \otimes_H p)_{\{-1\}} \otimes \psi_{2,N,P}(m_{(1)} \cdot (n \otimes_H p)_{\{0\}}) \\
&= (\mathrm{Id}_{\mathscr{U}(M)} \otimes \psi_{2,N,P})\psi_{2,M,N \otimes_H P}(m \otimes_H n \otimes_H p),
\end{aligned}$$

for all $m \in M$, $n \in N$ and $p \in P$, as required. We leave it to the reader to check that $\psi$ makes the two corresponding square diagrams in (1.3.2) commutative. □

At this point we can prove one of the main results of this section.

**Theorem 9.16** *Let $H$ be a quasi-bialgebra. Then giving a coalgebra in ${}_H^H\mathscr{M}_H^H$ is equivalent to giving a pair $(C, \pi)$ consisting of an $H$-bimodule coalgebra $C$ and an $H$-bimodule coalgebra morphism $\pi : C \to H$.*

*Proof* Let $(C,\underline{\Delta},\underline{\varepsilon})$ be a coalgebra in ${}^H_H\mathscr{M}^H_H$. The forgetful functor $\mathscr{U}$ in Lemma 9.15 is opmonoidal, so it follows that $C$ is an $H$-bimodule coalgebra via the original $H$-bimodule structure, but with comultiplication $\Delta_C$ and counit $\varepsilon_C$ defined by

$$\Delta_C : C = \mathscr{U}(C) \overset{\mathscr{U}(\underline{\Delta})}{\longrightarrow} \mathscr{U}(C \otimes_H C) \overset{\psi_{2,C,C}}{\longrightarrow} \mathscr{U}(C) \otimes \mathscr{U}(C) = C \otimes C$$

and $\varepsilon_C : C = \mathscr{U}(C) \overset{\mathscr{U}(\underline{\varepsilon})}{\longrightarrow} \mathscr{U}(H) \overset{\psi_0}{\longrightarrow} k$. Explicitly, for all $c \in C$ we have

$$\Delta_C(c) = c_{\underline{1}(0)} \cdot c_{\underline{2}\{-1\}} \otimes c_{\underline{1}(1)} \cdot c_{\underline{2}\{0\}} \text{ and } \varepsilon_C = \varepsilon\underline{\varepsilon} : C \to k, \tag{9.5.1}$$

where we denote $\underline{\Delta}(c) := c_{\underline{1}} \otimes c_{\underline{2}}$. If we take $\pi = \underline{\varepsilon} : C \to H$, then $\pi$ is a morphism in ${}^H_H\mathscr{M}^H_H$, and so in particular $H$-bilinear. The left and right $H$-colinearity of $\pi$ read

$$\Delta(\pi(c)) = c_{\{-1\}} \otimes \pi(c_{\{0\}}) = \pi(c_{(0)}) \otimes c_{(1)},$$

for all $c \in C$. These equalities allow us to compute that

$$\begin{aligned}(\pi \otimes \pi)\Delta_C(c) &= \pi(c_{\underline{1}(0)} \cdot c_{\underline{2}\{-1\}}) \otimes \pi(c_{\underline{1}(1)} \cdot c_{\underline{2}\{0\}}) \\ &= \pi(c_{\underline{1}(0)})c_{\underline{2}\{-1\}} \otimes c_{\underline{1}(1)}\pi(c_{\underline{2}\{0\}}) \\ &= \pi(c_{\underline{1}})_1\pi(c_{\underline{2}})_1 \otimes \pi(c_{\underline{1}})_2\pi(c_{\underline{2}})_2 \\ &= \Delta(\pi(c_{\underline{1}})\pi(c_{\underline{2}})) \\ &= \Delta(\pi(\pi(c_{\underline{1}})c_{\underline{2}})) = \Delta(\pi(c)),\end{aligned}$$

for all $c \in C$, where we freely used that $\pi$ is an $H$-bimodule morphism and the counit of $\underline{\Delta}$. Hence we have shown that $C$ is a coalgebra in ${}_H\overline{\mathscr{M}}_H$, and that $\pi : C \to H$ is a morphism of coalgebras within ${}_H\overline{\mathscr{M}}_H$.

Conversely, let $(C,\pi)$ be a pair consisting of an $H$-bimodule coalgebra $C$ and an $H$-bimodule coalgebra morphism $\pi : C \to H$. As above, denote by $(\Delta_C, \varepsilon_C)$ the coalgebra structure of $C$ in ${}_H\overline{\mathscr{M}}_H$. We claim that $C$ becomes a coalgebra in ${}^H_H\mathscr{M}^H_H$ via the original $H$-bimodule structure of $C$, $H$-coactions given by

$$\lambda_C : C \ni c \mapsto c_{\{-1\}} \otimes c_{\{0\}} := \pi(c_1) \otimes c_2 \in H \otimes C\,, \tag{9.5.2}$$

$$\rho_C : C \ni c \mapsto c_{(0)} \otimes c_{(1)} := c_1 \otimes \pi(c_2) \in C \otimes H, \tag{9.5.3}$$

for all $c \in C$, and coalgebra structure determined by

$$\underline{\Delta}(c) = E(c_1) \otimes_H c_2 \quad \text{and } \underline{\varepsilon} = \pi, \tag{9.5.4}$$

for all $c \in C$, where $E$ is the projection in (6.4.1) specialized for the object $C$, considered in ${}_H\mathscr{M}^H_H$ with the above structure.

Indeed, the fact that $C$ is an object in ${}^H_H\mathscr{M}^H_H$ via its regular $H$-actions and $H$-coactions (9.5.2), (9.5.3) follows easily from the defining properties of the pair $(C,\pi)$, as well as the fact that $\pi : C \to H$ becomes a morphism in ${}^H_H\mathscr{M}^H_H$. The comultiplication $\underline{\Delta}$ in (9.5.4) is an $H$-bimodule morphisms since

$$\underline{\Delta}(h \cdot c) = E(h_1 \cdot c_1) \otimes_H h_2 \cdot c_2 = E(h_1 \cdot c_1) \cdot h_2 \otimes_H c_2 \overset{(6.4.3)}{=} h \cdot E(c_1) \otimes_H c_2 = h \cdot \underline{\Delta}(c)$$

and $\underline{\Delta}(c\otimes h)=E(c_1\cdot h_1)\otimes_H c_2\cdot h_2\overset{(6.4.2)}{=}E(c_1)\otimes_H c_2\cdot h=\underline{\Delta}(c)\cdot h$, for all $c\in C$ and $h\in H$. The computation

$$\begin{aligned}\rho_{C\otimes_H C}\underline{\Delta}(c) &= E(c_1)_{(0)}\otimes_H c_{(2,1)}\otimes E(c_1)_{(1)}\pi(c_{(2,2)})\\ &\overset{(6.4.11)}{=} E(x^1\cdot E(c_1))\otimes_H c_{(2,1)}\otimes x^3\pi(c_{(2,2)})\\ &\overset{(6.4.2)}{=} E(x^1\cdot c_1)\otimes_H x^2\cdot c_{(2,1)}\otimes \pi(x^3\cdot c_{(2,2)})\\ &= E(c_{(1,1)}\cdot x^1)\otimes_H c_{(1,2)}\cdot x^2\otimes \pi(c_2\cdot x^3)\\ &\overset{(6.4.2)}{=} E(c_{(1,1)})\otimes_H c_{(1,2)}\otimes \pi(c_2)\\ &= (\underline{\Delta}\otimes \mathrm{Id}_H)\rho_C(c),\end{aligned}$$

valid for all $c\in C$, shows that $\underline{\Delta}$ in (9.5.4) is right $H$-colinear. It is also left $H$-colinear. To see this, observe that, for all $c\in C$, we have

$$\begin{aligned}&E(X^1\cdot c_1)_1\cdot X^2\pi(c_2)\otimes E(X^1\cdot c_1)_2\cdot X^3\\ &\qquad = q^1_1\cdot c_{(1,1)}\cdot p^1\otimes q^1_2\cdot c_{(1,2)}\cdot p^2S(q^2\pi(c_2)).\end{aligned}\tag{9.5.5}$$

Indeed, since

$$\begin{aligned}q^1\cdot c_{(1,1)}\otimes S(q^2\pi(c_{(1,2)}))\pi(c_2) &= c_1\cdot X^1\otimes S(\pi(c_{(2,1)})X^2)\alpha\pi(c_{(2,2)})X^3\\ &= c\cdot q^1\otimes S(q^2),\end{aligned}$$

for all $c\in C$ we compute that

$$\begin{aligned}&E(X^1\cdot c_1)_1\cdot X^2\pi(c_2)\otimes E(X^1\cdot c_1)_2\cdot X^3\\ &= (q^1X^1_1\cdot c_{(1,1)}\cdot\beta S(q^2X^1_2\pi(c_{(1,2)})))_1\cdot X^2\pi(c_2)\\ &\qquad\otimes (q^1X^1_1\cdot c_{(1,1)}\cdot\beta S(q^2X^1_2\pi(c_{(1,2)})))_2\cdot X^3\\ &\overset{(3.2.13)}{\underset{(3.2.14)}{=}} (q^1X^1_1\cdot c_{(1,1)})_1\cdot\delta^1S(q^2_2X^1_{(2,2)}\pi(c_{(1,2)_2}))f^1X^2\pi(c_2)\\ &\qquad\otimes (q^1X^1_1\cdot c_{(1,1)})_2\cdot\delta^2S(q^2_1X^1_{(2,1)}\pi(c_{(1,2)_1}))f^2X^3\\ &\overset{(3.2.26)}{=} (q^1\cdot(Q^1\cdot c_{(1,1)})_1)_1\cdot x^1_1\delta^1S(Q^2\pi(c_{(1,2)})x^3)\pi(c_2)\\ &\qquad\otimes (q^1\cdot(Q^1\cdot c_{(1,1)})_1)_2\cdot x^1_2\delta^2S(q^2\pi((Q^1\cdot c_{(1,1)})_2)x^2)\\ &\overset{(3.2.6)}{=} (q^1\cdot c_1)_1\cdot Q^1_{(1,1)}p^1\beta S(Q^2)\otimes (q^1\cdot c_1)_2\cdot Q^1_{(1,2)}p^2S(q^2\pi(c_2)Q^1_2)\\ &\overset{(3.2.21),(3.2.2)}{=} q^1_1\cdot c_{(1,1)}\cdot p^1\otimes q^1_2\cdot c_{(1,2)}\cdot p^2S(q^2\pi(c_2)),\end{aligned}$$

as desired. With the help of this relation we have that

$$\begin{aligned}\lambda_{C\otimes_H C}\underline{\Delta}(c) &= \lambda_{C\otimes_H C}(E(c_1)\otimes_H c_2)\\ &= \pi(E(c_1)_1)\pi(c_{(2,1)})\otimes E(c_1)_2\otimes_H c_{(2,2)}\\ &= \pi(E(X^1\cdot c_{(1,1)}\cdot x^1)_1\cdot\pi(X^2\cdot c_{(1,2)}\cdot x^2))\\ &\qquad\otimes E(X^1\cdot c_{(1,1)}\cdot x^1)_2\otimes_H X^3\cdot c_2\cdot x^3\\ &\overset{(6.4.2)}{=} \pi(E(X^1\cdot c_{(1,1)})_1\cdot X^2\pi(c_{(1,2)}))\otimes E(X^1\cdot c_{(1,1)})_2\cdot X^3\otimes_H c_2\\ &= \pi(q^1_1\cdot(c_1)_{(1,1)}\cdot p^1)\otimes q^1_2\cdot(c_1)_{(1,2)}\cdot p^2S(q^2\pi((c_1)_2))\otimes_H c_2\end{aligned}$$

$$\overset{(3.2.23)}{=} \pi(q^1_1x^1\cdot(c_1)_1)\otimes q^1_2x^2\cdot(c_1)_{(2,1)}\cdot\beta S(q^2x^3\pi((c_1)_{(2,2)}))\otimes_H c_2$$
$$\overset{(5.5.17)}{=} \pi(X^1\cdot c_{(1,1)})\otimes E(X^2\cdot c_{(1,2)})\otimes_H X^3\cdot c_2$$
$$\overset{(6.4.2)}{=} \pi(c_1)\otimes E(c_{(2,1)})\otimes_H c_{(2,2)} = (\mathrm{Id}_H\otimes\underline{\Delta})\lambda_C(c),$$

for all $c\in C$, and therefore $\underline{\Delta}$ in (9.5.4) is left $H$-colinear, as stated. So it remains to show that $\underline{\Delta}$ is coassociative in ${}^H_H\mathscr{M}^H_H$, and that $\underline{\varepsilon}$ is a counit for it. To this end, note that, for all $c\in C$,

$$\begin{aligned} E(E(c)_1)\otimes E(c)_2 &= E(q^1_1\cdot c_{(1,1)}\cdot(\beta S(q^2\pi(c_2)))_1)\otimes q^1_2\cdot c_{(1,2)}\cdot(\beta S(q^2\pi(c_2)))_2\\ &\overset{(6.4.2)}{=} E(q^1_1\cdot c_{(1,1)})\otimes q^1_2\cdot c_{(1,2)}\cdot\beta S(q^2\pi(c_2)). \end{aligned}$$

Therefore, we get that, for all $c\in C$,

$$\begin{aligned} &(\underline{\Delta}\otimes\mathrm{Id}_C)\underline{\Delta}(c)\\ &= E(E(c_1)_1)\otimes_H E(c_1)_2\otimes_H c_2\\ &= E(q^1_1\cdot(c_1)_{(1,1)})\otimes_H q^1_2\cdot(c_1)_{(1,2)}\cdot\beta S(q^2\pi((c_1)_2))\otimes_H c_2\\ &\overset{(6.4.2)}{=} E(q^1_1x^1\cdot(c_1)_1)\otimes_H q^1_2x^2\cdot(c_1)_{(2,1)}\cdot\beta S(q^2x^3\cdot\pi((c_1)_{(2,2)}))\otimes_H c_2\\ &\overset{(5.5.17)}{=} E(X^1\cdot c_{(1,1)})\otimes_H E(X^2\cdot c_{(1,2)})\otimes_H X^3\cdot c_2\\ &\overset{(6.4.2)}{=} E(c_1)\otimes_H E(c_{(2,1)})\otimes_H c_{(2,2)} = (\mathrm{Id}_C\otimes\underline{\Delta})\underline{\Delta}(c), \end{aligned}$$

that is, $\underline{\Delta}$ is coassociative in ${}^H_H\mathscr{M}^H_H$, as required. Finally, $\pi$ is a counit for $\underline{\Delta}$ since

$$\begin{aligned} &E(c_1)\cdot\pi(c_2) = E(c_{(0)})\cdot c_{(1)} \overset{(6.4.4)}{=} c,\\ &\pi(E(c_1))\cdot c_2 = q^1\pi(c_{(1,1)})\beta S(q^2\pi(c_{(1,2)}))\cdot c_2\\ &\qquad = X^1\beta S(X^2)\alpha X^3\cdot\varepsilon_C(c_1)c_2 \overset{(3.2.2)}{=} c, \end{aligned}$$

for all $c\in C$. One can check that the two correspondences defined above are inverses of each other, so we are done. □

Denote by $H$–BimCoalg$(\pi)$ the category whose objects are pairs $(C,\pi)$ consisting of an $H$-bimodule coalgebra $C$ and an $H$-bimodule coalgebra morphism $\pi: C\to H$. A morphism $\tau:(C,\pi)\to(C',\pi')$ in $H$–BimCoalg$(\pi)$ is a morphism of coalgebras $\tau: C\to C'$ within ${}_H\overline{\mathscr{M}}_H$ such that $\pi'\tau=\pi$. Also, by Coalg$({}^H_H\mathscr{M}^H_H)$ we denote the category of coalgebras and coalgebra morphisms within ${}^H_H\mathscr{M}^H_H$.

**Corollary 9.17** *$H$–BimCoalg$(\pi)$ and Coalg$({}^H_H\mathscr{M}^H_H)$ are isomorphic categories.*

*Proof* By Theorem 9.16, the desired isomorphism is given by the functors $\mathscr{T}$ : $H$–BimCoalg$(\pi)\to$ Coalg$({}^H_H\mathscr{M}^H_H)$ and $\mathscr{V}$ : Coalg$({}^H_H\mathscr{M}^H_H)\to H$–BimCoalg$(\pi)$ defined as follows. $\mathscr{T}$ sends $(C,\pi)$ to $C$, viewed as coalgebra in ${}^H_H\mathscr{M}^H_H$ under the structure given by (9.5.2) and (9.5.4). $\mathscr{T}$ sends a morphism to itself. If $(C,\underline{\Delta},\underline{\varepsilon})$ is a coalgebra in ${}^H_H\mathscr{M}^H_H$ then $\mathscr{V}(C)=C$, considered as a coalgebra in ${}_H\overline{\mathscr{M}}_H$ with the structure in (9.5.1). $\mathscr{V}$ acts as identity on morphisms.

We leave the verification of all these details to the reader. □

**Definition 9.18** For a coalgebra $B$ in ${}_H^H\mathscr{YD}$ denote by $B \bowtie H$ the $k$-vector space $B \otimes H$ endowed with the comultiplication

$$\Delta(b \bowtie h) = y^1X^1 \cdot b_1 \bowtie y^2Y^1(x^1X^2 \cdot b_2)_{[-1]}x^2X_1^3h_1 \otimes y_1^3Y^2 \cdot (x^1X^2 \cdot b_2)_{[0]} \bowtie y_2^3Y^3x^3X_2^3h_2, \tag{9.5.6}$$

and counit $\varepsilon(b \bowtie h) = \varepsilon_B(b)\varepsilon(h)$, for all $b \in B$ and $h \in H$. As before, $b \mapsto b_{[-1]} \otimes b_{[0]}$ is the left coaction of $H$ on $B$, $\Delta_B(b) = b_1 \otimes b_2$ is the comultiplication of $B$ in ${}_H^H\mathscr{YD}$ and $\varepsilon_B$ is its counit. We call $B \bowtie H$ the smash product coalgebra of $B$ and $H$.

We now have all the necessary ingredients for the proof of the main result of this section. In particular, the result says that a smash product coalgebra is indeed a coalgebra, but within ${}_H\overline{\mathscr{M}}_H$. Note that in the Hopf case we do not need the $H$-module structure on $B$, and that $B \bowtie H$ is an ordinary $k$-coalgebra, too.

**Theorem 9.19** *Let $H$ be a quasi-Hopf algebra with bijective antipode, $C$ an $H$-bimodule coalgebra and $\pi : C \to H$ an $H$-bimodule coalgebra morphism. Then there exists a coalgebra $B$ in ${}_H^H\mathscr{YD}$ such that $C \simeq B \bowtie H$ as $H$-bimodule coalgebras.*

*Proof* Consider $C = \mathscr{T}((C,\pi))$ as a coalgebra in ${}_H^H\mathscr{M}_H^H$ with the structure given by (9.5.2) and (9.5.4). Then $B = C^{\mathrm{co}(H)}$ is a coalgebra in ${}_H^H\mathscr{YD}$ and $C$ is isomorphic to $B \otimes H$ as coalgebras in ${}_H^H\mathscr{M}_H^H$. The fact that $C$ and $B \otimes H$ are isomorphic objects in ${}_H^H\mathscr{M}_H^H$ follows from the structure theorem for two-sided two-cosided Hopf modules over $H$. That they are, moreover, isomorphic as coalgebras in ${}_H^H\mathscr{M}_H^H$ is a consequence of a more general result. Namely, if the functors $\mathscr{S} : \mathscr{C} \to \mathscr{D}$ and $\mathscr{R} : \mathscr{D} \to \mathscr{C}$ define a monoidal category equivalence then $\mathscr{R}\mathscr{S}(\mathbf{C}) \cong \mathbf{C}$ is a coalgebra isomorphism in $\mathscr{C}$, for any coalgebra $\mathbf{C}$ within $\mathscr{C}$, where $\mathscr{R}\mathscr{S}(\mathbf{C})$ has the coalgebra structure provided by the monoidal structure of $\mathscr{R}\mathscr{S}$ and the coalgebra structure of $\mathbf{C}$.

The structure that makes $B \otimes H$ an object in ${}_H^H\mathscr{M}_H^H$ is the one in (9.2.10)–(9.2.12), while the coalgebra structure of $B \otimes H$ in ${}_H^H\mathscr{M}_H^H$ is obtained from (6.6.2). With these structures, $\mathscr{T}((C,\pi))$ and $B \otimes H$ are isomorphic as coalgebras in ${}_H^H\mathscr{M}_H^H$. By Corollary 9.17 we deduce that $(C,\pi) = \mathscr{V}\mathscr{T}(C)$ is isomorphic to $\mathscr{V}(B \otimes H)$ as objects in $H$–BimCoalg$(\pi)$, and consequently as $H$-bimodule coalgebras. To end the proof it suffices to show that $\mathscr{V}(B \otimes H) = (B \bowtie H, \varepsilon_B \otimes \mathrm{Id}_H)$. As a byproduct, we get that $B \bowtie H$ is indeed a coalgebra in ${}_H\overline{\mathscr{M}}_H$, as claimed.

The latest assertion follows from the following computation:

$$\begin{aligned}
\Delta(b \otimes h) &\overset{(9.5.1)}{=} (b \otimes h)_{\underline{1}_{(0)}} \cdot (b \otimes h)_{\underline{2}_{\{-1\}}} \otimes (b \otimes h)_{\underline{1}_{(1)}} \cdot (b \otimes h)_{\underline{2}_{\{-1\}}} \\
&\overset{(6.6.2)}{=} (X^1 \cdot b_1 \otimes 1_H)_{(0)} \cdot (X^2 \cdot b_2 \otimes X^3h)_{\{-1\}} \\
&\qquad \otimes (X^1b_1 \otimes 1_H)_{(1)} \cdot (X^3 \cdot b_2 \otimes X^3h)_{\{0\}} \\
&\underset{(9.2.12)}{\overset{(9.2.11)}{=}} (y^1X^1 \cdot b_1 \otimes y^2) \cdot Y^1(x^1X^2 \cdot b_2)_{[-1]}x^2X_1^3h_1 \\
&\qquad \otimes y^3 \cdot (Y^2 \cdot (x^1X^2 \cdot b_2)_{[0]} \otimes Y^3x^3X_2^3h_2) \\
&\overset{(9.2.10)}{=} (y^1X^1 \cdot b_1 \otimes y^2Y^1(x^1X^2 \cdot b_2)_{[-1]}x^2X_1^3h_1)
\end{aligned}$$

$$\otimes(y_1^3Y^2\cdot(x^1X^2\cdot b_2)_{[0]}\otimes y_2^3Y^3x^3X_2^3h_2),$$

valid for any $b\in B$ and $h\in H$, and the fact that $\varepsilon(b\otimes h)=\varepsilon\underline{\varepsilon}(b\otimes h)=\varepsilon_B(b)\varepsilon(h)$. This finishes the proof of the theorem. □

## 9.6 A Braided Monoidal Structure on ${}_H^H\mathcal{M}_H^H$

We show that ${}_H^H\mathcal{M}_H^H$ is braided monoidally equivalent to ${}_H^H\mathcal{YD}$. To this end, we apply Proposition 1.56 to the strong monoidal equivalence in Proposition 9.9. Recall that the category ${}_H^H\mathcal{YD}$ is braided via the braiding given by (8.2.15).

If $H$ is a quasi-Hopf algebra with bijective antipode and $M\in{}_H^H\mathcal{M}_H^H$, then by $E_M: M\to M$ we denote the projection to the space of coinvariants of $M$ of the second type, defined by $E_M(m)=X^1\cdot m_{(0)}\cdot\beta S(X^2m_{(1)})\alpha X^3=q^1\cdot\overline{E}_M(m)\cdot S(q^2)$, for all $m\in M$. Here $M\ni m\mapsto\rho_M(m):=m_{(0)}\otimes m_{(1)}\in M\otimes H$ denotes the right coaction of $H$ on $M$ and $q_R=q^1\otimes q^2$.

In what follows, we need the following property of $\overline{E}_M$.

**Lemma 9.20** *Let $H$ be a quasi-Hopf algebra with bijective antipode and $M$ a two-sided two-cosided Hopf module over $H$. Then, for all $m\in M$, we have that*

$$m_{\{-1\}}\otimes\overline{E}_M(m_{\{0\}})=X^1Y_1^1\overline{E}_M(m_{(0)})_{\{-1\}}g^1S(q^2Y_2^2)Y^3m_{(1)}$$
$$\otimes X^2Y_2^1\cdot\overline{E}_M(m_{(0)})_{\{0\}}\cdot g^2S(X^3q^1Y_1^2). \qquad (9.6.1)$$

*Proof* For all $m\in M\in{}_H^H\mathcal{M}_H^H$ we have

$$X^1Y_1^1\overline{E}_M(m_{(0)})_{\{-1\}}g^1S(q^2Y_2^2)Y^3m_{(1)}\otimes X^2Y_2^1\cdot\overline{E}_M(m_{(0)})_{\{0\}}\cdot g^2S(X^3q^1Y_1^2)$$
$$\overset{(3.2.13)}{=} X^1Y_1^1m_{(0,0)_{\{-1\}}}\beta_1g^1S(q^2Y_2^2m_{(0,1)_2})Y^3m_{(1)}$$
$$\otimes X^2Y_2^1\cdot m_{(0,0)_{\{0\}}}\cdot\beta_2g^2S(X^3q^1Y_1^2m_{(0,1)_1})$$
$$\overset{(3.2.14)}{=} X^1m_{(0)_{\{-1\}}}Y_1^1\delta^1S(q^2m_{(1)_{(1,2)}}Y_2^2)m_{(1)_2}Y^3$$
$$\otimes X^2\cdot m_{(0)_{\{0\}}}\cdot Y_2^1\delta^2S(X^3q^1m_{(1)_{(1,1)}}Y_1^2)$$
$$\overset{(3.2.21)}{=} m_{\{-1\}}X^1Y_1^1\delta^1S(q^2Y_2^2)Y^3\otimes m_{\{0\}_{(0)}}\cdot X^2Y_2^1\delta^2S(m_{\{0\}_{(1)}}X^3q^1Y_1^2)$$
$$\overset{(5.5.17)}{=} m_{\{-1\}}X^1q^1_{(1,1)}x_1^1\delta^1S(q^2x^3)\otimes m_{\{0\}_{(0)}}\cdot X^2q^1_{(1,2)}x_2^1\delta^2S(m_{\{0\}_{(1)}}X^3q_2^1x^2)$$
$$\overset{(3.2.6),(3.1.7)}{=} m_{\{-1\}}q_1^1\beta S(q^2)\otimes m_{\{0\}_{(0)}}\cdot q^1_{(2,1)}\beta S(m_{\{0\}_{(1)}}q^1_{(2,2)})$$
$$\overset{(3.2.1),(3.2.2)}{=} m_{\{-1\}}\otimes m_{\{0\}_{(0)}}\cdot\beta S(m_{\{0\}_{(1)}})=m_{\{-1\}}\otimes\overline{E}_M(m_{\{0\}}),$$

as needed. □

We are now able to define a braiding for ${}_H^H\mathcal{M}_H^H$.

**Theorem 9.21** *If $H$ is a quasi-Hopf algebra with bijective antipode then ${}_H^H\mathcal{M}_H^H$ is a braided category with the braiding defined by*

$$d_{M,N}: M\otimes_HN\ni m\otimes_Hn\mapsto E_N(m_{\{-1\}}\cdot n_{(0)})\otimes_Hm_{\{0\}}\cdot n_{(1)}\in N\otimes_HM, \qquad (9.6.2)$$

*for all* $M,N \in {}_H^H\mathscr{M}_H^H$.

*Furthermore, if we consider* ${}_H^H\mathscr{M}_H^H$ *as a braided category with the braiding d, then* ${}_H^H\mathscr{M}_H^H$ *is braided monoidally equivalent to* ${}_H^H\mathscr{YD}$*, where the braiding on* ${}_H^H\mathscr{YD}$ *is c as in* (8.2.15).

*Proof* Let ${}_H^H\mathscr{YD} \underset{\mathscr{G}}{\overset{\mathscr{F}}{\rightleftarrows}} {}_H^H\mathscr{M}_H^H$ be the strong monoidal equivalence functors defined in Proposition 9.9. We have that $\overline{\nu}$ defined, for all $M \in {}_H^H\mathscr{M}_H^H$, by

$$\overline{\nu}_M : M^{\overline{\mathrm{co}(H)}} \otimes H \ni m \otimes h \mapsto X^1 \cdot m \cdot S(X^2)\alpha X^3 h \in M, \tag{9.6.3}$$

is a natural monoidal isomorphism between $\mathscr{F}\mathscr{G}$ and $\mathrm{Id}_{{}_H^H\mathscr{M}_H^H}$, while $\zeta$ given, for all $M \in {}_H^H\mathscr{YD}$, by

$$\zeta_M : (M \otimes H)^{\overline{\mathrm{co}(H)}} \ni m \otimes h \mapsto \varepsilon(h) m \in M \tag{9.6.4}$$

is a natural monoidal isomorphism between $\mathscr{G}\mathscr{F}$ and $\mathrm{Id}_{{}_H^H\mathscr{YD}}$.

We show that $(\mathscr{F}, \overline{\nu}, \zeta)$ obeys the condition in (1.5.12), that is, for all $M \in {}_H^H\mathscr{YD}$,

$$\overline{\nu}_{\mathscr{F}(M)} = \mathscr{F}(\zeta_M) : (M \otimes H)^{\overline{\mathrm{co}(H)}} \otimes H \to M \otimes H.$$

To this end, by Remark 6.17 we have $(M \otimes H)^{\overline{\mathrm{co}(H)}} = \{p^1 \cdot m \otimes p^2 \mid m \in M\}$, where $p_R = p^1 \otimes p^2$ is the element defined in (3.2.19). Thus $\zeta_M(p^1 \cdot m \otimes p^2) = m$, for all $m \in M$, and therefore $\mathscr{F}(\zeta_M)((p^1 \cdot m \otimes p^2) \otimes h) = m \otimes h$, for all $m \in M$ and $h \in H$.

If $q_R$ is the element in (3.2.19) we compute that

$$\begin{aligned}
\overline{\nu}_{\mathscr{F}(M)}((p^1 \cdot m \otimes p^2) \otimes h) &\overset{(9.6.3),(3.2.19)}{=} q^1 \cdot (p^1 \cdot m \otimes p^2) \cdot S(q^2) h \\
&\overset{(9.2.10)}{=} q_1^1 p^1 \cdot m \otimes q_2^1 p^2 S(q^2) h \\
&\overset{(3.2.23)}{=} m \otimes h,
\end{aligned}$$

for all $m \in M$ and $h \in H$. We conclude that $\overline{\nu}_{\mathscr{F}(M)} = \mathscr{F}(\zeta_M)$, as stated.

It follows by Proposition 1.56 and Corollary 1.57 that the braiding $c$ for ${}_H^H\mathscr{YD}$ can be transferred along $\mathscr{F}$ to a braiding $d$ on ${}_H^H\mathscr{M}_H^H$ such that $\mathscr{F}$ becomes a braided monoidal equivalence. It only remains to show that $d$ is as in (9.6.2). Using (1.5.13), we see that

$$\begin{aligned}
d_{M,N} &= \Big( m \otimes_H n \overset{\overline{\nu}_M^{-1} \otimes_H \overline{\nu}_N^{-1}}{\longrightarrow} (\overline{E}_M(m_{(0)}) \otimes m_{(1)}) \otimes_H (\overline{E}_N(n_{(0)}) \otimes n_{(1)}) \\
&\overset{\varphi_{2,\mathscr{G}(M),\mathscr{G}(N)}}{\longrightarrow} (x^1 \triangleright \overline{E}_M(m_{(0)}) \otimes x^2 m_{(1)_1} \triangleright \overline{E}_N(n_{(0)})) \otimes x^3 m_{(1)_2} n_{(1)} \\
&= (\overline{E}_M(x^1 \cdot m_{(0)}) \otimes x^2 m_{(1)_1} \triangleright \overline{E}_N(n_{(0)})) \otimes x^3 m_{(1)_2} n_{(1)} \\
&\overset{(6.4.2)}{=} (\overline{E}_M(m_{(0,0)}) \otimes m_{(0,1)} \triangleright \overline{E}_N(n_{(0)})) \otimes m_{(1)} n_{(1)} \\
&\overset{c_{\mathscr{G}(M),\mathscr{G}(N)} \otimes \mathrm{Id}_H}{\longrightarrow} (\overline{E}_M(m_{(0,0)})_{[-1]} m_{(0,1)} \triangleright \overline{E}_N(n_{(0)}) \otimes \overline{E}_M(m_{(0,0)})_{[0]}) \otimes m_{(1)} n_{(1)} \\
&\overset{(9.2.14)}{=} (X^1 Y_1^1 \overline{E}_M(m_{(0,0)})_{\{-1\}} g^1 S(q^2 Y_2^2) Y^3 m_{(0,1)} \triangleright \overline{E}_N(n_{(0)}) \\
&\qquad \otimes X^2 Y_2^1 \cdot \overline{E}_M(m_{(0,0)})_{\{0\}} \cdot g^2 S(X^3 q^1 Y_1^2)) \otimes m_{(1)} n_{(1)}
\end{aligned}$$

$$
\begin{aligned}
&\stackrel{(9.6.1)}{=} (m_{(0)_{\{-1\}}} \triangleright \overline{E}_N(n_{(0)}) \otimes \overline{E}_N(m_{(0)_{\{0\}}})) \otimes m_{(1)}n_{(1)}\\
&\stackrel{\varphi^{-1}_{2,\mathscr{G}(N),\mathscr{G}(M)}}{\longrightarrow} (X^1 m_{(0)_{\{-1\}}} \triangleright \overline{E}_N(n_{(0)}) \otimes 1_H) \otimes_H (X^2 \triangleright \overline{E}_N(m_{(0)_{\{0\}}}) \otimes X^3 m_{(1)}n_{(1)})\\
&\stackrel{(6.4.2)}{=} (m_{\{-1\}} \triangleright \overline{E}_N(n_{(0)}) \otimes 1_H) \otimes_H (\overline{E}_M(m_{\{0\}_{(0)}}) \otimes m_{\{0\}_{(1)}} n_{(1)})\\
&\stackrel{\overline{v}_N \otimes \overline{v}_M}{\longrightarrow} q^1 \cdot \overline{E}_N(m_{\{-1\}} \cdot n_{(0)}) \cdot S(q^2) \otimes_H Q^1 \cdot \overline{E}_M(m_{\{0\}_{(0)}}) \cdot S(Q^2) m_{\{0\}_{(1)}} n_{(1)}\\
&\stackrel{(3.2.19)}{=} X^1 m_{\{-1\}_1} \cdot \overline{E}_N(n_{(0)}) \cdot S(X^2 m_{\{-1\}_2})\alpha\\
&\qquad\qquad \otimes_H X^3 Q^1 \cdot \overline{E}_M(m_{\{0\}_{(0)}}) \cdot S(Q^2) m_{\{0\}_{(1)}} n_{(1)}\\
&\stackrel{(3.2.21)}{=} X^1 m_{\{-1\}_1} \cdot \overline{E}_N(n_{(0)}) \cdot S(X^2 m_{\{-1\}_2})\alpha\\
&\qquad\qquad \otimes_H Q^1 \cdot \overline{E}_M(X^3_1 \cdot m_{\{0\}_{(0)}}) \cdot S(Q^2) X^3_2 m_{\{0\}_{(1)}} n_{(1)}\\
&\stackrel{(6.4.2)}{=} m_{\{-1\}} X^1 \cdot \overline{E}_N(n_{(0)}) \cdot S(m_{\{0,-1\}} X^2)\alpha \otimes_H E_M(m_{\{0,0\}_{(0)}}) \cdot m_{\{0,0\}_{(1)}} X^3 n_{(1)}\\
&\stackrel{(6.4.4)}{=} m_{\{-1\}} X^1 \cdot \overline{E}_N(n_{(0)}) \cdot S(m_{\{0,-1\}} X^2)\alpha \otimes_H m_{\{0,0\}} \cdot X^3 n_{(1)}\\
&= q^1 \cdot \overline{E}_N(m_{\{-1\}} \cdot n_{(0)}) \cdot S(q^2) \otimes_H m_{\{0\}} \cdot n_{(1)}\\
&= \left. E_N(m_{\{-1\}} \cdot n_{(0)}) \otimes_H m_{\{0\}} \cdot n_{(1)} \right),
\end{aligned}
$$

for all $m \in M$ and $n \in N$, as desired. □

## 9.7 Hopf Algebras within ${}_H^H\mathscr{M}_H^H$

Let $H$ be a quasi-Hopf algebra with bijective antipode. The aim of this section is to characterize the bialgebras and the Hopf algebras in ${}_H^H\mathscr{M}_H^H$. We show that giving a Hopf algebra in ${}_H^H\mathscr{M}_H^H$ is equivalent to giving a quasi-Hopf algebra projection for $H$. Later, we will obtain that quasi-Hopf algebra projections are characterized by the biproduct quasi-Hopf algebras, and therefore by Hopf algebras in ${}_H^H\mathscr{YD}$, too.

We denote by $\underline{H\text{–qBialgProj}}$ (resp. $\underline{H\text{–qHopfProj}}$) the category whose objects are triples $(A, i, \pi)$ consisting of a quasi-bialgebra (resp. quasi-Hopf algebra) $A$ and two quasi-bialgebra (resp. quasi-Hopf algebra) morphisms $H \underset{\pi}{\overset{i}{\rightleftarrows}} A$ such that $\pi i = \mathrm{Id}_H$. A morphism in $\underline{H\text{–qBialgProj}}$ (resp. $\underline{H\text{–qHopfProj}}$) between $(A, i, \pi)$ and $(A', i', \pi')$ is a quasi-bialgebra (resp. quasi-Hopf algebra) morphism $\tau : A \to A'$ such that $\tau i = i'$ and $\pi\tau = \pi'$. The objects of $\underline{H\text{–qBialgProj}}$ (resp. $\underline{H\text{–qHopfProj}}$) will be called quasi-bialgebra (resp. quasi-Hopf algebra) projections for $H$.

We also denote by $\mathrm{Bialg}({}_H^H\mathscr{M}_H^H)$ (resp. $\mathrm{Hopf}({}_H^H\mathscr{M}_H^H)$) the category of bialgebras (resp. Hopf algebras) and bialgebra morphisms within ${}_H^H\mathscr{M}_H^H$.

As announced, we next prove that the categories $\mathrm{Bialg}({}_H^H\mathscr{M}_H^H)$ and $\underline{H\text{–qBialgProj}}$ (resp. $\mathrm{Hopf}({}_H^H\mathscr{M}_H^H)$ and $\underline{H\text{–qHopfProj}}$) are isomorphic. We first need some lemmas.

**Lemma 9.22** *Take $M, N \in {}_H^H\mathscr{M}_H^H$, and the elements $m, m' \in M$ and $n, n' \in N$. Then*

$$m \otimes_H n = m' \otimes_H n' \;\Leftrightarrow\; E(m_{(0)}) \otimes m_{(1)} \cdot n = E(m'_{(0)}) \otimes m'_{(1)} \cdot n'. \tag{9.7.1}$$

*Proof* From Section 6.4 we know that $\nu_M^{-1} : M \ni m \mapsto E_M(m_{(0)}) \otimes m_{(1)} \in M^{\mathrm{co}(H)} \otimes H$ is an isomorphism in ${}_H^H\mathscr{M}_H^H$, for all $M \in {}_H^H\mathscr{M}_H^H$. Here $M^{\mathrm{co}(H)}$ is the image of $E_M$, a left $H$-module via the structure given by $h \neg E_M(m) = E_M(h \cdot m)$, for all $h \in H$ and $m \in M$. For a $k$-vector space $U$ and $V \in {}_H\mathscr{M}$ denote by $\Upsilon_{U,V} : (U \otimes H) \otimes_H V \to U \otimes V$ the canonical isomorphism. We then have that $m \otimes_H n = m' \otimes_H n'$ if and only if

$$\Upsilon_{M^{\mathrm{co}(H)}, N^{\mathrm{co}(H)} \otimes H}(\nu_M^{-1} \otimes_H \nu_N^{-1})(m \otimes_H n) = \Upsilon_{M^{\mathrm{co}(H)}, N^{\mathrm{co}(H)} \otimes H}(\nu_M^{-1} \otimes_H \nu_N^{-1})(m' \otimes_H n'),$$

if and only if

$$E_M(m_{(0)}) \otimes E_N(m_{(1)_1} \cdot n_{(0)}) \otimes m_{(1)_2} n_{(1)} = E_M(m'_{(0)}) \otimes E_N(m'_{(1)_1} \cdot n'_{(0)}) \otimes m'_{(1)_2} n'_{(1)}.$$

Thus, if $m \otimes_H n = m' \otimes_H n'$ then

$$\begin{aligned} E_M(m_{(0)}) \otimes E_N(m_{(1)_1} \cdot n_{(0)}) \cdot m_{(1)_2} n_{(1)} &= E_M(m'_{(0)}) \otimes E_N(m'_{(1)_1} \cdot n'_{(0)}) \cdot m'_{(1)_2} n'_{(1)} \\ \overset{(6.4.3)}{\Leftrightarrow} E_M(m_{(0)}) \otimes m_{(1)} \cdot E_N(n_{(0)}) \cdot n_{(1)} &= E_M(m'_{(0)}) \otimes m'_{(1)} \cdot E_N(n'_{(0)}) \cdot n'_{(1)} \\ \overset{(6.4.4)}{\Leftrightarrow} E_M(m_{(0)}) \otimes m_{(1)} \cdot n &= E_M(m'_{(0)}) \otimes m'_{(1)} \cdot n'. \end{aligned}$$

The converse follows easily from (6.4.4), and we are done. □

Now we construct the functor that gives the desired isomorphism of categories.

**Proposition 9.23** *Let $H$ be a quasi-Hopf algebra with bijective antipode. Then there is a functor*

$$\mathscr{V} : \mathrm{Bialg}({}_H^H\mathscr{M}_H^H) \to H\text{–}\underline{\mathrm{qBialgProj}}.$$

*On objects, $\mathscr{V}$ sends a bialgebra $(B, \underline{m}_B, i : H \to B, \underline{\Delta}_B, \pi : B \to H)$ in ${}_H^H\mathscr{M}_H^H$ to the triple $(B, i, \pi)$, where $B$ is considered as a quasi-bialgebra via $m_B := \underline{m}_B q_{B,B}$ (where $q_{B,B} : B \otimes B \to B \otimes_H B$ is the canonical surjection), $1_B = i(1_H)$,*

$$\Delta_B(b) = b_{\underline{1}_{(0)}} \cdot b_{\underline{2}_{\{-1\}}} \otimes b_{\underline{1}_{(1)}} \cdot b_{\underline{2}_{\{0\}}} \quad \textit{and} \quad \varepsilon_B = \varepsilon\pi : B \to k$$

*as in (9.5.1), and $\Phi_B = (i \otimes i \otimes i)(\Phi)$. $\mathscr{V}$ acts as identity on morphisms.*

*Proof* We must check that $(B, m_B, 1_B, \Delta_B, \varepsilon_B, \Phi_B)$ is indeed a quasi-bialgebra and, moreover, that $i, \pi$ become quasi-bialgebra morphisms.

We know that $(B, \underline{m}_B, i)$ is an algebra in ${}_H^H\mathscr{M}_H^H$ if and only if $(B, m_B, 1_B)$ is a $k$-algebra and at the same time an $H$-bicomodule algebra via the original left and right $H$-coactions and reassociators $\Phi_\lambda = X^1 \otimes X^2 \otimes i(X^3)$, $\Phi_\rho = i(X^1) \otimes X^2 \otimes X^3$ and $\Phi_{\lambda,\rho} = X^1 \otimes i(X^2) \otimes X^3$, such that, for all $h \in H$,

$$\lambda(i(h)) = h_1 \otimes i(h_2) \quad \text{and} \quad \rho(i(h)) = i(h_1) \otimes h_2. \tag{9.7.2}$$

In other words, $i$ is an $H$-bicomodule algebra morphism. Furthermore, the $H$-bimodule structure on $B$ is nothing but the one induced by the restriction of scalars functor defined by $i$.

Similarly, we have that $(B, \Delta_B, \varepsilon_B = \varepsilon\pi)$ is a coalgebra within the monoidal category ${}_H\overline{\mathcal{M}}_H := ({}_H\mathcal{M}_H, \otimes, k, a', l', r')$, that is, an $H$-bimodule coalgebra, and $\pi : B \to H$ is a coalgebra morphism in ${}_H\overline{\mathcal{M}}_H$. If we denote $\Delta_B(b) = b_1 \otimes b_2$ we have

$$i(X^1)b_{(1,1)}i(x^1) \otimes i(X^2)b_{(1,2)}i(x^2) \otimes i(X^3)b_2 i(x^3) = b_1 \otimes b_{(2,1)} \otimes b_{(2,2)}, \qquad (9.7.3)$$

for all $b \in B$, $\varepsilon\pi = \varepsilon_B$ and $\Delta(\pi(b)) = \pi(b_1) \otimes \pi(b_2)$, for all $b \in B$.

The left and right $H$-coactions on $B$ can be recovered from $\Delta_B$ and $\pi$ as

$$\lambda(b) = \pi(b_1) \otimes b_2 \quad \text{and} \quad \rho(b) = b_1 \otimes \pi(b_2), \ \ \forall\, b \in B. \qquad (9.7.4)$$

Since $i$ is the unit and $\pi$ is the counit of the bialgebra $B$ within ${}^H_H\mathcal{M}^H_H$ it follows that $\pi i = \mathrm{Id}_H$, and therefore $\pi$ is surjective. Furthermore, $\pi : B \to H$ is an algebra morphism in ${}^H_H\mathcal{M}^H_H$, so $\pi$ is a $k$-algebra morphism as well. As we have seen, $\pi$ intertwines the comultiplications $\Delta_B$ and $\Delta$ of $B$ and $H$, too. If we define $\Phi_B := (i \otimes i \otimes i)(\Phi)$, it is clear that $(\pi \otimes \pi \otimes \pi)(\Phi_B) = \Phi$.

Combining (9.7.2) and (9.7.4) we get

$$\lambda(i(h)) = \pi(i(h)_1) \otimes i(h)_2 = h_1 \otimes i(h_2), \ \ \forall\, h \in H,$$

and therefore $\pi(i(h)_1) \otimes i(h)_2 = \pi(i(h_1)) \otimes i(h_2)$, for all $h \in H$. As $\pi$ is surjective, we obtain that $i$ intertwines the comultiplications $\Delta$ and $\Delta_B$ of $H$ and $B$, and so $\Delta_B(1_B) = \Delta_B(i(1_H)) = i(1_H) \otimes i(1_H) = 1_B \otimes 1_B$. It is also an algebra morphism such that $(i \otimes i \otimes i)(\Phi) = \Phi_B$ and $\varepsilon_B i = \varepsilon$.

The most difficult part is to show that $\Delta_B$ is multiplicative, that is,

$$\Delta_B(bb') = (b_{\underline{1}_{(0)}} \cdot b_{\underline{2}_{\{-1\}}})(b'_{\underline{1}_{(0)}} \cdot b'_{\underline{2}_{\{-1\}}}) \otimes (b_{\underline{1}_{(1)}} \cdot b_{\underline{2}_{\{0\}}})(b'_{\underline{1}_{(1)}} \cdot b'_{\underline{2}_{\{0\}}}), \qquad (9.7.5)$$

for all $b, b' \in B$. Towards this end, observe first that by (9.6.2) and (9.7.1) we have that $\underline{\Delta}_B$ is multiplicative in ${}^H_H\mathcal{M}^H_H$ if and only if

$$\begin{aligned} & E((bb')_{\underline{1}_{(0)}}) \otimes (bb')_{\underline{1}_{(1)}} \cdot (bb')_{\underline{2}} \\ & \quad = E(b_{\underline{1}_{(0)}} E(b_{\underline{2}_{\{-1\}}} \cdot b'_{\underline{1}_{(0)}})_{(0)}) \otimes b_{\underline{1}_{(1)}} E(b_{\underline{2}_{\{-1\}}} \cdot b'_{\underline{1}_{(0)}})_{(1)} \cdot (b_{\underline{2}_{\{0\}}} \cdot b'_{\underline{1}_{(1)}}) b'_{\underline{2}}, \qquad (9.7.6) \end{aligned}$$

for all $b, b' \in B$, where, for simplicity, from now on we denote $E_B$ by $E$. This allows us to compute that

$$\begin{aligned} & \Delta_B(bb') \\ & = (bb')_{\underline{1}_{(0)}} \cdot (bb')_{\underline{2}_{\{-1\}}} \otimes (bb')_{\underline{1}_{(1)}} \cdot (bb')_{\underline{2}_{\{0\}}} \\ & \overset{(6.4.4)}{=} E((bb')_{\underline{1}_{(0,0)}}) \cdot (bb')_{\underline{1}_{(0,1)}} (bb')_{\underline{2}_{\{-1\}}} \otimes (bb')_{\underline{1}_{(1)}} \cdot (bb')_{\underline{2}_{\{0\}}} \\ & \overset{(6.4.2)}{=} x^1 \neg E((bb')_{\underline{1}_{(0)}}) \cdot x^2((bb')_{\underline{1}_{(1)}} \cdot (bb')_{\underline{2}})_{\{-1\}} \otimes x^3 \cdot ((bb')_{\underline{1}_{(1)}} \cdot (bb')_{\underline{2}})_{\{0\}} \\ & \overset{(9.7.6)}{=} E(x^1 \cdot b_{\underline{1}_{(0)}} E(b_{\underline{2}_{\{-1\}}} \cdot b'_{\underline{1}_{(0)}})_{(0)}) \cdot x^2 b_{\underline{1}_{(1)_1}} E(b_{\underline{2}_{\{-1\}}} \cdot b'_{\underline{1}_{(0)}})_{(1)_1} \\ & \qquad ((b_{\underline{2}_{\{0\}}} \cdot b'_{\underline{1}_{(1)}}) b'_{\underline{2}})_{\{-1\}} \otimes x^3 b_{\underline{1}_{(1)_2}} E(b_{\underline{2}_{\{-1\}}} \cdot b'_{\underline{1}_{(0)}})_{(1)_2} \cdot ((b_{\underline{2}_{\{0\}}} \cdot b'_{\underline{1}_{(1)}}) b'_{\underline{2}})_{\{0\}} \\ & \overset{(6.4.2)}{=} E(b_{\underline{1}_{(0,0)}} E(b_{\underline{2}_{\{-1\}}} \cdot b'_{\underline{1}_{(0)}})_{(0,0)}) \cdot b_{\underline{1}_{(0,1)}} E(b_{\underline{2}_{\{-1\}}} \cdot b'_{\underline{1}_{(0)}})_{(0,1)} \\ & \qquad (b_{\underline{2}_{\{0\}}} \cdot b'_{\underline{1}_{(1)}})_{\{-1\}} b'_{\underline{2}_{\{-1\}}} \otimes b_{\underline{1}_{(1)}} E(b_{\underline{2}_{\{-1\}}} \cdot b'_{\underline{1}_{(0)}})_{(1)} \cdot (b_{\underline{2}_{\{0\}}} \cdot b'_{\underline{1}_{(1)}})_{\{0\}} b'_{\underline{2}_{\{0\}}} \end{aligned}$$

$$
\begin{aligned}
&\overset{(6.4.4)}{=} b_{\underline{1}_{(0)}} E(b_{\underline{\underline{2}}_{\{-1\}}} \cdot b'_{\underline{1}_{(0)}})_{(0)} \cdot b_{\underline{\underline{2}}_{\{0,-1\}}} b'_{\underline{1}_{(1)_1}} b'_{\underline{\underline{2}}_{\{-1\}}} \\
&\qquad \otimes b_{\underline{1}_{(1)}} E(b_{\underline{\underline{2}}_{\{-1\}}} \cdot b'_{\underline{1}_{(0)}})_{(1)} \cdot (b_{\underline{\underline{2}}_{\{0,0\}}} \cdot b'_{\underline{1}_{(1)_2}}) b'_{\underline{\underline{2}}_{\{0\}}} \\
&\overset{(6.4.11)}{=} b_{\underline{1}_{(0)}} E(x^1 b_{\underline{\underline{2}}_{\{-1\}}} \cdot b'_{\underline{1}_{(0)}}) \cdot x^2 b_{\underline{\underline{2}}_{\{0,-1\}}} b'_{\underline{1}_{(1)_1}} b'_{\underline{\underline{2}}_{\{-1\}}} \otimes (b_{\underline{1}_{(1)}} x^3 \cdot b_{\underline{\underline{2}}_{\{0,0\}}} \cdot b'_{\underline{1}_{(1)_2}}) b'_{\underline{\underline{2}}_{\{0\}}} \\
&\overset{(6.4.2)}{=} b_{\underline{1}_{(0)}} E(b_{\underline{\underline{2}}_{\{-1\}_1}} \cdot b'_{\underline{1}_{(0,0)}}) \cdot b_{\underline{\underline{2}}_{\{-1\}_2}} b'_{\underline{1}_{(0,1)}} b'_{\underline{\underline{2}}_{\{-1\}}} \otimes b_{\underline{1}_{(1)}} \cdot (b_{\underline{\underline{2}}_{\{0\}}} \cdot b'_{\underline{1}_{(1)}}) b'_{\underline{\underline{2}}_{\{0\}}} \\
&\overset{(6.4.3)}{=} (b_{\underline{1}_{(0)}} \cdot b_{\underline{\underline{2}}_{\{-1\}}})(E(b'_{\underline{1}_{(0,0)}}) \cdot b'_{\underline{1}_{(0,1)}} b'_{\underline{\underline{2}}_{\{-1\}}}) \otimes (b_{\underline{1}_{(1)}} \cdot b_{\underline{\underline{2}}_{\{0\}}})(b'_{\underline{1}_{(1)}} \cdot b'_{\underline{\underline{2}}_{\{0\}}}) \\
&\overset{(6.4.4)}{=} (b_{\underline{1}_{(0)}} \cdot b_{\underline{\underline{2}}_{\{-1\}}})(b'_{\underline{1}_{(0)}} \cdot b'_{\underline{\underline{2}}_{\{-1\}}}) \otimes (b_{\underline{1}_{(1)}} \cdot b_{\underline{\underline{2}}_{\{0\}}})(b'_{\underline{1}_{(1)}} \cdot b'_{\underline{\underline{2}}_{\{0\}}}),
\end{aligned}
$$

for all $b, b' \in B$, as needed. The remaining details are left to the reader. □

We can construct an inverse for $\mathscr{V}$ as follows.

**Proposition 9.24** *Let $H$ be a quasi-Hopf algebra and $(B, i, \pi)$ a quasi-bialgebra projection for it. If the antipode of $H$ is bijective then $B$ is a bialgebra in ${}_H^H\mathscr{M}_H^H$ with the structure given, for all $h, h' \in H$ and $b, b' \in B$, by*

$$h \cdot b \cdot h' = i(h) b i(h'); \tag{9.7.7}$$

$$\lambda : B \ni b \mapsto \pi(b_1) \otimes b_2 \in H \otimes B, \quad \rho : B \ni b \mapsto b_1 \otimes \pi(b_2) \in B \otimes H; \tag{9.7.8}$$

$$\underline{m}_B(b \otimes_H b') = bb', \quad i : H \to B; \tag{9.7.9}$$

$$\underline{\Delta}_B(b) = E(b_1) \otimes_H b_2 \quad \text{and} \quad \underline{\varepsilon}_B = \pi. \tag{9.7.10}$$

*In this way we have a well-defined functor $\mathscr{T} : H$–$\underline{\text{qBialgProj}} \to \text{Bialg}({}_H^H\mathscr{M}_H^H)$. $\mathscr{T}$ acts as identity on morphisms.*

*Proof* It is easy to see that $B$ is an object in ${}_H^H\mathscr{M}_H^H$ with the structure as in (9.7.7) and (9.7.8). Since $(b \cdot h)b' = b(h \cdot b')$, for all $b, b' \in B$ and $h \in H$, it follows that $\underline{m}_B : B \otimes_H B \to B$ given by $\underline{m}_B(b \otimes_H b') = bb'$, for all $b, b' \in B$, is well defined. Thus $(B, \underline{m}_B, i)$ is an algebra in ${}_H^H\mathscr{M}_H^H$, since

$$\lambda(i(h)) = \pi(i(h)_1) \otimes i(h)_2 = h_1 \otimes i(h_2) \ \text{ and } \ \rho(i(h)) = i(h)_1 \otimes \pi(i(h)_2) = i(h_1) \otimes h_2,$$

for all $h \in H$, that is, $i$ is an $H$-bicomodule morphism, where the $H$-bicomodule structure of $B$ is $(B, \lambda, \rho, \Phi_\lambda = X^1 \otimes X^2 \otimes i(X^3), \Phi_\rho = i(X^1) \otimes X^2 \otimes X^3, \Phi_{\lambda,\rho} = X^1 \otimes i(X^2) \otimes X^3)$. We should point out that all these facts follow because $i : H \to B$ is a quasi-bialgebra morphism.

We proved that $B$ is a coalgebra in ${}_H^H\mathscr{M}_H^H$ with the structure in (9.7.10). It remains to show that $\underline{\Delta}_B$ is an algebra morphism, where the algebra structure on $B \otimes_H B$ is the $d$-tensor product algebra one, via the braiding $d$ in (9.6.2). We compute

$$
\begin{aligned}
\underline{\Delta}_B i(h) &= E(i(h)_1) \otimes_H i(h)_2 \\
&= q^1 \cdot i(h_1)_{(0)} \cdot \beta S(q^2 i(h_1)_{(1)}) \otimes i(h_2) \\
&= i(q^1 h_{(1,1)} \beta S(q^2 h_{(1,2)}) h_2) \otimes_H 1_H \\
&\overset{(3.2.21)}{=} i(hq^1 \beta S(q^2)) \otimes_H 1_H \overset{(3.2.2)}{=} i(h) \otimes_H 1_H,
\end{aligned}
$$

and this shows that, up to the identification given by the unit constraints of the monoidal category $({}_H\mathcal{M}_H, \otimes_H, H)$, $\underline{\Delta}_B i = i \otimes_H i$.

Owing to (9.7.6) and (9.7.10), the fact that $\underline{\Delta}_B$ is multiplicative is equivalent to

$$\begin{aligned}\underline{\Delta}_B(bb') &= E(b_1)E(b_{2_{\{-1\}}} \cdot E(b'_1)_{(0)}) \otimes_H (b_{2_{\{0\}}} \cdot E(b'_1)_{(1)})b'_2 \\ &\overset{(6.4.11)}{=} E(b_1)E(\pi(b_{(2,1)}) \cdot E(x^1 \cdot b'_1) \cdot x^2) \otimes_H (b_{(2,2)} \cdot x^3)b'_2 \\ &\overset{(6.4.2)}{=} E(X^1 \cdot b_{(1,1)})E(X^2\pi(b_{(1,2)}) \cdot b'_1) \cdot X^3 \otimes_H b_2 b'_2,\end{aligned}$$

for all $b, b' \in B$. Since, for all $b, b' \in B$, we have that

$$\begin{aligned}&E(X^1 \cdot b_1)E(X^2\pi(b_2) \cdot b') \cdot X^3 \\ &= i(q^1X^1_1)b_{(1,1)}i(\beta S(q^2X^1_2\pi(b_{(1,2)})Q^1X^2_1\pi(b_{(2,1)}))b'_1 \\ &\qquad i(\beta S(Q^2X^2_2\pi(b_{(2,2)})\pi(b'_2))X^3) \\ &\overset{(5.5.17)}{=} i(q^1Q^1_{(1,1)})(x^1 \cdot b_1)_1 i(\beta S(q^2Q^1_{(1,2)}\pi((x^1 \cdot b_1)_2))Q^1_2\pi(x^2 \cdot b_{(2,1)})) \\ &\qquad b'_1 i(\beta S(Q^2\pi(x^3 \cdot b_{(2,2)})\pi(b'_2)) \\ &\overset{(3.2.21)}{=} i(Q^1q^1)(b_1)_{(1,1)}i(\beta S(q^2\pi((b_1)_{(1,2)}))\pi((b_1)_2))b'_1 i(\beta S(Q^2\pi(b_2)\pi(b'_2))) \\ &= i(Q^1)b_1 i(q^1\beta S(q^2))b'_1 i(\beta S(Q^2\pi(b_2b'_2))) \overset{(3.2.2)}{=} E(bb'),\end{aligned}$$

it follows that $\underline{\Delta}_B$ is multiplicative if and only if

$$E((bb')_1) \otimes_H (bb')_2 = E(b_1b'_1) \otimes_H b_2b'_2, \ \ \forall\, b,\ b' \in B.$$

The last equivalence is immediate since $\Delta_B$ is multiplicative. □

At this point we can prove the main result of this section.

**Theorem 9.25** *Let H be a quasi-Hopf algebra with bijective antipode. Then*

$$\mathrm{Bialg}({}_H^H\mathcal{M}_H^H) \underset{\mathscr{T}}{\overset{\mathscr{V}}{\rightleftarrows}} H\text{–}\underline{\mathrm{qBialgProj}}$$

*define an isomorphism of categories. They also produce an isomorphism of categories between* $\mathrm{Hopf}({}_H^H\mathcal{M}_H^H)$ *and* $H$–$\underline{\mathrm{qHopfProj}}$.

*Proof* One can check directly that $\mathscr{V}$ and $\mathscr{T}$ are inverse to each other; it is a straightforward computation left to the reader.

Take $(B, i, \pi) \in H$–$\underline{\mathrm{qHopfProj}}$, and denote by $S_B$ the antipode of $B$. We claim that $\mathscr{T}((B, i, \pi)) = B$ is a Hopf algebra in ${}_H^H\mathcal{M}_H^H$ with antipode determined by

$$\underline{S}(b) = q^1\pi(b_{(1,1)})\beta \cdot S_B(q^2 \cdot b_{(1,2)}) \cdot \pi(b_2), \ \ \forall\, b \in B.$$

A technical but straightforward computation ensures that $\underline{S}$ is a morphism in ${}_H^H\mathcal{M}_H^H$. Then one can check that $\underline{S}(E(b)) = q^1\pi(b_1)\beta \cdot S_B(q^2 \cdot b_2)$, for all $b \in B$, and this fact allows us to compute, for all $b \in B$, that

$$\begin{aligned}\underline{S}(b_{\underline{1}})b_{\underline{2}} &= \underline{S}(E(b_1))b_2 \\ &= i(\pi(X^1 \cdot b_{(1,1)})\beta)S_B(X^2 \cdot b_{(1,2)})i(\alpha)(X^3 \cdot b_2)\end{aligned}$$

$$= i\pi(b)i(X^1\beta S(X^2)\alpha X^3) = i\pi(b),$$

as required. Similarly, one can see that $i(S(\pi(b_1))\alpha)\underline{S}(b_2) = \varepsilon_B(b)i(\alpha)$, for all $b \in B$, and from here we get that

$$\begin{aligned} b_{\underline{1}}\underline{S}(b_{\underline{2}}) &= E(b_1)\underline{S}_B(b_2) \\ &= i(\pi(X^1 \cdot b_{(1,1)})\beta S(\pi(X^2 \cdot b_{(1,2)}))\alpha)\underline{S}(X^3 \cdot b_2) \\ &= i(\pi(b_1)X^1\beta S(\pi(b_{(2,1)})X^2)\alpha)\underline{S}(b_{(2,2)})i(X^3) \\ &= \varepsilon_B(b_2)i(\pi(b_1)i(X^1\beta S(X^2)\alpha X^3) = i\pi(b), \end{aligned}$$

for all $b \in B$. Hence our claim is proved.

In a similar manner one can prove that if $\underline{S}$ is the antipode for the bialgebra $B$ in ${}_H^H\mathcal{M}_H^H$ then the quasi-bialgebra $\mathcal{V}(B)$ is actually a quasi-Hopf algebra with antipode determined by

$$S_B(b) = S(b_{(0)_{\{-1\}}}p^1)\alpha \cdot \underline{S}(b_{(0)_{\{0\}}}) \cdot p^2 S(b_{(1)}), \ \forall\, b \in B, \tag{9.7.11}$$

and distinguished elements $\alpha_B = i(\alpha)$ and $\beta_B = i(\beta)$. □

## 9.8 Biproduct Quasi-Hopf Algebras

Let $H$ be a quasi-Hopf algebra with bijecive antipode. We keep the same notation for the various functors that appear in the previous sections of this chapter.

We present a second characterization for bialgebras and Hopf algebras in ${}_H^H\mathcal{M}_H^H$.

By Theorem 9.21, the categories ${}_H^H\mathcal{M}_H^H$ and ${}_H^H\mathcal{YD}$ are braided monoidally equivalent. Therefore bialgebras (resp. Hopf algebras) in ${}_H^H\mathcal{M}_H^H$ are in a one-to-one correspondence to bialgebras (resp. Hopf algebras) in ${}_H^H\mathcal{YD}$. More precisely, if $B$ is a bialgebra (resp. Hopf algebra) in ${}_H^H\mathcal{M}_H^H$ then $A := \overline{B^{\mathrm{co}(H)}}$ is a bialgebra (resp. Hopf algebra) in ${}_H^H\mathcal{YD}$. The inverse of this correspondence associates to any bialgebra (resp. Hopf algebra) $A$ in ${}_H^H\mathcal{YD}$ the bialgebra (resp. Hopf algebra) $\mathscr{F}(A) = A \otimes H$ in ${}_H^H\mathcal{M}_H^H$. Thus, $B$ and $A \otimes H$ are isomorphic as bialgebras (resp. Hopf algebras) in ${}_H^H\mathcal{M}_H^H$. Consequently, $\mathcal{V}(B)$ and $\mathcal{V}(A \otimes H)$ are isomorphic as objects in $H$–$\underline{\text{qBialgProj}}$ (resp. $H$–$\underline{\text{qHopfProj}}$).

Firstly, $A \otimes H$ is an object in ${}_H^H\mathcal{M}_H^H$ with the structure as in (9.2.10)–(9.2.12). As an algebra $\mathcal{V}(A \otimes H) = A\#H$, the smash product algebra of $A$ and $H$. Recall that the multiplication of $A\#H$ is given by

$$(a\#h)(a'\#h') = (x^1 \cdot a)(x^2h_1 \cdot a')\#x^3h_2h', \tag{9.8.1}$$

for all $a, a' \in A$ and $h, h' \in H$, and its unit is $1_A \otimes 1_H$. This contributes to the structure of $\mathcal{V}(A \otimes H)$ with $j : H \ni h \mapsto 1_A \otimes h \in A\#H$, so far an $H$-bicomodule algebra morphism, provided that $A$ is an algebra in ${}_H^H\mathcal{YD}$.

Secondly, as a coalgebra $\mathcal{V}(A \otimes H) = A \ltimes H$, the smash product coalgebra of $A$

and $H$. More precisely, the comultiplication is defined by

$$\Delta(a \bowtie h) = (y^1X^1 \cdot a_1 \bowtie y^2Y^1(x^1X^2 \cdot a_2)_{[-1]}x^2X_1^3h_1) \\ \otimes (y_1^3Y^2 \cdot (x^1X^2 \cdot a_2)_{[0]} \bowtie y_2^3Y^3x^3X_2^3h_2), \tag{9.8.2}$$

and the counit is $\varepsilon(a \otimes h) = \varepsilon_A(a)\varepsilon(h)$, for all $a \in A$ and $h \in H$. This contributes to the structure of $\mathscr{V}(A \otimes H)$ with $p : A \bowtie H \ni a \bowtie h \mapsto \varepsilon_A(a)h \in H$, so far an $H$-bimodule coalgebra morphism, provided that $A$ is a coalgebra in ${}_H^H\mathscr{YD}$. As before, $a \mapsto a_{[-1]} \otimes a_{[0]}$ is the left coaction of $H$ on $A$, $\Delta_A(a) = a_1 \otimes a_2$ is the comultiplication of $A$ in ${}_H^H\mathscr{YD}$ and $\varepsilon_A$ is its counit.

Summing up, we denote $\mathscr{V}(\mathscr{F}(A)) = (A \times H, j, p)$, where $A \times H$ is the $k$-vector space $A \otimes H$ endowed with multiplication and comultiplication defined by (9.8.1) and (9.8.2).

**Proposition 9.26** *Let $H$ be a quasi-Hopf algebra with bijective antipode and $B$ an object of ${}_H^H\mathscr{YD}$ which is at the same time an algebra and a coalgebra in ${}_H^H\mathscr{YD}$. Then the smash product algebra and the smash product coalgebra afford a quasi-bialgebra (resp. quasi-Hopf algebra) structure on $A \otimes H$ if and only if $A$ is a bialgebra (resp. Hopf algebra) in ${}_H^H\mathscr{YD}$.*

*Proof* Everything follows from the above comments and the fact that $\mathscr{F} : {}_H^H\mathscr{YD} \to {}_H^H\mathscr{M}_H^H$ is a braided monoidal equivalence, and that $\mathscr{T}$, $\mathscr{V}$ are inverse isomorphism functors.

Note that the antipode $s$ of the quasi-Hopf algebra $A \times H$ can be obtained from the antipode $S_A$ of $A$ in ${}_H^H\mathscr{YD}$ and the antipode $S$ of $H$ as follows. The antipode $\underline{S}$ of $\mathscr{F}(A)$ in ${}_H^H\mathscr{M}_H^H$ is $\mathscr{F}(S_A) = S_A \otimes \mathrm{Id}_H$, and so we have that

$$\begin{aligned}
s(a \times h) &\overset{(9.7.11)}{=} S((a \times h)_{(0)_{\{-1\}}}p^1)\alpha \cdot \underline{S}((a \times h)_{(0)_{\{0\}}}) \cdot p^2S((a \times h)_{(1)}) \\
&\overset{(9.2.12)}{=} S((x^1 \cdot a \times x^2h_1)_{\{-1\}}p^1)\alpha \cdot \underline{S}((x^1 \cdot a \times x^2h_1)_{\{0\}}) \cdot p^2S(x^3h_2) \\
&\overset{(9.2.10),(9.2.11)}{=} S(X^1(y^1x^1 \cdot a)_{[-1]}y^2x_1^2h_{(1,1)}p^1)\alpha \\
&\qquad \cdot \underline{S}(X^2 \cdot (y^1x^1 \cdot a)_{[0]} \times X^3y^3x_2^2h_{(1,2)}p^2S(x^3h_2)) \\
&\overset{(3.2.21),(5.5.16)}{=} S(X^1(p_1^1 \cdot a)_{[-1]}p_2^1h)\alpha \cdot \underline{S}(X^2 \cdot (p_1^1 \cdot a)_{[0]} \times X^3p^2) \\
&\overset{(8.2.3)}{=} S(X^1p_1^1a_{[-1]}h)\alpha \cdot (S_A(X^2p_2^1 \cdot a_{[0]}) \times X^3p^2) \\
&\overset{(9.2.10)}{=} (1_A \times S(X^1p_1^1a_{[-1]}h)\alpha)(X^2p_2^1 \cdot S_A(a_{[0]}) \times X^3p^2),
\end{aligned}$$

for all $a \in A$ and $h \in H$. Clearly, the distinguished elements that together with $s$ define the antipode for $A \times H$ are $j(\alpha) = 1_A \times \alpha$ and $j(\beta) = 1_A \times \beta$. In the above computation we wrote $a \times h$ in place of $a \otimes h$ in order to distinguish the quasi-bialgebra structure on $A \otimes H$ defined above. □

**Definition 9.27** $\mathscr{V}(\mathscr{F}(A)) = (A \times H, j, p)$ will be called in what follows the biproduct quasi-bialgebra (resp. quasi-Hopf algebra) between a bialgebra (resp. Hopf algebra) $A$ in ${}_H^H\mathscr{YD}$ and $H$.

**Remark 9.28** The formulas (9.8.1) and (9.8.2) define a quasi-bialgebra structure on $A \otimes H$ even if $H$ is only a quasi-bialgebra, not necessarily a quasi-Hopf algebra.

Collecting the results proved so far we get the following.

**Theorem 9.29** *Let $H$ be a quasi-Hopf algebra with bijective antipode. Then there is a one-to-one correspondence between:*

- *bialgebras (resp. Hopf algebras) in ${}_H^H\mathscr{M}_H^H$;*
- *quasi-bialgebra (resp. quasi-Hopf algebra) projections for $H$;*
- *bialgebras (resp. Hopf algebras) in ${}_H^H\mathscr{YD}$;*
- *biproduct quasi-bialgebra (resp. quasi-Hopf algebra) structures for $H$.*

We end this section by studying the invariance under twisting of the biproduct.

If $F = F^1 \otimes F^2$ is a gauge transformation for a quasi-bialgebra or quasi-Hopf algebra $H$ then there is a monoidal isomorphism between the tensor categories ${}_H\mathscr{M}$ and ${}_{H_F}\mathscr{M}$; see Proposition 3.5. This functor is the identity on objects and morphisms with the monoidal structure given by multiplication by $F^{-1}$ (the inverse of $F$), that is, for any two left $H$-modules $V, W$, $\varphi_{2,V,W} : V \otimes W \to V \otimes W$, $\varphi_{2,V,W}(v \otimes w) = G^1 \cdot v \otimes G^2 \cdot w$, where $v \in V$, $w \in W$ and $F^{-1} = G^1 \otimes G^2$. Moreover, this functor induces a monoidal isomorphism between the pre-braided categories ${}_H^H\mathscr{YD}$ and ${}_{H_F}^{H_F}\mathscr{YD}$ as follows: it is the identity on objects and morphisms and if $M \in {}_H^H\mathscr{YD}$ with $h \otimes m \mapsto h \cdot m$ and $\lambda_M(m) = m_{(-1)} \otimes m_{(0)}$, $h \in H$, $m \in M$, then $M$ becomes an object in ${}_{H_F}^{H_F}\mathscr{YD}$ with the same $H$-action and coaction given by:

$$\lambda_M^F(m) = F^1(G^1 \cdot m)_{(-1)}G^2 \otimes F^2 \cdot (G^1 \cdot m)_{(0)}, \quad \forall\, m \in M. \tag{9.8.3}$$

The above assertion follows easily from the quasi-bialgebra (quasi-Hopf algebra) structure of $H_F$ and Definition 8.7; the details are left to the reader. Under this isomorphism a (co)algebra, bialgebra, Hopf algebra object $B$ corresponds to a (co)algebra etc. object $B_F$. Note that if $B$ is a bialgebra (or Hopf algebra) in the first category then $B_F$ is a bialgebra (Hopf algebra) in the second category with the $H_F$-coaction (9.8.3), multiplication and comultiplication as follows:

$$b \diamond b' = (G^1 \cdot b)(G^2 \cdot b')\,, \quad \tilde{\Delta}(b) = F^1 \cdot b_1 \otimes F^2 \cdot b_2, \tag{9.8.4}$$

and with the same unit and counit as $B$, where $\Delta(b) = b_1 \otimes b_2$ is the comultiplication of $B$ in the first category (with the same antipode as $B$ in case that $B$ is a Hopf algebra). Therefore, we have two biproducts $B \times H$ and $B_F \times H_F$ which are quasi-bialgebras (quasi-Hopf algebras). We will prove that these biproducts are isomorphic in the following sense:

**Theorem 9.30** *Let $H$ be a quasi-bialgebra, $F$ a gauge transformation for $H$ and $B$ a bialgebra in ${}_H^H\mathscr{YD}$. If we denote by $\mathscr{F} = 1_B \times F^1 \otimes 1_B \times F^2$ the gauge transformation for $B \times H$ induced by $F$, then the map $\nu : (B \times H)_{\mathscr{F}} \to B_F \times H_F$ given by*

$$\nu(b \times h) = F^1 \cdot b \times F^2 h\,, \quad \forall\, b \in B,\ h \in H \tag{9.8.5}$$

*is a quasi-bialgebra isomorphism. Moreover, if $H$ is a quasi-Hopf algebra and $B$ is a Hopf algebra in ${}_H^H\mathscr{YD}$ then $\nu$ is a quasi-Hopf algebra isomorphism.*

*Proof* We only describe the structures involved and leave the verification of some details to the reader.

Because the biproduct considered as an algebra is a smash product and the new quasi-bialgebra $H_F$ has the same algebra structure as $H$, we know by Proposition 5.10 that $\nu$ defined above is an algebra isomorphism with inverse given by:

$$\nu^{-1}(b\times h)=G^1\cdot b\times G^2h\,,\quad \forall\, b\in B,\ h\in H.$$

Because $\nu(1_B\times h)=1_B\times h$, for all $h\in H$, it follows that $(\nu\otimes\nu\otimes\nu)(\Phi_{(B\times H)_{\mathscr{F}}})=\Phi_{B_F\times H_F}$. Thus, for the first statement, we only have to show that $\nu$ respects the comultiplications. This follows from the quasi-bialgebra structures of a biproduct and of a twisted quasi-bialgebra. Now, for the second assertion observe that the elements $\alpha$, $\beta$ for $(B\times H)_{\mathscr{F}}$, denoted by $\alpha_{\mathscr{F}}$ and $\beta_{\mathscr{F}}$, respectively, are in fact $\alpha_{\mathscr{F}}=1_B\times\alpha_F$, $\beta_{\mathscr{F}}=1_B\times\beta_F$, so $\nu(\alpha_{\mathscr{F}})=\alpha_{B_F\times H_F}$ and $\nu(\beta_{\mathscr{F}})=\beta_{B_H\times H_F}$. Thus the proof will be complete once we show that $s_{B_F\times H_F}\circ\nu=\nu\circ s$, because the antipode for $(B\times H)_{\mathscr{F}}$ is the same as the antipode for $B\times H$, say $s$. This equation follows mostly from

$$(1_B\times h)(b\times h')=F^1h_1G^1\cdot b\times F^2h_2G^2h',\quad \forall\, b\in B \text{ and } h,\ h'\in H,$$

which holds in the smash product $B_F\times H_F$. □

## 9.9 Notes

Two-sided two-cosided Hopf modules over a Hopf algebra $H$ were introduced by Woronowicz [220] under the name of bicovariant bimodules, as a tool in the study of non-commutative differential calculus on quantum groups. He also extended the structure theorem for Hopf modules to the category of Hopf bimodules ${}_H\mathscr{M}_H^H$ and two-sided two-cosided Hopf modules ${}_H^H\mathscr{M}_H^H$ over $H$. Later on, Schauenburg proved in [194] that the structure theorems provide the classification of Hopf bimodules and two-sided two-cosided Hopf modules in the form of category equivalences ${}_H\mathscr{M}_H^H\cong{}_H\mathscr{M}$ and ${}_H^H\mathscr{M}_H^H\cong{}_H^H\mathscr{YD}$. These equivalences are even monoidal and they can be regarded as a coordinate-free version of the classifications in [220]. Using categorical techniques, Schauenburg [195] also proved that all the results mentioned above remain valid in the setting provided by quasi-Hopf algebras.

The content of this chapter is based on the following papers. The connection between the categories of Yetter–Drinfeld modules and two-sided two-cosided Hopf modules is from [63, 195], the monoidal equivalence between them is from [195, 44], while the fact that they are, moreover, equivalent as braided monoidal categories is taken from [45]. The content of Section 9.4 is from [44, 75], while that of Section 9.5 is from [44].

The structure of a Hopf algebra with a projection is due to Radford [186]. A second characterization of Hopf algebras with a projection is due to Bespalov and Drabant [31], where Hopf algebras with a projection are identified with Hopf algebras within ${}_H^H\mathcal{M}_H^H$. Their techniques were adapted to quasi-Hopf algebras in [45], which was our source of inspiration for Section 9.7. Note that the comultiplication on $B \otimes H$ defined in (9.5.6) and its counit appeared for the first time in [54] as the quasi-coalgebra part of the Radford biproduct construction for quasi-Hopf algebras. At that time there was no clue how to introduce a smash product coalgebra, and by hard computations it was proved in [54] that the comultiplication of the biproduct is coassociative up to conjugation by an invertible element. At this point it is clear that this quasi-coassociativity of the biproduct is nothing but a reformulation of the fact that $B \ltimes H$ is a coalgebra in ${}_H\overline{\mathcal{M}}_H$, assuming that $B$ is a coalgebra in ${}_H^H\mathcal{YD}$. A proof for the assertion in Remark 9.28 can be found in [54].

# 10

# Quasitriangular Quasi-Hopf Algebras

By using categorical tools, we introduce the concept of a quasitriangular (QT for short) quasi-bialgebra. For QT quasi-Hopf algebras we show that the square of the antipode is an inner automorphism, and therefore bijective. We uncover the QT structure of the quantum double $D(H)$ of a finite-dimensional quasi-Hopf algebra $H$, and characterize $D(H)$ as a biproduct quasi-Hopf algebra in the case when $H$ itself is QT.

## 10.1 Quasitriangular Quasi-bialgebras and Quasi-Hopf Algebras

We introduced the notion of quasi-bialgebra by investigating when the forgetful functor $F$ to the category of vector spaces is quasi-monoidal. This led to the monoidal structure on the category of representations over a quasi-bialgebra, say $H$. We now go further: we will investigate when ${}_H\mathcal{M}$ is a braided category.

**Proposition 10.1** *Let $H$ be a quasi-bialgebra, so ${}_H\mathcal{M}$ is a monoidal category. Then ${}_H\mathcal{M}$ is braided if and only if there exists an invertible element $R = R^1 \otimes R^2 = r^1 \otimes r^2 \in H \otimes H$ (formal notation, summation implicitly understood) such that the following relations hold:*

$$(\Delta \otimes \mathrm{Id}_H)(R) = X^2R^1x^1Y^1 \otimes X^3x^3r^1Y^2 \otimes X^1R^2x^2r^2Y^3, \tag{10.1.1}$$

$$(\mathrm{Id}_H \otimes \Delta)(R) = x^3R^1X^2r^1y^1 \otimes x^1X^1r^2y^2 \otimes x^2R^2X^3y^3, \tag{10.1.2}$$

$$\Delta^{\mathrm{cop}}(h)R = R\Delta(h), \ \ \forall\, h \in H. \tag{10.1.3}$$

*Proof* Suppose that ${}_H\mathcal{M}$ is a braided category. Let $c$ be a braiding for ${}_H\mathcal{M}$ and regard $H \in {}_H\mathcal{M}$ via its multiplication. If $x \in X \in {}_H\mathcal{M}$ we define $\varphi_x : H \to X$ by $\varphi_x(h) = h \cdot x$, for all $h \in H$, a left $H$-linear morphism.

Since $c$ is a natural transformation we have

$$c_{X,Y}(\varphi_x \otimes \varphi_y) = (\varphi_y \otimes \varphi_x)c_{H,H},$$

for all $x \in X \in {}_H\mathcal{M}$ and $y \in Y \in {}_H\mathcal{M}$. By evaluating both sides of the above equality on $1_H \otimes 1_H$ we obtain that $c_{X,Y}(x \otimes y) = (\varphi_y \otimes \varphi_x)c_{H,H}(1_H \otimes 1_H)$. Thus $c_{H,H}(1_H \otimes 1_H) := R^2 \otimes R^1 \in H \otimes H$ determines completely the braiding $c$ since

$$c_{X,Y}(x \otimes y) = R^2 \cdot y \otimes R^1 \cdot x, \ \ \forall\, x \in X \in {}_H\mathcal{M}, \ \ y \in Y \in {}_H\mathcal{M}. \tag{10.1.4}$$

Now, it can be easily checked that $c_{X,Y}$ is a morphism in ${}_H\mathcal{M}$ if and only if (10.1.3) holds, (1.5.2) is equivalent to (10.1.1) and (1.5.1) is equivalent to (10.1.2).

For instance, (10.1.3) is the consequence of the following facts: for all $x \in X \in {}_H\mathcal{M}$ and $y \in Y \in {}_H\mathcal{M}$ we have

$$c_{X,Y}(h\cdot(x\otimes y)) = c_{X,Y}(h_1\cdot x\otimes h_2\cdot y) = R^2h_2\cdot y\otimes R^1h_1\cdot x,$$
$$h\cdot c_{X,Y}(x\otimes y) = h\cdot(R^2\cdot y\otimes R^1\cdot x) = h_1R^2\cdot y\otimes h_2R^1\cdot x,$$

so $c_{X,Y}$ is a morphism in ${}_H\mathcal{M}$ if and only (10.1.3) is fulfilled (for the direct implication take $x = y = 1_H \in X = Y = H \in {}_H\mathcal{M}$).

Since $c$ is a natural isomorphism it follows that $R := R^1 \otimes R^2$ is an invertible element in $H\otimes H$ with inverse $R^{-1} = c^{-1}_{H,H}(1_H\otimes 1_H)$.

Conversely, if there is an $R$ as in the statement then $c$ defined by (10.1.4) is a braiding for ${}_H\mathcal{M}$, we leave the verification of this fact to the reader. □

**Definition 10.2** A quasi-bialgebra or a quasi-Hopf algebra $H$ is called quasitriangular (QT for short) if there exists an invertible element $R \in H\otimes H$ obeying (10.1.1), (10.1.2) and (10.1.3) (such an element is called an $R$-matrix). In other words, $H$ is quasitriangular if and only if ${}_H\mathcal{M}$ has a braided structure such that the forgetful functor $F : {}_H\mathcal{M} \to {}_k\mathcal{M}$ is quasi-monoidal.

When we refer to a QT quasi-bialgebra or quasi-Hopf algebra we always indicate the $R$-matrix $R$ that produces the QT structure, by pointing out the pair $(H,R)$.

By applying $\varepsilon\otimes\varepsilon\otimes \mathrm{Id}_H$ to both sides of (10.1.1) we get that $\varepsilon(R^1)R^2$ is an invertible idempotent of $H$, so it must be equal to $1_H$. Similarly, by using (10.1.2) we deduce that $\varepsilon(R^2)R^1 = 1_H$, and so we have

$$(\varepsilon\otimes \mathrm{Id}_H)(R) = (\mathrm{Id}_H\otimes\varepsilon)(R) = 1_H, \tag{10.1.5}$$

in any QT quasi-bialgebra or quasi-Hopf algebra $(H,R)$.

**Definition 10.3** A morphism of QT quasi-bialgebras $\varphi : (H,R) \to (H',R')$ is a morphism of the underlying quasi-bialgebras such that $(\varphi\otimes\varphi)(R) = R'$.

**Remarks 10.4** (1) Let $H$ be a quasi-bialgebra. Then the pre-braided structures $c$ on ${}_H\mathcal{M}$ are given by (10.1.4), where $R \in H\otimes H$ is such that (10.1.1), (10.1.2) and (10.1.3) are satisfied. The invertibility of $R$ is necessary only to turn $c$ into a braiding on ${}_H\mathcal{M}$.

(2) If $\mathscr{C}$ is braided then $\mathscr{C}^{\mathrm{in}}$ is braided as well via $c^{\mathrm{in}}_{X,Y} = c^{-1}_{Y,X}$. This says that if $(H,R)$ is QT then it admits a second QT structure, namely $(H,R^{-1}_{21})$.

(3) Let $(H,R)$ be a QT quasi-bialgebra. If $t$ denotes a permutation of $\{1,2,3\}$, then we set $\Phi_{t(1)t(2)t(3)} = X^{t^{-1}(1)}\otimes X^{t^{-1}(2)}\otimes X^{t^{-1}(3)}$, and by $R_{ij}$ we denote the element obtained by acting with $R$ non-trivially in the $i$th and $j$th positions of $H\otimes H\otimes H$. Then an immediate consequence of Proposition 1.51 and Proposition 10.1 is the fact that $R$ satisfies the so-called quasi-Yang–Baxter equation:

$$R_{12}\Phi_{312}R_{13}\Phi^{-1}_{132}R_{23}\Phi = \Phi_{321}R_{23}\Phi^{-1}_{231}R_{13}\Phi_{213}R_{12}. \tag{10.1.6}$$

**Definition 10.5** A QT quasi-bialgebra or quasi-Hopf algebra $(H,R)$ is called triangular if $R^{-1}=R_{21}$. In other words, $(H,R)$ is triangular if and only if ${}_H\mathcal{M}$ admits a symmetric structure such that the forgetful functor $F:{}_H\mathcal{M}\to{}_k\mathcal{M}$ is quasi-monoidal.

**Example 10.6** We have shown in Proposition 1.44 that for an abelian group $G$ the braided structures on $\mathrm{Vect}^G$ are given by the abelian 3-cocycles $(\phi,\mathscr{R})$ on $G$. Furthermore, if $G$ is finite then the category $\mathrm{Vect}^G_\phi$ identifies as a monoidal category with the category of left representations over the quasi-Hopf algebra $k_\phi[G]^*$; see Proposition 3.51 and Example 3.50. Since $\mathrm{Vect}^G_{(\phi,\mathscr{R})}$ is a braided category it follows that so is ${}_{k_\phi[G]^*}\mathcal{M}$. Combined with Proposition 10.1 this tells us that $k_\phi[G]^*$ is a QT quasi-Hopf algebra, provided that $(\phi,\mathscr{R})$ is an abelian 3-cocycle of a finite abelian group $G$. Concrete examples of this type can be obtained by using Example 1.48.

In the next result we describe all the QT structures of $H(2)$.

**Example 10.7** Suppose that $k$ is a field of characteristic different from 2 containing a primitive fourth root of unity $i$ and let $H(2)$ be the quasi-Hopf algebra constructed in Example 3.26. Then there are exactly two different $R$-matrices for $H(2)$, namely $R_\pm=1-(1\pm i)p_-\otimes p_-$.

*Proof* As we have seen before, $H(2)=k_\phi[C_2]^*$, where $C_2$ is the cyclic group of order 2 and $\phi$ is the unique non-trivial normalized 3-cocycle on $C_2$ as defined in Example 1.12. So by Example 10.6 and Example 1.48 we know that $H(2)$ has at least two QT structures. We will prove that it has precisely two, namely the ones stated above.

Suppose that $R=a1\otimes 1+b1\otimes g+cg\otimes 1+dg\otimes g$ is an $R$-matrix for $H(2)$, where $a,b,c,d\in k$. By (10.1.5) we have that $a+c=a+b=1$ and $b+d=c+d=0$, and therefore $b=c=-d$ and $a=1-b$. Hence, $R$ must be of the form

$$R=(1-b)1\otimes 1+b1\otimes g+bg\otimes 1-bg\otimes g=1-b(1-g)\otimes(1-g)=1-\omega p_-\otimes p_-,$$

where we denote $4b=\omega$. Now, one can easily see that

$$\Phi^{-1}=\Phi=1-2p_-\otimes p_-\otimes p_-, \tag{10.1.7}$$

and since $X^2\otimes X^3\otimes X^1=\Phi$, the above relation implies

$$X^2R^1x^1\otimes X^3x^3\otimes X^1R^2x^2=1-\omega p_-\otimes 1\otimes p_-,$$

and therefore, after some computations, we get

$$\begin{aligned}&X^2R^1x^1Y^1\otimes X^3x^3r^1Y^2\otimes X^1R^2x^2r^2Y^3\\&=1-\omega p_-\otimes p_+\otimes p_- -\omega p_+\otimes p_-\otimes p_- -(2-2\omega+\omega^2)p_-\otimes p_-\otimes p_-.\end{aligned}$$

On the other hand, we have $\Delta(p_-)=p_-\otimes p_+ + p_+\otimes p_-$, so

$$(\Delta\otimes\mathrm{Id})(R)=1-\omega p_-\otimes p_+\otimes p_- -\omega p_+\otimes p_-\otimes p_-.$$

We conclude that (10.1.1) holds if and only if $2-2\omega+\omega^2=0$, and this is equivalent to $\omega=1\pm i$.

By using (10.1.7) for $\Phi^{-1}$, we obtain in a similar way that

$$x^3R^1X^2\otimes x^1X^1\otimes x^2R^2X^3=1-\omega p_-\otimes 1\otimes p_-.$$

By using this formula, it can be proved that

$$x^3R^1X^2r^1y^1\otimes x^1X^1r^2y^2\otimes x^2R^2X^3y^3$$
$$=1-\omega p_-\otimes p_-\otimes p_+-\omega p_-\otimes p_+\otimes p_- -(2-2\omega+\omega^2)p_-\otimes p_-\otimes p_-.$$

It is easy to see that

$$(\mathrm{Id}\otimes\Delta)(R)=1-\omega p_-\otimes p_-\otimes p_+-\omega p_-\otimes p_+\otimes p_-,$$

so the relation in (10.1.2) holds if and only if $2-2\omega+\omega^2=0$. The relation in (10.1.3) is automatically satisfied because of the commutativity and cocommutativity of $H(2)$. Thus the $R$-matrices for $H(2)$ are in bijective correspondence with the solutions of the equation $2-2\omega+\omega^2=0$, from where we deduce that $R_\pm=1-(1\pm i)p_-\otimes p_-$ are the only quasi-triangular structures on $H(2)$. □

**Remark 10.8** It is not difficult to show that $H(2)_+=(H(2),R_+)$ and $H(2)_-=(H(2),R_-)$ are non-isomorphic QT quasi-Hopf algebras, that is, there is no quasi-Hopf algebra isomorphism $\nu:\ H(2)\to H(2)$ satisfying $(\nu\otimes\nu)(R_+)=R_-$. Indeed, if such a $\nu$ exists then $(1+i)\nu(p_-)\otimes\nu(p_-)=(1-i)p_-\otimes p_-$. If we write $\nu(p_-)=ap_-+bp_+$, for some scalars $a,b\in k$, then from the above relation we obtain that $a^2=-i$ and $b=0$. Since $p_\pm^2=p_\pm$ and $\nu$ is an algebra map we get that $ap_-=\nu(p_-)=\nu(p_-^2)=(ap_-)^2=-ip_-$, and we conclude that $a=-i$. But $a^2=-i$, so $i\in\{-1,0\}$, a contradiction.

A remarkable fact is that, in the quasi-Hopf case, the conditions in (10.1.5) can replace the invertibility of $R$ in the definition of a QT quasi-Hopf algebra $H$. As we shall see, this follows basically from the properties of a left rigid (pre-)braided category.

**Lemma 10.9** *Let $H$ be a quasi-Hopf algebra with antipode $S$ and $R\in H\otimes H$ such that* (10.1.1), (10.1.2), (10.1.3) *and* (10.1.5) *are satisfied. Then $R$ is invertible with inverse given by*

$$R^{-1}=X^1\beta S(Y^2R^1x^1X^2)\alpha Y^3x^3X_2^3\otimes Y^1R^2x^2X_1^3. \tag{10.1.8}$$

*Proof* The category ${}_H\mathcal{M}^{\mathrm{fd}}$ is pre-braided and left rigid; see Remarks 10.4 and Proposition 3.33. By Theorem 1.77 we get that ${}_H\mathcal{M}^{\mathrm{fd}}$ is braided, and so $c$ defined by (10.1.4) is a natural isomorphism. Furthermore, by (1.8.1) and the left rigid monoidal structure on ${}_H\mathcal{M}^{\mathrm{fd}}$ we have that the inverse of $c_{X,Y}$ on objects of ${}_H\mathcal{M}^{\mathrm{fd}}$ is given by

$$c_{X,Y}^{-1}=\Big(y\otimes x\xrightarrow{\mathrm{coev}_X\otimes\mathrm{Id}}(\beta\cdot x_i\otimes x^i)\otimes(y\otimes x)$$
$$\xrightarrow{a^{-1}_{X,X^*,Y\otimes X}}X^1\beta\cdot x_i\otimes(X^2\cdot x^i\otimes(X_1^3\cdot y\otimes X_2^3\cdot x))$$

$$\overset{\mathrm{Id}_X\otimes a^{-1}_{X^*,Y,X}}{\longrightarrow} X^1\beta\cdot x_i\otimes((x^1X^2\cdot x^i\otimes x^2X^3_1\cdot y)\otimes x^3X^3_2\cdot x)$$
$$\overset{\mathrm{Id}_X\otimes(c_{X^*,Y}\otimes\mathrm{Id}_X)}{\longrightarrow} X^1\beta\cdot x_i\otimes((R^2x^2X^3_1\cdot y\otimes R^1x^1X^2\cdot x^i)\otimes x^3X^3_2\cdot x)$$
$$\overset{\mathrm{Id}_X\otimes a_{Y,X^*,X}}{\longrightarrow} X^1\beta\cdot x_i\otimes(Y^1R^2x^2X^3_1\cdot y\otimes(Y^2R^1x^1X^2\cdot x^i\otimes Y^3x^3X^3_2\cdot x))$$
$$\overset{\mathrm{Id}_X\otimes(\mathrm{Id}_Y\otimes\mathrm{ev}_X)}{\longrightarrow} X^1\beta S(Y^2R^1x^1X^2)\alpha Y^3x^3X^3_2\cdot x\otimes Y^1R^2x^2X^3_1\cdot y).$$

When $H$ is finite dimensional this gives the definition of $R^{-1}$ in (10.1.8) since, as we have seen in the proof of Proposition 10.1, $R^{-1}=c^{-1}_{H,H}(1_H\otimes 1_H)$.

In the general case (i.e. $H$ not necessarily finite dimensional), the fact that $R^{-1}$ and $R$ are inverse to each other can be proved by direct computations. On the one hand, if we denote by $r=r^1\otimes r^2$ another copy of $R$ then we have:

$$\begin{aligned}
R^{-1}R &= X^1\beta S(Y^2R^1y^1X^2)\alpha Y^3y^3X^3_2r^1\otimes Y^1R^2y^2X^3_1r^2\\
&\overset{(10.1.3)}{=} X^1\beta S(Y^2R^1y^1X^2)\alpha Y^3y^3r^1X^3_1\otimes Y^1R^2y^2r^2X^3_2\\
&\overset{(10.1.1)}{=} X^1\beta S(R^1_1y^1X^2)\alpha R^1_2y^2X^3_1\otimes R^2y^3X^3_2\\
&\overset{(3.2.1),(10.1.2),(10.1.5)}{=} X^1\beta S(y^1X^2)\alpha y^2X^3_1\otimes y^3X^3_2\\
&\overset{(3.1.9)}{=} Y^1Z^1y^1_1\beta S(Y^2_1Z^2y^1_2)\alpha Y^2_2Z^3y^2\otimes Y^3y^3\\
&\overset{(3.2.1),(3.1.10),(3.1.11)}{=} Z^1\beta S(Z^2)\alpha Z^3\otimes 1_H\overset{(3.2.2)}{=}1_H\otimes 1_H.
\end{aligned}$$

On the other hand, to show that $RR^{-1}=1_H\otimes 1_H$, we first observe that (10.1.8), (3.1.9), (3.2.1) and (3.1.11) imply

$$R^{-1}=X^1Z^1\beta S(Y^2X^2_2R^1Z^2)\alpha Y^3X^3\otimes Y^1X^2_1R^2Z^3, \tag{10.1.9}$$

and by again using (3.1.9), (3.2.1) and (3.1.11) we obtain that

$$R^{-1}=X^1_1x^1Y^1\beta S(X^2x^3R^1Y^2)\alpha X^3\otimes X^1_2x^2R^2Y^3. \tag{10.1.10}$$

Now, we calculate:

$$\begin{aligned}
RR^{-1} &= r^1X^1_1x^1Y^1\beta S(X^2x^3R^1Y^2)\alpha X^3\otimes r^2X^1_2x^2R^2Y^3\\
&\overset{(10.1.3)}{=} X^1_2r^1x^1Y^1\beta S(X^2x^3R^1Y^2)\alpha X^3\otimes X^1_1r^2x^2R^2Y^3\\
&\overset{(10.1.1)}{=} X^1_2x^2R^1_1\beta S(X^2x^3R^1_2)\alpha X^3\otimes X^1_1x^1R^2\\
&\overset{(3.2.1),(10.1.2),(10.1.5)}{=} X^1_2x^2\beta S(X^2x^3)\alpha X^3\otimes X^1_1x^1\\
&\overset{(3.1.9)}{=} x^2X^1Y^2_1\beta S(x^3_1X^2Y^2_2)\alpha x^3_2X^3Y^3\otimes x^1Y^1\\
&\overset{(3.2.1),(3.1.10),(3.1.11)}{=} X^1\beta S(X^2)\alpha X^3\otimes 1_H\overset{(3.2.2)}{=}1_H\otimes 1_H,
\end{aligned}$$

and this finishes the proof of the lemma. □

We show now that a twisting of a QT quasi-bialgebra by a gauge transformation produces another QT quasi-bialgebra.

**Proposition 10.10** *Let $(H,R)$ be a QT quasi-bialgebra and $F\in H\otimes H$ a gauge transformation. Define $R_F:=F_{21}RF^{-1}\in H\otimes H$, where, if $F=F^1\otimes F^2$, then $F_{21}:=$*

*$F^2\otimes F^1$. Then $(H_F,R_F)$ is a QT quasi-bialgebra as well, and the categories ${}_H\mathcal{M}$ and ${}_{H_F}\mathcal{M}$ are braided monoidally isomorphic.*

*Proof* We know from Proposition 3.5 that we have a monoidal isomorphism between ${}_H\mathcal{M}$ and ${}_{H_F}\mathcal{M}$, given by the identity functor ${}_H\mathcal{M}\to{}_{H_F}\mathcal{M}$ with (strong) monoidal structure defined by $\varphi_0=\mathrm{Id}_k$ and $\varphi_{2,X,Y}:X\otimes Y\to X\otimes Y$, $\varphi_{2,X,Y}(x\otimes y)=G^1\cdot x\otimes G^2\cdot y$, where $F^{-1}=G^1\otimes G^2$. Since $H$ is QT, we know that ${}_H\mathcal{M}$ is braided with braiding $c_{X,Y}:X\otimes Y\to Y\otimes X$, $c_{X,Y}(x\otimes y)=R^2\cdot y\otimes R^1\cdot x$. Thus, there exists a unique braiding on ${}_{H_F}\mathcal{M}$ such that the above functor becomes braided monoidal and, in view of Definition 1.52, this braiding, denoted by $c^F_{X,Y}$, is defined by $c^F_{X,Y}:X\otimes Y\to Y\otimes X$, $c^F_{X,Y}=\varphi^{-1}_{2,X,Y}\circ c_{X,Y}\circ\varphi_{2,X,Y}$, that is $c^F_{X,Y}(x\otimes y)=F^1R^2G^2\cdot y\otimes F^2R^1G^1\cdot x$, for all $x\in X\in{}_{H_F}\mathcal{M}$ and $y\in Y\in{}_{H_F}\mathcal{M}$. But then, in view of Proposition 10.1 and its proof, $H_F$ has to be quasitriangular and its $R$-matrix has to be defined by $R_F=F^2R^1G^1\otimes F^1R^2G^2$, that is, $R_F=F_{21}RF^{-1}$. □

## 10.2 Further Examples of Monoidal Algebras

Let $H$ be a quasi-bialgebra and denote by ${}_H\mathcal{M}_{H^{\mathrm{cop}}}$ the category of $H$-bimodules. In this category we introduce a tensor product, as follows. If $V,W\in{}_H\mathcal{M}_{H^{\mathrm{cop}}}$ then $V\otimes W\in{}_H\mathcal{M}_{H^{\mathrm{cop}}}$ with $h\cdot(v\otimes w)\cdot h'=\Delta(h)\cdot(v\otimes w)\cdot\Delta^{\mathrm{cop}}(h')=h_1\cdot v\cdot h'_2\otimes h_2\cdot w\cdot h'_1$. ${}_H\mathcal{M}_{H^{\mathrm{cop}}}$ becomes a monoidal category, with associativity constraint

$$\begin{aligned}a_{U,V,W}((u\otimes v)\otimes w)&=\Phi\cdot(u\otimes(v\otimes w))\cdot\Phi_{321}\\&=X^1\cdot u\cdot Y^3\otimes(X^2\cdot v\cdot Y^2\otimes X^3\cdot w\cdot Y^1),\end{aligned}$$

for $U,V,W\in{}_H\mathcal{M}_{H^{\mathrm{cop}}}$ (the unit constraints are the usual ones).

Suppose now that $(H,R)$ is a QT quasi-bialgebra. One can easily see that ${}_H\mathcal{M}_{H^{\mathrm{cop}}}$ becomes a braided category, the braiding being given by

$$c_{V,W}(v\otimes w)=R^2\cdot w\cdot U^1\otimes R^1\cdot v\cdot U^2,$$

for $V,W\in{}_H\mathcal{M}_{H^{\mathrm{cop}}}$, where $U=U^1\otimes U^2$ is the inverse of $R$.

Suppose again that $(H,R)$ is a QT quasi-bialgebra and consider the left and right regular actions of $H$ on $H^*$, that is, $(h\rightharpoonup p)(h')=p(h'h)$, $(p\leftharpoonup h)(h')=p(hh')$, for $p\in H^*$ and $h,h'\in H$, turning $H^*$ into an $H$-bimodule. On $H^*$ we can consider the convolution product, given by $(fg)(h)=f(h_1)g(h_2)$, for all $f,g\in H^*$ and $h\in H$. We introduce another product on $H^*$, by

$$f\cdot g=(R^2\rightharpoonup g)(R^1\rightharpoonup f),\ \ \forall\, f,\,g\in H^*. \tag{10.2.1}$$

Denote by $H^*_R$ the pair $(H^*,\cdot)$. Then we have the following result:

**Theorem 10.11** *Let $(H,R)$ be a QT quasi-bialgebra. Then:*

*(i) $H^*_R$ is an algebra in the monoidal category ${}_H\mathcal{M}_{H^{\mathrm{cop}}}$ (we say that it is an $H$–$H^{\mathrm{cop}}$-bimodule algebra), that is, for all $f,g,l\in H^*$ and $h,h'\in H$ we have*

$$h\rightharpoonup(f\cdot g)\leftharpoonup h'=(h_1\rightharpoonup f\leftharpoonup h'_2)\cdot(h_2\rightharpoonup g\leftharpoonup h'_1),$$

$$(f\cdot g)\cdot l=(X^1\rightharpoonup f\leftharpoonup Y^3)\cdot((X^2\rightharpoonup g\leftharpoonup Y^2)\cdot(X^3\rightharpoonup l\leftharpoonup Y^1)),$$
$$\varepsilon\cdot f=f\cdot\varepsilon=f,\quad h\rightharpoonup\varepsilon\leftharpoonup h'=\varepsilon(h)\varepsilon(h')\varepsilon.$$

*(ii) $H_R^*$ is commutative as an algebra in the braided monoidal category ${}_H\mathscr{M}_{H^{\rm cop}}$, that is, for all $f,g\in H^*$, we have*

$$f\cdot g=(R^2\rightharpoonup g\leftharpoonup U^1)\cdot(R^1\rightharpoonup f\leftharpoonup U^2).$$

*Proof* We denote by $R=R^1\otimes R^2=r^1\otimes r^2=\rho^1\otimes\rho^2$ several copies of $R$. For $f,g\in H^*$, the product $f\cdot g$ is given by $(f\cdot g)(h)=g(h_1R^2)f(h_2R^1)$, for all $h\in H$. First we prove (ii). We compute:

$$\begin{aligned}
&((R^2\rightharpoonup g\leftharpoonup U^1)\cdot(R^1\rightharpoonup f\leftharpoonup U^2))(h)\\
&\quad= (R^1\rightharpoonup f\leftharpoonup U^2)(h_1r^2)(R^2\rightharpoonup g\leftharpoonup U^1)(h_2r^1)\\
&\quad= f(U^2h_1r^2R^1)g(U^1h_2r^1R^2)\\
&\quad\overset{(10.1.3)}{=} f(U^2r^2h_2R^1)g(U^1r^1h_1R^2)\\
&\quad= f(h_2R^1)g(h_1R^2)=(f\cdot g)(h),
\end{aligned}$$

as desired. Now we prove (i); we have:

$$\begin{aligned}
&((h_1\rightharpoonup f\leftharpoonup h'_2)\cdot(h_2\rightharpoonup g\leftharpoonup h'_1))(h'')\\
&\quad= (h_2\rightharpoonup g\leftharpoonup h'_1)(h''_1R^2)(h_1\rightharpoonup f\leftharpoonup h'_2)(h''_2R^1)\\
&\quad= g(h'_1h''_1R^2h_2)f(h'_2h''_2R^1h_1)\\
&\quad\overset{(10.1.3)}{=} g(h'_1h''_1h_1R^2)f(h'_2h''_2h_2R^1)\\
&\quad= (f\cdot g)(h'h''h)=(h\rightharpoonup f\cdot g\leftharpoonup h')(h'').
\end{aligned}$$

For the second relation, we compute:

$$\begin{aligned}
&((f\cdot g)\cdot l)(h)\\
&\quad= l(h_1R^2)(f\cdot g)(h_2R^1)\\
&\quad= l(h_1R^2)g(h_{(2,1)}R^1_1r^2)f(h_{(2,2)}R^1_2r^1)\\
&\quad\overset{(10.1.1)}{=} l(h_1X^1R^2x^2\rho^2Y^3)g(h_{(2,1)}X^2R^1x^1Y^1r^2)f(h_{(2,2)}X^3x^3\rho^1Y^2r^1)\\
&\quad\overset{(10.1.6)}{=} l(h_1X^1r^2x^2\rho^2Y^3)g(h_{(2,1)}R^2X^3x^3\rho^1Y^2)f(h_{(2,2)}R^1X^2r^1x^1Y^1).
\end{aligned}$$

We now compute the right-hand side evaluated in $h$:

$$\begin{aligned}
&((X^1\rightharpoonup f\leftharpoonup Y^3)\cdot((X^2\rightharpoonup g\leftharpoonup Y^2)\cdot(X^3\rightharpoonup l\leftharpoonup Y^1)))(h)\\
&\quad= ((X^2\rightharpoonup g\leftharpoonup Y^2)\cdot(X^3\rightharpoonup l\leftharpoonup Y^1))(h_1R^2)(X^1\rightharpoonup f\leftharpoonup Y^3)(h_2R^1)\\
&\quad= l(Y^1h_{(1,1)}R^2_1r^2X^3)g(Y^2h_{(1,2)}R^2_2r^1X^2)f(Y^3h_2R^1X^1)\\
&\quad\overset{(3.1.7)}{=} l(h_1Y^1R^2_1r^2X^3)g(h_{(2,1)}Y^2R^2_2r^1X^2)f(h_{(2,2)}Y^3R^1X^1)\\
&\quad\overset{(10.1.2)}{=} l(h_1Y^1y^1T^1R^2x^2r^2X^3)g(h_{(2,1)}Y^2y^2\rho^2T^3x^3r^1X^2)\\
&\qquad f(h_{(2,2)}Y^3y^3\rho^1T^2R^1x^1X^1)\\
&\quad= l(h_1T^1R^2x^2r^2X^3)g(h_{(2,1)}\rho^2T^3x^3r^1X^2)f(h_{(2,2)}\rho^1T^2R^1x^1X^1),
\end{aligned}$$

and this is obviously equal to the expression obtained for $((f\cdot g)\cdot l)(h)$.

The relations $\varepsilon\cdot f = f\cdot\varepsilon = f$ and $h \rightharpoonup \varepsilon \leftharpoonup h' = \varepsilon(h)\varepsilon(h')\varepsilon$ are obvious, using the fact that $\varepsilon(h_1)h_2 = h = h_1\varepsilon(h_2)$ and $\varepsilon(R^1)R^2 = R^1\varepsilon(R^2) = 1_H$. □

## 10.3 The Square of the Antipode of a QT Quasi-Hopf Algebra

We show that the square of the antipode $S$ of a QT quasi-Hopf algebra $(H,R)$ is an inner automorphism of $H$. At first sight the formula of the invertible element $u$ that defines $S^2$ as an inner automorphism of $H$ looks "unnatural". But as in the case of the inverse of $R$, we shall see that it comes naturally from some canonical isomorphisms in a rigid braided category, as well as the formula for its inverse.

**Proposition 10.12** *Let $(H,R)$ be a finite-dimensional QT quasi-Hopf algebra. If we define the elements*

$$u := S^2(\tilde{q}^2\overline{R}^2\tilde{p}^1)\tilde{q}^1\overline{R}^1\tilde{p}^2 \quad \textit{and}\ u^{-1} := S^2(\tilde{q}^2R^1\tilde{p}^1)\tilde{q}^1R^2\tilde{p}^2,$$

*where $q_L = \tilde{q}^1\otimes\tilde{q}^2$ and $p_L = \tilde{p}^1\otimes\tilde{p}^2$ are as in* (3.2.20)*, and $R = R^1\otimes R^2$ and $R^{-1} = \overline{R}^1\otimes\overline{R}^2$, then for any finite-dimensional left $H$-module $V$ the map $V^*\ni v^*\mapsto u^{-1}\succ v^*\in {}^*V$ is a left $H$-linear isomorphism with inverse given by ${}^*V\ni {}^*v\mapsto u\succ {}^*v\in V^*$. Here, in order to avoid ambiguities, we have denoted by $\succ$ the left $H$-module structures of both left and right duals $V^*$ and ${}^*V$ of $V$ in ${}_H\mathcal{M}^{\rm fd}$. Consequently, $S^2(h) = uhu^{-1}$, for all $h\in H$.*

*Proof* Everything follows from the fact that in a braided rigid category the left and right dual objects are isomorphic; see Corollary 1.76.

More precisely, for $\mathscr{C} = {}_H\mathcal{M}^{\rm fd}$, a rigid braided category, we have an isomorphism $\Theta'_V : V^*\to {}^*V$ in ${}_H\mathcal{M}$, for any finite-dimensional left $H$-module $V$. As was explained in the proof of Corollary 1.76, $\Theta'_V$ is defined by the composition:

$$\begin{aligned}
\Theta'_V = \big(v^* &\xrightarrow{\mathrm{Id}_{V^*}\otimes\mathrm{coev}'_V} v^*\otimes({}^iv\otimes S^{-1}(\beta)\cdot{}_iv) \xrightarrow{a^{-1}_{V^*,{}^*V,V}} (x^1\cdot v^*\otimes x^2\succ{}^iv)\otimes x^3S^{-1}(\beta)\cdot{}_iv\\
&\xrightarrow{c_{V^*,{}^*V}\otimes\mathrm{Id}_V} (R^2x^2\succ{}^iv\otimes R^1x^1\cdot v^*)\otimes x^3S^{-1}(\beta)\cdot{}_iv\\
&\xrightarrow{a_{{}^*V,V^*,V}} X^1R^2x^2\succ{}^iv\otimes(X^2R^1x^1\cdot v^*\otimes X^3x^3S^{-1}(\beta)\cdot{}_iv)\\
&\xrightarrow{\mathrm{Id}_{{}^*V}\otimes\mathrm{ev}_V} v^*\big(S(X^2R^1x^1)\alpha X^3x^3S^{-1}(\beta)\cdot{}_iv\big)X^1R^2x^2\succ{}^iv\big),
\end{aligned}$$

where $\{{}_iv,{}^iv\}_i$ are dual bases in $V$ and $V^*$. In other words, we have

$$\begin{aligned}
\Theta'_V(v^*) &= v^*(S(X^2R^1x^1)\alpha X^3x^3S^{-1}(X^1R^2x^2\beta)\cdot v_i){}^iv\\
&= v^*(S^{-1}(u^{-1})\cdot{}_iv){}^iv = u^{-1}\succ v^*,
\end{aligned}$$

for all $v^*\in V^*$. In a similar manner, by the proof of Corollary 1.76 we have that its inverse ${\Theta'}_V^{-1}$ is given by

$$
\begin{aligned}
{\Theta'}_V^{-1} = \Big( {}^*v &\overset{\mathrm{coev}_V\otimes \mathrm{Id}_{{}^*V}}{\longrightarrow} (\beta\cdot {}_iv\otimes {}^iv)\otimes {}^*v \overset{a_{V,V^*,{}^*V}}{\longrightarrow} X^1\beta\cdot {}_iv\otimes (X^2\cdot {}^iv\otimes X^3\succ {}^*v)\\
&\overset{\mathrm{Id}_V\otimes c^{-1}_{V^*,{}^*V}}{\longrightarrow} X^1\beta\cdot {}_iv\otimes (\overline{R}^1X^3\succ {}^*v\otimes \overline{R}^2X^2\cdot {}^iv)\\
&\overset{a^{-1}_{V,{}^*V,V^*}}{\longrightarrow} (x^1X^1\beta\cdot {}_iv\otimes x^2\overline{R}^1X^3\succ {}^*v)\otimes x^3\overline{R}^2X^2\cdot {}^iv\\
&\overset{\mathrm{ev}'_V\otimes \mathrm{Id}_{V^*}}{\longrightarrow} {}^*v(S^{-1}(\alpha x^2\overline{R}^1X^3)x^1X^1\beta\cdot {}_iv)x^3\overline{R}^2X^2\cdot {}^iv\Big),
\end{aligned}
$$

for all ${}^*v\in {}^*V$. Thus we have shown that

$$
{\Theta'}_V^{-1}({}^*v) = {}^*v(S^{-1}(\tilde{q}^1\overline{R}^1\tilde{p}^2)S(\tilde{q}^2\overline{R}^2\tilde{p}^1)\cdot {}_iv)^iv = {}^*v(S^{-1}(u)\cdot {}_iv)^iv = u\succ {}^*v,
$$

for all ${}^*v\in {}^*V$. Hence, by using the fact that $\Theta_V$ and ${\Theta'}_V^{-1}$ are inverses of each other we get that $uu^{-1}\succ v^* = u^{-1}u\succ v^* = v^*$, for all $v^*\in V^*$. Since $H$ is finite dimensional this applies to $V=H$, and therefore

$$
S^{-1}(uu^{-1}) = \sum_i h^i(S^{-1}(uu^{-1}))h_i = \sum_i (uu^{-1}\succ h^i)(1_H)h_i = \sum_i h^i(1_H)h_i = 1_H,
$$

where $\{h_i, h^i\}_i$ are dual bases in $H$ and $H^*$. The antipode $S$ is bijective, so we conclude that $uu^{-1} = 1_H$. In a similar manner one can show that $u^{-1}u = 1_H$, and this finishes the first part of the proof.

Now, to prove that $u$ defines $S^2$ as an inner automorphism of $H$, observe that, for all $h\in H$ and $v^*\in V^*$,

$$
\begin{aligned}
\Theta'_V(h\succ v^*) &= u^{-1}\succ (h\succ v^*)\\
&= \sum_i u^{-1}\succ (h\succ v^*)({}_iv)^iv\\
&= \sum_i v^*(S(h)\cdot {}_iv)u^{-1}\succ {}^iv\\
&= u^{-1}S^2(h)\succ v^*,
\end{aligned}
$$

and similarly $h\succ \Theta'_V(v^*) = hu^{-1}\succ v^*$. Taking $V=H$, by arguments similar to the ones above we get that $u^{-1}S^2(h) = hu^{-1}$, for all $h\in H$, or equivalently $S^2(h) = uhu^{-1}$, for all $h\in H$. □

We next see that the results in Proposition 10.12 remain valid for any QT quasi-Hopf algebra, not necessarily finite dimensional. As we have explained, we gave a detailed proof for Proposition 10.12 in order to obtain, in a canonical way, the form of the element $u$ that defines $S^2$ as an inner automorphism of $H$, as well as the form of its inverse. Furthermore, in the forthcoming proofs we will not make use of the bijectivity of the antipode. Consequently, we will get that the antipode of a QT quasi-Hopf algebra is always bijective.

We start by proving a second formula for $u$. Note that other equivalent definitions for $u$ and $u^{-1}$ can be obtained by interchanging the definitions of the maps $\Theta_V$, $\Theta_V^{-1}$

and $\Theta'_V$, $\Theta'^{-1}_V$ respectively, given in the proof of Corollary 1.76, specialized of course for $\mathscr{C} = {}_H\mathscr{M}^{\rm fd}$.

**Lemma 10.13** *Let $(H,R)$ be a finite-dimensional QT quasi-Hopf algebra. If we denote $R^{-1} = \overline{R}^1 \otimes \overline{R}^2$ then the following relations hold:*

$$R = q^1\overline{R}^1x^2\tilde{p}^2_1 \otimes S(q^2\overline{R}^2x^1\tilde{p}^1)x^3\tilde{p}^2_2, \tag{10.3.1}$$

$$u = S(R^2p^2)\alpha R^1p^1. \tag{10.3.2}$$

*Proof* By Remarks 10.4 we have that $(H, R_{21}^{-1})$ is a QT quasi-Hopf algebra. By Lemma 10.9 it then follows that

$$R_{21} = S(q^2\overline{R}^2x^1\tilde{p}^1)x^3\tilde{p}^2_2 \otimes q^1\overline{R}^1x^2\tilde{p}^2_1.$$

Switching the order of the factors in the tensor product we get the first formula stated above. We can use it together with

$$x^1\tilde{p}^1 \otimes x^2\tilde{p}^2_1 \otimes x^3\tilde{p}^2_2 = X^2_1\tilde{p}^1S^{-1}(X^1) \otimes X^2_2\tilde{p}^2 \otimes X^3, \tag{10.3.3}$$

which can be viewed as the "op"-version of (5.2.7), to prove the second one as follows:

$$\begin{aligned}
S(R^2p^2)\alpha R^1p^1 &= S(x^3\tilde{p}^2_2p^2)S^2(q^2\overline{R}^2x^1\tilde{p}^1)\alpha q^1\overline{R}^1x^2\tilde{p}^2_1p^1\\
&\overset{(10.3.3)}{=} S(X^3p^2)S^2(q^2\overline{R}^2X^2_1\tilde{p}^1S^{-1}(X^1))\alpha q^1\overline{R}^1X^2_2\tilde{p}^2p^1\\
&\overset{(10.1.3)}{=} S(p^2)S^2(S^{-1}(X^3)q^2X^2_2\overline{R}^2\tilde{p}^1)S(X^1)\alpha q^1X^2_1\overline{R}^1\tilde{p}^2p^1\\
&\overset{(5.5.17)}{=} S(p^2)S^2(q^2x^3\overline{R}^2\tilde{p}^1)S(q^1_1x^1)\alpha q^1_2x^2\overline{R}^1\tilde{p}^2p^1\\
&\overset{(3.2.1),(3.2.20)}{=} S(\alpha p^2)S^2(\tilde{q}^2\overline{R}^2\tilde{p}^1)\tilde{q}^1\overline{R}^1\tilde{p}^2p^1\\
&= S(\alpha p^2)up^1 = S(S(p^1)\alpha p^2)u \overset{(3.2.2)}{=} u.
\end{aligned}$$

Note that in the penultimate equality we used the fact that $S^2$ is inner via $u$. So our proof is complete. □

From now on $(H,R)$ is an arbitrary QT quasi-Hopf algebra, so neither finite dimensional nor with bijective antipode. Also, $u$ is the element of $H$ defined by

$$u = S(R^2x^2\beta S(x^3))\alpha R^1x^1. \tag{10.3.4}$$

We record the obvious fact that $\varepsilon(u) = 1$.

In what follows we prove formulas that connect the $R$-matrix $R$ with the quasi-Hopf algebra structure of $H$.

**Lemma 10.14** *Let $(H,R)$ be a QT quasi-Hopf algebra with antipode $S$. Then the following relations hold:*

$$(S\otimes S)(R)\gamma = \gamma_{21}R, \tag{10.3.5}$$

*where $\gamma = \gamma^1 \otimes \gamma^2$ is the element defined by (3.2.5) and $\gamma_{21} = \gamma^2 \otimes \gamma^1$, and*

$$f_{21}Rf^{-1} = (S\otimes S)(R), \tag{10.3.6}$$

*where* $f = f^1 \otimes f^2$ *is the element defined in* (3.2.15) *with its inverse* $f^{-1}$ *as in* (3.2.16) *and* $f_{21} = f^2 \otimes f^1$.

*Proof* If $R = r^1 \otimes r^2$ is another copy of $R$, then by (10.1.9) and (3.2.5) we have:

$$
\begin{aligned}
&(S \otimes S)(R)\gamma R^{-1}\\
&\quad = S(T^2 y_2^1 R^1)\alpha T^3 y^2 X^1 Z^1 \beta S(Y^2 X_2^2 r^1 Z^2)\alpha Y^3 X^3 \otimes\\
&\qquad S(T^1 y_1^1 R^2)\alpha y^3 Y^1 X_1^2 r^2 Z^3\\
&\quad \overset{(3.2.1)}{=} S(T^2 y_2^1 R^1)\alpha T^3 y^2 X^1 Z^1 \beta S(y_{(2,1)}^3 Y^2 X_2^2 r^1 Z^2)\alpha y_{(2,2)}^3 Y^3 X^3\\
&\qquad \otimes S(T^1 y_1^1 R^2)\alpha y_1^3 Y^1 X_1^2 r^2 Z^3\\
&\quad \overset{(3.1.7)}{=} S(T^2 y_2^1 R^1)\alpha T^3 y^2 X^1 Z^1 \beta S(Y^2 (y_1^3 X^2)_2 r^1 Z^2)\alpha Y^3 y_2^3 X^3\\
&\qquad \otimes S(T^1 y_1^1 R^2)\alpha Y^1 (y_1^3 X^2)_1 r^2 Z^3\\
&\quad \overset{(3.1.9)}{=} S(T^2 X_{(1,2)}^1 z_2^1 y_2^1 R^1)\alpha T^3 X_2^1 z^2 y_1^2 Z^1 \beta S(Y^2 X_2^2 z_2^3 y_{(2,2)}^2 r^1 Z^2)\alpha Y^3 X^3 y^3\\
&\qquad \otimes S(T^1 X_{(1,1)}^1 z_1^1 y_1^1 R^2)\alpha Y^1 X_1^2 z_1^3 y_{(2,1)}^2 r^2 Z^3\\
&\quad \overset{(3.1.7)}{\underset{(3.2.1)}{=}} S(T^2 z_2^1 y_2^1 R^1)\alpha T^3 z^2 y_1^2 Z^1 \beta S(Y^2 X_2^2 z_2^3 y_{(2,2)}^2 r^1 Z^2)\alpha Y^3 X^3 y^3\\
&\qquad \otimes S(X^1 T^1 z_1^1 y_1^1 R^2)\alpha Y^1 X_1^2 z_1^3 y_{(2,1)}^2 r^2 Z^3\\
&\quad \overset{(10.1.3)}{=} S(T^2 z_2^1 y_2^1 R^1)\alpha T^3 z^2 y_1^2 Z^1 \beta S(Y^2 X_2^2 z_2^3 r^1 y_{(2,1)}^2 Z^2)\alpha Y^3 X^3 y^3\\
&\qquad \otimes S(X^1 T^1 z_1^1 y_1^1 R^2)\alpha Y^1 X_1^2 z_1^3 r^2 y_{(2,2)}^2 Z^3\\
&\quad \overset{(3.1.7)}{\underset{(3.2.1)}{=}} S(T^2 z_2^1 y_2^1 R^1)\alpha T^3 z^2 Z^1 \beta S(Y^2 X_2^2 z_2^3 r^1 Z^2)\alpha Y^3 X^3 y^3\\
&\qquad \otimes S(X^1 T^1 z_1^1 y_1^1 R^2)\alpha Y^1 X_1^2 z_1^3 r^2 Z^3 y^2\\
&\quad \overset{(3.1.9)}{=} S(T^2 z_2^1 y_2^1 R^1)\alpha T^3 z^2 Z^1 \beta S(X_1^3 Y^2 x^3 z_2^3 r^1 Z^2)\alpha X_2^3 Y^3 y^3\\
&\qquad \otimes S(X^1 Y_1^1 x^1 T^1 z_1^1 y_1^1 R^2)\alpha X^2 Y_2^1 x^2 z_1^3 r^2 Z^3 y^2\\
&\quad \overset{(3.2.1),(3.1.11)}{\underset{(10.1.3)}{=}} S(T^2 z_2^1 y_2^1 R^1)\alpha T^3 z^2 Z^1 \beta S(x^3 r^1 z_1^3 Z^2)\alpha y^3 \otimes S(x^1 T^1 z_1^1 y_1^1 R^2)\alpha x^2\\
&\qquad r^2 z_2^3 Z^3 y^2\\
&\quad \overset{(3.1.9)}{=} S(T^2 X_{(1,2)}^1 z_2^1 t_2^1 y_2^1 R^1)\alpha T^3 X_2^1 z^2 t_1^2 \beta S(x^3 r^1 X^2 z^3 t_2^2)\alpha y^3\\
&\qquad \otimes S(x^1 T^1 X_{(1,1)}^1 z_1^1 t_1^1 y_1^1 R^2)\alpha x^2 r^2 X^3 t^3 y^2\\
&\quad \overset{(3.1.7),(3.2.1)}{\underset{(3.1.10)}{=}} S(T^2 z_2^1 y_2^1 R^1)\alpha T^3 z^2 \beta S(x^3 r^1 X^2 z^3)\alpha y^3 \otimes S(x^1 X^1 T^1 z_1^1 y_1^1 R^2)\alpha x^2\\
&\qquad r^2 X^3 y^2\\
&\quad \overset{(3.1.9)}{\underset{(10.1.3)}{=}} S(z_1^2 t^1 T^2 R^1 y_1^1)\alpha z_2^2 t^2 T_1^3 \beta S(x^3 r^1 X^2 z^3 t^3 T_2^3)\alpha y^3\\
&\qquad \otimes S(x^1 X^1 z^1 T^1 R^2 y_2^1)\alpha x^2 r^2 X^3 y^2\\
&\quad \overset{(3.2.1),(3.1.10)}{\underset{(3.1.11)}{=}} S(R^1 y_1^1) S(t^1)\alpha t^2 \beta S(t^3) S(x^3 r^1 X^2)\alpha y^3\\
&\qquad \otimes S(x^1 X^1 R^2 y_2^1)\alpha x^2 r^2 X^3 y^2\\
&\quad \overset{(3.2.2)}{=} S(x^3 r^1 X^2 R^1 y_1^1)\alpha y^3 \otimes S(x^1 X^1 R^2 y_2^1)\alpha x^2 r^2 X^3 y^2\\
&\quad \overset{(10.1.2)}{=} S(R^1 X^1 y_1^1)\alpha y^3 \otimes S(R_1^2 X^2 y_2^1)\alpha R_2^2 X^3 y^2\\
&\quad \overset{(3.2.1),(10.1.5)}{=} S(X^1 y_1^1)\alpha y^3 \otimes S(X^2 y_2^1)\alpha X^3 y^2 \overset{(3.2.5)}{=} \gamma_{21}.
\end{aligned}
$$

The proof of the relation (10.3.6) is now immediate since by (10.3.5), (3.2.15) and (10.1.3) we have

$$\begin{aligned}(S\otimes S)(R)f &= (S\otimes S)(\Delta^{\rm cop}(x^1)R)\gamma\Delta(x^2\beta S(x^3))\\ &= (S\otimes S)(\Delta(x^1))(S\otimes S)(R)\gamma\Delta(x^2\beta S(x^3))\\ &= (S\otimes S)(\Delta(x^1))\gamma_{21}R\Delta(x^2\beta S(x^3))\\ &= (S\otimes S)(\Delta(x^1))\gamma_{21}\Delta^{\rm cop}(x^2\beta S(x^3))R = f_{21}R,\end{aligned}$$

as stated. This completes the proof of the lemma. □

**Lemma 10.15** *Let $(H,R)$ be a QT quasi-Hopf algebra with antipode $S$, and $u$ the element defined by* (10.3.4). *Then the following relations hold:*

$$S^2(h)u = uh,\ \ \forall\, h\in H, \tag{10.3.7}$$

$$S(\alpha)u = S(R^2)\alpha R^1. \tag{10.3.8}$$

*Proof* By (10.1.3) and (3.2.1) it is not hard to see that for all $h\in H$ we have

$$S^2(h)u = S(R^2h_{(1,2)}x^2\beta S(h_2x^3))\alpha R^1h_{(1,1)}x^1,$$

and then, by (3.1.7) and (3.2.1), it follows that $S^2(h)u = uh$. To prove (10.3.8), one performs the following substitution in $u$

$$\begin{aligned}x^1\otimes x^2\otimes x^3\otimes 1_H = (\Delta\otimes {\rm Id}_H\otimes {\rm Id}_H)(\Phi^{-1})({\rm Id}_H\otimes {\rm Id}_H\otimes\Delta)(\Phi^{-1})\\ (1_H\otimes\Phi)({\rm Id}_H\otimes\Delta\otimes {\rm Id}_H)(\Phi)\end{aligned}$$

and simplifies in several steps the resulting expression for $S(\alpha)u$ by using (3.2.1), (3.1.10), (3.1.11) and (3.2.2). □

A last preliminary result that we need is the following:

**Lemma 10.16** *Let $(H,R)$ be a QT quasi-Hopf algebra with antipode $S$. If $u$ is the element defined by* (10.3.4), *then $S^2(u) = u$.*

*Proof* We set $p_R = p^1\otimes p^2 = x^1\otimes x^2\beta S(x^3)$ and denote by $F^1\otimes F^2$ another copy of $f$. Then we compute:

$$\begin{aligned}S^2(u) &= S^2(S(R^2p^2)\alpha R^1p^1)\\ &= S(S(p^1)S(R^1)S(\alpha)S(S(p^2)S(R^2)))\\ &\overset{(10.3.6)}{=} S(S(p^1)f^2R^1g^1S(\alpha)S(S(p^2)f^1R^2g^2))\\ &\overset{(8.7.7)}{=} S(S(p^1)f^2R^1\beta S(S(p^2)f^1R^2))\\ &= S(S(R^2)S(f^1)S^2(p^2))S(\beta)S(R^1)S(f^2)S^2(p^1)\\ &\overset{(10.3.6)}{=} S(F^1R^2g^2S(f^1)S^2(p^2))S(\beta)F^2R^1g^1S(f^2)S^2(p^1)\\ &\overset{(8.7.7)}{=} S(R^2g^2S(f^1)S^2(p^2))\alpha R^1g^1S(f^2)S^2(p^1)\\ &\overset{(10.3.8)}{=} S(g^2S(f^1)S^2(p^2))S(\alpha)ug^1S(f^2)S^2(p^1)\\ &\overset{(10.3.7)}{=} S(g^2S(f^1)S^2(p^2))S(\alpha)S^2(g^1)uS(f^2)S^2(p^1)\end{aligned}$$

$$\begin{aligned}
&\overset{(10.3.7)}{=} S(S(g^1)\alpha g^2 S(f^1)S^2(p^2))S^3(f^2)uS^2(p^1)\\
&\overset{(8.7.7)}{=} S(S(f^1\beta S(f^2))S^2(p^2))uS^2(p^1)\\
&\overset{(8.7.7)}{=} S^2(S(\alpha p^2))uS^2(p^1)\overset{(10.3.7)}{=} uS(\alpha p^2)S^2(p^1)\\
&= uS(S(x^1)\alpha x^2\beta S(x^3))\overset{(3.2.2)}{=} u,
\end{aligned}$$

as needed. □

Now, we can prove the main result of this section.

**Theorem 10.17** *Let $H$ be a QT quasi-Hopf algebra with antipode $S$. If $u$ is the element defined by* (10.3.4) *then $u$ is invertible with inverse given by*

$$u^{-1} = X^1R^2p^2S(S(X^2R^1p^1)\alpha X^3), \tag{10.3.9}$$

*where $p_R = p^1 \otimes p^2 = x^1 \otimes x^2\beta S(x^3)$, and*

$$S^2(h) = uhu^{-1}, \ \forall\, h \in H. \tag{10.3.10}$$

*In particular, the antipode $S$ is bijective.*

*Proof* By the previous results we only have to check that $u$ and $u^{-1}$ are inverses. If $r^1 \otimes r^2$ is another copy of $R$, we calculate:

$$\begin{aligned}
u^{-1}u &\overset{(10.3.9)}{=} X^1R^2p^2S(\alpha X^3)S^2(X^2R^1p^1)u\\
&\overset{(10.3.7)}{=} X^1R^2p^2S(\alpha X^3)uX^2R^1p^1\\
&\overset{(10.3.8)}{=} X^1R^2p^2S(r^2X^3)\alpha r^1X^2R^1p^1\\
&= X^1R^2x^2\beta S(r^2X^3x^3)\alpha r^1X^2R^1x^1\\
&\overset{(10.1.2)}{=} X^1R^2_1\beta S(X^2R^2_2)\alpha X^3R^1\\
&\overset{(3.2.1),(10.1.5)}{=} X^1\beta S(X^2)\alpha X^3 \overset{(3.2.2)}{=} 1_H,
\end{aligned}$$

and thus $u^{-1}$ is a left inverse for $u$. It is also a right inverse. Indeed, Lemma 10.16, (10.3.7) and the fact that $S^2$ is an algebra morphism imply

$$uu^{-1} = S^2(u^{-1})u = S^2(u^{-1})S^2(u) = S^2(u^{-1}u) = S^2(1_H) = 1_H,$$

and the proof is complete. □

Since $S$ is always bijective we can give a second formula for the inverse of $R$. Namely, as in the proof of Lemma 10.9 (see also (10.3.6)) we have that

$$R^{-1} = x^3_1X^2R^1p^1 \otimes x^3_2X^3S^{-1}(S(x^1)\alpha x^2X^1R^2p^2). \tag{10.3.11}$$

Furthermore, by (3.1.9), (3.2.1) and (3.1.11) it follows that

$$R^{-1} = x^3y^2_2R^1p^1 \otimes y^3S^{-1}(S(x^1y^1)\alpha x^2y^2_1R^2p^2). \tag{10.3.12}$$

We leave the verification of all these details to the reader.

**Corollary 10.18** *Let $(H,R)$ be a QT quasi-Hopf algebra with antipode $S$ and $u \in H$ defined in* (10.3.4). *Then $uS(u) = S(u)u$ and this element is central in $H$.*

*Proof* Let $h \in H$. We apply $S$ to the equality $uh = S^2(h)u$ and we obtain $S(h)S(u) = S(u)S^3(h)$. By replacing $h$ with $S^{-1}(h)$ we obtain $hS(u) = S(u)S^2(h) = S(u)uhu^{-1}$, so $hS(u)u = S(u)uh$, which means that $S(u)u$ is central in $H$. By taking $h = u$, we obtain $uS(u) = S(u)u$. □

We now prove that the canonical element $u$ of a QT quasi-Hopf algebra is invariant under twisting.

**Proposition 10.19** *Let $(H, R)$ be a QT quasi-Hopf algebra and $F \in H \otimes H$ a gauge transformation. If we denote by $u$ and $u_F$ the canonical elements of the QT quasi-Hopf algebras $(H, R)$ and $(H_F, R_F)$, respectively, then $u = u_F$.*

*Proof* We denote $F = F^1 \otimes F^2 = \mathscr{F}^1 \otimes \mathscr{F}^2 = \overline{F}^1 \otimes \overline{F}^2$ and $F^{-1} = G^1 \otimes G^2 = \mathscr{G}^1 \otimes \mathscr{G}^2$. We have

$$\Phi_F = F_{23}(\mathrm{Id}_H \otimes \Delta)(F)\Phi(\Delta \otimes \mathrm{Id}_H)(F^{-1})F_{12}^{-1},$$

so with the above notation we obtain

$$\Phi_F^{-1} = F^1 \mathscr{F}_1^1 x^1 G^1 \otimes F^2 \mathscr{F}_2^1 x^2 G_1^2 \mathscr{G}^1 \otimes \mathscr{F}^2 x^3 G_2^2 \mathscr{G}^2.$$

Similarly, $\alpha_F = S(G^1)\alpha G^2$, $\beta_F = F^1 \beta S(F^2)$, $R_F = F^2 R^1 G^1 \otimes F^1 R^2 G^2$. Thus, we can compute the element $u_F$ given by (10.3.4) for $H_F$:

$$\begin{aligned}
u_F &= S(R_F^2 F^2 \mathscr{F}_2^1 x^2 G_1^2 \mathscr{G}^1 \overline{F}^1 \beta S(\overline{F}^2) S(\mathscr{F}^2 x^3 G_2^2 \mathscr{G}^2))\alpha_F R_F^1 F^1 \mathscr{F}_1^1 x^1 G^1 \\
&= S(R_F^2 F^2 \mathscr{F}_2^1 x^2 G_1^2 \beta S(\mathscr{F}^2 x^3 G_2^2))\alpha_F R_F^1 F^1 \mathscr{F}_1^1 x^1 G^1 \\
&= S(R_F^2 F^2 \mathscr{F}_2^1 x^2 G_1^2 \beta S(G_2^2) S(\mathscr{F}^2 x^3))\alpha_F R_F^1 F^1 \mathscr{F}_1^1 x^1 G^1 \\
&\overset{(3.2.1)}{=} S(R_F^2 F^2 \mathscr{F}_2^1 x^2 \beta S(\mathscr{F}^2 x^3))\alpha_F R_F^1 F^1 \mathscr{F}_1^1 x^1 \\
&= S(\overline{F}^1 R^2 G^2 F^2 \mathscr{F}_2^1 x^2 \beta S(\mathscr{F}^2 x^3)) S(\mathscr{G}^1)\alpha \mathscr{G}^2 \overline{F}^2 R^1 G^1 F^1 \mathscr{F}_1^1 x^1 \\
&= S(R^2 \mathscr{F}_2^1 x^2 \beta S(\mathscr{F}^2 x^3))\alpha R^1 \mathscr{F}_1^1 x^1 \\
&\overset{(10.1.3)}{=} S(\mathscr{F}_1^1 R^2 x^2 \beta S(\mathscr{F}^2 x^3))\alpha \mathscr{F}_2^1 R^1 x^1 \\
&\overset{(3.2.1)}{=} S(R^2 x^2 \beta S(x^3))\alpha R^1 x^1 = u,
\end{aligned}$$

finishing the proof. □

## 10.4 The QT Structure of the Quantum Double

Throughout this section $H$ is a finite-dimensional quasi-Hopf algebra and $\{e_i, e^i\}_i$ are dual bases in $H$ and $H^*$. Also, $D(H)$ is the quantum double of $H$, the quasi-Hopf algebra considered in Theorem 8.33. We recall that the quasi-Hopf algebra structure of $D(H)$ was obtained by using the reconstruction theorem for quasi-Hopf algebras and the category isomorphism ${}_H\mathscr{YD}^H \cong {}_{D(H)}\mathscr{M}$ proved in Corollary 8.31.

Since ${}_H\mathscr{YD}^H \cong \mathscr{Z}_r({}_H\mathscr{M})$ is a braided category, by Proposition 10.1 it follows that $D(H)$ has a unique QT structure such that the monoidal category isomorphism

${}_H\mathscr{YD}^H \cong {}_{D(H)}\mathscr{M}$ becomes braided monoidal. We compute this QT structure of $D(H)$ as follows.

**Theorem 10.20** *Let $H$ be a finite-dimensional quasi-Hopf algebra. Then $D(H)$, the quantum double of $H$, is a QT quasi-Hopf algebra with the R-matrix given by*

$$\mathscr{R}_D = \sum_{i=1}^{n} (\varepsilon \bowtie S^{-1}(p^2)e_i p_1^1) \otimes (e^i \bowtie p_2^1), \tag{10.4.1}$$

*where $p_R = p^1 \otimes p^2$ is as in (3.2.19). This QT structure of $D(H)$ turns the isomorphism ${}_H\mathscr{YD}^H \cong {}_{D(H)}\mathscr{M}$ from Corollary 8.31 into a braided monoidal one.*

*Proof* From the above comments we only have to transfer the braided structure of ${}_H\mathscr{YD}^H$ to the category ${}_{D(H)}\mathscr{M}$ through the categorical isomorphism in Corollary 8.31, and then to apply Proposition 10.1 in order to get the explicit form of the $R$-matrix of $D(H)$.

We have a strict monoidal functor $F : {}_H\mathscr{YD}^H \to {}_{D(H)}\mathscr{M}$ that acts as identity on objects and morphisms; see (8.5.6) for the explicit definition of $F$. Note also that any left $D(H)$-module $M$ is a left-right Yetter–Drinfeld module via the structure given by

$$h \cdot m = (\varepsilon \bowtie h)m \text{ and } M \ni m \mapsto \sum_i (e^i \bowtie p_2^1)m \otimes S^{-1}(p^2)e_i p_1^1 \in M \otimes H.$$

For the general case see Lemmas 8.27 and 8.28.

By using these correspondences we get that the braided structure of ${}_{D(H)}\mathscr{M}$ is given by the braiding $c$ defined, for all $M, N \in {}_{D(H)}\mathscr{M}$, by

$$\begin{aligned} c_{M,N}(m \otimes n) &= \mathrm{c}_{M,N}(m \otimes n) \\ &\overset{(8.2.18)}{=} n_{(0)} \otimes n_{(1)} \cdot m \\ &= \sum_i (e^i \bowtie p_2^1)n \otimes S^{-1}(p^2)e_i p_1^1 \cdot m \\ &= \sum_i (e^i \bowtie p_2^1)n \otimes (\varepsilon \bowtie S^{-1}(p^2)e_i p_1^1)m, \end{aligned}$$

for all $m \in M$ and $n \in N$. Now, by the proof of Proposition 10.1 we have that $(\mathscr{R}_D)_{21} = c_{D(H),D(H)}(1_D \otimes 1_D)$ and this leads to the formula for $\mathscr{R}_D$ in (10.4.1). □

As an application of the quantum double construction we next describe the QT quasi-Hopf algebra structure of $D(H(2))$.

**Proposition 10.21** *The quantum double of $H(2)$ is the associative unital algebra generated by $X$ and $Y$ with relations*

$$X^2 = 1, \; Y^2 = X, \; XY = YX.$$

*The quasi-coalgebra structure on $D(H(2))$ is given by the formulas:*

$$\begin{aligned} &\Delta(X) = X \otimes X, \quad \varepsilon(X) = 1, \\ &\Delta(Y) = -\frac{1}{2}(Y \otimes Y + XY \otimes Y + Y \otimes XY - XY \otimes XY), \quad \varepsilon(Y) = -1. \end{aligned}$$

*If we denote* $p^X_\pm := \frac{1}{2}(1 \pm X)$ *then the reassociator, the distinguished elements* $\alpha$ *and* $\beta$ *and the antipode are given by*

$$\Phi_X = 1 - 2p^X_- \otimes p^X_- \otimes p^X_-, \ \ \alpha = X, \ \ \beta = 1, \ \ S(X) = X, \ \ S(Y) = Y,$$

*respectively. Moreover,* $D(H(2))$ *is a QT quasi-Hopf algebra with R-matrix given by*

$$R = p^X_+ \otimes 1 - p^X_- \otimes XY.$$

*Proof* By using the commutativity and cocommutativity of $H(2)$, (3.2.1) and the fact that $\beta = 1$, we find that the multiplication rule (8.5.2) takes the following form on $D(H(2))$:

$$(\varphi \bowtie h)(\varphi' \bowtie h') = (\Omega^1\Omega^5 \rightharpoonup \varphi)(\Omega^2\Omega^4 \rightharpoonup \varphi') \bowtie \Omega^3 hh',$$

for all $\varphi, \varphi' \in H(2)^*$ and $h, h' \in H(2)$. From the definition (8.5.1) of $\Omega$ we find

$$\Omega^1\Omega^5 \otimes \Omega^2\Omega^4 \otimes \Omega^3 = X^1_{(1,1)}X^3 y^1 x^1 f^2 \otimes X^1_{(1,2)}X^2 y^2 x^2_1 x^3 f^1 \otimes X^1_2 y^3 x^2_2.$$

By using the expressions of $\Phi$ and $\Phi^{-1}$ in (10.1.7) we easily compute that

$$\begin{aligned} X^1_{(1,1)}X^3 \otimes X^1_{(1,2)}X^2 \otimes X^1_2 &= 1 - 2p_- \otimes p_- \otimes p_-, \\ x^1 \otimes x^2_1 x^3 \otimes x^2_2 &= 1 - 2p_- \otimes p_- \otimes p_+, \\ \Phi^{-1}(f^2 \otimes f^1 \otimes 1) &= p_- \otimes 1 \otimes p_- + p_- \otimes g \otimes p_+ + p_+ \otimes 1 \otimes 1, \end{aligned}$$

where $f = g \otimes p_- + 1 \otimes p_+$ is the Drinfeld twist of $H(2)$. By the above relations the multiplication of $D(H(2))$ comes out explicitly as

$$(\varphi \bowtie h)(\varphi' \bowtie h') = \varphi\varphi' \bowtie hh' - 2(p_- \rightharpoonup \varphi)(p_- \rightharpoonup \varphi') \bowtie p_- hh'.$$

Let $\{P_1, P_g\}$ be the dual basis of $H(2)^*$ corresponding to the basis $\{1, g\}$ of $H(2)$. Then $\varepsilon = P_1 + P_g$ and $\{\varepsilon, \mu = P_1 - P_g\}$ is clearly a basis for $H(2)^*$. Now let $X = \varepsilon \bowtie g$ and $Y = \mu \bowtie 1$. Since $p_- \rightharpoonup P_1 = \frac{1}{2}\mu$ and $p_- \rightharpoonup P_g = -\frac{1}{2}\mu$, we obtain

$$\begin{aligned} X^2 &= (\varepsilon \bowtie g)(\varepsilon \bowtie g) = \varepsilon \bowtie 1 - 2(p_- \rightharpoonup \varepsilon)(p_- \rightharpoonup \varepsilon) = 1, \\ XY &= YX = \mu \bowtie g, \\ Y^2 &= \mu^2 \bowtie 1 - 2(p_- \rightharpoonup \mu)^2 \bowtie p_- = \varepsilon \bowtie 1 - \varepsilon \bowtie 2p_- = \varepsilon \bowtie g = X, \end{aligned}$$

which are the multiplication rules that we stated. A direct computation ensures that the reassociator of $D(H(2))$ has the desired form.

Since $H(2)$ is commutative and cocommutative, $\beta = 1$ and $\Phi^{-1} = \Phi = Y^2 \otimes Y^1 \otimes Y^3$, by (8.5.9) we have

$$\Delta(\varphi \bowtie h) = p^1_1 p^2 \rightharpoonup \varphi_2 \bowtie p^1_2 X^1 h_1 \otimes X^2_1 X^3 \rightharpoonup \varphi_1 \bowtie X^2_2 h_2,$$

for all $\varphi \in H(2)^*$ and $h \in H(2)$. On the other hand,

$$\begin{aligned} &p^1_1 p^2 \otimes p^1_2 X^1 \otimes X^2_1 X^3 \otimes X^2_2 \\ &\quad = (p_+ \otimes 1 \otimes 1 \otimes 1 - p_- \otimes g \otimes 1 \otimes 1)(1 - 2 \otimes p_- \otimes p_- \otimes p_+) \\ &\quad = p_+ \otimes 1 \otimes 1 \otimes 1 - p_- \otimes g \otimes 1 \otimes 1 - 2 \otimes p_- \otimes p_- \otimes p_+. \end{aligned}$$

Then we have

$$\Delta(X)=\Delta(\varepsilon\bowtie g)=\varepsilon\bowtie g\otimes\varepsilon\bowtie g=X\otimes X,$$

and since $\Delta(P_1)=P_1\otimes P_1+P_g\otimes P_g$ and $\Delta(P_g)=P_1\otimes P_g+P_g\otimes P_1$ we get that $\Delta(\mu)=\Delta(P_1-P_g)=(P_1-P_g)\otimes(P_1-P_g)=\mu\otimes\mu$, and therefore

$$\begin{aligned}\Delta(Y)&=p_+\rightharpoonup\mu\bowtie 1\otimes\mu\bowtie 1-p_-\rightharpoonup\mu\bowtie g\otimes\mu\bowtie 1-2\mu\bowtie p_-\otimes p_-\rightharpoonup\mu\bowtie p_+\\&=-XY\otimes Y-\frac{1}{2}(Y-XY)\otimes(Y+XY)\\&=-\frac{1}{2}(Y\otimes Y+XY\otimes Y+Y\otimes XY-XY\otimes XY),\end{aligned}$$

as needed. It follows from (8.5.10) that $\varepsilon(X)=1$ and $\varepsilon(Y)=-1$.

In our particular situation, (8.5.14) takes the form

$$S(\varphi\bowtie h)=p_1^1p^2U^1f^2\rightharpoonup\varphi\bowtie p_2^1U^2f^1h$$

for all $\varphi\in H(2)^*$ and $h\in H$. But $p_1^1p^2f^2\otimes p_2^1f^1=U^1\otimes U^2=g\otimes 1$, so the antipode for $D(H(2))$ is the identity map. Obviously, $\alpha=\varepsilon\bowtie g=X$, $\beta=\varepsilon\bowtie 1=1$.

Finally, since $p_1^1p^2\otimes p_2^1=p_+\otimes 1-p_-\otimes g$ we have from (10.4.1) that the canonical $R$-matrix for $D(H(2))$ is $R=p_+^X\otimes 1-p_-^X\otimes XY$, so our proof is complete. □

The quantum double construction yields also a QT structure on the quasi-Hopf algebras $D^\omega(H)$ and $D^\omega(G)$ constructed in Section 8.6.

**Proposition 10.22** *The quasi-Hopf algebra $D^\omega(H)$ is QT with R-matrix*

$$R=\sum_{i=1}^{n}(e^i\#1)\otimes(\varepsilon\#e_i)\in D^\omega(H)\otimes D^\omega(H),$$

*where $\{e_i,e^i\}_i$ are dual bases in $H$ and $H^*$. The inverse of $R$ is*

$$R^{-1}=\sum_{i=1}^{n}(e^i\#1_H)\otimes(\sigma^{-1}(S(e_i),e_i)\#S(e_i)),$$

*where $\sigma^{-1}$ is the convolution inverse of $\sigma$.*

*Proof* This is a direct consequence of Theorem 8.34. □

**Remarks 10.23** (1) It follows that $D^\omega(G)$ is a QT quasi-Hopf algebra with $R$-matrix

$$R=\sum_{g\in G}(p_g\otimes e)\otimes\left(\sum_{h\in G}p_h\otimes g\right).$$

(2) The map $T$ described in Proposition 8.36 is an isomorphism of QT quasi-bialgebras between $D^{\omega'}(H)$ and $D^\omega(H)_{F^{-1}}$.

In the rest of this section we specialize Theorem 10.17 for the QT quasi-Hopf algebra $D(H)$. We prove first an explicit formula for the element $u_D$ of $D(H)$. In what follows we denote by $\overline{S}$ the antipode of the dual quasi-Hopf algebra $H^*$, that is, the endomorphism of $H^*$ defined by $\overline{S}(\varphi)=\varphi\circ S$, for all $\varphi\in H^*$, and by $\overline{S}^{-1}$ its composition inverse.

**Proposition 10.24** *Let $H$ be a finite-dimensional quasi-Hopf algebra and $u_D$ the corresponding element $u$ for $D(H)$, the quantum double of $H$, as in* (10.3.4). *Then*

$$u_D = \sum_{i=1}^{n} \beta \rightharpoonup \overline{S}^{-1}(e^i) \bowtie e_i. \tag{10.4.2}$$

*Proof* Let us start by noting that (3.2.17), (8.7.7) and (3.2.1) imply

$$f_1^1 p^1 \otimes f_2^1 p^2 S(f^2) = g^1 S(\tilde{q}^2) \otimes g^2 S(\tilde{q}^1). \tag{10.4.3}$$

Secondly, observe that the definition (8.5.14) of the antipode $S_D$ of $D(H)$ can be reformulated as follows:

$$\begin{aligned}
&S_D(\varphi \bowtie h)\\
&\overset{(8.5.2)}{=} (\varepsilon \bowtie S(h))\left((f_1^1 p^1)_1 U^1 \rightharpoonup \overline{S}^{-1}(\varphi) \leftharpoonup f^2 S^{-1}(f_2^1 p^2) \bowtie (f_1^1 p^1)_2 U^2\right)\\
&\overset{(10.4.3)}{=} (\varepsilon \bowtie S(h))\left(g_1^1 S(\tilde{q}^2)_1 U^1 \rightharpoonup \overline{S}^{-1}(\varphi) \leftharpoonup \tilde{q}^1 S^{-1}(g^2) \bowtie g_2^1 S(\tilde{q}^2)_2 U^2\right)\\
&\overset{(7.3.1),(3.2.13)}{=} (\varepsilon \bowtie S(h))\left(g_1^1 G^1 S(q^2 \tilde{q}_2^2) \rightharpoonup \overline{S}^{-1}(\varphi) \leftharpoonup \tilde{q}^1 S^{-1}(g^2) \bowtie g_2^1 G^2 S(q^1 \tilde{q}_1^2)\right),
\end{aligned}$$

where we denote by $G^1 \otimes G^2$ another copy of $f^{-1}$. Now, we claim that

$$S_D(\mathscr{R}^2)\alpha_D \mathscr{R}^1 = \sum_{i=1}^{n} \beta \rightharpoonup \overline{S}^{-1}(e^i) \leftharpoonup \alpha \bowtie e_i, \tag{10.4.4}$$

where $R_D = \mathscr{R}^1 \otimes \mathscr{R}^2$ is the $R$-matrix of $D(H)$ defined in (10.4.1). Indeed, one can easily check that

$$\overline{S}^{-1}(h \rightharpoonup \varphi) = \overline{S}^{-1}(\varphi) \leftharpoonup S(h) \text{ and } \overline{S}^{-1}(\varphi \leftharpoonup h) = S(h) \rightharpoonup \overline{S}^{-1}(\varphi), \tag{10.4.5}$$

for all $\varphi \in H^*$ and $h \in H$. Now, we calculate:

$$\begin{aligned}
&S_D(\mathscr{R}^2)\alpha_D \mathscr{R}^1\\
&\overset{(10.4.1),(8.5.13)}{=} \sum_{i=1}^{n} S_D(e^i \bowtie p_2^1)(\varepsilon \bowtie \alpha)(\varepsilon \bowtie S^{-1}(p^2) e_i p_1^1)\\
&\overset{(8.5.2)}{=} \sum_{i=1}^{n} \left(S(p_2^1)_1 g^1\right)_1 G^1 S(q^2 \tilde{q}_2^2) \rightharpoonup \overline{S}^{-1}(e^i) \leftharpoonup \tilde{q}^1 S^{-1}(S(p_2^1)_2 g^2)\\
&\qquad \bowtie \left(S(p_2^1)_1 g^1\right)_2 G^2 S(q^1 \tilde{q}_1^2) \alpha S^{-1}(p^2) e_i p_1^1\\
&\overset{(3.2.13)}{=} \sum_{i=1}^{n} g_1^1 G^1 S\left(q^2 (\tilde{q}^2 p_{(2,2)}^1)_2\right) \rightharpoonup \overline{S}^{-1}(p_1^1 \rightharpoonup e^i) \leftharpoonup \tilde{q}^1 p_{(2,1)}^1 S^{-1}(g^2)\\
&\qquad \bowtie g_2^1 G^2 S\left(q^1 (\tilde{q}^2 p_{(2,2)}^1)_1\right) \alpha S^{-1}(p^2) e_i\\
&\overset{(10.4.5),(3.2.22)}{=} \sum_{i=1}^{n} g_1^1 G^1 S(q^2 p_2^1 \tilde{q}_2^2) \rightharpoonup \overline{S}^{-1}(e^i \leftharpoonup S^{-1}(p^2)) \leftharpoonup \tilde{q}^1 S^{-1}(g^2)\\
&\qquad \bowtie g_2^1 G^2 S(q^1 p_1^1 \tilde{q}_1^2) \alpha e_i\\
&\overset{(10.4.5),(3.2.23)}{=} \sum_{i=1}^{n} g_1^1 G^1 S(\tilde{q}_2^2) \rightharpoonup \overline{S}^{-1}(e^i) \leftharpoonup \tilde{q}^1 S^{-1}(g^2) \bowtie g_2^1 G^2 S(\tilde{q}_1^2) \alpha e_i
\end{aligned}$$

$$\overset{(10.4.5)}{=} \sum_{i=1}^{n} g_1^1 G^1 \rightharpoonup \overline{S}^{-1}(e^i) \leftharpoonup \tilde{q}^1 S^{-1}(g^2) \bowtie g_2^1 G^2 S(\tilde{q}_1^2)\alpha \tilde{q}_2^2 e_i$$

$$\overset{(3.2.1),(3.2.20)}{=} \sum_{i=1}^{n} e^i \bowtie g_2^1 G^2 \alpha S^{-1}(\alpha S^{-1}(g^2) e_i g_1^1 G^1)$$

$$\overset{(8.7.7),(3.2.1)}{=} \sum_{i=1}^{n} e^i \bowtie S^{-1}(\alpha e_i \beta) = \sum_{i=1}^{n} \beta \rightharpoonup \overline{S}^{-1}(e^i) \leftharpoonup \alpha \bowtie e_i.$$

We are now able to calculate the element $u_D$. Since $H$ can be viewed as a quasi-Hopf subalgebra of $D(H)$ via the morphism $i_D$ it follows that the corresponding element $p_R$ for $D(H)$ is $(p_R)_D = p_D^1 \otimes p_D^2 = \varepsilon \bowtie p^1 \otimes \varepsilon \bowtie p^2$. Therefore:

$$u_D \overset{(10.3.4)}{=} S_D(\mathscr{R}^2 p_D^2)\alpha_D \mathscr{R}^1 p_D^1 = (\varepsilon \bowtie S(p^2)) S_D(\mathscr{R}^2)\alpha_D \mathscr{R}^1 (\varepsilon \bowtie p^1)$$

$$\overset{(10.4.4),(8.5.2)}{=} \sum_{i=1}^{n} S(p^2)_{(1,1)}\beta \rightharpoonup \overline{S}^{-1}(e^i) \leftharpoonup \alpha S^{-1}(S(p^2)_2) \bowtie S(p^2)_{(1,2)} e_i p^1$$

$$= \sum_{i=1}^{n} e^i \bowtie S(p^2)_{(1,2)} S^{-1}\left(\alpha S^{-1}(S(p^2)_2) e_i S(p^2)_{(1,1)}\beta\right) p^1$$

$$\overset{(3.2.1)}{=} \sum_{i=1}^{n} e^i \bowtie S^{-1}(S(p^1)\alpha p^2 e_i \beta)$$

$$\overset{(3.2.19),(3.2.2)}{=} \sum_{i=1}^{n} e^i \bowtie S^{-1}(e_i \beta) = \sum_{i=1}^{n} \beta \rightharpoonup \overline{S}^{-1}(e^i) \bowtie e_i,$$

as claimed. □

For the proof of the next proposition we need the formula

$$S(U^1)\tilde{q}^1 U_1^2 \otimes \tilde{q}^2 U_2^2 = f, \tag{10.4.6}$$

where $q_L = \tilde{q}^1 \otimes \tilde{q}^2$ and $U = U^1 \otimes U^2$ are the elements defined in (3.2.20) and (7.3.1), and $f = f^1 \otimes f^2$ is the Drinfeld's twist defined in (3.2.15). The formula (10.4.6) follows easily from the axioms and the basic properties of a quasi-Hopf algebra.

**Proposition 10.25** *Let $H$ be a finite-dimensional quasi-Hopf algebra and $D(H)$ its quantum double. Then*

$$S_D^2(\varphi \bowtie h) = g_1^1 G^1 S(f^2 F_2^2) \rightharpoonup \overline{S}^{-2}(\varphi) \leftharpoonup F^1 S^{-1}(g^2) \bowtie g_2^1 G^2 S(f^1 F_1^2) S^2(h), \tag{10.4.7}$$

*for all $\varphi \in H^*$ and $h \in H$, where $f = f^1 \otimes f^2 = F^1 \otimes F^2$ is the Drinfeld element defined in* (3.2.15) *and $f^{-1} = g^1 \otimes g^2 = G^1 \otimes G^2$ is its inverse as in* (3.2.16)*. Furthermore, the element $u_D$ in* (10.4.2) *defines $S_D^2$ as an inner automorphism of $D(H)$.*

*Proof* For all $\varphi \in H^*$ and $h \in H$ we compute:

$$S_D^2(\varphi \bowtie h)$$

$$= S_D(p_1^1 U^1 \rightharpoonup \overline{S}^{-1}(\varphi) \leftharpoonup f^2 S^{-1}(p^2) \bowtie p_2^1 U^2)(\varepsilon \bowtie S(S(h) f^1))$$

$$\overset{(8.5.14)}{=} (\varepsilon \bowtie S(p_2^1 U^2) F^1)(P_1^1 \mathscr{U}^1 \rightharpoonup \overline{S}^{-1}(p_1^1 U^1 \rightharpoonup \overline{S}^{-1}(\varphi) \leftharpoonup f^2 S^{-1}(p^2))$$

$$\leftharpoonup F^2S^{-1}(P^2) \bowtie P_2^1\mathscr{U}^2)(\varepsilon \bowtie S(f^1))(\varepsilon \bowtie S^2(h))$$

$$\overset{(8.5.2)}{\underset{(10.4.3)}{=}} (S(p_2^1U^2)_1g^1S(\tilde{q}^2))_1\mathscr{U}^1p^2S(f^2) \rightharpoonup \overline{S}^{-2}(\varphi) \leftharpoonup S(p_1^1U^1)\tilde{q}^1$$

$$S^{-1}(S(p_2^1U^2)_2g^2) \bowtie (S(p_2^1U^2)_1g^1S(\tilde{q}^2))_2\mathscr{U}^2S(f^1))(\varepsilon \bowtie S^2(h))$$

$$\overset{(3.2.13)}{\underset{(3.2.22)}{=}} (g_1^1S(p^1\tilde{q}^2U_2^2)_1\mathscr{U}^1p^2S(f^2) \rightharpoonup \overline{S}^{-2}(\varphi) \leftharpoonup S(U^1)\tilde{q}^1U_1^2S^{-1}(g^2)$$

$$\bowtie g_2^1S(p^1\tilde{q}^2U_2^2)_2\mathscr{U}^2S(f^1))(\varepsilon \bowtie S^2(h))$$

$$\overset{(7.3.1),(3.2.13)}{\underset{(3.2.21)}{=}} (g_1^1G^1S(f^2\tilde{q}_2^2U_{(2,2)}^2) \rightharpoonup \overline{S}^{-2}(\varphi) \leftharpoonup S(U^1)\tilde{q}^1U_1^2S^{-1}(g^2)$$

$$\bowtie g_2^1G^2S(f^1\tilde{q}_1^2U_{(2,1)}^2))(\varepsilon \bowtie S^2(h))$$

$$\overset{(10.4.6)}{=} g_1^1G^1S(f^2F_2^2) \rightharpoonup \overline{S}^{-2}(\varphi) \leftharpoonup F^1S^{-1}(g^2) \bowtie g_2^1G^2S(f^1F_1^2)S^2(h),$$

where $\mathscr{U}^1 \otimes \mathscr{U}^2$ is a second copy of the element $U$ defined in (7.3.1). Everything follows now from Theorem 10.17 and Proposition 10.24. □

## 10.5 The Quantum Double $D(H)$ when $H$ is Quasitriangular

We show that a finite-dimensional quasi-Hopf algebra $H$ is QT if and only if its quantum double $D(H)$ is a quasi-Hopf algebra with a projection. In this case, $D(H)$ is isomorphic to a biproduct quasi-Hopf algebra between a certain Hopf algebra $B^i$ in the braided category ${}_H^H\mathscr{YD}$ and $H$. Also, we will show that $B^i$ equals $H^*$ as a vector space, but with a different multiplication and comultiplication: the structures of $H^*$ in ${}_H^H\mathscr{YD}$ are induced by the $R$-matrix and the quasi-Hopf algebra structure of $H$.

**Lemma 10.26** *Let $H$ be a finite-dimensional quasi-Hopf algebra. Then there exists a quasi-Hopf algebra projection $\pi: D(H) \to H$ covering the canonical inclusion $i_D: H \to D(H)$ if and only if $H$ is QT.*

*Proof* First assume that there is a quasi-Hopf algebra morphism $\pi: D(H) \to H$ such that $\pi \circ i_D = \mathrm{Id}_H$. Then it is not hard to see that $R = \pi(\mathscr{R}_D^1) \otimes \pi(\mathscr{R}_D^2)$ is an $R$-matrix for $H$, where $\mathscr{R}_D = \mathscr{R}_D^1 \otimes \mathscr{R}_D^2$ is the canonical $R$-matrix of $D(H)$ defined in (10.4.1). Thus $H$ is QT.

Conversely, let $H$ be a QT quasi-Hopf algebra with $R$-matrix $R = R^1 \otimes R^2$, and define $\pi: D(H) \to H$ by

$$\pi(\varphi \bowtie h) = \varphi(q^2R^1)q^1R^2h, \tag{10.5.1}$$

where $q_R = q^1 \otimes q^2$ is as in (3.2.19). We have to show that $\pi$ is a quasi-Hopf algebra morphism and $\pi \circ i_D = \mathrm{Id}_H$.

As before, we write $q_R = Q^1 \otimes Q^2$ and $R = r^1 \otimes r^2$, and then compute for all $\varphi, \psi \in H^*$ and $h, h' \in H$ that

$$\pi((\varphi \bowtie h)(\psi \bowtie h'))$$

$$\overset{(8.5.2)}{=} \pi((\Omega^1 \rightharpoonup \varphi \leftharpoonup \Omega^5)(\Omega^2h_{(1,1)} \rightharpoonup \psi \leftharpoonup S^{-1}(h_2)\Omega^4) \bowtie \Omega^3h_{(1,2)}h')$$

$$= \varphi(\Omega^5q_1^2R_1^1\Omega^1)\psi(S^{-1}(h_2)\Omega^4q_2^2R_2^1\Omega^2h_{(1,1)})q^1R^2\Omega^3h_{(1,2)}h'$$

$$
\begin{aligned}
&\overset{(8.5.1)}{=} \varphi(S^{-1}(f^2X^3)q_1^2R_1^1X^1_{(1,1)}y^1x^1)\psi(S^{-1}(f^1X^2x^3h_2)q_2^2R_2^1X^1_{(1,2)}y^2x_1^2h_{(1,1)})\\
&\qquad q^1R^2X_2^1y^3x_2^2h_{(1,2)}h'\\
&\overset{(10.1.3),(3.2.26)}{=} \varphi(q^2Q_2^1z^2R_1^1y^1x^1)\psi(S^{-1}(x^3h_2)Q^2z^3R_2^1y^2x_1^2h_{(1,1)})\\
&\qquad q^1Q_1^1z^1R^2y^3x_2^2h_{(1,2)}h'\\
&\overset{(10.1.1)}{=} \varphi(q^2Q_2^1R^1y^1x^1)\psi(S^{-1}(x^3h_2)Q^2y^3r^1x_1^2h_{(1,1)})q^1Q_1^1R^2y^2r^2x_2^2h_{(1,2)}h'\\
&\overset{(3.2.19),(10.1.3)}{=} \varphi(q^2R^1X_1^1y^1x^1)\psi(S^{-1}(\alpha X^3x^3h_2)X^2y^3x_2^2h_{(1,2)}r^1)q^1R^2X_2^1y^2x_1^2h_{(1,1)}r^2h'\\
&\overset{(3.1.9),(3.2.19)}{=} \varphi(q^2R^1)\psi(S^{-1}(h_2)Q^2h_{(1,2)}r^1)q^1R^2Q^1h_{(1,1)}r^2h'\\
&\overset{(3.2.21)}{=} \pi(\varphi\bowtie h)\pi(\psi\bowtie h').
\end{aligned}
$$

From (10.1.5) and (3.2.19), it follows that $\pi(\varepsilon\bowtie h)=h$, for any $h\in H$. Thus we have shown that $\pi$ is an algebra map, and that $\pi\circ i_D=\mathrm{Id}_H$.

Since $\pi\circ i_D=\mathrm{Id}_H$, we have that $(\pi\otimes\pi\otimes\pi)(\Phi_D)=\Phi$. $\pi$ preserves the comultiplication, since

$$
\begin{aligned}
&\pi((\varphi\bowtie h)_1)\otimes\pi((\varphi\bowtie h)_2)\\
&\overset{(8.5.9)}{=} \pi((\varepsilon\bowtie X^1Y^1)(p_1^1x^1\rightharpoonup\varphi_2\leftharpoonup Y^2S^{-1}(p^2)\bowtie p_2^1x^2h_1))\\
&\qquad\otimes\pi(X_1^2\rightharpoonup\varphi_1\leftharpoonup S^{-1}(X^3)\bowtie X_2^2Y^3x^3h_2)\\
&= \varphi_2(Y^2S^{-1}(p^2)q^2R^1p_1^1x^1)\varphi_1(S^{-1}(X^3)Q^2r^1X_1^2)X^1Y^1q^1R^2p_2^1x^2h_1\\
&\qquad\otimes Q^1r^2X_2^2Y^3x^3h_2\\
&\overset{(10.1.3)}{=} \varphi(S^{-1}(X^3)Q^2r^1X_1^2Y^2S^{-1}(p^2)q^2p_2^1R^1x^1)\\
&\qquad X^1Y^1q^1p_1^1R^2x^2h_1\otimes Q^1r^2X_2^2Y^3x^3h_2\\
&\overset{(3.2.23),(10.1.3)}{=} \varphi(S^{-1}(X^3)Q^2X_2^2r^1Y^2R^1x^1)X^1Y^1R^2x^2h_1\otimes Q^1X_1^2r^2Y^3x^3h_2\\
&\overset{(5.5.17)}{=} \varphi(q^2y^3r^1Y^2R^1x^1)q_1^1y^1Y^1R^2x^2h_1\otimes q_2^1y^2r^2Y^3x^3h_2\\
&\overset{(10.1.2)}{=} \varphi(q^2R^1)\Delta(q^1R^2h)=\Delta(\pi(\varphi\bowtie h)).
\end{aligned}
$$

$\pi$ also preserves the counit, since $(\varepsilon\circ\pi)(\varphi\bowtie h)=\varphi(S^{-1}(\alpha))\varepsilon(h)=\varepsilon_D(\varphi\bowtie h)$.

One can easily see that $\pi(\alpha_D)=\alpha$ and $\pi(\beta_D)=\beta$, so we are done if we can show that $S\circ\pi=\pi\circ S_D$. This follows from the next computation:

$$
\begin{aligned}
&(\pi\circ S_D)(\varphi\bowtie h)\\
&\overset{(8.5.14)}{=} \pi((\varepsilon\bowtie S(h)f^1)(p_1^1U^1\rightharpoonup\overline{S}^{-1}(\varphi)\leftharpoonup f^2S^{-1}(p^2)\bowtie p_2^1U^2))\\
&= \langle\overline{S}^{-1}(\varphi),f^2S^{-1}(p^2)q^2R^1p_1^1U^1\rangle S(h)f^1q^1R^2p_2^1U^2\\
&\overset{(10.1.3),(3.2.23),(7.3.1)}{=} \langle\overline{S}^{-1}(\varphi),f^2R^1g^1S(q^2)\rangle S(h)f^1R^2g^2S(q^1)\\
&\overset{(10.3.6)}{=} \langle\overline{S}^{-1}(\varphi),S(q^2R^1)\rangle S(q^1R^2h)=(S\circ\pi)(\varphi\bowtie h),
\end{aligned}
$$

and this completes our proof. □

**Remark 10.27** Observe that, in the case where $(H,R)$ is QT, the map $\pi$ given by

(10.5.1) is a quasitriangular morphism, that is, $(\pi\otimes\pi)(R_D)=R$. Indeed,

$$(\pi\otimes\pi)(\mathscr{R}_D)\overset{(10.4.1)}{=}\sum_{i=1}^{n}\langle\pi\otimes\pi,\varepsilon\bowtie S^{-1}(p^2)e_ip_1^1\otimes e^i\bowtie p_2^1\rangle$$
$$\overset{(10.5.1)}{=}\sum_{i=1}^{n}e^i(q^2R^1)S^{-1}(p^2)e_ip_1^1\otimes q^1R^2p_2^1$$
$$\overset{(10.1.3)}{=}S^{-1}(p^2)q^2p_2^1R^1\otimes q^1p_1^1R^2\overset{(3.2.23)}{=}R.$$

Now we will apply the structure theorem of a quasi-Hopf algebra with a projection (Theorem 9.29) in our case, namely when $(H,R)$ is a finite-dimensional QT quasi-Hopf algebra, $B=D(H)$, $i=i_D$ is the canonical inclusion and $\pi$ is the map defined by (10.5.1). For this, we show first that $A:=B^{\mathrm{co}(H)}=\mathrm{Im}(\overline{E}_{D(H)})$ is isomorphic to $H^*$ as $k$-vector spaces, where $\overline{E}_{D(H)}$ is the projection associated to $D(H)$ as in (6.3.7). This follows from

$$\begin{aligned}\overline{E}_{D(H)}(\varphi\bowtie h)&=(\varphi\bowtie h)_1(\varepsilon\bowtie\beta S(\pi((\varphi\bowtie h)_2)))\\ &\overset{(8.5.9)}{\underset{(10.5.1)}{=}}\varphi_1(S^{-1}(X^3)q^2R^1X_1^2)(\varepsilon\bowtie X^1Y^1)(p_1^1y^1\rightharpoonup\varphi_2\\ &\qquad\leftharpoonup Y^2S^{-1}(p^2)\bowtie p_2^1y^2h_1)(\varepsilon\bowtie\beta S(q^1R^2X_2^2Y^3y^3h_2))\\ &\overset{(8.5.2)}{\underset{(3.2.1)}{=}}\varepsilon(h)\varphi_1(S^{-1}(X^3)q^2R^1X_1^2)(\varepsilon\bowtie X^1Y^1)(p_1^1y^1\rightharpoonup\varphi_2\\ &\qquad\leftharpoonup Y^2S^{-1}(p^2)\bowtie p_2^1y^2\beta S(q^1R^2X_2^2Y^3y^3))\\ &=\varepsilon(h)\overline{E}_{D(H)}(\varphi\bowtie 1_H),\end{aligned}$$

for all $\varphi\in H^*$ and $h\in H$, so $A=\overline{E}_{D(H)}(D(H))=\overline{E}_{D(H)}(H^*\bowtie 1_H)$; this means that we have a surjective $k$-linear map $H^*\to A$, $\varphi\mapsto\overline{E}_{D(H)}(\varphi\bowtie 1_H)$. This map is also injective since

$$\begin{aligned}&\langle\mathrm{Id}\otimes\varepsilon,\overline{E}_{D(H)}(\varphi\bowtie 1_H)\rangle\\ &\quad=\varphi_1(S^{-1}(X^3)q^2R^1X_1^2)\langle\mathrm{Id}\otimes\varepsilon,(\varepsilon\bowtie X^1Y^1)\\ &\qquad\qquad(p_1^1y^1\rightharpoonup\varphi_2\leftharpoonup Y^2S^{-1}(p^2)\bowtie p_2^1y^2\beta S(q^1R^2X_2^2Y^3y^3))\rangle\\ &\quad=\langle\mathrm{Id}\otimes\varepsilon,(\varepsilon\bowtie X^1Y^1)(p_1^1y^1\rightharpoonup\varphi\leftharpoonup S^{-1}(X^3)q^2R^1X_1^2Y^2S^{-1}(p^2)\\ &\qquad\bowtie p_2^1y^2\beta S(q^1R^2X_2^2Y^3y^3)))\rangle\\ &\quad\overset{(8.5.2)}{=}\langle\mathrm{Id}\otimes\varepsilon,(X^1Y^1)_{(1,1)}p_1^1y^1\rightharpoonup\varphi\leftharpoonup S^{-1}(X^3)q^2R^1X_1^2Y^2S^{-1}((X^1Y^1)_2p^2)\\ &\qquad\bowtie(X^1Y^1)_{(1,2)}p_2^1y^2\beta S(q^1R^2X_2^2Y^3y^3))\rangle\\ &\quad\overset{(10.1.5)}{=}X_1^1p^1\rightharpoonup\varphi\leftharpoonup S^{-1}(\alpha X^3)X^2S^{-1}(X_2^1p^2)\overset{(3.2.19),(3.2.23)}{=}\varphi.\end{aligned}$$

In fact we have shown that the map

$$\mu:\ A=\overline{E}_{D(H)}(D(H))\to H^*,\quad \mu(\overline{E}_{D(H)}(\varphi\bowtie h))=\varepsilon(h)\varphi$$

is an isomorphism of $k$-vector spaces, with inverse $\mu^{-1}(\varphi)=\overline{E}_{D(H)}(\varphi\bowtie 1_H)$. From now on, $\underline{H}^*$ will be the $k$-vector space $H^*$, with the structure of Hopf algebra in the braided category ${}_H^H\mathscr{YD}$ induced from $A$ via $\mu$. Let us compute the structure maps of $\underline{H}^*$ in ${}_H^H\mathscr{YD}$.

**Proposition 10.28** *The structure of $\underline{H}^*$ as an object in ${}_H^H\mathcal{YD}$ is given by the formulas*

$$h\cdot\varphi = h_1 \rightharpoonup \varphi \leftharpoonup S^{-1}(h_2), \tag{10.5.2}$$

$$\lambda_{\underline{H}^*}(\varphi) = R^2\otimes R^1\cdot\varphi. \tag{10.5.3}$$

*Proof* (10.5.2) is easy, and left to the reader. Observe that (5.5.16) yields

$$\begin{aligned}
&\overline{E}_{D(H)}(\varphi\otimes 1_H)\\
&= (\varepsilon\bowtie X^1Y^1)(p_1^1y^1\rightharpoonup\varphi\leftharpoonup S^{-1}(X^3)q^2R^1X_1^2Y^2S^{-1}(p^2)\\
&\qquad\bowtie p_2^1y^2\beta S(q^1R^2X_2^2Y^3y^3))\\
&\overset{(10.1.3),(5.5.17)}{\underset{(3.2.19)}{=}} (\varepsilon\bowtie q_1^1x^1Y^1)(p_1^1P^1\rightharpoonup\varphi\leftharpoonup q^2x^3R^1Y^2S^{-1}(p^2)\\
&\qquad\bowtie p_2^1P^2S(q_2^1x^2R^2Y^3))\\
&\overset{(8.5.2),(3.2.25)}{=} (\varepsilon\bowtie q_1^1x^1)\\
&\qquad(y^1p^1\rightharpoonup\varphi\leftharpoonup q^2x^3R^1S^{-1}(y^3p_2^2g^2)\bowtie y^2p_1^2g^1S(q_2^1x^2R^2))\\
&\overset{(10.3.6),(10.1.3)}{\underset{(8.5.2)}{=}} (q_1^1x^1)_{(1,1)}y^1p^1\rightharpoonup\varphi\leftharpoonup q^2x^3S^{-1}((q_1^1x^1)_2y^3R^1p_1^2g^1)\\
&\qquad\bowtie(q_1^1x^1)_{(1,2)}y^2R^2p_2^2g^2S(q_2^1x^2)\\
&\overset{(3.1.7),(10.1.3)}{\underset{(3.2.25)}{=}} y^1(q_1^1)_1X^1p_1^1P^1\rightharpoonup\varphi\leftharpoonup q^2S^{-1}(y^3R^1(q_1^1)_{(2,1)}X^2p_2^1P^2)\\
&\qquad\bowtie y^2R^2(q_1^1)_{(2,2)}X^3p^2S(q_2^1)\\
&\overset{(3.1.7),(3.2.21)}{=} y^1X^1p_1^1q_1^1P^1\rightharpoonup\varphi\leftharpoonup S^{-1}(y^3R^1X^2p_2^1q_2^1P^2S(q^2))\bowtie y^2R^2X^3p^2\\
&\overset{(3.2.23),(5.5.16)}{=} y^1x^1\rightharpoonup\varphi\leftharpoonup S^{-1}(y^3R^1x_1^2p^1)\bowtie y^2R^2x_2^2p^2S(x^3).
\end{aligned}\tag{10.5.4}$$

A similar computation, using (10.5.4), leads to

$$\begin{aligned}
&\pi(\overline{E}_{D(H)}(\varphi\bowtie 1_H)_1)\otimes\overline{E}_{D(H)}(\varphi\bowtie 1_H)_2 = X^1Y^1r^2z^2y_1^2R_1^2x_{(2,1)}^2p_1^2S(x^3)_1\\
&\otimes X_1^2Y^2r^1z^1y^1x^1\rightharpoonup\varphi\leftharpoonup S^{-1}(X^3y^3R^1x_1^2p^1)\bowtie X_2^2Y^3z^3y_2^2R_2^2x_{(2,2)}^2p_2^2S(x^3)_2.
\end{aligned}\tag{10.5.5}$$

Here $R = r^1\otimes r^2$ is another copy of $R$. Since

$$\begin{aligned}
&\langle \mathrm{Id}\otimes\varepsilon, \overline{E}_{D(H)}((\varphi\bowtie h)(\Psi\bowtie h'))\rangle = \langle\mathrm{Id}\otimes\varepsilon,(\varphi\bowtie h)(\Psi\bowtie h')\rangle\\
&\quad\varepsilon(h')(X_1^1x^1\rightharpoonup\varphi\leftharpoonup S^{-1}(f^2X^3))(X_2^1x^2h_1\rightharpoonup\Psi\leftharpoonup S^{-1}(f^1X^2x^3h_2)),
\end{aligned}\tag{10.5.6}$$

for all $\varphi,\Psi\in H^*$ and $h,h'\in H$, we obtain, using (9.2.13), that

$$\begin{aligned}
\lambda_{\underline{H}^*}(\varphi) &= q_1^1X^1r^2y^2R^2x_2^2p^2S(q^2x^3)\\
&\qquad q_{(2,1)}^1X^2r^1y^1x^1\rightharpoonup\varphi\leftharpoonup S^{-1}(q_{(2,2)}^1X^3y^3R^1x_1^2p^1)\\
&\overset{(3.1.7),(10.1.3)}{=} X^1r^2q_{(1,2)}^1y^2R^2x_2^2p^2S(q^2x^3)\\
&\qquad\otimes X^2r^1q_{(1,1)}^1y^1x^1\rightharpoonup\varphi\leftharpoonup S^{-1}(X^3q_2^1y^3R^1x_1^2p^1)\\
&\overset{(3.1.7),(10.1.3)}{\underset{(5.5.17)}{=}} X^1r^2y^2R^2q_2^1Y_{(1,2)}^2p^2S(q^2Y_2^2)Y^3\\
&\qquad\otimes X^2r^1y^1Y^1\rightharpoonup\varphi\leftharpoonup S^{-1}(X^3y^3R^1q_1^1Y_{(1,1)}^2p^1)\\
&\overset{(3.2.21),(3.2.23)}{=} X^1r^2y^2R^2Y^3\otimes X^2r^1y^1Y^1\rightharpoonup\varphi\leftharpoonup S^{-1}(X^3y^3R^1Y^2)\\
&\overset{(10.1.1),(10.5.2)}{=} R^2\otimes R^1\cdot\varphi,
\end{aligned}$$

as stated. □

Our next goal is to compute the algebra structure of $\underline{H}^*$.

**Lemma 10.29** *Let $H$ be a finite-dimensional quasi-Hopf algebra and $A$ a quasi-bialgebra. Let $\pi : A \to H$ be a quasi-bialgebra map that is a left inverse of the quasi-bialgebra map $i : H \to A$. Then $\overline{E}_{D(H)}$ satisfies the equation*

$$\overline{E}_{D(H)}(\overline{E}_{D(H)}(a) \circ a') = \overline{E}_{D(H)}(a) \circ \overline{E}_{D(H)}(a'), \tag{10.5.7}$$

*for all $a, a' \in D(H)$, where $\triangleright_i$ and $\circ$ are as in Proposition 4.3, specialized to $i_D$.*

*Proof* From the definition of an alternative projection we have that $\overline{E}_{D(H)}(ai(h)) = \varepsilon(h)\overline{E}_{D(H)}(a)$, $\overline{E}_{D(H)}(i(h)a) = h \triangleright_i \overline{E}_{D(H)}(a)$ and

$$\overline{E}_{D(H)}(aa') = a_1\overline{E}_{D(H)}(a')i(S(\pi(a_2))),$$

for all $a, a' \in D(H)$, $h \in H$. Then we compute:

$$\begin{aligned}
&\overline{E}_{D(H)}(\overline{E}_{D(H)}(a) \circ a') \\
&\quad = \overline{E}_{D(H)}(i(X^1)\Pi(a)i(S(x^1X^2)\alpha x^2X^3_1)a'i(S(x^3X^3_2))) \\
&\quad \overset{(3.2.19)}{=} q^1 \triangleright_i \overline{E}_{D(H)}(\overline{E}_{D(H)}(a)i(S(q^2))a') \\
&\quad = q^1 \triangleright_i [\overline{E}_{D(H)}(a)_1\overline{E}_{D(H)}(i(S(q^2))a')i(S(\pi(\overline{E}_{D(H)}(a)_2)))] \\
&\quad \overset{(*)}{=} q^1 \triangleright_i [i(x^1)\overline{E}_{D(H)}(a)i(S(x^3_2X^3)f^1)\overline{E}_{D(H)}(i(S(q^2))a') \\
&\qquad\qquad i(S(x^2X^1\beta S(x^3_1X^2)f^2))] \\
&\quad \overset{(3.2.13)}{=} q^1 \triangleright_i [i(x^1)\overline{E}_{D(H)}(a)i(S(q^2_2x^3_2X^3)f^1)\overline{E}_{D(H)}(a') \\
&\qquad\qquad i(S(x^2X^1\beta S(q^2_1x^3_1X^2)f^2))] \\
&\quad \overset{(3.1.9),(3.2.1)}{=} q^1X^1 \triangleright_i [i(x^1)\overline{E}_{D(H)}(a)i(S(q^2_2X^3)f^1)\overline{E}_{D(H)}(a') \\
&\qquad\qquad i(S(x^2\beta S(q^2_1X^2x^3)f^2))] \\
&\quad \overset{(3.2.26),(3.2.19)}{=} q^1Q^1_1y^1_{(1,1)} \triangleright_i [i(p^1)\overline{E}_{D(H)}(a)i(S(Q^2y^1_2)y^2)\overline{E}_{D(H)}(a') \\
&\qquad\qquad i(S(p^2S(q^2Q^1_2y^1_{(1,2)})y^3))] \\
&\quad \overset{(3.2.21)}{=} i(q^1_1p^1Q^1y^1_1)\overline{E}_{D(H)}(a)i(S(Q^2y^1_2)y^2)\overline{E}_{D(H)}(a')i(S(q^1_2p^2S(q^2)y^3)) \\
&\quad \overset{(3.2.23)}{=} i(Q^1y^1_1)\overline{E}_{D(H)}(a)i(S(Q^2y^1_2)y^2)\Pi(a')i(S(y^3)) \\
&\quad \underset{(3.2.1)}{\overset{(3.2.19),(3.1.9)}{=}} \overline{E}_{D(H)}(a) \circ \overline{E}_{D(H)}(a'),
\end{aligned}$$

as desired, where (*) refers to the fact that $\overline{E}_{D(H)}(a) \in D(H)^{\overline{\mathrm{co}(H)}}$. □

**Proposition 10.30** *The structure of $\underline{H}^*$ as a Hopf algebra in ${}_H^H\mathcal{YD}$ is defined by*

$$\begin{aligned}
\varphi \circ \Psi = (x^1X^1 \rightharpoonup \varphi \leftharpoonup S^{-1}(f^2x^3_2Y^3R^1X^2)) \\
(x^2Y^1R^2_1X^3_1 \rightharpoonup \Psi \leftharpoonup S^{-1}(f^1x^3_1Y^2R^2_2X^3_2)),
\end{aligned} \tag{10.5.8}$$

$$\underline{\Delta}(\varphi) = X^1_1p^1 \rightharpoonup \varphi_2 \leftharpoonup S^{-1}(X^1_2p^2) \otimes X^2 \rightharpoonup \varphi_1 \leftharpoonup S^{-1}(X^3), \tag{10.5.9}$$

$$\underline{\varepsilon}(\varphi) = \varphi(S^{-1}(\alpha)), \tag{10.5.10}$$

$$\underline{S}(\varphi) = (R^1 \rightharpoonup \varphi \leftharpoonup u^{-1}S(R^2)) \circ S. \tag{10.5.11}$$

*The unit element is $\varepsilon$.*

*Proof* It follows from (10.5.7) that the multiplication $\circ$ on $\underline{H}^*$ is

$$\begin{aligned}\varphi \circ \Psi &= \langle \mathrm{Id} \otimes \varepsilon, \overline{E}_{D(H)}(\varphi \bowtie 1_H) \circ \overline{E}_{D(H)}(\Psi \bowtie 1_H)\rangle \\ &= \langle \mathrm{Id} \otimes \varepsilon, \overline{E}_{D(H)}(\overline{E}_{D(H)}(\varphi \bowtie 1_H) \circ (\Psi \bowtie 1_H))\rangle.\end{aligned}$$

Now we can see that

$$\begin{aligned}
&\overline{E}_{D(H)}(\varphi \bowtie 1_H) \circ (\Psi \bowtie 1_H) \\
&\quad = i(X^1)\overline{E}_{D(H)}(\varphi \bowtie 1_H) i(S(z^1X^2)\alpha z^2X_1^3)(\Psi \bowtie 1_H) i(S(z^3X_2^3)) \\
&\overset{(3.1.9),(3.2.1)}{\underset{(10.5.4)}{=}} i(q^1z_1^1)(y^1x^1 \rightharpoonup \varphi \leftharpoonup S^{-1}(y^3R^1x_1^2p^1) \bowtie y^2R^2x_2^2p^2S(x^3)) \\
&\qquad i(S(q^2z_2^1)z^2)(\Psi \bowtie S(z^3)) \\
&\overset{(8.5.2),(3.1.7)}{\underset{(10.1.3)}{=}} (y^1q_1^1z_{(1,1)}^1x^1 \rightharpoonup \varphi \leftharpoonup S^{-1}(y^3R^1q_{(2,1)}^1(z_{(1,2)}^1x^2)_1p^1) \bowtie y^2R^2q_{(2,2)}^1 \\
&\qquad (z_{(1,2)}^1x^2)_2p^2S(q^2z_2^1x^3)z^2)(\Psi \bowtie S(z^3)) \\
&\overset{(3.1.7),(3.2.21)}{=} [y^1q_1^1x^1z_1^1 \rightharpoonup \varphi \leftharpoonup S^{-1}(y^3R^1q_{(2,1)}^1x_1^2p^1z_2^1) \bowtie y^2R^2q_{(2,2)}^1x_2^2p^2 \\
&\qquad S(q^2x^3)z^2](\Psi \bowtie S(z^3)) \\
&\overset{(5.5.17)}{=} [y^1X^1z_1^1 \rightharpoonup \varphi \leftharpoonup S^{-1}(y^3R^1q_1^1X_{(1,1)}^2p^1z_2^1) \bowtie y^2R^2q_2^1X_{(1,2)}^2p^2 \\
&\qquad S(q^2X_2^2)X^3z^2](\Psi \bowtie S(z^3)) \\
&\overset{(3.2.21),(3.2.23)}{=} y^1X^1z_1^1 \rightharpoonup \varphi \leftharpoonup S^{-1}(y^3R^1X^2z_2^1) \bowtie y^2R^2X^3z^2)(\Psi \bowtie S(z^3)).
\end{aligned}$$

By using (10.5.6) and (3.1.9) we obtain (10.5.8). It is easy to prove that $\varepsilon$ is the unit element of $\underline{H}^*$.

(10.5.9–10.5.10) follow from the following formula for comultiplication in $D(H)$ and the axioms of a quasi-bialgebra (we leave the details to the reader):

$$\underline{\Delta}(\overline{E}_{D(H)}(\varphi \bowtie 1_H)) = \overline{E}_{D(H)}(i(X^1)(\varphi \bowtie 1_H)_1) \otimes i(X^2)(\varphi \bowtie 1_H)_2 i(S(X^3)).$$

After a straightforward, but long and tedious, computation using (10.5.5), (8.5.14) and (10.5.6), we get that the antipode $\underline{S}$ of $\underline{H}^*$ is given by

$$\underline{S}(\varphi) = Q^1q^1R^2x^2 \cdot [p^1P^2S(Q^2) \rightharpoonup \overline{S}^{-1}(\varphi) \leftharpoonup S(q^2R^1x^1P^1)x^3S^{-1}(p^2)],$$

for all $\varphi \in H^*$. This formula is equivalent to the one in (10.5.11) since it implies

$$\begin{aligned}
\underline{S}(\varphi S) &\overset{(10.5.2)}{=} Q_1^1q_1^1R_1^2x_1^2p^1P^2S(Q^2) \rightharpoonup \varphi \leftharpoonup S(q^2R^1x^1P^1) \\
&\qquad x^3S^{-1}(Q_2^1q_2^1R_2^2x_2^2p^2) \\
&\overset{(10.1.2)}{=} Q_1^1q_1^1y^1X^1r^2z^2x_1^2p^1P^2S(Q^2) \rightharpoonup \varphi \\
&\qquad \leftharpoonup S(q^2y^3R^1X^2r^1z^1x^1P^1)x^3S^{-1}(Q_2^1q_2^1y^2R^2X^3z^3x_2^2p^2) \\
&\overset{(5.5.16),(5.5.17)}{\underset{(10.1.3),(3.2.19)}{=}} Q_1^1Y^1X^1y_1^1r^2P^2S(Q^2) \rightharpoonup \varphi \\
&\qquad \leftharpoonup S(q^2R^1Y_1^2X^2y_2^1r^1P^1)Y^3y^3S^{-1}(Q_2^1q^1R^2Y_2^2X^3y^2\beta) \\
&\overset{(3.1.9),(3.2.1)}{=} Q_1^1r^2P^2S(Q^2) \rightharpoonup \varphi \leftharpoonup S(q^2R^1x^1r^1P^1)S^{-1}(Q_2^1q^1R^2x^2\beta S(x^3))
\end{aligned}$$

$$\stackrel{(3.2.19),(10.3.9)}{=} Q_1^1 r^2 P^2 S(Q^2) \rightharpoonup \varphi \leftharpoonup S^{-1}(Q_2^1 u^{-1} S^2(r^1 P^1))$$
$$\stackrel{(10.3.10),(10.1.3)}{=} r^2 Q_2^1 P^2 S(Q^2) \rightharpoonup \varphi \leftharpoonup S^{-1}(r^1 Q_1^1 P^1 u^{-1})$$
$$\stackrel{(3.2.23),(10.3.10)}{=} r^2 \rightharpoonup \varphi \leftharpoonup S(r^1) S^{-1}(u^{-1}) = (r^1 \rightharpoonup \varphi S \leftharpoonup u^{-1} S(r^2)) \circ S,$$

for all $\varphi \in H^*$, where the last equality is based on the fact that $S^2(u^{-1}) = u^{-1}$ (see Lemma 10.16) and on the following computation, for $h \in H$:

$$\begin{aligned}\langle \varphi, S(r^1) S^{-1}(u^{-1}) h r^2 \rangle &\stackrel{(10.3.10)}{=} \langle \varphi, S(r^1) S^2(h r^2) S^{-1}(u^{-1}) \rangle \\ &= \langle \varphi S, S^{-2}(u^{-1}) S(h r^2) r^1 \rangle.\end{aligned}$$

Now (10.5.11) follows from the bijectivity of the antipode $S$ of $H$. □

**Remarks 10.31** (1) The quasi-Hopf algebra isomorphism $\chi : \underline{H}^* \times H \to D(H)$ is

$$\chi(\varphi \times h) = x^1 X^1 \rightharpoonup \varphi \leftharpoonup S^{-1}(x^3 R^1 X^2) \bowtie x^2 R^2 X^3 h. \tag{10.5.12}$$

(2) Let $(H, R)$ be a QT quasi-Hopf algebra. By the left-handed version of Corollary 8.4 we have a braided monoidal functor $\mathscr{Z}_l : {}_H\mathscr{M} \to \mathscr{Z}_l({}_H\mathscr{M}) \equiv {}_H^H\mathscr{YD}$ which sends algebras, coalgebras, bialgebras, etc. in ${}_H\mathscr{M}$ to the corresponding objects in ${}_H^H\mathscr{YD}$. If $M \in {}_H\mathscr{M}$ then $\mathscr{Z}_l(M) = M$ as left $H$-module, and together with the left $H$-coaction given by

$$\lambda_M : M \to H \otimes M, \quad \lambda_M(m) := R^2 \otimes R^1 \cdot m, \ \forall\, m \in M, \tag{10.5.13}$$

it becomes a left Yetter–Drinfeld module over $H$. $\mathscr{Z}_l$ maps a morphism to itself.

We observe that $\underline{H}^*$ lies in the image of $\mathscr{Z}_l$, that is, the left $H$-coaction is given by (10.5.13), which is exactly what we proved in Proposition 10.28.

## 10.6 Notes

QT quasi-Hopf algebras were introduced by Drinfeld in [82]. That the square of the antipode of a QT quasi-Hopf algebra $(H, R)$ is inner was proved for the first time in [9], but under the condition $S$ bijective and $R$ invertible. An alternative definition for a QT quasi-Hopf algebra $(H, R)$ was given in [55], and it was shown that with this definition $R$ is invertible and the square of the antipode $S$ of $H$ is inner, so in particular bijective; therefore it is equivalent to the initial one given by Drinfeld. The techniques used in [55] are similar to those used in the Hopf case [81, 187].

The content of Section 10.2 is taken from [173]. The presentation in Section 10.4 is from [64], although the $R$-matrix of $D(H)$ was found earlier by Hausser and Nill [108]. In Section 10.5 we presented the content of [48], were it was proved a quasi-Hopf algebra version of a result of Majid from [144].

# 11

# Factorizable Quasi-Hopf Algebras

We introduce the notion of factorizable quasi-Hopf algebras by using a categorical point of view. We show that the quantum double $D(H)$ of any finite-dimensional quasi-Hopf algebra $H$ is factorizable, and we characterize $D(H)$ when $H$ itself is factorizable. Finally, we prove that any finite-dimensional factorizable quasi-Hopf algebra is unimodular. In particular, we obtain that the quantum double $D(H)$ is a unimodular quasi-Hopf algebra.

## 11.1 Reconstruction in Rigid Monoidal Categories

Throughout this section, except in Proposition 11.1, $F$ is a strong monoidal functor between the monoidal categories $\mathscr{C}$ and $\mathscr{D}$. We further assume that the functor $\mathrm{Nat}(-\otimes F, F) : \mathscr{C}^{\mathrm{opp}} \to \mathrm{Set}$ is representable, where $\mathrm{Nat}(-\otimes F, F)$ stands for the set of natural transformations $\xi : -\otimes F \to F$, and where for any object $M$ of $\mathscr{D}$ we denote by $M\otimes F : \mathscr{C} \to \mathscr{D}$ the functor that sends $N \in \mathrm{Ob}\mathscr{C}$ to $(M\otimes F)(N) = M\otimes F(N)$; if $f : N \to N'$ is a morphism in $\mathscr{C}$ then $(M\otimes F)(f) = \mathrm{Id}_M \otimes F(f)$, a morphism in $\mathscr{D}$. In other words, there exists an object $B$ of $\mathscr{D}$ such that

$$\theta_M : \mathrm{Hom}_{\mathscr{D}}(M, B) \xrightarrow{\cong} \mathrm{Nat}(M\otimes F, F), \quad \forall\, M \in \mathrm{Ob}(\mathscr{D}), \tag{11.1.1}$$

by functorial bijections. If this is the case then we say that $F$ satisfies the representability assumption for modules.

We show that under some suitable conditions $B$ has a Hopf algebra structure in $\mathscr{D}$, and $F$ factors as $F : \mathscr{C} \to {}_B\mathscr{D} \xrightarrow{U} \mathscr{D}$, where $U$ is the forgetful functor. We start by showing that $B$ has an algebra structure in $\mathscr{D}$.

**Proposition 11.1** *Let $F : \mathscr{C} \to \mathscr{D}$ be a functor between two monoidal categories, satisfying the representability assumption for modules* (11.1.1). *Then $B$ becomes an algebra in $\mathscr{D}$, and $F$ factors as $F : \mathscr{C} \to {}_B\mathscr{D} \xrightarrow{U} \mathscr{D}$, where $U$ is the forgetful functor.*

*Proof* Take $\mu = \theta_B(\mathrm{Id}_B)$, a natural transformation between $B\otimes F$ and $F$. Then $\mu$ determines completely $\theta_M$ since, by the functoriality of $\theta_-$, for any morphism

$f : M \to B$ in $\mathscr{D}$ the diagram

$$\begin{array}{ccc} \mathrm{Hom}_{\mathscr{D}}(B,B) & \xrightarrow{\theta_B} & \mathrm{Nat}(B\otimes F,F) \\ \downarrow{\scriptstyle \mathrm{Hom}_{\mathscr{D}}(f,B)} & & \downarrow{\scriptstyle \mathrm{Nat}(f\otimes F,F)} \\ \mathrm{Hom}_{\mathscr{D}}(M,B) & \xrightarrow{\theta_M} & \mathrm{Nat}(M\otimes F,F) \end{array}$$

is commutative, and so $\theta_M(f) = (\mu_X(f\otimes \mathrm{Id}_{F(X)}))_{X\in \mathrm{Ob}(\mathscr{C})}$. Thus, if we denote $\mu_X =$

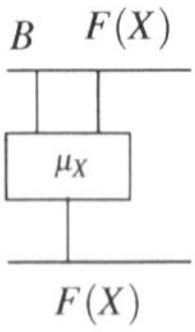


then, for any $M \in \mathrm{Ob}(\mathscr{D})$,

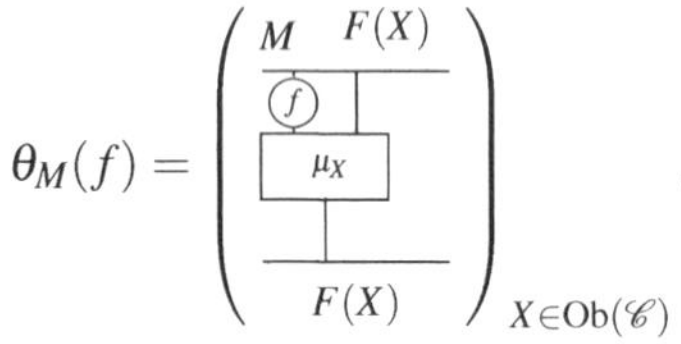


$$, \quad \forall\, f \in \mathrm{Hom}_{\mathscr{D}}(M,B). \tag{11.1.2}$$

Define $\xi := (\mu_X(\mathrm{Id}_B \otimes \mu_X))_{X\in\mathrm{Ob}(\mathscr{C})}$, a natural transformation from $B\otimes B\otimes F$ to $F$. Then there exists a unique morphism $\underline{m}_B : B\otimes B \to B$ such that $\theta_{B\otimes B}(\underline{m}_B) = \xi$. Hence $\underline{m}_B$ is the unique morphism in $\mathscr{D}$ obeying

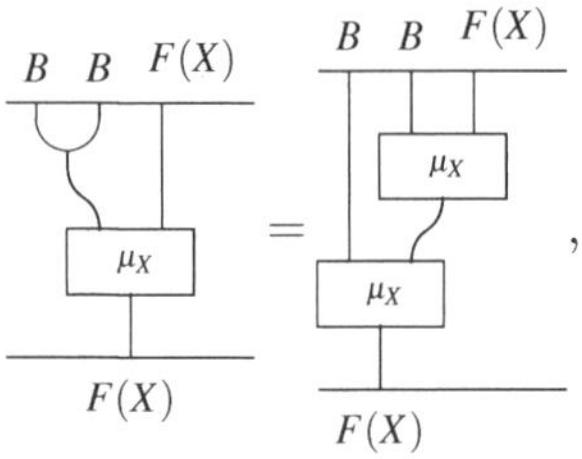


$$, \quad \forall\, X \in \mathrm{Ob}(\mathscr{C}). \tag{11.1.3}$$

It follows that

$$\theta_{B\otimes B\otimes B}(\underline{m}_B(\underline{m}_B \otimes \mathrm{Id}_B)) = \theta_{B\otimes B\otimes B}(\underline{m}_B(\mathrm{Id}_B \otimes \underline{m}_B)) =$$

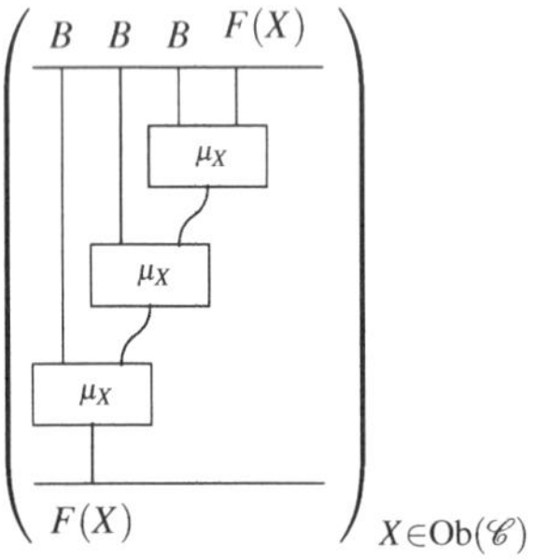


.

By our assumption we get that $\underline{m}_B$ is associative in $\mathscr{D}$. A unit for it is the unique morphism $\underline{\eta}_B : \underline{1} \to B$ such that $\theta_{\underline{1}}(\underline{\eta}_B) = 1_F$, the identity natural transformation of $F$. In other words, $\underline{\eta}_B$ is the unique morphism from $\underline{1}$ to $B$ in $\mathscr{D}$ such that

$$\mu_X(\underline{\eta}_B \otimes \mathrm{Id}_{F(X)}) = \mathrm{Id}_{F(X)}, \quad \forall\, X \in \mathrm{Ob}(\mathscr{C}). \tag{11.1.4}$$

By the above definitions one can see that $\theta_B(\underline{m}_B(\underline{\eta}_B \otimes \mathrm{Id}_B)) = \theta_B(\underline{m}_B(\mathrm{Id}_B \otimes \underline{\eta}_B)) = \mu = \theta_B(\mathrm{Id}_B)$, hence $\underline{\eta}_B$ is indeed a unit for $\underline{m}_B$. So we have proved that $(B, \underline{m}_B, \underline{\eta}_B)$ is an algebra in $\mathscr{D}$.

Finally, observe that the definition of $\mu$ was designed in such a way that each $\mu_X : B \otimes F(X) \to F(X)$ provides a left $B$-module structure on $F(X)$ in $\mathscr{D}$. Therefore $F$ factors as $F : \mathscr{C} \to {}_B\mathscr{D} \overset{U}{\to} \mathscr{D}$, as stated. □

In order to build a coalgebra structure on $B$ in $\mathscr{D}$ we need extra assumptions on $F$. Namely, we ask that $\mathscr{D}$ is braided and that for any $M \in \mathrm{Ob}(\mathscr{C})$ the maps $\theta^s_M : \mathrm{Hom}_{\mathscr{D}}(M, B^{\otimes s}) \to \mathrm{Nat}(M \otimes F^{\otimes s}, F^{\otimes s})$, $s \in \{2,3\}$, given by

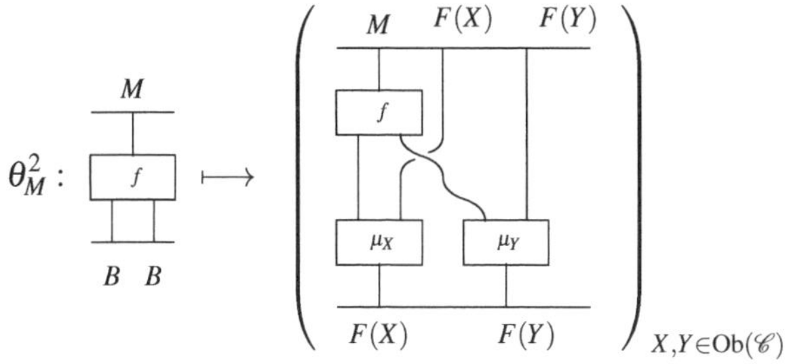


and by

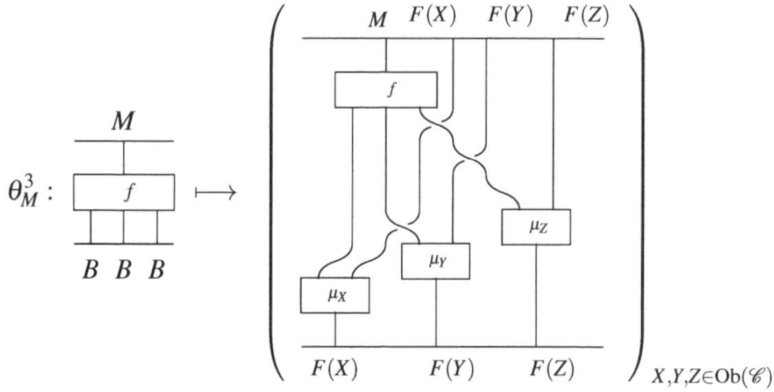


are bijections. This will be called the higher representability assumption for modules.

Under these extra assumptions, the coalgebra structure on $B$ can be defined as follows. Without loss of generality, we can assume (as before) that $\mathscr{C}, \mathscr{D}$ are strict monoidal categories and that $F$ is a strict monoidal functor, thus $F(X \otimes Y)$ and $F(X) \otimes F(Y)$ identify as objects of $\mathscr{D}$, for all $X, Y \in \mathrm{Ob}(\mathscr{C})$, and $F(\underline{1})$ is just the unit object of $\mathscr{D}$. Consequently, we can see $\mu_{X \otimes Y}$ either as a morphism from $B \otimes F(X \otimes Y)$ to $F(X \otimes Y)$ or as a morphism from $B \otimes F(X) \otimes F(Y) \to F(Y) \otimes F(Y)$ in $\mathscr{D}$, for all $X, Y \in \mathrm{Ob}(\mathscr{C})$, and $\mu_{\underline{1}}$ as a morphism from $B$ to $\underline{1}$ in $\mathscr{D}$.

**Proposition 11.2** *Let $F : \mathscr{C} \to \mathscr{D}$ be a strong monoidal functor satisfying the representability assumption for modules* (11.1.1). *If $\mathscr{D}$ is braided and $F$ satisfies the higher representability assumption for modules then $B$ has a bialgebra structure in $\mathscr{D}$, and $F$ factors, as a monoidal functor, as $F : \mathscr{C} \to {}_B\mathscr{D} \overset{U}{\to} \mathscr{D}$.*

*Proof* By the above comments we have that

$$\xi = (\mu_{X\otimes Y} : B\otimes F(X)\otimes F(Y) \to F(X)\otimes F(Y))_{X,Y\in \mathrm{Ob}(\mathscr{C})}$$

is a natural transformation between $B\otimes F^{\otimes 2}$ and $F^{\otimes 2}$. So by our higher assumption there exists a unique morphism $\underline{\Delta}_B : B \to B\otimes B$ in $\mathscr{D}$ such that $\theta_B^2(\underline{\Delta}_B) = \xi$. Equivalently, $\underline{\Delta}_B =$ 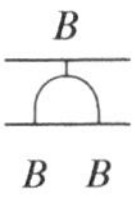 is the unique morphism in $\mathscr{D}$ from $B$ to $B\otimes B$ such that

$\mu_{X\otimes Y} =$ 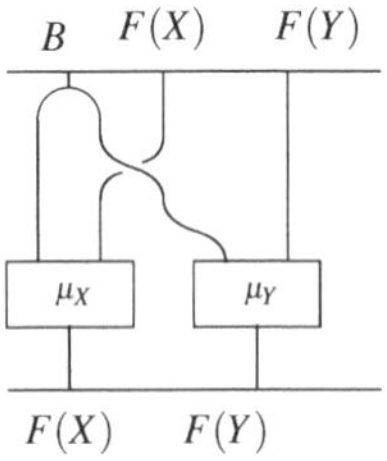 , $\forall\, X,\ Y \in \mathrm{Ob}(\mathscr{C})$. (11.1.5)

We have that, for all $X,Y,Z\in \mathrm{Ob}(\mathscr{C})$,

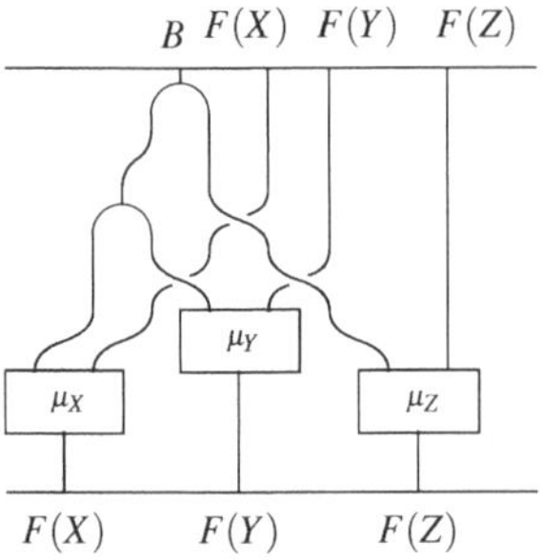

$=$

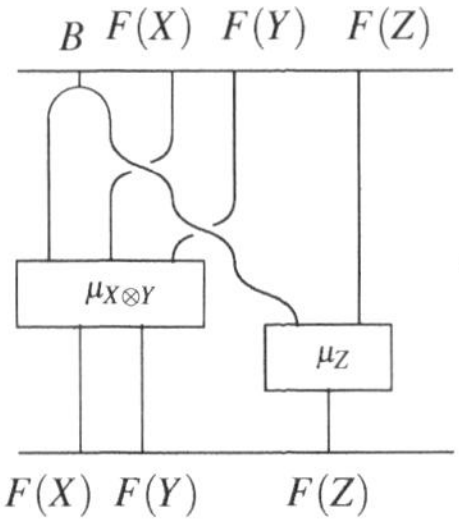

$= \mu_{(X\otimes Y)\otimes Z}$

and similarly

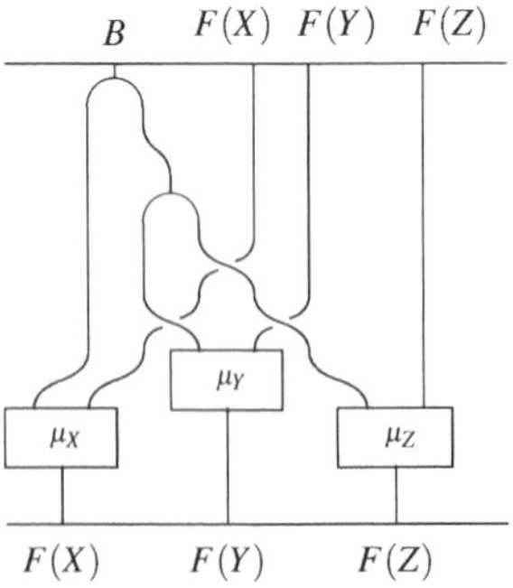

$\overset{(1.5.6)}{=}$

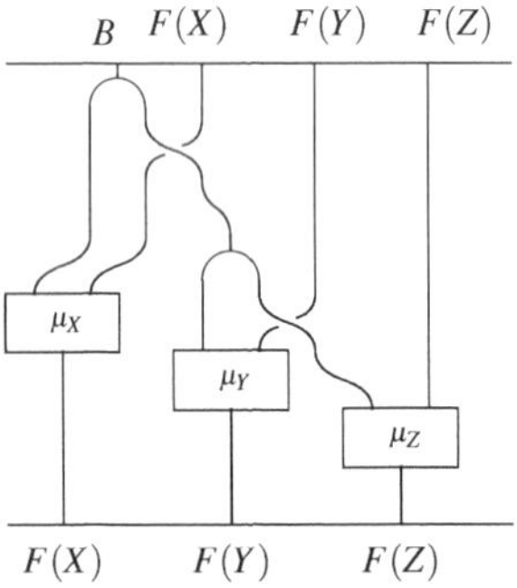

$=$

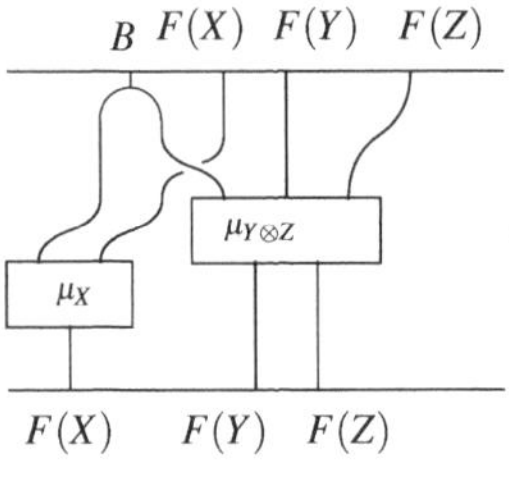

$= \mu_{X\otimes(Y\otimes Z)}$.

Since we assumed $\mathscr{C}$ strict monoidal we have $\theta_B^3((\underline{\Delta}_B \otimes \mathrm{Id}_B)\underline{\Delta}_B) = \theta_B^3((\mathrm{Id}_B \otimes \underline{\Delta}_B)\underline{\Delta}_B)$, and so $\underline{\Delta}_B$ is coassociative in $\mathscr{D}$.

We claim that $\underline{\varepsilon}_B := \mu_{\underline{1}} : \underline{1} \to B$ is a counit for $\underline{\Delta}_B$. One can see that $\theta_B((\underline{\varepsilon}_B \otimes \mathrm{Id}_B)\underline{\Delta}_B) = (\mu_{\underline{1}\otimes X})_{X\in \mathrm{Ob}(\mathscr{C})} = \mu = \theta_B(\mathrm{Id}_B)$ and $\theta_B((\mathrm{Id}_B \otimes \underline{\varepsilon}_B)\underline{\Delta}_B) = (\mu_{X\otimes \underline{1}})_{X\in \mathrm{Ob}(\mathscr{C})} = \mu = \theta_B(\mathrm{Id}_B)$. By our assumptions we get $(\underline{\varepsilon}_B \otimes \mathrm{Id}_B)\underline{\Delta}_B = \mathrm{Id}_B = (\mathrm{Id}_B \otimes \underline{\varepsilon}_B)\underline{\Delta}_B$.

For any two objects $X,Y$ of $\mathscr{D}$ we compute that

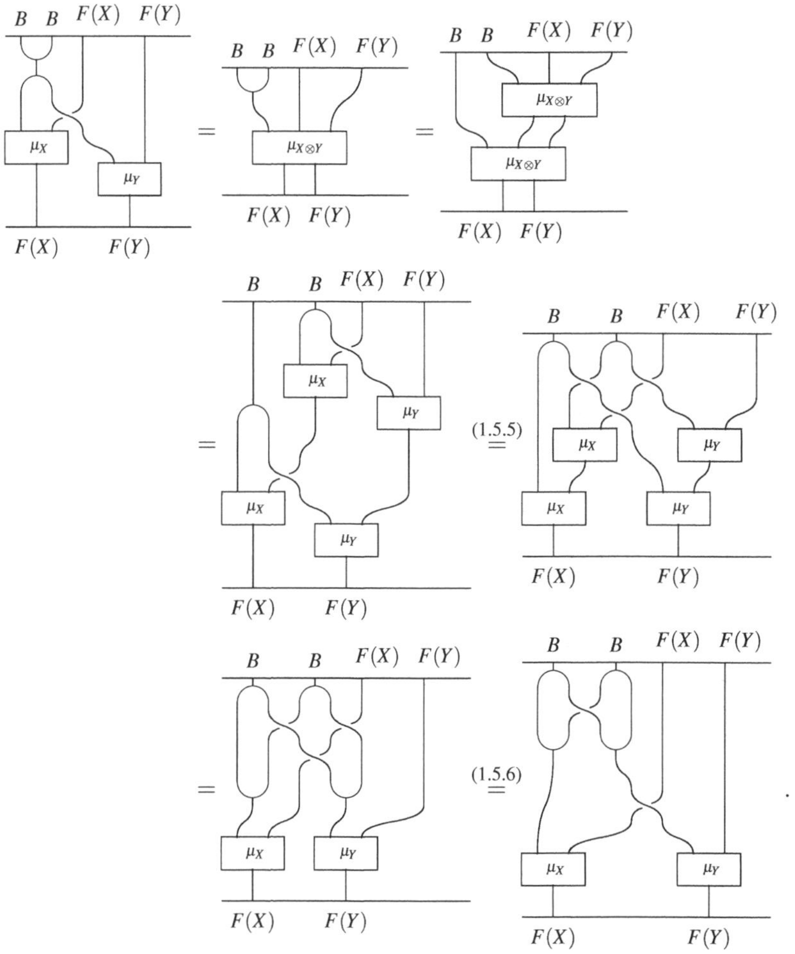


Together with the fact that $\theta_{B\otimes B}^2$ is a bijection this implies that $B$ is a bialgebra in $\mathscr{D}$. By (11.1.5) it follows that the bialgebra structure on $B$ is given in such a way that $F$ factors, as a monoidal functor, as $F : \mathscr{C} \to {}_B\mathscr{D} \overset{U}{\to} \mathscr{D}$. So our proof is finished. □

As in Section 3.5, to have an antipode for $B$ we need further to assume that $\mathscr{C}$ is a left rigid monoidal category. Consequently, since $F$ is strong monoidal, by Proposition 1.67 if follows that $F(X^*)$ is a left dual object for $F(X)$ in $\mathscr{D}$ whenever $X^*$ is a left dual object for $X$ in $\mathscr{C}$. Furthermore, if we assume $F$ strict monoidal then $F(X \otimes Y)$ and $F(X) \otimes F(Y)$ can be identified, for all $X,Y \in \mathrm{Ob}(\mathscr{C})$, hence in this case $(F(\mathrm{coev}_X), F(\mathrm{ev}_X)) : F(X^*) \dashv F(X)$ is an adjunction in $\mathscr{D}$. In other words, the corestriction of $F$ at its image is a left rigid monoidal functor, that is, we have a natural isomorphism $F(X^*) \cong F(X)^*$, indexed by $X \in \mathrm{Ob}(\mathscr{C})$.

For the adjunction $(F(\mathrm{coev}_X), F(\mathrm{ev}_X)) : F(X^*) \dashv F(X)$ we keep the same notation as in Section 1.6.

**Theorem 11.3** *Let $F : \mathscr{C} \to \mathscr{D}$ be a strong monoidal functor, where $\mathscr{C}$ is a left rigid monoidal category and $(\mathscr{D}, d)$ is a braided category. If $F$ satisfies the representability assumption* (11.1.1) *and the higher representability assumption for modules then $B$ is a Hopf algebra in $\mathscr{D}$ and $F$ factors, as a left rigid monoidal functor, as $F : \mathscr{C} \to {}_B\mathscr{D} \overset{U}{\to} \mathscr{D}$.*

*Proof* From the above results and comments, we only have to show that $B$ admits an antipode $\underline{S}$. To this end, define

$$\zeta = \big(F(\mathrm{Id}_X \otimes \mathrm{ev}_X)(\mathrm{Id}_{F(X)} \otimes \mu_{X^*} \otimes \mathrm{Id}_{F(X)})\\ (d_{B,F(X)} \otimes \mathrm{Id}_{F(X^* \otimes X)})(\mathrm{Id}_B \otimes F(\mathrm{coev}_X \otimes \mathrm{Id}_X))\big)_{X \in \mathrm{Ob}(\mathscr{C})}.$$

Clearly, $\zeta$ is a natural transformation between $B \otimes F$ and $F$, and since $\theta_B$ is a bijection there exists $\underline{S} : B \to B$ such that $\theta_B(\underline{S}) = \zeta$. Equivalently, $\underline{S}$ is uniquely determined by the equations

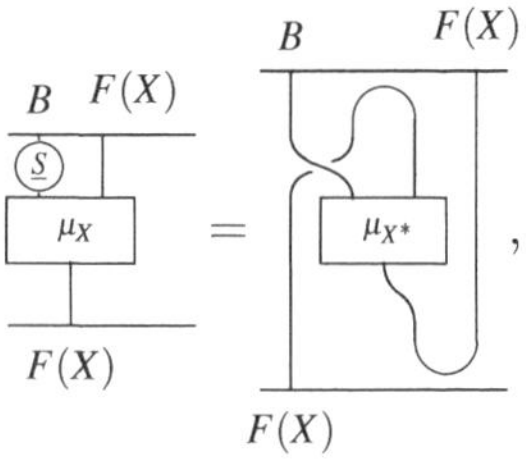


$, \ \forall X \in \mathrm{Ob}(\mathscr{C}).$ (11.1.6)

We end by proving that $\underline{S}$ is an antipode for the bialgebra $B$ in $\mathscr{D}$. Observe that the naturality of $\mu$ and the fact that $\mathrm{coev}_X : \underline{1} \to X \otimes X^*$ is a morphism in $\mathscr{C}$ imply

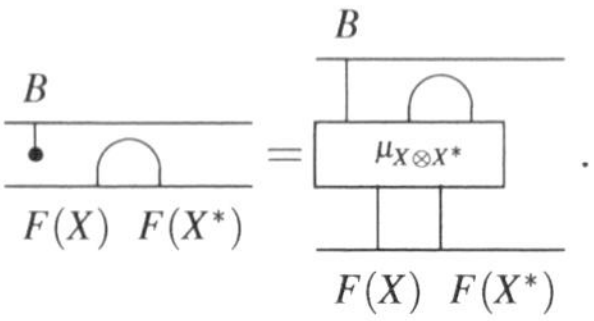


This allows us to compute that

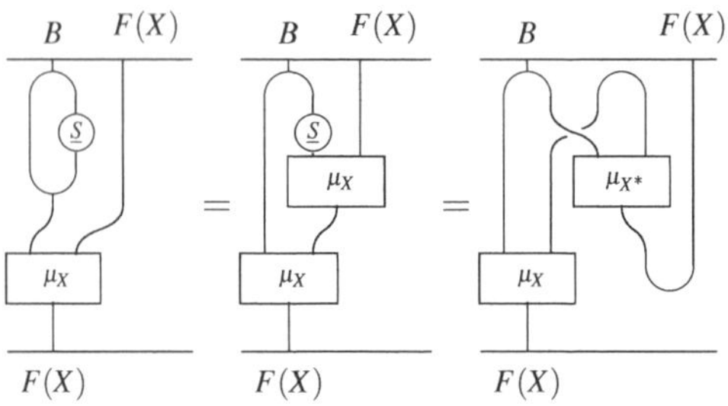

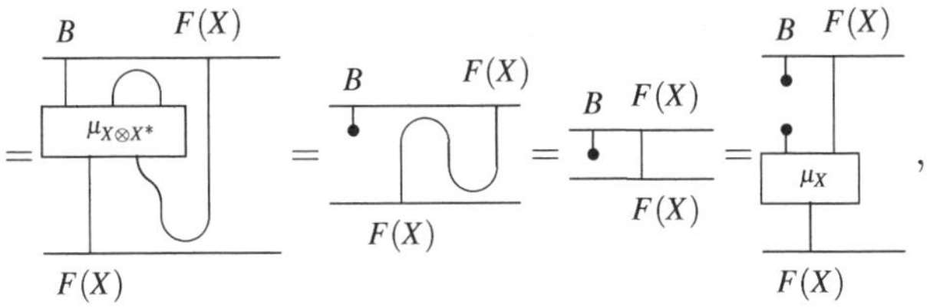


for all $X \in \mathrm{Ob}(\mathscr{C})$. In a similar manner, by using the equality

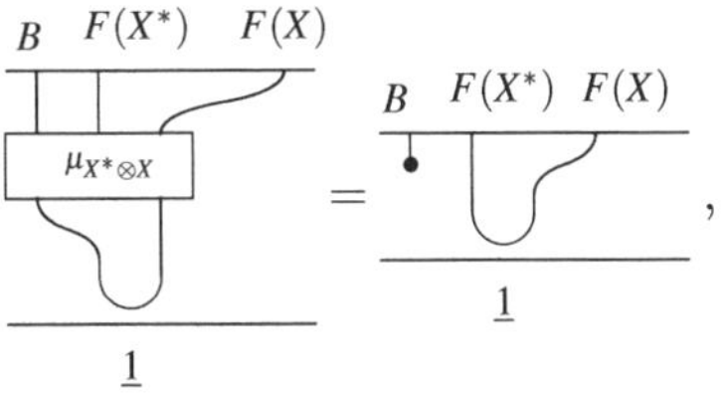


that is a consequence of the fact that $\mu$ is a natural transformation and $\mathrm{ev}_X : X^* \otimes X \to \underline{1}$ is a morphism in $\mathscr{C}$, we deduce that

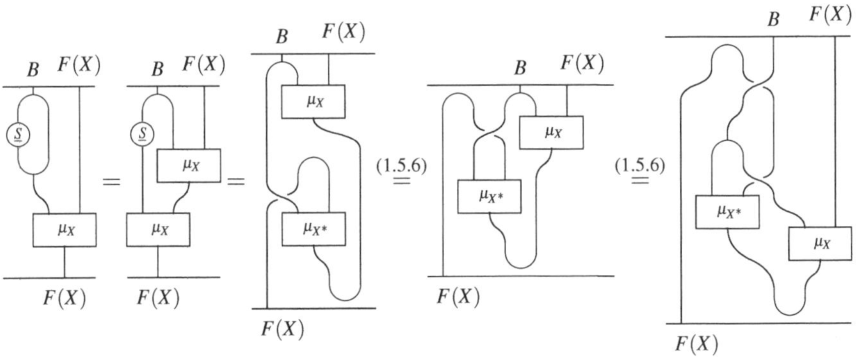


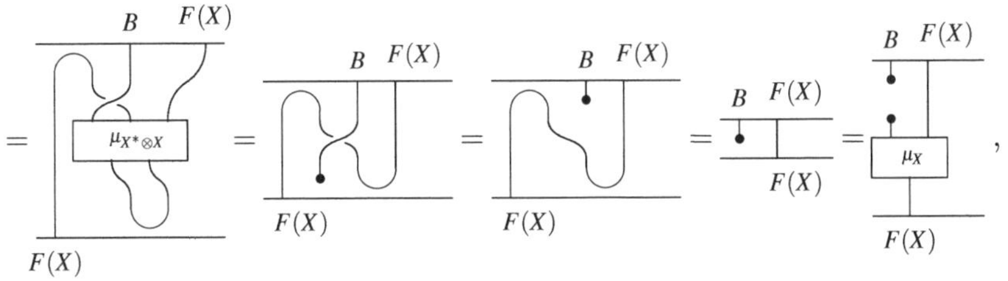

for all $X \in \mathrm{Ob}(\mathscr{C})$. Since $\theta_B$ is bijective, $\underline{S}$ satisfies (2.7.1), as needed. □

As an application of Theorem 11.3, in the next section we will give the explicit structure of $B$ as a braided Hopf algebra, if $(H,R)$ is a QT quasi-Hopf algebra, $\mathscr{C} = {}_H\mathscr{M}^{\mathrm{fd}}$ and $F = \mathrm{Id}_{\mathscr{C}}$. As we shall see next, in such a particular case the comultiplication of $B$ has an extra property, which reduces to the braided cocommutativity relation in the case when $\mathscr{C}$ is, moreover, a symmetric monoidal category.

**Corollary 11.4** *Let $(\mathscr{C},c)$ be a braided category and assume that the identity functor of $\mathscr{C}$ satisfies the representability assumption* (11.1.1) *and the higher representability assumption for modules. Then B admits a bialgebra structure in $\mathscr{C}$ such* *that its comultiplication is weakly braided cocommutative, in the sense that*

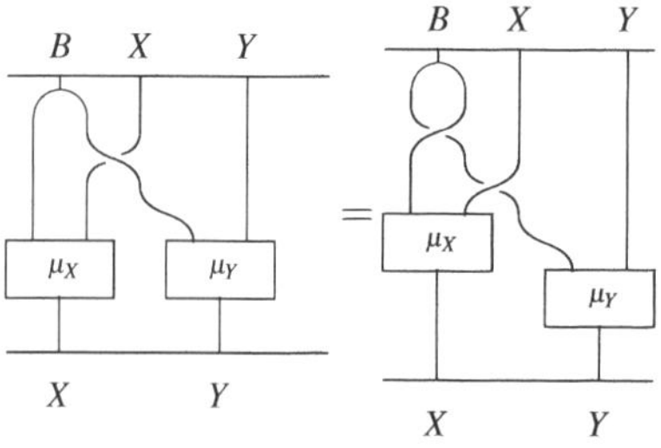


$$, \ \forall X,\, Y \in \mathrm{Ob}(\mathscr{C}). \qquad (11.1.7)$$

*Consequently, if $\mathscr{C}$ is a symmetric (resp. symmetric and rigid) category then $B$ is a braided cocommutative bialgebra (resp. Hopf algebra) in $\mathscr{C}$.*

*Proof* The relations in (11.1.7) follow from the following computation:

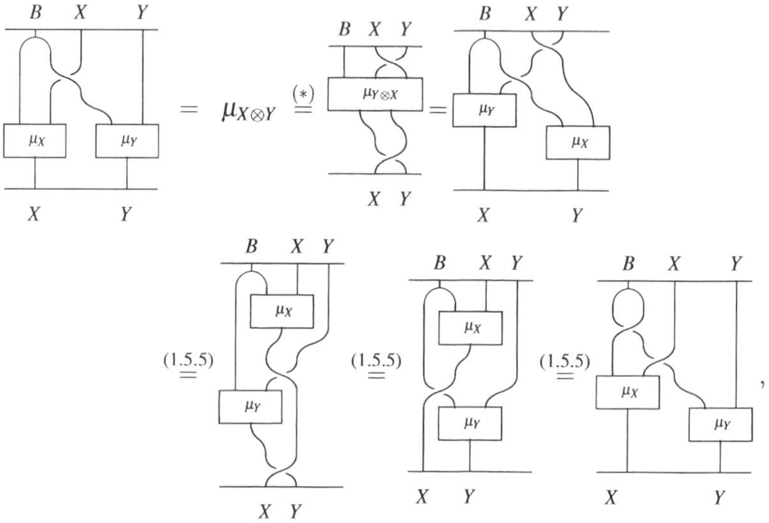


where in (*) we used that $\mu$ is a natural transformation and $c_{X,Y} : X \otimes Y \to Y \otimes X$ is a morphism in $\mathscr{C}$. □

Thus, the braided reconstruction theorem associates to any braided rigid category $\mathscr{C}$ a braided Hopf algebra $B$ which acts on each object of $\mathscr{C}$ and is weakly braided cocommutative. We call $B$ the automorphism braided group of $\mathscr{C}$.

## 11.2 The Enveloping Braided Group of a QT Quasi-Hopf Algebra

Throughout this section, $(H,R)$ is a QT quasi-Hopf algebra with $R$-matrix $R = R^1 \otimes R^2$, $\mathscr{C}$ is the braided category ${}_H\mathscr{M}$, $H_L$ is the vector space $H$ with the left regular action and $B$ is the same vector space $H$ but with the left adjoint action defined by (8.7.1). If $M, N \in \mathscr{C}$, let $\mathrm{Hom}_H(M,N)$ be the set of $H$-linear maps between $M$ and $N$.

Our goal is to apply Corollary 11.4 to the above setting. The first step is to show that $B$ as above is the object we need.

**Lemma 11.5** *Let $H$ be a quasi-Hopf algebra with bijective antipode, $p_R = p^1 \otimes p^2$ and $q_R = q^1 \otimes q^2$ the elements defined by* (3.2.19) *and $M \in \mathscr{C}$. Then, if we define for all $\xi \in \mathrm{Nat}(M \otimes \mathrm{Id}, \mathrm{Id})$*

$$\sigma_M : \mathrm{Nat}(M \otimes \mathrm{Id}, \mathrm{Id}) \to \mathrm{Hom}_H(M, B),\ \sigma_M(\xi)(m) = \xi_{H_L}(p^1 \cdot m \otimes p^2),\ \ \forall\, m \in M,$$

*then $\sigma_M$ is well defined and a functorial bijection with inverse $\theta_M$ given for all $N \in \mathscr{C}$, $\lambda \in \mathrm{Hom}_H(M, B)$, $m \in M$ and $n \in N$ by:*

$$\theta_M(\lambda)_N : M \otimes N \to N, \quad \theta_M(\lambda)_N(m \otimes n) = q^1 \lambda(m) S(q^2) \cdot n.$$

*Proof* We first have to check that $\sigma_M(\xi)$ is $H$-linear. For all $h \in H$ we have:

$$\begin{aligned} h \triangleright \sigma_M(\xi)(m) &= h_1 \xi_{H_L}(p^1 \cdot m \otimes p^2) S(h_2) \\ &= \xi_{H_L}(h_{(1,1)} p^1 \cdot m \otimes h_{(1,2)} p^2 S(h_2)) \\ &\overset{(3.2.21)}{=} \xi_{H_L}(p^1 h \cdot m \otimes p^2) = \sigma_M(\xi)(h \cdot m), \end{aligned}$$

where the second equality uses that $\xi_{H_L}$ is $H$-linear and $\xi$ is functorial under the morphism $H_L \to H_L$ defined by right multiplication. It follows that $\sigma$ is functorial and $\theta_M(\lambda)$ is a natural transformation. To see that $\theta_M$ is well defined we have to check that each $\theta_M(\lambda)_N$ is $H$-linear. That is true since for all $h \in H$

$$\begin{aligned} \theta_M(\lambda)_N(h \cdot (m \otimes n)) &= \theta_M(\lambda)_N(h_1 \cdot m \otimes h_2 \cdot n) \\ &= q^1 \lambda(h_1 \cdot m) S(q^2) h_2 \cdot n \\ &\overset{(*)}{=} q^1 h_{(1,1)} \lambda(m) S(q^2 h_{(1,2)}) h_2 \cdot n \\ &\overset{(3.2.21)}{=} h q^1 \lambda(m) S(q^2) \cdot n \\ &= h \cdot \theta_M(\lambda)_N(m \otimes n), \end{aligned}$$

where in (*) we used the fact that $\lambda$ is $H$-linear. Thus we only have to show that $\sigma_M$ and $\theta_M$ are inverses. Now,

$$\begin{aligned} (\sigma_M \circ \theta_M)(\lambda)(m) &= \theta_M(\lambda)_{H_L}(p^1 \cdot m \otimes p^2) \\ &= q^1 \lambda(p^1 \cdot m) S(q^2) p^2 \\ &\overset{(*)}{=} q^1 (p^1 \triangleright \lambda(m)) S(q^2) p^2 \\ &= q^1 p^1_1 \lambda(m) S(q^2 p^1_2) p^2 \overset{(3.2.23)}{=} \lambda(m), \end{aligned}$$

where in (*) we again used the fact that $\lambda$ is $H$-linear. Similarly,

$$\begin{aligned} (\theta_M \circ \sigma_M)(\xi)_N(m \otimes n) &= q^1 \sigma_M(\xi)(m) S(q^2) \cdot n \\ &= q^1 \xi_{H_L}(p^1 \cdot m \otimes p^2) S(q^2) \cdot n \\ &= \xi_{H_L}(q^1_1 p^1 \cdot m \otimes q^1_2 p^2 S(q^2)) \cdot n \\ &\overset{(3.2.23)}{=} \xi_{H_L}(m \otimes 1) \cdot n = \xi_N(m \otimes n), \end{aligned}$$

where the last equality uses that $\xi$ is functorial under $H_L \ni h \mapsto h \cdot n \in N$. □

By the above lemma, the natural transformation $\mu$ corresponding to the identity morphism $\mathrm{Id}_B$ is $\mu = \{\mu_N \mid N \in \mathscr{C}\}$, where each $\mu_N : B\otimes N \to N$ is given by

$$\mu_N(h\otimes n) = \theta_B(\mathrm{Id}_B)_N(h\otimes n) = q^1 hS(q^2)\cdot n,\ \ \forall\, h\in H,\ n\in N. \tag{11.2.1}$$

Now, we are able to begin our reconstruction. The reconstruction of the algebra structure on $B$ does not involve the braiding and is given as follows.

**Lemma 11.6** *$B = H_0$ as an algebra, where the algebra structure of $H_0$ in $\mathscr{C}$ is given in Definition 4.4.*

*Proof* Following the proof of Proposition 11.1, the multiplication on $B$ is obtained as $m_B = \theta_{B\otimes B}(\xi)$, where for all $N \in \mathscr{C}$

$$\xi_N : (B\otimes B)\otimes N \to N, \quad \xi_N = \mu_N \circ (\mathrm{Id}_B \otimes \mu_N)\circ a_{B,B,N}.$$

More precisely, $\xi_N((h\otimes h')\otimes n) = Q^1(X^1\triangleright h)S(Q^2)q^1(X^2\triangleright h')S(q^2)X^3\cdot n$, where we denote by $Q^1\otimes Q^2$ another copy of $q_R$, and therefore:

$$\begin{aligned}
m_B(h\otimes h') &= \xi_{H_L}(p^1\triangleright(h\otimes h')\otimes p^2)\\
&= Q^1(X^1p^1_1\triangleright h)S(Q^2)q^1(X^2p^1_2\triangleright h')S(q^2)X^3p^2\\
&\overset{(3.2.19)}{=} Q^1(X^1x^1_1\triangleright h)S(Q^2)q^1(X^2x^1_2\triangleright h')S(q^2)X^3x^2\beta S(x^3)\\
&\overset{(3.2.19),(3.2.1)}{=} Q^1(y^1\triangleright h)S(Q^2)q^1y^2_{(1,1)}(x^1\triangleright h')S(q^2y^2_{(1,2)})y^2_2x^2\beta S(y^3x^3)\\
&\overset{(3.2.21),(3.2.19)}{=} X^1y^1_1hS(X^2y^1_2)\alpha X^3y^2q^1(x^1\triangleright h')S(q^2)x^2\beta S(y^3x^3)\\
&\overset{(3.1.9),(3.2.19)}{=} X^1hS(y^1X^2)\alpha y^2X^3_1q^1p^1_1h'S(q^2p^1_2)p^2S(y^3X^3_2)\\
&\overset{(3.2.23),(4.1.9)}{=} h\circ h'.
\end{aligned}$$

Finally, the unit for $B$ is $\eta(1) = \theta_k(r)(1) = r_{H_L}(p^1\cdot 1\otimes p^2) = r_{H_L}(1\otimes\beta) = \beta$, where $r$ denotes the right unit constraint, and this finishes the proof. □

We now assume that the higher representability assumption for modules holds. By the proof of Proposition 11.2, the coproduct $\underline{\Delta}$ of $B$ is characterized as being the unique morphism $\underline{\Delta} : B \to B\otimes B$ in $\mathscr{C}$ such that:

$$\begin{aligned}
\mu_{M\otimes N} = (\mu_M\otimes\mu_N)\circ a^{-1}_{B,M,B\otimes N}\circ(\mathrm{Id}_B\otimes a_{M,B,N})\circ(\mathrm{Id}_B\otimes(c_{B,M}\otimes\mathrm{Id}_N))\\
\circ(\mathrm{Id}_B\otimes a^{-1}_{B,M,N})\circ a_{B,B,M\otimes N}\circ(\underline{\Delta}\otimes\mathrm{Id}_{M\otimes N}),
\end{aligned}$$

for all $M,N \in \mathscr{C}$. Explicitly, the map $\underline{\Delta}$ is characterized by

$$\begin{aligned}
(q^1hS(q^2))_1\cdot m\otimes(q^1hS(q^2))_2\cdot n = q^1(y^1X^1\triangleright h_{\underline{1}})S(q^2)y^2Y^1R^2x^2X^3_1\cdot m\\
\otimes Q^1(y^3_1Y^2R^1x^1X^2\triangleright h_{\underline{2}})S(Q^2)y^3_2Y^3x^3X^3_2\cdot n,
\end{aligned}$$

for all $m\in M$, $n\in N$, where $q_R = q^1\otimes q^2 = Q^1\otimes Q^2$. We used the braiding (10.1.4), computed $h\cdot(m\otimes n)$ in the usual way and set $\underline{\Delta}(h) = h_{\underline{1}}\otimes h_{\underline{2}}$. Since the above equality is true for all $m\in M$, $n\in N$, we conclude that $\underline{\Delta}$ is the unique morphism in $\mathscr{C}$ satisfying the following equality, for all $h\in H$:

$$\underline{\Delta}(q^1hS(q^2)) = q^1(y^1X^1\triangleright h_{\underline{1}})S(q^2)y^2Y^1R^2x^2X^3_1$$

$$\otimes Q^1(y_1^3Y^2R^1x^1X^2 \triangleright h_{\underline{2}})S(Q^2)y_2^3Y^3x^3X_2^3. \tag{11.2.2}$$

Now, under our assumption, the explicit formula for $\underline{\Delta}$ is given by the following:

**Lemma 11.7** *$B$ is a coalgebra in $\mathscr{C}$ with comultiplication*

$$\underline{\Delta}(h) = x^1X^1h_1g^1S(x^2R^2y^3X_2^3) \otimes x^3R^1 \triangleright y^1X^2h_2g^2S(y^2X_1^3), \tag{11.2.3}$$

*for all $h \in H$, and counit $\underline{\varepsilon} = \varepsilon$.*

*Proof* We prove that $\underline{\Delta}$ defined by (11.2.3) is a morphism in $\mathscr{C}$ which satisfies (11.2.2). The fact that $\underline{\Delta}$ is a $H$-linear map follows by applying (3.2.13), (3.1.7) and (10.1.3) several times. If we denote by $r^1 \otimes r^2$ another copy of $R$ then, for all $h \in H$, we have:

$$\begin{aligned}
&q^1(y^1X^1 \triangleright h_{\underline{1}})S(q^2)y^2Y^1R^2x^2X_1^3 \otimes Q^1(y_1^3Y^2R^1x^1X^2 \triangleright h_{\underline{2}})S(Q^2)y_2^3Y^3x^3X_2^3\\
&\quad = q^1y_1^1X_1^1z^1Z^1h_1g^1S(q^2y_2^1X_2^1z^2r^2t^3Z_2^3)y^2Y^1R^2x^2X_1^3\\
&\qquad \otimes Q^1y^3_{(1,1)}(Y^2R^1x^1X^2z^3r^1 \triangleright t^1Z^2h_2g^2S(t^2Z_1^3))S(Q^2y^3_{(1,2)})y_2^3Y^3x^3X_2^3\\
&\overset{(3.1.9),(3.2.21)}{\underset{(3.1.7),(10.1.3)}{=}} q^1y_1^1z^1X^1Z^1h_1g^1S(q^2y_2^1z^2T^1r^2X_2^2t^3Z_2^3)y^2Y^1R^2z^3_{(1,2)}x^2T_1^3X_1^3\\
&\qquad \otimes y^3Q^1(Y^2R^1z^3_{(1,1)}x^1T^2r^1X_1^2 \triangleright t^1Z^2h_2g^2S(t^2Z_1^3))S(Q^2)Y^3z_2^3x^3T_2^3X_2^3\\
&\overset{(3.1.9),(10.1.3)}{\underset{(3.1.7),(10.1.3)}{=}} q^1y_1^1z^1X^1Z^1h_1g^1S(q^2y_2^1z^2T^1V^1r^2v_2^1X_2^2t^3Z_2^3)y^2z_1^3Y^1T_1^2R^2V^3v^2X_1^3\\
&\qquad \otimes y^3Q^1(z^3_{(2,1)}Y^2T_2^2R^1V^2r^1v_1^1X_1^2 \triangleright t^1Z^2h_2g^2S(t^2Z_1^3))\\
&\qquad S(Q^2)z^3_{(2,2)}Y^3T^3v^3X_2^3\\
&\overset{(10.1.1),(3.2.21)}{\underset{(3.2.19),(3.1.9)}{=}} Y^1y_1^1z^1X^1Z^1h_1g^1S(Y^2y_2^1z^2V^1T_1^1R_1^2W^2v_2^1X_2^2t^3Z_2^3)\alpha Y^3y^2z_1^3V^2T_2^1R_2^2W^3\\
&\qquad v^2X_1^3 \otimes y^3z_2^3Q^1(V_1^3T^2R^1W^1v_1^1X_1^2 \triangleright t^1Z^2h_2g^2S(t^2Z_1^3))S(Q^2)V_2^3T^3v^3X_2^3\\
&\overset{(3.2.21),(3.1.9)}{\underset{(3.2.1),(10.1.5)}{=}} X^1Z^1h_1g^1S(W^2v_2^1X_2^2t^3Z_2^3)\alpha W^3v^2X_1^3\\
&\qquad \otimes Q^1(W^1v_1^1 \triangleright X^2_{(1,1)}t^1Z^2h_2g^2S(X^2_{(1,2)}t^2Z_1^3))S(Q^2)v^3X_2^3\\
&\overset{(3.1.7),(3.1.9)}{\underset{(3.2.1),(3.1.9)}{=}} X^1Z_1^1h_1g^1S(y^1Y^2t^3x_2^2X^3_{(1,2)}Z_2^2)\alpha y^2(Y^3x^3X_2^3Z^3)_1\\
&\qquad \otimes Q^1Y_1^1t^1x^1X^2Z_2^1h_2g^2S(Q^2Y_2^1t^2x_1^2X^3_{(1,1)}z_1^2)y^3(Y^3x^3X_2^3Z^3)_2\\
&\overset{(3.1.9),(3.1.7)}{\underset{(3.2.1)}{=}} X^1Z_1^1h_1g^1S(y^1Y^2X^3_{(1,2)}Z_2^2)\alpha y^2Y_1^3(X_2^3Z^3)_1\\
&\qquad \otimes Q^1x^1X^2Z_2^1h_2g^2S(Q^2x^2Y^1X^3_{(1,1)}Z_1^2)x^3y^3Y_2^3(X_2^3Z^3)_2\\
&\overset{(3.2.19),(3.1.9)}{\underset{(3.2.5),(3.2.13)}{=}} X^1Z_1^1h_1S(X_1^3Z^2)_1g^1\gamma^1(X_2^3Z^3)_1\\
&\qquad \otimes X^2Z_2^1h_2S(X_1^3Z^2)_2g^2\gamma^2(X_2^3Z^3)_2\\
&\overset{(3.2.14)}{=} X^1(Z^1hS(X_1^3Z^2)\alpha X_2^3Z^3)_1 \otimes X^2(Z^1hS(X_1^3Z^2)\alpha X_2^3Z^3)_2\\
&\overset{(3.2.1),(3.2.19)}{=} \Delta(q^1hS(q^2)).
\end{aligned}$$

Finally, by the proof of Proposition 11.2, the counit for $B$ is $\underline{\varepsilon}(h) = \mu_k(h \otimes 1) = q^1hS(q^2) \cdot 1 = \varepsilon(h)$, for all $h \in H$, where we used that $\varepsilon(\alpha) = 1$. □

Suppose now that $M \in \mathscr{C}$ is finite dimensional and let $M^*$ be its left dual in $\mathscr{C}$.

Recall that $M^* \in \mathscr{C}$ via $\langle h\cdot m^*, m\rangle = m^*(S(h)\cdot m)$, for all $h\in H$, $m^*\in M^*$, $m\in M$, and the morphisms $\mathrm{ev}_M$ and $\mathrm{coev}_M$ are the ones defined in Proposition 3.33.

According to the proof of Theorem 11.3, the reconstructed $\underline{S}$ is characterized by

$$\begin{aligned}\mu_M\circ(\underline{S}\otimes \mathrm{Id}_M) = l_M^{-1}&\circ(\mathrm{Id}_M\otimes \mathrm{ev}_M)\circ a_{M,M^*,M}\circ((\mathrm{Id}_M\otimes\mu_{M^*})\otimes\mathrm{Id}_M)\\ &\circ(a_{M,B,M^*}\otimes\mathrm{Id}_M)\circ((c_{B,M}\otimes\mathrm{Id}_{M^*})\otimes\mathrm{Id}_M)\circ(a^{-1}_{B,M,M^*}\otimes\mathrm{Id}_M)\\ &\circ((\mathrm{Id}_B\otimes\mathrm{coev}_M)\otimes\mathrm{Id}_M)\circ(l_B\otimes\mathrm{Id}_M),\end{aligned}$$

for all $M\in\mathscr{C}$ finite dimensional, where $l$, $a$ and $c$ are the left unit constraint, the associativity constraint and the braiding for $\mathscr{C}$, respectively. Explicitly, $\underline{S}$ is characterized by the following equality that holds for all $h\in H$, $m\in M$:

$$q^1\underline{S}(h)S(q^2)\cdot m = Y^1X^1R^2x^2\beta S(Y^2q^1(X^2R^1x^1\triangleright h)S(q^2)X^3x^3)\alpha Y^3\cdot m.$$

So we conclude that $\underline{S}$ is the unique morphism in $\mathscr{C}$ which satisfies

$$q^1\underline{S}(h)S(q^2) = Q^1X^1R^2x^2\beta S(q^1(X^2R^1x^1\triangleright h)S(q^2)X^3x^3)S(Q^2),$$

for all $h\in H$. Thus, by our assumption, the desired formula for $\underline{S}$ is

$$\underline{S}(h) = X^1R^2x^2\beta S(q^1(X^2R^1x^1\triangleright h)S(q^2)X^3x^3), \tag{11.2.4}$$

for all $h\in H$ (it is not hard to see that $\underline{S}$ as above is $H$-linear). Conversely, given these formulas for the braided group $B$ (denoted in what follows by $\underline{H}$), one can also check directly (by technical and tedious computations) that it satisfies the axioms for a Hopf algebra in a braided category. We call $\underline{H}$ the associated enveloping algebra braided group of $H$.

We summarize all these facts in the following:

**Theorem 11.8** *Let $(H,R)$ be a QT quasi-Hopf algebra and denote by $\mathscr{C}$ the braided category of left $H$-modules, ${}_H\mathscr{M}$. Then $H$ gives a braided Hopf algebra $\underline{H}$ in $\mathscr{C}$, considering the left adjoint action* (8.7.1), *the same algebra structure as $H_0$ defined in Definition 4.4, coproduct and counit as in Lemma 11.7 and antipode $\underline{S}$ as in* (11.2.4). *Moreover, $\underline{H}$ is weakly braided cocommutative, in the sense that*

$$\begin{aligned}&q^1(y^1X^1\triangleright h_{\underline{1}})S(q^2)y^2Y^1R^2x^2X^3_1\otimes y^3Q^1(Y^2R^1x^1X^2\triangleright h_{\underline{2}})S(Q^2)Y^3x^3X^3_2\\ &= Y^1\overline{R}^1y^2q^1(x^1X^2\triangleright h_{\underline{2}})S(q^2)x^2X^3_1\otimes Q^1(Y^2\overline{R}^2y^1X^1\triangleright h_{\underline{1}})S(Q^2)Y^3y^3x^3X^3_2,\end{aligned} \tag{11.2.5}$$

*where $q_R = q^1\otimes q^2 = Q^1\otimes Q^2$, $R = R^1\otimes R^2$ and $R^{-1} = \overline{R}^1\otimes\overline{R}^2$.*

*Proof* The relation (11.2.5) follows easily by using the equality between the first and the last but one terms of the computation performed in the proof of Corollary 11.4; we leave the details to the reader. □

An equivalent definition for $\underline{S}$ can be obtained with the help of the element $u$ defined in (10.3.4).

**Proposition 11.9** *Let $(H,R)$ be a QT quasi-Hopf algebra with R-matrix R and let u be the element given by* (10.3.4). *Then the antipode $\underline{S}$ defined by* (11.2.4) *admits a second description:*

$$\underline{S}(h)=u^{-1}S(R^1hS(R^2)),\ \ \forall\, h\in H.$$

*Proof* It is based on the following computation:

$$\begin{aligned}
\underline{S}(h) &= X^1R^2p^2S(q^1(X^2R^1p^1\triangleright h)S(q^2)X^3)\\
&= X^1R^2p^2S(q^1X^2_1R^1_1p^1_1hS(q^2X^2_2R^1_2p^1_2)X^3)\\
&\overset{(10.1.1)}{=} X^1Y^1R^2x^2r^2Z^3p^2S(T^1X^2_1Y^2R^1x^1Z^1\\
&\qquad\qquad p^1_1hS(T^2X^2_2Y^3x^3r^1Z^2p^1_2)\alpha T^3X^3)\\
&\overset{(3.1.9)}{=} Y^1_1R^2x^2r^2Z^3p^2S(Y^1_2R^1x^1Z^1p^1_1hS(Y^2x^3r^1Z^2p^1_2)\alpha Y^3)\\
&\overset{(10.1.3),(3.1.9)}{=} R^2q^1r^2X^2_2Z^3p^2S(R^1X^1Z^1p^1_1hS(q^2r^1X^2_1Z^2p^1_2)X^3)\\
&\overset{(3.1.9)}{=} R^2q^1r^2p^2S^2(q^2r^1p^1)S(R^1h)\\
&\overset{(10.3.9)}{=} R^2u^{-1}S(R^1h)\\
&\overset{(10.3.10)}{=} u^{-1}S(R^1hS(R^2)),
\end{aligned}$$

valid for all $h\in H$. Here $r^1\otimes r^2$ is another copy of $R$. □

**Remark 11.10** The braided antipode $\underline{S}$ is bijective: if we define $\underline{S}^{-1}:H\to H$ by

$$\underline{S}^{-1}(h)=\overline{R}^1S^{-1}(uh)S(\overline{R}^2),\ \ \forall\, h\in H, \tag{11.2.6}$$

then by using (10.3.10) several times as well as the second definition of $\underline{S}$ proved above, it is not hard to see that $\underline{S}$ and $\underline{S}^{-1}$ are inverses.

## 11.3 Bosonisation for Quasi-Hopf Algebras

In this section we introduce the bosonisation process. Let $(H,R)$ be a QT quasi-Hopf algebra, so the category ${}_H\mathcal{M}$ is braided. By Remark 10.31(2) we have a braided monoidal functor $\mathscr{Z}_l:{}_H\mathcal{M}\to\mathscr{Z}_l({}_H\mathcal{M})\equiv{}_H^H\mathcal{YD}$. Because $\mathscr{Z}_l$ is a braided monoidal functor, a braided bialgebra or Hopf algebra $B\in{}_H\mathcal{M}$ can also be regarded in ${}_H^H\mathcal{YD}$ and then we can consider the biproduct quasi-Hopf algebra $B\times H$ (see Section 9.8); it will be denoted by $\mathrm{bos}(B)$ and called the bosonisation of $B$.

Specializing the biproduct construction from Section 9.8 to the above setting we obtain the following result.

**Proposition 11.11** *Under the above hypothesis,* $\mathrm{bos}(B)$ *is a quasi-bialgebra with algebra structure given by the smash product*

$$(b\times h)(b'\times h')=(x^1\cdot b)(x^2h_1\cdot b')\times x^3h_2h', \tag{11.3.1}$$

*for all $b,b'\in B$, $h,h'\in H$, with unit $1_B\times 1_H$, with comultiplication*

$$\Delta(b\times h)=y^1X^1\cdot b_{\underline{1}}\times y^2Y^1R^2x^2X^3_1h_1\otimes y^3_1Y^2R^1x^1X^2\cdot b_{\underline{2}}\times y^3_2Y^3x^3X^3_2h_2 \tag{11.3.2}$$

*and counit $\varepsilon(b \times h) = \underline{\varepsilon}_B(b)\varepsilon(h)$, for all $b \in B$, $h \in H$, and with reassociator*

$$\Phi_{\mathrm{bos}(B)} = 1_B \times X^1 \otimes 1_B \times X^2 \otimes 1_B \times X^3. \tag{11.3.3}$$

*Moreover, if $H$ is a quasi-Hopf algebra and $B$ is a Hopf algebra in the braided category ${}_H^H\mathscr{YD}$ with antipode $\underline{S}_B$, then* bos$(B)$ *is a quasi-Hopf algebra with the antipode $s$ given by*

$$s(b \times h) = (1_B \times S(X^1 x_1^1 R^2 h)\alpha)(X^2 x_2^1 R^1 \cdot \underline{S}_B(b) \times X^3 x^2 \beta S(x^3)), \tag{11.3.4}$$

*for all $b \in B$, $h \in H$, and distinguished elements $1_B \times \alpha$ and $1_B \times \beta$.*

Because as an algebra bos$(B)$ is a smash product, from Proposition 5.7 it follows that the modules over $B$ in the braided category ${}_H\mathscr{M}$ correspond to the ordinary modules over bos$(B)$. Also, by Theorem 9.30 the bosonisation construction is invariant under a twist.

**Corollary 11.12** *If $H$ is a QT quasi-Hopf algebra then the smash product defined by the left adjoint action of $H$ on $H_0$ has a structure of a quasi-Hopf algebra.*

*Proof* Since $H$ is a QT quasi-Hopf algebra, by Theorem 11.8 we can associate to $H$ the braided Hopf algebra $\underline{H}$ in ${}_H\mathscr{M}$. Then bos$(\underline{H}) = H_0 \# H$ as an algebra, and is a quasi-Hopf algebra via the structure given by the bosonisation process. □

Our next goal is to compute the quasi-Hopf algebra structure on bos$(\underline{H})$, in the case where $H = H(2)_+$ or $H = H(2)_-$, the QT quasi-Hopf algebras considered in Example 10.6. To this end, note that, in general, bos$(\underline{H})$ is the $k$-vector space $H \otimes H$ with the following quasi-Hopf algebra structure:

$$(b \times h)(b' \times h') = (x^1 \triangleright b) \circ (x^2 h_1 \triangleright b') \times x^3 h_2 h', \tag{11.3.5}$$

$$\begin{aligned}\Delta(b \times h)& \\ &= y^1 X^1 \triangleright b_{\underline{1}} \times y^2 Y^1 R^2 x^2 X_1^3 h_1 \otimes y_1^3 Y^2 R^1 x^1 X^2 \triangleright b_{\underline{2}} \times y_2^3 Y^3 x^3 X_2^3 h_2,\end{aligned} \tag{11.3.6}$$

$$\Phi_{\mathrm{bos}(H_0)} = \beta \times X^1 \otimes \beta \times X^2 \otimes \beta \times X^3, \tag{11.3.7}$$

$$s(b \times h) = (\beta \times S(X^1 x_1^1 R^2 h)\alpha)(X^2 x_2^1 R^1 \triangleright S_{H_0}(b) \times X^3 x^2 \beta S(x^3)), \tag{11.3.8}$$

for all $b, b', h, h' \in H$, where $\triangleright$ stands for the left adjoint action of $H$ on itself, and

$$\underline{\Delta}(b) = b_{\underline{1}} \otimes b_{\underline{2}} := x^1 X^1 b_1 g^1 S(x^2 R^2 y^3 X_2^3) \otimes x^3 R^1 \triangleright y^1 X^2 b_2 g^2 S(y^2 X_1^3), \tag{11.3.9}$$

$$S_{H_0}(b) = X^1 R^2 p^2 S(q^1 (X^2 R^1 p^1 \triangleright b) S(q^2) X^3), \tag{11.3.10}$$

for all $b \in H$. Here $R = R^1 \otimes R^2$, while $f^{-1} = g^1 \otimes g^2$, $p_R = p^1 \otimes p^2$ and $q_R = q^1 \otimes q^2$ are the elements defined by (3.2.16) and (3.2.19), respectively.

The unit for bos$(\underline{H})$ is $\beta \times 1_H$, the counit is $\varepsilon(b \times h) = \varepsilon(b)\varepsilon(h)$, for all $b, h \in H$, and the distinguished elements $\alpha$ and $\beta$ are given by $\beta \times \alpha$ and $\beta \times \beta$, respectively.

**Example 11.13** bos$(H(2)_+) = $ bos$(H(2)_-) = k[C_2 \times C_2]$ as bialgebras, viewed as a quasi-Hopf algebra via the non-trivial reassociator $\Phi_x := 1 - 2p_-^x \otimes p_-^x \otimes p_-^x$, where $x$ is one of the generators of $C_2 \times C_2$, and where $p_-^x := \frac{1}{2}(1 - x)$. The antipode is the identity map and the distinguished elements $\alpha$ and $\beta$ are given by $\alpha = x$ and $\beta = 1$.

*Proof* Since $H := H(2)_\pm$ are commutative algebras and $\beta = 1$, it follows from (3.2.1) and (3.2.2) that the multiplication $\circ$ defined in (4.1.9) coincides with the original multiplication of $H(2)$. Also, from the definition of $H(2)$ it follows that the action $\triangleright$ is trivial, that is, $h \triangleright h' = \varepsilon(h)h'$, for all $h, h' \in H(2)$. Using (11.3.5) we obtain that the multiplication on $\mathrm{bos}(\underline{H})$ is the componentwise multiplication, and by (11.3.9) we get that the comultiplication $\underline{\Delta}$ on $H$ reduces to

$$\underline{\Delta}(b) = X^1 b_1 g^1 S(y^3 X_2^3) \otimes y^1 X^2 b_2 g^2 S(y^2 X_1^3) = \Delta(b)(X^1 X_2^3 y^3 \otimes X^2 X_1^3 y^1 y^2) f^{-1}.$$

We have that

$$(\mathrm{Id} \otimes \mathrm{Id} \otimes \Delta)(\Phi) = 1 - 2p_- \otimes p_- \otimes p_+ \otimes p_- - 2p_- \otimes p_- \otimes p_- \otimes p_+,$$

and therefore $X^1 X_2^3 \otimes X^2 X_1^3 = 1$. Also, by (10.1.7) we have

$$y^3 \otimes y^1 y^2 = 1 - 2p_- \otimes p_- = 1 \otimes p_+ + g \otimes p_-,$$

and a straightforward computation ensures that the Drinfeld twist $f$ and its inverse $f^{-1}$ for $H(2)$ are given by $f = f^{-1} = g \otimes p_- + 1 \otimes p_+$.

Combining all these facts we get $\underline{\Delta} = \Delta$, and keeping in mind that the action $\triangleright$ is trivial we conclude that the comultiplication in (11.3.6) is the componentwise comultiplication on $H(2) \otimes H(2)$. Thus $\mathrm{bos}(\underline{H(2)_+}) = \mathrm{bos}(\underline{H(2)_-}) = H(2) \otimes H(2)$ as bialgebras. Hence $\mathrm{bos}(\underline{H(2)_+}) = \mathrm{bos}(\underline{H(2)_-})$ is generated as an algebra by $x = 1 \times g$ and $y = g \times 1$, with relations $x^2 = y^2 = 1$ and $xy = yx$. The elements $x$ and $y$ are grouplike elements, so $\mathrm{bos}(\underline{H(2)_+}) = \mathrm{bos}(\underline{H(2)_-}) = k[C_2 \times C_2]$ as bialgebras. According to (11.3.7) the reassociator of $\mathrm{bos}(\underline{H(2)_+}) = \mathrm{bos}(\underline{H(2)_-})$ is given by

$$\begin{aligned}\Phi_x &= 1 \times X^1 \otimes 1 \times X^2 \otimes 1 \times X^3 \\ &= 1 \times 1 \otimes 1 \times 1 \otimes 1 \times 1 - 2 \times p_- \otimes 1 \times p_- \otimes 1 \times p_- \\ &= 1 - 2p_-^x \otimes p_-^x \otimes p_-^x,\end{aligned}$$

since $1 \times p_- = \frac{1}{2}(1 \times 1 - 1 \times g) = \frac{1}{2}(1 - x) = p_-^x$. Finally, using that $\triangleright$ is trivial, $\beta = 1$ and the axiom (3.2.2), we obtain $S_{\underline{H}} = S$, the antipode of $H(2)$. From (11.3.8) and (3.2.2) we deduce that the antipode of $\mathrm{bos}(\underline{H(2)_+}) = \mathrm{bos}(\underline{H(2)_-})$ is the identity map. Clearly, the elements $\alpha$ and $\beta$ are $1 \times g = x$ and $1 \times 1 = 1$, respectively. □

## 11.4 The Function Algebra Braided Group

We present a dual version of Theorem 11.3 and use it to associate to any co-quasitriangular (CQT for short) dual quasi-Hopf algebra $A$ a braided weakly commutative Hopf algebra $\underline{A}$ in the category of right $A$-comodules. We will call $\underline{A}$ the function algebra braided group associated to $A$. This procedure is the formal dual of the one performed in Section 11.2 where to any QT quasi-Hopf algebra $H$ is associated a weakly cocommutative braided group $\underline{H}$ in the braided category of left $H$-modules. We notice that, in the finite-dimensional case, $\underline{A}$ cannot be obtained from $\underline{H}$ by (usual) dualisation. In fact, we show that if $H$ is finite dimensional then the function algebra braided group $\underline{H^*}$ associated to $H^*$ is always isomorphic to the categorical left op-cop dual of $\underline{H}$ as braided Hopf algebra, and this will uncover the true meaning of the factorizable notion in the quasi-Hopf setting.

Throughout this section, $A$ will be a dual quasi-bialgebra or a dual quasi-Hopf algebra with structure as in Section 3.7. For a right $A$-comodule $M$ with structure morphism $\rho_M : M \to M \otimes A$ we denote $\rho_M(m) = m_{(0)} \otimes m_{(1)}$. Recall from Section 3.7 that $\mathscr{M}^A$ is a monoidal category. As we shall see, the braided structures on $\mathscr{M}^A$ are in a one-to-one correspondence with the CQT structures on $A$.

**Definition 11.14** A dual quasi-bialgebra or dual quasi-Hopf algebra is called CQT if there exists a $k$-bilinear form $\sigma : A \times A \to k$ such that the following relations hold:

$$\sigma(ab,c) = \varphi(c_1,a_1,b_1)\sigma(a_2,c_2)\varphi^{-1}(a_3,c_3,b_2)\sigma(b_3,c_4)\varphi(a_4,b_4,c_5), \qquad (11.4.1)$$

$$\sigma(a,bc) = \varphi^{-1}(b_1,c_1,a_1)\sigma(a_2,c_2)\varphi(b_2,a_3,c_3)\sigma(a_4,b_3)\varphi^{-1}(a_5,b_4,c_4), \qquad (11.4.2)$$

$$\sigma(a_1,b_1)a_2b_2 = b_1a_1\sigma(a_2,b_2), \qquad (11.4.3)$$

$$\sigma(a,1_A) = \sigma(1_A,a) = \varepsilon(a), \qquad (11.4.4)$$

for all $a,b,c \in A$.

If $(A,\sigma)$ is a CQT dual quasi-bialgebra or dual quasi-Hopf algebra then $\mathscr{M}^A$ is a (pre-)braided category. For any $M,N \in \mathscr{M}^A$ and $m \in M$, $n \in N$, the (pre-)braiding is

$$c_{M,N}(m \otimes n) = \sigma(m_{(1)},n_{(1)})n_{(0)} \otimes m_{(0)}.$$

Dual to the quasi-Hopf case, if $A$ is a CQT dual quasi-Hopf algebra we can prove that the bilinear form $\sigma$ is convolution invertible and the antipode $S$ is bijective.

**Proposition 11.15** *Let $(A,\sigma)$ be a CQT dual quasi-Hopf algebra. Then:*

*(i) $\sigma$ is convolution invertible. More precisely, its inverse $\sigma^{-1}$ is given by*

$$\sigma^{-1}(a,b) = \varphi(a_1,S(a_3),b_4a_{10})\beta(a_2)\varphi(b_1,S(a_6),a_8)$$
$$\sigma(S(a_5),b_2)\varphi^{-1}(S(a_4),b_3,a_9)\alpha(a_7), \qquad (11.4.5)$$

*for all $a,b \in A$.*

*(ii) The element $u \in A^*$, given by*

$$u(a) = \varphi^{-1}(a_7,S(a_3),S^2(a_1))\sigma(a_6,S(a_4))\alpha(a_5)\beta(S(a_2)) \qquad (11.4.6)$$

*for all $a \in A$, is invertible. Its inverse is given, for all $a \in A$, by*

$$u^{-1}(a) = \varphi(a_1,S^2(a_8),S(a_6))\beta(a_4)\sigma(S^2(a_9),a_2)\alpha(S(a_7))\varphi^{-1}(S^2(a_{10}),a_3,S(a_5)). \qquad (11.4.7)$$

*(iii) For all $a \in A$, the following relation holds:*

$$S^2(a) = u(a_1)a_2u^{-1}(a_3). \qquad (11.4.8)$$

*In particular, the antipode $S$ is bijective.*

*Proof* If $A$ is finite dimensional then the proof follows mostly from the results in Sections 10.1 and 10.3, by duality. This is why we give only a sketch of the proof, leaving other details to the reader.

(i) This follows by Lemma 10.9, by duality.

(ii) First, one can prove that

$$\sigma(S(a_1),S(b_1))\gamma(a_2,b_2)=\gamma(b_1,a_1)\sigma(a_2,b_2), \tag{11.4.9}$$

for all $a,b\in A$, and then that

$$f(b_1,a_1)\sigma(a_2,b_2)f^{-1}(a_3,b_3)=\sigma(S(a),S(b)). \tag{11.4.10}$$

Note that these formulas are the formal duals of Lemma 10.14. Secondly, by using (11.4.10) and the equalities

$$u(a_2)S^2(a_1)=u(a_1)a_2 \quad \text{and} \quad \alpha(S(a_1))u(a_2)=\sigma(a_3,S(a_1))\alpha(a_2) \tag{11.4.11}$$

one can show that $u\circ S^2=u$ (see Lemma 10.15 for the dual case). Now, by using (11.4.11), (11.4.2) and (11.4.4) it can be proved that $u^{-1}$ defined by (11.4.7) is a left inverse of $u$. It is also a right inverse since

$$u(a_1)u^{-1}(a_2)=u^{-1}(S^2(a_1))u(a_2)=u^{-1}(S^2(a_1))u(S^2(a_2))=\varepsilon(S^2(a))=\varepsilon(a),$$

because of (11.4.11) and $u\circ S^2=u$ (for the dual case see Theorem 10.17). □

We will now use the dual braided reconstruction theorem in order to obtain the structure of $\underline{A}$ as a braided Hopf algebra in $\mathcal{M}^A$. Since it is the formal dual result of Theorem 11.3 we restrict ourselves to presenting the concepts and results, leaving the verification of the details to the reader.

Let $\mathcal{C}$ and $\mathcal{D}$ be two monoidal categories with $\mathcal{D}$ braided. If $\mathbb{F}:\mathcal{C}\to\mathcal{D}$ is a functor then for any $M\in\mathcal{D}$ we denote by $\mathbb{F}\otimes M:\mathcal{C}\to\mathcal{D}$ the functor given by $(\mathbb{F}\otimes M)(N)=\mathbb{F}(N)\otimes M$, for all $N\in\mathcal{C}$. If $N\to N'$ is a morphism in $\mathcal{C}$ then $(\mathbb{F}\otimes M)(f)=\mathbb{F}(f)\otimes \mathrm{Id}_M$.

Suppose that there is an object $B\in\mathcal{D}$ such that for all $M\in\mathcal{D}$

$$\mathrm{Hom}_{\mathcal{D}}(B,M)\cong \mathrm{Nat}(\mathbb{F},\mathbb{F}\otimes M)$$

by functorial bijections $\theta_M$. This is the representability assumption for comodules.

Let $\mu=\{\mu_N:\mathbb{F}(N)\to\mathbb{F}(N)\otimes B\mid N\in\mathcal{C}\}$ be the natural transformation corresponding to the identity morphism $\mathrm{Id}_B$. Then, by using $\mu$ and the braiding in $\mathcal{D}$ we have induced maps, for $s\in\{2,3\}$:

$$\Theta^s_M:\mathrm{Hom}_{\mathcal{D}}(B^{\otimes s},M)\cong \mathrm{Nat}(\mathbb{F}^s,\mathbb{F}^s\otimes M),$$

and we assume that these are bijections. This is the higher representability assumption for comodules.

By using $(\Theta^2_B)^{-1}$, $\mu_{\underline{1}}$, $\theta^{-1}_{B\otimes B}$ and $\theta^{-1}_{\underline{1}}$ we can define a multiplication, a unit, a comultiplication and a counit for $B$. The next result is the formal dual of the one in Section 11.1, and this is why we will omit its proof. Also, the weak commutativity of $B$ will be uncovered at the end of this section.

**Theorem 11.16** *Let $\mathscr{C}$ and $\mathscr{D}$ be monoidal categories with $\mathscr{D}$ braided and $\mathbb{F}: \mathscr{C} \to \mathscr{D}$ a strong monoidal functor satisfying the representability and the higher representability assumptions for comodules. Then $B$ as above is a bialgebra in $\mathscr{D}$. If $\mathscr{C}$ is left rigid then $B$ is a Hopf algebra in $\mathscr{D}$.*

Let now $(A, \sigma)$ be a CQT dual quasi-Hopf algebra, $A_R$ the $k$-vector space $A$ viewed as a right $A$-comodule via $\Delta$ and $\underline{A}$ the same $k$-vector space $A$ but viewed now as an object of $\mathscr{M}^A$ via the right coadjoint coaction:

$$\rho_{\underline{A}}(a) = a_2 \otimes S(a_1)a_3, \tag{11.4.12}$$

for all $a \in A$. We apply Theorem 11.16 for $\mathscr{C} = \mathscr{D} = \mathscr{M}^A$ and $\mathbb{F} = \mathrm{Id}_{\mathscr{C}}$. The first step is to show that $\underline{A}$ is the representability object we need.

Dual to the quasi-Hopf case, since the antipode $S$ is bijective, we define the elements $p_L, q_L \in (A \otimes A)^*$ by

$$p_L(a,b) = \varphi(S^{-1}(a_3), a_1, b)\beta(S^{-1}(a_2)), \quad q_L(a,b) = \varphi^{-1}(S(a_1), a_3, b)\alpha(a_2), \tag{11.4.13}$$

for all $a, b \in A$. Then, for all $a, b \in A$, the following relations hold:

$$p_L(a_2, b_2)S^{-1}(a_3)(a_1b_1) = p_L(a, b_1)b_2, \quad q_L(a_2, b_1)S(a_1)(a_3b_2) = q_L(a, b_2)b_1, \tag{11.4.14}$$

$$p_L(S(a_1), a_3b_2)q_L(a_2, b_1) = \varepsilon(a)\varepsilon(b), \quad q_L(S^{-1}(a_3), a_1b_1)p_L(a_2, b_2) = \varepsilon(a)\varepsilon(b). \tag{11.4.15}$$

In what follows by $\mathrm{Hom}^A(N, M)$ we denote the space of right $A$-comodule morphisms between $N, M \in \mathscr{M}^A$.

**Lemma 11.17** *Let $A$ be a dual quasi-Hopf algebra and $M \in \mathscr{M}^A$. If we define*

$$\begin{aligned} &\theta_M : \mathrm{Hom}^A(\underline{A}, M) \to \mathrm{Nat}(\mathrm{Id}, \mathrm{Id} \otimes M), \\ &\theta_M(\chi)_N(n) = p_L(S(n_{(1)}), n_{(3)})n_{(0)} \otimes \chi(n_{(2)}), \end{aligned} \tag{11.4.16}$$

*for all $\chi \in Hom(\underline{A}, M)$, $N \in \mathscr{M}^A$ and $n \in N$, then $\theta_M$ is well defined and a bijection. Its inverse, $\theta_M^{-1}: \mathrm{Nat}(\mathrm{Id}, \mathrm{Id} \otimes M) \to \mathrm{Hom}^A(\underline{A}, M)$, is given by*

$$\theta_M^{-1}(\xi)(a) = q_L(a_1, (a_2)_{\langle 1\rangle_{(1)}})\varepsilon((a_2)_{\langle 0\rangle})(a_2)_{\langle 1\rangle_{(0)}}, \tag{11.4.17}$$

*for all $\xi \in \mathrm{Nat}(\mathrm{Id}, \mathrm{Id} \otimes M)$ and $a \in A$, where we denote $\xi_{A_R}(a) = a_{\langle 0\rangle} \otimes a_{\langle 1\rangle}$.*

*Proof* We have to prove first that $\theta_M$ is well defined, meaning that $\theta_M(\chi)_N$ is a right $A$-colinear map and $\theta_M(\chi)$ is a natural transformation. Since $\chi : \underline{A} \to M$ is a morphism in $\mathscr{M}^A$ we have

$$\chi(a)_{(0)} \otimes \chi(a)_{(1)} = \chi(a_2) \otimes S(a_1)a_3, \tag{11.4.18}$$

for all $a \in A$. Now, if $n \in N$ then:

$$\begin{aligned} \rho_{N\otimes M}(\theta_M(\chi)_N(n)) &= p_L(S(n_{(1)}), n_{(3)})\rho_{N\otimes M}(n_{(0)} \otimes \chi(n_{(2)})) \\ &\overset{(3.7.6)}{=} p_L(S(n_{(2)}), n_{(4)})n_{(0)} \otimes \chi(n_{(3)})_{(0)} \otimes n_{(1)}\chi(n_{(3)})_{(1)} \end{aligned}$$

$$\overset{(11.4.18)}{=} p_L(S(n_{(2)}), n_{(6)})n_{(0)} \otimes \chi(n_{(4)}) \otimes n_{(1)}(S(n_{(3)})n_{(5)})$$
$$\overset{(11.4.14)}{=} p_L(S(n_{(1)}), n_{(3)})n_{(0)} \otimes \chi(n_{(2)}) \otimes n_{(4)}$$
$$\overset{(11.4.16)}{=} \theta_M(\chi)_N(n_{(0)}) \otimes n_{(1)} = (\theta_M(\chi)_N \otimes id_A)(\rho_N(n)),$$

as needed. It is not hard to see that $\theta_M(\chi)$ is a natural transformation, so we are left to show that $\theta_M^{-1}$ is also well defined, and that $\theta_M$ and $\theta_M^{-1}$ are inverses. The first assertion follows from the following. Since $\xi_{A_R}$ is a right $A$-comodule map we have

$$(a_1)_{\langle 0\rangle} \otimes (a_1)_{\langle 1\rangle} \otimes a_2 = a_{\langle 0\rangle_1} \otimes a_{\langle 1\rangle_{(0)}} \otimes a_{\langle 0\rangle_2} a_{\langle 1\rangle_{(1)}}, \tag{11.4.19}$$

for all $a \in A$. On the other hand, for all $a^* \in A^*$ the map $\lambda_{a^*} : A_R \to A_R$, $\lambda_{a^*}(a) := a^*(a_1)a_2$, is right $A$-colinear. Since $\xi$ is functorial under the morphism $\lambda_{a^*}$ we get

$$a^*(a_{\langle 0\rangle_1})a_{\langle 0\rangle_2} \otimes a_{\langle 1\rangle} = a^*(a_1)(a_2)_{\langle 0\rangle} \otimes (a_2)_{\langle 1\rangle},$$

for all $a^* \in A^*$ and $a \in A$, and this is equivalent to

$$a_{\langle 0\rangle_1} \otimes a_{\langle 0\rangle_2} \otimes a_{\langle 1\rangle} = a_1 \otimes (a_2)_{\langle 0\rangle} \otimes (a_2)_{\langle 1\rangle}, \tag{11.4.20}$$

for all $a \in A$. Then for $a \in A$ we have

$$\begin{aligned}
&(\theta_M^{-1}(\xi) \otimes \mathrm{Id}_A)(\rho_{\underline{A}}(a)) \\
&= \theta_M^{-1}(\xi)(a_2) \otimes S(a_1)a_3 \\
&= q_L(a_2, (a_3)_{\langle 1\rangle_{(1)}})\varepsilon((a_3)_{\langle 0\rangle})(a_3)_{\langle 1\rangle_{(0)}} \otimes S(a_1)a_4 \\
&\overset{(11.4.19)}{=} q_L(a_2, (a_3)_{\langle 1\rangle_{(1)}})\varepsilon((a_3)_{\langle 0\rangle_1})(a_3)_{\langle 1\rangle_{(0)}} \otimes S(a_1)((a_3)_{\langle 0\rangle_2}(a_3)_{\langle 1\rangle_{(2)}}) \\
&= q_L(a_2, (a_3)_{\langle 1\rangle_{(1)}})\varepsilon((a_3)_{\langle 0\rangle_2})(a_3)_{\langle 1\rangle_{(0)}} \otimes S(a_1)((a_3)_{\langle 0\rangle_1}(a_3)_{\langle 1\rangle_{(2)}}) \\
&\overset{(11.4.20)}{=} q_L(a_2, (a_4)_{\langle 1\rangle_{(1)}})\varepsilon((a_4)_{\langle 0\rangle})(a_4)_{\langle 1\rangle_{(0)}} \otimes S(a_1)(a_3(a_4)_{\langle 1\rangle_{(2)}}) \\
&\overset{(11.4.14)}{=} q_L(a_1, (a_2)_{\langle 1\rangle_{(2)}})\varepsilon((a_2)_{\langle 0\rangle})(a_2)_{\langle 1\rangle_{(0)}} \otimes (a_2)_{\langle 1\rangle_{(1)}} \\
&= (\rho_M \circ \theta_M^{-1}(\xi))(a),
\end{aligned}$$

so $\theta_M^{-1}(\xi)$ is a right $A$-comodule map. We show now that $\theta_M^{-1}$ is a left inverse for $\theta_M$. Indeed, from the definitions we have

$$\theta_M(\xi)_{A_R}(a) = p_L(S(a_2), a_4)a_1 \otimes \xi(a_3) := a_{\langle 0\rangle} \otimes a_{\langle 1\rangle},$$

for all $a \in A$, and therefore

$$\begin{aligned}
(\theta_M^{-1} \circ \theta_M)(\chi)(a) &= q_L(a_1, \chi(a_4)_{(1)})\varepsilon(a_2)p_L(S(a_3), a_5)\chi(a_4)_{(0)} \\
&= q_L(a_1, S(a_3)a_4)p_L(S(a_2), a_5)\chi(a_2) \\
&\overset{(11.4.15)}{=} \varepsilon(a_1)\varepsilon(a_3)\chi(a_2) = \chi(a),
\end{aligned}$$

for all $\chi \in \mathrm{Hom}(\underline{A}, M)$ and $a \in A$. In order to prove that $\theta_M^{-1}$ is a right inverse for $\theta_M$ observe first that for any $N \in \mathscr{M}^A$ and $n^* \in N^*$, the map $\lambda_{n^*} : N \to A_R$, $\lambda_{n^*}(n) = n^*(n_{(0)})n_{(1)}$, is right $A$-colinear. That $\xi$ is functorial under the morphism $\lambda_{n^*}$ means

$$n^*(n_{[0]_{(0)}})n_{[0]_{(1)}} \otimes n_{[1]} = n^*(n_{(0)})n_{(1)_{\langle 0\rangle}} \otimes n_{(1)_{\langle 1\rangle}},$$

where we denote $\xi_N(n) := n_{[0]} \otimes n_{[1]}$. Since this is true for any $n^* \in N^*$ we obtain

$$n_{[0]_{(0)}} \otimes n_{[0]_{(1)}} \otimes n_{[1]} = n_{(0)} \otimes n_{(1)_{\langle 0\rangle}} \otimes n_{(1)_{\langle 1\rangle}} \tag{11.4.21}$$

for all $n \in N$. Now, again for all $n \in N$ we compute:

$$\begin{aligned}
&(\theta_M \circ \theta_M^{-1})(\xi)_N(n)\\
&= \theta_M(\theta_M^{-1}(\xi))_N(n) = p_L(S(n_{(1)}), n_{(3)})n_{(0)} \otimes \theta_M^{-1}(\xi)(n_{(2)})\\
&= p_L(S(n_{(1)}), n_{(4)})q_L(n_{(2)}, (n_{(3)})_{\langle 1\rangle_{(1)}})\varepsilon((n_{(3)})_{\langle 0\rangle})n_{(0)} \otimes (n_{(3)})_{\langle 1\rangle_{(0)}}\\
&\overset{(11.4.19)}{=} p_L(S(n_{(1)}), (n_{(3)})_{\langle 0\rangle_2}(n_{(3)})_{\langle 1\rangle_{(2)}})q_L(n_{(2)}, (n_{(3)})_{\langle 1\rangle_{(1)}})\\
&\qquad \varepsilon((n_{(3)})_{\langle 0\rangle_1})n_{(0)} \otimes (n_{(3)})_{\langle 1\rangle_{(0)}}\\
&\overset{(11.4.21)}{=} p_L(S(n_{[0]_{(1)}}), n_{[0]_{(3)}}n_{[1]_{(2)}})q_L(n_{[0]_{(2)}}, n_{[1]_{(1)}})n_{[0]_{(0)}} \otimes n_{[1]_{(0)}}\\
&\overset{(11.4.15)}{=} \varepsilon(n_{[0]_{(1)}})\varepsilon(n_{[1]_{(1)}})n_{[0]_{(0)}} \otimes n_{[1]_{(0)}} = n_{[0]} \otimes n_{[1]} = \xi_N(n),
\end{aligned}$$

as needed, and this finishes the proof. □

We are now able to begin our reconstruction. The natural transformation $\mu \in \mathrm{Nat}(\mathrm{Id}, \mathrm{Id} \otimes \underline{A})$ corresponding to the identity morphism $\mathrm{Id}_{\underline{A}}$ is given by

$$\mu_N(n) = \theta_{\underline{A}}(\mathrm{Id}_{\underline{A}})_N(n) = p_L(S(n_{(1)}), n_{(3)})n_{(0)} \otimes n_{(2)}$$

for all $N \in \mathscr{M}^A$ and $n \in N$. By the dual version of Proposition 11.1 the multiplication of $\underline{A}$ is characterized as being the unique morphism $\underline{m}:\ \underline{A} \otimes \underline{A} \to \underline{A}$ in $\mathscr{M}^A$ such that

$$\begin{aligned}
\mu_{M\otimes N} &= (\mathrm{Id}_{M\otimes N} \otimes \underline{m}) \circ a^{-1}_{M,N,\underline{A}\otimes\underline{A}} \circ (\mathrm{Id}_M \otimes a_{N,\underline{A},\underline{A}}) \circ (\mathrm{Id}_M \otimes (c_{\underline{A},N} \otimes \mathrm{Id}_{\underline{A}}))\\
&\quad \circ (\mathrm{Id}_M \otimes a^{-1}_{\underline{A},N,\underline{A}}) \circ a_{M,\underline{A},N\otimes\underline{A}} \circ (\mu_M \otimes \mu_N),
\end{aligned}$$

for any $M, N \in \mathscr{M}^A$. By using the braided categorical structure of $\mathscr{M}^A$ and the definition of $\mu$ it is not hard to see that $\underline{m}$ is the unique morphism in $\mathscr{M}^A$ which satisfies

$$\begin{aligned}
&p_L(S(m_{(1)}n_{(1)}), m_{(3)}n_{(3)})(m_{(0)} \otimes n_{(0)}) \otimes m_{(2)}n_{(2)}\\
&\quad = p_L(S(m_{(3)}), m_{(15)})p_L(S(n_{(5)}), n_{(13)})\varphi(m_{(2)}, S(m_{(4)})m_{(14)}, n_{(14)})\\
&\qquad \varphi^{-1}(S(m_{(5)})m_{(13)}, n_{(4)}, S(n_{(6)})n_{(12)})\sigma(S(m_{(6)})m_{(12)}, n_{(3)})\\
&\qquad \varphi(n_{(2)}, S(m_{(7)})m_{(11)}, S(n_{(7)})n_{(11)})\varphi^{-1}(m_{(1)}, n_{(1)}, m_{(9)}n_{(9)})\\
&\qquad (m_{(0)} \otimes n_{(0)}) \otimes (S(m_{(8)})m_{(10)})\underline{\cdot}(S(n_{(8)})n_{(10)})
\end{aligned}$$

for all $M, N \in \mathscr{M}^A$ and $m \in M$, $n \in N$, where we denote by $a\underline{\cdot}b := \underline{m}(a \otimes b)$. One can easily check that the above equality is equivalent to

$$\begin{aligned}
p_L(S(a_1b_1), a_3b_3)a_2b_2 &= p_L(S(a_3), a_{15})p_L(S(b_5), b_{13})\varphi(a_2, S(a_4)a_{14}, b_{14})\\
&\quad \varphi^{-1}(S(a_5)a_{13}, b_4, S(b_6)b_{12})\sigma(S(a_6)a_{12}, b_3)\varphi(b_2, S(a_7)a_{11}, S(b_7)b_{11})\\
&\quad \varphi^{-1}(a_1, b_1, a_9b_9)(S(a_8)a_{10})\underline{\cdot}(S(b_8)b_{10})
\end{aligned} \tag{11.4.22}$$

for all $a, b \in A$. Now, the explicit formula for the multiplication $\underline{\cdot}$ is the following:

$$a\underline{\cdot}b = \varphi(S(a_1), a_{10}, S(b_1)b_{12})f(b_6, a_3)\sigma(a_8, S(b_3))\varphi^{-1}(S(a_2), S(b_5), a_6b_9)$$

$$\sigma(a_4,b_7)\varphi^{-1}(a_9,S(b_2),b_{11})\varphi(S(b_4),a_7,b_{10})a_5b_8, \tag{11.4.23}$$

for all $a,b\in A$. Indeed, it is easy to see that the multiplication $\underline{\cdot}$ defined by (11.4.23) is a right $A$-colinear map. A straightforward but tedious computation ensures that $\underline{\cdot}$ satisfies the relation (11.4.22); we leave all these details to the reader. It is not hard to see that the unit of $\underline{A}$ is $1_A$, the unit of $A$.

Following the dual version of Proposition 11.2, the comultiplication of $\underline{A}$ is obtained as $\underline{\Delta}=\theta^{-1}_{\underline{A}\otimes\underline{A}}(\xi)$, where $\xi$ is defined by the following composition:

$$\xi_N:\ N\xrightarrow{\mu_N}N\otimes\underline{A}\xrightarrow{\mu_N}(N\otimes\underline{A})\otimes\underline{A}\xrightarrow{a_{N,\underline{A},\underline{A}}}N\otimes(\underline{A}\otimes\underline{A}),$$

for all $N\in\mathscr{M}^A$. Explicitly, for all $n\in N$,

$$\begin{aligned}\xi_N(n)=\varphi(n_{(1)},S(n_{(3)})n_{(5)},S(n_{(8)})n_{(10)})p_L(S(n_{(7)}),n_{(11)})\\ p_L(S(n_{(2)}),n_{(6)})n_{(0)}\otimes(n_{(4)}\otimes n_{(9)}).\end{aligned} \tag{11.4.24}$$

The counit $\underline{\varepsilon}$ is obtained as $\underline{\varepsilon}(a)=\theta_k^{-1}(l)(a)$, where $l$ is the left unit constraint.

**Proposition 11.18** *Let $A$ be a dual quasi-Hopf algebra. Then the comultiplication of $\underline{A}$ is given for all $a\in A$ by*

$$\underline{\Delta}(a)=\varphi^{-1}(S(a_1),a_5,S(a_7))\beta(a_6)\varphi(S(a_2)a_4,S(a_8),a_{10})a_3\otimes a_9. \tag{11.4.25}$$

*The counit of $\underline{\Delta}$ is $\underline{\varepsilon}=\alpha$.*

*Proof* Notice first that (3.7.3) and the definitions (11.4.13) of $p_L$ and $q_L$ imply:

$$\begin{aligned}&q_L(a_1,b_1c_1)\varphi(a_2,b_2,c_2)\\ &\qquad=\alpha(a_3)\varphi^{-1}(S(a_2),a_4,b_1)\varphi^{-1}(S(a_1),a_5b_2,c),\end{aligned} \tag{11.4.26}$$

$$\begin{aligned}&\varphi^{-1}(a,b_1,S(b_3)c_1)p_L(S(b_2),c_2)\\ &\qquad=\varphi(a_1b_1,S(b_5),c)\varphi^{-1}(a_2,b_2,S(b_4))\beta(b_3),\end{aligned} \tag{11.4.27}$$

for all $a,b,c\in A$. On the other hand, from (11.4.24) we can easily see that

$$\begin{aligned}\xi_{A_R}(a)=a_{\langle 0\rangle}\otimes a_{\langle 1\rangle}=p_L(S(a_3),a_7)p_L(S(a_8),a_{12})\\ \varphi(a_2,S(a_4)a_6,S(a_9)a_{11})a_1\otimes(a_5\otimes a_{10}).\end{aligned} \tag{11.4.28}$$

Now, for all $a\in A$ we compute:

$$\begin{aligned}\underline{\Delta}_{\underline{A}}(a)&=\theta^{-1}_{\underline{A}\otimes\underline{A}}(\xi)\\ &=q_L(a_1,(a_2)_{\langle 1\rangle_{(1)}})\varepsilon((a_2)_{\langle 0\rangle})(a_2)_{\langle 1\rangle_{(0)}}\\ &\overset{(11.4.28)}{=}q_L(a_1,(a_5\otimes a_{10})_{(1)})p_L(S(a_8),a_{12})p_L(S(a_3),a_7)\\ &\qquad\varphi(a_2,S(a_4)a_6,S(a_9)a_{11})(a_5\otimes a_{10})_{(0)}\\ &\overset{(3.7.6)}{=}q_L(a_1,(S(a_5)a_7)(S(a_{12})a_{14}))p_L(S(a_{10}),a_{16})p_L(S(a_3),a_9)\\ &\qquad\varphi(a_2,S(a_4)a_8,S(a_{11})a_{15})a_6\otimes a_{13}\\ &\overset{(11.4.26)}{=}\alpha(a_3)\varphi^{-1}(S(a_2),a_4,S(a_8)a_{10})\varphi^{-1}(S(a_1),a_5(S(a_7)a_{11}),S(a_{14})a_{16})\\ &\qquad p_L(S(a_{13}),a_{17})p_L(S(a_6),a_{12})a_9\otimes a_{15}\end{aligned}$$

$$\begin{aligned}
&\overset{(11.4.14)}{=} \alpha(a_3)\varphi^{-1}(S(a_2),a_4,S(a_6)a_8)\varphi^{-1}(S(a_1),a_{10},S(a_{12})a_{14})\\
&\qquad p_L(S(a_{11}),a_{15})p_L(S(a_5),a_9)a_7\otimes a_{13}\\
&\overset{(11.4.27)}{=} \alpha(a_4)\varphi^{-1}(S(a_3),a_5,S(a_7)a_9)\varphi(S(a_2)a_{11},S(a_{15}),a_{17})\\
&\qquad \varphi^{-1}(S(a_1),a_{12},S(a_{14}))\beta(a_{13})p_L(S(a_6),a_{10})a_8\otimes a_{16}\\
&\overset{(11.4.13)}{=} q_L(a_3,S(a_5)a_7)\varphi(S(a_2)a_9,S(a_{13}),a_{15})\varphi^{-1}(S(a_1),a_{10},S(a_{12}))\\
&\qquad \beta(a_{11})p_L(S(a_4),a_8)a_6\otimes a_{14}\\
&\overset{(11.4.15)}{=} \varphi^{-1}(S(a_1),a_5,S(a_7))\beta(a_6)\varphi(S(a_2)a_4,S(a_8),a_{10})a_3\otimes a_9.
\end{aligned}$$

The counit of $\underline{\Delta}$ is $\underline{\varepsilon}(a)=\theta_k^{-1}(l)(a)=q_L(a,1)=\alpha(a)$ for all $a\in A$, so $\underline{\varepsilon}=\alpha$. □

Let $M$ be a finite dimensional right $A$-comodule and $M^*$ its left dual object as in Proposition 3.52. By the dual version of Proposition 11.2, the reconstructed antipode $\underline{S}$ of $\underline{A}$ is characterized as being the unique morphism in $\mathscr{C}$ satisfying

$$\begin{aligned}
(\mathrm{Id}_M\circ\underline{S})\circ\mu_M &= l^{-1}_{M\otimes\underline{A}}\circ(\mathrm{Id}_{M\otimes\underline{A}}\otimes\mathrm{ev}_M)\circ(a_{M,\underline{A},M^*}\otimes\mathrm{Id}_M)\circ(a^{-1}_{M,\underline{A},M^*}\otimes\mathrm{Id}_M)\\
&\quad\circ((\mathrm{Id}_M\otimes c^{-1}_{\underline{A},M^*})\otimes\mathrm{Id}_M)\circ((\mathrm{Id}_M\otimes\mu_{M^*})\otimes\mathrm{Id}_M)\circ(\mathrm{coev}_M\otimes\mathrm{Id}_M)\circ r_M,
\end{aligned}$$

for any finite-dimensional object $M$ of $\mathscr{M}^A$, where $l$, $r$, $a$, $c$, ev and coev are the left unit constraint, the right unit constraint, the associativity constraint, the braiding of $\mathscr{M}^A$, and the evaluation and coevaluation maps, respectively. This reads

$$\begin{aligned}
p_L(S(m_{(1)}),m_{(3)})m_{(0)}\otimes\underline{S}(m_{(2)}) &= \beta(m_{(3)})p_L(S^2(m_{(12)}),S(m_{(4)}))\\
&\quad \sigma^{-1}(S^2(m_{(11)})S(m_{(5)}),S(m_{(13)}))\varphi^{-1}(m_{(2)},S^2(m_{(10)})S(m_{(6)}),S(m_{(14)}))\\
&\quad \varphi(m_{(1)}[S^2(m_{(9)})S(m_{(7)})],S(m_{(15)}),m_{(17)})\alpha(m_{(16)})m_{(0)}\otimes S(m_{(8)}),
\end{aligned}$$

for any finite-dimensional right $A$-comodule $M$ and $m\in M$. It follows that the above relation is equivalent to

$$\begin{aligned}
&p_L(S(a_1),a_3)\underline{S}(a_2)\\
&= \beta(a_3)p_L(S^2(a_{12}),S(a_4))\sigma^{-1}(S^2(a_{11})S(a_5),S(a_{13}))\\
&\qquad \varphi^{-1}(a_2,S^2(a_{10})S(a_6),S(a_{14}))\varphi(a_1[S^2(a_9)S(a_7)],S(a_{15}),a_{17})\alpha(a_{16})S(a_8)\\
&\overset{(3.7.3)}{\underset{(3.7.8)}{=}} \beta(a_2)p_L(S^2(a_{11}),S(a_3))\sigma^{-1}(S^2(a_{10})S(a_4),S(a_{12}))\\
&\qquad \varphi(S^2(a_8)S(a_6),S(a_{14}),a_{16})\varphi(a_1,[S^2(a_9)S(a_5)]S(a_{13}),a_{17})\alpha(a_{15})S(a_7)\\
&\overset{(11.4.14)}{\underset{(11.4.13)}{=}} p_L(S(a_1),a_{13})p_L(S^2(a_8),S(a_2))\sigma^{-1}(S^2(a_7)S(a_3),S(a_9))\\
&\qquad \varphi(S^2(a_6)S(a_4),S(a_{10}),a_{12})\alpha(a_{11})S(a_5)
\end{aligned}$$

for all $a\in A$, and therefore

$$\begin{aligned}
\underline{S}(a) &= p_L(S^2(a_7),S(a_1))\sigma^{-1}(S^2(a_6)S(a_2),S(a_8))\\
&\quad \varphi(S^2(a_5)S(a_3),S(a_9),a_{11})\alpha(a_{10})S(a_4), \qquad (11.4.29)
\end{aligned}$$

for all $a\in A$ (it is not hard to see that $\underline{S}$ defined above is right $A$-colinear).

We summarize all these facts in the following:

**Theorem 11.19** *Let $(A, \sigma)$ be a CQT dual quasi-Hopf algebra. Then there is a braided Hopf algebra $\underline{A}$ in the category $\mathcal{M}^A$. $\underline{A}$ coincides with $A$ as a k-linear space, and it is an object in $\mathcal{M}^A$ by the right coadjoint coaction*

$$\rho_{\underline{A}}(a) = a_2 \otimes S(a_1)a_3.$$

*The algebra structure, the coalgebra structure and the antipode are*

$$\begin{aligned}
a\underline{\cdot}b &= \varphi(S(a_1), a_{10}, S(b_1)b_{12})f(b_6, a_3)\sigma(a_8, S(b_3))\varphi^{-1}(S(a_2), S(b_5), a_6b_9)\\
&\qquad \sigma(a_4, b_7)\varphi^{-1}(a_9, S(b_2), b_{11})\varphi(S(b_4), a_7, b_{10})a_5b_8,\\
\underline{\Delta}(a) &= \varphi^{-1}(S(a_1), a_5, S(a_7))\beta(a_6)\varphi(S(a_2)a_4, S(a_8), a_{10})a_3 \otimes a_9,\\
\underline{S}(a) &= p_L(S^2(a_7), S(a_1))\sigma^{-1}(S^2(a_6)S(a_2), S(a_8))\\
&\qquad \varphi(S^2(a_5)S(a_3), S(a_9), a_{11})\alpha(a_{10})S(a_4),
\end{aligned}$$

*for all $a, b \in \underline{A}$. The unit element is $1_A$, the unit of $A$, and the counit is $\underline{\varepsilon} = \alpha$. We will call $\underline{A}$ the associated function algebra braided group of $A$.*

**Remark 11.20** The braided group $\underline{A}$ is weakly braided commutative in the following sense. The multiplication $\underline{m}_A$ satisfies the equality

$$\begin{array}{llcll}
M\otimes N \overset{\mu_M\otimes\mu_N}{\longrightarrow} & (M\otimes\underline{A})\otimes(N\otimes\underline{A}) & & M\otimes N \overset{\mu_M\otimes \mathrm{Id}_N}{\longrightarrow} & (M\otimes\underline{A})\otimes N\\
\overset{a_{M,\underline{A},N\otimes\underline{A}}}{\longrightarrow} & M\otimes(\underline{A}\otimes(N\otimes\underline{A})) & & \overset{a_{M,\underline{A},N}}{\longrightarrow} & M\otimes(\underline{A}\otimes N)\\
\overset{\mathrm{Id}_M\otimes a^{-1}_{\underline{A},N,\underline{A}}}{\longrightarrow} & M\otimes((\underline{A}\otimes N)\otimes\underline{A}) & & \overset{\mathrm{Id}_M\otimes c^{-1}_{N,\underline{A}}}{\longrightarrow} & M\otimes(N\otimes\underline{A})\\
\overset{\mathrm{Id}_M\otimes(c_{\underline{A},N}\otimes\mathrm{Id}_{\underline{A}})}{\longrightarrow} & M\otimes((N\otimes\underline{A})\otimes\underline{A}) & = & \overset{\mathrm{Id}_M\otimes(\mu_N\otimes\mathrm{Id}_{\underline{A}})}{\longrightarrow} & M\otimes((N\otimes\underline{A})\otimes\underline{A})\\
\overset{\mathrm{Id}_M\otimes a_{N,\underline{A},\underline{A}}}{\longrightarrow} & M\otimes(N\otimes(\underline{A}\otimes\underline{A})) & & \overset{\mathrm{Id}_M\otimes a_{N,\underline{A},\underline{A}}}{\longrightarrow} & M\otimes(N\otimes(\underline{A}\otimes\underline{A}))\\
\overset{a^{-1}_{M,N,\underline{A}\otimes\underline{A}}}{\longrightarrow} & (M\otimes N)\otimes(\underline{A}\otimes\underline{A}) & & \overset{a^{-1}_{M,N,\underline{A}\otimes\underline{A}}}{\longrightarrow} & (M\otimes N)\otimes(\underline{A}\otimes\underline{A})\\
\overset{\mathrm{Id}_{M\otimes N}\otimes\underline{m}_A}{\longrightarrow} & (M\otimes N)\otimes\underline{A} & & \overset{\mathrm{Id}_{M\otimes N}\otimes\underline{m}_A}{\longrightarrow} & (M\otimes N)\otimes\underline{A}\,,
\end{array}$$

for all $M, N \in \mathcal{M}^A$. Note that by writing down explicitly the above equality we get a relation that is dual to the one in (11.2.5); the details are left to the reader.

Suppose now that $(H, R)$ is a finite-dimensional QT quasi-Hopf algebra. Then $H^*$, the linear dual of $H$, is in an obvious way a CQT dual quasi-Hopf algebra, so it makes sense to consider $\underline{H^*}$, the function algebra braided group associated to $H^*$. It is a Hopf algebra in the category of right $H^*$-comodules, hence a Hopf algebra in the category of left $H$-modules. From (11.4.12), $\underline{H^*}$ is a left $H$-module via

$$h \blacktriangleright \chi = h_2 \rightharpoonup \chi \leftharpoonup S(h_1), \tag{11.4.30}$$

for all $h \in H$ and $\chi \in H^*$. By Theorem 11.19, the structure of $\underline{H^*}$ as a Hopf algebra in ${}_H\mathcal{M}$ is given by:

$$\begin{aligned}
\chi\underline{\cdot}\psi &= [x_1^3Y^2r^1y^1X^2 \rightharpoonup \chi \leftharpoonup S(x^1X^1)f^2R^1]\\
&\qquad [x_2^3Y^3y^3X_2^3 \rightharpoonup \psi \leftharpoonup S(x^2Y^1r^2y^2X_1^3)f^1R^2],
\end{aligned} \tag{11.4.31}$$

$$1_{\underline{H^*}} = \varepsilon, \tag{11.4.32}$$

$$\underline{\Delta}_{\underline{H^*}}(\chi) = \chi_1 \leftharpoonup S(x^1) \otimes x_2^3 X^3 \rightharpoonup \chi_2 \leftharpoonup x^2 X^1 \beta S(x_1^3 X^2), \tag{11.4.33}$$
$$\underline{\varepsilon}_{\underline{H^*}}(\chi) = \chi(\alpha), \tag{11.4.34}$$
$$\underline{S}(\chi) = q_2^1 \overline{R}_2^1 \tilde{p}^2 \rightharpoonup \chi S \leftharpoonup q^2 \overline{R}^2 S(q_1^1 \overline{R}_1^1 \tilde{p}^1), \tag{11.4.35}$$

for all $\chi, \psi \in H^*$. Here $p_R = p^1 \otimes p^2$ and $q_R = q^1 \otimes q^2$ are the elements defined by (3.2.19), $f = f^1 \otimes f^2$ is the Drinfeld's twist defined by (3.2.15), $R^{-1} = \overline{R}^1 \otimes \overline{R}^2$, and $q_L = \tilde{q}^1 \otimes \tilde{q}^2$ is the element given by (3.2.20).

On the other hand, since $(H,R)$ is finite dimensional, the categorical left dual of $\underline{H}$ has a braided Hopf algebra structure in ${}_H\mathcal{M}$. We have denoted $\underline{H}^*$ with this Hopf algebra structure by $(\underline{H})^\star$. By the above, $(\underline{H})^\star$ is a left $H$-module via

$$(h \succ \chi)(h') = \chi(S(h) \triangleright h'), \ \ \forall\, h, h' \in H, \ \chi \in H^*. \tag{11.4.36}$$

By Proposition 2.54 the structure of $(\underline{H})^\star$ as a bialgebra in ${}_H\mathcal{M}$ is given by the formulas

$$(\chi * \psi)(h) = \langle \chi, f^2 \triangleright h_{\underline{2}} \rangle \langle \psi, f^1 \triangleright h_{\underline{1}} \rangle, \tag{11.4.37}$$
$$1_{(\underline{H})^\star} = \varepsilon, \tag{11.4.38}$$
$$\underline{\Delta}_{(\underline{H})^\star}(\chi) = \langle \chi, (g^1 \triangleright {}_i e) \bullet (g^2 \triangleright {}_j e) \rangle {}^j e \otimes {}^i e, \tag{11.4.39}$$
$$\underline{\varepsilon}_{(\underline{H})^\star}(\chi) = \chi(\beta), \tag{11.4.40}$$

where $\{{}_i e\}_{i=\overline{1,n}}$ and $\{{}^i e\}_{i=\overline{1,n}}$ are dual bases in $H$ and $H^*$. Furthermore, by Proposition 2.66 we have that $(\underline{H})^\star$ is a Hopf algebra in ${}_H\mathcal{M}$ with antipode given by

$$\underline{S}_{(\underline{H})^\star}(\chi) = \chi \circ \underline{S}. \tag{11.4.41}$$

We show that, up to a braided Hopf algebra isomorphism, $\underline{H^*}$ is nothing but $(\underline{H})^\star$.

**Proposition 11.21** *Let $(H,R)$ be a finite-dimensional QT quasi-Hopf algebra, $\underline{H}$ the associated enveloping algebra braided group of $H$, $(\underline{H})^\star$ the left op-cop dual Hopf algebra structure of $\underline{H}$ in ${}_H\mathcal{M}$, and $\underline{H^*}$ the function algebra braided group associated to $H^*$. Then the map $\lambda : (\underline{H})^\star \to \underline{H^*}$ given for all $\chi \in H^*$ by*

$$\lambda(\chi) = S^{-1}(g^1) \rightharpoonup \chi \circ S \leftharpoonup g^2 \tag{11.4.42}$$

*is a braided Hopf algebra isomorphism. Here $g^1 \otimes g^2$ is the inverse of the Drinfeld twist $f$; see* (3.2.16).

*Proof* The map $\lambda$ is left $H$-linear since

$$\begin{aligned}
(h \blacktriangleright \lambda(\chi))(h') &= \langle \lambda(\chi), S(h_1) h' h_2 \rangle \\
&= \langle \chi, g^1 S(h_2) S(h') S(g^2 S(h_1)) \rangle \\
&\overset{(3.2.13),(8.7.1)}{=} \langle \chi, S(h) \triangleright (g^1 S(g^2 h')) \rangle \\
&\overset{(11.4.36)}{=} \langle (h \succ \chi) \circ S, g^2 h' S^{-1}(g^1) \rangle = \lambda(h \succ \chi)(h')
\end{aligned}$$

for all $h, h' \in H$ and $\chi \in H^*$. Next, we show that $\lambda$ is an algebra and coalgebra morphism, and that it is bijective. Firstly, for all $\chi, \psi \in H^*$ and $h \in H$ we compute

$$\begin{aligned}
\lambda(\chi * \psi)(h) &= \langle \chi, f^2 \triangleright (g^1 S(g^2 h))_{\underline{2}} \rangle \langle \psi, f^1 \triangleright (g^1 S(g^2 h))_{\underline{1}} \rangle \\
&\overset{(11.2.3),(8.7.1)}{\underset{(3.2.13)}{=}} \langle \chi, f^2 x^3 R^1 \triangleright y^1 X^2 g_2^1 G^2 S(y^2 X_1^3 g_1^2 h_1) \rangle \\
&\quad \langle \psi, f_1^1 x^1 X^1 g_1^1 G^1 S(f_2^1 x^2 R^2 y^3 X_2^3 g_2^2 h_2) \rangle \\
&\overset{(3.2.17),(3.1.7)}{\underset{(3.2.13)}{=}} \langle \chi, f^2 x^3 R^1 \triangleright G^2_{(1,1)} y^1 g^1 S(G^2_{(1,2)} y^2 g_1^2 \mathfrak{G}^1 S(X_2^1) F^1 h_1 X^2) \rangle \\
&\quad \langle \psi, f_1^1 x^1 G^1 S(f_2^1 x^2 R^2 G_2^2 y^3 g_2^2 \mathfrak{G}^2 S(X_1^1) F^2 h_2 X^3) \rangle \\
&\overset{(3.2.17),(8.7.1)}{\underset{(10.1.3)}{=}} \langle \chi, f^2 x^3 G_2^2 R^1 \mathfrak{G}^1 \triangleright g^1 S(g^2 S(X_2^1 y^2) F^1 h_1 X^2 y^3) \rangle \\
&\quad \langle \psi, f_1^1 x^1 G^1 S(f_2^1 x^2 G_1^2 R^2 \mathfrak{G}^2 S(X_1^1 y^1) F^2 h_2 X^3) \rangle \\
&\overset{(10.3.6),(3.2.17)}{\underset{(8.7.1),(3.2.13)}{=}} \langle \chi, g^1 S(g^2 S(X_2^1 y^2 R_1^1 x_1^1) F^1 h_1 X^2 y^3 R_2^1 x_2^1) \rangle \\
&\quad \langle \psi, G^1 S(G^2 S(X_1^1 y^1 R^2 x^2) F^2 h_2 X^3 x^3) \rangle,
\end{aligned}$$

and, on the other hand, by (11.4.31) we have

$$\begin{aligned}
(\lambda(\chi) \underline{\cdot} \lambda(\psi))(h) &= \langle \chi, g^1 S(g^2 S(x^1 X^1) f^2 R^1 h_1 x_1^3 Y^2 r^1 y^1 X^2) \rangle \\
&\quad \langle \psi, G^1 S(G^2 S(x^2 Y^1 r^2 y^2 X_1^3) f^1 R^2 h_2 x_2^3 Y^3 y^3 X_2^3) \rangle \\
&\overset{(10.3.6),(3.1.9)}{\underset{(10.1.3)}{=}} \langle \chi, g^1 S(g^2 S(Y_2^1 R^1 z^1 x^1 X^1) f^1 h_1 Y^2 z^3 r^1 x_1^2 y^1 X^2) \rangle \\
&\quad \langle \psi, G^1 S(G^2 S(Y_1^1 R^2 z^2 r^2 x_2^2 y^2 X_1^3) f^2 h_2 Y^3 x^3 y^3 X_2^3) \rangle \\
&\overset{(3.1.9),(10.1.1)}{=} \langle \chi, g^1 S(g^2 S(Y_2^1 z^2 R_1^1 x_1^1) f^1 h_1 Y^2 z^3 R_2^1 x_2^1) \rangle \\
&\quad \langle \psi, G^1 S(G^2 S(Y_1^1 z^1 R^2 y^2) f^2 h_2 Y^3 y^3) \rangle,
\end{aligned}$$

as needed. It is not hard to see that $\lambda(1_{(\underline{H})^*}) = 1_{\underline{H^*}}$, so $\lambda$ is an algebra morphism.

Now, $\lambda$ is a coalgebra morphism since

$$\begin{aligned}
&(\lambda \otimes \lambda)(\underline{\Delta}_{(\underline{H})^*}(\chi)) \\
&= \langle \chi, (g^1 \triangleright {}_i e) \bullet (g^2 \triangleright {}_j e) \rangle \lambda({}^j e) \otimes \lambda({}^i e) \\
&\overset{(11.4.42),(4.1.9),(8.7.1)}{=} \langle \chi, X^1 g_1^1 \mathfrak{G}^1 S(x^1 X^2 g_2^1 \mathfrak{G}^2 {}_i e) \alpha x^2 X_1^3 g_1^2 G^1 S(x^3 X_2^3 g_2^2 G^2 {}_j e) \rangle {}^j e \otimes {}^i e \\
&\overset{(3.2.17),(3.1.7),(3.2.1)}{=} \langle \chi, \mathfrak{G}^1 S(g^1 S(X^2 x^3)_i e X^3) \alpha g^2 S(\mathfrak{G}^2 S(X_1^1 x^1)_j e X_2^1 x^2) \rangle {}^j e \otimes {}^i e \\
&\overset{(8.7.7),(11.4.42)}{=} \langle \lambda(\chi), S(X_1^1 x^1)_j e X_2^1 x^2 \beta S(X^2 x^3)_i e X^3) \rangle {}^j e \otimes {}^i e \\
&\overset{(3.1.9),(11.4.33)}{=} \lambda(\chi)_1 \leftharpoonup S(X_1^1 x^1) \otimes X^3 \rightharpoonup \lambda(\chi)_2 \leftharpoonup X_2^1 x^2 \beta S(X^2 x^3) \\
&= \underline{\Delta}_{\underline{H^*}}(\lambda(\chi)),
\end{aligned}$$

for all $\chi \in H^*$, and since the definitions of counits imply $\underline{\varepsilon}_{\underline{H^*}} \circ \lambda = \underline{\varepsilon}_{(\underline{H})^*}$. It is easy to see that $\lambda$ is bijective with inverse $\lambda^{-1}(\chi) = S(f^2) \rightharpoonup \chi \circ S^{-1} \leftharpoonup f^1$, for all $\chi \in H^*$. Thus, the proof is complete. □

We have seen two processes that associate to a finite-dimensional QT quasi-Hopf algebra $H$ a braided Hopf algebra structure on $H^*$ within ${}_H\mathscr{M}$. The first one associates what we called the function algebra braided group on $H^*$, which was denoted by $\underline{H^*}$, while the second one associates $\underline{H}^*$ as in Proposition 10.30 (see also Remark 10.31(2). Actually, up to an isomorphism, these two processes coincide.

**Proposition 11.22** *Let $H$ be a finite-dimensional QT quasi-Hopf algebra, $\underline{H^*}$ the function algebra braided group on $H^*$ and $\underline{H}^*$ the Hopf algebra in ${}_H\mathcal{M}$ as in Proposition 10.30. Then $\overline{S}$, the antipode of $H^*$, yields a braided Hopf algebra isomorphism between $\underline{H^*}$ and $\underline{H}^*$.*

*Proof* One can see easily that

$$(h \rightharpoonup \chi \leftharpoonup h')\circ S = S^{-1}(h') \rightharpoonup \chi S \leftharpoonup S^{-1}(h)\ ,\ \ \forall\, h,\, h' \in\ H \text{ and } \chi \in H^*,$$

where, as before, $\rightharpoonup$ and $\leftharpoonup$ are the left and right regular actions of the algebra $H$ on its dual space $H^*$. The above formula together with (11.4.30) and (10.5.2) implies that $\overline{S} : \underline{H^*} \to \underline{H}^*$ is an isomorphism in ${}_H\mathcal{M}$. Also, it can be easily checked that $\overline{S}$ behaves well with respect to the units and counits of $\underline{H^*}$ and $\underline{H}^*$. It is also a multiplicative morphism since

$$\begin{aligned}
\langle \overline{S}(\chi \underline{\cdot} \psi), h\rangle &\overset{(11.4.31)}{=} \langle (x_1^3Y^2r^1y^1X^2 \rightharpoonup \chi \leftharpoonup S(x^1X^1)f^2R^1)\\
&\qquad (x_2^3Y^3y^3X_2^3 \rightharpoonup \psi \leftharpoonup S(x^2Y^1r^2y^2X_1^3)f^1R^2), S(h)\rangle\\
&\overset{(3.2.13),(10.1.3)}{=} \langle \chi, S(h_1x^1X^1)f^2x_2^3R^1Y^2r^1y^1X^2\rangle\\
&\qquad \langle \psi, S(h_2x^2Y^1r^2y^2X_1^3)f^1x_1^3R^2Y^3y^3X_2^3\rangle\\
&= \langle \chi S, S^{-1}(f^2x_2^3R^1Y^2r^1y^1X^2)h_1x^1X^1\rangle\\
&\qquad \langle \psi S, S^{-1}(f^1x_1^3R^2Y^3y^3X_2^3)h_2x^2Y^1r^2y^2X_1^3\rangle\\
&\overset{(10.1.2)}{=} \langle (x^1X^1 \rightharpoonup \chi S \leftharpoonup S^{-1}(f^2x_2^3Y^3R^1X^2))\\
&\qquad (x^2Y^1R_1^2X_1^3 \rightharpoonup \psi S \leftharpoonup S^{-1}(f^1x_1^3Y^2R_2^2X_2^3)), h\rangle\\
&\overset{(10.5.8)}{=} \langle (\chi S)\circ(\psi S), h\rangle,
\end{aligned}$$

for all $h \in H$, as required. Finally, $\overline{S}$ respects the comultiplications of $\underline{H^*}$ and $\underline{H}^*$ since $\Delta_{H^*}(\chi S) = \chi_2 S \otimes \chi_1 S$, for all $\chi \in H^*$, and therefore

$$\begin{aligned}
\underline{\Delta}_{\underline{H}^*}(\chi S) &\overset{(10.5.9)}{=} X_1^1p^1 \rightharpoonup \chi_1 S \leftharpoonup S^{-1}(X_2^1p^2) \otimes X^2 \rightharpoonup \chi_2 S \leftharpoonup S^{-1}(X^3)\\
&= (X_2^1p^2 \rightharpoonup \chi_1 \leftharpoonup S(X_1^1p^1))\circ S \otimes (X^3 \rightharpoonup \chi_2 \leftharpoonup S(X^2))\circ S\\
&= (\chi_1 \leftharpoonup S(X_1^1p^1))\circ S \otimes (X^3 \rightharpoonup \chi_2 \leftharpoonup X_2^1p^2S(X^2))\circ S\\
&\overset{(3.2.20),(11.5.7)}{=} (\chi_1 \leftharpoonup S(x^1))\circ S \otimes (x_2^3\tilde{p}^2 \rightharpoonup \chi_2 \leftharpoonup x^2S(x_1^3\tilde{p}^1))\circ S\\
&\overset{(3.2.20),(11.4.33)}{=} (\overline{S}\otimes\overline{S})\underline{\Delta}_{\underline{H^*}}(\chi),
\end{aligned}$$

for all $\chi \in H^*$, as desired. This finishes the proof. □

Consequently, in the QT case we have the following description for the quantum double quasi-Hopf algebra.

**Theorem 11.23** *The quantum double $D(H)$ of a finite-dimensional QT quasi-Hopf algebra $H$ can be characterized as follows:*

*(i) $D(H)$ is a biproduct between $\underline{H^*}$, the function algebra braided group, and $H$;*
*(ii) $D(H)$ is a biproduct between $(\underline{H})^\star$, the categorical left op-cop dual of the associated enveloping algebra braided group of $H$, and $H$.*

*Proof* The assertion (i) follows from Remark 10.31(1) and Proposition 11.22, while (ii) is a consequence of (i) and Proposition 11.21. □

## 11.5 Factorizable QT Quasi-Hopf Algebras

In this section we will introduce the notion of factorizable quasi-Hopf algebra and we will show that the quantum double is an example of this type.

If $(H,R)$ is a QT quasi-Hopf algebra we consider the $k$-linear map $\mathscr{Q}: H^* \to H$, given for all $\chi \in H^*$ by

$$\mathscr{Q}(\chi) = \langle \chi, S(X_2^2\tilde{p}^2)f^1R^2r^1U^1X^3\rangle X^1S(X_1^2\tilde{p}^1)f^2R^1r^2U^2, \tag{11.5.1}$$

where $r^1 \otimes r^2$ is another copy of $R$, $p_L = \tilde{p}^1 \otimes \tilde{p}^2$ is the element considered in (3.2.20) and $U = U^1 \otimes U^2$ is the element defined in (7.3.1).

**Definition 11.24** A QT quasi-Hopf algebra $(H,R)$ is called factorizable if the map $\mathscr{Q}$ defined by (11.5.1) is bijective.

In the second part of this section we will uncover the monoidal categorical interpretation for the definition of $\mathscr{Q}$.

**Example 11.25** For $(H(2),R_\pm)$ with $R_\pm$ as in Example 10.7, the map $\mathscr{Q}$ from (11.5.1) has the following form, for all $\chi \in H(2)^*$:

$$\mathscr{Q}(\chi) = \chi(1)p_- + \chi(g)p_+.$$

Since $\{p_-,p_+\}$ and $\{1,g\}$ are bases for $H(2)$ it follows that $\mathscr{Q}$ is bijective, so $(H(2),R_\pm)$ are factorizable quasi-Hopf algebras.

*Proof* For $H(2)$ the element $p_L$ has the form

$$p_L = X^1X^2 \otimes X^3 = 1 - 2p_- \otimes p_- = 1 - (1-g)\otimes p_- = 1\otimes p_+ + g\otimes p_- = f.$$

Also, one can easily see that $X^1X_1^2 \otimes X_2^2X^3 = 1$ and since $f = f^{-1}$ we conclude that

$$X_2^2X^3\tilde{p}^2f^1 \otimes X^1X_1^2\tilde{p}^1f^2 = 1.$$

On the other hand, since $\omega^2 - 2\omega = -2$ it follows that $R^2r^1 \otimes R^1r^2 = (1-\omega p_- \otimes p_-)^2 = 1 - 2p_- \otimes p_-$. We have already seen that $U = g\otimes 1$, and therefore

$$\begin{aligned} &S(X_2^2\tilde{p}^2)f^1R^2r^1U^1X^3 \otimes X^1S(X_1^2\tilde{p}^1)f^2R^1r^2U^2 \\ &\quad = (1-2p_- \otimes p_-)(g\otimes 1) = 1\otimes p_- + g\otimes p_+. \end{aligned}$$

It is now clear that $\mathscr{Q}(\chi) = \chi(1)p_- + \chi(g)p_+$, for all $\chi \in H(2)^*$. □

In what follows we will need a second formula for the map $\mathscr{Q}$ in (11.5.1). Also, another $k$-linear map $\overline{\mathscr{Q}}: H^* \to H$ is required.

**Proposition 11.26** *Let $(H,R)$ be a QT quasi-Hopf algebra.*

*(i) The map $\mathscr{Q}$ defined by* (11.5.1) *has a second formula given for all $\chi \in H^*$ by*

$$\mathscr{Q}(\chi) = \langle \chi, \tilde{q}^1 X^1 R^2 r^1 p^1 \rangle \tilde{q}^2_1 X^2 R^1 r^2 p^2 S(\tilde{q}^2_2 X^3), \tag{11.5.2}$$

*where $q_L = \tilde{q}^1 \otimes \tilde{q}^2$ and $p_R = p^1 \otimes p^2$ are the elements defined by* (3.2.20) *and* (3.2.19), *respectively.*

*(ii) Let $\overline{\mathscr{Q}} : H^* \to H$ be the k-linear map defined for all $\chi \in H^*$ by*

$$\overline{\mathscr{Q}}(\chi) = \langle \chi, S^{-1}(X^3) q^2 R^1 r^2 X^2_2 \tilde{p}^2 \rangle q^1 R^2 r^1 X^2_1 \tilde{p}^1 S^{-1}(X^1), \tag{11.5.3}$$

*where $q_R = q^1 \otimes q^2$ and $p_L = \tilde{p}^1 \otimes \tilde{p}^2$ are the elements defined by* (3.2.19) *and* (3.2.20), *respectively. Then $\mathscr{Q}$ is bijective if and only if $\overline{\mathscr{Q}}$ is bijective.*

*Proof* (i) We claim that

$$R^1 U^1 \otimes R^2 U^2 = \tilde{q}^1_2 R^1 p^1 \otimes \tilde{q}^1_1 R^2 p^2 S(\tilde{q}^2). \tag{11.5.4}$$

Indeed, we calculate:

$$\begin{aligned}
\tilde{q}^1_2 R^1 p^1 \otimes \tilde{q}^1_1 R^2 p^2 S(\tilde{q}^2) &\overset{(10.1.3)}{=} R^1 \tilde{q}^1_1 p^1 \otimes R^2 \tilde{q}^1_2 p^2 S(\tilde{q}^2)\\
&\overset{(7.3.5)}{=} R^1 (\tilde{q}^1 S(\tilde{p}^1))_1 U^1 \tilde{p}^2 \otimes R^2 (\tilde{q}^1 S(\tilde{p}^1))_2 U^2 S(\tilde{q}^2)\\
&\overset{(7.3.2)}{=} R^1 (\tilde{q}^1 S(\tilde{q}^2_1 \tilde{p}^1))_1 U^1 \tilde{q}^2_2 \tilde{p}^2 \otimes R^2 (\tilde{q}^1 S(\tilde{q}^2_1 \tilde{p}^1))_2 U^2\\
&\overset{(3.2.24)}{=} R^1 U^1 \otimes R^2 U^2,
\end{aligned}$$

as needed. Now, if we denote by $\tilde{Q}^1 \otimes \tilde{Q}^2$ another copy of $q_L$ we have

$$\begin{aligned}
\mathscr{Q}(\chi) &\overset{(11.5.1),(10.3.3)}{=} \langle \chi, S(x^2 \tilde{p}^2_1) f^1 R^2 r^1 U^1 x^3 \tilde{p}^2_2 \rangle S(x^1 \tilde{p}^1) f^2 R^1 r^2 U^2\\
&\overset{(11.5.4),(3.2.21)}{\underset{(10.1.3)}{=}} \langle \chi, S(x^2 \tilde{p}^2_1) f^1 \tilde{q}^1_1 (x^3 \tilde{p}^2_2)_{(1,1)} R^2 r^1 p^1 \rangle\\
&\qquad S(x^1 \tilde{p}^1) f^2 \tilde{q}^1_2 (x^3 \tilde{p}^2_2)_{(1,2)} R^1 r^2 p^2 S(\tilde{q}^2 (x^3 \tilde{p}^2_2)_2)\\
&\overset{(3.2.28)}{=} \langle \chi, S(\tilde{p}^2_1) \tilde{Q}^1 X^1 (\tilde{p}^2_2)_{(1,1)} R^2 r^1 p^1 \rangle\\
&\qquad S(\tilde{p}^1) \tilde{q}^1 \tilde{Q}^2_1 X^2 (\tilde{p}^2_2)_{(1,2)} R^1 r^2 p^2 S(\tilde{q}^2 \tilde{Q}^2_2 X^3 (\tilde{p}^2_2)_2)\\
&\overset{(3.1.7),(3.2.22)}{=} \langle \chi, \tilde{Q}^1 X^1 R^2 r^1 p^1 \rangle S(\tilde{p}^1) \tilde{q}^1 \tilde{p}^2_1 \tilde{Q}^2_1 X^2 R^1 r^2 p^2 S(\tilde{q}^2 \tilde{p}^2_2 \tilde{Q}^2_2 X^3)\\
&\overset{(3.2.24)}{=} \langle \chi, \tilde{Q}^1 X^1 R^2 r^1 p^1 \rangle \tilde{Q}^2_1 X^2 R^1 r^2 p^2 S(\tilde{Q}^2_2 X^3),
\end{aligned}$$

for all $\chi \in H^*$. So we have proved the relation (11.5.2).

(ii) For all $\chi \in H^*$ we have

$$\begin{aligned}
\mathscr{Q}(\chi) &\overset{(11.5.1),(7.3.1)}{=} \langle \chi, S(X^2_2 \tilde{p}^2) f^1 R^2 r^1 g^1 S(q^2) X^3 \rangle X^1 S(X^2_1 \tilde{p}^1) f^2 R^1 r^2 g^2 S(q^1)\\
&\overset{(10.3.6)\times 2}{=} \langle \chi, S(q^2 r^1 R^2 X^2_2 \tilde{p}^2) X^3 \rangle X^1 S(q^1 r^2 R^1 X^2_1 \tilde{p}^1)\\
&\overset{(11.5.2)}{=} S(\overline{\mathscr{Q}}(\chi \circ S)).
\end{aligned}$$

Since the antipode $S$ is bijective we conclude that $\mathscr{Q}$ is bijective if and only if $\overline{\mathscr{Q}}$ is bijective, so our proof is complete. □

We provide an important family of factorizable QT quasi-Hopf algebras.

**Proposition 11.27** *Let $H$ be a finite-dimensional quasi-Hopf algebra and $D(H)$ its quantum double. Then $D(H)$ is a factorizable quasi-Hopf algebra.*

*Proof* We will show that in the quantum double case the map $\overline{\mathscr{Q}}$ defined by (11.5.3) is bijective, so by Proposition 11.26 it follows that $D(H)$ is factorizable. For this we will compute first the element $\mathscr{R}^2\mathbf{R}^1\otimes\mathscr{R}^1\mathbf{R}^2$, where we denote by $\mathbf{R}^1\otimes\mathbf{R}^2$ another copy of the $R$-matrix $\mathscr{R}$ of $D(H)$. In fact, if we denote by $P^1\otimes P^2$ another copy of the element $p_R$ then we compute:

$$
\begin{aligned}
&\mathscr{R}^2\mathbf{R}^1\otimes\mathscr{R}^1\mathbf{R}^2\\
&\quad = ({}^ie\bowtie p^1_2)(\varepsilon\bowtie S^{-1}(P^2)_jeP^1_1)\otimes(\varepsilon\bowtie S^{-1}(p^2)_iep^1_1)({}^je\bowtie P^1_2)\\
&\quad \overset{(8.5.3)}{=} {}^ie\bowtie p^1_2S^{-1}(P^2)_jeP^1_1\otimes(S^{-1}(p^2)_iep^1_1)_{(1,1)}\rightharpoonup{}^je\leftharpoonup S^{-1}((S^{-1}(p^2)_iep^1_1)_2)\\
&\qquad\qquad \bowtie(S^{-1}(p^2)_iep^1_1)_{(1,2)}P^1_2\\
&\quad = {}^ie\bowtie p^1_2S^{-1}((S^{-1}(p^2)_iep^1_1)_2P^2)_je(S^{-1}(p^2)_iep^1_1)_{(1,1)}P^1_1\\
&\qquad\qquad \otimes{}^je\bowtie(S^{-1}(p^2)_iep^1_1)_{(1,2)}P^1_2\\
&\quad \overset{(3.2.21)}{=} {}^ie\bowtie S^{-1}((S^{-1}(p^2)_ie)_2P^2)_je(S^{-1}(p^2)_ie)_{(1,1)}P^1_1p^1_1\\
&\qquad\qquad \otimes{}^je\bowtie(S^{-1}(p^2)_ie)_{(1,2)}P^1_2p^1_2.
\end{aligned}
$$

Now, $H$ is a quasi-Hopf subalgebra of $D(H)$, so we have to calculate the element

$$b^1\otimes b^2:=(\varepsilon\bowtie S^{-1}(X^3)q^2)\mathscr{R}^1\mathbf{R}^2(\varepsilon\bowtie X^2_2\tilde{p}^2)\otimes(\varepsilon\bowtie q^1)\mathscr{R}^2\mathbf{R}^1(\varepsilon\bowtie X^2_1\tilde{p}^1S^{-1}(X^1)).$$

By dual bases and (8.5.3) we have

$$
\begin{aligned}
b^1\otimes b^2 &= (\varepsilon\bowtie S^{-1}(X^3))({}^je\bowtie(q^2S^{-1}(q^1_2p^2)_ie)_{(1,2)}((q^1_1)_{(1,1)}P^1)_2p^1_2)\\
&\quad (\varepsilon\bowtie X^2_2\tilde{p}^2)\otimes{}^ie\bowtie S^{-1}((q^2S^{-1}(q^1_2p^2)_ie)_2(q^1_1)_{(1,2)}P^2S((q^1_1)_2))_je\\
&\quad (q^2S^{-1}(q^1_2p^2)_ie)_{(1,1)}((q^1_1)_{(1,1)}P^1)_1p^1_1)(\varepsilon\bowtie X^2_1\tilde{p}^1S^{-1}(X^1))\\
&\overset{(3.2.21)}{\underset{(3.2.23)}{=}} (\varepsilon\bowtie S^{-1}(X^3))({}^je\bowtie({}_ie)_{(1,2)}P^1_2)(\varepsilon\bowtie X^2_2\tilde{p}^2)\\
&\quad \otimes({}^ie\bowtie S^{-1}(({}_ie)_2P^2)_je({}_ie)_{(1,1)}P^1_1)(\varepsilon\bowtie X^2_1\tilde{p}^1S^{-1}(X^1))\\
&\overset{(8.5.3)}{\underset{(10.3.3)}{=}} (\varepsilon\bowtie S^{-1}(x^3\tilde{p}^2_2)_ie)({}^je\bowtie P^1_2x^2\tilde{p}^2_1)\otimes({}^ie\bowtie S^{-1}(P^2)_jeP^1_1x^1\tilde{p}^1).
\end{aligned}
$$

Now we want an explicit formula for the element $S_D(b^1)\otimes b^2$. To this end we need the following relations:

$$
\begin{aligned}
&S(P^1_2x^2\tilde{p}^2_1)_1f^1_1p^1\otimes S(P^1_2x^2\tilde{p}^2_1)_2f^1_2p^2S(f^2)S^2(P^1_1x^1\tilde{p}^1)\\
&\qquad = g^1S(P^1y^3x^2_2\tilde{p}_{(1,2)})\otimes g^2S(S(y^1x^1\tilde{p}^1)\alpha y^2x^2_1\tilde{p}^2_{(1,1)}), \qquad (11.5.5)\\
&S(P^1)_2U^2\otimes S(P^1)_1U^1P^2=g^2\otimes g^1. \qquad (11.5.6)
\end{aligned}
$$

The first one follows by applying (3.2.13), (3.2.17), (3.1.7), (3.2.1) and then the formula $f^1\beta S(f^2)=S(\alpha)$ and (3.1.7), (3.2.1). The second one can be proved more easily by using (7.3.1), (3.2.13) and (3.2.23); we leave the details to the reader.

Therefore, if we denote by $G^1\otimes G^2$ another copy of $f^{-1}$ then from the definition

(8.5.14) of $S_D$, (11.5.5), (11.5.6) and the axioms of a quasi-Hopf algebra we obtain

$$
\begin{aligned}
&S_D(b^1)\otimes b^2\\
&= (\varepsilon\bowtie S(P^1_2x^2\tilde{p}^2_1))S_D(e^j\bowtie 1)(\varepsilon\bowtie S(e_i)x^3\tilde{p}^2_2)\\
&\quad\otimes(e^i\bowtie S^{-1}(P^2)e_jP^1_1x^1\tilde{p}^1)\\
&= (\varepsilon\bowtie S(P^1_2x^2\tilde{p}^2_1)f^1)(p^1_1U^1\rightharpoonup\overline{S}^{-1}(e^j)\leftharpoonup f^2S^{-1}(p^2)\\
&\quad\bowtie p^1_2U^2S(e_i)x^3\tilde{p}^2_2)\otimes(e^i\bowtie S^{-1}(P^2)e_jP^1_1x^1\tilde{p}^1)\\
&= (\varepsilon\bowtie S(P^1_2x^2\tilde{p}^2_1)f^1)(\overline{S}^{-1}(e^j)\bowtie p^1_2U^2S(e_i)x^3\tilde{p}^2_2)\\
&\quad\otimes(e^i\bowtie S^{-1}(p^1_1U^1P^2)e_jS^{-1}(f^2S^{-1}(p^2))P^1_1x^1\tilde{p}^1)\\
&\overset{(8.5.3)}{=} (\overline{S}^{-1}(e^j)\bowtie S(P^1_2x^2\tilde{p}^2_1)_{(1,2)}f^1_{(1,2)}p^1_2U^2S(e_i)x^3\tilde{p}^2_2)\otimes(e^i\bowtie S^{-1}(U^1P^2)\\
&\quad S^{-1}(S(P^1_2x^2\tilde{p}^2_1)_{(1,1)}f^1_{(1,1)}p^1_1)e_jS^{-1}(f^2S^{-1}(S(P^1_2x^2\tilde{p}^2_1)_2f^1_2p^2))P^1_1x^1\tilde{p}^1)\\
&\overset{(11.5.5)}{=} (\overline{S}^{-1}(e^j)\bowtie g^1_2S(P^1y^3x^2_2\tilde{p}^2_{(1,2)})_2U^2S(e_i)x^3\tilde{p}^2_2)\otimes(e^i\bowtie S^{-1}(U^1P^2)\\
&\quad S^{-1}(g^1_1S(P^1y^3x^2_2\tilde{p}^2_{(1,2)})_1)e_jS^{-2}(g^2S(S(y^1x^1\tilde{p}^1)\alpha y^2x^2_1\tilde{p}^2_{(1,1)})\\
&\overset{(10.3.3),(11.5.6)}{\underset{(3.2.20)}{=}} (\overline{S}^{-1}(e^j)\bowtie g^1_2S(\tilde{q}^2X^2_{(2,2)}\tilde{p}^2_2)_2G^2S(e_i)X^3)\\
&\quad\otimes(e^i\bowtie S^{-1}(g^1_1S(\tilde{q}^2X^2_{(2,2)}\tilde{p}^2_2)_1G^1)e_jS^{-2}(g^2S(X^1S(X^2_1\tilde{p}^1)\tilde{q}^1X^2_{(2,1)}\tilde{p}^2_1))\\
&\overset{(3.2.22),(3.2.24)}{=} (\overline{S}^{-1}(e^j)\bowtie g^1_2S(X^2)_2G^2S(e_i)X^3)\\
&\quad\otimes(e^i\bowtie S^{-1}(g^1_1S(X^2)_1G^1)e_jS^{-2}(g^2S(X^1))\\
&\overset{(3.2.13)}{=} (\overline{S}^{-1}({}^je)\bowtie S({}_ie))\otimes(X^2_1S^{-1}(g^1_2G^2)\rightharpoonup{}^ie\leftharpoonup S^{-1}(X^3)\\
&\quad\bowtie X^2_2S^{-1}(g^1_1G^1)_jeS^{-2}(g^2S(X^1))).
\end{aligned}
$$

We prove that $\overline{\mathscr{Q}}$ is injective. Let $\mathbf{D}\in(D(H))^*$ be such that $\overline{\mathscr{Q}}(\mathbf{D}\circ S_D)=0$. This means $\mathbf{D}(S_D(b^1))b^2=0$, which is equivalent to

$$
\begin{aligned}
\mathbf{D}(\overline{S}^{-1}({}^je)\bowtie S({}_ie))\langle{}^ie,S^{-1}(X^3)hX^2_1S^{-1}(g^1_2G^2)\rangle&\\
\langle\chi,X^2_2S^{-1}(g^1_1G^1)_jeS^{-2}(g^2S(X^1))\rangle&=0,
\end{aligned}
$$

for all $h\in H$ and $\chi\in H^*$. In particular,

$$
\begin{aligned}
&\mathbf{D}(\overline{S}^{-1}({}^je)\bowtie S({}_ie))\langle{}^ie,S^{-1}(X^3)(S^{-1}(x^3)hS^{-1}(F^2f^1_2)x^2_1)X^2_1S^{-1}(g^1_2G^2)\rangle\\
&\quad\langle S^{-2}(S(x^1)f^2)\rightharpoonup\chi\leftharpoonup S^{-1}(F^1f^1_1)x^2_2,X^2_2S^{-1}(g^1_1G^1)_jeS^{-2}(g^2S(X^1))\rangle=0,
\end{aligned}
$$

for all $h\in H$ and $\chi\in H^*$, and therefore

$$\mathbf{D}(\overline{S}^{-1}(\chi)\bowtie S(h))=0,\ \ \forall\,\chi\in H^*\ \text{ and }\ h\in H.$$

Since the antipode $S$ is bijective ($H$ is finite dimensional) we conclude that $\mathbf{D}=0$ and by using the bijectivity of $S_D$ it follows that $\overline{\mathscr{Q}}$ is injective. Finally, $\overline{\mathscr{Q}}$ is bijective because $D(H)$ is finite dimensional, so the proof is finished. □

We would like to stress that formula (11.5.1) was chosen in such a way that it provides a left $H$-module morphism from $\underline{H^*}$, the function algebra braided group associated to $H^*$, to $\underline{H}$, the enveloping braided group $\underline{H}$ of $(H,R)$. Indeed, for all $\chi \in H^*$ and $h \in H$ we have:

$$\begin{aligned}
h \triangleright \mathscr{Q}(\chi) &\overset{(11.5.1)}{=} \langle \chi, S(X_2^2\tilde{p}^2)f^1R^2r^1U^1X^3\rangle h_1X^1S(X_1^2\tilde{p}^1)f^2R^1r^2U^2S(h_2) \\
&\underset{(3.2.13)}{\overset{(7.3.2),(10.1.3)}{=}} \langle \chi, S((h_{(2,1)}X^2)_2\tilde{p}^2)f^1R^2r^1U^1h_{(2,2)}X^3\rangle \\
&\qquad h_1X^1S((h_{(2,1)}X^2)_1\tilde{p}^1)f^2R^1r^2U^2 \\
&\overset{(3.1.7),(3.2.22)}{=} \langle \chi, S(X_2^2\tilde{p}^2h_1)f^1R^2r^1U^1X^3h_2\rangle X^1S(X_1^2\tilde{p}^1)f^2R^1r^2U^2 \\
&\overset{(11.5.1),(11.4.30)}{=} \mathscr{Q}(h_2 \rightharpoonup \chi \leftharpoonup S(h_1)) = \mathscr{Q}(h \blacktriangleright \chi).
\end{aligned}$$

It is quite remarkable that (11.5.1) is a braided Hopf algebra morphism, too.

**Proposition 11.28** *Let $(H,R)$ be a finite-dimensional QT quasi-Hopf algebra, $\underline{H}$ the associated enveloping algebra braided group of $H$ and $\underline{H^*}$ the function algebra braided group associated to $H^*$. Then the map $\mathscr{Q}$ defined by* (11.5.1) *is a braided Hopf algebra morphism in ${}_H\mathscr{M}$ from $\underline{H^*}$ to $\underline{H}$.*

*Proof* We have already seen that $\mathscr{Q}$ is a morphism in ${}_H\mathscr{M}$. Hence, it remains to show that $\mathscr{Q}$ is an algebra and a coalgebra morphism. To this end, we will use the second formula (11.5.2) for the map $\mathscr{Q}$. We set $R = R^1 \otimes R^2 = r^1 \otimes r^2 = \mathbf{R}^1 \otimes \mathbf{R}^2 = \mathfrak{R}^1 \otimes \mathfrak{R}^2 = \mathfrak{r}^1 \otimes \mathfrak{r}^2 = \mathscr{R}^1 \otimes \mathscr{R}^2$, $q_L = \tilde{q}^1 \otimes \tilde{q}^2 = \tilde{q}^1 \otimes \tilde{q}^2$ and $p_R = p^1 \otimes p^2 = P^1 \otimes P^2$. Now, for all $\chi, \psi \in H^*$ we compute:

$$\begin{aligned}
&\mathscr{Q}(\chi\underline{:}\psi) \\
&= \langle \chi, S(x^1X^1)f^2R^1\tilde{q}_1^1Z_1^1\mathfrak{R}_1^2\mathfrak{r}_1^1p_1^1x_1^3Y^2r^1y^1X^2\rangle \\
&\qquad \langle \psi, S(x^2Y^1r^2y^2X_1^3)f^1R^2\tilde{q}_2^1Z_2^1\mathfrak{R}_2^2\mathfrak{r}_2^1p_2^1x_2^3Y^3y^3X_2^3\rangle \tilde{q}_1^2Z^2\mathfrak{R}^1\mathfrak{r}^2p^2S(\tilde{q}_2^2Z^3) \\
&\overset{(10.1.3),(5.5.16)}{=} \langle \chi, S(x^1X^1)f^2\tilde{q}_2^1[Z^1\mathfrak{R}^2\mathfrak{r}^1x_{(1,1)}^3p^1]_2R^1Y^2r^1y^1X^2\rangle \\
&\qquad \langle \psi, S(x^2Y^1r^2y^2X_1^3)f^1\tilde{q}_1^1[Z^1\mathfrak{R}^2\mathfrak{r}^1x_{(1,1)}^3p^1]_1R^2Y^3y^3X_2^3\rangle \\
&\qquad \tilde{q}_1^2Z^2\mathfrak{R}^1\mathfrak{r}^2x_{(1,2)}^3p^2S(\tilde{q}_2^2Z^3x_2^3) \\
&\underset{(3.2.28)}{\overset{(10.1.3),(3.1.7)}{=}} \langle \chi, S(X^1)\tilde{q}^1\widetilde{Q}_1^2T^2Z_2^1\mathfrak{R}_2^2\mathfrak{r}_2^1p_2^1R^1Y^2r^1y^1X^2\rangle \\
&\qquad \langle \psi, S(Y^1r^2y^2X_1^3)\tilde{q}^1T^1Z_1^1\mathfrak{R}_1^2\mathfrak{r}_1^1p_1^1R^2Y^3y^3X_2^3\rangle \\
&\qquad \tilde{q}_1^2\widetilde{Q}_{(2,1)}^2T_1^3Z^2\mathfrak{R}^1\mathfrak{r}^2p^2S(\tilde{q}_2^2\widetilde{Q}_{(2,2)}T_2^3Z^3) \\
&\underset{(5.2.7),(10.1.3)}{\overset{(3.1.9),(3.1.7)}{=}} \langle \chi, S(X^1)\tilde{q}^1V^1\widetilde{Q}_1^2x_{(2,1)}^2Z^2\mathfrak{R}_2^2R^1\mathfrak{r}_1^1p_1^1Y^2r^1y^1X^2\rangle \\
&\qquad \langle \psi, S(x^1Y^1r^2y^2X_1^3)\tilde{q}^1x_1^2Z^1\mathfrak{R}_1^2R^2\mathfrak{r}_2^1p_2^1Y^3y^3X_2^3\rangle \\
&\qquad \tilde{q}_1^2V^2\widetilde{Q}_2^2x_{(2,2)}^2Z^3\mathfrak{R}^1\mathfrak{r}^2p^2S(\tilde{q}_2^2V^3x^3) \\
&\underset{(5.5.16)}{\overset{(3.1.7),(10.1.3)}{=}} \langle \chi, S(X^1)\tilde{q}^1V^1\widetilde{Q}_1^2Z^2\mathfrak{R}_2^2R^1\mathfrak{r}_1^1T_1^2Y^2r^1y^1p_1^1X^2\rangle \\
&\qquad \langle \psi, S(T^1Y^1r^2y^2(p_2^1X^3)_1)\tilde{q}^1Z^1\mathfrak{R}_1^2R^2\mathfrak{r}_2^1T_2^2Y^3y^3(p_2^1X^3)_2\rangle \\
&\qquad \tilde{q}_1^2V^2\widetilde{Q}_2^2Z^3\mathfrak{R}^1\mathfrak{r}^2T^3p^2S(\tilde{q}_2^2V^3) \\
&\underset{(10.1.1),(10.1.2)}{\overset{(3.1.9),(10.1.3)}{=}} \langle \chi, S(X^1)\tilde{q}^1V^1\widetilde{Q}_1^2\mathfrak{R}^2Z^3x^3R^1W^2\mathfrak{r}^1z^1T^2r^1Y_1^1y^1p_1^1X^2\rangle \\
&\qquad \langle \psi, S(T^1r^2Y_2^1y^2(p_2^1X^3)_1)\tilde{q}^1Z^1\mathbf{R}^2x^2R^2W^3z^3\mathscr{R}^1T_1^3Y^2y^3(p_2^1X^3)_2\rangle
\end{aligned}$$

$$
\begin{aligned}
&\tilde{q}^2_1V^2\widetilde{Q}^2_2\mathfrak{R}^1Z^2\mathbf{R}^1x^1W^1\mathfrak{r}^2z^2\mathscr{R}^2T^3_2Y^3p^2S(\tilde{q}^2_2V^3)\\
\underset{(10.1.2)}{\overset{(3.1.9),(10.1.3)}{=}}\ &\langle\chi,S(X^1)\tilde{q}^1V^1\widetilde{Q}^2_1\mathfrak{R}^2Z^3x^3W^3_2R^1T^2r^1D^1z^1_1Y^1_1y^1p^1_1X^2\rangle\\
&\langle\psi,S(W^1T^1_1r^2_1D^2z^1_2Y^1_2y^2(p^1_2X^3)_1)\tilde{q}^1Z^1\mathbf{R}^2x^2W^3_1R^2T^3z^3\mathscr{R}^1\\
&Y^2y^3(p^1_2X^3)_2\rangle\tilde{q}^2_1V^2\widetilde{Q}^2_2\mathfrak{R}^1Z^2\mathbf{R}^1x^1W^2T^1_2r^2_2D^3z^2\mathscr{R}^2Y^3p^2S(\tilde{q}^2_2V^3)\\
\underset{(3.1.7)}{\overset{(10.1.2),(3.1.9)}{=}}\ &\langle\chi,S(X^1)\tilde{q}^1V^1\widetilde{Q}^2_1\mathfrak{R}^2Z^3T^3r^1Y^1p^1_1X^2\rangle\\
&\langle\psi,S(T^1r^2_1C^1Y^2_1(p^1_2X^3)_1)\tilde{q}^1Z^1\mathbf{R}^2T^2_2r^2_{(2,2)}\mathscr{R}^1C^2Y^2_2(p^1_2X^3)_2\rangle\\
&\tilde{q}^2_1V^2\widetilde{Q}^2_2\mathfrak{R}^1Z^2\mathbf{R}^1T^2_1r^2_{(2,1)}\mathscr{R}^2C^3Y^3p^2S(\tilde{q}^2_2V^3)\\
\underset{(3.1.7),(3.2.1)}{\overset{(5.2.7),(10.1.3)}{=}}\ &\langle\chi,S(y^1X^1)\tilde{q}^1\mathfrak{R}^2r^1y^2_1Y^1p^1_1X^2\rangle\\
&\langle\psi,S(x^1C^1(Y^2p^1_2)_1X^3_1)\alpha x^2\mathbf{R}^2\mathscr{R}^1C^2(Y^2p^1_2)_2X^3_2\rangle\\
&\tilde{q}^2\mathfrak{R}^1r^2y^2_2x^3\mathbf{R}^1\mathscr{R}^2C^3Y^3p^2S(y^3)\\
\underset{(3.1.7),(3.2.1)}{\overset{(5.5.16),(10.1.3)}{=}}\ &\langle\chi,S(y^1X^1)\tilde{q}^1\mathfrak{R}^2r^1y^2_1z^1X^2\rangle\\
&\langle\psi,S(C^1p^1_1X^3_1)\tilde{q}^1\mathbf{R}^2\mathscr{R}^1C^2p^1_2X^3_2\rangle\tilde{q}^2\mathfrak{R}^1r^2y^2_2z^2\tilde{q}^2\mathbf{R}^1\mathscr{R}^2C^3p^2S(y^3z^3).
\end{aligned}
$$

On the other hand, if we denote by $P^1\otimes P^2$ another copy of $p_R$ then by (4.1.9), (3.2.19), (11.5.2) we have:

$$
\begin{aligned}
&\mathscr{Q}(\chi)\circ\mathscr{Q}(\psi)\\
=\ &\langle\chi,\tilde{q}^1Y^1R^2r^1P^1\rangle\langle\psi,\tilde{q}^1Z^1\mathfrak{R}^2\mathfrak{r}^1p^1\rangle\\
&q^1y^1_1\tilde{q}^2_1Y^2R^1r^2P^2S(q^2y^1_2\tilde{q}^2_2Y^3)y^2\widetilde{Q}^2_1Z^2\mathfrak{R}^1\mathfrak{r}^2p^2S(y^3\widetilde{Q}^2_2Z^3)\\
\underset{(5.5.16)}{\overset{(5.2.7),(10.1.3)}{=}}\ &\langle\chi,S(X^1P^1_1)\tilde{q}^1R^2r^1X^2P^1_2\rangle\langle\psi,\tilde{q}^1Z^1\mathfrak{R}^2\mathfrak{r}^1p^1\rangle\\
&q^1y^1_1\tilde{q}^2R^1r^2X^3P^2S(q^2y^1_2)y^2\widetilde{Q}^2_1Z^2\mathfrak{R}^1\mathfrak{r}^2p^2S(y^3\widetilde{Q}^2_2Z^3)\\
\underset{(3.1.7),(3.2.21)}{\overset{(3.2.22),(10.1.3)}{=}}\ &\langle\chi,S(X^1(q^1_1P^1)_1y^1_1)\tilde{q}^1R^2r^1X^2(q^1_1P^1)_2y^1_2\rangle\\
&\langle\psi,\tilde{q}^1Z^1\mathfrak{R}^2\mathfrak{r}^1p^1\rangle\tilde{q}^2R^1r^2X^3q^1_2P^2S(q^2)y^2\widetilde{Q}^2_1Z^2\\
&\mathfrak{R}^1\mathfrak{r}^2p^2S(y^3\widetilde{Q}^2_2Z^3)\\
\underset{(10.1.3),(5.5.16)}{\overset{(3.2.23),(5.2.7),(3.1.9)}{=}}\ &\langle\chi,S(y^1X^1)\tilde{q}^1R^2r^1y^2_1x^1X^2\rangle\langle\psi,S(Y^1p^1_1)\tilde{q}^1\mathfrak{R}^2\mathfrak{r}^1Y^2p^1_2\rangle\\
&\tilde{q}^2R^1r^2y^2_2x^2X^3_1\tilde{q}^2\mathfrak{R}^1\mathfrak{r}^2Y^3p^2S(y^3x^3X^3_2)\\
\underset{(3.1.7),(3.2.21)}{\overset{(3.2.22),(10.1.3)}{=}}\ &\langle\chi,S(y^1X^1)\tilde{q}^1R^2r^1y^2_1x^1X^2\rangle\\
&\langle\psi,S(Y^1p^1_1X^3_1)\tilde{q}^1\mathfrak{R}^2\mathfrak{r}^1Y^2p^1_2X^3_2\rangle\\
&\tilde{q}^2R^1r^2y^2_2x^2\tilde{q}^2\mathfrak{R}^1\mathfrak{r}^2Y^3p^2S(y^3x^3).
\end{aligned}
$$

By the above it follows that $\mathscr{Q}$ is multiplicative. Since $\mathscr{Q}(1_{\underline{H}^*})=\mathscr{Q}(\varepsilon)=\beta=1_{\underline{H}}$, we conclude that $\mathscr{Q}$ is an algebra map. Thus, one has only to show that $\mathscr{Q}$ is a coalgebra map. To this end, observe first that (3.1.9), (3.2.1) imply

$$X^1_1p^1\otimes X^1_2p^2S(X^2)\otimes X^3=x^1\otimes x^2S(x^3_1\tilde{p}^1)\otimes x^3_2\tilde{p}^2. \tag{11.5.7}$$

Also, it is not hard to see that (11.2.3), (8.7.1), (10.1.1), (10.1.3) and (10.3.6) imply

$$\underline{\Delta}_{\underline{H}}(h)=x^1X^1h_1r^2g^2S(x^2Y^1R^2y^2X^3_1)\otimes x^3_1Y^2R^1y^1X^2h_2r^1g^1S(x^3_2Y^3y^3X^3_2). \tag{11.5.8}$$

Therefore, by (11.5.8) and (11.5.2), for any $\chi \in H^*$ we have

$$
\begin{aligned}
&\underline{\Delta_H}(\mathscr{Q}(\chi))\\
&= \langle \chi, \tilde{q}^1 Z^1 \mathfrak{R}^2 \mathfrak{r}^1 p^1 \rangle x^1 X^1 \tilde{q}^2_{(1,1)} Z^2_1 \mathfrak{R}^1_1 \mathfrak{r}^2_1 p^2_1 S(\tilde{q}^2_2 Z^3)_1 r^2 g^2 S(x^2 Y^1 R^2 y^2 X^3_1)\\
&\quad \otimes x^3_1 Y^2 R^1 y^1 X^2 \tilde{q}^2_{(1,2)} Z^2_2 \mathfrak{R}^1_2 \mathfrak{r}^2_2 p^2_2 S(\tilde{q}^2_2 Z^3)_2 r^1 g^1 S(x^3_2 Y^3 y^3 X^3_2)\\
&\overset{(10.1.3),(3.2.13)}{\underset{(3.2.25),(3.1.7)}{=}} \langle \chi, \tilde{q}^1 Z^1 \mathfrak{R}^2 \mathfrak{r}^1 V^1 (T^1_1 p^1)_1 P^1 \rangle\\
&\quad x^1 X^1 (\tilde{q}^2_1 Z^2)_1 \mathfrak{R}^1_1 r^2 \mathfrak{r}^2_2 V^3 T^1_2 p^2 S(x^2 Y^1 R^2 y^2 (X^3 \tilde{q}^2_2)_1 Z^3_1 T^2)\\
&\quad \otimes x^3_1 Y^2 R^1 y^1 X^2 (\tilde{q}^2_1 Z^2)_2 \mathfrak{R}^1_2 r^1 \mathfrak{r}^2_1 V^2 (T^1_1 p^1)_2 P^2 S(x^3_2 Y^3 y^3 (X^3 \tilde{q}^2_2)_2 Z^3_2 T^3)\\
&\overset{(11.5.7),(10.1.2)}{\underset{(10.1.1),(5.2.7)}{=}} \langle \chi, S(v^1) \tilde{q}^1 v^2_1 \mathfrak{R}^2 t^3 \mathfrak{r}^1_2 \mathbf{R}^1 z^1_1 P^1 \rangle\\
&\quad x^1 X^1 \tilde{q}^2_1 v^2_{(2,1)} \mathfrak{R}^1_1 t^1 \mathfrak{r}^2 z^2 S(x^2 Y^1 R^2 y^2 X^3_1 v^3_1 z^3_1 \tilde{p}^1)\\
&\quad \otimes x^3_1 Y^2 R^1 y^1 X^2 \tilde{q}^2_2 v^2_{(2,2)} \mathfrak{R}^1_2 t^2 \mathfrak{r}^1_1 \mathbf{R}^2 z^1_2 P^2 S(x^3_2 Y^3 y^3 X^3_2 v^3_2 z^3_2 \tilde{p}^2)\\
&\overset{(10.1.1),(10.1.3)}{\underset{(10.1.1)}{=}} \langle \chi, S(v^1) \tilde{q}^1 v^2_1 T^1 \mathfrak{R}^2 \mathfrak{r}^1 t^1 V^1 \mathscr{R}^2 \mathbf{R}^1 z^1_1 P^1 \rangle\\
&\quad x^1 X^1 \tilde{q}^2_1 v^2_{(2,1)} T^2 \mathfrak{R}^1 \mathfrak{r}^2 t^2 r^2 V^3 z^2 S(x^2 Y^1 R^2 y^2 X^3_1 v^3_1 z^3_1 \tilde{p}^1)\\
&\quad \otimes x^3_1 Y^2 R^1 y^1 X^2 \tilde{q}^2_2 v^2_{(2,2)} T^3 t^3 r^1 V^2 \mathscr{R}^1 \mathbf{R}^2 z^1_2 P^2 S(x^3_2 Y^3 y^3 X^3_2 v^3_2 z^3_2 \tilde{p}^2)\\
&\overset{(3.1.7),(5.2.7)}{\underset{(3.2.22),(3.1.9)}{=}} \langle \chi, S(v^1) \tilde{q}^1 v^2_1 X^1_1 \mathfrak{R}^2 \mathfrak{r}^1 t^1 V^1 \mathscr{R}^2 \mathbf{R}^1 z^1_1 P^1 \rangle\\
&\quad x^1 \tilde{q}^2 v^2_2 X^1_2 \mathfrak{R}^1 \mathfrak{r}^2 t^2 r^2 V^3 z^2 S(x^2 Y^1 R^2 y^2 v^3_{(2,1)} X^3_1 z^3_1 \tilde{p}^1)\\
&\quad \otimes x^3_1 Y^2 R^1 y^1 v^3_1 X^2 t^3 r^1 V^2 \mathscr{R}^1 \mathbf{R}^2 z^1_2 P^2 S(x^3_2 Y^3 y^3 v^3_{(2,2)} X^3_2 z^3_2 \tilde{p}^2)\\
&\overset{(10.1.3),(3.1.9)}{\underset{(3.2.22)}{=}} \langle \chi, S(v^1) \tilde{q}^1 v^2_1 \mathfrak{R}^2 \mathfrak{r}^1 t^1 Z^1 \mathscr{R}^2 \mathbf{R}^1 P^1 \rangle\\
&\quad x^1 \tilde{q}^2 v^2_2 \mathfrak{R}^1 \mathfrak{r}^2 t^2 X^1 r^2 z^2 S(x^2 Y^1 R^2 y^2 (v^3 t^3)_{(2,1)} X^3_1 z^3_1 \tilde{p}^1)\\
&\quad \otimes x^3_1 Y^2 R^1 y^1 (v^3 t^3)_1 X^2 r^1 z^1 Z^2 \mathscr{R}^1 \mathbf{R}^2 P^2 S(x^3_2 Y^3 y^3 (v^3 t^3)_{(2,2)} X^3_2 z^3_2 \tilde{p}^2 Z^3)\\
&\overset{(3.1.7),(10.1.3)}{\underset{(3.1.9),(10.1.2)}{=}} \langle \chi, S(v^1) \tilde{q}^1 v^2_1 \mathfrak{R}^2 \mathfrak{r}^1 t^1 Z^1 \mathscr{R}^2 \mathbf{R}^1 P^1 \rangle x^1 \tilde{q}^2 v^2_2 \mathfrak{R}^1 \mathfrak{r}^2 t^2 T^1 X^1 R^2_1 V^2 y^1_2 z^2\\
&\quad S(x^2 (v^3 t^3)_1 Y^1 T^2_1 X^2 R^2_2 V^3 y^2 z^3_1 \tilde{p}^1) \otimes x^3_1 (v^3 t^3)_{(2,1)}\\
&\quad Y^2 T^2_2 X^3 R^1 V^1 y^1_1 z^1 Z^2 \mathscr{R}^1 \mathbf{R}^2 P^2 S(x^3_2 (v^3 t^3)_{(2,2)} Y^3 T^3 y^3 z^3_2 \tilde{p}^2 Z^3)\\
&\overset{(3.1.9),(3.2.20)}{\underset{(3.2.1)}{=}} \langle \chi, S(v^1) \tilde{q}^1 v^2_1 \mathfrak{R}^2 \mathfrak{r}^1 t^1 Z^1 \mathscr{R}^2 \mathbf{R}^1 P^1 \rangle x^1 \tilde{q}^2 v^2_2 \mathfrak{R}^1 \mathfrak{r}^2 t^2 X^1 \beta S(x^2 (v^3 t^3)_1 X^2)\\
&\quad \otimes x^3_1 (v^3 t^3)_{(2,1)} X^3_1 Z^2 \mathscr{R}^1 \mathbf{R}^2 P^2 S(x^3_2 (v^3 t^3)_{(2,2)} X^3_2 Z^3)\\
&\overset{(3.1.9),(3.2.1)}{\underset{(3.2.22),(10.1.3)}{=}} \langle \chi, S(t^1 x^1) \tilde{q}^1 \mathfrak{R}^2 \mathfrak{r}^1 t^2_1 x^2_{(1,1)} z^1 Z^1 \mathscr{R}^2 \mathbf{R}^1 P^1 \rangle \tilde{q}^2 \mathfrak{R}^1 \mathfrak{r}^2 t^2_2 x^2_{(1,2)} z^2 \beta S(t^3 x^2_2 z^3)\\
&\quad \otimes x^3_1 Z^2 \mathscr{R}^1 \mathbf{R}^2 P^2 S(x^3_2 Z^3)\\
&\overset{(3.1.7),(3.2.1)}{\underset{(10.1.3),(11.5.2)}{=}} \mathscr{Q}(\chi_1 \leftharpoonup S(x^1)) \otimes \chi_2(x^2 Z^1 \mathscr{R}^2 \mathbf{R}^1 P^1) x^3_1 Z^2 \mathscr{R}^1 \mathbf{R}^2 P^2 S(x^3_2 Z^3).
\end{aligned}
$$

On the other hand, by (11.4.33) we have

$$
\begin{aligned}
&(\mathscr{Q} \otimes \mathscr{Q})(\underline{\Delta_{H^*}}(\chi))\\
&= \mathscr{Q}(\chi_1 \leftharpoonup S(x^1)) \otimes \mathscr{Q}(x^3_2 X^3 \rightharpoonup \chi_2 \leftharpoonup x^2 X^1 \beta S(x^3_1 X^2))\\
&\overset{(11.5.2),(3.2.21)}{\underset{(10.1.3)}{=}} \mathscr{Q}(\chi_1 \leftharpoonup S(x^1)) \otimes \langle \chi_2, x^2 X^1 \beta S(x^3_1 X^2) \tilde{q}^1 Z^1 (x^3_2 X^3)_{(1,1)} \mathscr{R}^2 \mathbf{R}^1 P^1 \rangle\\
&\quad \widetilde{Q}^2_1 Z^2 (x^3_2 X^3)_{(1,2)} \mathscr{R}^1 \mathbf{R}^2 P^2 S(\widetilde{Q}^2_2 Z^3 (x^3_2 X^3)_2)\\
&\overset{(3.1.7),(3.2.22)}{\underset{(3.2.20)}{=}} \mathscr{Q}(\chi_1 \leftharpoonup S(x^1)) \otimes \langle \chi_2, x^2 S(\tilde{p}^1) \tilde{q}^1 \tilde{p}^2_1 Z^1 \mathscr{R}^2 \mathbf{R}^1 P^1 \rangle
\end{aligned}
$$

$$
\begin{aligned}
& x_1^3(\widetilde{Q}^2\tilde{p}_2^2)_1Z^2\mathscr{R}^1\mathbf{R}^2P^2S(x_2^3(\widetilde{Q}^2\tilde{p}_2^2)_2Z^3)\\
\overset{(3.2.24)}{=} & \mathscr{Q}(\chi_1 \leftharpoonup S(x^1))\otimes\langle\chi_2, x^2Z^1\mathscr{R}^2\mathbf{R}^1P^1\rangle x_1^3Z^2\mathscr{R}^1\mathbf{R}^2P^2S(x_2^3Z^3).
\end{aligned}
$$

So $\mathscr{Q}$ is a coalgebra map since $(\underline{\varepsilon_H}\circ\mathscr{Q})(\chi)=\chi(\alpha)=\underline{\varepsilon_{H^*}}$. This ends the proof. □

Summarizing, we can now present the true meaning of the map $\mathscr{Q}: H^*\to H$ defined in (11.5.1). It is a morphism of braided groups from $\underline{H^*}$, the function algebra braided group associated to $H^*$, to $\underline{H}$, the associated enveloping algebra braided group of $H$. When $H$ is factorizable in the sense that the map $\mathscr{Q}$ is bijective then $\mathscr{Q}: \underline{H^*}\cong\underline{H}$ as braided Hopf algebras. In other words, the function algebra braided group associated to $H^*$ and the associated enveloping algebra braided group of $H$ are categorical self dual, cf. Proposition 11.21.

## 11.6 Factorizable Implies Unimodular

We show that any finite-dimensional factorizable QT quasi-Hopf algebra is unimodular. In particular, we obtain that for any finite-dimensional quasi-Hopf algebra $H$ its quantum double $D(H)$ is always a unimodular quasi-Hopf algebra.

Throughout this section, $H$ is a finite-dimensional quasi-Hopf algebra, $t\in H$ is a non-zero left integral in $H$ and $\mu$ is the modular element of $H^*$; see (7.2.6). We also consider $\lambda$ a non-zero left cointegral on $H$, $r$ a non-zero right integral in $H$ such that $\lambda(r)=1$ and $\underline{g}$ the modular element of $H$ with inverse $\underline{g}^{-1}$ as in (7.6.2).

**Remark 11.29** Let $H$ be a finite-dimensional quasi-Hopf algebra, $t$ a non-zero left integral in $H$ and $\mu$ the modular element of $H^*$. Then the relation (7.5.16) can be rewritten in the form

$$t_1p^1\otimes t_2p^2S(h\leftharpoonup\mu)=t_1p^1h\otimes t_2p^2, \tag{11.6.1}$$

for all $h\in H$, where for all $\chi\in H^*$ we denote $h\leftharpoonup\chi=\chi(h_1)h_2$.

We can now prove the main result of this section.

**Theorem 11.30** *Let $(H,R)$ be a finite-dimensional QT quasi-Hopf algebra and $\mu$ the modular element of $H^*$. Then the following assertions hold:*

*(i) If $q_R=q^1\otimes q^2=Q^1\otimes Q^2$ and $p_R=p^1\otimes p^2=P^1\otimes P^2$ are the elements defined by* (3.2.19) *then*

$$\mu(Q^1)q^2t_2p^2S(Q^2(R^2P^2\leftharpoonup\mu))R^1P^1\otimes q^1t_1p^1=S(u)q^1t_1p^1\otimes q^2t_2p^2, \tag{11.6.2}$$

*where $R=R^1\otimes R^2$ is the R-matrix of $H$ and $u$ is the element defined in* (10.3.4).

*(ii) If $(H,R)$ is factorizable then $H$ is unimodular.*

*Proof* (i) Note that $g^1S(g^2\alpha)=\beta$, (10.3.6), (10.3.8) and (10.3.10) imply

$$R^1\beta S(R^2)=S(\beta u). \tag{11.6.3}$$

Now, from (11.6.1) we have

$$\begin{aligned}
&\mu(Q^1)q^2t_2p^2S(Q^2(R^2P^2\leftharpoonup\mu))R^1P^1\otimes q^1t_1p^1\\
&\quad = \mu(Q^1)q^2t_2p^2S(Q^2)R^1P^1\otimes q^1t_1p^1R^2P^2\\
&\quad \overset{(7.2.6)}{=} q^2t_2Q^1_2p^2S(Q^2)R^1P^1\otimes q^1t_1Q^1_1p^1R^2P^2\\
&\quad \overset{(3.2.23)}{=} q^2t_2R^1P^1\otimes q^1t_1R^2P^2\\
&\quad \overset{(10.1.3)}{=} q^2R^1t_1P^1\otimes q^1R^2t_2P^2\\
&\quad \overset{(7.2.3)}{=} q^2R^1\beta Q^1t_1P^1\otimes q^1R^2Q^2t_2P^2\\
&\quad \overset{(7.2.2)}{=} q^2R^1\beta S(q^1R^2)Q^1t_1P^1\otimes Q^2t_2P^2\\
&\quad \overset{(11.6.3)}{=} S(q^1\beta uS^{-1}(q^2))Q^1t_1P^1\otimes Q^2t_2P^2\\
&\quad \overset{(10.3.10),(3.2.19),(3.2.2)}{=} S(u)Q^1t_1P^1\otimes Q^2t_2P^2,
\end{aligned}$$

and this proves the first assertion.

(ii) Let $(\lambda,t)\in\mathscr{L}\times\int_l^H$ be as in Proposition 7.59, that is, $\lambda$ is a non-zero left cointegral on $H$ and $t$ is a non-zero left integral in $H$ such that $\lambda(S^{-1}(t))=1$ and $\lambda(q^2t_2p^2)q^1t_1p^1=1_H$. So the definition (7.6.2) applies.

By applying $\mathrm{Id}_H\otimes\lambda$ to the equality (11.6.2) we obtain

$$\mu(Q^1)S^{-1}(\underline{g}^{-1})S(Q^2(R^2P^2\leftharpoonup\mu))R^1P^1=S(u),$$

and since $S^{-1}(\underline{g})S(u)=S(uS^{-2}(\underline{g}))=S(\underline{g}u)$, it follows that the above relation is equivalent to

$$\mu(Q^1)S(Q^2(R^2P^2\leftharpoonup\mu))R^1P^1=S(u)S(\underline{g}). \tag{11.6.4}$$

On the other hand, if we denote by $r^1\otimes r^2$ another copy of $R$, we have

$$\begin{aligned}
&\mu(Q^1)S(Q^2(R^2P^2\leftharpoonup\mu))R^1P^1\\
&\quad \overset{(3.2.19)}{=} \mu(X^1R^2_1P^2_1)S(X^2R^2_2P^2_2)\alpha X^3R^1P^1\\
&\quad \overset{(10.1.2)}{=} \mu(X^1R^2y^2P^2_1)S(r^2X^3y^3P^2_2)\alpha r^1X^2R^1y^1P^1\\
&\quad \overset{(10.3.8),(10.3.10),(3.2.19)}{=} \mu(q^1R^2y^2P^2_1)S(S(q^2)y^3P^2_2)uR^1y^1P^1\\
&\quad \overset{(3.2.25),(10.1.3)}{=} \mu(q^1X^1_{(1,1)}p^1_1R^2P^2S(X^3)f^1)\\
&\qquad\qquad S(S(q^2)X^1_2p^2S(X^2)f^2)uX^1_{(1,2)}p^1_2R^1P^1\\
&\quad \overset{(10.3.10),(3.2.21)}{=} \mu(X^1q^1p^1_1R^2P^2S(X^3)f^1)S(S(q^2p^1_2)p^2S(X^2)f^2)uR^1P^1\\
&\quad \overset{(3.2.23),(10.3.10)}{=} \mu(X^1R^2P^2S(X^3)f^1)uS^{-1}(S(X^2)f^2)R^1P^1.
\end{aligned}$$

From the above computation and (11.6.4) we obtain

$$\mu(X^1R^2P^2S(X^3)f^1)S^{-1}(S(X^2)f^2)R^1P^1=u^{-1}S(u)S(\underline{g}). \tag{11.6.5}$$

But, as we know, if $(H,R)$ is QT then $\tilde{R}=R_{21}^{-1}=\overline{R}^2\otimes\overline{R}^1$ is another $R$-matrix for $H$. Repeating the above computations for $(H,\tilde{R})$ instead of $(H,R)$, we find

$$\mu(X^1\overline{r}^1P^2S(X^3)f^1)S^{-1}(S(X^2)f^2)\overline{r}^2P^1=\tilde{u}^{-1}S(\tilde{u})S(\underline{g}), \tag{11.6.6}$$

where we denote by $\tilde{u}$ the element defined as in (10.3.4) for $(H,\tilde{R})$ instead of $(H,R)$, and where $\bar{r}^1\otimes\bar{r}^2$ is another copy of $R^{-1}$. More precisely, we have that

$$\tilde{u}=S(u^{-1}). \tag{11.6.7}$$

Indeed, one can easily see that (11.6.3) and (10.3.10) imply

$$\bar{r}^2\beta S(\bar{r}^1)=S^{-1}(\beta)u^{-1}=u^{-1}S(\beta). \tag{11.6.8}$$

Now, we compute:

$$\begin{aligned}
\tilde{u} &= S(\bar{r}^1x^2\beta S(x^3))\alpha\bar{r}^2x^1\\
&\overset{(*)}{=} S(\beta f^1\bar{r}^1x^2\beta S(x^3))f^2\bar{r}^2x^1\\
&\overset{(10.3.8),(3.2.19)}{=} S(\bar{r}^2\beta S(\bar{r}^1)f^2p^2)f^1p^1\\
&\overset{(11.6.8)}{=} S(S^{-1}(f^1p^1)u^{-1}S(\beta)f^2p^2)\\
&\overset{(10.3.10),(*),(3.2.19),(3.2.2)}{=} S(u^{-1}S(p^1)\alpha p^2)=S(u^{-1}),
\end{aligned}$$

where (*) means that we use the relation $S(\beta f^1)f^2=\alpha$. Now, since $S^2(u)=u$ the relation (11.6.6) becomes

$$\mu(X^1\bar{r}^1P^2S(X^3)f^1)S^{-1}(S(X^2)f^2)\bar{r}^2P^1=S(u)u^{-1}S(\underline{g}). \tag{11.6.9}$$

From Corollary 10.18 we know that $uS(u)=S(u)u$, so $u^{-1}S(u)=S(u)u^{-1}$. Hence, by (11.6.5) and (11.6.9) we obtain

$$\mu(X^1R^2P^2S(X^3)f^1)S^{-1}(S(X^2)f^2)R^1P^1=\mu(X^1\bar{r}^1P^2S(X^3)f^1)S^{-1}(S(X^2)f^2)\bar{r}^2P^1.$$

This comes out explicitly as $\mu(R^2P^2)R^1P^1=\mu(\bar{r}^1P^2)\bar{r}^2P^1$, and implies

$$\mu(Q_1^1R^2P^2S(Q^2))Q_2^1R^1P^1=\mu(Q_1^1\bar{r}^1P^2S(Q^2))Q_2^1\bar{r}^2P^1.$$

From (10.1.3) and (3.2.23) we deduce that

$$\mu(R^2)R^1=\mu(\bar{r}^1)\bar{r}^2\Leftrightarrow\mu(R^2r^1)R^1r^2=1_H. \tag{11.6.10}$$

Finally, the above relation allows to compute:

$$\begin{aligned}
\mathscr{Q}(\mu) &= \mu(\tilde{q}^1X^1R^2r^1p^1)\tilde{q}_1^2X^2R^1r^2p^2S(\tilde{q}_2^2X^3)\\
&\overset{(11.6.10),(3.2.19)}{=} \mu(\tilde{q}^1X^1x^1)\tilde{q}_1^2X^2x^2\beta S(\tilde{q}_2^2X^3x^3)\\
&\overset{(3.2.1),(3.2.20)}{=} \mu(\alpha)\beta=\mathscr{Q}(\mu(\alpha)\varepsilon).
\end{aligned}$$

If $(H,R)$ is factorizable then $\mathscr{Q}$ is bijective, so $\mu=\mu(\alpha)\varepsilon$. In particular, $1=\mu(1_H)=\mu(\alpha)\varepsilon(1_H)=\mu(\alpha)$. Hence $\mu=\varepsilon$, and this means that $H$ is unimodular. □

**Theorem 11.31** *Let H be a finite-dimensional quasi-Hopf algebra. Then the quantum double D(H) of H is a unimodular quasi-Hopf algebra.*

*Proof* It is a consequence of Proposition 11.27 and Theorem 11.30. □

**Remark 11.32** Let $(H,R)$ be a finite-dimensional QT quasi-Hopf algebra. By the proof of Theorem 11.30 we can derive a nicer formula for the antipode $\underline{S}$ of the function algebra braided group associated to $H^*$; see (11.4.35). In fact, since $(H^{\rm cop},R^{-1})$ is a QT quasi-Hopf algebra as well, by (10.1.1) applied to $(H^{\rm cop},R^{-1})$ (which is actually equivalent to (10.1.1) for $(H,R)$) we have that

$$\begin{aligned}
&q^1_2\overline{R}^1_2\tilde{p}^2\otimes q^2\overline{R}^2S(q^1_1\overline{R}^1_1\tilde{p}^1)\\
&\quad= q^1_2x^2\overline{R}^1X^3y^2\tilde{p}^2\otimes q^2x^3\overline{R}^2X^2\overline{r}^2y^1S(q^1_1x^1X^1\overline{r}^1y^2\tilde{p}^1)\\
&\overset{(3.2.20),(5.5.17)}{=} q^1Y^2_1\overline{R}^1X^3\otimes S^{-1}(Y^3)q^2Y^2_2\overline{R}^2X^2\overline{r}^2\beta S(Y^1x^1X^1\overline{r}^1)\\
&\overset{(10.1.3),(11.6.8)}{=} q^1\overline{R}^1Y^2_2X^3\otimes S^{-1}(Y^3)q^2\overline{R}^2Y^2_1X^2u^{-1}S(Y^1X^1\beta)\\
&\overset{(10.3.10),(3.2.20)}{=} q^1\overline{R}^1Y^2_2\tilde{p}^2\otimes u^{-1}S(Y^1S(q^2\overline{R}^2Y^2_1\tilde{p}^1)Y^3)\\
&\overset{(10.3.3)}{=} q^1\overline{R}^1x^2\tilde{p}^2_1\otimes u^{-1}S(S(q^2\overline{R}^2x^1\tilde{p}^1)x^3\tilde{p}^2_2)\\
&\overset{(10.3.1)}{=} R^1\otimes u^{-1}S(R^2).
\end{aligned}$$

Thus, by (11.4.35) we get that $\underline{S}(\chi)=R^1\rightharpoonup\chi S\leftharpoonup u^{-1}S(R^2)$, for all $\chi\in H^*$.

## 11.7 The Quantum Double of a Factorizable Quasi-Hopf Algebra

Throughout this section, $(H,R)$ will be a finite-dimensional QT quasi-Hopf algebra and $D(H)$ its quantum double. Thus $D(H)$ is a biproduct quasi-Hopf algebra. The goal of this section is to show that when $H$ is, moreover, factorizable then $D(H)$ is nothing but a twist deformation of $H\otimes H$.

When there is no danger of confusion, the elements $\chi\bowtie h$ of $D(H)$ will be simply denoted by $\chi h$. Since $H^*$ can be viewed only as a $k$-linear subspace of $D(H)$, we will denote by

$$\begin{aligned}
\chi_{(1)}\otimes\chi_{(2)} &= \Delta_D(\chi\bowtie 1_H)\\
&\overset{(8.5.9)}{=} (\varepsilon\bowtie X^1Y^1)(p^1_1x^1\rightharpoonup\chi_2\leftharpoonup Y^2S^{-1}(p^2)\bowtie p^1_2x^2)\\
&\qquad\otimes X^2_1\rightharpoonup\chi_1\leftharpoonup S^{-1}(X^3)\bowtie X^2_2Y^3x^3.
\end{aligned}$$

In this notation, for all $\chi\in H^*$ and $h\in H$, the comultiplication $\Delta_D$ of $D(H)$ is

$$\Delta_D(\chi h)=\chi_{(1)}h_1\otimes\chi_{(2)}h_2.$$

By Lemma 10.26, there exists a quasi-Hopf algebra projection $\pi: D(H)\to H$ covering the canonical inclusion $i_D: H\to D(H)$; see (10.5.1). Recall that $\pi$ is a quasi-Hopf algebra morphism and $\pi\circ i_D=\mathrm{Id}_H$.

As we have seen before, $\tilde{R}:=R^{-1}_{21}=\overline{R}^2\otimes\overline{R}^1$ is another $R$-matrix for $H$. So there is always a second projection $\tilde{\pi}: D(H)\to H$ covering the canonical inclusion $i_D$. Explicitly, the morphism $\tilde{\pi}$ is given, for all $\chi\in H^*$ and $h\in H$, by

$$\tilde{\pi}(\chi\bowtie h)=\chi(q^2\overline{R}^2)q^1\overline{R}^1h. \tag{11.7.1}$$

For $H$ a quasi-Hopf algebra, $A$ a quasi-bialgebra and $\nu : H \to A$ a quasi-bialgebra morphism, we denote by

$$H^{\mathrm{co}(\nu)} = \{h \in H \mid h_1 \otimes \nu(h_2) = x^1 h S(x_2^3 X^3) f^1 \otimes \nu(x^2 X^1 \beta S(x_1^3 X^2) f^2)\}, \quad (11.7.2)$$

the set of alternative coinvariants of $H$ relative to its structure in ${}_H\mathscr{M}_H^H$ and the morphism $\nu$.

**Lemma 11.33** *Let $(H,R)$ be a finite-dimensional QT quasi-Hopf algebra, and $\pi$ and $\widetilde{\pi}$ the quasi-Hopf algebra morphisms defined by* (10.5.1) *and* (11.7.1), *respectively. Let $j : D(H)^{\mathrm{co}(\pi)} \to D(H)$ be the inclusion map and $\Psi : H^* \to D(H)^{\mathrm{co}(\pi)}$ defined by*

$$\Psi(\chi) = \chi_{(1)} \beta S(\pi(\chi_{(2)})) \quad (11.7.3)$$

*for all $\chi \in H^*$. Then the following assertions hold:*

*(1) $\Psi$ is well defined and bijective.*

*(2) If $\overline{\mathscr{Q}}$ is the map defined by* (11.5.3) *then $S \circ \overline{\mathscr{Q}} = \widetilde{\pi} \circ j \circ \Psi$. In particular, $\overline{\mathscr{Q}}$ is bijective if and only if $\widetilde{\pi}\,|_{D(H)^{\mathrm{co}(\pi)}}$ is bijective.*

*Proof* (1) For all $\chi \in H^*$ we have

$$(\mathrm{Id} \otimes \pi)\Delta_D(\Psi(\chi)) = \chi_{((1),(1))} \beta_1 S(\pi(\chi_{(2)}))_1 \otimes \chi_{((1),(2))} \beta_2 S(\pi(\chi_{(2)}))_2,$$

where we use the Sweedler type notation

$$(\Delta_D \otimes \mathrm{Id})(\Delta_D(\chi)) = \chi_{((1),(1))} \otimes \chi_{((1),(2))} \otimes \chi_{(2)},$$
$$(\mathrm{Id} \otimes \Delta_D)(\Delta_D(\chi)) = \chi_{(1)} \otimes \chi_{((2),(1))} \otimes \chi_{((2),(2))}.$$

Now, since $H$ is a quasi-Hopf subalgebra of $D(H)$ and $\pi$ is a quasi-Hopf algebra morphism such that $\pi(h) = h$ for any $h \in H$, by a direct computation one can show that $\Psi(\chi) \in D(H)^{\mathrm{co}(\pi)}$, so $\Psi$ is well defined. We claim that the inverse of $\Psi$, $\Psi^{-1} : D(H)^{\mathrm{co}(\pi)} \to H^*$, is given for all $\mathbf{D} \in D(H)^{\mathrm{co}(\pi)}$ by the formula

$$\Psi^{-1}(\mathbf{D}) = (\mathrm{Id} \otimes \varepsilon)(\mathbf{D}).$$

Indeed, $\Psi^{-1}$ is a left inverse since

$$\begin{aligned}(\Psi^{-1} \circ \Psi)(\chi) &= \langle \mathrm{Id} \otimes \varepsilon, \chi_{(1)} \beta S(\pi(\chi_{(2)})) \rangle \\ &= (\mathrm{Id} \otimes \varepsilon)(\chi_{(1)}) \varepsilon_D(\chi_{(2)}) \\ &= (\mathrm{Id} \otimes \varepsilon)(\chi \otimes 1) = \chi,\end{aligned}$$

for all $\chi \in H^*$. It is also a right inverse. If $\mathbf{D} = {}_i\chi_i h \in D(H)^{\mathrm{co}(\pi)}$ then

$${}_i\chi_{(1)i}h_1 \otimes \pi({}_i\chi_{(2)})_i h_2 = x^1 [{}_i\chi_i h] S(x_2^3 X^3) f^1 \otimes x^2 X^1 \beta S(x_1^3 X^2) f^2$$

in $D(H) \otimes H$. Therefore,

$$\begin{aligned}(\Psi \circ \Psi^{-1})(\mathbf{D}) &= \varepsilon({}_i h) \Psi({}_i\chi) = \varepsilon({}_i h)\, {}_i\chi_{(1)} \beta S(\pi({}_i\chi_{(2)})) \\ &= {}_i\chi_{(1)i} h_1 \beta S(\pi({}_i\chi_{(2)})_i h_2)\end{aligned}$$

$$
\begin{aligned}
&= x^1[{}_i\chi_i h]S(x_2^3X^3)f^1\beta S(x^2X^1\beta S(x_1^3X^2)f^2)\\
&= {}_i\chi_i h = \mathbf{D},
\end{aligned}
$$

because of $f^1\beta S(f^2) = S(\alpha)$, and (3.2.1) and (3.2.2).

(2) By (11.7.3), (8.5.9) and (10.5.1), for any $\chi \in H^*$ we find that

$$
\begin{aligned}
\Psi(\chi) = (X^1Y^1)_{(1,1)}p_1^1x^1 &\rightharpoonup \chi \leftharpoonup S^{-1}(X^3)q^2R^1X_1^2Y^2S^{-1}((X^1Y^1)_2p^2)\\
&\bowtie (X^1Y^1)_{(1,2)}p_2^1x^2\beta S(q^1R^2X_2^2Y^3x^3)
\end{aligned}
$$

and, if we denote by $Q^1\otimes Q^2$ another copy of $q_R$, and by $r^1\otimes r^2$ another copy of $R$, then by (11.7.1) we compute that

$$
\begin{aligned}
&(\widetilde{\pi}\circ j\circ\Psi)(\chi)\\
&= \langle\chi, S^{-1}(X^3)q^2R^1X_1^2Y^2S^{-1}((X^1Y^1)_2p^2)Q^2\overline{R}^2(X^1Y^1)_{(1,1)}p_1^1x^1\rangle\\
&\qquad Q^1\overline{R}^1(X^1Y^1)_{(1,2)}p_2^1x^2\beta S(q^1R^2X_2^2Y^3x^3)\\
&\overset{(10.1.3),(3.2.21)}{=} \langle\chi, S^{-1}(X^3)q^2R^1X_1^2Y^2S^{-1}(p^2)Q^2\overline{R}^2p_1^1x^1\rangle\\
&\qquad X^1Y^1Q^1\overline{R}^1p_2^1x^2\beta S(q^1R^2X_2^2Y^3x^3)\\
&\overset{(10.1.3),(3.2.23)}{=} \langle\chi, S^{-1}(X^3)q^2R^1X_1^2Y^2\overline{R}^2x^1\rangle X^1Y^1\overline{R}^1x^2\beta S(q^1R^2X_2^2Y^3x^3)\\
&\overset{(10.1.3),(10.1.1)}{=} \langle\chi, S^{-1}(X^3)q^2X_2^2R^1r^2Y^3\overline{R}^2\rangle X^1Y^1\overline{R}_1^1\beta S(q^1X_1^2R^2r^1Y^2\overline{R}_2^1)\\
&\overset{(3.2.1),(10.1.5),(10.1.3)}{=} \langle\chi, S^{-1}(X^3)q^2R^1r^2X_2^2Y^3\rangle S(q^1R^2r^1X_1^2Y^2S^{-1}(X^1Y^1\beta))\\
&\overset{(3.2.20),(11.5.3)}{=} (S\circ\overline{\mathscr{Q}})(\chi),
\end{aligned}
$$

as needed. Since $H$ is finite dimensional the antipode $S$ is bijective, so $\overline{\mathscr{Q}}$ is bijective if and only if $\widetilde{\pi}\circ j$ is bijective. Thus, the proof is complete. □

The notion of right quasi-Hopf bimodule was introduced in Definition 6.1. For $M$ a right quasi-Hopf bimodule we have denoted by $\overline{M^{\mathrm{co}(H)}}$ the set of alternative coinvariants of $M$, see Remark 6.17 for the explicit definition of $\overline{M^{\mathrm{co}(H)}}$.

**Lemma 11.34** *Let $D$, $A$ and $B$ be quasi-bialgebras and $\vartheta, \upsilon, \kappa$ quasi-bialgebra morphisms as in the diagram below:*

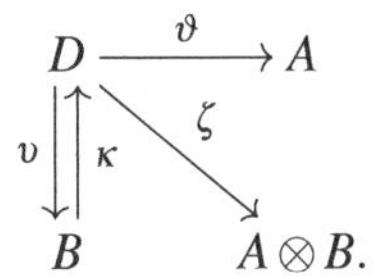


*Suppose $\upsilon\circ\kappa = \mathrm{Id}_B$ and let $\zeta := (\vartheta\otimes\upsilon)\circ\Delta_D$. The following assertions hold:*

*(i) $D$ and $A\otimes B$ are right quasi-Hopf $B$-bimodules via the following structures:*

$$
D \in {}_B\mathscr{M}_B^B : \begin{cases} b\cdot d\cdot b' = \kappa(b)d\kappa(b'),\\ \rho_D(d) = d_1\otimes\upsilon(d_2),\end{cases}
$$

$$
A\otimes B \in {}_B\mathscr{M}_B^B : \begin{cases} b'\cdot(a\otimes b)\cdot b'' = \vartheta(\kappa(b'_1))a\vartheta(\kappa(b''{}_1))\otimes b'_2bb''{}_2,\\ \rho_{A\otimes B}(a\otimes b) = \vartheta(\kappa(x^1))a\vartheta(\kappa(X^1))\otimes x^2b_1X^2\otimes x^3b_2X^3,\end{cases}
$$

*for $a \in A$, $b, b', b'' \in B$, $d \in D$, and $\zeta$ becomes a quasi-Hopf B-bimodule morphism.*

*(ii) If D, A and B are, moreover, quasi-Hopf algebras and $\vartheta, \upsilon$ and $\kappa$ are, moreover, quasi-Hopf algebra maps then $D^{\overline{\mathrm{co}(B)}} = D^{\mathrm{co}(\upsilon)}$ and*

$$(A \otimes B)^{\overline{\mathrm{co}(B)}} = \{\vartheta(\kappa(x^1))a\vartheta(\kappa(S(x_2^3X^3)f^1)) \otimes x^2X^1\beta S(x_1^3X^2)f^2 \mid a \in A\}.$$

*Proof* Since no confusion is possible we will write without subscripts $D, A$ or $B$ in the tensor components of the reassociators of $D, A$ or $B$, respectively. The same thing will be done when we write their inverses.

(i) This follows from the first part of the proof of Proposition 9.24. Also, one can check directly that $\zeta$ becomes a morphism in ${}_B\mathscr{M}_B^B$; details are left to the reader.

(ii) By definitions we have

$$\begin{aligned} D^{\overline{\mathrm{co}(B)}} &= \{d \in D \mid \rho_D(d) = x^1 \cdot d \cdot S(x_2^3X^3)f^1 \otimes x^2X^1\beta S(x_1^3X^2)f^2\} \\ &= \{d \in D \mid d_1 \otimes \upsilon(d_2) = \kappa(x^1)d\kappa(S(x_2^3X^3)f^1) \otimes \upsilon(\kappa(x^2X^1\beta S(x_1^3X^2)f^2))\} \\ &= D^{\mathrm{co}(\upsilon)}. \end{aligned}$$

Recall that $D^{\overline{\mathrm{co}(B)}} = \mathrm{Im}(\overline{E}_{A\otimes B})$, where $\overline{E}_{A\otimes B}$ is the projection on the space of alternative coinvariants of $A \otimes B$ defined in (6.3.7). Having in mind the structure of $A \otimes B$ as a right quasi-Hopf $B$-bimodule, we have that

$$\begin{aligned} \overline{E}_{A\otimes B}(a \otimes b) &= (\vartheta(\kappa(x^1))a\vartheta(\kappa(X^1)) \otimes x^2b_1X^2) \cdot \beta S(x^3b_2X^3) \\ &\overset{(3.2.13)}{\underset{(3.2.14)}{=}} \vartheta(\kappa(x^1))a\vartheta\kappa(X^1\delta^1S(x_2^3b_{(2,2)}X_2^3)f^1) \\ &\qquad \otimes x^2b_1X^2\delta^2S(x_1^3b_{(2,1)}X_1^3)f^2 \\ &\overset{(3.2.6)}{=} \vartheta(\kappa(x^1))a\vartheta\kappa(\beta S(x_2^3b_{(2,2)}X^3)f^1) \otimes x^2b_1X^1\beta S(x_1^3b_{(2,1)}X^2)f^2 \\ &\overset{(3.1.7)}{\underset{(3.2.1)}{=}} \vartheta(\kappa(x^1))a\vartheta\kappa(\beta S(b))\vartheta\kappa(S(x_2^3X^3)f^1) \otimes x^2X^1\beta S(x_1^3X^2)f^2, \end{aligned}$$

for all $a \in A$, $b \in B$. It follows that

$$D^{\overline{\mathrm{co}(B)}} = \{\vartheta(\kappa(x^1))a\vartheta(\kappa(S(x_2^3X^3)f^1)) \otimes x^2X^1\beta S(x_1^3X^2)f^2 \mid a \in A\},$$

since $A = \{a\vartheta\kappa(\beta S(b)) \mid a \in A, b \in B\}$. This finishes our proof. □

**Proposition 11.35** *Let D be a quasi-Hopf algebra, A and B two quasi-bialgebras and $\vartheta : D \to A$, $\upsilon : D \to B$ two quasi-bialgebra maps. Consider $\zeta : D \to A \otimes B$ given by $\zeta(d) = \vartheta(d_1) \otimes \upsilon(d_2)$, for all $d \in D$.*

*(1) Suppose that $(D, R)$ is QT and define $\mathfrak{F} = \mathfrak{F}^1 \otimes \mathfrak{F}^2 \in (A \otimes B)^{\otimes 2}$ by*

$$\mathfrak{F} = \vartheta(Y_1^1x^1X^1y_1^1) \otimes \upsilon(Y_2^1x^2\overline{R}^1X^3y^2) \otimes \vartheta(Y^2x^3\overline{R}^2X^2y_2^1) \otimes \upsilon(Y^3y^3), \qquad (11.7.4)$$

*where, as usual, $\overline{R}^1 \otimes \overline{R}^2$ is the inverse of the R-matrix R of D. Then $\mathfrak{F}$ is a twist on $A \otimes B$ (here $A \otimes B$ has the componentwise quasi-bialgebra structure) and $\zeta : D \to (A \otimes B)_{\mathfrak{F}}$ is a quasi-bialgebra morphism. Moreover, if A and B are quasi-Hopf algebras and $\vartheta$ and $\upsilon$ are quasi-Hopf algebra morphisms, then $\zeta : D \to (A \otimes B)_{\mathfrak{F}}^{\mathfrak{U}}$ is a quasi-Hopf algebra morphism, where $\mathfrak{U} = \vartheta(R^2g^2) \otimes \upsilon(R^1g^1)$.*

*(2) Suppose that A and B are quasi-Hopf algebras, $\vartheta$ and $\upsilon$ are quasi-Hopf algebra morphisms, and that there exists a quasi-Hopf algebra map $\kappa : B \to D$ such that $\upsilon \circ \kappa = \mathrm{Id}_B$. Then $\zeta$ is a bijective map if and only if the restriction of $\vartheta$ provides a bijection from $D^{\mathrm{co}(\vartheta)}$ to A.*

*Proof* (1) We have that $\zeta = (\vartheta \otimes \upsilon) \circ \Delta_D$, so clearly $\zeta$ is an algebra map. It also respects the comultiplications. Indeed, by applying (3.1.3), (3.1.7) twice, (10.1.3), and then again (3.1.7) twice, it is not hard to see that, for all $d \in D$,

$$(\Delta_{(A\otimes B)_{\mathfrak{F}}} \circ \zeta)(d) = ((\zeta \otimes \zeta) \circ \Delta_D)(d).$$

Obviously, $\varepsilon_{A\otimes B} \circ \zeta = \varepsilon_D$, so $\zeta$ respects the counits. It remains to show that

$$(\zeta \otimes \zeta \otimes \zeta)(\Phi_D) = \Phi_{(A\otimes B)_{\mathfrak{F}}}.$$

This follows from a long, technical but straightforward computation; we leave the details to the reader.

Suppose that $A$, $B$ are quasi-Hopf algebras and that $\vartheta$ and $\upsilon$ are quasi-Hopf algebra morphisms. In this case, $\zeta : D \to (A \otimes B)_{\mathfrak{F}}^{\mathfrak{U}}$ is also a quasi-bialgebra morphism since $(A\otimes B)_{\mathfrak{F}}^{\mathfrak{U}} = (A \otimes B)_{\mathfrak{F}}$ as quasi-bialgebras. Thus, we are left to show that

$$\zeta(\alpha) = \mathfrak{U}\alpha_{(A\otimes B)_{\mathfrak{F}}}, \quad \zeta(\beta) = \beta_{(A\otimes B)_{\mathfrak{F}}}\mathfrak{U}^{-1}, \quad (\zeta \circ S_D)(d) = \mathfrak{U}S_{A\otimes B}(\zeta(d))\mathfrak{U}^{-1}$$

for all $d \in D$. Take $\mathfrak{F}^{-1} = \mathfrak{G}^1 \otimes \mathfrak{G}^2$ as being the inverse of the twist $\mathfrak{F}$. By (3.2.4) and (11.7.4) we compute:

$$\begin{aligned}
\alpha_{(A\otimes B)_{\mathfrak{F}}} &= S_{A\otimes B}(\mathfrak{G}^1)\alpha_{A\otimes B}\mathfrak{G}^2\\
&= \vartheta(S(Y_1^1x^1X^1y_1^1)\alpha Y_2^1x^2R^2X^3y^2)\\
&\quad \otimes \upsilon(S(Y^2x^3R^1X^2y_2^1)\alpha Y^3y^3)\\
&\overset{(3.2.1),(10.1.2)}{=} \vartheta(S(R_1^2X^2\overline{R}^2y_1^1)\alpha R_2^2X^3y^2) \otimes \upsilon(S(R^1X^1\overline{R}^1y_2^1)\alpha y^3)\\
&\overset{(3.2.1),(10.1.5),(10.1.3)}{=} \vartheta(S(X^2y_2^1\overline{R}^2)\alpha X^3y^2) \otimes \upsilon(S(X^1y_1^1\overline{R}^1)\alpha y^3)\\
&\overset{(3.2.5),(3.2.14)}{=} \vartheta(S(\overline{R}^2)\gamma^1) \otimes \upsilon(S(\overline{R}^1)\gamma^2)\\
&= \vartheta(S(\overline{R}^2)f^1\alpha_1) \otimes \upsilon(S(\overline{R}^1)f^2\alpha_2)\\
&\overset{(10.3.6)}{=} \vartheta(f^2\overline{R}^2\alpha_1) \otimes \upsilon(f^1\overline{R}^1\alpha_2) = \mathfrak{U}^{-1}\zeta(\alpha),
\end{aligned}$$

as needed. In a similar manner one can prove that $\beta_{(A\otimes B)_{\mathfrak{F}}} = \zeta(\beta)\mathfrak{U}$; the details are left to the reader. Finally, for all $d \in D$ we have

$$\begin{aligned}
\mathfrak{U}S_{A\otimes B}(\zeta(d))\mathfrak{U}^{-1} &= \vartheta(R^2g^2S(d_1)f^2\overline{R}^2) \otimes \upsilon(R^1g^1S(d_2)f^1\overline{R}^1)\\
&\overset{(3.2.13),(10.1.3)}{=} \vartheta(S(d)_1) \otimes \upsilon(S(d)_2) = \zeta(S(d)).
\end{aligned}$$

(2) We are in the same hypothesis as in Lemma 11.34, so $\zeta : D \to A \otimes B$ is a right quasi-Hopf $B$-bimodule morphism. As we explained before Lemma 11.34, the morphism $\zeta$ is bijective if and only if $\zeta_0$, the restriction of $\zeta$, defines an isomorphism between $D^{\mathrm{co}(B)}$ and $(A\otimes B)^{\mathrm{co}(B)}$. But $D^{\mathrm{co}(B)} = D^{\mathrm{co}(\upsilon)}$, so if $d \in D^{\mathrm{co}(B)}$ then

$$\zeta(d) = \vartheta(d_1) \otimes \upsilon(d_2)$$

$$= \vartheta(\kappa(x^1))\vartheta(d)\vartheta(\kappa(S(x_2^3X^3)f^1)) \otimes x^2X^1\beta S(x_1^3X^2)f^2,$$

because of $\upsilon \circ \kappa = \mathrm{Id}_B$. Hence, by Lemma 11.34, $\zeta$ is bijective if and only if the map

$$\zeta_0 : D^{\mathrm{co}(\upsilon)} \to \{\vartheta(\kappa(x^1))a\vartheta(\kappa(S(x_2^3X^3)f^1)) \otimes x^2X^1\beta S(x_1^3X^2)f^2 \mid a \in A\},$$
$$\zeta_0(d) = \vartheta(\kappa(x^1))\vartheta(d)\vartheta(\kappa(S(x_2^3X^3)f^1)) \otimes x^2X^1\beta S(x_1^3X^2)f^2$$

is bijective. Now, it follows that $\zeta$ is bijective if and only if the restriction of $\vartheta$ defines a bijection between $D^{\mathrm{co}(\upsilon)}$ and $A$. □

We can now state the structure theorem of $D(H)$ when $H$ is factorizable.

**Theorem 11.36** *Let $(H,R)$ be a finite-dimensional QT quasi-Hopf algebra and $\pi, \widetilde{\pi} : D(H) \to H$ the quasi-Hopf algebra morphisms given by* (10.5.1) *and* (11.7.1). *Define $\zeta : D(H) \to H \otimes H$ by $\zeta(\mathbf{D}) = \widetilde{\pi}(\mathbf{D}_1) \otimes \pi(\mathbf{D}_2)$, for all $\mathbf{D} \in D(H)$, and*

$$\mathbf{F} = Y_1^1x^1X^1y_1^1 \otimes Y_2^1x^2R^2X^3y^2 \otimes Y^2x^3R^1X^2y_2^1 \otimes Y^3y^3, \tag{11.7.5}$$

*where $R^1 \otimes R^2$ is the R-matrix $R$ of $H$. Then the following assertions hold:*

*(1) $\zeta : D(H) \to (H \otimes H)_{\mathbf{F}}^{\mathbf{U}}$ is a quasi-Hopf algebra morphism, where we define $\mathbf{U} := \overline{R}^1g^2 \otimes \overline{R}^2g^1$.*

*(2) $\zeta$ is bijective if and only if $(H,R)$ is factorizable.*

*Proof* We consider in Proposition 11.35 $D = D(H)$, $A = B = H$, $\vartheta = \widetilde{\pi}$, $\upsilon = \pi$ and $\kappa = i_D$. So the map $\zeta$ in the statement is the map $\zeta$ in Proposition 11.35 specialized for our case. Moreover, from definition (10.4.1) of the $R$-matrix $\mathscr{R}$ of $D(H)$ we have

$$\begin{aligned}
\pi(\mathscr{R}^1) \otimes \widetilde{\pi}(\mathscr{R}^2) &= S^{-1}(p^2)_i e p_1^1 \otimes \langle {}^i e, q^2\overline{R}^2 \rangle q^1\overline{R}^1 p_2^1 \\
&= S^{-1}(p^2)q^2\overline{R}^2 p_1^1 \otimes q^1\overline{R}^1 p_2^1 \\
&\overset{(10.1.3),(3.2.23)}{=} S^{-1}(p^2)q^2p_2^1\overline{R}^2 \otimes q^1p_1^1\overline{R}^1 = \overline{R}^2 \otimes \overline{R}^1.
\end{aligned}$$

Since $\pi$ and $\widetilde{\pi}$ are algebra maps we obtain that $\pi(\overline{\mathscr{R}}^1) \otimes \widetilde{\pi}(\overline{\mathscr{R}}^2) = R^2 \otimes R^1$, so the twist (11.7.5) is the twist $\mathfrak{F}$ defined in (11.7.4) specialized for our situation. Also, the element $\mathbf{U}$ is the element $\mathfrak{U}$ defined in Proposition 11.35 specialized for our context and this proves the first assertion.

By applying again Proposition 11.35 we have that $\zeta$ is bijective if and only if the restriction of $\widetilde{\pi}$ provides a bijection from $D(H)^{\mathrm{co}(\pi)}$ to $H$. By Lemma 11.33 this is equivalent to $\overline{\mathscr{Q}}$ bijective. Finally, by Proposition 11.26 we obtain that $\zeta$ is bijective if and only if $(H,R)$ is factorizable, and this finishes our proof. □

We end this section with an application.

At first sight there is no relationship between $D(H(2))$ described in Proposition 10.21 and $H(2) \otimes H(2)$, so it comes as a surprise that these two quasi-Hopf algebras are twist equivalent. To show this, we will use Theorem 11.36.

**Example 11.37** Let $X, Y$ be the algebra generators of $D(H(2))$ defined in Proposition 10.21, and $x, y$ the generators of $H(2) \otimes H(2) \cong k[C_2 \times C_2]$, the tensor product quasi-Hopf algebra. Let $\omega_\pm = 1 \pm i$ and consider the elements

$$\mathbf{U}_\pm = p_+ \otimes 1 + p_- \otimes g + \omega_\pm p_- \otimes p_-,$$
$$\mathbf{F}_\pm = 1 - 2p^x_- p^y_- \otimes p^x_- p^y_+ - 2p^x_+ p^y_- \otimes p^x_- p^y_- - \omega_\pm p^x_- \otimes p^y_-,$$

where $p^x_\pm = \frac{1}{2}(1 \pm x)$ and $p^y_\pm = \frac{1}{2}(1 \pm y)$. Then the maps $\zeta_\pm : D(H(2)) \to (H(2) \otimes H(2))^{\mathbf{U}_\pm}_{\mathbf{F}_\pm}$, given by

$$\zeta_\pm(X) = xy, \quad \zeta_\pm(Y) = -1 + \omega_\pm p^x_- + \omega_\mp p^y_- = -\frac{1}{2}(\omega_\pm x + \omega_\mp y),$$

are quasi-Hopf algebra isomorphisms.

*Proof* By Example 11.25, $H(2)$ is a factorizable quasi-Hopf algebra. Then everything will follow from the general isomorphism presented in Theorem 11.36.

For $H(2)$ we have $q_R = 1 \otimes p_+ - g \otimes p_-$. Also, it is easy to see that the inverse of $R_\pm = 1 - \omega_\pm p_- \otimes p_-$ is $R_\mp = 1 - \omega_\mp p_- \otimes p_-$, and therefore

$$q^2 R^1_\pm \otimes q^1 R^2_\pm = p_+ \otimes 1 - p_- \otimes g - \omega_\pm p_- \otimes p_-,$$
$$q^2 \overline{R}^2_\pm \otimes q^1 \overline{R}^1_\pm = q^2 R^1_\mp \otimes q^1 R^2_\mp = p_+ \otimes 1 - p_- \otimes g - \omega_\mp p_- \otimes p_-.$$

From the structure of $D(H(2))$ in Proposition 10.21 we see that $\pi(X) = \tilde{\pi}(X) = g$,

$$\pi(Y) = \pi(\mu \bowtie 1) = \mu(p_+)1 - \mu(p_-)g - \omega_\pm \mu(p_-)p_- = -g - \omega_\pm p_-,$$

and, in a similar way, $\tilde{\pi}(Y) = -g - \omega_\mp p_-$. We get that $\pi(XY) = -1 + \omega_\pm p_-$ and $\tilde{\pi}(XY) = -1 + \omega_\mp p_-$, so $\zeta_\pm(X) = g \otimes g = xy$ and

$$\zeta_\pm(Y) = -\frac{1}{2}(\pi \otimes \tilde{\pi})\left(Y \otimes Y + XY \otimes Y + Y \otimes XY - XY \otimes XY\right).$$

After some straightforward computations we obtain

$$\pi(Y) \otimes \tilde{\pi}(Y) = xy + 2p^x_- p^y_- + \omega_\pm x p^y_- + \omega_\mp y p^x_-,$$
$$\pi(XY) \otimes \tilde{\pi}(Y) = x - 2p^x_- p^y_- - \omega_\pm x p^y_- + \omega_\mp p^x_-,$$
$$\pi(Y) \otimes \tilde{\pi}(XY) = y - 2p^x_- p^y_- + \omega_\pm p^y_- - \omega_\mp y p^x_-,$$
$$\pi(XY) \otimes \tilde{\pi}(XY) = 1 + 2p^x_- p^y_- - \omega_\pm p^y_- - \omega_\mp p^x_-.$$

Thus, we can compute:

$$\begin{aligned}
\zeta_\pm(Y) &= -\frac{1}{2}(xy + x + y - 1 - 4p^x_- p^y_- + 2\omega_\pm p^y_- + 2\omega_\mp p^x_-) \\
&= 1 - x - y - \omega_\pm p^y_- - \omega_\mp p^x_- \\
&= -1 + (2 - \omega_\mp)p^x_- + (2 - \omega\pm)p^y_- = -1 + \omega_\pm p^x_- + \omega_\mp p^y_-,
\end{aligned}$$

and this is exactly what we need. Finally, one can easily see that the corresponding elements $\mathbf{U}_\pm$ and $\mathbf{F}_\pm$ for $(H(2), R_\pm)$ are exactly the ones defined in the statement; we leave the details to the reader. □

**Remarks 11.38** (1) Keeping the notation used in Example 11.37, we have that

$$(\zeta_\pm \otimes \zeta_\pm \otimes \zeta_\pm)(\Phi_X) = \Phi_{xy} := 1 - 2p_-^{xy} \otimes p_-^{xy} \otimes p_-^{xy},$$

where $\Phi_X$ is the reassociator of $D(H(2))$ and $p_-^{xy} := \frac{1}{2}(1 - xy)$. Thus, $\Phi_{xy}$ is a 3-cocycle for $k[C_2 \times C_2]$ and $\Phi_{xy} = (\Phi_{x,y})_{\mathbf{F}} = (\Phi_x \Phi_y)_{\mathbf{F}}$, because of (3.1.5). Here $\Phi_y = 1 - 2p_-^y \otimes p_-^y \otimes p_-^y$ is the 3-cocycle on $k[C_2 \times C_2]$ corresponding to $y$. In other words we have proved that the 3-cocycles $\Phi_{xy}$ and $\Phi_x \Phi_y$ are equivalent.

(2) It follows from Example 11.37 that $k[C_4]$ and $k[C_2 \times C_2]$ are isomorphic as algebras if $\mathrm{char}(k) \neq 2$ and $k$ contains a primitive fourth root of unity. This is well known and can be easily seen directly: $k[C_4]$ and $k[C_2 \times C_2]$ are isomorphic (even as Hopf algebras) to their duals and the two duals are both isomorphic to $k^4$ as algebras. More explicitly, we have the following: $\zeta_+ = \gamma \circ \beta \circ \alpha$, where $\alpha, \beta, \gamma$ are the following three algebra isomorphisms ($\{e_1, e_2, e_3, e_4\}$ is the standard basis of $k^4$):

$$\alpha : k[C_4] = k[Y]/(Y^4 - 1) \to k^4, \;\; \alpha(Y) = e_1 + ie_2 - e_3 - ie_4;$$
$$\beta : k^4 \to k[C_2 \times C_2] = k[x, y],$$
$$\beta(e_1) = p_+^x p_+^y, \;\; \beta(e_2) = p_-^x p_+^y, \;\; \beta(e_3) = p_+^x p_-^y, \;\; \beta(e_4) = p_-^x p_-^y;$$
$$\gamma : k[C_2 \times C_2] \to k[C_2 \times C_2], \;\; \gamma(x) = xy, \;\; \gamma(y) = -x, \;\; \gamma(xy) = -y.$$

## 11.8 Notes

Factorizable (quasi-)Hopf algebras provide a special class of QT (quasi-)Hopf algebras, and in their theory an important role is played by the quantum double $D(H)$. Factorizable Hopf algebras were introduced and studied by Reshetikhin and Semenov-Tian-Shansky [190]. They are important in Hennings' investigation of 3-manifold invariants [110]. Afterwards, Kauffman reworked Hennings' construction; see [129] or [188] for more details.

Factorizable quasi-Hopf algebras were introduced in [61] by using a categorical point of view due to Majid [148]. An alternative definition was recently introduced in [98]. The process that associates to a QT quasi-Hopf algebra a braided Hopf algebra was introduced in [54], and the dual case in [61]. The bosonisation for quasi-Hopf algebras was also taken from [54]. That $D(H)$ is always factorizable, and therefore unimodular, was taken from [61], as well as the structure of $D(H)$ when $H$ itself is factorizable. Example 11.37 is from [52].

# 12

# The Quantum Dimension and Involutory Quasi-Hopf Algebras

We compute the quantum dimension of a finite dimensional quasi-Hopf algebra $H$ and of its quantum double $D(H)$, within the rigid braided category of finite dimensional left $D(H)$-modules. As we will see, this involves the semisimplicity of $D(H)$ and leads to the notion of involutory quasi-Hopf algebra, a concept that will be studied at the end of this chapter.

## 12.1 The Integrals of a Quantum Double

We provide explicit formulas for the integrals in the quantum double $D(H)$ of a finite-dimensional quasi-Hopf algebra $H$. Our aim is to apply Theorem 7.28 in order to see when $D(H)$ is a semisimple algebra, as this is important in computing the quantum dimension of $H$ and $D(H)$.

In what follows, $\{e_i\}_i$ is a basis of $H$ and $\{e^i\}_i$ is the corresponding dual basis of $H^*$. $\Omega$ is the element of $H^{\otimes 5}$ defined in (8.5.1), $\lambda$ is a non-zero left cointegral on $H$, $r$ is a non-zero right integral in $H$ and $\mu$ is the modular element of $H^*$.

If $\delta$ is the element defined in (3.2.6) and $T := \mu^{-1}(\delta^2)\delta^1 \rightharpoonup \lambda \in H^*$, we claim that $T \bowtie r$ is a non-zero left and right integral in $D(H)$. That it is non-zero follows easily from the fact that $r$ is non-zero and

$$T(r) = \mu^{-1}(\delta^2)\lambda(r\delta^1) = \mu^{-1}(\varepsilon(\delta^1)\delta^2)\lambda(r) = \mu^{-1}(\beta)\lambda(r) \neq 0,$$

as $\mathscr{L} \times \int_r^H \ni (\lambda', r') \mapsto \lambda'(r') \in k$ is non-degenerate.

The difficult part is to show that $T \bowtie r$ is a left and right integral in $D(H)$. To prove that it is a left integral we need the formula

$$Y^1\delta^1 S(Y_2^3) \otimes Y^2\delta^2 S(Y_1^3) = \beta S(\tilde{p}^2) \otimes S(\tilde{p}^1), \tag{12.1.1}$$

which can be deduced from the definitions of $\delta$ and $p_L$, and (3.1.9) and (3.2.1).

**Proposition 12.1** *With the above notation, $T \bowtie r$ is a left integral in $D(H)$.*

*Proof* We check this assertion by direct computation. If $\varphi \in H^*$ and $h \in H$ then

$$\begin{aligned}(\varphi \bowtie h)(T \bowtie r) \\ = \ & \mu^{-1}(\delta^2)(\Omega^1 \rightharpoonup \varphi \leftharpoonup \Omega^5)(\Omega^2 h_{(1,1)}\delta^1 \rightharpoonup \lambda \leftharpoonup S^{-1}(h_2)\Omega^4) \bowtie \Omega^3 h_{(1,2)} r\end{aligned}$$

$$
\begin{aligned}
&\overset{(7.2.7)}{=} \mu^{-1}(\Omega^3 h_{(1,2)}\delta^2)(\Omega^1 \rightharpoonup \varphi \leftharpoonup \Omega^5)(\Omega^2 h_{(1,1)}\delta^1 \rightharpoonup \lambda \leftharpoonup S^{-1}(h_2)\Omega^4) \bowtie r\\
&\overset{(7.5.11)}{=} \mu^{-1}(\Omega^3 h_{(1,2)}\delta^2 S((h_2)_1))(\Omega^2 h_{(1,1)}\delta^1 S((h_2)_2) \rightharpoonup \lambda \leftharpoonup \Omega^4) \bowtie \Omega^3 h_{(1,2)} r\\
&\overset{(3.2.8)}{=} \varepsilon(h)\mu^{-1}(\Omega^3\delta^2)(\Omega^1 \rightharpoonup \varphi \leftharpoonup \Omega^5)(\Omega^2\delta^1 \rightharpoonup \lambda \leftharpoonup \Omega^4) \bowtie r.
\end{aligned}
$$

Therefore it suffices to show that

$$
\mu^{-1}(\Omega^3\delta^2)\lambda(\Omega^4 h_2\Omega^2\delta^1)\Omega^5 h_1\Omega^1 = \mu^{-1}(\delta^2)\lambda(h\delta^1)S^{-1}(\alpha),
$$

for all $h \in H$. To this end, we compute

$$
\begin{aligned}
&\mu^{-1}(\Omega^3\delta^2)\lambda(\Omega^4 h_2\Omega^2\delta^1)\Omega^5 h_1\Omega^1\\
&\overset{(8.5.1)}{=} \mu^{-1}(X_2^1 y^3 x_2^2\delta^2)\lambda(S^{-1}(f^1X^2x^3)h_2X^1_{(1,2)}y^2x_1^2\delta^1)\\
&\qquad S^{-1}(f^2X^3)h_1X^1_{(1,1)}y^1x^1\\
&\overset{(3.1.9)}{\underset{(12.1.1)}{=}} \mu^{-1}(X_2^1y^2x_1^3S(\tilde{p}^1))\lambda(S^{-1}(f^1X^2y^3x_2^3)h_2X^1_{(1,2)}y_2^1x^2\beta S(\tilde{p}^2))\\
&\qquad S^{-1}(f^2X^3)h_1X^1_{(1,1)}y_1^1x^1\\
&\overset{(7.5.11)}{\underset{(3.2.22)}{=}} \mu^{-1}(X_2^1y^2S(\tilde{p}^1))\lambda(S^{-1}(f^1X^2y^3)h_2X^1_{(1,2)}y_2^1x^2\beta S(\tilde{p}^2x^3))\\
&\qquad S^{-1}(f^2X^3)h_1X^1_{(1,1)}y_1^1x^1\\
&\overset{(3.2.19)}{\underset{(7.3.5)}{=}} \mu^{-1}(X_2^1y^2S(\tilde{p}^1))\lambda(S^{-1}(f^1X^2y^3)(hX_1^1y^1S(\tilde{P}^1))_2U^2S(\tilde{p}^2))\\
&\qquad S^{-1}(f^2X^3)(hX_1^1y^1S(\tilde{P}^1))_1U^1\tilde{P}^2\\
&\overset{(7.5.11)}{\underset{(7.3.2)}{=}} \mu^{-1}(X_2^1y^2S(\tilde{p}^1))\mu(X_1^2y_1^3)\langle\lambda, S^{-1}(f^1)(hX_1^1y^1S(X^2_{(2,1)}y^3_{(2,1)}\tilde{p}_1^2\tilde{P}^1))_2U^2\rangle\\
&\qquad S^{-1}(f^2X^3)(hX_1^1y^1S(X^2_{(2,1)}y^3_{(2,1)}\tilde{p}_1^2\tilde{P}^1))_1U^1X^2_{(2,2)}y^3_{(2,2)}\tilde{p}_2^2\tilde{P}^2\\
&\overset{(7.55)}{=} \mu^{-1}(X_2^1y^2S(\tilde{p}^1))\mu(X_1^2y_1^3)\mu(q_1^1z^1)\langle\lambda, hX_1^1y^1S(q_2^1z^2X^2_{(2,1)}y^3_{(2,1)}\tilde{p}_1^2\tilde{P}^1)\rangle\\
&\qquad S^{-1}(X^3)q^2z^3X^2_{(2,2)}y^3_{(2,2)}\tilde{p}_2^2\tilde{P}^2\\
&\overset{(3.1.7)}{=} \mu^{-1}(X_2^1y^2)\mu(q_1^1X^2_{(1,1)}y^3_{(1,1)}z^1\tilde{p}^1)\langle\lambda, hX_1^1y^1S(q_2^1X^2_{(1,2)}y^3_{(1,2)}z^2\tilde{p}_1^2\tilde{P}^1\rangle\\
&\qquad S^{-1}(X^3)q^2X_2^2y_2^3z^3\tilde{p}_2^2\tilde{P}^2\\
&\overset{(10.3.3)}{\underset{(5.5.17)}{=}} \mu((q_2^1)_1(x^2y_1^3Y^2)_1Z^2S^{-1}(q^1_{(1,2)}x_2^1y^2Y^1Z^1\beta))\\
&\qquad \langle\lambda, hq^1_{(1,1)}x_1^1y^1S((q_2^1)_2(x^2y_1^3Y^2)_2Z^3\tilde{P}^1)\rangle q^2x^3y_2^3Y^3\tilde{P}^2\\
&\overset{(3.1.9),(3.1.7)}{\underset{(3.2.1)}{=}} \mu((q_2^1z^3)_1Z^2S^{-1}(q^1_{(1,2)}z^2Z^1\beta))\\
&\qquad \langle\lambda, hq^1_{(1,1)}z^1y^1S((q_2^1z^3)_2Z^3y^2\tilde{P}^1)\rangle q^2y^3\tilde{P}^2\\
&\overset{(3.1.7)}{=} \mu^{-1}(z^2(q_2^1)_1Z^1\beta S(z_1^3(q_2^1)_{(2,1)}Z^2))\\
&\qquad \langle\lambda, hz^1q_1^1y^1S(z_2^3(q_2^1)_{(2,2)}Z^3y^2\tilde{P}^1)\rangle q^2y^3\tilde{P}^2\\
&\overset{(3.1.7),(3.2.1)}{=} \mu^{-1}(z^2Z^1\beta S(z_1^3Z^2))\langle\lambda, hz^1q_1^1y^1S(z_2^3Z^3q_2^1y^2\tilde{P}^1)\rangle q^2y^3\tilde{P}^2\\
&\overset{(3.2.20),(3.2.1)}{=} \mu^{-1}(z^2Z^1\beta S(z_1^3Z^2))\langle\lambda, hz^1\beta S(z_2^3Z^3)\rangle S^{-1}(\alpha)\\
&\overset{(3.2.6)}{=} \mu^{-1}(\delta^2)\lambda(h\delta^1)S^{-1}(\alpha),
\end{aligned}
$$

for all $h \in H$, and this finishes the proof. □

**Corollary 12.2** *The quantum double $D(H)$ is a semisimple algebra if and only if $H$ is semisimple and admits a normalized left cointegral, that is, a left cointegral $\lambda$ satisfying $\lambda(S^{-1}(\alpha)\beta) \neq 0$.*

*Proof* This is an immediate consequence of Theorem 7.28. Note that, for the non-zero left integral $\mathbb{T} = \mu^{-1}(\delta^2)\delta^1 \rightharpoonup \lambda \bowtie r$ in $D(H)$, we have

$$\varepsilon_D(\mathbb{T}) = \varepsilon(r)\mu^{-1}(\delta^2)\lambda(S^{-1}(\alpha)\delta^1),$$

and so $\varepsilon_D(\mathbb{T}) \neq 0$ if and only if $\varepsilon(r) \neq 0$ and $\mu^{-1}(\delta^2)\lambda(S^{-1}(\alpha)\delta^1) \neq 0$. But $\varepsilon(r) \neq 0$ implies $H$ semisimple, and therefore unimodular, in which case $\mu^{-1}(\delta^2)\delta^1 = \beta$. □

**Examples 12.3** (1) $D(H(2))$ is semisimple because $H(2)$ is semisimple and the left cointegral $P_g$ on $H(2)$ found in Example 7.53 satisfies $P_g(S^{-1}(g)) = P_g(g) = 1$.
(2) $D(H_\pm(8))$ is not semisimple as $H_\pm(8)$ is not semisimple by Examples 7.31 (2).

From Theorem 11.31 we know that $\mathbb{T} = T \bowtie r$ is a right integral in $D(H)$, too. We will present a direct proof now, and for this we need first some technical results.

For Hopf algebras we have $\underline{g}^{-1} = S(\underline{g})$. Indeed, in this case $\mathscr{L}$ is an ideal of $H^*$ and since $\dim_k \mathscr{L} = 1$ it follows that for any $h^* \in H^*$ there is a $c_{h^*} \in k$ such that $\lambda h^* = c_{h^*}\lambda$. Evaluating in $t$ we obtain $c_{h^*} = h^*(S^{-1}(\underline{g}^{-1}))$, and so $\lambda(h_1)h_2 = \lambda(h)S^{-1}(\underline{g}^{-1})$, for all $h \in H$. From here we get that $\Delta(S^{-1}(\underline{g}^{-1})) = S^{-1}(\underline{g}^{-1}) \otimes S^{-1}(\underline{g}^{-1})$, that is, $S^{-1}(\underline{g}^{-1})$ is a grouplike element of $H$. In particular this implies that $S^{-1}(\underline{g}^{-1})^{-1} = S(S^{-1}(\underline{g}^{-1}))$, which is clearly equivalent to $\underline{g}^{-1} = S^{-1}(\underline{g})$, and so to $\underline{g}^{-1} = S(\underline{g})$, too. As a consequence, $\lambda(h_1)h_2 = \lambda(h)\underline{g}$, for all $h \in H$.

For quasi-Hopf algebras the above argument does not apply. Nevertheless, in this case we can prove the following.

**Proposition 12.4** *If $\lambda$ is a left cointegral on a finite-dimensional quasi-Hopf algebra $H$ then*

$$\begin{aligned}\lambda(S^{-1}(f^2)h_1g^1S(h'))S^{-1}(f^1)h_2g^2 = {} & \mu(f^1)\mu^{-1}(U_2^2\mathbf{U}^2\alpha)\mu(\beta) \\ & \mu(U^1y_2^1x^2)\lambda(hS(y^3x_2^3h'_2\tilde{p}^2)) \\ & S^{-1}(\underline{g}^{-1}y_1^1x^1)S(S(U_1^2\mathbf{U}^1y^2x_1^3h'_1\tilde{p}^1)f^2) \quad (12.1.2)\end{aligned}$$

*for all $h, h' \in H$, where $U = U^1 \otimes U^2 = \mathbf{U}^1 \otimes \mathbf{U}^2$ is the element defined in (7.3.1) and $\mu$ is the modular element of $H^*$.*

*Proof* The Nakayama isomorphism $\xi_{\rm cop}$ for $H^{\rm cop}$ is given by

$$\xi_{\rm cop} : H^* \to H, \quad \xi_{\rm cop}(h^*) = h^*(S^{-1}(\tilde{q}^1t_1\tilde{p}^1))\tilde{q}^2t_2\tilde{p}^2,$$

and is an isomorphism of left $H$-modules with inverse $\xi_{\rm cop}^{-1}(h) = h \rightharpoonup \Lambda \circ S$, where $\Lambda$ is a right cointegral on $H$ satisfying $\Lambda(S(t)) = 1$; see Remark 7.60.

Now take $h \in H$ and $h^* = \xi_{\rm cop}^{-1}(h)$. If $q^* := h^* \circ S^{-1}$ then

$$h = q^*(\tilde{q}^1t_1\tilde{p}^1)\tilde{q}^2t_2\tilde{p}^2 \overset{(7.6.4)}{=} q^*(q^1t_1p^1)q^2t_2p^2S^{-1}(u),$$

where $u$ is the element introduced in Proposition 7.61. Set $p_R = p^1 \otimes p^2 = P^1 \otimes P^2$

and $q_R = q^1 \otimes q^2 = Q^1 \otimes Q^2$ and compute

$$
\begin{aligned}
&\Delta(hS^{-1}(u^{-1}))\\
&\quad = q^*(q^1t_1p^1)q_1^2t_{(2,1)}p_1^2 \otimes q_2^2t_{(2,2)}p_2^2\\
&\quad \overset{(3.2.26)}{=} q^*(q^1Q_1^1x^1t_1p^1)S^{-1}(g^2)q^2Q_2^1x^2t_{(2,1)}p_1^2 \otimes S^{-1}(g^1)Q^2x^3t_{(2,2)}p_2^2\\
&\quad \overset{(3.2.25),(7.2.6)}{=} \mu(X^1)q^*(q^1(Q^1t_1P^1)_1p^1)S^{-1}(g^2)q^2(Q^1t_1P^1)_2p^2S(X^3)f^1\\
&\qquad \otimes S^{-1}(g^1)Q^2t_2P^2S(X^2)f^2.
\end{aligned}
$$

This equality is equivalent to

$$
\begin{aligned}
&(S^{-1}(f^2) \otimes S^{-1}(f^1))\Delta(hS^{-1}(u^{-1}))(g^1 \otimes g^2)\\
&\quad = \mu(X^1)q^*(q^1(Q^1t_1P^1)_1p^1)q^2(Q^1t_1P^1)_2p^2S(X^3) \otimes Q^2t_2P^2S(X^2).
\end{aligned}
$$

By applying $\lambda \otimes \mathrm{Id}_H$ to this formula, we find

$$
\begin{aligned}
&\lambda(S^{-1}(f^2)(hS^{-1}(u^{-1}))_1g^1S(h'))S^{-1}(f^1)(hS^{-1}(u^{-1}))_2g^2\\
&\quad = \mu(X^1)q^*(q^1(Q^1t_1P^1)_1p^1)\lambda(q^2(Q^1t_1P^1)_2p^2S(h'X^3))Q^2t_2P^2S(X^2)\\
&\quad \overset{(7.5.15)}{=} \mu(x^1X^1)q^*(\tilde{q}^2x^3h_2'X_2^3\tilde{p}^2)\lambda(S^{-1}(\tilde{q}^1)Q^1t_1P^1S(x^2h_1'X_1^3\tilde{p}^1))Q^2t_2P^2S(X^2).
\end{aligned}
$$

We know that $h^* = \xi_{\mathrm{cop}}^{-1}(h) = h \rightharpoonup \Lambda \circ S$, hence $q^* = (h \rightharpoonup \Lambda \circ S) \circ S^{-1} = \Lambda \leftharpoonup S(h)$. By Proposition 7.61 we have that $\Lambda = (\lambda \circ S^{-1}) \leftharpoonup u^{-1}$, hence $q^* = (\lambda \circ S^{-1}) \leftharpoonup u^{-1}S(h)$, and this implies that

$$
\begin{aligned}
&\lambda(S^{-1}(f^2)(hS^{-1}(u^{-1}))_1g^1S(h'))S^{-1}(f^1)(hS^{-1}(u^{-1}))_2g^2 = \mu(x^1X^1)\\
&\quad \lambda(S^{-1}(\tilde{q}^2x^3h_2'X_2^3\tilde{p}^2)hS^{-1}(u^{-1}))\lambda(S^{-1}(\tilde{q}^1)Q^1t_1p^1S(x^2h_1'X_1^3\tilde{p}^1))Q^2t_2p^2S(X^2).
\end{aligned}
$$

Since $u$ is invertible it follows that

$$
\begin{aligned}
&\lambda(S^{-1}(f^2)h_1g^1S(h'))S^{-1}(f^1)h_2g^2 = \mu(x^1X^1)\lambda(S^{-1}(\tilde{q}^2x^3h_2'X_2^3\tilde{p}^2)h)\\
&\qquad \lambda(S^{-1}(\tilde{q}^1)Q^1t_1p^1S(x^2h_1'X_1^3\tilde{p}^1))Q^2t_2p^2S(X^2)\\
&\overset{(7.5.16),(3.2.13)}{=} \mu(x^1X^1g^1S(x_2^2h'_{(1,2)}X^3_{(1,2)}\tilde{p}_2^1)f^1)\lambda(S^{-1}(\tilde{q}^2x^3h_2'X_2^3\tilde{p}^2)h)\\
&\qquad \lambda(S^{-1}(\tilde{q}^1)Q^1t_1p^1)Q^2t_2p^2S(X^2g^2S(x_1^2h'_{(1,1)}X^3_{(1,1)}\tilde{p}_1^1)f^2)\\
&\underset{(3.1.7),(7.5.16)}{\overset{(3.2.27),(7.5.11)}{=}} \mu(\tilde{q}_1^1x^1)\mu(S((\tilde{q}_2^1x^2)_2y^2(h_2'\tilde{p}^2)_1\tilde{P}^1)f^1)\lambda(S^{-1}(\tilde{q}^2x^3y^3(h_2'\tilde{p}^2)_2\tilde{P}^2)h)\\
&\qquad \lambda(Q^1t_1p^1)Q^2t_2p^2S(S((\tilde{q}_2^1x^2)_1y^1h_1'\tilde{p}^1)f^2)\\
&\underset{(6.5.1)}{\overset{(7.6.2),(3.2.28)}{=}} \lambda(S^{-1}(\tilde{q}^2y^3(x_2^3h_2'\tilde{p}^2)_2\tilde{P}^2)h)\mu(S(G_2^2S(q^1x_1^1)_2\tilde{q}_2^1y^2(x_2^3h_2'\tilde{p}^2)_1\tilde{P}^1)f^1)\\
&\qquad \mu(G^1S(q^2x_2^1)x^2)S^{-1}(\underline{g}^{-1})S(S(G_1^2S(q^1x_1^1)_1\tilde{q}_1^1y^1x_1^3h_1'\tilde{p}^1)f^2)\\
&\overset{(3.2.13),(3.2.28)}{=} \mu(G^1S(q^2x_2^1)x^2)\mu(S(G_2^2g^2S(X^1q_1^1x^1_{(1,1)})\tilde{q}^1(\tilde{Q}^2X_2^3x_2^3h_2'\tilde{p}^2)_1\tilde{P}^1)f^1)\\
&\qquad \lambda(S^{-1}(\tilde{q}^2(\tilde{Q}^2X_2^3x_2^3h_2'\tilde{p}^2)_2\tilde{P}^2)h)S^{-1}(\underline{g}^{-1})\\
&\qquad S(S(G_1^2g^1S(X^2q_2^1x^1_{(1,2)})\tilde{Q}^1X_1^3x_1^3h_1'\tilde{p}^1)f^2)\\
&\underset{(3.2.21)}{\overset{(6.5.1),(7.6.7)}{=}} \mu^{-1}(\alpha)\mu(\beta)\mu(G^1S(q^2(y_1^1x^1)_2)y_2^1x^2)\mu(S(G_2^2g^2S(Q^1q_1^1(y_1^1x^1)_{(1,1)}))f^1)\\
&\qquad \lambda(hS(y^3x_2^3h_2'\tilde{p}^2))S^{-1}(\underline{g}^{-1})S(S(G_1^2g^1S(Q^2q_2^1(y_1^1x^1)_{(1,2)})y^2x_1^3h_1'\tilde{p}^1)f^2)
\end{aligned}
$$

$$\overset{(3.2.26),(3.2.17)}{\underset{(3.1.7),(3.1.9)}{=}} \mu^{-1}(\alpha)\mu(\beta)\mu(Y^1g_1^1G^1S((q^2y_2^1)_2)F^1y_1^2x^1)\mu(S(Y^3g^2S(q^1y_1^1))f^1)$$
$$\lambda(hS(y^3x^3h_2'\tilde{p}^2))S^{-1}(\underline{g}^{-1})S(S(Y^2g_2^1G^2S((q^2y_2^1)_1)F^2y_2^2x^2h_1'\tilde{p}^1)f^2)$$
$$\overset{(3.2.13),(3.1.7)}{\underset{(7.3.1)}{=}} \mu^{-1}(\alpha)\mu(\beta)\mu(f^1)\mu(S(y^1)_1Y^1U_1^1y_1^2x^1)\mu^{-1}(S(y^1)_{(2,2)}Y^3U^2)$$
$$\lambda(hS(y^3x^3h_2'\tilde{p}^2))S^{-1}(\underline{g}^{-1})S(S(S(y^1)_{(2,1)}Y^2U_2^1y_2^2x^2h_1'\tilde{p}^1)f^2)$$
$$\overset{(7.6.13)}{=} \mu(\beta f^1)\mu^{-1}(Y^3U^2\alpha)\mu(Y^1U_1^1y_1^2x^1)\lambda(hS(y^3x^3h_2'\tilde{p}^2))$$
$$S^{-1}(\underline{g}^{-1}y^1)S(S(Y^2U_2^1y_2^2x^2h_1'\tilde{p}^1)f^2).$$

Thus we have shown that

$$\lambda(S^{-1}(f^2)h_1g^1S(h'))S^{-1}(f^1)h_2g^2 = \mu(\beta f^1)\mu^{-1}(Y^3U^2\alpha)$$
$$\mu(Y^1U_1^1y_1^2x^1)\lambda(hS(y^3x^3h_2'\tilde{p}^2))S^{-1}(\underline{g}^{-1}y^1)S(S(Y^2U_2^1y_2^2x^2h_1'\tilde{p}^1)f^2), \quad (12.1.3)$$

for all $h, h' \in H$. Finally, substituting (7.5.1) in (12.1.3) and applying (7.6.13), (3.1.9), we easily obtain (12.1.2), as desired. □

To prove that $T \bowtie r$ is a right integral in $D(H)$ we need the following formulas.

**Lemma 12.5** *Let $H$ be a finite-dimensional quasi-Hopf algebra, $h \in H$ and $r \in \int_r^H$. Then the following relations hold:*

$$X_1^1x^1\delta^1S(X_2^3) \otimes X_2^1x^2\delta_1^2S(X_1^3)_1 \otimes X^2x^3\delta_2^2S(X_1^3)_2$$
$$= (\beta S(X^3))_1g^1S(x^3) \otimes (\beta S(X^3))_2g^2S(x^2)f^1 \otimes x^2X^1\beta S(x^1X^2)f^2, \quad (12.1.4)$$
$$f^2V^1S^{-1}(f^1)_1 \otimes V^2S^{-1}(f^1)_2 = q_L, \quad (12.1.5)$$
$$S(p^1)F^2f_2^2X^3 \otimes S(p^2f^1X^1)F^1f_1^2X^2 = 1_H \otimes \alpha, \quad (12.1.6)$$
$$V^1r_1 \otimes \underline{g}^{-1}V^2r_2 = V^2r_2p^2 \otimes S^2(V^1r_1p^1)\alpha, \quad (12.1.7)$$
$$V^1r_1 \otimes S^{-1}(h)V^2r_2 = \mu(h_1)h_2V^1r_1 \otimes V^2r_2, \quad (12.1.8)$$
$$S(\tilde{p}^2)f^1r_1 \otimes \underline{g}^{-1}S(\tilde{p}^1)f^2r_2$$
$$= \mu(S(p^2)f^1)S(p^1)f^2V^2r_2P^2 \otimes S^2(V^1r_1P^1)\alpha. \quad (12.1.9)$$

*Proof* We have

$$X_1^1x^1\delta^1S(X_2^3) \otimes X_2^1x^2\delta_1^2S(X_1^3)_1 \otimes X^2x^3\delta_2^2S(X_1^3)_2$$
$$\overset{(3.2.14),(3.2.13)}{\underset{(3.1.7)}{=}} (X^1\beta_1)_1x^1g^1S(X_2^3) \otimes (X^1\beta_1)_2x^2g_1^2G^1S(X_{(1,2)}^3)f^1$$
$$\otimes X^2\beta_2x^3g_2^2G^2S(X_{(1,1)}^3)f^2$$
$$\overset{(3.2.17),(3.1.7)}{=} (X^1\beta_1g^1)_1G^1S(x^3X_{(2,2)}^3) \otimes (X^1\beta_1g^1)_2G^2S(x^2X_{(2,1)}^3)f^1$$
$$\otimes X^2\beta_2g^2S(x^2X_1^3)f^2$$
$$\overset{(3.2.14),(3.2.13)}{=} (X^1\delta^1S(X_2^3))_1G^1S(x^3) \otimes (X^1\delta^1S(X_2^3))_2G^2S(x^2)f^1$$
$$\otimes X^2\delta^2S(x^1X_1^3)f^2$$
$$\overset{(3.2.6)}{=} (\beta S(X^3))_1g^1S(x^3) \otimes (\beta S(X^3))_2g^2S(x^2)f^1 \otimes x^2X^1\beta S(x^1X^2)f^2,$$

and this proves (12.1.4). The equalities (12.1.5) and (12.1.6) follow from the definitions of $V$ and $p_R$, and from (3.2.17), (7.5.12). (12.1.8) can be proved with the help of (7.3.3). The verification of all these details is left to the reader.

In order to show (12.1.7), notice that (7.6.8) and (7.6.1) imply that

$$V^1 r_1 U^1 \otimes \underline{g}^{-1} V^2 r_2 U^2 = V^2 r_2 U^2 \otimes S^2(V^1 r_1 U^1).$$

As we have already observed, (7.6.10) guarantees that $\Delta(r)U = \Delta(r)p_R$, and so

$$\begin{aligned} V^1 r_1 \otimes \underline{g}^{-1} V^2 r_2 &\overset{(3.2.23)}{=} V^1 r_1 q_1^1 p^1 \otimes \underline{g}^{-1} V^2 r_2 q_2^1 p^2 S(q^2) \\ &= V^1 r_1 p^1 \otimes \underline{g}^{-1} V^2 r_2 p^2 \alpha = V^2 r_2 p^2 \otimes S^2(V^1 r_1 p^1)\alpha\ , \end{aligned}$$

as required. Finally, we have

$$\begin{aligned} &S(\tilde{p}^2) f^1 r_1 \otimes \underline{g}^{-1} S(\tilde{p}^1) f^2 r_2 \\ &\overset{(7.3.8)}{=} \mu^{-1}(g^1) q^1 r_1 \otimes S^{-1}(g^2) q^2 r_2 \\ &\overset{(7.3.4)}{=} \mu^{-1}(g^1)\mu(\tilde{q}^1)\tilde{q}^2 V^1 r_1 \otimes S^{-1}(g^2) V^2 r_2 \\ &\overset{(12.1.8)}{=} \mu(S(g^1)\tilde{q}^1 g_1^2)\tilde{q}^2 g_2^2 V^1 r_1 \otimes \underline{g}^{-1} V^2 r_2 \\ &\overset{(12.1.7),(7.6.12)}{=} \mu(S(p^2) f^1) S(p^1) f^2 V^2 r_2 P^2 \otimes S^2(V^1 r_1 P^1)\alpha, \end{aligned}$$

proving (12.1.9). This makes the proof complete. □

We can now (re)prove that a quantum double quasi-Hopf algebra is unimodular.

**Theorem 12.6** *If $H$ is a finite-dimensional quasi-Hopf algebra, $0 \neq r \in \int_r^H$ and $0 \neq \lambda \in \mathscr{L}$ then $\mathbb{T} = \mu^{-1}(\delta^2)\delta^1 \rightharpoonup \lambda \bowtie r$ is a non-zero right integral in $D(H)$. Consequently, $D(H)$ is a unimodular quasi-Hopf algebra.*

*Proof* For $\varphi \in H^*$ and $h \in H$, we compute that

$$\begin{aligned} &\mathbb{T}(\varphi \bowtie h) \\ &\underset{(7.5.11),(7.2.7)}{\overset{(8.5.2),(8.5.1)}{=}} (y^1 X_1^1 x^1 \delta^1 S(X_2^3) \rightharpoonup \lambda \leftharpoonup S^{-1}(f^2))(y^2 (X_2^1 x^2 \delta_1^2 S(X_1^3)_1)_1 \\ &\quad r_{(1,1)} \rightharpoonup \varphi \leftharpoonup S^{-1}(f^1 X^2 x^3 \delta_2^2 S(X_1^3)_2)) \bowtie y^3 (X_2^1 x^2 \delta_1^2 S(X_1^3)_1)_2 r_{(1,2)} h \\ &\underset{(3.2.13),(3.2.17)}{\overset{(12.1.4),(3.1.7)}{=}} (\beta S(X^3))_{(1,1)} g_1^1 G^1 S(x^3 y^3) \rightharpoonup \lambda \leftharpoonup S^{-1}(\mathbb{F}^2))((\beta S(X^3))_{(1,2)} \\ &\quad g_2^1 G^2 S(x_2^2 y^2) F^1 f_1^1 r_{(1,1)} \rightharpoonup \varphi \leftharpoonup S^{-1}(\mathbb{F}^1 X^1 \beta S(x^1 X^2) f^2 r_2)) \\ &\quad \bowtie (\beta S(X^3))_2 g^2 S(x_1^2 y^1) F^2 f_2^1 r_{(1,2)} h \\ &\underset{(3.2.14)}{\overset{(3.2.13)}{=}} ((\delta^1 S(X_2^3))_1 G^1 S(x^3 y^3) \rightharpoonup \lambda \leftharpoonup S^{-1}(\mathbb{F}^2))((\delta^1 S(X_2^3))_2 G^2 \\ &\quad S(x_2^2 y^2) F^1 f_1^1 r_{(1,1)} \rightharpoonup \varphi \leftharpoonup S^{-1}(\mathbb{F}^1 X^1 \beta S(x^1 X^2) f^2 r_2)) \\ &\quad \bowtie \delta^2 S(x_1^2 y^1 X_1^3) F^2 f_2^1 r_{(1,2)} h \\ &\overset{(12.1.2)}{=} \varphi\left(S^{-1}(\underline{g}^{-1} z_1^1 t^1 X^1 \beta S(x^1 X^2) f^2 r_2) S(x_2^2 y^2 S(U_1^2 \mathbf{U}^1 z^2 t_1^3 x_1^3 y_1^3 \tilde{p}^1)\mathbb{F}^2)\right. \\ &\quad \left. F^1 f_1^1 r_{(1,1)}\right) \mu(\beta \mathbb{F}^1)\mu^{-1}(U_2^2 \mathbf{U}^2 \alpha)\mu(U^1 z_2^1 t^2) \\ &\quad \delta^1 S(z^3 t_2^3 x_2^3 y_2^3 \tilde{p}^2 X_2^3) \rightharpoonup \lambda \bowtie \delta^2 S(x_1^2 y^1 X_1^3) F^2 f_2^1 r_{(1,2)} h \\ &\underset{(3.2.22)}{\overset{(3.1.9)}{=}} \varphi\left(S^{-1}(\underline{g}^{-1} z_1^1 t^1 X^1 \beta S(Y^1 y_1^1 x^1 X^2) f^2 r_2) S(Y^3 y^2 S(U_1^2 \mathbf{U}^1 z^2 t_1^3 y_1^3 \tilde{p}^1)\mathbb{F}^2)\right. \\ &\quad \left. F^1 f_1^1 r_{(1,1)}\right) \mu(\beta \mathbb{F}^1)\mu^{-1}(U_2^2 \mathbf{U}^2 \alpha)\mu(U^1 z_2^1 t^2) \end{aligned}$$

$$
\begin{aligned}
&\delta^1 S(z^3 t^3_2 y^3_2 \tilde{p}^2 x^3 X^3_2) \rightharpoonup \lambda \bowtie \delta^2 S(Y^2 y^1_2 x^2 X^3_1) F^2 f^1_2 r_{(1,2)} h \\
\overset{(11.5.7),(3.1.7)}{\underset{(7.2.7),(3.1.9)}{=}}\; & \varphi\big(S^{-1}(\underline{g}^{-1} z^1 t^1 X^1 \beta S(Y^1 p^1_1 x^1 X^2) f^2 r_2) S(Y^3 p^2 S(U^2_1 \mathbf{U}^1 z^3 t^2_2)\mathbb{F}^2) \\
& F^1 f^1_1 r_{(1,1)}\big)\, \mu(\beta \mathbb{F}^1) \mu^{-1}(U^2_2 \mathbf{U}^2 \alpha) \mu(U^1 z^2 t^2_1)\; \delta^1 S(t^3 x^3 X^3_2) \rightharpoonup \lambda \\
& \bowtie \delta^2 S(Y^2 p^1_2 x^2 X^3_1) F^2 f^1_2 r_{(1,2)} h \\
\overset{(7.2.7),(3.2.13)}{\underset{(3.1.7)}{=}}\; & \varphi\Big(S^{-1}(\underline{g}^{-1} z^1 t^1 X^1 \beta S(Y^1 (z^2_1 t^2_{(1,1)} p^1)_1 x^1 X^2) f^2 r_2) \mu(\beta \mathbb{F}^1) \mu(U^1) \\
& S(Y^3 z^2_2 t^2_{(1,2)} p^2 S(U^2_1 \mathbf{U}^1 z^3 t^2_2)\mathbb{F}^2) F^1 f^1_1 r_{(1,1)}\Big)\, \mu^{-1}(U^2_2 \mathbf{U}^2 \alpha) \\
& \delta^1 S(t^3 x^3 X^3_2) \rightharpoonup \lambda \bowtie \delta^2 S(Y^2 (z^2_1 t^2_{(1,1)} p^1)_2 x^2 X^3_1) F^2 f^1_2 r_{(1,2)} h \\
\overset{(3.2.21),(3.1.9)}{=}\; & \varphi\big(S^{-1}(\underline{g}^{-1} z^1 X^1 \beta S(Y^1 (z^2_1 p^1)_1 X^2) f^2 r_2) S(Y^3 z^2_2 p^2 S(U^2_1 \mathbf{U}^1 z^3)\mathbb{F}^2) \\
& F^1 f^1_1 r_{(1,1)}\big)\, \mu(\beta \mathbb{F}^1) \mu^{-1}(U^2_2 \mathbf{U}^2 \alpha) \mu(U^1)\; \delta^1 \rightharpoonup \lambda \\
& \bowtie \delta^2 S(Y^2 (z^2_1 p^1)_2 X^3) F^2 f^1_2 r_{(1,2)} h \\
\overset{(3.1.7),(5.5.16)}{\underset{(3.2.21)}{=}}\; & \varphi\big(S^{-1}(\underline{g}^{-1} z^1 X^1 \beta S(z^2_1 t^1 X^2) f^2 r_2) S((z^2_2 t^2 X^3_1)_2 p^2 S(U^2_1 \mathbf{U}^1 z^3 t^3 X^3_2)\mathbb{F}^2) \\
& F^1 f^1_1 r_{(1,1)}\big)\, \mu(\beta \mathbb{F}^1) \mu^{-1}(U^2_2 \mathbf{U}^2 \alpha) \mu(U^1)\; \delta^1 \rightharpoonup \lambda \\
& \bowtie \delta^2 S((z^2_2 t^2 X^3_1)_1 p^1) F^2 f^1_2 r_{(1,2)} h \\
\overset{(3.1.9),(3.2.20)}{\underset{(3.2.13)}{=}}\; & \varphi\big(S^{-1}(\underline{g}^{-1} S(\tilde{p}^1) f^2 r_2) S(p^2 S(U^2_1 \mathbf{U}^1)\mathbb{F}^2) F^1 (S(\tilde{p}^2) f^1 r_1)_1\big)\, \mu(\beta \mathbb{F}^1) \\
& \mu^{-1}(U^2_2 \mathbf{U}^2 \alpha) \mu(U^1)\; \delta^1 \rightharpoonup \lambda \bowtie \delta^2 S(p^1) F^2 (S(\tilde{p}^2) f^1 r_1)_2 h \\
\overset{(12.1.9),(3.2.13)}{=}\; & \varphi\big(S^{-1}(\alpha) S(p^2 S(\mathbf{U}^1)\mathbb{F}^2 S(U^2)_2 V^1 r_1 \mathbb{P}^1) F^1 (S(P^1) f^2 V^2 r_2 \mathbb{P}^2)_1\big) \\
& \mu(S(P^2) f^1) \mu(\beta \mathbb{F}^1) \mu^{-1}(U^2 \alpha) \mu(S(U^2)_1)\; \delta^1 \rightharpoonup \lambda \\
& \bowtie \delta^2 S(p^1) F^2 (S(P^1) f^2 V^2 r_2 \mathbb{P}^2)_2 h \\
\overset{(7.2.7),(12.1.8)}{\underset{(7.3.1),(3.2.19)}{=}}\; & \mu^{-1}(\beta) \mu(\alpha) \varphi\big(S^{-1}(\alpha) S(p^2 S(\mathbf{U}^1)\mathbb{F}^2 V^1 r_1 \mathbb{P}^1) F^1 (V^2 r_2 \mathbb{P}^2)_1\big) \\
& \mu(\beta \mathbb{F}^1) \mu^{-1}(\mathbf{U}^2 \alpha)\; \delta^1 \rightharpoonup \lambda \bowtie \delta^2 S(p^1) F^2 (V^2 r_2 \mathbb{P}^2)_2 h \\
\overset{(7.2.7),(12.1.5)}{\underset{(7.2.8)}{=}}\; & \mu^{-1}(\mathbf{U}^2) \varphi\big(S^{-1}(\alpha) S(p^2 S(\mathbf{U}^1) \tilde{q}^1 r_1 \mathbb{P}^1) F^1 (\tilde{q}^2 r_2 \mathbb{P}^2)_2\big) \\
& \delta^1 \rightharpoonup \lambda \bowtie \delta^2 S(p^1) F^2 (\tilde{q}^2 r_2 \mathbb{P}^2)_2 h \\
\overset{(7.2.7),(10.4.6)}{\underset{(3.2.25),(3.1.7)}{=}}\; & \varphi\big(S^{-1}(\alpha) S(p^2 f^1 X^1 (r_1 p^1)_1 P^1) F^1 f^2_1 X^2 (r_1 p^1)_2 P^2\big) \\
& \delta^1 \rightharpoonup \lambda \bowtie \delta^2 S(p^1) F^2 f^2_2 X^3 r_2 p^2 h \\
\overset{(12.1.6)}{=}\; & \varphi\big(S^{-1}(\alpha) S((r_1 p^1)_1 P^1) \alpha (r_1 p^1)_2 P^2\big)\; \delta^1 \rightharpoonup \lambda \bowtie \delta^2 r_2 p^2 h \\
\overset{(3.2.1),(3.2.2)}{=}\; & \varphi(S^{-1}(\alpha))\; \delta^1 \rightharpoonup \lambda \bowtie \delta^2 r \beta h \\
=\; & \varphi(S^{-1}(\alpha)) \varepsilon(h) \mu^{-1}(\delta^2) \rightharpoonup \lambda \bowtie r = \varepsilon_D(\varphi \bowtie h) T \bowtie r,
\end{aligned}
$$

as required. This finishes the proof. □

## 12.2 The Cointegrals of a Quantum Double

In what follows, we will identify $D(H)^* \cong H \otimes H^*$.

**Proposition 12.7** *Take non-zero elements $\lambda \in \mathscr{L}$ and $r \in \int_r^H$. Then*

$$\Gamma = r \bowtie \mu(\tilde{p}^1)S(\tilde{p}^2) \rightharpoonup \lambda \leftharpoonup \mu^{-1}(f^1)S^{-1}(f^2) \in D(H)^* \tag{12.2.1}$$

*is a non-zero left cointegral on $D(H)$.*

*Proof* It is clear that $\Gamma \neq 0$. Since $H$ is a quasi-Hopf subalgebra of $D(H)$ via $i_D$ it follows that the elements $U$ and $V$ for $D(H)$ are $U_D = \varepsilon \bowtie U^1 \otimes \varepsilon \bowtie U^2$ and $V_D = \varepsilon \bowtie V^1 \otimes \varepsilon \bowtie V^2$. Identifying $D(H)^* \cong H \otimes H^*$, we compute

$$\begin{aligned}
&\Gamma((\varepsilon \bowtie V^2)(X_1^2 \rightharpoonup \varphi_1 \leftharpoonup S^{-1}(X^3) \bowtie X_2^2 Y^3 x^3 h_2)(\varepsilon \bowtie U^2))\\
&\quad(\varepsilon \bowtie V^1)(\varepsilon \bowtie X^1Y^1)(p_1^1 x^1 \rightharpoonup \varphi_2 \leftharpoonup Y^2 S^{-1}(p^2) \bowtie p_2^1 x^2 h_1)(\varepsilon \bowtie U^1)\\
&\overset{(8.5.2)}{=} \varphi_1(S^{-1}(V_2^2 X^3) r V_{(1,1)}^2 X_1^2)\lambda(S^{-1}(f^2)V_{(1,2)}^2 X_2^2 Y^3 x^3 h_2 U^2 S(\tilde{p}^2))\mu(\tilde{p}^1)\\
&\qquad \mu^{-1}(f^1)(\varepsilon \bowtie V^1 X^1 Y^1)(p_1^1 x^1 \rightharpoonup \varphi_2 \leftharpoonup Y^2 S^{-1}(p^2) \bowtie p_2^1 x^2 h_1 U^1)\\
&= \varphi(r)\mu(S^{-1}(f^1)V_2^2 X^3)\mu(\tilde{p}^1)\lambda(S^{-1}(f^2)V_1^2 X^2 h_2 U^2 S(\tilde{p}^2))\\
&\qquad \varepsilon \bowtie V^1 X^1 h_1 U^1\\
&\overset{(7.3.1),(3.2.13)}{=} \varphi(r)\mu^{-1}(S(X^3)F^1 f_1^1 p_1^1)\mu(\tilde{p}^1)\lambda(S^{-1}(S(X^2)F^2 f_2^1 p_2^1)h_2 U^2 S(\tilde{p}^2))\\
&\qquad \varepsilon \bowtie S^{-1}(S(X^1)f^2 p^2)h_1 U^1\\
&\overset{(3.2.17),(5.5.16)}{=} \varphi(r)\mu^{-1}(f^1 x^1)\mu(\tilde{p}^1)\lambda(S^{-1}(F^1 f_1^2 x_1^2 p^1)h_2 U^2 S(\tilde{p}^2))\\
&\qquad \varepsilon \bowtie S^{-1}(F^2 f_2^2 x_2^2 p^2 S(x^3))h_1 U^1\\
&\overset{(3.2.13),(7.3.1)}{\underset{(7.3.2)}{=}} \varphi(r)\mu^{-1}(f^1 x^1)\mu(\tilde{p}^1)\lambda(V^2(S^{-1}(f^2 x^2)hS(\tilde{p}_1^2))_2 U^2)\\
&\qquad \varepsilon \bowtie x^3 V^1(S^{-1}(f^2x^2)hS(\tilde{p}_1^2))_1 U^1 \tilde{p}_2^2\\
&\overset{(7.5.13),(7.5.11)}{=} \varphi(r)\mu^{-1}(f^1 x^1)\mu(x_1^2 y^1 \tilde{p}^1)\lambda(S^{-1}(f^2)hS(x_2^2 y^2 \tilde{p}_1^2))\varepsilon \bowtie x^3 y^3 \tilde{p}_2^2\\
&\overset{(3.2.20),(3.1.9)}{=} \varphi(r)\mu(\tilde{p}^1)\mu^{-1}(f^1)\lambda(S^{-1}(f^2)hS(\tilde{p}^2))\varepsilon \bowtie 1 = \Gamma(\varphi \bowtie h)\varepsilon \bowtie 1_H.
\end{aligned}$$

As $D(H)$ is unimodular, it follows that $\Gamma$ is a left cointegral on $D(H)$. □

**Corollary 12.8** *$D(H)$ admits a normalized left cointegral if and only if $D(H)$ is a semisimple algebra.*

*Proof* For the non-zero left cointegral $\Gamma$ defined in (12.2.1) we have

$$\begin{aligned}
\Gamma(S_D^{-1}(\varepsilon \bowtie \alpha)(\varepsilon \bowtie \beta)) &= \Gamma(\varepsilon \bowtie S^{-1}(\alpha)\beta)\\
&= \varepsilon(r)\mu(\tilde{p}^1)\mu^{-1}(f^1)\lambda(S^{-1}(\alpha f^2)\beta\tilde{p}^2).
\end{aligned}$$

This is non-zero if and only if $\varepsilon(r) \neq 0$ and $\mu(\tilde{p}^1)\mu^{-1}(f^1)\lambda(S^{-1}(\alpha f^2)\beta\tilde{p}^2) \neq 0$. But, as we have already mentioned, $\varepsilon(r) \neq 0$ implies $H$ unimodular, and in this case

$$\mu(\tilde{p}^1)\mu^{-1}(f^1)\lambda(S^{-1}(\alpha f^2)\beta\tilde{p}^2) = \lambda(S^{-1}(\alpha)\beta).$$

Then the result follows from Corollary 12.2. □

Now we describe the space of right cointegrals on $D(H)$.

**Proposition 12.9** *If $t \in \int_l^H$ and $\lambda \in \mathscr{L}$ are non-zero then $t \bowtie \lambda \circ S$ is a non-zero right cointegral on $D(H)$.*

*Proof* Since $D(H)$ is unimodular, $\Gamma \circ S_D$ is a non-zero right cointegral on $D(H)$; see Corollary 7.69. So it suffices to show that $\Gamma \circ S_D = S(r) \bowtie \lambda \circ S$. By applying $\mu$ to both sides of (12.1.2) we obtain after a straightforward computation that

$$\mu(S^{-1}(f^1)h_2g^2)\lambda(S^{-1}(f^2)h_1g^1S(h'))$$
$$= \mu^{-1}(\alpha \underline{g}^{-1})\mu(\tilde{q}^1)\mu(\tilde{q}^2_1h'_1\tilde{p}^1)\lambda(hS(\tilde{q}^2_2h'_2\tilde{p}^2)),$$

for all $h, h' \in H$. Consequently, by (3.2.24) we obtain that

$$\mu^{-1}(\tilde{q}^1)\mu(S^{-1}(f^1)S(h)_2g^2)\lambda(S^{-1}(f^1)S(h)_1g^1S(h'\tilde{q}^2))$$
$$= \mu^{-1}(\alpha \underline{g}^{-1})\mu(\tilde{q}^1)\mu(\tilde{q}^2_1h'_1)\lambda(S(\tilde{q}^2_2h'_2h)), \tag{12.2.2}$$

for all $h, h' \in H$. Then we compute:

$$\begin{aligned}
&\Gamma \circ S_D(\varphi \bowtie h)\\
&= \Gamma((\varepsilon \bowtie S(h)f^1)(p^1_1U^1 \rightharpoonup \varphi \circ S^{-1} \leftharpoonup f^2S^{-1}(p^2) \bowtie p^1_2U^2))\\
&= \varphi \circ S^{-1}(f^2S^{-1}(S(h)_2f^1_2p^2)rS(h)_{(1,1)}f^1_{(1,1)}p^1_1U^1)\\
&\quad \mu(\tilde{p}^1)\mu^{-1}(F^1)\lambda(S^{-1}(F^2)S(h)_{(1,2)}f^1_{(1,2)}p^1_2U^2S(\tilde{p}^2))\\
&\overset{(7.6.12)}{=} \varphi(S^{-1}(r))\mu(S^{-1}(F^1)S(h)_2g^2S(\tilde{q}^1))\mu(\tilde{p}^1)\lambda(S^{-1}(F^2)S(h)_1g^1S(\tilde{p}^2\tilde{q}^2))\\
&\overset{(12.2.2)}{=} \mu^{-1}(\alpha \underline{g}^{-1})\varphi(S^{-1}(r))\mu((\tilde{q}^2\tilde{p}^2)_1)\mu(\tilde{q}^1\tilde{p}^1)\lambda(S((\tilde{q}^2\tilde{p}^2)_2h))\\
&\overset{(3.2.20)}{=} \mu^{-1}(\underline{g}^{-1})\varphi(S^{-1}(r))\mu^{-1}(\alpha)\mu(\alpha)\mu^{-1}(\beta)\lambda(S(h))\\
&\overset{(7.2.8)}{=} \varphi((\mu^{-1}(\underline{g})\mu(\beta))^{-1}S^{-1}(r))\lambda(S(h)) = (S(r) \bowtie \lambda \circ S)(\varphi \bowtie h),
\end{aligned}$$

for all $\varphi \in H^*$ and $h \in H$, where in the last equality we used Corollary 7.73. □

The modular element of $D(H)^*$ is $\mu_D = \varepsilon_D$. Our next aim is to compute the modular element $\underline{g}_D$ of $D(H)$. To this end, we will need an explicit formula for the inverse of the antipode $S_D$ of $D(H)$ and a lemma.

**Proposition 12.10** *The composition inverse $S_D^{-1}$ of the antipode of $D(H)$ is given by the following formula, for all $\varphi \in H^*$ and $h \in H$:*

$$S_D^{-1}(\varphi \bowtie h) = (\varepsilon \bowtie S^{-1}(f^2h))(p^1_1S^{-1}(q^2g^2) \rightharpoonup \varphi \circ S \leftharpoonup S^{-1}(p^2f^1) \bowtie p^1_2S^{-1}(q^1g^1)).$$

*Proof* We first observe that (3.2.21) and (3.2.23) imply that

$$(\varepsilon \bowtie q^1h_1)(p^1_1 \rightharpoonup \varphi \leftharpoonup q^2h_2S^{-1}(p^2) \bowtie p^1_2) = h_1 \rightharpoonup \varphi \bowtie h_2,$$

for all $\varphi \in H^*$ and $h \in H$. Consequently,

$$\begin{aligned}
&(\varepsilon \bowtie q^1S(P^1)_1)(p^1_1U^1P^2f^1 \rightharpoonup \varphi \leftharpoonup q^2S(P^1)_2S^{-1}(p^2) \bowtie p^1_2U^2)(\varepsilon \bowtie f^2)\\
&= (S(P^1)_1U^1P^2f^1 \rightharpoonup \varphi \bowtie S(P^1)_2U^2)(\varepsilon \bowtie f^2)\\
&\overset{(7.3.1),(3.2.13)}{=} (g^1S(q^2P^1_2)P^2f^1 \rightharpoonup \varphi \bowtie g^2S(q^1p^1_1))(\varepsilon \bowtie f^2) \overset{(3.2.23)}{=} \varphi \bowtie 1_H,
\end{aligned}$$

for all $\varphi \in H^*$. By the definition of $S_D$ we have

$$S_D(S^{-1}(g^2) \rightharpoonup \varphi \bowtie S^{-1}(g^1)) = p^1_1U^1 \rightharpoonup \varphi \circ S^{-1} \leftharpoonup S^{-1}(p^2) \bowtie p^1_2U^2,$$

and combining these two relations we find for all $\varphi \in H^*$ that $\varphi \bowtie 1_H$ equals

$$(\varepsilon \bowtie q^1S(P^1)_1)S_D(S^{-1}(g^2) \rightharpoonup (P^2f^1 \rightharpoonup \varphi \leftharpoonup q^2S(P^1)_2) \circ S \bowtie S^{-1}(g^1))(\varepsilon \bowtie f^2).$$

As $H$ is a quasi-Hopf subalgebra of $D(H)$, $S_D^{-1}(\varepsilon \bowtie h) = \varepsilon \bowtie S^{-1}(h)$, for all $h \in H$. This and the fact that $S_D^{-1}$ is an anti-algebra morphism imply

$$\begin{aligned} S_D^{-1}(\varphi \bowtie h) &= S_D^{-1}(\varepsilon \bowtie h)S_D^{-1}(\varphi \bowtie 1_H) \\ &= (\varepsilon \bowtie S^{-1}(f^2h))\,(S^{-1}(q^2S(P^1)_2g^2) \rightharpoonup \varphi \circ S \\ &\qquad \leftharpoonup S^{-1}(P^2f^1) \bowtie S^{-1}(q^1S(P^1)_1g^1)) \\ &\overset{(3.2.13)}{=} (\varepsilon \bowtie S^{-1}(f^2h))(P_1^1S^{-1}(q^2g^2) \rightharpoonup \varphi \circ S \\ &\qquad \leftharpoonup S^{-1}(P^2f^1) \bowtie P_2^1S^{-1}(q^1g^1)), \end{aligned}$$

for all $\varphi \in H^*$ and $h \in H$. □

**Lemma 12.11** *Let $H$ be a finite-dimensional quasi-Hopf algebra. Then for all $r \in \int_r^H$ and $h \in H$ the following equalities hold:*

$$hr_1 \otimes r_2 = \mu^{-1}(h_1p^1)q^1r_1 \otimes S^{-1}(h_2p^2)q^2r_2, \tag{12.2.3}$$
$$r_1U^1 \otimes r_2U^2S(h) = r_1U^1h \otimes r_2U^2. \tag{12.2.4}$$

*Proof* (12.2.3) follows since

$$\begin{aligned} hr_1 \otimes r_2 &\overset{(3.2.23)}{=} hq^1p_1^1r_1 \otimes S^{-1}(p^2)q^2p_2^1r_2 \\ &\overset{(3.2.21)}{=} q^1(h_1p^1r)_1 \otimes S^{-1}(h_2p^2)q^2(h_1p^1r)_2 \\ &= \mu^{-1}(h_1p^1)q^1r_1 \otimes S^{-1}(h_2p^2)q^2r_2. \end{aligned}$$

(12.2.4) is a direct consequence of (7.3.2). □

In order to compute the modular element $\underline{g}_D$ of $D(H)$ we need a left integral $\mathbb{T}$ in $D(H)$ and a left cointegral $\Gamma$ on $D(H)$ such that $\Gamma(S_D^{-1}(\mathbb{T})) = 1$. Since $\mu_D = \varepsilon_D$ it turns out that this is equivalent to $\Gamma(\mathbb{T}) = 1$. Also note that the unimodularity of $D(H)$ implies that $\Gamma \circ S_D = \Gamma \circ S_D^{-1}$.

Now take $\mathbb{T} = \mu^{-1}(\delta^2) \rightharpoonup \lambda' \bowtie r'$ for some $0 \neq \lambda' \in \mathscr{L}$ and $0 \neq r' \in \int_r^H$, and let $\Gamma$ be defined as in (12.2.1). A simple inspection ensures that

$$\Gamma \circ S_D^{-1}(\mathbb{T}) = \mu(\delta^1)\mu^{-1}(\delta^2)\lambda'(S(r))\lambda(S(r'))$$

and since $S(\delta^1)\alpha\delta^2 = S(\beta)$ and $\varepsilon(\underline{g}) = \mu(\beta)$, by Remark 7.60 we conclude that $\Gamma \circ S_D^{-1}(\mathbb{T}) = \mu^{-1}(\alpha)^{-1}\lambda'(S(r))\lambda(r')$. Thus we have to consider $\lambda, \lambda'$ and $r, r'$ such that $\lambda'(S(r))\lambda(r') = \mu^{-1}(\alpha)$.

**Proposition 12.12** *The modular element $\underline{g}_D$ of $D(H)$ is given by*

$$\begin{aligned} \underline{g}_D &= \mu(g_1^1)\mu^{-1}(g^2)S_D^{-1}(\mu \bowtie g_2^1S^{-2}(\underline{g}^{-1})) \\ &= \mu(\tilde{q}^1g^1)\mu^{-1}(\tilde{p}^1)(\varepsilon \bowtie S^{-3}(\underline{g}^{-1}))(\mu^{-1} \bowtie (S^{-1}(\tilde{q}^2g^2) \leftharpoonup \mu^{-1})\tilde{p}^2). \end{aligned}$$

*Proof* Let $\lambda, \lambda' \in \mathscr{L}$ and $r, r' \in \int_r^H$ be such that $\lambda'(S(r))\lambda(r') = \mu^{-1}(\alpha)$. By the above comments, (7.6.2) and (8.5.9), the modular element $\underline{g}_D$ can be computed as:

$$\begin{aligned}
\underline{g}_D &= \mu^{-1}(\delta^2)\Gamma \circ S_D^{-1}((q_1^2X^2)_1 \rightharpoonup (\delta^1 \rightharpoonup \lambda')_1 \leftharpoonup S^{-1}(X^3) \bowtie (q_1^2X^2)_2Y^3x^3r_2'P^2)\\
&\quad S_D^{-1}((\varepsilon \bowtie q^1X_1^1Y^1)(p_1^1x^1 \rightharpoonup (\delta^1 \rightharpoonup \lambda')_2 \leftharpoonup Y^2S^{-1}(p^2) \bowtie p_2^1x^2r_1'P^1))\\
&= \mu(q_1^2)\lambda(S(q_2^2Y^3x^3r_2'P^2))\lambda'(S(r))\mu(Y^2S^{-1}(p^2))\mu(p_1^1x^1)\\
&\quad \mu^{-1}(\delta^2)\mu(\delta^1)S_D^{-1}((\varepsilon \bowtie q^1Y^1)(\mu \bowtie p_2^1x^2r_1'P^1))\\
&\overset{(12.2.3)}{=} \mu^{-1}(p_{(2,1)}^1x_1^2\mathbb{P}^1)\lambda'(S(r))\mu(q_1^2)\lambda(S(q_2^2Y^3x^3S^{-1}(p_{(2,2)}^1x_2^2\mathbb{P}^2)Q^2r_2'P^2))\\
&\quad \mu^{-1}(\alpha)^{-1}\mu(\beta)\mu(Y^2S^{-1}(p^2))\mu(p_1^1x^1)S_D^{-1}((\varepsilon \bowtie q^1Y^1)(\mu \bowtie Q^1r_1'P^1))\\
&\underset{(3.1.7)}{\overset{(5.5.16)}{=}} \mu^{-1}(X^2S^{-1}(X^1\beta))\mu^{-1}(\alpha)^{-1}\lambda'(S(r))\mu(Y^2S^{-1}(p^2))\mu(q_1^2)\\
&\quad \lambda(S(q_2^2Y^3S^{-1}(X^3p^1\beta)Q^2r_2'P^2))S_D^{-1}((\varepsilon \bowtie q^1Y^1)(\mu \bowtie Q^1r_1'P^1))\\
&\underset{(7.6.10)}{\overset{(7.2.7)}{=}} \mu^{-1}(\alpha)^{-1}\lambda'(S(r))\mu(q_1^2)\mu(Y^2S^{-1}(p^2))\lambda(S(q_2^2Y^3S^{-1}(p^1\beta)V^2r_2'U^2))\\
&\quad S_D^{-1}((\varepsilon \bowtie q^1Y^1)(\mu \bowtie V^1r_1'U^1))\\
&\overset{(12.1.8)}{=} \mu^{-1}(q_2^1Y_2^1p^2S(Y^2))\lambda'(S(r))\mu(q_1^2)\lambda(S(q_2^2Y^3S^{-1}(q_1^1Y_1^1p^1\beta)V^2r_2'U^2))\\
&\quad \mu^{-1}(\alpha)^{-1}S_D^{-1}(\mu \bowtie V^1r_1'U^1)\\
&\overset{(3.2.6)}{=} \mu^{-1}(q_2^1\delta^2S(q_1^2))\lambda'(S(r))\lambda \circ S(S^{-1}(q_1^1\delta^1S(q_2^2))V^2r_2'U^2)\\
&\quad \mu^{-1}(\alpha)^{-1}S_D^{-1}(\mu \bowtie V^1r_1'U^1)\\
&\underset{(3.2.13)}{\overset{(3.2.14)}{=}} \mu^{-1}((q^1\beta S(q^2))_2g^2)\lambda'(S(r))\lambda \circ S(S^{-1}((q^1\beta S(q^2))_1g^1)V^2r_2'U^2)\\
&\quad \mu^{-1}(\alpha)^{-1}S_D^{-1}(\mu \bowtie V^1r_1'U^1)\\
&\overset{(12.1.8)}{=} \mu(g_1^1)\mu^{-1}(g^2)\mu^{-1}(\alpha)^{-1}\lambda'(S(r))\lambda \circ S(V^2r_2'U^2)S_D^{-1}(\mu \bowtie g_2^1V^1r_1'U^1)\\
&\overset{(7.5.13)}{=} \mu(g_1^1)\mu^{-1}(g^2)S_D^{-1}(\mu \bowtie g_2^1S^{-2}(\underline{g}^{-1})).
\end{aligned}$$

In the second equality we used Proposition 12.9 and the fact that $\Gamma \circ S_D^{-1} = \Gamma \circ S_D$, in the third one we used the properties $S(r) \in \int_l^H$ and $\mu$ is an algebra map, and in the last equality Remark 7.60 and (12.2.4). We have also denoted by $\mathbb{P}^1 \otimes \mathbb{P}^2$ another copy of $p_R$. This proves the first formula for $\underline{g}_D$. For the second one we use the form of $S_D^{-1}$ found above to compute

$$\begin{aligned}
\underline{g}_D &= \mu(g_1^1)\mu^{-1}(g^2)(\varepsilon \bowtie S^{-3}(\underline{g}^{-1})S^{-1}(f^2g_2^1))\\
&\quad (p_1^1S^{-1}(q^2G^2) \rightharpoonup \mu^{-1} \leftharpoonup S^{-1}(p^2f^1) \bowtie p_2^1S^{-1}(q^1G^1))\\
&\overset{(8.5.2)}{=} \mu^{-1}((S^{-1}(f^2g_2^1)_1p^1)_1S^{-1}(q^2G^2))\mu(S^{-1}(f^2g_2^1)_2p^2f^1)\mu(g_1^1)\\
&\quad \mu^{-1}(g^2)(\varepsilon \bowtie S^{-3}(\underline{g}^{-1}))(\mu^{-1} \bowtie (S^{-1}((f^2g_2^1)_1p^1)_2S^{-1}(q^1G^1))\\
&\overset{(3.2.13),(3.2.17)}{=} \mu^{-1}(S^{-1}(f^2x^3g_{(2,2)}^1\mathbb{G}^2)_1S^{-1}(q^2G^2))\mu^{-1}(g^2)\\
&\quad \mu(S^{-1}(F^2f_2^1x^2g_{(2,1)}^1\mathbb{G}^1)\beta F^1f_1^1x^1g_1^1)\\
&\quad (\varepsilon \bowtie S^{-3}(\underline{g}^{-1}))(\mu^{-1} \bowtie S^{-1}(f^2x^3g_{(2,2)}^1\mathbb{G}^2)_2S^{-1}(q^1G^1))\\
&\underset{(3.2.1)}{\overset{(7.5.12),(3.1.7)}{=}} \mu^{-1}(S^{-1}(\alpha x^2\mathbb{G}^1)x^1)\mu^{-1}(S^{-1}(g^1x^3\mathbb{G}^2)_1S^{-1}(q^2G^2))\mu^{-1}(g^2)\\
&\quad (\varepsilon \bowtie S^{-3}(\underline{g}^{-1}))(\mu^{-1} \bowtie S^{-1}(g^1x^3\mathbb{G}^2)_2S^{-1}(q^1G^1))
\end{aligned}$$

$$\overset{(3.2.20),(3.2.13)}{\underset{(7.3.8)}{=}} \mu(\tilde{q}^1\mathbb{G}^1)\mu(S(\tilde{p}^1)f^2\tilde{q}^2_2\mathbb{G}^2_2G^2)$$
$$(\varepsilon \bowtie S^{-3}(\underline{g}^{-1}))(\mu^{-1} \bowtie S^{-1}(S(\tilde{p}^2)f^1\tilde{q}^2_1\mathbb{G}^2_1G^1))$$
$$\overset{(3.2.13)}{=} \mu(\tilde{q}^1\mathbb{G}^1)\mu^{-1}(\tilde{p}^1)(\varepsilon \bowtie S^{-3}(\underline{g}^{-1}))(\mu^{-1} \bowtie (S^{-1}(\tilde{q}^2\mathbb{G}^2) \leftharpoonup \mu^{-1})\tilde{p}^2),$$

and this completes the proof. □

## 12.3 The Quantum Dimension

Let $\mathscr{C}$ be a braided category which is left rigid. If $V$ is an object of $\mathscr{C}$, and $\mathrm{ev}_V$ and $\mathrm{coev}_V$ are the evaluation and coevaluation morphisms associated to V, we define the quantum dimension (or representation-theoretic rank) of $V$ as follows:

$$\underline{\dim}(V) = \mathrm{ev}_V \circ c_{V,V^*} \circ \mathrm{coev}_V.$$

If $H$ is a quasi-Hopf algebra then the category ${}_H\mathscr{M}^{\mathrm{fd}}$ of finite-dimensional modules over $H$ is left rigid. Therefore, if $H$ is a QT quasi-Hopf algebra and $V$ a finite-dimensional left $H$-module it makes sense to consider the representation-theoretic rank of $V$. If $R = R^1 \otimes R^2$ is an $R$-matrix for $H$ then

$$\underline{\dim}(V) = \sum_{i=1}^{n} v^i(S(R^2)\alpha R^1\beta \cdot v_i) = \mathrm{Tr}(\eta), \tag{12.3.1}$$

where $\eta := S(R^2)\alpha R^1\beta$. Here $\mathrm{Tr}(\eta)$ is the trace of the linear endomorphism of $V$ defined by $v \mapsto \eta \cdot v$, and $\{v_i, v^i\}_i$ are dual bases in $V$ and $V^*$.

Let $u$ be the element defined in (10.3.4). By (10.3.8) we have that $S(R^2)\alpha R^1 = S(\alpha)u$, so by (10.3.7) we obtain

$$\eta = S(S(\beta)\alpha)u = uS^{-1}(\alpha)\beta. \tag{12.3.2}$$

The aim of this section is to compute the quantum dimension of $H$ and $D(H)$ within the braided rigid monoidal category ${}_{D(H)}\mathscr{M}^{\mathrm{fd}}$. If $\{e_i, e^i\}_i$ are dual bases in $H$ and $H^*$, by (10.4.2) and the fact that $H$ is a quasi-Hopf subalgebra of $D(H)$ we get that $\eta_D$, the corresponding element $\eta$ for $D(H)$, is given by

$$\eta_D = \sum_{i=1}^{n} \beta \rightharpoonup \overline{S}^{-1}(e^i) \bowtie e_i S^{-1}(\alpha)\beta.$$

### 12.3.1 The Quantum Dimension of $H$

We first compute the quantum dimension (or representation-theoretic rank) of $H$ within the braided rigid category ${}_{D(H)}\mathscr{M}^{\mathrm{fd}}$. Recall that $H$ is a left $D(H)$-module via the action $\rightarrow$ defined in (8.7.11). It can be rewritten as follows:

$$(\varphi \bowtie h) \rightarrow h' \overset{(3.1.9)}{\underset{(3.2.1)}{=}} \langle \varphi, S^{-1}(Y^3)q^2Y^2_2y^3_2S^{-1}(\tilde{q}^1y^2_1(h \triangleright h')_1g^1)y^1\rangle$$
$$Y^1\tilde{q}^2y^2_2(h \triangleright h')_2g^2S(q^1Y^2_1y^3_1)$$
$$\overset{(3.2.13)}{\underset{(7.3.1)}{=}} \langle \varphi, S^{-1}\left(\tilde{q}^1(y^2(h \triangleright h')S(Y^2y^3))_1U^1Y^3\right)y^1\rangle$$

$$
\begin{aligned}
&Y^1\tilde{q}^2(y^2(h\triangleright h')S(Y^2y^3))_2U^2\\
&\overset{(3.2.22)}{=}\langle\varphi,S^{-1}\left(\tilde{q}^1(Y_2^1y^2(h\triangleright h')S(Y^2y^3))_1U^1Y^3\right)Y_1^1y^1\rangle\\
&\quad\tilde{q}^2(Y_2^1y^2(h\triangleright h')S(Y^2y^3))_2U^2.
\end{aligned}
$$

Hence we have shown that for all $\varphi\in H^*$ and $h,h'\in H$ we have

$$
\begin{aligned}
(\varphi\bowtie h)\rightarrow h'=\langle\varphi,S^{-1}&\left(\tilde{q}^1(Y_2^1y^2(h\triangleright h')S(Y^2y^3))_1U^1Y^3\right)Y_1^1y^1\rangle\\
&\tilde{q}^2(Y_2^1y^2(h\triangleright h')S(Y^2y^3))_2U^2.
\end{aligned}
\tag{12.3.3}
$$

So this action defines on $H$ a left $D(H)$-module structure, and on $H_0$ a left $D(H)$-module algebra structure; see Proposition 8.42.

For the computation of $\underline{\dim}(H)$ we need the following formulas.

**Lemma 12.13** *Let $H$ be a finite-dimensional quasi-Hopf algebra and $\{e_i\}_i$ a basis in $H$ with dual basis $\{e^i\}_i$. Then for all $h,h',h''\in H$ the following relations hold:*

$$
\begin{aligned}
&\sum_{i=1}^n\langle e^i,S^{-1}(\beta)S^{-2}(\tilde{Q}^1(e_i)_1h')h\tilde{q}^2\tilde{Q}_2^2(e_i)_{(2,2)}h''S^{-1}(\tilde{q}^1\tilde{Q}_1^2(e_i)_{(2,1)})\rangle\\
&\qquad=\sum_{i=1}^n\langle e^i,S^{-1}(\beta)S^{-2}(\tilde{Q}^1(e_i)_1h')\tilde{q}^2\tilde{Q}_2^2(e_i)_{(2,2)}h''S^{-1}(\tilde{q}^1\tilde{Q}_1^2(e_i)_{(2,1)})h\rangle,
\end{aligned}
\tag{12.3.4}
$$

$$
\begin{aligned}
&\sum_{i=1}^n\langle e^i,S^{-1}(\beta)S^{-2}(\tilde{Q}^1(e_i)_1X^1p_1^1h')h_1\tilde{q}^2\tilde{Q}_2^2(e_i)_{(2,2)}X^3p^2S(h_2)h''\\
&\qquad S^{-1}(\tilde{q}^1\tilde{Q}_1^2(e_i)_{(2,1)}X^2p_2^1)\rangle=\sum_{i=1}^n\langle e^i,S^{-1}(\beta)S^{-2}(\tilde{Q}^1(e_i)_1X^1p_1^1h_1h')\\
&\qquad\qquad\tilde{q}^2\tilde{Q}_2^2(e_i)_{(2,2)}X^3p^2h''S^{-1}(\tilde{q}^1\tilde{Q}_1^2(e_i)_{(2,1)}X^2p_2^1h_2)\rangle,
\end{aligned}
\tag{12.3.5}
$$

*where we denoted $q_L=\tilde{q}^1\otimes\tilde{q}^2=\tilde{Q}^1\otimes\tilde{Q}^2$ and $p_R=p^1\otimes p^2$.*

*Proof* In order to prove (12.3.4) we will apply (3.2.22) twice, and then the properties of dual bases and (3.2.1). Explicitly,

$$
\begin{aligned}
&\sum_{i=1}^n\langle e^i,S^{-1}(\beta)S^{-2}(\tilde{Q}^1(e_i)_1h')h\tilde{q}^2\tilde{Q}_2^2(e_i)_{(2,2)}h''S^{-1}(\tilde{q}^1\tilde{Q}_1^2(e_i)_{(2,1)})\rangle\\
&\quad=\sum_{i=1}^n\langle e^i,S^{-1}(\beta)S^{-2}(\tilde{Q}^1(e_i)_1h')\tilde{q}^2(h_2\tilde{Q}^2)_2(e_i)_{(2,2)}h''\\
&\qquad\quad S^{-1}(\tilde{q}^1(h_2\tilde{Q}^2)_1(e_i)_{(2,1)})h_1\rangle\\
&\quad=\sum_{i=1}^n\langle e^i,S^{-1}(\beta)S^{-2}(S(h_{(2,1)})\tilde{Q}^1(h_{(2,2)}e_i)_1h')\tilde{q}^2\tilde{Q}_2^2(h_{(2,2)}e_i)_{(2,2)}h''\\
&\qquad\quad S^{-1}(\tilde{q}^1\tilde{Q}_1^2(h_{(2,2)}e_i)_{(2,1)})h_1\rangle\\
&\quad=\sum_{i=1}^n\langle e^i,h_{(2,2)}S^{-1}(h_{(2,1)}\beta)S^{-2}(\tilde{Q}^1(e_i)_1h')\tilde{q}^2\tilde{Q}_2^2(e_i)_{(2,2)}h''\\
&\qquad\quad S^{-1}(\tilde{q}^1\tilde{Q}_1^2(e_i)_{(2,1)})h_1\rangle
\end{aligned}
$$

$$= \sum_{i=1}^{n} \langle e^i, S^{-1}(\beta)S^{-2}(\tilde{Q}^1(e_i)_1h')\tilde{q}^2\tilde{Q}^2_2(e_i)_{(2,2)}h''S^{-1}(\tilde{q}^1\tilde{Q}^2_1(e_i)_{(2,1)})h\rangle.$$

In a similar manner one can prove (12.3.5). It follows by applying (12.3.4), dual bases, (3.1.7) and (3.2.21); we leave the details to the reader. □

We are now able to compute $\underline{\dim}(H)$.

**Proposition 12.14** *Let $H$ be a finite-dimensional quasi-Hopf algebra. Then the quantum dimension of $H$ within ${}_{D(H)}\mathcal{M}^{\mathrm{fd}}$ is*

$$\underline{\dim}(H) = \mathrm{Tr}\left(h \mapsto S^{-2}(S(\beta)\alpha h\beta S(\alpha))\right). \tag{12.3.6}$$

*Proof* We set $p_R = p^1 \otimes p^2 = P^1 \otimes P^2$, $q_L = \tilde{q}^1 \otimes \tilde{q}^2 = \tilde{Q}^1 \otimes \tilde{Q}^2$ and $f = f^1 \otimes f^2 = F^1 \otimes F^2$. Then by (12.3.1) and the above expression of $\eta_D$ we have:

$$\begin{aligned}
&\underline{\dim}(H)\\
&= \sum_{i,j=1}^{n} \langle e^j, (e^i \bowtie S^{-1}(\alpha e_i\beta)\beta) \rightarrow e_j\rangle\\
&\overset{(12.3.3)}{=} \sum_{i,j=1}^{n} \langle e^i, S^{-1}\left(\tilde{q}^1(Y^1_2y^2(S^{-1}(\alpha e_i\beta)\beta \triangleright e_j)S(Y^2y^3))_1U^1Y^3\right)Y^1_1y^1\rangle\\
&\qquad \langle e^j, \tilde{q}^2(Y^1_2y^2(S^{-1}(\alpha e_i\beta)\beta \triangleright e_j)S(Y^2y^3))_2U^2\rangle\\
&= \sum_{i,j,k=1}^{n} \langle e^k, Y^1_2y^2(S^{-1}(\alpha e_i\beta)\beta \triangleright e_j)S(Y^2y^3)\rangle\\
&\qquad \langle Y^1_1y^1 \rightharpoonup e^i, S^{-1}(\tilde{q}^1(e_k)_1U^1Y^3)\rangle\langle e^j, \tilde{q}^2(e_k)_2U^2\rangle\\
&= \sum_{i,k=1}^{n} \langle e^k, Y^1_2y^2\left(S^{-1}(\alpha e_iY^1_1y^1\beta)\beta \triangleright \tilde{q}^2(e_k)_2U^2\right)S(Y^2y^3)\rangle\\
&\qquad \langle e^i, S^{-1}(\tilde{q}^1(e_k)_1U^1Y^3)\rangle\\
&\overset{(8.7.1),(3.2.13)}{\underset{(3.2.14)}{=}} \sum_{i,k=1}^{n} \langle e^k, Y^1_2y^2S^{-1}(f^2Y^1_{(1,2)}y^1_2\delta^2)\left(S^{-1}(\alpha e_i)\beta \triangleright \tilde{q}^2(e_k)_2U^2\right)\\
&\qquad f^1Y^1_{(1,1)}y^1_1\delta^1S(Y^2y^3)\rangle\langle e^i, S^{-1}(\tilde{q}^1(e_k)_1U^1Y^3)\rangle\\
&\overset{(3.2.6),(3.1.9)}{\underset{(3.2.19)}{=}} \sum_{i,k=1}^{n} \langle e^k, Y^1_2S^{-1}(f^2Y^1_{(1,2)}p^2)\left(S^{-1}(\alpha e_i)\beta \triangleright \tilde{q}^2(e_k)_2U^2\right)\\
&\qquad f^1Y^1_{(1,1)}p^1\beta S(Y^2)\rangle\langle e^i, S^{-1}(\tilde{q}^1(e_k)_1U^1Y^3)\rangle\\
&\overset{(3.2.21),(3.2.20)}{=} \sum_{i,k=1}^{n} \langle S(\tilde{p}^1) \rightharpoonup e^k, S^{-1}(f^2p^2)\left(S^{-1}(\alpha e_i)\beta \triangleright \tilde{q}^2(e_k)_2U^2\right)f^1p^1\rangle\\
&\qquad \langle e^i, S^{-1}(\tilde{q}^1(e_k)_1U^1\tilde{p}^2)\rangle\\
&= \sum_{i,k=1}^{n} \langle e^k, S^{-1}(f^2p^2)\left(S^{-1}(\alpha e_i)\beta \triangleright \tilde{q}^2(e_k)_2S(\tilde{p}^1)_2U^2\right)f^1p^1\rangle\\
&\qquad \langle e^i, S^{-1}(\tilde{q}^1(e_k)_1S(\tilde{p}^1)_1U^1\tilde{p}^2)\rangle
\end{aligned}$$

$$
\begin{aligned}
&\overset{(7.3.5)}{=} \sum_{k=1}^{n} \langle e^k, S^{-1}(f^2p^2)\big(S^{-1}(\alpha S^{-1}(\tilde{q}^1(e_k)_1P^1))\beta \triangleright \tilde{q}^2(e_k)_2P^2\big) f^1p^1\rangle \\
&\underset{(3.2.19)}{\overset{(8.7.1),(3.2.13)}{=}} \sum_{k=1}^{n} \langle e^k \leftharpoonup x^3, S^{-1}(f^2S^{-1}(F^1\tilde{q}^1_1(e_k)_{(1,1)}p^1_1g^1)x^2\beta) \\
&\qquad \big(S^{-1}(\alpha)\beta \triangleright \tilde{q}^2(e_k)_2p^2\big) f^1S^{-1}(F^2\tilde{q}^1_2(e_k)_{(1,2)}p^1_2g^2)x^1\rangle \\
&= \sum_{k=1}^{n} \langle e^k, S^{-1}(f^2S^{-1}(F^1\tilde{q}^1_1x^3_{(1,1)}(e_k)_{(1,1)}p^1_1g^1)x^2\beta)\big(S^{-1}(\alpha)\beta \\
&\qquad \triangleright \tilde{q}^2x^3_2(e_k)_2p^2\big) f^1S^{-1}(F^2\tilde{q}^1_2x^3_{(1,2)}(e_k)_{(1,2)}p^1_2g^2)x^1\rangle \\
&\underset{(3.2.14)}{\overset{(3.2.20),(3.2.13)}{=}} \sum_{k=1}^{n} \langle e^k, S^{-1}(\gamma^2S^{-1}(\tilde{Q}^1X^1(e_k)_{(1,1)}p^1_1g^1)\beta) \\
&\qquad \beta_1\tilde{q}^2\tilde{Q}^2_2X^3(e_k)_2p^2S(\beta_2)\gamma^1S^{-1}(\tilde{q}^1\tilde{Q}^2_1X^2(e_k)_{(1,2)}p^1_2g^2)\rangle \\
&\overset{(3.1.7),(12.3.4)}{=} \sum_{k=1}^{n} \langle e^k, S^{-1}(\gamma^2S^{-1}(\tilde{Q}^1(e_k)_1X^1p^1_1g^1)\beta)\tilde{q}^2\tilde{Q}^2_2 \\
&\qquad (e_k)_{(2,2)}X^3p^2S(\beta_2)\gamma^1S^{-1}(\tilde{q}^1\tilde{Q}^2_1(e_k)_{(2,1)}X^2p^1_2g^2)\beta_1\rangle \\
&\overset{(12.3.5),(3.2.14)}{=} \sum_{k=1}^{n} \langle e^k, S^{-1}(\gamma^2S^{-1}(\tilde{Q}^1(e_k)_1X^1p^1_1\delta^1)\beta) \\
&\qquad \tilde{q}^2\tilde{Q}^2_2(e_k)_{(2,2)}X^3p^2\gamma^1S^{-1}(\tilde{q}^1\tilde{Q}^2_1(e_k)_{(2,1)}X^2p^1_2\delta^2)\rangle \\
&\overset{(3.2.5),(3.1.5)}{=} \sum_{k=1}^{n} \langle e^k, S^{-1}(\beta)S^{-2}(\tilde{Q}^1(e_k)_1X^1p^1_1Y^1_1x^1\beta S(S(Z^1)\alpha y^3Z^3_2Y^3)) \\
&\qquad \tilde{q}^2\tilde{Q}^2_2(e_k)_{(2,2)}X^3p^2S^{-1}(\tilde{q}^1\tilde{Q}^2_1(e_k)_{(2,1)}X^2p^1_2Y^1_2 \\
&\qquad x^2\beta S(S(y^1Z^2)\alpha y^2Z^3_1Y^2x^3))\rangle \\
&\overset{(3.2.21),(3.1.7)}{=} \sum_{k=1}^{n} \langle e^k, S^{-1}(\beta)S^{-2}(\tilde{Q}^1(e_k)_1X^1p^1_1x^1\beta S(S(Z^1)\alpha y^3Z^3_2Y^3)) \\
&\qquad \tilde{q}^2\tilde{Q}^2_2(e_k)_{(2,2)}X^3p^2S^{-1}(\tilde{q}^1\tilde{Q}^2_1(e_k)_{(2,1)}X^2p^1_2 \\
&\qquad x^2\beta S(S(y^1Z^2Y^1_2)\alpha y^2Z^3_1Y^2x^3)Y^1_1\rangle \\
&\underset{(3.2.1)}{\overset{(12.3.4),(3.1.9)}{=}} \sum_{k=1}^{n} \langle e^k, S^{-1}(\beta)S^{-2}(\tilde{Q}^1(e_k)_1X^1p^1_1x^1\beta S(\alpha))Y^1\tilde{q}^2\tilde{Q}^2_2(e_k)_{(2,2)} \\
&\qquad X^3p^2S^{-1}(\tilde{q}^1\tilde{Q}^2_1(e_k)_{(2,1)}X^2p^1_2x^2\beta S(S(Y^2)\alpha Y^3x^3)))\rangle \\
&\overset{(3.2.19),(12.3.4)}{=} \sum_{k=1}^{n} \langle q^1 \rightharpoonup e^k, S^{-1}(\beta)S^{-2}(\tilde{Q}^1(e_k)_1X^1p^1_1P^1\beta S(\alpha))\tilde{q}^2\tilde{Q}^2_2 \\
&\qquad (e_k)_{(2,2)}X^3p^2S(q^2)S^{-1}(\tilde{q}^1\tilde{Q}^2_1(e_k)_{(2,1)}X^2p^1_2P^2))\rangle \\
&\overset{(3.1.7)}{=} \sum_{k=1}^{n} \langle e^k, S^{-1}(\beta)S^{-2}(\tilde{Q}^1(e_k)_1X^1(q^1_1p^1)_1P^1\beta S(\alpha))\tilde{q}^2\tilde{Q}^2_2 \\
&\qquad (e_k)_{(2,2)}X^3q^1_2p^2S(q^2)S^{-1}(\tilde{q}^1\tilde{Q}^2_1(e_k)_{(2,1)}X^2(q^1_1p^1)_2P^2))\rangle
\end{aligned}
$$

$$\overset{(3.2.23),(3.2.19)}{=} \sum_{k=1}^{n} \langle e^k, S^{-1}(\beta)S^{-2}(\tilde{Q}^1(e_k)_1\beta S(\alpha))\tilde{q}^2$$
$$\tilde{Q}^2_2(e_k)_{(2,2)}S^{-1}(\tilde{q}^1\tilde{Q}^2_1(e_k)_{(2,1)}\beta)\rangle$$
$$\overset{(3.2.1),(3.2.20)}{=} \sum_{k=1}^{n} \langle e^k, S^{-1}(\beta)S^{-2}(\alpha e_k\beta S(\alpha))\tilde{q}^2 S^{-1}(\tilde{q}^1\beta)\rangle$$
$$\overset{(3.2.20),(3.2.2)}{=} \sum_{k=1}^{n} \langle e^k, S^{-2}(S(\beta)\alpha e_k\beta S(\alpha))\rangle$$
$$= \mathrm{Tr}\left(h\mapsto S^{-2}(S(\beta)\alpha h\beta S(\alpha))\right),$$

so the proof is finished. □

Further on in this chapter we shall see that the quantum dimension of $H$ is closely connected with what will be called the trace formula for quasi-Hopf algebras.

### 12.3.2 The Quantum Dimension of $D(H)$

We show that $\underline{\dim}(D(H)) = \underline{\dim}(H)$ within ${}_{D(H)}\mathscr{M}^{\mathrm{fd}}$.

**Lemma 12.15** *In a quasi-Hopf algebra H the following relations hold:*

$$\Omega^1_1\delta^1S^2(\Omega^4)\otimes\Omega^1_{(2,1)}\delta^2_1 g^1S(\Omega^3)\otimes\Omega^1_{(2,2)}\delta^2_2 g^2S(\Omega^2)\otimes\Omega^5$$
$$= X^1p^1_1P^1S(f^1\tilde{p}^1)\otimes X^2p^1_2P^2\otimes X^3p^2\otimes S^{-1}(f^2\tilde{p}^2), \tag{12.3.7}$$
$$\gamma^1X^1\otimes f^1\gamma^2_1X^2\otimes f^2\gamma^2_2X^3 = S(X^3)f^1\gamma^1_1\otimes S(X^2)f^2\gamma^1_2\otimes S(X^1)\gamma^2. \tag{12.3.8}$$

*Here* $\Omega = \Omega^1\otimes\cdots\otimes\Omega^5$, $\delta = \delta^1\otimes\delta^2$, $\gamma = \gamma^1\otimes\gamma^2$, $f = f^1\otimes f^2$, $f^{-1} = g^1\otimes g^2$, $q_R = q^1\otimes q^2$, $p_R = p^1\otimes p^2 = P^1\otimes P^2$ *and* $q_L = \tilde{q}^1\otimes\tilde{q}^2$ *are the elements defined in* (8.5.1), (3.2.5), (3.2.15), (3.2.16), (3.2.19) *and* (3.2.20), *respectively.*

*Proof* Using the definition of $\delta$ and $\Omega$ we compute:

$$\Omega^1_1\delta^1S^2(\Omega^4)\otimes\Omega^1_{(2,1)}\delta^2_1 g^1S(\Omega^3)\otimes\Omega^1_{(2,2)}\delta^2_2 g^2S(\Omega^2)\otimes\Omega^5$$
$$\overset{(3.2.13)}{=} X^1_{(1,1)_1}y^1_1p^1\beta S(f^1X^2)\otimes X^1_{(1,1)_{(2,1)}}y^1_{(2,1)}p^2_1g^1S(X^1_2y^3)$$
$$\otimes X^1_{(1,1)_{(2,2)}}y^1_{(2,2)}p^2_2g^2S(X^1_{(1,2)}y^2)\otimes S^{-1}(f^2X^3)$$
$$\overset{(3.2.19),(3.1.7)}{=} Y^1\left((X^1_1)_{(1,1)}p^1\right)_1P^1\beta S(f^1X^2)\otimes Y^2\left((X^1_1)_{(1,1)}p^1\right)_2P^2S(X^1_2)$$
$$\otimes Y^3(X^1_1)_{(1,2)}p^2S((X^1_1)_2)\otimes S^{-1}(f^2X^3)$$
$$\overset{(3.2.21),(3.2.20)}{=} Y^1p^1_1P^1S(f^1\tilde{p}^1)\otimes Y^2p^1_2P^2\otimes Y^3p^2\otimes S^{-1}(f^2\tilde{p}^2),$$

so (12.3.7) is proved. The relation in (12.3.8) follows more easily since

$$\gamma^1X^1\otimes f^1\gamma^2_1X^2\otimes f^2\gamma^2_2X^3$$
$$\overset{(3.2.14)}{=} F^1\alpha_1X^1\otimes f^1F^2_1\alpha_{(2,1)}X^2\otimes f^2F^2_2\alpha_{(2,2)}X^3$$
$$\overset{(3.1.7),(3.2.17)}{=} S(X^3)f^1F^1_1\alpha_{(1,1)}\otimes S(X^2)f^2F^1_2\alpha_{(1,2)}\otimes S(X^1)F^2\alpha_2$$
$$\overset{(3.2.14)}{=} S(X^3)f^1\gamma^1_1\otimes S(X^2)f^2\gamma^1_2\otimes S(X^1)\gamma^2,$$

where we denoted by $F^1 \otimes F^2$ another copy of $f$. □

In (7.2.1) we have constructed a projection onto the space of left integrals of a finite-dimensional quasi-Hopf algebra $H$. Replacing the quasi-Hopf algebra $H$ by $H^{\rm cop}$ we obtain a second projection onto the space of left integrals, denoted in what follows by $\tilde{\mathfrak{P}}$. Since in $H^{\rm cop}$ we have $(q_R)_{\rm cop} = \tilde{q}^2 \otimes \tilde{q}^1$ we obtain

$$\tilde{\mathfrak{P}}(h) = \sum_{i=1}^{n} \langle e^i, S^{-1}(\beta) S^{-2}(\tilde{q}^1 (e_i)_1) h \rangle \tilde{q}^2 (e_i)_2 \in \int_l^H, \quad \forall\, h \in H. \tag{12.3.9}$$

We can now compute the representation-theoretic rank of $D(H)$.

**Proposition 12.16** *Let $H$ be a finite-dimensional quasi-Hopf algebra and $D(H)$ its quantum double. Then*

$$\underline{\dim}(D(H)) = \underline{\dim}(H) = \mathrm{Tr}\left(h \mapsto S^{-2}(S(\beta)\alpha h \beta S(\alpha))\right).$$

*Proof* We set $p_R = p^1 \otimes p^2 = P^1 \otimes P^2$, $q_R = q^1 \otimes q^2 = Q^1 \otimes Q^2$ and $f = f^1 \otimes f^2 = F^1 \otimes F^2 = \mathscr{F}^1 \otimes \mathscr{F}^2$. The expression of $\eta_D$ allows to compute:

$$\begin{aligned}
&\underline{\dim}(D(H)) \\
&\quad = \sum_{i,j=1}^{n} \langle e_i \bowtie e^j, \eta_D(e^i \bowtie e_j) \rangle \\
&\quad = \sum_{i,j,k=1}^{n} \langle e_i \bowtie e^j, (\beta \rightharpoonup \overline{S}^{-1}(e^k) \leftharpoonup S(\beta)\alpha \bowtie e_k)(e^i \bowtie e_j) \rangle \\
&\overset{(8.5.2)}{=} \sum_{i,j,k=1}^{n} \langle \overline{S}^{-1}(e^k), S(\beta)\alpha \Omega^5 (e_i)_1 \Omega^1 \beta \rangle \\
&\qquad\qquad \langle e^i, S^{-1}((e_k)_2) \Omega^4 (e_i)_2 \Omega^2 (e_k)_{(1,1)} \rangle \langle e^j, \Omega^3 (e_k)_{(1,2)} e_j \rangle \\
&\overset{(3.2.13),(3.2.14)}{=} \sum_{i,j=1}^{n} \langle e^j, \Omega^3 S^{-1}\left(S(\beta_1)\gamma^2 \Omega_2^5 (e_i)_{(1,2)} \Omega_2^1 \delta^2\right)_2 e_j \rangle \\
&\qquad\qquad \langle e^i, S^{-2}(S(\beta_2)\gamma^1 \Omega_1^5 (e_i)_{(1,1)} \Omega_1^1 \delta^1) \Omega^4 (e_i)_2 \Omega^2 \\
&\qquad\qquad S^{-1}\left(S(\beta_1)\gamma^2 \Omega_2^5 (e_i)_{(1,2)} \Omega_2^1 \delta^2\right)_1 \rangle \\
&\overset{(3.2.13),(12.3.7)}{=} \sum_{i,j=1}^{n} \langle e^i, S^{-2}\left(S(\beta_2)\gamma^1 S^{-1}(F^2 \tilde{p}^2)_1 ((e_i)_1)_1 Y^1 p_1^1 P^1 S(F^1 \tilde{p}^1)\right) (e_i)_2 \\
&\qquad\qquad S^{-1}\left(f^2 S(\beta_1)_2 \gamma_2^2 S^{-1}(F^2 \tilde{p}^2)_{(2,2)} ((e_i)_1)_{(2,2)} Y^3 p^2\right) \rangle \\
&\qquad\qquad \langle e^j, S^{-1}\left(f^1 S(\beta_1)_1 \gamma_1^2 S^{-1}(F^2 \tilde{p}^2)_{(2,1)} ((e_i)_1)_{(2,1)} Y^2 p_2^1 P^2\right) e_j \rangle \\
&\overset{(3.1.7),(3.2.21)}{=} \sum_{i,j=1}^{n} \langle e^i, S^{-2}\left(S(\beta_2)\gamma^1 Y^1 (S^{-1}(F^2 \tilde{p}^2)_1 p^1)_1 (e_i)_1 P^1 S(F^1 \tilde{p}^1)\right) \\
&\qquad\qquad S^{-1}\left(f^2 S(\beta_1)_2 \gamma_2^2 Y^3 S^{-1}(F^2 \tilde{p}^2)_2 p^2\right) \rangle \\
&\qquad\qquad \langle e^j, S^{-1}\left(f^1 S(\beta_1)_1 \gamma_1^2 Y^2 (S^{-1}(F^2 \tilde{p}^2)_1 p^1)_2 (e_i)_2 P^2\right) e_j \rangle
\end{aligned}$$

$$\overset{(3.2.13),(3.2.19)}{\underset{(3.2.17)}{=}} \sum_{i,j=1}^{n} \langle e^i, S^{-2}\left(S(\beta_2)\gamma^1 Y^1 S^{-1}(F^2 x^3 \tilde{p}^2_2 g^2)_1 (e_i)_1 P^1 S(\mathscr{F}^1 F^1_1 x^1 \tilde{p}^1)\right)$$
$$S^{-1}\left(f^2 S(\beta_1)_2 \gamma^2_2 Y^3 S^{-1}(\mathscr{F}^2 F^1_2 x^2 \tilde{p}^2_1 g^1)\beta\right)\rangle$$
$$\langle e^j, S^{-1}\left(f^1 S(\beta_1)_1 \gamma^2_1 Y^2 S^{-1}(F^2 x^3 \tilde{p}^2_2 g^2)_2 (e_i)_2 P^2\right) e_j\rangle$$
$$\overset{(7.5.12),(3.2.1)}{\underset{(3.2.20)}{=}} \sum_{i,j=1}^{n} \langle e^i, S^{-2}(S(\beta_2)\gamma^1 Y^1 S^{-1}(\tilde{q}^2 \tilde{p}^2_2 g^2)_1 (e_i)_1 P^1 S(\tilde{p}^1)\tilde{q}^1 \tilde{p}^2_1 g^1$$
$$S(f^2 S(\beta_1)_2 \gamma^2_2 Y^3))\rangle\langle e^j, S^{-1}(f^1 S(\beta_1)_1 \gamma^2_1 Y^2$$
$$S^{-1}(\tilde{q}^2 \tilde{p}^2_2 g^2)_2 (e_i)_2 P^2) e_j\rangle$$
$$\overset{(3.2.24),(3.2.13)}{=} \sum_{i,j=1}^{n} \langle e^i, S^{-2}\left(S(\beta_2 g^2)\gamma^1 Y^1 (e_i)_1 P^1 g^1 S(f^2 \gamma^2_2 Y^3)\right) \beta_{(1,1)}\rangle$$
$$\langle e^j, S^{-1}(f^1 \gamma^2_1 Y^2 (e_i)_2 P^2)\beta_{(1,2)} e_j\rangle$$
$$\overset{(12.3.8)}{=} \sum_{i,j=1}^{n} \langle e^i, S^{-2}\left(S(Y^3 \beta_2 g^2) f^1 \gamma^1_1 (e_i)_1 (\beta_1)_{(1,1)} P^1 g^1 S(S(Y^1)\gamma^2)\right)\rangle$$
$$\langle e^j, (\beta_1)_2 S^{-1}(S(Y^2) f^2 \gamma^1_2 (e_i)_2 (\beta_1)_{(1,2)} P^2) e_j\rangle$$
$$\overset{(3.2.21),(3.2.14)}{=} \sum_{i,j=1}^{n} \langle e^i, \gamma^1 S^{-2}\left(S(Y^3 \delta^2) f^1 (e_i)_1 P^1 \delta^1 S(S(Y^1)\gamma^2)\right)\rangle$$
$$\langle e^j, S^{-1}(S(Y^2) f^2 (e_i)_2 P^2) e_j\rangle$$
$$\overset{(3.2.5),(3.2.6)}{=} \sum_{i,j=1}^{n} \langle \overline{S}^{-2}(e^i), S(X^1 x^1_1 Y^1)\alpha x^3 y^3_2 Z^3 S^{-1}(f^1 P^1 y^1 \beta)(e_i)_2 Y^3 y^2 Z^1 \beta$$
$$S(S(X^2)\alpha X^3 x^2 y^3_1 Z^2) S^2(x^1_2)\rangle\langle e^j, S^{-1}(f^2 P^2)(e_i)_1 Y^2 e_j\rangle$$
$$\overset{(3.2.19)}{=} \sum_{i,j=1}^{n} \langle \overline{S}^{-2}(e^i), S(q^1 x^1_1 Y^1)\alpha x^3 y^3_2 Z^3 S^{-1}(f^1 P^1 y^1 \beta)(e_i)_2 x^1_{(2,2)} Y^3$$
$$y^2 Z^1 \beta S(S(q^2) x^2 y^3_1 Z^2)\rangle\langle e^j, S^{-1}(f^2 P^2)(e_i)_1 x^1_{(2,1)} Y^2 e_j\rangle$$
$$\overset{(3.1.7),(3.2.13)}{=} \sum_{i,j=1}^{n} \langle \overline{S}^{-2}(e^i), S(q^1 Y^1)\alpha x^3 y^3_2 Z^3 S^{-1}(f^1 (x^1_1)_{(1,1)} P^1 y^1 \beta)(e_i)_2 Y^3 x^1_2$$
$$y^2 Z^1 \beta S(S(q^2) x^2 y^3_1 Z^2)\rangle$$
$$\langle e^j, (x^1_1)_2 S^{-1}(f^2 (x^1_1)_{(1,2)} P^2)(e_i)_1 Y^2 e_j\rangle$$
$$\overset{(3.2.21),(3.1.9)}{\underset{(3.2.1)}{=}} \sum_{i,j=1}^{n} \langle \overline{S}^{-2}(e^i), S(q^1 Y^1)\alpha S^{-1}(f^1 P^1 p^1 \beta)(e_i)_2 Y^3 p^2 S^2(q^2)\rangle$$
$$\langle e^j, S^{-1}(f^2 P^2)(e_i)_1 Y^2 e_j\rangle$$
$$\overset{(3.2.26)}{=} \sum_{i,j=1}^{n} \langle \overline{S}^{-2}(e^i), S(q^1 Q^1_1 x^1_{(1,1)})\alpha S^{-1}(f^1 P^1 p^1 \beta)(e_i)_2 S^{-1}(x^2 g^1)$$
$$Q^2 x^1_2 p^2\rangle\langle e^j, S^{-1}(f^2 P^2)(e_i)_1 S^{-1}(x^3 g^2) q^2 Q^1_2 x^1_{(1,2)} e_j\rangle$$

$$
\begin{aligned}
&\overset{(3.2.13)}{=} \sum_{i,j=1}^{n} \langle \overline{S}^{-1}(e^i), S(q^1Q^1_1x^1_{(1,1)})\alpha S^{-1}(x^2(e_i)_1P^1p^1\beta)Q^2x^1_2p^2\rangle \\
&\qquad \langle e^j, S^{-1}(x^3(e_i)_2P^2)q^2Q^1_2x^1_{(1,2)}e_j\rangle \\
&= \sum_{i,j=1}^{n} \langle \overline{S}^{-1}(e^i), \alpha S^{-1}\left(x^2(e_i)_1q^1_1(Q^1x^1_1)_{(1,1)}P^1p^1\beta\right) Q^2x^1_2p^2\rangle \\
&\qquad \langle e^j, q^2(Q^1x^1_1)_2S^{-1}\left(x^3(e_i)_2q^1_2(Q^1x^1_1)_{(1,2)}P^2\right) e_j\rangle \\
&\overset{(3.2.21),(3.2.23)}{=} \sum_{i,j=1}^{n} \langle \overline{S}^{-1}(e^i), \alpha S^{-1}(x^2(e_i)_1Q^1x^1_1p^1\beta)Q^2x^1_2p^2\rangle\langle e^j, S^{-1}(x^3(e_i)_2)e_j\rangle \\
&\overset{(6.5.1)}{=} \sum_{i,j=1}^{n} \langle \overline{S}^{-1}(e^i), \alpha S^{-1}(\tilde{q}^1(e_i)_1X^1p^1\beta)X^2p^2S(X^3)\rangle\langle e^j, S^{-1}(\tilde{q}^2(e_i)_2)e_j\rangle \\
&\overset{(3.2.19),(12.3.9)}{=} \sum_{j=1}^{n}\langle e^j, S^{-1}\left(\tilde{\mathfrak{P}}(S^{-2}(\beta S(\alpha)))\right) e_j\rangle \\
&\overset{(*)}{=} \varepsilon\left(\tilde{\mathfrak{P}}(S^{-2}(\beta S(\alpha)))\right) \\
&\overset{(12.3.9)}{=} \sum_{i=1}^{n}\langle e^i, S^{-2}(S(\beta)\alpha e_i\beta S(\alpha))\rangle \\
&= \mathrm{Tr}\left(h \mapsto S^{-2}(S(\beta)\alpha h\beta S(\alpha))\right),
\end{aligned}
$$

where in (*) we used the fact that $S^{-1}\left(\tilde{\mathfrak{P}}(S^{-2}(\beta S(\alpha)))\right) \in \int_r^H$. □

## 12.4 The Trace Formula for Quasi-Hopf Algebras

We will show that the quantum dimension $\underline{\dim}(H) = \underline{\dim}(D(H))$ computed in the previous section is zero unless $D(H)$ is a semisimple algebra. To this end, we need a trace formula for quasi-Hopf algebras. As before, by $\mathrm{Tr}(\chi)$ we denote the trace of an endomorphism $\chi$ of a finite-dimensional $k$-vector space $V$. This means $\mathrm{Tr}(\chi) = \sum_{i=1}^{n} v^i(\chi(v_i))$, where $\{v_i, v^i\}_i$ are dual bases in $V$ and $V^*$.

Recall also that, by the proof of Theorem 7.48, we have

$$
v : \mathscr{L} \otimes H \ni \lambda \otimes h \mapsto \lambda \cdot h = S(h) \rightharpoonup \lambda \in H^*,
$$

an isomorphism of right quasi-Hopf $H$-bimodules. The structure of $H^*$ in ${}_H\mathscr{M}_H^H$ is the one in Proposition 7.46, while $\mathscr{L} \otimes H$ is a right quasi-Hopf $H$-bimodules via the structure given, for all $\lambda \in \mathscr{L}$ and $h, h', h'' \in H$, by

$$
h' \cdot (\lambda \otimes h) \cdot h'' = \mu(h'_1)\lambda \otimes h'_2hh'', \tag{12.4.1}
$$

$$
\lambda \otimes h \mapsto \mu(x^1)\lambda \otimes x^2h_1 \otimes x^3h_2. \tag{12.4.2}
$$

**Theorem 12.17** *Let $H$ be a finite-dimensional quasi-Hopf algebra, $\mu$ the modular element of $H^*$, $\lambda$ a non-zero left cointegral on $H$ and $r$ a right integral in $H$ such that $\lambda(S(r)) = 1$. Then:*

*(i) For any linear map* $\chi : H \to H$ *we have that*

$$\mathrm{Tr}(\chi) = \mu(q_1^1 x^1)\lambda\left(\chi(q^2 x^3 r_2 p^2)S(q_2^1 x^2 r_1 p^1)\right);$$

*(ii)* $\mathrm{Tr}\left(h \mapsto \beta S(\alpha)S^2(h)S(\beta)\alpha\right) = \varepsilon(r)\lambda(S^{-1}(\alpha)\beta)$. *In particular, H is semisimple and admits a normalized left cointegral if and only if*

$$\mathrm{Tr}\left(h \mapsto \beta S(\alpha)S^2(h)S(\beta)\alpha\right) \neq 0.$$

*Proof* For any linear map $\chi : H \to H$ we denote by $\chi^* : H^* \to H^*$ the transpose of $\chi$. We also denote by $\vartheta : H^* \otimes H \to \mathrm{End}(H^*)$ the linear map defined for all $\varphi, \psi \in H^*$ and $h \in H$ by $\vartheta(\varphi \otimes h)(\psi) = \psi(h)\varphi$. Then one can see that

$$\vartheta(\varphi \otimes h) \circ \chi^* = \vartheta(\varphi \otimes \chi(h)), \tag{12.4.3}$$

$$\mathrm{Tr}(\vartheta(\varphi \otimes h)) = \varphi(h), \tag{12.4.4}$$

for all $\varphi \in H^*$, $h \in H$ and $\chi \in \mathrm{End}(H)$.

(i) The fact that $v$ is right $H$-colinear shows, by using (12.4.2) and (7.5.5), that

$$\varphi(V^1 h_1 U^1)\lambda(V^2 h_2 U^2 S(h')) = \mu(x^1)\varphi(x^3 h_2')\lambda(hS(x^2 h_1')),$$

for all $\varphi \in H^*$ and $h, h' \in H$. If we write the above equation for $h' = r$ and use the fact that $S(r) \in \int_l$ such that $\lambda(S(r)) = 1$, we obtain

$$\varphi(S^{-1}(\beta)h\alpha) = \mu(x^1)\varphi(x^3 r_2)\lambda(hS(x^2 r_1)),$$

for all $\varphi \in H^*$ and $h \in H$. In particular, we have that

$$\begin{aligned}&\langle p^2 \rightharpoonup \varphi \leftharpoonup q^2, S^{-1}(\beta)S^{-1}(q^1)hS(p^1)\alpha\rangle \\ &\quad = \mu(x^1)\langle p^2 \rightharpoonup \varphi \leftharpoonup q^2, x^3 r_2\rangle\lambda(S^{-1}(q^1)hS(p^1)S(x^2 r_1)),\end{aligned}$$

and this comes out explicitly as $\varphi(h) = \mu(q_1^1 x^1)\varphi(q^2 x^3 r_2 p^2)\lambda(hS(q_2^1 x^2 r_1 p^1))$, for all $\varphi \in H^*$ and $h \in H$, where we used (7.5.10). In other words we obtained

$$\vartheta(\lambda \leftharpoondown q_2^1 x^2 r_1 p^1 \otimes \mu(q_1^1 x^1)q^2 x^3 r_2 p^2) = \mathrm{Id}_{H^*}, \tag{12.4.5}$$

where for $h^* \in H^*$ and $h \in H$ we denote $h^* \leftharpoondown h = S(h) \rightharpoonup h^*$. Now, by using (12.4.3), (12.4.4) and the fact that $\mathrm{Tr}(\chi) = \mathrm{Tr}(\chi^*)$ we conclude that

$$\begin{aligned}\mathrm{Tr}(\chi) &= \mathrm{Tr}(\chi^*) = \mathrm{Tr}(\mathrm{Id}_{H^*} \circ \chi^*) \\ &= \mathrm{Tr}\left(\eta(\lambda \leftharpoondown q_2^1 x^2 r_1 p^1 \otimes \mu(q_1^1 x^1)q^2 x^3 r_2 p^2) \circ \chi^*\right) \\ &= \mathrm{Tr}\left(\eta(\lambda \leftharpoondown q_2^1 x^2 r_1 p^1 \otimes \mu(q_1^1 x^1)\chi(q^2 x^3 r_2 p^2)\right) \\ &= \mu(q_1^1 x^1)\lambda\left(\chi(q^2 x^3 r_2 p^2)S(q_2^1 x^2 r_1 p^1)\right).\end{aligned}$$

(ii) Combining (3.2.23) and (3.2.21) we obtain

$$r_1 \otimes r_2 = r_1 p^1 \otimes r_2 p^2 \alpha = r_1 p^1 S^{-1}(\alpha) \otimes r_2 p^2. \tag{12.4.6}$$

Now, by part (i) we have

$$\mathrm{Tr}\left(h \mapsto \beta S(\alpha)S^2(h)S(\beta)\alpha\right)$$

$$
\begin{aligned}
&= \mu(q_1^1x^1)\lambda\left(\beta S(\alpha)S^2(q^2x^3r_2p^2)S(\beta)\alpha S(q_2^1x^2r_1p^1)\right)\\
&\overset{(12.4.6)}{=} \mu(q_1^1x^1)\lambda\left(\beta S(\alpha)S(q_2^1x^2r_1\beta S(q^2x^3r_2))\right)\\
&\overset{(3.2.1),(3.2.19)}{=} \varepsilon(r)\mu(q_1^1p^1)\lambda\left(\beta S(\alpha)S(q_2^1p^2S(q^2))\right)\\
&\overset{(3.2.23)}{=} \varepsilon(r)\lambda(\beta S(\alpha)).
\end{aligned}
$$

Next, we claim that $\varepsilon(r)\lambda(\beta S(\alpha)) = \varepsilon(r)\lambda(S^{-1}(\alpha)\beta)$. Indeed, if $H$ is not semisimple then by Theorem 7.28 we have $\varepsilon(r) = 0$, and therefore

$$\varepsilon(r)\lambda(\beta S(\alpha)) = \varepsilon(r)\lambda(S^{-1}(\alpha)\beta) = 0.$$

On the other hand, if $H$ is semisimple then by the same theorem we have that $\varepsilon(\int_l) = \varepsilon(\int_r) \neq 0$. In this situation $H$ is unimodular, so $\mu = \varepsilon$. Finally, by (7.5.10) we get

$$\lambda(S^{-1}(\alpha)\beta) = \mu(\alpha_1)\lambda(\beta S(\alpha_2)) = \varepsilon(\alpha_1)\lambda(\beta S(\alpha_2)) = \lambda(\beta S(\alpha)),$$

as claimed. Thus the proof is finished. □

As a consequence of Proposition 12.16 and Theorem 12.17 we obtain the following formula for the representation-theoretic ranks of $H$ and $D(H)$.

**Theorem 12.18** *Let $H$ be a finite-dimensional quasi-Hopf algebra, $\lambda$ a left cointegral on $H$ and $r$ a right integral in $H$ such that $\lambda(r) = 1$. Then*

$$\underline{\dim}(H) = \underline{\dim}(D(H)) = \varepsilon(r)\lambda(S^{-1}(\alpha)\beta) = \varepsilon_D(\beta \rightharpoonup \lambda \bowtie r).$$

*In particular, if $H$ is not semisimple or it does not admit a normalized left cointegral, then $\underline{\dim}(H) = \underline{\dim}(D(H)) = 0$.*

*Proof* By $\lambda_{\rm op}$ we denote a left cointegral on $H^{\rm op}$. It is straightforward to check that in $H^{\rm op}$ we have $\mu_{\rm op} = \mu^{-1} := \mu \circ S$, and that the roles of $U$ and $V$ interchange. So $\lambda_{\rm op}$ is an element of $H^*$ satisfying

$$\lambda_{\rm op}(V^2h_2U^2)V^1h_1U^1 = \mu^{-1}(X^1)\lambda_{\rm op}(S^{-1}(X^2)h)X^3, \quad \forall\, h \in H.$$

Note that, if $H$ is unimodular, then $\mu = \varepsilon$ and therefore a left cointegral on $H^{\rm op}$ is nothing else than a left cointegral on $H$.

By applying Theorem 12.17 to the quasi-Hopf algebra $H^{\rm op}$ we obtain

$$\mathrm{Tr}\left(h \mapsto S^{-2}(S(\beta)\alpha h\beta S(\alpha)\right) = \varepsilon(t)\lambda_{\rm op}(S^{-1}(\alpha)\beta),$$

where $t$ is a left integral in $H$ such that $\lambda_{\rm op}(S^{-1}(t)) = 1$. If we denote $r = S^{-1}(t)$ we get that $r$ is a right integral in $H$ such that $\lambda_{\rm op}(r) = 1$. It follows that $\varepsilon(t) = \varepsilon(r)$, and

$$\underline{\dim}(H) = \underline{\dim}(D(H)) = \mathrm{Tr}\left(h \mapsto S^{-2}(S(\beta)\alpha h\beta S(\alpha)\right) = \varepsilon(r)\lambda_{\rm op}(S^{-1}(\alpha)\beta).$$

Finally, we apply the same trick as in the proof of the previous theorem. Namely, if $H$ is not semisimple then $\varepsilon(r) = 0$ and we are done. If $H$ is semisimple then it is unimodular. In this case we have seen that $\lambda_{\rm op}$ is a left cointegral on $H$ and since $\lambda_{\rm op}(r) = 1$ the above equality finishes the proof. □

## 12.5 Involutory Quasi-Hopf Algebras

The aim of this section is to introduce and study involutory quasi-Hopf algebras.

**Definition 12.19** A quasi-Hopf algebra is called involutory if

$$S^2(h) = S(\beta)\alpha h\beta S(\alpha), \ \forall\, h \in H. \tag{12.5.1}$$

The definition of an involutory quasi-Hopf algebra occurs as a consequence of the formula for the quantum dimension of $H$ and $D(H)$, the quantum double of $H$, and this is $h \mapsto S^{-2}(S(\beta)\alpha h\beta S(\alpha)) = \mathrm{Id}_H$; clearly, it is equivalent to the relation (12.5.1). A second way to introduce this notion is by using the trace formula for quasi-Hopf algebras, namely, $h \mapsto \beta S(\alpha)S^2(h)S(\beta)\alpha = \mathrm{Id}_H$. The result below says that the two ways above are equivalent.

**Lemma 12.20** *Let $H$ be an involutory quasi-Hopf algebra. Then $S(\beta)\alpha$ is an invertible element and $(S(\beta)\alpha)^{-1} = \beta S(\alpha)$. In particular, $S^2$ is inner and therefore $S$ is bijective. Moreover, $\alpha$ and $\beta$ are invertible elements and*

$$\alpha^{-1} = S^{-1}(\alpha\beta)\beta = \beta S(\beta\alpha), \tag{12.5.2}$$
$$\beta^{-1} = S(\beta\alpha)\alpha = \alpha S^{-1}(\alpha\beta). \tag{12.5.3}$$

*Proof* For simplicity denote $\mathfrak{U} = S(\beta)\alpha$ and $\mathfrak{V} = \beta S(\alpha)$. Then $S^2(h) = \mathfrak{U}h\mathfrak{V}$, for all $h \in H$. Since $S^2$ is an algebra map we get that $H \ni h \mapsto \mathfrak{U}h\mathfrak{V} \in H$ is an algebra endomorphism of $H$. Hence $\mathfrak{U}$ is invertible and $\mathfrak{U}^{-1} = \mathfrak{V}$. In other words we have proved that $S(\beta)\alpha$ is invertible and $(S(\beta)\alpha)^{-1} = \beta S(\alpha)$, as claimed.

The relation $\mathfrak{V}\mathfrak{U} = 1_H$ comes out as $\beta S(\beta\alpha)\alpha = 1_H$. Thus $\beta$ has a right inverse, namely $S(\beta\alpha)\alpha$. Similarly, from $\mathfrak{U}\mathfrak{V} = 1_H$ we obtain that $S(\beta)\alpha\beta S(\alpha) = 1_H$. Since $S$ is bijective this is equivalent to $\alpha S^{-1}(\alpha\beta)\beta = 1_H$, so $\beta$ has also a left inverse, namely $\alpha S^{-1}(\alpha\beta)$. Thus $\beta$ is invertible and $\beta^{-1} = S(\beta\alpha)\alpha = \alpha S^{-1}(\alpha\beta)$.

The relations in (12.5.2) follow from the fact that if in an algebra $A$ two elements $a$ and $b$ are such that $ab$ and $a$ are invertible, then $b$ is also invertible and $b^{-1} = (ab)^{-1}a$. In fact, we apply this elementary result to $A = H$, $a = S(\beta)$ and $b = \alpha$. □

One can easily verify that $H$ is involutory if and only if $H^{\mathrm{op}}$ is involutory, if and only if $H^{\mathrm{cop}}$ is involutory, if and only if $H^{\mathrm{op,cop}}$ is involutory.

**Proposition 12.21** *If $H$ is a finite-dimensional involutory quasi-Hopf algebra and* $\dim(H) \neq 0$ *in $k$ then $H$ is semisimple and admits a normalized left cointegral. Consequently, the quantum double of $H$ is a semisimple quasi-Hopf algebra admitting a normalized left cointegral.*

*Proof* By the comments made before Lemma 12.20, if $H$ is as in the statement then, on the one hand, $\underline{\dim}(H) = \mathrm{Tr}(\mathrm{Id}_H) = \dim(H) \neq 0$; see (12.3.6). On the other hand, $\underline{\dim}(H) = \varepsilon(r)\lambda(S^{-1}(\alpha)\beta)$; see Theorem 12.18, where $r$ is a non-zero right integral in $H$ and $\lambda$ is a non-zero left cointegral on $H$ such that $\lambda(r) = 1$. Thus $\varepsilon(r)\lambda(S^{-1}(\alpha)\beta) \neq 0$, and this is equivalent to the fact that $H$ is semisimple and admits a normalized left cointegral. The last assertion in the statement follows from Corollary 12.2 and Corollary 12.8. □

For an involutory quasi-Hopf algebra we have a kind of skew-antipode property, in the following sense.

**Proposition 12.22** *Let $H$ be an involutory quasi-Hopf algebra. Then for all $h \in H$ the following relations hold:*

$$S(h_2)\beta^{-1}h_1 = \varepsilon(h)\beta^{-1} \text{ and } h_2\alpha^{-1}S(h_1) = \varepsilon(h)\alpha^{-1}. \tag{12.5.4}$$

*Proof* As we have seen in Remark 3.16(2), if $\mathbb{U} \in H$ is invertible then we can define a new quasi-Hopf algebra $H^{\mathbb{U}} = (H, \Delta, \varepsilon, \Phi, S_{\mathbb{U}}, \alpha_{\mathbb{U}}, \beta_{\mathbb{U}})$, where $\alpha_{\mathbb{U}} := \mathbb{U}\alpha$, $\beta_{\mathbb{U}} := \beta\mathbb{U}^{-1}$ and $S_{\mathbb{U}}(h) := \mathbb{U}S(h)\mathbb{U}^{-1}$.

Now, consider $\mathbb{U} = \alpha$. We know from Lemma 12.20 that $\mathbb{U}$ is invertible, so it makes sense to consider the quasi-Hopf algebra $H^{\mathbb{U}}$. In this particular case we have that $\alpha_{\mathbb{U}} = 1_H$, $\beta_{\mathbb{U}} = \beta\alpha$ and

$$S_{\mathbb{U}}(h) = \alpha^{-1}S(h)\alpha = \beta S(\beta\alpha)S(h)\alpha = \beta S(\alpha)S(\beta^{-1}h\beta)S(\beta)\alpha = S^{-1}(\beta^{-1}h\beta).$$

Since $S_{\mathbb{U}}(h_1)\alpha_{\mathbb{U}}h_2 = \varepsilon(h)\alpha_{\mathbb{U}}$ for all $h \in H$, we get that $S^{-1}(\beta^{-1}h_1\beta)h_2 = \varepsilon(h)1_H$, and this is equivalent to $S(h_2)\beta^{-1}h_1\beta = \varepsilon(h)1_H$, for all $h \in H$. It follows now that $S(h_2)\beta^{-1}h_1 = \varepsilon(h)\beta^{-1}$ for all $h \in H$, as needed.

Similarly, by using the fact that $h_1\beta_{\mathbb{U}}S(h_2) = \varepsilon(h)\beta_{\mathbb{U}}$ for all $h \in H$, one can prove that $h_2\alpha^{-1}S(h_1) = \varepsilon(h)\alpha^{-1}$ for all $h \in H$; the details are left to the reader. □

We end this section by presenting examples of involutory quasi-Hopf algebras.

**Examples 12.23** (1) The two-dimensional quasi-Hopf algebra $H(2)$ described in Example 3.26 is an involutory quasi-Hopf algebra since $\alpha = g$ has order 2, $\beta = 1$ and $S$ is the identity map.

(2) The quasi-Hopf algebra $D^{\omega}(H)$ considered in Section 8.6 is an involutory quasi-Hopf algebra since for $D^{\omega}(H)$ we have $\alpha_{D^{\omega}(H)} = 1$, the unit of $D^{\omega}(H)$, $\beta^{-1}_{D^{\omega}(H)} = s(\beta_{D^{\omega}(H)})$ and $s^2(\varphi\#h) = \beta^{-1}_{D^{\omega}(H)}(\varphi\#h)\beta_{D^{\omega}(H)}$, for all $\varphi\#h \in D^{\omega}(H)$.

(3) For the quasi-Hopf algebra $\mathrm{bos}(H(2)_+) = \mathrm{bos}(H(2)_-)$ in Example 11.13 we have that the distinguished elements $\alpha$ and $\beta$ are given by $\alpha = x$ and $\beta = 1$, and the antipode is the identity map. Thus $\mathrm{bos}(H(2)_+) = \mathrm{bos}(H(2)_-)$ is an involutory quasi-Hopf algebra.

(4) Let $D(H(2))$ be the quantum double of $H(2)$ with the quasi-Hopf algebra structure computed in Proposition 10.21. We have that the antipode for $D(H(2))$ is the identity map, $\alpha = X$ and $\beta = 1$, so $D(H(2))$ is an involutory quasi-Hopf algebra.

The involutory property on $H$ transfers on $D(H)$ if a certain condition is fulfilled.

**Proposition 12.24** *Let $H$ be an involutory quasi-Hopf algebra such that*

$$\Delta(S(\beta)\alpha) = f^{-1}(S \otimes S)(f_{21})(S(\beta)\alpha \otimes S(\beta)\alpha), \tag{12.5.5}$$

*where $f_{21} = f^2 \otimes f^1$. Then $D(H)$ is an involutory quasi-Hopf algebra.*

*Proof* From (10.4.7) we know that

$$S_D^2(\varphi \bowtie h) = g_1^1 G^1 S(f^2 F_2^2) \rightharpoonup \overline{S}^{-2}(\varphi) \leftharpoonup F^1 S^{-1}(g^2) \bowtie g_2^1 G^2 S(f^1 F_1^2) S^2(h),$$

for all $\varphi \in H^*$ and $h \in H$. By using (12.5.5) twice and (3.2.13) we obtain

$$\begin{aligned}&(\Delta \otimes \mathrm{Id}_H)(\Delta(S(\beta)\alpha))\\ &\qquad = g_1^1 G^1 S(f^2 F_2^2) S(\beta)\alpha \otimes g_2^1 G^2 S(f^1 F_1^2) S(\beta)\alpha \otimes g^2 S(F^1) S(\beta)\alpha.\end{aligned}$$

By (12.5.1) and (12.5.2) we have

$$S(\beta S(\alpha)) = S^2(\alpha) S(\beta) = S(\beta)\alpha^2 \beta S(\beta\alpha) = S(\beta)\alpha^2\alpha^{-1} = S(\beta)\alpha,$$

or, equivalently, $S^{-1}(S(\beta)\alpha) = \beta S(\alpha)$. Then we have, for all $\varphi \in H^*$ and $h \in H$,

$$\begin{aligned}&(\varepsilon \bowtie S(\beta)\alpha)(\varphi \bowtie h)(\varepsilon \bowtie \alpha S(\beta))\\ &\overset{(8.5.2)}{=} g_1^1 G^1 S(f^2 F_2^2) S(\beta)\alpha \rightharpoonup \varphi \leftharpoonup \beta S(\alpha) S^{-1}(g^2 S(F^1)) \bowtie g_2^1 G^2 S(F^1 f_1^2) S^2(h)\\ &= g_1^1 G^1 S(f^2 F_2^2) \rightharpoonup \overline{S}^{-2}(\varphi) \leftharpoonup F^1 S^{-1}(g^2) \bowtie g_2^1 G^2 S(f^1 F_1^2) S^2(h)\\ &= S_D^2(\varphi \bowtie h).\end{aligned}$$

This means that $D(H)$ is an involutory quasi-Hopf algebra. □

Notice that for $H(2)$ the condition in (12.5.5) is $\Delta(g) = g \otimes g$, which is just part of the definition of $H(2)$. So Proposition 12.24 gives us a direct argument for the fact that $D(H(2))$ is an involutory quasi-Hopf algebra. It is still an open problem if (12.5.5) is automatic for an involutory quasi-Hopf algebra $H$, and so to have $H$ involutory if and only if $D(H)$ is involutory.

## 12.6 Representations of Involutory Quasi-Hopf Algebras

The goal of this section is to study the representations of an involutory quasi-Hopf algebra $H$ over a field $k$. We will prove that if $H$ is semisimple then the characteristic of $k$ does not divide the dimension of any finite-dimensional absolutely simple $H$-module. Recall that a left $H$-module $V$ is absolutely simple if for every field extension $k \subseteq K$, $K \otimes V$ is a simple $K \otimes H$-module or, equivalently, if every $H$-endomorphism of $V$ is of the form $c\mathrm{Id}_V$ for some scalar $c \in k$.

The case when $H$ is not semisimple is treated as well; in this case the characteristic of $k$ divides the dimension of any finite-dimensional projective $H$-module.

Now, in order to prove these results for involutory quasi-Hopf algebras we need some preliminary results.

Let $V$ and $W$ be two left $H$-modules. Then one can easily see that the set of $k$-linear maps from $V$ to $W$, $\mathrm{Hom}_k(V,W)$, has a left $H$-module structure defined by

$$(h \cdot \psi)(v) = h_1 \cdot \psi(S(h_2) \cdot v), \ \ \forall\, h \in H, \ \psi \in \mathrm{Hom}_k(V,W) \ \text{ and } \ v \in V.$$

Consequently, if $V = W$ then $\mathrm{End}_k(V) := \mathrm{Hom}_k(V,V)$ is a left $H$-module.

For any left $H$-module $V$ we define $V^H$, the set of $H$-invariants of $V$, as follows:

$$V^H = \{v \in V \mid h \cdot v = \varepsilon(h)v, \ \forall\, h \in H\}.$$

In particular, we have that the set of $H$-invariants of $\mathrm{Hom}_k(V,W)$ is

$$\mathrm{Hom}_k(V,W)^H = \{\psi \in \mathrm{Hom}_k(V,W) \mid h_1 \cdot \psi(S(h_2)\cdot v) = \varepsilon(h)\psi(v), \ \forall\, h \in H, \ v \in V\}. \tag{12.6.1}$$

We can characterize $\mathrm{Hom}_k(V,W)^H$ more precisely.

**Lemma 12.25** *Let $H$ be a quasi-Hopf algebra with bijective antipode and $V$, $W$ two left $H$-modules. If we denote by $\mathrm{Hom}_H(V,W)$ the set of left $H$-linear maps from $V$ to $W$ then*

$$\nu : \mathrm{Hom}_k(V,W)^H \to \mathrm{Hom}_H(V,W), \quad \nu(\psi)(v) = q^1 \cdot \psi(S(q^2)\cdot v),$$

*for all $\psi \in \mathrm{Hom}_k(V,W)$ and $v \in V$, is bijective. Its inverse is*

$$\nu^{-1} : \mathrm{Hom}_H(V,W) \to \mathrm{Hom}_k(V,W)^H, \quad \nu^{-1}(\chi)(v) = \chi(\beta \cdot v),$$

*for all $\chi \in \mathrm{Hom}_H(V,W)$ and $v \in V$, where $q_R = q^1 \otimes q^2$ is as defined in* (3.2.19).

*Proof* We first show that $\nu$ is well defined, that is, $\nu(\psi)$ is $H$-linear for any $\psi \in \mathrm{Hom}_k(V,W)^H$. Indeed, for any $\psi \in \mathrm{Hom}_k(V,W)^H$, $h \in H$ and $v \in V$, we have

$$\begin{aligned} h \cdot \nu(\psi)(v) &= hq^1 \cdot \psi(S(q^2)\cdot v) \\ &\overset{(3.2.21)}{=} q^1 h_{(1,1)} \cdot \psi(S(h_{(1,2)})S(q^2)h_2 \cdot v) \\ &\overset{(12.6.1),(3.1.8)}{=} q^1 \cdot \psi(S(q^2)h \cdot v) = \nu(\psi)(h \cdot v), \end{aligned}$$

as needed. Now, since for all $\chi \in \mathrm{Hom}_H(V,W)$, $h \in H$ and $v \in V$, we have

$$\begin{aligned} h_1 \cdot \nu^{-1}(\chi)(S(h_2)\cdot v) &= h_1 \cdot \chi(\beta S(h_2)\cdot v) \\ &= \chi(h_1 \beta S(h_2)\cdot v) \\ &\overset{(3.2.1)}{=} \varepsilon(h)\chi(\beta \cdot v) = \varepsilon(h)\nu^{-1}(\chi)(v), \end{aligned}$$

from (12.6.1) we deduce that $\nu^{-1}$ is well defined, too. So it remains to show that $\nu$ and $\nu^{-1}$ are inverses. Indeed, for all $\chi \in \mathrm{Hom}_H(V,W)$ and $v \in V$ we have

$$\begin{aligned} (\nu \circ \nu^{-1})(\chi)(v) &= q^1 \cdot \nu^{-1}(\chi)(S(q^2)\cdot v) \\ &= q^1 \cdot \chi(\beta S(q^2)\cdot v) \\ &= \chi(q^1 \beta S(q^2)\cdot v) = \chi(v), \end{aligned}$$

where in the third equality we used the fact that $\chi$ is $H$-linear and in the fourth equality we used the definition (3.2.19) of $q_R$ and (3.2.2).

Now let $p^1 \otimes p^2$ be the element $p_R$ defined in (3.2.19), $\psi \in \mathrm{Hom}_k(V,W)^H$ and $v \in V$. We have

$$\begin{aligned} (\nu^{-1} \circ \nu)(\psi)(v) &= \nu(\psi)(\beta \cdot v) \\ &= q^1 \cdot \psi(S(q^2)\beta \cdot v) \end{aligned}$$

$$\overset{(3.2.19),(3.1.11)}{=} \varepsilon(p^1)q^1\cdot\psi(S(q^2)p^2\cdot v)$$
$$\overset{(12.6.1)}{=} q^1p^1_1\cdot\psi(S(q^2p^1_2)p^2\cdot v)\overset{(3.2.23)}{=}\psi(v),$$

and this finishes the proof. □

**Lemma 12.26** *Let $H$ be an involutory quasi-Hopf algebra, $V$ a finite-dimensional left $H$-module and $\{v_i\}_{i=\overline{1,n}}$ a basis in $V$ with dual basis $\{v^i\}_{i=\overline{1,n}}$. Then the map*

$$\overline{\mathrm{Tr}}:\mathrm{End}_k(V)\ni\zeta\mapsto\sum_{i=1}^{n}\langle v^i,\beta^{-1}\cdot\zeta(v_i)\rangle\in k$$

*is $H$-linear. Moreover, the relation between $\overline{\mathrm{Tr}}$ and the classical trace function* Tr *is*

$$\overline{\mathrm{Tr}}(\zeta)=\mathrm{Tr}\left(v\mapsto\zeta(\beta^{-1}\cdot v)\right),\ \ \forall\,\zeta\in\mathrm{End}_k(V).$$

*Proof* Indeed, for any $h\in H$ and $\zeta\in\mathrm{End}_k(V)$ we have

$$\begin{aligned}
\overline{\mathrm{Tr}}(h\cdot\zeta) &= \sum_{i=1}^{n}\langle v^i,\beta^{-1}\cdot(h\cdot\zeta)(v_i)\rangle\\
&= \sum_{i=1}^{n}\langle v^i,\beta^{-1}h_1\cdot\zeta(S(h_2)\cdot v_i)\rangle\\
&= \sum_{i,j=1}^{n}\langle v^j,\zeta(S(h_2)\cdot v_i)\rangle\langle v^i,\beta^{-1}h_1\cdot v_j\rangle\\
&= \sum_{j=1}^{n}\langle v^j,\zeta(S(h_2)\beta^{-1}h_1\cdot v_j)\rangle\\
&\overset{(12.5.4)}{=} \varepsilon(h)\sum_{j=1}^{n}\langle v^j,\zeta(\beta^{-1}\cdot v_j)\rangle=\varepsilon(h)\mathrm{Tr}\left(v\mapsto\zeta(\beta^{-1}\cdot v)\right).
\end{aligned}$$

On the other hand, by using dual bases we have that

$$\begin{aligned}
\mathrm{Tr}\left(v\mapsto\zeta(\beta^{-1}\cdot v)\right) &= \sum_{j=1}^{n}\langle v^j,\zeta(\beta^{-1}\cdot v_j)\rangle\\
&= \sum_{i,j=1}^{n}\langle v^i,\beta^{-1}\cdot v_j\rangle\langle v^j,\zeta(v_i)\rangle\\
&= \sum_{i=1}^{n}\langle v^i,\beta^{-1}\cdot\zeta(v_i)\rangle=\overline{\mathrm{Tr}}(\zeta),
\end{aligned}$$

which is the second assertion in the statement. The first one follows now from the two computations above. □

By using Lemma 12.25 one can associate to any linear map an $H$-linear one. This can be done by using the integrals in $H$. Let $V$ and $W$ be two left $H$-modules and $\zeta\in\mathrm{Hom}_k(V,W)$. Since $\mathrm{Hom}_k(V,W)$ is a left $H$-module, by the definition of a left integral $t$ in $H$ it follows that $t\cdot\zeta$ belongs to $\mathrm{Hom}_k(V,W)^H$. Keeping the same notation as in Lemma 12.25, we get that $\tilde{\zeta} := v(t \cdot \zeta)$ is an $H$-linear map. Explicitly, to any $\zeta \in \mathrm{Hom}_k(V,W)$ we associated the map $\tilde{\zeta} \in \mathrm{Hom}_H(V,W)$ defined by

$$\tilde{\zeta}(v) = q^1 t_1 \cdot \zeta(S(q^2 t_2) \cdot v), \tag{12.6.2}$$

for all $v \in V$, where $t$ is a left integral in $H$.

We are now able to prove one of the main results of this section.

**Theorem 12.27** *Let $H$ be a semisimple involutory quasi-Hopf algebra over a field $k$ of characteristic $p \geq 0$. Then $p$ does not divide the dimension of any finite-dimensional absolutely simple $H$-module.*

*Proof* The assertion follows from the fact that the map $\overline{\mathrm{Tr}}$ defined in Lemma 12.26 is $H$-linear.

Actually, since $H$ is semisimple we know from Theorem 7.28 that there is a left integral $t$ in $H$ such that $\varepsilon(t) = 1$. Let now $V$ be a finite-dimensional absolutely simple $H$-module, that is, $V$ is finite dimensional and any element of $\mathrm{End}_H(V) := \mathrm{Hom}_H(V,V)$ is of the form $c\mathrm{Id}_V$, for some scalar $c \in k$. Therefore, for any $\zeta \in \mathrm{End}_k(V)$ there exists a scalar $c_\zeta \in k$ such that $\tilde{\zeta} = c_\zeta \mathrm{Id}_V$, where $\tilde{\zeta} \in \mathrm{End}_H(V)$ is the associated $H$-linear map of $\zeta$ as in (12.6.2). In other words, for any $\zeta \in \mathrm{End}_k(V)$ there is a scalar $c_\zeta$ such that $q^1 t_1 \cdot \zeta(S(q^2 t_2) \cdot v) = c_\zeta v$, for all $v \in V$. By (7.2.3) we have $t_1 \otimes S(t_2) = q^1 t_1 \otimes S(q^2 t_2)\beta$, so the above relation is equivalent to $t_1 \cdot \zeta(S(t_2)\beta^{-1} \cdot v) = c_\zeta v$, for all $v \in V$, and this implies:

$$\begin{aligned} c_\zeta \dim_k(V) &= \mathrm{Tr}(v \mapsto c_\zeta v) = \mathrm{Tr}\left(v \mapsto t_1 \cdot \zeta(S(t_2)\beta^{-1} \cdot v)\right) \\ &= \mathrm{Tr}\left(v \mapsto (t \cdot \zeta)(\beta^{-1} \cdot v)\right) = \overline{\mathrm{Tr}}(t \cdot \zeta) = \varepsilon(t)\overline{\mathrm{Tr}}(\zeta) = \overline{\mathrm{Tr}}(\zeta). \end{aligned}$$

Clearly, we can choose a map $\zeta \in \mathrm{End}_k(V)$ such that $\overline{\mathrm{Tr}}(\zeta) = \mathrm{Tr}\left(v \mapsto \zeta(\beta^{-1} \cdot v)\right) = 1$, so we conclude that $\dim_k(V) \neq 0$ in $k$. □

**Corollary 12.28** *Let $H$ be a semisimple involutory quasi-Hopf algebra over an algebraically closed field of characteristic $p \geq 0$. Then $p$ does not divide the dimension of any finite-dimensional simple $H$-module.*

*Proof* This follows from Schur's lemma and Theorem 12.27. □

We will focus now on the case when $H$ is not semisimple. In order to simplify the proof of the next theorem we first show the following result:

**Proposition 12.29** *Let $H$ be a finite-dimensional quasi-Hopf algebra and $P, Q$ finite-dimensional projective left $H$-modules. Then:*

*(i) $P^* = \mathrm{Hom}_k(P,k)$ is a projective left $H$-module;*

*(ii) $P \otimes Q$ is a projective left $H$-module, where the $H$-module structure of $P \otimes Q$ is defined by the comultiplication $\Delta$ of $H$.*

*Consequently, we obtain that $\mathrm{End}_k(P)$ is a projective left $H$-module.*

*Proof* (i) Recall that, if $V$ is a left $H$-module then $V^*$, the linear dual of V, is a left $H$-module with $H$-action $(h \cdot v^*)(v) = v^*(S(h) \cdot v)$, for all $v^* \in V^*$, $h \in H$, $v \in V$.

Since $P$ is finite dimensional it follows that $P$ is a finitely generated projective left $H$-module. Therefore, there exist a natural number $n$ and a left $H$-module $P'$ such that $P\oplus P'\cong H^n$ as left $H$-modules. Thus

$$\begin{aligned}(H^n)^* = \mathrm{Hom}_k(H^n,k) &\cong \mathrm{Hom}_k(P\oplus P',k)\\ &\cong \mathrm{Hom}_k(P,k)\oplus \mathrm{Hom}_k(P',k) = P^*\oplus P'^*,\end{aligned}$$

as left $H$-modules. Now, $H$ is finite dimensional, so from the proof of Theorem 7.48 we know that the application $H\ni h\mapsto (h'\mapsto \lambda(h'S(h)))\in H^*$ is bijective, where $\lambda$ is a non-zero left cointegral on $H$. Replacing $H$ by $H^{\mathrm{op,cop}}$ we get that $H\cong H^*$ as left $H$-modules, where the left $H$-module structures on $H$ and $H^*$ are given by the regular multiplication on $H$ and by the corresponding left $H$-module structure induced on its dual, respectively, namely $(h\cdot h^*)(h') = h^*(S(h)h')$, for $h,h'\in H$ and $h^*\in H^*$.

From the above we obtain that $H^n\cong (H^*)^n\cong (H^n)^*\cong P^*\oplus P'^*$, as left $H$-modules, so $P^*$ is a projective left $H$-module, too.

(ii) We follow the same line as above. There exist two natural numbers $n$ and $m$ and two left $H$-modules $P'$ and $Q'$ such that $P\oplus P'\cong H^n$ and $Q\otimes Q'\cong H^m$, as left $H$-modules. We then have

$$(H\otimes H)^{nm}\cong H^n\otimes H^m\cong (P\otimes Q)\oplus(P\otimes Q')\oplus(P'\otimes Q)\oplus(P'\otimes Q'),$$

as left $H$-modules. We now prove that $H\otimes H$ with the diagonal $H$-module structure is a free left $H$-module. Thus, as a consequence, we will obtain that $P\otimes Q$ is a projective left $H$-module. To this end, we claim that the map

$$\mu : .H\otimes .H\to .H\otimes H,\quad \mu(h\otimes h') = \tilde{q}^2h'_2\otimes S^{-1}(\tilde{q}^1h'_1)h,$$

for all $h,h'\in H$, is a left $H$-linear isomorphism. Here we denote by $.H\otimes .H$ and $.H\otimes H$ the $k$-vector space $H\otimes H$ endowed with the left $H$-module structure given by $\Delta$ and by the left regular multiplication on $H$, respectively. In addition, $q_L = \tilde{q}^1\otimes\tilde{q}^2$ is the element defined in (3.2.20). Indeed, for all $h,h',h''\in H$ we have

$$\begin{aligned}\mu(h''\cdot(h\otimes h')) &= \mu(h''_1h\otimes h''_2h')\\ &= \tilde{q}^2h''_{(2,2)}h'_2\otimes S^{-1}(\tilde{q}^1h''_{(2,1)}h'_1)h''_1h\\ &\overset{(3.2.22)}{=} h''\tilde{q}^2h'_2\otimes S^{-1}(\tilde{q}^1h'_1)h\\ &= h''\mu(h\otimes h'),\end{aligned}$$

and this means that $\mu$ is left $H$-linear. Next, one can easily check that the map $\mu^{-1}$ defined for all $h,h'\in H$ by

$$\mu^{-1}: .H\otimes H\to .H\otimes .H,\quad \mu^{-1}(h\otimes h') = h_1\tilde{p}^1h'\otimes h_2\tilde{p}^2$$

is the inverse of $\mu$. More precisely, (3.2.22) and (3.2.24) imply that $\mu^{-1}\circ\mu = \mathrm{Id}$, while (3.2.22) and (3.2.24) imply that $\mu\circ\mu^{-1} = \mathrm{Id}$, we leave the verification of the details to the reader.

Finally, if $V$ is a finite-dimensional $k$-vector space then $\mathrm{End}_k(V) \cong V \otimes V^*$ as $H$-modules. Indeed, one can prove that the maps

$$\mathrm{End}_k(V) \ni \chi \mapsto \sum_{i=1}^{n} \chi(v_i) \otimes v^i \in V \otimes V^*,$$
$$V \otimes V^* \ni v \otimes v^* \mapsto \big(v' \mapsto v^*(v')v\big) \in \mathrm{End}_k(V)$$

are $H$-linear and inverse to each other (the details are left to the reader).

Thus, if $P$ is a finite-dimensional projective left $H$-module then by part (i) $P^*$ is a projective left $H$-module, and then we deduce that $\mathrm{End}_k(P) \cong P \otimes P^*$ is a projective left $H$-module as well; see (ii). □

We are now able to prove the second important result of this section.

**Theorem 12.30** *Let $H$ be a finite-dimensional non-semisimple involutory quasi-Hopf algebra over a field of characteristic $p \geq 0$. Then $p$ divides the dimension of any finite-dimensional projective left $H$-module.*

*Proof* Let $P$ be a finite-dimensional projective left $H$-module and suppose that $p$ does not divide $\dim_k(P)$. Then the map

$$k \ni c \mapsto (\dim_k(P))^{-1} c\,(v \mapsto \beta \cdot v) \in \mathrm{End}_k(P)$$

is $H$-linear; see (3.2.1). Obviously, the map above is a section for the $H$-linear map $\overline{\mathrm{Tr}}$ defined in Lemma 12.26, specialized for $V = P$. So $k$ is isomorphic to a direct summand of $\mathrm{End}_k(P)$, which is a projective left $H$-module; see Proposition 12.29. Thus $k$ is a projective left $H$-module. By Theorem 7.28 we obtain that $H$ is semisimple, a contradiction. □

## 12.7 Notes

Section 12.1 is taken from [49], where an answer to a conjecture raised by Hausser and Nill in [109] was given. The form of the cointegrals on $D(H)$ is also taken from [49], as well as the explicit form of the modular element of $D(H)$. The explicit form of the (co)integrals in and on $D(H)$ leads to a characterization of the semisimplicity of $D(H)$, a property closely related to its representation-theoretic rank.

The computation of the representation-theoretic rank for $H$ and $D(H)$ was performed in [62]. The goal was to find a plausible definition for the involutory notion in the quasi-Hopf case, a notion that in the Hopf case is closely related to the fifth conjecture of Kaplansky [126]: a Hopf algebra $H$ is semisimple as an algebra if and only if it is involutory. It is well known that, over a field of characteristic zero, $H$ is semisimple if and only if it is cosemisimple, if and only if it is involutory, that is, $S^2 = \mathrm{Id}_H$. These remarkable results were proved by Larson and Radford in [133, 134], answering in the positive, in characteristic zero, the fifth conjecture of Kaplansky. They have also proved that in characteristic $p$ sufficiently large a semisimple cosemisimple Hopf algebra is involutory. Afterwards, using this result and a lifting theorem, Etingof and Gelaki proved in [90] that the antipode of a semisimple cosemisimple Hopf algebra over any field is an involution. Trying to generalize the above results for quasi-Hopf algebras, the first problem which occurs is: what could be an involutory quasi-Hopf algebra? According to Majid [143], $\mathrm{Tr}(S^2)$ arises in a very natural way as the representation-theoretic rank of the Schrödinger representation of $H$, $\underline{\dim}(H)$, or as the representation-theoretic rank of the canonical representation of the quantum double, $\underline{\dim}(D(H))$, and this stimulated performing similar computations in the quasi-Hopf case. This also led to the involutory notion in Section 12.5, a definition that agrees with the point of view in [96, Prop. 8.24 and 8.23] or [153]. We end by pointing out that the fifth conjecture of Kaplansky is still an open problem in quasi-Hopf algebra theory.

Apart from [49, 62], in the presentation of this chapter we also used [52].

# 13
# Ribbon Quasi-Hopf Algebras

We define and characterize ribbon quasi-Hopf algebras by using properties of a ribbon category. Consequently, we have a one-to-one correspondence between ribbon elements for a quasi-Hopf algebra and its grouplike elements. We also construct ribbon categories from left or right rigid monoidal categories and use this construction to introduce a special class of ribbon quasi-Hopf algebras.

## 13.1 Ribbon Categories

The concept of ribbon category is defined as follows.

**Definition 13.1** Let $(\mathscr{C}, c)$ be a braided category.

(1) $(\mathscr{C}, c)$ is called balanced if there exists a natural isomorphism $\eta = (\eta_V : V \to V)_{V \in \mathrm{Ob}(\mathscr{C})}$ such that, for all $V, W \in \mathscr{C}$,

$$\eta_{V\otimes W} = (\eta_V \otimes \eta_W) \circ c_{W,V} \circ c_{V,W}. \tag{13.1.1}$$

(2) A balanced category $(\mathscr{C}, c, \eta)$ is called ribbon if, in addition, $\mathscr{C}$ is left rigid and

$$\eta_{V^*} = (\eta_V)^* \tag{13.1.2}$$

for all $V \in \mathscr{C}$. If this is the case then $\eta$ is called a twist on $\mathscr{C}$.

**Examples 13.2** (1) Any symmetric category $\mathscr{C}$ is ribbon with $\eta$ defined as the identity natural transformation of $\mathscr{C}$. In particular, the category of representations of a triangular quasi-Hopf algebra is ribbon.

(2) If $(\mathscr{C}, c, \eta)$ is balanced then so are $\overline{\mathscr{C}} = (\mathscr{C}, \overline{c}, \eta)$, $\mathscr{C}^{\mathrm{in}} = (\mathscr{C}, \underline{c}, \eta^{-1})$ and $\mathscr{C}^{\mathrm{opp}} = (\mathscr{C}^{\mathrm{opp}}, c^{\mathrm{opp}}, \eta^{-1})$.

(3) Let $k$ be a field, $G$ a multiplicative finite abelian group and $\mathscr{R} : G \times G \to k^*$ a bilinear form. Then, endowed with the strict monoidal structure and with the braiding $c$ defined by (1.5.9), the category of $G$-graded vector spaces $\mathrm{Vect}^G$ is braided; see Proposition 1.44. Furthermore, if we restrict to the finite-dimensional case, then $\mathrm{vect}^G$ is ribbon. The braided structure on $\mathrm{vect}^G$ is induced by that of $\mathrm{Vect}^G$ described above, and a twist on it is given by $\eta_V : V \ni v \mapsto \mathscr{R}(|v|, |v|)v \in V$, extended by linearity, where we assumed that $v$ is a homogenous element in $V$ of degree $|v|$.

Indeed, one can see that the bilinearity of $\mathscr{R}$ implies (13.1.1). Also, by Example 1.65 the category $\mathrm{vect}^G$ is rigid, and (13.1.2) is satisfied by $\eta$ since, on the one hand, for all $v^* \in V^*$ homogenous of degree $g$ we have $(\eta_V)^*(v^*)(v) = v^*(\eta_V(v)) = \mathscr{R}(g^{-1}, g^{-1})v^*(v_{g^{-1}})$ and $\eta_{V^*}(v^*)(v) = \mathscr{R}(g,g)v^*(v) = \mathscr{R}(g,g)v^*(v_{g^{-1}})$, for all $v = \sum_{x\in G} v_x \in V$. On the other hand, $\mathscr{R}(g,g) = \mathscr{R}(g^{-1},g)^{-1} = \mathscr{R}(g^{-1},g^{-1})$, for all $g \in G$.

More examples of balanced categories can be obtained as follows.

**Proposition 13.3** *Let $(\mathscr{C}, c)$ be a braided category. Denote by $\mathscr{B}(\mathscr{C}, c)$ the category whose*

- *objects are pairs $(V, \eta_V)$ consisting of an object $V \in \mathscr{C}$ and an automorphism $\eta_V : V \to V$ of $V$ in $\mathscr{C}$;*
- *morphisms between $(V, \eta_V)$ and $(W, \eta_W)$ are morphisms $f : V \to W$ in $\mathscr{C}$ fulfilling $\eta_W f = f\eta_V$.*

*Then $\mathscr{B}(\mathscr{C}, c)$ is a balanced category via the following structure:*

*(i) The tensor product on $\mathscr{B}(\mathscr{C}, c)$ is given by $(V, \eta_V) \otimes (W, \eta_W) = (V \otimes W, \eta_{V\otimes W})$, with $\eta_{V\otimes W} = (\eta_V \otimes \eta_W)c_{W,V}c_{V,W}$; on morphisms it acts as the tensor product of $\mathscr{C}$. The unit object in $\mathscr{B}(\mathscr{C}, c)$ is $(\underline{1}, \eta_{\underline{1}} := \mathrm{Id}_{\underline{1}})$. Together with the associativity and the left and right unit constraints of $\mathscr{C}$ these give the monoidal structure on $\mathscr{B}(\mathscr{C}, c)$;*

*(ii) The braiding on $\mathscr{B}(\mathscr{C}, c)$ is determined by the braiding $c$ of $\mathscr{C}$;*

*(iii) The balancing on $\mathscr{B}(\mathscr{C}, c)$ is produced by $\eta := (\eta_V)_{(V,\eta_V)\in\mathscr{B}(\mathscr{C},c)}$.*

*Proof* We check that the associativity constraint of $\mathscr{C}$ is a morphism in $\mathscr{B}(\mathscr{C}, c)$; the remaining details are trivial. Assuming $\mathscr{C}$ is strict monoidal, this reduces to the equality $\eta_{(V\otimes W)\otimes T} = \eta_{V\otimes(W\otimes T)}$, for all $V, W, T \in \mathscr{C}$. In diagrammatic notation, the latter comes out as

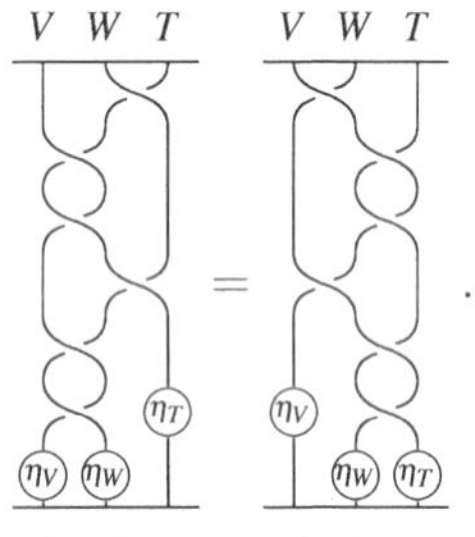

To prove the above equality we compute the right-hand side of it as follows: apply (1.5.10) twice, and then apply the naturality of the braiding $c$ to the morphism $c_{W,V}c_{V,W} : V \otimes W \to V \otimes W$. In this way we get the left-hand side of the equality. □

**Remark 13.4** By Proposition 13.3 we can associate to any monoidal category $\mathscr{C}$ two balanced ones: namely, $\mathfrak{B}_l(\mathscr{C}) := \mathscr{B}(\mathscr{Z}_l(\mathscr{C}), c)$ and $\mathfrak{B}_r(\mathscr{C}) := \mathscr{B}(\mathscr{Z}_r(\mathscr{C}), c)$, where $\mathscr{Z}_{l/r}(\mathscr{C})$ is the left/right center of $\mathscr{C}$ as in Section 8.1.

As a concrete example, we can take $\mathscr{C} = {}_H\mathscr{M}$, $H$ a quasi-Hopf algebra with bijective antipode, in which case we have $\mathfrak{B}_l(\mathscr{C}) = \mathscr{B}({}_H^H\mathscr{YD}, c)$ with $c$ as in (8.2.15), and $\mathfrak{B}_r(\mathscr{C}) = \mathscr{B}({}_H\mathscr{YD}^H, \mathfrak{c})$ with $\mathfrak{c}$ given by (8.2.18).

Note that Proposition 13.3 can also be applied to the category $_H\mathcal{M}$, where $(H,R)$ is a QT quasi-Hopf algebra. We will exploit this fact in the forthcoming sections.

The following result says that the inverse of the square of the twist is completely determined by the rigid braided structure of the category.

**Proposition 13.5** *Let* $(\mathscr{C}, c, \eta)$ *be a left rigid balanced category. Then*

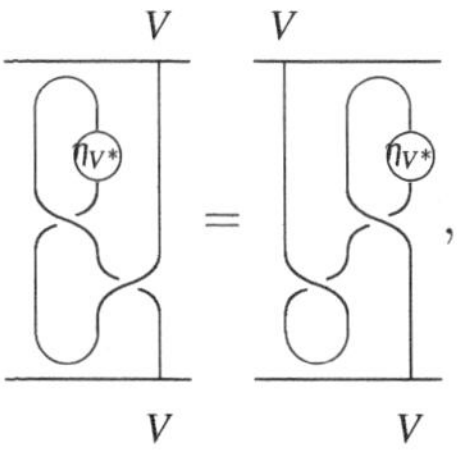

$$\eta_V^{-1} = \quad = \quad , \quad \forall\, V \in \mathscr{C}. \tag{13.1.3}$$

*Consequently, a left rigid balanced category* $(\mathscr{C}, c, \eta)$ *is ribbon if and only if*

$$\eta_V^{-2} = (\mathrm{ev}_V \otimes \mathrm{Id}_V)(\mathrm{Id}_{V^*} \otimes c_{V,V}^{-1})(c_{V,V^*}\mathrm{coev}_V \otimes \mathrm{Id}_V), \quad \forall\, V \in \mathscr{C}. \tag{13.1.4}$$

*Proof* Assume that $\mathscr{C}$ is strict monoidal. By taking $V = W = \underline{1}$ in (13.1.1), by Proposition 1.49 we get that $\eta_{\underline{1}\otimes\underline{1}} = \eta_{\underline{1}} \otimes \eta_{\underline{1}}$. By the naturality of $\eta$ and $l$ we see that

$$\eta_{\underline{1}} l_{\underline{1}} = l_{\underline{1}} \eta_{\underline{1}\otimes\underline{1}} = l_{\underline{1}}(\mathrm{Id}_{\underline{1}} \otimes \eta_{\underline{1}})(\eta_{\underline{1}} \otimes \mathrm{Id}_{\underline{1}}) = \eta_{\underline{1}} r_{\underline{1}}(\eta_{\underline{1}} \otimes \mathrm{Id}_{\underline{1}}) = \eta_{\underline{1}}^2 l_{\underline{1}},$$

where in the last two equalities we used the naturality of $r$ and the fact that $l_{\underline{1}} = r_{\underline{1}}$; see Proposition 1.5. We conclude that $\eta_{\underline{1}} = \mathrm{Id}_{\underline{1}}$, and so by the the naturality of $\eta$ we get that $\eta_{V\otimes V^*}\mathrm{coev}_V = \mathrm{coev}_V$, for any object $V$ of $\mathscr{C}$. Therefore,

$$\mathrm{Id}_V = \quad = \boxed{\eta_{V\otimes V^*}} \overset{(13.1.1)}{=} \quad \Leftrightarrow \eta_V^{-1} = \overset{(*)}{=} \quad ;$$

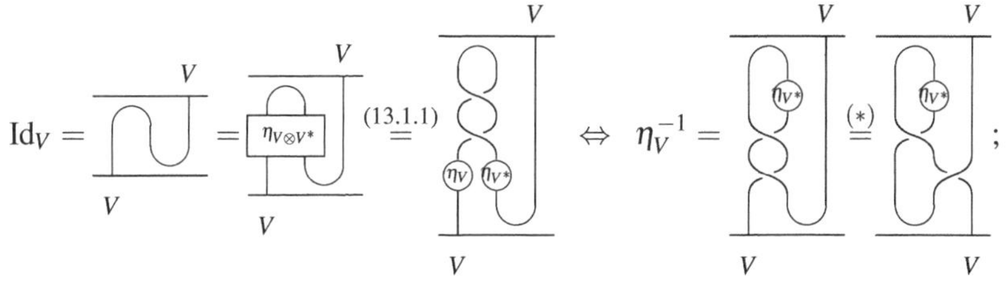

in (*) we applied the naturality of $c$ to $\mathrm{ev}_V$. The second equality in (13.1.3) follows from the naturality of $c$ applied to $c_{V,V^*}(\mathrm{Id}_V \otimes \eta_{V^*})\mathrm{coev}_V : \underline{1} \to V^* \otimes V$.

If $\eta_{V^*} = (\eta_V)^*$ then $\quad = \quad$, and therefore by (13.1.3) we have that

$$\eta_V^{-1} = \quad \Leftrightarrow \eta_V^{-2} = \quad = \quad , \tag{13.1.5}$$

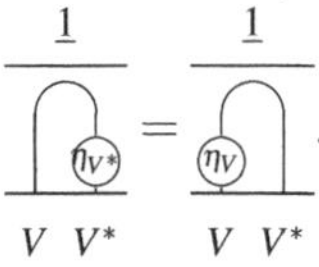

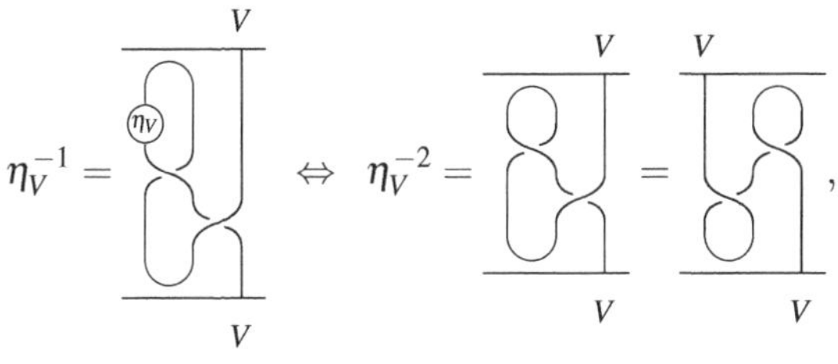

as desired. For the converse, if (13.1.4) holds then again by (13.1.3) we have that

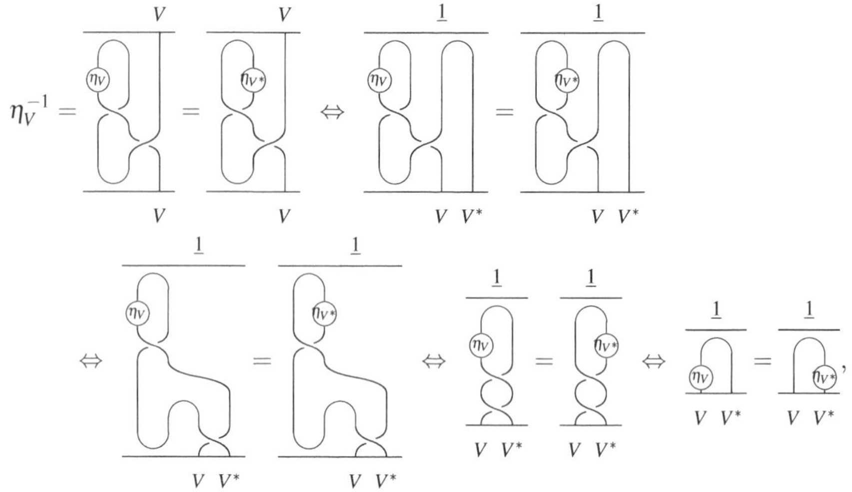

which is equivalent to $\eta_{V^*} = (\eta_V)^*$, as needed. So our proof is complete. □

Any ribbon category is rigid; see Proposition 1.74. Furthermore, the next results say that the right rigid structure can be constructed from the left rigid structure, braiding and twisting in such a way that the left and right dual functors coincide. Thus the choice of the left duals in the definition of a ribbon category is irrelevant.

**Proposition 13.6** *Let $(\mathscr{C}, c, \eta)$ be a left rigid balanced category. If $V$ is an object of $\mathscr{C}$ and $V^*$ is the left dual of $V$ in $\mathscr{C}$ with evaluation and coevaluation morphisms* $\mathrm{ev}_V : V^* \otimes V \to \underline{1}$ *and* $\mathrm{coev}_V : \underline{1} \to V \otimes V^*$, *then*

$$(V^*, \mathrm{ev}'_V := \mathrm{ev}_V c_{V,V^*}(\eta_V \otimes \mathrm{Id}_{V^*}), \mathrm{coev}'_V := (\eta_{V^*} \otimes \mathrm{Id}_V) c_{V,V^*} \mathrm{coev}_V) \qquad (13.1.6)$$

*is a right dual for $V$ in $\mathscr{C}$. Furthermore, with respect to it the left and right dual functors coincide as strong monoidal functors. Consequently, any left rigid balanced category is sovereign.*

*Proof* By using the equalities in (13.1.3) one can easily verify that $V^*$ with $\mathrm{ev}'$ and $\mathrm{coev}'$ as in (13.1.6) is a right dual for $V$ in $\mathscr{C}$. Also, with respect to this right duality we have ${}^*f = f^*$, for any morphism $f : V \to W$ in $\mathscr{C}$, since

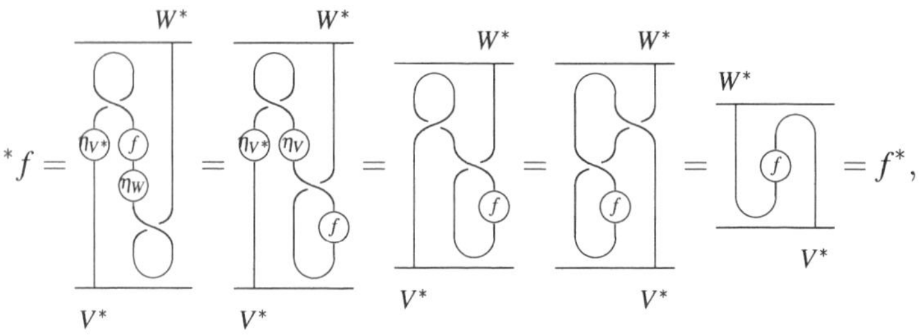

because of (13.1.1) and the naturality of $c$ and $\eta$, respectively.

The only thing left to show is that $\lambda$ in (1.7.1) equals $\lambda'$, its right-handed version. With notation as in Section 1.7 we have that

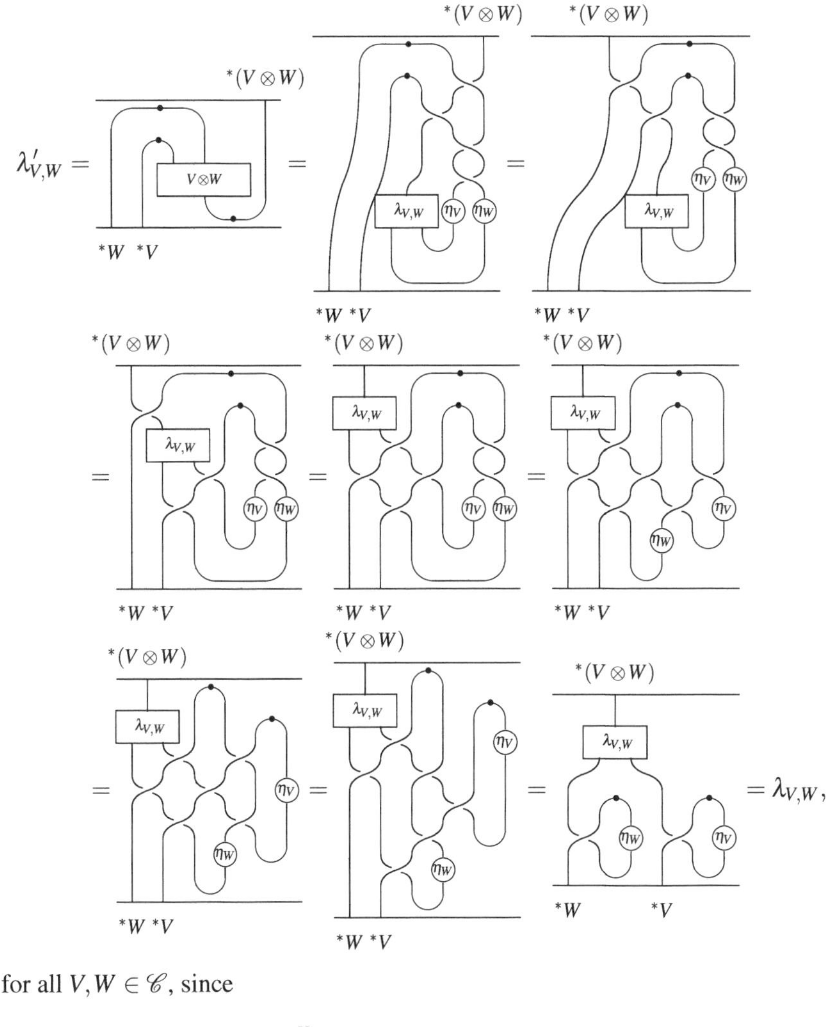


for all $V, W \in \mathscr{C}$, since

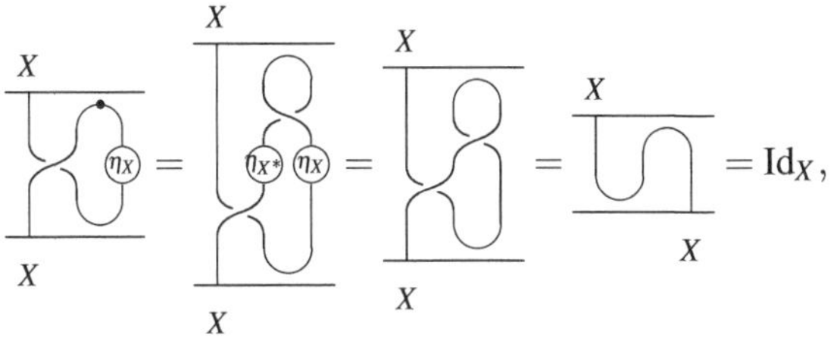


for all $X \in \mathscr{C}$. This ends the proof. □

**Corollary 13.7** *Any ribbon category is sovereign.*

**Remark 13.8** If $(\mathscr{C}, c, \eta)$ is ribbon, from $\eta_{V^*} = (\eta_V)^*$ it follows that $\mathrm{coev}'_V$ in

(13.1.6) can be restated as

$$\mathrm{coev}'_V = (\mathrm{Id}_{V^*} \otimes \eta_V) c_{V,V^*} \mathrm{coev}_V. \tag{13.1.7}$$

**Corollary 13.9** *A braided category $(\mathscr{C}, c)$ is ribbon if and only if it is right rigid and there exists a natural isomorphism $\theta = (\theta_V : V \to V)_{V \in \mathrm{Ob}(\mathscr{C})}$ such that*

$$\theta_{V \otimes W} = (\theta_V \otimes \theta_W) \circ c_{W,V} \circ c_{V,W}, \ \forall\, V, W \in \mathscr{C},$$
$$\theta_{^*V} = {}^*(\theta_V), \ \forall\, V \in \mathscr{C}.$$

*Proof* The direct implication follows from Proposition 13.6. Conversely, we have that $\overline{\mathscr{C}}$ is ribbon and by applying again Proposition 13.6 (this time to $\overline{\mathscr{C}}$, the reverse braided category associated to $(\mathscr{C}, c)$) the converse follows. □

The result in Proposition 13.6 has a converse. In particular, we can characterize ribbon categories in terms of sovereign categories.

**Theorem 13.10** *Let $(\mathscr{C}, c)$ be a left rigid braided category. Then*

*(i) $(\mathscr{C}, c)$ is balanced if and only if $\mathscr{C}$ is sovereign;*

*(ii) $(\mathscr{C}, c)$ is ribbon if and only if it is sovereign and, with respect to the rigid structure given by the fact that $\mathscr{C}$ is sovereign, we have*

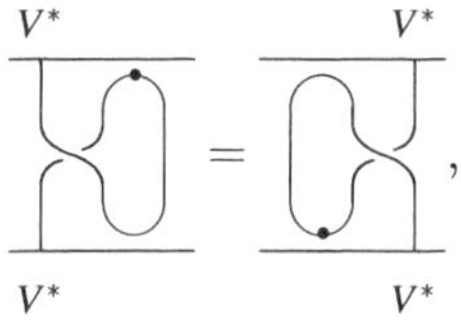


$$, \ \forall\, V \in \mathscr{C}. \tag{13.1.8}$$

*Proof* (i) By Proposition 13.6, a left rigid balanced category is sovereign.

Let $(\mathscr{C}, c)$ be a braided sovereign category. We claim that, for all $V \in \mathscr{C}$,

$\eta_V :=$ 

and $\eta_V^{-1} :=$ 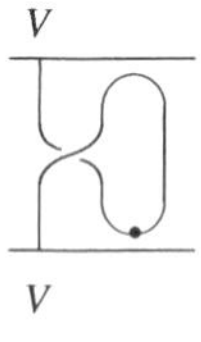

(13.1.9)

are inverses of each other in $\mathscr{C}$ and, moreover, $(\mathscr{C}, c)$ with $\eta := (\eta_V)_{V \in \mathrm{Ob}(\mathscr{C})}$ becomes a balanced category. As before, the graphical notation is as in Sections 1.6 and 1.8. Indeed, we use the naturality of $c^{-1}$ to compute that

$\eta_V \eta_V^{-1} =$ 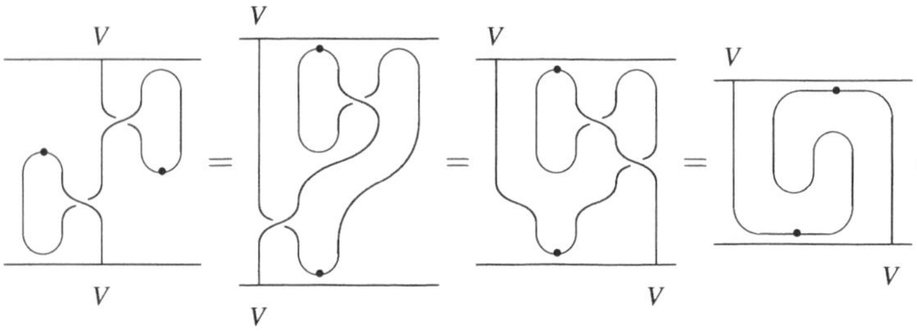

$= \mathrm{Id}_V.$

Likewise, by the naturality of $c$ and its inverse $c^{-1}$ we have

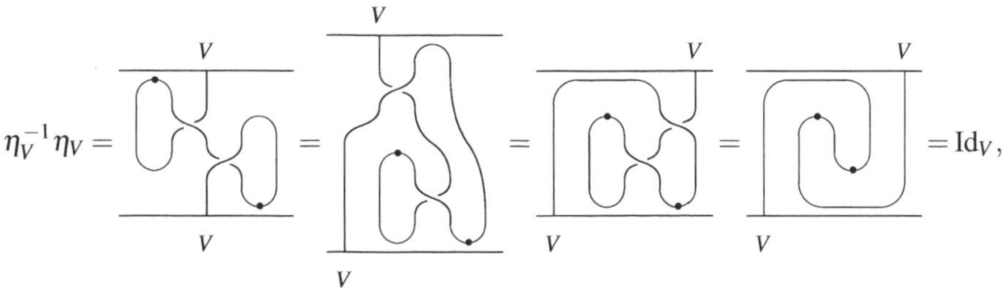


and so $\eta_V$ is an isomorphism in $\mathscr{C}$ with inverse $\eta_V^{-1}$, as claimed.

For $f : V \to W$ a morphism in $\mathscr{C}$, by the naturality of $c$ and $f^* = {}^*f$, we get that

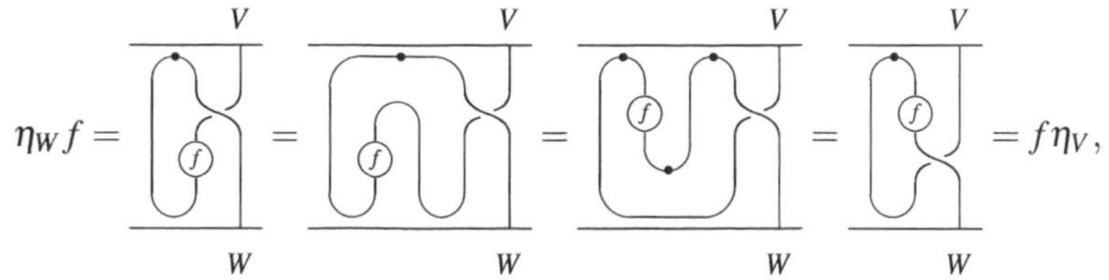


and therefore $\eta = (\eta_V)_{V \in \mathrm{Ob}(\mathscr{C})}$ is a natural isomorphism.

Since ${}^*(-)$ and $(-)^*$ are equal as strong monoidal functors, we calculate

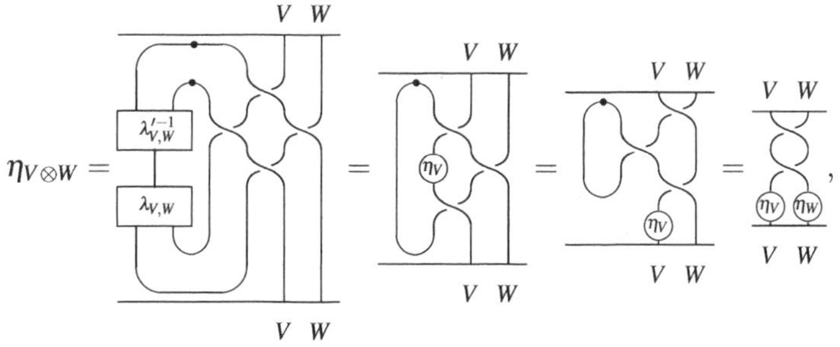


for all $V, W \in \mathscr{C}$, as needed.

(ii) By part (i), we have to show that, for all $V \in \mathscr{C}$, $\eta_{V^*} = (\eta_V)^*$ if and only if (13.1.8) holds. To this end, note that (13.1.6) and the naturality of $c$ imply

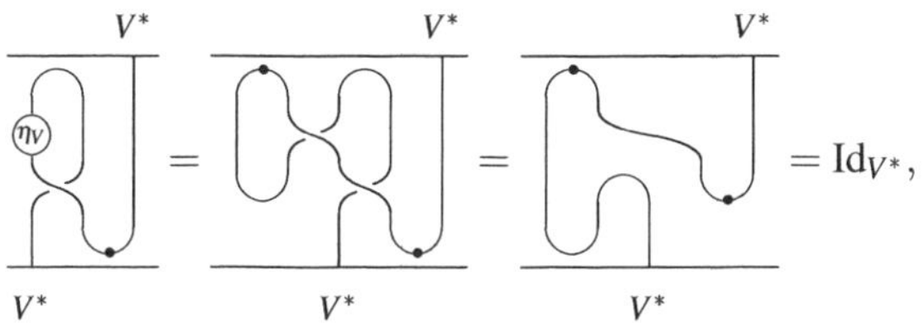

and therefore

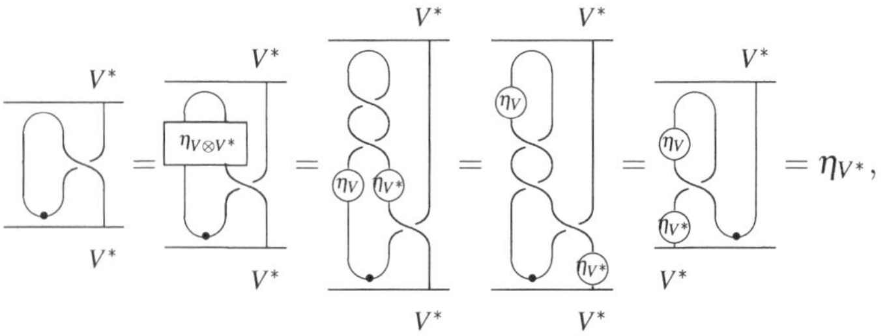


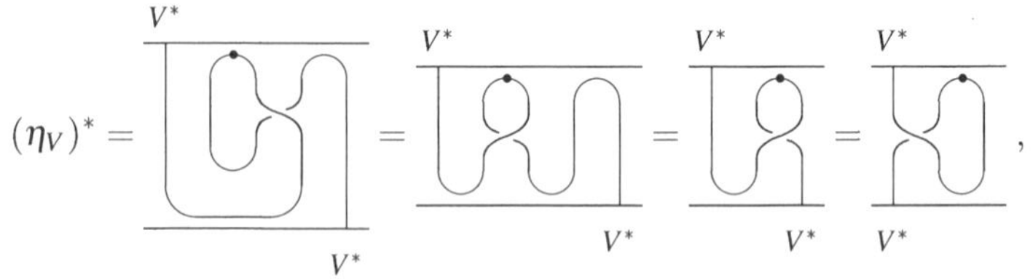


and this finishes the proof. □

**Corollary 13.11** *If $H$ is a sovereign quasi-Hopf algebra then the category of finite-dimensional left-right Yetter–Drinfeld modules ${}_H\mathscr{YD}^{H^{\rm fd}}$ is balanced.*

*Proof* The category ${}_H\mathscr{YD}^{H^{\rm fd}}$ is sovereign; see Theorem 8.25. As it is braided, the result follows from part (i) of Theorem 13.10. □

## 13.2 Ribbon Categories Obtained from Rigid Monoidal Categories

To any left rigid braided category $(\mathscr{C},c)$ (which is consequently rigid) we assign a ribbon category $\mathscr{R}(\mathscr{C},c)$. Thus, to any left (resp. right) rigid monoidal category $\mathscr{C}$ we can associate a ribbon monoidal one, that will be denoted by $\mathfrak{R}_l(\mathscr{C})$ (resp. $\mathfrak{R}_r(\mathscr{C})$). The latter are possible due to the left and right center constructions presented in Section 8.1.

Inspired by the formula in (13.1.4) we introduce the following category, which will turn out to be a ribbon category.

**Definition 13.12** If $(\mathscr{C},c)$ is a left rigid braided (strict) monoidal category then $\mathscr{R}(\mathscr{C},c)$ is the category whose

- objects are pairs $(V,\eta_V)$ consisting of an object $V$ of $\mathscr{C}$ and an automorphism $\eta_V$ of $V$ in $\mathscr{C}$ satisfying

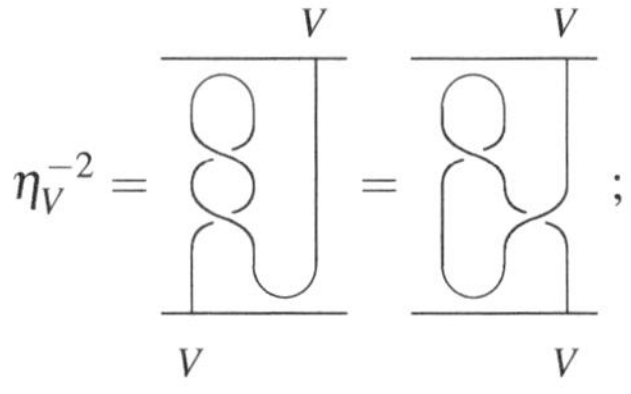


(13.2.1)

- morphisms $f : (V, \eta_V) \to (W, \eta_W)$ are morphisms $f : V \to W$ in $\mathscr{C}$ such that $\eta_W f = f\eta_V$.

The composition in $\mathscr{R}(\mathscr{C})$ is given by the composition in $\mathscr{C}$, and the identity morphism of an object $(V, \eta_V)$ is $\mathrm{Id}_V$.

In other words, $\mathscr{R}(\mathscr{C}, c)$ is the full subcategory of $\mathscr{B}(\mathscr{C}, c)$ considered in Proposition 13.3 determined by those objects $(V, \eta_V)$ of $\mathscr{B}(\mathscr{C}, c)$ for which $\eta_V$ obeys (13.2.1). As we pointed out in the second part of Proposition 13.5, this is the necessary and sufficient condition that turns $\mathscr{B}(\mathscr{C}, c)$ into a ribbon category. Actually, the ribbon structure of $\mathscr{R}(\mathscr{C}, c)$ is encoded in the following result.

**Theorem 13.13** *Let $(\mathscr{C}, c)$ be a left rigid braided (strict) monoidal category. Then $\mathscr{R}(\mathscr{C}, c)$ is a ribbon monoidal category as follows:*

- *if $(V, \eta_V)$, $(W, \eta_W) \in \mathscr{R}(\mathscr{C}, c)$ then $(V, \eta_V) \otimes (W, \eta_W) = (V \otimes W, \eta_{V\otimes W})$, where*

$$\eta_{V\otimes W} = (\eta_V \otimes \eta_W) c_{W,V} c_{V,W}; \tag{13.2.2}$$

- *the unit object is $(\underline{1}, \eta_{\underline{1}} = \mathrm{Id}_{\underline{1}})$, and the associativity and the left and right unit constraints are the same as those of $\mathscr{C}$;*
- *the braiding equals c, regarded as an isomorphism in $\mathscr{R}(\mathscr{C}, c)$;*
- *for $(V, \eta_V)$ an object in $\mathscr{R}(\mathscr{C}, c)$, a left dual object for it is $(V^*, \eta_{V^*})$, where*

$$\eta_{V^*} = (\eta_V)^*, \tag{13.2.3}$$

  *with evaluation and coevaluation morphisms equal to* $\mathrm{ev}_V$ *and* $\mathrm{coev}_V$, *viewed now as morphisms in $\mathscr{R}(\mathscr{C}, c)$;*
- *the twist is given by*

$$\eta_V : (V, \eta_V) \to (V, \eta_V), \tag{13.2.4}$$

  *an automorphism in $\mathscr{R}(\mathscr{C}, c)$.*

*Proof* We start by proving that $(V \otimes W, \eta_{V\otimes W})$ is an object of $\mathscr{R}(\mathscr{C}, c)$, that is, $\eta_{V\otimes W}$ in (13.2.2) obeys (13.2.1). To this end, we need the equalities

(a) 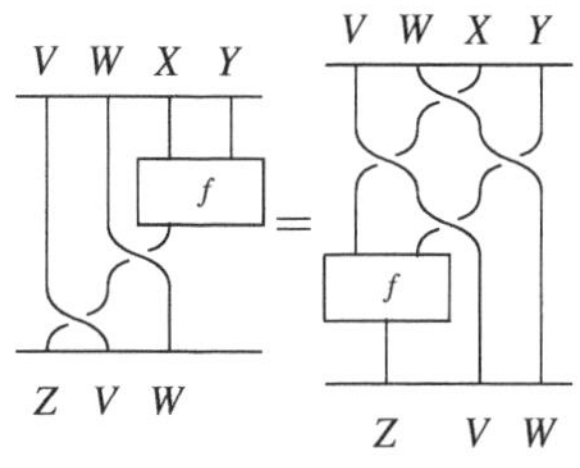

, (b) 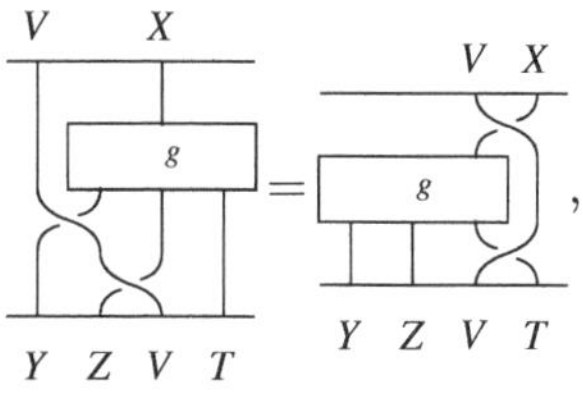

, (13.2.5)

for any morphisms $f : X \otimes Y \to Z$ and $g : X \to Y \otimes Z \otimes T$ in $\mathscr{C}$, and

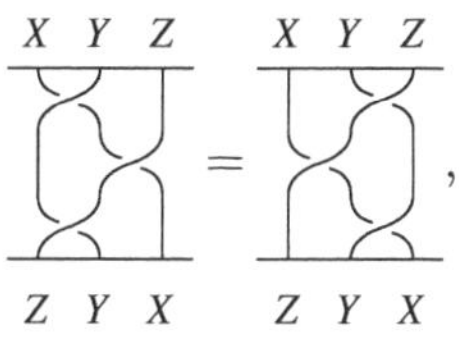

, (13.2.6)

valid for all $X, Y, Z \in \mathscr{C}$, respectively, which follow from the fact that $c_{-,-}$ is a natural isomorphism. Note that (13.2.6) is nothing but an equivalent form of the categorical version of the Yang–Baxter equation (1.5.10).

Now, if $\lambda_{V,W} : (V \otimes W)^* \to W^* \otimes V^*$ is the isomorphism in $\mathscr{C}$ defined in (1.7.1) and $\lambda_{V,W}^{-1}$ is its inverse as in (1.7.2), then we compute

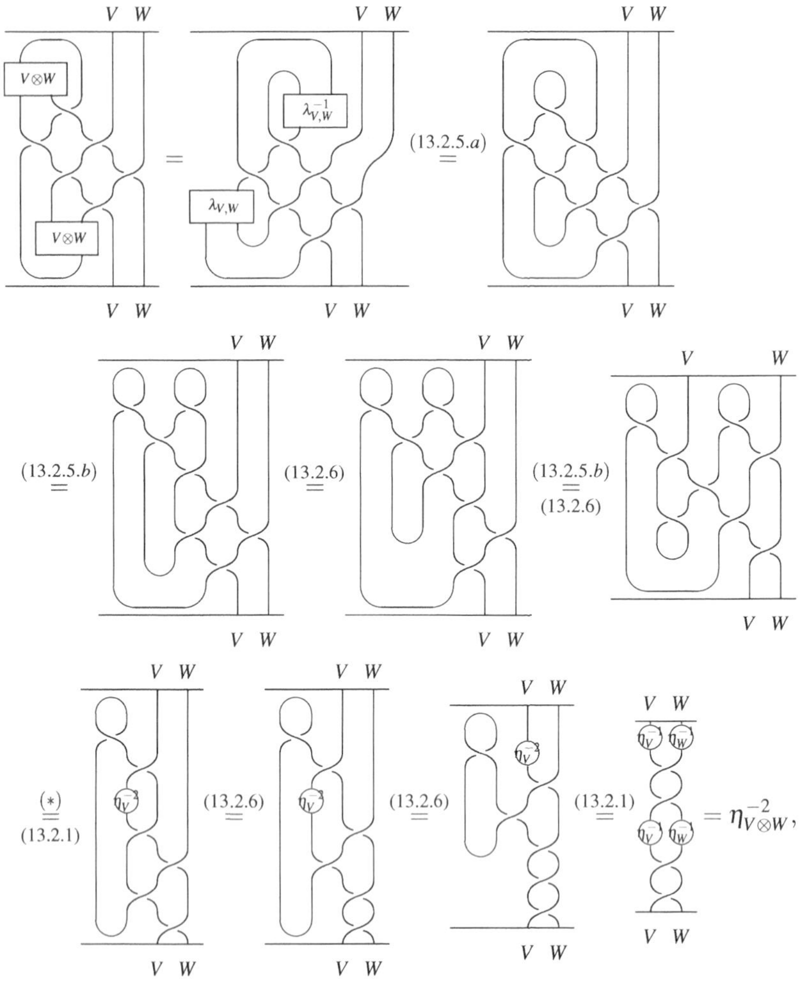


where (*) is

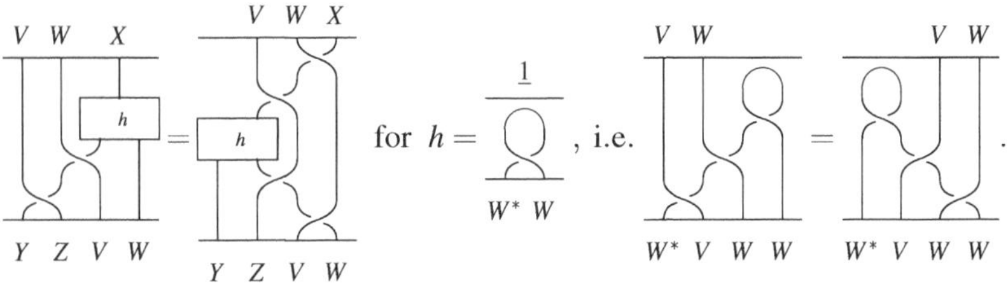

This ends the proof of the fact that the tensor product of $\mathscr{R}(\mathscr{C},c)$ is well defined at the level of objects. It is easy to see that for any two morphisms $f, f'$ in $\mathscr{R}(\mathscr{C},c)$ their tensor product $f \otimes f'$ in $\mathscr{C}$ is actually a morphism in $\mathscr{R}(\mathscr{C},c)$. Therefore, we have a tensor product functor on $\mathscr{R}(\mathscr{C},c)$ that together with the associativity and the left and right unit constraints of $\mathscr{C}$ defines on $\mathscr{R}(\mathscr{C},c)$ a monoidal structure; the unchecked details are left to the reader.

We next show that $c$ provides a braiding on $\mathscr{R}(\mathscr{C},c)$. The only thing that must be verified is that $c_{V,W}$ is a morphism in $\mathscr{R}(\mathscr{C},c)$, for any objects $(V,\eta_V)$, $(W,\eta_W)$ of $\mathscr{R}(\mathscr{C},c)$. This follows directly from the definitions and from the naturality of $c$.

Let $(V^*, \eta_{V^*})$ be as in (13.2.3). By using the naturality of $c$ one sees that

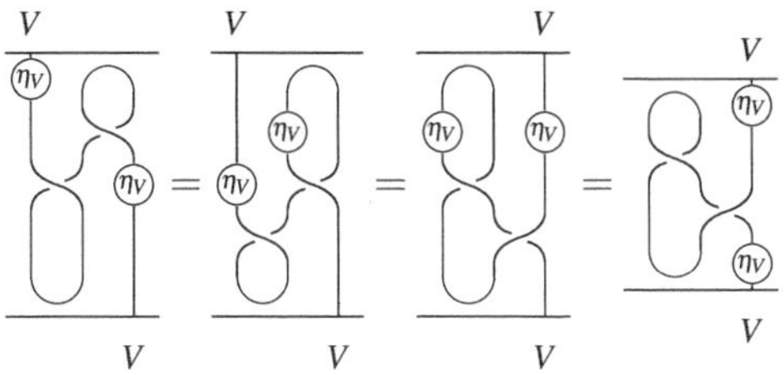


$= \mathrm{Id}_V,$ (13.2.7)

$\eta_V^{-2} =$ 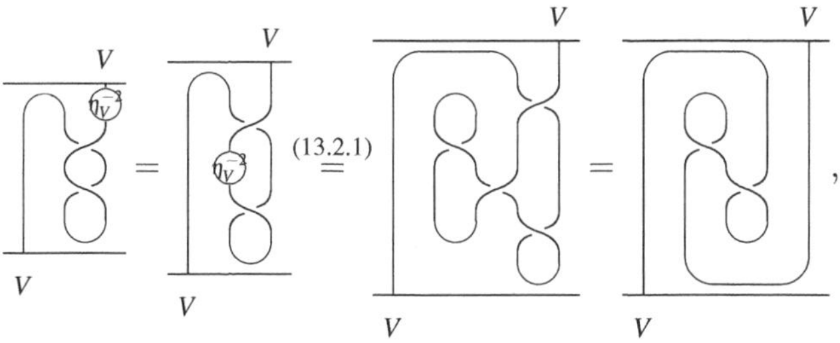


, (13.2.8)

where in the last equality of the second computation we applied the naturality of $c_{V,-}^{-1}$ to the morphism $\mathrm{ev}_V c_{V,V^*} : V \otimes V^* \to \underline{1}$. These formulas allow us to prove that

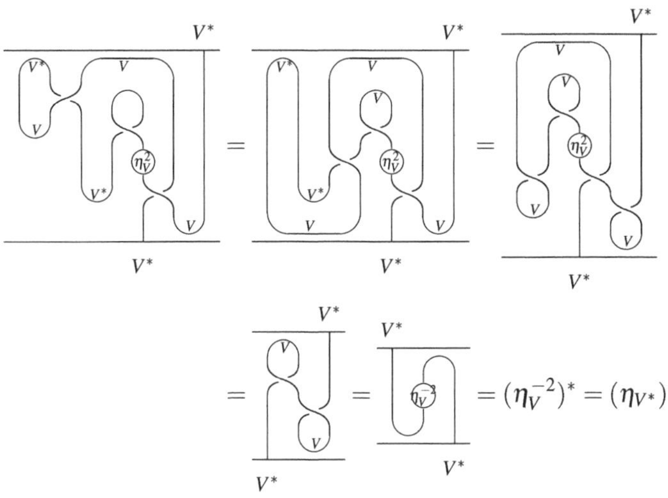


$= (\eta_V^{-2})^* = (\eta_{V^*})^{-2}.$

In the first equality we used an equivalent form of the naturality of $c_{V,-}^{-1}$ applied to $\mathrm{ev}_{V^*}$, in the second equality (1.6.6), in the third equality (13.2.7) and in the fourth equality (1.6.6) and (13.2.8). In other words, the relation in (13.2.1) is satisfied by $\eta_{V^*}$, and thus $(V^*, \eta_{V^*})$ is an object of $\mathscr{R}(\mathscr{C}, c)$. That it is a left dual of $(V, \eta_V)$ in $\mathscr{R}(\mathscr{C}, c)$ reduces to the fact that $\mathrm{ev}_V$ and $\mathrm{coev}_V$ are morphisms in $\mathscr{R}(\mathscr{C}, c)$. Towards this end, we need the equivalence

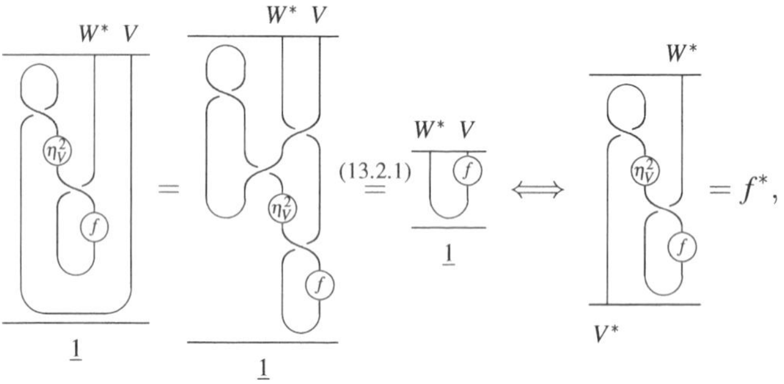


(13.2.9)

true for any morphism $f : V \to W$ in $\mathscr{C}$ with $(V, \eta_V) \in \mathscr{R}(\mathscr{C}, c)$, which follows from the naturality of $c_{V,-}^{-1}$ applied to $\mathrm{ev}_V(\mathrm{Id}_{V^*} \otimes f\eta_V^2)c_{V,W^*} : V \otimes W^* \to \underline{1}$ and the definition of $f^*$.

Now, that $\mathrm{ev}_V$ is a morphism in $\mathscr{R}(\mathscr{C}, c)$ is a consequence of the computation

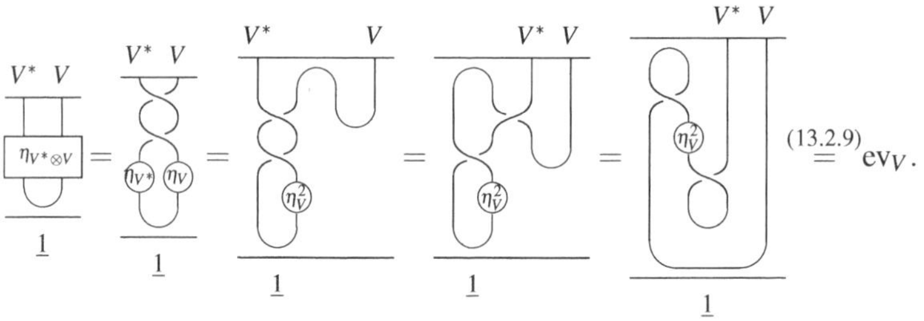


Likewise, we compute that

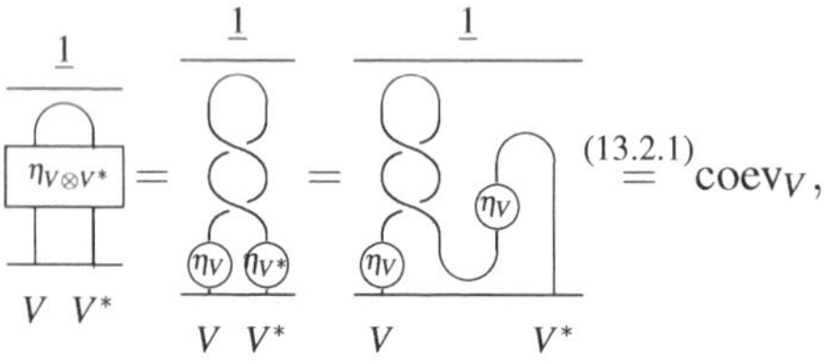


and so $\mathrm{coev}_V$ is a morphism in $\mathscr{R}(\mathscr{C}, c)$, as stated.

Finally, from the left rigid monoidal structure of $\mathscr{R}(\mathscr{C}, c)$ we get that $\eta := (\eta_V)_V$ is a twist on $\mathscr{R}(\mathscr{C}, c)$, and so the proof is finished. □

More generally, Theorem 13.13 allows us to construct ribbon categories from left or right rigid monoidal categories.

**Proposition 13.14** *Let $\mathscr{C}$ be a left rigid (strict) monoidal category. Then $\mathfrak{R}_l(\mathscr{C})$ is a category whose*

- *objects are triples $(V, c_{V,-}, \eta_V)$ consisting of an object $V$ of $\mathscr{C}$, a natural isomorphism $c_{V,-} = (c_{V,X} : V \otimes X \to X \otimes V)_{X \in \mathrm{Ob}(\mathscr{C})}$ and an automorphism $\eta_V$ of $V$ in $\mathscr{C}$ such that* (8.1.4) *and* (13.1.4) *hold, and*

$$(\mathrm{Id}_X \otimes \eta_V)c_{V,X} = c_{V,X}(\eta_V \otimes \mathrm{Id}_X), \quad \forall\, X \in \mathrm{Ob}(\mathscr{C}); \tag{13.2.10}$$

- *morphisms $f : (V, c_{V,-}, \eta_V) \to (V', c_{V',-}, \eta_{V'})$ are morphisms $f : V \to V'$ in $\mathscr{C}$ obeying $(\mathrm{Id}_X \otimes f)c_{V,X} = c_{V',X}(f \otimes \mathrm{Id}_X)$, for any object $X$ of $\mathscr{C}$, and $f\eta_V = \eta_{V'}f$.*

*Proof* One can easily check that $\mathfrak{R}_l(\mathscr{C}) = \mathscr{R}(\mathscr{Z}_l(\mathscr{C}), c)$, where $\mathscr{Z}_l(\mathscr{C})$ is the left center of $\mathscr{C}$ as in Section 8.1, a braided category. We need only note that (13.1.4) is nothing but the second equality in (13.2.1). □

We now uncover the ribbon structure of $\mathscr{R}_l(\mathscr{C})$. For the choice of the natural isomorphism $c_{V^*,-}$ below see (1.8.1) or Theorem 8.17.

**Theorem 13.15** *Let $\mathscr{C}$ be a left rigid monoidal category. Then $\mathscr{R}_l(\mathscr{C})$ is a ribbon category with the following structure:*

- *if $(V, c_{V,-}, \eta_V), (W, c_{W,-}, \eta_W) \in \mathscr{R}_l(\mathscr{C})$ then*

$$(V, c_{V,-}, \eta_V) \otimes (W, c_{W,-}, \eta_W) = (V \otimes W, c_{V\otimes W,-}, \eta_{V\otimes W}), \tag{13.2.11}$$

*where $c_{V\otimes W,-}$ is as in* (8.1.5) *and*

$$\eta_{V\otimes W} = (\eta_V \otimes \eta_W)c_{W,V}c_{V,W}; \tag{13.2.12}$$

- *the unit object is $(\underline{1}, c_{\underline{1},-} = (r_X^{-1} l_X)_{X\in\mathscr{C}} \equiv \mathrm{Id}, \eta_{\underline{1}} = \mathrm{Id}_{\underline{1}})$, and the associativity and the left and right unit constraints are the same as those of $\mathscr{C}$;*
- *the braiding $\mathfrak{c}$ is determined by*

$$c_{V,W} : (V, c_{V,-}, \eta_V) \otimes (W, c_{W,-}, \eta_W) \to (W, c_{W,-}, \eta_W) \otimes (V, c_{V,-}, \eta_V), \tag{13.2.13}$$

*an isomorphism in $\mathscr{R}_l(\mathscr{C})$;*
- *for $(V, c_{V,-}, \eta_V) \in \mathscr{R}_l(\mathscr{C})$, a left dual object for it is $(V^*, c_{V^*,-}, \eta_{V^*})$, where*

$$c_{V^*,X} = (\mathrm{ev}_V \otimes \mathrm{Id}_{X\otimes V^*})(\mathrm{Id}_{V^*} \otimes c_{V,X}^{-1} \otimes \mathrm{Id}_{V^*})(\mathrm{Id}_{V^*\otimes X} \otimes \mathrm{coev}_V), \tag{13.2.14}$$

*for all $X \in \mathrm{Ob}(\mathscr{C})$, and*

$$\eta_{V^*} = (\eta_V)^*, \tag{13.2.15}$$

*and the evaluation and coevaluation morphisms are precisely $\mathrm{ev}_V$ and $\mathrm{coev}_V$, viewed now as morphisms in $\mathscr{R}_l(\mathscr{C})$;*
- *the twist is given by*

$$\eta_V : (V, c_{V,-}, \eta_V) \to (V, c_{V,-}, \eta_V), \tag{13.2.16}$$

*an automorphism in $\mathscr{R}_l(\mathscr{C})$.*

*Proof* Since $\mathfrak{R}_l(\mathscr{C}) = \mathscr{R}(\mathscr{Z}_l(\mathscr{C}), c)$, the only thing we must check is the fact that $c_{V^*,X}$ in (13.2.14) is an isomorphism in $\mathscr{C}$, for all $X \in \mathrm{Ob}(\mathscr{C})$. In particular, Theorem 8.17 applies.

In the computations below, we use graphical notations similar to those used in Section 8.1. Namely, we denote

$c_{V,-} :=$ 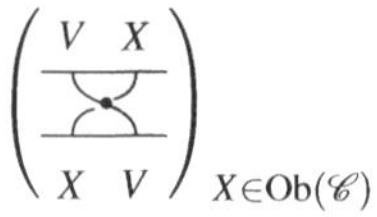

and $c_{V,-}^{-1} :=$ 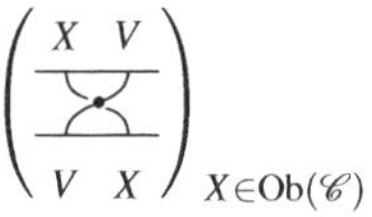

.

Let $c_{V^*,-}$ be defined by (13.2.14). We claim that, for all $X \in \mathrm{Ob}(\mathscr{C})$, $c_{V^*,X}$ is an isomorphism in $\mathscr{C}$ with inverse given by

$c_{V^*,X}^{-1} =$ 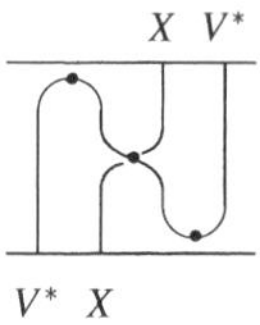

$=$ 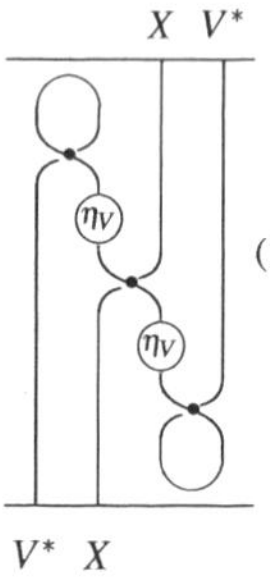

$\overset{(13.2.10)}{=}$ 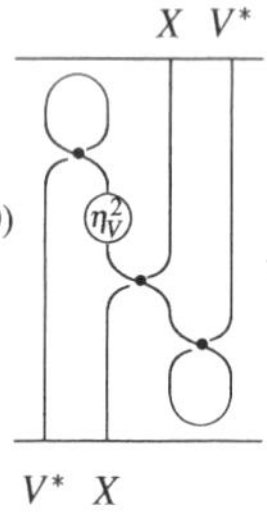

,

where, as the graphical notation suggests, the evaluation and coevaluation morphisms with a black dot are $\mathrm{ev}'_V$ and $\mathrm{coev}'_V$ defined as in (13.1.6). Since $\eta_{V^*} = (\eta_V)^*$, it follows that $\mathrm{coev}'_V$ can be restated as in (13.1.7).

In fact, (13.2.9) still holds if we replace 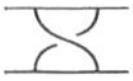 with 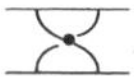. Specializing it for $f = \mathrm{Id}_V$, we get that

$c_{V^*,X} \circ c_{V^*,X}^{-1} =$ 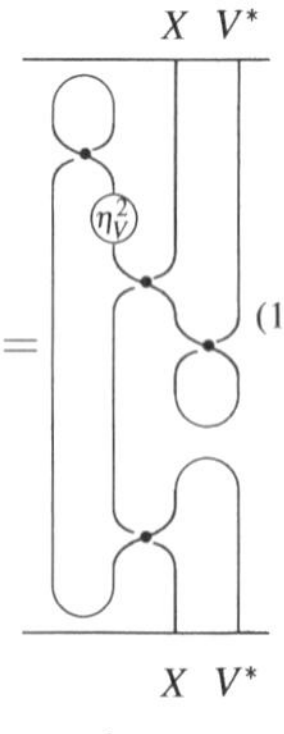

$\underset{(*)}{\overset{(13.2.10)}{=}}$ 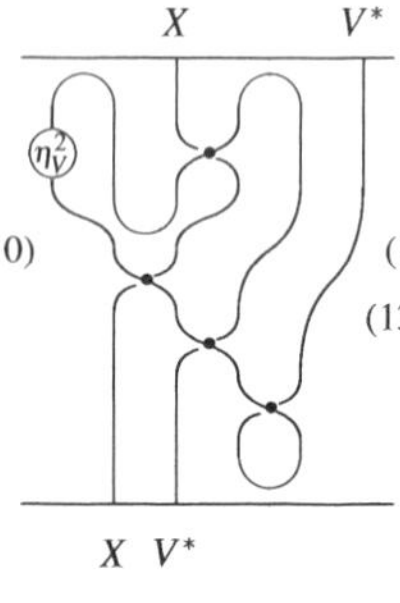

$\underset{(13.2.10)}{\overset{(1.6.6)}{=}}$ 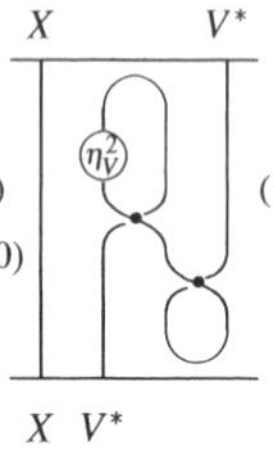

$\overset{(13.2.9)}{=} \mathrm{Id}_{X \otimes V^*}$,

where (*) refers to the naturality of $c_{V,-}$ applied to $c_{V^*,X}$. Likewise, again using the naturality of $c_{V,-}$ applied to $c_{V^*,X}$, one can show that $c_{V^*,X}^{-1} \circ c_{V^*,X} = \mathrm{Id}_{V^* \otimes X}$. □

**Definition 13.16** If $(\mathscr{C}, c, \eta)$ and $(\mathscr{D}, d, \theta)$ are ribbon categories then a functor $F : \mathscr{C} \to \mathscr{D}$ is called a ribbon functor if it is a braided monoidal functor such that $F(\eta_V) = \theta_{F(V)}$, for all $V \in \mathrm{Ob}(\mathscr{C})$, and is compatible with the duality.

The left-handed version of the Universal Property in Proposition 8.3 leads to a similar property for $\mathscr{R}_l(\mathscr{C})$.

**Proposition 13.17** *Let $(\mathscr{D},d,\theta)$ be a ribbon category, $\mathscr{C}$ a left rigid monoidal category and $F:\mathscr{D}\to\mathscr{C}$ a strong monoidal functor which is bijective on objects and surjective on morphisms. Then there exists a unique ribbon functor $\mathscr{R}_l(F):\mathscr{D}\to\mathscr{R}_l(\mathscr{C})$ such that $\Pi^l\circ\mathscr{R}_l(F)=F$, where $\Pi^l:\mathscr{R}_l(\mathscr{C})\to\mathscr{C}$ is the forgetful functor.*

*Proof* If $\overline{\Pi}:\mathscr{R}_l(\mathscr{C})\to\mathscr{Z}_l(\mathscr{C})$ is the functor that forgets the twist then $\overline{\Pi}$ is a braided monoidal functor and $\Pi\circ\overline{\Pi}=\Pi^l$, where $\Pi:\mathscr{Z}_l(\mathscr{C})\to\mathscr{C}$ is the strict monoidal functor that forgets the natural isomorphism.

According to the left-handed version of Proposition 8.3, there exists a unique braided monoidal functor $\mathscr{Z}_l(F):\mathscr{D}\to\mathscr{Z}_l(\mathscr{C})$ such that $\Pi\circ\mathscr{Z}_l(F)=F$. Then $\mathscr{R}_l(F):\mathscr{D}\to\mathscr{R}_l(\mathscr{C})$ given by $\mathscr{R}_l(F)(Y)=(\mathscr{Z}_l(F)(Y),F(\theta_Y))$, for all $Y\in\mathscr{D}$, and $\mathscr{R}_l(F)(f)=F(f)$, for any morphism $f$ in $\mathscr{D}$, is a well-defined functor. It follows that $\mathscr{R}_l(F)$ is a ribbon functor satisfying $\Pi^l\circ\mathscr{R}_l(F)=F$; it is, moreover, unique with these properties because of the uniqueness of $\mathscr{Z}_l(F)$ and of its ribbon property. □

**Corollary 13.18** *If $(\mathscr{C},c,\eta)$ is a ribbon category then there exists a unique ribbon functor $\mathscr{R}_l:\mathscr{C}\to\mathscr{R}_l(\mathscr{C})$ such that $\Pi^l\circ\mathscr{R}_l=\mathrm{Id}_{\mathscr{C}}$.*

*Proof* In Proposition 13.17, take $\mathscr{D}=\mathscr{C}$ and $F=\mathrm{Id}_{\mathscr{C}}$. Then $\mathscr{R}_l=\mathscr{R}_l(\mathrm{Id}_{\mathscr{C}})$. □

In what follows we also need the right-handed version of $\mathscr{R}_l(\mathscr{C})$. In fact, if we start with a right rigid monoidal category then $\overline{\mathscr{C}}$ is a left rigid monoidal category, and so we can consider $\mathscr{R}_l(\overline{\mathscr{C}})$. Thus $\mathscr{R}_r(\mathscr{C}):=\overline{\mathscr{R}_l(\overline{\mathscr{C}})}$ is a ribbon category, too. More precisely, we have the following.

**Proposition 13.19** *Let $\mathscr{C}$ be a right rigid monoidal category. Then the objects of $\mathscr{R}_r(\mathscr{C})$ are triples $(V,c_{-,V},\theta_V)$ consisting of an object $V$ of $\mathscr{C}$, a natural isomorphism $c_{-,V}=(c_{X,V}:X\otimes V\to V\otimes X)_{X\in\mathrm{Ob}(\mathscr{C})}$ and a morphism $\theta_V:V\to V$ in $\mathscr{C}$, subject to the following conditions:*

- $c_{\underline{1},V}\equiv\mathrm{Id}_V$ *and* $c_{X\otimes Y,V}=(c_{X,V}\otimes\mathrm{Id}_Y)(\mathrm{Id}_X\otimes c_{Y,V})$, *for all* $X,Y\in\mathscr{C}$;
- $\theta_V$ *is an automorphism of* $V$ *in* $\mathscr{C}$ *obeying* $c_{X,V}(\mathrm{Id}_X\otimes\theta_V)=(\theta_V\otimes\mathrm{Id}_X)c_{X,V}$, *for all* $X\in\mathscr{C}$, *and*

$$\theta_V^{-2}:=(\mathrm{Id}_V\otimes\mathrm{ev}'_V)(c_{V,V}^{-1}\otimes\mathrm{Id}_{{}^*V})(\mathrm{Id}_V\otimes c_{{}^*V,V})(\mathrm{Id}_V\otimes\mathrm{coev}'_V). \tag{13.2.17}$$

*A morphism $f:(V,c_{-,V},\theta_V)\to(W,c_{-,W},\theta_W)$ in $\mathscr{Z}_r(\mathscr{C})$ is a morphism $f:V\to W$ in $\mathscr{C}$ such that $c_{X,W}(f\otimes\mathrm{Id}_X)=(\mathrm{Id}_X\otimes f)c_{X,V}$, for all $X\in\mathscr{C}$.*

*The category $\mathscr{R}_r(\mathscr{C})$ is ribbon via the following structure:*

- *the tensor product of two objects $(V,c_{-,V},\theta_V)$ and $(W,c_{-,W},\theta_W)$ in $\mathscr{R}_r(\mathscr{C})$ is $(V\otimes W,c_{-,V\otimes W},\theta_{V\otimes W})$, where $c_{-,V\otimes W}$ is defined by* (8.1.1) *and $\theta_{V\otimes W}=(\theta_V\otimes\theta_W)c_{W,V}c_{V,W}$, and the tensor product of two morphisms in $\mathscr{R}_r(\mathscr{C})$ is their tensor product in $\mathscr{C}$, while the associativity and the left and right unit constraints are the same as those of $\mathscr{C}$;*

- *the braiding between two objects* $(V, c_{-,V}, \theta_V)$ *and* $(W, c_{-,W}, \theta_W)$ *in* $\mathscr{R}_r(\mathscr{C})$ *is given by* $c_{V,W}$;
- *the right dual object of* $(V, c_{-,V}, \theta_V)$ *in* $\mathscr{R}_r(\mathscr{C})$ *is* $({}^*V, c_{-,{}^*V}, \theta_{{}^*V})$ *determined by*

$$c_{X,{}^*V} = (\mathrm{Id}_{{}^*V\otimes X} \otimes \mathrm{ev}'_V)(\mathrm{Id}_{V^*} \otimes c^{-1}_{V,X} \otimes \mathrm{Id}_{{}^*V})(\mathrm{coev}'_V \otimes \mathrm{Id}_{X\otimes {}^*V}),$$

*for all* $X \in \mathscr{C}$, $\theta_{{}^*V} = {}^*(\theta_V)$, *and the evaluation and coevaluation morphisms equal* $\mathrm{ev}'_V$ *and* $\mathrm{coev}'_V$, *respectively;*
- *the twist is given by* $\theta_V : (V, c_{-,V}, \theta_V) \to (V, c_{-,V}, \theta_V)$, *for all* $V \in \mathscr{C}$.

*Proof* As $\mathscr{R}_r(\mathscr{C}) := \overline{\mathscr{R}_l(\overline{\mathscr{C}})}$, everything follows by applying Corollary 13.9, after we specialize Proposition 13.14 and Theorem 13.15 to $\overline{\mathscr{C}}$. □

## 13.3 Ribbon Quasi-Hopf Algebras

This section can be viewed as an extension of Section 10.1, as we continue to investigate when the category of representations over a QT quasi-bialgebra is a balanced or a ribbon category.

Let $H$ be a quasi-bialgebra, so ${}_H\mathscr{M}$ is monoidal. Then we have seen that ${}_H\mathscr{M}^{\mathrm{fd}}$ is left rigid braided monoidal (and so right rigid as well; see Theorem 1.77), provided that $H$ is a QT quasi-Hopf algebra. Furthermore, the converse is also true if we assume $H$ is finite dimensional; see the results in Section 3.5. Hence, in the finite-dimensional case, to see when ${}_H\mathscr{M}^{\mathrm{fd}}$ is a balanced/ribbon category such that the forgetful functor $F : {}_H\mathscr{M}^{\mathrm{fd}} \to {}_k\mathscr{M}^{\mathrm{fd}}$ is a left rigid quasi-monoidal functor reduces to the following problem: for a finite-dimensional QT quasi-Hopf algebra $(H, R)$ describe all the balanced structures/twists on ${}_H\mathscr{M}^{\mathrm{fd}}$.

The answer to the above question is given by the result below.

**Proposition 13.20** *Let* $(H, R)$ *be a finite-dimensional QT quasi-Hopf algebra, so* ${}_H\mathscr{M}^{\mathrm{fd}}$ *is a rigid braided category. Then* ${}_H\mathscr{M}^{\mathrm{fd}}$ *is, moreover,*

*(i) a balanced category if and only if there exists an invertible central element* $\eta \in H$ *such that*

$$\Delta(\eta) = (\eta \otimes \eta)R_{21}R, \tag{13.3.1}$$

*where, if* $R = R^1 \otimes R^2$ *then* $R_{21} = R^2 \otimes R^1$;

*(ii) a ribbon category if and only if there exists an invertible central element* $\eta \in H$ *obeying* (13.3.1) *and such that*

$$S(\eta) = \eta. \tag{13.3.2}$$

*Proof* Since $H$ is finite dimensional we can regard $H \in {}_H\mathscr{M}^{\mathrm{fd}}$ via its multiplication.

Suppose that $(\eta_V : V \to V)_V$, indexed by $V \in {}_H\mathscr{M}^{\mathrm{fd}}$, defines a balanced structure on ${}_H\mathscr{M}^{\mathrm{fd}}$. Then $\eta_H$ is completely determined by $\eta := \eta(1_H)$. Actually, $\eta$ describes $\eta_V$ completely, for any $V \in {}_H\mathscr{M}^{\mathrm{fd}}$. To see this, take $v \in V \in {}_H\mathscr{M}^{\mathrm{fd}}$ and define $\varphi_v : H \to V$ by $\varphi_v(h) = h \cdot v$, for all $h \in H$. Since $\varphi_v$ is left $H$-linear, by the naturality of the family $(\eta_V : V \to V)_V$ we have $\eta_V \circ \varphi_v = \varphi_v \circ \eta_H$. By evaluating both sides of this relation on $1_H$ we deduce that

$$\eta_V(v) = \eta \cdot v, \ \forall\, v \in V. \tag{13.3.3}$$

Since $\eta_V$ is an isomorphism for any $V \in {}_H\mathscr{M}^{\rm fd}$ it follows that $\eta$ is invertible (this follows from the fact that $\eta_H$ is a left $H$-linear isomorphism). Also, it can be easily seen that $\eta_V$ is left $H$-linear, for all $V \in {}_H\mathscr{M}^{\rm fd}$, if and only if $\eta$ is a central element of $H$. Furthermore, for all $v \in V \in {}_H\mathscr{M}^{\rm fd}$ and $w \in W \in {}_H\mathscr{M}^{\rm fd}$ we have

$$\begin{aligned}(\eta_V \otimes \eta_W) \circ c_{W,V} \circ c_{V,W}(v \otimes w) &= (\eta_V \otimes \eta_W)(r^2R^1 \cdot v \otimes r^1R^2 \cdot w)\\ &= \eta r^2R^1 \cdot v \otimes \eta r^1R^2 \cdot w\end{aligned}$$

and $\eta_{V\otimes W}(v \otimes w) = \eta \cdot (v \otimes w) = \eta_1 \cdot v \otimes \eta_2 \cdot w$. Thus (13.1.1) is satisfied for all $V, W \in {}_H\mathscr{M}^{\rm fd}$ if and only if (13.3.1) holds (for the direct implication take $v = w = 1_H \in V = W = H$).

For the ribbon case, we have $\eta_{V^*}(v^*)(v) = (\eta \cdot v^*)(v) = v^*(S(\eta) \cdot v)$, for all $v^* \in V^*$, $v \in V$. By Proposition 3.34 the left transpose of $\eta_V$ in ${}_H\mathscr{M}^{\rm fd}$ coincides with the usual transpose map of $\eta_V$ in ${}_k\mathscr{M}^{\rm fd}$, and so $(\eta_V)^*(v^*)(v) = (v^* \circ \eta_V)(v) = v^*(\eta \cdot v)$. We conclude that (13.1.2) is equivalent to (13.3.2), and this ends the direct assertion in both (i) and (ii).

Conversely, from the above computation it follows that $(\eta_V)_V$, with $\eta_V$ defined by an invertible central element $\eta$ of $H$ as in (13.3.3) that also obeys (13.3.1), provides a balanced structure on ${}_H\mathscr{M}^{\rm fd}$. If (13.3.2) is satisfied as well then $(\eta_V)_V$ is a twist on ${}_H\mathscr{M}^{\rm fd}$. □

The next definitions are imposed by the characterization that we have just proved in Proposition 13.20.

**Definition 13.21** (i) We call a QT quasi-bialgebra (resp. quasi-Hopf algebra) $(H,R)$ a balanced quasi-bialgebra (resp. quasi-Hopf algebra) if there exists an invertible central element $\eta \in H$ satisfying (13.3.1).

(ii) A QT quasi-Hopf algebra $(H,R)$ is called a ribbon quasi-Hopf algebra if there exists an invertible central element $\eta \in H$ satisfying (13.3.1) and (13.3.2).

By Theorem 3.38 and Proposition 13.20 we have the following consequence.

**Corollary 13.22** *For a finite-dimensional k-algebra H there exists a bijective correspondence between*

- *balanced/ribbon structures on* ${}_H\mathscr{M}^{\rm fd}$ *such that the forgetful functor* $F : {}_H\mathscr{M}^{\rm fd} \to {}_k\mathscr{M}^{\rm fd}$ *is a left rigid quasi-monoidal functor;*
- *balanced/ribbon quasi-Hopf algebra structures on H.*

According to Theorem 13.10, a ribbon category $\mathscr{C}$ is sovereign. When $\mathscr{C} = {}_H\mathscr{M}^{\rm fd}$, with $H$ a ribbon quasi-Hopf algebra, this implies the existence of an invertible element $\mathfrak{g} \in H$ fulfilling (3.6.3) and (3.6.4); see Proposition 3.41. In other words, we can construct sovereign/pivotal elements for $H$ out of ribbon elements. In what follows, we will give a concrete description for this correspondence. In particular, we will measure how far an element $\eta$ as in Proposition 13.20 is from being $u$, the element considered in (10.3.4). To see all these, we need the following technical result.

**Lemma 13.23** *Let $(H,R)$ be a QT quasi-Hopf algebra and $u$ the element defined in* (10.3.4). *Then*

$$\Delta(u) = f^{-1}(S\otimes S)(f_{21})(u\otimes u)(R_{21}R)^{-1}, \tag{13.3.4}$$

*where, as always, $f = f^1\otimes f^2$ is the element defined in* (3.2.15) *and $f^{-1}$ is its inverse as in* (3.2.16).

*Proof* We set $R = R^1\otimes R^2 = r^1\otimes r^2 = \mathscr{R}^1\otimes\mathscr{R}^2$, $R^{-1} = \overline{R}^1\otimes\overline{R}^2 = \overline{r}^1\otimes\overline{r}^2$, $\gamma = \gamma^1\otimes\gamma^2$ and $\delta = \delta^1\otimes\delta^2$ as in (3.2.5) and (3.2.6), and $f^{-1} = g^1\otimes g^2$. Then the formula in (13.3.4) is a consequence of the following technical computation:

$$\begin{aligned}
&\Delta(u)\\
&\overset{(10.3.4),(3.2.14)}{=} S(R^2p^2)_1g^1\gamma^1R^1_1p^1_1\otimes S(R^2p^2)_2g^2\gamma^2R^1_2p^1_2\\
&\overset{(3.2.13)}{=} g^1S(R^2_2p^2_2)\gamma^1R^1_1p^1_1\otimes g^2S(R^2_1p^2_1)\gamma^2R^1_2p^1_2\\
&\overset{(3.2.5),(3.1.9)}{\underset{(3.2.1),(3.2.1)}{=}} g^1S(x^1X^2R^2_2p^2_2)\alpha x^2(X^3R^1)_1p^1_1\\
&\qquad\otimes g^2S(X^1R^2_1p^2_1)\alpha x^3(X^3R^1)_2p^1_2\\
&\overset{(10.1.2)}{=} g^1S(x^1R^2X^3y^3p^2_2)\alpha x^2R^1_1X^2_1r^1_1y^1_1p^1_1\\
&\qquad\otimes g^2S(X^1r^2y^2p^2_1)\alpha x^3R^1_2X^2_2r^1_2y^1_2p^1_2\\
&\overset{(10.1.1)}{=} g^1S(\mathscr{R}^2x^2R^2Y^3X^3y^3p^2_2)\alpha\mathscr{R}^1x^1Y^1X^2_1r^1_1y^1_1p^1_1\\
&\qquad\otimes g^2S(X^1r^2y^2p^2_1)\alpha x^3R^1Y^2X^2_2r^1_2y^1_2p^1_2\\
&\overset{(10.3.8)}{=} g^1S(\alpha x^2R^2Y^3X^3y^3p^2_2)ux^1Y^1X^2_1r^1_1y^1_1p^1_1\\
&\qquad\otimes g^2S(X^1r^2y^2p^2_1)\alpha x^3R^1Y^2X^2_2r^1_2y^1_2p^1_2\\
&\overset{(10.3.10)}{=} g^1S(S(x^1)\alpha x^2R^2Y^3X^3y^3p^2_2)uY^1X^2_1r^1_1y^1_1p^1_1\\
&\qquad\otimes g^2S(X^1r^2y^2p^2_1)\alpha x^3R^1Y^2X^2_2r^1_2y^1_2p^1_2\\
&\overset{(3.2.14),(3.2.13)}{=} f^{-1}(S\otimes S)(f_{21})[S(S(x^1)\alpha x^2R^2Y^3X^3y^3z^2_2\delta^2S(z^3_1))u\\
&\qquad Y^1X^2_1r^1_1y^1_1z^1_1\otimes S(X^1r^2y^2z^2_1\delta^1S(z^3_2))\alpha x^3R^1Y^2X^2_2r^1_2y^1_2z^1_2]\\
&\overset{(3.2.6),(3.1.9)}{\underset{(3.2.1),(3.1.10)}{=}} f^{-1}(S\otimes S)(f_{21})[S(S(x^1)\alpha x^2R^2Y^3X^3y^3z^2_2t^2Z^1\beta S(z^3_1t^3_1Z^2))u\\
&\qquad Y^1X^2_1r^1_1y^1_1z^1_1\otimes S(X^1r^2y^2z^2_1t^1\beta S(z^3_2t^3_2Z^3))\alpha x^3R^1Y^2X^2_2r^1_2y^1_2z^1_2]\\
&\overset{(3.1.9)}{=} f^{-1}(S\otimes S)(f_{21})[S(S(x^1)\alpha x^2R^2Y^3X^3y^2z^3_1Z^1\beta S(y^3_1z^3_{(2,1)}Z^2))uY^1X^2_1\\
&\qquad (r^1y^1_1)_1z^1_1\otimes S(X^1r^2y^1_2z^2\beta S(y^3_2z^3_{(2,2)}Z^3))\alpha x^3R^1Y^2X^2_2(r^1y^1_1)_2z^1_2]\\
&\overset{(3.1.7),(3.2.1)}{\underset{(10.1.3)}{=}} f^{-1}(S\otimes S)(f_{21})[S(S(x^1)\alpha x^2R^2Y^3X^3y^2Z^1\beta S(y^3_1Z^2))uY^1\\
&\qquad (X^2y^1_2)_1r^1_1z^1_1\otimes S(X^1y^1_1r^2z^2\beta S(y^3_2Z^3z^3))\alpha x^3R^1Y^2(X^2y^1_2)_2r^1_2z^1_2]\\
&\overset{(3.1.9)}{=} f^{-1}(S\otimes S)(f_{21})[S(S(x^1)\alpha x^2R^2Y^3y^2_2t^2X^3_1Z^1\beta S((y^3t^3X^3_2)_1Z^2))uY^1\\
&\qquad (y^2_1t^1X^2r^1z^1)_1\otimes S(y^1X^1r^2z^2\beta S((y^3t^3X^3_2)_2Z^3z^3))\alpha x^3R^1Y^2\\
&\qquad (y^2_1t^1X^2r^1z^1)_2]
\end{aligned}$$

$$
\begin{aligned}
&\overset{(3.1.7),(3.2.1)}{=} f^{-1}(S\otimes S)(f_{21})[S(S(x^1)\alpha x^2R^2y^2_{(2,2)}Y^3t^2Z^1\beta S(y^3_1t^3_1Z^2))uy^2_1Y^1\\
&\qquad t^1_1X^2_1r^1_1z^1_1\otimes S(y^1X^1r^2z^2\beta S(y^3_2t^3_2Z^3X^3z^3))\alpha x^3R^1y^2_{(2,1)}Y^2t^1_2X^2_2r^1_2z^1_2]\\
&\overset{(10.1.3),(10.3.10)}{=} f^{-1}(S\otimes S)(f_{21})[S(S(x^1y^2_1)\alpha x^2y^2_{(2,1)}R^2Y^3t^2Z^1\beta S(y^3_1t^3_1Z^2))uY^1t^1_1\\
&\qquad X^2_1r^1_1z^1_1\otimes S(y^1X^1r^2z^2\beta S(y^3_2t^3_2Z^3X^3z^3))\alpha x^3y^2_{(2,2)}R^1Y^2t^1_2X^2_2r^1_2z^1_2]\\
&\underset{(3.1.9)}{\overset{(3.1.7),(3.2.1)}{=}} f^{-1}(S\otimes S)(f_{21})[S(S(x^1)\alpha x^2R^2t^2_2w^2Y^3_1Z^1\beta S((y^3t^3w^3Y^3_2)_1Z^2))u\\
&\qquad t^1Y^1X^2_1r^1_1z^1_1\otimes S(y^1X^1r^2z^2\beta S((y^3t^3w^3Y^3_2)_2Z^3X^3z^3))\alpha y^2x^3R^1\\
&\qquad t^2_1w^1Y^2X^2_2r^1_2z^1_2]\\
&\overset{(3.1.7),(3.2.1)}{=} f^{-1}(S\otimes S)(f_{21})[S(S(x^1)\alpha x^2R^2t^2_2w^2Z^1\beta S(y^3_1t^3_1w^3_1Z^2))ut^1Y^1X^2_1r^1_1z^1_1\\
&\qquad \otimes S(y^1X^1r^2z^2\beta S(y^3_2t^3_2w^3_2Z^3Y^3X^3z^3))\alpha y^2x^3R^1t^2_1w^1Y^2X^2_2r^1_2z^1_2]\\
&\underset{(3.1.10),(10.3.10)}{\overset{(3.1.9),(3.2.1)}{=}} f^{-1}(S\otimes S)(f_{21})[S(S(x^1t^1)\alpha x^2R^2(t^2Z^1)_2w^2\beta S(y^3_1t^3_1Z^2w^3))uY^1X^2_1\\
&\qquad r^1z^1_1\otimes S(y^1X^1r^2z^2\beta S(y^3_2t^3_2Z^3Y^3X^3z^3))\alpha y^2x^3R^1(t^2Z^1)_1\\
&\qquad w^1Y^2X^2_2r^1_2z^1_2]\\
&\overset{(3.1.9),(10.1.3)}{=} f^{-1}(S\otimes S)(f_{21})[S(S(x^1Z^1_1t^1v^1)\alpha x^2Z^1_{(2,1)}t^2_1R^2v^2_{(1,2)}w^2\beta\\
&\qquad S(y^3_1Z^2t^3v^2_2w^3))uY^1X^2_1r^1_1z^1_1\otimes S(y^1X^1r^2z^2\beta S(y^3_2Z^3v^3Y^3X^3z^3))\alpha\\
&\qquad y^2x^3Z^1_{(2,2)}t^2_2R^1v^2_{(1,1)}w^1Y^2X^2_2r^1_2z^1_2]\\
&\overset{(3.1.7),(3.2.1)}{=} f^{-1}(S\otimes S)(f_{21})[S(S(x^1t^1v^1)\alpha x^2t^2_1R^2w^2\beta S(y^3_1Z^2t^3w^3))uY^1X^2_1\\
&\qquad\qquad r^1_1z^1_1\otimes S(y^1X^1r^2z^2\beta S(y^3_2Z^3v^3Y^3X^3z^3))\alpha y^2Z^1x^3t^2_2R^1\\
&\qquad\qquad w^1v^2Y^2X^2_2r^1_2z^1_2]\\
&\overset{(10.3.12)}{=} f^{-1}(S\otimes S)(f_{21})[S(S(v^1)S(\overline{R}^2)S(y^3_1Z^2))uY^1X^2_1r^1_1z^1_1\otimes S(y^1X^1r^2\\
&\qquad\qquad z^2\beta S(y^3_2Z^3v^3Y^3X^3z^3))\alpha y^2Z^1\overline{R}^1v^2Y^2X^2_2r^1_2z^1_2]\\
&\overset{(10.3.10),(10.1.3)}{=} f^{-1}(S\otimes S)(f_{21})[uy^3_1Z^2X^2_2\overline{R}^2r^1_1z^1_1\\
&\qquad\qquad \otimes S(y^1X^1r^2z^2\beta S(y^3_2Z^3X^3z^3))\alpha y^2Z^1X^2_1\overline{R}^1r^1_2z^1_2]\\
&\underset{(3.1.11),(10.1.3)}{\overset{(3.1.9),(3.2.1)}{=}} f^{-1}(S\otimes S)(f_{21})[uy^3r^1_2z^1_2\overline{R}^2\\
&\qquad\qquad \otimes S(y^1r^2z^2\beta S(z^3))\alpha y^2r^1_1z^1_1\overline{R}^1]\\
&\overset{(10.1.1)}{=} f^{-1}(S\otimes S)(f_{21})[ux^3r^1Y^2z^1_2\overline{R}^2\\
&\qquad\qquad \otimes S(R^2x^2r^2Y^3z^2\beta S(z^3))\alpha R^1x^1Y^1z^1_1\overline{R}^1]\\
&\underset{(3.1.11),(10.3.8)}{\overset{(3.1.9),(3.2.1)}{=}} f^{-1}(S\otimes S)(f_{21})[ux^3r^1y^2_1z^1\overline{R}^2\\
&\qquad\qquad \otimes S(\alpha x^2r^2y^2_2z^2\beta S(y^3z^3))ux^1y^1\overline{R}^1\\
&\overset{(10.3.10),(10.1.3)}{=} f^{-1}(S\otimes S)(f_{21})(u\otimes u)[x^3y^2_2r^1z^1\overline{R}^2\\
&\qquad\qquad \otimes S^{-1}(S(x^1y^1)\alpha x^2y^2_1r^2z^2\beta S(y^3z^3))\overline{R}^1]\\
&\overset{(10.3.12)}{=} f^{-1}(S\otimes S)(f_{21})(u\otimes u)\overline{r}^1\overline{R}^2\otimes\overline{r}^2\overline{R}^1\\
&= f^{-1}(S\otimes S)(f_{21})(u\otimes u)(R_{21}R)^{-1},
\end{aligned}
$$

as stated. □

From the formula of $\Delta(u)$ we can derive other useful formulas.

**Corollary 13.24** *In the hypotheses of Lemma 13.23 we have:*

$$\Delta(u) = (R_{21}R)^{-1}f^{-1}(S\otimes S)(f_{21})(u\otimes u), \tag{13.3.5}$$

$$\Delta(S(u)) = (R_{21}R)^{-1}(S(u)\otimes S(u))(S\otimes S)(f_{21}^{-1})f, \tag{13.3.6}$$

$$\Delta(S(u)^{-1}u) = f^{-1}(S\otimes S)(f_{21})(S^2\otimes S^2)(f^{-1})(S^3\otimes S^3)(f_{21})(S(u)^{-1}u\otimes S(u)^{-1}u), \tag{13.3.7}$$

$$S^4(h) = (S(u)^{-1}u)h(S(u)^{-1}u)^{-1}, \ \ \forall\, h\in H. \tag{13.3.8}$$

*Proof* Let us start by noting that (10.3.6) implies

$$(S\otimes S)(RR_{21}) = fR_{21}Rf^{-1}. \tag{13.3.9}$$

By (13.3.9) we have $(S\otimes S)((R_{21}R)^{-1}) = f_{21}(RR_{21})^{-1}f_{21}^{-1}$. By applying $S\otimes S$ to both sides of this relation and taking into account (13.3.9) we deduce that

$$(S^2\otimes S^2)((R_{21}R)^{-1}) = (S\otimes S)(f_{21}^{-1})f(R_{21}R)^{-1}f^{-1}(S\otimes S)(f_{21}).$$

This together with (10.3.10) and (13.3.4) guarantees the fact that

$$\begin{aligned}\Delta(u) &= f^{-1}(S\otimes S)(f_{21})(S^2\otimes S^2)((R_{21}R)^{-1})(u\otimes u)\\ &= (R_{21}R)^{-1}f^{-1}(S\otimes S)(f_{21})(u\otimes u),\end{aligned}$$

proving (13.3.5). By using (13.3.9), (3.2.13) and (13.3.4), we compute:

$$\begin{aligned}\Delta(S(u)) &= f^{-1}(S\otimes S)(\Delta^{\rm cop}(u))f\\ &= f^{-1}(S\otimes S)((RR_{21})^{-1})(S(u)\otimes S(u))(S^2\otimes S^2)(f)(S\otimes S)(f_{21}^{-1})f\\ &= (R_{21}R)^{-1}f^{-1}(S(u)\otimes S(u))(S^2\otimes S^2)(f)(S\otimes S)(f_{21}^{-1})f\\ &= (R_{21}R)^{-1}\left(S(S(f^1)uS^{-1}(g^1))\otimes S(S(f^2)uS^{-1}(g^2))\right)(S\otimes S)(f_{21}^{-1})f\\ &\overset{(10.3.10)}{=} (R_{21}R)^{-1}(S(u)\otimes S(u))(S\otimes S)(f_{21}^{-1})f,\end{aligned}$$

and so (13.3.6) is proved. Now (13.3.7) follows easily from (13.3.6), (13.3.5) and (10.3.10); the details are left to the reader.

Finally, for all $h\in H$, we have:

$$(S(u)^{-1}u)h(S(u)^{-1}u)^{-1} = S(u)^{-1}S^2(h)S(u) = S(uS(h)u^{-1}) = S^4(h).$$

This finishes the proof of the corollary. □

We can prove the following characterization for ribbon quasi-Hopf algebras.

**Theorem 13.25** *A QT quasi-Hopf algebra $(H,R)$ is ribbon if and only if there exists a central element $\nu\in H$ such that*

$$\nu^2 = uS(u), \tag{13.3.10}$$

$$\Delta(\nu) = (\nu\otimes\nu)(R_{21}R)^{-1}, \tag{13.3.11}$$

$$S(\nu) = \nu, \tag{13.3.12}$$

*where, as before, $u$ is the element defined in* (10.3.4).

*Proof* Suppose that $(H,R)$ is ribbon and let $\eta \in H$ be an invertible central element such that (13.3.1) and (13.3.2) are satisfied. It follows that $\nu = \eta^{-1}$ is a central element in $H$, and it satisfies (13.3.11) and (13.3.12). To see that (13.3.10) is satisfied as well, observe first that by applying $\varepsilon \otimes \varepsilon$ to both sides of (13.3.1) we get $\varepsilon(\eta) = 1$; see (10.1.5). Thus, by using (13.3.1) again and the fact that $\eta$ is a central element in $H$, we deduce that

$$\begin{aligned}
\alpha &= \varepsilon(\eta)\alpha \\
&= S(\eta_1)\alpha\eta_2 \\
&= S(\eta R^2 r^1)\alpha\eta R^1 r^2 \\
&\overset{(13.3.2)}{=} \eta^2 S(R^2 r^1)\alpha R^1 r^2 \\
&\overset{(10.3.8)}{=} \eta^2 S(\alpha r^1) u r^2 \\
&\overset{(10.3.10)}{=} \eta^2 S(S(r^2)\alpha r^1) u \\
&\overset{(10.3.8)}{=} \eta^2 S(S(\alpha)u)u \\
&\overset{(10.3.10)}{=} \eta^2 S(u)u\alpha.
\end{aligned}$$

This fact allows to compute, for all $A, B \in H$:

$$\begin{aligned}
A\alpha B &= A\eta^2 S(u)u\alpha B \\
&= \eta^2 S(uS^{-1}(A))u\alpha B \\
&\overset{(10.3.10)}{=} \eta^2 S(u)S^2(A)u\alpha B \\
&\overset{(10.3.10)}{=} \eta^2 S(u)uA\alpha B.
\end{aligned}$$

In particular, by taking $A \otimes B = S(x^1) \otimes x^2\beta S(x^3)$, by (3.2.2) we conclude that $1_H = \eta^2 S(u)u$, and therefore $\nu^2 = S(u)u$, as needed. Note that $S(u)u = uS(u)$ by Corollary 10.18.

Conversely, let $\nu$ be a central element of $H$ such that (13.3.10), (13.3.11) and (13.3.12) are fulfilled. Then $\nu$ is invertible because $u$ is, and therefore $\eta := \nu^{-1}$ is an invertible central element of $H$. It follows easily that (13.3.2) and (13.3.1) are satisfied, so $(H,R)$ is indeed a ribbon quasi-Hopf algebra. □

**Definition 13.26** An element $\nu$ of a QT quasi-Hopf algebra $(H,R)$ is called a quasi-ribbon element if $\nu$ satisfies the equations (13.3.10), (13.3.11) and (13.3.12). If, moreover, $\nu$ is central then it is called a ribbon element of $H$.

**Remark 13.27** As in the proof of Theorem 13.25 we get that any (quasi-)ribbon element $\nu$ of $(H,R)$ also has the property that $\varepsilon(\nu) = 1$.

In order to achieve our main goal we need one more result. For a quasi-Hopf algebra $H$ denote

$$G(H) = \{l \in H \mid l \text{ is invertible with } l^{-1} = S(l) \text{ and } \Delta(l) = (l \otimes l)(S \otimes S)(f_{21}^{-1})f\}.$$

For $l \in G(H)$ one obtains $\varepsilon(l) = 1$, by applying $\varepsilon \otimes \varepsilon$ to both sides of the relation $\Delta(l) = (l \otimes l)(S \otimes S)(f_{21}^{-1})f$.

For a QT quasi-Hopf algebra $(H,R)$ we prove that there exists a one-to-one correspondence between quasi-ribbon (or ribbon) elements of $H$ and certain elements of $G(H)$.

**Lemma 13.28** *Let $(H,R)$ be a QT quasi-Hopf algebra and $u$ the element defined by (10.3.4). Suppose that $\nu$ is a quasi-ribbon element of $H$. If we set $\hbar = u^{-1}S(u)$ and $l = u^{-1}\nu$, then $l^2 = \hbar$ and $l \in G(H)$.*

*Proof* By (13.3.12) it follows that $S^2(\nu) = \nu$. Hence, by applying (10.3.10), we obtain that $u$ and $\nu$ commute. By (13.3.10) we have $\nu^2 = uS(u) = u^2\hbar$. Thus $l^2 = u^{-2}\nu^2 = \hbar$ since $u$ and $\nu$ commute.

Because $uS(u) = S(u)u$ is central we have

$$S(l)l = S(u^{-1}\nu)u^{-1}\nu = S(\nu)S(u^{-1})u^{-1}\nu = S(\nu)(uS(u))^{-1}\nu = (uS(u))^{-1}S(\nu)\nu.$$

Now, by (13.3.10) and (13.3.12) it follows that $S(l)l = 1_H$. By again using (13.3.10) and (13.3.12) we obtain that

$$lS(l) = u^{-1}\nu S(\nu)S(u^{-1}) = u^{-1}\nu^2 S(u)^{-1} = u^{-1}uS(u)S(u)^{-1} = 1_H.$$

Hence, $l$ is invertible and $l^{-1} = S(l)$.

By (13.3.11) and because $u$ and $\nu$ commute, and by using also (13.3.4), it follows that $\Delta(l) = (l\otimes l)(S\otimes S)(f_{21}^{-1})f$, and therefore $l \in G(H)$. □

The next result says that any (quasi-)ribbon element of $(H,R)$ comes from a deformation of $u$ by a suitable element of $G(H)$. It also says that $H$ is sovereign via $l = u^{-1}\nu$, if $\nu$ is a ribbon element for it.

**Theorem 13.29** *Let $(H,R)$ be a QT quasi-Hopf algebra and let $u$ and $\hbar$ be as in Lemma 13.28. Then the following hold.*

*(a) $l \mapsto ul$ defines a one-to-one correspondence*

$$\{l \in G(H) \mid l^2 = \hbar\} \longleftrightarrow \{\textit{quasi-ribbon elements of } H\}.$$

*(b) Suppose that $l \in G(H)$ satisfies $l^2 = \hbar$. Then $\nu = ul$ is a ribbon element of $H$ if and only if $S^2(h) = l^{-1}hl$, for all $h \in H$.*

*Proof* Let $l \in G(H)$ with $l^2 = \hbar$. By Lemma 13.28, to prove part (a) we need only to show that $\nu = ul$ is a quasi-ribbon element of $H$.

By $S(l) = l^{-1}$, we get $S^2(l) = l$, hence, by (10.3.10), it follows that $u$ and $l$ commute. Thus $\nu^2 = u^2l^2 = u^2\hbar = u^2u^{-1}S(u) = uS(u)$, so (13.3.10) holds for $\nu$.

To show that (13.3.12) holds for $\nu$, we first note that $\hbar$ is invertible with $\hbar^{-1} = S(u)^{-1}u$. Now $l^{-1} = \hbar^{-1}l$, which follows from the equation $l^2 = \hbar$. Therefore

$$S(\nu) = S(lu) = S(u)S(l) = S(u)l^{-1} = S(u)\hbar^{-1}l = S(u)S(u)^{-1}ul = ul = \nu.$$

Since $l \in G(H)$ and $u$ and $\nu$ commute (because $u$ and $l$ commute) it follows that

$$\begin{aligned}\Delta(u\nu^{-1}) &= \Delta(l^{-1})\\ &= f^{-1}(S\otimes S)(f_{21})(l^{-1}\otimes l^{-1})\end{aligned}$$

$$= f^{-1}(S\otimes S)(f_{21})(uv^{-1}\otimes uv^{-1}),$$

and thus (13.3.11) holds since

$$\begin{aligned}\Delta(v) &= \Delta(vu^{-1})\Delta(u)\\ &= (vu^{-1}\otimes vu^{-1})(S\otimes S)(f_{21}^{-1})f\Delta(u)\\ &\overset{(13.3.4)}{=} (v\otimes v)(R_{21}R)^{-1},\end{aligned}$$

as needed. Since $v = ul$ is central if and only if $l^{-1}hl = uhu^{-1}$ for all $h\in H$, part (b) follows by part (a) and (10.3.10). □

We end by showing that in the ribbon case the condition $l^{-1} = S(l)$ in the definition of an element $l\in G(H)$ is redundant.

**Corollary 13.30** *Let $(H,R)$ be a QT quasi-Hopf algebra and u the element defined in* (10.3.4). *If $\hbar = u^{-1}S(u)$ then $l\mapsto ul$ defines a one-to-one correspondence between*

$$R(H) := \{l\in H \mid l^2 = \hbar\ ,\ \Delta(l) = (l\otimes l)(S\otimes S)(f_{21}^{-1})f \text{ and } lS^2(h) = hl,\ \ \forall\, h\in H\}$$

*and ribbon elements of H.*

*Proof* According to Theorem 13.29 we only have to prove that

$$R(H) = \{l\in G(H) \mid l^2 = \hbar \text{ and } lS^2(h) = hl,\ \ \forall\, h\in H\}.$$

We will show this by double inclusion.

By the definitions of $R(H)$ and $G(H)$ it follows that

$$\{l\in G(H) \mid l^2 = \hbar \text{ and } lS^2(h) = hl,\ \ \forall\, h\in H\} \subseteq R(H).$$

To show the converse inclusion it suffices to show that any element $l\in R(H)$ is invertible and $l^{-1} = S(l)$. Indeed, if $l\in R(H)$ then by $l^2 = \hbar = u^{-1}S(u)$ and from the fact that $u$ is invertible it follows that $l$ is invertible, too. We claim now that $lS(l) = 1_H$, and this will end the proof.

To prove the claim, observe that by applying $\varepsilon\otimes\varepsilon$ to both sides of $\Delta(l) = (l\otimes l)(S\otimes S)(f_{21}^{-1})f$ we get $\varepsilon(l) = \varepsilon(l)^2$ in $k$, and since $l$ is invertible we deduce that $\varepsilon(l) = 1$. By the same equality we have now that

$$\begin{aligned}\beta &= \varepsilon(l)\beta = l_1\beta S(l_2) = lS(g^2)f^1\beta S(lS(g^1)f^2)\\ &\overset{(8.7.7)}{=} lS(g^2)S(\alpha)S(lS(g^1)) \overset{(8.7.7)}{=} lS^2(\beta)S(l) = \beta lS(l),\end{aligned}$$

where for the last equality we used the identity $lS^2(h) = hl$, applied to $h = \beta$. By using again the equality $lS^2(h) = hl$, valid for all $h\in H$, we compute that

$$A\beta B = A\beta lS(l)B = A\beta lS(S^{-1}(B)l) = A\beta lS(lS(B)) = A\beta lS^2(B)S(l) = A\beta BlS(l),$$

for all $A\otimes B\in H\otimes H$. By taking $A\otimes B = X^1\otimes S(X^2)\alpha X^3$ in the above equality, by (3.2.2) we conclude that $1_H = lS(l)$. So our proof is finished. □

**Corollary 13.31** *If $(H,R)$ is a unimodular QT quasi-Hopf algebra and $u$ is as in* (10.3.4) *then $l \mapsto ul$ defines a one-to-one correspondence between*

$$\{l \in H \mid l^2 = \underline{g}\,,\ \Delta(l) = (l \otimes l)(S \otimes S)(f_{21}^{-1})f \text{ and } lS^2(h) = hl,\ \ \forall\, h \in H\}$$

*and ribbon elements of $H$. Here $\underline{g}$ is the modular element of $H$; see* (7.6.2).

*Proof* When $H$ is unimodular, that is, $\mu = \varepsilon$, the formula in (11.6.5) gives the equality $u^{-1}S(u)S(\underline{g}) = 1_H$ or, equivalently, $u^{-1}S(u) = S(\underline{g}^{-1})$. As $uS(u) = S(u)u$ and $S^2(u) = u$ this implies

$$S^2(\underline{g}^{-1}) = S(u^{-1}S(u)) = S^2(u)S(u^{-1}) = uS(u)^{-1} = S(u)^{-1}u = (u^{-1}S(u))^{-1} = S(\underline{g}).$$

By using the bijectivity of $S$, we obtain that $S(\underline{g}^{-1}) = \underline{g}$, and so $\hbar = u^{-1}S(u) = \underline{g}$. Now everything follows from Corollary 13.30. □

With the help of Corollary 13.30 we can compute all the (quasi-)ribbon elements of $H(2)_\pm$, the QT quasi-Hopf algebras constructed in Example 10.7.

**Example 13.32** $H(2)_\pm$ has exactly four ribbon elements: namely, $v_1^\pm = \frac{1}{2}(1 \mp ig)$, $v_2^\pm = -\frac{1}{2}(1 \mp ig)$, $v_3^\pm = \frac{\mp i}{2}(1 \pm ig)$ and $v_4^\pm = \frac{\pm i}{2}(1 \pm ig)$.

*Proof* We have seen in Example 3.44 that in $H(2)$ we have $(S \otimes S)(f_{21})f = 1$. Also, as $S$ is the identity morphism of $H(2)$, we have $\hbar = u_\pm^{-1}S(u_\pm) = 1$ in $H(2)$. Therefore

$$R(H(2)_\pm) = \{l \in H(2) \mid l^2 = 1 \ \text{ and } \ \Delta(l) = l \otimes l\} = \{-1, 1, g, -g\}.$$

By (10.3.8) and the commutativity of $H(2)$ it follows that

$$u_\pm = S(R_\pm^2)R_\pm^1 = R_\pm^2 R_\pm^1 = 1 - (1 \pm i)p_- = \frac{1}{2}(1 \mp ig).$$

Corollary 13.30 now says that $lu_\pm$ with $l \in \{-1, 1, g, -g\}$ are all the (quasi-)ribbon elements of $H(2)_\pm$, so we are done. □

We prove that a twisting of a QT quasi-Hopf algebra preserves ribbon elements.

**Proposition 13.33** *Let $(H,R)$ be a QT quasi-Hopf algebra, $F \in H \otimes H$ a gauge transformation for $H$ and $\nu \in H$ a ribbon element for $(H,R)$. Then $\nu$ is a ribbon element for $(H_F, R_F)$ as well.*

*Proof* Since $u = u_F$ by Proposition 10.19 and $S_F = S$, the only thing we need to prove is that $\Delta_F(\nu) = (\nu \otimes \nu)((R_F)_{21}R_F)^{-1}$ (see Proposition 10.10 for the definition of $R_F$). One immediately sees that $(R_F)_{21}R_F = F(R_{21}R)F^{-1}$, so we can compute:

$$\begin{aligned}(\nu \otimes \nu)((R_F)_{21}R_F)^{-1} &= (\nu \otimes \nu)F(R_{21}R)^{-1}F^{-1}\\ &= F(\nu \otimes \nu)(R_{21}R)^{-1}F^{-1}\\ &\overset{(13.3.11)}{=} F\Delta(\nu)F^{-1} = \Delta_F(\nu),\end{aligned}$$

where for the second equality we used the fact that $\nu$ is central. □

## 13.4 A Class of Ribbon Quasi-Hopf Algebras

We have seen that to any finite-dimensional quasi-Hopf algebra we can associate a QT one, its quantum double. In this section we will go further by showing that to any QT quasi-bialgebra (resp. quasi-Hopf algebra) we can associate a balanced (resp. ribbon) one.

We start with the quasi-bialgebra case. In what follows, for $H$ a quasi-bialgebra, we denote by $H[\theta,\theta^{-1}]$ the free $k$-algebra generated by $H$ and $\theta$, with relations $h\theta=\theta h$, for all $h\in H$, and $\theta\theta^{-1}=\theta^{-1}\theta=1$; by analogy with the commutative case, we call $H[\theta,\theta^{-1}]$ the Laurent polynomial algebra over $H$.

**Proposition 13.34** *Let $(H,R)$ be a QT quasi-bialgebra. Then $H[\theta,\theta^{-1}]$ is a balanced quasi-bialgebra with structure given by*

$$\Delta\,|_H=\Delta_H\ ,\ \Delta(\theta^{\pm})=(R_{21}R)^{\mp}(\theta^{\pm}\otimes\theta^{\pm})\ ,\qquad \varepsilon\,|_H=\varepsilon_H\ ,\ \varepsilon(\theta^{\pm})=1,$$

*where, for simplicity, we denote $\theta=\theta^{+}$ and $\theta^{-}=\theta^{-1}$, and similarly for $(R_{21}R)^{\pm}$.*

*Proof* The only thing we have to prove is the quasi-coassociativity of $\Delta$ on $\theta$. It will then follow that the natural inclusion of $H$ into $H[\theta,\theta^{-1}]$ is a quasi-bialgebra morphism, and that $H[\theta,\theta^{-1}]$ is balanced via $\theta$.

Now, since $\Delta(h)R_{21}R=R_{21}R\Delta(h)$, for all $h\in H$, the quasi-coassociativity of $\Delta$ on $\theta$ reduces to

$$\Phi(R_{21}R\otimes 1_H)(\Delta_H\otimes \mathrm{Id}_H)(R_{21}R)\Phi^{-1}=(1_H\otimes R_{21}R)(\mathrm{Id}_H\otimes\Delta_H)(R_{21}R).$$

With notation as in Remark 10.4(3), this can be restated as

$$\Phi(\Delta_H\otimes \mathrm{Id}_H)(R_{21}R)R_{21}R_{12}\Phi^{-1}=R_{32}R_{23}(\mathrm{Id}_H\otimes\Delta_H)(R_{21}R).$$

By using (10.1.1) and (10.1.2), the left-hand side of the above equality equals

$$\Phi(\Delta(R^2)\otimes R^1)(\Delta(r^1)\otimes r^2)R_{21}R_{12}\Phi^{-1}=R_{32}\Phi_{132}R_{31}R_{13}\Phi^{-1}_{132}R_{23}\Phi R_{21}R_{12}\Phi^{-1},$$

while its right-hand side is equal to

$$R_{32}R_{23}(R^2\otimes\Delta_H(R^1))(r^1\otimes\Delta_H(r^2))=R_{32}R_{23}\Phi R_{21}\Phi^{-1}_{213}R_{31}R_{13}\Phi_{213}R_{12}\Phi^{-1}.$$

Thus we must show that

$$\Phi_{132}R_{31}R_{13}\Phi^{-1}_{132}R_{23}\Phi R_{21}=R_{23}\Phi R_{21}\Phi^{-1}_{213}R_{31}R_{13}\Phi_{213}.$$

If $\tau$ is the usual switch for the category of $k$-vector spaces, this follows from

$$\begin{aligned}
&R_{23}\Phi R_{21}\Phi^{-1}_{213}R_{31}R_{13}\Phi_{213}\\
&\quad= (\tau\otimes\mathrm{Id}_H)(R_{13}\Phi_{213}R_{12}\Phi^{-1})(\tau\otimes\mathrm{Id}_H)(1_H\otimes R_{21}R)\Phi_{213}\\
&\quad\overset{(10.1.2)}{=} (\tau\otimes\mathrm{Id}_H)(\Phi_{231}(\mathrm{Id}_H\otimes\Delta_H)(R)(1_H\otimes R_{21}R))\Phi_{231}\\
&\quad= \Phi_{132}(\tau\otimes\mathrm{Id}_H)(1_H\otimes R_{21}R)(\tau\otimes\mathrm{Id}_H)(\mathrm{Id}_H\otimes\Delta(R))\Phi_{213}\\
&\quad\overset{(10.1.2)}{=} \Phi_{132}R_{31}R_{13}(\tau\otimes\mathrm{Id}_H)(\Phi^{-1}_{231}R_{13}\Phi_{213}R_{12}\Phi^{-1})\Phi_{213}\\
&\quad= \Phi_{132}R_{31}R_{13}\Phi^{-1}_{132}R_{23}\Phi R_{21},
\end{aligned}$$

as desired. □

**Remark 13.35** Let $(H,R)$ be a QT quasi-bialgebra and $\mathscr{C} = {}_H\mathscr{M}$. Then $\mathscr{B}(\mathscr{C},c)$ and ${}_{H[\theta,\theta^{-1}]}\mathscr{M}$ are isomorphic as balanced categories, where $c$ is as in (10.1.4) and $\mathscr{B}(\mathscr{C},c)$ is as in the last part of Remark 13.4.

Indeed, the desired isomorphism is produced by the following correspondence. To $(V,\eta_V)$ in $\mathscr{B}(\mathscr{C},c)$ we associate $V$ regarded as a left $H[\theta,\theta^{-1}]$-module via the $H$-module structure of it and $\theta^{\pm}\cdot v = \eta_V^{\mp}(v)$, for all $v\in V$. In this way a morphism in $\mathscr{B}(\mathscr{C},c)$ becomes a morphism in ${}_{H[\theta,\theta^{-1}]}\mathscr{M}$.

In general, $H[\theta,\theta^{-1}]$ is not a quasi-Hopf algebra, and so not a ribbon quasi-Hopf algebra. To "make" it ribbon, we have to consider a quotient of it. Actually, we have to consider the $k$-algebra $H(\theta) := \frac{H[\theta]}{\langle\theta^2 - uS(u)\rangle}$ instead of $H[\theta,\theta^{-1}]$, where $H[\theta]$ is the free $k$-algebra generated by $H$ and $\theta$ with relations $h\theta = \theta h$, for all $h\in H$, and $u$ is as in (10.3.4).

In what follows, we still denote by $\theta$ the class of $\theta$ in $H(\theta)$.

**Proposition 13.36** *If $(H,R)$ is a QT quasi-Hopf algebra then $H(\theta)$ is a quasi-Hopf algebra with structure determined by $\Delta\mid_H = \Delta_H$, $\varepsilon\mid_H = \varepsilon_H$, $S\mid_H = S$,*

$$\Delta(\theta) = (\theta\otimes\theta)(R_{21}R)^{-1},\quad \varepsilon(\theta) = 1,\ S(\theta) = \theta,$$

*and the reassociator and distinguished elements that define the antipode equal to those of $H$. Furthermore, $(H(\theta),R)$ is QT and $\theta$ is a ribbon element for it.*

*Proof* As we have already pointed out, $\Delta(h)R_{21}R = R_{21}R\Delta(h)$, for all $h\in H$. So by (13.3.5) and (13.3.6) one can see that

$$\Delta(uS(u)) = f^{-1}(S\otimes S)(f_{21})(uS(u)\otimes uS(u))(S\otimes S)(f_{21}^{-1})f(R_{21}R)^{-2},$$

and because $uS(u)$ is central in $H$ we get $\Delta(uS(u)) = (uS(u)\otimes uS(u))(R_{21}R)^{-2}$. This shows that $\Delta$ is well defined on $\theta$. Also, it follows from Proposition 13.34 that $H(\theta)$ is a quasi-bialgebra.

Furthermore, since $S(\theta^2 - uS(u)) = S(\theta)^2 - S^2(u)S(u) = \theta^2 - uS(u)$ we deduce that $S$ is well defined on $\theta$, too. So it remains to show the equalities

$$S(\theta_1)\alpha\theta_2 = \alpha \quad\text{and}\quad \theta_1\beta S(\theta_2) = \beta\ .$$

The equality in (10.3.8) can be rewritten as $S(\alpha\overline{R}^2)u\overline{R}^1 = \alpha$ or, equivalently, as $S(\overline{R}^1)\alpha\overline{R}^2 = S^{-1}(\alpha u^{-1}) = S(u^{-1}\alpha)$; see (10.3.10). Hence

$$\begin{aligned} S(\theta_1)\alpha\theta_2 &= S(\overline{R}^1\overline{r}^2)\alpha\overline{R}^2\overline{r}^1\theta^2 \\ &= S(u^{-1}\alpha\overline{r}^2)\overline{r}^1\theta^2 \\ &\overset{(10.3.10)}{=} S(u^{-1}S(\overline{r}^1)\alpha\overline{r}^2)\theta^2 \\ &= S^2(u^{-1}\alpha)S(u^{-1})\theta^2 \\ &\overset{(10.3.10)}{=} \alpha u^{-1}S(u^{-1})\theta^2 = \alpha, \end{aligned}$$

as required; in the last equality we used that $uS(u) = S(u)u$.

Notice that the formula in (11.6.3) is equivalent to $\overline{R}^1 S(\overline{R}^2\beta u) = \beta$, and therefore to $\overline{R}^2\beta S(\overline{R}^1) = u^{-1}S(\beta)$, because of (10.3.10). This gives us

$$\begin{aligned}
\theta_1\beta S(\theta_2) &= \overline{R}^1\overline{r}^2\beta S(\overline{R}^2\overline{r}^1)\theta^2\\
&= \overline{R}^1 u^{-1}S(\overline{R}^2\beta)\theta^2\\
&\overset{(10.3.10)}{=} u^{-1}S(\overline{R}^2\beta S(\overline{R}^1))\theta^2\\
&= u^{-1}S(u^{-1}S(\beta))\theta^2\\
&\overset{(10.3.10)}{=} u^{-1}S(u^{-1})\theta^2\beta = \beta,
\end{aligned}$$

since $uS(u) = S(u)u$. Finally, $R$ is an $R$-matrix for $H(\theta)$ because it is for $H$ and

$$\begin{aligned}
R\Delta(\theta) &= R(R_{21}R)^{-1}(\theta\otimes\theta)\\
&= R_{21}^{-1}(\theta\otimes\theta)\\
&= (RR_{21})^{-1}R(\theta\otimes\theta) = \Delta^{\mathrm{cop}}(\theta)R.
\end{aligned}$$

Clearly $\theta$ is a ribbon element for $(H(\theta),R)$, and this completes the proof. □

Our next goal is to show that the category of finite-dimensional left modules over $H(\theta)$ can be identified with the category $\mathscr{R}({}_H\mathscr{M}^{\mathrm{fd}},c)$, where $c$ is defined by (10.1.4).

**Theorem 13.37** *Let $(H,R)$ be a QT quasi-Hopf algebra, so ${}_H\mathscr{M}^{\mathrm{fd}}$ is a rigid braided category. Then $\mathscr{R}({}_H\mathscr{M}^{\mathrm{fd}},c)$ identifies as a ribbon category with ${}_{H(\theta)}\mathscr{M}^{\mathrm{fd}}$.*

*Proof* Specializing Definition 13.12 for $\mathscr{C} = {}_H\mathscr{M}^{\mathrm{fd}}$, we deduce that an object of the category $\mathscr{R}({}_H\mathscr{M}^{\mathrm{fd}},c)$ is a pair $(V,\eta_V)$ consisting of a finite-dimensional left $H$-module $V$ and an $H$-automorphism $\eta_V$ of $V$ such that

$$\begin{aligned}
\eta^{-2}(v) &= X^1r^2R^1\beta S(X^2r^1R^2)\alpha X^3\cdot v\\
&= q^1r^2R^1\beta S(q^2r^1R^2)\cdot v\\
&\overset{(11.6.3)}{=} S(q^2r^1\beta uS^{-1}(q^1r^2))\cdot v\\
&\overset{(10.3.10)}{=} S(q^2r^1\beta S(q^1r^2)u)\cdot v\\
&\overset{(11.6.3)}{=} S(q^2S(q^1\beta u)u)\cdot v\\
&\overset{(10.3.10)}{=} S(u)S^2(q^1\beta S(q^2)u)\cdot v\\
&\overset{(3.2.2)}{=} S(u)u\cdot v = uS(u)\cdot v,
\end{aligned}$$

for all $v\in V$. Thus objects of $\mathscr{R}({}_H\mathscr{M}^{\mathrm{fd}},c)$ are pairs $(V,\eta_V)$ consisting of a finite-dimensional left $H$-module and an $H$-automorphism $\eta_V$ of $V$ satisfying $\eta^2(v) = (uS(u))^{-1}\cdot v$, for all $v\in V$.

Clearly, a morphism $f:(V,\eta_V)\to(W,\eta_W)$ is a left $H$-linear morphism $f:V\to W$ such that $\eta_W f = f\eta_V$.

Let $F:\mathscr{R}({}_H\mathscr{M}^{\mathrm{fd}},c)\to {}_{H(\theta)}\mathscr{M}^{\mathrm{fd}}$ be the functor defined as follows: $F((V,\eta_V)) = V$ regarded as an $H(\theta)$-module via the $H$-action on $V$ and $\theta\cdot v = \eta_V^{-1}(v)$; $F$ acts as identity on morphisms.

We can easily see that $\theta^2 = uS(u)$ together with $\eta^{-2}(v) = (uS(u))^{-1}\cdot v$, for all $v\in$

$V$, implies that $F$ is a well-defined functor. It provides an isomorphism of categories, its inverse being the functor $G : {}_{H(\theta)}\mathcal{M}^{\rm fd} \to \mathcal{R}({}_H\mathcal{M}^{\rm fd}, c)$ given by $G(V) = (V, \eta_V : V \ni v \mapsto \theta^{-1} \cdot v \in V)$, for all $V \in {}_{H(\theta)}\mathcal{M}^{\rm fd}$.

The functor $F$ is monoidal since

$$\begin{aligned}\theta \cdot (v \otimes w) &= \eta_{V\otimes W}^{-1}(v \otimes w)\\ &= (R_{21}R)^{-1}(\eta_V^{-1}(v) \otimes \eta_W^{-1}(w))\\ &= (R_{21}R)^{-1}(\theta \cdot v \otimes \theta \cdot w) = \theta_1 \cdot v \otimes \theta_2 \cdot w,\end{aligned}$$

for all $V, W \in \mathcal{R}({}_H\mathcal{M}^{\rm fd}, c)$, $v \in V$ and $w \in W$. Furthermore, $F$ is braided because for both categories the braiding is defined by $c$.

The ribbon structure $\widetilde{\eta}$ of ${}_{H(\theta)}\mathcal{M}^{\rm fd}$ is induced by the element $\theta^{-1}$, in the sense that $\widetilde{\eta}_V(v) = \theta^{-1} \cdot v = \eta_V(v)$, for all $v \in V \in {}_{H(\theta)}\mathcal{M}^{\rm fd}$. In other words, the two categories have the same ribbon structure, and therefore $F$ is a ribbon isomorphism functor. Observe also that $F$ is compatible with the left rigid monoidal structures on $\mathcal{R}({}_H\mathcal{M}^{\rm fd}, c)$ and ${}_{H(\theta)}\mathcal{M}^{\rm fd}$. More precisely, we have $F((V^*, \eta_{V^*})) = V^*$ with left $H(\theta)$-module structure induced by that of the left $H$-module and $\theta \cdot v^* = \eta_{V^*}^{-1}(v^*) = (\eta_V^{-1})^*(v^*) = v^* \circ \eta_V^{-1} = v^*(\theta \cdot) = v^*(S(\theta) \cdot)$, for all $v^* \in V^*$, as required. □

**Remark 13.38** As in the proof of Theorem 11.30, if $(H, R)$ is QT then $\tilde{R} = R_{21}^{-1} = \overline{R}^2 \otimes \overline{R}^1$ is another $R$-matrix for $H$. In addition, if $\tilde{u}$ is the element (10.3.4) corresponding to $(H, \tilde{R})$ then we have seen that $\tilde{u} = S(u^{-1})$, so $\tilde{u}S(\tilde{u}) = (uS(u))^{-1}$. Therefore, if $\tilde{c}$ is the braiding on ${}_H\mathcal{M}^{\rm fd}$ defined by $\tilde{R}$ then $\mathcal{R}({}_H\mathcal{M}^{\rm fd}, \tilde{c})$ is a ribbon category that is isomorphic to ${}_{H(\theta)}\mathcal{M}^{\rm fd}$, too. To see this, observe that an object of $\mathcal{R}({}_H\mathcal{M}^{\rm fd}, \tilde{c})$ is a pair $(V, \theta_V)$ consisting of $V \in {}_H\mathcal{M}^{\rm fd}$ and an automorphism $\theta_V$ of the left $H$-module $V$ such that $\theta_V^2(v) = uS(u) \cdot v$, for all $v \in V$. It is clear at this point that $(V, \eta_V) \mapsto (V, \theta_V := \eta_V^{-1})$ defines the desired isomorphism of categories.

**Corollary 13.39** *Let $H$ be a finite-dimensional quasi-Hopf algebra and $D(H)$ its quantum double. If $D(H, \theta) := D(H)(\theta)$ then*

$$\mathfrak{R}_r({}_H\mathcal{M}^{\rm fd}) \cong \mathcal{R}({}_H\mathcal{YD}^{H\rm fd}, c) \cong {}_{D(H,\theta)}\mathcal{M}^{\rm fd}$$

*as ribbon categories, where $c$ is the braiding defined in* (8.2.18).

*Proof* The first isomorphism can be deduced from the definition $\mathfrak{R}_r({}_H\mathcal{M}^{\rm fd}) = \mathcal{R}(\mathcal{Z}_r({}_H\mathcal{M}^{\rm fd}), c)$ and the braided isomorphism between $\mathcal{Z}_r({}_H\mathcal{M}^{\rm fd})$ and ${}_H\mathcal{YD}^{H\rm fd}$ induced by Theorem 8.8 and Proposition 8.12. The second one follows from the braided isomorphism $({}_H\mathcal{YD}^{H\rm fd}, c) \cong ({}_{D(H)}\mathcal{M}^{\rm fd}, \mathcal{R}_D)$ established in Theorem 10.20, and Theorem 13.37. □

## 13.5 Some Ribbon Elements for $D^\omega(H)$ and $D^\omega(G)$

Let $H$ be a finite-dimensional cocommutative Hopf algebra, $\omega$ a normalized 3-cocycle on $H$ and $D^\omega(H)$ the quasi-Hopf algebra constructed in Section 8.6. Recall from

Proposition 10.22 that $D^{\omega}(H)$ is a QT quasi-Hopf algebra isomorphic to the quantum double $D(H^*_{\omega})$.

In what follows, in order to avoid any confusion, we denote by $\mu_H \in H^*$ and $\underline{\mathfrak{g}}_H \in H$ the modular elements of $H$ as a Hopf algebra. Similar notation, $\mu_{H^*_{\omega}} \in H$ and $\underline{\mathfrak{g}}_{H^*_{\omega}} \in H^*$, is used for the modular elements of the quasi-Hopf algebra $H^*_{\omega}$. As $H$ is cocommutative it follows that $\underline{\mathfrak{g}}_H = \mu_{H^*_{\omega}} = 1_H$, that is, $H^*$ and $H^*_{\omega}$ are unimodular as Hopf and quasi-Hopf algebras, respectively. Also, it is clear that a left (and at the same time right) integral in the quasi-Hopf algebra $H^*_{\omega}$ is nothing but a left (and at the same time right) integral on the Hopf algebra $H$, that is, an element $\lambda \in H^*$ obeying $\lambda(h)h = \lambda(h)1_H$, for all $h \in H$.

In this section we show that the element $\nu = u(\zeta \# 1_H)\beta_{D^{\omega}(H)} = u(\zeta\beta \# 1_H)$ is a ribbon element for $D^{\omega}(H)$, provided that $\zeta : H \to k$ is an algebra map such that $\zeta^2 = \mu_H$; here, as before, $u$ is the element in (10.3.4) corresponding to $D^{\omega}(H)$ and the other notation is as in Section 8.6. Note that when $\dim_k H \neq 0$ in $k$ or $H$ is unimodular we can take $\zeta = \varepsilon$, and therefore $D^{\omega}(H)$ is ribbon with ribbon element $\nu = u(\beta \# 1_H)$. This applies for instance to a finite-dimensional Hopf group algebra $H = k[G]$, and so $D^{\omega}(G)$ is always a ribbon quasi-Hopf algebra.

To this end, we start by describing the space of left cointegrals for $H^*_{\omega}$.

**Lemma 13.40** *Let $t \in H$ be a non-zero left integral in $H$, that is, $ht = \varepsilon(h)t$, for all $h \in H$. Then $\mathfrak{t} := \beta(S(t))t \in H$ is a non-zero left cointegral for $H^*_{\omega}$.*

*Proof* Let $\lambda \in H^*$ be a non-zero (left and right) integral for $H^*_{\omega}$. As $\alpha, \beta$ are invertible in $H^*_{\omega}$, by the comments made before Example 7.53 we have that a non-zero left cointegral for $H^*_{\omega}$ is a non-zero element $\mathfrak{t} \in H$ satisfying $\lambda(h\mathfrak{t}) = \lambda(\mathfrak{t})\beta(h)$, for all $h \in H$.

Since the antipode of $H$ is bijective, by using the Hopf version of the isomorphism (7.2.11) we find a unique element $h^* \in H^*$ such that $h^*(t)t = \mathfrak{t}$; in particular, $h^*(t) \neq 0$. We have, for all $h \in H$, that $h^*(t)\lambda(ht) = h^*(t)\lambda(t)\beta(h)$, which is equivalent to $h^*(S(h)ht)\lambda(ht) = \lambda(t)h^*(1_H)\beta(h)$, which in turn is equivalent to $\lambda(t)h^*(S(h)) = \lambda(t)h^*(1_H)\beta(h)$.

As $\lambda(t) \neq 0$, it follows that $h^* = h^*(1_H)\beta \circ S$, which implies $\beta(S(t)) \neq 0$. Therefore, by rescaling, we can assume without loss of generality that $\mathfrak{t} = \beta(S(t))t$, as stated.

Conversely, $\mathfrak{t} = \beta(S(t))t$ is a left cointegral for $H^*_{\omega}$ since

$$\lambda(h\mathfrak{t}) = \beta(S(t))\lambda(ht) = \beta(S(ht)h)\lambda(ht) = \beta(h)\lambda(ht) = \beta(h)\lambda(t),$$

and this is equal to $\lambda(\mathfrak{t})\beta(h)$ because

$$\lambda(\mathfrak{t})\beta(h) = \beta(S(t))\lambda(t)\beta(h) = \lambda(t)\beta(S(1_H))\beta(h) = \lambda(t)\beta(h).$$

So our proof ends. □

**Corollary 13.41** *We have that $\underline{\mathfrak{g}}_{H^*_{\omega}} = \beta^2\mu_H$. Consequently, the modular element $\underline{\mathfrak{g}}_{D(H^*_{\omega})}$ of the quantum double $D(H^*_{\omega})$ equals $1_H \bowtie \beta^2\mu_H$.*

*Proof* Recall that $\mu_H$ is defined by $th = \mu_H(h)t$, for all $h \in H$ and $t \in H$ a non-zero left integral. Also, since $H^*$ is unimodular we have $\lambda \circ S = \lambda$. Thus, by specializing (7.6.2) to $H^*_\omega$ we compute, for all $h \in H$, that

$$\begin{aligned}
\underline{\mathfrak{g}}_{H^*_\omega} &= q^2(S(\mathfrak{t}))p^2(S(\mathfrak{t}))\lambda(S(\mathfrak{t}h))q^1(S(h))p^1(S(h)) \\
&= \omega(S(h), S(\mathfrak{t}), \mathfrak{t})\beta(S(\mathfrak{t}))\omega^{-1}(S(h), \mathfrak{t}, S(\mathfrak{t}))\lambda(S(\mathfrak{t}h)) \\
&= \omega(S(h), hS(\mathfrak{t}h), \mathfrak{t}hS(h))\beta(hS(\mathfrak{t}h))\omega^{-1}(S(h), \mathfrak{t}hS(h), hS(\mathfrak{t}h))\lambda(\mathfrak{t}h) \\
&= \omega(S(h), h, S(h))\beta(h)\omega^{-1}(S(h), h, S(h))\lambda(\mathfrak{t}h) \\
&= \beta(h)\beta(S(t))\lambda(th) \\
&= \beta(h)\beta(hS(th))\lambda(th) \\
&= \beta(h)^2\mu_H(h)\lambda(t).
\end{aligned}$$

But the pair $(\lambda, \mathfrak{t})$ obeys $\lambda(\mathfrak{t}) = 1$ or, equivalently, $\beta(S(t))\lambda(t) = 1$. The latter is equivalent to $\lambda(t) = 1$, and so $\underline{\mathfrak{g}}_{H^*_\omega} = \beta^2\mu_H$, as desired.

Finally, we have $\mu_{H^*_\omega} = 1_H$, hence the formula in Proposition 12.12 yields

$$\underline{\mathfrak{g}}_{D(H^*_\omega)} = 1_H \bowtie \underline{\mathfrak{g}}^{-1}_{H^*_\omega} \circ S^{-3} = 1_H \bowtie \underline{\mathfrak{g}}^{-1}_{H^*_\omega} \circ S = 1_H \bowtie \underline{\mathfrak{g}}_{H^*_\omega},$$

since $\underline{\mathfrak{g}}^{-1}_{H^*_\omega} = \beta^{-2}\mu_H^{-1}$ with $\beta^{-1} = \beta \circ S$ and $\mu_H^{-1} = \mu_H \circ S$, and combined with $S^2 = \mathrm{Id}_H$ this gives $\underline{\mathfrak{g}}^{-1}_{H^*_\omega} \circ S = \underline{\mathfrak{g}}_{H^*_\omega}$. □

Another preliminary result that we need is the following.

**Lemma 13.42** *The distinguished element $\beta_{D^\omega(H)} = \beta \# 1_H \in D^\omega(H)$ satisfies*

$$\Delta(\beta_{D^\omega(H)}) = (\beta_{D^\omega(H)} \otimes \beta_{D^\omega(H)})(s \otimes s)(\mathfrak{f}_{21}^{-1})\mathfrak{f}, \tag{13.5.1}$$

*where $\mathfrak{f} \in D^\omega(H) \otimes D^\omega(H)$ is the Drinfeld element. Consequently, the same relation is satisfied by any element of the form $\zeta\beta \# 1_H \in D^\omega(H)$, provided that $\zeta : H \to k$ is an algebra map.*

*Proof* By (3.2.14), the relation in (13.5.1) is equivalent to

$$\delta_{D^\omega(H)}(s \otimes s)(\mathfrak{f}_{21}) = \beta_{D^\omega(H)} \otimes \beta_{D^\omega(H)}, \tag{13.5.2}$$

where $\delta_{D^\omega(H)}$ is the element defined in (3.2.6), specialized for the quasi-Hopf algebra $D^\omega(H)$. To prove this relation we proceed as follows.

Since for $D^\omega(H)$ we have $\alpha_{D^\omega(H)} = \varepsilon \# 1_H$, the unit of $D^\omega(H)$, by the relation (3.2.14) we get $\gamma_{D^\omega(H)} = \mathfrak{f}$. It follows that $\mathfrak{f}$ and $\delta := \delta_{D^\omega(H)}$ can be written as:

$$\mathfrak{f} = (f_1 \# 1_H) \otimes (f_2 \# 1_H), \quad \delta = (\delta_1 \# 1_H) \otimes (\delta_2 \# 1_H), \tag{13.5.3}$$

with

$$f_1(x)f_2(y) = \omega(S(x), x, y)\omega^{-1}(S(y), S(x), xy), \tag{13.5.4}$$

$$\delta_1(x)\delta_2(y) = \beta(x)\beta(y)\omega(x, y, S(xy))\omega^{-1}(y, S(y), S(x)), \tag{13.5.5}$$

for all $x, y \in H$. We can now see that

$$\begin{aligned}
\delta_1 S(f_2)(x)\delta_2 S(f_1)(y) &= \delta_1(x)f_2(S(x))\delta_2(y)f_1(S(y)) \\
&\overset{(13.5.4)}{\underset{(13.5.5)}{=}} \beta(x)\beta(y)\omega(x,y,S(xy))\omega^{-1}(y,S(y),S(x)) \\
&\quad \omega(y,S(y),S(x))\omega^{-1}(x,y,S(xy)) \\
&= \beta(x)\beta(y),
\end{aligned}$$

for all $x, y \in H$, and so (13.5.2) is proved. The second assertion in the statement follows easily from the fact that $\Delta(\zeta\#1_H) = \zeta\#1_H \otimes \zeta\#1_H$ and from the fact that the comultiplication of $D^{\omega}(H)$ is multiplicative. □

We now have all the necessary ingredients in order to prove the following:

**Theorem 13.43** *Let $H$ be a finite-dimensional cocommutative Hopf algebra $H$, $\omega$ a normalized 3-cocycle on $H$ and $\zeta : H \to k$ an algebra map. Then the element $v = u(\zeta\beta\#1_H)$ is ribbon if and only if $\zeta^2 = \mu_H$.*

*Proof* The quasi-Hopf algebra $D^{\omega}(H)$ is unimodular; this follows from Theorems 8.34 and 12.6. Furthermore, by Corollary 13.41 and the definition of the isomorphism $w$ in (8.6.7) we can see that the modular element $\mathfrak{g}_{D^{\omega}(H)}$ of $D^{\omega}(H)$ is

$$\mathfrak{g}_{D^{\omega}(H)} = w(\mathfrak{g}_{D(H^*_{\omega})}) = \mathfrak{g}_{D(H^*_{\omega})}\#1_H = \beta^2\mu_H\#1_H.$$

So, according to Corollary 13.31, it suffices to see when the element $\mathfrak{l} := \zeta\beta\#1_H \in D^{\omega}(H)$ satisfies the relations

$$\mathfrak{l}^2 = \beta^2\mu_H\#1_H, \tag{13.5.6}$$

$$\Delta(\mathfrak{l}) = (\mathfrak{l}\otimes\mathfrak{l})(s\otimes s)(\mathfrak{f}_{21}^{-1})\mathfrak{f}, \tag{13.5.7}$$

$$\mathfrak{l}S^2(\varphi\#h) = (\varphi\#h)\mathfrak{l}\,,\ \ \forall\ \varphi\#h \in D^{\omega}(H). \tag{13.5.8}$$

It can be easily checked that (13.5.6) is equivalent to $\zeta^2\beta^2 = \beta^2\mu_H$, and the latter is equivalent to $\zeta^2 = \mu_H$, because $H^*$ is commutative and $\beta$ is convolution invertible. Also, Lemma 13.42 guarantees that (13.5.7) is always satisfied.

We look at (13.5.8). By the last paragraph of the proof of Theorem 8.35 we have that $s^2(\varphi\#h) = (\beta^{-1}\#1_H)(\varphi\#h)(\beta\#1_H)$, for all $\varphi\#h \in D^{\omega}(H)$. Hence (13.5.8) is equivalent to $(\zeta\#1_H)(\varphi\#h) = (\varphi\#h)(\zeta\#1_H)$, for all $\varphi\#h \in D^{\omega}(H)$. The last equation becomes $\zeta\varphi\#h = \varphi(h \rightharpoonup \zeta \leftharpoonup S(h))\#h$, and it holds for any $\varphi\#h \in D^{\omega}(H)$ since $\zeta$ is an algebra map and $H^*$ is commutative. □

We end this section with some concrete examples. By

$$G(H^*) := \{\zeta : H \to k \mid \zeta \ \text{ is an algebra map}\}$$

we denote the set of grouplike elements of $H^*$, assuming, as before, that $H$ is a finite-dimensional cocommutative Hopf algebra. $G(H^*)$ is a group under convolution, and so the group Hopf algebra $k[G(H^*)]$ is a Hopf subalgebra of $H^*$. By the Hopf algebraic version of the freeness theorem proved in Section 7.7 it follows that $\mid G(H^*) \mid$ divides $\dim_k(H)$.

**Example 13.44** Assume that $\mu_H$ has odd order in $G(H^*)$. Then $D^\omega(H)$ is a ribbon quasi-Hopf algebra.

*Proof* If $\mu_H$ has order $2m+1$ in $G(H^*)$ then $\zeta := \mu_H^{m+1} : H \to k$ is an algebra map such that $\zeta^2 = \mu_H$. By Theorem 13.43 we obtain that $u(\mu_H^{m+1}\beta\#1_H)$ is a ribbon element for $D^\omega(H)$. □

**Example 13.45** Suppose that either $\dim_k(H)$ or $|G(H^*)|$ is an odd number. Then $D^\omega(H)$ is a ribbon quasi-Hopf algebra.

*Proof* Since $|G(H^*)|$ divides $\dim_k(H)$, in either case we get that $\mu_H$ has odd order, and so Example 13.44 applies. □

**Example 13.46** The element $v = u(\beta\#1_H)$ is a ribbon element for $D^\omega(H)$ if and only if $H$ is unimodular.

*Proof* Take $\zeta = \varepsilon$ in Theorem 13.43; we obtain that $v = u(\beta\#1_H)$ is a ribbon element if and only if $\mu_H = \varepsilon$, i.e. $H$ is unimodular. □

**Example 13.47** The quasi-Hopf algebra $D^\omega(G)$ is ribbon.

*Proof* We have $D^\omega(G) = D^\omega(k[G])$ and $k[G]$ is unimodular. Indeed, $t = \sum_{g\in G} g$ is a left and right integral in $k[G]$. □

**Example 13.48** Suppose that $\dim_k(H) \neq 0$ in $k$ (this happens for instance when $k$ is of characteristic zero). Then $v = u(\beta\#1_H)$ is a ribbon element for $D^\omega(H)$.

*Proof* By part (ii) of Theorem 12.17, specialized to the Hopf algebra $H$, we get that $H$ is semisimple, and so unimodular, too. It then follows that $v = u(\beta\#1_H)$ is a ribbon element for $D^\omega(H)$. □

## 13.6 Notes

Ribbon (quasi-)Hopf algebras were introduced for topological reasons: they give rise to a topological invariant of knots and links in the 3-sphere; see [191]. From the algebraic point of view, ribbon quasi-Hopf algebras are Hopf-like objects defined by algebras for which their categories of representations are ribbon in the sense of Turaev [213], that is, are braided and have a twist which makes the braiding involutive (the balanced property) and at the same time relates it to duality. Note that ribbon categories were also introduced by Joyal and Street [120, 121] but under the name of tortile categories.

That rigid braided balanced categories are actually braided sovereign categories was proved by Deligne [74]; see also [149, 216]. Kassel and Turaev [128] associate to any rigid monoidal category a ribbon category and specialize this construction to the category of representations of a Hopf algebra. The slightly more general construction in Section 13.2 was done in [65], and has the advantage that in the quasi-Hopf case it leads to a more conceptual and less computational proof for the ribbon isomorphisms in Corollary 13.39. Note that, following the idea of Kassel and Turaev, balanced categories from monoidal ones were obtained by Drabant in [78]; his construction generalizes in the spirit of [65]; see Proposition 13.3 and Remark 13.4.

When $(H,R)$ is a ribbon quasi-Hopf algebra, the element responsible for the fact that ${}_H\mathscr{M}^{\rm fd}$ is sovereign was found in [56, 65, 204], by using the ideas of Kauffman and Radford in [130]; this is the content of Section 13.3. Section 13.4 uses the ideas in [191, 78] to construct $H[\theta,\theta^{-1}]$ and $H(\theta)$ in the quasi-Hopf setting. Proposition 13.33 is from [137].

Finally, the particular ribbon elements for $D^{\omega}(H)$ presented in Section 13.5 are taken from [57], which had as main sources of inspiration [56] and [130].

# Bibliography


[1] Abe, E. 1977. *Hopf Algebras*. Cambridge: Cambridge University Press.

[2] Aguiar, M. and Mahajan, S. 2010. Monoidal functors, species and Hopf algebras. CRM Monograph Series **29**, Providence, RI: American Mathematical Society.

[3] Akrami, S. E. and Majid, S. 2004. Braided cyclic cocycles and nonassociative geometry. *J. Math. Phys.*, **45** (10), 3883–3911.

[4] Albuquerque, H., Elduque, A. and Pérez-Izquierdo, J. M. 2002. $\mathbb{Z}_2$-quasialgebras. *Comm. Algebra*, **30** (2), 2161–2174.

[5] Albuquerque, H. and Majid, S. 1999. $\mathbb{Z}_n$-quasialgebras. In A. P. Santana, A. L. Duarte and J. F. Queiro (eds), *Matrices and Group Representations* (Coimbra, 1998). Textos Mat. Sér. B **19**, Coimbra: Universidade de Coimbra, pp. 57–64.

[6] Albuquerque, H. and Majid, S. 1999. Quasialgebra structure of the octonions. *J. Algebra*, **220**, 247–264.

[7] Albuquerque, H. and Majid, S. 2002. Clifford algebras obtained by twisting of group algebras. *J. Pure Appl. Algebra*, **171**, 133–148.

[8] Albuquerque, H. and Panaite, F. 2009. On quasi-Hopf smash products and twisted tensor products of quasialgebras. *Algebr. Represent. Theory*, **12**, 199–234.

[9] Altschuler, D. and Coste, A. 1992. Quasi-quantum groups, knots, three manifolds and topological field theory. *Comm. Math. Phys.*, **150**, 83–107.

[10] Anderson, F. W. and Fuller, K. R. 1992. *Rings and Categories of Modules*. Graduate Texts in Mathematics **13**, 2nd edition. New York: Springer-Verlag.

[11] Andruskiewitsch, N. and Dăscălescu, S. 2003. Co-Frobenius Hopf algebras and the coradical filtration. *Math. Z.*, **243** (1), 145–154.

[12] Andruskiewitsch, N. and Enriquez, B. 1992. Examples of compact matrix pseudogroups arising from the twisting operation. *Comm. Math. Phys.*, **149** (2), 195–207.

[13] Angiono, I. E. 2010. Basic quasi-Hopf algebras over cyclic groups. *Adv. Math.*, **225** (6), 3545–3575.

[14] Angiono, I. E., Ardizzoni, A. and Menini, C. 2017. Cohomology and coquasi-bialgebras in the category of Yetter-Drinfeld modules. *Ann. Sc. Norm. Super. Pisa Cl. Sci.*, **17** (2), 609–653.

[15] Ardizzoni, A., Beattie, M. and Menini, C. 2015. Quantum lines for dual quasi-bialgebras. *Algebr. Represent. Theory*, **18** (1), 35–64.

[16] Ardizzoni, A., Bulacu, D. and Menini, C. 2013. Quasi-bialgebra structures and torsion-free abelian groups. *Bull. Math. Soc. Sci. Math. Roumanie (N.S.)*, **56(104)**, 247–265.

[17] Ardizzoni, A., El Kaoutit, L. and Saracco, P. 2016. Functorial constructions for non-associative algebras with applications to quasi-bialgebras. *J. Algebra*, **449**, 460–496.

[18] Ardizzoni, A., Menini, C. and Ştefan, D. 2007. A monoidal approach to splitting morphisms of bialgebras. *Trans. Amer. Math. Soc.*, **359** (3), 991–1044.

[19] Ardizzoni, A. and Pavarin, A. 2012. Preantipodes for dual quasi-bialgebras. *Israel J. Math.*, **192** (1), 281–295.

[20] Ardizzoni, A. and Pavarin, A. 2013. Bosonization for dual quasi-bialgebras and preantipode. *J. Algebra*, **390**, 126–159.

[21] Babelon, O., Bernard, D. and Billey, E. 1996. A quasi-Hopf algebra interpretation of quantum 3-j and 6-j symbols and difference equations. *Phys. Lett. B*, **375** (1-4), 89–97.

[22] Bagheri, S. 2014. Adjunctions of Hom and tensor as endofunctors of (bi-)module categories over quasi-Hopf algebras. *Comm. Algebra*, **42** (2), 488–510.

[23] Bagheri, S. and Wisbauer, R. 2012. Hom-tensor relations for two-sided Hopf modules over quasi-Hopf algebras. *Comm. Algebra*, **40** (9), 3257–3287.

[24] Balan, A. 2009. A Morita context and Galois extensions for quasi-Hopf algebras. *Comm. Algebra*, **37** (4), 1129–1150.

[25] Balan, A. 2010. Galois extensions for coquasi-Hopf algebras. *Comm. Algebra*, **38** (4), 1491–1525.

[26] Balodi, M, Huang, H.-L. and Kumar, S. D. 2017. On the classification of finite quasi-quantum groups. *Rev. Math. Phys.*, **29** (10), 1730003, 20pp.

[27] Beattie, M., Bulacu, D. and Torrecillas, B. 2007. Radford's $S^4$ formula for co-Frobenius Hopf algebras. *J. Algebra*, **307** (1), 330–342.

[28] Beattie, M., Dăscălescu, S. and Grünenfelder, L. 1999. On the number of types of finite-dimensional Hopf algebras. *Invent. Math.*, **136** (1), 1–7.

[29] Beattie, M., Iovanov, M. and Raianu, Ş. 2009. The antipode of a dual quasi-Hopf algebra with nonzero integrals is bijective. *Algebr. Represent. Theory*, **12** (2-5), 251–255.

[30] Bénabou, J. 1963. Catégories avec multiplication. *C. R. Acad. Sci. Paris Sér. I Math.*, **256**, 1887–1890.

[31] Bespalov, Y. and Drabant, B. 1998. Hopf (bi-)modules and crossed modules in braided monoidal categories. *J. Pure Appl. Algebra*, **123**, 105–129.

[32] Bespalov, Y. and Drabant, B. 1999. Cross product bialgebras I. *J. Algebra*, **219**, 466–505.

[33] Bespalov, Y. and Drabant, B. 1999. Cross product bialgebras II. *J. Algebra*, **240**, 445–504.

[34] Boboc, C., Dăscălescu, S. and Van Wyk, L. 2012. Isomorphisms between Morita context rings. *Linear & Multilinear Algebra*, **60** (5), 545–563.

[35] Böhm, G. and Lack, S. 2016. Hopf comonads on naturally Frobenius map-monoidales. *J. Pure Appl. Algebra*, **220** (6), 2177–2213.

[36] Böhm, G. and Ştefan, D. 2012. A categorical approach to cyclic duality. *J. Noncommut. Geom.*, **6** (3), 481–538.

[37] Brauer, R. and Nesbitt, C. 1937. On the regular representations of algebras. *Proc. Nat. Acad. Sci. USA*, **23**, 236–240.

[38] Brzeziński, T. 2005. Galois comodules. *J. Algebra*, **290** (2), 503–537.

[39] Brzeziński, T. and Majid, S. 2000. Quantum geometry of algebra factorisations and coalgebra bundles. *Comm. Math. Phys.*, **213** (3), 491–521.

[40] Brzeziński, T., Marquez, A. V. and Vercruysse, J. 2011. The Eilenberg-Moore category and a Beck-type theorem for a Morita context. *Appl. Categ. Structures*, **19** (5), 821–858.

[41] Bulacu, D. 1999. On the antipode of semi-Hopf algebras and braided semi-Hopf algebras. *Rev. Roum. Math. Pures Appl.*, **44**, 329–340.

[42] Bulacu, D. 2009. The weak braided Hopf algebra structure of some Cayley-Dickson algebras. *J. Algebra*, **322**, 2404–2427.

[43] Bulacu, D. 2011. A Clifford algebra is a weak Hopf algebra in a suitable symmetric monoidal category. *J. Algebra*, **332**, 244–284.

[44] Bulacu, D. 2017. A structure theorem for quasi-Hopf bimodule coalgebras. *Theory Appl. Categ. (TAC)*, **32** (1), 1–30.

[45] Bulacu, D. 2018. Quasi-quantum groups obtained from tensor braided Hopf algebras. Preprint, submitted.

[46] Bulacu, D. and Caenepeel, S. 2003. Two-sided two-cosided Hopf modules and Doi-Hopf modules for quasi-Hopf algebras. *J. Algebra*, **270**, 55–95.

[47] Bulacu, D. and Caenepeel, S. 2003. Integrals for (dual) quasi-Hopf algebras: Applications. *J. Algebra*, **266**, 552–583.

[48] Bulacu, D. and Caenepeel, S. 2003. The quantum double for quasitriangular quasi-Hopf algebras. *Comm. Algebra*, **31** (3), 1403–1425.

[49] Bulacu, D. and Caenepeel, S. 2012. On integrals and cointegrals for quasi-Hopf algebras. *J. Algebra*, **351**, 390–425.

[50] Bulacu, D., Caenepeel, S. and Panaite, F. 2005. More properties of Yetter-Drinfeld modules over quasi-Hopf algebras, in S. Caenepeel and F. Van Oystaeyen (eds.), *Hopf Algebras in Noncommutative Geometry and Physics*. Lecture Notes in Pure and Applied Mathematics **239**. New York: Marcel Dekker, pp. 89–112.

[51] Bulacu, D., Caenepeel, S. and Panaite, F. 2006. Yetter-Drinfeld categories for quasi-Hopf algebras. *Comm. Algebra*, **34**, 1–35.

[52] Bulacu, D., Caenepeel, S. and Torrecillas, B. 2009. Involutory quasi-Hopf algebras. *Algebr. Represent. Theory*, **12**, 257–285.

[53] Bulacu, D., Caenepeel, S. and Torrecillas, B. 2011. The braided monoidal structures on the category of vector spaces graded by the Klein group. *Proc. Edinburgh Math. Soc.*, **54**, 613–641.

[54] Bulacu, D. and Nauwelaerts, E. 2002. Radford's biproduct for quasi-Hopf algebras and bosonization. *J. Pure Appl. Algebra*, **174**, 1–42.

[55] Bulacu D. and Nauwelaerts, E. 2003. Quasitriangular and ribbon quasi-Hopf algebras. *Comm. Algebra*, **31** (2), 1–16.

[56] Bulacu, D. and Panaite, F. 1998. A generalization of the quasi-Hopf algebra $D^{\omega}(G)$. *Comm. Algebra*, **26**, 4125–4141.

[57] Bulacu, D. and Panaite, F. 2018. Some ribbon elements for the quasi-Hopf algebra $D^{\omega}(H)$. Preprint, submitted.

[58] Bulacu, D., Panaite, F. and Van Oystaeyen, F. 1999. Quantum traces and quantum dimensions for quasi-Hopf algebras. *Comm. Algebra*, **27**, 6103–6122.

[59] Bulacu, D., Panaite, F. and Van Oystaeyen, F. 2000. Quasi-Hopf algebra actions and smash products. *Comm. Algebra*, **28**, 631–651.

[60] Bulacu, D., Panaite, F. and Van Oystaeyen, F. 2006. Generalized diagonal crossed products and smash products for quasi-Hopf algebras. Applications. *Comm. Math. Phys.*, **266**, 355–399.

[61] Bulacu, D. and Torrecillas, B. 2004. Factorizable quasi-Hopf algebras: Applications. *J. Pure Appl. Algebra*, **194** (1-2), 39–84.

[62] Bulacu, D. and Torrecillas, B. 2008. The representation-theoretic rank of the doubles of quasi-quantum groups. *J. Pure Appl. Algebra*, **212**, 919–940.

[63] Bulacu, D. and Torrecillas, B. 2006. Two-sided two-cosided Hopf modules and Yetter-Drinfeld modules for quasi-Hopf algebras. *Appl. Categor. Structures*, **28**, 503–530.

[64] Bulacu, D. and Torrecillas, B. 2018. Quasi-quantum groups obtained from the Tannaka-Krein reconstruction theorem. Preprint, submitted.

[65] Bulacu, D. and Torrecillas, B. 2018. On sovereign, balanced and ribbon quasi-Hopf algebras. Preprint, submitted.

[66] Burciu, S. and Natale, S. 2013. Fusion rules of equivariantizations of fusion categories. *J. Math. Phys.*, **54** (1), 013511, 21pp.

[67] Caenepeel, S., Ion, B., Militaru, G. and Zhu, S. 2000. The factorization problem and the smash biproduct of algebras and coalgebras. *Algebr. Represent. Theory*, **3**, 19–42.

[68] Caenepeel, S., Militaru, G. and Zhu, S. 2002. *Frobenius and Separable Functors for Generalized Module Categories and Nonlinear Equations*. Lecture Notes in Mathematics **1787**. Berlin: Springer-Verlag.

[69] Cap, A., Schichl, H. and Vanžura, J. 1995. On twisted tensor products of algebras. *Comm. Algebra*, **23**, 4701–4735.

[70] Curtis, C. W. and Reiner, I. 1962. *Representation Theory of Finite Groups and Associative Algebras*. Pure and Applied Mathematics **11**. London–New York: Interscience.

[71] Dăscălescu, S. 2008. Group gradings on diagonal algebras. *Arch. Math. (Bassel)*, **91** (3), 212–217.

[72] Dăscălescu, S., Iovanov, M. C. and Năstăsescu, C. 2013. Path subcoalgebras, finiteness properties and quantum groups. *J. Noncommut. Geom.*, **7** (3), 737–766.

[73] Dăscălescu, S., Năstăsescu, C. and Raianu, Ş. 2001. *Hopf Algebras: An Introduction*. Monographs Textbooks in Pure and Applied Mathematics **235**, Marcel Dekker, New York: Marcel Dekker.

[74] Deligne, P. 1990. Catégories Tannakiennes, in P. Cartier et al. (eds.), *The Grothendieck Festschrift*, Volume **2**. Basel: Birkhäuser, pp. 111–195.

[75] Dello, J., Panaite, F., Van Oystaeyen, F. and Zhang, Y. 2016. Structure theorems for bicomodule algebras over quasi-Hopf algebras, weak Hopf algebras and braided Hopf algebras. *Comm. Algebra*, **44**, 4609–4636.

[76] Dieudonné, J. 1958. Remarks on quasi-Frobenius rings. *Illinois J. Math.*, **2**, 346–354.

[77] Dijkgraaf, R., Pasquier, V. and Roche, P. 1990. Quasi-Hopf algebras, group cohomology and orbifold models. *Nuclear Phys. B Proc. Suppl.*, **18 B**, 60–72.

[78] Drabant, B. 2001. Notes on balanced categories and Hopf algebras, in A. Pressley (ed.), *Quantum Groups and Lie Theory* (Durham, 1999). Cambridge: Cambridge University Press, pp. 63–88.

[79] Drinfeld, V. G. 1989. Quasi-Hopf algebras and Knizhnik-Zamolodchikov equations, in A. A. Belavin and P. Kofen (eds.), *Problems of Modern Quantum Field Theory*. Research Report in Physics. Berlin: Springer, pp. 11–13.

[80] Drinfeld, V. G. 1989. Quasi-Hopf algebras. *Leningrad Math. J.*, **1** (6), 1419–1457.

[81] Drinfeld, V. G. 1990. On almost cocommutative Hopf algebras. *Leningrad Math. J.*, **1**, 321–342.

[82] Drinfeld, V. G. 1991. On quasitriangular quasi-Hopf algebras and a group closely connected with $\mathrm{Gal}(\overline{\mathbf{Q}}/\mathbf{Q})$. *Leningrad Math. J.*, **2** (4), 829–860.

[83] Drinfeld, V. G. 1992. On the structure of quasitriangular quasi-Hopf algebras. *Funct. Anal. Appl.*, **26** (1), 63–65.

[84] Eilenberg, S. and Kelly, G. M. 1966. Closed categories, in *Proceedings of the Conference on Categorical Algebra* (La Jolla, 1965). New York: Springer, pp. 421–562.

[85] Elhamdadi, M. and Makhlouf, A. 2013. Hom-quasi-bialgebras, in N. Andruskiewitsch, J. Cuadra and B. Torrecillas (eds.), *Hopf Algebras and Tensor Categories*, Contemporary Mathematics **585**. Providence, RI: American Mathematical Society, pp. 227–245.

[86] Enriquez, B. 2003. Quasi-Hopf algebras associated with semisimple Lie algebras and complex curves. *Selecta Math. (N.S.)*, **9** (1), 1–61.

[87] Enriquez, B. and Felder, G. 1998. Elliptic quantum groups $E_{\tau,\eta}(sl_2)$ and quasi-Hopf algebras. *Comm. Math. Phys.*, **195** (3), 651–689.

[88] Enriquez, B. and Halbout, G. 2004. Poisson algebras associated to quasi-Hopf algebras. *Adv. Math.*, **186** (2), 363–395.

[89] Enriquez, B. and Rubtsov, V. 1999. Quasi-Hopf algebras associated with $sl_2$ and complex curves. *Israel J. Math.*, **112**, 61–108.

[90] Etingof, P. and Gelaki, S. 1998. On finite dimensional semisimple and cosemisimple Hopf algebras in positive characteristic. *Internat. Math. Res. Notices*, **16**, 851–864.

[91] Etingof, P. and Gelaki, S. 2004. Finite dimensional quasi-Hopf algebras with radical of codimension 2. *Math. Res. Lett.*, **11**, 685–696.

[92] Etingof, P. and Gelaki, S. 2005. On radically graded finite-dimensional quasi-Hopf algebras. *Mosc. Math. J.*, **5** (2), 371–378.

[93] Etingof, P. and Gelaki, S. 2006. Liftings of graded quasi-Hopf algebras with radical of prime codimension. *J. Pure Appl. Algebra*, **205** (2), 310–322.

[94] Etingof, P. and Gelaki, S. 2009. The small quantum group as a quantum double. *J. Algebra*, **322** (7), 2580–2585.

[95] Etingof, P., Gelaki, S., Nikshych, D. and Ostrik, V. 2015. *Tensor Categories*. Mathematical Surveys and Monographs **205**. Providence, RI: American Mathematical Society.

[96] Etingof, P., Nikshych, D. and Ostrik, V. 2005. On fusion categories. *Ann. of Math.*, **162** (2), 581–642.

[97] Etingof, P., Rowell, E. and Witherspoon, S. 2008. Braid group representations from twisted quantum doubles of finite groups. *Pacific J. Math.*, **234** (1), 33–41.

[98] Farsad, V., Gainutdinov, A.M. and Runkel, I. 2018. SL$(2,Z)$-action for ribbon quasi-Hopf algebras. Preprint, arXiv:1702.01086v3.

[99] Freyd, P. and Yetter, D. 1989. Braided compact closed categories with applications to low dimensional topology. *Adv. Math.*, **77**, 156–182.

[100] Freyd, P. and Yetter, D. 1992. Coherence theorems via knot theory. *J. Pure Appl. Algebra*, **78**, 49–76.

[101] Gelaki, S. 2005. Basic quasi-Hopf algebras of dimension $n^3$. *J. Pure Appl. Algebra*, **198** (1-3), 165–174.

[102] Goff, C., Mason, G. and Ng, S.-H. 2007. On the gauge equivalence of twisted quantum doubles of elementary abelian and extra-special 2-groups. *J. Algebra*, **312** (2), 849–875.

[103] Gould, M. D. and Lekatsas, T. 2005. Some twisted results. *J. Phys. A*, **38** (47), 10123–10144.

[104] Gould, M. D. and Lekatsas, T. 2006. Quasi-Hopf *-algebras. Unpublished, arXiv:0604520.

[105] Gould, M. D. Zhang, Y.-Z. and Isaac, P. S. 2000. Casimir invariants from quasi-Hopf (super)algebras. *J. Math. Phys.*, **41** (1), 547–568.

[106] Gould, M. D., Zhang, Y.-Z. and Isaac, P. S. 2001. On quasi-Hopf superalgebras. *Comm. Math. Phys.*, **224** (2), 341–372.

[107] Hausser, F. and Nill, F. 1999. Diagonal crossed products by duals of quasi-quantum groups. *Rev. Math. Phys.*, **11**, 553–629.

[108] Hausser, F. and Nill, F. 1999. Doubles of quasi-quantum groups. *Comm. Math. Phys.*, **199**, 547–589.

[109] Hausser, F. and Nill, F. 1999. Integral theory for quasi-Hopf algebras. Unpublished, arXiv:9904164.

[110] Hennings, M. A. 1996. Invariants of links and 3-manifolds obtained from Hopf algebras. *J. Lond. Math. Soc. (2)*, **54** (3), 594–624.

[111] Huang, H.-L. 2009. Quiver approaches to quasi-Hopf algebras. *J. Math. Phys.*, **50** (4), 043501, 9 pp.

[112] Huang, H.-L. 2012. From projective representations to quasi-quantum groups, *Sci. China Math.*, **55** (10), 2067–2080.

[113] Huang, H.-L., Liu, G. and Ye, Y. 2011. Quivers, quasi-quantum groups and finite tensor categories. *Comm. Math. Phys.*, **303** (3), 595–612.

[114] Huang, H.-L., Liu, G., Ye, Y. 2015. Graded elementary quasi-Hopf algebras of tame representation type. *Israel J. Math.*, **209** (1), 157–186.

[115] Huang, H.-L. and Yang, Y. 2015. Quasi-quantum linear spaces. *J. Noncommut. Geom.*, **9** (4), 1227–1259.

[116] Huang, H.-L. and Yang, Y. 2015. Quasi-quantum planes and quasi-quantum groups of dimension $p^3$ and $p^4$. *Proc. Amer. Math. Soc.*, **143** (10), 4245–4260.

[117] Iglesias, F. C., Năstăsescu, C. and Vercruysse, J. 2010. Quasi-Frobenius functors: Applications. *Comm. Algebra*, **38** (8), 3057–3077.

[118] Iovanov, M. C. and Vercruysse, J. 2008. Co-Frobenius corings and adjoint functors. *J. Pure Appl. Algebra*, **212** (9), 2027–2058.

[119] Joyal, A. and Street, R. 1986. *Braided Monoidal Categories*. Mathematics Reports 860081. North Ryde: Macquarie University.

[120] Joyal, A. and Street, R. 1991. Tortile Yang-Baxter operators in tensor categories. *J. Pure Appl. Algebra*, **71**, 43–51.

[121] Joyal, A. and Street, R. 1993. Braided tensor categories. *Adv. Math.*, **102**, 20–78.

[122] Kadison, L. 2006. An approach to quasi-Hopf algebras via Frobenius coordinates. *J. Algebra*, **295** (1), 27–43.

[123] Kadison, L. 1999. *New Examples of Frobenius Extensions*. University Lecture Series **14**. Providence, RI: American Mathematical Society.

[124] Kadison, L. and Stolin, A. A. 2001. An approach to Hopf algebras via Frobenius coordinates. *Beiträge Algebra Geom.*, **42** (2), 359–384.

[125] Kadison, L. and Stolin, A. A. 2002. An approach to Hopf algebras via Frobenius coordinates. II. *J. Pure Appl. Algebra*, **176** (2-3), 127–152.

[126] Kaplansky, I. 1975. Bialgebras. *Lecture Notes in Mathematics*. University of Chicago.

[127] Kassel, C. 1995. *Quantum Groups*. Graduate Texts in Mathematics **155**. Berlin: Springer-Verlag.

[128] Kassel, C. and Turaev, V. G. 1995. Double construction for monoidal categories. *Acta Math.*, **175**, 1–48.

[129] Kauffman, L. H. 1993. Gauss codes, quantum groups and ribbon Hopf algebras. *Rev. Math. Phys.*, **5** (4), 735–773.

[130] Kauffman, L. H. and Radford, D. E. 1993. A necessary and sufficient condition for a finite-dimensional Drinfeld double to be a ribbon Hopf algebra. *J. Algebra*, **159**, 98–114.

[131] Lam, T. Y. 1999. *Lectures on Modules and Rings*. Graduate Texts in Mathematics **189**. New York: Springer-Verlag.

[132] Larson, R. G. 1971. Characters of Hopf algebras. *J. Algebra*, **17**, 352–368.

[133] Larson, R. G. and Radford, D. E. 1998. Finite-dimensional cosemisimple Hopf algebras in characteristic 0 are semisimple. *J. Algebra*, **117**, 267–289.

[134] Larson, R. G. and Radford, D. E. 1998. Semisimple cosemisimple Hopf algebras. *Amer. J. Math.*, **110**, 187–195.

[135] Larson, R. G. and Sweedler, M. E. 1969. An associative orthogonal bilinear form for Hopf algebras. *Amer. J. Math.*, **91**, 75–94.

[136] Laugwitz, R. 2015. Braided Drinfeld and Heisenberg doubles. *J. Pure Appl. Algebra*, **19** (10), 4541–4596.

[137] Links, J. R., Gould, M. D. and Zhang, Y.-Z. 2000. Twisting invariance of link polynomials derived from ribbon quasi-Hopf algebras. *J. Math. Phys.*, **41**, 5020–5032.

[138] Liu, G. 2014. The quasi-Hopf analogue of $u_q(sl_2)$. *Math. Res. Lett.*, **21** (3), 585–603.

[139] Liu, G., Van Oystaeyen, F. and Zhang, Y. 2016. Representations of the small quasi-quantum group $Qu_q(sl_2)$. *Bull. Belg. Math. Soc. Simon Stevin*, **23** (5), 779–800.

[140] Liu, G., Van Oystaeyen, F. and Zhang, Y. 2017. Quasi-Frobenius-Lusztig kernels for simple Lie algebras. *Trans. Amer. Math. Soc.*, **369** (3), 2049–2086.

[141] Mac Lane, S. 1971. Categories for the Working Mathematician. Graduate Texts in Mathematics **5**. New York: Springer-Verlag.

[142] Mack, G. and Schomerus, V. 1992. Quasi-Hopf quantum symmetry in quantum theory. *Nuclear Phys. B*, **370** (1), 185–230.

[143] Majid, S. 1990. Representation-theoretic rank and double Hopf algebras. *Comm. Algebra*, **18** (11), 3705–3712.

[144] Majid, S. 1991. Doubles of quasitriangular Hopf algebras. *Comm. Algebra*, **19** (11), 3061–3073.

[145] Majid, S. 1991. Representations, duals and quantum doubles of monoidal categories. *Rend. Circ. Math. Palermo (2) Suppl.*, **26**, 197–206.

[146] Majid, S. 1994. Algebras and Hopf algebras in braided categories. In J. Bergen and S. Montgomery (eds.), *Advances in Hopf Algebras* (Chicago, IL, 1992) Lecture Notes in Pure and Applied Mathematics **158**. New York: Dekker, pp. 55–105.

[147] Majid, S. 1998. Quantum double for quasi-Hopf algebras. *Lett. Math. Phys.*, **45**, 1–9.

[148] Majid, S. 1995. *Foundations of Quantum Group Theory*. Cambridge: Cambridge University Press.

[149] Maltsiniotis, G. 1995. Traces dans les catégories monoïdales, dualité et catégories monoïdales fibrées. *Cahiers Topologie Géom. Différentielle Catég.*, **36** (3), 195–288.

[150] Markl, M. and Shnider, S. 1996. Cohomology of Drinfeld algebras: a homological algebra approach. *Int. Math. Res. Not.*, **9**, 431–445.

[151] Markl, M. and Shnider, S. 1996. Drinfeld algebra deformations, homotopy comodules and the associahedra. *Trans. Amer. Math. Soc.*, **348** (9), 3505–3547.

[152] Mason, G. and Ng, S.-H. 2001. Group cohomology and gauge equivalence of some twisted quantum doubles. *Trans. Amer. Math. Soc.*, **353** (9), 3465–3509.

[153] Mason, G. and Ng, S.-H. 2005. Central invariants and Frobenius-Schur indicators for semisimple quasi-Hopf algebras. *Adv. Math.*, **190** (1), 161–195.

[154] Mason, G. and Ng, S.-H. 2014. Cleft extensions and quotients of twisted quantum doubles, in G. Mason, I. Penkov and J. A. Wolf (eds.), *Developments and Retrospectives in Lie Theory*. Developments in Mathematics **38**. Cham: Springer.

[155] Masuoka, A. 2003. Cohomology and coquasi-bialgebra extensions associated to a matched pair of bialgebras. *Adv. Math.*, **173** (2), 262–315.

[156] Milnor, J. and Moore, J. C. 1965. On the structure of Hopf algebras. *Ann. of Math.*, **81**, 211–264.

[157] Montgomery, S. 1993. Hopf Algebras and their Actions on Rings. CBMS Regional Conference Series in Mathematics **82**. Providence, RI: American Mathematical Society.

[158] Naidu, D. and Nikshych, D. 2008. Lagrangian subcategories and braided tensor equivalences of twisted quantum doubles of finite groups. *Comm. Math. Phys.*, **279**, 845–872.

[159] Naidu, D., Nikshych, D. and Witherspoon, S. 2009. Fusion subcategories of representation categories of twisted quantum doubles of finite groups. *Int. Math. Res. Not.*, **22**, 4183–4219.

[160] Nakayama, T. 1938. Note on symmetric algebras. *Ann. of Math.*, **39**, 659–668.

[161] Nakayama, T. 1939. On Frobeniusean algebras. *Ann. of Math.* (2), **40**, 611–633.

[162] Nakayama, T. 1941. On Frobeniusean algebras. II. *Ann. of Math.* (2), **42**, 1–21.

[163] Natale, S. 2003. On group theoretical Hopf algebras and exact factorizations of finite groups. *J. Algebra*, **270** (1), 199–211.

[164] Năstăsescu, C., Raianu, Ş. and Van Oystaeyen, F. 1990. Modules graded by $G$-sets. *Math. Z.*, **203** (4), 605–627.

[165] Năstăsescu, C. and Torrecillas, B. 2005. Morita duality for Grothendieck categories with applications to coalgebras. *Comm. Algebra*, **33** (11), 4083–4096.

[166] Năstăsescu, C., Van Den Bergh, M. and Van Oystaeyen, F. 1989. Separable functors applied to graded rings. *J. Algebra*, **123** (2), 397–413.

[167] Năstăsescu, C. and Van Oystaeyen, F. 1982. *Graded Ring Theory*. North-Holland Mathematical Library **28**. Amsterdam–New York: North Holland.

[168] Ng, S.-H. and Schauenburg, P. 2008. Central invariants and higher indicators for semisimple quasi-Hopf algebras. *Trans. Amer. Math. Soc.*, **360** (4), 1839–1860.

[169] Nichols, W. D. and Zoeller, M. B. 1989. A Hopf algebra freeness theorem. *Amer. J. Math.*, **111**, 381–385.

[170] Oberst, U. and Schneider, H. -J. 1974, Untergruppen formeller Gruppen von endlichem Index. *J. Algebra*, **31**, 10–44.

[171] Ospel, C. 2002. Cohomological properties of the quantum shuffle product and application to the construction of quasi-Hopf algebras. *J. Pure Appl. Algebra*, **173** (3), 315–337.

[172] Panaite, F. 1998. A Maschke-type theorem for quasi-Hopf algebras, in A. Verschoren and S. Caenepeel (eds.), *Rings, Hopf Algebras and Brauer Groups* (Antwerp/Brussels, 1996). Lecture Notes in Pure and Applied Mathematics **197**. New York: Marcel Dekker, pp. 201–207.

[173] Panaite, F. 2007. Doubles of (quasi) Hopf algebras and some examples of quantum groupoids and vertex groups related to them, in L. H. Kauffman, D. E. Radford and F. J. O. Souza (eds), *Hopf Algebras and Generalizations*. Contemporary Mathematics **441**. Providence, RI: American Mathematical Society, pp. 91–115.

[174] Panaite, F., Staic, M. D. and Van Oystaeyen, F. 2010. Pseudosymmetric braidings, twines and twisted algebras. *J. Pure Appl. Algebra*, **214** (6), 867–884.

[175] Panaite, F. and Ştefan, D. 1997. When is the category of comodules a braided tensor category? *Rev. Roum. Math. Pures Appl.*, **42**, 107–119.

[176] Panaite, F. and Van Oystaeyen, F. 2000. Existence of integrals for finite dimensional quasi-Hopf algebras. *Bull. Belg. Math. Soc. Simon Stevin*, **7**, 261–264.

[177] Panaite, F. and Van Oystaeyen, F. 2000. Quasi-Hopf algebras and the centre of a tensor category, in S. Caenepeel and F. Van Oystaeyen (eds.), *Hopf Algebras and Quantum Groups*. Lecture Notes in Pure and Applied Mathematics **209**. New York: Marcel Dekker, pp. 221–235.

[178] Panaite, F. and Van Oystaeyen, F. 2004. Quasi-Hopf algebras and representations of octonions and other quasialgebras. *J. Math. Phys.*, **45**, 3912–3929.

[179] Panaite, F. and Van Oystaeyen, F. 2007. L-R-smash product for (quasi-) Hopf algebras. *J. Algebra*, **309**, 168–191.

[180] Panaite, F. and Van Oystaeyen, F. 2007. A structure theorem for quasi-Hopf comodule algebras. *Proc. Amer. Math. Soc.*, **135**, 1669–1677.

[181] Pareigis, B. 1971. When Hopf algebras are Frobenius algebras. *J. Algebra*, **18**, 588–596.

[182] Pareigis, B. 1977. Non-additive ring and module theory I. General theory of monoids. *Publ. Math. Debrecen*, **24**, 189–204.

[183] Radford, D. E. 1976. The order of the antipode of a finite-dimensional Hopf algebra is finite. *Amer. J. Math.*, **98**, 333–355.

[184] Radford, D. E. 1977. Pointed Hopf algebras are free over Hopf subalgebras. *J. Algebra*, **45**, 266–273.

[185] Radford, D. E. 1977. Freeness (projectivity) criteria for Hopf algebras over Hopf subalgebras. *J. Pure Appl. Algebra*, **11**, 15–28.

[186] Radford, D. E. 1985. The structure of Hopf algebras with a projection. *J. Algebra*, **92** (2), 322–347.

[187] Radford, D. E. 1992. On the antipode of a quasitriangular Hopf algebra. *J. Algebra*, **151**, 1–11.

[188] Radford, D. E. 1994. On Kauffman's knot invariants arising from finite dimensional Hopf algebras, in J. Bergen and S. Montgomery (eds.), *Advances in Hopf algebras* (Chicago, IL, 1992), Lecture Notes in Pure and Applied Mathematics **158**, Dekker, New York: Marcel Dekker, pp. 205–266.

[189] Radford, D. E. 2012. *Hopf Algebras*. Series on Knots and Everything **49**. Hackensack, NJ: World Scientific.

[190] Reshetikhin, N. Y. and Semenov-Tian-Shansky, M. A. 1988. Quantum $R$-matrices and factorization problems. *J. Geom. Phys.*, **5**, 533–550.

[191] Reshetikhin, N. Y. and Turaev, V. G. 1990. Ribbon graphs and their invariants derived from quantum groups. *Comm. Math. Phys.*, **127**, 1–26.

[192] Rotman, J. 1979. *An Introduction to Homological Algebra*. New York: Academic Press.

[193] Sakáloš, Š. 2017. On categories associated to a quasi-Hopf algebra. *Comm. Algebra*, **45** (2), 722–748.

[194] Schauenburg, P. 1994. Hopf modules and Yetter-Drinfel'd modules. *J. Algebra*, **169**, 874–890.

[195] Schauenburg, P. 2002. Hopf modules and the double of a quasi-Hopf algebra. *Trans. Amer. Math. Soc.*, **354**, 3349–3378.

[196] Schauenburg, P. 2002. Hopf bimodules, coquasibialgebras, and an exact sequence of Kac. *Adv. Math.*, **165** (2), 194–263.

[197] Schauenburg, P. 2003. Actions of monoidal categories and generalized Hopf smash products. *J. Algebra*, **270** (2), 521–563.

[198] Schauenburg, P. 2004. On the Frobenius-Schur indicators for quasi-Hopf algebras. *J. Algebra*, **282** (1), 129–139.

[199] Schauenburg, P. 2004. Two characterizations of finite quasi-Hopf algebras. *J. Algebra*, **273** (2), 538–550.

[200] Schauenburg, P. 2004. A quasi-Hopf algebra freeness theorem. *Proc. Amer. Math. Soc.*, **132** (4), 965–972.

[201] Schauenburg, P. 2005. Quotients of finite quasi-Hopf algebras, in S. Caenepeel and F. Van Oysaeyen (eds.), *Hopf Algebras in Noncommutative Geometry and Physics*. Lecture Notes in Pure and Applied Mathematics **239**. New York: Marcel Dekker, pp. 281–290.

[202] Schneider, H.-J. 1995. *Lectures on Hopf Algebras*. Trabajos de Matematica **31**, University of Córdoba.

[203] Shnider, S. and Sternberg, S. 1993. *Quantum Groups: From Coalgebras to Drinfeld Algebras*. Graduate Texts in Mathematical Physics II. Cambridge, MA: International Press.

[204] Sommerhäuser, Y. 2010. On the notion of a ribbon quasi-Hopf algebra. *Rev. Unión Mat. Argent.*, **51**, 177–192.

[205] Sommerhäuser, Y. and Zhu, Y. 2013. On the central charge of a factorizable Hopf algebra. *Adv. Math.*, **236**, 158–223.

[206] Stasheff, J. 1992. Drinfeld's quasi-Hopf algebras and beyond, in J. Stasheff and M. Gerstenhaber (eds.), *Deformation Theory and Quantum Groups with Applications to Mathematical Physics* (Amherst, MA, 1990). Contemporary Mathematics **134**. Providence, RI: American Mathematical Society, pp. 297–307.

[207] Street, R. 1999. The quantum double and related constructions. *J. Pure Appl. Algebra*, **132**, 195–206.

[208] Street, R. 2004. Frobenius algebras and monoidal categories. Annual Meeting Aust. Math. Soc. Macquarie University.

[209] Sweedler, M. E. 1969. *Hopf Algebras*. New York: Benjamin.

[210] Sweedler, M. E. 1968. Cohomology of algebras over Hopf algebras. *Trans. Amer. Math. Soc.*, **133**, 205–239.

[211] Ştefan, D. 1997. The set of types of $n$-dimensional semisimple and cosemisimple Hopf algebras is finite. *J. Algebra*, **193** (2), 571–580.

[212] Takeuchi, M. 1999. Finite Hopf algebras in braided tensor categories. *J. Pure Appl. Algebra*, **138**, 59–82.

[213] Turaev, V. G. 1992. Modular categories and 3-manifold invariants. *Int. J. Modern Phys. B*, **6**, 1807–1824.

[214] Van Daele, A. 1997. The Haar measure on finite quantum groups. *Proc. Amer. Math. Soc.*, **125**, 3489–3500.

[215] Van Daele, A. and Van Keer, S. 1994. The Yang–Baxter and Pentagon equation. *Compositio Math.*, **91**, 201–221.

[216] Yetter, D. 1992. Framed tangles and a theorem of Deligne on braided deformations of Tannakian categories, in J. Stasheff and M. Gerstenhaber (eds.), *Deformation Theory and Quantum Groups with Applications to Mathematical Physics* (Amherst, MA, 1990). Contemporary Mathematics **134**. Providence, RI: American Mathematical Society, pp. 325–350.

[217] Yetter, D. 2001. *Functorial Knot Theory: Categories of Tangles, Coherence, Categorical Deformations, and Topological Invariance*. Series on Knots and Everything **26**. River Edge, NJ: World Scientific.

[218] Wang, Y., Zhang, L. Y. and Niu, R. F. 2013. Structure theorem for dual quasi-Hopf bicomodules and its application. *Math. Notes*, **94** (3-4), 470–481.

[219] Weibel, C. 1994. *An Introduction to Homological Algebra*. Cambridge Studies in Advanced Mathematics **38**. Cambridge: Cambridge University Press.

[220] Woronowicz, S. 1989. Differential calculus on compact matrix pseudogroups (quantum groups). *Comm. Math. Phys.*, **122**, 125–170.

# Index